Student's Solutions Manual
to accompany Jon Rogawski's

Multivariable
CALCULUS

SECOND EDITION

GREGORY P. DRESDEN

Washington and Lee University

JENNIFER BOWEN

The College of Wooster

RANDALL PAUL

Oregon Institute of Technology

W. H. FREEMAN AND COMPANY
NEW YORK

© 2012 by W. H. Freeman and Company

ISBN-13: 978-1-4292-5508-0
ISBN-10: 1-4292-5508-0

All rights reserved

Printed in the United States of America

First Printing

W. H. Freeman and Company, 41 Madison Avenue, New York, NY 10010
Houndmills, Basingstoke RG21 6XS, England
www.whfreeman.com

| CONTENTS

10 INFINITE SERIES

10.1 Sequences (LT Section 11.1)

Preliminary Questions

1. What is a_4 for the sequence $a_n = n^2 - n$?

SOLUTION Substituting $n = 4$ in the expression for a_n gives

$$a_4 = 4^2 - 4 = 12.$$

2. Which of the following sequences converge to zero?

(a) $\dfrac{n^2}{n^2 + 1}$ **(b)** 2^n **(c)** $\left(\dfrac{-1}{2}\right)^n$

SOLUTION

(a) This sequence does not converge to zero:

$$\lim_{n\to\infty} \frac{n^2}{n^2+1} = \lim_{x\to\infty} \frac{x^2}{x^2+1} = \lim_{x\to\infty} \frac{1}{1+\frac{1}{x^2}} = \frac{1}{1+0} = 1.$$

(b) This sequence does not converge to zero: this is a geometric sequence with $r = 2 > 1$; hence, the sequence diverges to ∞.

(c) Recall that if $|a_n|$ converges to 0, then a_n must also converge to zero. Here,

$$\left|\left(-\frac{1}{2}\right)^n\right| = \left(\frac{1}{2}\right)^n,$$

which is a geometric sequence with $0 < r < 1$; hence, $\left(\frac{1}{2}\right)^n$ converges to zero. It therefore follows that $\left(-\frac{1}{2}\right)^n$ converges to zero.

3. Let a_n be the nth decimal approximation to $\sqrt{2}$. That is, $a_1 = 1, a_2 = 1.4, a_3 = 1.41$, etc. What is $\lim_{n\to\infty} a_n$?

SOLUTION $\lim_{n\to\infty} a_n = \sqrt{2}$.

4. Which of the following sequences is defined recursively?

(a) $a_n = \sqrt{4+n}$ **(b)** $b_n = \sqrt{4 + b_{n-1}}$

SOLUTION

(a) a_n can be computed directly, since it depends on n only and not on preceding terms. Therefore a_n is defined explicitly and not recursively.

(b) b_n is computed in terms of the preceding term b_{n-1}, hence the sequence $\{b_n\}$ is defined recursively.

5. Theorem 5 says that every convergent sequence is bounded. Determine if the following statements are true or false and if false, give a counterexample.

(a) If $\{a_n\}$ is bounded, then it converges.

(b) If $\{a_n\}$ is not bounded, then it diverges.

(c) If $\{a_n\}$ diverges, then it is not bounded.

SOLUTION

(a) This statement is false. The sequence $a_n = \cos \pi n$ is bounded since $-1 \le \cos \pi n \le 1$ for all n, but it does not converge: since $a_n = \cos n\pi = (-1)^n$, the terms assume the two values 1 and -1 alternately, hence they do not approach one value.

(b) By Theorem 5, a converging sequence must be bounded. Therefore, if a sequence is not bounded, it certainly does not converge.

(c) The statement is false. The sequence $a_n = (-1)^n$ is bounded, but it does not approach one limit.

Exercises

1. Match each sequence with its general term:

$a_1, a_2, a_3, a_4, \ldots$	General term
(a) $\frac{1}{2}, \frac{2}{3}, \frac{3}{4}, \frac{4}{5}, \ldots$	(i) $\cos \pi n$
(b) $-1, 1, -1, 1, \ldots$	(ii) $\dfrac{n!}{2^n}$
(c) $1, -1, 1, -1, \ldots$	(iii) $(-1)^{n+1}$
(d) $\frac{1}{2}, \frac{2}{4}, \frac{6}{8}, \frac{24}{16} \ldots$	(iv) $\dfrac{n}{n+1}$

SOLUTION

(a) The numerator of each term is the same as the index of the term, and the denominator is one more than the numerator; hence $a_n = \frac{n}{n+1}$, $n = 1, 2, 3, \ldots$.

(b) The terms of this sequence are alternating between -1 and 1 so that the positive terms are in the even places. Since $\cos \pi n = 1$ for even n and $\cos \pi n = -1$ for odd n, we have $a_n = \cos \pi n$, $n = 1, 2, \ldots$.

(c) The terms a_n are 1 for odd n and -1 for even n. Hence, $a_n = (-1)^{n+1}$, $n = 1, 2, \ldots$

(d) The numerator of each term is $n!$, and the denominator is 2^n; hence, $a_n = \frac{n!}{2^n}$, $n = 1, 2, 3, \ldots$.

In Exercises 3–12, calculate the first four terms of the sequence, starting with $n = 1$.

3. $c_n = \dfrac{3^n}{n!}$

SOLUTION Setting $n = 1, 2, 3, 4$ in the formula for c_n gives

$$c_1 = \frac{3^1}{1!} = \frac{3}{1} = 3, \qquad c_2 = \frac{3^2}{2!} = \frac{9}{2},$$

$$c_3 = \frac{3^3}{3!} = \frac{27}{6} = \frac{9}{2}, \qquad c_4 = \frac{3^4}{4!} = \frac{81}{24} = \frac{27}{8}.$$

5. $a_1 = 2$, $\quad a_{n+1} = 2a_n^2 - 3$

SOLUTION For $n = 1, 2, 3$ we have:

$$a_2 = a_{1+1} = 2a_1^2 - 3 = 2 \cdot 4 - 3 = 5;$$

$$a_3 = a_{2+1} = 2a_2^2 - 3 = 2 \cdot 25 - 3 = 47;$$

$$a_4 = a_{3+1} = 2a_3^2 - 3 = 2 \cdot 2209 - 3 = 4415.$$

The first four terms of $\{a_n\}$ are $2, 5, 47, 4415$.

7. $b_n = 5 + \cos \pi n$

SOLUTION For $n = 1, 2, 3, 4$ we have

$$b_1 = 5 + \cos \pi = 4;$$

$$b_2 = 5 + \cos 2\pi = 6;$$

$$b_3 = 5 + \cos 3\pi = 4;$$

$$b_4 = 5 + \cos 4\pi = 6.$$

The first four terms of $\{b_n\}$ are $4, 6, 4, 6$.

9. $c_n = 1 + \dfrac{1}{2} + \dfrac{1}{3} + \cdots + \dfrac{1}{n}$

SOLUTION

$$c_1 = 1;$$

$$c_2 = 1 + \frac{1}{2} = \frac{3}{2};$$

$$c_3 = 1 + \frac{1}{2} + \frac{1}{3} = \frac{3}{2} + \frac{1}{3} = \frac{11}{6};$$

$$c_4 = 1 + \frac{1}{2} + \frac{1}{3} + \frac{1}{4} = \frac{11}{6} + \frac{1}{4} = \frac{25}{12}.$$

11. $b_1 = 2,\quad b_2 = 3,\quad b_n = 2b_{n-1} + b_{n-2}$

SOLUTION We need to find b_3 and b_4. Setting $n = 3$ and $n = 4$ and using the given values for b_1 and b_2 we obtain:

$$b_3 = 2b_{3-1} + b_{3-2} = 2b_2 + b_1 = 2 \cdot 3 + 2 = 8;$$

$$b_4 = 2b_{4-1} + b_{4-2} = 2b_3 + b_2 = 2 \cdot 8 + 3 = 19.$$

The first four terms of the sequence $\{b_n\}$ are 2, 3, 8, 19.

13. Find a formula for the nth term of each sequence.

(a) $\dfrac{1}{1}, \dfrac{-1}{8}, \dfrac{1}{27}, \cdots$

(b) $\dfrac{2}{6}, \dfrac{3}{7}, \dfrac{4}{8}, \cdots$

SOLUTION

(a) The denominators are the third powers of the positive integers starting with $n = 1$. Also, the sign of the terms is alternating with the sign of the first term being positive. Thus,

$$a_1 = \frac{1}{1^3} = \frac{(-1)^{1+1}}{1^3};\quad a_2 = -\frac{1}{2^3} = \frac{(-1)^{2+1}}{2^3};\quad a_3 = \frac{1}{3^3} = \frac{(-1)^{3+1}}{3^3}.$$

This rule leads to the following formula for the nth term:

$$a_n = \frac{(-1)^{n+1}}{n^3}.$$

(b) Assuming a starting index of $n = 1$, we see that each numerator is one more than the index and the denominator is four more than the numerator. Thus, the general term a_n is

$$a_n = \frac{n+1}{n+5}.$$

In Exercises 15–26, use Theorem 1 to determine the limit of the sequence or state that the sequence diverges.

15. $a_n = 12$

SOLUTION We have $a_n = f(n)$ where $f(x) = 12$; thus,

$$\lim_{n \to \infty} a_n = \lim_{x \to \infty} f(x) = \lim_{x \to \infty} 12 = 12.$$

17. $b_n = \dfrac{5n - 1}{12n + 9}$

SOLUTION We have $b_n = f(n)$ where $f(x) = \dfrac{5x - 1}{12x + 9}$; thus,

$$\lim_{n \to \infty} \frac{5n - 1}{12n + 9} = \lim_{x \to \infty} \frac{5x - 1}{12x + 9} = \frac{5}{12}.$$

19. $c_n = -2^{-n}$

SOLUTION We have $c_n = f(n)$ where $f(x) = -2^{-x}$; thus,

$$\lim_{n \to \infty} \left(-2^{-n}\right) = \lim_{x \to \infty} -2^{-x} = \lim_{x \to \infty} -\frac{1}{2^x} = 0.$$

21. $c_n = 9^n$

SOLUTION We have $c_n = f(n)$ where $f(x) = 9^x$; thus,

$$\lim_{n \to \infty} 9^n = \lim_{x \to \infty} 9^x = \infty$$

Thus, the sequence 9^n diverges.

23. $a_n = \dfrac{n}{\sqrt{n^2 + 1}}$

SOLUTION We have $a_n = f(n)$ where $f(x) = \dfrac{x}{\sqrt{x^2 + 1}}$; thus,

$$\lim_{n \to \infty} \frac{n}{\sqrt{n^2 + 1}} = \lim_{x \to \infty} \frac{x}{\sqrt{x^2 + 1}} = \lim_{x \to \infty} \frac{\frac{x}{x}}{\frac{\sqrt{x^2+1}}{x}} = \lim_{x \to \infty} \frac{1}{\sqrt{\frac{x^2+1}{x^2}}} = \lim_{x \to \infty} \frac{1}{\sqrt{1 + \frac{1}{x^2}}} = \frac{1}{\sqrt{1 + 0}} = 1.$$

25. $a_n = \ln\left(\dfrac{12n + 2}{-9 + 4n}\right)$

SOLUTION We have $a_n = f(n)$ where $f(x) = \ln\left(\dfrac{12x + 2}{-9 + 4x}\right)$; thus,

$$\lim_{n\to\infty} \ln\left(\frac{12n + 2}{-9 + 4n}\right) = \lim_{x\to\infty} \ln\left(\frac{12x + 2}{-9 + 4x}\right) = \ln \lim_{x\to\infty}\left(\frac{12x + 2}{-9 + 4x}\right) = \ln 3$$

In Exercises 27–30, use Theorem 4 to determine the limit of the sequence.

27. $a_n = \sqrt{4 + \dfrac{1}{n}}$

SOLUTION We have

$$\lim_{n\to\infty} 4 + \frac{1}{n} = \lim_{x\to\infty} 4 + \frac{1}{x} = 4$$

Since $\sqrt{x}$ is a continuous function for $x > 0$, Theorem 4 tells us that

$$\lim_{n\to\infty} \sqrt{4 + \frac{1}{n}} = \sqrt{\lim_{n\to\infty} 4 + \frac{1}{n}} = \sqrt{4} = 2$$

29. $a_n = \cos^{-1}\left(\dfrac{n^3}{2n^3 + 1}\right)$

SOLUTION We have

$$\lim_{n\to\infty} \frac{n^3}{2n^3 + 1} = \frac{1}{2}$$

Since $\cos^{-1}(x)$ is continuous for all x, Theorem 4 tells us that

$$\lim_{n\to\infty} \cos^{-1}\left(\frac{n^3}{2n^3 + 1}\right) = \cos^{-1}\left(\lim_{n\to\infty} \frac{n^3}{2n^3 + 1}\right) = \cos^{-1}(1/2) = \frac{\pi}{3}$$

31. Let $a_n = \dfrac{n}{n + 1}$. Find a number M such that:

(a) $|a_n - 1| \leq 0.001$ for $n \geq M$.

(b) $|a_n - 1| \leq 0.00001$ for $n \geq M$.

Then use the limit definition to prove that $\lim\limits_{n\to\infty} a_n = 1$.

SOLUTION

(a) We have

$$|a_n - 1| = \left|\frac{n}{n + 1} - 1\right| = \left|\frac{n - (n + 1)}{n + 1}\right| = \left|\frac{-1}{n + 1}\right| = \frac{1}{n + 1}.$$

Therefore $|a_n - 1| \leq 0.001$ provided $\frac{1}{n+1} \leq 0.001$, that is, $n \geq 999$. It follows that we can take $M = 999$.

(b) By part (a), $|a_n - 1| \leq 0.00001$ provided $\frac{1}{n+1} \leq 0.00001$, that is, $n \geq 99999$. It follows that we can take $M = 99999$.

We now prove formally that $\lim\limits_{n\to\infty} a_n = 1$. Using part (a), we know that

$$|a_n - 1| = \frac{1}{n + 1} < \epsilon,$$

provided $n > \frac{1}{\epsilon} - 1$. Thus, Let $\epsilon > 0$ and take $M = \frac{1}{\epsilon} - 1$. Then, for $n > M$, we have

$$|a_n - 1| = \frac{1}{n + 1} < \frac{1}{M + 1} = \epsilon.$$

33. Use the limit definition to prove that $\lim\limits_{n\to\infty} n^{-2} = 0$.

SOLUTION We see that

$$|n^{-2} - 0| = \left|\frac{1}{n^2}\right| = \frac{1}{n^2} < \epsilon$$

provided

$$n > \frac{1}{\sqrt{\epsilon}}.$$

Thus, let $\epsilon > 0$ and take $M = \frac{1}{\sqrt{\epsilon}}$. Then, for $n > M$, we have

$$|n^{-2} - 0| = \left| \frac{1}{n^2} \right| = \frac{1}{n^2} < \frac{1}{M^2} = \epsilon.$$

In Exercises 35–62, use the appropriate limit laws and theorems to determine the limit of the sequence or show that it diverges.

35. $a_n = 10 + \left(-\frac{1}{9} \right)^n$

SOLUTION By the Limit Laws for Sequences we have:

$$\lim_{n\to\infty} \left(10 + \left(-\frac{1}{9} \right)^n \right) = \lim_{n\to\infty} 10 + \lim_{n\to\infty} \left(-\frac{1}{9} \right)^n = 10 + \lim_{n\to\infty} \left(-\frac{1}{9} \right)^n.$$

Now,

$$-\left(\frac{1}{9} \right)^n \le \left(-\frac{1}{9} \right)^n \le \left(\frac{1}{9} \right)^n.$$

Because

$$\lim_{n\to\infty} \left(\frac{1}{9} \right)^n = 0,$$

by the Limit Laws for Sequences,

$$\lim_{n\to\infty} -\left(\frac{1}{9} \right)^n = -\lim_{n\to\infty} \left(\frac{1}{9} \right)^n = 0.$$

Thus, we have

$$\lim_{n\to\infty} \left(-\frac{1}{9} \right)^n = 0,$$

and

$$\lim_{n\to\infty} \left(10 + \left(-\frac{1}{9} \right)^n \right) = 10 + 0 = 10.$$

37. $c_n = 1.01^n$

SOLUTION Since $c_n = f(n)$ where $f(x) = 1.01^x$, we have

$$\lim_{n\to\infty} 1.01^n = \lim_{x\to\infty} 1.01^x = \infty$$

so that the sequence diverges.

39. $a_n = 2^{1/n}$

SOLUTION Because 2^x is a continuous function,

$$\lim_{n\to\infty} 2^{1/n} = \lim_{x\to\infty} 2^{1/x} = 2^{\lim_{x\to\infty}(1/x)} = 2^0 = 1.$$

41. $c_n = \dfrac{9^n}{n!}$

SOLUTION For $n \ge 9$, write

$$c_n = \frac{9^n}{n!} = \underbrace{\frac{9}{1} \cdot \frac{9}{2} \cdots \frac{9}{9}}_{\text{call this } C} \cdot \underbrace{\frac{9}{10} \cdot \frac{9}{11} \cdots \frac{9}{n-1} \cdot \frac{9}{n}}_{\text{Each factor is less than 1}}$$

Then clearly

$$0 \le \frac{9^n}{n!} \le C\frac{9}{n}$$

since each factor after the first nine is < 1. The squeeze theorem tells us that

$$\lim_{n\to\infty} 0 \le \lim_{n\to\infty} \frac{9^n}{n!} \le \lim_{n\to\infty} C\frac{9}{n} = C \lim_{n\to\infty} \frac{9}{n} = C \cdot 0 = 0$$

so that $\lim_{n\to\infty} c_n = 0$ as well.

43. $a_n = \dfrac{3n^2 + n + 2}{2n^2 - 3}$

SOLUTION

$$\lim_{n \to \infty} \frac{3n^2 + n + 2}{2n^2 - 3} = \lim_{x \to \infty} \frac{3x^2 + x + 2}{2x^2 - 3} = \frac{3}{2}.$$

45. $a_n = \dfrac{\cos n}{n}$

SOLUTION Since $-1 \le \cos n \le 1$ the following holds:

$$-\frac{1}{n} \le \frac{\cos n}{n} \le \frac{1}{n}.$$

We now apply the Squeeze Theorem for Sequences and the limits

$$\lim_{n \to \infty} -\frac{1}{n} = \lim_{n \to \infty} \frac{1}{n} = 0$$

to conclude that $\lim_{n \to \infty} \frac{\cos n}{n} = 0$.

47. $d_n = \ln 5^n - \ln n!$

SOLUTION Note that

$$d_n = \ln \frac{5^n}{n!}$$

so that

$$e^{d_n} = \frac{5^n}{n!} \quad \text{so} \quad \lim_{n \to \infty} e^{d_n} = \lim_{n \to \infty} \frac{5^n}{n!} = 0$$

by the method of Exercise 41. If d_n converged, we could, since $f(x) = e^x$ is continuous, then write

$$\lim_{n \to \infty} e^{d_n} = e^{\lim_{n \to \infty} d_n} = 0$$

which is impossible. Thus $\{d_n\}$ diverges.

49. $a_n = \left(2 + \dfrac{4}{n^2}\right)^{1/3}$

SOLUTION Let $a_n = \left(2 + \frac{4}{n^2}\right)^{1/3}$. Taking the natural logarithm of both sides of this expression yields

$$\ln a_n = \ln \left(2 + \frac{4}{n^2}\right)^{1/3} = \frac{1}{3} \ln \left(2 + \frac{4}{n^2}\right).$$

Thus,

$$\lim_{n \to \infty} \ln a_n = \lim_{n \to \infty} \frac{1}{3} \ln \left(2 + \frac{4}{n^2}\right)^{1/3} = \frac{1}{3} \lim_{x \to \infty} \ln \left(2 + \frac{4}{x^2}\right) = \frac{1}{3} \ln \left(\lim_{x \to \infty} \left(2 + \frac{4}{x^2}\right)\right)$$

$$= \frac{1}{3} \ln (2 + 0) = \frac{1}{3} \ln 2 = \ln 2^{1/3}.$$

Because $f(x) = e^x$ is a continuous function, it follows that

$$\lim_{n \to \infty} a_n = \lim_{n \to \infty} e^{\ln a_n} = e^{\lim_{n \to \infty} (\ln a_n)} = e^{\ln 2^{1/3}} = 2^{1/3}.$$

51. $c_n = \ln \left(\dfrac{2n + 1}{3n + 4}\right)$

SOLUTION Because $f(x) = \ln x$ is a continuous function, it follows that

$$\lim_{n \to \infty} c_n = \lim_{x \to \infty} \ln \left(\frac{2x + 1}{3x + 4}\right) = \ln \left(\lim_{x \to \infty} \frac{2x + 1}{3x + 4}\right) = \ln \frac{2}{3}.$$

53. $y_n = \dfrac{e^n}{2^n}$

SOLUTION $\frac{e^n}{2^n} = \left(\frac{e}{2}\right)^n$ and $\frac{e}{2} > 1$. By the Limit of Geometric Sequences, we conclude that $\lim_{n \to \infty} \left(\frac{e}{2}\right)^n = \infty$. Thus, the given sequence diverges.

55. $y_n = \dfrac{e^n + (-3)^n}{5^n}$

SOLUTION

$$\lim_{n \to \infty} \frac{e^n + (-3)^n}{5^n} = \lim_{n \to \infty} \left(\frac{e}{5}\right)^n + \lim_{n \to \infty} \left(\frac{-3}{5}\right)^n$$

assuming both limits on the right-hand side exist. But by the Limit of Geometric Sequences, since

$$-1 < \frac{-3}{5} < 0 < \frac{e}{5} < 1$$

both limits on the right-hand side are 0, so that y_n converges to 0.

57. $a_n = n \sin \dfrac{\pi}{n}$

SOLUTION By the Theorem on Sequences Defined by a Function, we have

$$\lim_{n \to \infty} n \sin \frac{\pi}{n} = \lim_{x \to \infty} x \sin \frac{\pi}{x}.$$

Now,

$$\lim_{x \to \infty} x \sin \frac{\pi}{x} = \lim_{x \to \infty} \frac{\sin \frac{\pi}{x}}{\frac{1}{x}} = \lim_{x \to \infty} \frac{\left(\cos \frac{\pi}{x}\right)\left(-\frac{\pi}{x^2}\right)}{-\frac{1}{x^2}} = \lim_{x \to \infty} \left(\pi \cos \frac{\pi}{x}\right)$$

$$= \pi \lim_{x \to \infty} \cos \frac{\pi}{x} = \pi \cos 0 = \pi \cdot 1 = \pi.$$

Thus,

$$\lim_{n \to \infty} n \sin \frac{\pi}{n} = \pi.$$

59. $b_n = \dfrac{3 - 4^n}{2 + 7 \cdot 4^n}$

SOLUTION Divide the numerator and denominator by 4^n to obtain

$$a_n = \frac{3 - 4^n}{2 + 7 \cdot 4^n} = \frac{\frac{3}{4^n} - \frac{4^n}{4^n}}{\frac{2}{4^n} + \frac{7 \cdot 4^n}{4^n}} = \frac{\frac{3}{4^n} - 1}{\frac{2}{4^n} + 7},$$

Thus,

$$\lim_{n \to \infty} a_n = \lim_{x \to \infty} \frac{\frac{3}{4^x} - 1}{\frac{2}{4^x} + 7} = \frac{\lim_{x \to \infty}\left(\frac{3}{4^x} - 1\right)}{\lim_{x \to \infty}\left(\frac{2}{4^x} + 7\right)} = \frac{3 \lim_{x \to \infty}\frac{1}{4^x} - \lim_{x \to \infty} 1}{2 \lim_{x \to \infty}\frac{1}{4^x} - \lim_{x \to \infty} 7} = \frac{3 \cdot 0 - 1}{2 \cdot 0 + 7} = -\frac{1}{7}.$$

61. $a_n = \left(1 + \dfrac{1}{n}\right)^n$

SOLUTION Taking the natural logarithm of both sides of this expression yields

$$\ln a_n = \ln \left(1 + \frac{1}{n}\right)^n = n \ln \left(1 + \frac{1}{n}\right) = \frac{\ln \left(1 + \frac{1}{n}\right)}{\frac{1}{n}}.$$

Thus,

$$\lim_{n \to \infty} (\ln a_n) = \lim_{x \to \infty} \frac{\ln \left(1 + \frac{1}{x}\right)}{\frac{1}{x}} = \lim_{x \to \infty} \frac{\frac{d}{dx}\left(\ln\left(1 + \frac{1}{x}\right)\right)}{\frac{d}{dx}\left(\frac{1}{x}\right)} = \lim_{x \to \infty} \frac{\frac{1}{1 + \frac{1}{x}} \cdot \left(-\frac{1}{x^2}\right)}{-\frac{1}{x^2}} = \lim_{x \to \infty} \frac{1}{1 + \frac{1}{x}} = \frac{1}{1 + 0} = 1.$$

Because $f(x) = e^x$ is a continuous function, it follows that

$$\lim_{n \to \infty} a_n = \lim_{n \to \infty} e^{\ln a_n} = e^{\lim_{n \to \infty}(\ln a_n)} = e^1 = e.$$

In Exercises 63–66, find the limit of the sequence using L'Hôpital's Rule.

63. $a_n = \dfrac{(\ln n)^2}{n}$

SOLUTION

$$\lim_{n\to\infty} \frac{(\ln n)^2}{n} = \lim_{x\to\infty} \frac{(\ln x)^2}{x} = \lim_{x\to\infty} \frac{\frac{d}{dx}(\ln x)^2}{\frac{d}{dx}x} = \lim_{x\to\infty} \frac{\frac{2\ln x}{x}}{1} = \lim_{x\to\infty} \frac{2\ln x}{x}$$

$$= \lim_{x\to\infty} \frac{\frac{d}{dx}2\ln x}{\frac{d}{dx}x} = \lim_{x\to\infty} \frac{\frac{2}{x}}{1} = \lim_{x\to\infty} \frac{2}{x} = 0$$

65. $c_n = n\left(\sqrt{n^2+1} - n\right)$

SOLUTION

$$\lim_{n\to\infty} n\left(\sqrt{n^2+1} - n\right) = \lim_{x\to\infty} x\left(\sqrt{x^2+1} - x\right) = \lim_{x\to\infty} \frac{x\left(\sqrt{x^2+1} - x\right)\left(\sqrt{x^2+1} + x\right)}{\sqrt{x^2+1} + x}$$

$$= \lim_{x\to\infty} \frac{x}{\sqrt{x^2+1} + x} = \lim_{x\to\infty} \frac{\frac{d}{dx}x}{\frac{d}{dx}\sqrt{x^2+1} + x} = \lim_{x\to\infty} \frac{1}{1 + \frac{x}{\sqrt{x^2+1}}}$$

$$= \lim_{x\to\infty} \frac{1}{1 + \sqrt{\frac{x^2}{x^2+1}}} = \lim_{x\to\infty} \frac{1}{1 + \sqrt{\frac{1}{1+(1/x^2)}}} = \frac{1}{2}$$

In Exercises 67–70, use the Squeeze Theorem to evaluate $\lim\limits_{n\to\infty} a_n$ by verifying the given inequality.

67. $a_n = \dfrac{1}{\sqrt{n^4+n^8}}, \quad \dfrac{1}{\sqrt{2}n^4} \le a_n \le \dfrac{1}{\sqrt{2}n^2}$

SOLUTION For all $n > 1$ we have $n^4 < n^8$, so the quotient $\dfrac{1}{\sqrt{n^4+n^8}}$ is smaller than $\dfrac{1}{\sqrt{n^4+n^4}}$ and larger than $\dfrac{1}{\sqrt{n^8+n^8}}$. That is,

$$a_n < \frac{1}{\sqrt{n^4+n^4}} = \frac{1}{\sqrt{n^4 \cdot 2}} = \frac{1}{\sqrt{2}n^2}; \text{ and}$$

$$a_n > \frac{1}{\sqrt{n^8+n^8}} = \frac{1}{\sqrt{2n^8}} = \frac{1}{\sqrt{2}n^4}.$$

Now, since $\lim\limits_{n\to\infty} \dfrac{1}{\sqrt{2}n^4} = \lim\limits_{n\to\infty} \dfrac{1}{\sqrt{2}n^2} = 0$, the Squeeze Theorem for Sequences implies that $\lim\limits_{n\to\infty} a_n = 0$.

69. $a_n = (2^n + 3^n)^{1/n}, \quad 3 \le a_n \le (2 \cdot 3^n)^{1/n} = 2^{1/n} \cdot 3$

SOLUTION Clearly $2^n + 3^n \ge 3^n$ for all $n \ge 1$. Therefore:

$$(2^n + 3^n)^{1/n} \ge (3^n)^{1/n} = 3.$$

Also $2^n + 3^n \le 3^n + 3^n = 2 \cdot 3^n$, so

$$(2^n + 3^n)^{1/n} \le (2 \cdot 3^n)^{1/n} = 2^{1/n} \cdot 3.$$

Thus,

$$3 \le (2^n + 3^n)^{1/n} \le 2^{1/n} \cdot 3.$$

Because

$$\lim_{n\to\infty} 2^{1/n} \cdot 3 = 3 \lim_{n\to\infty} 2^{1/n} = 3 \cdot 1 = 3$$

and $\lim_{n\to\infty} 3 = 3$, the Squeeze Theorem for Sequences guarantees

$$\lim_{n\to\infty} (2^n + 3^n)^{1/n} = 3.$$

71. Which of the following statements is equivalent to the assertion $\lim\limits_{n\to\infty} a_n = L$? Explain.

(a) For every $\epsilon > 0$, the interval $(L - \epsilon, L + \epsilon)$ contains at least one element of the sequence $\{a_n\}$.

(b) For every $\epsilon > 0$, the interval $(L - \epsilon, L + \epsilon)$ contains all but at most finitely many elements of the sequence $\{a_n\}$.

SOLUTION   Statement (b) is equivalent to Definition 1 of the limit, since the assertion "$|a_n - L| < \epsilon$ for all $n > M$" means that $L - \epsilon < a_n < L + \epsilon$ for all $n > M$; that is, the interval $(L - \epsilon, L + \epsilon)$ contains all the elements a_n except (maybe) the finite number of elements $a_1, a_2, \ldots, a_M$.

Statement (a) is not equivalent to the assertion $\lim_{n\to\infty} a_n = L$. We show this, by considering the following sequence:

$$a_n = \begin{cases} \dfrac{1}{n} & \text{for odd } n \\[2mm] 1 + \dfrac{1}{n} & \text{for even } n \end{cases}$$

Clearly for every $\epsilon > 0$, the interval $(-\epsilon, \epsilon) = (L - \epsilon, L + \epsilon)$ for $L = 0$ contains at least one element of $\{a_n\}$, but the sequence diverges (rather than converges to $L = 0$). Since the terms in the odd places converge to 0 and the terms in the even places converge to 1. Hence, a_n does not approach one limit.

73. Show that $a_n = \dfrac{3n^2}{n^2 + 2}$ is increasing. Find an upper bound.

SOLUTION   Let $f(x) = \frac{3x^2}{x^2+2}$. Then

$$f'(x) = \frac{6x(x^2 + 2) - 3x^2 \cdot 2x}{(x^2 + 2)^2} = \frac{12x}{(x^2 + 2)^2}.$$

$f'(x) > 0$ for $x > 0$, hence f is increasing on this interval. It follows that $a_n = f(n)$ is also increasing. We now show that $M = 3$ is an upper bound for a_n, by writing:

$$a_n = \frac{3n^2}{n^2 + 2} \le \frac{3n^2 + 6}{n^2 + 2} = \frac{3(n^2 + 2)}{n^2 + 2} = 3.$$

That is, $a_n \le 3$ for all n.

75. Give an example of a divergent sequence $\{a_n\}$ such that $\lim_{n\to\infty} |a_n|$ converges.

SOLUTION   Let $a_n = (-1)^n$. The sequence $\{a_n\}$ diverges because the terms alternate between $+1$ and -1; however, the sequence $\{|a_n|\}$ converges because it is a constant sequence, all of whose terms are equal to 1.

77. Using the limit definition, prove that if $\{a_n\}$ converges and $\{b_n\}$ diverges, then $\{a_n + b_n\}$ diverges.

SOLUTION   We will prove this result by contradiction. Suppose $\lim_{n\to\infty} a_n = L_1$ and that $\{a_n + b_n\}$ converges to a limit L_2. Now, let $\epsilon > 0$. Because $\{a_n\}$ converges to L_1 and $\{a_n + b_n\}$ converges to L_2, it follows that there exist numbers M_1 and M_2 such that:

$$|a_n - L_1| < \frac{\epsilon}{2} \qquad \text{for all } n > M_1,$$

$$|(a_n + b_n) - L_2| < \frac{\epsilon}{2} \qquad \text{for all } n > M_2.$$

Thus, for $n > M = \max\{M_1, M_2\}$,

$$|a_n - L_1| < \frac{\epsilon}{2} \quad \text{and} \quad |(a_n + b_n) - L_2| < \frac{\epsilon}{2}.$$

By the triangle inequality,

$$|b_n - (L_2 - L_1)| = |a_n + b_n - a_n - (L_2 - L_1)| = |(-a_n + L_1) + (a_n + b_n - L_2)|$$

$$\le |L_1 - a_n| + |a_n + b_n - L_2|.$$

Thus, for $n > M$,

$$|b_n - (L_2 - L_1)| < \frac{\epsilon}{2} + \frac{\epsilon}{2} = \epsilon;$$

that is, $\{b_n\}$ converges to $L_2 - L_1$, in contradiction to the given data. Thus, $\{a_n + b_n\}$ must diverge.

79. Theorem 1 states that if $\lim_{x\to\infty} f(x) = L$, then the sequence $a_n = f(n)$ converges and $\lim_{n\to\infty} a_n = L$. Show that the *converse* is false. In other words, find a function $f(x)$ such that $a_n = f(n)$ converges but $\lim_{x\to\infty} f(x)$ does not exist.

SOLUTION   Let $f(x) = \sin \pi x$ and $a_n = \sin \pi n$. Then $a_n = f(n)$. Since $\sin \pi x$ is oscillating between -1 and 1 the limit $\lim_{x\to\infty} f(x)$ does not exist. However, the sequence $\{a_n\}$ is the constant sequence in which $a_n = \sin \pi n = 0$ for all n, hence it converges to zero.

81. Let $b_n = a_{n+1}$. Use the limit definition to prove that if $\{a_n\}$ converges, then $\{b_n\}$ also converges and $\lim_{n\to\infty} a_n = \lim_{n\to\infty} b_n$.

SOLUTION Suppose $\{a_n\}$ converges to L. Let $b_n = a_{n+1}$, and let $\epsilon > 0$. Because $\{a_n\}$ converges to L, there exists an M' such that $|a_n - L| < \epsilon$ for $n > M'$. Now, let $M = M' - 1$. Then, whenever $n > M$, $n + 1 > M + 1 = M'$. Thus, for $n > M$,

$$|b_n - L| = |a_{n+1} - L| < \epsilon.$$

Hence, $\{b_n\}$ converges to L.

83. Proceed as in Example 12 to show that the sequence $\sqrt{3}, \sqrt{3\sqrt{3}}, \sqrt{3\sqrt{3\sqrt{3}}}, \ldots$ is increasing and bounded above by $M = 3$. Then prove that the limit exists and find its value.

SOLUTION This sequence is defined recursively by the formula:

$$a_{n+1} = \sqrt{3a_n}, \qquad a_1 = \sqrt{3}.$$

Consider the following inequalities:

$$a_2 = \sqrt{3a_1} = \sqrt{3\sqrt{3}} > \sqrt{3} = a_1 \quad \Rightarrow \quad a_2 > a_1;$$
$$a_3 = \sqrt{3a_2} > \sqrt{3a_1} = a_2 \quad \Rightarrow \quad a_3 > a_2;$$
$$a_4 = \sqrt{3a_3} > \sqrt{3a_2} = a_3 \quad \Rightarrow \quad a_4 > a_3.$$

In general, if we assume that $a_k > a_{k-1}$, then

$$a_{k+1} = \sqrt{3a_k} > \sqrt{3a_{k-1}} = a_k.$$

Hence, by mathematical induction, $a_{n+1} > a_n$ for all n; that is, the sequence $\{a_n\}$ is increasing.

Because $a_{n+1} = \sqrt{3a_n}$, it follows that $a_n \geq 0$ for all n. Now, $a_1 = \sqrt{3} < 3$. If $a_k \leq 3$, then

$$a_{k+1} = \sqrt{3a_k} \leq \sqrt{3 \cdot 3} = 3.$$

Thus, by mathematical induction, $a_n \leq 3$ for all n.

Since $\{a_n\}$ is increasing and bounded, it follows by the Theorem on Bounded Monotonic Sequences that this sequence is converging. Denote the limit by $L = \lim_{n\to\infty} a_n$. Using Exercise 81, it follows that

$$L = \lim_{n\to\infty} a_{n+1} = \lim_{n\to\infty} \sqrt{3a_n} = \sqrt{3 \lim_{n\to\infty} a_n} = \sqrt{3L}.$$

Thus, $L^2 = 3L$, so $L = 0$ or $L = 3$. Because the sequence is increasing, we have $a_n \geq a_1 = \sqrt{3}$ for all n. Hence, the limit also satisfies $L \geq \sqrt{3}$. We conclude that the appropriate solution is $L = 3$; that is, $\lim_{n\to\infty} a_n = 3$.

Further Insights and Challenges

85. Show that $\lim_{n\to\infty} \sqrt[n]{n!} = \infty$. *Hint:* Verify that $n! \geq (n/2)^{n/2}$ by observing that half of the factors of $n!$ are greater than or equal to $n/2$.

SOLUTION We show that $n! \geq \left(\frac{n}{2}\right)^{n/2}$. For $n \geq 4$ even, we have:

$$n! = \underbrace{1 \cdots \cdots \frac{n}{2}}_{\frac{n}{2} \text{ factors}} \cdot \underbrace{\left(\frac{n}{2} + 1\right) \cdots \cdots n}_{\frac{n}{2} \text{ factors}} \geq \underbrace{\left(\frac{n}{2} + 1\right) \cdots \cdots n}_{\frac{n}{2} \text{ factors}}.$$

Since each one of the $\frac{n}{2}$ factors is greater than $\frac{n}{2}$, we have:

$$n! \geq \underbrace{\left(\frac{n}{2} + 1\right) \cdots \cdots n}_{\frac{n}{2} \text{ factors}} \geq \underbrace{\frac{n}{2} \cdots \cdots \frac{n}{2}}_{\frac{n}{2} \text{ factors}} = \left(\frac{n}{2}\right)^{n/2}.$$

For $n \geq 3$ odd, we have:

$$n! = \underbrace{1 \cdots \cdots \frac{n-1}{2}}_{\frac{n-1}{2} \text{ factors}} \cdot \underbrace{\frac{n+1}{2} \cdots \cdots n}_{\frac{n+1}{2} \text{ factors}} \geq \underbrace{\frac{n+1}{2} \cdots \cdots n}_{\frac{n+1}{2} \text{ factors}}.$$

Since each one of the $\frac{n+1}{2}$ factors is greater than $\frac{n}{2}$, we have:

$$n! \geq \underbrace{\frac{n+1}{2} \cdot \ldots \cdot n}_{\frac{n+1}{2} \text{ factors}} \geq \underbrace{\frac{n}{2} \cdot \ldots \cdot \frac{n}{2}}_{\frac{n+1}{2} \text{ factors}} = \left(\frac{n}{2}\right)^{(n+1)/2} = \left(\frac{n}{2}\right)^{n/2} \sqrt{\frac{n}{2}} \geq \left(\frac{n}{2}\right)^{n/2}.$$

In either case we have $n! \geq \left(\frac{n}{2}\right)^{n/2}$. Thus,

$$\sqrt[n]{n!} \geq \sqrt{\frac{n}{2}}.$$

Since $\lim_{n\to\infty} \sqrt{\frac{n}{2}} = \infty$, it follows that $\lim_{n\to\infty} \sqrt[n]{n!} = \infty$. Thus, the sequence $a_n = \sqrt[n]{n!}$ diverges.

87. Given positive numbers $a_1 < b_1$, define two sequences recursively by

$$a_{n+1} = \sqrt{a_n b_n}, \qquad b_{n+1} = \frac{a_n + b_n}{2}$$

(a) Show that $a_n \leq b_n$ for all n (Figure 13).
(b) Show that $\{a_n\}$ is increasing and $\{b_n\}$ is decreasing.
(c) Show that $b_{n+1} - a_{n+1} \leq \dfrac{b_n - a_n}{2}$.
(d) Prove that both $\{a_n\}$ and $\{b_n\}$ converge and have the same limit. This limit, denoted $\mathrm{AGM}(a_1, b_1)$, is called the **arithmetic-geometric mean** of a_1 and b_1.
(e) Estimate $\mathrm{AGM}(1, \sqrt{2})$ to three decimal places.

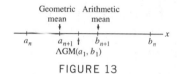

FIGURE 13

SOLUTION
(a) Examine the following:

$$b_{n+1} - a_{n+1} = \frac{a_n + b_n}{2} - \sqrt{a_n b_n} = \frac{a_n + b_n - 2\sqrt{a_n b_n}}{2} = \frac{\left(\sqrt{a_n}\right)^2 - 2\sqrt{a_n}\sqrt{b_n} + \left(\sqrt{b_n}\right)^2}{2}$$

$$= \frac{\left(\sqrt{a_n} - \sqrt{b_n}\right)^2}{2} \geq 0.$$

We conclude that $b_{n+1} \geq a_{n+1}$ for all $n > 1$. By the given information $b_1 > a_1$; hence, $b_n \geq a_n$ for all n.
(b) By part (a), $b_n \geq a_n$ for all n, so

$$a_{n+1} = \sqrt{a_n b_n} \geq \sqrt{a_n \cdot a_n} = \sqrt{a_n^2} = a_n$$

for all n. Hence, the sequence $\{a_n\}$ is increasing. Moreover, since $a_n \leq b_n$ for all n,

$$b_{n+1} = \frac{a_n + b_n}{2} \leq \frac{b_n + b_n}{2} = \frac{2b_n}{2} = b_n$$

for all n; that is, the sequence $\{b_n\}$ is decreasing.
(c) Since $\{a_n\}$ is increasing, $a_{n+1} \geq a_n$. Thus,

$$b_{n+1} - a_{n+1} \leq b_{n+1} - a_n = \frac{a_n + b_n}{2} - a_n = \frac{a_n + b_n - 2a_n}{2} = \frac{b_n - a_n}{2}.$$

Now, by part (a), $a_n \leq b_n$ for all n. By part (b), $\{b_n\}$ is decreasing. Hence $b_n \leq b_1$ for all n. Combining the two inequalities we conclude that $a_n \leq b_1$ for all n. That is, the sequence $\{a_n\}$ is increasing and bounded ($0 \leq a_n \leq b_1$). By the Theorem on Bounded Monotonic Sequences we conclude that $\{a_n\}$ converges. Similarly, since $\{a_n\}$ is increasing, $a_n \geq a_1$ for all n. We combine this inequality with $b_n \geq a_n$ to conclude that $b_n \geq a_1$ for all n. Thus, $\{b_n\}$ is decreasing and bounded ($a_1 \leq b_n \leq b_1$); hence this sequence converges.
   To show that $\{a_n\}$ and $\{b_n\}$ converge to the same limit, note that

$$b_n - a_n \leq \frac{b_{n-1} - a_{n-1}}{2} \leq \frac{b_{n-2} - a_{n-2}}{2^2} \leq \cdots \leq \frac{b_1 - a_1}{2^{n-1}}.$$

Thus,

$$\lim_{n\to\infty}(b_n - a_n) = (b_1 - a_1)\lim_{n\to\infty}\frac{1}{2^{n-1}} = 0.$$

(d) We have

$$a_{n+1} = \sqrt{a_n b_n}, \quad a_1 = 1; \quad b_{n+1} = \frac{a_n + b_n}{2}, \quad b_1 = \sqrt{2}$$

Computing the values of a_n and b_n until the first three decimal digits are equal in successive terms, we obtain:

$$a_2 = \sqrt{a_1 b_1} = \sqrt{1 \cdot \sqrt{2}} = 1.1892$$

$$b_2 = \frac{a_1 + b_1}{2} = \frac{1 + \sqrt{2}}{2} = 1.2071$$

$$a_3 = \sqrt{a_2 b_2} = \sqrt{1.1892 \cdot 1.2071} = 1.1981$$

$$b_3 = \frac{a_2 + b_2}{2} = \frac{1.1892 \cdot 1.2071}{2} = 1.1981$$

$$a_4 = \sqrt{a_3 b_3} = 1.1981$$

$$b_4 = \frac{a_3 + b_3}{2} = 1.1981$$

Thus,

$$AGM\left(1, \sqrt{2}\right) \approx 1.198.$$

89. Let $a_n = H_n - \ln n$, where H_n is the nth harmonic number

$$H_n = 1 + \frac{1}{2} + \frac{1}{3} + \cdots + \frac{1}{n}$$

(a) Show that $a_n \geq 0$ for $n \geq 1$. *Hint:* Show that $H_n \geq \int_1^{n+1} \frac{dx}{x}$.

(b) Show that $\{a_n\}$ is decreasing by interpreting $a_n - a_{n+1}$ as an area.

(c) Prove that $\lim_{n \to \infty} a_n$ exists.

This limit, denoted γ, is known as *Euler's Constant*. It appears in many areas of mathematics, including analysis and number theory, and has been calculated to more than 100 million decimal places, but it is still not known whether γ is an irrational number. The first 10 digits are $\gamma \approx 0.5772156649$.

SOLUTION

(a) Since the function $y = \frac{1}{x}$ is decreasing, the left endpoint approximation to the integral $\int_1^{n+1} \frac{dx}{x}$ is greater than this integral; that is,

$$1 \cdot 1 + \frac{1}{2} \cdot 1 + \frac{1}{3} \cdot 1 + \cdots + \frac{1}{n} \cdot 1 \geq \int_1^{n+1} \frac{dx}{x}$$

or

$$H_n \geq \int_1^{n+1} \frac{dx}{x}.$$

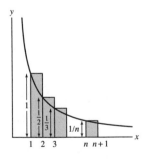

Moreover, since the function $y = \frac{1}{x}$ is positive for $x > 0$, we have:

$$\int_1^{n+1} \frac{dx}{x} \geq \int_1^n \frac{dx}{x}.$$

Thus,

$$H_n \geq \int_1^n \frac{dx}{x} = \ln x \Big|_1^n = \ln n - \ln 1 = \ln n,$$

and

$$a_n = H_n - \ln n \geq 0 \qquad \text{for all } n \geq 1.$$

(b) To show that $\{a_n\}$ is decreasing, we consider the difference $a_n - a_{n+1}$:

$$a_n - a_{n+1} = H_n - \ln n - \left(H_{n+1} - \ln(n+1)\right) = H_n - H_{n+1} + \ln(n+1) - \ln n$$

$$= 1 + \frac{1}{2} + \cdots + \frac{1}{n} - \left(1 + \frac{1}{2} + \cdots + \frac{1}{n} + \frac{1}{n+1}\right) + \ln(n+1) - \ln n$$

$$= -\frac{1}{n+1} + \ln(n+1) - \ln n.$$

Now, $\ln(n+1) - \ln n = \int_n^{n+1} \frac{dx}{x}$, whereas $\frac{1}{n+1}$ is the right endpoint approximation to the integral $\int_n^{n+1} \frac{dx}{x}$. Recalling $y = \frac{1}{x}$ is decreasing, it follows that

$$\int_n^{n+1} \frac{dx}{x} \geq \frac{1}{n+1}$$

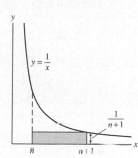

so

$$a_n - a_{n+1} \geq 0.$$

(c) By parts (a) and (b), $\{a_n\}$ is decreasing and 0 is a lower bound for this sequence. Hence $0 \leq a_n \leq a_1$ for all n. A monotonic and bounded sequence is convergent, so $\lim_{n \to \infty} a_n$ exists.

10.2 Summing an Infinite Series (LT Section 11.2)

Preliminary Questions

1. What role do partial sums play in defining the sum of an infinite series?

SOLUTION The sum of an infinite series is defined as the limit of the sequence of partial sums. If the limit of this sequence does not exist, the series is said to diverge.

2. What is the sum of the following infinite series?

$$\frac{1}{4} + \frac{1}{8} + \frac{1}{16} + \frac{1}{32} + \frac{1}{64} + \cdots$$

SOLUTION This is a geometric series with $c = \frac{1}{4}$ and $r = \frac{1}{2}$. The sum of the series is therefore

$$\frac{\frac{1}{4}}{1 - \frac{1}{2}} = \frac{\frac{1}{4}}{\frac{1}{2}} = \frac{1}{2}.$$

3. What happens if you apply the formula for the sum of a geometric series to the following series? Is the formula valid?

$$1 + 3 + 3^2 + 3^3 + 3^4 + \cdots$$

SOLUTION This is a geometric series with $c = 1$ and $r = 3$. Applying the formula for the sum of a geometric series then gives

$$\sum_{n=0}^{\infty} 3^n = \frac{1}{1-3} = -\frac{1}{2}.$$

Clearly, this is not valid: a series with all positive terms cannot have a negative sum. The formula is not valid in this case because a geometric series with $r = 3$ diverges.

4. Arvind asserts that $\sum_{n=1}^{\infty} \frac{1}{n^2} = 0$ because $\frac{1}{n^2}$ tends to zero. Is this valid reasoning?

SOLUTION Arvind's reasoning is not valid. Though the terms in the series do tend to zero, the general term in the sequence of partial sums,

$$S_n = 1 + \frac{1}{2^2} + \frac{1}{3^2} + \cdots + \frac{1}{n^2},$$

is clearly larger than 1. The sum of the series therefore cannot be zero.

5. Colleen claims that $\sum_{n=1}^{\infty} \frac{1}{\sqrt{n}}$ converges because

$$\lim_{n \to \infty} \frac{1}{\sqrt{n}} = 0$$

Is this valid reasoning?

SOLUTION Colleen's reasoning is not valid. Although the general term of a convergent series must tend to zero, a series whose general term tends to zero need not converge. In the case of $\sum_{n=1}^{\infty} \frac{1}{\sqrt{n}}$, the series diverges even though its general term tends to zero.

6. Find an N such that $S_N > 25$ for the series $\sum_{n=1}^{\infty} 2$.

SOLUTION The Nth partial sum of the series is:

$$S_N = \sum_{n=1}^{N} 2 = \underbrace{2 + \cdots + 2}_{N} = 2N.$$

7. Does there exist an N such that $S_N > 25$ for the series $\sum_{n=1}^{\infty} 2^{-n}$? Explain.

SOLUTION The series $\sum_{n=1}^{\infty} 2^{-n}$ is a convergent geometric series with the common ratio $r = \frac{1}{2}$. The sum of the series is:

$$S = \frac{\frac{1}{2}}{1 - \frac{1}{2}} = 1.$$

Notice that the sequence of partial sums $\{S_N\}$ is increasing and converges to 1; therefore $S_N \leq 1$ for all N. Thus, there does not exist an N such that $S_N > 25$.

8. Give an example of a divergent infinite series whose general term tends to zero.

SOLUTION Consider the series $\sum_{n=1}^{\infty} \frac{1}{n^{\frac{9}{10}}}$. The general term tends to zero, since $\lim_{n \to \infty} \frac{1}{n^{\frac{9}{10}}} = 0$. However, the Nth partial sum satisfies the following inequality:

$$S_N = \frac{1}{1^{\frac{9}{10}}} + \frac{1}{2^{\frac{9}{10}}} + \cdots + \frac{1}{N^{\frac{9}{10}}} \geq \frac{N}{N^{\frac{9}{10}}} = N^{1 - \frac{9}{10}} = N^{\frac{1}{10}}.$$

That is, $S_N \geq N^{\frac{1}{10}}$ for all N. Since $\lim_{N \to \infty} N^{\frac{1}{10}} = \infty$, the sequence of partial sums S_n diverges; hence, the series $\sum_{n=1}^{\infty} \frac{1}{n^{\frac{9}{10}}}$ diverges.

Exercises

1. Find a formula for the general term a_n (not the partial sum) of the infinite series.

(a) $\frac{1}{3} + \frac{1}{9} + \frac{1}{27} + \frac{1}{81} + \cdots$

(b) $\frac{1}{1} + \frac{5}{2} + \frac{25}{4} + \frac{125}{8} + \cdots$

(c) $\frac{1}{1} - \frac{2^2}{2 \cdot 1} + \frac{3^3}{3 \cdot 2 \cdot 1} - \frac{4^4}{4 \cdot 3 \cdot 2 \cdot 1} + \cdots$

(d) $\frac{2}{1^2 + 1} + \frac{1}{2^2 + 1} + \frac{2}{3^2 + 1} + \frac{1}{4^2 + 1} + \cdots$

SOLUTION

(a) The denominators of the terms are powers of 3, starting with the first power. Hence, the general term is:

$$a_n = \frac{1}{3^n}.$$

(b) The numerators are powers of 5, and the denominators are the same powers of 2. The first term is $a_1 = 1$ so,

$$a_n = \left(\frac{5}{2}\right)^{n-1}.$$

(c) The general term of this series is,

$$a_n = (-1)^{n+1}\frac{n^n}{n!}.$$

(d) Notice that the numerators of a_n equal 2 for odd values of n and 1 for even values of n. Thus,

$$a_n = \begin{cases} \dfrac{2}{n^2 + 1} & \text{odd } n \\[2mm] \dfrac{1}{n^2 + 1} & \text{even } n \end{cases}$$

The formula can also be rewritten as follows:

$$a_n = \frac{1 + \frac{(-1)^{n+1}+1}{2}}{n^2 + 1}.$$

In Exercises 3–6, compute the partial sums S_2, S_4, and S_6.

3. $1 + \dfrac{1}{2^2} + \dfrac{1}{3^2} + \dfrac{1}{4^2} + \cdots$

SOLUTION

$$S_2 = 1 + \frac{1}{2^2} = \frac{5}{4};$$

$$S_4 = 1 + \frac{1}{2^2} + \frac{1}{3^2} + \frac{1}{4^2} = \frac{205}{144};$$

$$S_6 = 1 + \frac{1}{2^2} + \frac{1}{3^2} + \frac{1}{4^2} + \frac{1}{5^2} + \frac{1}{6^2} = \frac{5369}{3600}.$$

5. $\dfrac{1}{1 \cdot 2} + \dfrac{1}{2 \cdot 3} + \dfrac{1}{3 \cdot 4} + \cdots$

SOLUTION

$$S_2 = \frac{1}{1 \cdot 2} + \frac{1}{2 \cdot 3} = \frac{1}{2} + \frac{1}{6} = \frac{4}{6} = \frac{2}{3};$$

$$S_4 = S_2 + a_3 + a_4 = \frac{2}{3} + \frac{1}{3 \cdot 4} + \frac{1}{4 \cdot 5} = \frac{2}{3} + \frac{1}{12} + \frac{1}{20} = \frac{4}{5};$$

$$S_6 = S_4 + a_5 + a_6 = \frac{4}{5} + \frac{1}{5 \cdot 6} + \frac{1}{6 \cdot 7} = \frac{4}{5} + \frac{1}{30} + \frac{1}{42} = \frac{6}{7}.$$

7. The series $S = 1 + \left(\frac{1}{5}\right) + \left(\frac{1}{5}\right)^2 + \left(\frac{1}{5}\right)^3 + \cdots$ converges to $\frac{5}{4}$. Calculate S_N for $N = 1, 2, \ldots$ until you find an S_N that approximates $\frac{5}{4}$ with an error less than 0.0001.

SOLUTION

$$S_1 = 1$$

$$S_2 = 1 + \frac{1}{5} = \frac{6}{5} = 1.2$$

$$S_3 = 1 + \frac{1}{5} + \frac{1}{25} = \frac{31}{25} = 1.24$$

$$S_3 = 1 + \frac{1}{5} + \frac{1}{25} + \frac{1}{125} = \frac{156}{125} = 1.248$$

$$S_4 = 1 + \frac{1}{5} + \frac{1}{25} + \frac{1}{125} + \frac{1}{625} = \frac{781}{625} = 1.2496$$

$$S_5 = 1 + \frac{1}{5} + \frac{1}{25} + \frac{1}{125} + \frac{1}{625} + \frac{1}{3125} = \frac{3906}{3125} = 1.24992$$

Note that

$$1.25 - S_5 = 1.25 - 1.24992 = 0.00008 < 0.0001$$

In Exercises 9 and 10, use a computer algebra system to compute S_{10}, S_{100}, S_{500}, and S_{1000} for the series. Do these values suggest convergence to the given value?

9. *CAS*

$$\frac{\pi - 3}{4} = \frac{1}{2 \cdot 3 \cdot 4} - \frac{1}{4 \cdot 5 \cdot 6} + \frac{1}{6 \cdot 7 \cdot 8} - \frac{1}{8 \cdot 9 \cdot 10} + \cdots$$

SOLUTION Write

$$a_n = \frac{(-1)^{n+1}}{2n \cdot (2n + 1) \cdot (2n + 2)}$$

Then

$$S_N = \sum_{i=1}^{N} a_n$$

Computing, we find

$$\frac{\pi - 3}{4} \approx 0.0353981635$$

$$S_{10} \approx 0.03535167962$$

$$S_{100} \approx 0.03539810274$$

$$S_{500} \approx 0.03539816290$$

$$S_{1000} \approx 0.03539816334$$

It appears that $S_N \to \frac{\pi - 3}{4}$.

11. Calculate S_3, S_4, and S_5 and then find the sum of the telescoping series

$$S = \sum_{n=1}^{\infty} \left(\frac{1}{n + 1} - \frac{1}{n + 2} \right)$$

SOLUTION

$$S_3 = \left(\frac{1}{2} - \frac{1}{3} \right) + \left(\frac{1}{3} - \frac{1}{4} \right) + \left(\frac{1}{4} - \frac{1}{5} \right) = \frac{1}{2} - \frac{1}{5} = \frac{3}{10};$$

$$S_4 = S_3 + \left(\frac{1}{5} - \frac{1}{6} \right) = \frac{1}{2} - \frac{1}{6} = \frac{1}{3};$$

$$S_5 = S_4 + \left(\frac{1}{6} - \frac{1}{7} \right) = \frac{1}{2} - \frac{1}{7} = \frac{5}{14}.$$

The general term in the sequence of partial sums is

$$S_N = \left(\frac{1}{2} - \frac{1}{3} \right) + \left(\frac{1}{3} - \frac{1}{4} \right) + \left(\frac{1}{4} - \frac{1}{5} \right) + \cdots + \left(\frac{1}{N + 1} - \frac{1}{N + 2} \right) = \frac{1}{2} - \frac{1}{N + 2};$$

thus,

$$S = \lim_{N \to \infty} S_N = \lim_{N \to \infty} \left(\frac{1}{2} - \frac{1}{N + 2} \right) = \frac{1}{2}.$$

The sum of the telescoping series is therefore $\frac{1}{2}$.

13. Calculate S_3, S_4, and S_5 and then find the sum $S = \sum_{n=1}^{\infty} \frac{1}{4n^2 - 1}$ using the identity

$$\frac{1}{4n^2 - 1} = \frac{1}{2} \left(\frac{1}{2n - 1} - \frac{1}{2n + 1} \right)$$

SOLUTION

$$S_3 = \frac{1}{2} \left(\frac{1}{1} - \frac{1}{3} \right) + \frac{1}{2} \left(\frac{1}{3} - \frac{1}{5} \right) + \frac{1}{2} \left(\frac{1}{5} - \frac{1}{7} \right) = \frac{1}{2} \left(1 - \frac{1}{7} \right) = \frac{3}{7};$$

$$S_4 = S_3 + \frac{1}{2} \left(\frac{1}{7} - \frac{1}{9} \right) = \frac{1}{2} \left(1 - \frac{1}{9} \right) = \frac{4}{9};$$

$$S_5 = S_4 + \frac{1}{2} \left(\frac{1}{9} - \frac{1}{11} \right) = \frac{1}{2} \left(1 - \frac{1}{11} \right) = \frac{5}{11}.$$

The general term in the sequence of partial sums is

$$S_N = \frac{1}{2}\left(\frac{1}{1} - \frac{1}{3}\right) + \frac{1}{2}\left(\frac{1}{3} - \frac{1}{5}\right) + \frac{1}{2}\left(\frac{1}{5} - \frac{1}{7}\right) + \cdots + \frac{1}{2}\left(\frac{1}{2N-1} - \frac{1}{2N+1}\right) = \frac{1}{2}\left(1 - \frac{1}{2N+1}\right);$$

thus,

$$S = \lim_{N\to\infty} S_N = \lim_{N\to\infty} \frac{1}{2}\left(1 - \frac{1}{2N+1}\right) = \frac{1}{2}.$$

15. Find the sum of $\dfrac{1}{1\cdot 3} + \dfrac{1}{3\cdot 5} + \dfrac{1}{5\cdot 7} + \cdots$.

SOLUTION We may write this sum as

$$\sum_{n=1}^{\infty} \frac{1}{(2n-1)(2n+1)} = \sum_{n=1}^{\infty} \frac{1}{2}\left(\frac{1}{2n-1} - \frac{1}{2n+1}\right).$$

The general term in the sequence of partial sums is

$$S_N = \frac{1}{2}\left(\frac{1}{1} - \frac{1}{3}\right) + \frac{1}{2}\left(\frac{1}{3} - \frac{1}{5}\right) + \frac{1}{2}\left(\frac{1}{5} - \frac{1}{7}\right) + \cdots + \frac{1}{2}\left(\frac{1}{2N-1} - \frac{1}{2N+1}\right) = \frac{1}{2}\left(1 - \frac{1}{2N+1}\right);$$

thus,

$$\lim_{N\to\infty} S_N = \lim_{N\to\infty} \frac{1}{2}\left(1 - \frac{1}{2N+1}\right) = \frac{1}{2},$$

and

$$\sum_{n=1}^{\infty} \frac{1}{(2n-1)(2n+1)} = \frac{1}{2}.$$

In Exercises 17–22, use Theorem 3 to prove that the following series diverge.

17. $\displaystyle\sum_{n=1}^{\infty} \frac{n}{10n+12}$

SOLUTION The general term, $\dfrac{n}{10n+12}$, has limit

$$\lim_{n\to\infty} \frac{n}{10n+12} = \lim_{n\to\infty} \frac{1}{10+(12/n)} = \frac{1}{10}$$

Since the general term does not tend to zero, the series diverges.

19. $\dfrac{0}{1} - \dfrac{1}{2} + \dfrac{2}{3} - \dfrac{3}{4} + \cdots$

SOLUTION The general term $a_n = (-1)^{n-1}\frac{n-1}{n}$ does not tend to zero. In fact, because $\lim_{n\to\infty} \frac{n-1}{n} = 1$, $\lim_{n\to\infty} a_n$ does not exist. By Theorem 3, we conclude that the given series diverges.

21. $\cos\dfrac{1}{2} + \cos\dfrac{1}{3} + \cos\dfrac{1}{4} + \cdots$

SOLUTION The general term $a_n = \cos\frac{1}{n+1}$ tends to 1, not zero. By Theorem 3, we conclude that the given series diverges.

In Exercises 23–36, use the formula for the sum of a geometric series to find the sum or state that the series diverges.

23. $\dfrac{1}{1} + \dfrac{1}{8} + \dfrac{1}{8^2} + \cdots$

SOLUTION This is a geometric series with $c = 1$ and $r = \frac{1}{8}$, so its sum is

$$\frac{1}{1 - \frac{1}{8}} = \frac{1}{7/8} = \frac{8}{7}$$

25. $\displaystyle\sum_{n=3}^{\infty} \left(\frac{3}{11}\right)^{-n}$

SOLUTION Rewrite this series as

$$\sum_{n=3}^{\infty} \left(\frac{11}{3}\right)^{n}$$

This is a geometric series with $r = \dfrac{11}{3} > 1$, so it is divergent.

27. $\displaystyle\sum_{n=-4}^{\infty}\left(-\frac{4}{9}\right)^{n}$

SOLUTION   This is a geometric series with $c = 1$ and $r = -\dfrac{4}{9}$, starting at $n = -4$. Its sum is thus

$$\frac{cr^{-4}}{1-r} = \frac{c}{r^4 - r^5} = \frac{1}{\frac{4^4}{9^4} + \frac{4^5}{9^5}} = \frac{9^5}{9 \cdot 4^4 + 4^5} = \frac{59{,}049}{3328}$$

29. $\displaystyle\sum_{n=1}^{\infty} e^{-n}$

SOLUTION   Rewrite the series as

$$\sum_{n=1}^{\infty}\left(\frac{1}{e}\right)^{n}$$

to recognize it as a geometric series with $c = \frac{1}{e}$ and $r = \frac{1}{e}$. Thus,

$$\sum_{n=1}^{\infty} e^{-n} = \frac{\frac{1}{e}}{1 - \frac{1}{e}} = \frac{1}{e - 1}.$$

31. $\displaystyle\sum_{n=0}^{\infty} \frac{8 + 2^n}{5^n}$

SOLUTION   Rewrite the series as

$$\sum_{n=0}^{\infty} \frac{8}{5^n} + \sum_{n=0}^{\infty} \frac{2^n}{5^n} = \sum_{n=0}^{\infty} 8 \cdot \left(\frac{1}{5}\right)^{n} + \sum_{n=0}^{\infty}\left(\frac{2}{5}\right)^{n},$$

which is a sum of two geometric series. The first series has $c = 8\left(\frac{1}{5}\right)^{0} = 8$ and $r = \frac{1}{5}$; the second has $c = \left(\frac{2}{5}\right)^{0} = 1$ and $r = \frac{2}{5}$. Thus,

$$\sum_{n=0}^{\infty} 8 \cdot \left(\frac{1}{5}\right)^{n} = \frac{8}{1 - \frac{1}{5}} = \frac{8}{\frac{4}{5}} = 10,$$

$$\sum_{n=0}^{\infty}\left(\frac{2}{5}\right)^{n} = \frac{1}{1 - \frac{2}{5}} = \frac{1}{\frac{3}{5}} = \frac{5}{3},$$

and

$$\sum_{n=0}^{\infty} \frac{8 + 2^n}{5^n} = 10 + \frac{5}{3} = \frac{35}{3}.$$

33. $5 - \dfrac{5}{4} + \dfrac{5}{4^2} - \dfrac{5}{4^3} + \cdots$

SOLUTION   This is a geometric series with $c = 5$ and $r = -\frac{1}{4}$. Thus,

$$\sum_{n=0}^{\infty} 5 \cdot \left(-\frac{1}{4}\right)^{n} = \frac{5}{1 - \left(-\frac{1}{4}\right)} = \frac{5}{1 + \frac{1}{4}} = \frac{5}{\frac{5}{4}} = 4.$$

35. $\dfrac{7}{8} - \dfrac{49}{64} + \dfrac{343}{512} - \dfrac{2401}{4096} + \cdots$

SOLUTION   This is a geometric series with $c = \frac{7}{8}$ and $r = -\frac{7}{8}$. Thus,

$$\sum_{n=0}^{\infty} \frac{7}{8} \cdot \left(-\frac{7}{8}\right)^{n} = \frac{\frac{7}{8}}{1 - \left(-\frac{7}{8}\right)} = \frac{\frac{7}{8}}{\frac{15}{8}} = \frac{7}{15}.$$

37. Which of the following are *not* geometric series?

(a) $\sum_{n=0}^{\infty} \dfrac{7^n}{29^n}$

(b) $\sum_{n=3}^{\infty} \dfrac{1}{n^4}$

(c) $\sum_{n=0}^{\infty} \dfrac{n^2}{2^n}$

(d) $\sum_{n=5}^{\infty} \pi^{-n}$

SOLUTION

(a) $\sum_{n=0}^{\infty} \dfrac{7^n}{29^n} = \sum_{n=0}^{\infty} \left(\dfrac{7}{29}\right)^n$: this is a geometric series with common ratio $r = \dfrac{7}{29}$.

(b) The ratio between two successive terms is

$$\frac{a_{n+1}}{a_n} = \frac{\frac{1}{(n+1)^4}}{\frac{1}{n^4}} = \frac{n^4}{(n+1)^4} = \left(\frac{n}{n+1}\right)^4.$$

This ratio is not constant since it depends on n. Hence, the series $\sum_{n=3}^{\infty} \dfrac{1}{n^4}$ is not a geometric series.

(c) The ratio between two successive terms is

$$\frac{a_{n+1}}{a_n} = \frac{\frac{(n+1)^2}{2^{n+1}}}{\frac{n^2}{2^n}} = \frac{(n+1)^2}{n^2} \cdot \frac{2^n}{2^{n+1}} = \left(1+\frac{1}{n}\right)^2 \cdot \frac{1}{2}.$$

This ratio is not constant since it depends on n. Hence, the series $\sum_{n=0}^{\infty} \dfrac{n^2}{2^n}$ is not a geometric series.

(d) $\sum_{n=5}^{\infty} \pi^{-n} = \sum_{n=5}^{\infty} \left(\dfrac{1}{\pi}\right)^n$: this is a geometric series with common ratio $r = \dfrac{1}{\pi}$.

39. Prove that if $\sum_{n-1}^{\infty} a_n$ converges and $\sum_{n=1}^{\infty} b_n$ diverges, then $\sum_{n=1}^{\infty} (a_n + b_n)$ diverges. *Hint:* If not, derive a contradiction by writing

$$\sum_{n=1}^{\infty} b_n = \sum_{n=1}^{\infty} (a_n + b_n) - \sum_{n=1}^{\infty} a_n$$

SOLUTION Suppose to the contrary that $\sum_{n=1}^{\infty} a_n$ converges, $\sum_{n=1}^{\infty} b_n$ diverges, but $\sum_{n=1}^{\infty} (a_n + b_n)$ converges. Then by the Linearity of Infinite Series, we have

$$\sum_{n=1}^{\infty} b_n = \sum_{n=1}^{\infty} (a_n + b_n) - \sum_{n=1}^{\infty} a_n$$

so that $\sum_{n=1}^{\infty} b_n$ converges, a contradiction.

41. Give a counterexample to show that each of the following statements is false.

(a) If the general term a_n tends to zero, then $\sum_{n=1}^{\infty} a_n = 0$.

(b) The Nth partial sum of the infinite series defined by $\{a_n\}$ is a_N.

(c) If a_n tends to zero, then $\sum_{n=1}^{\infty} a_n$ converges.

(d) If a_n tends to L, then $\sum_{n=1}^{\infty} a_n = L$.

SOLUTION

(a) Let $a_n = 2^{-n}$. Then $\lim_{n \to \infty} a_n = 0$, but a_n is a geometric series with $c = 2^0 = 1$ and $r = 1/2$, so its sum is $\dfrac{1}{1 - (1/2)} = 2$.

(b) Let $a_n = 1$. Then the nth partial sum is $a_1 + a_2 + \cdots + a_n = n$ while $a_n = 1$.

(c) Let $a_n = \dfrac{1}{\sqrt{n}}$. An example in the text shows that while a_n tends to zero, the sum $\sum_{n=1}^{\infty} a_n$ does not converge.

(d) Let $a_n = 1$. Then clearly a_n tends to $L = 1$, while the series $\sum_{n=1}^{\infty} a_n$ obviously diverges.

43. Compute the total area of the (infinitely many) triangles in Figure 4.

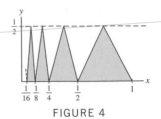

FIGURE 4

SOLUTION The area of a triangle with base B and height H is $A = \frac{1}{2}BH$. Because all of the triangles in Figure 4 have height $\frac{1}{2}$, the area of each triangle equals one-quarter of the base. Now, for $n \geq 0$, the nth triangle has a base which extends from $x = \frac{1}{2^{n+1}}$ to $x = \frac{1}{2^n}$. Thus,

$$B = \frac{1}{2^n} - \frac{1}{2^{n+1}} = \frac{1}{2^{n+1}} \quad \text{and} \quad A = \frac{1}{4}B = \frac{1}{2^{n+3}}.$$

The total area of the triangles is then given by the geometric series

$$\sum_{n=0}^{\infty} \frac{1}{2^{n+3}} = \sum_{n=0}^{\infty} \frac{1}{8}\left(\frac{1}{2}\right)^n = \frac{\frac{1}{8}}{1 - \frac{1}{2}} = \frac{1}{4}.$$

45. Find the total length of the infinite zigzag path in Figure 5 (each zag occurs at an angle of $\frac{\pi}{4}$).

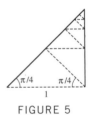

FIGURE 5

SOLUTION Because the angle at the lower left in Figure 5 has measure $\frac{\pi}{4}$ and each zag in the path occurs at an angle of $\frac{\pi}{4}$, every triangle in the figure is an isosceles right triangle. Accordingly, the length of each new segment in the path is $\frac{1}{\sqrt{2}}$ times the length of the previous segment. Since the first segment has length 1, the total length of the path is

$$\sum_{n=0}^{\infty} \left(\frac{1}{\sqrt{2}}\right)^n = \frac{1}{1 - \frac{1}{\sqrt{2}}} = \frac{\sqrt{2}}{\sqrt{2} - 1} = 2 + \sqrt{2}.$$

47. Show that if a is a positive integer, then

$$\sum_{n=1}^{\infty} \frac{1}{n(n+a)} = \frac{1}{a}\left(1 + \frac{1}{2} + \cdots + \frac{1}{a}\right).$$

SOLUTION By partial fraction decomposition

$$\frac{1}{n(n+a)} = \frac{A}{n} + \frac{B}{n+a};$$

clearing the denominators gives

$$1 = A(n+a) + Bn.$$

Setting $n = 0$ then yields $A = \frac{1}{a}$, while setting $n = -a$ yields $B = -\frac{1}{a}$. Thus,

$$\frac{1}{n(n+a)} = \frac{\frac{1}{a}}{n} - \frac{\frac{1}{a}}{n+a} = \frac{1}{a}\left(\frac{1}{n} - \frac{1}{n+a}\right),$$

and

$$\sum_{n=1}^{\infty} \frac{1}{n(n+a)} = \sum_{n=1}^{\infty} \frac{1}{a}\left(\frac{1}{n} - \frac{1}{n+a}\right).$$

For $N > a$, the Nth partial sum is

$$S_N = \frac{1}{a}\left(1 + \frac{1}{2} + \frac{1}{3} + \cdots + \frac{1}{a}\right) - \frac{1}{a}\left(\frac{1}{N+1} + \frac{1}{N+2} + \frac{1}{N+3} + \cdots + \frac{1}{N+a}\right).$$

Thus,

$$\sum_{n=1}^{\infty} \frac{1}{n(n+a)} = \lim_{N\to\infty} S_N = \frac{1}{a}\left(1 + \frac{1}{2} + \frac{1}{3} + \cdots + \frac{1}{a}\right).$$

49. Let $\{b_n\}$ be a sequence and let $a_n = b_n - b_{n-1}$. Show that $\displaystyle\sum_{n=1}^{\infty} a_n$ converges if and only if $\displaystyle\lim_{n\to\infty} b_n$ exists.

SOLUTION Let $a_n = b_n - b_{n-1}$. The general term in the sequence of partial sums for the series $\displaystyle\sum_{n=1}^{\infty} a_n$ is then

$$S_N = (b_1 - b_0) + (b_2 - b_1) + (b_3 - b_2) + \cdots + (b_N - b_{N-1}) = b_N - b_0.$$

Now, if $\displaystyle\lim_{N\to\infty} b_N$ exists, then so does $\displaystyle\lim_{N\to\infty} S_N$ and $\displaystyle\sum_{n=1}^{\infty} a_n$ converges. On the other hand, if $\displaystyle\sum_{n=1}^{\infty} a_n$ converges, then

$\displaystyle\lim_{N\to\infty} S_N$ exists, which implies that $\displaystyle\lim_{N\to\infty} b_N$ also exists. Thus, $\displaystyle\sum_{n=1}^{\infty} a_n$ converges if and only if $\displaystyle\lim_{n\to\infty} b_n$ exists.

Further Insights and Challenges

Exercises 51–53 use the formula

$$1 + r + r^2 + \cdots + r^{N-1} = \frac{1 - r^N}{1 - r} \qquad \boxed{7}$$

51. Professor George Andrews of Pennsylvania State University observed that we can use Eq. (7) to calculate the derivative of $f(x) = x^N$ (for $N \geq 0$). Assume that $a \neq 0$ and let $x = ra$. Show that

$$f'(a) = \lim_{x\to a} \frac{x^N - a^N}{x - a} = a^{N-1} \lim_{r\to 1} \frac{r^N - 1}{r - 1}$$

and evaluate the limit.

SOLUTION According to the definition of derivative of $f(x)$ at $x = a$

$$f'(a) = \lim_{x\to a} \frac{x^N - a^N}{x - a}.$$

Now, let $x = ra$. Then $x \to a$ if and only if $r \to 1$, and

$$f'(a) = \lim_{x\to a} \frac{x^N - a^N}{x - a} = \lim_{r\to 1} \frac{(ra)^N - a^N}{ra - a} = \lim_{r\to 1} \frac{a^N\left(r^N - 1\right)}{a(r - 1)} = a^{N-1} \lim_{r\to 1} \frac{r^N - 1}{r - 1}.$$

By Eq. (7) for a geometric sum,

$$\frac{1 - r^N}{1 - r} = \frac{r^N - 1}{r - 1} = 1 + r + r^2 + \cdots + r^{N-1},$$

so

$$\lim_{r\to 1} \frac{r^N - 1}{r - 1} = \lim_{r\to 1} \left(1 + r + r^2 + \cdots + r^{N-1}\right) = 1 + 1 + 1^2 + \cdots + 1^{N-1} = N.$$

Therefore, $f'(a) = a^{N-1} \cdot N = Na^{N-1}$.

53. Verify the Gregory–Leibniz formula as follows.

(a) Set $r = -x^2$ in Eq. (7) and rearrange to show that

$$\frac{1}{1 + x^2} = 1 - x^2 + x^4 - \cdots + (-1)^{N-1}x^{2N-2} + \frac{(-1)^N x^{2N}}{1 + x^2}$$

(b) Show, by integrating over [0, 1], that

$$\frac{\pi}{4} = 1 - \frac{1}{3} + \frac{1}{5} - \frac{1}{7} + \cdots + \frac{(-1)^{N-1}}{2N-1} + (-1)^N \int_0^1 \frac{x^{2N}\, dx}{1+x^2}$$

(c) Use the Comparison Theorem for integrals to prove that

$$0 \le \int_0^1 \frac{x^{2N}\, dx}{1+x^2} \le \frac{1}{2N+1}$$

Hint: Observe that the integrand is $\le x^{2N}$.

(d) Prove that

$$\frac{\pi}{4} = 1 - \frac{1}{3} + \frac{1}{5} - \frac{1}{7} + \frac{1}{9} - \cdots$$

Hint: Use (b) and (c) to show that the partial sums S_N of satisfy $\left| S_N - \frac{\pi}{4} \right| \le \frac{1}{2N+1}$, and thereby conclude that $\lim_{N \to \infty} S_N = \frac{\pi}{4}$.

SOLUTION

(a) Start with Eq. (7), and substitute $-x^2$ for r:

$$1 + r + r^2 + \cdots + r^{N-1} = \frac{1-r^N}{1-r}$$

$$1 - x^2 + x^4 + \cdots + (-1)^{N-1}x^{2N-2} = \frac{1-(-1)^N x^{2N}}{1-(-x^2)}$$

$$1 - x^2 + x^4 + \cdots + (-1)^{N-1}x^{2N-2} = \frac{1}{1+x^2} - \frac{(-1)^N x^{2N}}{1+x^2}$$

$$\frac{1}{1+x^2} = 1 - x^2 + x^4 + \cdots + (-1)^{N-1}x^{2N-2} + \frac{(-1)^N x^{2N}}{1+x^2}$$

(b) The integrals of both sides must be equal. Now,

$$\int_0^1 \frac{1}{1+x^2}\, dx = \tan^{-1} x \Big|_0^1 = \tan^{-1} 1 - \tan^{-1} 0 = \frac{\pi}{4}$$

while

$$\int_0^1 \left(1 - x^2 + x^4 + \cdots + (-1)^{N-1}x^{2N-2} + \frac{(-1)^N x^{2N}}{1+x^2} \right) dx$$

$$= \left(x - \frac{1}{3}x^3 + \frac{1}{5}x^5 + \cdots + (-1)^{N-1}\frac{1}{2N-1}x^{2N-1} \right) + (-1)^N \int_0^1 \frac{x^{2N}\, dx}{1+x^2}$$

$$= 1 - \frac{1}{3} + \frac{1}{5} + \cdots + (-1)^{N-1}\frac{1}{2N-1} + (-1)^N \int_0^1 \frac{x^{2N}\, dx}{1+x^2}$$

(c) Note that for $x \in [0, 1]$, we have $1 + x^2 \ge 1$, so that

$$0 \le \frac{x^{2N}}{1+x^2} \le x^{2N}$$

By the Comparison Theorem for integrals, we then see that

$$0 \le \int_0^1 \frac{x^{2N}\, dx}{1+x^2} \le \int_0^1 x^{2N}\, dx = \frac{1}{2N+1}x^{2N+1} \Big|_0^1 = \frac{1}{2N+1}$$

(d) Write

$$a_n = (-1)^n \frac{1}{2n-1}, \quad n \ge 1$$

and let S_N be the partial sums. Then

$$\left| S_N - \frac{\pi}{4} \right| = \left| (-1)^N \int_0^1 \frac{x^{2N}\, dx}{1+x^2} \right| = \int_0^1 \frac{x^{2N}\, dx}{1+x^2} \le \frac{1}{2N+1}$$

Thus $\lim_{N \to \infty} S_N = \dfrac{\pi}{4}$ so that

$$\frac{\pi}{4} = 1 - \frac{1}{3} + \frac{1}{5} - \frac{1}{7} + \frac{1}{9} - \cdots$$

55. The **Koch snowflake** (described in 1904 by Swedish mathematician Helge von Koch) is an infinitely jagged "fractal" curve obtained as a limit of polygonal curves (it is continuous but has no tangent line at any point). Begin with an equilateral triangle (stage 0) and produce stage 1 by replacing each edge with four edges of one-third the length, arranged as in Figure 8. Continue the process: At the nth stage, replace each edge with four edges of one-third the length.

(a) Show that the perimeter P_n of the polygon at the nth stage satisfies $P_n = \frac{4}{3} P_{n-1}$. Prove that $\lim_{n \to \infty} P_n = \infty$. The snowflake has infinite length.

(b) Let A_0 be the area of the original equilateral triangle. Show that $(3)4^{n-1}$ new triangles are added at the nth stage, each with area $A_0/9^n$ (for $n \geq 1$). Show that the total area of the Koch snowflake is $\frac{8}{5} A_0$.

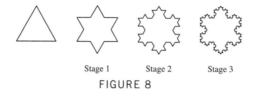

Stage 1   Stage 2   Stage 3

FIGURE 8

SOLUTION

(a) Each edge of the polygon at the $(n-1)$st stage is replaced by four edges of one-third the length; hence the perimeter of the polygon at the nth stage is $\frac{4}{3}$ times the perimeter of the polygon at the $(n-1)$th stage. That is, $P_n = \frac{4}{3} P_{n-1}$. Thus,

$$P_1 = \frac{4}{3} P_0; \quad P_2 = \frac{4}{3} P_1 = \left(\frac{4}{3}\right)^2 P_0, \quad P_3 = \frac{4}{3} P_2 = \left(\frac{4}{3}\right)^3 P_0,$$

and, in general, $P_n = \left(\frac{4}{3}\right)^n P_0$. As $n \to \infty$, it follows that

$$\lim_{n \to \infty} P_n = P_0 \lim_{n \to \infty} \left(\frac{4}{3}\right)^n = \infty.$$

(b) When each edge is replaced by four edges of one-third the length, one new triangle is created. At the $(n-1)$st stage, there are $3 \cdot 4^{n-1}$ edges in the snowflake, so $3 \cdot 4^{n-1}$ new triangles are generated at the nth stage. Because the area of an equilateral triangle is proportional to the square of its side length and the side length for each new triangle is one-third the side length of triangles from the previous stage, it follows that the area of the triangles added at each stage is reduced by a factor of $\frac{1}{9}$ from the area of the triangles added at the previous stage. Thus, each triangle added at the nth stage has an area of $A_0/9^n$. This means that the nth stage contributes

$$3 \cdot 4^{n-1} \cdot \frac{A_0}{9^n} = \frac{3}{4} A_0 \left(\frac{4}{9}\right)^n$$

to the area of the snowflake. The total area is therefore

$$A = A_0 + \frac{3}{4} A_0 \sum_{n=1}^{\infty} \left(\frac{4}{9}\right)^n = A_0 + \frac{3}{4} A_0 \frac{\frac{4}{9}}{1 - \frac{4}{9}} = A_0 + \frac{3}{4} A_0 \cdot \frac{4}{5} = \frac{8}{5} A_0.$$

10.3 Convergence of Series with Positive Terms (LT Section 11.3)

Preliminary Questions

1. Let $S = \sum_{n=1}^{\infty} a_n$. If the partial sums S_N are increasing, then (choose the correct conclusion):

(a) $\{a_n\}$ is an increasing sequence.

(b) $\{a_n\}$ is a positive sequence.

SOLUTION The correct response is **(b)**. Recall that $S_N = a_1 + a_2 + a_3 + \cdots + a_N$; thus, $S_N - S_{N-1} = a_N$. If S_N is increasing, then $S_N - S_{N-1} \geq 0$. It then follows that $a_N \geq 0$; that is, $\{a_n\}$ is a positive sequence.

2. What are the hypotheses of the Integral Test?

SOLUTION The hypotheses for the Integral Test are: A function $f(x)$ such that $a_n = f(n)$ must be positive, decreasing, and continuous for $x \geq 1$.

3. Which test would you use to determine whether $\sum\limits_{n=1}^{\infty} n^{-3.2}$ converges?

SOLUTION Because $n^{-3.2} = \frac{1}{n^{3.2}}$, we see that the indicated series is a p-series with $p = 3.2 > 1$. Therefore, the series converges.

4. Which test would you use to determine whether $\sum\limits_{n=1}^{\infty} \frac{1}{2^n + \sqrt{n}}$ converges?

SOLUTION Because

$$\frac{1}{2^n + \sqrt{n}} < \frac{1}{2^n} = \left(\frac{1}{2}\right)^n,$$

and

$$\sum_{n=1}^{\infty} \left(\frac{1}{2}\right)^n$$

is a convergent geometric series, the comparison test would be an appropriate choice to establish that the given series converges.

5. Ralph hopes to investigate the convergence of $\sum\limits_{n=1}^{\infty} \frac{e^{-n}}{n}$ by comparing it with $\sum\limits_{n=1}^{\infty} \frac{1}{n}$. Is Ralph on the right track?

SOLUTION No, Ralph is not on the right track. For $n \geq 1$,

$$\frac{e^{-n}}{n} < \frac{1}{n};$$

however, $\sum\limits_{n=1}^{\infty} \frac{1}{n}$ is a divergent series. The Comparison Test therefore does not allow us to draw a conclusion about the convergence or divergence of the series $\sum\limits_{n=1}^{\infty} \frac{e^{-n}}{n}$.

Exercises

In Exercises 1–14, use the Integral Test to determine whether the infinite series is convergent.

1. $\sum\limits_{n=1}^{\infty} \frac{1}{n^4}$

SOLUTION Let $f(x) = \frac{1}{x^4}$. This function is continuous, positive and decreasing on the interval $x \geq 1$, so the Integral Test applies. Moreover,

$$\int_1^{\infty} \frac{dx}{x^4} = \lim_{R \to \infty} \int_1^R x^{-4} \, dx = -\frac{1}{3} \lim_{R \to \infty} \left(\frac{1}{R^3} - 1\right) = \frac{1}{3}.$$

The integral converges; hence, the series $\sum\limits_{n=1}^{\infty} \frac{1}{n^4}$ also converges.

3. $\sum\limits_{n=1}^{\infty} n^{-1/3}$

SOLUTION Let $f(x) = x^{-\frac{1}{3}} = \frac{1}{\sqrt[3]{x}}$. This function is continuous, positive and decreasing on the interval $x \geq 1$, so the Integral Test applies. Moreover,

$$\int_1^{\infty} x^{-1/3} \, dx = \lim_{R \to \infty} \int_1^R x^{-1/3} \, dx = \frac{3}{2} \lim_{R \to \infty} \left(R^{2/3} - 1\right) = \infty.$$

The integral diverges; hence, the series $\sum\limits_{n=1}^{\infty} n^{-1/3}$ also diverges.

5. $\displaystyle\sum_{n=25}^{\infty} \frac{n^2}{(n^3+9)^{5/2}}$

SOLUTION   Let $f(x) = \dfrac{x^2}{\left(x^3+9\right)^{5/2}}$. This function is positive and continuous for $x \geq 25$. Moreover, because

$$f'(x) = \frac{2x(x^3+9)^{5/2} - x^2 \cdot \frac{5}{2}(x^3+9)^{3/2} \cdot 3x^2}{(x^3+9)^5} = \frac{x(36-11x^3)}{2(x^3+9)^{7/2}},$$

we see that $f'(x) < 0$ for $x \geq 25$, so f is decreasing on the interval $x \geq 25$. The Integral Test therefore applies. To evaluate the improper integral, we use the substitution $u = x^3+9$, $du = 3x^2 dx$. We then find

$$\int_{25}^{\infty} \frac{x^2}{(x^3+9)^{5/2}}\,dx = \lim_{R\to\infty} \int_{25}^{R} \frac{x^2}{(x^3+9)^{5/2}}\,dx = \frac{1}{3}\lim_{R\to\infty} \int_{15634}^{R^3+9} \frac{du}{u^{5/2}}$$

$$= -\frac{2}{9}\lim_{R\to\infty}\left(\frac{1}{(R^3+9)^{3/2}} - \frac{1}{15634^{3/2}}\right) = \frac{2}{9\cdot 15634^{3/2}}.$$

The integral converges; hence, the series $\displaystyle\sum_{n=25}^{\infty}\frac{n^2}{(n^3+9)^{5/2}}$ also converges.

7. $\displaystyle\sum_{n=1}^{\infty}\frac{1}{n^2+1}$

SOLUTION   Let $f(x) = \dfrac{1}{x^2+1}$. This function is positive, decreasing and continuous on the interval $x \geq 1$, hence the Integral Test applies. Moreover,

$$\int_{1}^{\infty} \frac{dx}{x^2+1} = \lim_{R\to\infty}\int_{1}^{R}\frac{dx}{x^2+1} = \lim_{R\to\infty}\left(\tan^{-1} R - \frac{\pi}{4}\right) - \frac{\pi}{2} - \frac{\pi}{4} = \frac{\pi}{4}.$$

The integral converges; hence, the series $\displaystyle\sum_{n=1}^{\infty}\frac{1}{n^2+1}$ also converges.

9. $\displaystyle\sum_{n=1}^{\infty}\frac{1}{n(n+1)}$

SOLUTION   Let $f(x) = \dfrac{1}{x(x+1)}$. This function is positive, continuous and decreasing on the interval $x \geq 1$, so the Integral Test applies. We compute the improper integral using partial fractions:

$$\int_{1}^{\infty}\frac{dx}{x(x+1)} = \lim_{R\to\infty}\int_{1}^{R}\left(\frac{1}{x} - \frac{1}{x+1}\right)dx = \lim_{R\to\infty} \ln\frac{x}{x+1}\Big|_{1}^{R} = \lim_{R\to\infty}\left(\ln\frac{R}{R+1} - \ln\frac{1}{2}\right) = \ln 1 - \ln\frac{1}{2} = \ln 2.$$

The integral converges; hence, the series $\displaystyle\sum_{n=1}^{\infty}\frac{1}{n(n+1)}$ converges.

11. $\displaystyle\sum_{n=2}^{\infty}\frac{1}{n(\ln n)^2}$

SOLUTION   Let $f(x) = \dfrac{1}{x(\ln x)^2}$. This function is positive and continuous for $x \geq 2$. Moreover,

$$f'(x) = -\frac{1}{x^2(\ln x)^4}\left(1\cdot(\ln x)^2 + x\cdot 2(\ln x)\cdot\frac{1}{x}\right) = -\frac{1}{x^2(\ln x)^4}\left((\ln x)^2 + 2\ln x\right).$$

Since $\ln x > 0$ for $x > 1$, $f'(x)$ is negative for $x > 1$; hence, f is decreasing for $x \geq 2$. To compute the improper integral, we make the substitution $u = \ln x$, $du = \dfrac{1}{x}\,dx$. We obtain:

$$\int_{2}^{\infty}\frac{1}{x(\ln x)^2}\,dx = \lim_{R\to\infty}\int_{2}^{R}\frac{1}{x(\ln x)^2}\,dx = \lim_{R\to\infty}\int_{\ln 2}^{\ln R}\frac{du}{u^2}$$

$$= -\lim_{R\to\infty}\left(\frac{1}{\ln R} - \frac{1}{\ln 2}\right) = \frac{1}{\ln 2}.$$

The integral converges; hence, the series $\displaystyle\sum_{n=2}^{\infty}\frac{1}{n(\ln n)^2}$ also converges.

13. $\displaystyle\sum_{n=1}^{\infty} \frac{1}{2^{\ln n}}$

SOLUTION Note that

$$2^{\ln n} = (e^{\ln 2})^{\ln n} = (e^{\ln n})^{\ln 2} = n^{\ln 2}.$$

Thus,

$$\sum_{n=1}^{\infty} \frac{1}{2^{\ln n}} = \sum_{n=1}^{\infty} \frac{1}{n^{\ln 2}}.$$

Now, let $f(x) = \dfrac{1}{x^{\ln 2}}$. This function is positive, continuous and decreasing on the interval $x \geq 1$; therefore, the Integral Test applies. Moreover,

$$\int_{1}^{\infty} \frac{dx}{x^{\ln 2}} = \lim_{R \to \infty} \int_{1}^{R} \frac{dx}{x^{\ln 2}} = \frac{1}{1 - \ln 2} \lim_{R \to \infty} (R^{1 - \ln 2} - 1) = \infty,$$

because $1 - \ln 2 > 0$. The integral diverges; hence, the series $\displaystyle\sum_{n=1}^{\infty} \frac{1}{2^{\ln n}}$ also diverges.

15. Show that $\displaystyle\sum_{n=1}^{\infty} \frac{1}{n^3 + 8n}$ converges by using the Comparison Test with $\displaystyle\sum_{n=1}^{\infty} n^{-3}$.

SOLUTION We compare the series with the p-series $\displaystyle\sum_{n=1}^{\infty} n^{-3}$. For $n \geq 1$,

$$\frac{1}{n^3 + 8n} \leq \frac{1}{n^3}.$$

Since $\displaystyle\sum_{n=1}^{\infty} \frac{1}{n^3}$ converges (it is a p-series with $p = 3 > 1$), the series $\displaystyle\sum_{n=1}^{\infty} \frac{1}{n^3 + 8n}$ also converges by the Comparison Test.

17. Let $S = \displaystyle\sum_{n=1}^{\infty} \frac{1}{n + \sqrt{n}}$. Verify that for $n \geq 1$,

$$\frac{1}{n + \sqrt{n}} \leq \frac{1}{n}, \qquad \frac{1}{n + \sqrt{n}} \leq \frac{1}{\sqrt{n}}$$

Can either inequality be used to show that S diverges? Show that $\dfrac{1}{n + \sqrt{n}} \geq \dfrac{1}{2n}$ and conclude that S diverges.

SOLUTION For $n \geq 1$, $n + \sqrt{n} \geq n$ and $n + \sqrt{n} \geq \sqrt{n}$. Taking the reciprocal of each of these inequalities yields

$$\frac{1}{n + \sqrt{n}} \leq \frac{1}{n} \quad \text{and} \quad \frac{1}{n + \sqrt{n}} \leq \frac{1}{\sqrt{n}}.$$

These inequalities indicate that the series $\displaystyle\sum_{n=1}^{\infty} \frac{1}{n + \sqrt{n}}$ is smaller than both $\displaystyle\sum_{n=1}^{\infty} \frac{1}{n}$ and $\displaystyle\sum_{n=1}^{\infty} \frac{1}{\sqrt{n}}$; however, $\displaystyle\sum_{n=1}^{\infty} \frac{1}{n}$ and $\displaystyle\sum_{n=1}^{\infty} \frac{1}{\sqrt{n}}$ both diverge so neither inequality allows us to show that S diverges.

On the other hand, for $n \geq 1$, $n \geq \sqrt{n}$, so $2n \geq n + \sqrt{n}$ and

$$\frac{1}{n + \sqrt{n}} \geq \frac{1}{2n}.$$

The series $\displaystyle\sum_{n=1}^{\infty} \frac{1}{2n} = 2 \sum_{n=1}^{\infty} \frac{1}{n}$ diverges, since the harmonic series diverges. The Comparison Test then lets us conclude that the larger series $\displaystyle\sum_{n=1}^{\infty} \frac{1}{n + \sqrt{n}}$ also diverges.

In Exercises 19–30, use the Comparison Test to determine whether the infinite series is convergent.

19. $\displaystyle\sum_{n=1}^{\infty} \frac{1}{n2^n}$

SOLUTION We compare with the geometric series $\displaystyle\sum_{n=1}^{\infty} \left(\frac{1}{2}\right)^n$. For $n \geq 1$,

$$\frac{1}{n2^n} \leq \frac{1}{2^n} = \left(\frac{1}{2}\right)^n.$$

Since $\displaystyle\sum_{n=1}^{\infty} \left(\frac{1}{2}\right)^n$ converges (it is a geometric series with $r = \frac{1}{2}$), we conclude by the Comparison Test that $\displaystyle\sum_{n=1}^{\infty} \frac{1}{n2^n}$ also converges.

21. $\displaystyle\sum_{n=1}^{\infty} \frac{1}{n^{1/3} + 2^n}$

SOLUTION For $n \geq 1$,

$$\frac{1}{n^{1/3} + 2^n} \leq \frac{1}{2^n}.$$

The series $\sum_{n=1}^{\infty} \frac{1}{2^n}$ is a geometric series with $r = \frac{1}{2}$, so it converges. By the Comparison test, so does $\displaystyle\sum_{n=1}^{\infty} \frac{1}{n^{1/3} + 2^n}$.

23. $\displaystyle\sum_{m=1}^{\infty} \frac{4}{m! + 4^m}$

SOLUTION For $m \geq 1$,

$$\frac{4}{m! + 4^m} \leq \frac{4}{4^m} = \left(\frac{1}{4}\right)^{m-1}.$$

The series $\displaystyle\sum_{m=1}^{\infty} \left(\frac{1}{4}\right)^{m-1}$ is a geometric series with $r = \frac{1}{4}$, so it converges. By the Comparison Test we can therefore conclude that the series $\displaystyle\sum_{m=1}^{\infty} \frac{4}{m! + 4^m}$ also converges.

25. $\displaystyle\sum_{k=1}^{\infty} \frac{\sin^2 k}{k^2}$

SOLUTION For $k \geq 1$, $0 \leq \sin^2 k \leq 1$, so

$$0 \leq \frac{\sin^2 k}{k^2} \leq \frac{1}{k^2}.$$

The series $\displaystyle\sum_{k=1}^{\infty} \frac{1}{k^2}$ is a p-series with $p = 2 > 1$, so it converges. By the Comparison Test we can therefore conclude that the series $\displaystyle\sum_{k=1}^{\infty} \frac{\sin^2 k}{k^2}$ also converges.

27. $\displaystyle\sum_{n=1}^{\infty} \frac{2}{3^n + 3^{-n}}$

SOLUTION Since $3^{-n} > 0$ for all n,

$$\frac{2}{3^n + 3^{-n}} \leq \frac{2}{3^n} = 2\left(\frac{1}{3}\right)^n.$$

The series $\displaystyle\sum_{n=1}^{\infty} 2\left(\frac{1}{3}\right)^n$ is a geometric series with $r = \frac{1}{3}$, so it converges. By the Comparison Theorem we can therefore conclude that the series $\displaystyle\sum_{n=1}^{\infty} \frac{2}{3^n + 3^{-n}}$ also converges.

29. $\sum_{n=1}^{\infty} \dfrac{1}{(n+1)!}$

SOLUTION Note that for $n \geq 2$,

$$(n+1)! = 1 \cdot \underbrace{2 \cdot 3 \cdots n \cdot (n+1)}_{n \text{ factors}} \leq 2^n$$

so that

$$\sum_{n=1}^{\infty} \frac{1}{(n+1)!} = 1 + \sum_{n=2}^{\infty} \frac{1}{(n+1)!} \leq 1 + \sum_{n=2}^{\infty} \frac{1}{2^n}$$

But $\sum_{n=2}^{\infty} \dfrac{1}{2^n}$ is a geometric series with ratio $r = \dfrac{1}{2}$, so it converges. By the comparison test, $\sum_{n=1}^{\infty} \dfrac{1}{(n+1)!}$ converges as well.

Exercise 31–36: For all $a > 0$ and $b > 1$, the inequalities

$$\ln n \leq n^a, \qquad n^a < b^n$$

are true for n sufficiently large (this can be proved using L'Hopital's Rule). Use this, together with the Comparison Theorem, to determine whether the series converges or diverges.

31. $\sum_{n=1}^{\infty} \dfrac{\ln n}{n^3}$

SOLUTION For n sufficiently large (say $n = k$, although in this case $n = 1$ suffices), we have $\ln n \leq n$, so that

$$\sum_{n=k}^{\infty} \frac{\ln n}{n^3} \leq \sum_{n=k}^{\infty} \frac{n}{n^3} = \sum_{n=k}^{\infty} \frac{1}{n^2}$$

This is a p-series with $p = 2 > 1$, so it converges. Thus $\sum_{n=k}^{\infty} \frac{\ln n}{n^3}$ also converges; adding back in the finite number of terms for $1 \leq n \leq k$ does not affect this result.

33. $\sum_{n=1}^{\infty} \dfrac{(\ln n)^{100}}{n^{1.1}}$

SOLUTION Choose N so that $\ln n \leq n^{0.0005}$ for $n \geq N$. Then also for $n > N$, $(\ln n)^{100} \leq (n^{0.0005})^{100} = n^{0.05}$. Then

$$\sum_{n=N}^{\infty} \frac{(\ln n)^{100}}{n^{1.1}} \leq \sum_{n=N}^{\infty} \frac{n^{0.05}}{n^{1.1}} = \sum_{n=N}^{\infty} \frac{1}{n^{1.05}}$$

But $\sum_{n=N}^{\infty} \dfrac{1}{n^{1.05}}$ is a p-series with $p = 1.05 > 1$, so is convergent. It follows that $\sum_{n=N}^{\infty} \frac{(\ln n)^{100}}{n^{1.1}}$ is also convergent;

adding back in the finite number of terms for $n = 1, 2, \ldots, N - 1$ shows that $\sum_{n=1}^{\infty} \dfrac{(\ln n)^{100}}{n^{1.1}}$ converges as well.

35. $\sum_{n=1}^{\infty} \dfrac{n}{3^n}$

SOLUTION Choose N such that $n \leq 2^n$ for $n \geq N$. Then

$$\sum_{n=N}^{\infty} \frac{n}{3^n} \leq \sum_{n=N}^{\infty} \left(\frac{2}{3}\right)^n$$

The latter sum is a geometric series with $r = \dfrac{2}{3} < 1$, so it converges. Thus the series on the left converges as well. Adding

back in the finite number of terms for $n < N$ shows that $\sum_{n=1}^{\infty} \dfrac{n}{3^n}$ converges.

37. Show that $\sum_{n=1}^{\infty} \sin \dfrac{1}{n^2}$ converges. *Hint:* Use the inequality $\sin x \leq x$ for $x \geq 0$.

SOLUTION For $n \geq 1$,

$$0 \leq \frac{1}{n^2} \leq 1 < \pi;$$

therefore, $\sin \frac{1}{n^2} > 0$ for $n \geq 1$. Moreover, for $n \geq 1$,

$$\sin \frac{1}{n^2} \leq \frac{1}{n^2}.$$

The series $\sum_{n=1}^{\infty} \frac{1}{n^2}$ is a p-series with $p = 2 > 1$, so it converges. By the Comparison Test we can therefore conclude that the series $\sum_{n=1}^{\infty} \sin \frac{1}{n^2}$ also converges.

In Exercises 39–48, use the Limit Comparison Test to prove convergence or divergence of the infinite series.

39. $\sum_{n=2}^{\infty} \frac{n^2}{n^4 - 1}$

SOLUTION Let $a_n = \frac{n^2}{n^4 - 1}$. For large n, $\frac{n^2}{n^4 - 1} \approx \frac{n^2}{n^4} = \frac{1}{n^2}$, so we apply the Limit Comparison Test with $b_n = \frac{1}{n^2}$. We find

$$L = \lim_{n \to \infty} \frac{a_n}{b_n} = \lim_{n \to \infty} \frac{\frac{n^2}{n^4 - 1}}{\frac{1}{n^2}} = \lim_{n \to \infty} \frac{n^4}{n^4 - 1} = 1.$$

The series $\sum_{n=1}^{\infty} \frac{1}{n^2}$ is a p-series with $p = 2 > 1$, so it converges; hence, $\sum_{n=2}^{\infty} \frac{1}{n^2}$ also converges. Because L exists, by the Limit Comparison Test we can conclude that the series $\sum_{n=2}^{\infty} \frac{n^2}{n^4 - 1}$ converges.

41. $\sum_{n=2}^{\infty} \frac{n}{\sqrt{n^3 + 1}}$

SOLUTION Let $a_n = \frac{n}{\sqrt{n^3 + 1}}$. For large n, $\frac{n}{\sqrt{n^3 + 1}} \approx \frac{n}{\sqrt{n^3}} = \frac{1}{\sqrt{n}}$, so we apply the Limit Comparison test with $b_n = \frac{1}{\sqrt{n}}$. We find

$$L = \lim_{n \to \infty} \frac{a_n}{b_n} = \lim_{n \to \infty} \frac{\frac{n}{\sqrt{n^3 + 1}}}{\frac{1}{\sqrt{n}}} = \lim_{n \to \infty} \frac{\sqrt{n^3}}{\sqrt{n^3 + 1}} = 1.$$

The series $\sum_{n=1}^{\infty} \frac{1}{\sqrt{n}}$ is a p-series with $p = \frac{1}{2} < 1$, so it diverges; hence, $\sum_{n=2}^{\infty} \frac{1}{\sqrt{n}}$ also diverges. Because $L > 0$, by the Limit Comparison Test we can conclude that the series $\sum_{n=2}^{\infty} \frac{n}{\sqrt{n^3 + 1}}$ diverges.

43. $\sum_{n=3}^{\infty} \frac{3n + 5}{n(n - 1)(n - 2)}$

SOLUTION Let $a_n = \frac{3n + 5}{n(n - 1)(n - 2)}$. For large n, $\frac{3n + 5}{n(n - 1)(n - 2)} \approx \frac{3n}{n^3} = \frac{3}{n^2}$, so we apply the Limit Comparison Test with $b_n = \frac{1}{n^2}$. We find

$$L = \lim_{n \to \infty} \frac{a_n}{b_n} = \lim_{n \to \infty} \frac{\frac{3n+5}{n(n+1)(n+2)}}{\frac{1}{n^2}} = \lim_{n \to \infty} \frac{3n^3 + 5n^2}{n(n + 1)(n + 2)} = 3.$$

The series $\sum_{n=1}^{\infty} \frac{1}{n^2}$ is a p-series with $p = 2 > 1$, so it converges; hence, the series $\sum_{n=3}^{\infty} \frac{1}{n^2}$ also converges. Because L exists, by the Limit Comparison Test we can conclude that the series $\sum_{n=3}^{\infty} \frac{3n + 5}{n(n - 1)(n - 2)}$ converges.

45. $\displaystyle\sum_{n=1}^{\infty} \frac{1}{\sqrt{n} + \ln n}$

SOLUTION Let

$$a_n = \frac{1}{\sqrt{n} + \ln n}$$

For large n, $\sqrt{n} + \ln n \approx \sqrt{n}$, so apply the Comparison Test with $b_n = \dfrac{1}{\sqrt{n}}$. We find

$$L = \lim_{n \to \infty} \frac{a_n}{b_n} = \lim_{n \to \infty} \frac{1}{\sqrt{n} + \ln n} \cdot \frac{\sqrt{n}}{1} = \lim_{n \to \infty} \frac{1}{1 + \frac{\ln n}{\sqrt{n}}} = 1$$

The series $\displaystyle\sum_{n=1}^{\infty} \frac{1}{\sqrt{n}}$ is a p-series with $p = \dfrac{1}{2} < 1$, so it diverges. Because L exists, the Limit Comparison Test tells us the the original series also diverges.

47. $\displaystyle\sum_{n=1}^{\infty} \left(1 - \cos \frac{1}{n}\right)$ *Hint:* Compare with $\displaystyle\sum_{n=1}^{\infty} n^{-2}$.

SOLUTION Let $a_n = 1 - \cos \dfrac{1}{n}$, and apply the Limit Comparison Test with $b_n = \dfrac{1}{n^2}$. We find

$$L = \lim_{n \to \infty} \frac{a_n}{b_n} = \lim_{n \to \infty} \frac{1 - \cos \frac{1}{n}}{\frac{1}{n^2}} = \lim_{x \to \infty} \frac{1 - \cos \frac{1}{x}}{\frac{1}{x^2}} = \lim_{x \to \infty} \frac{-\frac{1}{x^2} \sin \frac{1}{x}}{-\frac{2}{x^3}} = \frac{1}{2} \lim_{x \to \infty} \frac{\sin \frac{1}{x}}{\frac{1}{x}}.$$

As $x \to \infty$, $u = \frac{1}{x} \to 0$, so

$$L = \frac{1}{2} \lim_{x \to \infty} \frac{\sin \frac{1}{x}}{\frac{1}{x}} = \frac{1}{2} \lim_{u \to 0} \frac{\sin u}{u} = \frac{1}{2}.$$

The series $\displaystyle\sum_{n=1}^{\infty} \frac{1}{n^2}$ is a p-series with $p = 2 > 1$, so it converges. Because L exists, by the Limit Comparison Test we can

conclude that the series $\displaystyle\sum_{n=1}^{\infty} \left(1 - \cos \frac{1}{n}\right)$ also converges.

In Exercises 49–74, determine convergence or divergence using any method covered so far.

49. $\displaystyle\sum_{n=4}^{\infty} \frac{1}{n^2 - 9}$

SOLUTION Apply the Limit Comparison Test with $a_n = \dfrac{1}{n^2 - 9}$ and $b_n = \dfrac{1}{n^2}$:

$$L = \lim_{n \to \infty} \frac{a_n}{b_n} = \lim_{n \to \infty} \frac{\frac{1}{n^2 - 9}}{\frac{1}{n^2}} = \lim_{n \to \infty} \frac{n^2}{n^2 - 9} = 1.$$

Since the p-series $\displaystyle\sum_{n=1}^{\infty} \frac{1}{n^2}$ converges, the series $\displaystyle\sum_{n=4}^{\infty} \frac{1}{n^2}$ also converges. Because L exists, by the Limit Comparison Test we can conclude that the series $\displaystyle\sum_{n=4}^{\infty} \frac{1}{n^2 - 9}$ converges.

51. $\displaystyle\sum_{n=1}^{\infty} \frac{\sqrt{n}}{4n + 9}$

SOLUTION Apply the Limit Comparison Test with $a_n = \dfrac{\sqrt{n}}{4n + 9}$ and $b_n = \dfrac{1}{\sqrt{n}}$:

$$L = \lim_{n \to \infty} \frac{a_n}{b_n} = \lim_{n \to \infty} \frac{\frac{\sqrt{n}}{4n+9}}{\frac{1}{\sqrt{n}}} = \lim_{n \to \infty} \frac{n}{4n + 9} = \frac{1}{4}.$$

The series $\displaystyle\sum_{n=1}^{\infty} \frac{1}{\sqrt{n}}$ is a divergent p-series. Because $L > 0$, by the Limit Comparison Test we can conclude that the series $\displaystyle\sum_{n=1}^{\infty} \frac{\sqrt{n}}{4n + 9}$ also diverges.

53. $\displaystyle\sum_{n=1}^{\infty} \frac{n^2 - n}{n^5 + n}$

SOLUTION First rewrite $a_n = \dfrac{n^2 - n}{n^5 + n} = \dfrac{n(n-1)}{n(n^4 + 1)} = \dfrac{n - 1}{n^4 + 1}$ and observe

$$\frac{n - 1}{n^4 + 1} < \frac{n}{n^4} = \frac{1}{n^3}$$

for $n \geq 1$. The series $\displaystyle\sum_{n=1}^{\infty} \frac{1}{n^3}$ is a convergent p-series, so by the Comparison Test we can conclude that the series

$\displaystyle\sum_{n=1}^{\infty} \frac{n^2 - n}{n^5 + n}$ also converges.

55. $\displaystyle\sum_{n=5}^{\infty} (4/5)^{-n}$

SOLUTION

$$\sum_{n=5}^{\infty} \left(\frac{4}{5}\right)^{-n} = \sum_{n=5}^{\infty} \left(\frac{5}{4}\right)^{n}$$

which is a geometric series starting at $n = 5$ with ratio $r = \dfrac{5}{4} > 1$. Thus the series diverges.

57. $\displaystyle\sum_{n=2}^{\infty} \frac{1}{n^{3/2} \ln n}$

SOLUTION For $n \geq 3$, $\ln n > 1$, so $n^{3/2} \ln n > n^{3/2}$ and

$$\frac{1}{n^{3/2} \ln n} < \frac{1}{n^{3/2}}.$$

The series $\displaystyle\sum_{n=1}^{\infty} \frac{1}{n^{3/2}}$ is a convergent p-series, so the series $\displaystyle\sum_{n=3}^{\infty} \frac{1}{n^{3/2}}$ also converges. By the Comparison Test we can

therefore conclude that the series $\displaystyle\sum_{n=3}^{\infty} \frac{1}{n^{3/2} \ln n}$ converges. Hence, the series $\displaystyle\sum_{n=2}^{\infty} \frac{1}{n^{3/2} \ln n}$ also converges.

59. $\displaystyle\sum_{k=1}^{\infty} 4^{1/k}$

SOLUTION

$$\lim_{k \to \infty} a_k = \lim_{k \to \infty} 4^{1/k} = 4^0 = 1 \neq 0;$$

therefore, the series $\displaystyle\sum_{k=1}^{\infty} 4^{1/k}$ diverges by the Divergence Test.

61. $\displaystyle\sum_{n=2}^{\infty} \frac{1}{(\ln n)^4}$

SOLUTION By the comment preceding Exercise 31, we can choose N so that for $n \geq N$, we have $\ln n < n^{1/8}$, so that $(\ln n)^4 < n^{1/2}$. Then

$$\sum_{n=N}^{\infty} \frac{1}{(\ln n)^4} > \sum_{n=N}^{\infty} \frac{1}{n^{1/2}}$$

which is a divergent p-series. Thus the series on the left diverges as well, and adding back in the finite number of terms

for $n < N$ does not affect the result. Thus $\displaystyle\sum_{n=2}^{\infty} \frac{1}{(\ln n)^4}$ diverges.

63. $\displaystyle\sum_{n=1}^{\infty} \frac{1}{n \ln n - n}$

SOLUTION   For $n \geq 2$, $n \ln n - n \leq n \ln n$; therefore,

$$\frac{1}{n \ln n - n} \geq \frac{1}{n \ln n}.$$

Now, let $f(x) = \dfrac{1}{x \ln x}$. For $x \geq 2$, this function is continuous, positive and decreasing, so the Integral Test applies. Using the substitution $u = \ln x$, $du = \frac{1}{x} dx$, we find

$$\int_2^{\infty} \frac{dx}{x \ln x} = \lim_{R \to \infty} \int_2^R \frac{dx}{x \ln x} = \lim_{R \to \infty} \int_{\ln 2}^{\ln R} \frac{du}{u} = \lim_{R \to \infty} (\ln(\ln R) - \ln(\ln 2)) = \infty.$$

The integral diverges; hence, the series $\displaystyle\sum_{n=2}^{\infty} \frac{1}{n \ln n}$ also diverges. By the Comparison Test we can therefore conclude that the series $\displaystyle\sum_{n=2}^{\infty} \frac{1}{n \ln n - n}$ diverges.

65. $\displaystyle\sum_{n=1}^{\infty} \frac{1}{n^n}$

SOLUTION   For $n \geq 2$, $n^n \geq 2^n$; therefore,

$$\frac{1}{n^n} \leq \frac{1}{2^n} = \left(\frac{1}{2}\right)^n.$$

The series $\displaystyle\sum_{n=1}^{\infty} \left(\frac{1}{2}\right)^n$ is a convergent geometric series, so $\displaystyle\sum_{n=2}^{\infty} \left(\frac{1}{2}\right)^n$ also converges. By the Comparison Test we can therefore conclude that the series $\displaystyle\sum_{n=2}^{\infty} \frac{1}{n^n}$ converges. Hence, the series $\displaystyle\sum_{n=1}^{\infty} \frac{1}{n^n}$ converges.

67. $\displaystyle\sum_{n=1}^{\infty} \frac{1 + (-1)^n}{n}$

SOLUTION   Let

$$a_n = \frac{1 + (-1)^n}{n}$$

Then

$$a_n = \begin{cases} 0 & n \text{ odd} \\ \frac{2}{2k} = \frac{1}{k} & n = 2k \text{ even} \end{cases}$$

Therefore, $\{a_n\}$ consists of 0s in the odd places and the harmonic series in the even places, so $\sum_{i=1}^{\infty} a_n$ is just the sum of the harmonic series, which diverges. Thus $\sum_{i=1}^{\infty} a_n$ diverges as well.

69. $\displaystyle\sum_{n=1}^{\infty} \sin \frac{1}{n}$

SOLUTION   Apply the Limit Comparison Test with $a_n = \sin \dfrac{1}{n}$ and $b_n = \dfrac{1}{n}$:

$$L = \lim_{n \to \infty} \frac{\sin \frac{1}{n}}{\frac{1}{n}} = \lim_{u \to 0} \frac{\sin u}{u} = 1,$$

where $u = \frac{1}{n}$. The harmonic series diverges. Because $L > 0$, by the Limit Comparison Test we can conclude that the series $\displaystyle\sum_{n=1}^{\infty} \sin \frac{1}{n}$ also diverges.

71. $\displaystyle\sum_{n=1}^{\infty} \frac{2n + 1}{4^n}$

SOLUTION   For $n \geq 3$, $2n + 1 < 2^n$, so

$$\frac{2n + 1}{4^n} < \frac{2^n}{4^n} = \left(\frac{1}{2}\right)^n.$$

The series $\sum_{n=1}^{\infty} \left(\frac{1}{2}\right)^n$ is a convergent geometric series, so $\sum_{n=3}^{\infty} \left(\frac{1}{2}\right)^n$ also converges. By the Comparison Test we can therefore conclude that the series $\sum_{n=3}^{\infty} \frac{2n+1}{4^n}$ converges. Finally, the series $\sum_{n=1}^{\infty} \frac{2n+1}{4^n}$ converges.

73. $\sum_{n=4}^{\infty} \frac{\ln n}{n^2 - 3n}$

SOLUTION By the comment preceding Exercise 31, we can choose $N \geq 4$ so that for $n \geq N$, $\ln n < n^{1/2}$. Then

$$\sum_{n=N}^{\infty} \frac{\ln n}{n^2 - 3n} \leq \sum_{n=N}^{\infty} \frac{n^{1/2}}{n^2 - 3n} = \sum_{n=N}^{\infty} \frac{1}{n^{3/2} - 3n^{1/2}}$$

To evaluate convergence of the latter series, let $a_n = \frac{1}{n^{3/2} - 3n^{1/2}}$ and $b_n = \frac{1}{n^{3/2}}$, and apply the Limit Comparison Test:

$$L = \lim_{n \to \infty} \frac{a_n}{b_n} = \lim_{n \to \infty} \frac{1}{n^{3/2} - 3n^{1/2}} \cdot n^{3/2} = \lim_{n \to \infty} \frac{1}{1 - 3n^{-1}} = 0$$

Thus $\sum a_n$ converges if $\sum b_n$ does. But $\sum b_n$ is a convergent p-series. Thus $\sum a_n$ converges and, by the comparison test, so does the original series. Adding back in the finite number of terms for $n < N$ does not affect convergence.

75. $\sum_{n=2}^{\infty} \frac{1}{n^{1/2} \ln n}$

SOLUTION By the comment preceding Exercise 31, we can choose $N \geq 2$ so that for $n \geq N$, $\ln n < n^{1/4}$. Then

$$\sum_{n=N}^{\infty} \frac{1}{n^{1/2} \ln n} > \sum_{n=N}^{\infty} \frac{1}{n^{3/4}}$$

which is a divergent p-series. Thus the original series diverges as well - as usual, adding back in the finite number of terms for $n < N$ does not affect convergence.

77. $\sum_{n=1}^{\infty} \frac{4n^2 + 15n}{3n^4 - 5n^2 - 17}$

SOLUTION Apply the Limit Comparison Test with

$$a_n = \frac{4n^2 + 15n}{3n^4 - 5n^2 - 17}, \qquad b_n = \frac{4n^2}{3n^4} = \frac{4}{3n^2}$$

We have

$$L = \lim_{n \to \infty} \frac{a_n}{b_n} = \lim_{n \to \infty} \frac{4n^2 + 15n}{3n^4 - 5n^2 - 17} \cdot \frac{3n^2}{4} = \lim_{n \to \infty} \frac{12n^4 + 45n^3}{12n^4 - 20n^2 - 68} = \lim_{n \to \infty} \frac{12 + 45/n}{12 - 20/n^2 - 68/n^4} = 1$$

Now, $\sum_{n=1}^{\infty} b_n$ is a p-series with $p = 2 > 1$, so converges. Since $L = 1$, we see that $\sum_{n=1}^{\infty} \frac{4n^2 + 15n}{3n^4 - 5n^2 - 17}$ converges as well.

79. For which a does $\sum_{n=2}^{\infty} \frac{1}{n(\ln n)^a}$ converge?

SOLUTION First consider the case $a > 0$ but $a \neq 1$. Let $f(x) = \frac{1}{x(\ln x)^a}$. This function is continuous, positive and decreasing for $x \geq 2$, so the Integral Test applies. Now,

$$\int_2^{\infty} \frac{dx}{x(\ln x)^a} = \lim_{R \to \infty} \int_2^R \frac{dx}{x(\ln x)^a} = \lim_{R \to \infty} \int_{\ln 2}^{\ln R} \frac{du}{u^a} = \frac{1}{1-a} \lim_{R \to \infty} \left(\frac{1}{(\ln R)^{a-1}} - \frac{1}{(\ln 2)^{a-1}} \right).$$

Because

$$\lim_{R \to \infty} \frac{1}{(\ln R)^{a-1}} = \begin{cases} \infty, & 0 < a < 1 \\ 0, & a > 1 \end{cases}$$

we conclude the integral diverges when $0 < a < 1$ and converges when $a > 1$. Therefore

$$\sum_{n=2}^{\infty} \frac{1}{n(\ln n)^a} \text{ converges for } a > 1 \text{ and diverges for } 0 < a < 1.$$

Next, consider the case $a = 1$. The series becomes $\displaystyle\sum_{n=2}^{\infty} \frac{1}{n \ln n}$. Let $f(x) = \dfrac{1}{x \ln x}$. For $x \geq 2$, this function is continuous, positive and decreasing, so the Integral Test applies. Using the substitution $u = \ln x$, $du = \frac{1}{x} dx$, we find

$$\int_2^{\infty} \frac{dx}{x \ln x} = \lim_{R \to \infty} \int_2^R \frac{dx}{x \ln x} = \lim_{R \to \infty} \int_{\ln 2}^{\ln R} \frac{du}{u} = \lim_{R \to \infty} (\ln(\ln R) - \ln(\ln 2)) = \infty.$$

The integral diverges; hence, the series also diverges.

Finally, consider the case $a < 0$. Let $b = -a > 0$ so the series becomes $\displaystyle\sum_{n=2}^{\infty} \frac{(\ln n)^b}{n}$. Since $\ln n > 1$ for all $n \geq 3$, it follows that

$$(\ln n)^b > 1 \quad \text{so} \quad \frac{(\ln n)^b}{n} > \frac{1}{n}.$$

The series $\displaystyle\sum_{n=3}^{\infty} \frac{1}{n}$ diverges, so by the Comparison Test we can conclude that $\displaystyle\sum_{n=3}^{\infty} \frac{(\ln n)^b}{n}$ also diverges. Consequently, $\displaystyle\sum_{n=2}^{\infty} \frac{(\ln n)^b}{n}$ diverges. Thus,

$$\sum_{n=2}^{\infty} \frac{1}{n(\ln n)^a} \text{ diverges for } a < 0.$$

To summarize:

$$\sum_{n=2}^{\infty} \frac{1}{n(\ln n)^a} \text{ converges if } a > 1 \text{ and diverges if } a \leq 1.$$

Approximating Infinite Sums *In Exercises 81–83, let $a_n = f(n)$, where $f(x)$ is a continuous, decreasing function such that $f(x) \geq 0$ and $\int_1^{\infty} f(x)\, dx$ converges.*

81. Show that

$$\int_1^{\infty} f(x)\, dx \leq \sum_{n=1}^{\infty} a_n \leq a_1 + \int_1^{\infty} f(x)\, dx \qquad \boxed{3}$$

SOLUTION From the proof of the Integral Test, we know that

$$a_2 + a_3 + a_4 + \cdots + a_N \leq \int_1^N f(x)\, dx \leq \int_1^{\infty} f(x)\, dx;$$

that is,

$$S_N - a_1 \leq \int_1^{\infty} f(x)\, dx \quad \text{or} \quad S_N \leq a_1 + \int_1^{\infty} f(x)\, dx.$$

Also from the proof of the Integral test, we know that

$$\int_1^N f(x)\, dx \leq a_1 + a_2 + a_3 + \cdots + a_{N-1} = S_N - a_N \leq S_N.$$

Thus,

$$\int_1^N f(x)\, dx \leq S_N \leq a_1 + \int_1^{\infty} f(x)\, dx.$$

Taking the limit as $N \to \infty$ yields Eq. (3), as desired.

83. Let $S = \sum\limits_{n=1}^{\infty} a_n$. Arguing as in Exercise 81, show that

$$\sum_{n=1}^{M} a_n + \int_{M+1}^{\infty} f(x)\, dx \le S \le \sum_{n=1}^{M+1} a_n + \int_{M+1}^{\infty} f(x)\, dx \qquad \boxed{4}$$

Conclude that

$$0 \le S - \left(\sum_{n=1}^{M} a_n + \int_{M+1}^{\infty} f(x)\, dx \right) \le a_{M+1} \qquad \boxed{5}$$

This provides a method for approximating S with an error of at most a_{M+1}.

SOLUTION Following the proof of the Integral Test and the argument in Exercise 81, but starting with $n = M + 1$ rather than $n = 1$, we obtain

$$\int_{M+1}^{\infty} f(x)\, dx \le \sum_{n=M+1}^{\infty} a_n \le a_{M+1} + \int_{M+1}^{\infty} f(x)\, dx.$$

Adding $\sum\limits_{n=1}^{M} a_n$ to each part of this inequality yields

$$\sum_{n=1}^{M} a_n + \int_{M+1}^{\infty} f(x)\, dx \le \sum_{n=1}^{\infty} a_n = S \le \sum_{n=1}^{M+1} a_n + \int_{M+1}^{\infty} f(x)\, dx.$$

Subtracting $\sum\limits_{n=1}^{M} a_n + \int_{M+1}^{\infty} f(x)\, dx$ from each part of this last inequality then gives us

$$0 \le S - \left(\sum_{n=1}^{M} a_n + \int_{M+1}^{\infty} f(x)\, dx \right) \le a_{M+1}.$$

85. *CAS* Apply Eq. (4) with $M = 40,000$ to show that

$$1.644934066 \le \sum_{n=1}^{\infty} \frac{1}{n^2} \le 1.644934068$$

Is this consistent with Euler's result, according to which this infinite series has sum $\pi^2/6$?

SOLUTION Using Eq. (4) with $f(x) = \dfrac{1}{x^2}$, $a_n = \dfrac{1}{n^2}$ and $M = 40,000$, we find

$$S_{40,000} + \int_{40,001}^{\infty} \frac{dx}{x^2} \le \sum_{n=1}^{\infty} \frac{1}{n^2} \le S_{40,001} + \int_{40,001}^{\infty} \frac{dx}{x^2}.$$

Now,

$$S_{40,000} = 1.6449090672;$$

$$S_{40,001} = S_{40,000} + \frac{1}{40,001} = 1.6449090678;$$

and

$$\int_{40,001}^{\infty} \frac{dx}{x^2} = \lim_{R \to \infty} \int_{40,001}^{R} \frac{dx}{x^2} = - \lim_{R \to \infty} \left(\frac{1}{R} - \frac{1}{40,001} \right) = \frac{1}{40,001} = 0.0000249994.$$

Thus,

$$1.6449090672 + 0.0000249994 \le \sum_{n=1}^{\infty} \frac{1}{n^2} \le 1.6449090678 + 0.0000249994,$$

or

$$1.6449340665 \le \sum_{n=1}^{\infty} \frac{1}{n^2} \le 1.6449340672.$$

Since $\dfrac{\pi^2}{6} \approx 1.6449340668$, our approximation is consistent with Euler's result.

87. CAS Using a CAS and Eq. (5), determine the value of $\displaystyle\sum_{n=1}^{\infty} n^{-5}$ to within an error less than 10^{-4}.

SOLUTION Using Eq. (5) with $f(x) = x^{-5}$ and $a_n = n^{-5}$, we have

$$0 \le \sum_{n=1}^{\infty} n^{-5} - \left(\sum_{n=1}^{M+1} n^{-5} + \int_{M+1}^{\infty} x^{-5}\, dx \right) \le (M+1)^{-5}.$$

To guarantee an error less than 10^{-4}, we need $(M+1)^{-5} \le 10^{-4}$. This yields $M \ge 10^{4/5} - 1 \approx 5.3$, so we choose $M = 6$. Now,

$$\sum_{n=1}^{7} n^{-5} = 1.0368498887,$$

and

$$\int_{7}^{\infty} x^{-5}\, dx = \lim_{R \to \infty} \int_{7}^{R} x^{-5}\, dx = -\frac{1}{4} \lim_{R \to \infty} \left(R^{-4} - 7^{-4} \right) = \frac{1}{4 \cdot 7^4} = 0.0001041233.$$

Thus,

$$\sum_{n=1}^{\infty} n^{-5} \approx \sum_{n=1}^{7} n^{-5} + \int_{7}^{\infty} x^{-5}\, dx = 1.0368498887 + 0.0001041233 = 1.0369540120.$$

89. The following argument proves the divergence of the harmonic series $S = \displaystyle\sum_{n=1}^{\infty} 1/n$ without using the Integral Test.

Let

$$S_1 = 1 + \frac{1}{3} + \frac{1}{5} + \cdots, \qquad S_2 = \frac{1}{2} + \frac{1}{4} + \frac{1}{6} + \cdots$$

Show that if S converges, then

(a) S_1 and S_2 also converge and $S = S_1 + S_2$.

(b) $S_1 > S_2$ and $S_2 = \frac{1}{2}S$.

Observe that (b) contradicts (a), and conclude that S diverges.

SOLUTION Assume throughout that S converges; we will derive a contradiction. Write

$$a_n = \frac{1}{n}, \quad b_n = \frac{1}{2n-1}, \quad c_n = \frac{1}{2n}$$

for the nth terms in the series S, S_1, and S_2. Since $2n - 1 \ge n$ for $n \ge 1$, we have $b_n < a_n$. Since $S = \sum a_n$ converges, so does $S_1 = \sum b_n$ by the Comparison Test. Also, $c_n = \frac{1}{2}a_n$, so again by the Comparison Test, the convergence of S implies the convergence of $S_2 = \sum c_n$. Now, define two sequences

$$b'_n = \begin{cases} b_{(n+1)/2} & n \text{ odd} \\ 0 & n \text{ even} \end{cases}$$

$$c'_n = \begin{cases} 0 & n \text{ odd} \\ c_{n/2} & n \text{ even} \end{cases}$$

That is, b'_n and c'_n look like b_n and c_n, but have zeros inserted in the "missing" places compared to a_n. Then $a_n = b'_n + c'_n$; also $S_1 = \sum b_n = \sum b'_n$ and $S_2 = \sum c_n = \sum c'_n$. Finally, since S, S_1, and S_2 all converge, we have

$$S = \sum_{n=1}^{\infty} a_n = \sum_{n=1}^{\infty} (b'_n + c'_n) = \sum_{n=1}^{\infty} b'_n + \sum_{n=1}^{\infty} c'_n = \sum_{n=1}^{\infty} b_n + \sum_{n=1}^{\infty} c_n = S_1 + S_2$$

Now, $b_n > c_n$ for every n, so that $S_1 > S_2$. Also, we showed above that $c_n = \frac{1}{2}a_n$, so that $2S_2 = S$. Putting all this together gives

$$S = S_1 + S_2 > S_2 + S_2 = 2S_2 = S$$

so that $S > S$, a contradiction. Thus S must diverge.

Further Insights and Challenges

91. Kummer's Acceleration Method   Suppose we wish to approximate $S = \sum_{n=1}^{\infty} 1/n^2$. There is a similar telescoping series whose value can be computed exactly (Example 1 in Section 10.2):

$$\sum_{n=1}^{\infty} \frac{1}{n(n+1)} = 1$$

(a) Verify that

$$S = \sum_{n=1}^{\infty} \frac{1}{n(n+1)} + \sum_{n=1}^{\infty} \left(\frac{1}{n^2} - \frac{1}{n(n+1)} \right)$$

Thus for M large,

$$S \approx 1 + \sum_{n=1}^{M} \frac{1}{n^2(n+1)} \qquad \boxed{6}$$

(b) Explain what has been gained. Why is Eq. (6) a better approximation to S than is $\sum_{n=1}^{M} 1/n^2$?

(c) $\mathcal{CAS}$ Compute

$$\sum_{n=1}^{1000} \frac{1}{n^2}, \qquad 1 + \sum_{n=1}^{100} \frac{1}{n^2(n+1)}$$

Which is a better approximation to S, whose exact value is $\pi^2/6$?

SOLUTION

(a) Because the series $\sum_{n=1}^{\infty} \frac{1}{n^2}$ and $\sum_{n=1}^{\infty} \frac{1}{n(n+1)}$ both converge,

$$\sum_{n=1}^{\infty} \frac{1}{n(n+1)} + \sum_{n=1}^{\infty} \left(\frac{1}{n^2} - \frac{1}{n(n+1)} \right) = \sum_{n=1}^{\infty} \frac{1}{n(n+1)} + \sum_{n=1}^{\infty} \frac{1}{n^2} - \sum_{n=1}^{\infty} \frac{1}{n(n+1)} = \sum_{n=1}^{\infty} \frac{1}{n^2} = S.$$

Now,

$$\frac{1}{n^2} - \frac{1}{n(n+1)} = \frac{n+1}{n^2(n+1)} - \frac{n}{n^2(n+1)} = \frac{1}{n^2(n+1)},$$

so, for M large,

$$S \approx 1 + \sum_{n=1}^{M} \frac{1}{n^2(n+1)}.$$

(b) The series $\sum_{n=1}^{\infty} \frac{1}{n^2(n+1)}$ converges more rapidly than $\sum_{n=1}^{\infty} \frac{1}{n^2}$ since the degree of n in the denominator is larger.

(c) Using a computer algebra system, we find

$$\sum_{n=1}^{1000} \frac{1}{n^2} = 1.6439345667 \quad \text{and} \quad 1 + \sum_{n=1}^{100} \frac{1}{n^2(n+1)} = 1.6448848903.$$

The second sum is more accurate because it is closer to the exact solution $\dfrac{\pi^2}{6} \approx 1.6449340668$.

10.4 Absolute and Conditional Convergence   (LT Section 11.4)

Preliminary Questions

1. Give an example of a series such that $\sum a_n$ converges but $\sum |a_n|$ diverges.

SOLUTION   The series $\sum \frac{(-1)^n}{\sqrt[3]{n}}$ converges by the Leibniz Test, but the positive series $\sum \frac{1}{\sqrt[3]{n}}$ is a divergent p-series.

2. Which of the following statements is equivalent to Theorem 1?

(a) If $\sum_{n=0}^{\infty} |a_n|$ diverges, then $\sum_{n=0}^{\infty} a_n$ also diverges.

(b) If $\sum_{n=0}^{\infty} a_n$ diverges, then $\sum_{n=0}^{\infty} |a_n|$ also diverges.

(c) If $\sum_{n=0}^{\infty} a_n$ converges, then $\sum_{n=0}^{\infty} |a_n|$ also converges.

SOLUTION   The correct answer is **(b)**: If $\sum_{n=0}^{\infty} a_n$ diverges, then $\sum_{n=0}^{\infty} |a_n|$ also diverges. Take $a_n = (-1)^n \frac{1}{n}$ to see that statements **(a)** and **(c)** are not true in general.

3. Lathika argues that $\sum_{n=1}^{\infty} (-1)^n \sqrt{n}$ is an alternating series and therefore converges. Is Lathika right?

SOLUTION   No. Although $\sum_{n=1}^{\infty} (-1)^n \sqrt{n}$ is an alternating series, the terms $a_n = \sqrt{n}$ do not form a decreasing sequence that tends to zero. In fact, $a_n = \sqrt{n}$ is an increasing sequence that tends to ∞, so $\sum_{n=1}^{\infty} (-1)^n \sqrt{n}$ diverges by the Divergence Test.

4. Suppose that a_n is positive, decreasing, and tends to 0, and let $S = \sum_{n=1}^{\infty} (-1)^{n-1} a_n$. What can we say about $|S - S_{100}|$ if $a_{101} = 10^{-3}$? Is S larger or smaller than S_{100}?

SOLUTION   From the text, we know that $|S - S_{100}| < a_{101} = 10^{-3}$. Also, the Leibniz test tells us that $S_{2N} < S < S_{2N+1}$ for any $N \geq 1$, so that $S_{100} < S$.

Exercises

1. Show that

$$\sum_{n=0}^{\infty} \frac{(-1)^n}{2^n}$$

converges absolutely.

SOLUTION   The positive series $\sum_{n=0}^{\infty} \frac{1}{2^n}$ is a geometric series with $r = \frac{1}{2}$. Thus, the positive series converges, and the given series converges absolutely.

In Exercises 3–10, determine whether the series converges absolutely, conditionally, or not at all.

3. $\sum_{n=1}^{\infty} \frac{(-1)^{n-1}}{n^{1/3}}$

SOLUTION   The sequence $a_n = \frac{1}{n^{1/3}}$ is positive, decreasing, and tends to zero; hence, the series $\sum_{n=1}^{\infty} \frac{(-1)^{n-1}}{n^{1/3}}$ converges by the Leibniz Test. However, the positive series $\sum_{n=1}^{\infty} \frac{1}{n^{1/3}}$ is a divergent p-series, so the original series converges conditionally.

5. $\displaystyle\sum_{n=0}^{\infty} \frac{(-1)^{n-1}}{(1.1)^n}$

SOLUTION The positive series $\displaystyle\sum_{n=0}^{\infty} \left(\frac{1}{1.1}\right)^n$ is a convergent geometric series; thus, the original series converges absolutely.

7. $\displaystyle\sum_{n=2}^{\infty} \frac{(-1)^n}{n \ln n}$

SOLUTION Let $a_n = \frac{1}{n \ln n}$. Then a_n forms a decreasing sequence (note that n and $\ln n$ are both increasing functions of n) that tends to zero; hence, the series $\displaystyle\sum_{n=2}^{\infty} \frac{(-1)^n}{n \ln n}$ converges by the Leibniz Test. However, the positive series $\displaystyle\sum_{n=2}^{\infty} \frac{1}{n \ln n}$ diverges, so the original series converges conditionally.

9. $\displaystyle\sum_{n=2}^{\infty} \frac{\cos n\pi}{(\ln n)^2}$

SOLUTION Since $\cos n\pi$ alternates between $+1$ and -1,

$$\sum_{n=2}^{\infty} \frac{\cos n\pi}{(lnn)^2} = \sum_{n=2}^{\infty} \frac{(-1)^n}{(lnn)^2}$$

This is an alternating series whose general term decreases to zero, so it converges. The associated positive series,

$$\sum_{n=2}^{\infty} \frac{1}{(\ln n)^2}$$

is a divergent series, so the original series converges conditionally.

11. Let $S = \displaystyle\sum_{n=1}^{\infty} (-1)^{n+1} \frac{1}{n^3}$.

(a) Calculate S_n for $1 \le n \le 10$.

(b) Use Eq. (2) to show that $0.9 \le S \le 0.902$.

SOLUTION

(a)

$$S_1 = 1 \qquad\qquad S_6 = S_5 - \frac{1}{6^3} = 0.899782407$$

$$S_2 = 1 - \frac{1}{2^3} = \frac{7}{8} = 0.875 \qquad\qquad S_7 = S_6 + \frac{1}{7^3} = 0.902697859$$

$$S_3 = S_2 + \frac{1}{3^3} = 0.912037037 \qquad\qquad S_8 = S_7 - \frac{1}{8^3} = 0.900744734$$

$$S_4 = S_3 - \frac{1}{4^3} = 0.896412037 \qquad\qquad S_9 = S_8 + \frac{1}{9^3} = 0.902116476$$

$$S_5 = S_4 + \frac{1}{5^3} = 0.904412037 \qquad\qquad S_{10} = S_9 - \frac{1}{10^3} = 0.901116476$$

(b) By Eq. (2),

$$|S_{10} - S| \le a_{11} = \frac{1}{11^3},$$

so

$$S_{10} - \frac{1}{11^3} \le S \le S_{10} + \frac{1}{11^3},$$

or

$$0.900365161 \le S \le 0.901867791.$$

13. Approximate $\displaystyle\sum_{n=1}^{\infty} \frac{(-1)^{n+1}}{n^4}$ to three decimal places.

SOLUTION Let $S = \displaystyle\sum_{n=1}^{\infty} \frac{(-1)^{n+1}}{n^4}$, so that $a_n = \dfrac{1}{n^4}$. By Eq. (2),

$$|S_N - S| \le a_{N+1} = \frac{1}{(N+1)^4}.$$

To guarantee accuracy to three decimal places, we must choose N so that

$$\frac{1}{(N+1)^4} < 5 \times 10^{-4} \quad \text{or} \quad N > \sqrt[4]{2000} - 1 \approx 5.7.$$

The smallest value that satisfies the required inequality is then $N = 6$. Thus,

$$S \approx S_6 = 1 - \frac{1}{2^4} + \frac{1}{3^4} - \frac{1}{4^4} + \frac{1}{5^4} - \frac{1}{6^4} = 0.946767824.$$

In Exercises 15 and 16, find a value of N such that S_N approximates the series with an error of at most 10^{-5}. If you have a CAS, compute this value of S_N.

15. $\displaystyle\sum_{n=1}^{\infty} \frac{(-1)^{n+1}}{n(n+2)(n+3)}$

SOLUTION Let $S = \displaystyle\sum_{n=1}^{\infty} \frac{(-1)^{n+1}}{n\,(n+2)\,(n+3)}$, so that $a_n = \dfrac{1}{n\,(n+2)\,(n+3)}$. By Eq. (2),

$$|S_N - S| \le a_{N+1} = \frac{1}{(N+1)(N+3)(N+4)}.$$

We must choose N so that

$$\frac{1}{(N+1)(N+3)(N+4)} \le 10^{-5} \quad \text{or} \quad (N+1)(N+3)(N+4) \ge 10^5.$$

For $N = 43$, the product on the left hand side is $95{,}128$, while for $N = 44$ the product is $101{,}520$; hence, the smallest value of N which satisfies the required inequality is $N = 44$. Thus,

$$S \approx S_{44} = \sum_{n=1}^{44} \frac{(-1)^{n+1}}{n(n+2)(n+3)} = 0.0656746.$$

In Exercises 17–32, determine convergence or divergence by any method.

17. $\displaystyle\sum_{n=0}^{\infty} 7^{-n}$

SOLUTION This is a (positive) geometric series with $r = \dfrac{1}{7} < 1$, so it converges.

19. $\displaystyle\sum_{n=1}^{\infty} \frac{1}{5^n - 3^n}$

SOLUTION Use the Limit Comparison Test with $\dfrac{1}{5^n}$:

$$L = \lim_{n\to\infty} \frac{1/(5^n - 3^n)}{1/5^n} = \lim_{n\to\infty} \frac{5^n}{5^n - 3^n} = \lim_{n\to\infty} \frac{1}{1 - (3/5)^n} = 1.$$

But $\sum_{n=1}^{\infty} \dfrac{1}{5^n}$ is a convergent geometric series. Since $L = 1$, the Limit Comparison Test tells us that the original series converges as well.

21. $\sum_{n=1}^{\infty} \dfrac{1}{3n^4 + 12n}$

SOLUTION Use the Limit Comparison Test with $\dfrac{1}{3n^4}$:

$$L = \lim_{n\to\infty} \frac{(1/(3n^4 + 12n))}{1/3n^4} = \lim_{n\to\infty} \frac{3n^4}{3n^4 + 12n} = \lim_{n\to\infty} \frac{1}{1 + 4n^{-3}} = 1$$

But $\sum_{n=1}^{\infty} \dfrac{1}{3n^4} = \dfrac{1}{3}\sum_{n=1}^{\infty}\dfrac{1}{n^4}$ is a convergent p-series. Since $L = 1$, the Limit Comparison Test tells us that the original series converges as well.

23. $\sum_{n=1}^{\infty} \dfrac{1}{\sqrt{n^2 + 1}}$

SOLUTION Apply the Limit Comparison Test and compare the series with the divergent harmonic series:

$$L = \lim_{n\to\infty} \frac{\frac{1}{\sqrt{n^2+1}}}{\frac{1}{n}} = \lim_{n\to\infty} \frac{n}{\sqrt{n^2+1}} = 1.$$

Because $L > 0$, we conclude that the series $\sum_{n=1}^{\infty} \dfrac{1}{\sqrt{n^2+1}}$ diverges.

25. $\sum_{n=1}^{\infty} \dfrac{3^n + (-2)^n}{5^n}$

SOLUTION The series

$$\sum_{n=1}^{\infty} \frac{3^n}{5^n} = \sum_{n=1}^{\infty} \left(\frac{3}{5}\right)^n$$

is a convergent geometric series, as is the series

$$\sum_{n=1}^{\infty} \frac{(-1)^n\, 2^n}{5^n} = \sum_{n=1}^{\infty} \left(-\frac{2}{5}\right)^n.$$

Hence,

$$\sum_{n=1}^{\infty} \frac{3^n + (-1)^n 2^n}{5^n} = \sum_{n-1}^{\infty} \left(\frac{3}{5}\right)^n + \sum_{n=1}^{\infty} \left(-\frac{2}{5}\right)^n$$

also converges.

27. $\sum_{n=1}^{\infty} (-1)^n n^2 e^{-n^3/3}$

SOLUTION Consider the associated positive series $\sum_{n=1}^{\infty} n^2 e^{-n^3/3}$. This series can be seen to converge by the Integral Test:

$$\int_1^{\infty} x^2 e^{-x^3/3}\, dx = \lim_{R\to\infty} \int_1^R x^2 e^{-x^3/3}\, dx = -\lim_{R\to\infty} e^{-x^3/3}\Big|_1^R = e^{-1/3} + \lim_{R\to\infty} e^{-R^3/3} = e^{-1/3}.$$

The integral converges, so the original series converges absolutely.

29. $\sum_{n=2}^{\infty} \dfrac{(-1)^n}{n^{1/2}(\ln n)^2}$

SOLUTION This is an alternating series with $a_n = \dfrac{1}{n^{1/2}(\ln n)^2}$. Because a_n is a decreasing sequence which converges to zero, the series $\sum_{n=2}^{\infty} \dfrac{(-1)^n}{n^{1/2}(\ln n)^2}$ converges by the Leibniz Test. (Note that the series converges only conditionally, not absolutely; the associated positive series is eventually greater than $\dfrac{1}{n^{3/4}}$, which is a divergent p-series).

31. $\sum\limits_{n=1}^{\infty} \dfrac{\ln n}{n^{1.05}}$

SOLUTION Choose N so that for $n \geq N$ we have $\ln n \leq n^{0.01}$. Then

$$\sum_{n=N}^{\infty} \frac{\ln n}{n^{1.05}} \leq \sum_{n=N}^{\infty} \frac{n^{0.01}}{n^{1.05}} = \sum_{n=N}^{\infty} \frac{1}{n^{1.04}}$$

This is a convergent p-series, so by the Comparison Test, the original series converges as well.

33. Show that

$$S = \frac{1}{2} - \frac{1}{2} + \frac{1}{3} - \frac{1}{3} + \frac{1}{4} - \frac{1}{4} + \cdots$$

converges by computing the partial sums. Does it converge absolutely?

SOLUTION The sequence of partial sums is

$$S_1 = \frac{1}{2}$$

$$S_2 = S_1 - \frac{1}{2} = 0$$

$$S_3 = S_2 + \frac{1}{3} = \frac{1}{3}$$

$$S_4 = S_3 - \frac{1}{3} = 0$$

and, in general,

$$S_N = \begin{cases} \dfrac{1}{N}, & \text{for odd } N \\ 0, & \text{for even } N \end{cases}$$

Thus, $\lim\limits_{N \to \infty} S_N = 0$, and the series converges to 0. The positive series is

$$\frac{1}{2} + \frac{1}{2} + \frac{1}{3} + \frac{1}{3} + \frac{1}{4} + \frac{1}{4} + \cdots = 2\sum_{n=2}^{\infty} \frac{1}{n};$$

which diverges. Therefore, the original series converges conditionally, not absolutely.

35. 📖 **Assumptions Matter** Show by counterexample that the Leibniz Test does not remain true if the sequence a_n tends to zero but is not assumed nonincreasing. *Hint:* Consider

$$R = \frac{1}{2} - \frac{1}{4} + \frac{1}{3} - \frac{1}{8} + \frac{1}{4} - \frac{1}{16} + \cdots + \left(\frac{1}{n} - \frac{1}{2^n}\right) + \cdots$$

SOLUTION Let

$$R = \frac{1}{2} - \frac{1}{4} + \frac{1}{3} - \frac{1}{8} + \frac{1}{4} - \frac{1}{16} + \cdots + \left(\frac{1}{n+1} - \frac{1}{2^{n+1}}\right) + \cdots$$

This is an alternating series with

$$a_n = \begin{cases} \dfrac{1}{k+1}, & n = 2k-1 \\ \dfrac{1}{2^{k+1}}, & n = 2k \end{cases}$$

Note that $a_n \to 0$ as $n \to \infty$, but the sequence $\{a_n\}$ is not decreasing. We will now establish that R diverges.
For sake of contradiction, suppose that R converges. The geometric series

$$\sum_{n=1}^{\infty} \frac{1}{2^{n+1}}$$

converges, so the sum of R and this geometric series must also converge; however,

$$R + \sum_{n=1}^{\infty} \frac{1}{2^{n+1}} = \sum_{n=2}^{\infty} \frac{1}{n},$$

which diverges because the harmonic series diverges. Thus, the series R must diverge.

37. Prove that if $\sum a_n$ converges absolutely, then $\sum a_n^2$ also converges. Then give an example where $\sum a_n$ is only conditionally convergent and $\sum a_n^2$ diverges.

SOLUTION Suppose the series $\sum a_n$ converges absolutely. Because $\sum |a_n|$ converges, we know that

$$\lim_{n \to \infty} |a_n| = 0.$$

Therefore, there exists a positive integer N such that $|a_n| < 1$ for all $n \geq N$. It then follows that for $n \geq N$,

$$0 \leq a_n^2 = |a_n|^2 = |a_n| \cdot |a_n| < |a_n| \cdot 1 = |a_n|.$$

By the Comparison Test we can then conclude that $\sum a_n^2$ also converges.

Consider the series $\sum_{n=1}^{\infty} \frac{(-1)^n}{\sqrt{n}}$. This series converges by the Leibniz Test, but the corresponding positive series is a

divergent p-series; that is, $\sum_{n=1}^{\infty} \frac{(-1)^n}{\sqrt{n}}$ is conditionally convergent. Now, $\sum_{n=1}^{\infty} a_n^2$ is the divergent harmonic series $\sum_{n=1}^{\infty} \frac{1}{n}$.

Thus, $\sum a_n^2$ need not converge if $\sum a_n$ is only conditionally convergent.

Further Insights and Challenges

39. Use Exercise 38 to show that the following series converges:

$$S = \frac{1}{\ln 2} + \frac{1}{\ln 3} - \frac{2}{\ln 4} + \frac{1}{\ln 5} + \frac{1}{\ln 6} - \frac{2}{\ln 7} + \cdots$$

SOLUTION The given series has the structure of the generic series from Exercise 38 with $a_n = \frac{1}{\ln(n+1)}$. Because a_n is a positive, decreasing sequence with $\lim_{n \to \infty} a_n = 0$, we can conclude from Exercise 38 that the given series converges.

41. Show that the following series diverges:

$$S = 1 + \frac{1}{2} + \frac{1}{3} - \frac{2}{4} + \frac{1}{5} + \frac{1}{6} + \frac{1}{7} - \frac{2}{8} + \cdots$$

Hint: Use the result of Exercise 40 to write S as the sum of a convergent series and a divergent series.

SOLUTION Let

$$R = 1 + \frac{1}{2} + \frac{1}{3} - \frac{3}{4} + \frac{1}{5} + \frac{1}{6} + \frac{1}{7} - \frac{3}{8} + \cdots$$

and

$$S = 1 + \frac{1}{2} + \frac{1}{3} - \frac{2}{4} + \frac{1}{5} + \frac{1}{6} + \frac{1}{7} - \frac{2}{8} + \cdots$$

For sake of contradiction, suppose the series S converges. From Exercise 40, we know that the series R converges. Thus, the series $S - R$ must converge; however,

$$S - R = \frac{1}{4} + \frac{1}{8} + \frac{1}{12} + \cdots = \frac{1}{4} \sum_{k=1}^{\infty} \frac{1}{k},$$

which diverges because the harmonic series diverges. Thus, the series S must diverge.

43. We say that $\{b_n\}$ is a rearrangement of $\{a_n\}$ if $\{b_n\}$ has the same terms as $\{a_n\}$ but occurring in a different order. Show that if $\{b_n\}$ is a rearrangement of $\{a_n\}$ and $S = \sum_{n=1}^{\infty} a_n$ converges absolutely, then $T = \sum_{n=1}^{\infty} b_n$ also converges absolutely.

(This result does not hold if S is only conditionally convergent.) *Hint:* Prove that the partial sums $\sum_{n=1}^{N} |b_n|$ are bounded.

It can be shown further that $S = T$.

SOLUTION Suppose the series $S = \sum_{n=1}^{\infty} a_n$ converges absolutely and denote the corresponding positive series by

$$S^+ = \sum_{n=1}^{\infty} |a_n|.$$

Further, let $T_N = \sum_{n=1}^{N} |b_n|$ denote the Nth partial sum of the series $\sum_{n=1}^{\infty} |b_n|$. Because $\{b_n\}$ is a rearrangement of $\{a_n\}$, we know that

$$0 \le T_N \le \sum_{n=1}^{\infty} |a_n| = S^+;$$

that is, the sequence $\{T_N\}$ is bounded. Moreover,

$$T_{N+1} = \sum_{n=1}^{N+1} |b_n| = T_N + |b_{N+1}| \ge T_N;$$

that is, $\{T_N\}$ is increasing. It follows that $\{T_N\}$ converges, so the series $\sum_{n=1}^{\infty} |b_n|$ converges, which means the series $\sum_{n=1}^{\infty} b_n$ converges absolutely.

10.5 The Ratio and Root Tests (LT Section 11.5)

Preliminary Questions

1. In the Ratio Test, is ρ equal to $\lim_{n\to\infty} \left| \dfrac{a_{n+1}}{a_n} \right|$ or $\lim_{n\to\infty} \left| \dfrac{a_n}{a_{n+1}} \right|$?

SOLUTION In the Ratio Test ρ is the limit $\lim_{n\to\infty} \left| \dfrac{a_{n+1}}{a_n} \right|$.

2. Is the Ratio Test conclusive for $\sum_{n=1}^{\infty} \dfrac{1}{2^n}$? Is it conclusive for $\sum_{n=1}^{\infty} \dfrac{1}{n}$?

SOLUTION The general term of $\sum_{n=1}^{\infty} \dfrac{1}{2^n}$ is $a_n = \dfrac{1}{2^n}$; thus,

$$\left| \frac{a_{n+1}}{a_n} \right| = \frac{1}{2^{n+1}} \cdot \frac{2^n}{1} = \frac{1}{2},$$

and

$$\rho = \lim_{n\to\infty} \left| \frac{a_{n+1}}{a_n} \right| = \frac{1}{2} < 1.$$

Consequently, the Ratio Test guarantees that the series $\sum_{n=1}^{\infty} \dfrac{1}{2^n}$ converges.

The general term of $\sum_{n=1}^{\infty} \dfrac{1}{n}$ is $a_n = \dfrac{1}{n}$; thus,

$$\left| \frac{a_{n+1}}{a_n} \right| = \frac{1}{n+1} \cdot \frac{n}{1} = \frac{n}{n+1},$$

and

$$\rho = \lim_{n\to\infty} \left| \frac{a_{n+1}}{a_n} \right| = \lim_{n\to\infty} \frac{n}{n+1} = 1.$$

The Ratio Test is therefore inconclusive for the series $\sum_{n=1}^{\infty} \dfrac{1}{n}$.

3. Can the Ratio Test be used to show convergence if the series is only conditionally convergent?

SOLUTION No. The Ratio Test can only establish absolute convergence and divergence, not conditional convergence.

Exercises

In Exercises 1–20, apply the Ratio Test to determine convergence or divergence, or state that the Ratio Test is inconclusive.

1. $\displaystyle\sum_{n=1}^{\infty} \frac{1}{5^n}$

SOLUTION With $a_n = \frac{1}{5^n}$,

$$\left|\frac{a_{n+1}}{a_n}\right| = \frac{1}{5^{n+1}} \cdot \frac{5^n}{1} = \frac{1}{5} \quad \text{and} \quad \rho = \lim_{n\to\infty}\left|\frac{a_{n+1}}{a_n}\right| = \frac{1}{5} < 1.$$

Therefore, the series $\displaystyle\sum_{n=1}^{\infty} \frac{1}{5^n}$ converges by the Ratio Test.

3. $\displaystyle\sum_{n=1}^{\infty} \frac{1}{n^n}$

SOLUTION With $a_n = \frac{1}{n^n}$,

$$\left|\frac{a_{n+1}}{a_n}\right| = \frac{1}{(n+1)^{n+1}} \cdot \frac{n^n}{1} = \frac{1}{n+1}\left(\frac{n}{n+1}\right)^n = \frac{1}{n+1}\left(1+\frac{1}{n}\right)^{-n},$$

and

$$\rho = \lim_{n\to\infty}\left|\frac{a_{n+1}}{a_n}\right| = 0 \cdot \frac{1}{e} = 0 < 1.$$

Therefore, the series $\displaystyle\sum_{n=1}^{\infty} \frac{1}{n^n}$ converges by the Ratio Test.

5. $\displaystyle\sum_{n=1}^{\infty} \frac{n}{n^2+1}$

SOLUTION With $a_n = \frac{n}{n^2+1}$,

$$\left|\frac{a_{n+1}}{a_n}\right| = \frac{n+1}{(n+1)^2+1} \cdot \frac{n^2+1}{n} = \frac{n+1}{n} \cdot \frac{n^2+1}{n^2+2n+2},$$

and

$$\rho = \lim_{n\to\infty}\left|\frac{a_{n+1}}{a_n}\right| = 1 \cdot 1 = 1.$$

Therefore, for the series $\displaystyle\sum_{n=1}^{\infty} \frac{n}{n^2+1}$, the Ratio Test is inconclusive.

We can show that this series diverges by using the Limit Comparison Test and comparing with the divergent harmonic series.

7. $\displaystyle\sum_{n=1}^{\infty} \frac{2^n}{n^{100}}$

SOLUTION With $a_n = \frac{2^n}{n^{100}}$,

$$\left|\frac{a_{n+1}}{a_n}\right| = \frac{2^{n+1}}{(n+1)^{100}} \cdot \frac{n^{100}}{2^n} = 2\left(\frac{n}{n+1}\right)^{100} \quad \text{and} \quad \rho = \lim_{n\to\infty}\left|\frac{a_{n+1}}{a_n}\right| = 2 \cdot 1^{100} = 2 > 1.$$

Therefore, the series $\displaystyle\sum_{n=1}^{\infty} \frac{2^n}{n^{100}}$ diverges by the Ratio Test.

9. $\displaystyle\sum_{n=1}^{\infty} \frac{10^n}{2^{n^2}}$

SOLUTION With $a_n = \frac{10^n}{2^{n^2}}$,

$$\left|\frac{a_{n+1}}{a_n}\right| = \frac{10^{n+1}}{2^{(n+1)^2}} \cdot \frac{2^{n^2}}{10^n} = 10 \cdot \frac{1}{2^{2n+1}} \quad \text{and} \quad \rho = \lim_{n\to\infty}\left|\frac{a_{n+1}}{a_n}\right| = 10 \cdot 0 = 0 < 1.$$

Therefore, the series $\displaystyle\sum_{n=1}^{\infty} \frac{10^n}{2^{n^2}}$ converges by the Ratio Test.

11. $\displaystyle\sum_{n=1}^{\infty} \frac{e^n}{n^n}$

SOLUTION With $a_n = \frac{e^n}{n^n}$,

$$\left| \frac{a_{n+1}}{a_n} \right| = \frac{e^{n+1}}{(n+1)^{n+1}} \cdot \frac{n^n}{e^n} = \frac{e}{n+1} \left(\frac{n}{n+1} \right)^n = \frac{e}{n+1} \left(1 + \frac{1}{n} \right)^{-n},$$

and

$$\rho = \lim_{n \to \infty} \left| \frac{a_{n+1}}{a_n} \right| = 0 \cdot \frac{1}{e} = 0 < 1.$$

Therefore, the series $\displaystyle\sum_{n=1}^{\infty} \frac{e^n}{n^n}$ converges by the Ratio Test.

13. $\displaystyle\sum_{n=0}^{\infty} \frac{n!}{6^n}$

SOLUTION With $a_n = \frac{n!}{6^n}$,

$$\left| \frac{a_{n+1}}{a_n} \right| = \frac{(n+1)!}{6^{n+1}} \cdot \frac{6^n}{n!} = \frac{n+1}{6} \quad \text{and} \quad \rho = \lim_{n \to \infty} \left| \frac{a_{n+1}}{a_n} \right| = \infty > 1.$$

Therefore, the series $\displaystyle\sum_{n=0}^{\infty} \frac{n!}{6^n}$ diverges by the Ratio Test.

15. $\displaystyle\sum_{n=2}^{\infty} \frac{1}{n \ln n}$

SOLUTION With $a_n = \frac{1}{n \ln n}$,

$$\left| \frac{a_{n+1}}{a_n} \right| = \frac{1}{(n+1) \ln(n+1)} \cdot \frac{n \ln n}{1} = \frac{n}{n+1} \frac{\ln n}{\ln(n+1)},$$

and

$$\rho = \lim_{n \to \infty} \left| \frac{a_{n+1}}{a_n} \right| = 1 \cdot \lim_{n \to \infty} \frac{\ln n}{\ln(n+1)}.$$

Now,

$$\lim_{n \to \infty} \frac{\ln n}{\ln(n+1)} = \lim_{x \to \infty} \frac{\ln x}{\ln(x+1)} = \lim_{x \to \infty} \frac{1/(x+1)}{1/x} = \lim_{x \to \infty} \frac{x}{x+1} = 1.$$

Thus, $\rho = 1$, and the Ratio Test is inconclusive for the series $\displaystyle\sum_{n=2}^{\infty} \frac{1}{n \ln n}$.

Using the Integral Test, we can show that the series $\displaystyle\sum_{n=2}^{\infty} \frac{1}{n \ln n}$ diverges.

17. $\displaystyle\sum_{n=1}^{\infty} \frac{n^2}{(2n+1)!}$

SOLUTION With $a_n = \frac{n^2}{(2n+1)!}$,

$$\left| \frac{a_{n+1}}{a_n} \right| = \frac{(n+1)^2}{(2n+3)!} \cdot \frac{(2n+1)!}{n^2} = \left(\frac{n+1}{n} \right)^2 \frac{1}{(2n+3)(2n+2)},$$

and

$$\rho = \lim_{n \to \infty} \left| \frac{a_{n+1}}{a_n} \right| = 1^2 \cdot 0 = 0 < 1.$$

Therefore, the series $\displaystyle\sum_{n=1}^{\infty} \frac{n^2}{(2n+1)!}$ converges by the Ratio Test.

19. $\displaystyle\sum_{n=2}^{\infty} \frac{1}{2^n + 1}$

SOLUTION With $a_n = \dfrac{1}{2^n + 1}$,

$$\left| \frac{a_{n+1}}{a_n} \right| = \frac{1}{2^{n+1} + 1} \cdot \frac{2^n + 1}{1} = \frac{1 + 2^{-n}}{2 + 2^{-n}}$$

and

$$\rho = \lim_{n \to \infty} \left| \frac{a_{n+1}}{a_n} \right| = \frac{1}{2} < 1$$

Therefore, the series $\displaystyle\sum_{n=2}^{\infty} \frac{1}{2^n + 1}$ converges by the Ratio Test.

21. Show that $\displaystyle\sum_{n=1}^{\infty} n^k 3^{-n}$ converges for all exponents k.

SOLUTION With $a_n = n^k 3^{-n}$,

$$\left| \frac{a_{n+1}}{a_n} \right| = \frac{(n+1)^k 3^{-(n+1)}}{n^k 3^{-n}} = \frac{1}{3}\left(1 + \frac{1}{n}\right)^k,$$

and, for all k,

$$\rho = \lim_{n \to \infty} \left| \frac{a_{n+1}}{a_n} \right| = \frac{1}{3} \cdot 1 = \frac{1}{3} < 1.$$

Therefore, the series $\displaystyle\sum_{n=1}^{\infty} n^k 3^{-n}$ converges for all exponents k by the Ratio Test.

23. Show that $\displaystyle\sum_{n=1}^{\infty} 2^n x^n$ converges if $|x| < \frac{1}{2}$.

SOLUTION With $a_n = 2^n x^n$,

$$\left| \frac{a_{n+1}}{a_n} \right| = \frac{2^{n+1}|x|^{n+1}}{2^n|x|^n} = 2|x| \quad \text{and} \quad \rho = \lim_{n \to \infty} \left| \frac{a_{n+1}}{a_n} \right| = 2|x|.$$

Therefore, $\rho < 1$ and the series $\displaystyle\sum_{n=1}^{\infty} 2^n x^n$ converges by the Ratio Test provided $|x| < \frac{1}{2}$.

25. Show that $\displaystyle\sum_{n=1}^{\infty} \frac{r^n}{n}$ converges if $|r| < 1$.

SOLUTION With $a_n = \dfrac{r^n}{n}$,

$$\left| \frac{a_{n+1}}{a_n} \right| = \frac{|r|^{n+1}}{n+1} \cdot \frac{n}{|r|^n} = |r|\frac{n}{n+1} \quad \text{and} \quad \rho = \lim_{n \to \infty} \left| \frac{a_{n+1}}{a_n} \right| = 1 \cdot |r| = |r|.$$

Therefore, by the Ratio Test, the series $\displaystyle\sum_{n=1}^{\infty} \frac{r^n}{n}$ converges provided $|r| < 1$.

27. Show that $\displaystyle\sum_{n=1}^{\infty} \frac{n!}{n^n}$ converges. *Hint:* Use $\displaystyle\lim_{n \to \infty}\left(1 + \frac{1}{n}\right)^n = e$.

SOLUTION With $a_n = \dfrac{n!}{n^n}$,

$$\left| \frac{a_{n+1}}{a_n} \right| = \frac{(n+1)!}{(n+1)^{n+1}} \cdot \frac{n^n}{n!} = \left(\frac{n}{n+1}\right)^n = \left(1 + \frac{1}{n}\right)^{-n},$$

and

$$\rho = \lim_{n \to \infty} \left| \frac{a_{n+1}}{a_n} \right| = \frac{1}{e} < 1.$$

Therefore, the series $\sum\limits_{n=1}^{\infty} \dfrac{n!}{n^n}$ converges by the Ratio Test.

In Exercises 28–33, assume that $|a_{n+1}/a_n|$ converges to $\rho = \frac{1}{3}$. What can you say about the convergence of the given series?

29. $\sum\limits_{n=1}^{\infty} n^3 a_n$

SOLUTION   Let $b_n = n^3 a_n$. Then

$$\rho = \lim_{n \to \infty} \left| \frac{b_{n+1}}{b_n} \right| = \lim_{n \to \infty} \left(\frac{n+1}{n} \right)^3 \left| \frac{a_{n+1}}{a_n} \right| = 1^3 \cdot \frac{1}{3} = \frac{1}{3} < 1.$$

Therefore, the series $\sum\limits_{n=1}^{\infty} n^3 a_n$ converges by the Ratio Test.

31. $\sum\limits_{n=1}^{\infty} 3^n a_n$

SOLUTION   Let $b_n = 3^n a_n$. Then

$$\rho = \lim_{n \to \infty} \left| \frac{b_{n+1}}{b_n} \right| = \lim_{n \to \infty} \frac{3^{n+1}}{3^n} \left| \frac{a_{n+1}}{a_n} \right| = 3 \cdot \frac{1}{3} = 1.$$

Therefore, the Ratio Test is inconclusive for the series $\sum\limits_{n=1}^{\infty} 3^n a_n$.

33. $\sum\limits_{n=1}^{\infty} a_n^2$

SOLUTION   Let $b_n = a_n^2$. Then

$$\rho = \lim_{n \to \infty} \left| \frac{b_{n+1}}{b_n} \right| = \lim_{n \to \infty} \left| \frac{a_{n+1}}{a_n} \right|^2 = \left(\frac{1}{3} \right)^2 = \frac{1}{9} < 1.$$

Therefore, the series $\sum\limits_{n=1}^{\infty} a_n^2$ converges by the Ratio Test.

35. Is the Ratio Test conclusive for the p-series $\sum\limits_{n=1}^{\infty} \dfrac{1}{n^p}$?

SOLUTION   With $a_n = \frac{1}{n^p}$,

$$\left| \frac{a_{n+1}}{a_n} \right| = \frac{1}{(n+1)^p} \cdot \frac{n^p}{1} = \left(\frac{n}{n+1} \right)^p \quad \text{and} \quad \rho = \lim_{n \to \infty} \left| \frac{a_{n+1}}{a_n} \right| = 1^p = 1.$$

Therefore, the Ratio Test is inconclusive for the p-series $\sum\limits_{n=1}^{\infty} \dfrac{1}{n^p}$.

In Exercises 36–41, use the Root Test to determine convergence or divergence (or state that the test is inconclusive).

37. $\sum\limits_{n=1}^{\infty} \dfrac{1}{n^n}$

SOLUTION   With $a_n = \frac{1}{n^n}$,

$$\sqrt[n]{a_n} = \sqrt[n]{\frac{1}{n^n}} = \frac{1}{n} \quad \text{and} \quad \lim_{n \to \infty} \sqrt[n]{a_n} = 0 < 1.$$

Therefore, the series $\sum\limits_{n=1}^{\infty} \dfrac{1}{n^n}$ converges by the Root Test.

39. $\displaystyle\sum_{k=0}^{\infty} \left(\frac{k}{3k+1}\right)^k$

SOLUTION With $a_k = \left(\frac{k}{3k+1}\right)^k$,

$$\sqrt[k]{a_k} = \sqrt[k]{\left(\frac{k}{3k+1}\right)^k} = \frac{k}{3k+1} \quad \text{and} \quad \lim_{k\to\infty} \sqrt[k]{a_k} = \frac{1}{3} < 1.$$

Therefore, the series $\displaystyle\sum_{k=0}^{\infty} \left(\frac{k}{3k+1}\right)^k$ converges by the Root Test.

41. $\displaystyle\sum_{n=4}^{\infty} \left(1+\frac{1}{n}\right)^{-n^2}$

SOLUTION With $a_k = \left(1+\frac{1}{n}\right)^{-n^2}$,

$$\sqrt[n]{a_n} = \sqrt[n]{\left(1+\frac{1}{n}\right)^{-n^2}} = \left(1+\frac{1}{n}\right)^{-n} \quad \text{and} \quad \lim_{n\to\infty} \sqrt[n]{a_n} = e^{-1} < 1.$$

Therefore, the series $\displaystyle\sum_{n=4}^{\infty} \left(1+\frac{1}{n}\right)^{-n^2}$ converges by the Root Test.

In Exercises 43–56, determine convergence or divergence using any method covered in the text so far.

43. $\displaystyle\sum_{n=1}^{\infty} \frac{2^n + 4^n}{7^n}$

SOLUTION Because the series

$$\sum_{n=1}^{\infty} \frac{2^n}{7^n} = \sum_{n=1}^{\infty} \left(\frac{2}{7}\right)^n \quad \text{and} \quad \sum_{n=1}^{\infty} \frac{4^n}{7^n} = \sum_{n=1}^{\infty} \left(\frac{4}{7}\right)^n$$

are both convergent geometric series, it follows that

$$\sum_{n=1}^{\infty} \frac{2^n + 4^n}{7^n} = \sum_{n=1}^{\infty} \left(\frac{2}{7}\right)^n + \sum_{n=1}^{\infty} \left(\frac{4}{7}\right)^n$$

also converges.

45. $\displaystyle\sum_{n=1}^{\infty} \frac{n^3}{5^n}$

SOLUTION The presence of the exponential term suggests applying the Ratio Test. With $a_n = \frac{n^3}{5^n}$,

$$\left|\frac{a_{n+1}}{a_n}\right| = \frac{(n+1)^3}{5^{n+1}} \cdot \frac{5^n}{n^3} = \frac{1}{5}\left(1+\frac{1}{n}\right)^3 \quad \text{and} \quad \rho = \lim_{n\to\infty} \left|\frac{a_{n+1}}{a_n}\right| = \frac{1}{5} \cdot 1^3 = \frac{1}{5} < 1.$$

Therefore, the series $\displaystyle\sum_{n=1}^{\infty} \frac{n^3}{5^n}$ converges by the Ratio Test.

47. $\displaystyle\sum_{n=2}^{\infty} \frac{1}{\sqrt{n^3 - n^2}}$

SOLUTION This series is similar to a *p*-series; because

$$\frac{1}{\sqrt{n^3 - n^2}} \approx \frac{1}{\sqrt{n^3}} = \frac{1}{n^{3/2}}$$

for large n, we will apply the Limit Comparison Test comparing with the *p*-series with $p = \frac{3}{2}$. Now,

$$L = \lim_{n\to\infty} \frac{\frac{1}{\sqrt{n^3 - n^2}}}{\frac{1}{n^{3/2}}} = \lim_{n\to\infty} \sqrt{\frac{n^3}{n^3 - n^2}} = 1.$$

The *p*-series with $p = \frac{3}{2}$ converges and L exists; therefore, the series $\displaystyle\sum_{n=2}^{\infty} \frac{1}{\sqrt{n^3 - n^2}}$ also converges.

49. $\displaystyle\sum_{n=1}^{\infty} n^{-0.8}$

SOLUTION

$$\sum_{n=1}^{\infty} n^{-0.8} = \sum_{n=1}^{\infty} \frac{1}{n^{0.8}}$$

so that this is a divergent p-series.

51. $\displaystyle\sum_{n=1}^{\infty} 4^{-2n+1}$

SOLUTION   Observe

$$\sum_{n=1}^{\infty} 4^{-2n+1} = \sum_{n=1}^{\infty} 4 \cdot (4^{-2})^n = \sum_{n=1}^{\infty} 4 \left(\frac{1}{16}\right)^n$$

is a geometric series with $r = \frac{1}{16}$; therefore, this series converges.

53. $\displaystyle\sum_{n=1}^{\infty} \sin \frac{1}{n^2}$

SOLUTION   Here, we will apply the Limit Comparison Test, comparing with the p-series with $p = 2$. Now,

$$L = \lim_{n\to\infty} \frac{\sin \frac{1}{n^2}}{\frac{1}{n^2}} = \lim_{u\to 0} \frac{\sin u}{u} = 1,$$

where $u = \frac{1}{n^2}$. The p-series with $p = 2$ converges and L exists; therefore, the series $\displaystyle\sum_{n=1}^{\infty} \sin \frac{1}{n^2}$ also converges.

55. $\displaystyle\sum_{n=1}^{\infty} \frac{(-2)^n}{\sqrt{n}}$

SOLUTION   Because

$$\lim_{n\to\infty} \frac{2^n}{\sqrt{n}} = \lim_{x\to\infty} \frac{2^x}{\sqrt{x}} = \lim_{x\to\infty} \frac{2^x \ln 2}{\frac{1}{2\sqrt{x}}} = \lim_{x\to\infty} 2^{x+1} \sqrt{x} \ln 2 = \infty \neq 0,$$

the general term in the series $\displaystyle\sum_{n=1}^{\infty} \frac{(-2)^n}{\sqrt{n}}$ does not tend toward zero; therefore, the series diverges by the Divergence Test.

Further Insights and Challenges

57.   **Proof of the Root Test**   Let $S = \displaystyle\sum_{n=0}^{\infty} a_n$ be a positive series, and assume that $L = \lim_{n\to\infty} \sqrt[n]{a_n}$ exists.

(a) Show that S converges if $L < 1$. *Hint:* Choose R with $L < R < 1$ and show that $a_n \leq R^n$ for n sufficiently large. Then compare with the geometric series $\sum R^n$.

(b) Show that S diverges if $L > 1$.

SOLUTION   Suppose $\lim_{n\to\infty} \sqrt[n]{a_n} = L$ exists.

(a) If $L < 1$, let $\epsilon = \dfrac{1-L}{2}$. By the definition of a limit, there is a positive integer N such that

$$-\epsilon \leq \sqrt[n]{a_n} - L \leq \epsilon$$

for $n \geq N$. From this, we conclude that

$$0 \leq \sqrt[n]{a_n} \leq L + \epsilon$$

for $n \geq N$. Now, let $R = L + \epsilon$. Then

$$R = L + \frac{1-L}{2} = \frac{L+1}{2} < \frac{1+1}{2} = 1,$$

and

$$0 \leq \sqrt[n]{a_n} \leq R \quad \text{or} \quad 0 \leq a_n \leq R^n$$

for $n \geq N$. Because $0 \leq R < 1$, the series $\sum\limits_{n=N}^{\infty} R^n$ is a convergent geometric series, so the series $\sum\limits_{n=N}^{\infty} a_n$ converges by the Comparison Test. Therefore, the series $\sum\limits_{n=0}^{\infty} a_n$ also converges.

(b) If $L > 1$, let $\epsilon = \dfrac{L-1}{2}$. By the definition of a limit, there is a positive integer N such that

$$-\epsilon \leq \sqrt[n]{a_n} - L \leq \epsilon$$

for $n \geq N$. From this, we conclude that

$$L - \epsilon \leq \sqrt[n]{a_n}$$

for $n \geq N$. Now, let $R = L - \epsilon$. Then

$$R = L - \frac{L-1}{2} = \frac{L+1}{2} > \frac{1+1}{2} = 1,$$

and

$$R \leq \sqrt[n]{a_n} \quad \text{or} \quad R^n \leq a_n$$

for $n \geq N$. Because $R > 1$, the series $\sum\limits_{n=N}^{\infty} R^n$ is a divergent geometric series, so the series $\sum\limits_{n=N}^{\infty} a_n$ diverges by the Comparison Test. Therefore, the series $\sum\limits_{n=0}^{\infty} a_n$ also diverges.

59. Let $S = \sum\limits_{n=1}^{\infty} \dfrac{c^n n!}{n^n}$, where c is a constant.

(a) Prove that S converges absolutely if $|c| < e$ and diverges if $|c| > e$.

(b) It is known that $\lim\limits_{n\to\infty} \dfrac{e^n n!}{n^{n+1/2}} = \sqrt{2\pi}$. Verify this numerically.

(c) Use the Limit Comparison Test to prove that S diverges for $c = e$.

SOLUTION

(a) With $a_n = \dfrac{c^n n!}{n^n}$,

$$\left| \frac{a_{n+1}}{a_n} \right| = \frac{|c|^{n+1}(n+1)!}{(n+1)^{n+1}} \cdot \frac{n^n}{|c|^n n!} = |c|\left(\frac{n}{n+1}\right)^n = |c|\left(1 + \frac{1}{n}\right)^{-n},$$

and

$$\rho = \lim_{n\to\infty} \left| \frac{a_{n+1}}{a_n} \right| = |c|e^{-1}.$$

Thus, by the Ratio Test, the series $\sum\limits_{n=1}^{\infty} \dfrac{c^n n!}{n^n}$ converges when $|c|e^{-1} < 1$, or when $|c| < e$. The series diverges when $|c| > e$.

(b) The table below lists the value of $\dfrac{e^n n!}{n^{n+1/2}}$ for several increasing values of n. Since $\sqrt{2\pi} = 2.506628275$, the numerical evidence verifies that

$$\lim_{n\to\infty} \frac{e^n n!}{n^{n+1/2}} = \sqrt{2\pi}.$$

n	100	1000	10000	100000
$\dfrac{e^n n!}{n^{n+1/2}}$	2.508717995	2.506837169	2.506649163	2.506630363

(c) With $c = e$, the series S becomes $\sum\limits_{n=1}^{\infty} \dfrac{e^n n!}{n^n}$. Using the result from part (b),

$$L = \lim_{n\to\infty} \frac{\frac{e^n n!}{n^n}}{\sqrt{n}} = \lim_{n\to\infty} \frac{e^n n!}{n^{n+1/2}} = \sqrt{2\pi}.$$

Because the series $\sum\limits_{n=1}^{\infty} \sqrt{n}$ diverges by the Divergence Test and $L > 0$, we conclude that $\sum\limits_{n=1}^{\infty} \dfrac{e^n n!}{n^n}$ diverges by the Limit Comparison Test.

10.6 Power Series (LT Section 11.6)

Preliminary Questions

1. Suppose that $\sum a_n x^n$ converges for $x = 5$. Must it also converge for $x = 4$? What about $x = -3$?

SOLUTION The power series $\sum a_n x^n$ is centered at $x = 0$. Because the series converges for $x = 5$, the radius of convergence must be at least 5 and the series converges absolutely at least for the interval $|x| < 5$. Both $x = 4$ and $x = -3$ are inside this interval, so the series converges for $x = 4$ and for $x = -3$.

2. Suppose that $\sum a_n (x - 6)^n$ converges for $x = 10$. At which of the points (a)–(d) must it also converge?

(a) $x = 8$ **(b)** $x = 11$ **(c)** $x = 3$ **(d)** $x = 0$

SOLUTION The given power series is centered at $x = 6$. Because the series converges for $x = 10$, the radius of convergence must be at least $|10 - 6| = 4$ and the series converges absolutely at least for the interval $|x - 6| < 4$, or $2 < x < 10$.

(a) $x = 8$ is inside the interval $2 < x < 10$, so the series converges for $x = 8$.

(b) $x = 11$ is not inside the interval $2 < x < 10$, so the series may or may not converge for $x = 11$.

(c) $x = 3$ is inside the interval $2 < x < 10$, so the series converges for $x = 2$.

(d) $x = 0$ is not inside the interval $2 < x < 10$, so the series may or may not converge for $x = 0$.

3. What is the radius of convergence of $F(3x)$ if $F(x)$ is a power series with radius of convergence $R = 12$?

SOLUTION If the power series $F(x)$ has radius of convergence $R = 12$, then the power series $F(3x)$ has radius of convergence $R = \frac{12}{3} = 4$.

4. The power series $F(x) = \sum_{n=1}^{\infty} n x^n$ has radius of convergence $R = 1$. What is the power series expansion of $F'(x)$ and what is its radius of convergence?

SOLUTION We obtain the power series expansion for $F'(x)$ by differentiating the power series expansion for $F(x)$ term-by-term. Thus,

$$F'(x) = \sum_{n=1}^{\infty} n^2 x^{n-1}.$$

The radius of convergence for this series is $R = 1$, the same as the radius of convergence for the series expansion for $F(x)$.

Exercises

1. Use the Ratio Test to determine the radius of convergence R of $\sum_{n=0}^{\infty} \frac{x^n}{2^n}$. Does it converge at the endpoints $x = \pm R$?

SOLUTION With $a_n = \frac{x^n}{2^n}$,

$$\left| \frac{a_{n+1}}{a_n} \right| = \frac{|x|^{n+1}}{2^{n+1}} \cdot \frac{2^n}{|x|^n} = \frac{|x|}{2} \quad \text{and} \quad \rho = \lim_{n \to \infty} \left| \frac{a_{n+1}}{a_n} \right| = \frac{|x|}{2}.$$

By the Ratio Test, the series converges when $\rho = \frac{|x|}{2} < 1$, or $|x| < 2$, and diverges when $\rho = \frac{|x|}{2} > 1$, or $|x| > 2$. The radius of convergence is therefore $R = 2$. For $x = -2$, the left endpoint, the series becomes $\sum_{n=0}^{\infty} (-1)^n$, which is divergent. For $x = 2$, the right endpoint, the series becomes $\sum_{n=0}^{\infty} 1$, which is also divergent. Thus the series diverges at both endpoints.

3. Show that the power series (a)–(c) have the same radius of convergence. Then show that (a) diverges at both endpoints, (b) converges at one endpoint but diverges at the other, and (c) converges at both endpoints.

(a) $\sum_{n=1}^{\infty} \frac{x^n}{3^n}$ **(b)** $\sum_{n=1}^{\infty} \frac{x^n}{n 3^n}$ **(c)** $\sum_{n=1}^{\infty} \frac{x^n}{n^2 3^n}$

SOLUTION

(a) With $a_n = \frac{x^n}{3^n}$,

$$\rho = \lim_{n \to \infty} \left| \frac{a_{n+1}}{a_n} \right| = \lim_{n \to \infty} \left| \frac{x^{n+1}}{3^{n+1}} \cdot \frac{3^n}{x^n} \right| = \lim_{n \to \infty} \left| \frac{x}{3} \right| = \left| \frac{x}{3} \right|$$

Then $\rho < 1$ if $|x| < 3$, so that the radius of convergence is $R = 3$. For the endpoint $x = 3$, the series becomes

$$\sum_{n=1}^{\infty} \frac{3^n}{3^n} = \sum_{n=1}^{\infty} 1,$$

which diverges by the Divergence Test. For the endpoint $x = -3$, the series becomes

$$\sum_{n=1}^{\infty} \frac{(-3)^n}{3^n} = \sum_{n=1}^{\infty} (-1)^n,$$

which also diverges by the Divergence Test.

(b) With $a_n = \frac{x^n}{n3^n}$,

$$\rho = \lim_{n \to \infty} \left| \frac{a_{n+1}}{a_n} \right| = \lim_{n \to \infty} \left| \frac{x^{n+1}}{(n+1)3^{n+1}} \cdot \frac{n3^n}{x^n} \right| = \lim_{n \to \infty} \left| \frac{x}{3} \left(\frac{n}{n+1} \right) \right| = \left| \frac{x}{3} \right|.$$

Then $\rho < 1$ when $|x| < 3$, so that the radius of convergence is $R = 3$. For the endpoint $x = 3$, the series becomes

$$\sum_{n=1}^{\infty} \frac{3^n}{n3^n} = \sum_{n=1}^{\infty} \frac{1}{n},$$

which is the divergent harmonic series. For the endpoint $x = -3$, the series becomes

$$\sum_{n=1}^{\infty} \frac{(-3)^n}{n3^n} = \sum_{n=1}^{\infty} \frac{(-1)^n}{n},$$

which converges by the Leibniz Test.

(c) With $a_n = \frac{x^n}{n^2 3^n}$,

$$\rho = \lim_{n \to \infty} \left| \frac{a_{n+1}}{a_n} \right| = \lim_{n \to \infty} \left| \frac{x^{n+1}}{(n+1)^2 3^{n+1}} \cdot \frac{n^2 3^n}{x^n} \right| = \lim_{n \to \infty} \left| \frac{x}{3} \left(\frac{n}{n+1} \right)^2 \right| = \left| \frac{x}{3} \right|$$

Then $\rho < 1$ when $|x| < 3$, so that the radius of convergence is $R = 3$. For the endpoint $x = 3$, the series becomes

$$\sum_{n=1}^{\infty} \frac{3^n}{n^2 3^n} = \sum_{n=1}^{\infty} \frac{1}{n^2},$$

which is a convergent p-series. For the endpoint $x = -3$, the series becomes

$$\sum_{n=1}^{\infty} \frac{(-3)^n}{n^2 3^n} = \sum_{n=1}^{\infty} \frac{(-1)^n}{n^2},$$

which converges by the Leibniz Test.

5. Show that $\displaystyle\sum_{n=0}^{\infty} n^n x^n$ diverges for all $x \neq 0$.

SOLUTION With $a_n = n^n x^n$, and assuming $x \neq 0$,

$$\rho = \lim_{n \to \infty} \left| \frac{a_{n+1}}{a_n} \right| = \lim_{n \to \infty} \left| \frac{(n+1)^{n+1} x^{n+1}}{n^n x^n} \right| = \lim_{n \to \infty} \left| x \left(1 + \frac{1}{n} \right)^n (n+1) \right| = \infty$$

$\rho < 1$ only if $x = 0$, so that the radius of convergence is therefore $R = 0$. In other words, the power series converges only for $x = 0$.

7. Use the Ratio Test to show that $\displaystyle\sum_{n=0}^{\infty} \frac{x^{2n}}{3^n}$ has radius of convergence $R = \sqrt{3}$.

SOLUTION With $a_n = \frac{x^{2n}}{3^n}$,

$$\rho = \lim_{n \to \infty} \left| \frac{a_{n+1}}{a_n} \right| = \lim_{n \to \infty} \left| \frac{x^{2(n+1)}}{3^{n+1}} \cdot \frac{3^n}{x^{2n}} \right| = \lim_{n \to \infty} \left| \frac{x^2}{3} \right| = \left| \frac{x^2}{3} \right|$$

Then $\rho < 1$ when $|x^2| < 3$, or $x = \sqrt{3}$, so the radius of convergence is $R = \sqrt{3}$.

In Exercises 9–34, find the interval of convergence.

9. $\displaystyle\sum_{n=0}^{\infty} nx^n$

SOLUTION With $a_n = nx^n$,

$$\rho = \lim_{n\to\infty}\left|\frac{a_{n+1}}{a_n}\right| = \lim_{n\to\infty}\left|\frac{(n+1)x^{n+1}}{nx^n}\right| = \lim_{n\to\infty}\left|x\frac{n+1}{n}\right| = |x|$$

Then $\rho < 1$ when $|x| < 1$, so that the radius of convergence is $R = 1$, and the series converges absolutely on the interval $|x| < 1$, or $-1 < x < 1$. For the endpoint $x = 1$, the series becomes $\displaystyle\sum_{n=0}^{\infty} n$, which diverges by the Divergence Test.

For the endpoint $x = -1$, the series becomes $\displaystyle\sum_{n=1}^{\infty} (-1)^n n$, which also diverges by the Divergence Test. Thus, the series

$\displaystyle\sum_{n=0}^{\infty} nx^n$ converges for $-1 < x < 1$ and diverges elsewhere.

11. $\displaystyle\sum_{n=1}^{\infty} (-1)^n \frac{x^{2n+1}}{2^n n}$

SOLUTION With $a_n = (-1)^n \dfrac{x^{2n+1}}{2^n n}$,

$$\rho = \lim_{n\to\infty}\left|\frac{x^{2(n+1)+1}}{2^{n+1}(n+1)} \cdot \frac{2^n n}{x^{2n+1}}\right| = \lim_{n\to\infty}\left|\frac{x^2}{2} \cdot \frac{n}{n+1}\right| = \left|\frac{x^2}{2}\right|$$

Then $\rho < 1$ when $|x| < \sqrt{2}$, so the radius of convergence is $R = \sqrt{2}$, and the series converges absolutely on the interval $-\sqrt{2} < x < \sqrt{2}$. For the endpoint $x = -\sqrt{2}$, the series becomes $\displaystyle\sum_{n=1}^{\infty}(-1)^n\frac{-\sqrt{2}}{n} = \sum_{n=1}^{\infty}(-1)^{n+1}\frac{\sqrt{2}}{n}$, which converges

by the Leibniz test. For the endpoint $x = \sqrt{2}$, the series becomes $\displaystyle\sum_{n=1}^{\infty}(-1)^n\frac{\sqrt{2}}{n}$ which also converges by the Leibniz test.

Thus the series $\displaystyle\sum_{n=1}^{\infty}(-1)^n\frac{x^{2n+1}}{2^n n}$ converges for $-\sqrt{2} \le x \le \sqrt{2}$ and diverges elsewhere.

13. $\displaystyle\sum_{n=4}^{\infty} \frac{x^n}{n^5}$

SOLUTION With $a_n = \frac{x^n}{n^5}$,

$$\rho = \lim_{n\to\infty}\left|\frac{a_{n+1}}{a_n}\right| = \lim_{n\to\infty}\left|\frac{x^{n+1}}{(n+1)^5} \cdot \frac{n^5}{x^n}\right| = \lim_{n\to\infty}\left|x\left(\frac{n}{n+1}\right)^5\right| = |x|$$

Then $\rho < 1$ when $|x| < 1$, so the radius of convergence is $R = 1$, and the series converges absolutely on the interval $|x| < 1$, or $-1 < x < 1$. For the endpoint $x = 1$, the series becomes $\displaystyle\sum_{n=1}^{\infty}\frac{1}{n^5}$, which is a convergent p-series. For the

endpoint $x = -1$, the series becomes $\displaystyle\sum_{n=1}^{\infty}\frac{(-1)^n}{n^5}$, which converges by the Leibniz Test. Thus, the series $\displaystyle\sum_{n=4}^{\infty}\frac{x^n}{n^5}$ converges

for $-1 \le x \le 1$ and diverges elsewhere.

15. $\displaystyle\sum_{n=0}^{\infty} \frac{x^n}{(n!)^2}$

SOLUTION With $a_n = \frac{x^n}{(n!)^2}$,

$$\rho = \lim_{n\to\infty}\left|\frac{a_{n+1}}{a_n}\right| = \lim_{n\to\infty}\left|\frac{x^{n+1}}{((n+1)!)^2} \cdot \frac{(n!)^2}{x^n}\right| = \lim_{n\to\infty}\left|x\left(\frac{1}{n+1}\right)^2\right| = 0$$

$\rho < 1$ for all x, so the radius of convergence is $R = \infty$, and the series converges absolutely for all x.

17. $\sum_{n=0}^{\infty} \dfrac{(2n)!}{(n!)^3} x^n$

SOLUTION With $a_n = \dfrac{(2n)! x^n}{(n!)^3}$, and assuming $x \neq 0$,

$$\rho = \lim_{n \to \infty} \left| \frac{a_{n+1}}{a_n} \right| = \lim_{n \to \infty} \left| \frac{(2(n+1))! x^{n+1}}{((n+1)!)^3} \cdot \frac{(n!)^3}{(2n)! x^n} \right| = \lim_{n \to \infty} \left| x \frac{(2n+2)(2n+1)}{(n+1)^3} \right|$$

$$= \lim_{n \to \infty} \left| x \frac{4n^2 + 6n + 2}{n^3 + 3n^2 + 3n + 1} \right| = \lim_{n \to \infty} \left| x \frac{4n^{-1} + 6n^{-1} + 2n^{-3}}{1 + 3n^{-1} + 3n^{-2} + n^{-3}} \right| = 0$$

Then $\rho < 1$ for all x, so the radius of convergence is $R = \infty$, and the series converges absolutely for all x.

19. $\sum_{n=0}^{\infty} \dfrac{(-1)^n x^n}{\sqrt{n^2 + 1}}$

SOLUTION With $a_n = \dfrac{(-1)^n x^n}{\sqrt{n^2+1}}$,

$$\rho = \lim_{n \to \infty} \left| \frac{a_{n+1}}{a_n} \right| = \lim_{n \to \infty} \left| \frac{(-1)^{n+1} x^{n+1}}{\sqrt{n^2 + 2n + 2}} \cdot \frac{\sqrt{n^2 + 1}}{(-1)^n x^n} \right|$$

$$= \lim_{n \to \infty} \left| x \frac{\sqrt{n^2 + 1}}{\sqrt{n^2 + 2n + 2}} \right| = \lim_{n \to \infty} \left| x \sqrt{\frac{n^2 + 1}{n^2 + 2n + 2}} \right| = \lim_{n \to \infty} \left| x \sqrt{\frac{1 + 1/n^2}{1 + 2/n + 2/n^2}} \right|$$

$$= |x|$$

Then $\rho < 1$ when $|x| < 1$, so the radius of convergence is $R = 1$, and the series converges absolutely on the interval $-1 < x < 1$. For the endpoint $x = 1$, the series becomes $\sum_{n=1}^{\infty} \dfrac{(-1)^n}{\sqrt{n^2 + 1}}$, which converges by the Leibniz Test. For the endpoint $x = -1$, the series becomes $\sum_{n=1}^{\infty} \dfrac{1}{\sqrt{n^2 + 1}}$, which diverges by the Limit Comparison Test comparing with the divergent harmonic series. Thus, the series $\sum_{n=0}^{\infty} \dfrac{(-1)^n x^n}{\sqrt{n^2 + 1}}$ converges for $-1 < x \leq 1$ and diverges elsewhere.

21. $\sum_{n=15}^{\infty} \dfrac{x^{2n+1}}{3n + 1}$

SOLUTION With $a_n = \dfrac{x^{2n+1}}{3n+1}$,

$$\rho = \lim_{n \to \infty} \left| \frac{a_{n+1}}{a_n} \right| = \lim_{n \to \infty} \left| \frac{x^{2n+3}}{3n + 4} \cdot \frac{3n + 1}{x^{2n+1}} \right| = \lim_{n \to \infty} \left| x^2 \frac{3n + 1}{3n + 4} \right| = |x^2|$$

Then $\rho < 1$ when $|x^2| < 1$, so the radius of convergence is $R = 1$, and the series converges absolutely for $-1 < x < 1$. For the endpoint $x = 1$, the series becomes $\sum_{n=15}^{\infty} \dfrac{1}{3n + 1}$, which diverges by the Limit Comparison Test comparing with the divergent harmonic series. For the endpoint $x = -1$, the series becomes $\sum_{n=15}^{\infty} \dfrac{-1}{3n + 1}$, which also diverges by the Limit Comparison Test comparing with the divergent harmonic series. Thus, the series $\sum_{n=15}^{\infty} \dfrac{x^{2n+1}}{3n + 1}$ converges for $-1 < x < 1$ and diverges elsewhere.

23. $\sum_{n=2}^{\infty} \dfrac{x^n}{\ln n}$

SOLUTION With $a_n = \dfrac{x^n}{\ln n}$,

$$\rho = \lim_{n \to \infty} \left| \frac{a_{n+1}}{a_n} \right| = \lim_{n \to \infty} \left| \frac{x^{n+1}}{\ln(n+1)} \cdot \frac{\ln n}{x^n} \right| = \lim_{n \to \infty} \left| x \frac{\ln(n+1)}{\ln n} \right| = \lim_{n \to \infty} \left| x \frac{1/(n+1)}{1/n} \right| = \lim_{n \to \infty} \left| x \frac{n}{n + 1} \right| = |x|$$

using L'Hôpital's rule. Then $\rho < 1$ when $|x| < 1$, so the radius of convergence is 1, and the series converges absolutely on the interval $|x| < 1$, or $-1 < x < 1$. For the endpoint $x = 1$, the series becomes $\sum_{n=2}^{\infty} \frac{1}{\ln n}$. Because $\frac{1}{\ln n} > \frac{1}{n}$ and

$\sum_{n=2}^{\infty} \frac{1}{n}$ is the divergent harmonic series, the endpoint series diverges by the Comparison Test. For the endpoint $x = -1$,

the series becomes $\sum_{n=2}^{\infty} \frac{(-1)^n}{\ln n}$, which converges by the Leibniz Test. Thus, the series $\sum_{n=2}^{\infty} \frac{x^n}{\ln n}$ converges for $-1 \le x < 1$ and diverges elsewhere.

25. $\sum_{n=1}^{\infty} n(x - 3)^n$

SOLUTION With $a_n = n(x - 3)^n$,

$$\rho = \lim_{n \to \infty} \left| \frac{a_{n+1}}{a_n} \right| = \lim_{n \to \infty} \left| \frac{(n + 1)(x - 3)^{n+1}}{n(x - 3)^n} \right| = \lim_{n \to \infty} \left| (x - 3) \cdot \frac{n + 1}{n} \right| = |x - 3|$$

Then $\rho < 1$ when $|x - 3| < 1$, so the radius of convergence is 1, and the series converges absolutely on the interval $|x - 3| < 1$, or $2 < x < 4$. For the endpoint $x = 4$, the series becomes $\sum_{n=1}^{\infty} n$, which diverges by the Divergence Test.

For the endpoint $x = 2$, the series becomes $\sum_{n=1}^{\infty} (-1)^n n$, which also diverges by the Divergence Test. Thus, the series

$\sum_{n=1}^{\infty} n(x - 3)^n$ converges for $2 < x < 4$ and diverges elsewhere.

27. $\sum_{n=1}^{\infty} (-1)^n n^5 (x - 7)^n$

SOLUTION With $a_n = (-1)^n n^5 (x - 7)^n$,

$$\rho = \lim_{n \to \infty} \left| \frac{a_{n+1}}{a_n} \right| = \lim_{n \to \infty} \left| \frac{(-1)^{n+1} (n + 1)^5 (x - 7)^{n+1}}{(-1)^n n^5 (x - 7)^n} \right| = \lim_{n \to \infty} \left| (x - 7) \cdot \frac{(n + 1)^5}{n^5} \right|$$

$$= \lim_{n \to \infty} \left| (x - 7) \cdot \frac{n^5 + \cdots}{n^5} \right| = |x - 7|$$

Then $\rho < 1$ when $|x - 7| < 1$, so the radius of convergence is 1, and the series converges absolutely on the interval $|x - 7| < 1$, or $6 < x < 8$. For the endpoint $x = 6$, the series becomes $\sum_{n=1}^{\infty} (-1)^{2n} n^5 = \sum_{n=1}^{\infty} n^5$, which diverges by the

Divergence Test. For the endpoint $x = 8$, the series becomes $\sum_{n=1}^{\infty} (-1)^n n^5$, which also diverges by the Divergence Test.

Thus, the series $\sum_{n=1}^{\infty} (-1)^n n^5 (x - 7)^n$ converges for $6 < x < 8$ and diverges elsewhere.

29. $\sum_{n=1}^{\infty} \frac{2^n}{3n} (x + 3)^n$

SOLUTION With $a_n = \frac{2^n (x+3)^n}{3n}$,

$$\rho = \lim_{n \to \infty} \left| \frac{a_{n+1}}{a_n} \right| = \lim_{n \to \infty} \left| \frac{2^{n+1} (x + 3)^{n+1}}{3(n + 1)} \cdot \frac{3n}{2^n (x + 3)^n} \right| = \lim_{n \to \infty} \left| 2(x + 3) \cdot \frac{3n}{3n + 3} \right|$$

$$= \lim_{n \to \infty} \left| 2(x + 3) \cdot \frac{1}{1 + 1/n} \right| = |2(x + 3)|$$

Then $\rho < 1$ when $|2(x+3)| < 1$, so when $|x+3| < \frac{1}{2}$. Thus the radius of convergence is $\frac{1}{2}$, and the series converges absolutely on the interval $|x+3| < \frac{1}{2}$, or $-\frac{7}{2} < x < -\frac{5}{2}$. For the endpoint $x = -\frac{5}{2}$, the series becomes $\sum_{n=1}^{\infty} \frac{1}{3n}$, which diverges because it is a multiple of the divergent harmonic series. For the endpoint $x = -\frac{7}{2}$, the series becomes $\sum_{n=1}^{\infty} \frac{(-1)^n}{3n}$, which converges by the Leibniz Test. Thus, the series $\sum_{n=1}^{\infty} \frac{2^n}{3n}(x+3)^n$ converges for $-\frac{7}{2} \le x < -\frac{5}{2}$ and diverges elsewhere.

31. $\sum_{n=0}^{\infty} \frac{(-5)^n}{n!}(x+10)^n$

SOLUTION With $a_n = \frac{(-5)^n}{n!}(x+10)^n$,

$$\rho = \lim_{n\to\infty} \left| \frac{a_{n+1}}{a_n} \right| = \lim_{n\to\infty} \left| \frac{(-5)^{n+1}(x+10)^{n+1}}{(n+1)!} \cdot \frac{n!}{(-5)^n(x+10)^n} \right| = \lim_{n\to\infty} \left| 5(x+10)\frac{1}{n} \right| = 0$$

Thus $\rho < 1$ for all x, so the radius of convergence is infinite, and $\sum_{n=0}^{\infty} \frac{(-5)^n}{n!}(x+10)^n$ converges for all x.

33. $\sum_{n=12}^{\infty} e^n(x-2)^n$

SOLUTION With $a_n = e^n(x-2)^n$,

$$\rho = \lim_{n\to\infty} \left| \frac{a_{n+1}}{a_n} \right| = \lim_{n\to\infty} \left| \frac{e^{n+1}(x-2)^{n+1}}{e^n(x-2)^n} \right| = \lim_{n\to\infty} |e(x-2)| = |e(x-2)|$$

Thus $\rho < 1$ when $|e(x-2)| < 1$, so when $|x-2| < e^{-1}$. Thus the radius of convergence is e^{-1}, and the series converges absolutely on the interval $|x-2| < e^{-1}$, or $2 - e^{-1} < x < 2 + e^{-1}$. For the endpoint $x = 2 + e^{-1}$, the series becomes $\sum_{n=1}^{\infty} 1$, which diverges by the Divergence Test. For the endpoint $x = 2 - e^{-1}$, the series becomes $\sum_{n=1}^{\infty} (-1)^n$, which also diverges by the Divergence Test. Thus, the series $\sum_{n=12}^{\infty} e^n(x-2)^n$ converges for $2 - e^{-1} < x < 2 + e^{-1}$ and diverges elsewhere.

In Exercises 35–40, use Eq. (2) to expand the function in a power series with center $c = 0$ and determine the interval of convergence.

35. $f(x) = \dfrac{1}{1-3x}$

SOLUTION Substituting $3x$ for x in Eq. (2), we obtain

$$\frac{1}{1-3x} = \sum_{n=0}^{\infty} (3x)^n = \sum_{n=0}^{\infty} 3^n x^n.$$

This series is valid for $|3x| < 1$, or $|x| < \frac{1}{3}$.

37. $f(x) = \dfrac{1}{3-x}$

SOLUTION First write

$$\frac{1}{3-x} = \frac{1}{3} \cdot \frac{1}{1 - \frac{x}{3}}.$$

Substituting $\frac{x}{3}$ for x in Eq. (2), we obtain

$$\frac{1}{1 - \frac{x}{3}} = \sum_{n=0}^{\infty} \left(\frac{x}{3} \right)^n = \sum_{n=0}^{\infty} \frac{x^n}{3^n};$$

Thus,

$$\frac{1}{3-x} = \frac{1}{3} \sum_{n=0}^{\infty} \frac{x^n}{3^n} = \sum_{n=0}^{\infty} \frac{x^n}{3^{n+1}}.$$

This series is valid for $|x/3| < 1$, or $|x| < 3$.

39. $f(x) = \dfrac{1}{1 + x^2}$

SOLUTION   Substituting $-x^2$ for x in Eq. (2), we obtain

$$\frac{1}{1 + x^2} = \sum_{n=0}^{\infty} (-x^2)^n = \sum_{n=0}^{\infty} (-1)^n x^{2n}$$

This series is valid for $|x| < 1$.

41. Use the equalities

$$\frac{1}{1 - x} = \frac{1}{-3 - (x - 4)} = \frac{-\frac{1}{3}}{1 + \left(\frac{x-4}{3}\right)}$$

to show that for $|x - 4| < 3$,

$$\frac{1}{1 - x} = \sum_{n=0}^{\infty} (-1)^{n+1} \frac{(x - 4)^n}{3^{n+1}}$$

SOLUTION   Substituting $-\frac{x-4}{3}$ for x in Eq. (2), we obtain

$$\frac{1}{1 + \left(\frac{x-4}{3}\right)} = \sum_{n=0}^{\infty} \left(-\frac{x-4}{3}\right)^n = \sum_{n=0}^{\infty} (-1)^n \frac{(x - 4)^n}{3^n}.$$

Thus,

$$\frac{1}{1 - x} = -\frac{1}{3} \sum_{n=0}^{\infty} (-1)^n \frac{(x - 4)^n}{3^n} = \sum_{n=0}^{\infty} (-1)^{n+1} \frac{(x - 4)^n}{3^{n+1}}.$$

This series is valid for $\left| -\frac{x-4}{3} \right| < 1$, or $|x - 4| < 3$.

43. Use the method of Exercise 41 to expand $1/(4 - x)$ in a power series with center $c = 5$. Determine the interval of convergence.

SOLUTION   First write

$$\frac{1}{4 - x} = \frac{1}{-1 - (x - 5)} = -\frac{1}{1 + (x - 5)}.$$

Substituting $-(x - 5)$ for x in Eq. (2), we obtain

$$\frac{1}{1 + (x - 5)} = \sum_{n=0}^{\infty} (-(x - 5))^n = \sum_{n=0}^{\infty} (-1)^n (x - 5)^n.$$

Thus,

$$\frac{1}{4 - x} = -\sum_{n=0}^{\infty} (-1)^n (x - 5)^n = \sum_{n=0}^{\infty} (-1)^{n+1} (x - 5)^n.$$

This series is valid for $|-(x - 5)| < 1$, or $|x - 5| < 1$.

45. Apply integration to the expansion

$$\frac{1}{1 + x} = \sum_{n=0}^{\infty} (-1)^n x^n = 1 - x + x^2 - x^3 + \cdots$$

to prove that for $-1 < x < 1$,

$$\ln(1 + x) = \sum_{n=1}^{\infty} \frac{(-1)^{n-1} x^n}{n} = x - \frac{x^2}{2} + \frac{x^3}{3} - \frac{x^4}{4} + \cdots$$

SOLUTION   To obtain the first expansion, substitute $-x$ for x in Eq. (2):

$$\frac{1}{1 + x} = \sum_{n=0}^{\infty} (-x)^n = \sum_{n=0}^{\infty} (-1)^n x^n.$$

This expansion is valid for $|-x| < 1$, or $-1 < x < 1$.

Upon integrating both sides of the above equation, we find

$$\ln(1+x) = \int \frac{dx}{1+x} = \int \left(\sum_{n=0}^{\infty} (-1)^n x^n \right) dx.$$

Integrating the series term-by-term then yields

$$\ln(1+x) = C + \sum_{n=0}^{\infty} (-1)^n \frac{x^{n+1}}{n+1}.$$

To determine the constant C, set $x = 0$. Then $0 = \ln(1+0) = C$. Finally,

$$\ln(1+x) = \sum_{n=0}^{\infty} (-1)^n \frac{x^{n+1}}{n+1} = \sum_{n=1}^{\infty} (-1)^{n-1} \frac{x^n}{n}.$$

47. Let $F(x) = (x+1)\ln(1+x) - x$.

(a) Apply integration to the result of Exercise 45 to prove that for $-1 < x < 1$,

$$F(x) = \sum_{n=1}^{\infty} (-1)^{n+1} \frac{x^{n+1}}{n(n+1)}$$

(b) Evaluate at $x = \frac{1}{2}$ to prove

$$\frac{3}{2}\ln\frac{3}{2} - \frac{1}{2} = \frac{1}{1 \cdot 2 \cdot 2^2} - \frac{1}{2 \cdot 3 \cdot 2^3} + \frac{1}{3 \cdot 4 \cdot 2^4} - \frac{1}{4 \cdot 5 \cdot 2^5} + \cdots$$

(c) Use a calculator to verify that the partial sum S_4 approximates the left-hand side with an error no greater than the term a_5 of the series.

SOLUTION

(a) Note that

$$\int \ln(x+1)\, dx = (x+1)\ln(x+1) - x + C$$

Then integrating both sides of the result of Exercise 45 gives

$$(x+1)\ln(x+1) - x = \int \ln(x+1)\, dx = \int \sum_{n=1}^{\infty} \frac{(-1)^{n-1} x^n}{n}\, dx$$

For $-1 < x < 1$, which is the interval of convergence of the series in Exercise 45, therefore, we can integrate term by term to get

$$(x+1)\ln(x+1) - x = \sum_{n=1}^{\infty} \frac{(-1)^{n-1}}{n} \int x^n\, dx = \sum_{n=1}^{\infty} \frac{(-1)^{n-1}}{n} \cdot \frac{x^{n+1}}{n+1} + C = \sum_{n=1}^{\infty} (-1)^{n+1} \frac{x^{n+1}}{n(n+1)} + C$$

(noting that $(-1)^{n-1} = (-1)^{n+1}$). To determine C, evaluate both sides at $x = 0$ to get

$$0 = \ln 1 - 0 = 0 + C$$

so that $C = 0$ and we get finally

$$(x+1)\ln(x+1) - x = \sum_{n=1}^{\infty} (-1)^{n+1} \frac{x^{n+1}}{n(n+1)}$$

(b) Evaluating the result of part(a) at $x = \frac{1}{2}$ gives

$$\frac{3}{2}\ln\frac{3}{2} - \frac{1}{2} = \sum_{n=1}^{\infty} (-1)^{n+1} \frac{1}{n(n+1)2^{n+1}}$$

$$= \frac{1}{1 \cdot 2 \cdot 2^2} - \frac{1}{2 \cdot 3 \cdot 2^3} + \frac{1}{3 \cdot 4 \cdot 2^4} - \frac{1}{4 \cdot 5 \cdot 2^5} + \cdots$$

(c)

$$S_4 = \frac{1}{1 \cdot 2 \cdot 2^2} - \frac{1}{2 \cdot 3 \cdot 2^3} + \frac{1}{3 \cdot 4 \cdot 2^4} - \frac{1}{4 \cdot 5 \cdot 2^5} = 0.1078125$$

$$a_5 = \frac{1}{5 \cdot 6 \cdot 2^6} \approx 0.0005208$$

$$\frac{3}{2} \ln \frac{3}{2} - \frac{1}{2} \approx 0.10819766$$

and

$$\left| S_4 - \frac{3}{2} \ln \frac{3}{2} - \frac{1}{2} \right| \approx 0.0003852 < a_5$$

49. Use the result of Example 7 to show that

$$F(x) = \frac{x^2}{1 \cdot 2} - \frac{x^4}{3 \cdot 4} + \frac{x^6}{5 \cdot 6} - \frac{x^8}{7 \cdot 8} + \cdots$$

is an antiderivative of $f(x) = \tan^{-1} x$ satisfying $F(0) = 0$. What is the radius of convergence of this power series?

SOLUTION For $-1 < x < 1$, which is the interval of convergence for the power series for arctangent, we can integrate term-by-term, so integrate that power series to get

$$F(x) = \int \tan^{-1} x \, dx = \sum_{n=0}^{\infty} \int \frac{(-1)^n x^{2n+1}}{2n+1} \, dx = \sum_{n=0}^{\infty} (-1)^n \frac{x^{2n+2}}{(2n+1)(2n+2)}$$

$$= \frac{x^2}{1 \cdot 2} - \frac{x^4}{3 \cdot 4} + \frac{x^6}{5 \cdot 6} - \frac{x^8}{7 \cdot 8} + \cdots + C$$

If we assume $F(0) = 0$, then we have $C = 0$. The radius of convergence of this power series is the same as that of the original power series, which is 1.

51. Evaluate $\displaystyle\sum_{n=1}^{\infty} \frac{n}{2^n}$. *Hint:* Use differentiation to show that

$$(1-x)^{-2} = \sum_{n=1}^{\infty} nx^{n-1} \quad \text{(for } |x| < 1\text{)}$$

SOLUTION Differentiate both sides of Eq. (2) to obtain

$$\frac{1}{(1-x)^2} = \sum_{n=1}^{\infty} nx^{n-1}.$$

Setting $x = \frac{1}{2}$ then yields

$$\sum_{n=1}^{\infty} \frac{n}{2^{n-1}} = \frac{1}{\left(1 - \frac{1}{2}\right)^2} = 4.$$

Divide this equation by 2 to obtain

$$\sum_{n=1}^{\infty} \frac{n}{2^n} = 2.$$

53. Show that the following series converges absolutely for $|x| < 1$ and compute its sum:

$$F(x) = 1 - x - x^2 + x^3 - x^4 - x^5 + x^6 - x^7 - x^8 + \cdots$$

Hint: Write $F(x)$ as a sum of three geometric series with common ratio x^3.

SOLUTION Because the coefficients in the power series are all ± 1, we find

$$r = \lim_{n \to \infty} \left| \frac{a_{n+1}}{a_n} \right| = 1.$$

The radius of convergence is therefore $R = r^{-1} = 1$, and the series converges absolutely for $|x| < 1$.

By Exercise 43 of Section 10.4, any rearrangement of the terms of an absolutely convergent series yields another absolutely convergent series with the same sum as the original series. Following the hint, we now rearrange the terms of $F(x)$ as the sum of three geometric series:

$$F(x) = \left(1 + x^3 + x^6 + \cdots\right) - \left(x + x^4 + x^7 + \cdots\right) - \left(x^2 + x^5 + x^8 + \cdots\right)$$

$$= \sum_{n=0}^{\infty}(x^3)^n - \sum_{n=0}^{\infty}x(x^3)^n - \sum_{n=0}^{\infty}x^2(x^3)^n = \frac{1}{1-x^3} - \frac{x}{1-x^3} - \frac{x^2}{1-x^3} = \frac{1-x-x^2}{1-x^3}.$$

55. Find all values of x such that $\sum_{n=1}^{\infty}\frac{x^{n^2}}{n!}$ converges.

SOLUTION With $a_n = \frac{x^{n^2}}{n!}$,

$$\left|\frac{a_{n+1}}{a_n}\right| = \frac{|x|^{(n+1)^2}}{(n+1)!} \cdot \frac{n!}{|x|^{n^2}} = \frac{|x|^{2n+1}}{n+1}.$$

if $|x| \leq 1$, then

$$\lim_{n\to\infty}\frac{|x|^{2n+1}}{n+1} = 0,$$

and the series converges absolutely. On the other hand, if $|x| > 1$, then

$$\lim_{n\to\infty}\frac{|x|^{2n+1}}{n+1} = \infty,$$

and the series diverges. Thus, $\sum_{n=1}^{\infty}\frac{x^{n^2}}{n!}$ converges for $-1 \leq x \leq 1$ and diverges elsewhere.

57. Find a power series $P(x) = \sum_{n=0}^{\infty}a_n x^n$ satisfying the differential equation $y' = -y$ with initial condition $y(0) = 1$. Then use Theorem 1 of Section 5.8 to conclude that $P(x) = e^{-x}$.

SOLUTION Let $P(x) = \sum_{n=0}^{\infty}a_n x^n$ and note that $P(0) = a_0$; thus, to satisfy the initial condition $P(0) = 1$, we must take $a_0 = 1$. Now,

$$P'(x) = \sum_{n=1}^{\infty}na_n x^{n-1},$$

so

$$P'(x) + P(x) = \sum_{n=1}^{\infty}na_n x^{n-1} + \sum_{n=0}^{\infty}a_n x^n = \sum_{n=0}^{\infty}\left[(n+1)a_{n+1} + a_n\right]x^n.$$

In order for this series to be equal to zero, the coefficient of x^n must be equal to zero for each n; thus

$$(n+1)a_{n+1} + a_n = 0 \quad \text{or} \quad a_{n+1} = -\frac{a_n}{n+1}.$$

Starting from $a_0 = 1$, we then calculate

$$a_1 = -\frac{a_0}{1} = -1;$$

$$a_2 = -\frac{a_1}{2} = \frac{1}{2};$$

$$a_3 = -\frac{a_2}{3} = -\frac{1}{6} = -\frac{1}{3!};$$

and, in general,

$$a_n = (-1)^n\frac{1}{n!}.$$

Hence,

$$P(x) = \sum_{n=0}^{\infty} (-1)^n \frac{x^n}{n!}.$$

The solution to the initial value problem $y' = -y$, $y(0) = 1$ is $y = e^{-x}$. Because this solution is unique, it follows that

$$P(x) = \sum_{n=0}^{\infty} (-1)^n \frac{x^n}{n!} = e^{-x}.$$

59. Use the power series for $y = e^x$ to show that

$$\frac{1}{e} = \frac{1}{2!} - \frac{1}{3!} + \frac{1}{4!} - \cdots$$

Use your knowledge of alternating series to find an N such that the partial sum S_N approximates e^{-1} to within an error of at most 10^{-3}. Confirm this using a calculator to compute both S_N and e^{-1}.

SOLUTION Recall that the series for e^x is

$$\sum_{n=0}^{\infty} \frac{x^n}{n!} = 1 + x + \frac{x^2}{2!} + \frac{x^3}{3!} + \frac{x^4}{4!} + \cdots.$$

Setting $x = -1$ yields

$$e^{-1} = 1 - 1 + \frac{1}{2!} - \frac{1}{3!} + \frac{1}{4!} - + \cdots = \frac{1}{2!} - \frac{1}{3!} + \frac{1}{4!} - + \cdots.$$

This is an alternating series with $a_n = \frac{1}{(n+1)!}$. The error in approximating e^{-1} with the partial sum S_N is therefore bounded by

$$|S_N - e^{-1}| \le a_{N+1} = \frac{1}{(N+2)!}.$$

To make the error at most 10^{-3}, we must choose N such that

$$\frac{1}{(N+2)!} \le 10^{-3} \quad \text{or} \quad (N+2)! \ge 1000.$$

For $N = 4$, $(N+2)! = 6! = 720 < 1000$, but for $N = 5$, $(N+2)! = 7! = 5040$; hence, $N = 5$ is the smallest value that satisfies the error bound. The corresponding approximation is

$$S_5 = \frac{1}{2!} - \frac{1}{3!} + \frac{1}{4!} - \frac{1}{5!} + \frac{1}{6!} = 0.368055555$$

Now, $e^{-1} = 0.367879441$, so

$$|S_5 - e^{-1}| = 1.761 \times 10^{-4} < 10^{-3}.$$

61. Find a power series $P(x)$ satisfying the differential equation

$$y'' - xy' + y = 0 \qquad \boxed{9}$$

with initial condition $y(0) = 1$, $y'(0) = 0$. What is the radius of convergence of the power series?

SOLUTION Let $P(x) = \sum_{n=0}^{\infty} a_n x^n$. Then

$$P'(x) = \sum_{n=1}^{\infty} n a_n x^{n-1} \quad \text{and} \quad P''(x) = \sum_{n=2}^{\infty} n(n-1) a_n x^{n-2}.$$

Note that $P(0) = a_0$ and $P'(0) = a_1$; in order to satisfy the initial conditions $P(0) = 1$, $P'(0) = 0$, we must have $a_0 = 1$ and $a_1 = 0$. Now,

$$P''(x) - x P'(x) + P(x) = \sum_{n=2}^{\infty} n(n-1) a_n x^{n-2} - \sum_{n=1}^{\infty} n a_n x^n + \sum_{n=0}^{\infty} a_n x^n$$

$$= \sum_{n=0}^{\infty} (n+2)(n+1) a_{n+2} x^n - \sum_{n=1}^{\infty} n a_n x^n + \sum_{n=0}^{\infty} a_n x^n$$

$$= 2a_2 + a_0 + \sum_{n=1}^{\infty} \left[(n+2)(n+1) a_{n+2} - n a_n + a_n \right] x^n.$$

In order for this series to be equal to zero, the coefficient of x^n must be equal to zero for each n; thus, $2a_2 + a_0 = 0$ and $(n+2)(n+1)a_{n+2} - (n-1)a_n = 0$, or

$$a_2 = -\frac{1}{2}a_0 \quad \text{and} \quad a_{n+2} = \frac{n-1}{(n+2)(n+1)}a_n.$$

Starting from $a_1 = 0$, we calculate

$$a_3 = \frac{1-1}{(3)(2)}a_1 = 0;$$

$$a_5 = \frac{2}{(5)(4)}a_3 = 0;$$

$$a_7 = \frac{4}{(7)(6)}a_5 = 0;$$

and, in general, all of the odd coefficients are zero. As for the even coefficients, we have $a_0 = 1$, $a_2 = -\frac{1}{2}$,

$$a_4 = \frac{1}{(4)(3)}a_2 = -\frac{1}{4!};$$

$$a_6 = \frac{3}{(6)(5)}a_4 = -\frac{3}{6!};$$

$$a_8 = \frac{5}{(8)(7)}a_6 = -\frac{15}{8!}$$

and so on. Thus,

$$P(x) = 1 - \frac{1}{2}x^2 - \frac{1}{4!}x^4 - \frac{3}{6!}x^6 - \frac{15}{8!}x^8 - \cdots$$

To determine the radius of convergence, treat this as a series in the variable x^2, and observe that

$$r = \lim_{k\to\infty}\left|\frac{a_{2k+2}}{a_{2k}}\right| = \lim_{k\to\infty}\frac{2k-1}{(2k+2)(2k+1)} = 0.$$

Thus, the radius of convergence is $R = r^{-1} = \infty$.

63. Prove that

$$J_2(x) = \sum_{k=0}^{\infty}\frac{(-1)^k}{2^{2k+2}\,k!\,(k+3)!}x^{2k+2}$$

is a solution of the Bessel differential equation of order 2:

$$x^2 y'' + x y' + (x^2 - 4)y = 0$$

SOLUTION Let $J_2(x) = \sum_{k=0}^{\infty}\frac{(-1)^k}{2^{2k+2}\,k!\,(k+2)!}x^{2k+2}$. Then

$$J_2'(x) = \sum_{k=0}^{\infty}\frac{(-1)^k(k+1)}{2^{2k+1}\,k!\,(k+2)!}x^{2k+1}$$

$$J_2''(x) = \sum_{k=0}^{\infty}\frac{(-1)^k(k+1)(2k+1)}{2^{2k+1}\,k!\,(k+2)!}x^{2k}$$

and

$$x^2 J_2''(x) + x J_2'(x) + (x^2 - 4)J_2(x) = \sum_{k=0}^{\infty}\frac{(-1)^k(k+1)(2k+1)}{2^{2k+1}\,k!\,(k+2)!}x^{2k+2} + \sum_{k=0}^{\infty}\frac{(-1)^k(k+1)}{2^{2k+1}\,k!\,(k+2)!}x^{2k+2}$$

$$- \sum_{k=0}^{\infty}\frac{(-1)^k}{2^{2k+2}\,k!\,(k+2)!}x^{2k+4} - \sum_{k=0}^{\infty}\frac{(-1)^k}{2^{2k}\,k!\,(k+2)!}x^{2k+2}$$

$$= \sum_{k=0}^{\infty}\frac{(-1)^k k(k+2)}{2^{2k}k!(k+2)!}x^{2k+2} + \sum_{k=1}^{\infty}\frac{(-1)^{k-1}}{2^{2k}\,(k-1)!\,(k+1)!}x^{2k+2}$$

$$= \sum_{k=1}^{\infty}\frac{(-1)^k}{2^{2k}(k-1)!(k+1)!}x^{2k+2} - \sum_{k=1}^{\infty}\frac{(-1)^k}{2^{2k}(k-1)!(k+1)!}x^{2k+2} = 0.$$

Further Insights and Challenges

65. Suppose that the coefficients of $F(x) = \sum_{n=0}^{\infty} a_n x^n$ are *periodic*; that is, for some whole number $M > 0$, we have $a_{M+n} = a_n$. Prove that $F(x)$ converges absolutely for $|x| < 1$ and that

$$F(x) = \frac{a_0 + a_1 x + \cdots + a_{M-1} x^{M-1}}{1 - x^M}$$

Hint: Use the hint for Exercise 53.

SOLUTION Suppose the coefficients of $F(x)$ are periodic, with $a_{M+n} = a_n$ for some whole number M and all n. The $F(x)$ can be written as the sum of M geometric series:

$$F(x) = a_0 \left(1 + x^M + x^{2M} + \cdots \right) + a_1 \left(x + x^{M+1} + x^{2M+1} + \cdots \right) +$$

$$= a_2 \left(x^2 + x^{M+2} + x^{2M+2} + \cdots \right) + \cdots + a_{M-1} \left(x^{M-1} + x^{2M-1} + x^{3M-1} + \cdots \right)$$

$$= \frac{a_0}{1 - x^M} + \frac{a_1 x}{1 - x^M} + \frac{a_2 x^2}{1 - x^M} + \cdots + \frac{a_{M-1} x^{M-1}}{1 - x^M} = \frac{a_0 + a_1 x + a_2 x^2 + \cdots + a_{M-1} x^{M-1}}{1 - x^M}.$$

As each geometric series converges absolutely for $|x| < 1$, it follows that $F(x)$ also converges absolutely for $|x| < 1$.

10.7 Taylor Series (LT Section 11.7)

Preliminary Questions

1. Determine $f(0)$ and $f'''(0)$ for a function $f(x)$ with Maclaurin series

$$T(x) = 3 + 2x + 12x^2 + 5x^3 + \cdots$$

SOLUTION The Maclaurin series for a function f has the form

$$f(0) + \frac{f'(0)}{1!}x + \frac{f''(0)}{2!}x^2 + \frac{f'''(0)}{3!}x^3 + \cdots$$

Matching this general expression with the given series, we find $f(0) = 3$ and $\dfrac{f'''(0)}{3!} = 5$. From this latter equation, it follows that $f'''(0) = 30$.

2. Determine $f(-2)$ and $f^{(4)}(-2)$ for a function with Taylor series

$$T(x) = 3(x + 2) + (x + 2)^2 - 4(x + 2)^3 + 2(x + 2)^4 + \cdots$$

SOLUTION The Taylor series for a function f centered at $x = -2$ has the form

$$f(-2) + \frac{f'(-2)}{1!}(x + 2) + \frac{f''(-2)}{2!}(x + 2)^2 + \frac{f'''(-2)}{3!}(x + 2)^3 + \frac{f^{(4)}(-2)}{4!}(x + 2)^4 + \cdots$$

Matching this general expression with the given series, we find $f(-2) = 0$ and $\dfrac{f^{(4)}(-2)}{4!} = 2$. From this latter equation, it follows that $f^{(4)}(-2) = 48$.

3. What is the easiest way to find the Maclaurin series for the function $f(x) = \sin(x^2)$?

SOLUTION The easiest way to find the Maclaurin series for $\sin\left(x^2\right)$ is to substitute x^2 for x in the Maclaurin series for $\sin x$.

4. Find the Taylor series for $f(x)$ centered at $c = 3$ if $f(3) = 4$ and $f'(x)$ has a Taylor expansion

$$f'(x) = \sum_{n=1}^{\infty} \frac{(x - 3)^n}{n}$$

SOLUTION Integrating the series for $f'(x)$ term-by-term gives

$$f(x) = C + \sum_{n=1}^{\infty} \frac{(x - 3)^{n+1}}{n(n + 1)}.$$

Substituting $x = 3$ then yields

$$f(3) = C = 4;$$

so

$$f(x) = 4 + \sum_{n=1}^{\infty} \frac{(x-3)^{n+1}}{n(n+1)}.$$

5. Let $T(x)$ be the Maclaurin series of $f(x)$. Which of the following guarantees that $f(2) = T(2)$?

(a) $T(x)$ converges for $x = 2$.

(b) The remainder $R_k(2)$ approaches a limit as $k \to \infty$.

(c) The remainder $R_k(2)$ approaches zero as $k \to \infty$.

SOLUTION The correct response is **(c)**: $f(2) = T(2)$ if and only if the remainder $R_k(2)$ approaches zero as $k \to \infty$.

Exercises

1. Write out the first four terms of the Maclaurin series of $f(x)$ if

$$f(0) = 2, \quad f'(0) = 3, \quad f''(0) = 4, \quad f'''(0) = 12$$

SOLUTION The first four terms of the Maclaurin series of $f(x)$ are

$$f(0) + f'(0)x + \frac{f''(0)}{2!}x^2 + \frac{f'''(0)}{3!}x^3 = 2 + 3x + \frac{4}{2}x^2 + \frac{12}{6}x^3 = 2 + 3x + 2x^2 + 2x^3.$$

In Exercises 3–18, find the Maclaurin series and find the interval on which the expansion is valid.

3. $f(x) = \dfrac{1}{1 - 2x}$

SOLUTION Substituting $2x$ for x in the Maclaurin series for $\frac{1}{1-x}$ gives

$$\frac{1}{1 - 2x} = \sum_{n=0}^{\infty} (2x)^n = \sum_{n=0}^{\infty} 2^n x^n.$$

This series is valid for $|2x| < 1$, or $|x| < \frac{1}{2}$.

5. $f(x) = \cos 3x$

SOLUTION Substituting $3x$ for x in the Maclaurin series for $\cos x$ gives

$$\cos 3x = \sum_{n=0}^{\infty} (-1)^n \frac{(3x)^{2n}}{(2n)!} = \sum_{n=0}^{\infty} (-1)^n \frac{9^n x^{2n}}{(2n)!}.$$

This series is valid for all x.

7. $f(x) = \sin(x^2)$

SOLUTION Substituting x^2 for x in the Maclaurin series for $\sin x$ gives

$$\sin x^2 = \sum_{n=0}^{\infty} (-1)^n \frac{(x^2)^{2n+1}}{(2n+1)!} = \sum_{n=0}^{\infty} (-1)^n \frac{x^{4n+2}}{(2n+1)!}.$$

This series is valid for all x.

9. $f(x) = \ln(1 - x^2)$

SOLUTION Substituting $-x^2$ for x in the Maclaurin series for $\ln(1 + x)$ gives

$$\ln(1 - x^2) = \sum_{n=1}^{\infty} \frac{(-1)^{n-1}(-x^2)^n}{n} = \sum_{n=1}^{\infty} \frac{(-1)^{2n-1} x^{2n}}{n} = -\sum_{n=1}^{\infty} \frac{x^{2n}}{n}.$$

This series is valid for $|x| < 1$.

11. $f(x) = \tan^{-1}(x^2)$

SOLUTION Substituting x^2 for x in the Maclaurin series for $\tan^{-1} x$ gives

$$\tan^{-1}(x^2) = \sum_{n=0}^{\infty} (-1)^n \frac{(x^2)^{2n+1}}{2n + 1} = \sum_{n=0}^{\infty} (-1)^n \frac{x^{4n+2}}{2n + 1}.$$

This series is valid for $|x| \leq 1$.

13. $f(x) = e^{x-2}$

SOLUTION $e^{x-2} = e^{-2}e^x$; thus,

$$e^{x-2} = e^{-2} \sum_{n=0}^{\infty} \frac{x^n}{n!} = \sum_{n=0}^{\infty} \frac{x^n}{e^2 n!}.$$

This series is valid for all x.

15. $f(x) = \ln(1 - 5x)$

SOLUTION Substituting $-5x$ for x in the Maclaurin series for $\ln(1 + x)$ gives

$$\ln(1 - 5x) = \sum_{n=1}^{\infty} \frac{(-1)^{n-1}(-5x)^n}{n} = \sum_{n=1}^{\infty} \frac{(-1)^{2n-1}5^n x^n}{n} = -\sum_{n=1}^{\infty} \frac{5^n x^n}{n}.$$

This series is valid for $|5x| < 1$, or $|x| < \frac{1}{5}$, and for $x = -\frac{1}{5}$.

17. $f(x) = \sinh x$

SOLUTION Recall that

$$\sinh x = \frac{1}{2}(e^x - e^{-x}).$$

Therefore,

$$\sinh x = \frac{1}{2}\left(\sum_{n=0}^{\infty} \frac{x^n}{n!} - \sum_{n=0}^{\infty} \frac{(-x)^n}{n!} \right) = \sum_{n=0}^{\infty} \frac{x^n}{2(n!)}\left(1 - (-1)^n\right).$$

Now,

$$1 - (-1)^n = \begin{cases} 0, & n \text{ even} \\ 2, & n \text{ odd} \end{cases}$$

so

$$\sinh x = \sum_{k=0}^{\infty} 2\frac{x^{2k+1}}{2(2k+1)!} = \sum_{k=0}^{\infty} \frac{x^{2k+1}}{(2k+1)!}.$$

This series is valid for all x.

In Exercises 19–28, find the terms through degree four of the Maclaurin series of $f(x)$. Use multiplication and substitution as necessary.

19. $f(x) = e^x \sin x$

SOLUTION Multiply the fourth-order Taylor Polynomials for e^x and $\sin x$:

$$\left(1 + x + \frac{x^2}{2} + \frac{x^3}{6} + \frac{x^4}{24}\right)\left(x - \frac{x^3}{6}\right)$$

$$= x + x^2 - \frac{x^3}{6} + \frac{x^3}{2} - \frac{x^4}{6} + \frac{x^4}{6} + \text{higher-order terms}$$

$$= x + x^2 + \frac{x^3}{3} + \text{higher-order terms}.$$

The terms through degree four in the Maclaurin series for $f(x) = e^x \sin x$ are therefore

$$x + x^2 + \frac{x^3}{3}.$$

21. $f(x) = \dfrac{\sin x}{1 - x}$

SOLUTION Multiply the fourth order Taylor Polynomials for $\sin x$ and $\dfrac{1}{1 - x}$:

$$\left(x - \frac{x^3}{6}\right)\left(1 + x + x^2 + x^3 + x^4\right)$$

$$= x + x^2 - \frac{x^3}{6} + x^3 + x^4 - \frac{x^4}{6} + \text{higher-order terms}$$

$$= x + x^2 + \frac{5x^3}{6} + \frac{5x^4}{6} + \text{higher-order terms}.$$

The terms through order four of the Maclaurin series for $f(x) = \dfrac{\sin x}{1-x}$ are therefore

$$x + x^2 + \frac{5x^3}{6} + \frac{5x^4}{6}.$$

23. $f(x) = (1+x)^{1/4}$

SOLUTION The first five generalized binomial coefficients for $a = \frac{1}{4}$ are

$$1, \quad \frac{1}{4}, \quad \frac{\frac{1}{4}\left(\frac{-3}{4}\right)}{2!} = -\frac{3}{32}, \quad \frac{\frac{1}{4}\left(\frac{-3}{4}\right)\left(\frac{-7}{4}\right)}{3!} = \frac{7}{128}, \quad \frac{\frac{1}{4}\left(\frac{-3}{4}\right)\left(\frac{-7}{4}\right)\left(\frac{-11}{4}\right)}{4!} = \frac{-77}{2048}$$

Therefore, the first four terms in the binomial series for $(1+x)^{1/4}$ are

$$1 + \frac{1}{4}x - \frac{3}{32}x^2 + \frac{7}{128}x^3 - \frac{77}{2048}x^4$$

25. $f(x) = e^x \tan^{-1} x$

SOLUTION Using the Maclaurin series for e^x and $\tan^{-1} x$, we find

$$e^x \tan^{-1} x = \left(1 + x + \frac{x^2}{2} + \frac{x^3}{6} + \cdots\right)\left(x - \frac{x^3}{3} + \cdots\right) = x + x^2 - \frac{x^3}{3} + \frac{x^3}{2} + \frac{x^4}{6} - \frac{x^4}{3} + \cdots$$

$$= x + x^2 + \frac{1}{6}x^3 - \frac{1}{6}x^4 + \cdots.$$

27. $f(x) = e^{\sin x}$

SOLUTION Substituting $\sin x$ for x in the Maclaurin series for e^x and then using the Maclaurin series for $\sin x$, we find

$$e^{\sin x} = 1 + \sin x + \frac{\sin^2 x}{2} + \frac{\sin^3 x}{6} + \frac{\sin^4 x}{24} + \cdots$$

$$= 1 + \left(x - \frac{x^3}{6} + \cdots\right) + \frac{1}{2}\left(x - \frac{x^3}{6} + \cdots\right)^2 + \frac{1}{6}(x - \cdots)^3 + \frac{1}{24}(x - \cdots)^4$$

$$= 1 + x + \frac{1}{2}x^2 - \frac{1}{6}x^3 + \frac{1}{6}x^3 - \frac{1}{6}x^4 + \frac{1}{24}x^4 + \cdots$$

$$= 1 + x + \frac{1}{2}x^2 - \frac{1}{8}x^4 + \cdots.$$

In Exercises 29–38, find the Taylor series centered at c and find the interval on which the expansion is valid.

29. $f(x) = \dfrac{1}{x}, \quad c = 1$

SOLUTION Write

$$\frac{1}{x} = \frac{1}{1 + (x - 1)},$$

and then substitute $-(x - 1)$ for x in the Maclaurin series for $\frac{1}{1-x}$ to obtain

$$\frac{1}{x} = \sum_{n=0}^{\infty} [-(x-1)]^n = \sum_{n=0}^{\infty} (-1)^n (x-1)^n.$$

This series is valid for $|x - 1| < 1$.

31. $f(x) = \dfrac{1}{1-x}, \quad c = 5$

SOLUTION Write

$$\frac{1}{1-x} = \frac{1}{-4 - (x-5)} = -\frac{1}{4} \cdot \frac{1}{1 + \frac{x-5}{4}}.$$

Substituting $-\frac{x-5}{4}$ for x in the Maclaurin series for $\frac{1}{1-x}$ yields

$$\frac{1}{1+\frac{x-5}{4}} = \sum_{n=0}^{\infty}\left(-\frac{x-5}{4}\right)^n = \sum_{n=0}^{\infty}(-1)^n\frac{(x-5)^n}{4^n}.$$

Thus,

$$\frac{1}{1-x} = -\frac{1}{4}\sum_{n=0}^{\infty}(-1)^n\frac{(x-5)^n}{4^n} = \sum_{n=0}^{\infty}(-1)^{n+1}\frac{(x-5)^n}{4^{n+1}}.$$

This series is valid for $\left|\frac{x-5}{4}\right| < 1$, or $|x-5| < 4$.

33. $f(x) = x^4 + 3x - 1, \quad c = 2$

SOLUTION To determine the Taylor series with center $c = 2$, we compute

$$f'(x) = 4x^3 + 3, \quad f''(x) = 12x^2, \quad f'''(x) = 24x,$$

and $f^{(4)}(x) = 24$. All derivatives of order five and higher are zero. Now,

$$f(2) = 21, \quad f'(2) = 35, \quad f''(2) = 48, \quad f'''(2) = 48,$$

and $f^{(4)}(2) = 24$. Therefore, the Taylor series is

$$21 + 35(x-2) + \frac{48}{2}(x-2)^2 + \frac{48}{6}(x-2)^3 + \frac{24}{24}(x-2)^4,$$

or

$$21 + 35(x-2) + 24(x-2)^2 + 8(x-2)^3 + (x-2)^4.$$

35. $f(x) = \frac{1}{x^2}, \quad c = 4$

SOLUTION We will first find the Taylor series for $\frac{1}{x}$ and then differentiate to obtain the series for $\frac{1}{x^2}$. Write

$$\frac{1}{x} = \frac{1}{4 + (x-4)} = \frac{1}{4}\cdot\frac{1}{1 + \frac{x-4}{4}}.$$

Now substitute $-\frac{x-4}{4}$ for x in the Maclaurin series for $\frac{1}{1-x}$ to obtain

$$\frac{1}{x} = \frac{1}{4}\sum_{n=}^{\infty}\left(-\frac{x-4}{4}\right)^n = \sum_{n=0}^{\infty}(-1)^n\frac{(x-4)^n}{4^{n+1}}.$$

Differentiating term-by-term yields

$$-\frac{1}{x^2} = \sum_{n=1}^{\infty}(-1)^n n\frac{(x-4)^{n-1}}{4^{n+1}},$$

so that

$$\frac{1}{x^2} = \sum_{n=1}^{\infty}(-1)^{n-1}n\frac{(x-4)^{n-1}}{4^{n+1}} = \sum_{n=0}^{\infty}(-1)^n(n+1)\frac{(x-4)^n}{4^{n+2}}.$$

This series is valid for $\left|\frac{x-4}{4}\right| < 1$, or $|x-4| < 4$.

37. $f(x) = \frac{1}{1-x^2}, \quad c = 3$

SOLUTION By partial fraction decomposition

$$\frac{1}{1-x^2} = \frac{\frac{1}{2}}{1-x} + \frac{\frac{1}{2}}{1+x},$$

so

$$\frac{1}{1-x^2} = \frac{\frac{1}{2}}{-2-(x-3)} + \frac{\frac{1}{2}}{4+(x-3)} = -\frac{1}{4}\cdot\frac{1}{1+\frac{x-3}{2}} + \frac{1}{8}\cdot\frac{1}{1+\frac{x-3}{4}}.$$

Substituting $-\frac{x-3}{2}$ for x in the Maclaurin series for $\frac{1}{1-x}$ gives

$$\frac{1}{1 + \frac{x-3}{2}} = \sum_{n=0}^{\infty} \left(-\frac{x-3}{2}\right)^n = \sum_{n=0}^{\infty} \frac{(-1)^n}{2^n}(x-3)^n,$$

while substituting $-\frac{x-3}{4}$ for x in the same series gives

$$\frac{1}{1 + \frac{x-3}{4}} = \sum_{n=0}^{\infty} \left(-\frac{x-3}{4}\right)^n = \sum_{n=0}^{\infty} \frac{(-1)^n}{4^n}(x-3)^n.$$

Thus,

$$\frac{1}{1-x^2} = -\frac{1}{4}\sum_{n=0}^{\infty} \frac{(-1)^n}{2^n}(x-3)^n + \frac{1}{8}\sum_{n=0}^{\infty} \frac{(-1)^n}{4^n}(x-3)^n = \sum_{n=0}^{\infty} \frac{(-1)^{n+1}}{2^{n+2}}(x-3)^n + \sum_{n=0}^{\infty} \frac{(-1)^n}{2^{2n+3}}(x-3)^n$$

$$= \sum_{n=0}^{\infty} \left(\frac{(-1)^{n+1}}{2^{n+2}} + \frac{(-1)^n}{2^{2n+3}}\right)(x-3)^n = \sum_{n=0}^{\infty} \frac{(-1)^{n+1}(2^{n+1}-1)}{2^{2n+3}}(x-3)^n.$$

This series is valid for $|x-3| < 2$.

39. Use the identity $\cos^2 x = \frac{1}{2}(1 + \cos 2x)$ to find the Maclaurin series for $\cos^2 x$.

SOLUTION   The Maclaurin series for $\cos 2x$ is

$$\sum_{n=0}^{\infty} (-1)^n \frac{(2x)^{2n}}{(2n)!} = \sum_{n=0}^{\infty} (-1)^n \frac{2^{2n}x^{2n}}{(2n)!}$$

so the Maclaurin series for $\cos^2 x = \frac{1}{2}(1 + \cos 2x)$ is

$$\frac{1 + \left(1 + \sum_{n=1}^{\infty} (-1)^n \frac{2^{2n}x^{2n}}{(2n)!}\right)}{2} = 1 + \sum_{n=1}^{\infty} (-1)^n \frac{2^{2n-1}x^{2n}}{(2n)!}$$

41. Use the Maclaurin series for $\ln(1+x)$ and $\ln(1-x)$ to show that

$$\frac{1}{2}\ln\left(\frac{1+x}{1-x}\right) = x + \frac{x^3}{3} + \frac{x^5}{5} + \cdots$$

for $|x| < 1$. What can you conclude by comparing this result with that of Exercise 40?

SOLUTION   Using the Maclaurin series for $\ln(1+x)$ and $\ln(1-x)$, we have for $|x| < 1$

$$\ln(1+x) - \ln(1-x) = \sum_{n=1}^{\infty} \frac{(-1)^{n-1}}{n}x^n - \sum_{n=1}^{\infty} \frac{(-1)^{n-1}}{n}(-x)^n$$

$$= \sum_{n=1}^{\infty} \frac{(-1)^{n-1}}{n}x^n + \sum_{n=1}^{\infty} \frac{x^n}{n} = \sum_{n=1}^{\infty} \frac{1 + (-1)^{n-1}}{n}x^n.$$

Since $1 + (-1)^{n-1} = 0$ for even n and $1 + (-1)^{n-1} = 2$ for odd n,

$$\ln(1+x) - \ln(1-x) = \sum_{k=0}^{\infty} \frac{2}{2k+1}x^{2k+1}.$$

Thus,

$$\frac{1}{2}\ln\left(\frac{1+x}{1-x}\right) = \frac{1}{2}(\ln(1+x) - \ln(1-x)) = \frac{1}{2}\sum_{k=0}^{\infty} \frac{2}{2k+1}x^{2k+1} = \sum_{k=0}^{\infty} \frac{x^{2k+1}}{2k+1}.$$

Observe that this is the same series we found in Exercise 40; therefore,

$$\frac{1}{2}\ln\left(\frac{1+x}{1-x}\right) = \tanh^{-1} x.$$

43. Show, by integrating the Maclaurin series for $f(x) = \dfrac{1}{\sqrt{1-x^2}}$, that for $|x| < 1$,

$$\sin^{-1} x = x + \sum_{n=1}^{\infty} \frac{1 \cdot 3 \cdot 5 \cdots (2n-1)}{2 \cdot 4 \cdot 6 \cdots (2n)} \frac{x^{2n+1}}{2n+1}$$

SOLUTION   From Example 10, we know that for $|x| < 1$

$$\frac{1}{\sqrt{1-x^2}} = \sum_{n=0}^{\infty} \frac{1 \cdot 3 \cdot 5 \cdots (2n-1)}{2 \cdot 4 \cdot 6 \cdots (2n)} x^{2n} = 1 + \sum_{n=1}^{\infty} \frac{1 \cdot 3 \cdot 5 \cdots (2n-1)}{2 \cdot 4 \cdot 6 \cdots (2n)} x^{2n},$$

so, for $|x| < 1$,

$$\sin^{-1} x = \int \frac{dx}{\sqrt{1-x^2}} = C + x + \sum_{n=1}^{\infty} \frac{1 \cdot 3 \cdot 5 \cdots (2n-1)}{2 \cdot 4 \cdot 6 \cdots (2n)} \frac{x^{2n+1}}{2n+1}.$$

Since $\sin^{-1} 0 = 0$, we find that $C = 0$. Thus,

$$\sin^{-1} x = x + \sum_{n=1}^{\infty} \frac{1 \cdot 3 \cdot 5 \cdots (2n-1)}{2 \cdot 4 \cdot 6 \cdots (2n)} \frac{x^{2n+1}}{2n+1}.$$

45. How many terms of the Maclaurin series of $f(x) = \ln(1+x)$ are needed to compute $\ln 1.2$ to within an error of at most 0.0001? Make the computation and compare the result with the calculator value.

SOLUTION   Substitute $x = 0.2$ into the Maclaurin series for $\ln(1+x)$ to obtain:

$$\ln 1.2 = \sum_{n=1}^{\infty} (-1)^{n-1} \frac{(0.2)^n}{n} = \sum_{n=1}^{\infty} (-1)^{n-1} \frac{1}{5^n n}.$$

This is an alternating series with $a_n = \dfrac{1}{n \cdot 5^n}$. Using the error bound for alternating series

$$|\ln 1.2 - S_N| \le a_{N+1} = \frac{1}{(N+1)5^{N+1}},$$

so we must choose N so that

$$\frac{1}{(N+1)5^{N+1}} < 0.0001 \quad \text{or} \quad (N+1)5^{N+1} > 10,000.$$

For $N = 3$, $(N+1)5^{N+1} = 4 \cdot 5^4 = 2500 < 10,000$, and for $N = 4$, $(N+1)5^{N+1} = 5 \cdot 5^5 = 15,625 > 10,000$; thus, the smallest acceptable value for N is $N = 4$. The corresponding approximation is:

$$S_4 = \sum_{n=1}^{4} \frac{(-1)^{n-1}}{5^n \cdot n} = \frac{1}{5} - \frac{1}{5^2 \cdot 2} + \frac{1}{5^3 \cdot 3} - \frac{1}{5^4 \cdot 4} = 0.182266666.$$

Now, $\ln 1.2 = 0.182321556$, so

$$|\ln 1.2 - S_4| = 5.489 \times 10^{-5} < 0.0001.$$

47. Use the Maclaurin expansion for e^{-t^2} to express the function $F(x) = \int_0^x e^{-t^2}\,dt$ as an alternating power series in x (Figure 4).

(a) How many terms of the Maclaurin series are needed to approximate the integral for $x = 1$ to within an error of at most 0.001?

(b) *CAS*   Carry out the computation and check your answer using a computer algebra system.

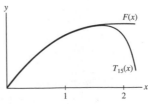

FIGURE 4   The Maclaurin polynomial $T_{15}(x)$ for $F(t) = \displaystyle\int_0^x e^{-t^2}\,dt$.

SOLUTION Substituting $-t^2$ for t in the Maclaurin series for e^t yields

$$e^{-t^2} = \sum_{n=0}^{\infty} \frac{(-t^2)^n}{n!} = \sum_{n=0}^{\infty} (-1)^n \frac{t^{2n}}{n!};$$

thus,

$$\int_0^x e^{-t^2}\, dt = \sum_{n=0}^{\infty} (-1)^n \frac{x^{2n+1}}{n!(2n+1)}.$$

(a) For $x = 1$,

$$\int_0^1 e^{-t^2}\, dt = \sum_{n=0}^{\infty} (-1)^n \frac{1}{n!(2n+1)}.$$

This is an alternating series with $a_n = \frac{1}{n!(2n+1)}$; therefore, the error incurred by using S_N to approximate the value of the definite integral is bounded by

$$\left| \int_0^1 e^{-t^2}\, dt - S_N \right| \le a_{N+1} = \frac{1}{(N+1)!(2N+3)}.$$

To guarantee the error is at most 0.001, we must choose N so that

$$\frac{1}{(N+1)!(2N+3)} < 0.001 \quad \text{or} \quad (N+1)!(2N+3) > 1000.$$

For $N = 3$, $(N+1)!(2N+3) = 4! \cdot 9 = 216 < 1000$ and for $N = 4$, $(N+1)!(2N+3) = 5! \cdot 11 = 1320 > 1000$; thus, the smallest acceptable value for N is $N = 4$. The corresponding approximation is

$$S_4 = \sum_{n=0}^{4} \frac{(-1)^n}{n!(2n+1)} = 1 - \frac{1}{3} + \frac{1}{2! \cdot 5} - \frac{1}{3! \cdot 7} + \frac{1}{4! \cdot 9} = 0.747486772.$$

(b) Using a computer algebra system, we find

$$\int_0^1 e^{-t^2}\, dt = 0.746824133;$$

therefore

$$\left| \int_0^1 e^{-t^2}\, dt - S_4 \right| = 6.626 \times 10^{-4} < 10^{-3}.$$

In Exercises 49–52, express the definite integral as an infinite series and find its value to within an error of at most 10^{-4}.

49. $\displaystyle \int_0^1 \cos(x^2)\, dx$

SOLUTION Substituting x^2 for x in the Maclaurin series for $\cos x$ yields

$$\cos(x^2) = \sum_{n=0}^{\infty} (-1)^n \frac{(x^2)^{2n}}{(2n)!} = \sum_{n=0}^{\infty} (-1)^n \frac{x^{4n}}{(2n)!};$$

therefore,

$$\int_0^1 \cos(x^2)\, dx = \sum_{n=0}^{\infty} (-1)^n \frac{x^{4n+1}}{(2n)!(4n+1)} \bigg|_0^1 = \sum_{n=0}^{\infty} \frac{(-1)^n}{(2n)!(4n+1)}.$$

This is an alternating series with $a_n = \frac{1}{(2n)!(4n+1)}$; therefore, the error incurred by using S_N to approximate the value of the definite integral is bounded by

$$\left| \int_0^1 \cos(x^2)\, dx - S_N \right| \le a_{N+1} = \frac{1}{(2N+2)!(4N+5)}.$$

To guarantee the error is at most 0.0001, we must choose N so that

$$\frac{1}{(2N+2)!(4N+5)} < 0.0001 \quad \text{or} \quad (2N+2)!(4N+5) > 10,000.$$

For $N = 2$, $(2N+2)!(4N+5) = 6! \cdot 13 = 9360 < 10,000$ and for $N = 3$, $(2N+2)!(4N+5) = 8! \cdot 17 = 685,440 > 10,000$; thus, the smallest acceptable value for N is $N = 3$. The corresponding approximation is

$$S_3 = \sum_{n=0}^{3} \frac{(-1)^n}{(2n)!(4n+1)} = 1 - \frac{1}{5 \cdot 2!} + \frac{1}{9 \cdot 4!} - \frac{1}{13 \cdot 6!} = 0.904522792.$$

51. $\displaystyle\int_0^1 e^{-x^3}\, dx$

SOLUTION   Substituting $-x^3$ for x in the Maclaurin series for e^x yields

$$e^{-x^3} = \sum_{n=0}^{\infty} \frac{(-x^3)^n}{n!} = \sum_{n=0}^{\infty} (-1)^n \frac{x^{3n}}{n!};$$

therefore,

$$\int_0^1 e^{-x^3}\, dx = \sum_{n=0}^{\infty} (-1)^n \frac{x^{3n+1}}{n!(3n+1)} \Bigg|_0^1 = \sum_{n=0}^{\infty} \frac{(-1)^n}{n!(3n+1)}.$$

This is an alternating series with $a_n = \frac{1}{n!(3n+1)}$; therefore, the error incurred by using S_N to approximate the value of the definite integral is bounded by

$$\left| \int_0^1 e^{-x^3}\, dx - S_N \right| \le a_{N+1} = \frac{1}{(N+1)!(3N+4)}.$$

To guarantee the error is at most 0.0001, we must choose N so that

$$\frac{1}{(N+1)!(3N+4)} < 0.0001 \quad \text{or} \quad (N+1)!(3N+4) > 10,000.$$

For $N = 4$, $(N+1)!(3N+4) = 5! \cdot 16 = 1920 < 10,000$ and for $N = 5$, $(N+1)!(3N+4) = 6! \cdot 19 = 13,680 > 10,000$; thus, the smallest acceptable value for N is $N = 5$. The corresponding approximation is

$$S_5 = \sum_{n=0}^{5} \frac{(-1)^n}{n!(3n+1)} = 0.807446200.$$

In Exercises 53–56, express the integral as an infinite series.

53. $\displaystyle\int_0^x \frac{1 - \cos(t)}{t}\, dt, \quad$ for all x

SOLUTION   The Maclaurin series for $\cos t$ is

$$\cos t = \sum_{n=0}^{\infty} (-1)^n \frac{t^{2n}}{(2n)!} = 1 + \sum_{n=1}^{\infty} (-1)^n \frac{t^{2n}}{(2n)!},$$

so

$$1 - \cos t = -\sum_{n=1}^{\infty} (-1)^n \frac{t^{2n}}{(2n)!} = \sum_{n=1}^{\infty} (-1)^{n+1} \frac{t^{2n}}{(2n)!},$$

and

$$\frac{1 - \cos t}{t} = \frac{1}{t} \sum_{n=1}^{\infty} (-1)^{n+1} \frac{t^{2n}}{(2n)!} = \sum_{n=1}^{\infty} (-1)^{n+1} \frac{t^{2n-1}}{(2n)!}.$$

Thus,

$$\int_0^x \frac{1 - \cos(t)}{t}\, dt = \sum_{n=1}^{\infty} (-1)^{n+1} \frac{t^{2n}}{(2n)!2n} \Bigg|_0^x = \sum_{n=1}^{\infty} (-1)^{n+1} \frac{x^{2n}}{(2n)!2n}.$$

55. $\displaystyle\int_0^x \ln(1 + t^2)\, dt,$ for $|x| < 1$

SOLUTION   Substituting t^2 for t in the Maclaurin series for $\ln(1 + t)$ yields

$$\ln(1 + t^2) = \sum_{n=1}^{\infty}(-1)^{n-1}\frac{(t^2)^n}{n} = \sum_{n=1}^{\infty}(-1)^n\frac{t^{2n}}{n}.$$

Thus,

$$\int_0^x \ln(1 + t^2)\, dt = \sum_{n=1}^{\infty}(-1)^n\frac{t^{2n+1}}{n(2n+1)}\bigg|_0^x = \sum_{n=1}^{\infty}(-1)^n\frac{x^{2n+1}}{n(2n+1)}.$$

57. Which function has Maclaurin series $\displaystyle\sum_{n=0}^{\infty}(-1)^n 2^n x^n$?

SOLUTION   We recognize that

$$\sum_{n=0}^{\infty}(-1)^n 2^n x^n = \sum_{n=0}^{\infty}(-2x)^n$$

is the Maclaurin series for $\frac{1}{1-x}$ with x replaced by $-2x$. Therefore,

$$\sum_{n=0}^{\infty}(-1)^n 2^n x^n = \frac{1}{1-(-2x)} = \frac{1}{1+2x}.$$

In Exercises 59–62, use Theorem 2 to prove that the $f(x)$ is represented by its Maclaurin series for all x.

59. $f(x) = \sin\left(\frac{x}{2}\right) + \cos\left(\frac{x}{3}\right),$

SOLUTION   All derivatives of $f(x)$ consist of sin or cos applied to each of $x/2$ and $x/3$ and added together, so each summand is bounded by 1. Thus $\left|f^{(n)}(x)\right| \le 2$ for all n and x. By Theorem 2, $f(x)$ is represented by its Taylor series for every x.

61. $f(x) = \sinh x,$

SOLUTION   By definition, $\sinh x = \frac{1}{2}(e^x - e^{-x})$, so if both e^x and e^{-x} are represented by their Taylor series centered at c, then so is $\sinh x$. But the previous exercise shows that e^{-x} is so represented, and the text shows that e^x is.

In Exercises 63–66, find the functions with the following Maclaurin series (refer to Table 1 on page 599).

63. $1 + x^3 + \dfrac{x^6}{2!} + \dfrac{x^9}{3!} + \dfrac{x^{12}}{4!} + \cdots$

SOLUTION   We recognize

$$1 + x^3 + \frac{x^6}{2!} + \frac{x^9}{3!} + \frac{x^{12}}{4!} + \cdots = \sum_{n=0}^{\infty}\frac{x^{3n}}{n!} = \sum_{n=0}^{\infty}\frac{(x^3)^n}{n!}$$

as the Maclaurin series for e^x with x replaced by x^3. Therefore,

$$1 + x^3 + \frac{x^6}{2!} + \frac{x^9}{3!} + \frac{x^{12}}{4!} + \cdots = e^{x^3}.$$

65. $1 - \dfrac{5^3 x^3}{3!} + \dfrac{5^5 x^5}{5!} - \dfrac{5^7 x^7}{7!} + \cdots$

SOLUTION   Note

$$1 - \frac{5^3 x^3}{3!} + \frac{5^5 x^5}{5!} - \frac{5^7 x^7}{7!} + \cdots = 1 - 5x + \left(5x - \frac{5^3 x^3}{3!} + \frac{5^5 x^5}{5!} - \frac{5^7 x^7}{7!} + \cdots\right)$$

$$= 1 - 5x + \sum_{n=0}^{\infty}(-1)^n\frac{(5x)^{2n+1}}{(2n+1)!}.$$

The series is the Maclaurin series for $\sin x$ with x replaced by $5x$, so

$$1 - \frac{5^3 x^3}{3!} + \frac{5^5 x^5}{5!} - \frac{5^7 x^7}{7!} + \cdots = 1 - 5x + \sin(5x).$$

In Exercises 67 and 68, let

$$f(x) = \frac{1}{(1-x)(1-2x)}$$

67. Find the Maclaurin series of $f(x)$ using the identity

$$f(x) = \frac{2}{1-2x} - \frac{1}{1-x}$$

SOLUTION   Substituting $2x$ for x in the Maclaurin series for $\dfrac{1}{1-x}$ gives

$$\frac{1}{1-2x} = \sum_{n=0}^{\infty}(2x)^n = \sum_{n=0}^{\infty}2^n x^n$$

which is valid for $|2x| < 1$, or $|x| < \frac{1}{2}$. Because the Maclaurin series for $\dfrac{1}{1-x}$ is valid for $|x| < 1$, the two series together are valid for $|x| < \frac{1}{2}$. Thus, for $|x| < \frac{1}{2}$,

$$\frac{1}{(1-2x)(1-x)} = \frac{2}{1-2x} - \frac{1}{1-x} = 2\sum_{n=0}^{\infty}2^n x^n - \sum_{n=0}^{\infty}x^n$$

$$= \sum_{n=0}^{\infty}2^{n+1}x^n - \sum_{n=0}^{\infty}x^n = \sum_{n=0}^{\infty}\left(2^{n+1}-1\right)x^n.$$

69. When a voltage V is applied to a series circuit consisting of a resistor R and an inductor L, the current at time t is

$$I(t) = \left(\frac{V}{R}\right)\left(1 - e^{-Rt/L}\right)$$

Expand $I(t)$ in a Maclaurin series. Show that $I(t) \approx \dfrac{Vt}{L}$ for small t.

SOLUTION   Substituting $-\frac{Rt}{L}$ for t in the Maclaurin series for e^t gives

$$e^{-Rt/L} = \sum_{n=0}^{\infty}\frac{\left(-\frac{Rt}{L}\right)^n}{n!} = \sum_{n=0}^{\infty}\frac{(-1)^n}{n!}\left(\frac{R}{L}\right)^n t^n = 1 + \sum_{n=1}^{\infty}\frac{(-1)^n}{n!}\left(\frac{R}{L}\right)^n t^n$$

Thus,

$$1 - e^{-Rt/L} = 1 - \left(1 + \sum_{n=1}^{\infty}\frac{(-1)^n}{n!}\left(\frac{R}{L}\right)^n t^n\right) = \sum_{n=1}^{\infty}\frac{(-1)^{n+1}}{n!}\left(\frac{Rt}{L}\right)^n,$$

and

$$I(t) = \frac{V}{R}\sum_{n=1}^{\infty}\frac{(-1)^{n+1}}{n!}\left(\frac{Rt}{L}\right)^n = \frac{Vt}{L} + \frac{V}{R}\sum_{n=2}^{\infty}\frac{(-1)^{n+1}}{n!}\left(\frac{Rt}{L}\right)^n.$$

If t is small, then we can approximate $I(t)$ by the first (linear) term, and ignore terms with higher powers of t; then we find

$$V(t) \approx \frac{Vt}{L}.$$

71. Find the Maclaurin series for $f(x) = \cos(x^3)$ and use it to determine $f^{(6)}(0)$.

SOLUTION   The Maclaurin series for $\cos x$ is

$$\cos x = \sum_{n=0}^{\infty}(-1)^n \frac{x^{2n}}{(2n)!}$$

Substituting x^3 for x gives

$$\cos(x^3) = \sum_{n=0}^{\infty}(-1)^n \frac{x^{6n}}{(2n)!}$$

Now, the coefficient of x^6 in this series is

$$-\frac{1}{2!} = -\frac{1}{2} = \frac{f^{(6)}(0)}{6!}$$

so

$$f^{(6)}(0) = -\frac{6!}{2} = -360$$

73. Use substitution to find the first three terms of the Maclaurin series for $f(x) = e^{x^{20}}$. How does the result show that $f^{(k)}(0) = 0$ for $1 \le k \le 19$?

SOLUTION Substituting x^{20} for x in the Maclaurin series for e^x yields

$$e^{x^{20}} = \sum_{n=0}^{\infty} \frac{(x^{20})^n}{n!} = \sum_{n=0}^{\infty} \frac{x^{20n}}{n!};$$

the first three terms in the series are then

$$1 + x^{20} + \frac{1}{2}x^{40}.$$

Recall that the coefficient of x^k in the Maclaurin series for f is $\frac{f^{(k)}(0)}{k!}$. For $1 \le k \le 19$, the coefficient of x^k in the Maclaurin series for $f(x) = e^{x^{20}}$ is zero; it therefore follows that

$$\frac{f^{(k)}(0)}{k!} = 0 \quad \text{or} \quad f^{(k)}(0) = 0$$

for $1 \le k \le 19$.

75. Does the Maclaurin series for $f(x) = (1+x)^{3/4}$ converge to $f(x)$ at $x = 2$? Give numerical evidence to support your answer.

SOLUTION The Taylor series for $f(x) = (1+x)^{3/4}$ converges to $f(x)$ for $|x| < 1$; because $x = 2$ is not contained on this interval, the series does not converge to $f(x)$ at $x = 2$. The graph below displays

$$S_N = \sum_{n=0}^{N} \binom{\frac{3}{4}}{n} 2^n$$

for $0 \le N \le 14$. The divergent nature of the sequence of partial sums is clear.

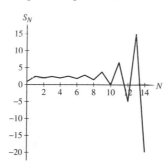

77. GU Let $f(x) = \sqrt{1+x}$.

(a) Use a graphing calculator to compare the graph of f with the graphs of the first five Taylor polynomials for f. What do they suggest about the interval of convergence of the Taylor series?

(b) Investigate numerically whether or not the Taylor expansion for f is valid for $x = 1$ and $x = -1$.

SOLUTION

(a) The five first terms of the Binomial series with $a = \frac{1}{2}$ are

$$\sqrt{1+x} = 1 + \frac{1}{2}x + \frac{\frac{1}{2}\left(\frac{1}{2}-1\right)}{2!}x^2 + \frac{\frac{1}{2}\left(\frac{1}{2}-1\right)\left(\frac{1}{2}-2\right)}{3!}x^3 + \frac{\frac{1}{2}\left(\frac{1}{2}-1\right)\left(\frac{1}{2}-2\right)\left(\frac{1}{2}-3\right)}{4!}x^4 + \cdots$$

$$= 1 + \frac{1}{2}x - \frac{1}{8}x^2 + \frac{9}{4}x^3 - \frac{45}{2}x^4 + \cdots$$

Therefore, the first five Taylor polynomials are

$$T_0(x) = 1;$$

$$T_1(x) = 1 + \frac{1}{2}x;$$

$$T_2(x) = 1 + \frac{1}{2}x - \frac{1}{8}x^2;$$

$$T_3(x) = 1 + \frac{1}{2}x - \frac{1}{8}x^2 + \frac{1}{8}x^3;$$

$$T_4(x) = 1 + \frac{1}{2}x - \frac{1}{8}x^2 + \frac{1}{8}x^3 - \frac{5}{128}x^4.$$

The figure displays the graphs of these Taylor polynomials, along with the graph of the function $f(x) = \sqrt{1+x}$, which is shown in red.

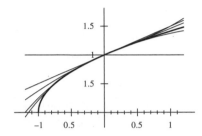

The graphs suggest that the interval of convergence for the Taylor series is $-1 < x < 1$.

(b) Using a computer algebra system to calculate $S_N = \sum_{n=0}^{N} \binom{\frac{1}{2}}{n} x^n$ for $x = 1$ we find

$$S_{10} = 1.409931183, \quad S_{100} = 1.414073048, \quad S_{1000} = 1.414209104,$$

which appears to be converging to $\sqrt{2}$ as expected. At $x = -1$ we calculate $S_N = \sum_{n=0}^{N} \binom{\frac{1}{2}}{n} \cdot (-1)^n$, and find

$$S_{10} = 0.176197052, \quad S_{100} = 0.056348479, \quad S_{1000} = 0.017839011,$$

which appears to be converging to zero, though slowly.

79. Use Example 11 and the approximation $\sin x \approx x$ to show that the period T of a pendulum released at an angle θ has the following second-order approximation:

$$T \approx 2\pi \sqrt{\frac{L}{g}} \left(1 + \frac{\theta^2}{16}\right)$$

SOLUTION   The period T of a pendulum of length L released from an angle θ is

$$T = 4\sqrt{\frac{L}{g}} E(k),$$

where $g \approx 9.8$ m/s^2 is the acceleration due to gravity, $E(k)$ is the elliptic function of the first kind and $k = \sin \frac{\theta}{2}$. From Example 11, we know that

$$E(k) = \frac{\pi}{2} \sum_{n=0}^{\infty} \left(\frac{1 \cdot 3 \cdot 5 \cdots (2n-1)}{2 \cdot 4 \cdot 6 \cdots (2n)}\right)^2 k^{2n}.$$

Using the approximation $\sin x \approx x$, we have

$$k = \sin \frac{\theta}{2} \approx \frac{\theta}{2};$$

moreover, using the first two terms of the series for $E(k)$, we find

$$E(k) \approx \frac{\pi}{2}\left[1 + \left(\frac{1}{2}\right)^2 \left(\frac{\theta}{2}\right)^2\right] = \frac{\pi}{2}\left(1 + \frac{\theta^2}{16}\right).$$

Therefore,

$$T = 4\sqrt{\frac{L}{g}}E(k) \approx 2\pi\sqrt{\frac{L}{g}}\left(1 + \frac{\theta^2}{16}\right).$$

In Exercises 80–83, find the Maclaurin series of the function and use it to calculate the limit.

81. $\displaystyle\lim_{x\to 0} \frac{\sin x - x + \frac{x^3}{6}}{x^5}$

SOLUTION Using the Maclaurin series for $\sin x$, we find

$$\sin x = \sum_{n=0}^{\infty}(-1)^n\frac{x^{2n+1}}{(2n+1)!} = x - \frac{x^3}{6} + \frac{x^5}{120} + \sum_{n=3}^{\infty}(-1)^n\frac{x^{2n+1}}{(2n+1)!}.$$

Thus,

$$\sin x - x + \frac{x^3}{6} = \frac{x^5}{120} + \sum_{n=3}^{\infty}(-1)^n\frac{x^{2n+1}}{(2n+1)!}$$

and

$$\frac{\sin x - x + \frac{x^3}{6}}{x^5} = \frac{1}{120} + \sum_{n=3}^{\infty}(-1)^n\frac{x^{2n-4}}{(2n+1)!}$$

Note that the radius of convergence for this series is infinite, and recall from the previous section that a convergent power series is continuous within its radius of convergence. Thus to calculate the limit of this power series as $x \to 0$ it suffices to evaluate it at $x = 0$:

$$\lim_{x\to 0}\frac{\sin x - x + \frac{x^3}{6}}{x^5} = \lim_{x\to 0}\left(\frac{1}{120} + \sum_{n=3}^{\infty}(-1)^n\frac{x^{2n-4}}{(2n+1)!}\right) = \frac{1}{120} + 0 = \frac{1}{120}$$

83. $\displaystyle\lim_{x\to 0}\left(\frac{\sin(x^2)}{x^4} - \frac{\cos x}{x^2}\right)$

SOLUTION We start with

$$\sin x = \sum_{n=0}^{\infty}(-1)^n\frac{x^{2n+1}}{(2n+1)!} \qquad \cos x = \sum_{n=0}^{\infty}(-1)^n\frac{x^{2n}}{(2n)!}$$

so that

$$\frac{\sin(x^2)}{x^4} = \sum_{n=0}^{\infty}(-1)^n\frac{x^{4n+2}}{(2n+1)!x^4} = \sum_{n=0}^{\infty}(-1)^n\frac{x^{4n-2}}{(2n+1)!}$$

$$\frac{\cos x}{x^2} = \sum_{n=0}^{\infty}(-1)^n\frac{x^{2n-2}}{(2n)!}$$

Expanding the first few terms gives

$$\frac{\sin(x^2)}{x^4} = \frac{1}{x^2} - \sum_{n=1}^{\infty}(-1)^n\frac{x^{4n-2}}{(2n+1)!}$$

$$\frac{\cos x}{x^2} = \frac{1}{x^2} - \frac{1}{2} + \sum_{n=2}^{\infty}(-1)^n\frac{x^{2n-2}}{(2n)!}$$

so that

$$\frac{\sin(x^2)}{x^4} - \frac{\cos x}{x^2} = \frac{1}{2} - \sum_{n=1}^{\infty}(-1)^n\frac{x^{4n-2}}{(2n+1)!} - \sum_{n=2}^{\infty}(-1)^n\frac{x^{2n-2}}{(2n)!}$$

Note that all terms under the summation signs have positive powers of x. Now, the radius of convergence of the series for both sin and cos is infinite, so the radius of convergence of this series is infinite. Recall from the previous section that a convergent power series is continuous within its radius of convergence. Thus to calculate the limit of this power series as $x \to 0$ it suffices to evaluate it at $x = 0$:

$$\lim_{x\to 0}\left(\frac{\sin(x^2)}{x^4} - \frac{\cos x}{x^2}\right) = \lim_{x\to 0}\left(\frac{1}{2} - \sum_{n=1}^{\infty}(-1)^n\frac{x^{4n-2}}{(2n+1)!} - \sum_{n=2}^{\infty}(-1)^n\frac{x^{2n-2}}{(2n)!}\right) = \frac{1}{2} + 0 = \frac{1}{2}$$

Further Insights and Challenges

85. Let $g(t) = \dfrac{1}{1+t^2} - \dfrac{t}{1+t^2}$.

(a) Show that $\displaystyle\int_0^1 g(t)\,dt = \frac{\pi}{4} - \frac{1}{2}\ln 2$.

(b) Show that $g(t) = 1 - t - t^2 + t^3 - t^4 - t^5 - t^6 + \cdots$

(c) Evaluate $S = 1 - \frac{1}{2} - \frac{1}{3} + \frac{1}{4} - \frac{1}{5} - \frac{1}{6} - \frac{1}{7} + \cdots$

SOLUTION

(a)

$$\int_0^1 g(t)\,dt = \left(\tan^{-1}t - \frac{1}{2}\ln(t^2+1)\right)\Big|_0^1 = \tan^{-1}1 - \frac{1}{2}\ln 2 = \frac{\pi}{4} - \frac{1}{2}\ln 2$$

(b) Start with the Taylor series for $\frac{1}{1+t}$:

$$\frac{1}{1+t} = \sum_{n=0}^{\infty}(-1)^n t^n$$

and substitute t^2 for t to get

$$\frac{1}{1+t^2} = \sum_{n=0}^{\infty}(-1)^n t^{2n} = 1 - t^2 + t^4 - t^6 + \cdots$$

so that

$$\frac{t}{1+t^2} = \sum_{n=0}^{\infty}(-1)^n t^{2n+1} = t - t^3 + t^5 - t^7 + \cdots$$

Finally,

$$g(t) = \frac{1}{1+t^2} - \frac{t}{1+t^2} = 1 - t - t^2 + t^3 + t^4 - t^5 - t^6 + t^7 + \cdots$$

(c) We have

$$\int g(t)\,dt = \int(1 - t - t^2 + t^3 + t^4 - t^5 - \cdots)\,dt = t - \frac{1}{2}t^2 - \frac{1}{3}t^3 + \frac{1}{4}t^4 + \frac{1}{5}t^5 - \frac{1}{6}t^6 - \cdots + C$$

The radius of convergence of the series for $g(t)$ is 1, so the radius of convergence of this series is also 1. However, this series converges at the right endpoint, $t = 1$, since

$$\left(1 - \frac{1}{2}\right) - \left(\frac{1}{3} - \frac{1}{4}\right) + \left(\frac{1}{5} - \frac{1}{6}\right) - \cdots$$

is an alternating series with general term decreasing to zero. Thus by part (a),

$$1 - \frac{1}{2} - \frac{1}{3} + \frac{1}{4} + \frac{1}{5} - \frac{1}{6} - \cdots = \frac{\pi}{4} - \frac{1}{2}\ln 2$$

In Exercises 86 and 87, we investigate the convergence of the binomial series

$$T_a(x) = \sum_{n=0}^{\infty}\binom{a}{n}x^n$$

87. By Exercise 86, $T_a(x)$ converges for $|x| < 1$, but we do not yet know whether $T_a(x) = (1+x)^a$.

(a) Verify the identity

$$a\binom{a}{n} = n\binom{a}{n} + (n+1)\binom{a}{n+1}$$

(b) Use (a) to show that $y = T_a(x)$ satisfies the differential equation $(1 + x)y' = ay$ with initial condition $y(0) = 1$.

(c) Prove that $T_a(x) = (1 + x)^a$ for $|x| < 1$ by showing that the derivative of the ratio $\dfrac{T_a(x)}{(1+x)^a}$ is zero.

SOLUTION

(a)

$$n\binom{a}{n} + (n+1)\binom{a}{n+1} = n \cdot \frac{a(a-1)\cdots(a-n+1)}{n!} + (n+1) \cdot \frac{a(a-1)\cdots(a-n+1)(a-n)}{(n+1)!}$$

$$= \frac{a(a-1)\cdots(a-n+1)}{(n-1)!} + \frac{a(a-1)\cdots(a-n+1)(a-n)}{n!}$$

$$= \frac{a(a-1)\cdots(a-n+1)(n+(a-n))}{n!} = a \cdot \binom{a}{n}$$

(b) Differentiating $T_a(x)$ term-by-term yields

$$T_a'(x) = \sum_{n=1}^{\infty} n\binom{a}{n} x^{n-1}.$$

Thus,

$$(1+x)T_a'(x) = \sum_{n=1}^{\infty} n\binom{a}{n} x^{n-1} + \sum_{n=1}^{\infty} n\binom{a}{n} x^n = \sum_{n=0}^{\infty} (n+1)\binom{a}{n+1} x^n + \sum_{n=0}^{\infty} n\binom{a}{n} x^n$$

$$= \sum_{n=0}^{\infty} \left[(n+1)\binom{a}{n+1} + n\binom{a}{n} \right] x^n = a \sum_{n=0}^{\infty} \binom{a}{n} x^n = aT_a(x).$$

Moreover,

$$T_a(0) = \binom{a}{0} = 1.$$

(c)

$$\frac{d}{dx}\left(\frac{T_a(x)}{(1+x)^a} \right) = \frac{(1+x)^a T_a'(x) - a(1+x)^{a-1}T_a(x)}{(1+x)^{2a}} = \frac{(1+x)T_a'(x) - aT_a(x)}{(1+x)^{a+1}} = 0.$$

Thus,

$$\frac{T_a(x)}{(1+x)^a} = C,$$

for some constant C. For $x = 0$,

$$\frac{T_a(0)}{(1+0)^a} = \frac{1}{1} = 1, \text{ so } C = 1.$$

Finally, $T_a(x) = (1+x)^a$.

89. Assume that $a < b$ and let L be the arc length (circumference) of the ellipse $\left(\frac{x}{a}\right)^2 + \left(\frac{y}{b}\right)^2 = 1$ shown in Figure 5. There is no explicit formula for L, but it is known that $L = 4bG(k)$, with $G(k)$ as in Exercise 88 and $k = \sqrt{1 - a^2/b^2}$. Use the first three terms of the expansion of Exercise 88 to estimate L when $a = 4$ and $b = 5$.

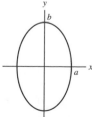

FIGURE 5 The ellipse $\left(\frac{x}{a}\right)^2 + \left(\frac{y}{b}\right)^2 = 1$.

SOLUTION   With $a = 4$ and $b = 5$,

$$k = \sqrt{1 - \frac{4^2}{5^2}} = \frac{3}{5},$$

and the arc length of the ellipse $\left(\frac{x}{4}\right)^2 + \left(\frac{y}{5}\right)^2 = 1$ is

$$L = 20G\left(\frac{3}{5}\right) = 20\left(\frac{\pi}{2} - \frac{\pi}{2}\sum_{n=1}^{\infty}\left(\frac{1 \cdot 3 \cdots (2n-1)}{2 \cdot 4 \cdots (2n)}\right)^2 \frac{\left(\frac{3}{5}\right)^{2n}}{2n-1}\right).$$

Using the first three terms in the series for $G(k)$ gives

$$L \approx 10\pi - 10\pi\left(\left(\frac{1}{2}\right)^2 \cdot \frac{(3/5)^2}{1} + \left(\frac{1 \cdot 3}{2 \cdot 4}\right)^2 \cdot \frac{(3/5)^4}{3}\right) = 10\pi\left(1 - \frac{9}{100} - \frac{243}{40,000}\right) = \frac{36,157\pi}{4000} \approx 28.398.$$

91. Irrationality of e   Prove that e is an irrational number using the following argument by contradiction. Suppose that $e = M/N$, where M, N are nonzero integers.

(a) Show that $M!\,e^{-1}$ is a whole number.

(b) Use the power series for e^x at $x = -1$ to show that there is an integer B such that $M!\,e^{-1}$ equals

$$B + (-1)^{M+1}\left(\frac{1}{M+1} - \frac{1}{(M+1)(M+2)} + \cdots\right)$$

(c) Use your knowledge of alternating series with decreasing terms to conclude that $0 < |M!\,e^{-1} - B| < 1$ and observe that this contradicts (a). Hence, e is not equal to M/N.

SOLUTION   Suppose that $e = M/N$, where M, N are nonzero integers.

(a) With $e = M/N$,

$$M!\,e^{-1} = M!\frac{N}{M} = (M-1)!N,$$

which is a whole number.

(b) Substituting $x = -1$ into the Maclaurin series for e^x and multiplying the resulting series by $M!$ yields

$$M!\,e^{-1} = M!\left(1 - 1 + \frac{1}{2!} - \frac{1}{3!} + \cdots + \frac{(-1)^k}{k!} + \cdots\right).$$

For all $k \le M$, $\dfrac{M!}{k!}$ is a whole number, so

$$M!\left(1 - 1 + \frac{1}{2!} - \frac{1}{3!} + \cdots + \frac{(-1)^k}{M!}\right)$$

is an integer. Denote this integer by B. Thus,

$$M!\,e^{-1} = B + M!\left(\frac{(-1)^{M+1}}{(M+1)!} + \frac{(-1)^{M+2}}{(M+2)!} + \cdots\right) = B + (-1)^{M+1}\left(\frac{1}{M+1} - \frac{1}{(M+1)(M+2)} + \cdots\right).$$

(c) The series for $M!\,e^{-1}$ obtained in part (b) is an alternating series with $a_n = \frac{M!}{n!}$. Using the error bound for an alternating series and noting that $B = S_M$, we have

$$\left|M!\,e^{-1} - B\right| \le a_{M+1} = \frac{1}{M+1} < 1.$$

This inequality implies that $M!\,e^{-1} - B$ is not a whole number; however, B is a whole number so $M!\,e^{-1}$ cannot be a whole number. We get a contradiction to the result in part (a), which proves that the original assumption that e is a rational number is false.

CHAPTER REVIEW EXERCISES

1. Let $a_n = \dfrac{n-3}{n!}$ and $b_n = a_{n+3}$. Calculate the first three terms in each sequence.

(a) a_n^2

(b) b_n

(c) $a_n b_n$

(d) $2a_{n+1} - 3a_n$

SOLUTION

(a)

$$a_1^2 = \left(\frac{1-3}{1!}\right)^2 = (-2)^2 = 4;$$

$$a_2^2 = \left(\frac{2-3}{2!}\right)^2 = \left(-\frac{1}{2}\right)^2 = \frac{1}{4};$$

$$a_3^2 = \left(\frac{3-3}{3!}\right)^2 = 0.$$

(b)

$$b_1 = a_4 = \frac{4-3}{4!} = \frac{1}{24};$$

$$b_2 = a_5 = \frac{5-3}{5!} = \frac{1}{60};$$

$$b_3 = a_6 = \frac{6-3}{6!} = \frac{1}{240}.$$

(c) Using the formula for a_n and the values in (b) we obtain:

$$a_1 b_1 = \frac{1-3}{1!} \cdot \frac{1}{24} = -\frac{1}{12};$$

$$a_2 b_2 = \frac{2-3}{2!} \cdot \frac{1}{60} = -\frac{1}{120};$$

$$a_3 b_3 = \frac{3-3}{3!} \cdot \frac{1}{240} = 0.$$

(d)

$$2a_2 - 3a_1 = 2\left(-\frac{1}{2}\right) - 3(-2) = 5;$$

$$2a_3 - 3a_2 = 2 \cdot 0 - 3\left(-\frac{1}{2}\right) = \frac{3}{2};$$

$$2a_4 - 3a_3 = 2 \cdot \frac{1}{24} - 3 \cdot 0 = \frac{1}{12}.$$

In Exercises 3–8, compute the limit (or state that it does not exist) assuming that $\lim\limits_{n\to\infty} a_n = 2$.

3. $\lim\limits_{n\to\infty} (5a_n - 2a_n^2)$

SOLUTION

$$\lim_{n\to\infty}\left(5a_n - 2a_n^2\right) = 5\lim_{n\to\infty} a_n - 2\lim_{n\to\infty} a_n^2 = 5\lim_{n\to\infty} a_n - 2\left(\lim_{n\to\infty} a_n\right)^2 = 5\cdot 2 - 2\cdot 2^2 = 2.$$

5. $\lim\limits_{n\to\infty} e^{a_n}$

SOLUTION   The function $f(x) = e^x$ is continuous, hence:

$$\lim_{n\to\infty} e^{a_n} = e^{\lim_{n\to\infty} a_n} = e^2.$$

7. $\lim\limits_{n\to\infty} (-1)^n a_n$

SOLUTION   Because $\lim\limits_{n\to\infty} a_n \neq 0$, it follows that $\lim\limits_{n\to\infty} (-1)^n a_n$ does not exist.

In Exercises 9–22, determine the limit of the sequence or show that the sequence diverges.

9. $a_n = \sqrt{n+5} - \sqrt{n+2}$

SOLUTION   First rewrite a_n as follows:

$$a_n = \frac{\left(\sqrt{n+5} - \sqrt{n+2}\right)\left(\sqrt{n+5} + \sqrt{n+2}\right)}{\sqrt{n+5} + \sqrt{n+2}} = \frac{(n+5) - (n+2)}{\sqrt{n+5} + \sqrt{n+2}} = \frac{3}{\sqrt{n+5} + \sqrt{n+2}}.$$

Thus,

$$\lim_{n \to \infty} a_n = \lim_{n \to \infty} \frac{3}{\sqrt{n+5} + \sqrt{n+2}} = 0.$$

11. $a_n = 2^{1/n^2}$

SOLUTION   The function $f(x) = 2^x$ is continuous, so

$$\lim_{n \to \infty} a_n = \lim_{n \to \infty} 2^{1/n^2} = 2^{\lim_{n \to \infty}(1/n^2)} = 2^0 = 1.$$

13. $b_m = 1 + (-1)^m$

SOLUTION   Because $1 + (-1)^m$ is equal to 0 for m odd and is equal to 2 for m even, the sequence $\{b_m\}$ does not approach one limit; hence this sequence diverges.

15. $b_n = \tan^{-1}\left(\dfrac{n+2}{n+5}\right)$

SOLUTION   The function $\tan^{-1} x$ is continuous, so

$$\lim_{n \to \infty} b_n = \lim_{n \to \infty} \tan^{-1}\left(\frac{n+2}{n+5}\right) = \tan^{-1}\left(\lim_{n \to \infty} \frac{n+2}{n+5}\right) = \tan^{-1} 1 = \frac{\pi}{4}.$$

17. $b_n = \sqrt{n^2 + n} - \sqrt{n^2 + 1}$

SOLUTION   Rewrite b_n as

$$b_n = \frac{\left(\sqrt{n^2+n} - \sqrt{n^2+1}\right)\left(\sqrt{n^2+n} + \sqrt{n^2+1}\right)}{\sqrt{n^2+n} + \sqrt{n^2+1}} = \frac{\left(n^2+n\right) - \left(n^2+1\right)}{\sqrt{n^2+n} + \sqrt{n^2+1}} = \frac{n-1}{\sqrt{n^2+n} + \sqrt{n^2+1}}.$$

Then

$$\lim_{n \to \infty} b_n = \lim_{n \to \infty} \frac{\frac{n}{n} - \frac{1}{n}}{\sqrt{\frac{n^2}{n^2} + \frac{n}{n^2}} + \sqrt{\frac{n^2}{n^2} + \frac{1}{n^2}}} = \lim_{n \to \infty} \frac{1 - \frac{1}{n}}{\sqrt{1 + \frac{1}{n}} + \sqrt{1 + \frac{1}{n^2}}} = \frac{1-0}{\sqrt{1+0} + \sqrt{1+0}} = \frac{1}{2}.$$

19. $b_m = \left(1 + \dfrac{1}{m}\right)^{3m}$

SOLUTION $\displaystyle\lim_{m \to \infty} b_m = \lim_{m \to \infty} \left(1 + \frac{1}{m}\right)^m = e.$

21. $b_n = n\big(\ln(n+1) - \ln n\big)$

SOLUTION   Write

$$b_n = n \ln\left(\frac{n+1}{n}\right) = \frac{\ln\left(1 + \frac{1}{n}\right)}{\frac{1}{n}}.$$

Using L'Hôpital's Rule, we find

$$\lim_{n \to \infty} b_n = \lim_{n \to \infty} \frac{\ln\left(1 + \frac{1}{n}\right)}{\frac{1}{n}} = \lim_{x \to \infty} \frac{\ln\left(1 + \frac{1}{x}\right)}{\frac{1}{x}} = \lim_{x \to \infty} \frac{\left(1 + \frac{1}{x}\right)^{-1} \cdot \left(-\frac{1}{x^2}\right)}{-\frac{1}{x^2}} = \lim_{x \to \infty} \left(1 + \frac{1}{x}\right)^{-1} = 1.$$

23. Use the Squeeze Theorem to show that $\lim\limits_{n\to\infty} \dfrac{\arctan(n^2)}{\sqrt{n}} = 0$.

SOLUTION   For all x,

$$-\frac{\pi}{2} < \arctan x < \frac{\pi}{2},$$

so

$$-\frac{\pi/2}{\sqrt{n}} < \frac{\arctan(n^2)}{\sqrt{n}} < \frac{\pi/2}{\sqrt{n}},$$

for all n. Because

$$\lim_{n\to\infty}\left(-\frac{\pi/2}{\sqrt{n}}\right) = \lim_{n\to\infty}\frac{\pi/2}{\sqrt{n}} = 0,$$

it follows by the Squeeze Theorem that

$$\lim_{n\to\infty}\frac{\arctan(n^2)}{\sqrt{n}} = 0.$$

25. Calculate $\lim\limits_{n\to\infty} \dfrac{a_{n+1}}{a_n}$, where $a_n = \dfrac{1}{2}3^n - \dfrac{1}{3}2^n$.

SOLUTION   Because

$$\frac{1}{2}3^n - \frac{1}{3}2^n \ge \frac{1}{2}3^n - \frac{1}{3}3^n = \frac{3^n}{6}$$

and

$$\lim_{n\to\infty}\frac{3^n}{6} = \infty,$$

we conclude that $\lim_{n\to\infty} a_n = \infty$, so L'Hôpital's rule may be used:

$$\lim_{n\to\infty}\frac{a_{n+1}}{a_n} = \lim_{n\to\infty}\frac{\frac{1}{2}3^{n+1} - \frac{1}{3}2^{n+1}}{\frac{1}{2}3^n - \frac{1}{3}2^n} = \lim_{n\to\infty}\frac{3^{n+2} - 2^{n+2}}{3^{n+1} - 2^{n+1}} = \lim_{n\to\infty}\frac{3 - 2\left(\frac{2}{3}\right)^{n+1}}{1 - \left(\frac{2}{3}\right)^{n+1}} = \frac{3 - 0}{1 - 0} = 3.$$

27. Calculate the partial sums S_4 and S_7 of the series $\displaystyle\sum_{n=1}^{\infty} \frac{n-2}{n^2 + 2n}$.

SOLUTION

$$S_4 = -\frac{1}{3} + 0 + \frac{1}{15} + \frac{2}{24} = -\frac{11}{60} = -0.183333;$$

$$S_7 = -\frac{1}{3} + 0 + \frac{1}{15} + \frac{2}{24} + \frac{3}{35} + \frac{4}{48} + \frac{5}{63} = \frac{287}{4410} = 0.065079.$$

29. Find the sum $\dfrac{4}{9} + \dfrac{8}{27} + \dfrac{16}{81} + \dfrac{32}{243} + \cdots$.

SOLUTION   This is a geometric series with common ratio $r = \frac{2}{3}$. Therefore,

$$\frac{4}{9} + \frac{8}{27} + \frac{16}{81} + \frac{32}{243} + \cdots = \frac{\frac{4}{9}}{1 - \frac{2}{3}} = \frac{4}{3}.$$

31. Find the sum $\displaystyle\sum_{n=-1}^{\infty} \frac{2^{n+3}}{3^n}$.

SOLUTION   Note

$$\sum_{n=-1}^{\infty}\frac{2^{n+3}}{3^n} = 2^3\sum_{n=-1}^{\infty}\frac{2^n}{3^n} = 8\sum_{n=-1}^{\infty}\left(\frac{2}{3}\right)^n;$$

therefore,

$$\sum_{n=-1}^{\infty}\frac{2^{n+3}}{3^n} = 8\cdot\frac{3}{2}\cdot\frac{1}{1-\frac{2}{3}} = 36.$$

33. Give an example of divergent series $\sum_{n=1}^{\infty} a_n$ and $\sum_{n=1}^{\infty} b_n$ such that $\sum_{n=1}^{\infty} (a_n + b_n) = 1$.

SOLUTION   Let $a_n = \left(\frac{1}{2}\right)^n + 1$, $b_n = -1$. The corresponding series diverge by the Divergence Test; however,

$$\sum_{n=1}^{\infty} (a_n + b_n) = \sum_{n=1}^{\infty} \left(\frac{1}{2}\right)^n = \frac{\frac{1}{2}}{1 - \frac{1}{2}} = 1.$$

35. Evaluate $S = \sum_{n=3}^{\infty} \frac{1}{n(n+3)}$.

SOLUTION   Note that

$$\frac{1}{n(n+3)} = \frac{1}{3}\left(\frac{1}{n} - \frac{1}{n+3}\right)$$

so that

$$\sum_{n=3}^{N} \frac{1}{n(n+3)} = \frac{1}{3} \sum_{n=3}^{N} \left(\frac{1}{n} - \frac{1}{n+3}\right)$$

$$= \frac{1}{3}\left(\left(\frac{1}{3} - \frac{1}{6}\right) + \left(\frac{1}{4} - \frac{1}{7}\right) + \left(\frac{1}{5} - \frac{1}{8}\right) \right.$$

$$\left. \left(\frac{1}{6} - \frac{1}{9}\right) + \cdots + \left(\frac{1}{N-1} - \frac{1}{N+2}\right) + \left(\frac{1}{N} - \frac{1}{N+3}\right)\right)$$

$$= \frac{1}{3}\left(\frac{1}{3} + \frac{1}{4} + \frac{1}{5} - \frac{1}{N+1} - \frac{1}{N+2} - \frac{1}{N+3}\right)$$

Thus

$$\sum_{n=3}^{\infty} \frac{1}{n(n+3)} = \frac{1}{3} \lim_{N \to \infty} \sum_{n=3}^{N} \left(\frac{1}{n} - \frac{1}{n+3}\right)$$

$$= \frac{1}{3}\left(\frac{1}{3} + \frac{1}{4} + \frac{1}{5} - \frac{1}{N+1} - \frac{1}{N+2} - \frac{1}{N+3}\right) = \frac{1}{3}\left(\frac{1}{3} + \frac{1}{4} + \frac{1}{5}\right) = \frac{47}{180}$$

In Exercises 37–40, use the Integral Test to determine whether the infinite series converges.

37. $\sum_{n=1}^{\infty} \frac{n^2}{n^3 + 1}$

SOLUTION   Let $f(x) = \frac{x^2}{x^3+1}$. This function is continuous and positive for $x \geq 1$. Because

$$f'(x) = \frac{(x^3 + 1)(2x) - x^2(3x^2)}{(x^3 + 1)^2} = \frac{x(2 - x^3)}{(x^3 + 1)^2},$$

we see that $f'(x) < 0$ and f is decreasing on the interval $x \geq 2$. Therefore, the Integral Test applies on the interval $x \geq 2$. Now,

$$\int_{2}^{\infty} \frac{x^2}{x^3 + 1}\, dx = \lim_{R \to \infty} \int_{2}^{R} \frac{x^2}{x^3 + 1}\, dx = \frac{1}{3} \lim_{R \to \infty} \left(\ln(R^3 + 1) - \ln 9\right) = \infty.$$

The integral diverges; hence, the series $\sum_{n=2}^{\infty} \frac{n^2}{n^3 + 1}$ diverges, as does the series $\sum_{n=1}^{\infty} \frac{n^2}{n^3 + 1}$.

39. $\sum_{n=1}^{\infty} \frac{1}{(n + 2)(\ln(n + 2))^3}$

SOLUTION   Let $f(x) = \frac{1}{(x+2)\ln^3(x+2)}$. Using the substitution $u = \ln(x + 2)$, so that $du = \frac{1}{x+2}\, dx$, we have

$$\int_{0}^{\infty} f(x)\, dx = \int_{\ln 2}^{\infty} \frac{1}{u^3}\, du = \lim_{R \to \infty} \int_{\ln 2}^{\infty} \frac{1}{u^3}\, du = \lim_{R \to \infty} \left(-\frac{1}{2u^2}\Big|_{\ln 2}^{R}\right)$$

$$= \lim_{R \to \infty} \left(\frac{1}{2(\ln 2)^2} - \frac{1}{2(\ln R)^2}\right) = \frac{1}{2(\ln 2)^2}$$

Since the integral of $f(x)$ converges, so does the series.

In Exercises 41–48, use the Comparison or Limit Comparison Test to determine whether the infinite series converges.

41. $\displaystyle\sum_{n=1}^{\infty} \frac{1}{(n+1)^2}$

SOLUTION For all $n \geq 1$,

$$0 < \frac{1}{n+1} < \frac{1}{n} \quad \text{so} \quad \frac{1}{(n+1)^2} < \frac{1}{n^2}.$$

The series $\displaystyle\sum_{n=1}^{\infty} \frac{1}{n^2}$ is a convergent p-series, so the series $\displaystyle\sum_{n=1}^{\infty} \frac{1}{(n+1)^2}$ converges by the Comparison Test.

43. $\displaystyle\sum_{n=2}^{\infty} \frac{n^2+1}{n^{3.5}-2}$

SOLUTION Apply the Limit Comparison Test with $a_n = \frac{n^2+1}{n^{3.5}-2}$ and $b_n = \frac{1}{n^{1.5}}$. Now,

$$L = \lim_{n\to\infty} \frac{\frac{n^2+1}{n^{3.5}-2}}{\frac{1}{n^{1.5}}} = \lim_{n\to\infty} \frac{n^{3.5}+n^{1.5}}{n^{3.5}-2} = 1.$$

Because L exists and $\displaystyle\sum_{n=1}^{\infty} \frac{1}{n^{1.5}}$ is a convergent p-series, we conclude by the Limit Comparison Test that the series

$\displaystyle\sum_{n=2}^{\infty} \frac{n^2+1}{n^{3.5}-2}$ also converges.

45. $\displaystyle\sum_{n=2}^{\infty} \frac{n}{\sqrt{n^5+5}}$

SOLUTION For all $n \geq 2$,

$$\frac{n}{\sqrt{n^5+5}} < \frac{n}{n^{5/2}} = \frac{1}{n^{3/2}}.$$

The series $\displaystyle\sum_{n=2}^{\infty} \frac{1}{n^{3/2}}$ is a convergent p-series, so the series $\displaystyle\sum_{n=2}^{\infty} \frac{n}{\sqrt{n^5+5}}$ converges by the Comparison Test.

47. $\displaystyle\sum_{n=1}^{\infty} \frac{n^{10}+10^n}{n^{11}+11^n}$

SOLUTION Apply the Limit Comparison Test with $a_n = \frac{n^{10}+10^n}{n^{11}+11^n}$ and $b_n = \left(\frac{10}{11}\right)^n$. Then,

$$L = \lim_{n\to\infty} \frac{a_n}{b_n} = \lim_{n\to\infty} \frac{\frac{n^{10}+10^n}{n^{11}+11^n}}{\left(\frac{10}{11}\right)^n} = \lim_{n\to\infty} \frac{\frac{n^{10}+10^n}{10^n}}{\frac{n^{11}+11^n}{11^n}} = \lim_{n\to\infty} \frac{\frac{n^{10}}{10^n}+1}{\frac{n^{11}}{11^n}+1} = 1.$$

The series $\displaystyle\sum_{n=1}^{\infty} \left(\frac{10}{11}\right)^n$ is a convergent geometric series; because L exists, we may therefore conclude by the Limit

Comparison Test that the series $\displaystyle\sum_{n=1}^{\infty} \frac{n^{10}+10^n}{n^{11}+11^n}$ also converges.

49. Determine the convergence of $\displaystyle\sum_{n=1}^{\infty} \frac{2^n+n}{3^n-2}$ using the Limit Comparison Test with $b_n = \left(\frac{2}{3}\right)^n$.

SOLUTION With $a_n = \frac{2^n+n}{3^n-2}$, we have

$$L = \lim_{n\to\infty} \frac{a_n}{b_n} = \lim_{n\to\infty} \frac{2^n+n}{3^n-2} \cdot \frac{3^n}{2^n} = \lim_{n\to\infty} \frac{6^n+n3^n}{6^n-2^{n+1}} = \lim_{n\to\infty} \frac{1+n\left(\frac{1}{2}\right)^n}{1-2\left(\frac{1}{3}\right)^n} = 1$$

Since $L = 1$, the two series either both converge or both diverge. Since $\displaystyle\sum_{n=1}^{\infty} \left(\frac{2}{3}\right)^n$ is a convergent geometric series, the

Limit Comparison Test tells us that $\displaystyle\sum_{n=1}^{\infty} \frac{2^n+n}{3^n-2}$ also converges.

51. Let $a_n = 1 - \sqrt{1 - \frac{1}{n}}$. Show that $\lim\limits_{n \to \infty} a_n = 0$ and that $\sum\limits_{n=1}^{\infty} a_n$ diverges. *Hint:* Show that $a_n \geq \frac{1}{2n}$.

SOLUTION

$$1 - \sqrt{1 - \frac{1}{n}} = 1 - \sqrt{\frac{n-1}{n}} = \frac{\sqrt{n} - \sqrt{n-1}}{\sqrt{n}} = \frac{n - (n-1)}{\sqrt{n}(\sqrt{n} + \sqrt{n-1})} = \frac{1}{n + \sqrt{n^2 - n}}$$

$$\geq \frac{1}{n + \sqrt{n^2}} = \frac{1}{2n}.$$

The series $\sum\limits_{n=2}^{\infty} \frac{1}{2n}$ diverges, so the series $\sum\limits_{n=2}^{\infty} \left(1 - \sqrt{1 - \frac{1}{n}}\right)$ also diverges by the Comparison Test.

53. Let $S = \sum\limits_{n=1}^{\infty} \frac{n}{(n^2 + 1)^2}$.

(a) Show that S converges.

(b) *CAS* Use Eq. (4) in Exercise 83 of Section 10.3 with $M = 99$ to approximate S. What is the maximum size of the error?

SOLUTION

(a) For $n \geq 1$,

$$\frac{n}{(n^2 + 1)^2} < \frac{n}{(n^2)^2} = \frac{1}{n^3}.$$

The series $\sum\limits_{n=1}^{\infty} \frac{1}{n^3}$ is a convergent p-series, so the series $\sum\limits_{n=1}^{\infty} \frac{n}{(n^2 + 1)^2}$ also converges by the Comparison Test.

(b) With $a_n = \frac{n}{(n^2+1)^2}$, $f(x) = \frac{x}{(x^2+1)^2}$ and $M = 99$, Eq. (4) in Exercise 83 of Section 10.3 becomes

$$\sum_{n=1}^{99} \frac{n}{(n^2 + 1)^2} + \int_{100}^{\infty} \frac{x}{(x^2 + 1)^2}\, dx \leq S \leq \sum_{n=1}^{100} \frac{n}{(n^2 + 1)^2} + \int_{100}^{\infty} \frac{x}{(x^2 + 1)^2}\, dx,$$

or

$$0 \leq S - \left(\sum_{n=1}^{99} \frac{n}{(n^2 + 1)^2} + \int_{100}^{\infty} \frac{x}{(x^2 + 1)^2}\, dx \right) \leq \frac{100}{(100^2 + 1)^2}.$$

Now,

$$\sum_{n=1}^{99} \frac{n}{(n^2 + 1)^2} = 0.397066274; \text{ and}$$

$$\int_{100}^{\infty} \frac{x}{(x^2 + 1)^2}\, dx = \lim_{R \to \infty} \int_{100}^{R} \frac{x}{(x^2 + 1)^2}\, dx = \frac{1}{2} \lim_{R \to \infty} \left(-\frac{1}{R^2 + 1} + \frac{1}{100^2 + 1} \right)$$

$$= \frac{1}{20002} = 0.000049995;$$

thus,

$$S \approx 0.397066274 + 0.000049995 = 0.397116269.$$

The bound on the error in this approximation is

$$\frac{100}{(100^2 + 1)^2} = 9.998 \times 10^{-7}.$$

In Exercises 54–57, determine whether the series converges absolutely. If it does not, determine whether it converges conditionally.

55. $\sum\limits_{n=1}^{\infty} \frac{(-1)^n}{n^{1.1} \ln(n + 1)}$

SOLUTION Consider the corresponding positive series $\sum\limits_{n=1}^{\infty} \frac{1}{n^{1.1} \ln(n + 1)}$. Because

$$\frac{1}{n^{1.1} \ln(n + 1)} < \frac{1}{n^{1.1}}$$

and $\displaystyle\sum_{n=1}^{\infty} \frac{1}{n^{1.1}}$ is a convergent p-series, we can conclude by the Comparison Test that $\displaystyle\sum_{n=1}^{\infty} \frac{(-1)^n}{n^{1.1}\ln(n+1)}$ also converges.

Thus, $\displaystyle\sum_{n=1}^{\infty} \frac{(-1)^n}{n^{1.1}\ln(n+1)}$ converges absolutely.

57. $\displaystyle\sum_{n=1}^{\infty} \frac{\cos\left(\frac{\pi}{4}+2\pi n\right)}{\sqrt{n}}$

SOLUTION $\cos\left(\frac{\pi}{4}+2\pi n\right) = \cos\frac{\pi}{4} = \frac{\sqrt{2}}{2}$, so

$$\sum_{n=1}^{\infty} \frac{\cos\left(\frac{\pi}{4}+2\pi n\right)}{\sqrt{n}} = \frac{\sqrt{2}}{2}\sum_{n=1}^{\infty} \frac{1}{\sqrt{n}}.$$

This is a divergent p-series, so the series $\displaystyle\sum_{n=1}^{\infty} \frac{\cos\left(\frac{\pi}{4}+2\pi n\right)}{\sqrt{n}}$ diverges.

59. Catalan's constant is defined by $K = \displaystyle\sum_{k=0}^{\infty} \frac{(-1)^k}{(2k+1)^2}$.

(a) How many terms of the series are needed to calculate K with an error of less than 10^{-6}?
(b) $\mathcal{CAS}$ Carry out the calculation.

SOLUTION   Using the error bound for an alternating series, we have

$$|S_N - K| \le \frac{1}{(2(N+1)+1)^2} = \frac{1}{(2N+3)^2}.$$

For accuracy to three decimal places, we must choose N so that

$$\frac{1}{(2N+3)^2} < 5\times 10^{-3} \quad\text{or}\quad (2N+3)^2 > 2000.$$

Solving for N yields

$$N > \frac{1}{2}\left(\sqrt{2000}-3\right) \approx 20.9.$$

Thus,

$$K \approx \sum_{k=0}^{21} \frac{(-1)^k}{(2k+1)^2} = 0.915707728.$$

61. Let $\displaystyle\sum_{n=1}^{\infty} a_n$ be an absolutely convergent series. Determine whether the following series are convergent or divergent:

(a) $\displaystyle\sum_{n=1}^{\infty}\left(a_n + \frac{1}{n^2}\right)$

(b) $\displaystyle\sum_{n=1}^{\infty}(-1)^n a_n$

(c) $\displaystyle\sum_{n=1}^{\infty} \frac{1}{1+a_n^2}$

(d) $\displaystyle\sum_{n=1}^{\infty} \frac{|a_n|}{n}$

SOLUTION   Because $\displaystyle\sum_{n=1}^{\infty} a_n$ converges absolutely, we know that $\displaystyle\sum_{n=1}^{\infty} a_n$ converges and that $\displaystyle\sum_{n=1}^{\infty} |a_n|$ converges.

(a) Because we know that $\displaystyle\sum_{n=1}^{\infty} a_n$ converges and the series $\displaystyle\sum_{n=1}^{\infty} \frac{1}{n^2}$ is a convergent p-series, the sum of these two series, $\displaystyle\sum_{n=1}^{\infty}\left(a_n + \frac{1}{n^2}\right)$ also converges.

(b) We have,

$$\sum_{n=1}^{\infty}\left|(-1)^n a_n\right| = \sum_{n=1}^{\infty} |a_n|$$

Because $\displaystyle\sum_{n=1}^{\infty} |a_n|$ converges, it follows that $\displaystyle\sum_{n=1}^{\infty}(-1)^n a_n$ converges absolutely, which implies that $\displaystyle\sum_{n=1}^{\infty}(-1)^n a_n$ converges.

(c) Because $\sum\limits_{n=1}^{\infty} a_n$ converges, $\lim_{n\to\infty} a_n = 0$. Therefore,

$$\lim_{n\to\infty} \frac{1}{1+a_n^2} = \frac{1}{1+0^2} = 1 \neq 0,$$

and the series $\sum\limits_{n=1}^{\infty} \frac{1}{1+a_n^2}$ diverges by the Divergence Test.

(d) $\frac{|a_n|}{n} \leq |a_n|$ and the series $\sum\limits_{n=1}^{\infty} |a_n|$ converges, so the series $\sum\limits_{n=1}^{\infty} \frac{|a_n|}{n}$ also converges by the Comparison Test.

In Exercises 63–70, apply the Ratio Test to determine convergence or divergence, or state that the Ratio Test is inconclusive.

63. $\sum\limits_{n=1}^{\infty} \frac{n^5}{5^n}$

SOLUTION   With $a_n = \frac{n^5}{5^n}$,

$$\left| \frac{a_{n+1}}{a_n} \right| = \frac{(n+1)^5}{5^{n+1}} \cdot \frac{5^n}{n^5} = \frac{1}{5}\left(1+\frac{1}{n}\right)^5,$$

and

$$\rho = \lim_{n\to\infty} \left| \frac{a_{n+1}}{a_n} \right| = \frac{1}{5}\lim_{n\to\infty}\left(1+\frac{1}{n}\right)^5 = \frac{1}{5}\cdot 1 = \frac{1}{5}.$$

Because $\rho < 1$, the series converges by the Ratio Test.

65. $\sum\limits_{n=1}^{\infty} \frac{1}{n2^n + n^3}$

SOLUTION   With $a_n = \frac{1}{n2^n+n^3}$,

$$\left| \frac{a_{n+1}}{a_n} \right| = \frac{n2^n+n^3}{(n+1)2^{n+1}+(n+1)^3} = \frac{n2^n\left(1+\frac{n^2}{2^n}\right)}{(n+1)2^{n+1}\left(1+\frac{(n+1)^2}{2^{n+1}}\right)} = \frac{1}{2}\cdot\frac{n}{n+1}\cdot\frac{1+\frac{n^2}{2^n}}{1+\frac{(n+1)^2}{2^{n+1}}},$$

and

$$\rho = \lim_{n\to\infty}\left| \frac{a_{n+1}}{a_n} \right| = \frac{1}{2}\cdot 1 \cdot 1 = \frac{1}{2}.$$

Because $\rho < 1$, the series converges by the Ratio Test.

67. $\sum\limits_{n=1}^{\infty} \frac{2^{n^2}}{n!}$

SOLUTION   With $a_n = \frac{2^{n^2}}{n!}$,

$$\left| \frac{a_{n+1}}{a_n} \right| = \frac{2^{(n+1)^2}}{(n+1)!}\cdot\frac{n!}{2^{n^2}} = \frac{2^{2n+1}}{n+1} \quad\text{and}\quad \rho = \lim_{n\to\infty}\left| \frac{a_{n+1}}{a_n} \right| = \infty.$$

Because $\rho > 1$, the series diverges by the Ratio Test.

69. $\sum\limits_{n=1}^{\infty} \left(\frac{n}{2}\right)^n \frac{1}{n!}$

SOLUTION   With $a_n = \left(\frac{n}{2}\right)^n \frac{1}{n!}$,

$$\left| \frac{a_{n+1}}{a_n} \right| = \left(\frac{n+1}{2}\right)^{n+1}\frac{1}{(n+1)!}\cdot\left(\frac{2}{n}\right)^n n! = \frac{1}{2}\left(\frac{n+1}{n}\right)^n = \frac{1}{2}\left(1+\frac{1}{n}\right)^n,$$

and

$$\rho = \lim_{n\to\infty}\left| \frac{a_{n+1}}{a_n} \right| = \frac{1}{2}e.$$

Because $\rho = \frac{e}{2} > 1$, the series diverges by the Ratio Test.

In Exercises 71–74, apply the Root Test to determine convergence or divergence, or state that the Root Test is inconclusive.

71. $\displaystyle\sum_{n=1}^{\infty} \frac{1}{4^n}$

SOLUTION   With $a_n = \frac{1}{4^n}$,

$$L = \lim_{n\to\infty} \sqrt[n]{a_n} = \lim_{n\to\infty} \sqrt[n]{\frac{1}{4^n}} = \frac{1}{4}.$$

Because $L < 1$, the series converges by the Root Test.

73. $\displaystyle\sum_{n=1}^{\infty} \left(\frac{3}{4n}\right)^n$

SOLUTION   With $a_n = \left(\frac{3}{4n}\right)^n$,

$$L = \lim_{n\to\infty} \sqrt[n]{a_n} = \lim_{n\to\infty} \sqrt[n]{\left(\frac{3}{4n}\right)^n} = \lim_{n\to\infty} \frac{3}{4n} = 0.$$

Because $L < 1$, the series converges by the Root Test.

In Exercises 75–92, determine convergence or divergence using any method covered in the text.

75. $\displaystyle\sum_{n=1}^{\infty} \left(\frac{2}{3}\right)^n$

SOLUTION   This is a geometric series with ratio $r = \frac{2}{3} < 1$; hence, the series converges.

77. $\displaystyle\sum_{n=1}^{\infty} e^{-0.02n}$

SOLUTION   This is a geometric series with common ratio $r = \frac{1}{e^{0.02}} \approx 0.98 < 1$; hence, the series converges.

79. $\displaystyle\sum_{n=1}^{\infty} \frac{(-1)^{n-1}}{\sqrt{n} + \sqrt{n+1}}$

SOLUTION   In this alternating series, $a_n = \frac{1}{\sqrt{n}+\sqrt{n+1}}$. The sequence $\{a_n\}$ is decreasing, and

$$\lim_{n\to\infty} a_n = 0;$$

therefore the series converges by the Leibniz Test.

81. $\displaystyle\sum_{n=2}^{\infty} \frac{(-1)^n}{\ln n}$

SOLUTION   The sequence $a_n = \frac{1}{\ln n}$ is decreasing for $n \geq 10$ and

$$\lim_{n\to\infty} a_n = 0;$$

therefore, the series converges by the Leibniz Test.

83. $\displaystyle\sum_{n=1}^{\infty} \frac{1}{n\sqrt{n} + \ln n}$

SOLUTION   For $n \geq 1$,

$$\frac{1}{n\sqrt{n} + \ln n} \leq \frac{1}{n\sqrt{n}} = \frac{1}{n^{3/2}}.$$

The series $\displaystyle\sum_{n=1}^{\infty} \frac{1}{n^{3/2}}$ is a convergent p-series, so the series $\displaystyle\sum_{n=1}^{\infty} \frac{1}{n\sqrt{n} + \ln n}$ converges by the Comparison Test.

85. $\displaystyle\sum_{n=1}^{\infty} \left(\frac{1}{\sqrt{n}} - \frac{1}{\sqrt{n+1}} \right)$

SOLUTION This series telescopes:

$$\sum_{n=1}^{\infty} \left(\frac{1}{\sqrt{n}} - \frac{1}{\sqrt{n+1}} \right) = \left(1 - \frac{1}{\sqrt{2}} \right) + \left(\frac{1}{\sqrt{2}} - \frac{1}{\sqrt{3}} \right) + \left(\frac{1}{\sqrt{3}} - \frac{1}{\sqrt{4}} \right) + \cdots$$

so that the nth partial sum S_n is

$$S_n = \left(1 - \frac{1}{\sqrt{2}} \right) + \left(\frac{1}{\sqrt{2}} - \frac{1}{\sqrt{3}} \right) + \left(\frac{1}{\sqrt{3}} - \frac{1}{\sqrt{4}} \right) + \cdots + \left(\frac{1}{\sqrt{n}} - \frac{1}{\sqrt{n+1}} \right) = 1 - \frac{1}{\sqrt{n+1}}$$

and then

$$\sum_{n=1}^{\infty} \left(\frac{1}{\sqrt{n}} - \frac{1}{\sqrt{n+1}} \right) = \lim_{n\to\infty} S_n = 1 - \lim_{n\to\infty} \frac{1}{\sqrt{n+1}} = 1$$

87. $\displaystyle\sum_{n=1}^{\infty} \frac{1}{n + \sqrt{n}}$

SOLUTION For $n \geq 1$, $\sqrt{n} \leq n$, so that

$$\sum_{n=1}^{\infty} \frac{1}{n + \sqrt{n}} \geq \sum_{n=1}^{\infty} \frac{1}{2n}$$

which diverges since it is a constant multiple of the harmonic series. Thus $\displaystyle\sum_{n=1}^{\infty} \frac{1}{n + \sqrt{n}}$ diverges as well, by the Comparison Test.

89. $\displaystyle\sum_{n=2}^{\infty} \frac{1}{n^{\ln n}}$

SOLUTION For $n \geq N$ large enough, $\ln n \geq 2$ so that

$$\sum_{n=N}^{\infty} \frac{1}{n^{\ln n}} \leq \sum_{n=N}^{\infty} \frac{1}{n^2}$$

which is a convergent p-series. Thus by the Comparison Test, $\displaystyle\sum_{n=N}^{\infty} \frac{1}{n^{\ln n}}$ also converges; adding back in the terms for $n < N$ does not affect convergence.

91. $\displaystyle\sum_{n=1}^{\infty} \sin^2 \frac{\pi}{n}$

SOLUTION For all $x > 0$, $\sin x < x$. Therefore, $\sin^2 x < x^2$, and for $x = \frac{\pi}{n}$,

$$\sin^2 \frac{\pi}{n} < \frac{\pi^2}{n^2} = \pi^2 \cdot \frac{1}{n^2}.$$

The series $\displaystyle\sum_{n=1}^{\infty} \frac{1}{n^2}$ is a convergent p-series, so the series $\displaystyle\sum_{n=1}^{\infty} \sin^2 \frac{\pi}{n}$ also converges by the Comparison Test.

In Exercises 93–98, find the interval of convergence of the power series.

93. $\displaystyle\sum_{n=0}^{\infty} \frac{2^n x^n}{n!}$

SOLUTION With $a_n = \frac{2^n x^n}{n!}$,

$$\rho = \lim_{n\to\infty} \left| \frac{a_{n+1}}{a_n} \right| = \lim_{n\to\infty} \left| \frac{2^{n+1} x^{n+1}}{(n+1)!} \cdot \frac{n!}{2^n x^n} \right| = \lim_{n\to\infty} \left| x \cdot \frac{2}{n} \right| = 0$$

Then $\rho < 1$ for all x, so that the radius of convergence is $R = \infty$, and the series converges for all x.

95. $\displaystyle\sum_{n=0}^{\infty} \frac{n^6}{n^8+1}(x-3)^n$

SOLUTION With $a_n = \frac{n^6(x-3)^n}{n^8+1}$,

$$\rho = \lim_{n\to\infty}\left|\frac{a_{n+1}}{a_n}\right| = \lim_{n\to\infty}\left|\frac{(n+1)^6(x-3)^{n+1}}{(n+1)^8-1}\cdot\frac{n^8+1}{n^6(x-3)^n}\right|$$

$$= \lim_{n\to\infty}\left|(x-3)\cdot\frac{(n+1)^6(n^8+1)}{n^6((n+1)^8+1)}\right|$$

$$= \lim_{n\to\infty}\left|(x-3)\cdot\frac{n^{14}+\text{terms of lower degree}}{n^{14}+\text{terms of lower degree}}\right| = |x-3|$$

Then $\rho < 1$ when $|x-3| < 1$, so the radius of convergence is 1, and the series converges absolutely for $|x-3| < 1$, or $2 < x < 4$. For the endpoint $x = 4$, the series becomes $\displaystyle\sum_{n=0}^{\infty}\frac{n^6}{n^8+1}$, which converges by the Comparison Test comparing with the convergent p-series $\displaystyle\sum_{n=1}^{\infty}\frac{1}{n^2}$. For the endpoint $x = 2$, the series becomes $\displaystyle\sum_{n=0}^{\infty}\frac{n^6(-1)^n}{n^8+1}$, which converges by the Leibniz Test. The series $\displaystyle\sum_{n-0}^{\infty}\frac{n^6(x-3)^n}{n^8+1}$ therefore converges for $2 \le x \le 4$.

97. $\displaystyle\sum_{n=0}^{\infty}(nx)^n$

SOLUTION With $a_n = n^n x^n$, and assuming $x \ne 0$,

$$\rho = \lim_{n\to\infty}\left|\frac{a_{n+1}}{a_n}\right| = \lim_{n\to\infty}\left|\frac{(n+1)^{n+1}x^{n+1}}{n^n x^n}\right| = \lim_{n\to\infty}\left|x(n+1)\cdot\left(\frac{n+1}{n}\right)^n\right| = \infty$$

since $\left(\frac{n+1}{n}\right)^n = \left(1+\frac{1}{n}\right)^n$ converges to e and the $(n+1)$ term diverges to ∞. Thus $\rho < 1$ only when $x = 0$, so the series converges only for $x = 0$.

99. Expand $f(x) = \dfrac{2}{4-3x}$ as a power series centered at $c = 0$. Determine the values of x for which the series converges.

SOLUTION Write

$$\frac{2}{4-3x} = \frac{1}{2}\frac{1}{1-\frac{3}{4}x}.$$

Substituting $\frac{3}{4}x$ for x in the Maclaurin series for $\frac{1}{1-x}$, we obtain

$$\frac{1}{1-\frac{3}{4}x} = \sum_{n=0}^{\infty}\left(\frac{3}{4}\right)^n x^n.$$

This series converges for $\left|\frac{3}{4}x\right| < 1$, or $|x| < \frac{4}{3}$. Hence, for $|x| < \frac{4}{3}$,

$$\frac{2}{4-3x} = \frac{1}{2}\sum_{n=0}^{\infty}\left(\frac{3}{4}\right)^n x^n.$$

101. Let $F(x) = \sum_{k=0}^{\infty} \frac{x^{2k}}{2^k \cdot k!}$.

(a) Show that $F(x)$ has infinite radius of convergence.

(b) Show that $y = F(x)$ is a solution of

$$y'' = xy' + y, \qquad y(0) = 1, \qquad y'(0) = 0$$

(c) $\mathcal{CAS}$ Plot the partial sums S_N for $N = 1, 3, 5, 7$ on the same set of axes.

SOLUTION

(a) With $a_k = \frac{x^{2k}}{2^k \cdot k!}$,

$$\left| \frac{a_{k+1}}{a_k} \right| = \frac{|x|^{2k+2}}{2^{k+1} \cdot (k+1)!} \cdot \frac{2^k \cdot k!}{|x|^{2k}} = \frac{x^2}{2(k+1)},$$

and

$$\rho = \lim_{k \to \infty} \left| \frac{a_{k+1}}{a_k} \right| = x^2 \cdot 0 = 0.$$

Because $\rho < 1$ for all x, we conclude that the series converges for all x; that is, $R = \infty$.

(b) Let

$$y = F(x) = \sum_{k=0}^{\infty} \frac{x^{2k}}{2^k \cdot k!}.$$

Then

$$y' = \sum_{k=1}^{\infty} \frac{2k x^{2k-1}}{2^k k!} = \sum_{k=1}^{\infty} \frac{x^{2k-1}}{2^{k-1}(k-1)!},$$

$$y'' = \sum_{k=1}^{\infty} \frac{(2k-1) x^{2k-2}}{2^{k-1}(k-1)!},$$

and

$$xy' + y = x \sum_{k=1}^{\infty} \frac{x^{2k-1}}{2^{k-1}(k-1)!} + \sum_{k=0}^{\infty} \frac{x^{2k}}{2^k k!} = \sum_{k=1}^{\infty} \frac{x^{2k}}{2^{k-1}(k-1)!} + 1 + \sum_{k=1}^{\infty} \frac{x^{2k}}{2^k k!}$$

$$= 1 + \sum_{k=1}^{\infty} \frac{(2k+1) x^{2k}}{2^k k!} = \sum_{k=0}^{\infty} \frac{(2k+1) x^{2k}}{2^k k!} = \sum_{k=1}^{\infty} \frac{(2k-1) x^{2k-2}}{2^{k-1}(k-1)!} = y''.$$

Moreover,

$$y(0) = 1 + \sum_{k=1}^{\infty} \frac{0^{2k}}{2^k k!} = 1 \quad \text{and} \quad y'(0) = \sum_{k=1}^{\infty} \frac{0^{2k-1}}{2^{k-1}(k-1)!} = 0.$$

Thus, $\sum_{k=0}^{\infty} \frac{x^{2k}}{2^k k!}$ is the solution to the equation $y'' = xy' + y$ satisfying $y(0) = 1$, $y'(0) = 0$.

(c) The partial sums S_1, S_3, S_5 and S_7 are plotted in the figure below.

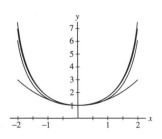

In Exercises 103–112, find the Taylor series centered at c.

103. $f(x) = e^{4x}, \quad c = 0$

SOLUTION Substituting $4x$ for x in the Maclaurin series for e^x yields

$$e^{4x} = \sum_{n=0}^{\infty} \frac{(4x)^n}{n!} = \sum_{n=0}^{\infty} \frac{4^n}{n!} x^n.$$

105. $f(x) = x^4, \quad c = 2$

SOLUTION We have

$$f'(x) = 4x^3 \quad f''(x) = 12x^2 \quad f'''(x) = 24x \quad f^{(4)}(x) = 24$$

and all higher derivatives are zero, so that

$$f(2) = 2^4 = 16 \quad f'(2) = 4 \cdot 2^3 = 32 \quad f''(2) = 12 \cdot 2^2 = 48 \quad f'''(2) = 24 \cdot 2 = 48 \quad f^{(4)}(2) = 24$$

Thus the Taylor series centered at $c = 2$ is

$$\sum_{n=0}^{4} \frac{f^{(n)}(2)}{n!} (x - 2)^n = 16 + \frac{32}{1!}(x - 2) + \frac{48}{2!}(x - 2)^2 + \frac{48}{3!}(x - 2)^3 + \frac{24}{4!}(x - 2)^4$$

$$= 16 + 32(x - 2) + 24(x - 2)^2 + 8(x - 2)^3 + (x - 2)^4$$

107. $f(x) = \sin x, \quad c = \pi$

SOLUTION We have

$$f^{(4n)}(x) = \sin x \quad f^{(4n+1)}(x) = \cos x \quad f^{(4n+2)}(x) = -\sin x \quad f^{(4n+3)}(x) = -\cos x$$

so that

$$f^{(4n)}(\pi) = \sin \pi = 0 \quad f^{(4n+1)}(\pi) = \cos \pi = -1 \quad f^{(4n+2)}(\pi) = -\sin \pi = 0 \quad f^{(4n+3)}(\pi) = -\cos \pi = 1$$

Then the Taylor series centered at $c = \pi$ is

$$\sum_{n=0}^{\infty} \frac{f^{(n)}(\pi)}{n!} (x - \pi)^n = \frac{-1}{1!}(x - \pi) + \frac{1}{3!}(x - \pi)^3 + \frac{-1}{5!}(x - \pi)^5 + \frac{1}{7!}(x - \pi)^7 - \dots$$

$$= -(x - \pi) + \frac{1}{6}(x - \pi)^3 - \frac{1}{120}(x - \pi)^5 + \frac{1}{5040}(x - \pi)^7 - \dots$$

109. $f(x) = \dfrac{1}{1 - 2x}, \quad c = -2$

SOLUTION Write

$$\frac{1}{1 - 2x} = \frac{1}{5 - 2(x + 2)} = \frac{1}{5} \frac{1}{1 - \frac{2}{5}(x + 2)}.$$

Substituting $\frac{2}{5}(x + 2)$ for x in the Maclaurin series for $\frac{1}{1-x}$ yields

$$\frac{1}{1 - \frac{2}{5}(x + 2)} = \sum_{n=0}^{\infty} \frac{2^n}{5^n}(x + 2)^n;$$

hence,

$$\frac{1}{1 - 2x} = \frac{1}{5} \sum_{n=0}^{\infty} \frac{2^n}{5^n}(x + 2)^n = \sum_{n=0}^{\infty} \frac{2^n}{5^{n+1}}(x + 2)^n.$$

111. $f(x) = \ln \dfrac{x}{2}, \quad c = 2$

SOLUTION   Write

$$\ln \frac{x}{2} = \ln \left(\frac{(x-2)+2}{2} \right) = \ln \left(1 + \frac{x-2}{2} \right).$$

Substituting $\frac{x-2}{2}$ for x in the Maclaurin series for $\ln(1+x)$ yields

$$\ln \frac{x}{2} = \sum_{n=1}^{\infty} \frac{(-1)^{n+1} \left(\frac{x-2}{2} \right)^n}{n} = \sum_{n=1}^{\infty} \frac{(-1)^{n+1}(x-2)^n}{n \cdot 2^n}.$$

This series is valid for $|x - 2| < 2$.

In Exercises 113–116, find the first three terms of the Maclaurin series of $f(x)$ and use it to calculate $f^{(3)}(0)$.

113. $f(x) = (x^2 - x)e^{x^2}$

SOLUTION   Substitute x^2 for x in the Maclaurin series for e^x to get

$$e^{x^2} = 1 + x^2 + \frac{1}{2}x^4 + \frac{1}{6}x^6 + \cdots$$

so that the Maclaurin series for $f(x)$ is

$$(x^2 - x)e^{x^2} = x^2 + x^4 + \frac{1}{2}x^6 + \cdots - x - x^3 - \frac{1}{2}x^5 - \cdots = -x + x^2 - x^3 + x^4 + \cdots$$

The coefficient of x^3 is

$$\frac{f'''(0)}{3!} = -1$$

so that $f'''(0) = -6$.

115. $f(x) = \dfrac{1}{1 + \tan x}$

SOLUTION   Substitute $-\tan x$ in the Maclaurin series for $\frac{1}{1-x}$ to get

$$\frac{1}{1 + \tan x} = 1 - \tan x + (\tan x)^2 - (\tan x)^3 + \cdots$$

We have not yet encountered the Maclaurin series for $\tan x$. We need only the terms up through x^3, so compute

$$\tan'(x) = \sec^2 x \quad \tan''(x) = 2(\tan x)\sec^2 x \quad \tan'''(x) = 2(1 + \tan^2 x)\sec^2 x + 4(\tan^2 x)\sec^2 x$$

so that

$$\tan'(0) = 1 \quad \tan''(0) = 0 \quad \tan'''(0) = 2$$

Then the Maclaurin series for $\tan x$ is

$$\tan x = \tan 0 + \frac{\tan'(0)}{1!}x + \frac{\tan''(0)}{2!}x^2 + \frac{\tan'''(0)}{3!}x^3 + \cdots = x + \frac{1}{3}x^3 + \cdots$$

Substitute these into the series above to get

$$\frac{1}{1 + \tan x} = 1 - \left(x + \frac{1}{3}x^3 \right) + \left(x + \frac{1}{3}x^3 \right)^2 - \left(x + \frac{1}{3}x^3 \right)^3 + \cdots$$

$$= 1 - x - \frac{1}{3}x^3 + x^2 - x^3 + \text{higher degree terms}$$

$$= 1 - x + x^2 - \frac{4}{3}x^3 + \text{higher degree terms}$$

The coefficient of x^3 is

$$\frac{f'''(0)}{3!} = -\frac{4}{3}$$

so that

$$f'''(0) = -6 \cdot \frac{4}{3} = -8$$

117. Calculate $\dfrac{\pi}{2} - \dfrac{\pi^3}{2^3 3!} + \dfrac{\pi^5}{2^5 5!} - \dfrac{\pi^7}{2^7 7!} + \cdots$.

SOLUTION We recognize that

$$\frac{\pi}{2} - \frac{\pi^3}{2^3 3!} + \frac{\pi^5}{2^5 5!} - \frac{\pi^7}{2^7 7!} + \cdots = \sum_{n=0}^{\infty} (-1)^n \frac{(\pi/2)^{2n+1}}{(2n+1)!}$$

is the Maclaurin series for $\sin x$ with x replaced by $\pi/2$. Therefore,

$$\frac{\pi}{2} - \frac{\pi^3}{2^3 3!} + \frac{\pi^5}{2^5 5!} - \frac{\pi^7}{2^7 7!} + \cdots = \sin \frac{\pi}{2} = 1.$$

11 PARAMETRIC EQUATIONS, POLAR COORDINATES, AND CONIC SECTIONS

11.1 Parametric Equations (LT Section 12.1)

Preliminary Questions

1. Describe the shape of the curve $x = 3\cos t$, $y = 3\sin t$.

SOLUTION For all t,

$$x^2 + y^2 = (3\cos t)^2 + (3\sin t)^2 = 9(\cos^2 t + \sin^2 t) = 9 \cdot 1 = 9,$$

therefore the curve is on the circle $x^2 + y^2 = 9$. Also, each point on the circle $x^2 + y^2 = 9$ can be represented in the form $(3\cos t, 3\sin t)$ for some value of t. We conclude that the curve $x = 3\cos t$, $y = 3\sin t$ is the circle of radius 3 centered at the origin.

2. How does $x = 4 + 3\cos t$, $y = 5 + 3\sin t$ differ from the curve in the previous question?

SOLUTION In this case we have

$$(x - 4)^2 + (y - 5)^2 = (3\cos t)^2 + (3\sin t)^2 = 9(\cos^2 t + \sin^2 t) = 9 \cdot 1 = 9$$

Therefore, the given equations parametrize the circle of radius 3 centered at the point $(4, 5)$.

3. What is the maximum height of a particle whose path has parametric equations $x = t^9$, $y = 4 - t^2$?

SOLUTION The particle's height is $y = 4 - t^2$. To find the maximum height we set the derivative equal to zero and solve:

$$\frac{dy}{dt} = \frac{d}{dt}(4 - t^2) = -2t = 0 \quad \text{or} \quad t = 0$$

The maximum height is $y(0) = 4 - 0^2 = 4$.

4. Can the parametric curve $(t, \sin t)$ be represented as a graph $y = f(x)$? What about $(\sin t, t)$?

SOLUTION In the parametric curve $(t, \sin t)$ we have $x = t$ and $y = \sin t$, therefore, $y = \sin x$. That is, the curve can be represented as a graph of a function. In the parametric curve $(\sin t, t)$ we have $x = \sin t$, $y = t$, therefore $x = \sin y$. This equation does not define y as a function of x, therefore the parametric curve $(\sin t, t)$ cannot be represented as a graph of a function $y = f(x)$.

5. Match the derivatives with a verbal description:

(a) $\dfrac{dx}{dt}$ **(b)** $\dfrac{dy}{dt}$ **(c)** $\dfrac{dy}{dx}$

(i) Slope of the tangent line to the curve

(ii) Vertical rate of change with respect to time

(iii) Horizontal rate of change with respect to time

SOLUTION

(a) The derivative $\dfrac{dx}{dt}$ is the horizontal rate of change with respect to time.

(b) The derivative $\dfrac{dy}{dt}$ is the vertical rate of change with respect to time.

(c) The derivative $\dfrac{dy}{dx}$ is the slope of the tangent line to the curve.

Hence, (a) $\leftrightarrow$ (iii), (b) $\leftrightarrow$ (ii), (c) $\leftrightarrow$ (i)

Exercises

1. Find the coordinates at times $t = 0, 2, 4$ of a particle following the path $x = 1 + t^3$, $y = 9 - 3t^2$.

SOLUTION Substituting $t = 0$, $t = 2$, and $t = 4$ into $x = 1 + t^3$, $y = 9 - 3t^2$ gives the coordinates of the particle at these times respectively. That is,

$$
\begin{aligned}
(t = 0) \quad & x = 1 + 0^3 = 1, \ y = 9 - 3 \cdot 0^2 = 9 && \Rightarrow (1, 9) \\
(t = 2) \quad & x = 1 + 2^3 = 9, \ y = 9 - 3 \cdot 2^2 = -3 && \Rightarrow (9, -3) \\
(t = 4) \quad & x = 1 + 4^3 = 65, \ y = 9 - 3 \cdot 4^2 = -39 && \Rightarrow (65, -39).
\end{aligned}
$$

3. Show that the path traced by the bullet in Example 3 is a parabola by eliminating the parameter.

SOLUTION The path traced by the bullet is given by the following parametric equations:

$$x = 200t, \ y = 400t - 16t^2$$

We eliminate the parameter. Since $x = 200t$, we have $t = \dfrac{x}{200}$. Substituting into the equation for y we obtain:

$$y = 400t - 16t^2 = 400 \cdot \frac{x}{200} - 16 \left(\frac{x}{200} \right)^2 = 2x - \frac{x^2}{2500}$$

The equation $y = -\dfrac{x^2}{2500} + 2x$ is the equation of a parabola.

5. Graph the parametric curves. Include arrows indicating the direction of motion.

(a) $(t, t), \quad -\infty < t < \infty$

(b) $(\sin t, \sin t), \quad 0 \le t \le 2\pi$

(c) $(e^t, e^t), \quad -\infty < t < \infty$

(d) $(t^3, t^3), \quad -1 \le t \le 1$

SOLUTION

(a) For the trajectory $c(t) = (t, t)$, $-\infty < t < \infty$ we have $y = x$. Also the two coordinates tend to ∞ and $-\infty$ as $t \to \infty$ and $t \to -\infty$ respectively. The graph is shown next:

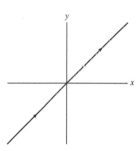

(b) For the curve $c(t) = (\sin t, \sin t)$, $0 \le t \le 2\pi$, we have $y = x$. $\sin t$ is increasing for $0 \le t \le \frac{\pi}{2}$, decreasing for $\frac{\pi}{2} \le t \le \frac{3\pi}{2}$ and increasing again for $\frac{3\pi}{2} \le t \le 2\pi$. Hence the particle moves from $c(0) = (0, 0)$ to $c(\frac{\pi}{2}) = (1, 1)$, then moves back to $c(\frac{3\pi}{2}) = (-1, -1)$ and then returns to $c(2\pi) = (0, 0)$. We obtain the following trajectory:

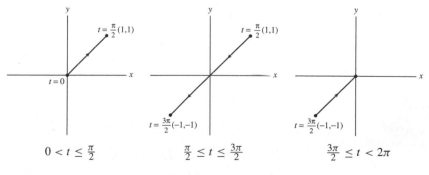

These three parts of the trajectory are shown together in the next figure:

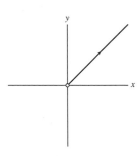

(c) For the trajectory $c(t) = (e^t, e^t)$, $-\infty < t < \infty$, we have $y = x$. However since $\lim_{t \to -\infty} e^t = 0$ and $\lim_{t \to \infty} e^t = \infty$, the trajectory is the part of the line $y = x$, $0 < x$.

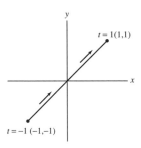

(d) For the trajectory $c(t) = (t^3, t^3)$, $-1 \le t \le 1$, we have again $y = x$. Since the function t^3 is increasing the particle moves in one direction starting at $((-1)^3, (-1)^3) = (-1, -1)$ and ending at $(1^3, 1^3) = (1, 1)$. The trajectory is shown next:

In Exercises 7–14, express in the form $y = f(x)$ by eliminating the parameter.

7. $x = t + 3$, $y = 4t$

SOLUTION We eliminate the parameter. Since $x = t + 3$, we have $t = x - 3$. Substituting into $y = 4t$ we obtain

$$y = 4t = 4(x - 3) \Rightarrow y = 4x - 12$$

9. $x = t$, $y = \tan^{-1}(t^3 + e^t)$

SOLUTION Replacing t by x in the equation for y we obtain $y = \tan^{-1}(x^3 + e^x)$.

11. $x = e^{-2t}$, $y = 6e^{4t}$

SOLUTION We eliminate the parameter. Since $x = e^{-2t}$, we have $-2t = \ln x$ or $t = -\frac{1}{2}\ln x$. Substituting in $y = 6e^{4t}$ we get

$$y = 6e^{4t} = 6e^{4 \cdot (-\frac{1}{2}\ln x)} = 6e^{-2\ln x} = 6e^{\ln x^{-2}} = 6x^{-2} \Rightarrow y = \frac{6}{x^2}, \quad x > 0.$$

13. $x = \ln t$, $y = 2 - t$

SOLUTION Since $x = \ln t$ we have $t = e^x$. Substituting in $y = 2 - t$ we obtain $y = 2 - e^x$.

In Exercises 15–18, graph the curve and draw an arrow specifying the direction corresponding to motion.

15. $x = \frac{1}{2}t$, $y = 2t^2$

SOLUTION Let $c(t) = (x(t), y(t)) = (\frac{1}{2}t, 2t^2)$. Then $c(-t) = (-x(t), y(t))$ so the curve is symmetric with respect to the y-axis. Also, the function $\frac{1}{2}t$ is increasing. Hence there is only one direction of motion on the curve. The corresponding function is the parabola $y = 2 \cdot (2x)^2 = 8x^2$. We obtain the following trajectory:

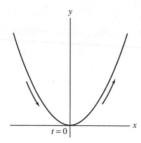

17. $x = \pi t, \quad y = \sin t$

SOLUTION We find the function by eliminating t. Since $x = \pi t$, we have $t = \frac{x}{\pi}$. Substituting $t = \frac{x}{\pi}$ into $y = \sin t$ we get $y = \sin \frac{x}{\pi}$. We obtain the following curve:

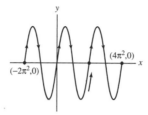

19. Match the parametrizations (a)–(d) below with their plots in Figure 14, and draw an arrow indicating the direction of motion.

(I)   (II)   (III)   (IV)

FIGURE 14

(a) $c(t) = (\sin t, -t)$ **(b)** $c(t) = (t^2 - 9, 8t - t^3)$

(c) $c(t) = (1 - t, t^2 - 9)$ **(d)** $c(t) = (4t + 2, 5 - 3t)$

SOLUTION

(a) In the curve $c(t) = (\sin t, -t)$ the x-coordinate is varying between -1 and 1 so this curve corresponds to plot IV. As t increases, the y-coordinate $y = -t$ is decreasing so the direction of motion is downward.

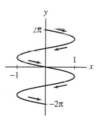

(IV) $c(t) = (\sin t, -t)$

(b) The curve $c(t) = (t^2 - 9, -t^3 - 8)$ intersects the x-axis where $y = -t^3 - 8 = 0$, or $t = -2$. The x-intercept is $(-5, 0)$. The y-intercepts are obtained where $x = t^2 - 9 = 0$, or $t = \pm 3$. The y-intercepts are $(0, -35)$ and $(0, 19)$. As t increases from $-\infty$ to 0, x and y decrease, and as t increases from 0 to ∞, x increases and y decreases. We obtain the following trajectory:

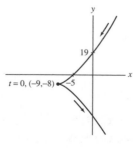

(II)

(c) The curve $c(t) = (1 - t, t^2 - 9)$ intersects the y-axis where $x = 1 - t = 0$, or $t = 1$. The y-intercept is $(0, -8)$. The x-intercepts are obtained where $t^2 - 9 = 0$ or $t = \pm 3$. These are the points $(-2, 0)$ and $(4, 0)$. Setting $t = 1 - x$ we get

$$y = t^2 - 9 = (1 - x)^2 - 9 = x^2 - 2x - 8.$$

As t increases the x coordinate decreases and we obtain the following trajectory:

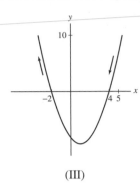

(III)

(d) The curve $c(t) = (4t + 2, 5 - 3t)$ is a straight line, since eliminating t in $x = 4t + 2$ and substituting in $y = 5 - 3t$ gives $y = 5 - 3 \cdot \frac{x-2}{4} = -\frac{3}{4}x + \frac{13}{2}$ which is the equation of a line. As t increases, the x coordinate $x = 4t + 2$ increases and the y-coordinate $y = 5 - 3t$ decreases. We obtain the following trajectory:

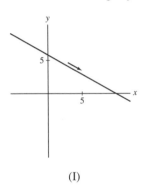

(I)

21. Find an interval of t-values such that $c(t) = (\cos t, \sin t)$ traces the lower half of the unit circle.

SOLUTION For $t = \pi$, we have $c(\pi) = (-1, 0)$. As t increases from π to 2π, the x-coordinate of $c(t)$ increases from -1 to 1, and the y-coordinate decreases from 0 to -1 (at $t = 3\pi/2$) and then returns to 0. Thus, for t in $[\pi, 2\pi]$, the equation traces the lower part of the circle.

In Exercises 23–38, find parametric equations for the given curve.

23. $y = 9 - 4x$

SOLUTION This is a line through $P = (0, 9)$ with slope $m = -4$. Using the parametric representation of a line, as given in Example 3, we obtain $c(t) = (t, 9 - 4t)$.

25. $4x - y^2 = 5$

SOLUTION We define the parameter $t = y$. Then, $x = \dfrac{5 + y^2}{4} = \dfrac{5 + t^2}{4}$, giving us the parametrization $c(t) = \left(\dfrac{5 + t^2}{4}, t \right)$.

27. $(x + 9)^2 + (y - 4)^2 = 49$

SOLUTION This is a circle of radius 7 centered at $(-9, 4)$. Using the parametric representation of a circle we get $c(t) = (-9 + 7 \cos t, 4 + 7 \sin t)$.

29. Line of slope 8 through $(-4, 9)$

SOLUTION Using the parametric representation of a line given in Example 3, we get the parametrization $c(t) = (-4 + t, 9 + 8t)$.

31. Line through $(3, 1)$ and $(-5, 4)$

SOLUTION We use the two-point parametrization of a line with $P = (a, b) = (3, 1)$ and $Q = (c, d) = (-5, 4)$. Then $c(t) = (3 - 8t, 1 + 3t)$ for $-\infty < t < \infty$.

33. Segment joining (1, 1) and (2, 3)

SOLUTION We use the two-point parametrization of a line with $P = (a, b) = (1, 1)$ and $Q = (c, d) = (2, 3)$. Then $c(t) = (1 + t, 1 + 2t)$; since we want only the segment joining the two points, we want $0 \le t \le 1$.

35. Circle of radius 4 with center (3, 9)

SOLUTION Substituting $(a, b) = (3, 9)$ and $R = 4$ in the parametric equation of the circle we get $c(t) = (3 + 4\cos t, 9 + 4\sin t)$.

37. $y = x^2$, translated so that the minimum occurs at $(-4, -8)$

SOLUTION We may parametrize $y = x^2$ by (t, t^2) for $-\infty < t < \infty$. The minimum of $y = x^2$ occurs at $(0, 0)$, so the desired curve is translated by $(-4, -8)$ from $y = x^2$. Thus a parametrization of the desired curve is $c(t) = (-4 + t, -8 + t^2)$.

In Exercises 39–42, find a parametrization $c(t)$ of the curve satisfying the given condition.

39. $y = 3x - 4$, $c(0) = (2, 2)$

SOLUTION Let $x(t) = t + a$ and $y(t) = 3x - 4 = 3(t + a) - 4$. We want $x(0) = 2$, thus we must use $a = 2$. Our line is $c(t) = (x(t), y(t)) = (t + 2, 3(t + 2) - 4) = (t + 2, 3t + 2)$.

41. $y = x^2$, $c(0) = (3, 9)$

SOLUTION Let $x(t) = t + a$ and $y(t) = x^2 = (t + a)^2$. We want $x(0) = 3$, thus we must use $a = 3$. Our curve is $c(t) = (x(t), y(t)) = (t + 3, (t + 3)^2) = (t + 3, t^2 + 6t + 9)$.

43. Describe $c(t) = (\sec t, \tan t)$ for $0 \le t < \frac{\pi}{2}$ in the form $y = f(x)$. Specify the domain of x.

SOLUTION The function $x = \sec t$ has period 2π and $y = \tan t$ has period π. The graphs of these functions in the interval $-\pi \le t \le \pi$, are shown below:

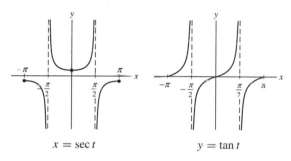

$$x = \sec t$$ $$y = \tan t$$

$$x = \sec t \Rightarrow x^2 = \sec^2 t$$

$$y = \tan t \Rightarrow y^2 = \tan^2 t = \frac{\sin^2 t}{\cos^2 t} = \frac{1 - \cos^2 t}{\cos^2 t} = \sec^2 t - 1 = x^2 - 1$$

Hence the graph of the curve is the hyperbola $x^2 - y^2 = 1$. The function $x = \sec t$ is an even function while $y = \tan t$ is odd. Also x has period 2π and y has period π. It follows that the intervals $-\pi \le t < -\frac{\pi}{2}$, $\frac{-\pi}{2} < t < \frac{\pi}{2}$ and $\frac{\pi}{2} < t < \pi$ trace the curve exactly once. The corresponding curve is shown next:

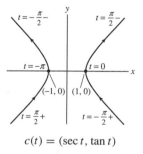

$$c(t) = (\sec t, \tan t)$$

45. The graphs of $x(t)$ and $y(t)$ as functions of t are shown in Figure 15(A). Which of (I)–(III) is the plot of $c(t) = (x(t), y(t))$? Explain.

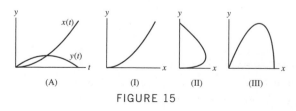

FIGURE 15

SOLUTION As seen in Figure 15(A), the x-coordinate is an increasing function of t, while $y(t)$ is first increasing and then decreasing. In Figure I, x and y are both increasing or both decreasing (depending on the direction on the curve). In Figure II, x does not maintain one tendency, rather, it is decreasing and increasing for certain values of t. The plot $c(t) = (x(t), y(t))$ is plot III.

47. Sketch $c(t) = (t^3 - 4t, t^2)$ following the steps in Example 7.

SOLUTION We note that $x(t) = t^3 - 4t$ is odd and $y(t) = t^2$ is even, hence $c(-t) = (x(-t), y(-t)) = (-x(t), y(t))$. It follows that $c(-t)$ is the reflection of $c(t)$ across y-axis. That is, $c(-t)$ and $c(t)$ are symmetric with respect to the y-axis; thus, it suffices to graph the curve for $t \geq 0$. For $t = 0$, we have $c(0) = (0, 0)$ and the y-coordinate $y(t) = t^2$ tends to ∞ as $t \to \infty$. To analyze the x-coordinate, we graph $x(t) = t^3 - 4t$ for $t \geq 0$:

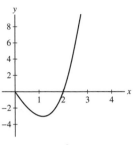

$$x = t^3 - 4t$$

We see that $x(t) < 0$ and decreasing for $0 < t < 2/\sqrt{3}$, $x(t) < 0$ and increasing for $2/\sqrt{3} < t < 2$ and $x(t) > 0$ and increasing for $t > 2$. Also $x(t)$ tends to ∞ as $t \to \infty$. Therefore, starting at the origin, the curve first directs to the left of the y-axis, then at $t = 2/\sqrt{3}$ it turns to the right, always keeping an upward direction. The part of the path for $t \leq 0$ is obtained by reflecting across the y-axis. We also use the points $c(0) = (0, 0)$, $c(1) = (-3, 1)$, $c(2) = (0, 4)$ to obtain the following graph for $c(t)$:

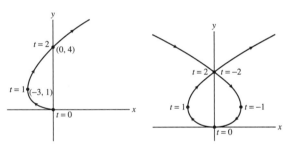

Graph of $c(t)$ for $t \geq 0$.   Graph of $c(t)$ for all t.

In Exercises 49–52, use Eq. (7) to find dy/dx at the given point.

49. $(t^3, t^2 - 1)$, $t = -4$

SOLUTION By Eq. (7) we have

$$\frac{dy}{dx} = \frac{y'(t)}{x'(t)} = \frac{(t^2 - 1)'}{(t^3)'} = \frac{2t}{3t^2} = \frac{2}{3t}$$

Substituting $t = -4$ we get

$$\frac{dy}{dx} = \frac{2}{3t}\bigg|_{t=-4} = \frac{2}{3 \cdot (-4)} = -\frac{1}{6}.$$

51. $(s^{-1} - 3s, s^3)$, $s = -1$

SOLUTION Using Eq. (7) we get

$$\frac{dy}{dx} = \frac{y'(s)}{x'(s)} = \frac{(s^3)'}{(s^{-1} - 3s)'} = \frac{3s^2}{-s^{-2} - 3} = \frac{3s^4}{-1 - 3s^2}$$

Substituting $s = -1$ we obtain

$$\frac{dy}{dx} = \frac{3s^4}{-1 - 3s^2}\bigg|_{s=-1} = \frac{3 \cdot (-1)^4}{-1 - 3 \cdot (-1)^2} = -\frac{3}{4}.$$

In Exercises 53–56, find an equation $y = f(x)$ for the parametric curve and compute dy/dx in two ways: using Eq. (7) and by differentiating $f(x)$.

53. $c(t) = (2t + 1, 1 - 9t)$

SOLUTION Since $x = 2t + 1$, we have $t = \dfrac{x - 1}{2}$. Substituting in $y = 1 - 9t$ we have

$$y = 1 - 9\left(\frac{x - 1}{2}\right) = -\frac{9}{2}x + \frac{11}{2}$$

Differentiating $y = -\dfrac{9}{2}x + \dfrac{11}{2}$ gives $\dfrac{dy}{dx} = -\dfrac{9}{2}$. We now find $\dfrac{dy}{dx}$ using Eq. (7):

$$\frac{dy}{dx} = \frac{y'(t)}{x'(t)} = \frac{(1 - 9t)'}{(2t + 1)'} = -\frac{9}{2}$$

55. $x = s^3, \quad y = s^6 + s^{-3}$

SOLUTION We find y as a function of x:

$$y = s^6 + s^{-3} = \left(s^3\right)^2 + \left(s^3\right)^{-1} = x^2 + x^{-1}.$$

We now differentiate $y = x^2 + x^{-1}$. This gives

$$\frac{dy}{dx} = 2x - x^{-2}.$$

Alternatively, we can use Eq. (7) to obtain the following derivative:

$$\frac{dy}{dx} = \frac{y'(s)}{x'(s)} = \frac{\left(s^6 + s^{-3}\right)'}{\left(s^3\right)'} = \frac{6s^5 - 3s^{-4}}{3s^2} = 2s^3 - s^{-6}.$$

Hence, since $x = s^3$,

$$\frac{dy}{dx} = 2x - x^{-2}.$$

57. Find the points on the curve $c(t) = (3t^2 - 2t, t^3 - 6t)$ where the tangent line has slope 3.

SOLUTION We solve

$$\frac{dy}{dx} = \frac{3t^2 - 6}{6t - 2} = 3$$

or $3t^2 - 6 = 18t - 6$, or $t^2 - 6t = 0$, so the slope is 3 at $t = 0, 6$ and the points are $(0, 0)$ and $(96, 180)$

In Exercises 59–62, let $c(t) = (t^2 - 9, t^2 - 8t)$ (see Figure 17).

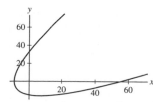

FIGURE 17 Plot of $c(t) = (t^2 - 9, t^2 - 8t)$.

59. Draw an arrow indicating the direction of motion, and determine the interval of t-values corresponding to the portion of the curve in each of the four quadrants.

SOLUTION We plot the functions $x(t) = t^2 - 9$ and $y(t) = t^2 - 8t$:

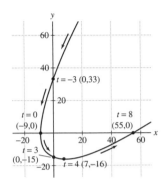

$$x = t^2 - 9 \qquad\qquad y = t^2 - 8t$$

We note carefully where each of these graphs are positive or negative, increasing or decreasing. In particular, $x(t)$ is decreasing for $t < 0$, increasing for $t > 0$, positive for $|t| > 3$, and negative for $|t| < 3$. Likewise, $y(t)$ is decreasing for $t < 4$, increasing for $t > 4$, positive for $t > 8$ or $t < 0$, and negative for $0 < t < 8$. We now draw arrows on the path following the decreasing/increasing behavior of the coordinates as indicated above. We obtain:

This plot also shows that:

- The graph is in the first quadrant for $t < -3$ or $t > 8$.
- The graph is in the second quadrant for $-3 < t < 0$.
- The graph is in the third quadrant for $0 < t < 3$.
- The graph is in the fourth quadrant for $3 < t < 8$.

61. Find the points where the tangent has slope $\frac{1}{2}$.

SOLUTION   The slope of the tangent at t is

$$\frac{dy}{dx} = \frac{\left(t^2 - 8t\right)'}{\left(t^2 - 9\right)'} = \frac{2t - 8}{2t} = 1 - \frac{4}{t}$$

The point where the tangent has slope $\frac{1}{2}$ corresponds to the value of t that satisfies

$$\frac{dy}{dx} = 1 - \frac{4}{t} = \frac{1}{2} \Rightarrow \frac{4}{t} = \frac{1}{2} \Rightarrow t = 8.$$

We substitute $t = 8$ in $x(t) = t^2 - 9$ and $y(t) = t^2 - 8t$ to obtain the following point:

$$\begin{array}{l} x(8) = 8^2 - 9 = 55 \\ y(8) = 8^2 - 8 \cdot 8 = 0 \end{array} \quad \Rightarrow \quad (55, 0)$$

63. Let A and B be the points where the ray of angle θ intersects the two concentric circles of radii $r < R$ centered at the origin (Figure 18). Let P be the point of intersection of the horizontal line through A and the vertical line through B. Express the coordinates of P as a function of θ and describe the curve traced by P for $0 \le \theta \le 2\pi$.

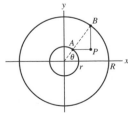

FIGURE 18

SOLUTION We use the parametric representation of a circle to determine the coordinates of the points A and B. That is,

$$A = (r\cos\theta, r\sin\theta), \quad B = (R\cos\theta, R\sin\theta)$$

The coordinates of P are therefore

$$P = (R\cos\theta, r\sin\theta)$$

In order to identify the curve traced by P, we notice that the x and y coordinates of P satisfy $\frac{x}{R} = \cos\theta$ and $\frac{y}{r} = \sin\theta$. Hence

$$\left(\frac{x}{R}\right)^2 + \left(\frac{y}{r}\right)^2 = \cos^2\theta + \sin^2\theta = 1.$$

The equation

$$\left(\frac{x}{R}\right)^2 + \left(\frac{y}{r}\right)^2 = 1$$

is the equation of ellipse. Hence, the coordinates of P, $(R\cos\theta, r\sin\theta)$ describe an ellipse for $0 \le \theta \le 2\pi$.

In Exercises 65–68, refer to the Bézier curve defined by Eqs. (8) and (9).

65. Show that the Bézier curve with control points

$$P_0 = (1, 4), \quad P_1 = (3, 12), \quad P_2 = (6, 15), \quad P_3 = (7, 4)$$

has parametrization

$$c(t) = (1 + 6t + 3t^2 - 3t^3, 4 + 24t - 15t^2 - 9t^3)$$

Verify that the slope at $t = 0$ is equal to the slope of the segment $\overline{P_0 P_1}$.

SOLUTION For the given Bézier curve we have $a_0 = 1, a_1 = 3, a_2 = 6, a_3 = 7$, and $b_0 = 4, b_1 = 12, b_2 = 15, b_3 = 4$. Substituting these values in Eq. (8)–(9) and simplifying gives

$$
\begin{aligned}
x(t) &= (1-t)^3 + 9t(1-t)^2 + 18t^2(1-t) + 7t^3 \\
&= 1 - 3t + 3t^2 - t^3 + 9t(1 - 2t + t^2) + 18t^2 - 18t^3 + 7t^3 \\
&= 1 - 3t + 3t^2 - t^3 + 9t - 18t^2 + 9t^3 + 18t^2 - 18t^3 + 7t^3 \\
&= -3t^3 + 3t^2 + 6t + 1 \\
y(t) &= 4(1-t)^3 + 36t(1-t)^2 + 45t^2(1-t) + 4t^3 \\
&= 4(1 - 3t + 3t^2 - t^3) + 36t(1 - 2t + t^2) + 45t^2 - 45t^3 + 4t^3 \\
&= 4 - 12t + 12t^2 - 4t^3 + 36t - 72t^2 + 36t^3 + 45t^2 - 45t^3 + 4t^3 \\
&= 4 + 24t - 15t^2 - 9t^3
\end{aligned}
$$

Then

$$c(t) = (1 + 6t + 3t^2 - 3t^3, 4 + 24t - 15t^2 - 9t^3), \quad 0 \le t \le 1.$$

We find the slope at $t = 0$. Using the formula for slope of the tangent line we get

$$\frac{dy}{dx} = \frac{(4 + 24t - 15t^2 - 9t^3)'}{(1 + 6t + 3t^2 - 3t^3)'} = \frac{24 - 30t - 27t^2}{6 + 6t - 9t^2} \Rightarrow \left.\frac{dy}{dx}\right|_{t=0} = \frac{24}{6} = 4.$$

The slope of the segment $\overline{P_0 P_1}$ is the slope of the line determined by the points $P_0 = (1, 4)$ and $P_1 = (3, 12)$. That is, $\frac{12-4}{3-1} = \frac{8}{2} = 4$. We see that the slope of the tangent line at $t = 0$ is equal to the slope of the segment $\overline{P_0 P_1}$, as expected.

67. *CAS* Find and plot the Bézier curve $c(t)$ passing through the control points

$$P_0 = (3, 2), \quad P_1 = (0, 2), \quad P_2 = (5, 4), \quad P_3 = (2, 4)$$

SOLUTION Setting $a_0 = 3, a_1 = 0, a_2 = 5, a_3 = 2$, and $b_0 = 2, b_1 = 2, b_2 = 4, b_3 = 4$ into Eq. (8)–(9) and simplifying gives

$$
\begin{aligned}
x(t) &= 3(1-t)^3 + 0 + 15t^2(1-t) + 2t^3 \\
&= 3(1 - 3t + 3t^2 - t^3) + 15t^2 - 15t^3 + 2t^3 = 3 - 9t + 24t^2 - 16t^3
\end{aligned}
$$

$$y(t) = 2(1-t)^3 + 6t(1-t)^2 + 12t^2(1-t) + 4t^3$$
$$= 2(1 - 3t + 3t^2 - t^3) + 6t(1 - 2t + t^2) + 12t^2 - 12t^3 + 4t^3$$
$$= 2 - 6t + 6t^2 - 2t^3 + 6t - 12t^2 + 6t^3 + 12t^2 - 12t^3 + 4t^3 = 2 + 6t^2 - 4t^3$$

We obtain the following equation

$$c(t) = (3 - 9t + 24t^2 - 16t^3, 2 + 6t^2 - 4t^3), \quad 0 \le t \le 1.$$

The graph of the Bézier curve is shown in the following figure:

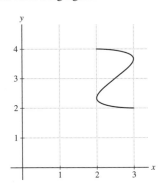

69. A bullet fired from a gun follows the trajectory

$$x = at, \qquad y = bt - 16t^2 \quad (a, b > 0)$$

Show that the bullet leaves the gun at an angle $\theta = \tan^{-1}\left(\frac{b}{a}\right)$ and lands at a distance $ab/16$ from the origin.

SOLUTION The height of the bullet equals the value of the y-coordinate. When the bullet leaves the gun, $y(t) = t(b - 16t) = 0$. The solutions to this equation are $t = 0$ and $t = \frac{b}{16}$, with $t = 0$ corresponding to the moment the bullet leaves the gun. We find the slope m of the tangent line at $t = 0$:

$$\frac{dy}{dx} = \frac{y'(t)}{x'(t)} = \frac{b - 32t}{a} \Rightarrow m = \left.\frac{b - 32t}{a}\right|_{t=0} = \frac{b}{a}$$

It follows that $\tan\theta = \frac{b}{a}$ or $\theta = \tan^{-1}\left(\frac{b}{a}\right)$. The bullet lands at $t = \frac{b}{16}$. We find the distance of the bullet from the origin at this time, by substituting $t = \frac{b}{16}$ in $x(t) = at$. This gives

$$x\left(\frac{b}{16}\right) = \frac{ab}{16}$$

71. **CAS** Plot the astroid $x = \cos^3\theta$, $y = \sin^3\theta$ and find the equation of the tangent line at $\theta = \frac{\pi}{3}$.

SOLUTION The graph of the astroid $x = \cos^3\theta$, $y = \sin^3\theta$ is shown in the following figure:

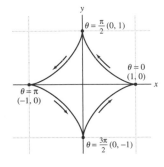

The slope of the tangent line at $\theta = \frac{\pi}{3}$ is

$$m = \left.\frac{dy}{dx}\right|_{\theta=\pi/3} = \left.\frac{(\sin^3\theta)'}{(\cos^3\theta)'}\right|_{\theta=\pi/3} = \left.\frac{3\sin^2\theta\cos\theta}{3\cos^2\theta(-\sin\theta)}\right|_{\theta=\pi/3} = \left.-\frac{\sin\theta}{\cos\theta}\right|_{\theta=\pi/3} = \left.-\tan\theta\right|_{\pi/3} = -\sqrt{3}$$

We find the point of tangency:

$$\left(x\left(\frac{\pi}{3}\right), y\left(\frac{\pi}{3}\right)\right) = \left(\cos^3\frac{\pi}{3}, \sin^3\frac{\pi}{3}\right) = \left(\frac{1}{8}, \frac{3\sqrt{3}}{8}\right)$$

The equation of the tangent line at $\theta = \frac{\pi}{3}$ is, thus,

$$y - \frac{3\sqrt{3}}{8} = -\sqrt{3}\left(x - \frac{1}{8}\right) \Rightarrow y = -\sqrt{3}x + \frac{\sqrt{3}}{2}$$

73. Find the points with horizontal tangent line on the cycloid with parametric equation (5).

SOLUTION The parametric equations of the cycloid are

$$x = t - \sin t, \quad y = 1 - \cos t$$

We find the slope of the tangent line at t:

$$\frac{dy}{dx} = \frac{(1 - \cos t)'}{(t - \sin t)'} = \frac{\sin t}{1 - \cos t}$$

The tangent line is horizontal where it has slope zero. That is,

$$\frac{dy}{dx} = \frac{\sin t}{1 - \cos t} = 0 \quad \Rightarrow \quad \begin{matrix} \sin t = 0 \\ \cos t \neq 1 \end{matrix} \quad \Rightarrow \quad t = (2k-1)\pi, \quad k = 0, \pm 1, \pm 2, \ldots$$

We find the coordinates of the points with horizontal tangent line, by substituting $t = (2k-1)\pi$ in $x(t)$ and $y(t)$. This gives

$$x = (2k-1)\pi - \sin(2k-1)\pi = (2k-1)\pi$$

$$y = 1 - \cos((2k-1)\pi) = 1 - (-1) = 2$$

The required points are

$$((2k-1)\pi, 2), \quad k = 0, \pm 1, \pm 2, \ldots$$

75. A *curtate cycloid* (Figure 21) is the curve traced by a point at a distance h from the center of a circle of radius R rolling along the x-axis where $h < R$. Show that this curve has parametric equations $x = Rt - h \sin t$, $y = R - h \cos t$.

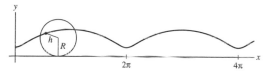

FIGURE 21 Curtate cycloid.

SOLUTION Let P be a point at a distance h from the center C of the circle. Assume that at $t = 0$, the line of CP is passing through the origin. When the circle rolls a distance Rt along the x-axis, the length of the arc $\overset{\frown}{SQ}$ (see figure) is also Rt and the angle $\angle SCQ$ has radian measure t. We compute the coordinates x and y of P.

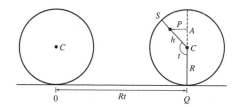

$$x = Rt - \overline{PA} = Rt - h\sin(\pi - t) = Rt - h\sin t$$

$$y = R + \overline{AC} = R + h\cos(\pi - t) = R - h\cos t$$

We obtain the following parametrization:

$$x = Rt - h\sin t, \quad y = R - h\cos t.$$

77. Show that the line of slope t through $(-1, 0)$ intersects the unit circle in the point with coordinates

$$x = \frac{1 - t^2}{t^2 + 1}, \qquad y = \frac{2t}{t^2 + 1}$$ **10**

Conclude that these equations parametrize the unit circle with the point $(-1, 0)$ excluded (Figure 22). Show further that $t = y/(x + 1)$.

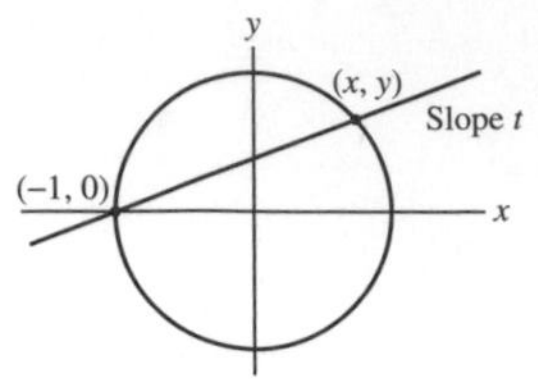

FIGURE 22 Unit circle.

SOLUTION The equation of the line of slope t through $(-1, 0)$ is $y = t(x + 1)$. The equation of the unit circle is $x^2 + y^2 = 1$. Hence, the line intersects the unit circle at the points (x, y) that satisfy the equations:

$$y = t(x + 1) \tag{1}$$

$$x^2 + y^2 = 1 \tag{2}$$

Substituting y from equation (1) into equation (2) and solving for x we obtain

$$x^2 + t^2(x + 1)^2 = 1$$
$$x^2 + t^2 x^2 + 2t^2 x + t^2 = 1$$
$$(1 + t^2)x^2 + 2t^2 x + (t^2 - 1) = 0$$

This gives

$$x_{1,2} = \frac{-2t^2 \pm \sqrt{4t^4 - 4(t^2 + 1)(t^2 - 1)}}{2(1 + t^2)} = \frac{-2t^2 \pm 2}{2(1 + t^2)} = \frac{\pm 1 - t^2}{1 + t^2}$$

So $x_1 = -1$ and $x_2 = \dfrac{1 - t^2}{t^2 + 1}$. The solution $x = -1$ corresponds to the point $(-1, 0)$. We are interested in the second point of intersection that is varying as t varies. Hence the appropriate solution is

$$x = \frac{1 - t^2}{t^2 + 1}$$

We find the y-coordinate by substituting x in equation (1). This gives

$$y = t(x + 1) = t\left(\frac{1 - t^2}{t^2 + 1} + 1\right) = t \cdot \frac{1 - t^2 + t^2 + 1}{t^2 + 1} = \frac{2t}{t^2 + 1}$$

We conclude that the line and the unit circle intersect, besides at $(-1, 0)$, at the point with the following coordinates:

$$x = \frac{1 - t^2}{t^2 + 1}, \quad y = \frac{2t}{t^2 + 1} \tag{3}$$

Since these points determine all the points on the unit circle except for $(-1, 0)$ and no other points, the equations in (3) parametrize the unit circle with the point $(-1, 0)$ excluded.

We show that $t = -\dfrac{y}{x + 1}$. Using (3) we have

$$\frac{y}{x + 1} = \frac{\frac{2t}{t^2 + 1}}{\frac{1 - t^2}{t^2 + 1} + 1} = \frac{\frac{2t}{t^2 + 1}}{\frac{1 - t^2 + t^2 + 1}{t^2 + 1}} = \frac{\frac{2t}{t^2 + 1}}{\frac{2}{t^2 + 1}} = \frac{2t}{2} = t.$$

79. Use the results of Exercise 78 to show that the asymptote of the folium is the line $x + y = -a$. *Hint:* Show that $\lim_{t \to -1} (x + y) = -a$.

SOLUTION We must show that as $x \to \infty$ or $x \to -\infty$ the graph of the folium is getting arbitrarily close to the line $x + y = -a$, and the derivative $\frac{dy}{dx}$ is approaching the slope -1 of the line.

In Exercise 78 we showed that $x \to \infty$ when $t \to (-1^-)$ and $x \to -\infty$ when $t \to (-1^+)$. We first show that the graph is approaching the line $x + y = -a$ as $x \to \infty$ or $x \to -\infty$, by showing that $\lim_{t \to -1^-} x + y = \lim_{t \to -1^+} x + y = -a$.

For $x(t) = \dfrac{3at}{1+t^3}$, $y(t) = \dfrac{3at^2}{1+t^3}$, $a > 0$, calculated in Exercise 78, we obtain using L'Hôpital's Rule:

$$\lim_{t \to -1-} (x+y) = \lim_{t \to -1-} \frac{3at + 3at^2}{1+t^3} = \lim_{t \to -1-} \frac{3a + 6at}{3t^2} = \frac{3a - 6a}{3} = -a$$

$$\lim_{t \to -1+} (x+y) = \lim_{t \to -1+} \frac{3at + 3at^2}{1+t^3} = \lim_{t \to -1+} \frac{3a + 6at}{3t^2} = \frac{3a - 6a}{3} = -a$$

We now show that $\dfrac{dy}{dx}$ is approaching -1 as $t \to -1-$ and as $t \to -1+$. We use $\dfrac{dy}{dx} = \dfrac{6at - 3at^4}{3a - 6at^3}$ computed in Exercise 78 to obtain

$$\lim_{t \to -1-} \frac{dy}{dx} = \lim_{t \to -1-} \frac{6at - 3at^4}{3a - 6at^3} = \frac{-9a}{9a} = -1$$

$$\lim_{t \to -1+} \frac{dy}{dx} = \lim_{t \to -1+} \frac{6at - 3at^4}{3a - 6at^3} = \frac{-9a}{9a} = -1$$

We conclude that the line $x + y = -a$ is an asymptote of the folium as $x \to \infty$ and as $x \to -\infty$.

81. Second Derivative for a Parametrized Curve Given a parametrized curve $c(t) = (x(t), y(t))$, show that

$$\frac{d}{dt}\left(\frac{dy}{dx}\right) = \frac{x'(t)y''(t) - y'(t)x''(t)}{x'(t)^2}$$

Use this to prove the formula

$$\boxed{\frac{d^2y}{dx^2} = \frac{x'(t)y''(t) - y'(t)x''(t)}{x'(t)^3}}$$

$\boxed{11}$

SOLUTION By the formula for the slope of the tangent line we have

$$\frac{dy}{dx} = \frac{y'(t)}{x'(t)}$$

Differentiating with respect to t, using the Quotient Rule, gives

$$\frac{d}{dt}\left(\frac{dy}{dx}\right) = \frac{d}{dt}\left(\frac{y'(t)}{x'(t)}\right) = \frac{x'(t)y''(t) - y'(t)x''(t)}{x'(t)^2}$$

By the Chain Rule we have

$$\frac{d^2y}{dx^2} = \frac{d}{dx}\left(\frac{dy}{dx}\right) = \frac{d}{dt}\left(\frac{dy}{dx}\right) \cdot \frac{dt}{dx}$$

Substituting into the above equation $\left(\text{and using } \dfrac{dt}{dx} = \dfrac{1}{dx/dt} = \dfrac{1}{x'(t)}\right)$ gives

$$\frac{d^2y}{dx^2} = \frac{x'(t)y''(t) - y'(t)x''(t)}{x'(t)^2} \cdot \frac{1}{x'(t)} = \frac{x'(t)y''(t) - y'(t)x''(t)}{x'(t)^3}$$

In Exercises 83–86, use Eq. (11) to find d^2y/dx^2.

83. $x = t^3 + t^2$, $y = 7t^2 - 4$, $t = 2$

SOLUTION We find the first and second derivatives of $x(t)$ and $y(t)$:

$$x'(t) = 3t^2 + 2t \Rightarrow x'(2) = 3 \cdot 2^2 + 2 \cdot 2 = 16$$

$$x''(t) = 6t + 2 \quad \Rightarrow x''(2) = 6 \cdot 2 + 2 = 14$$

$$y'(t) = 14t \qquad \Rightarrow y'(2) = 14 \cdot 2 = 28$$

$$y''(t) = 14 \qquad \Rightarrow y''(2) = 14$$

Using Eq. (11) we get

$$\frac{d^2y}{dx^2}\bigg|_{t=2} = \frac{x'(t)y''(t) - y'(t)x''(t)}{x'(t)^3}\bigg|_{t=2} = \frac{16 \cdot 14 - 28 \cdot 14}{16^3} = \frac{-21}{512}$$

85. $x = 8t + 9$, $y = 1 - 4t$, $t = -3$

SOLUTION   We compute the first and second derivatives of $x(t)$ and $y(t)$:

$$x'(t) = 8 \quad \Rightarrow x'(-3) = 8$$
$$x''(t) = 0 \quad \Rightarrow x''(-3) = 0$$
$$y'(t) = -4 \Rightarrow y'(-3) = -4$$
$$y''(t) = 0 \quad \Rightarrow y''(-3) = 0$$

Using Eq. (11) we get

$$\left.\frac{d^2 y}{dx^2}\right|_{t=-3} = \frac{x'(-3)y''(-3) - y'(-3)x''(-3)}{x'(-3)^3} = \frac{8 \cdot 0 - (-4) \cdot 0}{8^3} = 0$$

87. Use Eq. (11) to find the t-intervals on which $c(t) = (t^2, t^3 - 4t)$ is concave up.

SOLUTION   The curve is concave up where $\frac{d^2 y}{dx^2} > 0$. Thus,

$$\frac{x'(t)y''(t) - y'(t)x''(t)}{x'(t)^3} > 0 \tag{1}$$

We compute the first and second derivatives:

$$x'(t) = 2t, \qquad x''(t) = 2$$
$$y'(t) = 3t^2 - 4, \quad y''(t) = 6t$$

Substituting in (1) and solving for t gives

$$\frac{12t^2 - (6t^2 - 8)}{8t^3} = \frac{6t^2 + 8}{8t^3}$$

Since $6t^2 + 8 > 0$ for all t, the quotient is positive if $8t^3 > 0$. We conclude that the curve is concave up for $t > 0$.

89. Area Under a Parametrized Curve   Let $c(t) = (x(t), y(t))$, where $y(t) > 0$ and $x'(t) > 0$ (Figure 24). Show that the area A under $c(t)$ for $t_0 \leq t \leq t_1$ is

$$A = \int_{t_0}^{t_1} y(t)x'(t)\, dt \tag{12}$$

Hint: Because it is increasing, the function $x(t)$ has an inverse $t = g(x)$ and $c(t)$ is the graph of $y = y(g(x))$. Apply the change-of-variables formula to $A = \int_{x(t_0)}^{x(t_1)} y(g(x))\, dx$.

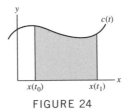

FIGURE 24

SOLUTION   Let $x_0 = x(t_0)$ and $x_1 = x(t_1)$. We are given that $x'(t) > 0$, hence $x = x(t)$ is an increasing function of t, so it has an inverse function $t = g(x)$. The area A is given by $\int_{x_0}^{x_1} y(g(x))\, dx$. Recall that y is a function of t and $t = g(x)$, so the height y at any point x is given by $y = y(g(x))$. We find the new limits of integration. Since $x_0 = x(t_0)$ and $x_1 = x(t_1)$, the limits for t are t_0 and t_1, respectively. Also since $x'(t) = \frac{dx}{dt}$, we have $dx = x'(t)dt$. Performing this substitution gives

$$A = \int_{x_0}^{x_1} y(g(x))\, dx = \int_{t_0}^{t_1} y(g(x))x'(t)\, dt.$$

Since $g(x) = t$, we have $A = \int_{t_0}^{t_1} y(t)x'(t)\, dt.$

91. What does Eq. (12) say if $c(t) = (t, f(t))$?

SOLUTION In the parametrization $x(t) = t$, $y(t) = f(t)$ we have $x'(t) = 1$, $t_0 = x(t_0)$, $t_1 = x(t_1)$. Hence Eq. (12) becomes

$$A = \int_{t_0}^{t_1} y(t)x'(t)\,dt = \int_{x(t_0)}^{x(t_1)} f(t)\,dt$$

We see that in this parametrization Eq. (12) is the familiar formula for the area under the graph of a positive function.

93. Galileo tried unsuccessfully to find the area under a cycloid. Around 1630, Gilles de Roberval proved that the area under one arch of the cycloid $c(t) = (Rt - R\sin t, R - R\cos t)$ generated by a circle of radius R is equal to three times the area of the circle (Figure 25). Verify Roberval's result using Eq. (12).

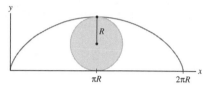

FIGURE 25 The area of one arch of the cycloid equals three times the area of the generating circle.

SOLUTION This reduces to

$$\int_0^{2\pi} (R - R\cos t)(Rt - R\sin t)'\,dt = \int_0^{2\pi} R^2(1 - \cos t)^2\,dt = 3\pi R^2.$$

Further Insights and Challenges

95. Derive the formula for the slope of the tangent line to a parametric curve $c(t) = (x(t), y(t))$ using a method different from that presented in the text. Assume that $x'(t_0)$ and $y'(t_0)$ exist and that $x'(t_0) \neq 0$. Show that

$$\lim_{h \to 0} \frac{y(t_0 + h) - y(t_0)}{x(t_0 + h) - x(t_0)} = \frac{y'(t_0)}{x'(t_0)}$$

Then explain why this limit is equal to the slope dy/dx. Draw a diagram showing that the ratio in the limit is the slope of a secant line.

SOLUTION Since $y'(t_0)$ and $x'(t_0)$ exist, we have the following limits:

$$\lim_{h \to 0} \frac{y(t_0 + h) - y(t_0)}{h} = y'(t_0), \quad \lim_{h \to 0} \frac{x(t_0 + h) - x(t_0)}{h} = x'(t_0) \tag{1}$$

We use Basic Limit Laws, the limits in (1) and the given data $x'(t_0) \neq 0$, to write

$$\lim_{h \to 0} \frac{y(t_0 + h) - y(t_0)}{x(t_0 + h) - x(t_0)} = \lim_{h \to 0} \frac{\frac{y(t_0+h) - y(t_0)}{h}}{\frac{x(t_0+h) - x(t_0)}{h}} = \frac{\lim_{h \to 0} \frac{y(t_0+h) - y(t_0)}{h}}{\lim_{h \to 0} \frac{x(t_0+h) - x(t_0)}{h}} = \frac{y'(t_0)}{x'(t_0)}$$

Notice that the quotient $\dfrac{y(t_0 + h) - y(t_0)}{x(t_0 + h) - x(t_0)}$ is the slope of the secant line determined by the points $P = (x(t_0), y(t_0))$ and $Q = (x(t_0 + h), y(t_0 + h))$. Hence, the limit of the quotient as $h \to 0$ is the slope of the tangent line at P, that is the derivative $\dfrac{dy}{dx}$.

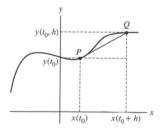

97. In Exercise 54 of Section 9.1 (LT Exercise 54 of Section 10.1), we described the tractrix by the differential equation

$$\frac{dy}{dx} = -\frac{y}{\sqrt{\ell^2 - y^2}}$$

Show that the curve $c(t)$ identified as the tractrix in Exercise 96 satisfies this differential equation. Note that the derivative on the left is taken with respect to x, not t.

SOLUTION   Note that $dx/dt = 1 - \text{sech}^2(t/\ell) = \tanh^2(t/\ell)$ and $dy/dt = -\text{sech}(t/\ell)\tanh(t/\ell)$. Thus,

$$\frac{dy}{dx} = \frac{dy/dt}{dx/dt} = \frac{-\text{sech}(t/\ell)}{\tanh(t/\ell)} = \frac{-y/\ell}{\sqrt{1 - y^2/\ell^2}}$$

Multiplying top and bottom by ℓ/ℓ gives

$$\frac{dy}{dx} = \frac{-y}{\sqrt{\ell^2 - y^2}}$$

In Exercises 98 and 99, refer to Figure 28.

99. Show that the parametrization of the ellipse by the angle θ is

$$x = \frac{ab\cos\theta}{\sqrt{a^2\sin^2\theta + b^2\cos^2\theta}}$$

$$y = \frac{ab\sin\theta}{\sqrt{a^2\sin^2\theta + b^2\cos^2\theta}}$$

SOLUTION   We consider the ellipse

$$\frac{x^2}{a^2} + \frac{y^2}{b^2} = 1.$$

For the angle θ we have $\tan\theta = \frac{y}{x}$, hence,

$$y = x\tan\theta \tag{1}$$

Substituting in the equation of the ellipse and solving for x we obtain

$$\frac{x^2}{a^2} + \frac{x^2\tan^2\theta}{b^2} = 1$$

$$b^2x^2 + a^2x^2\tan^2\theta = a^2b^2$$

$$(a^2\tan^2\theta + b^2)x^2 = a^2b^2$$

$$x^2 = \frac{a^2b^2}{a^2\tan^2\theta + b^2} = \frac{a^2b^2\cos^2\theta}{a^2\sin^2\theta + b^2\cos^2\theta}$$

We now take the square root. Since the sign of the x-coordinate is the same as the sign of $\cos\theta$, we take the positive root, obtaining

$$x = \frac{ab\cos\theta}{\sqrt{a^2\sin^2\theta + b^2\cos^2\theta}} \tag{2}$$

Hence by (1), the y-coordinate is

$$y = x\tan\theta = \frac{ab\cos\theta\tan\theta}{\sqrt{a^2\sin^2\theta + b^2\cos^2\theta}} = \frac{ab\sin\theta}{\sqrt{a^2\sin^2\theta + b^2\cos^2\theta}} \tag{3}$$

Equalities (2) and (3) give the following parametrization for the ellipse:

$$c_1(\theta) = \left(\frac{ab\cos\theta}{\sqrt{a^2\sin^2\theta + b^2\cos^2\theta}}, \frac{ab\sin\theta}{\sqrt{a^2\sin^2\theta + b^2\cos^2\theta}} \right)$$

11.2 Arc Length and Speed   (LT Section 12.2)

Preliminary Questions

1. What is the definition of arc length?

SOLUTION   A curve can be approximated by a polygonal path obtained by connecting points

$$p_0 = c(t_0),\ p_1 = c(t_1),\ \dots,\ p_N = c(t_N)$$

on the path with segments. One gets an approximation by summing the lengths of the segments. The definition of arc length is the limit of that approximation when increasing the number of points so that the lengths of the segments approach zero. In doing so, we obtain the following theorem for the arc length:

$$S = \int_a^b \sqrt{x'(t)^2 + y'(t)^2}\, dt,$$

which is the length of the curve $c(t) = (x(t), y(t))$ for $a \leq t \leq b$.

2. What is the interpretation of $\sqrt{x'(t)^2 + y'(t)^2}$ for a particle following the trajectory $(x(t), y(t))$?

SOLUTION The expression $\sqrt{x'(t)^2 + y'(t)^2}$ denotes the speed at time t of a particle following the trajectory $(x(t), y(t))$.

3. A particle travels along a path from $(0, 0)$ to $(3, 4)$. What is the displacement? Can the distance traveled be determined from the information given?

SOLUTION The net displacement is the distance between the initial point $(0, 0)$ and the endpoint $(3, 4)$. That is

$$\sqrt{(3-0)^2 + (4-0)^2} = \sqrt{25} = 5.$$

The distance traveled can be determined only if the trajectory $c(t) = (x(t), y(t))$ of the particle is known.

4. A particle traverses the parabola $y = x^2$ with constant speed 3 cm/s. What is the distance traveled during the first minute? *Hint:* No computation is necessary.

SOLUTION Since the speed is constant, the distance traveled is the following product: $L = st = 3 \cdot 60 = 180$ cm.

Exercises

In Exercises 1–10, use Eq. (3) to find the length of the path over the given interval.

1. $(3t + 1, 9 - 4t)$, $0 \leq t \leq 2$

SOLUTION Since $x = 3t + 1$ and $y = 9 - 4t$ we have $x' = 3$ and $y' = -4$. Hence, the length of the path is

$$S = \int_0^2 \sqrt{3^2 + (-4)^2}\, dt = 5 \int_0^2 dt = 10.$$

3. $(2t^2, 3t^2 - 1)$, $0 \leq t \leq 4$

SOLUTION Since $x = 2t^2$ and $y = 3t^2 - 1$, we have $x' = 4t$ and $y' = 6t$. By the formula for the arc length we get

$$S = \int_0^4 \sqrt{x'(t)^2 + y'(t)^2}\, dt = \int_0^4 \sqrt{16t^2 + 36t^2}\, dt = \sqrt{52} \int_0^4 t\, dt = \sqrt{52} \cdot \frac{t^2}{2}\Big|_0^4 = 16\sqrt{13}$$

5. $(3t^2, 4t^3)$, $1 \leq t \leq 4$

SOLUTION We have $x = 3t^2$ and $y = 4t^3$. Hence $x' = 6t$ and $y' = 12t^2$. By the formula for the arc length we get

$$S = \int_1^4 \sqrt{x'(t)^2 + y'(t)^2}\, dt = \int_1^4 \sqrt{36t^2 + 144t^4}\, dt = 6 \int_1^4 \sqrt{1 + 4t^2}\, t\, dt.$$

Using the substitution $u = 1 + 4t^2$, $du = 8t\, dt$ we obtain

$$S = \frac{6}{8} \int_5^{65} \sqrt{u}\, du = \frac{3}{4} \cdot \frac{2}{3} u^{3/2}\Big|_5^{65} = \frac{1}{2}(65^{3/2} - 5^{3/2}) \approx 256.43$$

7. $(\sin 3t, \cos 3t)$, $0 \leq t \leq \pi$

SOLUTION We have $x = \sin 3t$, $y = \cos 3t$, hence $x' = 3 \cos 3t$ and $y' = -3 \sin 3t$. By the formula for the arc length we obtain:

$$S = \int_0^\pi \sqrt{x'(t)^2 + y'(t)^2}\, dt = \int_0^\pi \sqrt{9 \cos^2 3t + 9 \sin^2 3t}\, dt = \int_0^\pi \sqrt{9}\, dt = 3\pi$$

In Exercises 9 and 10, use the identity

$$\frac{1 - \cos t}{2} = \sin^2 \frac{t}{2}$$

9. $(2 \cos t - \cos 2t, 2 \sin t - \sin 2t), \quad 0 \le t \le \frac{\pi}{2}$

SOLUTION　We have $x = 2 \cos t - \cos 2t$, $y = 2 \sin t - \sin 2t$. Thus, $x' = -2 \sin t + 2 \sin 2t$ and $y' = 2 \cos t - 2 \cos 2t$. We get

$$\begin{aligned}
x'(t)^2 + y'(t)^2 &= (-2 \sin t + 2 \sin 2t)^2 + (2 \cos t - 2 \cos 2t)^2 \\
&= 4 \sin^2 t - 8 \sin t \sin 2t + 4 \sin^2 2t + 4 \cos^2 t - 8 \cos t \cos 2t + 4 \cos^2 2t \\
&= 4(\sin^2 t + \cos^2 t) + 4(\sin^2 2t + \cos^2 2t) - 8(\sin t \sin 2t + \cos t \cos 2t) \\
&= 4 + 4 - 8 \cos(2t - t) = 8 - 8 \cos t = 8(1 - \cos t)
\end{aligned}$$

We now use the formula for the arc length to obtain

$$S = \int_0^{\pi/2} \sqrt{x'(t)^2 + y'(t)^2} = \int_0^{\pi/2} \sqrt{8(1 - \cos t)}\, dt = \int_0^{\pi/2} \sqrt{16 \sin^2 \frac{t}{2}}\, dt = 4 \int_0^{\pi/2} \sin \frac{t}{2}\, dt$$

$$= -8 \cos \frac{t}{2} \bigg|_0^{\pi/2} = -8 \left(\cos \frac{\pi}{4} - \cos 0 \right) = -8 \left(\frac{\sqrt{2}}{2} - 1 \right) \approx 2.34$$

11.　Show that one arch of a cycloid generated by a circle of radius R has length $8R$.

SOLUTION　Recall from earlier that the cycloid generated by a circle of radius R has parametric equations $x = Rt - R \sin t$, $y = R - R \cos t$. Hence, $x' = R - R \cos t$, $y' = R \sin t$. Using the identity $\sin^2 \frac{t}{2} = \frac{1 - \cos t}{2}$, we get

$$\begin{aligned}
x'(t)^2 + y'(t)^2 &= R^2(1 - \cos t)^2 + R^2 \sin^2 t = R^2(1 - 2 \cos t + \cos^2 t + \sin^2 t) \\
&= R^2(1 - 2 \cos t + 1) = 2R^2(1 - \cos t) = 4R^2 \sin^2 \frac{t}{2}
\end{aligned}$$

One arch of the cycloid is traced as t varies from 0 to 2π. Hence, using the formula for the arc length we obtain:

$$S = \int_0^{2\pi} \sqrt{x'(t)^2 + y'(t)^2}\, dt = \int_0^{2\pi} \sqrt{4R^2 \sin^2 \frac{t}{2}}\, dt = 2R \int_0^{2\pi} \sin \frac{t}{2}\, dt = 4R \int_0^{\pi} \sin u\, du$$

$$= -4R \cos u \bigg|_0^{\pi} = -4R(\cos \pi - \cos 0) = 8R$$

13.　Find the length of the tractrix (see Figure 6)

$$c(t) = (t - \tanh(t), \operatorname{sech}(t)), \qquad 0 \le t \le A$$

SOLUTION　Since $x = t - \tanh(t)$ and $y = \operatorname{sech}(t)$ we have $x' = 1 - \operatorname{sech}^2(t)$ and $y' = -\operatorname{sech}(t) \tanh(t)$. Hence,

$$\begin{aligned}
x'(t)^2 + y'(t)^2 &= \left(1 - \operatorname{sech}^2(t)\right)^2 + \operatorname{sech}^2(t)\tanh^2(t) \\
&= 1 - 2 \operatorname{sech}^2(t) + \operatorname{sech}^4(t) + \operatorname{sech}^2(t)\tanh^2(t) \\
&= 1 - 2 \operatorname{sech}^2(t) + \operatorname{sech}^2(t)(\operatorname{sech}^2(t) + \tanh^2(t)) \\
&= 1 - 2 \operatorname{sech}^2(t) + \operatorname{sech}^2(t) = 1 - \operatorname{sech}^2(t) = \tanh^2(t)
\end{aligned}$$

Hence, using the formula for the arc length we get:

$$S = \int_0^A \sqrt{x'(t)^2 + y'(t)^2}\, dt = \int_0^A \sqrt{\tanh^2(t)}\, dt = \int_0^A \tanh(t)\, dt = \ln(\cosh(t)) \bigg|_0^A$$

$$= \ln(\cosh(A)) - \ln(\cosh(0)) = \ln(\cosh(A)) - \ln 1 = \ln(\cosh(A))$$

In Exercises 15–18, determine the speed s at time t (assume units of meters and seconds).

15. $(t^3, t^2), \quad t = 2$

SOLUTION　We have $x(t) = t^3$, $y(t) = t^2$ hence $x'(t) = 3t^2$, $y'(t) = 2t$. The speed of the particle at time t is thus, $\frac{ds}{dt} = \sqrt{x'(t)^2 + y'(t)^2} = \sqrt{9t^4 + 4t^2} = t\sqrt{9t^2 + 4}$. At time $t = 2$ the speed is

$$\frac{ds}{dt} \bigg|_{t=2} = 2\sqrt{9 \cdot 2^2 + 4} = 2\sqrt{40} = 4\sqrt{10} \approx 12.65 \text{ m/s.}$$

17. $(5t + 1, 4t - 3)$, $t = 9$

SOLUTION Since $x = 5t + 1$, $y = 4t - 3$, we have $x' = 5$ and $y' = 4$. The speed of the particle at time t is

$$\frac{ds}{dt} = \sqrt{x'(t) + y'(t)} = \sqrt{5^2 + 4^2} = \sqrt{41} \approx 6.4 \text{ m/s}.$$

We conclude that the particle has constant speed of 6.4 m/s.

19. Find the minimum speed of a particle with trajectory $c(t) = (t^3 - 4t, t^2 + 1)$ for $t \geq 0$. *Hint:* It is easier to find the minimum of the square of the speed.

SOLUTION We first find the speed of the particle. We have $x(t) = t^3 - 4t$, $y(t) = t^2 + 1$, hence $x'(t) = 3t^2 - 4$ and $y'(t) = 2t$. The speed is thus

$$\frac{ds}{dt} = \sqrt{(3t^2 - 4)^2 + (2t)^2} = \sqrt{9t^4 - 24t^2 + 16 + 4t^2} = \sqrt{9t^4 - 20t^2 + 16}.$$

The square root function is an increasing function, hence the minimum speed occurs at the value of t where the function $f(t) = 9t^4 - 20t^2 + 16$ has minimum value. Since $\lim_{t \to \infty} f(t) = \infty$, f has a minimum value on the interval $0 \leq t < \infty$, and it occurs at a critical point or at the endpoint $t = 0$. We find the critical point of f on $t \geq 0$:

$$f'(t) = 36t^3 - 40t = 4t(9t^2 - 10) = 0 \Rightarrow t = 0, t = \sqrt{\frac{10}{9}}.$$

We compute the values of f at these points:

$$f(0) = 9 \cdot 0^4 - 20 \cdot 0^2 + 16 = 16$$

$$f\left(\sqrt{\frac{10}{9}}\right) = 9\left(\sqrt{\frac{10}{9}}\right)^4 - 20\left(\sqrt{\frac{10}{9}}\right)^2 + 16 = \frac{44}{9} \approx 4.89$$

We conclude that the minimum value of f on $t \geq 0$ is 4.89. The minimum speed is therefore

$$\left(\frac{ds}{dt}\right)_{\min} \approx \sqrt{4.89} \approx 2.21.$$

21. Find the speed of the cycloid $c(t) = (4t - 4\sin t, 4 - 4\cos t)$ at points where the tangent line is horizontal.

SOLUTION We first find the points where the tangent line is horizontal. The slope of the tangent line is the following quotient:

$$\frac{dy}{dx} = \frac{dy/dt}{dx/dt} = \frac{4\sin t}{4 - 4\cos t} = \frac{\sin t}{1 - \cos t}.$$

To find the points where the tangent line is horizontal we solve the following equation for $t \geq 0$:

$$\frac{dy}{dx} = 0, \quad \frac{\sin t}{1 - \cos t} = 0 \Rightarrow \sin t = 0 \quad \text{and} \quad \cos t \neq 1.$$

Now, $\sin t = 0$ and $t \geq 0$ at the points $t = \pi k$, $k = 0, 1, 2, \ldots$. Since $\cos \pi k = (-1)^k$, the points where $\cos t \neq 1$ are $t = \pi k$ for k odd. The points where the tangent line is horizontal are, therefore:

$$t = \pi(2k - 1), \quad k = 1, 2, 3, \ldots$$

The speed at time t is given by the following expression:

$$\frac{ds}{dt} = \sqrt{x'(t)^2 + y'(t)^2} = \sqrt{(4 - 4\cos t)^2 + (4\sin t)^2}$$

$$= \sqrt{16 - 32\cos t + 16\cos^2 t + 16\sin^2 t} = \sqrt{16 - 32\cos t + 16}$$

$$= \sqrt{32(1 - \cos t)} = \sqrt{32 \cdot 2\sin^2 \frac{t}{2}} = 8\left|\sin \frac{t}{2}\right|$$

That is, the speed of the cycloid at time t is

$$\frac{ds}{dt} = 8\left|\sin \frac{t}{2}\right|.$$

We now substitute

$$t = \pi(2k - 1), \quad k = 1, 2, 3, \ldots$$

to obtain

$$\frac{ds}{dt} = 8 \left| \sin \frac{\pi(2k-1)}{2} \right| = 8|(-1)^{k+1}| = 8$$

CAS *In Exercises 23–26, plot the curve and use the Midpoint Rule with $N = 10, 20, 30,$ and 50 to approximate its length.*

23. $c(t) = (\cos t, e^{\sin t})$ for $0 \le t \le 2\pi$

SOLUTION   The curve of $c(t) = (\cos t, e^{\sin t})$ for $0 \le t \le 2\pi$ is shown in the figure below:

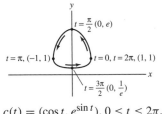

$$c(t) = (\cos t, e^{\sin t}), 0 \le t \le 2\pi.$$

The length of the curve is given by the following integral:

$$S = \int_0^{2\pi} \sqrt{x'(t)^2 + y'(t)^2} \, dt = \int_0^{2\pi} \sqrt{(-\sin t)^2 + (\cos t \, e^{\sin t})^2} \, dt.$$

That is, $S = \int_0^{2\pi} \sqrt{\sin^2 t + \cos^2 t \, e^{2\sin t}} \, dt$. We approximate the integral using the Mid-Point Rule with $N = 10, 20,$ $30, 50$. For $f(t) = \sqrt{\sin^2 t + \cos^2 t \, e^{2\sin t}}$ we obtain

$$(N = 10): \quad \Delta x = \frac{2\pi}{10} = \frac{\pi}{5}, c_i = \left(i - \frac{1}{2}\right) \cdot \frac{\pi}{5}$$

$$M_{10} = \frac{\pi}{5} \sum_{i=1}^{10} f(c_i) = 6.903734$$

$$(N = 20): \quad \Delta x = \frac{2\pi}{20} = \frac{\pi}{10}, c_i = \left(i - \frac{1}{2}\right) \cdot \frac{\pi}{10}$$

$$M_{20} = \frac{\pi}{10} \sum_{i=1}^{20} f(c_i) = 6.915035$$

$$(N = 30): \quad \Delta x = \frac{2\pi}{30} = \frac{\pi}{15}, c_i = \left(i - \frac{1}{2}\right) \cdot \frac{\pi}{15}$$

$$M_{30} = \frac{\pi}{15} \sum_{i=1}^{30} f(c_i) = 6.914949$$

$$(N = 50): \quad \Delta x = \frac{2\pi}{50} = \frac{\pi}{25}, c_i = \left(i - \frac{1}{2}\right) \cdot \frac{\pi}{25}$$

$$M_{50} = \frac{\pi}{25} \sum_{i=1}^{50} f(c_i) = 6.914951$$

25. The ellipse $\left(\dfrac{x}{5}\right)^2 + \left(\dfrac{y}{3}\right)^2 = 1$

SOLUTION   We use the parametrization given in Example 4, section 12.1, that is, $c(t) = (5\cos t, 3\sin t), 0 \le t \le 2\pi$. The curve is shown in the figure below:

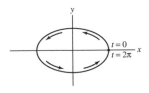

$$c(t) = (5\cos t, 3\sin t), 0 \le t \le 2\pi.$$

The length of the curve is given by the following integral:

$$S = \int_0^{2\pi} \sqrt{x'(t)^2 + y'(t)^2}\, dt = \int_0^{2\pi} \sqrt{(-5\sin t)^2 + (3\cos t)^2}\, dt$$

$$= \int_0^{2\pi} \sqrt{25\sin^2 t + 9\cos^2 t}\, dt = \int_0^{2\pi} \sqrt{9(\sin^2 t + \cos^2 t) + 16\sin^2 t}\, dt = \int_0^{2\pi} \sqrt{9 + 16\sin^2 t}\, dt.$$

That is,

$$S = \int_0^{2\pi} \sqrt{9 + 16\sin^2 t}\, dt.$$

We approximate the integral using the Mid-Point Rule with $N = 10, 20, 30, 50$, for $f(t) = \sqrt{9 + 16\sin^2 t}$. We obtain

$$(N = 10): \quad \Delta x = \frac{2\pi}{10} = \frac{\pi}{5}, c_i = \left(i - \frac{1}{2}\right) \cdot \frac{\pi}{5}$$

$$M_{10} = \frac{\pi}{5} \sum_{i=1}^{10} f(c_i) = 25.528309$$

$$(N = 20): \quad \Delta x = \frac{2\pi}{20} = \frac{\pi}{10}, c_i = \left(i - \frac{1}{2}\right) \cdot \frac{\pi}{10}$$

$$M_{20} = \frac{\pi}{10} \sum_{i=1}^{20} f(c_i) = 25.526999$$

$$(N = 30): \quad \Delta x = \frac{2\pi}{30} = \frac{\pi}{15}, c_i = \left(i - \frac{1}{2}\right) \cdot \frac{\pi}{15}$$

$$M_{30} = \frac{\pi}{15} \sum_{i=1}^{30} f(c_i) = 25.526999$$

$$(N = 50): \quad \Delta x = \frac{2\pi}{50} = \frac{\pi}{25}, c_i = \left(i - \frac{1}{2}\right) \cdot \frac{\pi}{25}$$

$$M_{50} = \frac{\pi}{25} \sum_{i=1}^{50} f(c_i) = 25.526999$$

27. If you unwind thread from a stationary circular spool, keeping the thread taut at all times, then the endpoint traces a curve C called the **involute** of the circle (Figure 9). Observe that $\overline{PQ}$ has length $R\theta$. Show that C is parametrized by

$$c(\theta) = \left(R(\cos\theta + \theta\sin\theta), R(\sin\theta - \theta\cos\theta)\right)$$

Then find the length of the involute for $0 \le \theta \le 2\pi$.

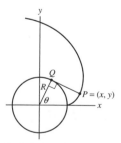

FIGURE 9 Involute of a circle.

SOLUTION Suppose that the arc $\overset{\frown}{QT}$ corresponding to the angle θ is unwound. Then the length of the segment $\overline{QP}$ equals the length of this arc. That is, $\overline{QP} = R\theta$. With the help of the figure we can see that

$$x = \overline{OA} + \overline{AB} = \overline{OA} + \overline{EP} = R\cos\theta + \overline{QP}\sin\theta = R\cos\theta + R\theta\sin\theta = R(\cos\theta + \theta\sin\theta).$$

Furthermore,

$$y = \overline{QA} - \overline{QE} = R\sin\theta - \overline{QP}\cos\theta = R\sin\theta - R\theta\cos\theta = R(\sin\theta - \theta\cos\theta)$$

The coordinates of P with respect to the parameter θ form the following parametrization of the curve:

$$c(\theta) = (R(\cos\theta + \theta\sin\theta), R(\sin\theta - \theta\cos\theta)), \qquad 0 \le \theta \le 2\pi.$$

We find the length of the involute for $0 \le \theta \le 2\pi$, using the formula for the arc length:

$$S = \int_0^{2\pi} \sqrt{x'(\theta)^2 + y'(\theta)^2}\, d\theta.$$

We compute the integrand:

$$x'(\theta) = \frac{d}{d\theta}(R(\cos\theta + \theta\sin\theta)) = R(-\sin\theta + \sin\theta + \theta\cos\theta) = R\theta\cos\theta$$

$$y'(\theta) = \frac{d}{d\theta}(R(\sin\theta - \theta\cos\theta)) = R(\cos\theta - (\cos\theta - \theta\sin\theta)) = R\theta\sin\theta$$

$$\sqrt{x'(\theta)^2 + y'(\theta)^2} = \sqrt{(R\theta\cos\theta)^2 + (R\theta\sin\theta)^2} = \sqrt{R^2\theta^2(\cos^2\theta + \sin^2\theta)} = \sqrt{R^2\theta^2} = R\theta$$

We now compute the arc length:

$$S = \int_0^{2\pi} R\theta\, d\theta = \frac{R\theta^2}{2}\Big|_0^{2\pi} = \frac{R\cdot(2\pi)^2}{2} = 2\pi^2 R.$$

In Exercises 29–32, use Eq. (4) to compute the surface area of the given surface.

29. The cone generated by revolving $c(t) = (t, mt)$ about the x-axis for $0 \le t \le A$

SOLUTION Substituting $y(t) = mt$, $y'(t) = m$, $x'(t) = 1$, $a = 0$, and $b = 0$ in the formula for the surface area, we get

$$S = 2\pi\int_0^A mt\sqrt{1+m^2}\, dt = 2\pi\sqrt{1+m^2}\, m\int_0^A t\, dt = 2\pi m\sqrt{1+m^2}\cdot\frac{t^2}{2}\Big|_0^A = m\sqrt{1+m^2}\,\pi A^2$$

31. The surface generated by revolving one arch of the cycloid $c(t) = (t - \sin t, 1 - \cos t)$ about the x-axis

SOLUTION One arch of the cycloid is traced as t varies from 0 to 2π. Since $x(t) = t - \sin t$ and $y(t) = 1 - \cos t$, we have $x'(t) = 1 - \cos t$ and $y'(t) = \sin t$. Hence, using the identity $1 - \cos t = 2\sin^2\frac{t}{2}$, we get

$$x'(t)^2 + y'(t)^2 = (1 - \cos t)^2 + \sin^2 t = 1 - 2\cos t + \cos^2 t + \sin^2 t = 2 - 2\cos t = 4\sin^2\frac{t}{2}$$

By the formula for the surface area we obtain:

$$S = 2\pi\int_0^{2\pi} y(t)\sqrt{x'(t)^2 + y'(t)^2}\, dt = 2\pi\int_0^{2\pi}(1 - \cos t)\cdot 2\sin\frac{t}{2}\, dt$$

$$= 2\pi\int_0^{2\pi} 2\sin^2\frac{t}{2}\cdot 2\sin\frac{t}{2}\, dt = 8\pi\int_0^{2\pi}\sin^3\frac{t}{2}\, dt = 16\pi\int_0^\pi \sin^3 u\, du$$

We use a reduction formula to compute this integral, obtaining

$$S = 16\pi\left[\frac{1}{3}\cos^3 u - \cos u\right]\Big|_0^\pi = 16\pi\left[\frac{4}{3}\right] = \frac{64\pi}{3}$$

Further Insights and Challenges

33. ⌂⌂⌂ Let $b(t)$ be the "Butterfly Curve":

$$x(t) = \sin t\left(e^{\cos t} - 2\cos 4t - \sin\left(\frac{t}{12}\right)^5\right)$$

$$y(t) = \cos t\left(e^{\cos t} - 2\cos 4t - \sin\left(\frac{t}{12}\right)^5\right)$$

(a) Use a computer algebra system to plot $b(t)$ and the speed $s'(t)$ for $0 \le t \le 12\pi$.

(b) Approximate the length $b(t)$ for $0 \le t \le 10\pi$.

SOLUTION

(a) Let $f(t) = e^{\cos t} - 2\cos 4t - \sin\left(\frac{t}{12}\right)^5$, then

$$x(t) = \sin t f(t)$$

$$y(t) = \cos t f(t)$$

and so

$$(x'(t))^2 + (y'(t))^2 = [\sin t f'(t) + \cos t f(t)]^2 + [\cos t f'(t) - \sin t f(t)]^2$$

Using the identity $\sin^2 t + \cos^2 t = 1$, we get

$$(x'(t))^2 + (y'(t))^2 = (f'(t))^2 + (f(t))^2.$$

Thus, $s'(t)$ is the following:

$$\sqrt{\left[e^{\cos t} - 2\cos 4t - \sin\left(\frac{t}{12}\right)^5\right]^2 + \left[-\sin t e^{\cos t} + 8\sin 4t - \frac{5}{12}\left(\frac{t}{12}\right)^4 \cos\left(\frac{t}{12}\right)^5\right]^2}.$$

The following figures show the curves of $b(t)$ and the speed $s'(t)$ for $0 \le t \le 10\pi$:

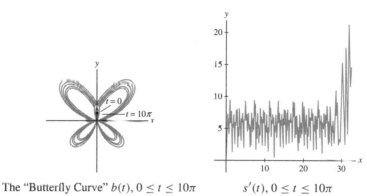

The "Butterfly Curve" $b(t), 0 \le t \le 10\pi$ $s'(t), 0 \le t \le 10\pi$

Looking at the graph, we see it would be difficult to compute the length using numeric integration; due to the high frequency oscillations, very small steps would be needed.

(b) The length of $b(t)$ for $0 \le t \le 10\pi$ is given by the integral: $L = \int_0^{10\pi} s'(t)\,dt$ where $s'(t)$ is given in part (a). We approximate the length using the Midpoint Rule with $N = 30$. The numerical methods in Mathematica approximate the answer by 211.952. Using the Midpoint Rule with $N = 50$, we get 204.48; with $N = 500$, we get 211.6; and with $N = 5000$, we get 212.09.

35. A satellite orbiting at a distance R from the center of the earth follows the circular path $x = R\cos\omega t$, $y = R\sin\omega t$.

(a) Show that the period T (the time of one revolution) is $T = 2\pi/\omega$.

(b) According to Newton's laws of motion and gravity,

$$x''(t) = -Gm_e\frac{x}{R^3}, \qquad y''(t) = -Gm_e\frac{y}{R^3}$$

where G is the universal gravitational constant and m_e is the mass of the earth. Prove that $R^3/T^2 = Gm_e/4\pi^2$. Thus, R^3/T^2 has the same value for all orbits (a special case of Kepler's Third Law).

SOLUTION

(a) As shown in Example 4, the circular path has constant speed of $\frac{ds}{dt} = \omega R$. Since the length of one revolution is $2\pi R$, the period T is

$$T = \frac{2\pi R}{\omega R} = \frac{2\pi}{\omega}.$$

(b) Differentiating $x = R\cos\omega t$ twice with respect to t gives

$$x'(t) = -R\omega \sin\omega t$$

$$x''(t) = -R\omega^2 \cos\omega t$$

Substituting $x(t)$ and $x''(t)$ in the equation $x''(t) = -Gm_e \dfrac{x}{R^3}$ and simplifying, we obtain

$$-R\omega^2 \cos \omega t = -Gm_e \cdot \frac{R \cos \omega t}{R^3}$$

$$-R\omega^2 = -\frac{Gm_e}{R^2} \Rightarrow R^3 = \frac{Gm_e}{\omega^2}$$

By part (a), $T = \dfrac{2\pi}{\omega}$. Hence, $\omega = \dfrac{2\pi}{T}$. Substituting yields

$$R^3 = \frac{Gm_e}{\dfrac{4\pi^2}{T^2}} = \frac{T^2 Gm_e}{4\pi^2} \Rightarrow \frac{R^3}{T^2} = \frac{Gm_e}{4\pi^2}$$

11.3 Polar Coordinates   (LT Section 12.3)

Preliminary Questions

1. Points P and Q with the same radial coordinate (choose the correct answer):

(a) Lie on the same circle with the center at the origin.

(b) Lie on the same ray based at the origin.

SOLUTION   Two points with the same radial coordinate are equidistant from the origin, therefore they lie on the same circle centered at the origin. The angular coordinate defines a ray based at the origin. Therefore, if the two points have the same angular coordinate, they lie on the same ray based at the origin.

2. Give two polar representations for the point $(x, y) = (0, 1)$, one with negative r and one with positive r.

SOLUTION   The point $(0, 1)$ is on the y-axis, distant one unit from the origin, hence the polar representation with positive r is $(r, \theta) = \left(1, \frac{\pi}{2}\right)$. The point $(r, \theta) = \left(-1, \frac{\pi}{2}\right)$ is the reflection of $(r, \theta) = \left(1, \frac{\pi}{2}\right)$ through the origin, hence we must add π to return to the original point.

We obtain the following polar representation of $(0, 1)$ with negative r:

$$(r, \theta) = \left(-1, \frac{\pi}{2} + \pi\right) = \left(-1, \frac{3\pi}{2}\right).$$

3. Describe each of the following curves:

(a) $r = 2$ **(b)** $r^2 = 2$ **(c)** $r \cos \theta = 2$

SOLUTION

(a) Converting to rectangular coordinates we get

$$\sqrt{x^2 + y^2} = 2 \quad \text{or} \quad x^2 + y^2 = 2^2.$$

This is the equation of the circle of radius 2 centered at the origin.

(b) We convert to rectangular coordinates, obtaining $x^2 + y^2 = 2$. This is the equation of the circle of radius $\sqrt{2}$, centered at the origin.

(c) We convert to rectangular coordinates. Since $x = r \cos \theta$ we obtain the following equation: $x = 2$. This is the equation of the vertical line through the point $(2, 0)$.

4. If $f(-\theta) = f(\theta)$, then the curve $r = f(\theta)$ is symmetric with respect to the (choose the correct answer):

(a) x-axis   **(b)** y-axis   **(c)** origin

SOLUTION   The equality $f(-\theta) = f(\theta)$ for all θ implies that whenever a point (r, θ) is on the curve, also the point $(r, -\theta)$ is on the curve. Since the point $(r, -\theta)$ is the reflection of (r, θ) with respect to the x-axis, we conclude that the curve is symmetric with respect to the x-axis.

Exercises

1. Find polar coordinates for each of the seven points plotted in Figure 16.

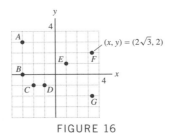

FIGURE 16

SOLUTION We mark the points as shown in the figure.

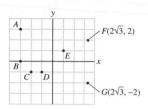

Using the data given in the figure for the x and y coordinates and the quadrants in which the point are located, we obtain:

(A), with rectangular coordinates $(-3, 4)$: $\quad \begin{array}{l} r = \sqrt{(-3)^2 + 3^2} = \sqrt{18} \\ \theta = \pi - \frac{\pi}{4} = \frac{3\pi}{4} \end{array} \Rightarrow (r, \theta) = \left(3\sqrt{2}, \frac{3\pi}{4}\right)$

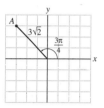

(B), with rectangular coordinates $(-3, 0)$: $\quad \begin{array}{l} r = 3 \\ \theta = \pi \end{array} \Rightarrow (r, \theta) = (3, \pi)$

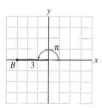

(C), with rectangular coordinates $(-2, -1)$:

$$r = \sqrt{2^2 + 1^2} = \sqrt{5} \approx 2.2$$
$$\theta = \tan^{-1}\left(\frac{-1}{-2}\right) = \tan^{-1}\left(\frac{1}{2}\right) = \pi + 0.46 \approx 3.6 \quad \Rightarrow (r, \theta) \approx \left(\sqrt{5}, 3.6\right)$$

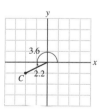

(D), with rectangular coordinates $(-1, -1)$: $\quad \begin{array}{l} r = \sqrt{1^2 + 1^2} = \sqrt{2} \approx 1.4 \\ \theta = \pi + \frac{\pi}{4} = \frac{5\pi}{4} \end{array} \Rightarrow (r, \theta) \approx \left(\sqrt{2}, \frac{5\pi}{4}\right)$

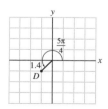

(E), with rectangular coordinates $(1, 1)$:
$$r = \sqrt{1^2 + 1^2} = \sqrt{2} \approx 1.4$$
$$\theta = \tan^{-1}\left(\tfrac{1}{1}\right) = \tfrac{\pi}{4}$$
$$\Rightarrow (r, \theta) \approx \left(\sqrt{2}, \tfrac{\pi}{4}\right)$$

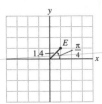

(F), with rectangular coordinates $(2\sqrt{3}, 2)$:
$$r = \sqrt{\left(2\sqrt{3}\right)^2 + 2^2} = \sqrt{16} = 4$$
$$\theta = \tan^{-1}\left(\tfrac{2}{2\sqrt{3}}\right) = \tan^{-1}\left(\tfrac{1}{\sqrt{3}}\right) = \tfrac{\pi}{6}$$
$$\Rightarrow (r, \theta) = \left(4, \tfrac{\pi}{6}\right)$$

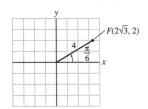

(G), with rectangular coordinates $(2\sqrt{3}, -2)$: G is the reflection of F about the x axis, hence the two points have equal radial coordinates, and the angular coordinate of G is obtained from the angular coordinate of F: $\theta = 2\pi - \tfrac{\pi}{6} = \tfrac{11\pi}{6}$.
Hence, the polar coordinates of G are $\left(4, \tfrac{11\pi}{6}\right)$.

3. Convert from rectangular to polar coordinates.

(a) $(1, 0)$ **(b)** $(3, \sqrt{3})$ **(c)** $(-2, 2)$ **(d)** $(-1, \sqrt{3})$

SOLUTION

(a) The point $(1, 0)$ is on the positive x axis distanced one unit from the origin. Hence, $r = 1$ and $\theta = 0$. Thus, $(r, \theta) = (1, 0)$.

(b) The point $\left(3, \sqrt{3}\right)$ is in the first quadrant so $\theta = \tan^{-1}\left(\tfrac{\sqrt{3}}{3}\right) = \tfrac{\pi}{6}$. Also, $r = \sqrt{3^2 + \left(\sqrt{3}\right)^2} = \sqrt{12}$. Hence, $(r, \theta) = \left(\sqrt{12}, \tfrac{\pi}{6}\right)$.

(c) The point $(-2, 2)$ is in the second quadrant. Hence,
$$\theta = \tan^{-1}\left(\frac{2}{-2}\right) = \tan^{-1}(-1) = \pi - \frac{\pi}{4} = \frac{3\pi}{4}.$$

Also, $r = \sqrt{(-2)^2 + 2^2} = \sqrt{8}$. Hence, $(r, \theta) = \left(\sqrt{8}, \tfrac{3\pi}{4}\right)$.

(d) The point $\left(-1, \sqrt{3}\right)$ is in the second quadrant, hence,
$$\theta = \tan^{-1}\left(\frac{\sqrt{3}}{-1}\right) = \tan^{-1}\left(-\sqrt{3}\right) = \pi - \frac{\pi}{3} = \frac{2\pi}{3}.$$

Also, $r = \sqrt{(-1)^2 + \left(\sqrt{3}\right)^2} = \sqrt{4} = 2$. Hence, $(r, \theta) = \left(2, \tfrac{2\pi}{3}\right)$.

5. Convert from polar to rectangular coordinates:

(a) $\left(3, \tfrac{\pi}{6}\right)$ **(b)** $\left(6, \tfrac{3\pi}{4}\right)$ **(c)** $\left(0, \tfrac{\pi}{5}\right)$ **(d)** $\left(5, -\tfrac{\pi}{2}\right)$

SOLUTION

(a) Since $r = 3$ and $\theta = \tfrac{\pi}{6}$, we have:

$$x = r\cos\theta = 3\cos\frac{\pi}{6} = 3 \cdot \frac{\sqrt{3}}{2} \approx 2.6$$
$$y = r\sin\theta = 3\sin\frac{\pi}{6} = 3 \cdot \frac{1}{2} = 1.5$$
$$\Rightarrow \quad (x, y) \approx (2.6, 1.5).$$

(b) For $\left(6, \frac{3\pi}{4}\right)$ we have $r = 6$ and $\theta = \frac{3\pi}{4}$. Hence,

$$x = r\cos\theta = 6\cos\frac{3\pi}{4} \approx -4.24$$

$$\Rightarrow \quad (x, y) \approx (-4.24, 4.24).$$

$$y = r\sin\theta = 6\sin\frac{3\pi}{4} \approx 4.24$$

(c) For $\left(0, \frac{\pi}{5}\right)$, we have $r = 0$, so that the rectangular coordinates are $(x, y) = (0, 0)$.

(d) Since $r = 5$ and $\theta = -\frac{\pi}{2}$ we have

$$x = r\cos\theta = 5\cos\left(-\frac{\pi}{2}\right) = 5 \cdot 0 = 0$$

$$\Rightarrow \quad (x, y) = (0, -5)$$

$$y = r\sin\theta = 5\sin\left(-\frac{\pi}{2}\right) = 5 \cdot (-1) = -5$$

7. Describe each shaded sector in Figure 17 by inequalities in r and θ.

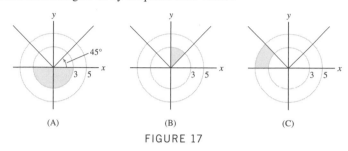

(A)   (B)   (C)

FIGURE 17

SOLUTION

(a) In the sector shown below r is varying between 0 and 3 and θ is varying between π and 2π. Hence the following inequalities describe the sector:

$$0 \leq r \leq 3$$

$$\pi \leq \theta \leq 2\pi$$

(b) In the sector shown below r is varying between 0 and 3 and θ is varying between $\frac{\pi}{4}$ and $\frac{\pi}{2}$. Hence, the inequalities for the sector are:

$$0 \leq r \leq 3$$

$$\frac{\pi}{4} \leq \theta \leq \frac{\pi}{2}$$

(c) In the sector shown below r is varying between 3 and 5 and θ is varying between $\frac{3\pi}{4}$ and π. Hence, the inequalities are:

$$3 \leq r \leq 5$$

$$\frac{3\pi}{4} \leq \theta \leq \pi$$

9. What is the slope of the line $\theta = \frac{3\pi}{5}$?

SOLUTION This line makes an angle $\theta_0 = \frac{3\pi}{5}$ with the positive x-axis, hence the slope of the line is $m = \tan\frac{3\pi}{5} \approx -3.1$.

In Exercises 11–16, convert to an equation in rectangular coordinates.

11. $r = 7$

SOLUTION $r = 7$ describes the points having distance 7 from the origin, that is, the circle with radius 7 centered at the origin. The equation of the circle in rectangular coordinates is

$$x^2 + y^2 = 7^2 = 49.$$

13. $r = 2\sin\theta$

SOLUTION We multiply the equation by r and substitute $r^2 = x^2 + y^2$, $r\sin\theta = y$. This gives

$$r^2 = 2r\sin\theta$$

$$x^2 + y^2 = 2y$$

Moving the $2y$ and completing the square yield: $x^2 + y^2 - 2y = 0$ and $x^2 + (y - 1)^2 = 1$. Thus, $r = 2\sin\theta$ is the equation of a circle of radius 1 centered at $(0, 1)$.

15. $r = \dfrac{1}{\cos\theta - \sin\theta}$

SOLUTION   We multiply the equation by $\cos\theta - \sin\theta$ and substitute $y = r\sin\theta$, $x = r\cos\theta$. This gives

$$r(\cos\theta - \sin\theta) = 1$$

$$r\cos\theta - r\sin\theta = 1$$

$x - y = 1 \Rightarrow y = x - 1$. Thus,

$$r = \dfrac{1}{\cos\theta - \sin\theta}$$

is the equation of the line $y = x - 1$.

In Exercises 17–20, convert to an equation in polar coordinates.

17. $x^2 + y^2 = 5$

SOLUTION   We make the substitution $x^2 + y^2 = r^2$ to obtain; $r^2 = 5$ or $r = \sqrt{5}$.

19. $y = x^2$

SOLUTION   Substituting $y = r\sin\theta$ and $x = r\cos\theta$ yields

$$r\sin\theta = r^2\cos^2\theta.$$

Then, dividing by $r\cos^2\theta$ we obtain,

$$\dfrac{\sin\theta}{\cos^2\theta} = r \qquad \text{so} \qquad r = \tan\theta\sec\theta$$

21. Match each equation with its description.

(a) $r = 2$ (i) Vertical line
(b) $\theta = 2$ (ii) Horizontal line
(c) $r = 2\sec\theta$ (iii) Circle
(d) $r = 2\csc\theta$ (iv) Line through origin

SOLUTION

(a) $r = 2$ describes the points 2 units from the origin. Hence, it is the equation of a circle.
(b) $\theta = 2$ describes the points P so that $\overline{OP}$ makes an angle of $\theta_0 = 2$ with the positive x-axis. Hence, it is the equation of a line through the origin.
(c) This is $r\cos\theta = 2$, which is $x = 2$, a vertical line.
(d) Converting to rectangular coordinates, we get $r = 2\csc\theta$, so $r\sin\theta = 2$ and $y = 2$. This is the equation of a horizontal line.

23. Suppose that $P = (x, y)$ has polar coordinates (r, θ). Find the polar coordinates for the points:

(a) $(x, -y)$ (b) $(-x, -y)$ (c) $(-x, y)$ (d) (y, x)

SOLUTION

(a) $(x, -y)$ is the symmetric point of (x, y) with respect to the x-axis, hence the two points have the same radial coordinate, and the angular coordinate of $(x, -y)$ is $2\pi - \theta$. Hence, $(x, -y) = (r, 2\pi - \theta)$.

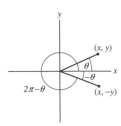

(b) $(-x, -y)$ is the symmetric point of (x, y) with respect to the origin. Hence, $(-x, -y) = (r, \theta + \pi)$.

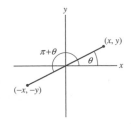

(c) $(-x, y)$ is the symmetric point of (x, y) with respect to the y-axis. Hence the two points have the same radial coordinates and the angular coordinate of $(-x, y)$ is $\pi - \theta$. Hence, $(-x, y) = (r, \pi - \theta)$.

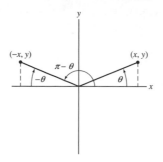

(d) Let (r_1, θ_1) denote the polar coordinates of (y, x). Hence,

$$r_1 = \sqrt{y^2 + x^2} = \sqrt{x^2 + y^2} = r$$

$$\tan \theta_1 = \frac{x}{y} = \frac{1}{y/x} = \frac{1}{\tan \theta} = \cot \theta = \tan\left(\frac{\pi}{2} - \theta\right)$$

Since the points (x, y) and (y, x) are in the same quadrant, the solution for θ_1 is $\theta_1 = \frac{\pi}{2} - \theta$. We obtain the following polar coordinates: $(y, x) = \left(r, \frac{\pi}{2} - \theta\right)$.

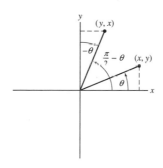

25. What are the polar equations of the lines parallel to the line $r \cos\left(\theta - \frac{\pi}{3}\right) = 1$?

SOLUTION The line $r \cos\left(\theta - \frac{\pi}{3}\right) = 1$, or $r = \sec\left(\theta - \frac{\pi}{3}\right)$, is perpendicular to the ray $\theta = \frac{\pi}{3}$ and at distance $d = 1$ from the origin. Hence, the lines parallel to this line are also perpendicular to the ray $\theta = \frac{\pi}{3}$, so the polar equations of these lines are $r = d \sec\left(\theta - \frac{\pi}{3}\right)$ or $r \cos\left(\theta - \frac{\pi}{3}\right) = d$.

27. Sketch the curve $r = \frac{1}{2}\theta$ (the spiral of Archimedes) for θ between 0 and 2π by plotting the points for $\theta = 0, \frac{\pi}{4}, \frac{\pi}{2}, \ldots, 2\pi$.

SOLUTION We first plot the following points (r, θ) on the spiral:

$$O = (0, 0), \quad A = \left(\frac{\pi}{8}, \frac{\pi}{4}\right), \quad B = \left(\frac{\pi}{4}, \frac{\pi}{2}\right), \quad C = \left(\frac{3\pi}{8}, \frac{3\pi}{4}\right), \quad D = \left(\frac{\pi}{2}, \pi\right),$$

$$E = \left(\frac{5\pi}{8}, \frac{5\pi}{4}\right), \quad F = \left(\frac{3\pi}{4}, \frac{3\pi}{2}\right), \quad G = \left(\frac{7\pi}{8}, \frac{7\pi}{4}\right), \quad H = (\pi, 2\pi).$$

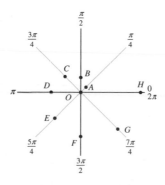

Since $r(0) = \frac{0}{2} = 0$, the graph begins at the origin and moves toward the points A, B, C, D, E, F, G and H as θ varies from $\theta = 0$ to the other values stated above. Connecting the points in this direction we obtain the following graph for $0 \le \theta \le 2\pi$:

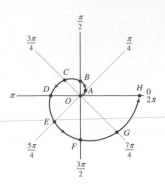

29. Sketch the cardioid curve $r = 1 + \cos\theta$.

SOLUTION   Since $\cos\theta$ is period with period 2π, the entire curve will be traced out as θ varies from 0 to 2π. Additionally, since $\cos(2\pi - \theta) = \cos(\theta)$, we can sketch the curve for θ between 0 and π and reflect the result through the x axis to obtain the whole curve. Use the values $\theta = 0, \frac{\pi}{6}, \frac{\pi}{4}, \frac{\pi}{3}, \frac{\pi}{2}, \frac{2\pi}{3}, \frac{3\pi}{4}, \frac{5\pi}{6}$, and π:

θ	r	point
0	$1 + \cos 0 = 2$	$(2, 0)$
$\frac{\pi}{6}$	$1 + \cos\frac{\pi}{6} = \frac{2+\sqrt{3}}{2}$	$\left(\frac{2+\sqrt{3}}{2}, \frac{\pi}{6}\right)$
$\frac{\pi}{4}$	$1 + \cos\frac{\pi}{4} = \frac{2+\sqrt{2}}{2}$	$\left(\frac{2+\sqrt{2}}{2}, \frac{\pi}{4}\right)$
$\frac{\pi}{3}$	$1 + \cos\frac{\pi}{3} = \frac{3}{2}$	$\left(\frac{3}{2}, \frac{\pi}{3}\right)$
$\frac{\pi}{2}$	$1 + \cos\frac{\pi}{2} = 1$	$\left(1, \frac{\pi}{2}\right)$
$\frac{2\pi}{3}$	$1 + \cos\frac{2\pi}{3} = \frac{1}{2}$	$\left(\frac{1}{2}, \frac{2\pi}{3}\right)$
$\frac{3\pi}{4}$	$1 + \cos\frac{3\pi}{4} = \frac{2-\sqrt{2}}{2}$	$\left(\frac{2-\sqrt{2}}{2}, \frac{3\pi}{4}\right)$
$\frac{5\pi}{6}$	$1 + \cos\frac{5\pi}{6} = \frac{2-\sqrt{3}}{2}$	$\left(\frac{2-\sqrt{3}}{2}, \frac{5\pi}{6}\right)$

$\theta = 0$ corresponds to the point $(2, 0)$, and the graph moves clockwise as θ increases from 0 to π. Thus the graph is

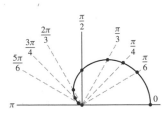

Reflecting through the x axis gives the other half of the curve:

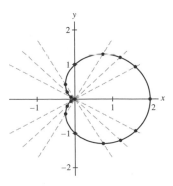

31. Figure 20 displays the graphs of $r = \sin 2\theta$ in rectangular coordinates and in polar coordinates, where it is a "rose with four petals." Identify:

(a) The points in (B) corresponding to points $A–I$ in (A).

(b) The parts of the curve in (B) corresponding to the angle intervals $\left[0, \frac{\pi}{2}\right]$, $\left[\frac{\pi}{2}, \pi\right]$, $\left[\pi, \frac{3\pi}{2}\right]$, and $\left[\frac{3\pi}{2}, 2\pi\right]$.

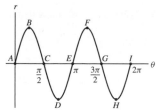
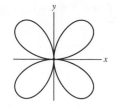

(A) Graph of r as a function
of θ, where $r = \sin 2\theta$

(B) Graph of $r = \sin 2\theta$
in polar coordinates

FIGURE 20

SOLUTION

(a) The graph (A) gives the following polar coordinates of the labeled points:

$$A: \quad \theta = 0, \quad r = 0$$

$$B: \quad \theta = \frac{\pi}{4}, \quad r = \sin\frac{2\pi}{4} = 1$$

$$C: \quad \theta = \frac{\pi}{2}, \quad r = 0$$

$$D: \quad \theta = \frac{3\pi}{4}, \quad r = \sin\frac{2\cdot 3\pi}{4} = -1$$

$$E: \quad \theta = \pi, \quad r = 0$$

$$F: \quad \theta = \frac{5\pi}{4}, \quad r = 1$$

$$G: \quad \theta = \frac{3\pi}{2}, \quad r = 0$$

$$H: \quad \theta = \frac{7\pi}{4}, \quad r = -1$$

$$I: \quad \theta = 2\pi, \quad r = 0.$$

Since the maximal value of $|r|$ is 1, the points with $r = 1$ or $r = -1$ are the furthest points from the origin. The corresponding quadrant is determined by the value of θ and the sign of r. If $r_0 < 0$, the point (r_0, θ_0) is on the ray $\theta = -\theta_0$. These considerations lead to the following identification of the points in the xy plane. Notice that A, C, G, E, and I are the same point.

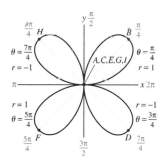

(b) We use the graph (A) to find the sign of $r = \sin 2\theta$: $0 \le \theta \le \frac{\pi}{2} \Rightarrow r \ge 0 \Rightarrow (r, \theta)$ is in the first quadrant. $\frac{\pi}{2} \le \theta \le \pi \Rightarrow r \le 0 \Rightarrow (r, \theta)$ is in the fourth quadrant. $\pi \le \theta \le \frac{3\pi}{2} \Rightarrow r \ge 0 \Rightarrow (r, \theta)$ is in the third quadrant. $\frac{3\pi}{2} \le \theta \le 2\pi \Rightarrow r \le 0 \Rightarrow (r, \theta)$ is in the second quadrant. That is,

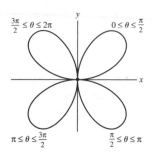

33. CAS   Plot the **cissoid** $r = 2\sin\theta\tan\theta$ and show that its equation in rectangular coordinates is

$$y^2 = \frac{x^3}{2-x}$$

SOLUTION   Using a CAS we obtain the following curve of the cissoid:

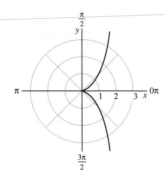

We substitute $\sin\theta = \frac{y}{r}$ and $\tan\theta = \frac{y}{x}$ in $r = 2\sin\theta\tan\theta$ to obtain

$$r = 2\frac{y}{r}\cdot\frac{y}{x}.$$

Multiplying by rx, setting $r^2 = x^2 + y^2$ and simplifying, yields

$$r^2 x = 2y^2$$
$$(x^2 + y^2)x = 2y^2$$
$$x^3 + y^2 x = 2y^2$$
$$y^2(2-x) = x^3$$

so

$$y^2 = \frac{x^3}{2-x}$$

35. Show that

$$r = a\cos\theta + b\sin\theta$$

is the equation of a circle passing through the origin. Express the radius and center (in rectangular coordinates) in terms of a and b.

SOLUTION   We multiply the equation by r and then make the substitution $x = r\cos\theta$, $y = r\sin\theta$, and $r^2 = x^2 + y^2$. This gives

$$r^2 = ar\cos\theta + br\sin\theta$$
$$x^2 + y^2 = ax + by$$

Transferring sides and completing the square yields

$$x^2 - ax + y^2 - by = 0$$
$$\left(x^2 - 2\cdot\frac{a}{2}x + \left(\frac{a}{2}\right)^2\right) + \left(y^2 - 2\cdot\frac{b}{2}y + \left(\frac{b}{2}\right)^2\right) = \left(\frac{a}{2}\right)^2 + \left(\frac{b}{2}\right)^2$$
$$\left(x - \frac{a}{2}\right)^2 + \left(y - \frac{b}{2}\right)^2 = \frac{a^2 + b^2}{4}$$

This is the equation of the circle with radius $\frac{\sqrt{a^2+b^2}}{2}$ centered at the point $\left(\frac{a}{2}, \frac{b}{2}\right)$. By plugging in $x = 0$ and $y = 0$ it is clear that the circle passes through the origin.

37. Use the identity $\cos 2\theta = \cos^2\theta - \sin^2\theta$ to find a polar equation of the hyperbola $x^2 - y^2 = 1$.

SOLUTION   We substitute $x = r\cos\theta$, $y = r\sin\theta$ in $x^2 - y^2 = 1$ to obtain

$$r^2\cos^2\theta - r^2\sin^2\theta = 1$$
$$r^2(\cos^2\theta - \sin^2\theta) = 1$$

Using the identity $\cos 2\theta = \cos^2 \theta - \sin^2 \theta$ we obtain the following equation of the hyperbola:

$$r^2 \cos 2\theta = 1 \quad \text{or} \quad r^2 = \sec 2\theta.$$

39. Show that $\cos 3\theta = \cos^3 \theta - 3 \cos \theta \sin^2 \theta$ and use this identity to find an equation in rectangular coordinates for the curve $r = \cos 3\theta$.

SOLUTION We use the identities $\cos(\alpha + \beta) = \cos \alpha \cos \beta - \sin \alpha \sin \beta$, $\cos 2\alpha = \cos^2 \alpha - \sin^2 \alpha$, and $\sin 2\alpha = 2 \sin \alpha \cos \alpha$ to write

$$\cos 3\theta = \cos(2\theta + \theta) = \cos 2\theta \cos \theta - \sin 2\theta \sin \theta$$

$$= (\cos^2 \theta - \sin^2 \theta) \cos \theta - 2 \sin \theta \cos \theta \sin \theta$$

$$= \cos^3 \theta - \sin^2 \theta \cos \theta - 2 \sin^2 \theta \cos \theta$$

$$= \cos^3 \theta - 3 \sin^2 \theta \cos \theta$$

Using this identity we may rewrite the equation $r = \cos 3\theta$ as follows:

$$r = \cos^3 \theta - 3 \sin^2 \theta \cos \theta \tag{1}$$

Since $x = r \cos \theta$ and $y = r \sin \theta$, we have $\cos \theta = \frac{x}{r}$ and $\sin \theta = \frac{y}{r}$. Substituting into (1) gives:

$$r = \left(\frac{x}{r}\right)^3 - 3\left(\frac{y}{r}\right)^2 \left(\frac{x}{r}\right)$$

$$r = \frac{x^3}{r^3} - \frac{3y^2 x}{r^3}$$

We now multiply by r^3 and make the substitution $r^2 = x^2 + y^2$ to obtain the following equation for the curve:

$$r^4 = x^3 - 3y^2 x$$

$$(x^2 + y^2)^2 = x^3 - 3y^2 x$$

In Exercises 41–44, find an equation in polar coordinates of the line $\mathcal{L}$ with the given description.

41. The point on $\mathcal{L}$ closest to the origin has polar coordinates $\left(2, \frac{\pi}{9}\right)$.

SOLUTION In Example 5, it is shown that the polar equation of the line where (r, α) is the point on the line closest to the origin is $r = d \sec(\theta - \alpha)$. Setting $(d, \alpha) = \left(2, \frac{\pi}{9}\right)$ we obtain the following equation of the line:

$$r = 2 \sec\left(\theta - \frac{\pi}{9}\right).$$

43. $\mathcal{L}$ is tangent to the circle $r = 2\sqrt{10}$ at the point with rectangular coordinates $(-2, -6)$.

SOLUTION

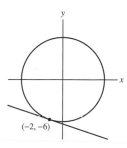

Since $\mathcal{L}$ is tangent to the circle at the point $(-2, -6)$, this is the point on $\mathcal{L}$ closest to the center of the circle which is at the origin. Therefore, we may use the polar coordinates (d, α) of this point in the equation of the line:

$$r = d \sec(\theta - \alpha) \tag{1}$$

We thus must convert the coordinates $(-2, -6)$ to polar coordinates. This point is in the third quadrant so $\pi < \alpha < \frac{3\pi}{2}$. We get

$$d = \sqrt{(-2)^2 + (-6)^2} = \sqrt{40} = 2\sqrt{10}$$

$$\alpha = \tan^{-1}\left(\frac{-6}{-2}\right) = \tan^{-1} 3 \approx \pi + 1.25 \approx 4.39$$

Substituting in (1) yields the following equation of the line:

$$r = 2\sqrt{10}\sec(\theta - 4.39).$$

45. Show that every line that does not pass through the origin has a polar equation of the form

$$r = \frac{b}{\sin\theta - a\cos\theta}$$

where $b \neq 0$.

SOLUTION   Write the equation of the line in rectangular coordinates as $y = ax + b$. Since the line does not pass through the origin, we have $b \neq 0$. Substitute for y and x to convert to polar coordinates, and simplify:

$$y = ax + b$$
$$r\sin\theta = ar\cos\theta + b$$
$$r(\sin\theta - a\cos\theta) = b$$
$$r = \frac{b}{\sin\theta - a\cos\theta}$$

47. For $a > 0$, a **lemniscate curve** is the set of points P such that the product of the distances from P to $(a, 0)$ and $(-a, 0)$ is a^2. Show that the equation of the lemniscate is

$$(x^2 + y^2)^2 = 2a^2(x^2 - y^2)$$

Then find the equation in polar coordinates. To obtain the simplest form of the equation, use the identity $\cos 2\theta = \cos^2\theta - \sin^2\theta$. Plot the lemniscate for $a = 2$ if you have a computer algebra system.

SOLUTION   We compute the distances d_1 and d_2 of $P(x, y)$ from the points $(a, 0)$ and $(-a, 0)$ respectively. We obtain:

$$d_1 = \sqrt{(x-a)^2 + (y-0)^2} = \sqrt{(x-a)^2 + y^2}$$
$$d_2 = \sqrt{(x+a)^2 + (y-0)^2} = \sqrt{(x+a)^2 + y^2}$$

For the points $P(x, y)$ on the lemniscate we have $d_1 d_2 = a^2$. That is,

$$a^2 = \sqrt{(x-a)^2 + y^2}\sqrt{(x+a)^2 + y^2} = \sqrt{\left[(x-a)^2 + y^2\right]\left[(x+a)^2 + y^2\right]}$$
$$= \sqrt{(x-a)^2(x+a)^2 + y^2(x-a)^2 + y^2(x+a)^2 + y^4}$$
$$= \sqrt{(x^2-a^2)^2 + y^2\left[(x-a)^2 + (x+a)^2\right] + y^4}$$
$$= \sqrt{x^4 - 2a^2x^2 + a^4 + y^2\left(x^2 - 2xa + a^2 + x^2 + 2xa + a^2\right) + y^4}$$
$$= \sqrt{x^4 - 2a^2x^2 + a^4 + 2y^2x^2 + 2y^2a^2 + y^4}$$
$$= \sqrt{x^4 + 2x^2y^2 + y^4 + 2a^2(y^2 - x^2) + a^4}$$
$$= \sqrt{(x^2 + y^2)^2 + 2a^2(y^2 - x^2) + a^4}.$$

Squaring both sides and simplifying yields

$$a^4 = (x^2 + y^2)^2 + 2a^2(y^2 - x^2) + a^4$$
$$0 = (x^2 + y^2)^2 + 2a^2(y^2 - x^2)$$

so

$$(x^2 + y^2)^2 = 2a^2(x^2 - y^2)$$

We now find the equation in polar coordinates. We substitute $x = r\cos\theta$, $y = r\sin\theta$ and $x^2 + y^2 = r^2$ into the equation of the lemniscate. This gives

$$(r^2)^2 = 2a^2(r^2\cos^2\theta - r^2\sin^2\theta) = 2a^2r^2(\cos^2\theta - \sin^2\theta) = 2a^2r^2\cos 2\theta$$

$$r^4 = 2a^2r^2\cos 2\theta$$

$r = 0$ is a solution, hence the origin is on the curve. For $r \neq 0$ we divide the equation by r^2 to obtain $r^2 = 2a^2\cos 2\theta$. This curve also includes the origin ($r = 0$ is obtained for $\theta = \frac{\pi}{4}$ for example), hence this is the polar equation of the lemniscate. Setting $a = 2$ we get $r^2 = 8\cos 2\theta$.

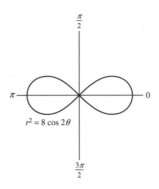

$r^2 = 8\cos 2\theta$

49. The Derivative in Polar Coordinates Show that a polar curve $r = f(\theta)$ has parametric equations

$$x = f(\theta)\cos\theta, \qquad y = f(\theta)\sin\theta$$

Then apply Theorem 2 of Section 11.1 to prove

$$\frac{dy}{dx} = \frac{f(\theta)\cos\theta + f'(\theta)\sin\theta}{-f(\theta)\sin\theta + f'(\theta)\cos\theta} \qquad \boxed{2}$$

where $f'(\theta) = df/d\theta$.

SOLUTION Multiplying both sides of the given equation by $\cos\theta$ yields $r\cos\theta = f(\theta)\cos\theta$; multiplying both sides by $\sin\theta$ yields $r\sin\theta = f(\theta)\sin\theta$. The left-hand sides of these two equations are the x and y coordinates in rectangular coordinates, so for any θ we have $x = f(\theta)\cos\theta$ and $y = f(\theta)\sin\theta$, showing that the parametric equations are as claimed. Now, by the formula for the derivative we have

$$\frac{dy}{dx} = \frac{y'(\theta)}{x'(\theta)} \qquad (1)$$

We differentiate the functions $x = f(\theta)\cos\theta$ and $y = f(\theta)\sin\theta$ using the Product Rule for differentiation. This gives

$$y'(\theta) = f'(\theta)\sin\theta + f(\theta)\cos\theta$$

$$x'(\theta) = f'(\theta)\cos\theta - f(\theta)\sin\theta$$

Substituting in (1) gives

$$\frac{dy}{dx} = \frac{f'(\theta)\sin\theta + f(\theta)\cos\theta}{f'(\theta)\cos\theta - f(\theta)\sin\theta} = \frac{f(\theta)\cos\theta + f'(\theta)\sin\theta}{-f(\theta)\sin\theta + f'(\theta)\cos\theta}.$$

51. Use Eq. (2) to find the slope of the tangent line to $r = \theta$ at $\theta = \frac{\pi}{2}$ and $\theta = \pi$.

SOLUTION In the given curve we have $r = f(\theta) = \theta$. Using Eq. (2) we obtain the following derivative, which is the slope of the tangent line at (r, θ).

$$\frac{dy}{dx} = \frac{f(\theta)\cos\theta + f'(\theta)\sin\theta}{-f(\theta)\sin\theta + f'(\theta)\cos\theta} = \frac{\theta\cos\theta + 1\cdot\sin\theta}{-\theta\sin\theta + 1\cdot\cos\theta} \qquad (1)$$

The slope, m, of the tangent line at $\theta = \frac{\pi}{2}$ and $\theta = \pi$ is obtained by substituting these values in (1). We get ($\theta = \frac{\pi}{2}$):

$$m = \frac{\frac{\pi}{2}\cos\frac{\pi}{2} + \sin\frac{\pi}{2}}{-\frac{\pi}{2}\sin\frac{\pi}{2} + \cos\frac{\pi}{2}} = \frac{\frac{\pi}{2}\cdot 0 + 1}{-\frac{\pi}{2}\cdot 1 + 0} = \frac{1}{-\frac{\pi}{2}} = -\frac{2}{\pi}.$$

($\theta = \pi$):

$$m = \frac{\pi\cos\pi + \sin\pi}{-\pi\sin\pi + \cos\pi} = \frac{-\pi}{-1} = \pi.$$

53. Find the polar coordinates of the points on the lemniscate $r^2 = \cos 2t$ in Figure 23 where the tangent line is horizontal.

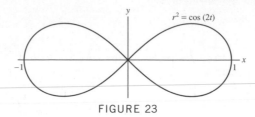

$r^2 = \cos(2t)$

FIGURE 23

SOLUTION   This curve is defined for $-\frac{\pi}{2} \le 2t \le \frac{\pi}{2}$ (where $\cos 2t \ge 0$), so for $-\frac{\pi}{4} \le t \le \frac{\pi}{4}$. For each θ in that range, there are two values of r satisfying the equation ($\pm\sqrt{\cos 2t}$). By symmetry, we need only calculate the coordinates of the points corresponding to the positive square root (i.e. to the right of the y axis). Then the equation becomes $r = \sqrt{\cos 2t}$. Now, by Eq. (2), with $f(t) = \sqrt{\cos(2t)}$ and $f'(t) = -\sin(2t)(\cos(2t))^{-1/2}$, we have

$$\frac{dy}{dx} = \frac{f(t)\cos t + f'(t)\sin t}{-f(t)\sin t + f'(t)\cos t} = \frac{\cos t\sqrt{\cos(2t)} - \sin(2t)\sin t(\cos(2t))^{-1/2}}{-\sin t\sqrt{\cos(2t)} - \sin(2t)\cos t(\cos(2t))^{-1/2}}$$

The tangent line is horizontal when this derivative is zero, which occurs when the numerator of the fraction is zero and the denominator is not. Multiply top and bottom of the fraction by $\sqrt{\cos(2t)}$, and use the identities $\cos 2t = \cos^2 t - \sin^2 t$, $\sin 2t = 2\sin t\cos t$ to get

$$-\frac{\cos t\cos 2t - \sin t\sin 2t}{\sin t\cos 2t + \cos t\sin 2t} = -\frac{\cos t(\cos^2 t - 3\sin^2 t)}{\sin t\cos 2t + \cos t\sin 2t}$$

The numerator is zero when $\cos t = 0$, so when $t = \frac{\pi}{2}$ or $t = \frac{3\pi}{2}$, or when $\tan t = \pm\frac{1}{\sqrt{3}}$, so when $t = \pm\frac{\pi}{6}$ or $t = \pm\frac{5\pi}{6}$. Of these possibilities, only $t = \pm\frac{\pi}{6}$ lie in the range $-\frac{\pi}{4} \le t \le \frac{\pi}{4}$. Note that the denominator is nonzero for $t = \pm\frac{\pi}{6}$, so these are the two values of t for which the tangent line is horizontal. The corresponding values of r are solutions to

$$r^2 = \cos\left(2 \cdot \frac{\pi}{6}\right) = \cos\left(\frac{\pi}{3}\right) = \frac{1}{2}$$

$$r^2 = \cos\left(2 \cdot \frac{-\pi}{6}\right) = \cos\left(-\frac{\pi}{3}\right) = \frac{1}{2}$$

Finally, the four points are $(r, t) =$

$$\left(\frac{1}{\sqrt{2}}, \frac{\pi}{6}\right), \qquad \left(-\frac{1}{\sqrt{2}}, \frac{\pi}{6}\right), \qquad \left(\frac{1}{\sqrt{2}}, -\frac{pi}{6}\right), \qquad \left(-\frac{1}{\sqrt{2}}, -\frac{\pi}{6}\right)$$

If desired, we can change the second and fourth points by adding π to the angle and making r positive, to get

$$\left(\frac{1}{\sqrt{2}}, \frac{\pi}{6}\right), \qquad \left(\frac{1}{\sqrt{2}}, \frac{7\pi}{6}\right), \qquad \left(\frac{1}{\sqrt{2}}, -\frac{pi}{6}\right), \qquad \left(\frac{1}{\sqrt{2}}, \frac{5\pi}{6}\right)$$

55. Use Eq. (2) to show that for $r = \sin\theta + \cos\theta$,

$$\frac{dy}{dx} = \frac{\cos 2\theta + \sin 2\theta}{\cos 2\theta - \sin 2\theta}$$

Then calculate the slopes of the tangent lines at points A, B, C in Figure 19.

SOLUTION   In Exercise 49 we proved that for a polar curve $r = f(\theta)$ the following formula holds:

$$\frac{dy}{dx} = \frac{f(\theta)\cos\theta + f'(\theta)\sin\theta}{-f(\theta)\sin\theta + f'(\theta)\cos\theta} \tag{1}$$

For the given circle we have $r = f(\theta) = \sin\theta + \cos\theta$, hence $f'(\theta) = \cos\theta - \sin\theta$. Substituting in (1) we have

$$\frac{dy}{dx} = \frac{(\sin\theta + \cos\theta)\cos\theta + (\cos\theta - \sin\theta)\sin\theta}{-(\sin\theta + \cos\theta)\sin\theta + (\cos\theta - \sin\theta)\cos\theta} = \frac{\sin\theta\cos\theta + \cos^2\theta + \cos\theta\sin\theta - \sin^2\theta}{-\sin^2\theta - \cos\theta\sin\theta + \cos^2\theta - \sin\theta\cos\theta}$$

$$= \frac{\cos^2\theta - \sin^2\theta + 2\sin\theta\cos\theta}{\cos^2\theta - \sin^2\theta - 2\sin\theta\cos\theta}$$

We use the identities $\cos^2\theta - \sin^2\theta = \cos 2\theta$ and $2\sin\theta\cos\theta = \sin 2\theta$ to obtain

$$\frac{dy}{dx} = \frac{\cos 2\theta + \sin 2\theta}{\cos 2\theta - \sin 2\theta} \tag{2}$$

The derivative $\frac{dy}{dx}$ is the slope of the tangent line at (r, θ). The slopes of the tangent lines at the points with polar coordinates $A = \left(1, \frac{\pi}{2}\right)$ $B = \left(0, \frac{3\pi}{4}\right)$ $C = (1, 0)$ are computed by substituting the values of θ in (2). This gives

$$\frac{dy}{dx}\bigg|_A = \frac{\cos\left(2 \cdot \frac{\pi}{2}\right) + \sin\left(2 \cdot \frac{\pi}{2}\right)}{\cos\left(2 \cdot \frac{\pi}{2}\right) - \sin\left(2 \cdot \frac{\pi}{2}\right)} = \frac{\cos \pi + \sin \pi}{\cos \pi - \sin \pi} = \frac{-1 + 0}{-1 - 0} = 1$$

$$\frac{dy}{dx}\bigg|_B = \frac{\cos\left(2 \cdot \frac{3\pi}{4}\right) + \sin\left(2 \cdot \frac{3\pi}{4}\right)}{\cos\left(2 \cdot \frac{3\pi}{4}\right) - \sin\left(2 \cdot \frac{3\pi}{4}\right)} = \frac{\cos \frac{3\pi}{2} + \sin \frac{3\pi}{2}}{\cos \frac{3\pi}{2} - \sin \frac{3\pi}{2}} = \frac{0 - 1}{0 + 1} = -1$$

$$\frac{dy}{dx}\bigg|_C = \frac{\cos (2 \cdot 0) + \sin (2 \cdot 0)}{\cos (2 \cdot 0) - \sin (2 \cdot 0)} = \frac{\cos 0 + \sin 0}{\cos 0 - \sin 0} = \frac{1 + 0}{1 - 0} = 1$$

Further Insights and Challenges

57. GU Use a graphing utility to convince yourself that the polar equations $r = f_1(\theta) = 2\cos\theta - 1$ and $r = f_2(\theta) = 2\cos\theta + 1$ have the same graph. Then explain why. *Hint:* Show that the points $(f_1(\theta + \pi), \theta + \pi)$ and $(f_2(\theta), \theta)$ coincide.

SOLUTION The graphs of $r = 2\cos\theta - 1$ and $r = 2\cos\theta + 1$ in the xy-plane coincide as shown in the graph obtained using a CAS.

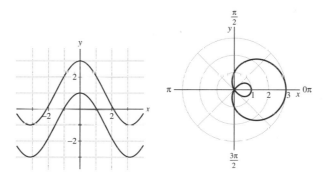

Recall that (r, θ) and $(-r, \theta + \pi)$ represent the same point. Replacing θ by $\theta + \pi$ and r by $(-r)$ in $r = 2\cos\theta - 1$ we obtain

$$-r = 2\cos(\theta + \pi) - 1$$
$$-r = -2\cos\theta - 1$$
$$r = 2\cos\theta + 1$$

Thus, the two equations define the same graph. (One could also convert both equations to rectangular coordinates and note that they come out identical.)

11.4 Area and Arc Length in Polar Coordinates (LT Section 12.4)

Preliminary Questions

1. Polar coordinates are suited to finding the area (choose one):

(a) Under a curve between $x = a$ and $x = b$.

(b) Bounded by a curve and two rays through the origin.

SOLUTION Polar coordinates are best suited to finding the area bounded by a curve and two rays through the origin. The formula for the area in polar coordinates gives the area of this region.

2. Is the formula for area in polar coordinates valid if $f(\theta)$ takes negative values?

SOLUTION The formula for the area

$$\frac{1}{2} \int_\alpha^\beta f(\theta)^2 \, d\theta$$

always gives the actual (positive) area, even if $f(\theta)$ takes on negative values.

3. The horizontal line $y = 1$ has polar equation $r = \csc \theta$. Which area is represented by the integral $\dfrac{1}{2} \displaystyle\int_{\pi/6}^{\pi/2} \csc^2 \theta \, d\theta$ (Figure 12)?

(a) $\square ABCD$ (b) $\triangle ABC$ (c) $\triangle ACD$

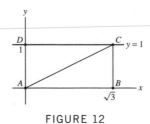

FIGURE 12

SOLUTION   This integral represents an area taken from $\theta = \pi/6$ to $\theta = \pi/2$, which can only be the triangle $\triangle ACD$, as seen in part (c).

Exercises

1. Sketch the area bounded by the circle $r = 5$ and the rays $\theta = \frac{\pi}{2}$ and $\theta = \pi$, and compute its area as an integral in polar coordinates.

SOLUTION   The region bounded by the circle $r = 5$ and the rays $\theta = \frac{\pi}{2}$ and $\theta = \pi$ is the shaded region in the figure. The area of the region is given by the following integral:

$$\frac{1}{2} \int_{\pi/2}^{\pi} r^2 \, d\theta = \frac{1}{2} \int_{\pi/2}^{\pi} 5^2 \, d\theta = \frac{25}{2} \left(\pi - \frac{\pi}{2} \right) = \frac{25\pi}{4}$$

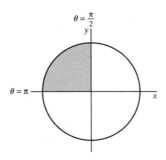

3. Calculate the area of the circle $r = 4 \sin \theta$ as an integral in polar coordinates (see Figure 4). Be careful to choose the correct limits of integration.

SOLUTION   The equation $r = 4 \sin \theta$ defines a circle of radius 2 tangent to the x-axis at the origin as shown in the figure:

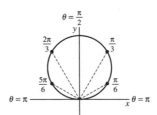

The circle is traced as θ varies from 0 to π. We use the area in polar coordinates and the identity

$$\sin^2 \theta = \frac{1}{2} (1 - \cos 2\theta)$$

to obtain the following area:

$$A = \frac{1}{2} \int_0^{\pi} r^2 \, d\theta = \frac{1}{2} \int_0^{\pi} (4 \sin \theta)^2 \, d\theta = 8 \int_0^{\pi} \sin^2 \theta \, d\theta = 4 \int_0^{\pi} (1 - \cos 2\theta) \, d\theta = 4 \left[\theta - \frac{\sin 2\theta}{2} \right]_0^{\pi}$$

$$= 4 \left(\left(\pi - \frac{\sin 2\pi}{2} \right) - 0 \right) = 4\pi.$$

5. Find the area of the shaded region in Figure 14. Note that θ varies from 0 to $\frac{\pi}{2}$.

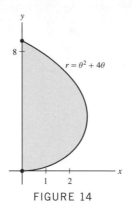

FIGURE 14

SOLUTION Since θ varies from 0 to $\frac{\pi}{2}$, the area is

$$\frac{1}{2}\int_0^{\pi/2} r^2\,d\theta = \frac{1}{2}\int_0^{\pi/2} (\theta^2 + 4\theta)^2\,d\theta = \frac{1}{2}\int_0^{\pi/2} \theta^4 + 8\theta^3 + 16\theta^2\,d\theta$$

$$= \frac{1}{2}\left(\frac{1}{5}\theta^5 + 2\theta^4 + \frac{16}{3}\theta^3\right)\Big|_0^{\pi/2} = \frac{\pi^5}{320} + \frac{\pi^4}{16} + \frac{\pi^2}{3}$$

7. Find the total area enclosed by the cardioid in Figure 16.

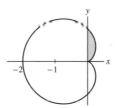

FIGURE 16 The cardioid $r = 1 - \cos\theta$.

SOLUTION We graph $r = 1 - \cos\theta$ in r and θ (cartesian, not polar, this time):

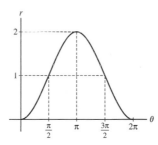

We see that as θ varies from 0 to π, the radius r increases from 0 to 2, so we get the upper half of the cardioid (the lower half is obtained as θ varies from π to 2π and consequently r decreases from 2 to 0). Since the cardioid is symmetric with respect to the x-axis we may compute the upper area and double the result. Using

$$\cos^2\theta = \frac{\cos 2\theta + 1}{2}$$

we get

$$A = 2 \cdot \frac{1}{2}\int_0^\pi r^2\,d\theta = \int_0^\pi (1 - \cos\theta)^2\,d\theta = \int_0^\pi \left(1 - 2\cos\theta + \cos^2\theta\right)d\theta$$

$$= \int_0^\pi \left(1 - 2\cos\theta + \frac{\cos 2\theta + 1}{2}\right)d\theta = \int_0^\pi \left(\frac{3}{2} - 2\cos\theta + \frac{1}{2}\cos 2\theta\right)d\theta$$

$$= \frac{3}{2}\theta - 2\sin\theta + \frac{1}{4}\sin 2\theta\Big|_0^\pi = \frac{3\pi}{2}$$

The total area enclosed by the cardioid is $A = \frac{3\pi}{2}$.

9. Find the area of one leaf of the "four-petaled rose" $r = \sin 2\theta$ (Figure 17). Then prove that the total area of the rose is equal to one-half the area of the circumscribed circle.

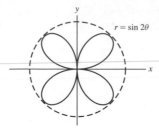

FIGURE 17 Four-petaled rose $r = \sin 2\theta$.

SOLUTION We consider the graph of $r = \sin 2\theta$ in cartesian and in polar coordinates:

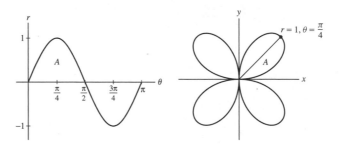

We see that as θ varies from 0 to $\frac{\pi}{4}$ the radius r is increasing from 0 to 1, and when θ varies from $\frac{\pi}{4}$ to $\frac{\pi}{2}$, r is decreasing back to zero. Hence, the leaf in the first quadrant is traced as θ varies from 0 to $\frac{\pi}{2}$. The area of the leaf (the four leaves have equal areas) is thus

$$A = \frac{1}{2} \int_0^{\pi/2} r^2 \, d\theta = \frac{1}{2} \int_0^{\pi/2} \sin^2 2\theta \, d\theta.$$

Using the identity

$$\sin^2 2\theta = \frac{1 - \cos 4\theta}{2}$$

we get

$$A = \frac{1}{2} \int_0^{\pi/2} \left(\frac{1}{2} - \frac{\cos 4\theta}{2} \right) d\theta = \frac{1}{2} \left(\frac{\theta}{2} - \frac{\sin 4\theta}{8} \right) \bigg|_0^{\pi/2} = \frac{1}{2} \left(\left(\frac{\pi}{4} - \frac{\sin 2\pi}{8} \right) - 0 \right) = \frac{\pi}{8}$$

The area of one leaf is $A = \frac{\pi}{8} \approx 0.39$. It follows that the area of the entire rose is $\frac{\pi}{2}$. Since the "radius" of the rose (the point where $\theta = \frac{\pi}{4}$) is 1, and the circumscribed circle is tangent there, the circumscribed circle has radius 1 and thus area π. Hence the area of the rose is half that of the circumscribed circle.

11. Sketch the spiral $r = \theta$ for $0 \le \theta \le 2\pi$ and find the area bounded by the curve and the first quadrant.

SOLUTION The spiral $r = \theta$ for $0 \le \theta \le 2\pi$ is shown in the following figure in the xy-plane:

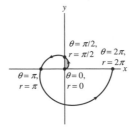

The spiral $r = \theta$

We must compute the area of the shaded region. This region is traced as θ varies from 0 to $\frac{\pi}{2}$. Using the formula for the area in polar coordinates we get

$$A = \frac{1}{2} \int_0^{\pi/2} r^2 \, d\theta = \frac{1}{2} \int_0^{\pi/2} \theta^2 \, d\theta = \frac{1}{2} \frac{\theta^3}{3} \bigg|_0^{\pi/2} = \frac{1}{6} \left(\frac{\pi}{2} \right)^3 = \frac{\pi^3}{48}$$

13. Find the area of region A in Figure 19.

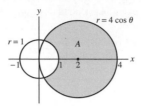

FIGURE 19

SOLUTION We first find the values of θ at the points of intersection of the two circles, by solving the following equation for $-\frac{\pi}{2} \leq x \leq \frac{\pi}{2}$:

$$4\cos\theta = 1 \Rightarrow \cos\theta = \frac{1}{4} \Rightarrow \theta_1 = \cos^{-1}\left(\frac{1}{4}\right)$$

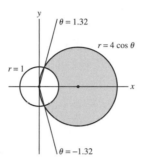

We now compute the area using the formula for the area between two curves:

$$A = \frac{1}{2}\int_{-\theta_1}^{\theta_1}\left((4\cos\theta)^2 - 1^2\right)d\theta = \frac{1}{2}\int_{-\theta_1}^{\theta_1}\left(16\cos^2\theta - 1\right)d\theta$$

Using the identity $\cos^2\theta = \frac{\cos 2\theta + 1}{2}$ we get

$$A = \frac{1}{2}\int_{-\theta_1}^{\theta_1}\left(\frac{16(\cos 2\theta + 1)}{2} - 1\right)d\theta = \frac{1}{2}\int_{-\theta_1}^{\theta_1}(8\cos 2\theta + 7)\,d\theta = \frac{1}{2}(4\sin 2\theta + 7\theta)\Big|_{-\theta_1}^{\theta_1}$$

$$= 4\sin 2\theta_1 + 7\theta_1 = 8\sin\theta_1\cos\theta_1 + 7\theta_1 = 8\sqrt{1 - \cos^2\theta_1}\cos\theta_1 + 7\theta_1$$

Using the fact that $\cos\theta_1 = \frac{1}{4}$ we get

$$A = \frac{\sqrt{15}}{2} + 7\cos^{-1}\left(\frac{1}{4}\right) \approx 11.163$$

15. Find the area of the inner loop of the limaçon with polar equation $r = 2\cos\theta - 1$ (Figure 21).

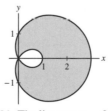

FIGURE 21 The limaçon $r = 2\cos\theta - 1$.

SOLUTION We consider the graph of $r = 2\cos\theta - 1$ in cartesian and in polar, for $-\frac{\pi}{2} \leq x \leq \frac{\pi}{2}$:

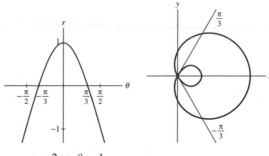

$r = 2\cos\theta - 1$

As θ varies from $-\frac{\pi}{3}$ to 0, r increases from 0 to 1. As θ varies from 0 to $\frac{\pi}{3}$, r decreases from 1 back to 0. Hence, the inner loop of the limaçon is traced as θ varies from $-\frac{\pi}{3}$ to $\frac{\pi}{3}$. The area of the shaded region is thus

$$A = \frac{1}{2} \int_{-\pi/3}^{\pi/3} r^2 \, d\theta = \frac{1}{2} \int_{-\pi/3}^{\pi/3} (2\cos\theta - 1)^2 \, d\theta = \frac{1}{2} \int_{-\pi/3}^{\pi/3} \left(4\cos^2\theta - 4\cos\theta + 1 \right) d\theta$$

$$= \frac{1}{2} \int_{-\pi/3}^{\pi/3} (2(\cos 2\theta + 1) - 4\cos\theta + 1) \, d\theta = \frac{1}{2} \int_{-\pi/3}^{\pi/3} (2\cos 2\theta - 4\cos\theta + 3) \, d\theta$$

$$= \frac{1}{2} (\sin 2\theta - 4\sin\theta + 3\theta) \Big|_{-\pi/3}^{\pi/3} = \frac{1}{2} \left(\left(\sin\frac{2\pi}{3} - 4\sin\frac{\pi}{3} + \pi \right) - \left(\sin\left(-\frac{2\pi}{3}\right) - 4\sin\left(-\frac{\pi}{3}\right) - \pi \right) \right)$$

$$= \frac{\sqrt{3}}{2} - \frac{4\sqrt{3}}{2} + \pi = \pi - \frac{3\sqrt{3}}{2} \approx 0.54$$

17. Find the area of the part of the circle $r = \sin\theta + \cos\theta$ in the fourth quadrant (see Exercise 26 in Section 11.3).

SOLUTION The value of θ corresponding to the point B is the solution of $r = \sin\theta + \cos\theta = 0$ for $-\pi \le \theta \le \pi$.

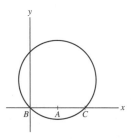

That is,

$$\sin\theta + \cos\theta = 0 \Rightarrow \sin\theta = -\cos\theta \Rightarrow \tan\theta = -1 \Rightarrow \theta = -\frac{\pi}{4}$$

At the point C, we have $\theta = 0$. The part of the circle in the fourth quadrant is traced if θ varies between $-\frac{\pi}{4}$ and 0. This leads to the following area:

$$A = \frac{1}{2} \int_{-\pi/4}^{0} r^2 \, d\theta = \frac{1}{2} \int_{-\pi/4}^{0} (\sin\theta + \cos\theta)^2 \, d\theta = \frac{1}{2} \int_{-\pi/4}^{0} \left(\sin^2\theta + 2\sin\theta\cos\theta + \cos^2\theta \right) d\theta$$

Using the identities $\sin^2\theta + \cos^2\theta = 1$ and $2\sin\theta\cos\theta = \sin 2\theta$ we get:

$$A = \frac{1}{2} \int_{-\pi/4}^{0} (1 + \sin 2\theta) \, d\theta = \frac{1}{2} \left(\theta - \frac{\cos 2\theta}{2} \right) \Big|_{-\pi/4}^{0}$$

$$= \frac{1}{2} \left(\left(0 - \frac{1}{2}\right) - \left(-\frac{\pi}{4} - \frac{\cos\left(\frac{-\pi}{2}\right)}{2}\right) \right) = \frac{1}{2} \left(\frac{\pi}{4} - \frac{1}{2} \right) = \frac{\pi}{8} - \frac{1}{4} \approx 0.14.$$

19. Find the area between the two curves in Figure 22(A).

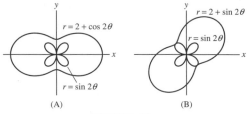

FIGURE 22

SOLUTION We compute the area A between the two curves as the difference between the area A_1 of the region enclosed in the outer curve $r = 2 + \cos 2\theta$ and the area A_2 of the region enclosed in the inner curve $r = \sin 2\theta$. That is,

$$A = A_1 - A_2.$$

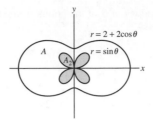

In Exercise 9 we showed that $A_2 = \frac{\pi}{2}$, hence,

$$A = A_1 - \frac{\pi}{2} \tag{1}$$

We compute the area A_1.

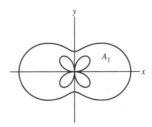

Using symmetry, the area is four times the area enclosed in the first quadrant. That is,

$$A_1 = 4 \cdot \frac{1}{2} \int_0^{\pi/2} r^2 \, d\theta = 2 \int_0^{\pi/2} (2 + \cos 2\theta)^2 \, d\theta = 2 \int_0^{\pi/2} \left(4 + 4\cos 2\theta + \cos^2 2\theta \right) d\theta$$

Using the identity $\cos^2 2\theta = \frac{1}{2}\cos 4\theta + \frac{1}{2}$ we get

$$A_1 = 2 \int_0^{\pi/2} \left(4 + 4\cos 2\theta + \frac{1}{2}\cos 4\theta + \frac{1}{2} \right) d\theta = 2 \int_0^{\pi/2} \left(\frac{9}{2} + \frac{1}{2}\cos 4\theta + 4\cos 2\theta \right) d\theta$$

$$= 2 \left(\frac{9\theta}{2} + \frac{\sin 4\theta}{8} + 2\sin 2\theta \right) \Big|_0^{\pi/2} = 2 \left(\left(\frac{9\pi}{4} + \frac{\sin 2\pi}{8} + 2\sin \pi \right) - 0 \right) = \frac{9\pi}{2} \tag{2}$$

Combining (1) and (2) we obtain

$$A = \frac{9\pi}{2} - \frac{\pi}{2} = 4\pi$$

21. Find the area inside both curves in Figure 23.

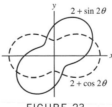

FIGURE 23

SOLUTION The area we need to find is the area of the shaded region in the figure.

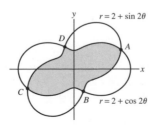

We first find the values of θ at the points of intersection A, B, C, and D of the two curves, by solving the following equation for $-\pi \leq \theta \leq \pi$:

$$2 + \cos 2\theta = 2 + \sin 2\theta$$

$$\cos 2\theta = \sin 2\theta$$

$$\tan 2\theta = 1 \Rightarrow 2\theta = \frac{\pi}{4} + \pi k \Rightarrow \theta = \frac{\pi}{8} + \frac{\pi k}{2}$$

The solutions for $-\pi \leq \theta \leq \pi$ are

$$A: \quad \theta = \frac{\pi}{8}.$$

$$B: \quad \theta = -\frac{3\pi}{8}.$$

$$C: \quad \theta = -\frac{7\pi}{8}.$$

$$D: \quad \theta = \frac{5\pi}{8}.$$

Using symmetry, we compute the shaded area in the figure below and multiply it by 4:

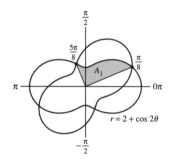

$$A = 4 \cdot A_1 = 4 \cdot \frac{1}{2} \cdot \int_{\pi/8}^{5\pi/8} (2 + \cos 2\theta)^2 \, d\theta = 2 \int_{\pi/8}^{5\pi/8} \left(4 + 4\cos 2\theta + \cos^2 2\theta \right) d\theta$$

$$= 2 \int_{\pi/8}^{5\pi/8} \left(4 + 4\cos 2\theta + \frac{1 + \cos 4\theta}{2} \right) d\theta = \int_{\pi/8}^{5\pi/8} (9 + 8\cos 2\theta + \cos 4\theta) \, d\theta$$

$$= 9\theta + 4\sin 2\theta + \frac{\sin 4\theta}{4} \bigg|_{\pi/8}^{5\pi/8} = 9 \left(\frac{5\pi}{8} - \frac{\pi}{8} \right) + 4 \left(\sin \frac{5\pi}{4} - \sin \frac{\pi}{4} \right) + \frac{1}{4} \left(\sin \frac{5\pi}{2} - \sin \frac{\pi}{2} \right) = \frac{9\pi}{2} - 4\sqrt{2}$$

23. Calculate the total length of the circle $r = 4\sin\theta$ as an integral in polar coordinates.

SOLUTION   We use the formula for the arc length:

$$S = \int_\alpha^\beta \sqrt{f(\theta)^2 + f'(\theta)^2} \, d\theta \tag{1}$$

In this case, $f(\theta) = 4\sin\theta$ and $f'(\theta) = 4\cos\theta$, hence

$$\sqrt{f(\theta)^2 + f'(\theta)^2} = \sqrt{(4\sin\theta)^2 + (4\cos\theta)^2} = \sqrt{16} = 4$$

The circle is traced as θ is varied from 0 to π. Substituting $\alpha = 0$, $\beta = \pi$ in (1) yields $S = \int_0^\pi 4 \, d\theta = 4\pi$.

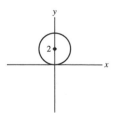

The circle $r = 4\sin\theta$

In Exercises 25–30, compute the length of the polar curve.

25. The length of $r = \theta^2$ for $0 \le \theta \le \pi$

SOLUTION We use the formula for the arc length. In this case $f(\theta) = \theta^2$, $f'(\theta) = 2\theta$, so we obtain

$$S = \int_0^\pi \sqrt{(\theta^2)^2 + (2\theta)^2}\, d\theta = \int_0^\pi \sqrt{\theta^4 + 4\theta^2}\, d\theta = \int_0^\pi \theta\sqrt{\theta^2 + 4}\, d\theta$$

We compute the integral using the substitution $u = \theta^2 + 4$, $du = 2\theta\, d\theta$. This gives

$$S = \frac{1}{2}\int_4^{\pi^2+4} \sqrt{u}\, du = \frac{1}{2}\cdot\frac{2}{3}u^{3/2}\Big|_4^{\pi^2+4} = \frac{1}{3}\left(\left(\pi^2+4\right)^{3/2} - 4^{3/2}\right) = \frac{1}{3}\left(\left(\pi^2+4\right)^{3/2} - 8\right) \approx 14.55$$

27. The equiangular spiral $r = e^\theta$ for $0 \le \theta \le 2\pi$

SOLUTION Since $f(\theta) = e^\theta$, by the formula for the arc length we have:

$$L = \int_0^{2\pi} \sqrt{f'(\theta)^2 + f(\theta)}\, d\theta + \int_0^{2\pi} \sqrt{(e^\theta)^2 + (e^\theta)^2}\, d\theta = \int_0^{2\pi} \sqrt{2e^{2\theta}}\, d\theta$$

$$= \sqrt{2}\int_0^{2\pi} e^\theta\, d\theta = \sqrt{2}e^\theta\Big|_0^{2\pi} = \sqrt{2}\left(e^{2\pi} - e^0\right) = \sqrt{2}\left(e^{2\pi} - 1\right) \approx 755.9$$

29. The cardioid $r = 1 - \cos\theta$ in Figure 16

SOLUTION In the equation of the cardioid, $f(\theta) = 1 - \cos\theta$. Using the formula for arc length in polar coordinates we have:

$$L = \int_\alpha^\beta \sqrt{f(\theta)^2 + f'(\theta)^2}\, d\theta \tag{1}$$

We compute the integrand:

$$\sqrt{f(\theta)^2 + f'(\theta)^2} = \sqrt{(1-\cos\theta)^2 + (\sin\theta)^2} = \sqrt{1 - 2\cos\theta + \cos^2\theta + \sin^2\theta} = \sqrt{2(1-\cos\theta)}$$

We identify the interval of θ. Since $-1 \le \cos\theta \le 1$, every $0 \le \theta \le 2\pi$ corresponds to a nonnegative value of r. Hence, θ varies from 0 to 2π. By (1) we obtain

$$L = \int_0^{2\pi} \sqrt{2(1-\cos\theta)}\, d\theta$$

Now, $1 - \cos\theta = 2\sin^2(\theta/2)$, and on the interval $0 \le \theta \le \pi$, $\sin(\theta/2)$ is nonnegative, so that $\sqrt{2(1-\cos\theta)} = \sqrt{4\sin^2(\theta/2)} = 2\sin(\theta/2)$ there. The graph is symmetric, so it suffices to compute the integral for $0 \le \theta \le \pi$, and we have

$$L = 2\int_0^\pi \sqrt{2(1-\cos\theta)}\, d\theta = 2\int_0^\pi 2\sin(\theta/2)\, d\theta = 8\sin\frac{\theta}{2}\Big|_0^\pi = 8$$

In Exercises 31 and 32, express the length of the curve as an integral but do not evaluate it.

31. $r = (2 - \cos\theta)^{-1}$, $\quad 0 \le \theta \le 2\pi$

SOLUTION We have $f(\theta) = (2 - \cos\theta)^{-1}$, $f'(\theta) = -(2 - \cos\theta)^{-2}\sin\theta$, hence,

$$\sqrt{f^2(\theta) + f'(\theta)^2} = \sqrt{(2-\cos\theta)^{-2} + (2-\cos\theta)^{-4}\sin^2\theta} = \sqrt{(2-\cos\theta)^{-4}\left((2-\cos\theta)^2 + \sin^2\theta\right)}$$

$$= (2-\cos\theta)^{-2}\sqrt{4 - 4\cos\theta + \cos^2\theta + \sin^2\theta} = (2-\cos\theta)^{-2}\sqrt{5 - 4\cos\theta}$$

Using the integral for the arc length we get

$$L = \int_0^{2\pi} \sqrt{5 - 4\cos\theta}\,(2 - \cos\theta)^{-2}\, d\theta.$$

In Exercises 33–36, use a computer algebra system to calculate the total length to two decimal places.

33. $\boxed{CAS}$ The three-petal rose $r = \cos 3\theta$ in Figure 20

SOLUTION   We have $f(\theta) = \cos 3\theta$, $f'(\theta) = -3\sin 3\theta$, so that

$$\sqrt{f(\theta)^2 + f'(\theta)^2} = \sqrt{\cos^2 3\theta + 9\sin^2 3\theta} = \sqrt{\cos^2 3\theta + \sin^2 3\theta + 8\sin^2 3\theta} = \sqrt{1 + 8\sin^2 3\theta}$$

Note that the curve is traversed completely for $0 \le \theta \le \pi$. Using the arc length formula and evaluating with Maple gives

$$L = \int_0^\pi \sqrt{f(\theta)^2 + f'(\theta)^2}\, d\theta = \int_0^\pi \sqrt{1 + 8\sin^2 3\theta}\, d\theta \approx 6.682446608$$

35. $\boxed{CAS}$ The curve $r = \theta \sin \theta$ in Figure 24 for $0 \le \theta \le 4\pi$

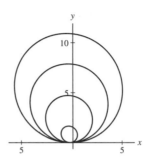

FIGURE 24 $r = \theta \sin \theta$ for $0 \le \theta \le 4\pi$.

SOLUTION   We have $f(\theta) = \theta \sin \theta$, $f'(\theta) = \sin \theta + \theta \cos \theta$, so that

$$\sqrt{f(\theta)^2 + f'(\theta)^2} = \sqrt{\theta^2 \sin^2 \theta + (\sin \theta + \theta \cos \theta)^2} = \sqrt{\theta^2 \sin^2 \theta + \sin^2 \theta + 2\theta \sin \theta \cos \theta + \theta^2 \cos^2 \theta}$$

$$= \sqrt{\theta^2 + \sin^2 \theta + \theta \sin 2\theta}$$

using the identities $\sin^2 \theta + \cos^2 \theta = 1$ and $2\sin \theta \cos \theta = \sin 2\theta$. Thus by the arc length formula and evaluating with Maple, we have

$$L = \int_0^{4\pi} \sqrt{f(\theta)^2 + f'(\theta)^2}\, d\theta = \int_0^{4\pi} \sqrt{\theta^2 + \sin^2 \theta + \theta \sin 2\theta}\, d\theta \approx 79.56423976$$

Further Insights and Challenges

37.   Suppose that the polar coordinates of a moving particle at time t are $(r(t), \theta(t))$. Prove that the particle's speed is equal to $\sqrt{(dr/dt)^2 + r^2(d\theta/dt)^2}$.

SOLUTION   The speed of the particle in rectangular coordinates is:

$$\frac{ds}{dt} = \sqrt{x'(t)^2 + y'(t)^2} \tag{1}$$

We need to express the speed in polar coordinates. The x and y coordinates of the moving particles as functions of t are

$$x(t) = r(t) \cos \theta(t), \quad y(t) = r(t) \sin \theta(t)$$

We differentiate $x(t)$ and $y(t)$, using the Product Rule for differentiation. We obtain (omitting the independent variable t)

$$x' = r' \cos \theta - r (\sin \theta)\, \theta'$$
$$y' = r' \sin \theta - r (\cos \theta)\, \theta'$$

Hence,

$$x'^2 + y'^2 = \left(r' \cos \theta - r\theta' \sin \theta \right)^2 + \left(r' \sin \theta + r\theta' \cos \theta \right)^2$$

$$= r'^2 \cos^2 \theta - 2r'r\theta' \cos \theta \sin \theta + r^2\theta'^2 \sin^2 \theta + r'^2 \sin^2 \theta + 2r'r\theta' \sin^2 \theta \cos \theta + r^2\theta'^2 \cos^2 \theta$$

$$= r'^2 \left(\cos^2 \theta + \sin^2 \theta \right) + r^2\theta'^2 \left(\sin^2 \theta + \cos^2 \theta \right) = r'^2 + r^2\theta'^2 \tag{2}$$

Substituting (2) into (1) we get

$$\frac{ds}{dt} = \sqrt{r'^2 + r^2\theta'^2} = \sqrt{\left(\frac{dr}{dt}\right)^2 + r^2\left(\frac{d\theta}{dt}\right)^2}$$

11.5 Conic Sections (LT Section 12.5)

Preliminary Questions

1. Which of the following equations defines an ellipse? Which does not define a conic section?

(a) $4x^2 - 9y^2 = 12$

(b) $-4x + 9y^2 = 0$

(c) $4y^2 + 9x^2 = 12$

(d) $4x^3 + 9y^3 = 12$

SOLUTION

(a) This is the equation of the hyperbola $\left(\frac{x}{\sqrt{3}}\right)^2 - \left(\frac{y}{\frac{2}{\sqrt{3}}}\right)^2 = 1$, which is a conic section.

(b) The equation $-4x + 9y^2 = 0$ can be rewritten as $x = \frac{9}{4}y^2$, which defines a parabola. This is a conic section.

(c) The equation $4y^2 + 9x^2 = 12$ can be rewritten in the form $\left(\frac{y}{\sqrt{3}}\right)^2 + \left(\frac{x}{\frac{2}{\sqrt{3}}}\right)^2 - 1$, hence it is the equation of an ellipse, which is a conic section

(d) This is not the equation of a conic section, since it is not an equation of degree two in x and y.

2. For which conic sections do the vertices lie between the foci?

SOLUTION If the vertices lie between the foci, the conic section is a hyperbola.

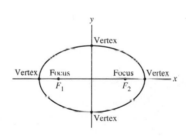
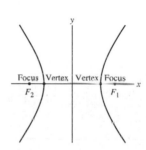

ellipse: foci between vertices   hyperbola: vertices between foci

3. What are the foci of

$$\left(\frac{x}{a}\right)^2 + \left(\frac{y}{b}\right)^2 = 1 \quad \text{if } a < b?$$

SOLUTION If $a < b$ the foci of the ellipse $\left(\frac{x}{a}\right)^2 + \left(\frac{y}{b}\right)^2 = 1$ are at the points $(0, c)$ and $(0, -c)$ on the y-axis, where $c = \sqrt{b^2 - a^2}$.

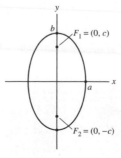

$$\left(\tfrac{x}{a}\right)^2 + \left(\tfrac{y}{b}\right)^2 = 1; a < b$$

4. What is the geometric interpretation of b/a in the equation of a hyperbola in standard position?

SOLUTION   The vertices, i.e., the points where the focal axis intersects the hyperbola, are at the points $(a, 0)$ and $(-a, 0)$. The values $\pm \frac{b}{a}$ are the slopes of the two asymptotes of the hyperbola.

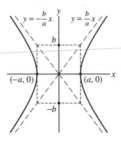

Hyperbola in standard position

Exercises

In Exercises 1–6, find the vertices and foci of the conic section.

1. $\left(\frac{x}{9}\right)^2 + \left(\frac{y}{4}\right)^2 = 1$

SOLUTION   This is an ellipse in standard position with $a = 9$ and $b = 4$. Hence, $c = \sqrt{9^2 - 4^2} = \sqrt{65} \approx 8.06$. The foci are at $F_1 = (-8.06, 0)$ and $F_2 = (8.06, 0)$, and the vertices are $(9, 0)$, $(-9, 0)$, $(0, 4)$, $(0, -4)$.

3. $\left(\frac{x}{4}\right)^2 - \left(\frac{y}{9}\right)^2 = 1$

SOLUTION   This is a hyperbola in standard position with $a = 4$ and $b = 9$. Hence, $c = \sqrt{a^2 + b^2} = \sqrt{97} \approx 9.85$. The foci are at $(\pm\sqrt{97}, 0)$ and the vertices are $(\pm 2, 0)$.

5. $\left(\frac{x-3}{7}\right)^2 - \left(\frac{y+1}{4}\right)^2 = 1$

SOLUTION   We first consider the hyperbola $\left(\frac{x}{7}\right)^2 - \left(\frac{y}{4}\right)^2 = 1$. For this hyperbola, $a = 7$, $b = 4$ and $c = \sqrt{7^2 + 4^2} \approx 8.06$. Hence, the foci are at $(8.06, 0)$ and $(-8.06, 0)$ and the vertices are at $(7, 0)$ and $(-7, 0)$. Since the given hyperbola is obtained by translating the center of the hyperbola $\left(\frac{x}{7}\right)^2 - \left(\frac{y}{4}\right)^2 = 1$ to the point $(3, -1)$, the foci are at $F_1 = (8.06 + 3, 0 - 1) = (11.06, -1)$ and $F_2 = (-8.06 + 3, 0 - 1) = (-5.06, -1)$ and the vertices are $A = (7 + 3, 0 - 1) = (10, -1)$ and $A' = (-7 + 3, 0 - 1) = (-4, -1)$.

In Exercises 7–10, find the equation of the ellipse obtained by translating (as indicated) the ellipse

$$\left(\frac{x-8}{6}\right)^2 + \left(\frac{y+4}{3}\right)^2 = 1$$

7. Translated with center at the origin

SOLUTION   Recall that the equation

$$\frac{(x-h)^2}{a^2} + \frac{(y-k)^2}{b^2} = 1$$

describes an ellipse with center (h, k). Thus, for our ellipse to be located at the origin, it must have equation

$$\frac{x^2}{6^2} + \frac{y^2}{3^2} = 1$$

9. Translated to the right six units

SOLUTION   Recall that the equation

$$\frac{(x-h)^2}{a^2} + \frac{(y-k)^2}{b^2} = 1$$

describes an ellipse with center (h, k). The original ellipse has center at $(8, -4)$, so we want an ellipse with center $(14, -4)$. Thus its equation is

$$\frac{(x-14)^2}{6^2} + \frac{(y+4)^2}{3^2} = 1$$

In Exercises 11–14, find the equation of the given ellipse.

11. Vertices ($\pm 5, 0$) and ($0, \pm 7$)

SOLUTION   Since both sets of vertices are symmetric around the origin, the center of the ellipse is at $(0, 0)$. We have $a = 5$ and $b = 7$, so the equation of the ellipse is

$$\left(\frac{x}{5}\right)^2 + \left(\frac{y}{7}\right)^2 = 1$$

13. Foci ($0, \pm 10$) and eccentricity $e = \frac{3}{5}$

SOLUTION   Since the foci are on the y axis, this ellipse has a vertical major axis with center $(0, 0)$, so its equation is

$$\left(\frac{x}{b}\right)^2 + \left(\frac{y}{a}\right)^2 = 1$$

We have $a = \frac{c}{e} = \frac{10}{3/5} = \frac{50}{3}$ and

$$b = \sqrt{a^2 - c^2} = \sqrt{\frac{2500}{9} - 100} = \frac{1}{3}\sqrt{2500 - 900} = \frac{40}{3}$$

Thus the equation of the ellipse is

$$\left(\frac{x}{40/3}\right)^2 + \left(\frac{y}{50/3}\right)^2 = 1$$

In Exercises 15–20, find the equation of the given hyperbola.

15. Vertices ($\pm 3, 0$) and foci ($\pm 5, 0$)

SOLUTION   The equation is $\left(\frac{x}{a}\right)^2 - \left(\frac{y}{b}\right)^2 = 1$. The vertices are $(\pm a, 0)$ with $a = 3$ and the foci $(\pm c, 0)$ with $c = 5$. We use the relation $c = \sqrt{a^2 + b^2}$ to find b:

$$b = \sqrt{c^2 - a^2} = \sqrt{5^2 - 3^2} = \sqrt{16} = 4$$

Therefore, the equation of the hyperbola is

$$\left(\frac{x}{3}\right)^2 - \left(\frac{y}{4}\right)^2 = 1.$$

17. Foci ($\pm 4, 0$) and eccentricity $e = 2$

SOLUTION   We have $c = 4$ and $e = 2$; from $c = ae$ we get $a = 2$, and then

$$b = \sqrt{c^2 - a^2} = \sqrt{4^2 - 2^2} = 2\sqrt{3}$$

The hyperbola has center at $(0, 0)$ and horizontal axis, so its equation is

$$\left(\frac{x}{2}\right)^2 - \left(\frac{y}{2\sqrt{3}}\right)^2 = 1$$

19. Vertices ($-3, 0$), ($7, 0$) and eccentricity $e = 3$

SOLUTION   The center is at $\frac{-3+7}{2} = 2$ with a horizontal focal axis, so the equation is

$$\left(\frac{x-2}{a}\right)^2 - \left(\frac{y}{b}\right)^2 = 1.$$

Then $a = 7 - 2 = 5$, and $c = ae = 5 \cdot 3 = 15$. Finally,

$$b = \sqrt{c^2 - a^2} = \sqrt{15^2 - 5^2} = 10\sqrt{2}$$

so that the equation of the hyperbola is

$$\left(\frac{x-2}{5}\right)^2 - \left(\frac{y}{10\sqrt{2}}\right)^2 = 1$$

In Exercises 21–28, find the equation of the parabola with the given properties.

21. Vertex $(0, 0)$, focus $\left(\frac{1}{12}, 0\right)$

SOLUTION Since the focus is on the x-axis rather than the y-axis, and the vertex is $(0, 0)$, the equation is $x = \frac{1}{4c} y^2$. The focus is $(0, c)$ with $c = \frac{1}{12}$, so the equation is

$$x = \frac{1}{4 \cdot \frac{1}{12}} y^2 = 3y^2$$

23. Vertex $(0, 0)$, directrix $y = -5$

SOLUTION The equation is $y = \frac{1}{4c} x^2$. The directrix is $y = -c$ with $c = 5$, hence $y = \frac{1}{20} x^2$.

25. Focus $(0, 4)$, directrix $y = -4$

SOLUTION The focus is $(0, c)$ with $c = 4$ and the directrix is $y = -c$ with $c = 4$, hence the equation of the parabola is

$$y = \frac{1}{4c} x^2 = \frac{x^2}{16}.$$

27. Focus $(2, 0)$, directrix $x = -2$

SOLUTION The focus is on the x-axis rather than on the y-axis and the directrix is a vertical line rather than horizontal as in the parabola in standard position. Therefore, the equation of the parabola is obtained by interchanging x and y in $y = \frac{1}{4c} x^2$. Also, by the given information $c = 2$. Hence, $x = \frac{1}{4c} y^2 = \frac{1}{4 \cdot 2} y^2$ or $x = \frac{y^2}{8}$.

In Exercises 29–38, find the vertices, foci, center (if an ellipse or a hyperbola), and asymptotes (if a hyperbola).

29. $x^2 + 4y^2 = 16$

SOLUTION We first divide the equation by 16 to convert it to the equation in standard form:

$$\frac{x^2}{16} + \frac{4y^2}{16} = 1 \Rightarrow \frac{x^2}{16} + \frac{y^2}{4} = 1 \Rightarrow \left(\frac{x}{4}\right)^2 + \left(\frac{y}{2}\right)^2 = 1$$

For this ellipse, $a = 4$ and $b = 2$ hence $c = \sqrt{4^2 - 2^2} = \sqrt{12} \approx 3.5$. Since $a > b$ we have:

- The vertices are at $(\pm 4, 0)$, $(0, \pm 2)$.
- The foci are $F_1 = (-3.5, 0)$ and $F_2 = (3.5, 0)$.
- The focal axis is the x-axis and the conjugate axis is the y-axis.
- The ellipse is centered at the origin.

31. $\left(\frac{x - 3}{4}\right)^2 - \left(\frac{y + 5}{7}\right)^2 = 1$

SOLUTION For this hyperbola $a = 4$ and $b = 7$ so $c = \sqrt{4^2 + 7^2} = \sqrt{65} \approx 8.06$. For the standard hyperbola $\left(\frac{x}{4}\right)^2 - \left(\frac{y}{7}\right)^2 = 1$, we have

- The vertices are $A = (4, 0)$ and $A' = (-4, 0)$.
- The foci are $F = (\sqrt{65}, 0)$ and $F' = (-\sqrt{65}, 0)$.
- The focal axis is the x-axis $y = 0$, and the conjugate axis is the y-axis $x = 0$.
- The center is at the midpoint of $\overline{FF'}$; that is, at the origin.
- The asymptotes $y = \pm \frac{b}{a} x$ are $y = \pm \frac{7}{4} x$.

The given hyperbola is a translation of the standard hyperbola, 3 units to the right and 5 units downward. Hence the following holds:

- The vertices are at $A = (7, -5)$ and $A' = (-1, -5)$.
- The foci are at $F = (3 + \sqrt{65}, -5)$ and $F' = (3 - \sqrt{65}, -5)$.
- The focal axis is $y = -5$ and the conjugate axis is $x = 3$.
- The center is at $(3, -5)$.
- The asymptotes are $y + 5 = \pm \frac{7}{4}(x - 3)$.

33. $4x^2 - 3y^2 + 8x + 30y = 215$

SOLUTION Since there is no cross term, we complete the square of the terms involving x and y separately:

$$4x^2 - 3y^2 + 8x + 30y = 4\left(x^2 + 2x\right) - 3\left(y^2 - 10y\right) = 4(x+1)^2 - 4 - 3(y-5)^2 + 75 = 215$$

Hence,

$$4(x+1)^2 - 3(y-5)^2 = 144$$

$$\frac{4(x+1)^2}{144} - \frac{3(y-5)^2}{144} = 1$$

$$\left(\frac{x+1}{6}\right)^2 - \left(\frac{y-5}{\sqrt{48}}\right)^2 = 1$$

This is the equation of the hyperbola obtained by translating the hyperbola $\left(\frac{x}{6}\right)^2 - \left(\frac{y}{\sqrt{48}}\right)^2 = 1$ one unit to the left and five units upwards. Since $a = 6, b = \sqrt{48}$, we have $c = \sqrt{36 + 48} = \sqrt{84} \sim 9.2$. We obtain the following table:

	Standard position	Translated hyperbola
vertices	$(6, 0), (-6, 0)$	$(5, 5), (-7, 5)$
foci	$(\pm 9.2, 0)$	$(8.2, 5), (-10.2, 5)$
focal axis	The x-axis	$y = 5$
conjugate axis	The y-axis	$x = -1$
center	The origin	$(-1, 5)$
asymptotes	$y = \pm 1.15x$	$y = 1.15x + 3.85$ $y = 1.15x + 6.15$

35. $y = 4(x - 4)^2$

SOLUTION By Exercise 34, the parabola $y = 4x^2$ has the vertex at the origin, the focus at $\left(0, \frac{1}{16}\right)$ and its axis is the y-axis. Our parabola is a translation of the standard parabola four units to the right. Hence its vertex is at $(4, 0)$, the focus is at $\left(4, \frac{1}{16}\right)$ and its axis is the vertical line $x = 4$.

37. $4x^2 + 25y^2 - 8x - 10y = 20$

SOLUTION Since there are no cross terms this conic section is obtained by translating a conic section in standard position. To identify the conic section we complete the square of the terms involving x and y separately:

$$4x^2 + 25y^2 - 8x - 10y = 4\left(x^2 - 2x\right) + 25\left(y^2 - \frac{2}{5}y\right)$$

$$= 4(x-1)^2 - 4 + 25\left(y - \frac{1}{5}\right)^2 - 1$$

$$= 4(x-1)^2 + 25\left(y - \frac{1}{5}\right)^2 - 5 = 20$$

Hence,

$$4(x-1)^2 + 25\left(y - \frac{1}{5}\right)^2 = 25$$

$$\frac{4}{25}(x-1)^2 + \left(y - \frac{1}{5}\right)^2 = 1$$

$$\left(\frac{x-1}{\frac{5}{2}}\right)^2 + \left(y - \frac{1}{5}\right)^2 = 1$$

This is the equation of the ellipse obtained by translating the ellipse in standard position $\left(\frac{x}{\frac{5}{2}}\right)^2 + y^2 = 1$ one unit to the

right and $\frac{1}{5}$ unit upward. Since $a = \frac{5}{2}, b = 1$ we have $c = \sqrt{\left(\frac{5}{2}\right)^2 - 1} \approx 2.3$, so we obtain the following table:

	Standard position	Translated ellipse
Vertices	$\left(\pm\frac{5}{2}, 0\right), (0, \pm 1)$	$\left(1 \pm \frac{5}{2}, \frac{1}{5}\right), \left(1, \frac{1}{5} \pm 1\right)$
Foci	$(-2.3, 0), (2.3, 0)$	$\left(-1.3, \frac{1}{5}\right), \left(3.3, \frac{1}{5}\right)$
Focal axis	The x-axis	$y = \frac{1}{5}$
Conjugate axis	The y-axis	$x = 1$
Center	The origin	$\left(1, \frac{1}{5}\right)$

In Exercises 39–42, use the Discriminant Test to determine the type of the conic section (in each case, the equation is nondegenerate). Plot the curve if you have a computer algebra system.

39. $4x^2 + 5xy + 7y^2 = 24$

SOLUTION   Here, $D = 25 - 4 \cdot 4 \cdot 7 = -87$, so the conic section is an ellipse.

41. $2x^2 - 8xy + 3y^2 - 4 = 0$

SOLUTION   Here, $D = 64 - 4 \cdot 2 \cdot 3 = 40$, giving us a hyperbola.

43. Show that the "conic" $x^2 + 3y^2 - 6x + 12y + 23 = 0$ has no points.

SOLUTION   Complete the square in each variable separately:

$$-23 = x^2 - 6x + 3y^2 + 12y = (x^2 - 6x + 9) + (3y^2 + 12y + 12) - 9 - 12 = (x - 3)^2 + 3(y + 2)^2 - 21$$

Collecting constants and reversing sides gives

$$(x - 3)^2 + 3(y + 2)^2 = -2$$

which has no solutions since the left-hand side is a sum of two squares so is always nonnegative.

45. Show that $\dfrac{b}{a} = \sqrt{1 - e^2}$ for a standard ellipse of eccentricity e.

SOLUTION   By the definition of eccentricity:

$$e = \frac{c}{a} \tag{1}$$

For the ellipse in standard position, $c = \sqrt{a^2 - b^2}$. Substituting into (1) and simplifying yields

$$e = \frac{\sqrt{a^2 - b^2}}{a} = \sqrt{\frac{a^2 - b^2}{a^2}} = \sqrt{1 - \left(\frac{b}{a}\right)^2}$$

We square the two sides and solve for $\frac{b}{a}$:

$$e^2 = 1 - \left(\frac{b}{a}\right)^2 \Rightarrow \left(\frac{b}{a}\right)^2 = 1 - e^2 \Rightarrow \frac{b}{a} = \sqrt{1 - e^2}$$

47. Explain why the dots in Figure 23 lie on a parabola. Where are the focus and directrix located?

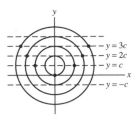

FIGURE 23

SOLUTION All the circles are centered at $(0, c)$ and the kth circle has radius kc. Hence the indicated point P_k on the kth circle has a distance kc from the point $F = (0, c)$. The point P_k also has distance kc from the line $y = -c$. That is, the indicated point on each circle is equidistant from the point $F = (0, c)$ and the line $y = -c$, hence it lies on the parabola with focus at $F = (0, c)$ and directrix $y = -c$.

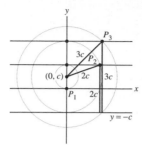

49. A **latus rectum** of a conic section is a chord through a focus parallel to the directrix. Find the area bounded by the parabola $y = x^2/(4c)$ and its latus rectum (refer to Figure 8).

SOLUTION The directrix is $y = -c$, and the focus is $(0, c)$. The chord through the focus parallel to $y = -c$ is clearly $y = c$; this line intersects the parabola when $c = x^2/(4c)$ or $4c^2 = x^2$, so when $x = \pm 2c$. The desired area is then

$$\int_{-2c}^{2c} c - \frac{1}{4c}x^2 \, dx = \left(cx - \frac{1}{12c}x^3 \right) \Big|_{-2c}^{2c}$$

$$= 2c^2 - \frac{8c^3}{12c} - \left(-2c^2 - \frac{(-2c)^3}{12c} \right) = 4c^2 - \frac{4}{3}c^2 = \frac{8}{3}c^2$$

In Exercises 51–54, find the polar equation of the conic with the given eccentricity and directrix, and focus at the origin.

51. $e = \frac{1}{2}$, $x = 3$

SOLUTION Substituting $e = \frac{1}{2}$ and $d = 3$ in the polar equation of a conic section we obtain

$$r = \frac{ed}{1 + e\cos\theta} = \frac{\frac{1}{2} \cdot 3}{1 + \frac{1}{2}\cos\theta} = \frac{3}{2 + \cos\theta} \Rightarrow r = \frac{3}{2 + \cos\theta}$$

53. $e = 1$, $x = 4$

SOLUTION We substitute $e = 1$ and $d = 4$ in the polar equation of a conic section to obtain

$$r = \frac{ed}{1 + e\cos\theta} = \frac{1 \cdot 4}{1 + 1 \cdot \cos\theta} = \frac{4}{1 + \cos\theta} \Rightarrow r = \frac{4}{1 + \cos\theta}$$

In Exercises 55–58, identify the type of conic, the eccentricity, and the equation of the directrix.

55. $r = \dfrac{8}{1 + 4\cos\theta}$

SOLUTION Matching with the polar equation $r = \frac{ed}{1 + e\cos\theta}$ we get $ed = 8$ and $e = 4$ yielding $d = 2$. Since $e > 1$, the conic section is a hyperbola, having eccentricity $e = 4$ and directrix $x = 2$ (referring to the focus-directrix definition (11)).

57. $r = \dfrac{8}{4 + 3\cos\theta}$

SOLUTION We first rewrite the equation in the form $r = \frac{ed}{1 + e\cos\theta}$, obtaining

$$r = \frac{2}{1 + \frac{3}{4}\cos\theta}$$

Hence, $ed = 2$ and $e = \frac{3}{4}$ yielding $d = \frac{8}{3}$. Since $e < 1$, the conic section is an ellipse, having eccentricity $e = \frac{3}{4}$ and directrix $x = \frac{8}{3}$.

59. Find a polar equation for the hyperbola with focus at the origin, directrix $x = -2$, and eccentricity $e = 1.2$.

SOLUTION We substitute $d = -2$ and $e = 1.2$ in the polar equation $r = \frac{ed}{1 + e\cos\theta}$ and use Exercise 40 to obtain

$$r = \frac{1.2 \cdot (-2)}{1 + 1.2\cos\theta} = \frac{-2.4}{1 + 1.2\cos\theta} = \frac{-12}{5 + 6\cos\theta} = \frac{12}{5 - 6\cos\theta}$$

61. Find an equation in rectangular coordinates of the conic

$$r = \frac{16}{5 + 3\cos\theta}$$

Hint: Use the results of Exercise 60.

SOLUTION   Put this equation in the form of the referenced exercise:

$$\frac{16}{5 + 3\cos\theta} = \frac{\frac{16}{5}}{1 + \frac{3}{5}\cos\theta} = \frac{\frac{16}{3}\cdot\frac{3}{5}}{1 + \frac{3}{5}\cos\theta}$$

so that $e = \frac{3}{5}$ and $d = \frac{16}{3}$. Then the center of the ellipse has x-coordinate

$$-\frac{de^2}{1 - e^2} = -\frac{\frac{16}{3}\cdot\frac{9}{25}}{1 - \frac{9}{25}} = -\frac{16}{3}\cdot\frac{9}{25}\cdot\frac{25}{16} = -3$$

and y-coordinate 0, and A' has x-coordinate

$$-\frac{de}{1 - e} = -\frac{\frac{16}{3}\cdot\frac{3}{5}}{1 - \frac{3}{5}} = -\frac{16}{3}\cdot\frac{3}{5}\cdot\frac{5}{2} = -8$$

and y-coordinate 0, so $a = -3 - (-8) = 5$, and the equation is

$$\left(\frac{x+3}{5}\right)^2 + \left(\frac{y}{b}\right)^2 = 1$$

To find b, set $\theta = \frac{\pi}{2}$; then $r = \frac{16}{5}$. But the point corresponding to $\theta = \frac{\pi}{2}$ lies on the y-axis, so has coordinates $\left(0, \frac{16}{5}\right)$. This point is on the ellipse, so that

$$\left(\frac{0+3}{5}\right)^2 + \left(\frac{\frac{16}{5}}{b}\right)^2 = 1 \quad \Rightarrow \quad \frac{256}{25\cdot b^2} = \frac{16}{25} \quad \Rightarrow \quad \frac{256}{b^2} = 16 \quad \Rightarrow \quad b = 4$$

and the equation is

$$\left(\frac{x+3}{5}\right)^2 + \left(\frac{y}{4}\right)^2 = 1$$

63. Kepler's First Law states that planetary orbits are ellipses with the sun at one focus. The orbit of Pluto has eccentricity $e \approx 0.25$. Its **perihelion** (closest distance to the sun) is approximately 2.7 billion miles. Find the **aphelion** (farthest distance from the sun).

SOLUTION   We define an xy-coordinate system so that the orbit is an ellipse in standard position, as shown in the figure.

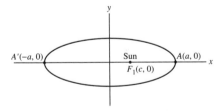

The aphelion is the length of $\overline{A'F_1}$, that is $a + c$. By the given data, we have

$$0.25 = e = \frac{c}{a} \Rightarrow c = 0.25a$$

$$a - c = 2.7 \Rightarrow c = a - 2.7$$

Equating the two expressions for c we get

$$0.25a = a - 2.7$$

$$0.75a = 2.7 \Rightarrow a = \frac{2.7}{0.75} = 3.6, \ c = 3.6 - 2.7 = 0.9$$

The aphelion is thus

$$\overline{A'F_0} = a + c = 3.6 + 0.9 = 4.5 \text{ billion miles.}$$

Further Insights and Challenges

65. Verify Theorem 2.

SOLUTION Let $F_1 = (c, 0)$ and $F_2 = (-c, 0)$ and let $P(x, y)$ be an arbitrary point on the hyperbola. Then for some constant a,

$$\overline{PF_1} - \overline{PF_2} = \pm 2a$$

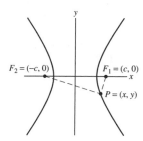

Using the distance formula we write this as

$$\sqrt{(x - c)^2 + y^2} - \sqrt{(x + c)^2 + y^2} = \pm 2a.$$

Moving the second term to the right and squaring both sides gives

$$\sqrt{(x - c)^2 + y^2} = \sqrt{(x + c)^2 + y^2} \pm 2a$$

$$(x - c)^2 + y^2 = (x + c)^2 + y^2 \pm 4a\sqrt{(x + c)^2 + y^2} + 4a^2$$

$$(x - c)^2 - (x + c)^2 - 4a^2 = \pm 4a\sqrt{(x + c)^2 + y^2}$$

$$xc + a^2 = \pm a\sqrt{(x + c)^2 + y^2}$$

We square and simplify to obtain

$$x^2 c^2 + 2xca^2 + a^4 = a^2 \left((x + c)^2 + y^2 \right)$$

$$= a^2 x^2 + 2a^2 xc + a^2 c^2 + a^2 y^2$$

$$\left(c^2 - a^2 \right) x^2 - a^2 y^2 = a^2 \left(c^2 - a^2 \right)$$

$$\frac{x^2}{a^2} - \frac{y^2}{c^2 - a^2} = 1$$

For $b = \sqrt{c^2 - a^2}$ (or $c = \sqrt{a^2 + b^2}$) we get

$$\frac{x^2}{a^2} - \frac{y^2}{b^2} = 1 \Rightarrow \left(\frac{x}{a} \right)^2 - \left(\frac{y}{b} \right)^2 = 1.$$

67. Verify that if $e > 1$, then Eq. (11) defines a hyperbola of eccentricity e, with its focus at the origin and directrix at $x = d$.

SOLUTION The points $P = (r, \theta)$ on the hyperbola satisfy $\overline{PF} = e\overline{PD}$, $e > 1$. Referring to the figure we see that

$$\overline{PF} = r, \overline{PD} = d - r \cos \theta \tag{1}$$

Hence

$$r = e(d - r \cos \theta)$$

$$r = ed - er \cos \theta$$

$$r(1 + e \cos \theta) = ed \Rightarrow r = \frac{ed}{1 + e \cos \theta}$$

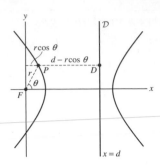

Remark: Equality (1) holds also for $\theta > \frac{\pi}{2}$. For example, in the following figure, we have

$$\overline{PD} = d + r\cos(\pi - \theta) = d - r\cos\theta$$

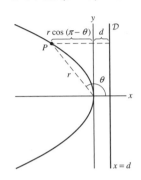

Reflective Property of the Ellipse   *In Exercises 68–70, we prove that the focal radii at a point on an ellipse make equal angles with the tangent line $\mathcal{L}$. Let $P = (x_0, y_0)$ be a point on the ellipse in Figure 25 with foci $F_1 = (-c, 0)$ and $F_2 = (c, 0)$, and eccentricity $e = c/a$.*

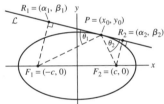

FIGURE 25 The ellipse $\left(\dfrac{x}{a}\right)^2 + \left(\dfrac{y}{b}\right)^2 = 1$.

69. Points R_1 and R_2 in Figure 25 are defined so that $\overline{F_1 R_1}$ and $\overline{F_2 R_2}$ are perpendicular to the tangent line.

(a) Show, with A and B as in Exercise 68, that

$$\frac{\alpha_1 + c}{\beta_1} = \frac{\alpha_2 - c}{\beta_2} = \frac{A}{B}$$

(b) Use (a) and the distance formula to show that

$$\frac{F_1 R_1}{F_2 R_2} = \frac{\beta_1}{\beta_2}$$

(c) Use (a) and the equation of the tangent line in Exercise 68 to show that

$$\beta_1 = \frac{B(1 + Ac)}{A^2 + B^2}, \qquad \beta_2 = \frac{B(1 - Ac)}{A^2 + B^2}$$

SOLUTION

(a) Since $R_1 = (\alpha_1, \beta_1)$ and $R_2 = (\alpha_2, \beta_2)$ lie on the tangent line at P, that is on the line $Ax + By = 1$, we have

$$A\alpha_1 + B\beta_1 = 1 \quad \text{and} \quad A\alpha_2 + B\beta_2 = 1$$

The slope of the line $R_1 F_1$ is $\frac{\beta_1}{\alpha_1 + c}$ and it is perpendicular to the tangent line having slope $-\frac{A}{B}$. Similarly, the slope of the line $R_2 F_2$ is $\frac{\beta_2}{\alpha_2 - c}$ and it is also perpendicular to the tangent line. Hence,

$$\frac{\alpha_1 + c}{\beta_1} = \frac{A}{B} \quad \text{and} \quad \frac{\alpha_2 - c}{\beta_2} = \frac{A}{B}.$$

(b) Using the distance formula, we have

$$\overline{R_1 F_1}^2 = (\alpha_1 + c)^2 + \beta_1^2$$

Thus,

$$\overline{R_1 F_1}^2 = \beta_1^2 \left(\left(\frac{\alpha_1 + c}{\beta_1} \right)^2 + 1 \right) \tag{1}$$

By part (a), $\frac{\alpha_1 + c}{\beta_1} = \frac{A}{B}$. Substituting in (1) gives

$$\overline{R_1 F_1}^2 = \beta_1^2 \left(\frac{A^2}{B^2} + 1 \right) \tag{2}$$

Likewise,

$$\overline{R_2 F_2}^2 = (\alpha_2 - c)^2 + \beta_2^2 = \beta_2^2 \left(\left(\frac{\alpha_2 - c}{\beta_2} \right)^2 + 1 \right) \tag{3}$$

but since $\frac{\alpha_2 - c}{\beta_2} = \frac{A}{B}$, substituting in (3) gives

$$\overline{R_2 F_2}^2 = \beta_2^2 \left(\frac{A^2}{B^2} + 1 \right). \tag{4}$$

Dividing, we find that

$$\frac{\overline{R_1 F_1}^2}{\overline{R_2 F_2}^2} = \frac{\beta_1^2}{\beta_2^2} \quad \text{so} \quad \frac{\overline{R_1 F_1}}{\overline{R_2 F_2}} = \frac{\beta_1}{\beta_2},$$

as desired.

(c) In part (a) we showed that

$$\begin{cases} A\alpha_1 + B\beta_1 = 1 \\ \dfrac{\beta_1}{\alpha_1 + c} = \dfrac{B}{A} \end{cases}$$

Eliminating α_1 and solving for β_1 gives

$$\beta_1 = \frac{B(1 + Ac)}{A^2 + B^2}. \tag{5}$$

Similarly, we have

$$\begin{cases} A\alpha_2 + B\beta_2 = 1 \\ \dfrac{\beta_2}{\alpha_2 - c} = \dfrac{B}{A} \end{cases}$$

Eliminating α_2 and solving for β_2 yields

$$\beta_2 = \frac{B(1 - Ac)}{A^2 + B^2} \tag{6}$$

71. Here is another proof of the Reflective Property.
(a) Figure 25 suggests that $\mathcal{L}$ is the unique line that intersects the ellipse only in the point P. Assuming this, prove that $QF_1 + QF_2 > PF_1 + PF_2$ for all points Q on the tangent line other than P.
(b) Use the Principle of Least Distance (Example 6 in Section 4.7) to prove that $\theta_1 = \theta_2$.

SOLUTION

(a) Consider a point $Q \neq P$ on the line $\mathcal{L}$ (see figure). Since $\mathcal{L}$ intersects the ellipse in only one point, the remainder of the line lies outside the ellipse, so that QR does not have zero length, and $F_2 QR$ is a triangle. Thus

$$QF_1 + QF_2 = QR + RF_1 + QF_2 = RF_1 + (QR + QF_2) > RF_1 + RF_2$$

since the sum of lengths of two sides of a triangle exceeds the length of the third side. But since point R lies on the ellipse, $RF_2 + RF_2 = PF_1 + PF_2$, and we are done.

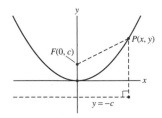

(b) Consider a beam of light traveling from F_1 to F_2 by reflection off of the line $\mathcal{L}$. By the principle of least distance, the light takes the shortest path, which by part (a) is the path through P. By Example 6 in Section 4.7, this shortest path has the property that the angle of incidence (θ_1) is equal to the angle of reflection (θ_2).

73. Show that $y = x^2/4c$ is the equation of a parabola with directrix $y = -c$, focus $(0, c)$, and the vertex at the origin, as stated in Theorem 3.

SOLUTION The points $P = (x, y)$ on the parabola are equidistant from $F = (0, c)$ and the line $y = -c$.

That is, by the distance formula, we have

$$\overline{PF} = \overline{PD}$$

$$\sqrt{x^2 + (y - c)^2} = |y + c|$$

Squaring and simplifying yields

$$x^2 + (y - c)^2 = (y + c)^2$$
$$x^2 + y^2 - 2yc + c^2 = y^2 + 2yc + c^2$$
$$x^2 - 2yc = 2yc$$

$$x^2 = 4yc \Rightarrow y = \frac{x^2}{4c}$$

Thus, we showed that the points that are equidistant from the focus $F = (0, c)$ and the directrix $y = -c$ satisfy the equation $y = \frac{x^2}{4c}$.

75. Derive Eqs. (13) and (14) in the text as follows. Write the coordinates of P with respect to the rotated axes in Figure 21 in polar form $x' = r \cos \alpha$, $y' = r \sin \alpha$. Explain why P has polar coordinates $(r, \alpha + \theta)$ with respect to the standard x and y-axes and derive Eqs. (13) and (14) using the addition formulas for cosine and sine.

SOLUTION If the polar coordinates of P with respect to the rotated axes are (r, α), then the line from the origin to P has length r and makes an angle of α with the rotated x-axis (the x'-axis). Since the x'-axis makes an angle of θ with the x-axis, it follows that the line from the origin to P makes an angle of $\alpha + \theta$ with the x-axis, so that the polar coordinates of P with respect to the standard axes are $(r, \alpha + \theta)$. Write (x', y') for the rectangular coordinates of P with respect to the rotated axes and (x, y) for the rectangular coordinates of P with respect to the standard axes. Then

$$x = r \cos(\alpha + \theta) = (r \cos \alpha) \cos \theta - (r \sin \alpha) \sin \theta = x' \cos \theta - y' \sin \theta$$

$$y = r \sin(\alpha + \theta) = r \sin \alpha \cos \theta + r \cos \alpha \sin \theta = (r \cos \alpha) \sin \theta + (r \sin \alpha) \cos \theta = x' \sin \theta + y' \cos \theta$$

CHAPTER REVIEW EXERCISES

1. Which of the following curves pass through the point $(1, 4)$?

(a) $c(t) = (t^2, t + 3)$ **(b)** $c(t) = (t^2, t - 3)$

(c) $c(t) = (t^2, 3 - t)$ **(d)** $c(t) = (t - 3, t^2)$

SOLUTION To check whether it passes through the point $(1, 4)$, we solve the equations $c(t) = (1, 4)$ for the given curves.

(a) Comparing the second coordinate of the curve and the point yields:

$$t + 3 = 4$$
$$t = 1$$

We substitute $t = 1$ in the first coordinate, to obtain

$$t^2 = 1^2 = 1$$

Hence the curve passes through $(1, 4)$.

(b) Comparing the second coordinate of the curve and the point yields:

$$t - 3 = 4$$
$$t = 7$$

We substitute $t = 7$ in the first coordinate to obtain

$$t^2 = 7^2 = 49 \neq 1$$

Hence the curve does not pass through $(1, 4)$.

(c) Comparing the second coordinate of the curve and the point yields

$$3 - t = 4$$
$$t = -1$$

We substitute $t = -1$ in the first coordinate, to obtain

$$t^2 = (-1)^2 = 1$$

Hence the curve passes through $(1, 4)$.

(d) Comparing the first coordinate of the curve and the point yields

$$t - 3 = 1$$
$$t = 4$$

We substitute $t = 4$ in the second coordinate, to obtain:

$$t^2 = 4^2 = 16 \neq 4$$

Hence the curve does not pass through $(1, 4)$.

3. Find parametric equations for the circle of radius 2 with center $(1, 1)$. Use the equations to find the points of intersection of the circle with the x- and y-axes.

SOLUTION Using the standard technique for parametric equations of curves, we obtain

$$c(t) = (1 + 2\cos t, 1 + 2\sin t)$$

We compare the x coordinate of $c(t)$ to 0:

$$1 + 2\cos t = 0$$
$$\cos t = -\frac{1}{2}$$
$$t = \pm\frac{2\pi}{3}$$

Substituting in the y coordinate yields

$$1 + 2\sin\left(\pm\frac{2\pi}{3}\right) = 1 \pm 2\frac{\sqrt{3}}{2} = 1 \pm \sqrt{3}$$

Hence, the intersection points with the y-axis are $(0, 1 \pm \sqrt{3})$. We compare the y coordinate of $c(t)$ to 0:

$$1 + 2\sin t = 0$$
$$\sin t = -\frac{1}{2}$$
$$t = -\frac{\pi}{6} \quad \text{or} \quad \frac{7}{6}\pi$$

Substituting in the x coordinates yields

$$1 + 2\cos\left(-\frac{\pi}{6}\right) = 1 + 2\frac{\sqrt{3}}{2} = 1 + \sqrt{3}$$

$$1 + 2\cos\left(\frac{7}{6}\pi\right) = 1 - 2\cos\left(\frac{\pi}{6}\right) = 1 - 2\frac{\sqrt{3}}{2} = 1 - \sqrt{3}$$

Hence, the intersection points with the x-axis are $(1 \pm \sqrt{3}, 0)$.

5. Find a parametrization $c(\theta)$ of the unit circle such that $c(0) = (-1, 0)$.

SOLUTION   The unit circle has the parametrization

$$c(t) = (\cos t, \sin t)$$

This parametrization does not satisfy $c(0) = (-1, 0)$. We replace the parameter t by a parameter θ so that $t = \theta + \alpha$, to obtain another parametrization for the circle:

$$c^*(\theta) = (\cos(\theta + \alpha), \sin(\theta + \alpha)) \tag{1}$$

We need that $c^*(0) = (1, 0)$, that is,

$$c^*(0) = (\cos\alpha, \sin\alpha) = (-1, 0)$$

Hence

$$\begin{matrix} \cos\alpha = -1 \\ \sin\alpha = 0 \end{matrix} \quad \Rightarrow \quad \alpha = \pi$$

Substituting in (1) we obtain the following parametrization:

$$c^*(\theta) = (\cos(\theta + \pi), \sin(\theta + \pi))$$

7. Find a path $c(t)$ that traces the line $y = 2x + 1$ from $(1, 3)$ to $(3, 7)$ for $0 \le t \le 1$.

SOLUTION   Solution 1: By one of the examples in section 12.1, the line through $P = (1, 3)$ with slope 2 has the parametrization

$$c(t) = (1 + t, 3 + 2t)$$

But this parametrization does not satisfy $c(1) = (3, 7)$. We replace the parameter t by a parameter s so that $t = \alpha s + \beta$. We get

$$c^*(s) = (1 + \alpha s + \beta, 3 + 2(\alpha s + \beta)) = (\alpha s + \beta + 1, 2\alpha s + 2\beta + 3)$$

We need that $c^*(0) = (1, 3)$ and $c^*(1) = (3, 7)$. Hence,

$$c^*(0) = (1 + \beta, 3 + 2\beta) = (1, 3)$$

$$c^*(1) = (\alpha + \beta + 1, 2\alpha + 2\beta + 3) = (3, 7)$$

We obtain the equations

$$\begin{matrix} 1 + \beta = 1 \\ 3 + 2\beta = 3 \\ \alpha + \beta + 1 = 3 \\ 2\alpha + 2\beta + 3 = 7 \end{matrix} \quad \Rightarrow \quad \beta = 0, \alpha = 2$$

Substituting in (1) gives

$$c^*(s) = (2s + 1, 4s + 3)$$

Solution 2: The segment from $(1, 3)$ to $(3, 7)$ has the following vector parametrization:

$$(1 - t)\langle 1, 3\rangle + t\langle 3, 7\rangle = \langle 1 - t + 3t, 3(1 - t) + 7t\rangle = \langle 1 + 2t, 3 + 4t\rangle$$

The parametrization is thus

$$c(t) = (1 + 2t, 3 + 4t)$$

In Exercises 9–12, express the parametric curve in the form $y = f(x)$.

9. $c(t) = (4t - 3, 10 - t)$

SOLUTION   We use the given equation to express t in terms of x.

$$x = 4t - 3$$

$$4t = x + 3$$

$$t = \frac{x + 3}{4}$$

Substituting in the equation of y yields

$$y = 10 - t = 10 - \frac{x + 3}{4} = -\frac{x}{4} + \frac{37}{4}$$

That is,

$$y = -\frac{x}{4} + \frac{37}{4}$$

11. $c(t) = \left(3 - \frac{2}{t}, t^3 + \frac{1}{t}\right)$

SOLUTION   We use the given equation to express t in terms of x:

$$x = 3 - \frac{2}{t}$$

$$\frac{2}{t} = 3 - x$$

$$t = \frac{2}{3 - x}$$

Substituting in the equation of y yields

$$y = \left(\frac{2}{3 - x}\right)^3 + \frac{1}{2/(3 - x)} = \frac{8}{(3 - x)^3} + \frac{3 - x}{2}$$

In Exercises 13–16, calculate dy/dx at the point indicated

13. $c(t) = (t^3 + t, t^2 - 1), \quad t = 3$

SOLUTION   The parametric equations are $x = t^3 + t$ and $y = t^2 - 1$. We use the theorem on the slope of the tangent line to find $\frac{dy}{dx}$:

$$\frac{dy}{dx} = \frac{\frac{dy}{dt}}{\frac{dx}{dt}} = \frac{2t}{3t^2 + 1}$$

We now substitute $t = 3$ to obtain

$$\frac{dy}{dx}\bigg|_{t=3} = \frac{2 \cdot 3}{3 \cdot 3^2 + 1} = \frac{3}{14}$$

15. $c(t) = (e^t - 1, \sin t), \quad t = 20$

SOLUTION   We use the theorem for the slope of the tangent line to find $\frac{dy}{dx}$:

$$\frac{dy}{dx} = \frac{\frac{dy}{dt}}{\frac{dx}{dt}} = \frac{(\sin t)'}{(e^t - 1)'} = \frac{\cos t}{e^t}$$

We now substitute $t = 20$:

$$\frac{dy}{dx}\bigg|_{t=0} = \frac{\cos 20}{e^{20}}$$

17. CAS   Find the point on the cycloid $c(t) = (t - \sin t, 1 - \cos t)$ where the tangent line has slope $\frac{1}{2}$.

SOLUTION   Since $x = t - \sin t$ and $y = 1 - \cos t$, the theorem on the slope of the tangent line gives

$$\frac{dy}{dx} = \frac{\frac{dy}{dt}}{\frac{dx}{dt}} = \frac{\sin t}{1 - \cos t}$$

The points where the tangent line has slope $\frac{1}{2}$ are those where $\frac{dy}{dx} = \frac{1}{2}$. We solve for t:

$$\frac{dy}{dx} = \frac{1}{2}$$

$$\frac{\sin t}{1 - \cos t} = \frac{1}{2} \tag{1}$$

$$2 \sin t = 1 - \cos t$$

We let $u = \sin t$. Then $\cos t = \pm\sqrt{1 - \sin^2 t} = \pm\sqrt{1 - u^2}$. Hence

$$2u = 1 \pm \sqrt{1 - u^2}$$

We transfer sides and square to obtain

$$\pm\sqrt{1 - u^2} = 2u - 1$$

$$1 - u^2 = 4u^2 - 4u + 1$$

$$5u^2 - 4u = u(5u - 4) = 0$$

$$u = 0, \ u = \frac{4}{5}$$

We find t by the relation $u = \sin t$:

$$u = 0: \quad \sin t = 0 \Rightarrow t = 0, t = \pi$$

$$u = \frac{4}{5}: \quad \sin t = \frac{4}{5} \Rightarrow t \approx 0.93, t \approx 2.21$$

These correspond to the points $(0, 1)$, $(\pi, 2)$, $(0.13, 0.40)$, and $(1.41, 1.60)$, respectively, for $0 < t < 2\pi$.

19. Find the equation of the Bézier curve with control points

$$P_0 = (-1, -1), \quad P_1 = (-1, 1), \quad P_2 = (1, 1), \quad P_3(1, -1)$$

SOLUTION We substitute the given points in the appropriate formulas in the text to find the parametric equations of the Bézier curve. We obtain

$$x(t) = -(1 - t)^3 - 3t(1 - t)^2 + t^2(1 - t) + t^3$$

$$= -(1 - 3t + 3t^2 - t^3) - (3t - 6t^2 + 3t^3) + (t^2 - t^3) + t^3$$

$$= (-2t^3 + 4t^2 - 1)$$

$$y(t) = -(1 - t)^3 + 3t(1 - t)^2 + t^2(1 - t) - t^3$$

$$= -(1 - 3t + 3t^2 - t^3) + (3t - 6t^2 + 3t^3) + (t^2 - t^3) - t^3$$

$$= (2t^3 - 8t^2 + 6t - 1)$$

21. Find the speed (as a function of t) of a particle whose position at time t seconds is $c(t) = (\sin t + t, \cos t + t)$. What is the particle's maximal speed?

SOLUTION We use the parametric definition to find the speed. We obtain

$$\frac{ds}{dt} = \sqrt{((\sin t + t)')^2 + ((\cos t + t)')^2} = \sqrt{(\cos t + 1)^2 + (1 - \sin t)^2}$$

$$= \sqrt{\cos^2 t + 2\cos t + 1 + 1 - 2\sin t + \sin^2 t} = \sqrt{3 + 2(\cos t - \sin t)}$$

We now differentiate the speed function to find its maximum:

$$\frac{d^2 s}{dt^2} = \left(\sqrt{3 + 2(\cos t - \sin t)}\right)' = \frac{-\sin t - \cos t}{\sqrt{3 + 2(\cos t - \sin t)}}$$

We equate the derivative to zero, to obtain the maximum point:

$$\frac{d^2s}{dt^2} = 0$$

$$\frac{-\sin t - \cos t}{\sqrt{3 + 2(\cos t - \sin t)}} = 0$$

$$-\sin t - \cos t = 0$$

$$-\sin t = \cos t$$

$$\sin(-t) = \cos(-t)$$

$$-t = \frac{\pi}{4} + \pi k$$

$$t = -\frac{\pi}{4} + \pi k$$

Substituting t in the function of speed we obtain the value of the maximal speed:

$$\sqrt{3 + 2\left(\cos -\frac{\pi}{4} - \sin -\frac{\pi}{4}\right)} = \sqrt{3 + 2\left(\frac{\sqrt{2}}{2} - \left(-\frac{\sqrt{2}}{2}\right)\right)} = \sqrt{3 + 2\sqrt{2}}$$

In Exercises 23 and 24, let $c(t) = (e^{-t}\cos t, e^{-t}\sin t)$.

23. Show that $c(t)$ for $0 \le t < \infty$ has finite length and calculate its value.

SOLUTION We use the formula for arc length, to obtain:

$$s = \int_0^\infty \sqrt{((e^{-t}\cos t)')^2 + ((e^{-t}\sin t)')^2}\,dt$$

$$= \int_0^\infty \sqrt{(-e^{-t}\cos t - e^{-t}\sin t)^2 + (-e^{-t}\sin t + e^{-t}\cos t)^2}\,dt$$

$$= \int_0^\infty \sqrt{e^{-2t}(\cos t + \sin t)^2 + e^{-2t}(\cos t - \sin t)^2}\,dt$$

$$= \int_0^\infty e^{-t}\sqrt{\cos^2 t + 2\sin t \cos t + \sin^2 t + \cos^2 t - 2\sin t \cos t + \sin^2 t}\,dt$$

$$= \int_0^\infty e^{-t}\sqrt{2}\,dt = \sqrt{2}(-e^{-t})\Big|_0^\infty = -\sqrt{2}\left(\lim_{t\to\infty} e^{-t} - e^0\right)$$

$$= -\sqrt{2}(0 - 1) = \sqrt{2}$$

25. **CAS** Plot $c(t) = (\sin 2t, 2\cos t)$ for $0 \le t \le \pi$. Express the length of the curve as a definite integral, and approximate it using a computer algebra system.

SOLUTION We use a CAS to plot the curve. The resulting graph is shown here.

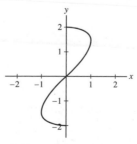

Plot of the curve $(\sin 2t, 2\cos t)$

To calculate the arc length we use the formula for the arc length to obtain

$$s = \int_0^\pi \sqrt{(2\cos 2t)^2 + (-2\sin t)^2}\, dt = 2\int_0^\pi \sqrt{\cos^2 2t + \sin^2 t}\, dt$$

We use a CAS to obtain $s = 6.0972$.

27. Convert the points $(r, \theta) = \left(1, \frac{\pi}{6}\right), \left(3, \frac{5\pi}{4}\right)$ from polar to rectangular coordinates.

SOLUTION We convert the points from polar coordinates to cartesian coordinates. For the first point we have

$$x = r\cos\theta = 1 \cdot \cos\frac{\pi}{6} = \frac{\sqrt{3}}{2}$$

$$y = r\sin\theta = 1 \cdot \sin\frac{\pi}{6} = \frac{1}{2}$$

For the second point we have

$$x = r\cos\theta = 3\cos\frac{5\pi}{4} = -\frac{3\sqrt{2}}{2}$$

$$y = r\sin\theta = 3\sin\frac{5\pi}{4} = -\frac{3\sqrt{2}}{2}$$

29. Write $r = \dfrac{2\cos\theta}{\cos\theta - \sin\theta}$ as an equation in rectangular coordinates.

SOLUTION We use the formula for converting from polar coordinates to cartesian coordinates to substitute x and y for r and θ:

$$r = \frac{2\cos\theta}{\cos\theta - \sin\theta}$$

$$\sqrt{x^2 + y^2} = \frac{2r\cos\theta}{r\cos\theta - r\sin\theta}$$

$$\sqrt{x^2 + y^2} = \frac{2x}{x - y}$$

31. $\boxed{\text{GU}}$ Convert the equation

$$9(x^2 + y^2) = (x^2 + y^2 - 2y)^2$$

to polar coordinates, and plot it with a graphing utility.

SOLUTION We use the formula for converting from cartesian coordinates to polar coordinates to substitute r and θ for x and y:

$$9(x^2 + y^2) = (x^2 + y^2 - 2y)^2$$

$$9r^2 = (r^2 - 2r\sin\theta)^2$$

$$3r = r^2 - 2r\sin\theta$$

$$3 = r - 2\sin\theta$$

$$r = 3 + 2\sin\theta$$

The plot of $r = 3 + 2\sin\theta$ is shown here:

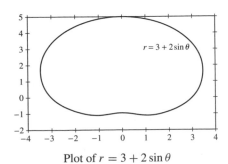

Plot of $r = 3 + 2\sin\theta$

33. Calculate the area of one petal of $r = \sin 4\theta$ (see Figure 1).

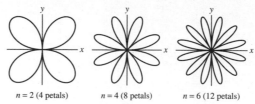

$n = 2$ (4 petals) $n = 4$ (8 petals) $n = 6$ (12 petals)

FIGURE 1 Plot of $r = \sin(n\theta)$.

SOLUTION   We use a CAS to generate the plot, as shown here.

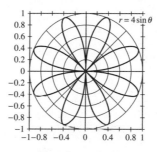

Plot of $r = \sin 4\theta$

We can see that one leaf lies between the rays $\theta = 0$ and $\theta = \dfrac{\theta}{4}$. We now use the formula for area in polar coordinates to obtain

$$A = \frac{1}{2} \int_0^{\pi/4} \sin^2 4\theta \, d\theta = \frac{1}{4} \int_0^{\pi/4} (1 - \cos 8\theta) \, d\theta = \frac{1}{4} \left(\theta - \frac{\sin 8\theta}{8} \Big|_0^{\pi/4} \right)$$

$$= \frac{\pi}{16} - \frac{1}{32} (\sin 2\pi - \sin 0) = \frac{\pi}{16}$$

35. Calculate the total area enclosed by the curve $r^2 = \cos\theta e^{\sin\theta}$ (Figure 2).

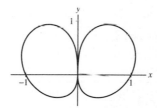

FIGURE 2 Graph of $r^2 = \cos\theta e^{\sin\theta}$.

SOLUTION   Note that this is defined only for θ between $-\pi/2$ and $\pi/2$. We use the formula for area in polar coordinates to obtain:

$$A = \frac{1}{2} \int_{-\pi/2}^{\pi/2} r^2 \, d\theta = \frac{1}{2} \int_{-\pi/2}^{\pi/2} \cos\theta e^{\sin\theta} \, d\theta$$

We evaluate the integral by making the substitution $x = \sin\theta \, dx = \cos\theta \, d\theta$:

$$A = \frac{1}{2} \int_{-\pi/2}^{\pi/2} \cos\theta e^{\sin\theta} \, d\theta = \frac{1}{2} e^x \Big|_{-1}^{1} = \frac{1}{2} \left(e - e^{-1} \right)$$

37. Find the area enclosed by the cardioid $r = a(1 + \cos\theta)$, where $a > 0$.

SOLUTION   The graph of $r = a(1 + \cos\theta)$ in the $r\theta$-plane for $0 \le \theta \le 2\pi$ and the cardioid in the xy-plane are shown in the following figures:

$$r = a\,(1 + \cos\theta)$$

The cardioid $r = a\,(1 + \cos\theta), a > 0$

As θ varies from 0 to π the radius r decreases from $2a$ to 0, and this gives the upper part of the cardioid.

The lower part is traced as θ varies from π to 2π and consequently r increases from 0 back to $2a$. We compute the area enclosed by the upper part of the cardioid and the x-axis, using the following integral (we use the identity $\cos^2\theta = \frac{1}{2} + \frac{1}{2}\cos 2\theta$):

$$\frac{1}{2}\int_0^\pi r^2\,d\theta = \frac{1}{2}\int_0^\pi a^2(1 + \cos\theta)^2\,d\theta = \frac{a^2}{2}\int_0^\pi \left(1 + 2\cos\theta + \cos^2\theta\right)\,d\theta$$

$$= \frac{a^2}{2}\int_0^\pi \left(1 + 2\cos\theta + \frac{1}{2} + \frac{1}{2}\cos 2\theta\right)\,d\theta = \frac{a^2}{2}\int_0^\pi \left(\frac{3}{2} + 2\cos\theta + \frac{1}{2}\cos 2\theta\right)\,d\theta$$

$$= \frac{a^2}{2}\left[\frac{3\theta}{2} + 2\sin\theta + \frac{1}{4}\sin 2\theta\right]\Big|_0^\pi = \frac{a^2}{2}\left[\frac{3\pi}{2} + 2\sin\pi + \frac{1}{4}\sin 2\pi - 0\right] = \frac{3\pi a^2}{4}$$

Using symmetry, the total area A enclosed by the cardioid is

$$A = 2\cdot\frac{3\pi a^2}{4} = \frac{3\pi a^2}{2}$$

39. CAS Figure 5 shows the graph of $r = e^{0.5\theta}\sin\theta$ for $0 \le \theta \le 2\pi$. Use a computer algebra system to approximate the difference in length between the outer and inner loops.

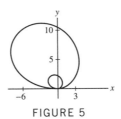

FIGURE 5

SOLUTION We note that the inner loop is the curve for $\theta \in [0, \pi]$, and the outer loop is the curve for $\theta \in [\pi, 2\pi]$. We express the length of these loops using the formula for the arc length. The length of the inner loop is

$$s_1 = \int_0^\pi \sqrt{(e^{0.5\theta}\sin\theta)^2 + ((e^{0.5\theta}\sin\theta)')^2}\,d\theta = \int_0^\pi \sqrt{e^\theta\sin^2\theta + \left(\frac{e^{0.5\theta}\sin\theta}{2} + e^{0.5\theta}\cos\theta\right)^2}\,d\theta$$

and the length of the outer loop is

$$s_2 = \int_\pi^{2\pi} \sqrt{e^\theta\sin^2\theta + \left(\frac{e^{0.5\theta}\sin\theta}{2} + e^{0.5\theta}\cos\theta\right)^2}\,d\theta$$

We now use the CAS to calculate the arc length of each of the loops. We obtain that the length of the inner loop is 7.5087 and the length of the outer loop is 36.121, hence the outer one is 4.81 times longer than the inner one.

In Exercises 41–44, identify the conic section. Find the vertices and foci.

41. $\left(\frac{x}{3}\right)^2 + \left(\frac{y}{2}\right)^2 = 1$

SOLUTION This is an ellipse in standard position. Its foci are $(\pm\sqrt{3^2 - 2^2}, 0) = (\pm\sqrt{5}, 0)$ and its vertices are $(\pm 3, 0), (0, \pm 2)$.

43. $\left(2x + \frac{1}{2}y\right)^2 = 4 - (x - y)^2$

SOLUTION We simplify the equation:

$$\left(2x + \frac{1}{2}y\right)^2 = 4 - (x - y)^2$$

$$4x^2 + 2xy + \frac{1}{4}y^2 = 4 - x^2 + 2xy - y^2$$

$$5x^2 + \frac{5}{4}y^2 = 4$$

$$\frac{5x^2}{4} + \frac{5y^2}{16} = 1$$

$$\left(\frac{x}{\frac{2}{\sqrt{5}}}\right)^2 + \left(\frac{y}{\frac{4}{\sqrt{5}}}\right)^2 = 1$$

This is an ellipse in standard position, with foci $\left(0, \pm\sqrt{\left(\frac{4}{\sqrt{5}}\right)^2 - \left(\frac{2}{\sqrt{5}}\right)^2}\right) = \left(0, \pm\sqrt{\frac{12}{5}}\right)$ and vertices $\left(\pm\frac{2}{\sqrt{5}}, 0\right)$, $\left(0, \pm\frac{4}{\sqrt{5}}\right)$.

In Exercises 45–50, find the equation of the conic section indicated.

45. Ellipse with vertices $(\pm 8, 0)$ and foci $(\pm\sqrt{3}, 0)$

SOLUTION Since the foci of the desired ellipse are on the x-axis, we conclude that $a > b$. We are given that the points $(\pm 8, 0)$ are vertices of the ellipse, and since they are on the x-axis, $a = 8$. We are given that the foci are $(\pm\sqrt{3}, 0)$ and we have shown that $a > b$, hence we have that $\sqrt{a^2 - b^2} = \sqrt{3}$. Solving for b yields

$$\sqrt{a^2 - b^2} = \sqrt{3}$$

$$a^2 - b^2 = 3$$

$$8^2 - b^2 = 3$$

$$b^2 = 61$$

$$b = \sqrt{61}$$

Next we use a and b to construct the equation of the ellipse:

$$\left(\frac{x}{8}\right)^2 + \left(\frac{y}{\sqrt{61}}\right)^2 = 1.$$

47. Hyperbola with vertices $(\pm 8, 0)$, asymptotes $y = \pm\frac{3}{4}x$

SOLUTION Since the asymptotes of the hyperbola are $y = \pm\frac{3}{4}x$, and the equation of the asymptotes for a general hyperbola in standard position is $y = \pm\frac{b}{a}x$, we conclude that $\frac{b}{a} = \frac{3}{4}$. We are given that the vertices are $(\pm 8, 0)$, thus $a = 8$. We substitute and solve for b:

$$\frac{b}{a} = \frac{3}{4}$$

$$\frac{b}{8} = \frac{3}{4}$$

$$b = 6$$

Next we use a and b to construct the equation of the hyperbola:

$$\left(\frac{x}{8}\right)^2 - \left(\frac{y}{6}\right)^2 = 1.$$

49. Parabola with focus $(8, 0)$, directrix $x = -8$

SOLUTION This is similar to the usual equation of a parabola, but we must use y as x, and x as y, to obtain

$$x = \frac{1}{32}y^2.$$

51. Find the asymptotes of the hyperbola $3x^2 + 6x - y^2 - 10y = 1$.

SOLUTION   We complete the squares and simplify:

$$3x^2 + 6x - y^2 - 10y = 1$$

$$3(x^2 + 2x) - (y^2 + 10y) = 1$$

$$3(x^2 + 2x + 1 - 1) - (y^2 + 10y + 25 - 25) = 1$$

$$3(x + 1)^2 - 3 - (y + 5)^2 + 25 = 1$$

$$3(x + 1)^2 - (y + 5)^2 = -21$$

$$\left(\frac{y + 5}{\sqrt{21}}\right)^2 - \left(\frac{x + 1}{\sqrt{7}}\right)^2 = 1$$

We obtained a hyperbola with focal axis that is parallel to the y-axis, and is shifted -5 units on the y-axis, and -1 units in the x-axis. Therefore, the asymptotes are

$$x + 1 = \pm \frac{\sqrt{7}}{\sqrt{21}}(y + 5) \quad \text{or} \quad y + 5 = \pm \sqrt{3}(x + 1).$$

53. Show that the relation $\frac{dy}{dx} = (e^2 - 1)\frac{x}{y}$ holds on a standard ellipse or hyperbola of eccentricity e.

SOLUTION   We differentiate the equations of the standard ellipse and the hyperbola with respect to x:

$$\text{Ellipse:} \qquad\qquad \text{Hyperbola:}$$

$$\frac{x^2}{a^2} + \frac{y^2}{b^2} = 1 \qquad\qquad \frac{x^2}{a^2} - \frac{y^2}{b^2} = 1$$

$$\frac{2x}{a^2} + \frac{2y}{b^2}\frac{dy}{dx} = 0 \qquad\qquad \frac{2x}{a^2} - \frac{2y}{b^2}\frac{dy}{dx} = 0$$

$$\frac{dy}{dx} = -\frac{b^2}{a^2}\frac{x}{y} \qquad\qquad \frac{dy}{dx} = \frac{b^2}{a^2}\frac{x}{y}$$

The eccentricity of the ellipse is $e = \frac{\sqrt{a^2 - b^2}}{a}$, hence $e^2 a^2 = a^2 - b^2$ or $e^2 = 1 - \frac{b^2}{a^2}$ yielding $\frac{b^2}{a^2} = 1 - e^2$.

The eccentricity of the hyperbola is $e = \frac{\sqrt{a^2 + b^2}}{a}$, hence $e^2 a^2 = a^2 + b^2$ or $e^2 = 1 + \frac{b^2}{a^2}$, giving $\frac{b^2}{a^2} = e^2 - 1$. Combining with the expressions for $\frac{dy}{dx}$ we get:

$$\text{Ellipse:} \qquad\qquad \text{Hyperbola:}$$

$$\frac{dy}{dx} = -(1 - e^2)\frac{x}{y} = (e^2 - 1)\frac{x}{y} \qquad \frac{dy}{dx} = (e^2 - 1)\frac{x}{y}$$

We, thus, proved that the relation $\frac{dy}{dx} = (e^2 - 1)\frac{x}{y}$ holds on a standard ellipse or hyperbola of eccentricity e.

55. Refer to Figure 25 in Section 11.5. Prove that the product of the perpendicular distances $F_1 R_1$ and $F_2 R_2$ from the foci to a tangent line of an ellipse is equal to the square b^2 of the semiminor axes.

SOLUTION   We first consider the ellipse in standard position:

$$\frac{x^2}{a^2} + \frac{y^2}{b^2} = 1$$

The equation of the tangent line at $P = (x_0, y_0)$ is

$$\frac{x_0 x}{a^2} + \frac{y_0 y}{b^2} = 1$$

or

$$b^2 x_0 x + a^2 y_0 y - a^2 b^2 = 0$$

The distances of the foci $F_1 = (c, 0)$ and $F_2 = (-c, 0)$ from the tangent line are

$$\overline{F_1 R_1} = \frac{|b^2 x_0 c - a^2 b^2|}{\sqrt{b^4 x_0^2 + a^4 y_0^2}}; \quad \overline{F_2 R_2} = \frac{|b^2 x_0 c + a^2 b^2|}{\sqrt{b^4 x_0^2 + a^4 y_0^2}}$$

We compute the product of the distances:

$$\overline{F_1 R_1} \cdot \overline{F_2 R_2} = \left| \frac{\left(b^2 x_0 c - a^2 b^2\right)\left(b^2 x_0 c + a^2 b^2\right)}{b^4 x_0^2 + a^4 y_0^2} \right| = \left| \frac{b^4 x_0^2 c^2 - a^4 b^4}{b^4 x_0^2 + a^4 y_0^2} \right| \tag{1}$$

The point $P = (x_0, y_0)$ lies on the ellipse, hence:

$$\frac{x_0^2}{a^2} + \frac{y_0^2}{b^2} = 1 \Rightarrow a^4 y_0^2 = a^4 b^2 - a^2 b^2 x_0^2$$

We substitute in (1) to obtain (notice that $b^2 - a^2 = -c^2$)

$$\overline{F_1 R_1} \cdot \overline{F_2 R_2} = \frac{|b^4 x_0^2 c^2 - a^4 b^4|}{|b^4 x_0^2 + a^4 b^2 - a^2 b^2 x_0^2|} = \frac{|b^4 x_0^2 c^2 - a^4 b^4|}{|b^2 (b^2 - a^2) x_0^2 + a^4 b^2|}$$

$$= \frac{|b^4 x_0^2 c^2 - a^4 b^4|}{|-b^2 x_0^2 c^2 + a^4 b^2|} = \frac{|b^2 (x_0^2 c^2 - a^4)|}{|-(x_0^2 c^2 - a^4)|} = |-b^2| = b^2$$

The product $\overline{F_1 R_1} \cdot \overline{F_2 R_2}$ remains unchanged if we translate the standard ellipse.

12 VECTOR GEOMETRY

12.1 Vectors in the Plane (LT Section 13.1)

Preliminary Questions

1. Answer true or false. Every nonzero vector is:

(a) Equivalent to a vector based at the origin.

(b) Equivalent to a unit vector based at the origin.

(c) Parallel to a vector based at the origin.

(d) Parallel to a unit vector based at the origin.

SOLUTION

(a) This statement is true. Translating the vector so that it is based on the origin, we get an equivalent vector based at the origin.

(b) Equivalent vectors have equal lengths, hence vectors that are not unit vectors, are not equivalent to a unit vector.

(c) This statement is true. A vector based at the origin such that the line through this vector is parallel to the line through the given vector, is parallel to the given vector.

(d) Since parallel vectors do not necessarily have equal lengths, the statement is true by the same reasoning as in (c).

2. What is the length of $-3\mathbf{a}$ if $\|\mathbf{a}\| = 5$?

SOLUTION Using properties of the length we get

$$\|-3\mathbf{a}\| = |-3|\|\mathbf{a}\| = 3\|\mathbf{a}\| = 3 \cdot 5 = 15$$

3. Suppose that $\mathbf{v}$ has components $\langle 3, 1 \rangle$. How, if at all, do the components change if you translate $\mathbf{v}$ horizontally two units to the left?

SOLUTION Translating $\mathbf{v} = \langle 3, 1 \rangle$ yields an equivalent vector, hence the components are not changed.

4. What are the components of the zero vector based at $P = (3, 5)$?

SOLUTION The components of the zero vector are always $\langle 0, 0 \rangle$, no matter where it is based.

5. True or false?

(a) The vectors $\mathbf{v}$ and $-2\mathbf{v}$ are parallel.

(b) The vectors $\mathbf{v}$ and $-2\mathbf{v}$ point in the same direction.

SOLUTION

(a) The lines through $\mathbf{v}$ and $-2\mathbf{v}$ are parallel, therefore these vectors are parallel.

(b) The vector $-2\mathbf{v}$ is a scalar multiple of $\mathbf{v}$, where the scalar is negative. Therefore $-2\mathbf{v}$ points in the opposite direction as $\mathbf{v}$.

6. Explain the commutativity of vector addition in terms of the Parallelogram Law.

SOLUTION To determine the vector $\mathbf{v} + \mathbf{w}$, we translate $\mathbf{w}$ to the equivalent vector $\mathbf{w}'$ whose tail coincides with the head of $\mathbf{v}$. The vector $\mathbf{v} + \mathbf{w}$ is the vector pointing from the tail of $\mathbf{v}$ to the head of $\mathbf{w}'$.

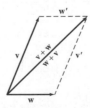

To determine the vector $\mathbf{w} + \mathbf{v}$, we translate $\mathbf{v}$ to the equivalent vector $\mathbf{v}'$ whose tail coincides with the head of $\mathbf{w}$. Then $\mathbf{w} + \mathbf{v}$ is the vector pointing from the tail of $\mathbf{w}$ to the head of $\mathbf{v}'$. In either case, the resulting vector is the vector with the tail at the basepoint of $\mathbf{v}$ and $\mathbf{w}$, and head at the opposite vertex of the parallelogram. Therefore $\mathbf{v} + \mathbf{w} = \mathbf{w} + \mathbf{v}$.

Exercises

1. Sketch the vectors $\mathbf{v}_1$, $\mathbf{v}_2$, $\mathbf{v}_3$, $\mathbf{v}_4$ with tail P and head Q, and compute their lengths. Are any two of these vectors equivalent?

	$\mathbf{v}_1$	$\mathbf{v}_2$	$\mathbf{v}_3$	$\mathbf{v}_4$
P	$(2, 4)$	$(-1, 3)$	$(-1, 3)$	$(4, 1)$
Q	$(4, 4)$	$(1, 3)$	$(2, 4)$	$(6, 3)$

SOLUTION Using the definitions we obtain the following answers:

$$\mathbf{v}_1 = \overrightarrow{PQ} = \langle 4 - 2, 4 - 4 \rangle = \langle 2, 0 \rangle \qquad \mathbf{v}_2 = \langle 1 - (-1), 3 - 3 \rangle = \langle 2, 0 \rangle$$

$$\|\mathbf{v}_1\| = \sqrt{2^2 + 0^2} = 2 \qquad\qquad \|\mathbf{v}_2\| = \sqrt{2^2 + 0^2} = 2$$

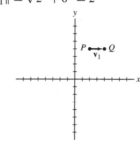
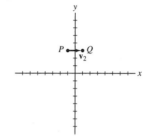

$$\mathbf{v}_3 = \langle 2 - (-1), 4 - 3 \rangle = \langle 3, 1 \rangle \qquad \mathbf{v}_4 = \langle 6 - 4, 3 - 1 \rangle = \langle 2, 2 \rangle$$

$$\|\mathbf{v}_3\| = \sqrt{3^2 + 1^2} = \sqrt{10} \qquad\qquad \|\mathbf{v}_4\| = \sqrt{2^2 + 2^2} = \sqrt{8} = 2\sqrt{2}$$

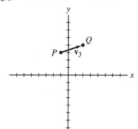
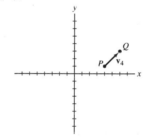

$\mathbf{v}_1$ and $\mathbf{v}_2$ are parallel and have the same length, hence they are equivalent.

3. What is the terminal point of the vector $\mathbf{a} = \langle 1, 3 \rangle$ based at $P = (2, 2)$? Sketch $\mathbf{a}$ and the vector $\mathbf{a}_0$ based at the origin and equivalent to $\mathbf{a}$.

SOLUTION The terminal point Q of the vector $\mathbf{a}$ is located 1 unit to the right and 3 units up from $P = (2, 2)$. Therefore, $Q = (2 + 1, 2 + 3) = (3, 5)$. The vector $\mathbf{a}_0$ equivalent to $\mathbf{a}$ based at the origin is shown in the figure, along with the vector $\mathbf{a}$.

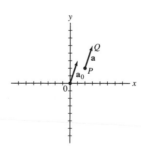

In Exercises 5–8, find the components of $\overrightarrow{PQ}$.

5. $P = (3, 2)$, $\quad Q = (2, 7)$

SOLUTION Using the definition of the components of a vector we have $\overrightarrow{PQ} = \langle 2 - 3, 7 - 2 \rangle = \langle -1, 5 \rangle$.

7. $P = (3, 5)$, $\quad Q = (1, -4)$

SOLUTION By the definition of the components of a vector, we obtain $\overrightarrow{PQ} = \langle 1 - 3, -4 - 5 \rangle = \langle -2, -9 \rangle$.

In Exercises 9–14, calculate.

9. $\langle 2, 1 \rangle + \langle 3, 4 \rangle$

SOLUTION Using vector algebra we have $\langle 2, 1 \rangle + \langle 3, 4 \rangle = \langle 2 + 3, 1 + 4 \rangle = \langle 5, 5 \rangle$.

11. $5 \langle 6, 2 \rangle$

SOLUTION $5\langle 6, 2 \rangle = \langle 5 \cdot 6, 5 \cdot 2 \rangle = \langle 30, 10 \rangle$

13. $\left\langle -\frac{1}{2}, \frac{5}{3} \right\rangle + \left\langle 3, \frac{10}{3} \right\rangle$

SOLUTION The vector sum is $\left\langle -\frac{1}{2}, \frac{5}{3} \right\rangle + \left\langle 3, \frac{10}{3} \right\rangle = \left\langle -\frac{1}{2} + 3, \frac{5}{3} + \frac{10}{3} \right\rangle = \left\langle \frac{5}{2}, 5 \right\rangle$.

15. Which of the vectors (A)–(C) in Figure 21 is equivalent to $\mathbf{v} - \mathbf{w}$?

(A)   **(B)**   **(C)**

FIGURE 21

SOLUTION The vector $-\mathbf{w}$ has the same length as $\mathbf{w}$ but points in the opposite direction. The sum $\mathbf{v} + (-\mathbf{w})$, which is the difference $\mathbf{v} - \mathbf{w}$, is obtained by the parallelogram law. This vector is the vector shown in (b).

17. Sketch $2\mathbf{v}$, $-\mathbf{w}$, $\mathbf{v} + \mathbf{w}$, and $2\mathbf{v} - \mathbf{w}$ for the vectors in Figure 23.

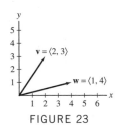

FIGURE 23

SOLUTION The scalar multiple $2\mathbf{v}$ points in the same direction as $\mathbf{v}$ and its length is twice the length of $\mathbf{v}$. It is the vector $2\mathbf{v} = \langle 4, 6 \rangle$.

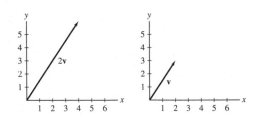

$-\mathbf{w}$ has the same length as $\mathbf{w}$ but points to the opposite direction. It is the vector $-\mathbf{w} = \langle -4, -1 \rangle$.

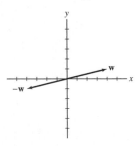

The vector sum $\mathbf{v} + \mathbf{w}$ is the vector:

$$\mathbf{v} + \mathbf{w} = \langle 2, 3 \rangle + \langle 4, 1 \rangle = \langle 6, 4 \rangle.$$

This vector is shown in the following figure:

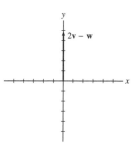

The vector $2\mathbf{v} - \mathbf{w}$ is

$$2\mathbf{v} - \mathbf{w} = 2\langle 2, 3\rangle - \langle 4, 1\rangle = \langle 4, 6\rangle - \langle 4, 1\rangle = \langle 0, 5\rangle$$

It is shown next:

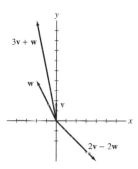

19. Sketch $\mathbf{v} = \langle 0, 2\rangle$, $\mathbf{w} = \langle -2, 4\rangle$, $3\mathbf{v} + \mathbf{w}$, $2\mathbf{v} - 2\mathbf{w}$.

SOLUTION We compute the vectors and then sketch them:

$$3\mathbf{v} + \mathbf{w} = 3\langle 0, 2\rangle + \langle -2, 4\rangle = \langle 0, 6\rangle + \langle -2, 4\rangle = \langle -2, 10\rangle$$

$$2\mathbf{v} - 2\mathbf{w} = 2\langle 0, 2\rangle - 2\langle -2, 4\rangle = \langle 0, 4\rangle - \langle -4, 8\rangle = \langle 4, -4\rangle$$

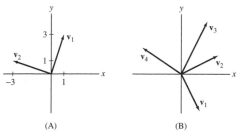

21. Sketch the vector $\mathbf{v}$ such that $\mathbf{v} + \mathbf{v}_1 + \mathbf{v}_2 = \mathbf{0}$ for $\mathbf{v}_1$ and $\mathbf{v}_2$ in Figure 24(A).

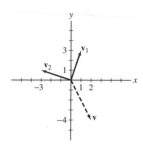

FIGURE 24

SOLUTION Since $\mathbf{v} + \mathbf{v}_1 + \mathbf{v}_2 = \mathbf{0}$, we have that $\mathbf{v} = -\mathbf{v}_1 - \mathbf{v}_2$, and since $\mathbf{v}_1 = \langle 1, 3\rangle$ and $\mathbf{v}_2 = \langle -3, 1\rangle$, then $\mathbf{v} = -\mathbf{v}_1 - \mathbf{v}_2 = \langle 2, -4\rangle$, as seen in this picture.

23. Let $\mathbf{v} = \overrightarrow{PQ}$, where $P = (-2, 5)$, $Q = (1, -2)$. Which of the following vectors with the given tails and heads are equivalent to $\mathbf{v}$?

(a) $(-3, 3)$, $(0, 4)$

(b) $(0, 0)$, $(3, -7)$

(c) $(-1, 2)$, $(2, -5)$

(d) $(4, -5)$, $(1, 4)$

SOLUTION   Two vectors are equivalent if they have the same components. We thus compute the vectors and check whether this condition is satisfied.

$$\mathbf{v} = \overrightarrow{PQ} = \langle 1 - (-2), -2 - 5 \rangle = \langle 3, -7 \rangle$$

(a) $\langle 0 - (-3), 4 - 3 \rangle = \langle 3, 1 \rangle$

(b) $\langle 3 - 0, -7 - 0 \rangle = \langle 3, -7 \rangle$

(c) $\langle 2 - (-1), -5 - 2 \rangle = \langle 3, -7 \rangle$

(d) $\langle 1 - 4, 4 - (-5) \rangle = \langle -3, 9 \rangle$

We see that the vectors in (b) and (c) are equivalent to $\mathbf{v}$.

In Exercises 25–28, sketch the vectors $\overrightarrow{AB}$ and $\overrightarrow{PQ}$, and determine whether they are equivalent.

25. $A = (1, 1)$, $B = (3, 7)$, $P = (4, -1)$, $Q = (6, 5)$

SOLUTION   We compute the vectors and check whether they have the same components:

$$\overrightarrow{AB} = \langle 3 - 1, 7 - 1 \rangle = \langle 2, 6 \rangle$$
$$\overrightarrow{PQ} = \langle 6 - 4, 5 - (-1) \rangle = \langle 2, 6 \rangle$$

$\Rightarrow$ The vectors are equivalent.

27. $A = (-3, 2)$, $B = (0, 0)$, $P = (0, 0)$, $Q = (3, -2)$

SOLUTION   We compute the vectors $\overrightarrow{AB}$ and $\overrightarrow{PQ}$:

$$\overrightarrow{AB} = \langle 0 - (-3), 0 - 2 \rangle = \langle 3, -2 \rangle$$
$$\overrightarrow{PQ} = \langle 3 - 0, -2 - 0 \rangle = \langle 3, -2 \rangle$$

$\Rightarrow$ The vectors are equivalent.

In Exercises 29–32, are $\overrightarrow{AB}$ and $\overrightarrow{PQ}$ parallel? And if so, do they point in the same direction?

29. $A = (1, 1)$, $B = (3, 4)$, $P = (1, 1)$, $Q = (7, 10)$

SOLUTION   We compute the vectors $\overrightarrow{AB}$ and $\overrightarrow{PQ}$:

$$\overrightarrow{AB} = \langle 3 - 1, 4 - 1 \rangle = \langle 2, 3 \rangle$$
$$\overrightarrow{PQ} = \langle 7 - 1, 10 - 1 \rangle = \langle 6, 9 \rangle$$

Since $\overrightarrow{AB} = \frac{1}{3} \langle 6, 9 \rangle$, the vectors are parallel and point in the same direction.

31. $A = (2, 2)$, $B = (-6, 3)$, $P = (9, 5)$, $Q = (17, 4)$

SOLUTION   We compute the vectors $\overrightarrow{AB}$ and $\overrightarrow{PQ}$:

$$\overrightarrow{AB} = \langle -6 - 2, 3 - 2 \rangle = \langle -8, 1 \rangle$$
$$\overrightarrow{PQ} = \langle 17 - 9, 4 - 5 \rangle = \langle 8, -1 \rangle$$

Since $\overrightarrow{AB} = -\overrightarrow{PQ}$, the vectors are parallel and point in opposite directions.

In Exercises 33–36, let $R = (-2, 7)$. Calculate the following.

33. The length of $\overrightarrow{OR}$

SOLUTION   Since $\overrightarrow{OR} = \langle -2, 7 \rangle$, the length of the vector is $\|\overrightarrow{OR}\| = \sqrt{(-2)^2 + 7^2} = \sqrt{53}$.

35. The point P such that $\overrightarrow{PR}$ has components $\langle -2, 7 \rangle$

SOLUTION   Denoting $P = (x_0, y_0)$ we have:

$$\overrightarrow{PR} = \langle -2 - x_0, 7 - y_0 \rangle = \langle -2, 7 \rangle$$

Equating corresponding components yields:

$$\begin{aligned} -2 - x_0 &= -2 \\ 7 - y_0 &= 7 \end{aligned}$$

$\Rightarrow$ $x_0 = 0, \ y_0 = 0$ $\Rightarrow$ $P = (0, 0)$

In Exercises 37–42, find the given vector.

37. Unit vector $\mathbf{e_v}$ where $\mathbf{v} = \langle 3, 4 \rangle$

SOLUTION   The unit vector $\mathbf{e_v}$ is the following vector:

$$\mathbf{e_v} = \frac{1}{\|\mathbf{v}\|} \mathbf{v}$$

We find the length of $\mathbf{v} = \langle 3, 4 \rangle$:

$$\|\mathbf{v}\| = \sqrt{3^2 + 4^2} = \sqrt{25} = 5$$

Thus

$$\mathbf{e_v} = \frac{1}{5} \langle 3, 4 \rangle = \left\langle \frac{3}{5}, \frac{4}{5} \right\rangle.$$

39. Vector of length 4 in the direction of $\mathbf{u} = \langle -1, -1 \rangle$

SOLUTION   Since $\|u\| = \sqrt{(-1)^2 + (-1)^2} = \sqrt{2}$, the unit vector in the direction of $\mathbf{u}$ is $\mathbf{e_u} = \left\langle -\frac{1}{\sqrt{2}}, -\frac{1}{\sqrt{2}} \right\rangle$. We multiply $\mathbf{e_u}$ by 4 to obtain the desired vector:

$$4\mathbf{e_u} = 4 \left\langle -\frac{1}{\sqrt{2}}, -\frac{1}{\sqrt{2}} \right\rangle = \left\langle -2\sqrt{2}, -2\sqrt{2} \right\rangle$$

41. Unit vector $\mathbf{e}$ making an angle of $\frac{4\pi}{7}$ with the x-axis

SOLUTION   The unit vector $\mathbf{e}$ is the following vector:

$$\mathbf{e} = \left\langle \cos \frac{4\pi}{7}, \sin \frac{4\pi}{7} \right\rangle = \langle -0.22, 0.97 \rangle.$$

43. Find all scalars λ such that $\lambda \langle 2, 3 \rangle$ has length 1.

SOLUTION   We have:

$$\|\lambda \langle 2, 3 \rangle\| = |\lambda| \|\langle 2, 3 \rangle\| = |\lambda| \sqrt{2^2 + 3^2} = |\lambda| \sqrt{13}$$

The scalar λ must satisfy

$$|\lambda| \sqrt{13} = 1$$
$$|\lambda| = \frac{1}{\sqrt{13}} \quad \Rightarrow \quad \lambda_1 = \frac{1}{\sqrt{13}}, \ \lambda_2 = -\frac{1}{\sqrt{13}}$$

45. What are the coordinates of the point P in the parallelogram in Figure 25(A)?

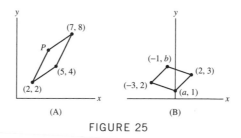

(A)   (B)

FIGURE 25

SOLUTION   We denote by A, B, C the points in the figure.

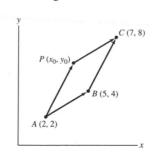

Let $P = (x_0, y_0)$. We compute the following vectors:

$$\overrightarrow{PC} = \langle 7 - x_0, 8 - y_0 \rangle$$

$$\overrightarrow{AB} = \langle 5 - 2, 4 - 2 \rangle = \langle 3, 2 \rangle$$

The vectors $\overrightarrow{PC}$ and $\overrightarrow{AB}$ are equivalent, hence they have the same components. That is:

$$\begin{aligned} 7 - x_0 &= 3 \\ 8 - y_0 &= 2 \end{aligned} \quad \Rightarrow \quad x_0 = 4, \ y_0 = 6 \quad \Rightarrow \quad P = (4, 6)$$

47. Let $\mathbf{v} = \overrightarrow{AB}$ and $\mathbf{w} = \overrightarrow{AC}$, where A, B, C are three distinct points in the plane. Match (a)–(d) with (i)–(iv). (*Hint:* Draw a picture.)

(a) $-\mathbf{w}$ (b) $-\mathbf{v}$ (c) $\mathbf{w} - \mathbf{v}$ (d) $\mathbf{v} - \mathbf{w}$

(i) $\overrightarrow{CB}$ (ii) $\overrightarrow{CA}$ (iii) $\overrightarrow{BC}$ (iv) $\overrightarrow{BA}$

SOLUTION

(a) $-\mathbf{w}$ has the same length as $\mathbf{w}$ and points in the opposite direction. Hence: $-\mathbf{w} = \overrightarrow{CA}$.

(b) $-\mathbf{v}$ has the same length as $\mathbf{v}$ and points in the opposite direction. Hence: $-\mathbf{v} = \overrightarrow{BA}$.

(c) By the parallelogram law we have:

$$\overrightarrow{BC} = \overrightarrow{BA} + \overrightarrow{AC} = -\mathbf{v} + \mathbf{w} = \mathbf{w} - \mathbf{v}$$

That is,

$$\mathbf{w} - \mathbf{v} = \overrightarrow{BC}$$

(d) By the parallelogram law we have:

$$\overrightarrow{CB} = \overrightarrow{CA} + \overrightarrow{AB} = -\mathbf{w} + \mathbf{v} = \mathbf{v} - \mathbf{w}$$

That is,

$$\mathbf{v} - \mathbf{w} = \overrightarrow{CB}.$$

In Exercises 49–52, calculate the linear combination.

49. $3\mathbf{j} + (9\mathbf{i} + 4\mathbf{j})$

SOLUTION   We have:

$$3\mathbf{j} + (9\mathbf{i} + 4\mathbf{j}) = 3 \langle 0, 1 \rangle + 9 \langle 1, 0 \rangle + 4 \langle 0, 1 \rangle = \langle 9, 7 \rangle$$

51. $(3\mathbf{i} + \mathbf{j}) - 6\mathbf{j} + 2(\mathbf{j} - 4\mathbf{i})$

SOLUTION We have:

$$(3\mathbf{i} + \mathbf{j}) - 6\mathbf{j} + 2(\mathbf{j} - 4\mathbf{i}) = (\langle 3, 0 \rangle + \langle 0, 1 \rangle) - \langle 0, 6 \rangle + 2(\langle 0, 1 \rangle - \langle 4, 0 \rangle) = \langle -5, -3 \rangle$$

53. For each of the position vectors **u** with endpoints A, B, and C in Figure 26, indicate with a diagram the multiples $r\mathbf{v}$ and $s\mathbf{w}$ such that $\mathbf{u} = r\mathbf{v} + s\mathbf{w}$. A sample is shown for $\mathbf{u} = \overrightarrow{OQ}$.

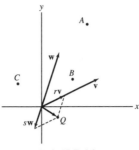

FIGURE 26

SOLUTION See the following three figures:

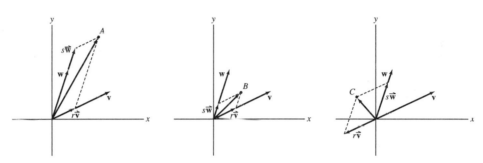

In Exercises 55 and 56, express **u** *as a linear combination* $\mathbf{u} = r\mathbf{v} + s\mathbf{w}$. *Then sketch* **u**, **v**, **w**, *and the parallelogram formed by* $r\mathbf{v}$ *and* $s\mathbf{w}$.

55. $\mathbf{u} = \langle 3, -1 \rangle$; $\mathbf{v} = \langle 2, 1 \rangle$, $\mathbf{w} = \langle 1, 3 \rangle$

SOLUTION We have

$$\mathbf{u} = \langle 3, -1 \rangle = r\mathbf{v} + s\mathbf{w} = r\langle 2, 1 \rangle + s\langle 1, 3 \rangle$$

which becomes the two equations

$$3 = 2r + s$$

$$-1 = r + 3s$$

Solving the second equation for r gives $r = -1 - 3s$, and substituting that into the first equation gives $3 = 2(-1 - 3s) + s = -2 - 6s + s$, so $5 = -5s$, so $s = -1$, and thus $r = 2$. In other words,

$$\mathbf{u} = \langle 3, -1 \rangle = 2\langle 2, 1 \rangle - 1\langle 1, 3 \rangle$$

as seen in this sketch:

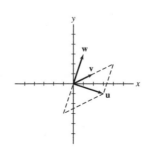

57. Calculate the magnitude of the force on cables 1 and 2 in Figure 27.

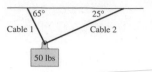

FIGURE 27

SOLUTION   The three forces acting on the point P are:

- The force $\mathbf{F}$ of magnitude 50 lb that acts vertically downward.
- The forces $\mathbf{F}_1$ and $\mathbf{F}_2$ that act through cables 1 and 2 respectively.

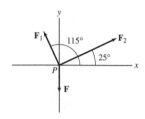

Since the point P is not in motion we have

$$\mathbf{F}_1 + \mathbf{F}_2 + \mathbf{F} = 0 \tag{1}$$

We compute the forces. Letting $\|\mathbf{F}_1\| = f_1$ and $\|\mathbf{F}_2\| = f_2$ we have:

$$\mathbf{F}_1 = f_1 \langle \cos 115°, \sin 115° \rangle = f_1 \langle -0.423, 0.906 \rangle$$
$$\mathbf{F}_2 = f_2 \langle \cos 25°, \sin 25° \rangle = f_2 \langle 0.906, 0.423 \rangle$$
$$\mathbf{F} = \langle 0, -50 \rangle$$

Substituting the forces in (1) gives

$$f_1 \langle -0.423, 0.906 \rangle + f_2 \langle 0.906, 0.423 \rangle + \langle 0, -50 \rangle = \langle 0, 0 \rangle$$
$$\langle -0.423 f_1 + 0.906 f_2, 0.906 f_1 + 0.423 f_2 - 50 \rangle = \langle 0, 0 \rangle$$

We equate corresponding components and get

$$-0.423 f_1 + 0.906 f_2 = 0$$
$$0.906 f_1 + 0.423 f_2 - 50 = 0$$

By the first equation, $f_2 = 0.467 f_1$. Substituting in the second equation and solving for f_1 yields

$$0.906 f_1 + 0.423 \cdot 0.467 f_1 - 50 = 0$$
$$1.104 f_1 = 50 \quad \Rightarrow \quad f_1 = 45.29, \ f_2 = 0.467 f_1 = 21.15$$

We conclude that the magnitude of the force on cable 1 is $f_1 = 45.29$ lb and the magnitude of the force on cable 2 is $f_2 = 21.15$ lb.

59. A plane flying due east at 200 km/h encounters a 40-km/h wind blowing in the north-east direction. The resultant velocity of the plane is the vector sum $\mathbf{v} = \mathbf{v}_1 + \mathbf{v}_2$, where $\mathbf{v}_1$ is the velocity vector of the plane and $\mathbf{v}_2$ is the velocity vector of the wind (Figure 29). The angle between $\mathbf{v}_1$ and $\mathbf{v}_2$ is $\frac{\pi}{4}$. Determine the resultant *speed* of the plane (the length of the vector $\mathbf{v}$).

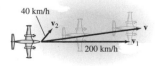

FIGURE 29

SOLUTION The resultant speed of the plane is the length of the sum vector $\mathbf{v} = \mathbf{v}_1 + \mathbf{v}_2$. We place the xy-coordinate system as shown in the figure, and compute the components of the vectors $\mathbf{v}_1$ and $\mathbf{v}_2$. This gives

$$\mathbf{v}_1 = \langle v_1, 0 \rangle$$

$$\mathbf{v}_2 = \left\langle v_2 \cos \frac{\pi}{4}, v_2 \sin \frac{\pi}{4} \right\rangle = \left\langle v_2 \cdot \frac{\sqrt{2}}{2}, v_2 \cdot \frac{\sqrt{2}}{2} \right\rangle$$

We now compute the sum $\mathbf{v} = \mathbf{v}_1 + \mathbf{v}_2$:

$$\mathbf{v} = \langle v_1, 0 \rangle + \left\langle \frac{\sqrt{2}v_2}{2}, \frac{\sqrt{2}v_2}{2} \right\rangle = \left\langle \frac{\sqrt{2}}{2} v_2 + v_1, \frac{\sqrt{2}}{2} v_2 \right\rangle$$

The resultant speed is the length of $\mathbf{v}$, that is,

$$v = \|\mathbf{v}\| = \sqrt{\left(\frac{\sqrt{2}v_2}{2}\right)^2 + \left(v_1 + \frac{\sqrt{2}v_2}{2}\right)^2} = \sqrt{\frac{v_2^2}{2} + v_1^2 + 2 \cdot \frac{\sqrt{2}}{2} v_2 v_1 + \frac{v_2^2}{2}} = \sqrt{v_1^2 + v_2^2 + \sqrt{2}v_1 v_2}$$

Finally, we substitute the given information $v_1 = 200$ and $v_2 = 40$ in the equation above, to obtain

$$v = \sqrt{200^2 + 40^2 + \sqrt{2} \cdot 200 \cdot 40} \approx 230 \text{ km/hr}$$

Further Insights and Challenges

In Exercises 60–62, refer to Figure 30, which shows a robotic arm consisting of two segments of lengths L_1 and L_2.

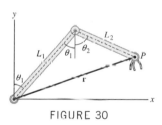

FIGURE 30

61. Let $L_1 = 5$ and $L_2 = 3$. Find $\mathbf{r}$ for $\theta_1 = \frac{\pi}{3}, \theta_2 = \frac{\pi}{4}$.

SOLUTION In Exercise 60 we showed that

$$\mathbf{r} = \langle L_1 \sin \theta_1 + L_2 \sin \theta_2, L_1 \cos \theta_1 - L_2 \cos \theta_2 \rangle$$

Substituting the given information we obtain

$$\mathbf{r} = \left\langle 5 \sin \frac{\pi}{3} + 3 \sin \frac{\pi}{4}, 5 \cos \frac{\pi}{3} - 3 \cos \frac{\pi}{4} \right\rangle = \left\langle \frac{5\sqrt{3}}{2} + \frac{3\sqrt{2}}{2}, \frac{5}{2} - \frac{3\sqrt{2}}{2} \right\rangle \approx \langle 6.45, 0.38 \rangle$$

63. Use vectors to prove that the diagonals $\overline{AC}$ and $\overline{BD}$ of a parallelogram bisect each other (Figure 31). *Hint:* Observe that the midpoint of $\overline{BD}$ is the terminal point of $\mathbf{w} + \frac{1}{2}(\mathbf{v} - \mathbf{w})$.

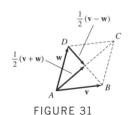

FIGURE 31

SOLUTION   We denote by O the midpoint of $\overline{BD}$. Hence,

$$\overrightarrow{DO} = \frac{1}{2}\overrightarrow{DB}$$

Using the Parallelogram Law we have

$$\overrightarrow{AO} = \overrightarrow{AD} + \overrightarrow{DO} = \overrightarrow{AD} + \frac{1}{2}\overrightarrow{DB}$$

Since $\overrightarrow{AD} = \mathbf{w}$ and $\overrightarrow{DB} = \mathbf{v} - \mathbf{w}$ we get

$$\overrightarrow{AO} = \mathbf{w} + \frac{1}{2}(\mathbf{v} - \mathbf{w}) = \frac{\mathbf{w} + \mathbf{v}}{2} \tag{1}$$

On the other hand, $\overrightarrow{AC} = \overrightarrow{AD} + \overrightarrow{DC} = \mathbf{w} + \mathbf{v}$, hence the midpoint O' of the diagonal $\overline{AC}$ is the terminal point of $\frac{\mathbf{w}+\mathbf{v}}{2}$. That is,

$$\overrightarrow{AO'} = \frac{\mathbf{w} + \mathbf{v}}{2} \tag{2}$$

We combine (1) and (2) to conclude that O and O' are the same point. That is, the diagonal $\overline{AC}$ and $\overline{BD}$ bisect each other.

65. Prove that two vectors $\mathbf{v} = \langle a, b \rangle$ and $\mathbf{w} = \langle c, d \rangle$ are perpendicular if and only if

$$ac + bd = 0$$

SOLUTION   Suppose that the vectors $\mathbf{v}$ and $\mathbf{w}$ make angles θ_1 and θ_2, which are not $\frac{\pi}{2}$ or $\frac{3\pi}{2}$, respectively, with the positive x-axis. Then their components satisfy

$$\begin{aligned} a &= \|\mathbf{v}\| \cos\theta_1 \\ b &= \|\mathbf{v}\| \sin\theta_1 \end{aligned} \quad\Rightarrow\quad \frac{b}{a} = \frac{\sin\theta_1}{\cos\theta_1} = \tan\theta_1$$

$$\begin{aligned} c &= \|\mathbf{w}\| \cos\theta_2 \\ d &= \|\mathbf{w}\| \sin\theta_2 \end{aligned} \quad\Rightarrow\quad \frac{d}{c} = \frac{\sin\theta_2}{\cos\theta_2} = \tan\theta_2$$

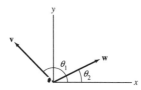

That is, the vectors $\mathbf{v}$ and $\mathbf{w}$ are on the lines with slopes $\frac{b}{a}$ and $\frac{d}{c}$, respectively. The lines are perpendicular if and only if their slopes satisfy

$$\frac{b}{a} \cdot \frac{d}{c} = -1 \quad\Rightarrow\quad bd = -ac \quad\Rightarrow\quad ac + bd = 0$$

We now consider the case where one of the vectors, say $\mathbf{v}$, is perpendicular to the x-axis. In this case $a = 0$, and the vectors are perpendicular if and only if $\mathbf{w}$ is parallel to the x-axis, that is, $d = 0$. So $ac + bd = 0 \cdot c + b \cdot 0 = 0$.

12.2 Vectors in Three Dimensions   (LT Section 13.2)

Preliminary Questions

1. What is the terminal point of the vector $\mathbf{v} = \langle 3, 2, 1 \rangle$ based at the point $P = (1, 1, 1)$?

SOLUTION   We denote the terminal point by $Q = (a, b, c)$. Then by the definition of components of a vector, we have

$$\langle 3, 2, 1 \rangle = \langle a - 1, b - 1, c - 1 \rangle$$

Equivalent vectors have equal components respectively, thus,

$$3 = a - 1 \qquad a = 4$$
$$2 = b - 1 \quad \Rightarrow \quad b = 3$$
$$1 = c - 1 \qquad c = 2$$

The terminal point of **v** is thus $Q = (4, 3, 2)$.

2. What are the components of the vector $\mathbf{v} = \langle 3, 2, 1 \rangle$ based at the point $P = (1, 1, 1)$?

SOLUTION The component of $\mathbf{v} = \langle 3, 2, 1 \rangle$ are $\langle 3, 2, 1 \rangle$ regardless of the base point. The component of **v** and the base point $P = (1, 1, 1)$ determine the head $Q = (a, b, c)$ of the vector, as found in the previous exercise.

3. If $\mathbf{v} = -3\mathbf{w}$, then (choose the correct answer):

(a) **v** and **w** are parallel.

(b) **v** and **w** point in the same direction.

SOLUTION The vectors **v** and **w** lie on parallel lines, hence these vectors are parallel. Since **v** is a scalar multiple of **w** by a negative scalar, **v** and **w** point in opposite directions. Thus, (a) is correct and (b) is not.

4. Which of the following is a direction vector for the line through $P = (3, 2, 1)$ and $Q = (1, 1, 1)$?

(a) $\langle 3, 2, 1 \rangle$ **(b)** $\langle 1, 1, 1 \rangle$ **(c)** $\langle 2, 1, 0 \rangle$

SOLUTION Any vector that is parallel to the vector $\overrightarrow{PQ}$ is a direction vector for the line through P and Q. We compute the vector $\overrightarrow{PQ}$:

$$\overrightarrow{PQ} = \langle 1 - 3, 1 - 2, 1 - 1 \rangle = \langle -2, -1, 0 \rangle.$$

The vectors $\langle 3, 2, 1 \rangle$ and $\langle 1, 1, 1 \rangle$ are not constant multiples of $\overrightarrow{PQ}$, hence they are not parallel to $\overrightarrow{PQ}$. However $\langle 2, 1, 0 \rangle = -1\langle -2, -1, 0 \rangle = -\overrightarrow{PQ}$, hence the vector $\langle 2, 1, 0 \rangle$ is parallel to $\overrightarrow{PQ}$ Therefore, the vector $\langle 2, 1, 0 \rangle$ is a direction vector for the line through P and Q.

5. How many different direction vectors does a line have?

SOLUTION All the vectors that are parallel to a line are also direction vectors for that line. Therefore, there are infinitely many direction vectors for a line.

6. True or false? If **v** is a direction vector for a line $\mathcal{L}$, then $-\mathbf{v}$ is also a direction vector for $\mathcal{L}$.

SOLUTION True. Every vector that is parallel to **v** is a direction vector for the line L. Since $-\mathbf{v}$ is parallel to **v**, it is also a direction vector for L.

Exercises

1. Sketch the vector $\mathbf{v} = \langle 1, 3, 2 \rangle$ and compute its length.

SOLUTION The vector $\mathbf{v} = \langle 1, 3, 2 \rangle$ is shown in the following figure:

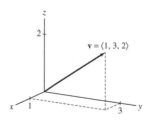

The length of **v** is

$$\|\mathbf{v}\| = \sqrt{1^2 + 3^2 + 2^2} = \sqrt{14}$$

3. Sketch the vector $\mathbf{v} = \langle 1, 1, 0 \rangle$ based at $P = (0, 1, 1)$. Describe this vector in the form $\overrightarrow{PQ}$ for some point Q, and sketch the vector $\mathbf{v}_0$ based at the origin equivalent to **v**.

SOLUTION The vector $\mathbf{v} = \langle 1, 1, 0 \rangle$ based at $P = (0, 1, 1)$ is shown in the figure:

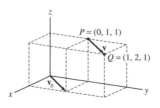

The head Q of the vector $\mathbf{v} = \overrightarrow{PQ}$ is at the point $Q = (0 + 1, 1 + 1, 1 + 0) = (1, 2, 1)$.

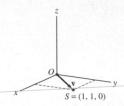

The vector $\mathbf{v}_0$ based at the origin and equivalent to $\mathbf{v}$ is

$$\mathbf{v}_0 = \langle 1, 1, 0 \rangle = \overrightarrow{OS}, \text{ where } S = (1, 1, 0).$$

In Exercises 5–8, find the components of the vector $\overrightarrow{PQ}$.

5. $P = (1, 0, 1), \quad Q = (2, 1, 0)$

SOLUTION   By the definition of the vector components we have

$$\overrightarrow{PQ} = \langle 2 - 1, 1 - 0, 0 - 1 \rangle = \langle 1, 1, -1 \rangle$$

7. $P = (4, 6, 0), \quad Q = \left(-\frac{1}{2}, \frac{9}{2}, 1\right)$

SOLUTION   Using the definition of vector components we have

$$\overrightarrow{PQ} = \left\langle -\frac{1}{2} - 4, \frac{9}{2} - 6, 1 - 0 \right\rangle = \left\langle -\frac{9}{2}, -\frac{3}{2}, 1 \right\rangle$$

In Exercises 9–12, let $R = (1, 4, 3)$.

9. Calculate the length of $\overrightarrow{OR}$.

SOLUTION   The length of $\overrightarrow{OR}$ is the distance from $R = (1, 4, 3)$ to the origin. That is,

$$\|\overrightarrow{OR}\| = \sqrt{(1 - 0)^2 + (4 - 0)^2 + (3 - 0)^2} = \sqrt{26} \approx 5.1.$$

11. Find the point P such that $\mathbf{w} = \overrightarrow{PR}$ has components $\langle 3, -2, 3 \rangle$, and sketch $\mathbf{w}$.

SOLUTION   Denoting $P = (x_0, y_0, z_0)$ we get

$$\overrightarrow{PR} = \langle 1 - x_0, 4 - y_0, 3 - z_0 \rangle = \langle 3, -2, 3 \rangle$$

Equating corresponding components gives

$$\begin{aligned} 1 - x_0 &= 3 \\ 4 - y_0 &= -2 \quad \Rightarrow \quad x_0 = -2, \ y_0 = 6, \ z_0 = 0 \\ 3 - z_0 &= 3 \end{aligned}$$

The point P is, thus, $P = (-2, 6, 0)$.

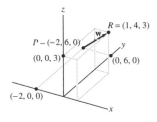

13. Let $\mathbf{v} = \langle 4, 8, 12 \rangle$. Which of the following vectors is parallel to $\mathbf{v}$? Which point in the same direction?
(a) $\langle 2, 4, 6 \rangle$ **(b)** $\langle -1, -2, 3 \rangle$
(c) $\langle -7, -14, -21 \rangle$ **(d)** $\langle 6, 10, 14 \rangle$

SOLUTION   A vector is parallel to $\mathbf{v}$ if it is a scalar multiple of $\mathbf{v}$. It points in the same direction if the multiplying scalar is positive. Using these properties we obtain the following answer:
(a) $\langle 2, 4, 6 \rangle = \frac{1}{2}\mathbf{v} \Rightarrow$ The vectors are parallel and point in the same direction.
(b) $\langle -1, -2, 3 \rangle$ is not a scalar multiple of $\mathbf{v}$, hence these vectors are not parallel.
(c) $\langle -7, -14, -21 \rangle = -\frac{7}{4}\mathbf{v} \Rightarrow$ The vectors are parallel but point in opposite directions.
(d) $\langle 6, 10, 14 \rangle$ is not a constant multiple of $\mathbf{v}$, hence these vectors are not parallel.

In Exercises 14–17, determine whether $\overrightarrow{AB}$ is equivalent to $\overrightarrow{PQ}$.

15. $A = (1, 4, 1) \quad B = (-2, 2, 0)$
$P = (2, 5, 7) \quad Q = (-3, 2, 1)$

SOLUTION We compute the two vectors:

$$\overrightarrow{AB} = \langle -2 - 1, 2 - 4, 0 - 1 \rangle = \langle -3, -2, -1 \rangle$$
$$\overrightarrow{PQ} = \langle -3 - 2, 2 - 5, 1 - 7 \rangle = \langle -5, -3, -6 \rangle$$

The components of $\overrightarrow{AB}$ and $\overrightarrow{PQ}$ are not equal, hence they are not a translate of each other, that is, the vectors are not equivalent.

17. $A = (1, 1, 0) \quad B = (3, 3, 5)$
$P = (2, -9, 7) \quad Q = (4, -7, 13)$

SOLUTION The vectors $\overrightarrow{AB}$ and $\overrightarrow{PQ}$ are the following vectors:

$$\overrightarrow{AB} = \langle 3 - 1, 3 - 1, 5 - 0 \rangle = \langle 2, 2, 5 \rangle$$
$$\overrightarrow{PQ} = \langle 4 - 2, -7 - (-9), 13 - 7 \rangle = \langle 2, 2, 6 \rangle$$

The z-coordinates of the vectors are not equal, hence the vectors are not equivalent.

In Exercises 18–23, calculate the linear combinations.

19. $-2 \langle 8, 11, 3 \rangle + 4 \langle 2, 1, 1 \rangle$

SOLUTION Using the operations of vector addition and scalar multiplication we have

$$-2\langle 8, 11, 3 \rangle + 4\langle 2, 1, 1 \rangle = \langle -16, -22, -6 \rangle + \langle 8, 4, 4 \rangle = \langle -8, -18, -2 \rangle.$$

21. $\frac{1}{2} \langle 4, -2, 8 \rangle - \frac{1}{3} \langle 12, 3, 3 \rangle$

SOLUTION Using the operations on vectors we have

$$\frac{1}{2} \langle 4, -2, 8 \rangle - \frac{1}{3} \langle 12, 3, 3 \rangle = \langle 2, -1, 4 \rangle - \langle 4, 1, 1 \rangle = \langle -2, -2, 3 \rangle.$$

23. $4 \langle 6, -1, 1 \rangle - 2 \langle 1, 0, -1 \rangle + 3 \langle -2, 1, 1 \rangle$

SOLUTION Using the operations of vector addition and scalar multiplication we have

$$4 \langle 6, -1, 1 \rangle - 2 \langle 1, 0, -1 \rangle + 3 \langle -2, 1, 1 \rangle = \langle 24, -4, 4 \rangle + \langle -2, 0, 2 \rangle + \langle -6, 3, 3 \rangle$$
$$= \langle 16, -1, 9 \rangle.$$

In Exercises 24–27, find the given vector.

25. $\mathbf{e_w}$, where $\mathbf{w} = \langle 4, -2, -1 \rangle$

SOLUTION We first find the length of $\mathbf{w}$:

$$\|\mathbf{w}\| = \sqrt{4^2 + (-2)^2 + 1^2} = \sqrt{21}$$

Hence,

$$\mathbf{e_w} = \frac{1}{\|\mathbf{w}\|} \mathbf{w} = \left\langle \frac{4}{\sqrt{21}}, \frac{-2}{\sqrt{21}}, \frac{-1}{\sqrt{21}} \right\rangle$$

27. Unit vector in the direction opposite to $\mathbf{v} = \langle -4, 4, 2 \rangle$

SOLUTION A unit vector in the direction opposite to $\mathbf{v} = \langle -4, 4, 2 \rangle$ is the following vector:

$$-\mathbf{e_v} = -\frac{1}{\|\mathbf{v}\|} \mathbf{v}$$

We compute the length of $\mathbf{v}$:

$$\|\mathbf{v}\| = \sqrt{(-4)^2 + 4^2 + 2^2} = 6$$

The desired vector is, thus,

$$-\mathbf{e_v} = -\frac{1}{6}\langle -4, 4, 2 \rangle = \left\langle \frac{-4}{-6}, \frac{4}{-6}, \frac{2}{-6} \right\rangle = \left\langle \frac{2}{3}, -\frac{2}{3}, -\frac{1}{3} \right\rangle$$

In Exercises 29–36, find a vector parametrization for the line with the given description.

29. Passes through $P = (1, 2, -8)$, direction vector $\mathbf{v} = \langle 2, 1, 3 \rangle$

SOLUTION The vector parametrization for the line is

$$\mathbf{r}(t) = \overrightarrow{OP} + t\mathbf{v}$$

Inserting the given data we get

$$\mathbf{r}(t) = \langle 1, 2, -8 \rangle + t \langle 2, 1, 3 \rangle = \langle 1 + 2t, 2 + t, -8 + 3t \rangle$$

31. Passes through $P = (4, 0, 8)$, direction vector $\mathbf{v} = 7\mathbf{i} + 4\mathbf{k}$

SOLUTION Since $\mathbf{v} = 7\mathbf{i} + 4\mathbf{k} = \langle 7, 0, 4 \rangle$ we obtain the following parametrization:

$$\mathbf{r}(t) = \overrightarrow{OP} + t\mathbf{v} = \langle 4, 0, 8 \rangle + t \langle 7, 0, 4 \rangle = \langle 4 + 7t, 0, 8 + 4t \rangle$$

33. Passes through $(1, 1, 1)$ and $(3, -5, 2)$

SOLUTION We use the equation of the line through two points P and Q:

$$\mathbf{r}(t) = (1 - t)\overrightarrow{OP} + t\overrightarrow{OQ}$$

Since $\overrightarrow{OP} = \langle 1, 1, 1 \rangle$ and $\overrightarrow{OQ} = \langle 3, -5, 2 \rangle$ we obtain

$$\mathbf{r}(t) = (1 - t)\langle 1, 1, 1 \rangle + t \langle 3, -5, 2 \rangle = \langle 1 - t, 1 - t, 1 - t \rangle + \langle 3t, -5t, 2t \rangle = \langle 1 + 2t, 1 - 6t, 1 + t \rangle$$

35. Passes through O and $(4, 1, 1)$

SOLUTION By the equation of the line through two points we get

$$\mathbf{r}(t) = (1 - t)\langle 0, 0, 0 \rangle + t \langle 4, 1, 1 \rangle = \langle 0, 0, 0 \rangle + \langle 4t, t, t \rangle = \langle 4t, t, t \rangle$$

In Exercises 37–40, find parametric equations for the lines with the given description.

37. Perpendicular to the xy-plane, passes through the origin

SOLUTION A direction vector for the line is a vector parallel to the z-axis, for instance, we may choose $\mathbf{v} = \langle 0, 0, 1 \rangle$. The line passes through the origin $(0, 0, 0)$, hence we obtain the following parametrization:

$$\mathbf{r}(t) = \langle 0, 0, 0 \rangle + t \langle 0, 0, 1 \rangle = \langle 0, 0, t \rangle$$

or $x = 0$, $y = 0$, $z = t$.

39. Parallel to the line through $(1, 1, 0)$ and $(0, -1, -2)$, passes through $(0, 0, 4)$

SOLUTION The direction vector is $\mathbf{v} = \langle 0 - 1, -1 - 1, -2 - 0 \rangle = \langle -1, -2, -2 \rangle$. Hence, using the equation of a line we obtain

$$\mathbf{r}(t) = \langle 0, 0, 4 \rangle + t \langle -1, -2, -2 \rangle = \langle -t, -2t, 4 - 2t \rangle$$

41. Which of the following is a parametrization of the line through $P = (4, 9, 8)$ perpendicular to the xz-plane (Figure 18)?

(a) $\mathbf{r}(t) = \langle 4, 9, 8 \rangle + t \langle 1, 0, 1 \rangle$ **(b)** $\mathbf{r}(t) = \langle 4, 9, 8 \rangle + t \langle 0, 0, 1 \rangle$

(c) $\mathbf{r}(t) = \langle 4, 9, 8 \rangle + t \langle 0, 1, 0 \rangle$ **(d)** $\mathbf{r}(t) = \langle 4, 9, 8 \rangle + t \langle 1, 1, 0 \rangle$

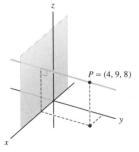

FIGURE 18

SOLUTION Since the direction vector must be perpendicular to the xz-plane, then the direction vector for the line must be parallel to $\mathbf{j}$, which is only satisfied by solution (c).

In Exercises 43–46, let $P = (2, 1, -1)$ and $Q = (4, 7, 7)$. Find the coordinates of each of the following.

43. The midpoint of $\overline{PQ}$

SOLUTION We first parametrize the line through $P = (2, 1, -1)$ and $Q = (4, 7, 7)$:

$$\mathbf{r}(t) = (1 - t)\langle 2, 1, -1\rangle + t\langle 4, 7, 7\rangle = \langle 2 + 2t, 1 + 6t, -1 + 8t\rangle$$

The midpoint of $\overline{PQ}$ occurs at $t = \frac{1}{2}$, that is,

$$\text{midpoint} = \mathbf{r}\left(\frac{1}{2}\right) = \left\langle 2 + 2 \cdot \frac{1}{2}, 1 + 6 \cdot \frac{1}{2}, -1 + 8 \cdot \frac{1}{2}\right\rangle = \langle 3, 4, 3\rangle$$

The midpoint of $\overline{PQ}$ is the terminal point of the vector $\mathbf{r}(t)$, that is, $(3, 4, 3)$. (One could also use the midpoint formula to arrive at the same solution.)

45. The point R such that Q is the midpoint of $\overline{PR}$

SOLUTION We denote $R = (x_0, y_0, z_0)$. By the formula for the midpoint of a segment we have

$$\langle 4, 7, 7\rangle = \left\langle\frac{2 + x_0}{2}, \frac{1 + y_0}{2}, \frac{-1 + z_0}{2}\right\rangle$$

Equating corresponding components we get

$$4 = \frac{2 + x_0}{2}$$
$$7 = \frac{1 + y_0}{2} \quad \Rightarrow \quad x_0 = 6, \; y_0 = 13, \; z_0 = 15 \quad \Rightarrow \quad R = (6, 13, 15)$$
$$7 = \frac{-1 + z_0}{2}$$

47. Show that $\mathbf{r}_1(t)$ and $\mathbf{r}_2(t)$ define the same line, where

$$\mathbf{r}_1(t) = \langle 3, -1, 4\rangle + t\langle 8, 12, -6\rangle$$
$$\mathbf{r}_2(t) = \langle 11, 11, -2\rangle + t\langle 4, 6, -3\rangle$$

Hint: Show that $\mathbf{r}_2$ passes through $(3, -1, 4)$ and that the direction vectors for $\mathbf{r}_1$ and $\mathbf{r}_2$ are parallel.

SOLUTION We observe first that the direction vectors of $\mathbf{r}_1(t)$ and $\mathbf{r}_2(t)$ are multiples of each other:

$$\langle 8, 12, -6\rangle = 2\langle 4, 6, -3\rangle$$

Therefore $\mathbf{r}_1(t)$ and $\mathbf{r}_2(t)$ are parallel. To show they coincide, it suffices to prove that they share a point in common, so we verify that $\mathbf{r}_1(0) = \langle 3, -1, 4\rangle$ lies on $\mathbf{r}_2(t)$ by solving for t:

$$\langle 3, -1, 4\rangle = \langle 11, 11, -2\rangle + t\langle 4, 6, -3\rangle$$
$$\langle 3, -1, 4\rangle - \langle 11, 11, -2\rangle = t\langle 4, 6, -3\rangle$$
$$\langle -8, -12, 6\rangle = t\langle 4, 6, -3\rangle$$

This equation is satisfied for $t = -2$, so $\mathbf{r}_1$ and $\mathbf{r}_2$ coincide.

49. Find two different vector parametrizations of the line through $P = (5, 5, 2)$ with direction vector $\mathbf{v} = \langle 0, -2, 1\rangle$.

SOLUTION Two different parameterizations are

$$\mathbf{r}_1(t) = \langle 5, 5, 2\rangle + t\langle 0, -2, 1\rangle$$
$$\mathbf{r}_2(t) = \langle 5, 5, 2\rangle + t\langle 0, -20, 10\rangle$$

51. Show that the lines $\mathbf{r}_1(t) = \langle -1, 2, 2\rangle + t\langle 4, -2, 1\rangle$ and $\mathbf{r}_2(t) = \langle 0, 1, 1\rangle + t\langle 2, 0, 1\rangle$ do not intersect.

SOLUTION The two lines intersect if there exist parameter values t_1 and t_2 such that

$$\langle -1, 2, 2\rangle + t_1\langle 4, -2, 1\rangle = \langle 0, 1, 1\rangle + t_2\langle 2, 0, 1\rangle$$
$$\langle -1 + 4t_1, 2 - 2t_1, 2 + t_1\rangle = \langle 2t_2, 1, 1 + t_2\rangle$$

Equating corresponding components yields

$$-1 + 4t_1 = 2t_2$$

$$2 - 2t_1 = 1$$

$$2 + t_1 = 1 + t_2$$

The second equation implies $t_1 = \frac{1}{2}$. Substituting into the first and third equations we get

$$-1 + 4 \cdot \frac{1}{2} = 2t_2 \quad \Rightarrow \quad t_2 = \frac{1}{2}$$

$$2 + \frac{1}{2} = 1 + t_2 \quad \Rightarrow \quad t_2 = \frac{3}{2}$$

We conclude that the equations do not have solutions, which means that the two lines do not intersect.

53. Determine whether the lines $\mathbf{r}_1(t) = \langle 0, 1, 1 \rangle + t \langle 1, 1, 2 \rangle$ and $\mathbf{r}_2(s) = \langle 2, 0, 3 \rangle + s \langle 1, 4, 4 \rangle$ intersect, and if so, find the point of intersection.

SOLUTION   The lines intersect if there exist parameter values t and s such that

$$\langle 0, 1, 1 \rangle + t \langle 1, 1, 2 \rangle = \langle 2, 0, 3 \rangle + s \langle 1, 4, 4 \rangle$$

$$\langle t, 1 + t, 1 + 2t \rangle = \langle 2 + s, 4s, 3 + 4s \rangle \tag{1}$$

Equating corresponding components we get

$$t = 2 + s$$

$$1 + t = 4s$$

$$1 + 2t = 3 + 4s$$

Substituting t from the first equation into the second equation we get

$$\begin{matrix} 1 + 2 + s = 4s \\ 3s = 3 \end{matrix} \quad \Rightarrow \quad s = 1, \; t = 2 + s = 3$$

We now check whether $s = 1$, $t = 3$ satisfy the third equation:

$$1 + 2 \cdot 3 = 3 + 4 \cdot 1$$

$$7 = 7$$

We conclude that $s = 1$, $t = 3$ is the solution of (1), hence the two lines intersect. To find the point of intersection we substitute $s = 1$ in the right-hand side of (1) to obtain

$$\langle 2 + 1, 4 \cdot 1, 3 + 4 \cdot 1 \rangle = \langle 3, 4, 7 \rangle$$

The point of intersection is the terminal point of this vector, that is, $(3, 4, 7)$.

55. Find the components of the vector $\mathbf{v}$ whose tail and head are the midpoints of segments $\overline{AC}$ and $\overline{BC}$ in Figure 19.

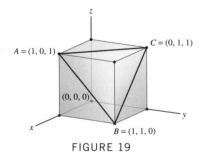

FIGURE 19

SOLUTION   We denote by P and Q the midpoints of the segments $\overline{AC}$ and $\overline{BC}$ respectively. Thus,

$$\mathbf{v} = \overrightarrow{PQ} \tag{1}$$

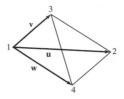

We use the formula for the midpoint of a segment to find the coordinates of the points P and Q. This gives

$$P = \left(\frac{1+0}{2}, \frac{0+1}{2}, \frac{1+1}{2}\right) = \left(\frac{1}{2}, \frac{1}{2}, 1\right)$$

$$Q = \left(\frac{1+0}{2}, \frac{1+1}{2}, \frac{0+1}{2}\right) = \left(\frac{1}{2}, 1, \frac{1}{2}\right)$$

Substituting in (1) yields the following vector:

$$\mathbf{v} = \overrightarrow{PQ} = \left\langle \frac{1}{2} - \frac{1}{2}, 1 - \frac{1}{2}, \frac{1}{2} - 1 \right\rangle = \left\langle 0, \frac{1}{2}, -\frac{1}{2} \right\rangle.$$

Further Insights and Challenges

*In Exercises 57–63, we consider the equations of a line in **symmetric form**, when $a \neq 0$, $b \neq 0$, $c \neq 0$.*

$$\frac{x - x_0}{a} = \frac{y - y_0}{b} = \frac{z - z_0}{c} \qquad \boxed{12}$$

57. Let $\mathcal{L}$ be the line through $P_0 = (x_0, y_0, c_0)$ with direction vector $\mathbf{v} = \langle a, b, c \rangle$. Show that $\mathcal{L}$ is defined by the symmetric Eq. (12). *Hint:* Use the vector parametrization to show that every point on $\mathcal{L}$ satisfies Eq. (12).

SOLUTION $\mathcal{L}$ is given by vector parametrization

$$\mathbf{r}(t) = \langle x_0, y_0, z_0 \rangle + t \langle a, b, c \rangle$$

which gives us the equations

$$x - x_0 \mid at$$
$$y = y_0 + bt$$
$$z = z_0 + ct.$$

Solving for t gives

$$t = \frac{x - x_0}{a}$$
$$t = \frac{y - y_0}{b}$$
$$t = \frac{z - z_0}{c}$$

Setting each equation equal to the other gives Eq. (12).

59. Find the symmetric equations of the line through $P = (1, 1, 2)$ and $Q = (-2, 4, 0)$.

SOLUTION This line has direction vector $\overrightarrow{PQ} = \langle -3, 3, -2 \rangle$. Using $(x_0, y_0, z_0) = P = (1, 1, 2)$ and $\langle a, b, c \rangle = \overrightarrow{PQ} = \langle -3, 3, -2 \rangle$ in Eq. (12) gives

$$\frac{x - 1}{-3} = \frac{y - 1}{3} = \frac{z - 2}{-2}$$

61. Find a vector parametrization for the line

$$\frac{x - 5}{9} = \frac{y + 3}{7} = z - 10$$

SOLUTION Using $(x_0, y_0, z_0) = (5, -3, 10)$ and $\langle a, b, c \rangle = \langle 9, 7, 1 \rangle$ gives

$$\mathbf{r}(t) = \langle 5, -3, 10 \rangle + t \langle 9, 7, 1 \rangle$$

63. Show that the line in the plane through (x_0, y_0) of slope m has symmetric equations

$$x - x_0 = \frac{y - y_0}{m}$$

SOLUTION The line through (x_0, y_0) of slope m has equation $y - y_0 = m(x - x_0)$, which becomes $x - x_0 = \frac{1}{m}(y - y_0)$, which becomes

$$\frac{x - x_0}{1} = \frac{y - y_0}{m}$$

65. A median of a tetrahedron is a segment joining a vertex to the centroid of the opposite face. The tetrahedron in Figure 20(B) has vertices at the origin and at the terminal points of vectors $\mathbf{u}$, $\mathbf{v}$, and $\mathbf{w}$. Show that the medians intersect at the terminal point of $\frac{1}{4}(\mathbf{u} + \mathbf{v} + \mathbf{w})$.

SOLUTION We first find vectors from the origin to the centroids of the four faces (labelled 1,2,3,4 after their opposite vertices, also labelled 1,2,3,4). Now, by the previous problem (Exercise 64), a vector from the origin (vertex 1) to the centroid of the opposite face (face 1) is $\frac{1}{3}(\mathbf{u} + \mathbf{v} + \mathbf{w})$. As for face 2, a vector from vertex 2 to the centroid of face 2 is $\frac{1}{3}(-\mathbf{u} + (\mathbf{v} - \mathbf{u}) + (\mathbf{w} - \mathbf{u}))$, but since vertex 2 is at the head of vector $\mathbf{u}$, then a vector from the origin to the centroid of face 2 is $\mathbf{u} + \frac{1}{3}(-\mathbf{u} + (\mathbf{v} - \mathbf{u}) + (\mathbf{w} - \mathbf{u})) = \frac{1}{3}(\mathbf{v} + \mathbf{w})$. Similarly, a vector from the origin to the centroid of face 3 is $\mathbf{v} + \frac{1}{3}(-\mathbf{v} + (\mathbf{u} - \mathbf{v}) + (\mathbf{w} - \mathbf{v})) = \frac{1}{3}(\mathbf{u} + \mathbf{w})$, and from the origin to the centroid of face 4 is $\frac{1}{3}(\mathbf{u} + \mathbf{v})$.

We now find the paramentric equations of four lines $\ell_1, \ldots, \ell_4$, each from vertex i to the centroid of the (opposite) face i.

$$\ell_1(t) = t\mathbf{0} + (1 - t)\frac{1}{3}(\mathbf{u} + \mathbf{v} + \mathbf{w})$$

$$\ell_2(t) = t\mathbf{u} + (1 - t)\frac{1}{3}(\mathbf{v} + \mathbf{w})$$

$$\ell_3(t) = t\mathbf{v} + (1 - t)\frac{1}{3}(\mathbf{u} + \mathbf{w})$$

$$\ell_4(t) = t\mathbf{w} + (1 - t)\frac{1}{3}(\mathbf{u} + \mathbf{v})$$

By substituting $t = 1/4$ into each line, we find that they all intersect in the same point:

$$\ell_1(1/4) = 1/4\mathbf{0} + (1 - 1/4)\frac{1}{3}(\mathbf{u} + \mathbf{v} + \mathbf{w}) = 1/4(\mathbf{u} + \mathbf{v} + \mathbf{w})$$

$$\ell_2(1/4) = 1/4\mathbf{u} + (1 - 1/4)\frac{1}{3}(\mathbf{v} + \mathbf{w}) = 1/4(\mathbf{u} + \mathbf{v} + \mathbf{w})$$

$$\ell_3(1/4) = 1/4\mathbf{v} + (1 - 1/4)\frac{1}{3}(\mathbf{u} + \mathbf{w}) = 1/4(\mathbf{u} + \mathbf{v} + \mathbf{w})$$

$$\ell_4(1/4) = 1/4\mathbf{w} + (1 - 1/4)\frac{1}{3}(\mathbf{u} + \mathbf{v}) = 1/4(\mathbf{u} + \mathbf{v} + \mathbf{w})$$

We conclude that all four lines intersect at the terminal point of the vector $1/4(\mathbf{u} + \mathbf{v} + \mathbf{w})$, as desired.

12.3 Dot Product and the Angle between Two Vectors   (LT Section 13.3)

Preliminary Questions

1. Is the dot product of two vectors a scalar or a vector?

SOLUTION The dot product of two vectors is the sum of products of scalars, hence it is a scalar.

2. What can you say about the angle between $\mathbf{a}$ and $\mathbf{b}$ if $\mathbf{a} \cdot \mathbf{b} < 0$?

SOLUTION Since the cosine of the angle between $\mathbf{a}$ and $\mathbf{b}$ satisfies $\cos\theta = \frac{\mathbf{a} \cdot \mathbf{b}}{\|\mathbf{a}\| \|\mathbf{b}\|}$, also $\cos\theta < 0$. By definition $0 \le \theta \le \pi$, but since $\cos\theta < 0$ then θ is in $(\pi/2, \pi]$. In other words, the angle between $\mathbf{a}$ and $\mathbf{b}$ is obtuse.

3. Which property of dot products allows us to conclude that if $\mathbf{v}$ is orthogonal to both $\mathbf{u}$ and $\mathbf{w}$, then $\mathbf{v}$ is orthogonal to $\mathbf{u} + \mathbf{w}$?

SOLUTION One property is that two vectors are orthogonal if and only if the dot product of the two vectors is zero. The second property is the Distributive Law. Since $\mathbf{v}$ is orthogonal to $\mathbf{u}$ and $\mathbf{w}$, we have $\mathbf{v} \cdot \mathbf{u} = 0$ and $\mathbf{v} \cdot \mathbf{w} = 0$. Therefore,

$$\mathbf{v} \cdot (\mathbf{u} + \mathbf{w}) = \mathbf{v} \cdot \mathbf{u} + \mathbf{v} \cdot \mathbf{w} = 0 + 0 = 0$$

We conclude that $\mathbf{v}$ is orthogonal to $\mathbf{u} + \mathbf{w}$.

4. Which is the projection of $\mathbf{v}$ along $\mathbf{v}$: (a) $\mathbf{v}$ or (b) $\mathbf{e_v}$?

SOLUTION The projection of $\mathbf{v}$ along itself is $\mathbf{v}$, since

$$\mathbf{v}_{||} = \left(\frac{\mathbf{v} \cdot \mathbf{v}}{\mathbf{v} \cdot \mathbf{v}}\right)\mathbf{v} = \mathbf{v}$$

Also, the projection of $\mathbf{v}$ along $\mathbf{e_v}$ is the same answer, $\mathbf{v}$, because

$$\mathbf{v}_{||} = \left(\frac{\mathbf{v} \cdot \mathbf{e_v}}{\mathbf{e_v} \cdot \mathbf{e_v}}\right)\mathbf{e_v} = \|\mathbf{v}\|\mathbf{e_v} = \mathbf{v}$$

5. Let $\mathbf{u}_{||}$ be the projection of $\mathbf{u}$ along $\mathbf{v}$. Which of the following is the projection $\mathbf{u}$ along the vector $2\mathbf{v}$ and which is the projection of $2\mathbf{u}$ along $\mathbf{v}$?

(a) $\frac{1}{2}\mathbf{u}_{||}$ (b) $\mathbf{u}_{||}$ (c) $2\mathbf{u}_{||}$

SOLUTION Since $\mathbf{u}_{||}$ is the projection of $\mathbf{u}$ along $\mathbf{v}$, we have,

$$\mathbf{u}_{||} = \left(\frac{\mathbf{u} \cdot \mathbf{v}}{\mathbf{v} \cdot \mathbf{v}}\right)\mathbf{v}$$

The projection of $\mathbf{u}$ along the vector $2\mathbf{v}$ is

$$\left(\frac{\mathbf{u} \cdot 2\mathbf{v}}{2\mathbf{v} \cdot 2\mathbf{v}}\right)2\mathbf{v} = \left(\frac{2\mathbf{u} \cdot \mathbf{v}}{4\mathbf{v} \cdot \mathbf{v}}\right)2\mathbf{v} = \left(\frac{4\mathbf{u} \cdot \mathbf{v}}{4\mathbf{v} \cdot \mathbf{v}}\right)\mathbf{v} = \left(\frac{\mathbf{u} \cdot \mathbf{v}}{\mathbf{v} \cdot \mathbf{v}}\right)\mathbf{v} = \mathbf{u}_{||}$$

That is, $\mathbf{u}_{||}$ is the projection of $\mathbf{u}$ along $2\mathbf{v}$, so our answer is (b) for the first part. Notice that the projection of $\mathbf{u}$ along $\mathbf{v}$ is the projection of $\mathbf{u}$ along the unit vector $\mathbf{e_v}$, hence it depends on the direction of $\mathbf{v}$ rather than on the length of $\mathbf{v}$. Therefore, the projection of $\mathbf{u}$ along $\mathbf{v}$ and along $2\mathbf{v}$ is the same vector.

On the other hand, the projection of $2\mathbf{u}$ along $\mathbf{v}$ is as follows:

$$\left(\frac{2\mathbf{u} \cdot \mathbf{v}}{\mathbf{v} \cdot \mathbf{v}}\right)\mathbf{v} = 2\left(\frac{\mathbf{u} \cdot \mathbf{v}}{\mathbf{v} \cdot \mathbf{v}}\right)\mathbf{v} = 2\mathbf{u}_{||}$$

giving us answer (c) for the second part.

6. Which of the following is equal to $\cos\theta$, where θ is the angle between $\mathbf{u}$ and $\mathbf{v}$?

(a) $\mathbf{u} \cdot \mathbf{v}$ (b) $\mathbf{u} \cdot \mathbf{e_v}$ (c) $\mathbf{e_u} \cdot \mathbf{e_v}$

SOLUTION By the Theorems on the Dot Product and the Angle Between Vectors, we have

$$\cos\theta = \frac{\mathbf{u} \cdot \mathbf{v}}{\|\mathbf{u}\|\|\mathbf{v}\|} = \frac{\mathbf{u}}{\|\mathbf{u}\|} \cdot \frac{\mathbf{v}}{\|\mathbf{v}\|} = \mathbf{e_u} \cdot \mathbf{e_v}$$

The correct answer is (c).

Exercises

In Exercises 1–12, compute the dot product.

1. $\langle 1, 2, 1 \rangle \cdot \langle 4, 3, 5 \rangle$

SOLUTION Using the definition of the dot product we obtain

$$\langle 1, 2, 1 \rangle \cdot \langle 4, 3, 5 \rangle = 1 \cdot 4 + 2 \cdot 3 + 1 \cdot 5 = 15$$

3. $\langle 0, 1, 0 \rangle \cdot \langle 7, 41, -3 \rangle$

SOLUTION The dot product is

$$\langle 0, 1, 0 \rangle \cdot \langle 7, 41, -3 \rangle = 0 \cdot 7 + 1 \cdot 41 + 0 \cdot (-3) = 41$$

5. $\langle 3, 1 \rangle \cdot \langle 4, -7 \rangle$

SOLUTION The dot product of the two vectors is the following scalar:

$$\langle 3, 1 \rangle \cdot \langle 4, -7 \rangle = 3 \cdot 4 + 1 \cdot (-7) = 5$$

7. $\mathbf{k} \cdot \mathbf{j}$

SOLUTION By the orthogonality of $\mathbf{j}$ and $\mathbf{k}$, we have $\mathbf{k} \cdot \mathbf{j} = 0$

9. $(\mathbf{i} + \mathbf{j}) \cdot (\mathbf{j} + \mathbf{k})$

SOLUTION By the distributive law and the orthogonality of $\mathbf{i}, \mathbf{j}$ and $\mathbf{k}$ we have

$$(\mathbf{i} + \mathbf{j}) \cdot (\mathbf{j} + \mathbf{k}) = \mathbf{i} \cdot \mathbf{j} + \mathbf{i} \cdot \mathbf{k} + \mathbf{j} \cdot \mathbf{j} + \mathbf{j} \cdot \mathbf{k} = 0 + 0 + 1 + 0 = 1$$

11. $(\mathbf{i} + \mathbf{j} + \mathbf{k}) \cdot (3\mathbf{i} + 2\mathbf{j} - 5\mathbf{k})$

SOLUTION We use properties of the dot product to obtain

$$(\mathbf{i} + \mathbf{j} + \mathbf{k}) \cdot (3\mathbf{i} + 2\mathbf{j} - 5\mathbf{k}) = 3\mathbf{i} \cdot \mathbf{i} + 2\mathbf{i} \cdot \mathbf{j} - 5\mathbf{i} \cdot \mathbf{k} + 3\mathbf{j} \cdot \mathbf{i} + 2\mathbf{j} \cdot \mathbf{j} - 5\mathbf{j} \cdot \mathbf{k} + 3\mathbf{k} \cdot \mathbf{i} + 2\mathbf{k} \cdot \mathbf{j} - 5\mathbf{k} \cdot \mathbf{k}$$
$$= 3\|\mathbf{i}\|^2 + 2\|\mathbf{j}\|^2 - 5\|\mathbf{k}\|^2 = 3 \cdot 1 + 2 \cdot 1 - 5 \cdot 1 = 0$$

In Exercises 13–18, determine whether the two vectors are orthogonal and, if not, whether the angle between them is acute or obtuse.

13. $\langle 1, 1, 1 \rangle, \quad \langle 1, -2, -2 \rangle$

SOLUTION We compute the dot product of the two vectors:

$$\langle 1, 1, 1 \rangle \cdot \langle 1, -2, -2 \rangle = 1 \cdot 1 + 1 \cdot (-2) + 1 \cdot (-2) = -3$$

Since the dot product is negative, the angle between the vectors is obtuse.

15. $\langle 1, 2, 1 \rangle, \quad \langle 7, -3, -1 \rangle$

SOLUTION We compute the dot product:

$$\langle 1, 2, 1 \rangle \cdot \langle 7, -3, -1 \rangle = 1 \cdot 7 + 2 \cdot (-3) + 1 \cdot (-1) = 0$$

The dot product is zero, hence the vectors are orthogonal.

17. $\left(\frac{12}{5}, -\frac{4}{5} \right), \left(\frac{1}{2}, -\frac{7}{4} \right)$

SOLUTION We find the dot product of the two vectors:

$$\left\langle \frac{12}{5}, -\frac{4}{5} \right\rangle \cdot \left\langle \frac{1}{2}, -\frac{7}{4} \right\rangle = \frac{12}{5} \cdot \frac{1}{2} + \left(-\frac{4}{5} \right) \cdot \left(-\frac{7}{4} \right) = \frac{12}{10} + \frac{28}{20} = \frac{13}{5}$$

The dot product is positive, hence the angle between the vectors is acute.

In Exercises 19–22, find the cosine of the angle between the vectors.

19. $\langle 0, 3, 1 \rangle, \quad \langle 4, 0, 0 \rangle$

SOLUTION Since $\langle 0, 3, 1 \rangle \cdot \langle 4, 0, 0 \rangle = 0 \cdot 4 + 3 \cdot 0 + 1 \cdot 0 = 0$, the vectors are orthogonal, that is, the angle between them is $\theta = 90°$ and $\cos \theta = 0$.

21. $\mathbf{i} + \mathbf{j}, \quad \mathbf{j} + 2\mathbf{k}$

SOLUTION We use the formula for the cosine of the angle between two vectors. Let $\mathbf{v} = \mathbf{i} + \mathbf{j}$ and $\mathbf{w} = \mathbf{j} + 2\mathbf{k}$. We compute the following values:

$$\|\mathbf{v}\| = \|\mathbf{i} + \mathbf{j}\| = \sqrt{1^2 + 1^2} = \sqrt{2}$$
$$\|\mathbf{w}\| = \|\mathbf{j} + 2\mathbf{k}\| = \sqrt{1^2 + 2^2} = \sqrt{5}$$
$$\mathbf{v} \cdot \mathbf{w} = (\mathbf{i} + \mathbf{j}) \cdot (\mathbf{j} + 2\mathbf{k}) = \mathbf{i} \cdot \mathbf{j} + 2\mathbf{i} \cdot \mathbf{k} + \mathbf{j} \cdot \mathbf{j} + 2\mathbf{j} \cdot \mathbf{k} = \|\mathbf{j}\|^2 = 1$$

Hence,

$$\cos \theta = \frac{\mathbf{v} \cdot \mathbf{w}}{\|\mathbf{v}\| \|\mathbf{w}\|} = \frac{1}{\sqrt{2}\sqrt{5}} = \frac{1}{\sqrt{10}}.$$

In Exercises 23–28, find the angle between the vectors. Use a calculator if necessary.

23. $\langle 2, \sqrt{2} \rangle, \quad \langle 1 + \sqrt{2}, 1 - \sqrt{2} \rangle$

SOLUTION We write $\mathbf{v} = \langle 2, \sqrt{2} \rangle$ and $\mathbf{w} = \langle 2, \sqrt{2} \rangle$. To use the formula for the cosine of the angle θ between two vectors we need to compute the following values:

$$\|\mathbf{v}\| = \sqrt{4 + 2} = \sqrt{6}$$
$$\|\mathbf{w}\| = \sqrt{(1 + \sqrt{2})^2 + (1 - \sqrt{2})^2} = \sqrt{6}$$
$$\mathbf{v} \cdot \mathbf{w} = 2 + 2\sqrt{2} + \sqrt{2} - 2 = 3\sqrt{2}$$

Hence,

$$\cos\theta = \frac{\mathbf{v}\cdot\mathbf{w}}{\|\mathbf{v}\|\|\mathbf{w}\|} = \frac{3\sqrt{2}}{\sqrt{6}\sqrt{6}} = \frac{\sqrt{2}}{2}$$

and so,

$$\theta = \cos^{-1}\frac{\sqrt{2}}{2} = \pi/4$$

25. $\langle 1, 1, 1\rangle$, $\langle 1, 0, 1\rangle$

SOLUTION We denote $\mathbf{v} = \langle 1, 1, 1\rangle$ and $\mathbf{w} = \langle 1, 0, 1\rangle$. To use the formula for the cosine of the angle θ between two vectors we need to compute the following values:

$$\|\mathbf{v}\| = \sqrt{1^2 + 1^2 + 1^2} = \sqrt{3}$$
$$\|\mathbf{w}\| = \sqrt{1^2 + 0^2 + 1^2} = \sqrt{2}$$
$$\mathbf{v}\cdot\mathbf{w} = 1 + 0 + 1 = 2$$

Hence,

$$\cos\theta = \frac{\mathbf{v}\cdot\mathbf{w}}{\|\mathbf{v}\|\|\mathbf{w}\|} = \frac{2}{\sqrt{3}\sqrt{2}} = \frac{\sqrt{6}}{3}$$

and so,

$$\theta = \cos^{-1}\frac{\sqrt{6}}{3} \approx 0.615$$

27. $\langle 0, 1, 1\rangle$, $\langle 1, -1, 0\rangle$

SOLUTION We denote $\mathbf{v} = \langle 0, 1, 1\rangle$ and $\mathbf{w} = \langle 1, -1, 0\rangle$. To use the formula for the cosine of the angle θ between two vectors we need to compute the following values:

$$\|\mathbf{v}\| = \sqrt{0^2 + 1^2 + 1^2} = \sqrt{2}$$
$$\|\mathbf{w}\| = \sqrt{1^2 + (-1)^2 + 0^2} = \sqrt{2}$$
$$\mathbf{v}\cdot\mathbf{w} = 0 + (-1) + 0 = -1$$

Hence,

$$\cos\theta = \frac{\mathbf{v}\cdot\mathbf{w}}{\|\mathbf{v}\|\|\mathbf{w}\|} = \frac{-1}{\sqrt{2}\sqrt{2}} = -\frac{1}{2}$$

and so,

$$\theta = \cos^{-1}-\frac{1}{2} = \frac{2\pi}{3}$$

29. Find all values of b for which the vectors are orthogonal.

(a) $\langle b, 3, 2\rangle$, $\langle 1, b, 1\rangle$ **(b)** $\langle 4, -2, 7\rangle$, $\langle b^2, b, 0\rangle$

SOLUTION

(a) The vectors are orthogonal if and only if the scalar product is zero. That is,

$$\langle b, 3, 2\rangle \cdot \langle 1, b, 1\rangle = 0$$
$$b\cdot 1 + 3\cdot b + 2\cdot 1 = 0$$
$$4b + 2 = 0 \quad \Rightarrow \quad b = -\frac{1}{2}$$

(b) We set the scalar product of the two vectors equal to zero and solve for b. This gives

$$\langle 4, -2, 7\rangle \cdot \langle b^2, b, 0\rangle = 0$$
$$4b^2 - 2b + 7\cdot 0 = 0$$
$$2b(2b - 1) = 0 \quad \Rightarrow \quad b = 0 \text{ or } b = \frac{1}{2}$$

31. Find two vectors that are not multiples of each other and are both orthogonal to $\langle 2, 0, -3 \rangle$.

SOLUTION We denote by $\langle a, b, c \rangle$, a vector orthogonal to $\langle 2, 0, -3 \rangle$. Hence,

$$\langle a, b, c \rangle \cdot \langle 2, 0, -3 \rangle = 0$$

$$2a + 0 - 3c = 0$$

$$2a - 3c = 0 \quad \Rightarrow \quad a = \frac{3}{2}c$$

Thus, the vectors orthogonal to $\langle 2, 0, -3 \rangle$ are of the form

$$\left\langle \frac{3}{2}c, b, c \right\rangle.$$

We may find two such vectors by setting $c = 0$, $b = 1$ and $c = 2$, $b = 2$. We obtain

$$\mathbf{v}_1 = \langle 0, 1, 0 \rangle, \quad \mathbf{v}_2 = \langle 3, 2, 2 \rangle.$$

33. Find $\mathbf{v} \cdot \mathbf{e}$ where $\|\mathbf{v}\| = 3$, $\mathbf{e}$ is a unit vector, and the angle between $\mathbf{e}$ and $\mathbf{v}$ is $\frac{2\pi}{3}$.

SOLUTION Since $\mathbf{v} \cdot \mathbf{e} = \|\mathbf{v}\| \|\mathbf{e}\| \cos 2\pi/3$, and $\|\mathbf{v}\| = 3$ and $\|\mathbf{e}\| = 1$, we have $\mathbf{v} \cdot \mathbf{e} = 3 \cdot 1 \cdot (-1/2) = -3/2$.

In Exercises 35–38, simplify the expression.

35. $(\mathbf{v} - \mathbf{w}) \cdot \mathbf{v} + \mathbf{v} \cdot \mathbf{w}$

SOLUTION By properties of the dot product we obtain

$$(\mathbf{v} - \mathbf{w}) \cdot \mathbf{v} + \mathbf{v} \cdot \mathbf{w} = \mathbf{v} \cdot \mathbf{v} - \mathbf{w} \cdot \mathbf{v} + \mathbf{v} \cdot \mathbf{w} = \|\mathbf{v}\|^2 - \mathbf{v} \cdot \mathbf{w} + \mathbf{v} \cdot \mathbf{w} = \|\mathbf{v}\|^2$$

37. $(\mathbf{v} + \mathbf{w}) \cdot \mathbf{v} - (\mathbf{v} + \mathbf{w}) \cdot \mathbf{w}$

SOLUTION We use properties of the dot product to write

$$(\mathbf{v} + \mathbf{w}) \cdot \mathbf{v} - (\mathbf{v} + \mathbf{w}) \cdot \mathbf{w} = \mathbf{v} \cdot \mathbf{v} + \mathbf{w} \cdot \mathbf{v} - \mathbf{v} \cdot \mathbf{w} - \mathbf{w} \cdot \mathbf{w}$$

$$= \|\mathbf{v}\|^2 + \mathbf{w} \cdot \mathbf{v} - \mathbf{w} \cdot \mathbf{v} - \|\mathbf{w}\|^2 = \|\mathbf{v}\|^2 - \|\mathbf{w}\|^2$$

In Exercises 39–42, use the properties of the dot product to evaluate the expression, assuming that $\mathbf{u} \cdot \mathbf{v} = 2$, $\|\mathbf{u}\| = 1$, *and* $\|\mathbf{v}\| = 3$.

39. $\mathbf{u} \cdot (4\mathbf{v})$

SOLUTION Using properties of the dot product we get

$$\mathbf{u} \cdot (4\mathbf{v}) = 4(\mathbf{u} \cdot \mathbf{v}) = 4 \cdot 2 = 8.$$

41. $2\mathbf{u} \cdot (3\mathbf{u} - \mathbf{v})$

SOLUTION By properties of the dot product we obtain

$$2\mathbf{u} \cdot (3\mathbf{u} - \mathbf{v}) = (2\mathbf{u}) \cdot (3\mathbf{u}) - (2\mathbf{u}) \cdot \mathbf{v} = 6(\mathbf{u} \cdot \mathbf{u}) - 2(\mathbf{u} \cdot \mathbf{v})$$

$$= 6\|\mathbf{u}\|^2 - 2(\mathbf{u} \cdot \mathbf{v}) = 6 \cdot 1^2 - 2 \cdot 2 = 2$$

43. Find the angle between $\mathbf{v}$ and $\mathbf{w}$ if $\mathbf{v} \cdot \mathbf{w} = -\|\mathbf{v}\| \|\mathbf{w}\|$.

SOLUTION Using the formula for dot product, and the given equation $\mathbf{v} \cdot \mathbf{w} = -\|\mathbf{v}\| \|\mathbf{w}\|$, we get:

$$\|\mathbf{v}\| \|\mathbf{w}\| \cos \theta = -\|\mathbf{v}\| \|\mathbf{w}\|,$$

which implies $\cos \theta = -1$, and so the angle between the two vectors is $\theta = \pi$.

45. Assume that $\|\mathbf{v}\| = 3$, $\|\mathbf{w}\| = 5$ and that the angle between $\mathbf{v}$ and $\mathbf{w}$ is $\theta = \frac{\pi}{3}$.

(a) Use the relation $\|\mathbf{v} + \mathbf{w}\|^2 = (\mathbf{v} + \mathbf{w}) \cdot (\mathbf{v} + \mathbf{w})$ to show that $\|\mathbf{v} + \mathbf{w}\|^2 = 3^2 + 5^2 + 2\mathbf{v} \cdot \mathbf{w}$.
(b) Find $\|\mathbf{v} + \mathbf{w}\|$.

SOLUTION For part (a), we use the distributive property to get:

$$\|\mathbf{v} + \mathbf{w}\|^2 = (\mathbf{v} + \mathbf{w}) \cdot (\mathbf{v} + \mathbf{w})$$

$$= \mathbf{v} \cdot \mathbf{v} + \mathbf{v} \cdot \mathbf{w} + \mathbf{w} \cdot \mathbf{v} + \mathbf{w} \cdot \mathbf{w}$$

$$= \|\mathbf{v}\|^2 + 2\mathbf{v} \cdot \mathbf{w} + \|\mathbf{w}\|^2$$

$$= 3^2 + 5^2 + 2\mathbf{v} \cdot \mathbf{w}$$

For part (b), we use the definition of dot product on the previous equation to get:

$$\|\mathbf{v} + \mathbf{w}\|^2 = 3^2 + 5^2 + 2\mathbf{v} \cdot \mathbf{w}$$
$$= 34 + 2 \cdot 3 \cdot 5 \cdot \cos \pi/3$$
$$= 34 + 15 = 49$$

Thus, $\|\mathbf{v} + \mathbf{w}\| = \sqrt{49} = 7$.

47. Show that if $\mathbf{e}$ and $\mathbf{f}$ are unit vectors such that $\|\mathbf{e} + \mathbf{f}\| = \frac{3}{2}$, then $\|\mathbf{e} - \mathbf{f}\| = \frac{\sqrt{7}}{2}$. *Hint:* Show that $\mathbf{e} \cdot \mathbf{f} = \frac{1}{8}$.

SOLUTION We use the relation of the dot product with length and properties of the dot product to write

$$9/4 = \|\mathbf{e} + \mathbf{f}\|^2 = (\mathbf{e} + \mathbf{f}) \cdot (\mathbf{e} + \mathbf{f}) = \mathbf{e} \cdot \mathbf{e} + \mathbf{e} \cdot \mathbf{f} + \mathbf{f} \cdot \mathbf{e} + \mathbf{f} \cdot \mathbf{f}$$
$$= \|\mathbf{e}\|^2 + 2\mathbf{e} \cdot \mathbf{f} + \|\mathbf{f}\|^2 = 1^2 + 2\mathbf{e} \cdot \mathbf{f} + 1^2 = 2 + 2\mathbf{e} \cdot \mathbf{f}$$

We now find $\mathbf{e} \cdot \mathbf{f}$:

$$9/4 = 2 + 2\mathbf{e} \cdot \mathbf{f} \quad \Rightarrow \quad \mathbf{e} \cdot \mathbf{f} = 1/8$$

Hence, using the same method as above, we have:

$$\|\mathbf{e} - \mathbf{f}\|^2 = (\mathbf{e} - \mathbf{f}) \cdot (\mathbf{e} - \mathbf{f}) = \mathbf{e} \cdot \mathbf{e} - \mathbf{e} \cdot \mathbf{f} - \mathbf{f} \cdot \mathbf{e} + \mathbf{f} \cdot \mathbf{f}$$
$$= \|\mathbf{e}\|^2 - 2\mathbf{e} \cdot \mathbf{f} + \|\mathbf{f}\|^2 = 1^2 - 2\mathbf{e} \cdot \mathbf{f} + 1^2 = 2 - 2\mathbf{e} \cdot \mathbf{f} = 2 - 2/8 = 7/4.$$

Taking square roots, we get:

$$\|\mathbf{e} - \mathbf{f}\| = \frac{\sqrt{7}}{2}$$

49. Find the angle θ in the triangle in Figure 12.

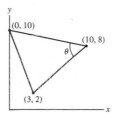

FIGURE 12

SOLUTION We denote by $\mathbf{u}$ and $\mathbf{v}$ the vectors in the figure.

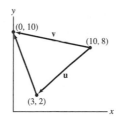

Hence,

$$\cos \theta = \frac{\mathbf{v} \cdot \mathbf{u}}{\|\mathbf{v}\| \|\mathbf{u}\|} \tag{1}$$

We find the vectors $\mathbf{v}$ and $\mathbf{u}$, and then compute their length and the dot product $\mathbf{v} \cdot \mathbf{u}$. This gives

$$\mathbf{v} = \langle 0 - 10, 10 - 8 \rangle = \langle -10, 2 \rangle$$
$$\mathbf{u} = \langle 3 - 10, 2 - 8 \rangle = \langle -7, -6 \rangle$$
$$\|\mathbf{v}\| = \sqrt{(-10)^2 + 2^2} = \sqrt{104}$$
$$\|\mathbf{u}\| = \sqrt{(-7)^2 + (-6)^2} = \sqrt{85}$$
$$\mathbf{v} \cdot \mathbf{u} = \langle -10, 2 \rangle \cdot \langle -7, -6 \rangle = (-10) \cdot (-7) + 2 \cdot (-6) = 58$$

Substituting these values in (1) yields

$$\cos\theta = \frac{58}{\sqrt{104}\sqrt{85}} \approx 0.617$$

Hence the angle of the triangle is $51.91°$.

*In Exercises 51–58, find the projection of **u** along **v**.*

51. $\mathbf{u} = \langle 2, 5 \rangle, \quad \mathbf{v} = \langle 1, 1 \rangle$

SOLUTION   We first compute the following dot products:

$$\mathbf{u} \cdot \mathbf{v} = \langle 2, 5 \rangle \cdot \langle 1, 1 \rangle = 7$$
$$\mathbf{v} \cdot \mathbf{v} = \|\mathbf{v}\|^2 = 1^2 + 1^2 = 2$$

The projection of **u** along **v** is the following vector:

$$\mathbf{u}_{\parallel} = \left(\frac{\mathbf{u} \cdot \mathbf{v}}{\mathbf{v} \cdot \mathbf{v}}\right) \mathbf{v} = \frac{7}{2}\mathbf{v} = \left\langle \frac{7}{2}, \frac{7}{2} \right\rangle$$

53. $\mathbf{u} = \langle -1, 2, 0 \rangle, \quad \mathbf{v} = \langle 2, 0, 1 \rangle$

SOLUTION   The projection of **u** along **v** is the following vector:

$$\mathbf{u}_{\parallel} = \left(\frac{\mathbf{u} \cdot \mathbf{v}}{\mathbf{v} \cdot \mathbf{v}}\right) \mathbf{v}$$

We compute the values in this expression:

$$\mathbf{u} \cdot \mathbf{v} = \langle -1, 2, 0 \rangle \cdot \langle 2, 0, 1 \rangle = -1 \cdot 2 + 2 \cdot 0 + 0 \cdot 1 = -2$$
$$\mathbf{v} \cdot \mathbf{v} = \|\mathbf{v}\|^2 = 2^2 + 0^2 + 1^2 = 5$$

Hence,

$$\mathbf{u}_{\parallel} = -\frac{2}{5}\langle 2, 0, 1 \rangle = \left\langle -\frac{4}{5}, 0, -\frac{2}{5} \right\rangle.$$

55. $\mathbf{u} = 5\mathbf{i} + 7\mathbf{j} - 4\mathbf{k}, \quad \mathbf{v} = \mathbf{k}$

SOLUTION   The projection of **u** along **v** is the following vector:

$$\mathbf{u}_{\parallel} = \left(\frac{\mathbf{u} \cdot \mathbf{v}}{\mathbf{v} \cdot \mathbf{v}}\right) \mathbf{v}$$

We compute the dot products:

$$\mathbf{u} \cdot \mathbf{v} = (5\mathbf{i} + 7\mathbf{j} - 4\mathbf{k}) \cdot \mathbf{k} = -4\mathbf{k} \cdot \mathbf{k} = -4$$
$$\mathbf{v} \cdot \mathbf{v} = \|\mathbf{v}\|^2 = \|\mathbf{k}\|^2 = 1$$

Hence,

$$\mathbf{u}_{\parallel} = \frac{-4}{1}\mathbf{k} = -4\mathbf{k}$$

57. $\mathbf{u} = \langle a, b, c \rangle, \quad \mathbf{v} = \mathbf{i}$

SOLUTION   The component of **u** along **v** is a, since

$$\mathbf{u} \cdot \mathbf{e_v} = (a\mathbf{i} + b\mathbf{j} + c\mathbf{k}) \cdot \mathbf{i} = a$$

Therefore, the projection of **u** along **v** is the vector

$$\mathbf{u}_{\parallel} = (\mathbf{u} \cdot \mathbf{e_v})\mathbf{e_v} = a\mathbf{i}$$

*In Exercises 59 and 60, compute the component of **u** along **v**.*

59. $\mathbf{u} = \langle 3, 2, 1 \rangle, \quad \mathbf{v} = \langle 1, 0, 1 \rangle$

SOLUTION   We first compute the following dot products:

$$\mathbf{u} \cdot \mathbf{v} = \langle 3, 2, 1 \rangle \cdot \langle 1, 0, 1 \rangle = 4$$
$$\mathbf{v} \cdot \mathbf{v} = \|\mathbf{v}\|^2 = 1^2 + 1^2 = 2$$

The component of **u** along **v** is the length of the projection of **u** along **v**

$$\left\| \left(\frac{\mathbf{u} \cdot \mathbf{v}}{\mathbf{v} \cdot \mathbf{v}} \right) \mathbf{v} \right\| = \frac{4}{2} \|\mathbf{v}\| = 2\|\mathbf{v}\| = 2\sqrt{2}$$

61. Find the length of $\overline{OP}$ in Figure 14.

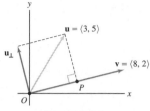

FIGURE 14

SOLUTION   This is just the component of $\mathbf{u} = \langle 3, 5 \rangle$ along $\mathbf{v} = \langle 8, 2 \rangle$. We first compute the following dot products:

$$\mathbf{u} \cdot \mathbf{v} = \langle 3, 5 \rangle \cdot \langle 8, 2 \rangle = 34$$

$$\mathbf{v} \cdot \mathbf{v} = \|\mathbf{v}\|^2 = 8^2 + 2^2 = 68$$

The component of **u** along **v** is the length of the projection of **u** along **v**

$$\left\| \left(\frac{\mathbf{u} \cdot \mathbf{v}}{\mathbf{v} \cdot \mathbf{v}} \right) \mathbf{v} \right\| = \frac{34}{68} \|\mathbf{v}\| = \frac{34}{68} \sqrt{68}$$

In Exercises 63–68, find the decomposition $\mathbf{a} = \mathbf{a}_{\|} + \mathbf{a}_{\perp}$ *with respect to* **b**.

63. $\mathbf{a} = \langle 1, 0 \rangle$, $\quad \mathbf{b} = \langle 1, 1 \rangle$

SOLUTION

Step 1. We compute $\mathbf{a} \cdot \mathbf{b}$ and $\mathbf{b} \cdot \mathbf{b}$

$$\mathbf{a} \cdot \mathbf{b} = \langle 1, 0 \rangle \cdot \langle 1, 1 \rangle = 1 \cdot 1 + 0 \cdot 1 = 1$$

$$\mathbf{b} \cdot \mathbf{b} = \|\mathbf{b}\|^2 = 1^2 + 1^2 = 2$$

Step 2. We find the projection of **a** along **b**:

$$\mathbf{a}_{\|} = \left(\frac{\mathbf{a} \cdot \mathbf{b}}{\mathbf{b} \cdot \mathbf{b}} \right) \mathbf{b} = \frac{1}{2} \langle 1, 1 \rangle = \left\langle \frac{1}{2}, \frac{1}{2} \right\rangle$$

Step 3. We find the orthogonal part as the difference:

$$\mathbf{a}_{\perp} = \mathbf{a} - \mathbf{a}_{\|} = \langle 1, 0 \rangle - \left\langle \frac{1}{2}, \frac{1}{2} \right\rangle = \left\langle \frac{1}{2}, -\frac{1}{2} \right\rangle$$

Hence,

$$\mathbf{a} = \mathbf{a}_{\|} + \mathbf{a}_{\perp} = \left\langle \frac{1}{2}, \frac{1}{2} \right\rangle + \left\langle \frac{1}{2}, -\frac{1}{2} \right\rangle.$$

65. $\mathbf{a} = \langle 4, -1, 0 \rangle$, $\quad \mathbf{b} = \langle 0, 1, 1 \rangle$

SOLUTION   We first compute $\mathbf{a} \cdot \mathbf{b}$ and $\mathbf{b} \cdot \mathbf{b}$ to find the projection of **a** along **b**:

$$\mathbf{a} \cdot \mathbf{b} = \langle 4, -1, 0 \rangle \cdot \langle 0, 1, 1 \rangle = 4 \cdot 0 + (-1) \cdot 1 + 0 \cdot 1 = -1$$

$$\mathbf{b} \cdot \mathbf{b} = \|\mathbf{b}\|^2 = 0^2 + 1^2 + 1^2 = 2$$

Hence,

$$\mathbf{a}_{\|} = \left(\frac{\mathbf{a} \cdot \mathbf{b}}{\mathbf{b} \cdot \mathbf{b}} \right) \mathbf{b} = \frac{-1}{2} \langle 0, 1, 1 \rangle = \left\langle 0, -\frac{1}{2}, -\frac{1}{2} \right\rangle$$

We now find the vector $\mathbf{a}_{\perp}$ orthogonal to **b** by computing the difference:

$$\mathbf{a} - \mathbf{a}_{\|} = \langle 4, -1, 0 \rangle - \left\langle 0, -\frac{1}{2}, -\frac{1}{2} \right\rangle = \left\langle 4, -\frac{1}{2}, \frac{1}{2} \right\rangle$$

Thus, we have

$$\mathbf{a} = \mathbf{a}_{\|} + \mathbf{a}_{\perp} = \left\langle 0, -\frac{1}{2}, -\frac{1}{2} \right\rangle + \left\langle 4, -\frac{1}{2}, \frac{1}{2} \right\rangle$$

67. $\mathbf{a} = \langle x, y \rangle$, $\mathbf{b} = \langle 1, -1 \rangle$

SOLUTION   We first compute $\mathbf{a} \cdot \mathbf{b}$ and $\mathbf{b} \cdot \mathbf{b}$ to find the projection of $\mathbf{a}$ along $\mathbf{b}$:

$$\mathbf{a} \cdot \mathbf{b} = \langle x, y \rangle \cdot \langle 1, -1 \rangle = x - y$$

$$\mathbf{b} \cdot \mathbf{b} = \|\mathbf{b}\|^2 = 1^2 + (-1)^2 = 2$$

Hence,

$$\mathbf{a}_{\|} = \left(\frac{\mathbf{a} \cdot \mathbf{b}}{\mathbf{b} \cdot \mathbf{b}} \right) \mathbf{b} = \frac{x-y}{2} \langle 1, -1 \rangle = \left\langle \frac{x-y}{2}, \frac{y-x}{2} \right\rangle$$

We now find the vector $\mathbf{a}_{\perp}$ orthogonal to $\mathbf{b}$ by computing the difference:

$$\mathbf{a} - \mathbf{a}_{\|} = \langle x, y \rangle - \left\langle \frac{x-y}{2}, \frac{y-x}{2} \right\rangle = \left\langle \frac{x+y}{2}, \frac{x+y}{2} \right\rangle$$

Thus, we have

$$\mathbf{a} = \mathbf{a}_{\|} + \mathbf{a}_{\perp} = \left\langle \frac{x-y}{2}, \frac{y-x}{2} \right\rangle + \left\langle \frac{x+y}{2}, \frac{x+y}{2} \right\rangle$$

69. Let $\mathbf{e}_{\theta} = \langle \cos \theta, \sin \theta \rangle$. Show that $\mathbf{e}_{\theta} \cdot \mathbf{e}_{\psi} = \cos(\theta - \psi)$ for any two angles θ and ψ.

SOLUTION   First, $\mathbf{e}_{\theta}$ is a unit vector since by a trigonometric identity we have

$$\|\mathbf{e}_{\theta}\| = \sqrt{\cos^2 \theta + \sin^2 \theta} = \sqrt{1} = 1$$

The cosine of the angle α between $\mathbf{e}_{\theta}$ and the vector $\mathbf{i}$ in the direction of the positive x-axis is

$$\cos \alpha = \frac{\mathbf{e}_{\theta} \cdot \mathbf{i}}{\|\mathbf{e}_{\theta}\| \cdot \|\mathbf{i}\|} = \mathbf{e}_{\theta} \cdot \mathbf{i} = ((\cos \theta)\mathbf{i} + (\sin \theta)\mathbf{j}) \cdot \mathbf{i} = \cos \theta$$

The solution of $\cos \alpha = \cos \theta$ for angles between 0 and π is $\alpha = \theta$. That is, the vector $\mathbf{e}_{\theta}$ makes an angle θ with the x-axis. We now use the trigonometric identity

$$\cos \theta \cos \psi + \sin \theta \sin \psi = \cos(\theta - \psi)$$

to obtain the following equality:

$$\mathbf{e}_{\theta} \cdot \mathbf{e}_{\psi} = \langle \cos \theta, \sin \theta \rangle \cdot \langle \cos \psi, \sin \psi \rangle = \cos \theta \cos \psi + \sin \theta \sin \psi = \cos(\theta - \psi)$$

In Exercises 71–74, refer to Figure 15.

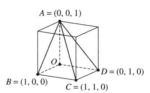

$A = (0, 0, 1)$

$B = (1, 0, 0)$ $C = (1, 1, 0)$ $D = (0, 1, 0)$

FIGURE 15   Unit cube in $\mathbf{R}^3$.

71. Find the angle between $\overline{AB}$ and $\overline{AC}$.

SOLUTION   The cosine of the angle α between the vectors $\overrightarrow{AB}$ and $\overrightarrow{AC}$ is

$$\cos \alpha = \frac{\overrightarrow{AB} \cdot \overrightarrow{AC}}{\|\overrightarrow{AB}\|\|\overrightarrow{AC}\|} \tag{1}$$

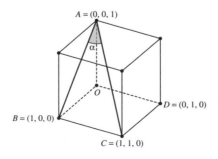

$A = (0, 0, 1)$

$B = (1, 0, 0)$ $D = (0, 1, 0)$

$C = (1, 1, 0)$

We compute the vectors $\overrightarrow{AB}$ and $\overrightarrow{AC}$ and then calculate their dot product and lengths. We get

$$\overrightarrow{AB} = \langle 1-0, 0-0, 0-1 \rangle = \langle 1, 0, -1 \rangle$$

$$\overrightarrow{AC} = \langle 1-0, 1-0, 0-1 \rangle = \langle 1, 1, -1 \rangle$$

$$\overrightarrow{AB} \cdot \overrightarrow{AC} = \langle 1, 0, -1 \rangle \cdot \langle 1, 1, -1 \rangle = 1 \cdot 1 + 0 \cdot 1 + (-1) \cdot (-1) = 2$$

$$\|\overrightarrow{AB}\| = \sqrt{1^2 + 0^2 + (-1)^2} = \sqrt{2}$$

$$\|\overrightarrow{AC}\| = \sqrt{1^2 + 1^2 + (-1)^2} = \sqrt{3}$$

Substituting in (1) and solving for $0 \le \alpha \le 90°$ gives

$$\cos \alpha = \frac{2}{\sqrt{2} \cdot \sqrt{3}} \approx 0.816 \quad \Rightarrow \quad \alpha \approx 35.31°.$$

73. Calculate the projection of $\overrightarrow{AC}$ along $\overrightarrow{AD}$.

SOLUTION $\overline{DC}$ is perpendicular to the face OAD of the cube. Hence, it is orthogonal to the segment $\overline{AD}$ on this face. Therefore, the projection of the vector $\overrightarrow{AC}$ along $\overrightarrow{AD}$ is the vector $\overrightarrow{AD}$ itself.

75. Let **v** and **w** be nonzero vectors and set $\mathbf{u} = \mathbf{e_v} + \mathbf{e_w}$. Use the dot product to show that the angle between **u** and **v** is equal to the angle between **u** and **w**. Explain this result geometrically with a diagram.

SOLUTION We denote by α the angle between **u** and **v** and by β the angle between **u** and **w**. Since $\mathbf{e_v}$ and $\mathbf{e_w}$ are vectors in the directions of **v** and **w** respectively, α is the angle between **u** and $\mathbf{e_v}$ and β is the angle between **u** and $\mathbf{e_w}$. The cosines of these angles are thus

$$\cos \alpha = \frac{\mathbf{u} \cdot \mathbf{e_v}}{\|\mathbf{u}\| \|\mathbf{e_v}\|} = \frac{\mathbf{u} \cdot \mathbf{e_v}}{\|\mathbf{u}\|}; \quad \cos \beta = \frac{\mathbf{u} \cdot \mathbf{e_w}}{\|\mathbf{u}\| \|\mathbf{e_w}\|} = \frac{\mathbf{u} \cdot \mathbf{e_w}}{\|\mathbf{u}\|}$$

To show that $\cos \alpha = \cos \beta$ (which implies that $\alpha = \beta$) we must show that

$$\mathbf{u} \cdot \mathbf{e_v} = \mathbf{u} \cdot \mathbf{e_w}.$$

We compute the two dot products:

$$\mathbf{u} \cdot \mathbf{e_v} = (\mathbf{e_v} + \mathbf{e_w}) \cdot \mathbf{e_v} = \mathbf{e_v} \cdot \mathbf{e_v} + \mathbf{e_w} \cdot \mathbf{e_v} = 1 + \mathbf{e_w} \cdot \mathbf{e_v}$$

$$\mathbf{u} \cdot \mathbf{e_w} = (\mathbf{e_v} + \mathbf{e_w}) \cdot \mathbf{e_w} = \mathbf{e_v} \cdot \mathbf{e_w} + \mathbf{e_w} \cdot \mathbf{e_w} = \mathbf{e_v} \cdot \mathbf{e_w} + 1$$

We see that $\mathbf{u} \cdot \mathbf{e_v} = \mathbf{u} \cdot \mathbf{e_w}$. We conclude that $\cos \alpha = \cos \beta$, hence $\alpha = \beta$. Geometrically, **u** is a diagonal in the rhombus $OABC$ (see figure), hence it bisects the angle $\sphericalangle AOC$ of the rhombus.

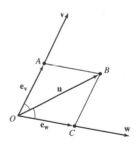

77. Calculate the force (in newtons) required to push a 40-kg wagon up a 10° incline (Figure 16).

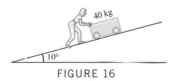

FIGURE 16

SOLUTION Gravity exerts a force $\mathbf{F}_g$ of magnitude $40g$ newtons where $g = 9.8$. The magnitude of the force required to push the wagon equals the component of the force $\mathbf{F}_g$ along the ramp. Resolving $\mathbf{F}_g$ into a sum $\mathbf{F}_g = \mathbf{F}_{\|} + \mathbf{F}_{\perp}$, where $\mathbf{F}_{\|}$ is the force along the ramp and $\mathbf{F}_{\perp}$ is the force orthogonal to the ramp, we need to find the magnitude of $\mathbf{F}_{\|}$. The angle between $\mathbf{F}_g$ and the ramp is $90° - 10° = 80°$. Hence,

$$\mathbf{F}_{\|} = \|\mathbf{F}_g\| \cos 80° = 40 \cdot 9.8 \cdot \cos 80° \approx 68.07 \text{ N}.$$

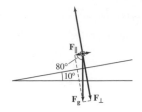

Therefore the minimum force required to push the wagon is 68.07 N. (Actually, this is the force required to keep the wagon from sliding down the hill; any slight amount greater than this force will serve to push it up the hill.)

79. A light beam travels along the ray determined by a unit vector **L**, strikes a flat surface at point P, and is reflected along the ray determined by a unit vector **R**, where $\theta_1 = \theta_2$ (Figure 18). Show that if **N** is the unit vector orthogonal to the surface, then

$$\mathbf{R} = 2(\mathbf{L} \cdot \mathbf{N})\mathbf{N} - \mathbf{L}$$

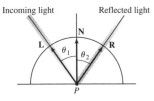

FIGURE 18

SOLUTION We denote by **W** a unit vector orthogonal to **N** in the direction shown in the figure, and let $\theta_1 = \theta_2 = \theta$.

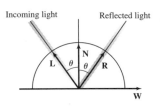

We resolve the unit vectors **R** and **L** into a sum of forces along **N** and **W**. This gives

$$\mathbf{R} = \cos(90 - \theta)\mathbf{W} + \cos\theta\mathbf{N} = \sin\theta\mathbf{W} + \cos\theta\mathbf{N}$$

$$\mathbf{L} = -\cos(90 - \theta)\mathbf{W} + \cos\theta\mathbf{N} = -\sin\theta\mathbf{W} + \cos\theta\mathbf{N} \tag{1}$$

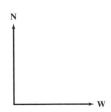

Now, since

$$\mathbf{L} \cdot \mathbf{N} = \|\mathbf{L}\|\|\mathbf{N}\| \cos\theta = 1 \cdot 1 \cos\theta = \cos\theta$$

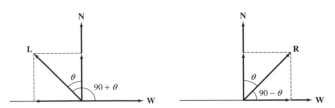

we have by (1):

$$2(\mathbf{L} \cdot \mathbf{N})\mathbf{N} - \mathbf{L} = (2\cos\theta)\mathbf{N} - \mathbf{L} = (2\cos\theta)\mathbf{N} - ((-\sin\theta)\mathbf{W} + (\cos\theta)\mathbf{N})$$

$$= (2\cos\theta)\mathbf{N} + (\sin\theta)\mathbf{W} - (\cos\theta)\mathbf{N} = (\sin\theta)\mathbf{W} + (\cos\theta)\mathbf{N} = \mathbf{R}$$

81. Prove that $\|\mathbf{v} + \mathbf{w}\|^2 - \|\mathbf{v} - \mathbf{w}\|^2 = 4\mathbf{v} \cdot \mathbf{w}$.

SOLUTION We compute the following values:

$$\|\mathbf{v} + \mathbf{w}\|^2 = (\mathbf{v} + \mathbf{w}) \cdot (\mathbf{v} + \mathbf{w}) = \mathbf{v} \cdot \mathbf{v} + \mathbf{v} \cdot \mathbf{w} + \mathbf{w} \cdot \mathbf{v} + \mathbf{w} \cdot \mathbf{w} = \|\mathbf{v}\|^2 + 2\mathbf{v} \cdot \mathbf{w} + \|\mathbf{w}\|^2$$

$$\|\mathbf{v} - \mathbf{w}\|^2 = (\mathbf{v} - \mathbf{w}) \cdot (\mathbf{v} - \mathbf{w}) = \mathbf{v} \cdot \mathbf{v} - \mathbf{v} \cdot \mathbf{w} - \mathbf{w} \cdot \mathbf{v} - \mathbf{w} \cdot \mathbf{w} = \|\mathbf{v}\|^2 - 2\mathbf{v} \cdot \mathbf{w} + \|\mathbf{w}\|^2$$

Hence,

$$\|\mathbf{v} + \mathbf{w}\|^2 - \|\mathbf{v} - \mathbf{w}\|^2 = (\|\mathbf{v}\|^2 + 2\mathbf{v} \cdot \mathbf{w} + \|\mathbf{w}\|^2) - (\|\mathbf{v}\|^2 - 2\mathbf{v} \cdot \mathbf{w} + \|\mathbf{w}\|^2) = 4\mathbf{v} \cdot \mathbf{w}$$

83. Show that the two diagonals of a parallelogram are perpendicular if and only if its sides have equal length. *Hint:* Use Exercise 82 to show that $\mathbf{v} - \mathbf{w}$ and $\mathbf{v} + \mathbf{w}$ are orthogonal if and only if $\|\mathbf{v}\| = \|\mathbf{w}\|$.

SOLUTION We denote the vectors $\overrightarrow{AB}$ and $\overrightarrow{AD}$ by

$$\mathbf{w} = \overrightarrow{AB}, \quad \mathbf{v} = \overrightarrow{AD}.$$

Then,

$$\overrightarrow{AC} = \mathbf{w} + \mathbf{v}, \quad \overrightarrow{BD} = -\mathbf{w} + \mathbf{v}.$$

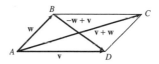

The diagonals are perpendicular if and only if the vectors $\mathbf{v} + \mathbf{w}$ and $\mathbf{v} - \mathbf{w}$ are orthogonal. By Exercise 82 these vectors are orthogonal if and only if the norms of the sum $(\mathbf{v} + \mathbf{w}) + (\mathbf{v} - \mathbf{w}) = 2\mathbf{v}$ and the difference $(\mathbf{v} + \mathbf{w}) - (\mathbf{v} - \mathbf{w}) = 2\mathbf{w}$ are equal, that is,

$$\|2\mathbf{v}\| = \|2\mathbf{w}\|$$

$$2\|\mathbf{v}\| - 2\|\mathbf{w}\| \quad \Rightarrow \quad \|\mathbf{v}\| = \|\mathbf{w}\|$$

85. Verify that $(\lambda\mathbf{v}) \cdot \mathbf{w} = \lambda(\mathbf{v} \cdot \mathbf{w})$ for any scalar λ.

SOLUTION We denote the components of the vectors $\mathbf{v}$ and $\mathbf{w}$ by

$$\mathbf{v} = \langle a_1, a_2, a_3 \rangle \quad \mathbf{w} = \langle b_1, b_2, b_3 \rangle$$

Thus,

$$(\lambda\mathbf{v}) \cdot \mathbf{w} = (\lambda\langle a_1, a_2, a_3 \rangle) \cdot \langle b_1, b_2, b_3 \rangle = \langle \lambda a_1, \lambda a_2, \lambda a_3 \rangle \cdot \langle b_1, b_2, b_3 \rangle$$

$$= \lambda a_1 b_1 + \lambda a_2 b_2 + \lambda a_3 b_3$$

Recalling that λ, a_i, and b_i are scalars and using the definitions of scalar multiples of vectors and the dot product, we get

$$(\lambda\mathbf{v}) \cdot \mathbf{w} = \lambda(a_1 b_1 + a_2 b_2 + a_3 b_3) = \lambda(\langle a_1, a_2, a_3 \rangle \cdot \langle b_1, b_2, b_3 \rangle) = \lambda(\mathbf{v} \cdot \mathbf{w})$$

Further Insights and Challenges

87. In this exercise, we prove the Cauchy–Schwarz inequality: If $\mathbf{v}$ and $\mathbf{w}$ are any two vectors, then

$$|\mathbf{v} \cdot \mathbf{w}| \leq \|\mathbf{v}\| \, \|\mathbf{w}\|$$

$$\boxed{6}$$

(a) Let $f(x) = \|x\mathbf{v} + \mathbf{w}\|^2$ for x a scalar. Show that $f(x) = ax^2 + bx + c$, where $a = \|\mathbf{v}\|^2$, $b = 2\mathbf{v} \cdot \mathbf{w}$, and $c = \|\mathbf{w}\|^2$.

(b) Conclude that $b^2 - 4ac \leq 0$. *Hint:* Observe that $f(x) \geq 0$ for all x.

SOLUTION

(a) We express the norm as a dot product and compute it:

$$f(x) = \|x\mathbf{v} + \mathbf{w}\|^2 = (x\mathbf{v} + \mathbf{w}) \cdot (x\mathbf{v} + \mathbf{w})$$

$$= x^2\mathbf{v} \cdot \mathbf{v} + x\mathbf{v} \cdot \mathbf{w} + x\mathbf{w} \cdot \mathbf{v} + \mathbf{w} \cdot \mathbf{w} = \|\mathbf{v}\|^2 x^2 + 2(\mathbf{v} \cdot \mathbf{w})x + \|\mathbf{w}\|^2$$

Hence, $f(x) = ax^2 + bx + c$, where $a = \|\mathbf{v}\|^2$, $b = 2\mathbf{v} \cdot \mathbf{w}$, and $c = \|\mathbf{w}\|^2$.

(b) If f has distinct real roots x_1 and x_2, then $f(x)$ is negative for x between x_1 and x_2, but this is impossible since f is the square of a length.

$$f(x) = ax^2 + bx + c, a > 0$$

Using properties of quadratic functions, it follows that f has a nonpositive discriminant. That is, $b^2 - 4ac \leq 0$. Substituting the values for a, b, and c, we get

$$4(\mathbf{v} \cdot \mathbf{w})^2 - 4\|\mathbf{v}\|^2\|\mathbf{w}\|^2 \leq 0$$

$$(\mathbf{v} \cdot \mathbf{w})^2 \leq \|\mathbf{v}\|^2\|\mathbf{w}\|^2$$

Taking the square root of both sides we obtain

$$|\mathbf{v} \cdot \mathbf{w}| \leq \|\mathbf{v}\|\|\mathbf{w}\|$$

89. This exercise gives another proof of the relation between the dot product and the angle θ between two vectors $\mathbf{v} = \langle a_1, b_1 \rangle$ and $\mathbf{w} = \langle a_2, b_2 \rangle$ in the plane. Observe that $\mathbf{v} = \|\mathbf{v}\| \langle \cos \theta_1, \sin \theta_1 \rangle$ and $\mathbf{w} = \|\mathbf{w}\| \langle \cos \theta_2, \sin \theta_2 \rangle$, with θ_1 and θ_2 as in Figure 21. Then use the addition formula for the cosine to show that

$$\mathbf{v} \cdot \mathbf{w} = \|\mathbf{v}\| \|\mathbf{w}\| \cos \theta$$

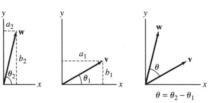

FIGURE 21

SOLUTION   Using the trigonometric function for angles in right triangles, we have

$$a_2 = \|\mathbf{v}\| \sin \theta_1, \qquad a_1 = \|\mathbf{v}\| \cos \theta_1$$

$$b_2 = \|\mathbf{w}\| \sin \theta_2, \qquad b_1 = \|\mathbf{w}\| \cos \theta_2$$

Hence, using the given identity we obtain

$$\mathbf{v} \cdot \mathbf{w} = \langle a_1, a_2 \rangle \cdot \langle b_1, b_2 \rangle = a_1 b_1 + a_2 b_2 = \|\mathbf{v}\| \cos \theta_1 \|\mathbf{w}\| \cos \theta_2 + \|\mathbf{v}\| \sin \theta_1 \|\mathbf{w}\| \sin \theta_2$$

$$= \|\mathbf{v}\|\|\mathbf{w}\|(\cos \theta_1 \cos \theta_2 + \sin \theta_1 \sin \theta_2) = \|\mathbf{v}\|\|\mathbf{w}\| \cos(\theta_1 - \theta_2)$$

That is,

$$\mathbf{v} \cdot \mathbf{w} = \|\mathbf{v}\|\|\mathbf{w}\| \cos(\theta)$$

91. Let $\mathbf{v}$ be a nonzero vector. The angles α, β, γ between $\mathbf{v}$ and the unit vectors $\mathbf{i}, \mathbf{j}, \mathbf{k}$ are called the direction angles of $\mathbf{v}$ (Figure 22). The cosines of these angles are called the **direction cosines** of $\mathbf{v}$. Prove that

$$\cos^2 \alpha + \cos^2 \beta + \cos^2 \gamma = 1$$

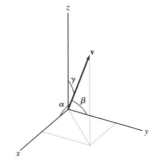

FIGURE 22 Direction angles of $\mathbf{v}$.

SOLUTION We use the relation between the dot product and the angle between two vectors to write

$$\cos \alpha = \frac{\mathbf{v} \cdot \mathbf{i}}{\|\mathbf{v}\|\|\mathbf{i}\|} = \frac{\mathbf{v} \cdot \mathbf{i}}{\|\mathbf{v}\|}$$

$$\cos \beta = \frac{\mathbf{v} \cdot \mathbf{j}}{\|\mathbf{v}\|\|\mathbf{j}\|} = \frac{\mathbf{v} \cdot \mathbf{j}}{\|\mathbf{v}\|} \tag{1}$$

$$\cos \gamma = \frac{\mathbf{v} \cdot \mathbf{k}}{\|\mathbf{v}\|\|\mathbf{k}\|} = \frac{\mathbf{v} \cdot \mathbf{k}}{\|\mathbf{v}\|}$$

We compute the values involved in (1). Letting $\mathbf{v} = \langle v_1, v_2, v_3 \rangle$ we get

$$\mathbf{v} \cdot \mathbf{i} = \langle v_1, v_2, v_3 \rangle \cdot \langle 1, 0, 0 \rangle = v_1$$

$$\mathbf{v} \cdot \mathbf{j} = \langle v_1, v_2, v_3 \rangle \cdot \langle 0, 1, 0 \rangle = v_2$$

$$\mathbf{v} \cdot \mathbf{k} = \langle v_1, v_2, v_3 \rangle \cdot \langle 0, 0, 1 \rangle = v_3$$

$$\|\mathbf{v}\| = \sqrt{v_1^2 + v_2^2 + v_3^2} \tag{2}$$

We now substitute (2) into (1) to obtain

$$\cos \alpha = \frac{v_1}{\|\mathbf{v}\|}, \quad \cos \beta = \frac{v_2}{\|\mathbf{v}\|}, \quad \cos \gamma = \frac{v_3}{\|\mathbf{v}\|}$$

Finally, we compute the sum of squares of the direction cosines:

$$\cos^2 \alpha + \cos^2 \beta + \cos^2 \gamma = \left(\frac{v_1}{\|\mathbf{v}\|}\right)^2 + \left(\frac{v_2}{\|\mathbf{v}\|}\right)^2 + \left(\frac{v_3}{\|\mathbf{v}\|}\right)^2 = \frac{1}{\|\mathbf{v}\|^2}(v_1^2 + v_2^2 + v_3^2) = \frac{1}{\|\mathbf{v}\|^2} \cdot \|\mathbf{v}\|^2 = 1$$

93. The set of all points $X = (x, y, z)$ equidistant from two points P, Q in $\mathbf{R}^3$ is a plane (Figure 23). Show that X lies on this plane if

$$\overrightarrow{PQ} \cdot \overrightarrow{OX} = \frac{1}{2}\left(\|\overrightarrow{OQ}\|^2 - \|\overrightarrow{OP}\|^2\right) \qquad \boxed{7}$$

Hint: If R is the midpoint of $\overline{PQ}$, then X is equidistant from P and Q if and only if $\overrightarrow{XR}$ is orthogonal to $\overrightarrow{PQ}$.

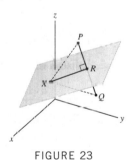

FIGURE 23

SOLUTION Let R be the midpoint of the segment $\overline{PQ}$. The points $X = (x, y, z)$ that are equidistant from P and Q are the points for which the vector $\overrightarrow{XR}$ is orthogonal to $\overrightarrow{PQ}$. That is,

$$\overrightarrow{XR} \cdot \overrightarrow{PQ} = 0 \tag{1}$$

Since $\overrightarrow{XR} = \overrightarrow{XO} + \overrightarrow{OR}$ we have by (1):

$$0 = \left(\overrightarrow{XO} + \overrightarrow{OR}\right) \cdot \overrightarrow{PQ} = \overrightarrow{XO} \cdot \overrightarrow{PQ} + \overrightarrow{OR} \cdot \overrightarrow{PQ} = -\overrightarrow{OX} \cdot \overrightarrow{PQ} + \overrightarrow{OR} \cdot \overrightarrow{PQ}$$

Transferring sides we get

$$\overrightarrow{OX} \cdot \overrightarrow{PQ} = \overrightarrow{OR} \cdot \overrightarrow{PQ} \tag{2}$$

We now write $\overrightarrow{PQ} = \overrightarrow{PO} + \overrightarrow{OQ}$ on the right-hand-side of (2), and $\overrightarrow{OR} = \dfrac{\overrightarrow{OP} + \overrightarrow{OQ}}{2}$. We get

$$\overrightarrow{OX} \cdot \overrightarrow{PQ} = \frac{1}{2}\left(\overrightarrow{OP} + \overrightarrow{OQ}\right) \cdot \left(\overrightarrow{PO} + \overrightarrow{OQ}\right) = \frac{1}{2}\left(\overrightarrow{OP} + \overrightarrow{OQ}\right) \cdot \left(\overrightarrow{OQ} - \overrightarrow{OP}\right)$$

$$= \frac{1}{2}\left(\overrightarrow{OP} \cdot \overrightarrow{OQ} - \overrightarrow{OP} \cdot \overrightarrow{OP} + \overrightarrow{OQ} \cdot \overrightarrow{OQ} - \overrightarrow{OQ} \cdot \overrightarrow{OP}\right) = \frac{1}{2}\left(\|\overrightarrow{OQ}\|^2 - \|\overrightarrow{OP}\|^2\right)$$

Thus, we showed that the vector equation of the plane is

$$\overrightarrow{OX} \cdot \overrightarrow{PQ} = \frac{1}{2}\left(\|\overrightarrow{OQ}\|^2 - \|\overrightarrow{OP}\|^2\right).$$

95. Use Eq. (7) to find the equation of the plane consisting of all points $X = (x, y, z)$ equidistant from $P = (2, 1, 1)$ and $Q = (1, 0, 2)$.

SOLUTION Using Eq. (7) with $X = (x, y, z)$, $P = (2, 1, 1)$, and $Q = (1, 0, 2)$ gives

$$\langle x, y, z\rangle \cdot \langle -1, -1, 1\rangle = \frac{1}{2}\left((\sqrt{5})^2 - (\sqrt{6})^2\right) = -\frac{1}{2}$$

This gives us $-1x - 1y + 1z = -\frac{1}{2}$, which leads to $2x + 2y - 2z = 1$.

12.4 The Cross Product   (LT Section 13.4)

Preliminary Questions

1. What is the $(1, 3)$ minor of the matrix $\begin{vmatrix} 3 & 4 & 2 \\ -5 & -1 & 1 \\ 4 & 0 & 3 \end{vmatrix}$?

SOLUTION The $(1, 3)$ minor is obtained by crossing out the first row and third column of the matrix. That is,

$$\begin{vmatrix} 3 & 4 & 2 \\ -5 & -1 & 1 \\ 4 & 0 & 3 \end{vmatrix} \quad \Rightarrow \quad \begin{vmatrix} -5 & -1 \\ 4 & 0 \end{vmatrix}$$

2. The angle between two unit vectors $\mathbf{e}$ and $\mathbf{f}$ is $\frac{\pi}{6}$. What is the length of $\mathbf{e} \times \mathbf{f}$?

SOLUTION We use the Formula for the Length of the Cross Product:

$$\|\mathbf{e} \times \mathbf{f}\| = \|\mathbf{e}\|\,\|\mathbf{f}\|\sin\theta$$

Since $\mathbf{e}$ and $\mathbf{f}$ are unit vectors, $\|\mathbf{e}\| = \|\mathbf{f}\| = 1$. Also $\theta = \frac{\pi}{6}$, therefore,

$$\|\mathbf{e} \times \mathbf{f}\| = 1 \cdot 1 \cdot \sin\frac{\pi}{6} = \frac{1}{2}$$

The length of $\mathbf{e} \times \mathbf{f}$ is $\frac{1}{2}$.

3. What is $\mathbf{u} \times \mathbf{w}$, assuming that $\mathbf{w} \times \mathbf{u} = \langle 2, 2, 1\rangle$?

SOLUTION By anti-commutativity of the cross product, we have

$$\mathbf{u} \times \mathbf{w} = -\mathbf{w} \times \mathbf{u} = -\langle 2, 2, 1\rangle = \langle -2, -2, -1\rangle$$

4. Find the cross product without using the formula:

(a) $\langle 4, 8, 2\rangle \times \langle 4, 8, 2\rangle$ **(b)** $\langle 4, 8, 2\rangle \times \langle 2, 4, 1\rangle$

SOLUTION By properties of the cross product, the cross product of parallel vectors is the zero vector. In particular, the cross product of a vector with itself is the zero vector. Since $\langle 4, 8, 2\rangle = 2\langle 2, 4, 1\rangle$, the vectors $\langle 4, 8, 2\rangle$ and $\langle 2, 4, 1\rangle$ are parallel. We conclude that

$$\langle 4, 8, 2\rangle \times \langle 4, 8, 2\rangle = \mathbf{0} \quad \text{and} \quad \langle 4, 8, 2\rangle \times \langle 2, 4, 1\rangle = \mathbf{0}.$$

5. What are $\mathbf{i} \times \mathbf{j}$ and $\mathbf{i} \times \mathbf{k}$?

SOLUTION The cross product $\mathbf{i} \times \mathbf{j}$ and $\mathbf{i} \times \mathbf{k}$ are determined by the right-hand rule. We can also use the following figure to determine these cross-products:

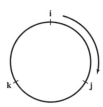

We get

$$\mathbf{i} \times \mathbf{j} = \mathbf{k} \text{ and } \mathbf{i} \times \mathbf{k} = -\mathbf{j}$$

6. When is the cross product $\mathbf{v} \times \mathbf{w}$ equal to zero?

SOLUTION The cross product $\mathbf{v} \times \mathbf{w}$ is equal to zero if one of the vectors $\mathbf{v}$ or $\mathbf{w}$ (or both) is the zero vector, or if $\mathbf{v}$ and $\mathbf{w}$ are parallel vectors.

Exercises

In Exercises 1–4, calculate the 2×2 determinant.

1. $\begin{vmatrix} 1 & 2 \\ 4 & 3 \end{vmatrix}$

SOLUTION Using the definition of 2×2 determinant we get

$$\begin{vmatrix} 1 & 2 \\ 4 & 3 \end{vmatrix} = 1 \cdot 3 - 2 \cdot 4 = -5$$

3. $\begin{vmatrix} -6 & 9 \\ 1 & 1 \end{vmatrix}$

SOLUTION We evaluate the determinant to obtain

$$\begin{vmatrix} -6 & 9 \\ 1 & 1 \end{vmatrix} = -6 \cdot 1 - 9 \cdot 1 = -15$$

In Exercises 5–8, calculate the 3×3 determinant.

5. $\begin{vmatrix} 1 & 2 & 1 \\ 4 & -3 & 0 \\ 1 & 0 & 1 \end{vmatrix}$

SOLUTION Using the definition of 3×3 determinant we obtain

$$\begin{vmatrix} 1 & 2 & 1 \\ 4 & -3 & 0 \\ 1 & 0 & 1 \end{vmatrix} = 1 \begin{vmatrix} -3 & 0 \\ 0 & 1 \end{vmatrix} - 2 \begin{vmatrix} 4 & 0 \\ 1 & 1 \end{vmatrix} + 1 \begin{vmatrix} 4 & -3 \\ 1 & 0 \end{vmatrix}$$

$$= 1 \cdot (-3 \cdot 1 - 0 \cdot 0) - 2 \cdot (4 \cdot 1 - 0 \cdot 1) + 1 \cdot (4 \cdot 0 - (-3) \cdot 1)$$

$$= -3 - 8 + 3 = -8$$

7. $\begin{vmatrix} 1 & 2 & 3 \\ 2 & 4 & 6 \\ -3 & -4 & 2 \end{vmatrix}$

SOLUTION We have

$$\begin{vmatrix} 1 & 2 & 3 \\ 2 & 4 & 6 \\ -3 & -4 & 2 \end{vmatrix} = 1 \begin{vmatrix} 4 & 6 \\ -4 & 2 \end{vmatrix} - 2 \begin{vmatrix} 2 & 6 \\ -3 & 2 \end{vmatrix} + 3 \begin{vmatrix} 2 & 4 \\ -3 & -4 \end{vmatrix}$$

$$= 1(4 \cdot 2 - 6 \cdot (-4)) - 2(2 \cdot 2 - 6 \cdot (-3)) + 3(2 \cdot (-4) - 4 \cdot (-3))$$

$$= 32 - 44 + 12 = 0$$

In Exercises 9–12, calculate $\mathbf{v} \times \mathbf{w}$.

9. $\mathbf{v} = \langle 1, 2, 1 \rangle, \quad \mathbf{w} = \langle 3, 1, 1 \rangle$

SOLUTION Using the definition of the cross product we get

$$\mathbf{v} \times \mathbf{w} = \begin{vmatrix} \mathbf{i} & \mathbf{j} & \mathbf{k} \\ 1 & 2 & 1 \\ 3 & 1 & 1 \end{vmatrix} = \begin{vmatrix} 2 & 1 \\ 1 & 1 \end{vmatrix} \mathbf{i} - \begin{vmatrix} 1 & 1 \\ 3 & 1 \end{vmatrix} \mathbf{j} + \begin{vmatrix} 1 & 2 \\ 3 & 1 \end{vmatrix} \mathbf{k}$$

$$= (2 - 1)\mathbf{i} - (1 - 3)\mathbf{j} + (1 - 6)\mathbf{k} = \mathbf{i} + 2\mathbf{j} - 5\mathbf{k}$$

11. $\mathbf{v} = \left\langle \frac{2}{3}, 1, \frac{1}{2} \right\rangle$, $\quad \mathbf{w} = \langle 4, -6, 3 \rangle$

SOLUTION   We have

$$\mathbf{v} \times \mathbf{w} = \begin{vmatrix} \mathbf{i} & \mathbf{j} & \mathbf{k} \\ \frac{2}{3} & 1 & \frac{1}{2} \\ 4 & -6 & 3 \end{vmatrix} = \begin{vmatrix} 1 & \frac{1}{2} \\ -6 & 3 \end{vmatrix} \mathbf{i} - \begin{vmatrix} \frac{2}{3} & \frac{1}{2} \\ 4 & 3 \end{vmatrix} \mathbf{j} + \begin{vmatrix} \frac{2}{3} & 1 \\ 4 & -6 \end{vmatrix} \mathbf{k}$$

$$= (3 + 3)\mathbf{i} - (2 - 2)\mathbf{j} + (-4 - 4)\mathbf{k} = 6\mathbf{i} - 8\mathbf{k}$$

In Exercises 13–16, use the relations in Eq. (5) to calculate the cross product.

13. $(\mathbf{i} + \mathbf{j}) \times \mathbf{k}$

SOLUTION   We use basic properties of the cross product to obtain

$$(\mathbf{i} + \mathbf{j}) \times \mathbf{k} = \mathbf{i} \times \mathbf{k} + \mathbf{j} \times \mathbf{k} = -\mathbf{j} + \mathbf{i}$$

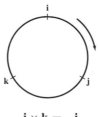

$$\mathbf{i} \times \mathbf{k} = -\mathbf{j}$$
$$\mathbf{j} \times \mathbf{k} = \mathbf{i}$$

15. $(\mathbf{i} - 3\mathbf{j} + 2\mathbf{k}) \times (\mathbf{j} - \mathbf{k})$

SOLUTION   Using the distributive law we obtain

$$(\mathbf{i} - 3\mathbf{j} + 2\mathbf{k}) \times (\mathbf{j} - \mathbf{k}) = (\mathbf{i} - 3\mathbf{j} + 2\mathbf{k}) \times \mathbf{j} - (\mathbf{i} - 3\mathbf{j} + 2\mathbf{k}) \times (\mathbf{k})$$

$$= \mathbf{i} \times \mathbf{j} + 2\mathbf{k} \times \mathbf{j} - \mathbf{i} \times \mathbf{k} - (-3\mathbf{j}) \times \mathbf{k}$$

$$= \mathbf{i} + \mathbf{j} + \mathbf{k}$$

In Exercises 17–22, calculate the cross product assuming that

$$\mathbf{u} \times \mathbf{v} = \langle 1, 1, 0 \rangle, \quad \mathbf{u} \times \mathbf{w} = \langle 0, 3, 1 \rangle, \quad \mathbf{v} \times \mathbf{w} = \langle 2, -1, 1 \rangle$$

17. $\mathbf{v} \times \mathbf{u}$

SOLUTION   Using the properties of the cross product we obtain

$$\mathbf{v} \times \mathbf{u} = -\mathbf{u} \times \mathbf{v} = \langle -1, -1, 0 \rangle$$

19. $\mathbf{w} \times (\mathbf{u} + \mathbf{v})$

SOLUTION   Using the properties of the cross product we obtain

$$\mathbf{w} \times (\mathbf{u} + \mathbf{v}) = \mathbf{w} \times \mathbf{u} + \mathbf{w} \times \mathbf{v} = -\mathbf{u} \times \mathbf{w} - \mathbf{v} \times \mathbf{w} = \langle -2, -2, -2 \rangle \,.$$

21. $(\mathbf{u} - 2\mathbf{v}) \times (\mathbf{u} + 2\mathbf{v})$

SOLUTION   Using the properties of the cross product we obtain

$$(\mathbf{u} - 2\mathbf{v}) \times (\mathbf{u} + 2\mathbf{v}) = (\mathbf{u} - 2\mathbf{v}) \times \mathbf{u} + (\mathbf{u} - 2\mathbf{v}) \times 2\mathbf{v} = \mathbf{u} \times \mathbf{u} - 2\mathbf{v} \times \mathbf{u} + \mathbf{u} \times 2\mathbf{v} - 4\mathbf{v} \times \mathbf{v}$$

$$= 0 + 2\mathbf{u} \times \mathbf{v} + 2\mathbf{u} \times \mathbf{v} - 0 = 0 + 4\mathbf{u} \times \mathbf{v} = \langle 4, 4, 0 \rangle$$

23. Let $\mathbf{v} = \langle a, b, c \rangle$. Calculate $\mathbf{v} \times \mathbf{i}$, $\mathbf{v} \times \mathbf{j}$, and $\mathbf{v} \times \mathbf{k}$.

SOLUTION   We write $\mathbf{v} = a\mathbf{i} + b\mathbf{j} + c\mathbf{k}$ and use the distributive law:

$$\mathbf{v} \times \mathbf{i} = (a\mathbf{i} + b\mathbf{j} + c\mathbf{k}) \times \mathbf{i} = a\mathbf{i} \times \mathbf{i} + b\mathbf{j} \times \mathbf{i} + c\mathbf{k} \times \mathbf{i} = a \cdot \mathbf{0} - b\mathbf{k} + c\mathbf{j} = -b\mathbf{k} + c\mathbf{j} = \langle 0, c, -b \rangle$$

$$\mathbf{v} \times \mathbf{j} = (a\mathbf{i} + b\mathbf{j} + c\mathbf{k}) \times \mathbf{j} = a\mathbf{i} \times \mathbf{j} + b\mathbf{j} \times \mathbf{j} + c\mathbf{k} \times \mathbf{j} = a\mathbf{k} + b\mathbf{0} - c\mathbf{i} = a\mathbf{k} - c\mathbf{i} = \langle -c, 0, a \rangle$$

$$\mathbf{v} \times \mathbf{k} = (a\mathbf{i} + b\mathbf{j} + c\mathbf{k}) \times \mathbf{k} = a\mathbf{i} \times \mathbf{k} + b\mathbf{j} \times \mathbf{k} + c\mathbf{k} \times \mathbf{k} = -a\mathbf{j} + b\mathbf{i} + c\mathbf{0} = -a\mathbf{j} + b\mathbf{i} = \langle b, -a, 0 \rangle$$

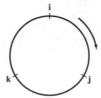

In Exercises 25 and 26, refer to Figure 16.

FIGURE 16

25. Which of **u** and −**u** is equal to **v** × **w**?

SOLUTION   The direction of **v** × **w** is determined by the right-hand rule, that is, our thumb points in the direction of **v** × **w** when the fingers of our right hand curl from **v** to **w**. Therefore **v** × **w** equals −**u** rather than **u**.

27. Let $\mathbf{v} = \langle 3, 0, 0 \rangle$ and $\mathbf{w} = \langle 0, 1, -1 \rangle$. Determine $\mathbf{u} = \mathbf{v} \times \mathbf{w}$ using the geometric properties of the cross product rather than the formula.

SOLUTION   The cross product $\mathbf{u} = \mathbf{v} \times \mathbf{w}$ is orthogonal to **v**.

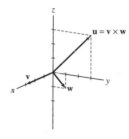

Since **v** lies along the x-axis, **u** lies in the yz-plane, therefore $\mathbf{u} = \langle 0, b, c \rangle$. **u** is also orthogonal to **w**, so $\mathbf{u} \cdot \mathbf{w} = 0$. This gives $\mathbf{u} \cdot \mathbf{w} = \langle 0, b, c \rangle \cdot \langle 0, 1, -1 \rangle = b - c = 0 \Rightarrow b = c$. Thus, $\mathbf{u} = \langle 0, b, b \rangle$. By the right-hand rule, **u** points to the positive z-direction so $b > 0$. We compute the length of **u**. Since $\mathbf{v} \cdot \mathbf{w} = \langle 3, 0, 0 \rangle \cdot \langle 0, 1, -1 \rangle = 0$, **v** and **w** are orthogonal. Hence,

$$\|\mathbf{v} \times \mathbf{w}\| = \|\mathbf{v}\|\|\mathbf{w}\| \sin \frac{\pi}{2} = \|\mathbf{v}\|\|\mathbf{w}\| = 3 \cdot \sqrt{2}.$$

Also since $b > 0$, we have

$$\|\mathbf{u}\| = \|\langle 0, b, b \rangle\| = \sqrt{2b^2} = b\sqrt{2}$$

Equating the lengths gives

$$b\sqrt{2} = 3\sqrt{2} \quad \Rightarrow \quad b = 3.$$

We conclude that $\mathbf{u} = \mathbf{v} \times \mathbf{w} = \langle 0, 3, 3 \rangle$.

29. Show that if **v** and **w** lie in the yz-plane, then **v** × **w** is a multiple of **i**.

SOLUTION   **v** × **w** is orthogonal to **v** and **w**. Since **v** and **w** lie in the yz-plane, **v** × **w** must lie along the x axis which is perpendicular to yz-plane. That is, **v** × **w** is a scalar multiple of the unit vector **i**.

31. Let **e** and $\mathbf{e}'$ be unit vectors in $\mathbf{R}^3$ such that $\mathbf{e} \perp \mathbf{e}'$. Use the geometric properties of the cross product to compute $\mathbf{e} \times (\mathbf{e}' \times \mathbf{e})$.

SOLUTION   Let $\mathbf{u} = \mathbf{e} \times (\mathbf{e}' \times \mathbf{e})$ and $\mathbf{v} = \mathbf{e}' \times \mathbf{e}$. The vector **v** is orthogonal to $\mathbf{e}'$ and **e**, hence **v** is orthogonal to the plane π defined by $\mathbf{e}'$ and **e**. Now **u** is orthogonal to **v**, hence **u** lies in the plane π orthogonal to **v**. **u** is orthogonal to **e**, which is in this plane, hence **u** is a multiple of $\mathbf{e}'$:

$$\mathbf{u} = \lambda \mathbf{e}' \tag{1}$$

The right-hand rule implies that $\mathbf{u}$ is in the direction of $\mathbf{e}'$, hence $\lambda > 0$. To find λ, we compute the length of $\mathbf{u}$:

$$\|\mathbf{v}\| = \|\mathbf{e}' \times \mathbf{e}\| = \|\mathbf{e}'\|\|\mathbf{e}\| \sin \frac{\pi}{2} = 1 \cdot 1 \cdot 1 = 1$$

$$\|\mathbf{u}\| = \|\mathbf{e} \times \mathbf{v}\| = \|\mathbf{e}\|\|\mathbf{v}\| \sin \frac{\pi}{2} = 1 \cdot 1 \cdot 1 = 1 \tag{2}$$

Combining (1), (2), and $\lambda > 0$ we conclude that

$$\mathbf{u} = \mathbf{e} \times \left(\mathbf{e}' \times \mathbf{e}\right) = \mathbf{e}'.$$

33. An electron moving with velocity $\mathbf{v}$ in the plane experiences a force $\mathbf{F} = q(\mathbf{v} \times \mathbf{B})$, where q is the charge on the electron and $\mathbf{B}$ is a uniform magnetic field pointing directly out of the page. Which of the two vectors $\mathbf{F}_1$ or $\mathbf{F}_2$ in Figure 17 represents the force on the electron? Remember that q is negative.

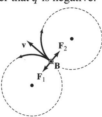

FIGURE 17 The magnetic field vector $\mathbf{B}$ points directly out of the page.

SOLUTION Since the magnetic field $\mathbf{B}$ points directly out of the page (toward us), the right-hand rule implies that the cross product $\mathbf{v} \times \mathbf{B}$ is in the direction of $\mathbf{F}_2$ (see figure).

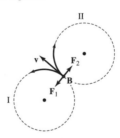

Since $\mathbf{F} = q\,(\mathbf{v} \times \mathbf{B})$ and $q < 0$, the force $\mathbf{F}$ on the electron is represented by the opposite vector $\mathbf{F}_1$.

35. Verify identity (10) for vectors $\mathbf{v} = \langle 3, -2, 2 \rangle$ and $\mathbf{w} = \langle 4, -1, 2 \rangle$.

SOLUTION We compute the cross product $\mathbf{v} \times \mathbf{w}$:

$$\mathbf{v} \times \mathbf{w} = \begin{vmatrix} \mathbf{i} & \mathbf{j} & \mathbf{k} \\ 3 & -2 & 2 \\ 4 & -1 & 2 \end{vmatrix} = \begin{vmatrix} -2 & 2 \\ -1 & 2 \end{vmatrix}\mathbf{i} - \begin{vmatrix} 3 & 2 \\ 4 & 2 \end{vmatrix}\mathbf{j} + \begin{vmatrix} 3 & -2 \\ 4 & -1 \end{vmatrix}\mathbf{k}$$

$$= (-4 + 2)\mathbf{i} - (6 - 8)\mathbf{j} + (-3 + 8)\mathbf{k} = -2\mathbf{i} + 2\mathbf{j} + 5\mathbf{k} = \langle -2, 2, 5 \rangle$$

We now find the dot product $\mathbf{v} \cdot \mathbf{w}$:

$$\mathbf{v} \cdot \mathbf{w} = \langle 3, -2, 2 \rangle \cdot \langle 4, -1, 2 \rangle = 3 \cdot 4 + (-2) \cdot (-1) + 2 \cdot 2 = 18$$

Finally we compute the squares of the lengths of $\mathbf{v}$, $\mathbf{w}$ and $\mathbf{v} \times \mathbf{w}$:

$$\|\mathbf{v}\|^2 = 3^2 + (-2)^2 + 2^2 = 17$$

$$\|\mathbf{w}\|^2 = 4^2 + (-1)^2 + 2^2 = 21$$

$$\|\mathbf{v} \times \mathbf{w}\|^2 = (-2)^2 + 2^2 + 5^2 = 33$$

We now verify the equality:

$$\|\mathbf{v}\|^2\|\mathbf{w}\|^2 - (\mathbf{v} \cdot \mathbf{w})^2 = 17 \cdot 21 - 18^2 = 33 = \|\mathbf{v} \times \mathbf{w}\|^2$$

37. Find the area of the parallelogram spanned by **v** and **w** in Figure 18.

SOLUTION The area of the parallelogram equals the length of the cross product of the two vectors $\mathbf{v} = \langle 1, 3, 1 \rangle$ and $\mathbf{w} = \langle -4, 2, 6 \rangle$. We calculate the cross product as follows:

$$\mathbf{v} \times \mathbf{w} = \begin{vmatrix} \mathbf{i} & \mathbf{j} & \mathbf{k} \\ 1 & 3 & 1 \\ -4 & 2 & 6 \end{vmatrix} = (18 - 2)\mathbf{i} - (6 + 4)\mathbf{j} + (2 + 12)\mathbf{k} = 16\mathbf{i} - 10\mathbf{j} + 14\mathbf{k}$$

The length of this vector $16\mathbf{i} - 10\mathbf{j} + 14\mathbf{k}$ is $\sqrt{16^2 + 10^2 + 14^2} = 2\sqrt{138}$. Thus, the area of the parallelogram is $2\sqrt{138}$.

39. Sketch and compute the volume of the parallelepiped spanned by

$$\mathbf{u} = \langle 1, 0, 0 \rangle, \qquad \mathbf{v} = \langle 0, 2, 0 \rangle, \qquad \mathbf{w} = \langle 1, 1, 2 \rangle$$

SOLUTION Using $\mathbf{u} = \langle 1, 0, 0 \rangle$, $\mathbf{v} = \langle 0, 2, 0 \rangle$, and $\mathbf{w} = \langle 1, 1, 2 \rangle$, the volume is given by the following scalar triple product:

$$\mathbf{u} \cdot (\mathbf{v} \times \mathbf{w}) = \begin{vmatrix} 1 & 0 & 0 \\ 0 & 2 & 0 \\ 1 & 1 & 2 \end{vmatrix} = 1(4 - 0) - 0 + 0 = 4.$$

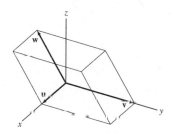

41. Calculate the area of the parallelogram spanned by $\mathbf{u} = \langle 1, 0, 3 \rangle$ and $\mathbf{v} = \langle 2, 1, 1 \rangle$.

SOLUTION The area of the parallelogram is the length of the vector $\mathbf{u} \times \mathbf{v}$. We first compute this vector:

$$\mathbf{u} \times \mathbf{v} = \begin{vmatrix} \mathbf{i} & \mathbf{j} & \mathbf{k} \\ 1 & 0 & 3 \\ 2 & 1 & 1 \end{vmatrix} = \begin{vmatrix} 0 & 3 \\ 1 & 1 \end{vmatrix} \mathbf{i} - \begin{vmatrix} 1 & 3 \\ 2 & 1 \end{vmatrix} \mathbf{j} + \begin{vmatrix} 1 & 0 \\ 2 & 1 \end{vmatrix} \mathbf{k} = -3\mathbf{i} - (1 - 6)\mathbf{j} + \mathbf{k} = -3\mathbf{i} + 5\mathbf{j} + \mathbf{k}$$

The area A is the length

$$A = \|\mathbf{u} \times \mathbf{v}\| = \sqrt{(-3)^2 + 5^2 + 1^2} = \sqrt{35} \approx 5.92.$$

43. Sketch the triangle with vertices at the origin O, $P = (3, 3, 0)$, and $Q = (0, 3, 3)$, and compute its area using cross products.

SOLUTION The triangle OPQ is shown in the following figure.

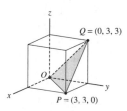

The area S of the triangle is half of the area of the parallelogram determined by the vectors $\overrightarrow{OP} = \langle 3, 3, 0 \rangle$ and $\overrightarrow{OQ} = \langle 0, 3, 3 \rangle$. Thus,

$$S = \frac{1}{2} \|\overrightarrow{OP} \times \overrightarrow{OQ}\| \tag{1}$$

We compute the cross product:

$$\overrightarrow{OP} \times \overrightarrow{OQ} = \begin{vmatrix} \mathbf{i} & \mathbf{j} & \mathbf{k} \\ 3 & 3 & 0 \\ 0 & 3 & 3 \end{vmatrix} = \begin{vmatrix} 3 & 0 \\ 3 & 3 \end{vmatrix} \mathbf{i} - \begin{vmatrix} 3 & 0 \\ 0 & 3 \end{vmatrix} \mathbf{j} + \begin{vmatrix} 3 & 3 \\ 0 & 3 \end{vmatrix} \mathbf{k}$$

$$= 9\mathbf{i} - 9\mathbf{j} + 9\mathbf{k} = 9\langle 1, -1, 1 \rangle$$

Substituting into (1) gives

$$S = \frac{1}{2}\|9\langle 1, -1, 1\rangle\| = \frac{9}{2}\|\langle 1, -1, 1\rangle\| = \frac{9}{2}\sqrt{1^2 + (-1)^2 + 1^2} = \frac{9\sqrt{3}}{2} \approx 7.8$$

The area of the triangle is $S = \frac{9\sqrt{3}}{2} \approx 7.8$.

In Exercises 45–47, verify the identity using the formula for the cross product.

45. $\mathbf{v} \times \mathbf{w} = -\mathbf{w} \times \mathbf{v}$

SOLUTION   Let $\mathbf{v} = \langle a, b, c\rangle$ and $\mathbf{w} = \langle d, e, f\rangle$. By the definition of the cross product we have

$$\mathbf{v} \times \mathbf{w} = \begin{vmatrix} \mathbf{i} & \mathbf{j} & \mathbf{k} \\ a & b & c \\ d & e & f \end{vmatrix} = \begin{vmatrix} b & c \\ e & f \end{vmatrix}\mathbf{i} - \begin{vmatrix} a & c \\ d & f \end{vmatrix}\mathbf{j} + \begin{vmatrix} a & b \\ d & e \end{vmatrix}\mathbf{k} = (bf - ec)\mathbf{i} - (af - dc)\mathbf{j} + (ae - db)\mathbf{k}$$

We also have

$$-\mathbf{w} \times \mathbf{v} = \begin{vmatrix} \mathbf{i} & \mathbf{j} & \mathbf{k} \\ -d & -e & -f \\ a & b & c \end{vmatrix} = (-ec + bf)\mathbf{i} - (-dc + af)\mathbf{j} + (-db + ea)\mathbf{k}$$

Thus, $\mathbf{v} \times \mathbf{w} = -\mathbf{w} \times \mathbf{v}$, as desired.

47. $(\mathbf{u} + \mathbf{v}) \times \mathbf{w} = \mathbf{u} \times \mathbf{w} + \mathbf{v} \times \mathbf{w}$

SOLUTION   We let $\mathbf{u} = \langle a_1, a_2, a_3\rangle$, $\mathbf{v} = \langle b_1, b_2, b_3\rangle$ and $\mathbf{w} = \langle c_1, c_2, c_3\rangle$. Computing the left-hand side gives

$$(\mathbf{u} + \mathbf{v}) \times \mathbf{w} = \langle a_1 + b_1, a_2 + b_2, a_3 + b_3\rangle \times \langle c_1, c_2, c_3\rangle = \begin{vmatrix} \mathbf{i} & \mathbf{j} & \mathbf{k} \\ a_1 + b_1 & a_2 + b_2 & a_3 + b_3 \\ c_1 & c_2 & c_3 \end{vmatrix}$$

$$= \begin{vmatrix} a_2 + b_2 & a_3 + b_3 \\ c_2 & c_3 \end{vmatrix}\mathbf{i} - \begin{vmatrix} a_1 + b_1 & a_3 + b_3 \\ c_1 & c_3 \end{vmatrix}\mathbf{j} + \begin{vmatrix} a_1 + b_1 & a_2 + b_2 \\ c_1 & c_2 \end{vmatrix}\mathbf{k}$$

$$= (c_3(a_2 + b_2) - c_2(a_3 + b_3))\mathbf{i} - (c_3(a_1 + b_1) - c_1(a_3 + b_3))\mathbf{j} + (c_2(a_1 + b_1) - c_1(a_2 + b_2))\mathbf{k}$$

We now compute the right-hand-side of the equality:

$$\mathbf{u} \times \mathbf{w} + \mathbf{v} \times \mathbf{w} = \begin{vmatrix} \mathbf{i} & \mathbf{j} & \mathbf{k} \\ a_1 & a_2 & a_3 \\ c_1 & c_2 & c_3 \end{vmatrix} + \begin{vmatrix} \mathbf{i} & \mathbf{j} & \mathbf{k} \\ b_1 & b_2 & b_3 \\ c_1 & c_2 & c_3 \end{vmatrix}$$

$$= \begin{vmatrix} a_2 & a_3 \\ c_2 & c_3 \end{vmatrix}\mathbf{i} - \begin{vmatrix} a_1 & a_3 \\ c_1 & c_3 \end{vmatrix}\mathbf{j} + \begin{vmatrix} a_1 & a_2 \\ c_1 & c_2 \end{vmatrix}\mathbf{k} + \begin{vmatrix} b_2 & b_3 \\ c_2 & c_3 \end{vmatrix}\mathbf{i} - \begin{vmatrix} b_1 & b_3 \\ c_1 & c_3 \end{vmatrix}\mathbf{j} + \begin{vmatrix} b_1 & b_2 \\ c_1 & c_2 \end{vmatrix}\mathbf{k}$$

$$= (a_2 c_3 - a_3 c_2)\mathbf{i} - (a_1 c_3 - a_3 c_1)\mathbf{j} + (a_1 c_2 - a_2 c_1)\mathbf{k}$$

$$\quad + (b_2 c_3 - b_3 c_2)\mathbf{i} - (b_1 c_3 - b_3 c_1)\mathbf{j} + (b_1 c_2 - b_2 c_1)\mathbf{k}$$

$$= (a_2 c_3 - a_3 c_2 + b_2 c_3 - b_3 c_2)\mathbf{i} - (a_1 c_3 - a_3 c_1 + b_1 c_3 - b_3 c_1)\mathbf{j} + (a_1 c_2 - a_2 c_1 + b_1 c_2 - b_2 c_1)\mathbf{k}$$

$$= (c_3(a_2 + b_2) - c_2(a_3 + b_3))\mathbf{i} - (c_3(a_1 + b_1) - c_1(a_3 + b_3))\mathbf{j} + (c_2(a_1 + b_1) - c_1(a_2 + b_2))\mathbf{k}$$

The results are the same. Hence,

$$(\mathbf{u} + \mathbf{v}) \times \mathbf{w} = \mathbf{u} \times \mathbf{w} + \mathbf{v} \times \mathbf{w}.$$

49. Verify the relations (5).

SOLUTION   We must verify the following relations:

$$\mathbf{i} \times \mathbf{j} = \mathbf{k}, \quad \mathbf{j} \times \mathbf{k} = \mathbf{i}, \quad \mathbf{k} \times \mathbf{i} = \mathbf{j}, \quad \mathbf{i} \times \mathbf{i} = \mathbf{j} \times \mathbf{j} = \mathbf{k} \times \mathbf{k} = \mathbf{0}$$

We compute the cross products using the definition of the cross product. This gives

$$\mathbf{i} \times \mathbf{j} = \begin{vmatrix} \mathbf{i} & \mathbf{j} & \mathbf{k} \\ 1 & 0 & 0 \\ 0 & 1 & 0 \end{vmatrix} = \begin{vmatrix} 0 & 0 \\ 1 & 0 \end{vmatrix}\mathbf{i} - \begin{vmatrix} 1 & 0 \\ 0 & 0 \end{vmatrix}\mathbf{j} + \begin{vmatrix} 1 & 0 \\ 0 & 1 \end{vmatrix}\mathbf{k} = \mathbf{k}$$

$$\mathbf{j} \times \mathbf{k} = \begin{vmatrix} \mathbf{i} & \mathbf{j} & \mathbf{k} \\ 0 & 1 & 0 \\ 0 & 0 & 1 \end{vmatrix} = \begin{vmatrix} 1 & 0 \\ 0 & 1 \end{vmatrix}\mathbf{i} - \begin{vmatrix} 0 & 0 \\ 0 & 1 \end{vmatrix}\mathbf{j} + \begin{vmatrix} 0 & 1 \\ 0 & 0 \end{vmatrix}\mathbf{k} = \mathbf{i}$$

$$\mathbf{k} \times \mathbf{i} = \begin{vmatrix} \mathbf{i} & \mathbf{j} & \mathbf{k} \\ 0 & 0 & 1 \\ 1 & 0 & 0 \end{vmatrix} = \begin{vmatrix} 0 & 1 \\ 0 & 0 \end{vmatrix}\mathbf{i} - \begin{vmatrix} 0 & 1 \\ 1 & 0 \end{vmatrix}\mathbf{j} + \begin{vmatrix} 0 & 0 \\ 1 & 0 \end{vmatrix}\mathbf{k} = \mathbf{j}$$

$$\mathbf{i} \times \mathbf{i} = \begin{vmatrix} \mathbf{i} & \mathbf{j} & \mathbf{k} \\ 1 & 0 & 0 \\ 1 & 0 & 0 \end{vmatrix} = \begin{vmatrix} 0 & 0 \\ 0 & 0 \end{vmatrix}\mathbf{i} - \begin{vmatrix} 1 & 0 \\ 1 & 0 \end{vmatrix}\mathbf{j} + \begin{vmatrix} 1 & 0 \\ 1 & 0 \end{vmatrix}\mathbf{k} = \mathbf{0}$$

$$\mathbf{j} \times \mathbf{j} = \begin{vmatrix} \mathbf{i} & \mathbf{j} & \mathbf{k} \\ 0 & 1 & 0 \\ 0 & 1 & 0 \end{vmatrix} = \begin{vmatrix} 1 & 0 \\ 1 & 0 \end{vmatrix}\mathbf{i} - \begin{vmatrix} 0 & 0 \\ 0 & 0 \end{vmatrix}\mathbf{j} + \begin{vmatrix} 0 & 1 \\ 0 & 1 \end{vmatrix}\mathbf{k} = \mathbf{0}$$

$$\mathbf{k} \times \mathbf{k} = \begin{vmatrix} \mathbf{i} & \mathbf{j} & \mathbf{k} \\ 0 & 0 & 1 \\ 0 & 0 & 1 \end{vmatrix} = \begin{vmatrix} 0 & 1 \\ 0 & 1 \end{vmatrix}\mathbf{i} - \begin{vmatrix} 0 & 1 \\ 0 & 1 \end{vmatrix}\mathbf{j} + \begin{vmatrix} 0 & 0 \\ 0 & 0 \end{vmatrix}\mathbf{k} = \mathbf{0}$$

51. The components of the cross product have a geometric interpretation. Show that the absolute value of the $\mathbf{k}$-component of $\mathbf{v} \times \mathbf{w}$ is equal to the area of the parallelogram spanned by the projections $\mathbf{v}_0$ and $\mathbf{w}_0$ onto the xy-plane (Figure 20).

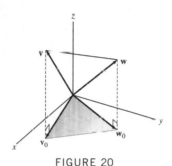

FIGURE 20

SOLUTION Let $\mathbf{v} = \langle a_1, a_2, a_3 \rangle$ and $\mathbf{w} = \langle b_1, b_2, b_3 \rangle$, hence, $\mathbf{v}_0 = \langle a_1, a_2, 0 \rangle$ and $\mathbf{w}_0 = \langle b_1, b_2, 0 \rangle$. The area S of the parallelogram spanned by $\mathbf{v}_0$ and $\mathbf{w}_0$ is the following value:

$$S = \|\mathbf{v}_0 \times \mathbf{w}_0\| \tag{1}$$

We compute the cross product:

$$\mathbf{v}_0 \times \mathbf{w}_0 = \begin{vmatrix} \mathbf{i} & \mathbf{j} & \mathbf{k} \\ a_1 & a_2 & 0 \\ b_1 & b_2 & 0 \end{vmatrix} = \begin{vmatrix} a_2 & 0 \\ b_2 & 0 \end{vmatrix}\mathbf{i} - \begin{vmatrix} a_1 & 0 \\ b_1 & 0 \end{vmatrix}\mathbf{j} + \begin{vmatrix} a_1 & a_2 \\ b_1 & b_2 \end{vmatrix}\mathbf{k}$$

$$= 0\mathbf{i} - 0\mathbf{j} + (a_1 b_2 - a_2 b_1)\mathbf{k} = \langle 0, 0, a_1 b_2 - a_2 b_1 \rangle$$

Using (1) we have

$$S = \sqrt{0^2 + 0^2 + (a_1 b_2 - a_2 b_1)^2} = |a_1 b_2 - a_2 b_1| \tag{2}$$

We now compute $\mathbf{v} \times \mathbf{w}$:

$$\mathbf{v} \times \mathbf{w} = \begin{vmatrix} \mathbf{i} & \mathbf{j} & \mathbf{k} \\ a_1 & a_2 & a_3 \\ b_1 & b_2 & b_3 \end{vmatrix} = \begin{vmatrix} a_2 & a_3 \\ b_2 & b_3 \end{vmatrix}\mathbf{i} - \begin{vmatrix} a_1 & a_3 \\ b_1 & b_3 \end{vmatrix}\mathbf{j} + \begin{vmatrix} a_1 & a_2 \\ b_1 & b_2 \end{vmatrix}\mathbf{k}$$

The $\mathbf{k}$-component of $\mathbf{v} \times \mathbf{w}$ is, thus,

$$\begin{vmatrix} a_1 & a_2 \\ b_1 & b_2 \end{vmatrix} = a_1 b_2 - a_2 b_1 \tag{3}$$

By (2) and (3) we obtain the desired result.

53. Show that three points P, Q, R are collinear (lie on a line) if and only if $\overrightarrow{PQ} \times \overrightarrow{PR} = \mathbf{0}$.

SOLUTION The points P, Q, and R lie on one line if and only if the vectors $\overrightarrow{PQ}$ and $\overrightarrow{PR}$ are parallel. By basic properties of the cross product this is equivalent to $\overrightarrow{PQ} \times \overrightarrow{PR} = \mathbf{0}$.

55. Solve the equation $\langle 1, 1, 1 \rangle \times \mathbf{X} = \langle 1, -1, 0 \rangle$, where $\mathbf{X} = \langle x, y, z \rangle$. *Note:* There are infinitely many solutions.

SOLUTION Let $\mathbf{X} = \langle a, b, c \rangle$. We compute the cross product:

$$\langle 1, 1, 1 \rangle \times \langle a, b, c \rangle = \begin{vmatrix} \mathbf{i} & \mathbf{j} & \mathbf{k} \\ 1 & 1 & 1 \\ a & b & c \end{vmatrix} = \begin{vmatrix} 1 & 1 \\ b & c \end{vmatrix} \mathbf{i} - \begin{vmatrix} 1 & 1 \\ a & c \end{vmatrix} \mathbf{j} + \begin{vmatrix} 1 & 1 \\ a & b \end{vmatrix} \mathbf{k}$$

$$= (c - b)\mathbf{i} - (c - a)\mathbf{j} + (b - a)\mathbf{k} = \langle c - b, a - c, b - a \rangle$$

The equation for $\mathbf{X}$ is, thus,

$$\langle c - b, a - c, b - a \rangle = \langle 1, -1, 0 \rangle$$

Equating corresponding components we get

$$c - b = 1$$
$$a - c = -1$$
$$b - a = 0$$

The third equation implies $a = b$. Substituting in the first and second equations gives

$$\begin{aligned} c - a &= 1 \\ a - c &= -1 \end{aligned} \quad \Rightarrow \quad c = a + 1$$

The solution is thus, $b = a$, $c = a + 1$. The corresponding solutions $\mathbf{X}$ are

$$\mathbf{X} = \langle a, b, c \rangle = \langle a, a, a + 1 \rangle$$

One possible solution is obtained for $a = 0$, that is, $\mathbf{X} = \langle 0, 0, 1 \rangle$.

57. Let $\mathbf{X} = \langle x, y, z \rangle$. Show that $\mathbf{i} \times \mathbf{X} = \mathbf{v}$ has a solution if and only if $\mathbf{v}$ is contained in the yz-plane (the $\mathbf{i}$-component is zero).

SOLUTION The cross product vector $\mathbf{i} \times \mathbf{X} = \mathbf{v}$ must be orthogonal to the vector $\mathbf{i} = \langle 1, 0, 0 \rangle$. This condition is true if and only if $\langle 1, 0, 0 \rangle \cdot \mathbf{v} = 0$, which is true if and only if the $\mathbf{i}$-component of $\mathbf{v}$ is zero (that is, $\mathbf{v}$ is in the yz-plane).

*In Exercises 59–62: The **torque** about the origin O due to a force $\mathbf{F}$ acting on an object with position vector $\mathbf{r}$ is the vector quantity $\tau = \mathbf{r} \times \mathbf{F}$. If several forces $\mathbf{F}_j$ act at positions $\mathbf{r}_j$, then the net torque (units: N-m or lb-ft) is the sum*

$$\tau = \sum \mathbf{r}_j \times \mathbf{F}_j$$

Torque measures how much the force causes the object to rotate. By Newton's Laws, τ is equal to the rate of change of angular momentum.

59. Calculate the torque τ about O acting at the point P on the mechanical arm in Figure 21(A), assuming that a 25-N force acts as indicated. Ignore the weight of the arm itself.

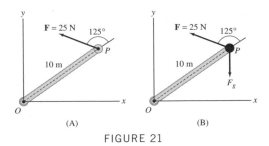

FIGURE 21

SOLUTION We denote by O and P the points shown in the figure and compute the position vector $\mathbf{r} = \overrightarrow{OP}$ and the force vector $\mathbf{F}$.

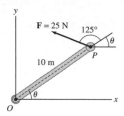

Denoting by θ the angle between the arm and the x-axis we have

$$\mathbf{r} = \overrightarrow{OP} = 10\,(\cos\theta\,\mathbf{i} + \sin\theta\,\mathbf{j})$$

The angle between the force vector $\mathbf{F}$ and the x-axis is $(\theta + 125°)$, hence,

$$\mathbf{F} = 25\,\left(\cos\left(\theta + 125°\right)\mathbf{i} + \sin\left(\theta + 125°\right)\mathbf{j}\right)$$

The torque τ about O acting at the point P is the cross product $\tau = \mathbf{r} \times \mathbf{F}$. We compute it using the cross products of the unit vectors $\mathbf{i}$ and $\mathbf{j}$:

$$
\begin{aligned}
\tau = \mathbf{r} \times \mathbf{F} &= 10\,(\cos\theta\,\mathbf{i} + \sin\theta\,\mathbf{j}) \times 25\,\left(\cos\left(\theta + 125°\right)\mathbf{i} + \sin\left(\theta + 125°\right)\mathbf{j}\right)\\
&= 250\,(\cos\theta\,\mathbf{i} + \sin\theta\,\mathbf{j}) \times \left(\cos\left(\theta + 125°\right)\mathbf{i} + \sin\left(\theta + 125°\right)\mathbf{j}\right)\\
&= 250\,\left(\cos\theta\,\sin\left(\theta + 125°\right)\mathbf{k} + \sin\theta\,\cos\left(\theta + 125°\right)(-\mathbf{k})\right)\\
&= 250\,\left(\sin\left(\theta + 125°\right)\cos\theta - \sin\theta\,\cos\left(\theta + 125°\right)\right)\mathbf{k}
\end{aligned}
$$

We now use the identity $\sin\alpha\cos\beta - \sin\beta\cos\alpha = \sin(\alpha - \beta)$ to obtain

$$\tau = 250\sin\left(\theta + 125° - \theta\right)\mathbf{k} = 250\sin 125°\mathbf{k} \approx 204.79\mathbf{k}$$

61. Let τ be the net torque about O acting on the robotic arm of Figure 22. Assume that the two segments of the arms have mass m_1 and m_2 (in kg) and that a weight of m_3 kg is located at the endpoint P. In calculating the torque, we may assume that the entire mass of each arm segment lies at the midpoint of the arm (its center of mass). Show that the position vectors of the masses m_1, m_2, and m_3 are

$$\mathbf{r}_1 = \frac{1}{2}L_1(\sin\theta_1\mathbf{i} + \cos\theta_1\mathbf{j})$$

$$\mathbf{r}_2 = L_1(\sin\theta_1\mathbf{i} + \cos\theta_1\mathbf{j}) + \frac{1}{2}L_2(\sin\theta_2\mathbf{i} - \cos\theta_2\mathbf{j})$$

$$\mathbf{r}_3 = L_1(\sin\theta_1\mathbf{i} + \cos\theta_1\mathbf{j}) + L_2(\sin\theta_2\mathbf{i} - \cos\theta_2\mathbf{j})$$

Then show that

$$\tau = -g\left(L_1\left(\frac{1}{2}m_1 + m_2 + m_3\right)\sin\theta_1 + L_2\left(\frac{1}{2}m_2 + m_3\right)\sin\theta_2\right)\mathbf{k}$$

where $g = 9.8 m/s^2$. To simplify the computation, note that all three gravitational forces act in the $-\mathbf{j}$ direction, so the $\mathbf{j}$-components of the position vectors $\mathbf{r}_i$ do not contribute to the torque.

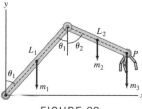

FIGURE 22

SOLUTION We denote by O, P, and Q the points shown in the figure.

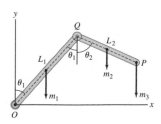

The coordinates of O and Q are

$$O = (0, 0), \quad Q = (L_1 \sin \theta_1, L_1 \cos \theta_1)$$

The midpoint of the segment OQ is, thus,

$$\left(\frac{0 + L_1 \sin \theta_1}{2}, \frac{0 + L_1 \cos \theta_1}{2} \right) = \left(\frac{L_1 \sin \theta_1}{2}, \frac{L_1 \cos \theta_1}{2} \right)$$

Since the mass m_1 is assumed to lie at the midpoint of the arm, the position vector of m_1 is

$$\mathbf{r}_1 = \frac{L_1}{2} \left(\sin \theta_1 \mathbf{i} + \cos \theta_1 \mathbf{j} \right) \tag{1}$$

We now find the position vector $\mathbf{r}_2$ of m_2. We have (see figure)

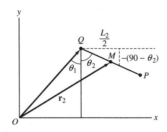

$$\mathbf{r}_2 = \overrightarrow{OQ} + \overrightarrow{QM} \tag{2}$$

$$\overrightarrow{OQ} = L_1 \sin \theta_1 \mathbf{i} + L_1 \cos \theta_1 \mathbf{j} = L_1 \left(\sin \theta_1 \mathbf{i} + \cos \theta_1 \mathbf{j} \right) \tag{3}$$

The vector $\overrightarrow{QM}$ makes an angle of $- \left(90° - \theta_2 \right)$ with the x axis and has length $\frac{L_2}{2}$, hence,

$$\overrightarrow{QM} = \frac{L_2}{2} \left(\cos \left(- \left(90° - \theta_2 \right) \right) \mathbf{i} + \sin \left(- \left(90° - \theta_2 \right) \right) \mathbf{j} \right) = \frac{L_2}{2} \left(\sin \theta_2 \mathbf{i} - \cos \theta_2 \mathbf{j} \right) \tag{4}$$

Combining (2), (3) and (4) we get

$$\mathbf{r}_2 = L_1 \left(\sin \theta_1 \mathbf{i} + \cos \theta_1 \mathbf{j} \right) + \frac{L_2}{2} \left(\sin \theta_2 \mathbf{i} - \cos \theta_2 \mathbf{j} \right) \tag{5}$$

Finally, we find the position vector $\mathbf{r}_3$:

$$\mathbf{r}_3 = \overrightarrow{OQ} + \overrightarrow{QP} = \overrightarrow{OQ} + 2\overrightarrow{QM}$$

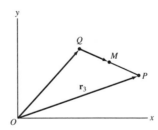

Substituting (3) and (4) we get

$$\mathbf{r}_3 = L_1 \left(\sin \theta_1 \mathbf{i} + \cos \theta_1 \mathbf{j} \right) + L_2 \left(\sin \theta_2 \mathbf{i} - \cos \theta_2 \mathbf{j} \right) \tag{6}$$

The net torque is the following vector:

$$\tau = \mathbf{r}_1 \times \left(-m_1 g \mathbf{j} \right) + \mathbf{r}_2 \times \left(-m_2 g \mathbf{j} \right) + \mathbf{r}_3 \times \left(-m_3 g \mathbf{j} \right)$$

In computing the cross products, the $\mathbf{j}$ components of $\mathbf{r}_1$, $\mathbf{r}_2$ and $\mathbf{r}_3$ do not contribute to the torque since $\mathbf{j} \times \mathbf{j} = \mathbf{0}$. We thus consider only the $\mathbf{i}$ components of $\mathbf{r}_1$, $\mathbf{r}_2$ and $\mathbf{r}_3$ in (1), (5) and (6). This gives

$$\tau = \frac{L_1}{2} \sin \theta_1 \mathbf{i} \times \left(-m_1 g \mathbf{j} \right) + \left(L_1 \sin \theta_1 + \frac{L_2}{2} \sin \theta_2 \right) \mathbf{i} \times \left(-m_2 g \mathbf{j} \right) + \left(L_1 \sin \theta_1 + L_2 \sin \theta_2 \right) \mathbf{i} \times \left(-m_3 g \mathbf{j} \right)$$

$$= -\frac{L_1 m_1 g \sin \theta_1}{2} \mathbf{k} - \left(L_1 m_2 g \sin \theta_1 + \frac{L_2 m_2 g}{2} \sin \theta_2 \right) \mathbf{k} - \left(L_1 m_3 g \sin \theta_1 + L_2 m_3 g \sin \theta_2 \right) \mathbf{k}$$

$$= -g \left(L_1 \left(\frac{1}{2} m_1 + m_2 + m_3 \right) \sin \theta_1 + L_2 \left(\frac{1}{2} m_2 + m_3 \right) \sin \theta_2 \right) \mathbf{k}$$

Further Insights and Challenges

63. Show that 3×3 determinants can be computed using the **diagonal rule**: Repeat the first two columns of the matrix and form the products of the numbers along the six diagonals indicated. Then add the products for the diagonals that slant from left to right and subtract the products for the diagonals that slant from right to left.

$$\det(A) = \begin{vmatrix} a_{11} & a_{12} & a_{13} \\ a_{21} & a_{22} & a_{23} \\ a_{31} & a_{32} & a_{33} \end{vmatrix} \begin{matrix} a_{11} & a_{12} \\ a_{21} & a_{22} \\ a_{31} & a_{32} \end{matrix}$$

$$= a_{11}a_{22}a_{33} + a_{12}a_{23}a_{31} + a_{13}a_{21}a_{32} - a_{13}a_{22}a_{31} - a_{11}a_{23}a_{32} - a_{12}a_{21}a_{33}$$

SOLUTION Using the definition of 3×3 determinants given in Eq. (2) we get

$$\det(A) = a_{11} \begin{vmatrix} a_{22} & a_{23} \\ a_{32} & a_{33} \end{vmatrix} - a_{12} \begin{vmatrix} a_{21} & a_{23} \\ a_{31} & a_{33} \end{vmatrix} + a_{13} \begin{vmatrix} a_{21} & a_{22} \\ a_{31} & a_{32} \end{vmatrix}$$

Using the definition of 2×2 determinants given in Eq. (1) we get

$$\det(A) = a_{11}(a_{22}a_{33} - a_{23}a_{32}) - a_{12}(a_{21}a_{33} - a_{23}a_{31}) + a_{13}(a_{21}a_{32} - a_{22}a_{31})$$

$$= a_{11}a_{22}a_{33} - a_{11}a_{23}a_{32} - a_{12}a_{21}a_{33} + a_{12}a_{23}a_{31} + a_{13}a_{21}a_{32} - a_{13}a_{22}a_{31}$$

$$= a_{11}a_{22}a_{33} + a_{12}a_{23}a_{31} + a_{13}a_{21}a_{32} - a_{13}a_{22}a_{31} - a_{11}a_{23}a_{32} - a_{12}a_{21}a_{33}$$

65. Prove that $\mathbf{v} \times \mathbf{w} = \mathbf{v} \times \mathbf{u}$ if and only if $\mathbf{u} = \mathbf{w} + \lambda\mathbf{v}$ for some scalar λ. Assume that $\mathbf{v} \neq \mathbf{0}$.

SOLUTION Transferring sides and using the distributive law and the property of parallel vectors, we obtain the following equivalent equalities:

$$\mathbf{v} \times \mathbf{w} = \mathbf{v} \times \mathbf{u}$$

$$\mathbf{0} = \mathbf{v} \times \mathbf{u} - \mathbf{v} \times \mathbf{w}$$

$$\mathbf{0} = \mathbf{v} \times (\mathbf{u} - \mathbf{w})$$

This holds if and only if there exists a scalar λ such that

$$\mathbf{u} - \mathbf{w} = \lambda\mathbf{v}$$

$$\mathbf{u} = \mathbf{w} + \lambda\mathbf{v}$$

67. Show that if $\mathbf{u}$, $\mathbf{v}$, and $\mathbf{w}$ are nonzero vectors and $(\mathbf{u} \times \mathbf{v}) \times \mathbf{w} = \mathbf{0}$, then either (i) $\mathbf{u}$ and $\mathbf{v}$ are parallel, or (ii) $\mathbf{w}$ is orthogonal to $\mathbf{u}$ and $\mathbf{v}$.

SOLUTION By the theorem on basic properties of the cross product, part (c), it follows that $(\mathbf{u} \times \mathbf{v}) \times \mathbf{w} = \mathbf{0}$ if and only if

- $\mathbf{u} \times \mathbf{v} = \mathbf{0}$ or
- $\mathbf{w} = \lambda(\mathbf{u} \times \mathbf{v})$

We consider the two possibilities.

1. $\mathbf{u} \times \mathbf{v} = \mathbf{0}$ is equivalent to $\mathbf{u}$ and $\mathbf{v}$ being parallel vectors or one of them being the zero vector.
2. The cross product $\mathbf{u} \times \mathbf{v}$ is orthogonal to $\mathbf{u}$ and $\mathbf{v}$, hence $\mathbf{w} = \lambda(\mathbf{u} \times \mathbf{v})$ implies that $\mathbf{w}$ is also orthogonal to $\mathbf{u}$ and $\mathbf{v}$ (for $\lambda \neq 0$) or $\mathbf{w} = \mathbf{0}$ (for $\lambda = 0$).

Conclusions: $(\mathbf{u} \times \mathbf{v}) \times \mathbf{w} = \mathbf{0}$ implies that either $\mathbf{u}$ and $\mathbf{v}$ are parallel, or $\mathbf{w}$ is orthogonal to $\mathbf{u}$ and $\mathbf{v}$, or one of the vectors $\mathbf{u}$, $\mathbf{v}$, $\mathbf{w}$ is the zero vector.

69. Let $\mathbf{a}$, $\mathbf{b}$, $\mathbf{c}$ be nonzero vectors. Assume that $\mathbf{b}$ and $\mathbf{c}$ are not parallel, and set

$$\mathbf{v} = \mathbf{a} \times (\mathbf{b} \times \mathbf{c}), \qquad \mathbf{w} = (\mathbf{a} \cdot \mathbf{c})\mathbf{b} - (\mathbf{a} \cdot \mathbf{b})\mathbf{c}$$

(a) Prove that

(i) $\mathbf{v}$ lies in the plane spanned by $\mathbf{b}$ and $\mathbf{c}$.

(ii) $\mathbf{v}$ is orthogonal to $\mathbf{a}$.

(b) Prove that $\mathbf{w}$ also satisfies (i) and (ii). Conclude that $\mathbf{v}$ and $\mathbf{w}$ are parallel.

(c) Show algebraically that $\mathbf{v} = \mathbf{w}$ (Figure 23).

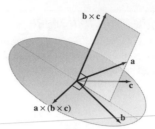

FIGURE 23

SOLUTION

(a) Since $\mathbf{v}$ is the cross product of $\mathbf{a}$ and another vector $(\mathbf{b} \times \mathbf{c})$, then $\mathbf{v}$ is orthogonal to $\mathbf{a}$. Furthermore, $\mathbf{v}$ is orthogonal to $(\mathbf{b} \times \mathbf{c})$, so it is orthogonal to the normal vector to the plane containing $\mathbf{b}$ and $\mathbf{c}$, so $\mathbf{v}$ must be in that plane.

(b) $\mathbf{w} \cdot \mathbf{a} = ((\mathbf{a} \cdot \mathbf{c})\mathbf{b} - (\mathbf{a} \cdot \mathbf{b})\mathbf{c}) \cdot \mathbf{a} = (\mathbf{a} \cdot \mathbf{c})(\mathbf{b} \cdot \mathbf{a}) - (\mathbf{a} \cdot \mathbf{b})(\mathbf{c} \cdot \mathbf{a}) = 0$ (since $\mathbf{a} \cdot \mathbf{c} = \mathbf{c} \cdot \mathbf{a}$ and $\mathbf{b} \cdot \mathbf{a} = \mathbf{a} \cdot \mathbf{b}$). Thus, $\mathbf{w}$ is orthogonal to $\mathbf{a}$. Also, $\mathbf{w}$ is a multiple of $\mathbf{b}$ and $\mathbf{c}$, so $\mathbf{w}$ must be in the plane containing $\mathbf{b}$ and $\mathbf{c}$.

Now, if $\mathbf{a}$ is perpendicular to the plane spanned by $\mathbf{b}$ and $\mathbf{c}$, then $\mathbf{a}$ is parallel to $\mathbf{b} \times \mathbf{c}$ and so $\mathbf{a} \times (\mathbf{b} \times \mathbf{c}) = 0$, which means $\mathbf{v} = 0$, but also $\mathbf{a} \cdot \mathbf{b} = \mathbf{a} \cdot \mathbf{c} = 0$ which means $\mathbf{w} = 0$. Thus, $\mathbf{v}$ and $\mathbf{w}$ are parallel (in fact, equal).

Now, if $\mathbf{a}$ is not perpendicular to the plane spanned by $\mathbf{b}$ and $\mathbf{c}$, then the set of vectors on that plane that are also perpendicular to $\mathbf{a}$ form a line, and thus all such vectors are parallel. We conclude that $\mathbf{v}$ and $\mathbf{w}$, being on that plane and perpendicular to $\mathbf{a}$, are parallel.

(c) On the one hand,

$$\mathbf{v} = \mathbf{a} \times (\mathbf{b} \times \mathbf{c}) = \langle a_1, a_2, a_3 \rangle \times \begin{vmatrix} \mathbf{i} & \mathbf{j} & \mathbf{k} \\ b_1 & b_2 & b_3 \\ c_1 & c_2 & c_3 \end{vmatrix}$$

$$= \begin{vmatrix} \mathbf{i} & \mathbf{j} & \mathbf{k} \\ a_1 & a_2 & a_3 \\ (b_2c_3 - b_3c_2) & (b_3c_1 - b_1c_3) & (b_1c_2 - b_2c_1) \end{vmatrix}$$

$$= \langle a_2(b_1c_2 - b_2c_1) - a_3(b_3c_1 - b_1c_3), a_3(b_2c_3 - b_3c_2) - a_1(b_1c_2 - b_2c_1),$$

$$a_1(b_3c_1 - b_1c_3) - a_2(b_2c_3 - b_3c_2) \rangle$$

but on the other hand,

$$\mathbf{w} = (\mathbf{a} \cdot \mathbf{c})\mathbf{b} - (\mathbf{a} \cdot \mathbf{b})\mathbf{c}$$

$$= (a_1c_1 + a_2c_2 + a_3c_3)\langle b_1, b_2, b_3 \rangle - (a_1b_1 + a_2b_2 + a_3b_3)\langle c_1, c_2, c_3 \rangle$$

$$= \langle a_2c_2b_1 + a_3c_3b_1 - a_2b_2c_1 - a_3b_3c_1, a_1c_1b_2 + a_3c_3b_2 - a_1b_1c_2 - a_3b_3c_2,$$

$$a_1c_1b_3 + a_2c_2b_3 - a_1b_1c_3 - a_2b_2c_3 \rangle$$

$$= \langle a_2(b_1c_2 - b_2c_1) - a_3(b_3c_1 - b_1c_3), a_3(b_2c_3 - b_3c_2) - a_1(b_1c_2 - b_2c_1),$$

$$a_1(b_3c_1 - b_1c_3) - a_2(b_2c_3 - b_3c_2) \rangle$$

which is the same as $\mathbf{v}$.

71. Show that if $\mathbf{a}, \mathbf{b}$ are nonzero vectors such that $\mathbf{a} \perp \mathbf{b}$, then there exists a vector $\mathbf{X}$ such that

$$\mathbf{a} \times \mathbf{X} = \mathbf{b} \qquad \boxed{13}$$

Hint: Show that if $\mathbf{X}$ is orthogonal to $\mathbf{b}$ and is not a multiple of $\mathbf{a}$, then $\mathbf{a} \times \mathbf{X}$ is a multiple of $\mathbf{b}$.

SOLUTION We define the following vectors:

$$\mathbf{X} = \frac{\mathbf{b} \times \mathbf{a}}{\|\mathbf{a}\|^2}, \quad \mathbf{c} = \mathbf{X} \times \mathbf{a} \qquad (1)$$

We show that $\mathbf{c} = \mathbf{b}$. Since $\mathbf{X}$ is orthogonal to $\mathbf{a}$ and $\mathbf{b}$, $\mathbf{X}$ is orthogonal to the plane of $\mathbf{a}$ and $\mathbf{b}$. But $\mathbf{c}$ is orthogonal to $\mathbf{X}$, hence $\mathbf{c}$ is contained in the plane of $\mathbf{a}$ and $\mathbf{b}$, that is, $\mathbf{a}, \mathbf{b}$ and $\mathbf{c}$ are in the same plane. Now the vectors $\mathbf{a}, \mathbf{b}$ and $\mathbf{c}$ are in one plane, and the vectors $\mathbf{c}$ and $\mathbf{b}$ are orthogonal to $\mathbf{a}$.

$$\text{It follows that } \mathbf{c} \text{ and } \mathbf{b} \text{ are parallel.} \qquad (2)$$

We now show that $\|\mathbf{c}\| = \|\mathbf{b}\|$. We use the cross-product identity to obtain

$$\|\mathbf{c}\|^2 = \|\mathbf{X} \times \mathbf{a}\|^2 = \|\mathbf{X}\|^2\|\mathbf{a}\|^2 - (\mathbf{X} \cdot \mathbf{a})^2$$

X is orthogonal to **a**, hence $\mathbf{X} \cdot \mathbf{a} = 0$, and we obtain

$$\|\mathbf{c}\|^2 = \|\mathbf{X}\|^2 \|\mathbf{a}\|^2 = \left\| \frac{\mathbf{b} \times \mathbf{a}}{\|\mathbf{a}\|^2} \right\|^2 \|\mathbf{a}\|^2 = \frac{1}{\|\mathbf{a}\|^4} \|\mathbf{b} \times \mathbf{a}\|^2 \|\mathbf{a}\|^2 = \frac{1}{\|\mathbf{a}\|^2} \|\mathbf{b} \times \mathbf{a}\|^2$$

By the given data, **a** and **b** are orthogonal vectors, so,

$$\|\mathbf{c}\|^2 = \frac{1}{\|\mathbf{a}\|^2} \left(\|\mathbf{b}\|^2 \|\mathbf{a}\|^2 \right) = \|\mathbf{b}\|^2 \Rightarrow \|\mathbf{c}\| = \|\mathbf{b}\| \tag{3}$$

By (2) and (3) it follows that $\mathbf{c} = \mathbf{b}$ or $\mathbf{c} = -\mathbf{b}$. We thus proved that the vector $\mathbf{X} = \dfrac{\mathbf{b} \times \mathbf{a}}{\|\mathbf{a}\|^2}$ satisfies $\mathbf{X} \times \mathbf{a} = \mathbf{b}$ or $\mathbf{X} \times \mathbf{a} = -\mathbf{b}$. If $\mathbf{X} \times \mathbf{a} = -\mathbf{b}$, then $(-\mathbf{X}) \times \mathbf{a} = \mathbf{b}$. Hence, there exists a vector **X** such that $\mathbf{X} \times \mathbf{a} = \mathbf{b}$.

73. Assume that **v** and **w** lie in the first quadrant in $\mathbf{R}^2$ as in Figure 24. Use geometry to prove that the area of the parallelogram is equal to $\det \begin{pmatrix} \mathbf{v} \\ \mathbf{w} \end{pmatrix}$.

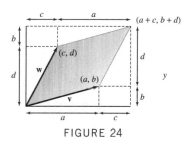

FIGURE 24

SOLUTION We denote the components of **u** and **v** by

$$\mathbf{u} = \langle c, d \rangle$$
$$\mathbf{v} = \langle a, b \rangle$$

We also denote by $O, A, B, C, D, E, F, G, H, K$ the points shown in the figure.

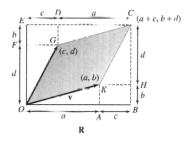

Since $OGCK$ is a parallelogram, it follows by geometrical properties that the triangles OFG and KHC and also the triangles DGC and AKO are congruent. It also follows that the rectangles $EFDG$ and $ABHK$ have equal areas. We use the following notation:

$A:$ The area of the parallelogram
$S:$ The area of the rectangle $OBCE$
$S_1:$ The area of the rectangle $EFDG$
$S_2:$ The area of the triangle OFG
$S_3:$ The area of the triangle DGC

Hence,

$$A = S - 2(S_1 + S_2 + S_3) \tag{1}$$

Using the formulas for the areas of rectangles and triangles we have (see figure)

$$S = OB \cdot OE = (a + c)(d + b)$$

$$S_1 = bc, \quad S_2 = \frac{cd}{2}, \quad S_3 = \frac{ab}{2}$$

Substituting into (1) we get

$$A = (a+c)(d+b) - 2\left(bc + \frac{cd}{2} + \frac{ab}{2}\right)$$

$$= ad + ab + cd + cb - 2bc - cd - ab \qquad (2)$$

$$= ad - bc$$

On the other hand,

$$\det\begin{pmatrix} \mathbf{v} \\ \mathbf{w} \end{pmatrix} = \begin{vmatrix} a & b \\ c & d \end{vmatrix} = ad - bc \qquad (3)$$

By (2) and (3) we obtain the desired result.

75. In the notation of Exercise 74, suppose that $\mathbf{a}, \mathbf{b}, \mathbf{c}$ are mutually perpendicular as in Figure 25(B). Let S_F be the area of face F. Prove the following three-dimensional version of the Pythagorean Theorem:

$$S_A^2 + S_B^2 + S_C^2 = S_D^2$$

SOLUTION   Since $\|\mathbf{v}_D\| = S_D$ then using Exercise 74 we obtain

$$S_D^2 = \|\mathbf{v}_D\|^2 = \mathbf{v}_D \cdot \mathbf{v}_D = (\mathbf{v}_A + \mathbf{v}_B + \mathbf{v}_C) \cdot (\mathbf{v}_A + \mathbf{v}_B + \mathbf{v}_C)$$

$$= \mathbf{v}_A \cdot \mathbf{v}_A + \mathbf{v}_A \cdot \mathbf{v}_B + \mathbf{v}_A \cdot \mathbf{v}_C + \mathbf{v}_B \cdot \mathbf{v}_A + \mathbf{v}_B \cdot \mathbf{v}_B + \mathbf{v}_B \cdot \mathbf{v}_C + \mathbf{v}_C \cdot \mathbf{v}_A + \mathbf{v}_C \cdot \mathbf{v}_B + \mathbf{v}_C \cdot \mathbf{v}_C$$

$$= \|\mathbf{v}_A\|^2 + \|\mathbf{v}_B\|^2 + \|\mathbf{v}_C\|^2 + 2\left(\mathbf{v}_A \cdot \mathbf{v}_B + \mathbf{v}_A \cdot \mathbf{v}_C + \mathbf{v}_B \cdot \mathbf{v}_C\right) \qquad (1)$$

Now, the normals $\mathbf{v}_A$, $\mathbf{v}_B$, and $\mathbf{v}_C$ to the coordinate planes are mutually orthogonal, hence,

$$\mathbf{v}_A \cdot \mathbf{v}_B = \mathbf{v}_A \cdot \mathbf{v}_C = \mathbf{v}_B \cdot \mathbf{v}_C = 0 \qquad (2)$$

Combining (1) and (2) and using the relations $\|\mathbf{v}_F\| = S_F$ we obtain

$$S_D^2 = S_A^2 + S_B^2 + S_C^2$$

12.5 Planes in Three-Space   (LT Section 13.5)

Preliminary Questions

1. What is the equation of the plane parallel to $3x + 4y - z = 5$ passing through the origin?

SOLUTION   The two planes are parallel, therefore the vector $\mathbf{n} = \langle 3, 4, -1 \rangle$ that is normal to the given plane is also normal to the plane we need to find. This plane is passing through the origin, hence we may substitute $\langle x_0, y_0, z_0 \rangle = \langle 0, 0, 0 \rangle$ in the vector form of the equation of the plane. This gives

$$\mathbf{n} \cdot \langle x, y, z \rangle = \mathbf{n} \cdot \langle x_0, y_0, z_0 \rangle$$

$$\langle 3, 4, -1 \rangle \cdot \langle x, y, z \rangle = \langle 3, 4, -1 \rangle \cdot \langle 0, 0, 0 \rangle = 0$$

or in scalar form

$$3x + 4y - z = 0$$

2. The vector $\mathbf{k}$ is normal to which of the following planes?

(a) $x = 1$ **(b)** $y = 1$ **(c)** $z = 1$

SOLUTION   The planes $x = 1$, $y = 1$, and $z = 1$ are orthogonal to the x, y, and z-axes respectively. Since the plane $z = 1$ is orthogonal to the z-axis, the vector $\mathbf{k}$ is normal to this plane.

3. Which of the following planes is not parallel to the plane $x + y + z = 1$?

(a) $2x + 2y + 2z = 1$ **(b)** $x + y + z = 3$
(c) $x - y + z = 0$

SOLUTION   The two planes are parallel if vectors that are normal to the planes are parallel. The vector $\mathbf{n} = \langle 1, 1, 1 \rangle$ is normal to the plane $x + y + z = 1$. We identify the following normals:

- $\mathbf{v} = \langle 2, 2, 2 \rangle$ is normal to plane (a)
- $\mathbf{u} = \langle 1, 1, 1 \rangle$ is normal to plane (b)
- $\mathbf{w} = \langle 1, -1, 1 \rangle$ is normal to plane (c)

The vectors $\mathbf{v}$ and $\mathbf{u}$ are parallel to $\mathbf{n}$, whereas $\mathbf{w}$ is not. (These vectors are not constant multiples of each other). Therefore, only plane (c) is not parallel to the plane $x + y + z = 1$.

4. To which coordinate plane is the plane $y = 1$ parallel?

SOLUTION The plane $y = 1$ is parallel to the xz-plane.

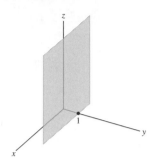

5. Which of the following planes contains the z-axis?

(a) $z = 1$ **(b)** $x + y = 1$ **(c)** $x + y = 0$

SOLUTION The points on the z-axis are the points with zero x and y coordinates. A plane contains the z-axis if and only if the points $(0, 0, c)$ satisfy the equation of the plane for all values of c.

(a) Plane (a) does not contain the z-axis, rather it is orthogonal to this axis. Only the point $(0, 0, 1)$ is on the plane.

(b) $x = 0$ and $y = 0$ do not satisfy the equation of the plane, since $0 + 0 \neq 1$. Therefore the plane does not contain the z-axis.

(c) The plane $x + y = 0$ contains the z-axis since $x = 0$ and $y = 0$ satisfy the equation of the plane.

6. Suppose that a plane $\mathcal{P}$ with normal vector $\mathbf{n}$ and a line $\mathcal{L}$ with direction vector $\mathbf{v}$ both pass through the origin and that $\mathbf{n} \cdot \mathbf{v} = 0$. Which of the following statements is correct?

(a) $\mathcal{L}$ is contained in $\mathcal{P}$.

(b) $\mathcal{L}$ is orthogonal to $\mathcal{P}$.

SOLUTION The direction vector of the line $\mathcal{L}$ is orthogonal to the vector $\mathbf{n}$ that is normal to the plane. Therefore, $\mathcal{L}$ is either parallel or contained in the plane. Since the origin is common to $\mathcal{L}$ and $\mathcal{P}$, the line is contained in the plane. That is, statement (a) is correct.

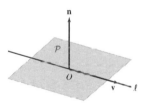

Exercises

In Exercises 1–8, write the equation of the plane with normal vector $\mathbf{n}$ *passing through the given point in each of the three forms (one vector form and two scalar forms).*

1. $\mathbf{n} = \langle 1, 3, 2 \rangle$, $(4, -1, 1)$

SOLUTION The vector equation is

$$\langle 1, 3, 2 \rangle \cdot \langle x, y, z \rangle = \langle 1, 3, 2 \rangle \cdot \langle 4, -1, 1 \rangle = 4 - 3 + 2 = 3$$

To obtain the scalar forms we compute the dot product on the left-hand side of the previous equation:

$$x + 3y + 2z = 3$$

or in the other scalar form:

$$(x - 4) + 3(y + 1) + 2(z - 1) + 4 - 3 + 2 = 3$$
$$(x - 4) + 3(y + 1) + 2(z - 1) = 0$$

3. $\mathbf{n} = \langle -1, 2, 1 \rangle$, $(4, 1, 5)$

SOLUTION The vector form is

$$\langle -1, 2, 1 \rangle \cdot \langle x, y, z \rangle = \langle -1, 2, 1 \rangle \cdot \langle 4, 1, 5 \rangle = -4 + 2 + 5 = 3$$

To obtain the scalar form we compute the dot product above:

$$-x + 2y + z = 3$$

or in the other scalar form:

$$-(x - 4) + 2(y - 1) + (z - 5) = 3 + 4 - 2 - 5 = 0$$

$$-(x - 4) + 2(y - 1) + (z - 5) = 0$$

5. $n = i$, $(3, 1, -9)$

SOLUTION   We find the vector form of the equation of the plane. We write the vector $n = i$ as $n = \langle 1, 0, 0 \rangle$ and obtain

$$\langle 1, 0, 0 \rangle \cdot \langle x, y, z \rangle = \langle 1, 0, 0 \rangle \cdot \langle 3, 1, -9 \rangle = 3 + 0 + 0 = 3$$

Computing the dot product above gives the scalar form:

$$x + 0 + 0 = 3$$

$$x = 3$$

Or in the other scalar form:

$$(x - 3) + 0 \cdot (y - 1) + 0 \cdot (z + 9) = 3 - 3 = 0$$

7. $n = k$, $(6, 7, 2)$

SOLUTION   We write the normal $n = k$ in the form $n = \langle 0, 0, 1 \rangle$ and obtain the following vector form of the equation of the plane:

$$\langle 0, 0, 1 \rangle \cdot \langle x, y, z \rangle = \langle 0, 0, 1 \rangle \cdot \langle 6, 7, 2 \rangle = 0 + 0 + 2 = 2$$

We compute the dot product to obtain the scalar form:

$$0x + 0y + 1z = 2$$

$$z = 2$$

or in the other scalar form:

$$0(x - 6) + 0(y - 7) + 1(z - 2) = 0$$

9. Write down the equation of any plane through the origin.

SOLUTION   We can use any equation $ax + by + cz = d$ which contains the point $(x, y, z) = (0, 0, 0)$. One solution (and there are many) is $x + y + z = 0$.

11. Which of the following statements are true of a plane that is parallel to the yz-plane?
(a) $n = \langle 0, 0, 1 \rangle$ is a normal vector.
(b) $n = \langle 1, 0, 0 \rangle$ is a normal vector.
(c) The equation has the form $ay + bz = d$
(d) The equation has the form $x = d$

SOLUTION
(a) For $n = \langle 0, 0, 1 \rangle$ a normal vector, the plane would be parallel to the xy-plane, not the yz-plane. This statement is false.
(b) For $n = \langle 1, 0, 0 \rangle$ a normal vector, the plane would be parallel to the yz-plane. This statement is true.
(c) For the equation $ay + bz = d$, this plane intersects the yz-plane at $y = 0, z = d/b$ or $y = d/a, z = 0$ depending on whether a or b is non-zero, but it is not equal to the yz-plane (which has equation $x = d$) Thus, it is not parallel to the yz-plane This statement is false.
(d) For the equation of the form $x = d$, this has $\langle 1, 0, 0 \rangle$ as a normal vector and is parallel to the yz-plane. This statement is true.

In Exercises 13–16, find a vector normal to the plane with the given equation.

13. $9x - 4y - 11z = 2$

SOLUTION   Using the scalar form of the equation of the plane, a vector normal to the plane is the coefficients vector:

$$n = \langle 9, -4, -11 \rangle$$

15. $3(x - 4) - 8(y - 1) + 11z = 0$

SOLUTION   Using the scalar form of the equation of the plane, $3x - 8y + 11z = 4$ a vector normal to the plane is the coefficients vector:

$$n = \langle 3, -8, 11 \rangle$$

In Exercises 17–20, find an equation of the plane passing through the three points given.

17. $P = (2, -1, 4)$, $Q = (1, 1, 1)$, $R = (3, 1, -2)$

SOLUTION We go through the steps below:

Step 1. Find the normal vector **n**. The vectors $\mathbf{a} = \overrightarrow{PQ}$ and $\mathbf{b} = \overrightarrow{PR}$ lie on the plane, hence the cross product $\mathbf{n} = \mathbf{a} \times \mathbf{b}$ is normal to the plane. We compute the cross product:

$$\mathbf{a} = \overrightarrow{PQ} = \langle 1 - 2, 1 - (-1), 1 - 4 \rangle = \langle -1, 2, -3 \rangle$$

$$\mathbf{b} = \overrightarrow{PR} = \langle 3 - 2, 1 - (-1), -2 - 4 \rangle = \langle 1, 2, -6 \rangle$$

$$\mathbf{n} = \mathbf{a} \times \mathbf{b} = \begin{vmatrix} \mathbf{i} & \mathbf{j} & \mathbf{k} \\ -1 & 2 & -3 \\ 1 & 2 & -6 \end{vmatrix} = \begin{vmatrix} 2 & -3 \\ 2 & -6 \end{vmatrix} \mathbf{i} - \begin{vmatrix} -1 & -3 \\ 1 & -6 \end{vmatrix} \mathbf{j} + \begin{vmatrix} -1 & 2 \\ 1 & 2 \end{vmatrix} \mathbf{k}$$

$$= -6\mathbf{i} - 9\mathbf{j} - 4\mathbf{k} = \langle -6, -9, -4 \rangle$$

Step 2. Choose a point on the plane. We choose any one of the three points on the plane, for instance $Q = (1, 1, 1)$. Using the vector form of the equation of the plane we get

$$\mathbf{n} \cdot \langle x, y, z \rangle = \mathbf{n} \cdot \langle x_0, y_0, z_0 \rangle$$

$$\langle -6, -9, -4 \rangle \cdot \langle x, y, z \rangle = \langle -6, -9, -4 \rangle \cdot \langle 1, 1, 1 \rangle$$

Computing the dot products we obtain the following equation:

$$-6x - 9y - 4z = -6 - 9 - 4 = -19$$

$$6x + 9y + 4z = 19$$

19. $P = (1, 0, 0)$, $Q = (0, 1, 1)$, $R = (2, 0, 1)$

SOLUTION We use the vector form of the equation of the plane:

$$\mathbf{n} \cdot \langle x, y, z \rangle = d \tag{1}$$

To find the normal vector to the plane, **n**, we first compute the vectors $\overrightarrow{PQ}$ and $\overrightarrow{PR}$ that lie in the plane, and then find the cross product of these vectors. This gives

$$\overrightarrow{PQ} = \langle 0, 1, 1 \rangle - \langle 1, 0, 0 \rangle = \langle -1, 1, 1 \rangle$$

$$\overrightarrow{PR} = \langle 2, 0, 1 \rangle - \langle 1, 0, 0 \rangle = \langle 1, 0, 1 \rangle$$

$$\mathbf{n} = \overrightarrow{PQ} \times \overrightarrow{PR} = \begin{vmatrix} \mathbf{i} & \mathbf{j} & \mathbf{k} \\ -1 & 1 & 1 \\ 1 & 0 & 1 \end{vmatrix} = \begin{vmatrix} 1 & 1 \\ 0 & 1 \end{vmatrix} \mathbf{i} - \begin{vmatrix} -1 & 1 \\ 1 & 1 \end{vmatrix} \mathbf{j} + \begin{vmatrix} -1 & 1 \\ 1 & 0 \end{vmatrix} \mathbf{k}$$

$$= \mathbf{i} + 2\mathbf{j} - \mathbf{k} = \langle 1, 2, -1 \rangle \tag{2}$$

We now choose any one of the three points in the plane, say $P = (1, 0, 0)$, and compute d:

$$d = \mathbf{n} \cdot \overrightarrow{OP} = \langle 1, 2, -1 \rangle \cdot \langle 1, 0, 0 \rangle = 1 \cdot 1 + 2 \cdot 0 + (-1) \cdot 0 = 1 \tag{3}$$

Finally we substitute (2) and (3) into (1) to obtain the following equation of the plane:

$$\langle 1, 2, -1 \rangle \cdot \langle x, y, z \rangle = 1$$

$$x + 2y - z = 1$$

In Exercises 21–28, find the equation of the plane with the given description.

21. Passes through O and is parallel to $4x - 9y + z = 3$

SOLUTION The vector $\mathbf{n} = \langle 4, -9, 1 \rangle$ is normal to the plane $4x - 9y + z = 3$, and so is also normal to the parallel plane. Setting $\mathbf{n} = \langle 4, -9, 1 \rangle$ and $(x_0, y_0, z_0) = (0, 0, 0)$ in the vector equation of the plane yields

$$\langle 4, -9, 1 \rangle \cdot \langle x, y, z \rangle = \langle 4, -9, 1 \rangle \cdot \langle 0, 0, 0 \rangle = 0$$

$$4x - 9y + z = 0$$

23. Passes through $(4, 1, 9)$ and is parallel to $x = 3$

SOLUTION The vector form of the plane $x = 3$ is

$$\langle 1, 0, 0 \rangle \cdot \langle x, y, z \rangle = 3$$

Hence, $\mathbf{n} = \langle 1, 0, 0 \rangle$ is normal to this plane. This vector is also normal to the parallel plane. Setting $(x_0, y_0, z_0) = (4, 1, 9)$ and $\mathbf{n} = \langle 1, 0, 0 \rangle$ in the vector equation of the plane yields

$$\langle 1, 0, 0 \rangle \cdot \langle x, y, z \rangle = \langle 1, 0, 0 \rangle \cdot \langle 4, 1, 9 \rangle = 4 + 0 + 0 = 4$$

or

$$x + 0 + 0 = 4 \quad \Rightarrow \quad x = 4$$

25. Passes through $(-2, -3, 5)$ and has normal vector $\mathbf{i} + \mathbf{k}$

SOLUTION   We substitute $\mathbf{n} = \langle 1, 0, 1 \rangle$ and $(x_0, y_0, z_0) = (-2, -3, 5)$ in the vector equation of the plane to obtain

$$\langle 1, 0, 1 \rangle \cdot \langle x, y, z \rangle = \langle 1, 0, 1 \rangle \cdot \langle -2, -3, 5 \rangle$$

or

$$x + 0 + z = -2 + 0 + 5 = 3$$
$$x + z = 3$$

27. Contains the lines $\mathbf{r}_1(t) = \langle 2, 1, 0 \rangle + \langle t, 2t, 3t \rangle$ and $\mathbf{r}_2(t) = \langle 2, 1, 0 \rangle + \langle 3t, t, 8t \rangle$

SOLUTION   Since the plane contains the lines $\mathbf{r}_1(t)$ and $\mathbf{r}_2(t)$, the direction vectors $\mathbf{v}_1 = \langle 1, 2, 3 \rangle$ and $\mathbf{v}_2 = \langle 3, 1, 8 \rangle$ of the lines lie in the plane. Therefore the cross product $\mathbf{n} = \mathbf{v}_1 \times \mathbf{v}_2$ is normal to the plane. We compute the cross product:

$$\mathbf{n} = \langle 1, 2, 3 \rangle \times \langle 3, 1, 8 \rangle = \begin{vmatrix} \mathbf{i} & \mathbf{j} & \mathbf{k} \\ 1 & 2 & 3 \\ 3 & 1 & 8 \end{vmatrix} = \begin{vmatrix} 2 & 3 \\ 1 & 8 \end{vmatrix} \mathbf{i} - \begin{vmatrix} 1 & 3 \\ 3 & 8 \end{vmatrix} \mathbf{j} + \begin{vmatrix} 1 & 2 \\ 3 & 1 \end{vmatrix} \mathbf{k}$$

$$= 13\mathbf{i} + \mathbf{j} - 5\mathbf{k} = \langle 13, 1, -5 \rangle$$

We now must choose a point on the plane. Since the line $\mathbf{r}_1(t) = \langle 2 + t, 1 + 2t, 3t \rangle$ is contained in the plane, all of its points are on the plane. We choose the point corresponding to $t = 0$, that is,

$$\langle x_0, y_0, z_0 \rangle = \langle 2, 1, 0 \rangle$$

We now use the vector equation of the plane to determine the equation of the desired plane:

$$\mathbf{n} \cdot \langle x, y, z \rangle = \mathbf{n} \cdot \langle x_0, y_0, z_0 \rangle$$
$$\langle 13, 1, -5 \rangle \cdot \langle x, y, z \rangle = \langle 13, 1, -5 \rangle \cdot \langle 2, 1, 0 \rangle$$
$$13x + y - 5z = 26 + 1 + 0 = 27$$
$$13x + y - 5z = 27$$

29. Are the planes $\frac{1}{2}x + 2x - y = 5$ and $3x + 12x - 6y = 1$ parallel?

SOLUTION   The planes $2\frac{1}{2}x - y = 5$ and $15x - 6y = 1$, are parallel if and only if the vectors $\mathbf{n}_1 = \langle 2\frac{1}{2}, -1, 0 \rangle$ and $\mathbf{n}_2 = \langle 15, -6, 0 \rangle$ normal to the planes are parallel. Since $\mathbf{n}_2 = 6\mathbf{n}_1$ the planes are parallel.

31. Find an equation of the plane $\mathcal{P}$ in Figure 8.

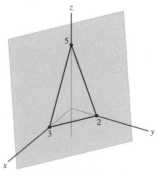

FIGURE 8

SOLUTION   We must find the equation of the plane passing though the points $P = (3, 0, 0)$, $Q = (0, 2, 0)$, and $R = (0, 0, 5)$.

We use the following steps:

Step 1. Find a normal vector $\mathbf{n}$. The vectors $\mathbf{a} = \overrightarrow{PQ}$ and $\mathbf{b} = \overrightarrow{PR}$ lie in the plane, hence the cross product $\mathbf{n} = \mathbf{a} \times \mathbf{b}$ is normal to the plane. We compute the cross product:

$$\mathbf{a} = \overrightarrow{PQ} = \langle 0 - 3, 2 - 0, 0 - 0 \rangle = \langle -3, 2, 0 \rangle$$

$$\mathbf{b} = \overrightarrow{PR} = \langle 0 - 3, 0 - 0, 5 - 0 \rangle = \langle -3, 0, 5 \rangle$$

$$\mathbf{n} = \mathbf{a} \times \mathbf{b} = \begin{vmatrix} \mathbf{i} & \mathbf{j} & \mathbf{k} \\ -3 & 2 & 0 \\ -3 & 0 & 5 \end{vmatrix} = \begin{vmatrix} 2 & 0 \\ 0 & 5 \end{vmatrix} \mathbf{i} - \begin{vmatrix} -3 & 0 \\ -3 & 5 \end{vmatrix} \mathbf{j} + \begin{vmatrix} -3 & 2 \\ -3 & 0 \end{vmatrix} \mathbf{k}$$

$$= 10\mathbf{i} + 15\mathbf{j} + 6\mathbf{k} = \langle 10, 15, 6 \rangle$$

Step 2. Choose a point on the plane We choose one of the points on the plane, say $P = (3, 0, 0)$. Substituting $\mathbf{n} = \langle 10, 15, 6 \rangle$ and $(x_0, y_0, z_0) = (3, 0, 0)$ in the vector form of the equation of the plane gives

$$\mathbf{n} \cdot \langle x, y, z \rangle = \mathbf{n} \cdot \langle x_0, y_0, z_0 \rangle$$

$$\langle 10, 15, 6 \rangle \cdot \langle x, y, z \rangle = \langle 10, 15, 6 \rangle \cdot \langle 3, 0, 0 \rangle$$

Computing the dot products we get the following scalar form of the equation of the plane:

$$10x + 15y + 6z = 10 \cdot 3 + 0 + 0 = 30$$

$$10x + 15y + 6z = 30$$

In Exercises 33–36, find the intersection of the line and the plane.

33. $x + y + z = 14$, $\mathbf{r}(t) = \langle 1, 1, 0 \rangle + t \langle 0, 2, 4 \rangle$

SOLUTION The line has parametric equations

$$x = 1, \quad y = 1 + 2t, \quad z = 4t$$

To find a value of t for which (x, y, z) lies on the plane, we substitute the parametric equations in the equation of the plane and solve for t:

$$x + y + z = 14$$

$$1 + (1 + 2t) + 4t = 14$$

$$6t = 12 \quad \Rightarrow \quad t = 2$$

The point P of intersection has coordinates

$$x = 1, \quad y = 1 + 2 \cdot 2 = 5, \quad z = 4 \cdot 2 = 8$$

That is, $P = (1, 5, 8)$.

35. $z = 12$, $\mathbf{r}(t) = t \langle -6, 9, 36 \rangle$

SOLUTION The parametric equations of the line are

$$x = -6t, \quad y = 9t, \quad z = 36t \tag{1}$$

We substitute the parametric equations in the equation of the plane and solve for t:

$$z = 12$$

$$36t = 12 \quad \Rightarrow \quad t = \frac{1}{3}$$

The value of the parameter at the point of intersection is $t = \frac{1}{3}$. Substituting into (1) gives the coordinates of the point P of intersection:

$$x = -6 \cdot \frac{1}{3} = -2, \quad y = 9 \cdot \frac{1}{3} = 3, \quad z = 36 \cdot \frac{1}{3} = 12$$

That is,

$$P = (-2, 3, 12).$$

In Exercises 37–42, find the trace of the plane in the given coordinate plane.

37. $3x - 9y + 4z = 5$, yz

SOLUTION   The yz-plane has the equation $x = 0$, hence the intersection of the plane with the yz-plane must satisfy both $x = 0$ and the equation of the plane $3x - 9y + 4z = 5$. That is, this is the set of all points $(0, y, z)$ in the yz-plane such that $-9y + 4z = 5$.

39. $3x + 4z = -2$, xy

SOLUTION   The trace of the plane $3x + 4z = -2$ in the xy coordinate plane is the set of all points that satisfy the equation of the plane and the equation $z = 0$ of the xy coordinate plane. Thus, we substitute $z = 0$ in $3x + 4z = -2$ to obtain the line $3x = -2$ or $x = -\frac{2}{3}$ in the xy-plane.

41. $-x + y = 4$, xz

SOLUTION   The trace of the plane $-x + y = 4$ on the xz-plane is the set of all points that satisfy both the equation of the given plane and the equation $y = 0$ of the xz-plane. That is, the set of all points $(x, 0, z)$ such that $-x + 0 = 4$, or $x = -4$. This is a vertical line in the xz-plane.

43. Does the plane $x = 5$ have a trace in the yz-plane? Explain.

SOLUTION   The yz-plane has the equation $x = 0$, hence the x-coordinates of the points in this plane are zero, whereas the x-coordinates of the points in the plane $x = 5$ are 5. Thus, the two planes have no common points.

45. Give equations for two distinct planes whose trace in the yz-plane has equation $y = 4z$.

SOLUTION   The yz-plane has the equation $x = 0$, hence the trace of a plane $ax + by + cz = 0$ in the yz-plane is obtained by substituting $x = 0$ in the equation of the plane. Therefore, the following two planes have trace $y = 4z$ (that is, $y - 4z = 0$) in the yz-plane:

$$x + y - 4z = 0; \quad 2x + y - 4z = 0$$

47. Find all planes in $\mathbf{R}^3$ whose intersection with the xz-plane is the line with equation $3x + 2z = 5$.

SOLUTION   The intersection of the plane $ax + by + cz = d$ with the xz-plane is obtained by substituting $y = 0$ in the equation of the plane. This gives the following line in the xz-plane:

$$ax + cz = d$$

This is the equation of the line $3x + 2z = 5$ if and only if for some $\lambda \neq 0$,

$$a = 3\lambda, \quad c = 2\lambda, \quad d = 5\lambda$$

Notice that b can have any value. The planes are thus

$$(3\lambda)x + by + (2\lambda)z = 5\lambda, \quad \lambda \neq 0.$$

In Exercises 49–54, compute the angle between the two planes, defined as the angle θ (between 0 and π) between their normal vectors (Figure 9).

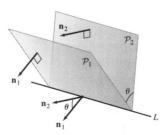

FIGURE 9 By definition, the angle between two planes is the angle between their normal vectors.

49. Planes with normals $\mathbf{n}_1 = \langle 1, 0, 1 \rangle$, $\mathbf{n}_2 = \langle -1, 1, 1 \rangle$

SOLUTION Using the formula for the angle between two vectors we get

$$\cos\theta = \frac{\mathbf{n}_1 \cdot \mathbf{n}_2}{\|\mathbf{n}_1\| \|\mathbf{n}_2\|} = \frac{\langle 1, 0, 1 \rangle \cdot \langle -1, 1, 1 \rangle}{\|\langle 1, 0, 1 \rangle\| \|\langle -1, 1, 1 \rangle\|} = \frac{-1 + 0 + 1}{\sqrt{1^2 + 0 + 1^2}\sqrt{(-1)^2 + 1^2 + 1^2}} = 0$$

The solution for $0 \le \theta < \pi$ is $\theta = \frac{\pi}{2}$.

51. $2x + 3y + 7z = 2$ and $4x - 2y + 2z = 4$

SOLUTION The planes $2x + 3y + 7z = 2$ and $4x - 2y + 2z = 4$ have the normals $\mathbf{n}_1 = \langle 2, 3, 7 \rangle$ and $\mathbf{n}_2 = \langle 4, -2, 2 \rangle$ respectively. The cosine of the angle between $\mathbf{n}_1$ and $\mathbf{n}_2$ is

$$\cos\theta = \frac{\mathbf{n}_1 \cdot \mathbf{n}_2}{\|\mathbf{n}_1\| \|\mathbf{n}_2\|} = \frac{\langle 2, 3, 7 \rangle \cdot \langle 4, -2, 2 \rangle}{\|\langle 2, 3, 7 \rangle\| \|\langle 4, -2, 2 \rangle\|} = \frac{8 - 6 + 14}{\sqrt{2^2 + 3^2 + 7^2}\sqrt{4^2 + (-2)^2 + 2^2}} = \frac{16}{\sqrt{62}\sqrt{24}} \approx 0.415$$

The solution for $0 \le \theta < \pi$ is $\theta = 1.143$ rad or $\theta = 65.49°$.

53. $3(x - 1) - 5y + 2(z - 12) = 0$ and the plane with normal $\mathbf{n} = \langle 1, 0, 1 \rangle$

SOLUTION The plane $3(x - 1) - 5y + 2(z - 12) = 0$ has the normal $\mathbf{n}_1 = \langle 3, -5, 2 \rangle$, and our second plane has given normal $\mathbf{n}_2 = \langle 1, 0, 1 \rangle$. We use the formula for the angle between two vectors:

$$\cos\theta = \frac{\mathbf{n}_1 \cdot \mathbf{n}_2}{\|\mathbf{n}_1\| \|\mathbf{n}_2\|} = \frac{\langle 3, -5, 2 \rangle \cdot \langle 1, 0, 1 \rangle}{\|\langle 3, -5, 2 \rangle\| \|\langle 1, 0, 1 \rangle\|} = \frac{3 + 0 + 2}{\sqrt{3^2 + (-5)^2 + 2^2}\sqrt{1^2 + 0 + 1^2}} = \frac{5}{\sqrt{38}\sqrt{2}} \approx 0.5735$$

The solution for $0 \le \theta < \pi$ is $\theta = 0.96$ rad or $\theta = 55°$.

55. Find an equation of a plane making an angle of $\frac{\pi}{2}$ with the plane $3x + y - 4z - 2$.

SOLUTION The angle θ between two planes (chosen so that $0 \le \theta < \pi$) is defined as the angle between their normal vectors. The following vector is normal to the plane $3x + y - 4z = 2$:

$$\mathbf{n}_1 = \langle 3, 1, -4 \rangle$$

Let $\mathbf{n} \cdot \langle x, y, z \rangle = d$ denote the equation of a plane making an angle of $\frac{\pi}{2}$ with the given plane, where $\mathbf{n} = \langle a, b, c \rangle$. Since the two planes are perpendicular, the dot product of their normal vectors is zero. That is,

$$\mathbf{n} \cdot \mathbf{n}_1 = \langle a, b, c \rangle \cdot \langle 3, 1, -4 \rangle = 3a + b - 4c = 0 \quad \Rightarrow \quad b = -3a + 4c$$

Thus, the required planes (there is more than one plane) have the following normal vector:

$$\mathbf{n} = \langle a, -3a + 4c, c \rangle$$

We obtain the following equation:

$$\mathbf{n} \cdot \langle x, y, c \rangle = d$$
$$\langle a, -3a + 4c, c \rangle \cdot \langle x, y, z \rangle = d$$
$$ax + (4c - 3a)y + cz = d$$

Every choice of the values of a, c and d yields a plane with the desired property. For example, we set $a = c = d = 1$ to obtain

$$x + y + z = 1$$

57. Find a plane that is perpendicular to the two planes $x + y = 3$ and $x + 2y - z = 4$.

SOLUTION The vector forms of the equations of the planes are $\langle 1, 1, 0 \rangle \cdot \langle x, y, z \rangle = 3$ and $\langle 1, 2, -1 \rangle \cdot \langle x, y, z \rangle = 4$, hence the vectors $\mathbf{n}_1 = \langle 1, 1, 0 \rangle$ and $\mathbf{n}_2 = \langle 1, 2, -1 \rangle$ are normal to the planes. We denote the equation of the planes which are perpendicular to the two planes by

$$ax + by + cz = d \qquad (1)$$

Then, the normal $\mathbf{n} = \langle a, b, c \rangle$ to the planes is orthogonal to the normals $\mathbf{n}_1$ and $\mathbf{n}_2$ of the given planes. Therefore, $\mathbf{n} \cdot \mathbf{n}_1 = 0$ and $\mathbf{n} \cdot \mathbf{n}_2 = 0$ which gives us

$$\langle a, b, c \rangle \cdot \langle 1, 1, 0 \rangle = 0, \quad \langle a, b, c \rangle \cdot \langle 1, 2, -1 \rangle = 0$$

We obtain the following equations:

$$\begin{cases} a+b=0 \\ a+2b-c=0 \end{cases}$$

The first equation implies that $b=-a$. Substituting in the second equation we get $a-2a-c=0$, or $c=-a$. Substituting $b=-a$ and $c=-a$ in (1) gives (for $a \neq 0$):

$$ax-ay-az=d \quad \Rightarrow \quad x-y-z=\frac{d}{a}$$

$\frac{d}{a}$ is an arbitrary constant which we denote by f. The planes which are perpendicular to the given planes are, therefore,

$$x-y-z=f$$

59. Let $\mathcal{L}$ denote the intersection of the planes $x-y-z=1$ and $2x+3y+z=2$. Find parametric equations for the line $\mathcal{L}$. *Hint:* To find a point on $\mathcal{L}$, substitute an arbitrary value for z (say, $z=2$) and then solve the resulting pair of equations for x and y.

SOLUTION We use Exercise 56 to find a direction vector for the line of intersection $\mathcal{L}$ of the planes $x-y-z=1$ and $2x+3y+z=2$. We identify the normals $\mathbf{n}_1=\langle 1,-1,-1 \rangle$ and $\mathbf{n}_2=\langle 2,3,1 \rangle$ to the two planes respectively. Hence, a direction vector for $\mathcal{L}$ is the cross product $\mathbf{v}=\mathbf{n}_1 \times \mathbf{n}_2$. We find it here:

$$\mathbf{v}=\mathbf{n}_1 \times \mathbf{n}_2=\begin{vmatrix} \mathbf{i} & \mathbf{j} & \mathbf{k} \\ 1 & -1 & -1 \\ 2 & 3 & 1 \end{vmatrix}=2\mathbf{i}-3\mathbf{j}+5\mathbf{k}=\langle 2,-3,5 \rangle$$

We now need to find a point on $\mathcal{L}$. We choose $z=2$, substitute in the equations of the planes and solve the resulting equations for x and y. This gives

$$\begin{array}{ccc} x-y-2=1 \\ 2x+3y+2=2 \end{array} \quad \text{or} \quad \begin{array}{c} x-y=3 \\ 2x+3y=0 \end{array}$$

The 1st equation implies that $y=x-3$. Substituting in the 2nd equation and solving for x gives

$$2x+3(x-3)=0$$

$$5x=9 \quad \Rightarrow \quad x=\frac{9}{5}, \quad y=\frac{9}{5}-3=-\frac{6}{5}$$

We conclude that the point $\left(\frac{9}{5},-\frac{6}{5},2 \right)$ is on $\mathcal{L}$. We now use the vector parametrization of a line to obtain the following parametrization for $\mathcal{L}$:

$$\mathbf{r}(t)=\left\langle \frac{9}{5},-\frac{6}{5},2 \right\rangle + t \langle 2,-3,5 \rangle$$

This yields the parametric equations

$$x=\frac{9}{5}+2t, \quad y=-\frac{6}{5}-3t, \quad z=2+5t$$

61. Two vectors $\mathbf{v}$ and $\mathbf{w}$, each of length 12, lie in the plane $x+2y-2z=0$. The angle between $\mathbf{v}$ and $\mathbf{w}$ is $\pi/6$. This information determines $\mathbf{v} \times \mathbf{w}$ up to a sign ± 1. What are the two possible values of $\mathbf{v} \times \mathbf{w}$?

SOLUTION The length of $\mathbf{v} \times \mathbf{w}$ is $\|\mathbf{v}\|\|\mathbf{w}\| \sin \theta$, but since both vectors have length 12 and since the angle between them is $\pi/6$, then the length of $\mathbf{v} \times \mathbf{w}$ is $12 \cdot 12 \cdot 1/2=72$. The direction of $\mathbf{v} \times \mathbf{w}$ is perpendicular to the plane containing them, which is the plane $x+2y-2z=0$, which has normal vector $\mathbf{n}=\langle 1,2,-2 \rangle$. Since $\mathbf{v} \times \mathbf{w}$ must have length 72 and must be parallel to $\langle 1,2,-2 \rangle$, then it must be ± 72 times the unit vector $\langle 1,2,-2 \rangle / \sqrt{1^2+2^2+(-2)^2}=\langle 1/3,2/3,-2/3 \rangle$. Thus,

$$\mathbf{v} \times \mathbf{w}=\pm 72 \cdot \langle 1/3,2/3,-2/3 \rangle=\pm 24 \cdot \langle 1,2,-2 \rangle$$

63. In this exercise, we show that the orthogonal distance D from the plane $\mathcal{P}$ with equation $ax+by+cz=d$ to the origin O is equal to (Figure 10)

$$D=\frac{|d|}{\sqrt{a^2+b^2+c^2}}$$

Let $\mathbf{n}=\langle a,b,c \rangle$, and let P be the point where the line through $\mathbf{n}$ intersects $\mathcal{P}$. By definition, the orthogonal distance from $\mathcal{P}$ to O is the distance from P to O.

(a) Show that P is the terminal point of $\mathbf{v} = \left(\dfrac{d}{\mathbf{n} \cdot \mathbf{n}}\right) \mathbf{n}$.

(b) Show that the distance from P to O is D.

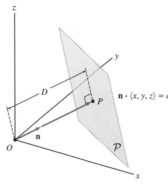

FIGURE 10

SOLUTION Let $\mathbf{v}$ be the vector $\mathbf{v} = \left(\dfrac{d}{\mathbf{n} \cdot \mathbf{n}}\right) \mathbf{n}$. Then $\mathbf{v}$ is parallel to $\mathbf{n}$ and the two vectors are on the same ray.

(a) First we must show that the terminal point of $\mathbf{v}$ lies on the plane $ax + by + cz = d$. Since the terminal point of $\mathbf{v}$ is the point

$$\left(\frac{d}{\mathbf{n} \cdot \mathbf{n}}\right)(a, b, c) = \left(\frac{da}{a^2 + b^2 + c^2}, \frac{db}{a^2 + b^2 + c^2}, \frac{dc}{a^2 + b^2 + c^2}\right)$$

then we need only show that this point satisfies $ax + by + cz = d$. Plugging in, we find:

$$ax + by + cz = a \cdot \frac{da}{a^2 + b^2 + c^2} + b \cdot \frac{db}{a^2 + b^2 + c^2} + c \cdot \frac{dc}{a^2 + b^2 + c^2} = \frac{a^2 d + b^2 d + c^2 d}{a^2 + b^2 + c^2} = d$$

(b) We now show that the distance from P to O is D. This distance is just the length of the vector $\mathbf{v}$, which is:

$$\|\mathbf{v}\| = \left(\frac{|d|}{\mathbf{n} \cdot \mathbf{n}}\right)\|\mathbf{n}\| = \frac{|d|}{\|\mathbf{n}\|} = \frac{|d|}{\sqrt{a^2 + b^2 + c^2}}$$

as desired.

Further Insights and Challenges

In Exercises 65 and 66, let $\mathcal{P}$ be a plane with equation

$$ax + by + cz = d$$

and normal vector $\mathbf{n} = \langle a, b, c \rangle$. For any point Q, there is a unique point P on $\mathcal{P}$ that is closest to Q, and is such that $\overline{PQ}$ is orthogonal to $\mathcal{P}$ (Figure 11).

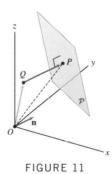

FIGURE 11

65. Show that the point P on $\mathcal{P}$ closest to Q is determined by the equation

$$\overrightarrow{OP} = \overrightarrow{OQ} + \left(\frac{d - \overrightarrow{OQ} \cdot \mathbf{n}}{\mathbf{n} \cdot \mathbf{n}}\right) \mathbf{n}$$

$\boxed{7}$

SOLUTION   Since $\overrightarrow{PQ}$ is orthogonal to the plane $\mathcal{P}$, it is parallel to the vector $\mathbf{n} = \langle a, b, c \rangle$ which is normal to the plane. Hence,

$$\overrightarrow{PQ} = \lambda \mathbf{n} \qquad (1)$$

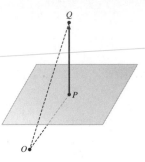

Since $\overrightarrow{OP} + \overrightarrow{PQ} = \overrightarrow{OQ}$, we have $\overrightarrow{PQ} = \overrightarrow{OQ} - \overrightarrow{OP}$. Thus, by (1) we get

$$\overrightarrow{OQ} - \overrightarrow{OP} = \lambda \mathbf{n} \quad \Rightarrow \quad \overrightarrow{OP} = \overrightarrow{OQ} - \lambda \mathbf{n} \qquad (2)$$

The point P is on the plane, hence $\overrightarrow{OP}$ satisfies the vector form of the equation of the plane, that is,

$$\mathbf{n} \cdot \overrightarrow{OP} = d \qquad (3)$$

Substituting (2) into (3) and solving for λ yields

$$\mathbf{n} \cdot \left(\overrightarrow{OQ} - \lambda \mathbf{n} \right) = d$$

$$\mathbf{n} \cdot \overrightarrow{OQ} - \lambda \mathbf{n} \cdot \mathbf{n} = d$$

$$\lambda \mathbf{n} \cdot \mathbf{n} = \mathbf{n} \cdot \overrightarrow{OQ} - d \quad \Rightarrow \quad \lambda = \frac{\mathbf{n} \cdot \overrightarrow{OQ} - d}{\mathbf{n} \cdot \mathbf{n}} \qquad (4)$$

Finally, we combine (2) and (4) to obtain

$$\overrightarrow{OP} = \overrightarrow{OQ} + \left(\frac{d - \mathbf{n} \cdot \overrightarrow{OQ}}{\mathbf{n} \cdot \mathbf{n}} \right) \mathbf{n}$$

67. Use Eq. (7) to find the point P nearest to $Q = (2, 1, 2)$ on the plane $x + y + z = 1$.

SOLUTION   We identify $\mathbf{n} = \langle 1, 1, 1 \rangle$ as a vector normal to the plane. By Eq. (7) the nearest point P to Q is determined by

$$\overrightarrow{OP} = \overrightarrow{OQ} + \left(\frac{d - \overrightarrow{OQ} \cdot \mathbf{n}}{\mathbf{n} \cdot \mathbf{n}} \right) \mathbf{n}$$

We substitute $\mathbf{n} = \langle 1, 1, 1 \rangle$, $\overrightarrow{OQ} = \langle 2, 1, 2 \rangle$ and $d = 1$ in this equation to obtain

$$\overrightarrow{OP} = \langle 2, 1, 2 \rangle + \frac{1 - \langle 2, 1, 2 \rangle \cdot \langle 1, 1, 1 \rangle}{\langle 1, 1, 1 \rangle \cdot \langle 1, 1, 1 \rangle} \langle 1, 1, 1 \rangle = \langle 2, 1, 2 \rangle + \frac{1 - (2 + 1 + 2)}{1 + 1 + 1} \langle 1, 1, 1 \rangle$$

$$= \langle 2, 1, 2 \rangle - \frac{4}{3} \langle 1, 1, 1 \rangle = \left\langle \frac{2}{3}, -\frac{1}{3}, \frac{2}{3} \right\rangle$$

The terminal point $P = \left(\frac{2}{3}, -\frac{1}{3}, \frac{2}{3} \right)$ of $\overrightarrow{OP}$ is the nearest point to $Q = (2, 1, 2)$ on the plane.

69. Use Eq. (8) to find the distance from $Q = (1, 1, 1)$ to the plane $2x + y + 5z = 2$.

SOLUTION   By Eq. (8), the distance from $Q = \langle x_1, y_1, z_1 \rangle$ to the plane $ax + by + cz = d$ is

$$\ell = \frac{|ax_1 + by_1 + cz_1 - d|}{\|\mathbf{n}\|} \qquad (1)$$

We identify the vector $\mathbf{n} = \langle 2, 1, 5 \rangle$ as a normal to the plane $2x + y + 5z = 2$. Also $a = 2$, $b = 1$, $c = 5$, $d = 2$, and $(x_1, y_1, z_1) = (1, 1, 1)$. Substituting in (1) above we get

$$\ell = \frac{|2 \cdot 1 + 1 \cdot 1 + 5 \cdot 1 - 2|}{\|\langle 2, 1, 5 \rangle\|} = \frac{6}{\sqrt{2^2 + 1^2 + 5^2}} = \frac{6}{\sqrt{30}} \approx 1.095$$

71. What is the distance from $Q = (a, b, c)$ to the plane $x = 0$? Visualize your answer geometrically and explain without computation. Then verify that Eq. (8) yields the same answer.

SOLUTION The plane $x = 0$ is the yz-coordinate plane. The nearest point to Q on the plane is the projection of Q on the plane, which is the point $Q' = (0, b, c)$.

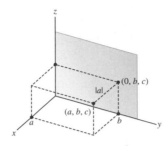

Hence, the distance from Q to the plane is the length of the vector $\overrightarrow{Q'Q} = \langle a, 0, 0 \rangle$ which is $|a|$. We now verify that Eq. (8) gives the same answer. The plane $x = 0$ has the vector parametrization $\langle 1, 0, 0 \rangle \cdot \langle x, y, z \rangle = 0$, hence $\mathbf{n} = \langle 1, 0, 0 \rangle$. The coefficients of the plane $x = 0$ are $A = 1$, $B = C = D = 0$. Also $(x_1, y_1, z_1) = (a, b, c)$. Substituting this value in Eq. (8) we get

$$\frac{|Ax_1 + By_1 + Cz_1 - D|}{\|\mathbf{n}\|} = \frac{|1 \cdot a + 0 + 0 - 0|}{\|\langle 1, 0, 0 \rangle\|} = \frac{|a|}{\sqrt{1^2 + 0^2 + 0^2}} = |a|$$

The two answers agree, as expected.

12.6 A Survey of Quadric Surfaces (LT Section 13.6)

Preliminary Questions

1. True or false? All traces of an ellipsoid are ellipses.

SOLUTION This statement is true, mostly. All traces of an ellipsoid $\left(\frac{x}{a}\right)^2 + \left(\frac{y}{b}\right)^2 + \left(\frac{z}{c}\right)^2 = 1$ are ellipses, except for the traces obtained by intersecting the ellipsoid with the planes $x = \pm a$, $y = \pm b$ and $z = \pm c$. These traces reduce to the single points $(\pm a, 0, 0)$, $(0, \pm b, 0)$ and $(0, 0, +c)$ respectively.

2. True or false? All traces of a hyperboloid are hyperbolas.

SOLUTION The statement is false. For a hyperbola in the standard orientation, the horizontal traces are ellipses (or perhaps empty for a hyperbola of two sheets), and the vertical traces are hyperbolas.

3. Which quadric surfaces have both hyperbolas and parabolas as traces?

SOLUTION The hyperbolic paraboloid $z = \left(\frac{x}{a}\right)^2 - \left(\frac{y}{b}\right)^2$ has vertical trace curves which are parabolas. If we set $x = x_0$ or $y = y_0$ we get

$$z = \left(\frac{x_0}{a}\right)^2 - \left(\frac{y}{b}\right)^2 \quad \Rightarrow \quad z = -\left(\frac{y}{b}\right)^2 + C$$

$$z = \left(\frac{x}{a}\right)^2 - \left(\frac{y_0}{b}\right)^2 \quad \Rightarrow \quad z = \left(\frac{x}{a}\right)^2 + C$$

The hyperbolic paraboloid has vertical traces which are hyperbolas, since for $z = z_0$, $(z_0 > 0)$, we get

$$z_0 = \left(\frac{x}{a}\right)^2 - \left(\frac{y}{b}\right)^2$$

4. Is there any quadric surface whose traces are all parabolas?

SOLUTION There is no quadric surface whose traces are all parabolas.

5. A surface is called **bounded** if there exists $M > 0$ such that every point on the surface lies at a distance of at most M from the origin. Which of the quadric surfaces are bounded?

SOLUTION The only quadric surface that is bounded is the ellipsoid

$$\left(\frac{x}{a}\right)^2 + \left(\frac{y}{b}\right)^2 + \left(\frac{z}{c}\right)^2 = 1.$$

All other quadric surfaces are not bounded, since at least one of the coordinates can increase or decrease without bound.

6. What is the definition of a parabolic cylinder?

SOLUTION A parabolic cylinder consists of all vertical lines passing through a parabola C in the xy-plane.

Exercises

In Exercises 1–6, state whether the given equation defines an ellipsoid or hyperboloid, and if a hyperboloid, whether it is of one or two sheets.

1. $\left(\frac{x}{2}\right)^2 + \left(\frac{y}{3}\right)^2 + \left(\frac{z}{5}\right)^2 = 1$

SOLUTION This equation is the equation of an ellipsoid.

3. $x^2 + 3y^2 + 9z^2 = 1$

SOLUTION We rewrite the equation as follows:

$$x^2 + \left(\frac{y}{\frac{1}{\sqrt{3}}}\right)^2 + \left(\frac{z}{\frac{1}{3}}\right)^2 = 1$$

This equation defines an ellipsoid.

5. $x^2 - 3y^2 + 9z^2 = 1$

SOLUTION We rewrite the equation in the form

$$x^2 - \left(\frac{y}{\frac{1}{\sqrt{3}}}\right)^2 + \left(\frac{z}{\frac{1}{3}}\right)^2 = 1$$

This is the equation of a hyperboloid of one sheet.

In Exercises 7–12, state whether the given equation defines an elliptic paraboloid, a hyperbolic paraboloid, or an elliptic cone.

7. $z = \left(\frac{x}{4}\right)^2 + \left(\frac{y}{3}\right)^2$

SOLUTION This equation defines an elliptic paraboloid.

9. $z = \left(\frac{x}{9}\right)^2 - \left(\frac{y}{12}\right)^2$

SOLUTION This equation defines a hyperbolic paraboloid.

11. $3x^2 - 7y^2 = z$

SOLUTION Rewriting the equation as

$$z = \left(\frac{x}{\frac{1}{\sqrt{3}}}\right)^2 - \left(\frac{y}{\frac{1}{\sqrt{7}}}\right)^2$$

we identify it as the equation of a hyperbolic paraboloid.

In Exercises 13–20, state the type of the quadric surface and describe the trace obtained by intersecting with the given plane.

13. $x^2 + \left(\frac{y}{4}\right)^2 + z^2 = 1, \quad y = 0$

SOLUTION The equation $x^2 + \left(\frac{y}{4}\right)^2 + z^2 = 1$ defines an ellipsoid. The xz-trace is obtained by substituting $y = 0$ in the equation of the ellipsoid. This gives the equation $x^2 + z^2 = 1$ which defines a circle in the xz-plane.

15. $x^2 + \left(\frac{y}{4}\right)^2 + z^2 = 1, \quad z = \frac{1}{4}$

SOLUTION The quadric surface is an ellipsoid, since its equation has the form $\left(\frac{x}{a}\right)^2 + \left(\frac{y}{b}\right)^2 + \left(\frac{z}{c}\right)^2 = 1$ for $a = 1$, $b = 4, c = 1$. To find the trace obtained by intersecting the ellipsoid with the plane $z = \frac{1}{4}$, we set $z = \frac{1}{4}$ in the equation of the ellipsoid. This gives

$$lx^2 + \left(\frac{y}{4}\right)^2 + \left(\frac{1}{4}\right)^2 = 1$$

$$x^2 + \frac{y^2}{16} = \frac{15}{16}$$

To get the standard form we divide by $\frac{15}{16}$ to obtain

$$\frac{x^2}{\frac{15}{16}} + \frac{y^2}{\frac{16 \cdot 15}{16}} = 1 \quad \Rightarrow \quad \left(\frac{x}{\frac{\sqrt{15}}{4}}\right)^2 + \left(\frac{y}{\sqrt{15}}\right)^2 = 1 \tag{1}$$

We conclude that the trace is an ellipse on the xy-plane, whose equation is given in (1).

17. $\left(\frac{x}{3}\right)^2 + \left(\frac{y}{5}\right)^2 - 5z^2 = 1, \quad y = 1$

SOLUTION Rewriting the equation in the form

$$\left(\frac{x}{3}\right)^2 + \left(\frac{y}{5}\right)^2 - \left(\frac{z}{\frac{1}{\sqrt{5}}}\right)^2 = 1$$

we identify it as the equation of a hyperboloid of one sheet. Substituting $y = 1$ we get

$$\frac{x^2}{9} + \frac{1}{25} - 5z^2 = 1$$

$$\frac{x^2}{9} - 5z^2 = \frac{24}{25}$$

$$\frac{25}{24 \cdot 9}x^2 - \frac{25 \cdot 5}{24}z^2 = 1$$

$$\left(\frac{x}{\frac{6\sqrt{6}}{5}}\right)^2 - \left(\frac{z}{\frac{2}{5}\sqrt{\frac{6}{5}}}\right)^2 = 1$$

Thus, the trace on the plane $y = 1$ is a hyperbola.

19. $y = 3x^2, \quad z = 27$

SOLUTION This equation defines a parabolic cylinder, consisting of all vertical lines passing through the parabola $y = 3x^2$ in the xy-plane. Hence, the trace of the cylinder on the plane $z = 27$ is the parabola $y = 3x^2$ on this plane, that is, the following set:

$$\{(x, y, z) : y = 3x^2, \ z = 27\}.$$

21. Match each of the ellipsoids in Figure 12 with the correct equation:
(a) $x^2 + 4y^2 + 4z^2 = 16$ **(b)** $4x^2 + y^2 + 4z^2 = 16$
(c) $4x^2 + 4y^2 + z^2 = 16$

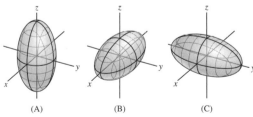

(A)   (B)   (C)

FIGURE 12

SOLUTION
(a) We rewrite the equation in the form

$$\left(\frac{x}{4}\right)^2 + \left(\frac{y}{2}\right)^2 + \left(\frac{z}{2}\right)^2 = 1$$

The ellipsoid intersects the x, y, and z axes at the points $(\pm 4, 0, 0)$, $(0, \pm 2, 0)$, and $(0, 0, \pm 2)$, hence (B) is the corresponding figure.
(b) We rewrite the equation in the form

$$\left(\frac{x}{2}\right)^2 + \left(\frac{y}{4}\right)^2 + \left(\frac{z}{2}\right)^2 = 1$$

The x, y, and z intercepts are $(\pm 2, 0, 0)$, $(0, \pm 4, 0)$, and $(0, 0, \pm 2)$ respectively, hence (C) is the correct figure.

(c) We write the equation in the form

$$\left(\frac{x}{2}\right)^2 + \left(\frac{y}{2}\right)^2 + \left(\frac{z}{4}\right)^2 = 1$$

The x, y, and z intercepts are $(\pm2, 0, 0)$, $(0, \pm2, 0)$, and $(0, 0, \pm4)$ respectively, hence the corresponding figure is (A).

23. What is the equation of the surface obtained when the elliptic paraboloid $z = \left(\frac{x}{2}\right)^2 + \left(\frac{y}{4}\right)^2$ is rotated about the x-axis by $90°$? Refer to Figure 13.

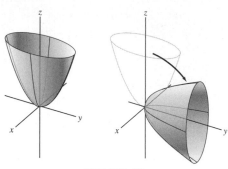

FIGURE 13

SOLUTION   The axis of symmetry of the resulting surface is the y-axis rather than the z-axis. Interchanging y and z in the given equation gives the following equation of the rotated paraboloid:

$$y = \left(\frac{x}{2}\right)^2 + \left(\frac{z}{4}\right)^2$$

In Exercises 25–30, sketch the given surface.

25. $x^2 + y^2 - z^2 = 1$

SOLUTION   This equation defines a hyperboloid of one sheet. The trace on the plane $z = z_0$ is the circle $x^2 + y^2 = 1 + z_0^2$. The trace on the plane $y = y_0$ is the hyperbola $x^2 - z^2 = 1 - y_0^2$ and the trace on the plane $x = x_0$ is the hyperbola $y^2 - z^2 = 1 - x_0^2$. We obtain the following surface:

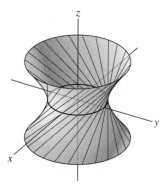

Graph of $x^2 + y^2 - z^2 = 1$

27. $z = \left(\frac{x}{4}\right)^2 + \left(\frac{y}{8}\right)^2$

SOLUTION   This equation defines an elliptic paraboloid, as shown in the following figure:

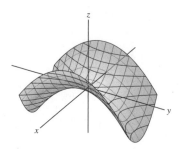

29. $z^2 = \left(\frac{x}{4}\right)^2 + \left(\frac{y}{8}\right)^2$

SOLUTION This equation defines the following elliptic cone:

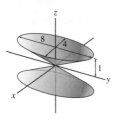

31. Find the equation of the ellipsoid passing through the points marked in Figure 14(A).

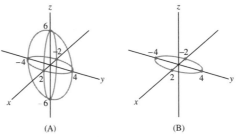

(A)   (B)

FIGURE 14

SOLUTION The equation of an ellipsoid is

$$\left(\frac{x}{a}\right)^2 + \left(\frac{y}{b}\right)^2 + \left(\frac{z}{c}\right)^2 = 1 \tag{1}$$

The x, y and z intercepts are $(\pm a, 0, 0)$, $(0, \pm b, 0)$ and $(0, 0, \pm c)$ respectively. The x, y and z intercepts of the desired ellipsoid are $(\pm 2, 0, 0)$, $(0, \pm 4, 0)$ and $(0, 0, \pm 6)$ respectively, hence $a = 2$, $b = 4$ and $c = 6$. Substituting into (1) we get

$$\left(\frac{x}{2}\right)^2 + \left(\frac{y}{4}\right)^2 + \left(\frac{z}{6}\right)^2 = 1.$$

33. Find the equation of the hyperboloid shown in Figure 15(A).

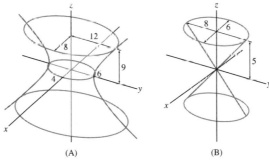

(A)   (B)

FIGURE 15

SOLUTION The hyperboloid in the figure is of one sheet and the intersections with the planes $z = z_0$ are ellipses. Hence, the equation of the hyperboloid has the form

$$\left(\frac{x}{a}\right)^2 + \left(\frac{y}{b}\right)^2 - \left(\frac{z}{c}\right)^2 = 1 \tag{1}$$

Substituting $z = 0$ we get

$$\left(\frac{x}{a}\right)^2 + \left(\frac{y}{b}\right)^2 = 1$$

By the given information this ellipse has x and y intercepts at the points $(\pm 4, 0)$ and $(0, \pm 6)$ hence $a = 4$, $b = 6$. Substituting in (1) we get

$$\left(\frac{x}{4}\right)^2 + \left(\frac{y}{6}\right)^2 - \left(\frac{z}{c}\right)^2 = 1 \tag{2}$$

Substituting $z = 9$ we get

$$\frac{x^2}{16} + \frac{y^2}{36} - \frac{9^2}{c^2} = 1$$

$$\frac{x^2}{16} + \frac{y^2}{36} = 1 + \frac{81}{c^2} = \frac{c^2 + 81}{c^2}$$

$$\frac{c^2 x^2}{16(81 + c^2)} + \frac{c^2 y^2}{36(81 + c^2)} = 1$$

$$\left(\frac{x}{\frac{4}{c}\sqrt{81 + c^2}}\right)^2 + \left(\frac{y}{\frac{6}{c}\sqrt{81 + c^2}}\right)^2 = 1$$

By the given information the following must hold:

$$\begin{aligned}\frac{4}{c}\sqrt{81 + c^2} = 8 \\ \frac{6}{c}\sqrt{81 + c^2} = 12\end{aligned} \quad \Rightarrow \quad \frac{\sqrt{81 + c^2}}{c} = 2 \quad \Rightarrow \quad 81 + c^2 = 4c^2 \quad \Rightarrow \quad 3c^2 = 81$$

Thus, $c = 3\sqrt{3}$, and by substituting in (2) we obtain the following equation:

$$\left(\frac{x}{4}\right)^2 + \left(\frac{y}{6}\right)^2 - \left(\frac{z}{3\sqrt{3}}\right)^2 = 1$$

35. Determine the vertical traces of elliptic and parabolic cylinders in standard form.

SOLUTION   The vertical traces of elliptic or parabolic cylinders are one or two vertical lines, or an empty set.

37. Let $\mathcal{C}$ be an ellipse in a horizontal plane lying above the xy-plane. Which type of quadric surface is made up of all lines passing through the origin and a point on $\mathcal{C}$?

SOLUTION   The quadric surface is the upper part of an elliptic cone.

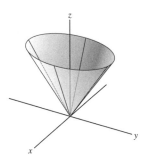

Further Insights and Challenges

39. Let $\mathcal{S}$ be the hyperboloid $x^2 + y^2 = z^2 + 1$ and let $P = (\alpha, \beta, 0)$ be a point on $\mathcal{S}$ in the (x, y)-plane. Show that there are precisely two lines through P entirely contained in $\mathcal{S}$ (Figure 16). *Hint:* Consider the line $\mathbf{r}(t) = \langle \alpha + at, \beta + bt, t \rangle$ through P. Show that $\mathbf{r}(t)$ is contained in $\mathcal{S}$ if (a, b) is one of the two points on the unit circle obtained by rotating (α, β) through $\pm\frac{\pi}{2}$. This proves that a hyperboloid of one sheet is a **doubly ruled surface**, which means that it can be swept out by moving a line in space in two different ways.

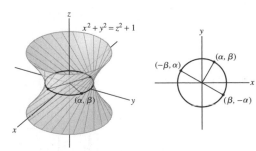

FIGURE 16

SOLUTION The parametric equations of the lines through $P = (\alpha, \beta, 0)$ have the form

$$x = \alpha + ks, \quad y = \beta + \ell s, \quad z = ms$$

Setting the parameter $t = ms$ and replacing $\frac{k}{m}$ and $\frac{\ell}{m}$ by a and b, respectively, we obtain the following (normalized) form

$$x = \alpha + at, \quad y = \beta + bt, \quad z = t$$

The line is entirely contained in S if and only if for all values of the parameter t, the following equality holds:

$$(\alpha + at)^2 + (\beta + bt)^2 = t^2 + 1$$

That is, for all t,

$$\alpha^2 + 2\alpha at + a^2 t^2 + \beta^2 + 2\beta bt + b^2 t^2 = t^2 + 1$$
$$(a^2 + b^2 - 1)t^2 + 2(\alpha a + \beta b)t + (\alpha^2 + \beta^2 - 1) = 0$$

This equality holds for all t if and only if all the coefficients are zero. That is, if and only if

$$\begin{cases} a^2 + b^2 - 1 = 0 \\ \alpha a + \beta b = 0 \\ \alpha^2 + \beta^2 - 1 = 0 \end{cases}$$

The first and the third equations imply that (a, b) and (α, β) are points on the unit circle $x^2 + y^2 = 1$. The second equation implies that the vector $\mathbf{u} = \langle a, b \rangle$ is orthogonal to the vector $\mathbf{v} = \langle \alpha, \beta \rangle$ (since $\mathbf{u} \cdot \mathbf{v} = a\alpha + b\beta = 0$).

Conclusions: There are precisely two lines through P entirely contained in S. For the direction vectors $(a, b, 1)$ of these lines, (a, b) is obtained by rotating (α, β) through $\pm\frac{\pi}{2}$ about the origin.

In Exercises 40 and 41, let C be a curve in $\mathbf{R}^3$ not passing through the origin. The cone on C is the surface consisting of all lines passing through the origin and a point on C. [Figure 17(A)].

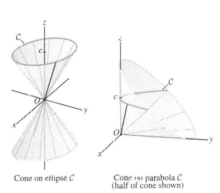

Cone on ellipse C Cone on parabola C
(half of cone shown)

FIGURE 17

41. Let a and c be nonzero constants and let C be the parabola at height c consisting of all points (x, ax^2, c) [Figure 17(B)]. Let S be the cone consisting of all lines passing through the origin and a point on C. This exercise shows that S is also an elliptic cone.

(a) Show that S has equation $yz = acx^2$.
(b) Show that under the change of variables $y = u + v$ and $z = u - v$, this equation becomes $acx^2 = u^2 - v^2$ or $u^2 = acx^2 + v^2$ (the equation of an elliptic cone in the variables x, v, u).

SOLUTION A point P on the parabola C has the form $P = \left(x_0, ax_0^2, c\right)$, hence the parametric equations of the line through the origin and P are

$$x = tx_0, \quad y = tax_0^2, \quad z = tc$$

To find a direct relation between xy and z we notice that

$$yz = tax_0^2 ct = ac(tx_0)^2 = acx^2$$

Now, defining new variables $z = u - v$ and $y = u + v$. This equation becomes

$$(u + v)(u - v) = acx^2$$
$$u^2 - v^2 = acx^2 \quad \Rightarrow \quad u^2 = acx^2 + v^2$$

This is the equation of an elliptic cone in the variables x, v, u. We, thus, showed that the cone on the parabola C is transformed to an elliptic cone by the transformation (change of variables) $y = u + v, z = u - v, x = x$.

12.7 Cylindrical and Spherical Coordinates   (LT Section 13.7)

Preliminary Questions

1. Describe the surfaces $r = R$ in cylindrical coordinates and $\rho = R$ in spherical coordinates.

SOLUTION   The surface $r = R$ consists of all points located at a distance R from the z-axis. This surface is the cylinder of radius R whose axis is the z-axis. The surface $\rho = R$ consists of all points located at a distance R from the origin. This is the sphere of radius R centered at the origin.

2. Which statement about cylindrical coordinates is correct?
(a) If $\theta = 0$, then P lies on the z-axis.
(b) If $\theta = 0$, then P lies in the xz-plane.

SOLUTION   The equation $\theta = 0$ defines the half-plane of all points that project onto the ray $\theta = 0$, that is, onto the nonnegative x-axis. This half plane is part of the (x, z)-plane, therefore if $\theta = 0$, then P lies in the (x, z)-plane.

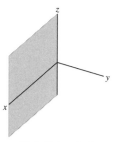

The half-plane $\theta = 0$

For instance, the point $P = (1, 0, 1)$ satisfies $\theta = 0$, but it does not lie on the z-axis. We conclude that statement (b) is correct and statement (a) is false.

3. Which statement about spherical coordinates is correct?
(a) If $\phi = 0$, then P lies on the z-axis.
(b) If $\phi = 0$, then P lies in the xy-plane.

SOLUTION   The equation $\phi = 0$ describes the nonnegative z-axis. Therefore, if $\phi = 0$, P lies on the z-axis as stated in (a). Statement (b) is false, since the point $(0, 0, 1)$ satisfies $\phi = 0$, but it does not lie in the (x, y)-plane.

4. The level surface $\phi = \phi_0$ in spherical coordinates, usually a cone, reduces to a half-line for two values of ϕ_0. Which two values?

SOLUTION   For $\phi_0 = 0$, the level surface $\phi = 0$ is the upper part of the z-axis. For $\phi_0 = \pi$, the level surface $\phi = \pi$ is the lower part of the z-axis. These are the two values of ϕ_0 where the level surface $\phi = \phi_0$ reduces to a half-line.

5. For which value of ϕ_0 is $\phi = \phi_0$ a plane? Which plane?

SOLUTION   For $\phi_0 = \frac{\pi}{2}$, the level surface $\phi = \frac{\pi}{2}$ is the xy-plane.

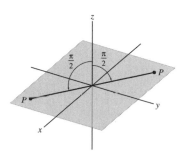

Exercises

In Exercises 1–4, convert from cylindrical to rectangular coordinates.

1. $(4, \pi, 4)$

SOLUTION   By the given data $r = 4$, $\theta = \pi$ and $z = 4$. Hence,

$$x = r \cos \theta = 4 \cos \pi = 4 \cdot (-1) = -4$$
$$y = r \sin \theta = 4 \sin \pi = 4 \cdot 0 \qquad \Rightarrow \qquad (x, y, z) = (-4, 0, 4)$$
$$z = 4$$

3. $\left(0, \frac{\pi}{5}, \frac{1}{2}\right)$

SOLUTION We have $r = 0, \theta = \frac{\pi}{5}, z = \frac{1}{2}$. Thus,

$$x = r \cos \theta = 0 \cdot \cos \frac{\pi}{5} = 0$$

$$y = r \sin \theta = 0 \cdot \sin \frac{\pi}{5} = 0 \quad \Rightarrow \quad (x, y, z) = \left(0, 0, \frac{1}{2}\right)$$

$$z = \frac{1}{2}$$

In Exercises 5–10, convert from rectangular to cylindrical coordinates.

5. $(1, -1, 1)$

SOLUTION We are given that $x = 1, y = -1, z = 1$. We find r:

$$r = \sqrt{x^2 + y^2} = \sqrt{1^2 + (-1)^2} = \sqrt{2}$$

Next we find θ. The point $(x, y) = (1, -1)$ lies in the fourth quadrant, hence,

$$\tan \theta = \frac{y}{x} = \frac{-1}{1} = -1, \quad \frac{3\pi}{2} \le \theta \le 2\pi \quad \Rightarrow \quad \theta = \frac{7\pi}{4}$$

We conclude that the cylindrical coordinates of the point are

$$(r, \theta, z) = \left(\sqrt{2}, \frac{7\pi}{4}, 1\right).$$

7. $(1, \sqrt{3}, 7)$

SOLUTION We have $x = 1, y = \sqrt{3}, z = 7$. We first find r:

$$r = \sqrt{x^2 + y^2} = \sqrt{1^2 + \left(\sqrt{3}\right)^2} = 2$$

Since the point $(x, y) = \left(1, \sqrt{3}\right)$ lies in the first quadrant, $0 \le \theta \le \frac{\pi}{2}$. Hence,

$$\tan \theta = \frac{y}{x} = \frac{\sqrt{3}}{1} = \sqrt{3}, \quad 0 \le \theta \le \frac{\pi}{2} \quad \Rightarrow \quad \theta = \frac{\pi}{3}$$

The cylindrical coordinates are thus

$$(r, \theta, z) = \left(2, \frac{\pi}{3}, 7\right).$$

9. $\left(\frac{5}{\sqrt{2}}, \frac{5}{\sqrt{2}}, 2\right)$

SOLUTION We have $x = \frac{5}{\sqrt{2}}, y = \frac{5}{\sqrt{2}}, z = 2$. We find r:

$$r = \sqrt{x^2 + y^2} = \sqrt{\left(\frac{5}{\sqrt{2}}\right)^2 + \left(\frac{5}{\sqrt{2}}\right)^2} = \sqrt{25} = 5$$

Since the point $(x, y) = \left(\frac{5}{\sqrt{2}}, \frac{5}{\sqrt{2}}\right)$ is in the first quadrant, $0 \le \theta \le \frac{\pi}{2}$, therefore,

$$\tan \theta = \frac{y}{x} = \frac{5/\sqrt{2}}{5/\sqrt{2}} = 1, \quad 0 \le \theta \le \frac{\pi}{2} \quad \Rightarrow \quad \theta = \frac{\pi}{4}$$

The corresponding cylindrical coordinates are

$$(r, \theta, z) = \left(5, \frac{\pi}{4}, 2\right).$$

In Exercises 11–16, describe the set in cylindrical coordinates.

11. $x^2 + y^2 \le 1$

SOLUTION The inequality describes a solid cylinder of radius 1 centered on the z-axis. Since $x^2 + y^2 = r^2$, this inequality can be written as $r^2 \le 1$.

13. $y^2 + z^2 \leq 4$, $x = 0$

SOLUTION The projection of the points in this set onto the xy-plane are points on the y axis, thus $\theta = \frac{\pi}{2}$ or $\theta = \frac{3\pi}{2}$. Therefore, $y = r \sin \frac{\pi}{2} = r \cdot 1 = r$ or $y = r \sin \left(\frac{3\pi}{2}\right) = -r$. In both cases, $y^2 = r^2$, thus the inequality $y^2 + z^2 \leq 4$ becomes $r^2 + z^2 \leq 4$. In cylindrical coordinates, we obtain the following inequality

$$r^2 + z^2 \leq 4, \quad \theta = \frac{\pi}{2} \quad \text{or} \quad \theta = \frac{3\pi}{2}$$

15. $x^2 + y^2 \leq 9$, $x \geq y$

SOLUTION The equation $x^2 + y^2 \leq 9$ in cylindrical coordinates becomes $r^2 \leq 9$, which becomes $r \leq 3$. However, we also have the restriction that $x \geq y$. This means that the projection of our set onto the xy plane is below and to the right of the line $y = x$. In other words, our θ is restricted to $-3\pi/4 \leq \theta \leq \pi/4$. In conclusion, the answer is:

$$r \leq 3, \qquad -3\pi/4 \leq \theta \leq \pi/4$$

In Exercises 17–24, sketch the set (described in cylindrical coordinates).

17. $r = 4$

SOLUTION The surface $r = 4$ consists of all points located at a distance 4 from the z-axis. It is a cylinder of radius 4 whose axis is the z-axis. The cylinder is shown in the following figure:

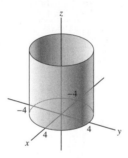

19. $z = -2$

SOLUTION $z = -2$ is the horizontal plane at height -2, shown in the following figure:

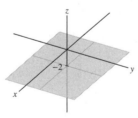

21. $1 \leq r \leq 3$, $0 \leq z \leq 4$

SOLUTION The region $1 \leq r \leq 3, 0 \leq z \leq 4$ is shown in the following figure:

23. $z^2 + r^2 \leq 4$

SOLUTION The region $z^2 + r^2 \leq 4$ is shown in the following figure:

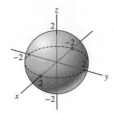

In rectangular coordinates the inequality is $z^2 + \left(x^2 + y^2\right) \le 4$, or $x^2 + y^2 + z^2 \le 4$, which is a ball of radius 2.

In Exercises 25–30, find an equation of the form $r = f(\theta, z)$ in cylindrical coordinates for the following surfaces.

25. $z = x + y$

SOLUTION We substitute $x = r\cos\theta$, $y = r\sin\theta$ to obtain the following equation in cylindrical coordinates:

$$z = r\cos\theta + r\sin\theta$$
$$z = r(\cos\theta + \sin\theta) \quad \Rightarrow \quad r = \frac{z}{\cos\theta + \sin\theta}.$$

27. $\dfrac{x^2}{yz} = 1$

SOLUTION We rewrite the equation in the form

$$\frac{x}{\frac{y}{x}z} = 1$$

Substituting $x = r\cos\theta$ and $\frac{y}{x} = \tan\theta$ we get

$$\frac{r\cos\theta}{(\tan\theta)z} - 1$$

$$r = \frac{z\tan\theta}{\cos\theta}$$

29. $x^2 + y^2 = 4$

SOLUTION Since $x^2 + y^2 = r^2$, the equation in cylindrical coordinates is, $r^2 = 4$ or $r = 2$.

In Exercises 31–36, convert from spherical to rectangular coordinates.

31. $\left(3, 0, \dfrac{\pi}{2}\right)$

SOLUTION We are given that $\rho = 3$, $\theta = 0$, $\phi = \frac{\pi}{2}$. Using the relations between spherical and rectangular coordinates we have

$$x = \rho\sin\phi\cos\theta = 3\sin\frac{\pi}{2}\cos 0 - 3\cdot 1\cdot 1 = 3$$

$$y = \rho\sin\phi\sin\theta = 3\sin\frac{\pi}{2}\sin 0 = 3\cdot 1\cdot 0 = 0 \quad \Rightarrow \quad (x, y, z) = (3, 0, 0)$$

$$z = \rho\cos\phi = 3\cos\frac{\pi}{2} = 3\cdot 0 = 0$$

33. $(3, \pi, 0)$

SOLUTION We have $\rho = 3$, $\theta = \pi$, $\phi = 0$. Hence,

$$x = \rho\sin\phi\cos\theta = 3\sin 0\cos\pi = 0$$
$$y = \rho\sin\phi\sin\theta = 3\sin 0\sin\pi = 0 \quad \Rightarrow \quad (x, y, z) = (0, 0, 3)$$
$$z = \rho\cos\phi = 3\cos 0 = 3$$

35. $\left(6, \dfrac{\pi}{6}, \dfrac{5\pi}{6}\right)$

SOLUTION Since $\rho = 6$, $\theta = \frac{\pi}{6}$, and $\phi = \frac{5\pi}{6}$ we get

$$x = \rho\sin\phi\cos\theta = 6\sin\frac{5\pi}{6}\cos\frac{\pi}{6} = 6\cdot\frac{1}{2}\cdot\frac{\sqrt{3}}{2} = \frac{3\sqrt{3}}{2}$$

$$y = \rho\sin\phi\sin\theta = 6\sin\frac{5\pi}{6}\sin\frac{\pi}{6} = 6\cdot\frac{1}{2}\cdot\frac{1}{2} = \frac{3}{2} \quad \Rightarrow \quad (x, y, z) = \left(\frac{3\sqrt{3}}{2}, \frac{3}{2}, -3\sqrt{3}\right)$$

$$z = \rho\cos\phi = 6\cos\frac{5\pi}{6} = 6\cdot\left(-\frac{\sqrt{3}}{2}\right) = -3\sqrt{3}$$

In Exercises 37–42, convert from rectangular to spherical coordinates.

37. $(\sqrt{3}, 0, 1)$

SOLUTION   By the given data $x = \sqrt{3}$, $y = 0$, and $z = 1$. We find the radial coordinate:

$$\rho = \sqrt{x^2 + y^2 + z^2} = \sqrt{\left(\sqrt{3}\right)^2 + 0^2 + 1^2} = 2$$

The angular coordinate θ satisfies

$$\tan \theta = \frac{y}{x} = \frac{0}{\sqrt{3}} = 0 \quad \Rightarrow \quad \theta = 0 \quad \text{or} \quad \theta = \pi$$

Since the point $(x, y) = \left(\sqrt{3}, 0\right)$ lies in the first quadrant, the correct choice is $\theta = 0$. The angle of declination ϕ satisfies

$$\cos \phi = \frac{z}{\rho} = \frac{1}{2}, \quad 0 \le \phi \le \pi \quad \Rightarrow \quad \phi = \frac{\pi}{3}$$

The spherical coordinates of the given points are thus

$$(\rho, \theta, \phi) = \left(2, 0, \frac{\pi}{3}\right)$$

39. $(1, 1, 1)$

SOLUTION   We have $x = y = z = 1$. The radial coordinate is

$$\rho = \sqrt{x^2 + y^2 + z^2} = \sqrt{1^2 + 1^2 + 1^2} = \sqrt{3}$$

The angular coordinate θ is determined by $\tan \theta = \frac{y}{x} = \frac{1}{1} = 1$ and by the quadrant of the point $(x, y) = (1, 1)$, that is, $\theta = \frac{\pi}{4}$. The angle of declination ϕ satisfies

$$\cos \phi = \frac{z}{\rho} = \frac{1}{\sqrt{3}}, \quad 0 \le \phi \le \pi \quad \Rightarrow \quad \phi = 0.955$$

The spherical coordinates are thus

$$\left(\sqrt{3}, \frac{\pi}{4}, 0.955\right)$$

41. $\left(\dfrac{1}{2}, \dfrac{\sqrt{3}}{2}, \sqrt{3}\right)$

SOLUTION   We have $x = \frac{1}{2}$, $y = \frac{\sqrt{3}}{2}$, and $z = \sqrt{3}$. Thus

$$\rho = \sqrt{x^2 + y^2 + z^2} = \sqrt{\left(\frac{1}{2}\right)^2 + \left(\frac{\sqrt{3}}{2}\right)^2 + \left(\sqrt{3}\right)^2} = 2$$

The angular coordinate θ satisfies $0 \le \theta \le \frac{\pi}{2}$, since the point $(x, y) = \left(\frac{1}{2}, \frac{\sqrt{3}}{2}\right)$ is in the first quadrant. Also $\tan \theta = \frac{y}{x} = \frac{\sqrt{3}/2}{1/2} = \sqrt{3}$, hence the angle is $\theta = \frac{\pi}{3}$. The angle of declination ϕ satisfies

$$\cos \phi = \frac{z}{\rho} = \frac{\sqrt{3}}{2}, \quad 0 \le \phi \le \pi \quad \Rightarrow \quad \phi = \frac{\pi}{6}$$

We conclude that

$$(\rho, \theta, \phi) = \left(2, \frac{\pi}{3}, \frac{\pi}{6}\right)$$

In Exercises 43 and 44, convert from cylindrical to spherical coordinates.

43. $(2, 0, 2)$

SOLUTION   We are given that $r = 2$, $\theta = 0$, $z = 2$. Using the conversion formulas, we have

$$\rho = \sqrt{x^2 + y^2 + z^2} = \sqrt{r^2 + z^2} = \sqrt{2^2 + 2^2} = 2\sqrt{2}$$

$$\theta = \theta = 0$$

$$\phi = \cos^{-1}(z/\rho) = \cos^{-1}(2/(2\sqrt{2})) = \pi/4$$

In Exercises 45 and 46, convert from spherical to cylindrical coordinates.

45. $\left(4, 0, \frac{\pi}{4}\right)$

SOLUTION We are given that $\rho = 4, \theta = 0$, and $\phi = \pi/4$. To find r, we use the formulas $x = r \cos \theta$ and $x = \rho \cos \theta \sin \phi$ to get $r \cos \theta = \rho \cos \theta \sin \phi$, and so

$$r = \rho \sin \phi = 4 \sin \pi/4 = 2\sqrt{2}$$

Clearly $\theta = 0$, and as for z,

$$z = \rho \cos \phi = 4 \cos \pi/4 = 2\sqrt{2}$$

So, in cylindrical coordinates, our point is $(2\sqrt{2}, 0, 2\sqrt{2})$

In Exercises 47–52, describe the given set in spherical coordinates.

47. $x^2 + y^2 + z^2 \le 1$

SOLUTION Substituting $\rho^2 = x^2 + y^2 + z^2$ we obtain $\rho^2 \le 1$ or $0 \le \rho \le 1$.

49. $x^2 + y^2 + z^2 = 1, \quad x \ge 0, \quad y \ge 0, \quad z \ge 0$

SOLUTION By $\rho^2 = x^2 + y^2 + z^2$, we get $\rho^2 = 1$ or $\rho = 1$. The inequalities $x \ge 0, y > 0$ determine the first quadrant, which is also determined by $0 \le \theta \le \frac{\pi}{2}$. Finally, $z \ge 0$ gives $\cos \phi = \frac{z}{\rho} \ge 0$. Also $0 \le \phi \le \pi$, hence $0 \le \phi \le \frac{\pi}{2}$. We obtain the following description:

$$\rho = 1, \quad 0 \le \theta < \frac{\pi}{2}, \quad 0 \le \phi \le \frac{\pi}{2}$$

51. $y^2 + z^2 \le 4, \quad x = 0$

SOLUTION We substitute $y = \rho \sin \theta \sin \phi$ and $z = \rho \cos \phi$ in the given inequality. This gives

$$4 \ge \rho^2 \sin^2 \theta \sin^2 \phi + \rho^2 \cos^2 \phi \tag{1}$$

The equality $x = 0$ determines that $\theta = \frac{\pi}{2}$ or $\theta = \frac{3\pi}{2}$ (and the origin). In both cases, $\sin^2 \theta = 1$. Hence by (1) we get

$$\rho^2 \sin^2 \phi + \rho^2 \cos^2 \phi \le 4$$
$$\rho^2 (1) \le 4$$
$$\rho \le 2$$

We obtain the following description:

$$\left\{ (\rho, \theta, \phi) : 0 \le \rho \le 2, \theta = \frac{\pi}{2} \text{ or } \theta = \frac{3\pi}{2} \right\}$$

In Exercises 53–60, sketch the set of points (described in spherical coordinates).

53. $\rho = 4$

SOLUTION $\rho = 4$ describes the sphere of radius 4. This is shown in the following figure:

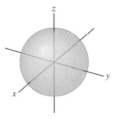

55. $\rho = 2, \quad \theta = \frac{\pi}{4}$

SOLUTION The equation $\rho = 2$ is a sphere of radius 2, and the equation $\theta = \frac{\pi}{4}$ is the vertical plane $y = x$. These two surfaces intersect in a (vertical) circle of radius 2, as seen here.

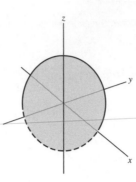

57. $\rho = 2$, $\quad 0 \le \phi \le \dfrac{\pi}{2}$

SOLUTION The set

$$\rho = 2, \quad 0 \le \phi \le \frac{\pi}{2}$$

is shown in the following figure:

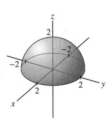

It is the upper half of the sphere with radius 2.

59. $\rho \le 2$, $\quad 0 \le \theta \le \dfrac{\pi}{2}$, $\quad \dfrac{\pi}{2} \le \phi \le \pi$

SOLUTION This set is the part of the ball of radius 2 which is below the first quadrant of the xy-plane, as shown in the following figure:

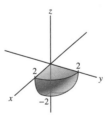

In Exercises 61–66, find an equation of the form $\rho = f(\theta, \phi)$ in spherical coordinates for the following surfaces.

61. $z = 2$

SOLUTION Since $z = \rho \cos \phi$, we have $\rho \cos \phi = 2$, or $\rho = \frac{2}{\cos \phi}$.

63. $x = z^2$

SOLUTION Substituting $x = \rho \cos \theta \sin \phi$ and $z = \rho \cos \phi$ we obtain

$$\rho \cos \theta \sin \phi = \rho^2 \cos^2 \phi$$

$$\cos \theta \sin \phi = \rho \cos^2 \phi$$

$$\rho = \frac{\cos \theta \sin \phi}{\cos^2 \phi} = \frac{\cos \theta \tan \phi}{\cos \phi}$$

65. $x^2 - y^2 = 4$

SOLUTION We substitute $x = \rho \cos \theta \sin \phi$ and $y = \rho \sin \theta \sin \phi$ to obtain

$$4 = \rho^2 \cos^2 \theta \sin^2 \phi - \rho^2 \sin^2 \theta \sin^2 \phi = \rho^2 \sin^2 \phi \left(\cos^2 \theta - \sin^2 \theta \right)$$

Using the identity $\cos^2\theta - \sin^2\theta = \cos 2\theta$ we get

$$4 = \rho^2 \sin^2\phi \cos 2\theta$$

$$\rho^2 = \frac{4}{\sin^2\phi \cos 2\theta}$$

We take the square root of both sides. Since $0 < \phi < \pi$ we have $\sin\phi > 0$, hence,

$$\rho = \frac{2}{\sin\phi\sqrt{\cos 2\theta}}$$

67. Which of (a)–(c) is the equation of the cylinder of radius R in spherical coordinates? Refer to Figure 15.

(a) $R\rho = \sin\phi$ **(b)** $\rho \sin\phi = R$ **(c)** $\rho = R\sin\phi$

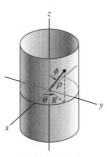

FIGURE 15

SOLUTION The equation of the cylinder of radius R in rectangular coordinates is $x^2 + y^2 = R^2$ (z is unlimited). Substituting the formulas for x and y in terms of ρ, θ and ϕ yields

$$R^2 = \rho^2 \cos^2\theta \sin^2\phi + \rho^2 \sin^2\theta \sin^2\phi = \rho^2 \sin^2\phi\left(\cos^2\theta + \sin^2\theta\right) = \rho^2 \sin^2\phi$$

Hence,

$$R^2 = \rho^2 \sin^2\phi$$

We take the square root of both sides. Since $0 \le \phi \le \pi$, we have $\sin\phi \ge 0$, therefore,

$$R = \rho \sin\phi$$

Equation (b) is the correct answer.

69. Find the spherical angles (θ, ϕ) for Helsinki, Finland (60.1° N, 25.0° E) and Sao Paulo, Brazil (23.52° S, 46.52° W).

SOLUTION For Helsinki, θ is 25° and ϕ is $90 - 60.1 = 29.9°$.
   For Sao Paulo, θ is $360 - 46.52 = 313.48°$ and ϕ is $90 + 23.52 = 113.52°$.

71. Consider a rectangular coordinate system with origin at the center of the earth, z-axis through the North Pole, and x-axis through the prime meridian. Find the rectangular coordinates of Sydney, Australia (34° S, 151° E), and Bogotá, Colombia (4° 32′ N, 74° 15′ W). A minute is $1/60°$. Assume that the earth is a sphere of radius $R = 6370$ km.

SOLUTION We first find the angle (θ, ϕ) for the two towns. For Sydney $\theta = 151°$, since its longitude lies to the east of Greenwich, that is, in the positive θ direction. Sydney's latitude is south of the equator, hence $\phi = 90 + 34 = 124°$.
   For Bogota, we have $\theta = 360° - 74°15′ = 285°45′$, since $74°15′W$ refers to $74°15′$ in the negative θ direction. The latitude is north of the equator hence $\phi = 90° - 4°32′ = 85°28′$.
   We now use the formulas of x, y and z in terms of ρ, θ, ϕ to find the rectangular coordinates of the two towns. (Notice that $285°45′ = 285.75°$ and $85°28′ = 85.47°$).
Sydney:

$$x = \rho\cos\theta \sin\phi = 6370\cos 151° \sin 124° = -4618.8$$

$$y = \rho\sin\theta \sin\phi = 6370\sin 151° \sin 124° = 2560$$

$$z = \rho\cos\phi = 6370\cos 124° = -3562.1$$

Bogota:

$$x = \rho\cos\theta \sin\phi = 6370\cos 285.75° \sin 85.47° = 1723.7$$

$$y = \rho\sin\theta \sin\phi = 6370\sin 285.75° \sin 85.47° = -6111.7$$

$$z = \rho\cos\phi = 6370\cos 85.47° = 503.1$$

73. Find an equation of the form $z = f(r, \theta)$ in cylindrical coordinates for $z^2 = x^2 - y^2$.

SOLUTION In cylindrical coordinates, $x = r \cos \theta$ and $y = r \sin \theta$. Hence,

$$z^2 = x^2 - y^2 = r^2 \cos^2 \theta - r^2 \sin^2 \theta$$

We use the identity $\cos^2 \theta - \sin^2 \theta = \cos 2\theta$ to obtain

$$z^2 = r^2 \cos 2\theta \quad \Rightarrow \quad z = \pm r \sqrt{\cos 2\theta}$$

75. Explain the following statement: If the equation of a surface in cylindrical or spherical coordinates does not involve the coordinate θ, then the surface is rotationally symmetric with respect to the z-axis.

SOLUTION Since the equation of the surface does not involve the coordinate θ, then for every point P on the surface ($P = (\rho_0, \theta_0, \phi_0)$ in spherical coordinates or $P = (r_0, \theta_0, z_0)$ in cylindrical coordinates) so also all the points (ρ_0, θ, ϕ_0) or (r_0, θ, z_0) are on the surface. That is, all the points obtained by rotating P around the z-axis are on the surface. Hence, the surface is rotationally symmetric with respect to the z-axis.

77. Find equations $r = g(\theta, z)$ (cylindrical) and $\rho = f(\theta, \phi)$ (spherical) for the hyperboloid $x^2 + y^2 = z^2 + 1$ (Figure 16). Do there exist points on the hyperboloid with $\phi = 0$ or π? Which values of ϕ occur for points on the hyperboloid?

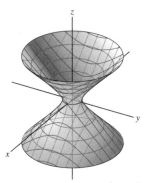

FIGURE 16 The hyperboloid $x^2 + y^2 = z^2 + 1$.

SOLUTION For the cylindrical coordinates (r, θ, z) we have $x^2 + y^2 = r^2$. Substituting into the equation $x^2 + y^2 = z^2 + 1$ gives

$$r^2 = z^2 + 1 \quad \Rightarrow \quad r = \sqrt{z^2 + 1}$$

For the spherical coordinates (ρ, θ, ϕ) we have $x = \rho \sin \phi \cos \theta$, $y = \rho \sin \phi \sin \theta$ and $z = \rho \cos \phi$. We substitute into the equation of the hyperboloid $x^2 + y^2 = z^2 + 1$ and simplify to obtain

$$\rho^2 \sin^2 \phi \cos^2 \theta + \rho^2 \sin^2 \phi \sin^2 \theta = \rho^2 \cos^2 \phi + 1$$
$$\rho^2 \sin^2 \phi (\cos^2 \theta + \sin^2 \theta) = \rho^2 \cos^2 \phi + 1$$
$$\rho^2 (\sin^2 \phi - \cos^2 \phi) = 1$$

Using the trigonometric identity $\cos 2\phi = \cos^2 \phi - \sin^2 \phi$ we get

$$\rho^2 \cdot (-\cos 2\phi) = 1 \quad \Rightarrow \quad \rho = \sqrt{-\frac{1}{\cos 2\phi}}$$

For $\phi = 0$ and $\phi = \pi$ we have $\cos 2 \cdot 0 = 1$ and $\cos 2\pi = 1$. In both cases $-\frac{1}{\cos 2\phi} = -1 < 0$, hence there is no real value of ρ satisfying $\rho = \sqrt{-\frac{1}{\cos 2\phi}}$. We conclude that there are no points on the hyperboloid with $\phi = 0$ or π.

To obtain a real ρ such that $\rho = \sqrt{-\frac{1}{\cos 2\phi}}$, we must have $-\frac{1}{\cos 2\phi} > 0$. That is, $\cos 2\phi < 0$ (and of course $0 \le \phi \le \pi$). The corresponding values of ϕ are

$$\frac{\pi}{2} < 2\phi \le \frac{3\pi}{2} \quad \Rightarrow \quad \frac{\pi}{4} < \phi \le \frac{3\pi}{4}$$

Further Insights and Challenges

*In Exercises 78–82, a **great circle** on a sphere S with center O is a circle obtained by intersecting S with a plane that passes through O (Figure 17). If P and Q are not antipodal (on opposite sides), there is a unique great circle through P and Q on S (intersect S with the plane through O, P, and Q). The geodesic distance from P to Q is defined as the length of the smaller of the two circular arcs of this great circle.*

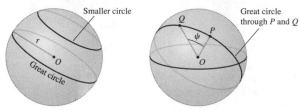

FIGURE 17

79. Show that the geodesic distance from $Q = (a, b, c)$ to the North Pole $P = (0, 0, R)$ is equal to $R \cos^{-1}\left(\dfrac{c}{R}\right)$.

SOLUTION Let ψ be the central angle between P and Q, that is, the angle between the vectors $\mathbf{v} = \overrightarrow{OP}$ and $\mathbf{u} = \overrightarrow{OQ}$. By Exercise 78 the geodesic distance from P to Q is $R\psi$. We find ψ. By the formula for the cosine of the angle between two vectors, we have

$$\cos\psi = \frac{\mathbf{u}\cdot\mathbf{v}}{\|\mathbf{u}\|\|\mathbf{v}\|} \tag{1}$$

We compute the values in this quotient·

$$\mathbf{u}\cdot\mathbf{v} = \langle 0, 0, R\rangle \cdot \langle a, b, c\rangle = 0 + 0 + Rc = Rc$$

$$\|\mathbf{v}\| = \|\overrightarrow{OP}\| = R$$

$$\|\mathbf{u}\| = \|\overrightarrow{OQ}\| = \sqrt{a^2 + b^2 + c^2} = R$$

Substituting in (1) we get

$$\cos\psi = \frac{Rc}{R^2} = \frac{c}{R} \quad \Rightarrow \quad \psi = \cos^{-1}\left(\frac{c}{R}\right)$$

The geodesic distance from Q to P is thus

$$R\psi = R\cos^{-1}\left(\frac{c}{R}\right)$$

81. Show that the central angle ψ between points P and Q on a sphere (of any radius) with angular coordinates (θ, ϕ) and (θ', ϕ') is equal to

$$\psi = \cos^{-1}\left(\sin\phi\sin\phi'\cos(\theta - \theta') + \cos\phi\cos\phi'\right)$$

Hint: Compute the dot product of $\overrightarrow{OP}$ and $\overrightarrow{OQ}$. Check this formula by computing the geodesic distance between the North and South Poles.

SOLUTION We denote the vectors $\mathbf{u} = \overrightarrow{OP}$ and $\mathbf{v} = \overrightarrow{OQ}$. By the formula for the angle between two vectors we have

$$\psi = \cos^{-1}\left(\frac{\mathbf{u}\cdot\mathbf{v}}{\|\mathbf{u}\|\|\mathbf{v}\|}\right)$$

Denoting by R the radius of the sphere, we have $\|\mathbf{u}\| = \|\mathbf{v}\| = R$, hence,

$$\psi = \cos^{-1}\left(\frac{\mathbf{u}\cdot\mathbf{v}}{R^2}\right) \tag{1}$$

The rectangular coordinates of $\mathbf{u}$ and $\mathbf{v}$ are

u	v
$x = R\sin\phi\cos\theta$	$x' = R\sin\phi'\cos\theta'$
$y = R\sin\phi\sin\theta$	$y' = R\sin\phi'\sin\theta'$
$z = R\cos\phi$	$z' = R\cos\phi'$

Hence,

$$\mathbf{u} \cdot \mathbf{v} = R^2 \sin\phi \cos\theta \sin\phi' \cos\theta' + R^2 \sin\phi \sin\theta \sin\phi' \sin\theta' + R^2 \cos\phi \cos\phi'$$

$$= R^2 \left[\sin\phi \sin\phi' \left(\cos\theta \cos\theta' + \sin\theta \sin\theta'\right) + \cos\phi \cos\phi'\right]$$

We use the identity $\cos(\alpha - \beta) = \cos\alpha \cos\beta + \sin\alpha \sin\beta$ to obtain

$$\mathbf{u} \cdot \mathbf{v} = R^2 \left(\sin\phi \sin\phi' \cos\left(\theta - \theta'\right) + \cos\phi \cos\phi'\right)$$

Substituting in (1) we obtain

$$\psi = \cos^{-1}\left(\sin\phi \sin\phi' \cos\left(\theta - \theta'\right) + \cos\phi \cos\phi'\right) \tag{2}$$

We now check this formula in the case where P and Q are the north and south poles respectively. In this case $\theta = \theta' = 0$, $\phi = 0$, $\phi' = \pi$. Substituting in (2) gives

$$\psi = \cos^{-1}\left(\sin 0 \sin\pi \cos 0 + \cos 0 \cos\pi\right) = \cos^{-1}(-1) = \pi$$

Using Exercise 78, the geodesic distance between the two poles is $R\psi = R\pi$, in accordance with the formula for the length of a semicircle.

CHAPTER REVIEW EXERCISES

In Exercises 1–6, let $\mathbf{v} = \langle -2, 5\rangle$ *and* $\mathbf{w} = \langle 3, -2\rangle$.

1. Calculate $5\mathbf{w} - 3\mathbf{v}$ and $5\mathbf{v} - 3\mathbf{w}$.

SOLUTION   We use the definition of basic vector operations to compute the two linear combinations:

$$5\mathbf{w} - 3\mathbf{v} = 5\langle 3, -2\rangle - 3\langle -2, 5\rangle = \langle 15, -10\rangle + \langle 6, -15\rangle = \langle 21, -25\rangle$$

$$5\mathbf{v} - 3\mathbf{w} = 5\langle -2, 5\rangle - 3\langle 3, -2\rangle = \langle -10, 25\rangle + \langle -9, 6\rangle = \langle -19, 31\rangle$$

3. Find the unit vector in the direction of $\mathbf{v}$.

SOLUTION   The unit vector in the direction of $\mathbf{v}$ is

$$\mathbf{e_v} = \frac{1}{\|\mathbf{v}\|}\mathbf{v}$$

We compute the length of $\mathbf{v}$:

$$\|\mathbf{v}\| = \sqrt{(-2)^2 + 5^2} = \sqrt{29}$$

Hence,

$$\mathbf{e_v} = \frac{\mathbf{v}}{\|\mathbf{v}\|} = \frac{\langle -2, 5\rangle}{\sqrt{29}} = \left\langle \frac{-2}{\sqrt{29}}, \frac{5}{\sqrt{29}}\right\rangle.$$

5. Express $\mathbf{i}$ as a linear combination $r\mathbf{v} + s\mathbf{w}$.

SOLUTION   We use basic properties of vector algebra to write

$$\mathbf{i} = r\mathbf{v} + s\mathbf{w} \tag{1}$$

$$\langle 1, 0\rangle = r\langle -2, 5\rangle + s\langle 3, -2\rangle = \langle -2r + 3s, 5r - 2s\rangle$$

The vector are equivalent, hence,

$$1 = -2r + 3s$$

$$0 = 5r - 2s$$

The second equation implies that $s = \frac{5}{2}r$. We substitute in the first equation and solve for r:

$$1 = -2r + 3 \cdot \frac{5}{2}r$$

$$1 = \frac{11}{2}r$$

$$r = \frac{2}{11} \quad \Rightarrow \quad s = \frac{5}{2} \cdot \frac{2}{11} = \frac{5}{11}$$

Substituting in (1) we obtain

$$i = \frac{2}{11}v + \frac{5}{11}w.$$

7. If $P = (1, 4)$ and $Q = (-3, 5)$, what are the components of $\overrightarrow{PQ}$? What is the length of $\overrightarrow{PQ}$?

SOLUTION By the Definition of Components of a Vector we have

$$\overrightarrow{PQ} = \langle -3 - 1, 5 - 4 \rangle = \langle -4, 1 \rangle$$

The length of $\overrightarrow{PQ}$ is

$$\|\overrightarrow{PQ}\| = \sqrt{(-4)^2 + 1^2} = \sqrt{17}.$$

9. Find the vector with length 3 making an angle of $\frac{7\pi}{4}$ with the positive x-axis.

SOLUTION We denote the vector by $v = \langle a, b \rangle$. v makes an angle $\theta = \frac{7\pi}{4}$ with the x-axis, and its length is 3, hence,

$$a = \|v\| \cos \theta = 3 \cos \frac{7\pi}{4} = \frac{3}{\sqrt{2}}$$

$$b = \|v\| \sin \theta = 3 \sin \frac{7\pi}{4} = -\frac{3}{\sqrt{2}}$$

That is,

$$v = \langle a, b \rangle = \left\langle \frac{3}{\sqrt{2}}, -\frac{3}{\sqrt{2}} \right\rangle.$$

11. Find the value of β for which $w = \langle -2, \beta \rangle$ is parallel to $v = \langle 4, -3 \rangle$.

SOLUTION If $v = \langle 4, -3 \rangle$ and $w = \langle -2, \beta \rangle$ are parallel, there exists a scalar λ such that $w = \lambda v$. That is,

$$\langle -2, \beta \rangle = \lambda \langle 4, -3 \rangle = \langle 4\lambda, -3\lambda \rangle$$

yielding

$$-2 = 4\lambda \quad \text{and} \quad \beta = -3\lambda$$

These equations imply that $\lambda = -\frac{1}{2}$ and $\lambda = -\frac{\beta}{3}$. Equating the two expressions for λ gives

$$-\frac{1}{2} = -\frac{\beta}{3} \quad \text{or} \quad \beta = \frac{3}{2}.$$

13. Let $w = \langle 2, -2, 1 \rangle$ and $v = \langle 4, 5, -4 \rangle$. Solve for u if $v + 5u = 3w - u$.

SOLUTION Using vector algebra we have

$$v + 5u = 3w - u$$

$$6u = 3w - v$$

$$u = \frac{1}{2}w - \frac{1}{6}v = \left\langle 1, -1, \frac{1}{2} \right\rangle - \left\langle \frac{4}{6}, \frac{5}{6}, -\frac{4}{6} \right\rangle = \left\langle \frac{1}{3}, -\frac{11}{6}, \frac{7}{6} \right\rangle$$

15. Find a parametrization $r_1(t)$ of the line passing through $(1, 4, 5)$ and $(-2, 3, -1)$. Then find a parametrization $r_2(t)$ of the line parallel to r_1 passing through $(1, 0, 0)$.

SOLUTION Since the points $P = (-2, 3, -1)$ and $Q = (1, 4, 5)$ are on the line l_1, the vector $\overrightarrow{PQ}$ is a direction vector for the line. We find this vector:

$$\overrightarrow{PQ} = \langle 1 - (-2), 4 - 3, 5 - (-1) \rangle = \langle 3, 1, 6 \rangle$$

Substituting $v = \langle 3, 1, 6 \rangle$ and $P_0 = \langle 1, 4, 5 \rangle$ in the vector parametrization of the line we obtain the following equation for l_1:

$$r_1(t) = \overrightarrow{OP_0} + tv$$

$$r_1(t) = \langle 1, 4, 5 \rangle + t\langle 3, 1, 6 \rangle = \langle 1 + 3t, 4 + t, 5 + 6t \rangle$$

The line l_2 is parallel to l_1, hence $\overrightarrow{PQ} = \langle 3, 1, 6 \rangle$ is also a direction vector for l_2. Substituting $\mathbf{v} = \langle 3, 1, 6 \rangle$ and $P_0 = (1, 0, 0)$ in the vector parametrization of the line we obtain the following equation for l_2:

$$\mathbf{r}_2(t) = \overrightarrow{OP_0} + t\mathbf{v}$$

$$\mathbf{r}_2(t) = \langle 1, 0, 0 \rangle + t \langle 3, 1, 6 \rangle = \langle 1 + 3t, t, 6t \rangle$$

17. Find a and b such that the lines $\mathbf{r}_1 = \langle 1, 2, 1 \rangle + t \langle 1, -1, 1 \rangle$ and $\mathbf{r}_2 = \langle 3, -1, 1 \rangle + t \langle a, b, -2 \rangle$ are parallel.

SOLUTION   The lines are parallel if and only if the direction vectors $\mathbf{v}_1 = \langle 1, -1, 1 \rangle$ and $\mathbf{v}_2 = \langle a, b, -2 \rangle$ are parallel. That is, if and only if there exists a scalar λ such that:

$$\mathbf{v}_2 = \lambda \mathbf{v}_1$$

$$\langle a, b, -2 \rangle = \lambda \langle 1, -1, 1 \rangle = \langle \lambda, -\lambda, \lambda \rangle$$

We obtain the following equations:

$$a = \lambda$$
$$b = -\lambda \quad \Rightarrow \quad a = -2, \quad b = 2$$
$$-2 = \lambda$$

19. Sketch the vector sum $\mathbf{v} = \mathbf{v}_1 - \mathbf{v}_2 + \mathbf{v}_3$ for the vectors in Figure 1(A).

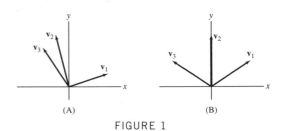

(A)   (B)

FIGURE 1

SOLUTION   Using the Parallelogram Law we obtain the vector sum shown in the figure.

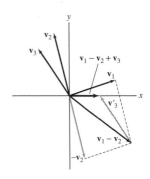

We first add $\mathbf{v}_1$ and $-\mathbf{v}_2$, then we add $\mathbf{v}_3$ to $\mathbf{v}_1 - \mathbf{v}_2$.

In Exercises 21–26, let $\mathbf{v} = \langle 1, 3, -2 \rangle$ and $\mathbf{w} = \langle 2, -1, 4 \rangle$.

21. Compute $\mathbf{v} \cdot \mathbf{w}$.

SOLUTION   Using the definition of the dot product we have

$$\mathbf{v} \cdot \mathbf{w} = \langle 1, 3, -2 \rangle \cdot \langle 2, -1, 4 \rangle = 1 \cdot 2 + 3 \cdot (-1) + (-2) \cdot 4 = 2 - 3 - 8 = -9$$

23. Compute $\mathbf{v} \times \mathbf{w}$.

SOLUTION   We use the definition of the cross product as a "determinant":

$$\mathbf{v} \times \mathbf{w} = \begin{vmatrix} \mathbf{i} & \mathbf{j} & \mathbf{k} \\ 1 & 3 & -2 \\ 2 & -1 & 4 \end{vmatrix} = \begin{vmatrix} 3 & -2 \\ -1 & 4 \end{vmatrix} \mathbf{i} - \begin{vmatrix} 1 & -2 \\ 2 & 4 \end{vmatrix} \mathbf{j} + \begin{vmatrix} 1 & 3 \\ 2 & -1 \end{vmatrix} \mathbf{k}$$

$$= (12 - 2)\mathbf{i} - (4 + 4)\mathbf{j} + (-1 - 6)\mathbf{k} = 10\mathbf{i} - 8\mathbf{j} - 7\mathbf{k} = \langle 10, -8, -7 \rangle$$

25. Find the volume of the parallelepiped spanned by $\mathbf{v}$, $\mathbf{w}$, and $\mathbf{u} = \langle 1, 2, 6 \rangle$.

SOLUTION   The volume V of the parallelepiped spanned by $\mathbf{v}$, $\mathbf{w}$ and $\mathbf{u}$ is the following determinant:

$$V = \left| \det \begin{pmatrix} \mathbf{v} \\ \mathbf{w} \\ \mathbf{u} \end{pmatrix} \right| = \begin{vmatrix} 1 & 3 & -2 \\ 2 & -1 & 4 \\ 1 & 2 & 6 \end{vmatrix} = \left| 1 \cdot \begin{vmatrix} -1 & 4 \\ 2 & 6 \end{vmatrix} - 3 \begin{vmatrix} 2 & 4 \\ 1 & 6 \end{vmatrix} - 2 \begin{vmatrix} 2 & -1 \\ 1 & 2 \end{vmatrix} \right|$$

$$= |1 \cdot (-6 - 8) - 3(12 - 4) - 2(4 + 1)| = 48$$

27. Use vectors to prove that the line connecting the midpoints of two sides of a triangle is parallel to the third side.

SOLUTION   Let E and F be the midpoints of sides AC and BC in a triangle ABC (see figure).

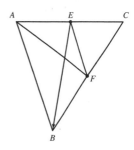

We must show that

$$\overrightarrow{EF} \parallel \overrightarrow{AB}$$

Using the Parallelogram Law we have

$$\overrightarrow{EF} = \overrightarrow{EA} + \overrightarrow{AB} + \overrightarrow{BF} \tag{1}$$

By the definition of the points E and F,

$$\overrightarrow{EA} = \frac{1}{2}\overrightarrow{CA}; \quad \overrightarrow{BF} = \frac{1}{2}\overrightarrow{BC}$$

We substitute (1) to obtain

$$\overrightarrow{EF} = \frac{1}{2}\overrightarrow{CA} + \overrightarrow{AB} + \frac{1}{2}\overrightarrow{BC} = \overrightarrow{AB} + \frac{1}{2}\left(\overrightarrow{CA} + \overrightarrow{BC}\right)$$

$$= \overrightarrow{AB} + \frac{1}{2}\left(\overrightarrow{BC} + \overrightarrow{CA}\right) = \overrightarrow{AB} + \frac{1}{2}\overrightarrow{BA} = \overrightarrow{AB} - \frac{1}{2}\overrightarrow{AB} = \frac{1}{2}\overrightarrow{AB}$$

Therefore, $\overrightarrow{EF}$ is a constant multiple of $\overrightarrow{AB}$, which implies that $\overrightarrow{EF}$ and $\overrightarrow{AB}$ are parallel vectors.

29. Calculate the component of $\mathbf{v} = \left\langle -2, \frac{1}{2}, 3 \right\rangle$ along $\mathbf{w} = \langle 1, 2, 2 \rangle$.

SOLUTION   We first compute the following dot products:

$$\mathbf{v} \cdot \mathbf{w} = \left\langle -2, \frac{1}{2}, 3 \right\rangle \cdot \langle 1, 2, 2 \rangle = 5$$

$$\mathbf{w} \cdot \mathbf{w} = \|\mathbf{w}\|^2 = 1^2 + 2^2 + 2^2 = 9$$

The component of $\mathbf{v}$ along $\mathbf{w}$ is the following number:

$$\left\| \left(\frac{\mathbf{v} \cdot \mathbf{w}}{\mathbf{w} \cdot \mathbf{w}} \right) \mathbf{w} \right\| = \frac{5}{9} \|\mathbf{w}\| = \frac{5}{9} \cdot 3 = \frac{5}{3}$$

31. A 50-kg wagon is pulled to the right by a force $\mathbf{F}_1$ making an angle of $30°$ with the ground. At the same time the wagon is pulled to the left by a horizontal force $\mathbf{F}_2$.

(a) Find the magnitude of $\mathbf{F}_1$ in terms of the magnitude of $\mathbf{F}_2$ if the wagon does not move.

(b) What is the maximal magnitude of $\mathbf{F}_1$ that can be applied to the wagon without lifting it?

SOLUTION

(a) By Newton's Law, at equilibrium, the total force acting on the wagon is zero.

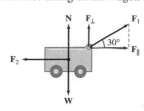

We resolve the force $\mathbf{F}_1$ into its components:

$$\mathbf{F}_1 = \mathbf{F}_{\parallel} + \mathbf{F}_{\perp}$$

where $\mathbf{F}_{\parallel}$ is the horizontal component and $\mathbf{F}_{\perp}$ is the vertical component. Since the wagon does not move, the magnitude of $\mathbf{F}_{\parallel}$ must be equal to the magnitude of $\mathbf{F}_2$. That is,

$$\|\mathbf{F}_{\parallel}\| = \|\mathbf{F}_1\| \cos 30° = \|\mathbf{F}_2\|$$

The above equation gives:

$$\|\mathbf{F}_1\| \frac{\sqrt{3}}{2} = \|\mathbf{F}_2\| \quad \Rightarrow \quad \|\mathbf{F}_1\| = \frac{2\|\mathbf{F}_2\|}{\sqrt{3}}$$

(b) The maximum magnitude of force $\mathbf{F}_1$ that can be applied to the wagon without lifting the wagon is found by comparing the vertical forces:

$$\|\mathbf{F}_1\| \sin 30° = 9.8 \cdot 50$$

$$\|\mathbf{F}_1\| \cdot \frac{1}{2} = 9.8 \cdot 50 \quad \Rightarrow \quad \|\mathbf{F}_1\| = 9.8 \cdot 100 = 980 \text{ N}$$

In Exercises 33–36, let $\mathbf{v} = \langle 1, 2, 4 \rangle$, $\mathbf{u} = \langle 6, -1, 2 \rangle$, *and* $\mathbf{w} = \langle 1, 0, -3 \rangle$. *Calculate the given quantity.*

33. $\mathbf{v} \times \mathbf{w}$

SOLUTION   We use the definition of the cross product as a determinant to compute $\mathbf{v} \times \mathbf{w}$:

$$\mathbf{v} \times \mathbf{w} = \begin{vmatrix} \mathbf{i} & \mathbf{j} & \mathbf{k} \\ 1 & 2 & 4 \\ 1 & 0 & -3 \end{vmatrix} = \begin{vmatrix} 2 & 4 \\ 0 & -3 \end{vmatrix} \mathbf{i} - \begin{vmatrix} 1 & 4 \\ 1 & -3 \end{vmatrix} \mathbf{j} + \begin{vmatrix} 1 & 2 \\ 1 & 0 \end{vmatrix} \mathbf{k}$$

$$= (-6 - 0)\mathbf{i} - (-3 - 4)\mathbf{j} + (0 - 2)\mathbf{k} = -6\mathbf{i} + 7\mathbf{j} - 2\mathbf{k} = \langle -6, 7, -2 \rangle$$

35. $\det \begin{pmatrix} \mathbf{u} \\ \mathbf{v} \\ \mathbf{w} \end{pmatrix}$

SOLUTION   We compute the determinant:

$$\det \begin{pmatrix} \mathbf{u} \\ \mathbf{v} \\ \mathbf{w} \end{pmatrix} = \begin{vmatrix} 6 & -1 & 2 \\ 1 & 2 & 4 \\ 1 & 0 & -3 \end{vmatrix} = 6 \cdot \begin{vmatrix} 2 & 4 \\ 0 & -3 \end{vmatrix} + 1 \cdot \begin{vmatrix} 1 & 4 \\ 1 & -3 \end{vmatrix} + 2 \begin{vmatrix} 1 & 2 \\ 1 & 0 \end{vmatrix}$$

$$= 6 \cdot (-6 - 0) + 1 \cdot (-3 - 4) + 2 \cdot (0 - 2) = -47$$

37. Use the cross product to find the area of the triangle whose vertices are $(1, 3, -1)$, $(2, -1, 3)$, and $(4, 1, 1)$.

SOLUTION   Let $A = (1, 3, -1)$, $B = (2, -1, 3)$ and $C = (4, 1, 1)$.

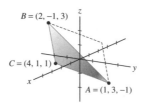

The area S of the triangle ABC is half the area of the parallelogram spanned by $\overrightarrow{AB}$ and $\overrightarrow{AC}$. Using the Formula for the Area of the Parallelogram, we conclude that the area of the triangle is:

$$S = \frac{1}{2} \left\| \overrightarrow{AB} \times \overrightarrow{AC} \right\| \tag{1}$$

We first compute the vectors $\overrightarrow{AB}$ and $\overrightarrow{AC}$:

$$\overrightarrow{AB} = \langle 2 - 1, -1 - 3, 3 - (-1) \rangle = \langle 1, -4, 4 \rangle$$

$$\overrightarrow{AC} = \langle 4 - 1, 1 - 3, 1 - (-1) \rangle = \langle 3, -2, 2 \rangle$$

We compute the cross product of the two vectors:

$$\overrightarrow{AB} \times \overrightarrow{AC} = \begin{vmatrix} \mathbf{i} & \mathbf{j} & \mathbf{k} \\ 1 & -4 & 4 \\ 3 & -2 & 2 \end{vmatrix} = \begin{vmatrix} -4 & 4 \\ -2 & 2 \end{vmatrix} \mathbf{i} - \begin{vmatrix} 1 & 4 \\ 3 & 2 \end{vmatrix} \mathbf{j} + \begin{vmatrix} 1 & -4 \\ 3 & -2 \end{vmatrix} \mathbf{k}$$

$$= (-8 - (-8))\mathbf{i} - (2 - 12)\mathbf{j} + (-2 - (-12))\mathbf{k}$$

$$= 10\mathbf{j} + 10\mathbf{k} = \langle 0, 10, 10 \rangle = 10\langle 0, 1, 1 \rangle$$

The length of $\overrightarrow{AB} \times \overrightarrow{AC}$ is, thus:

$$\|\overrightarrow{AB} \times \overrightarrow{AC}\| = \|10\langle 0, 1, 1 \rangle\| = 10\|\langle 0, 1, 1 \rangle\| = 10\sqrt{0^2 + 1^2 + 1^2} = 10\sqrt{2}$$

Substituting in (1) gives the following area:

$$S = \frac{1}{2} \cdot 10\sqrt{2} = 5\sqrt{2}.$$

39. Show that if the vectors $\mathbf{v}$, $\mathbf{w}$ are orthogonal, then $\|\mathbf{v} + \mathbf{w}\|^2 = \|\mathbf{v}\|^2 + \|\mathbf{w}\|^2$.

SOLUTION The vectors $\mathbf{v}$ and $\mathbf{w}$ are orthogonal, hence:

$$\mathbf{v} \cdot \mathbf{w} = 0 \tag{1}$$

Using the relation of the dot product with length and properties of the dot product we obtain:

$$\|\mathbf{v} + \mathbf{w}\|^2 = (\mathbf{v} + \mathbf{w}) \cdot (\mathbf{v} + \mathbf{w}) = \mathbf{v} \cdot (\mathbf{v} + \mathbf{w}) + \mathbf{w} \cdot (\mathbf{v} + \mathbf{w})$$

$$= \mathbf{v} \cdot \mathbf{v} + \mathbf{v} \cdot \mathbf{w} + \mathbf{w} \cdot \mathbf{v} + \mathbf{w} \cdot \mathbf{w} = \|\mathbf{v}\|^2 + 2\mathbf{v} \cdot \mathbf{w} + \|\mathbf{w}\|^2 \tag{2}$$

Combining (1) and (2) we get:

$$\|\mathbf{v} + \mathbf{w}\|^2 = \|\mathbf{v}\|^2 + \|\mathbf{w}\|^2.$$

41. Find $\|\mathbf{e} - 4\mathbf{f}\|$, assuming that $\mathbf{e}$ and $\mathbf{f}$ are unit vectors such that $\|\mathbf{e} + \mathbf{f}\| = \sqrt{3}$.

SOLUTION We use the relation of the dot product with length and properties of the dot product to write

$$3 = \|\mathbf{e} + \mathbf{f}\|^2 = (\mathbf{e} + \mathbf{f}) \cdot (\mathbf{e} + \mathbf{f}) = \mathbf{e} \cdot \mathbf{e} + \mathbf{e} \cdot \mathbf{f} + \mathbf{f} \cdot \mathbf{e} + \mathbf{f} \cdot \mathbf{f}$$

$$= \|\mathbf{e}\|^2 + 2\mathbf{e} \cdot \mathbf{f} + \|\mathbf{f}\|^2 = 1^2 + 2\mathbf{e} \cdot \mathbf{f} + 1^2 = 2 + 2\mathbf{e} \cdot \mathbf{f}$$

We now find $\mathbf{e} \cdot \mathbf{f}$.

$$3 = 2 + 2\mathbf{e} \cdot \mathbf{f} \quad \Rightarrow \quad \mathbf{e} \cdot \mathbf{f} = 1/2$$

Hence, using the same method as above, we have:

$$\|\mathbf{e} - 4\mathbf{f}\|^2 = (\mathbf{e} - 4\mathbf{f}) \cdot (\mathbf{e} - 4\mathbf{f})$$

$$= \|\mathbf{e}\|^2 - 2 \cdot \mathbf{e} \cdot 4\mathbf{f} + \|4\mathbf{f}\|^2 = 1^2 - 8\mathbf{e} \cdot \mathbf{f} + 4^2 = 17 - 4 = 13$$

Taking square roots, we get:

$$\|\mathbf{e} - 4\mathbf{f}\| = \sqrt{13}$$

43. Show that the equation $\langle 1, 2, 3 \rangle \times \mathbf{v} = \langle -1, 2, a \rangle$ has no solution for $a \neq -1$.

SOLUTION By properties of the cross product, the vector $\langle -1, 2, a \rangle$ is orthogonal to $\langle 1, 2, 3 \rangle$, hence the dot product of these vectors is zero. That is:

$$\langle -1, 2, a \rangle \cdot \langle 1, 2, 3 \rangle = 0$$

We compute the dot product and solve for a:

$$-1 + 4 + 3a = 0$$

$$3a = -3 \quad \Rightarrow \quad a = -1$$

We conclude that if the given equation is solvable, then $a = -1$.

45. Use the identity

$$\mathbf{u} \times (\mathbf{v} \times \mathbf{w}) = (\mathbf{u} \cdot \mathbf{w})\mathbf{v} - (\mathbf{u} \cdot \mathbf{v})\mathbf{w}$$

to prove that

$$\mathbf{u} \times (\mathbf{v} \times \mathbf{w}) + \mathbf{v} \times (\mathbf{w} \times \mathbf{u}) + \mathbf{w} \times (\mathbf{u} \times \mathbf{v}) = \mathbf{0}$$

SOLUTION   The given identity implies that:

$$\mathbf{u} \times (\mathbf{v} \times \mathbf{w}) = (\mathbf{u} \cdot \mathbf{w})\mathbf{v} - (\mathbf{u} \cdot \mathbf{v})\mathbf{w}$$

$$\mathbf{v} \times (\mathbf{w} \times \mathbf{u}) = (\mathbf{v} \cdot \mathbf{u})\mathbf{w} - (\mathbf{v} \cdot \mathbf{w})\mathbf{u}$$

$$\mathbf{w} \times (\mathbf{u} \times \mathbf{v}) = (\mathbf{w} \cdot \mathbf{v})\mathbf{u} - (\mathbf{w} \cdot \mathbf{u})\mathbf{v}$$

Adding the three equations and using the commutativity of the dot product we find that:

$$\mathbf{u} \times (\mathbf{v} \times \mathbf{w}) + \mathbf{v} \times (\mathbf{w} \times \mathbf{u}) + \mathbf{w} \times (\mathbf{u} \times \mathbf{v})$$

$$= (\mathbf{u} \cdot \mathbf{w} - \mathbf{w} \cdot \mathbf{u})\mathbf{v} + (\mathbf{v} \cdot \mathbf{u} - \mathbf{u} \cdot \mathbf{v})\mathbf{w} + (\mathbf{w} \cdot \mathbf{v} - \mathbf{v} \cdot \mathbf{w})\mathbf{u} = \mathbf{0}$$

47. Write the equation of the plane $\mathcal{P}$ with vector equation

$$\langle 1, 4, -3 \rangle \cdot \langle x, y, z \rangle = 7$$

in the form

$$a(x - x_0) + b(y - y_0) + c(z - z_0) = 0$$

Hint: You must find a point $P = (x_0, y_0, z_0)$ on $\mathcal{P}$.

SOLUTION   We identify the vector $\mathbf{n} = \langle a, b, c \rangle = \langle 1, 4, -3 \rangle$ that is normal to the plane, hence we may choose,

$$a = 1, \quad b = 4, \quad c = -3.$$

We now must find a point in the plane. The point $(x_0, y_0, z_0) = (0, 1, -1)$, for instance, satisfies the equation of the plane, therefore the equation may be written in the form:

$$1(x - 0) + 4(y - 1) - 3(z - (-1)) = 0$$

or

$$(x - 0) + 4(y - 1) - 3(z + 1) = 0$$

49. Find the plane through $P = (4, -1, 9)$ containing the line $\mathbf{r}(t) = \langle 1, 4, -3 \rangle + t\langle 2, 1, 1 \rangle$.

SOLUTION   Since the plane contains the line, the direction vector of the line, $\mathbf{v} = \langle 2, 1, 1 \rangle$, is in the plane. To find another vector in the plane, we use the points $A = (1, 4, -3)$ and $B = (4, -1, 9)$ that lie in the plane, and compute the vector $\mathbf{u} = \overrightarrow{AB}$:

$$\mathbf{u} = \overrightarrow{AB} = \langle 4 - 1, -1 - 4, 9 - (-3) \rangle = \langle 3, -5, 12 \rangle$$

We now compute the cross product $\mathbf{n} = \mathbf{v} \times \mathbf{u}$ that is normal to the plane:

$$\mathbf{n} = \mathbf{v} \times \mathbf{u} = \begin{vmatrix} \mathbf{i} & \mathbf{j} & \mathbf{k} \\ 2 & 1 & 1 \\ 3 & -5 & 12 \end{vmatrix} = \begin{vmatrix} 1 & 1 \\ -5 & 12 \end{vmatrix}\mathbf{i} - \begin{vmatrix} 2 & 1 \\ 3 & 12 \end{vmatrix}\mathbf{j} + \begin{vmatrix} 2 & 1 \\ 3 & -5 \end{vmatrix}\mathbf{k}$$

$$= (12 + 5)\mathbf{i} - (24 - 3)\mathbf{j} + (-10 - 3)\mathbf{k} = 17\mathbf{i} - 21\mathbf{j} - 13\mathbf{k} = \langle 17, -21, -13 \rangle$$

Finally, we use the vector form of the equation of the plane with $\mathbf{n} = \langle 17, -21, -13 \rangle$ and $P_0 = (4, -1, 9)$ to obtain the following equation:

$$\mathbf{n} \cdot \langle x, y, z \rangle = \mathbf{n} \cdot \langle x_0, y_0, z_0 \rangle$$

$$\langle 17, -21, -13 \rangle \cdot \langle x, y, z \rangle = \langle 17, -21, -13 \rangle \cdot \langle 4, -1, 9 \rangle$$

$$17x - 21y - 13z = 17 \cdot 4 + 21 - 13 \cdot 9 = -28$$

The equation of the plane is, thus,

$$17x - 21y - 13z = -28.$$

51. Find the trace of the plane $3x - 2y + 5z = 4$ in the xy-plane.

SOLUTION   The xy-plane has equation $z = 0$, therefore the intersection of the plane $3x - 2y + 5z = 4$ with the xy-plane must satisfy both $z = 0$ and the equation of the plane. Therefore the trace has the following equation:

$$3x - 2y + 5 \cdot 0 = 4 \quad \Rightarrow \quad 3x - 2y = 4$$

We conclude that the trace of the plane in the xy-plane is the line $3x - 2y = 4$ in the xy-plane.

In Exercises 53–58, determine the type of the quadric surface.

53. $\left(\dfrac{x}{3}\right)^2 + \left(\dfrac{y}{4}\right)^2 + 2z^2 = 1$

SOLUTION   Writing the equation in the form:

$$\left(\frac{x}{3}\right)^2 + \left(\frac{y}{4}\right)^2 + \left(\frac{z}{\frac{1}{\sqrt{2}}}\right)^2 = 1$$

we identify the quadric surface as an ellipsoid.

55. $\left(\dfrac{x}{3}\right)^2 + \left(\dfrac{y}{4}\right)^2 - 2z = 0$

SOLUTION   We rewrite this equation as:

$$2z = \left(\frac{x}{3}\right)^2 + \left(\frac{y}{4}\right)^2$$

or

$$z = \left(\frac{x}{3\sqrt{2}}\right)^2 + \left(\frac{y}{4\sqrt{2}}\right)^2$$

This is the equation of an elliptic paraboloid.

57. $\left(\dfrac{x}{3}\right)^2 - \left(\dfrac{y}{4}\right)^2 - 2z^2 = 0$

SOLUTION   This equation may be rewritten in the form

$$\left(\frac{x}{3}\right)^2 - \left(\frac{y}{4}\right)^2 - \left(\frac{z}{\frac{1}{\sqrt{2}}}\right)^2$$

we identify the quadric surface as an elliptic cone.

59. Determine the type of the quadric surface $ax^2 + by^2 - z^2 = 1$ if:
(a) $a < 0, \quad b < 0$
(b) $a > 0, \quad b > 0$
(c) $a > 0, \quad b < 0$

SOLUTION

(a) If $a < 0$, $b < 0$ then for all x, y and z we have $ax^2 + by^2 - z^2 < 0$, hence there are no points that satisfy $ax^2 + by^2 - z^2 = 1$. Therefore it is the empty set.
(b) For $a > 0$ and $b > 0$ we rewrite the equation as

$$\left(\frac{x}{\frac{1}{\sqrt{a}}}\right)^2 + \left(\frac{y}{\frac{1}{\sqrt{b}}}\right)^2 - z^2 = 1$$

which is the equation of a hyperboloid of one sheet.
(c) For $a > 0, b < 0$ we rewrite the equation in the form

$$\left(\frac{x}{\frac{1}{\sqrt{a}}}\right)^2 - \left(\frac{y}{\frac{1}{\sqrt{|b|}}}\right)^2 - z^2 = 1$$

which is the equation of a hyperboloid of two sheets.

61. Convert $(x, y, z) = (3, 4, -1)$ from rectangular to cylindrical and spherical coordinates.

SOLUTION   In cylindrical coordinates (r, θ, z) we have

$$r = \sqrt{x^2 + y^2}, \quad \tan \theta = \frac{y}{x}$$

Therefore, $r = \sqrt{3^2 + 4^2} = 5$ and $\tan\theta = \frac{4}{3}$. The projection of the point $(3, 4, -1)$ onto the xy-plane is the point $(3, 4)$, in the first quadrant. Therefore, the corresponding value of θ is $\tan^{-1}\frac{4}{3} \approx 0.93$ rad. The cylindrical coordinates are, thus,

$$(r, \theta, z) = \left(5, \tan^{-1}\frac{4}{3}, -1\right)$$

The spherical coordinates (ρ, θ, ϕ) satisfy

$$\rho = \sqrt{x^2 + y^2 + z^2}, \quad \tan\theta = \frac{y}{x}, \quad \cos\phi = \frac{z}{\rho}$$

Therefore,

$$\rho = \sqrt{3^2 + 4^2 + (-1)^2} = \sqrt{26}$$

$$\tan\theta = \frac{4}{3}$$

$$\cos\phi = \frac{-1}{\sqrt{26}}$$

The angle θ is the same as in the cylindrical coordinates, that is, $\theta = \tan^{-1}\frac{4}{3}$. The angle ϕ is the solution of $\cos\phi = \frac{-1}{\sqrt{26}}$ that satisfies $0 \le \phi \le \pi$, that is, $\phi = \cos^1\left(\frac{-1}{\sqrt{26}}\right) \approx 1.77$ rad. The spherical coordinates are, thus,

$$(\rho, \theta, \phi) = \left(\sqrt{26}, \tan^{-1}\frac{4}{3}, \cos^{-1}\left(\frac{-1}{\sqrt{26}}\right)\right).$$

63. Convert the point $(\rho, \theta, \phi) = \left(3, \frac{\pi}{6}, \frac{\pi}{3}\right)$ from spherical to cylindrical coordinates.

SOLUTION   By the given information, $\rho = 3$, $\theta = \frac{\pi}{6}$, and $\phi = \frac{\pi}{3}$. We must determine the cylindrical coordinates (r, θ, z). The angle θ is the same as in spherical coordinates. We find z using the relation $\cos\phi = \frac{z}{\rho}$, or $z = \rho\cos\phi$. We obtain

$$z = \rho\cos\phi = 3\cos\frac{\pi}{3} = 3 \cdot \frac{1}{2} = \frac{3}{2}$$

We find r using the relation $\rho^2 = x^2 + y^2 + z^2 = r^2 + z^2$, or $r = \sqrt{\rho^2 - z^2}$, we get

$$r = \sqrt{3^2 - \left(\frac{3}{2}\right)^2} = \sqrt{\frac{27}{4}} = \frac{3\sqrt{3}}{2}$$

Hence, in cylindrical coordinates we obtain the following description:

$$(r, \theta, z) = \left(\frac{3\sqrt{3}}{2}, \frac{\pi}{6}, \frac{3}{2}\right).$$

65. Sketch the graph of the cylindrical equation $z = 2r\cos\theta$ and write the equation in rectangular coordinates.

SOLUTION   To obtain the equation in rectangular coordinates, we substitute $x = r\cos\theta$ in the equation $z = 2r\cos\theta$:

$$z = 2r\cos\theta = 2x \quad \Rightarrow \quad z = 2x$$

This is the equation of a plane normal to the xz-plane, whose intersection with the xz-plane is the line $z = 2x$. The graph of the plane is shown in the following figure (the same plane drawn twice, using the cylindrical coordinates' equation and using the rectangular coordinates' equation):

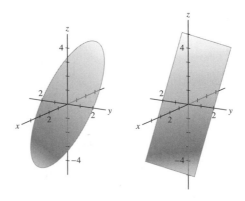

67. Show that the cylindrical equation

$$r^2(1 - 2\sin^2\theta) + z^2 = 1$$

is a hyperboloid of one sheet.

SOLUTION We rewrite the equation in the form

$$r^2 - 2(r\sin\theta)^2 + z^2 = 1$$

To write this equation in rectangular coordinates, we substitute $r^2 = x^2 + y^2$ and $r\sin\theta = y$. This gives

$$x^2 + y^2 - 2y^2 + z^2 = 1$$
$$x^2 - y^2 + z^2 = 1$$

We now can identify the surface as a hyperboloid of one sheet.

69. Describe how the surface with spherical equation

$$\rho^2(1 + A\cos^2\phi) = 1$$

depends on the constant A.

SOLUTION To identify the surface we convert the equation to rectangular coordinates. We write

$$\rho^2 + A\rho^2\cos^2\phi = 1$$

To obtain the following equation in terms of x, y, z only, we substitute $\rho^2 = x^2 + y^2 + z^2$ and $\rho\cos\phi = z$:

$$x^2 + y^2 + z^2 + Az^2 = 1$$
$$x^2 + y^2 + (1 + A)z^2 = 1 \tag{1}$$

Case 1: $A < -1$. Then $A + 1 < 0$ and the equation can be rewritten in the form

$$x^2 + y^2 - \left(\frac{z}{|1 + A|^{-1/2}}\right)^2 = 1$$

The corresponding surface is a hyperboloid of one sheet.
Case 2: $A = -1$. Equation (1) becomes:

$$x^2 + y^2 = 1$$

In R^3, this equation describes a cylinder with the z-axis as its central axis.
Case 3: $A > -1$. Then equation (1) can be rewritten as

$$x^2 + y^2 + \left(\frac{z}{(1 + A)^{1/2}}\right)^2 = 1$$

Then if $A = 0$ the equation $x^2 + y^2 + z^2 = 1$ describes the unit sphere in R^3. Otherwise, the surface is an ellipsoid.

71. Let c be a scalar, let $\mathbf{a}$ and $\mathbf{b}$ be vectors, and let $\mathbf{X} = \langle x, y, z \rangle$. Show that the equation $(\mathbf{X} - \mathbf{a}) \cdot (\mathbf{X} - \mathbf{b}) = c^2$ defines a sphere with center $\mathbf{m} = \frac{1}{2}(\mathbf{a} + \mathbf{b})$ and radius R, where $R^2 = c^2 + \left\|\frac{1}{2}(\mathbf{a} - \mathbf{b})\right\|^2$.

SOLUTION We evaluate the following length:

$$\|\mathbf{x} - \mathbf{m}\|^2 = \left\|\mathbf{x} - \frac{1}{2}(\mathbf{a} + \mathbf{b})\right\|^2 = \left((\mathbf{x} - \mathbf{a}) + \frac{1}{2}(\mathbf{a} - \mathbf{b})\right) \cdot \left((\mathbf{x} - \mathbf{b}) - \frac{1}{2}(\mathbf{a} - \mathbf{b})\right)$$

$$= (\mathbf{x} - \mathbf{a}) \cdot (\mathbf{x} - \mathbf{b}) - \frac{1}{2}(\mathbf{x} - \mathbf{a}) \cdot (\mathbf{a} - \mathbf{b}) + \frac{1}{2}(\mathbf{a} - \mathbf{b}) \cdot (\mathbf{x} - \mathbf{b}) - \frac{1}{4}(\mathbf{a} - \mathbf{b}) \cdot (\mathbf{a} - \mathbf{b})$$

$$= (\mathbf{x} - \mathbf{a}) \cdot (\mathbf{x} - \mathbf{b}) + \frac{1}{2}(\mathbf{a} - \mathbf{b}) \cdot (\mathbf{x} - \mathbf{b} - \mathbf{x} + \mathbf{a}) - \frac{1}{4}(\mathbf{a} - \mathbf{b}) \cdot (\mathbf{a} - \mathbf{b})$$

$$= (\mathbf{x} - \mathbf{a}) \cdot (\mathbf{x} - \mathbf{b}) + \frac{1}{2}(\mathbf{a} - \mathbf{b}) \cdot (\mathbf{a} - \mathbf{b}) - \frac{1}{4}(\mathbf{a} - \mathbf{b}) \cdot (\mathbf{a} - \mathbf{b})$$

$$= (\mathbf{x} - \mathbf{a}) \cdot (\mathbf{x} - \mathbf{b}) + \frac{1}{4}(\mathbf{a} - \mathbf{b}) \cdot (\mathbf{a} - \mathbf{b})$$

$$= (\mathbf{x} - \mathbf{a}) \cdot (\mathbf{x} - \mathbf{b}) + \left\|\frac{1}{2}(\mathbf{a} - \mathbf{b})\right\|^2$$

Since $R^2 = c^2 + \left\|\frac{1}{2}(\mathbf{a} - \mathbf{b})\right\|^2$ we get

$$\|\mathbf{x} - \mathbf{m}\|^2 = (\mathbf{x} - \mathbf{a}) \cdot (\mathbf{x} - \mathbf{b}) + R^2 - c^2$$

We conclude that if $(\mathbf{x} - \mathbf{a})(\mathbf{x} - \mathbf{b}) = c^2$ then $\|\mathbf{x} - \mathbf{m}\|^2 = R^2$. That is, the equation $(\mathbf{x} - \mathbf{a})(\mathbf{x} - \mathbf{b}) = c^2$ defines a sphere with center $\mathbf{m}$ and radius R.

13 CALCULUS OF VECTOR-VALUED FUNCTIONS

13.1 Vector-Valued Functions   (LT Section 14.1)

Preliminary Questions

1. Which one of the following does *not* parametrize a line?

(a) $\mathbf{r}_1(t) = \langle 8 - t, 2t, 3t \rangle$

(b) $\mathbf{r}_2(t) = t^3\mathbf{i} - 7t^3\mathbf{j} + t^3\mathbf{k}$

(c) $\mathbf{r}_3(t) = \langle 8 - 4t^3, 2 + 5t^2, 9t^3 \rangle$

SOLUTION

(a) This is a parametrization of the line passing through the point $(8, 0, 0)$ in the direction parallel to the vector $\langle -1, 2, 3 \rangle$, since:

$$\langle 8 - t, 2t, 3t \rangle = \langle 8, 0, 0 \rangle + t \langle -1, 2, 3 \rangle$$

(b) Using the parameter $s = t^3$ we get:

$$\langle t^3, -7t^3, t^3 \rangle = \langle s, -7s, s \rangle = s \langle 1, -7, 1 \rangle$$

This is a parametrization of the line through the origin, with the direction vector $\mathbf{v} = \langle -1, 7, 1 \rangle$.

(c) The parametrization $\langle 8 - 4t^3, 2 + 5t^2, 9t^3 \rangle$ does not parametrize a line. In particular, the points $(8, 2, 0)$ (at $t = 0$), $(4, 7, 9)$ (at $t = 1$), and $(-24, 22, 72)$ (at $t = 2$) are not collinear.

2. What is the projection of $\mathbf{r}(t) = t\mathbf{i} + t^4\mathbf{j} + e^t\mathbf{k}$ onto the xz-plane?

SOLUTION   The projection of the path onto the xz-plane is the curve traced by $t\mathbf{i} + e^t\mathbf{k} = \langle t, 0, e^t \rangle$. This is the curve $z = e^x$ in the xz-plane.

3. Which projection of $\langle \cos t, \cos 2t, \sin t \rangle$ is a circle?

SOLUTION   The parametric equations are

$$x = \cos t, \quad y = \cos 2t, \quad z = \sin t$$

The projection onto the xz-plane is $\langle \cos t, 0, \sin t \rangle$. Since $x^2 + z^2 = \cos^2 t + \sin^2 t = 1$, the projection is a circle in the xz-plane. The projection onto the xy-plane is traced by the curve $\langle \cos t, \cos 2t, 0 \rangle$. Therefore, $x = \cos t$ and $y = \cos 2t$. We express y in terms of x:

$$y = \cos 2t = 2\cos^2 t - 1 = 2x^2 - 1$$

The projection onto the xy-plane is a parabola. The projection onto the yz-plane is the curve $\langle 0, \cos 2t, \sin t \rangle$. Hence $y = \cos 2t$ and $z = \sin t$. We find y as a function of z:

$$y = \cos 2t = 1 - 2\sin^2 t = 1 - 2z^2$$

The projection onto the yz-plane is again a parabola.

4. What is the center of the circle with parametrization

$$\mathbf{r}(t) = (-2 + \cos t)\mathbf{i} + 2\mathbf{j} + (3 - \sin t)\mathbf{k}?$$

SOLUTION   The parametric equations are

$$x = -2 + \cos t, \quad y = 2, \quad z = 3 - \sin t$$

Therefore, the curve is contained in the plane $y = 2$, and the following holds:

$$(x + 2)^2 + (z - 3)^2 = \cos^2 t + \sin^2 t = 1$$

We conclude that the curve $\mathbf{r}(t)$ is the circle of radius 1 in the plane $y = 2$ centered at the point $(-2, 2, 3)$.

5. How do the paths $\mathbf{r}_1(t) = \langle \cos t, \sin t \rangle$ and $\mathbf{r}_2(t) = \langle \sin t, \cos t \rangle$ around the unit circle differ?

SOLUTION The two paths describe the unit circle. However, as t increases from 0 to 2π, the point on the path $\sin t\mathbf{i} + \cos t\mathbf{j}$ moves in a clockwise direction, whereas the point on the path $\cos t\mathbf{i} + \sin t\mathbf{j}$ moves in a counterclockwise direction.

6. Which three of the following vector-valued functions parametrize the same space curve?

(a) $(-2 + \cos t)\mathbf{i} + 9\mathbf{j} + (3 - \sin t)\mathbf{k}$ **(b)** $(2 + \cos t)\mathbf{i} - 9\mathbf{j} + (-3 - \sin t)\mathbf{k}$

(c) $(-2 + \cos 3t)\mathbf{i} + 9\mathbf{j} + (3 - \sin 3t)\mathbf{k}$ **(d)** $(-2 - \cos t)\mathbf{i} + 9\mathbf{j} + (3 + \sin t)\mathbf{k}$

(e) $(2 + \cos t)\mathbf{i} + 9\mathbf{j} + (3 + \sin t)\mathbf{k}$

SOLUTION All the curves except for (b) lie in the vertical plane $y = 9$. We identify each one of the curves (a), (c), (d) and (e).

(a) The parametric equations are:

$$x = -2 + \cos t, \quad y = 9, \quad z = 3 - \sin t$$

Hence,

$$(x + 2)^2 + (z - 3)^2 = (\cos t)^2 + (-\sin t)^2 = 1$$

This is the circle of radius 1 in the plane $y = 9$, centered at $(-2, 9, 3)$.

(c) The parametric equations are:

$$x = -2 + \cos 3t, \quad y = 9, \quad z = 3 - \sin 3t$$

Hence,

$$(x + 2)^2 + (z - 3)^2 = (\cos 3t)^2 + (-\sin 3t)^2 = 1$$

This is the circle of radius 1 in the plane $y = 9$, centered at $(-2, 9, 3)$.

(d) In this curve we have:

$$x = -2 - \cos t, \quad y = 9, \quad z = 3 + \sin t$$

Hence,

$$(x + 2)^2 + (z - 3)^2 = (-\cos t)^2 + (\sin t)^2 = 1$$

Again, the circle of radius 1 in the plane $y = 9$, centered at $(-2, 9, 3)$.

(e) In this parametrization we have:

$$x = 2 + \cos t, \quad y = 9, \quad z = 3 + \sin t$$

Hence,

$$(x - 2)^2 + (z - 3)^2 = (\cos t)^2 + (\sin t)^2 = 1$$

This is the circle of radius 1 in the plane $y = 9$, centered at $(2, 9, 3)$.

We conclude that (a), (c) and (d) parametrize the same circle whereas (b) and (e) are different curves.

Exercises

1. What is the domain of $\mathbf{r}(t) = e^t\mathbf{i} + \frac{1}{t}\mathbf{j} + (t + 1)^{-3}\mathbf{k}$?

SOLUTION $\mathbf{r}(t)$ is defined for $t \neq 0$ and $t \neq -1$, hence the domain of $\mathbf{r}(t)$ is:

$$D = \{t \in \mathbf{R} : t \neq 0, t \neq -1\}$$

3. Evaluate $\mathbf{r}(2)$ and $\mathbf{r}(-1)$ for $\mathbf{r}(t) = \left\langle \sin \frac{\pi}{2}t, t^2, (t^2 + 1)^{-1} \right\rangle$.

SOLUTION Since $\mathbf{r}(t) = \left\langle \sin \frac{\pi}{2}t, t^2, (t^2 + 1)^{-1} \right\rangle$, then

$$\mathbf{r}(2) = \left\langle \sin \pi, 4, 5^{-1} \right\rangle = \left\langle 0, 4, \frac{1}{5} \right\rangle$$

and

$$\mathbf{r}(-1) = \left\langle \sin \frac{-\pi}{2}, 1, 2^{-1} \right\rangle = \left\langle -1, 1, \frac{1}{2} \right\rangle$$

5. Find a vector parametrization of the line through $P = (3, -5, 7)$ in the direction $\mathbf{v} = \langle 3, 0, 1 \rangle$.

SOLUTION   We use the vector parametrization of the line to obtain:

$$\mathbf{r}(t) = \overrightarrow{OP} + t\mathbf{v} = \langle 3, -5, 7 \rangle + t \langle 3, 0, 1 \rangle = \langle 3 + 3t, -5, 7 + t \rangle$$

or in the form:

$$\mathbf{r}(t) = (3 + 3t)\mathbf{i} - 5\mathbf{j} + (7 + t)\mathbf{k}, \quad -\infty < t < \infty$$

7. Match the space curves in Figure 8 with their projections onto the xy-plane in Figure 9.

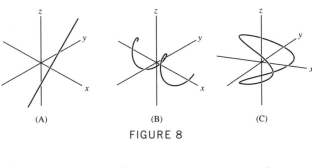

(A)   (B)   (C)

FIGURE 8

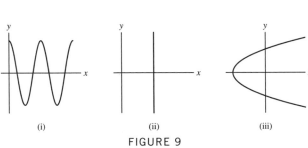

(i)   (ii)   (iii)

FIGURE 9

SOLUTION   The projection of curve (C) onto the xy-plane is neither a segment nor a periodic wave. Hence, the correct projection is (iii), rather than the two other graphs. The projection of curve (A) onto the xy-plane is a vertical line, hence the corresponding projection is (ii). The projection of curve (B) onto the xy-plane is a periodic wave as illustrated in (i).

9. Match the vector-valued functions (a)–(f) with the space curves (i)–(vi) in Figure 10.

(a) $\mathbf{r}(t) = \langle t + 15, e^{0.08t} \cos t, e^{0.08t} \sin t \rangle$ **(b)** $\mathbf{r}(t) = \langle \cos t, \sin t, \sin 12t \rangle$

(c) $\mathbf{r}(t) = \left\langle t, t, \dfrac{25t}{1 + t^2} \right\rangle$ **(d)** $\mathbf{r}(t) = \langle \cos^3 t, \sin^3 t, \sin 2t \rangle$

(e) $\mathbf{r}(t) = \langle t, t^2, 2t \rangle$ **(f)** $\mathbf{r}(t) = \langle \cos t, \sin t, \cos t \sin 12t \rangle$

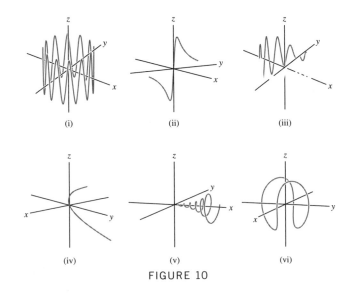

(i)   (ii)   (iii)

(iv)   (v)   (vi)

FIGURE 10

SOLUTION

(a) (v)   **(b)** (i)   **(c)** (ii)

(d) (vi)   **(e)** (iv)   **(f)** (iii)

11. Match the space curves (A)–(C) in Figure 11 with their projections (i)–(iii) onto the xy-plane.

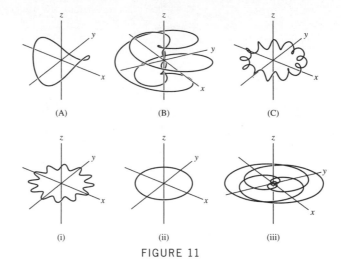

(A)   (B)   (C)

(i)   (ii)   (iii)

FIGURE 11

SOLUTION Observing the curves and the projections onto the xy-plane we conclude that: Projection (i) corresponds to curve (C); Projection (ii) corresponds to curve (A); Projection (iii) corresponds to curve (B).

In Exercises 13–16, the function $\mathbf{r}(t)$ *traces a circle. Determine the radius, center, and plane containing the circle.*

13. $\mathbf{r}(t) = (9\cos t)\mathbf{i} + (9\sin t)\mathbf{j}$

SOLUTION Since $x(t) = 9\cos t$, $y(t) = 9\sin t$ we have:

$$x^2 + y^2 = 81\cos^2 t + 81\sin^2 t = 81(\cos^2 t + \sin^2 t) = 81$$

This is the equation of a circle with radius 9 centered at the origin. The circle lies in the xy-plane.

15. $\mathbf{r}(t) = \langle \sin t, 0, 4 + \cos t \rangle$

SOLUTION $x(t) = \sin t$, $z(t) = 4 + \cos t$, hence:

$$x^2 + (z - 4)^2 = \sin^2 t + \cos^2 t = 1$$

$y = 0$ is the equation of the xz-plane. We conclude that the function traces the circle of radius 1, centered at the point $(0, 0, 4)$, and contained in the xz-plane.

17. Let $\mathcal{C}$ be the curve $\mathbf{r}(t) = \langle t\cos t, t\sin t, t \rangle$.
(a) Show that $\mathcal{C}$ lies on the cone $x^2 + y^2 = z^2$.
(b) Sketch the cone and make a rough sketch of $\mathcal{C}$ on the cone.

SOLUTION $x = t\cos t$, $y = t\sin t$ and $z = t$, hence:

$$x^2 + y^2 = t^2\cos^2 t + t^2\sin^2 t = t^2(\cos^2 t + \sin^2 t) = t^2 = z^2.$$

$x^2 + y^2 = z^2$ is the equation of a circular cone, hence the curve lies on a circular cone. As the height $z = t$ increases linearly with time, the x and y coordinates trace out points on the circles of increasing radius. We obtain the following curve:

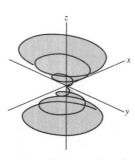

$$\mathbf{r}(t) = \langle t\cos t, t\sin t, t \rangle$$

In Exercises 19 and 20, let

$$\mathbf{r}(t) = \langle \sin t, \cos t, \sin t \cos 2t \rangle$$

as shown in Figure 12.

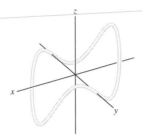

FIGURE 12

19. Find the points where $\mathbf{r}(t)$ intersects the xy-plane.

SOLUTION   The curve intersects the xy-plane at the points where $z = 0$. That is, $\sin t \cos 2t = 0$ and so either $\sin t = 0$ or $\cos 2t = 0$. The solutions are, thus:

$$t = \pi k \text{ or } t = \frac{\pi}{4} + \frac{\pi k}{2}, \quad k = 0, \pm 1, \pm 2, \dots$$

The values $t = \pi k$ yield the points: $(\sin \pi k, \cos \pi k, 0) = \left(0, (-1)^k, 0 \right)$. The values $t = \frac{\pi}{4} + \frac{\pi k}{2}$ yield the points:

$$k = 0: \left(\sin \frac{\pi}{4}, \cos \frac{\pi}{4}, 0 \right) = \left(\frac{1}{\sqrt{2}}, \frac{1}{\sqrt{2}}, 0 \right)$$

$$k = 1: \left(\sin \frac{3\pi}{4}, \cos \frac{3\pi}{4}, 0 \right) = \left(\frac{1}{\sqrt{2}}, -\frac{1}{\sqrt{2}}, 0 \right)$$

$$k = 2: \left(\sin \frac{5\pi}{4}, \cos \frac{5\pi}{4}, 0 \right) = \left(-\frac{1}{\sqrt{2}}, -\frac{1}{\sqrt{2}}, 0 \right)$$

$$k = 3: \left(\sin \frac{7\pi}{4}, \cos \frac{7\pi}{4}, 0 \right) = \left(-\frac{1}{\sqrt{2}}, \frac{1}{\sqrt{2}}, 0 \right)$$

(Other values of k do not provide new points). We conclude that the curve intersects the xy-plane at the following points:
$(0, 1, 0), (0, -1, 0), \left(\frac{1}{\sqrt{2}}, \frac{1}{\sqrt{2}}, 0 \right), \left(\frac{1}{\sqrt{2}}, -\frac{1}{\sqrt{2}}, 0 \right), \left(-\frac{1}{\sqrt{2}}, -\frac{1}{\sqrt{2}}, 0 \right), \left(-\frac{1}{\sqrt{2}}, \frac{1}{\sqrt{2}}, 0 \right)$

21. Parametrize the intersection of the surfaces

$$y^2 - z^2 = x - 2, \qquad y^2 + z^2 = 9$$

using $t = y$ as the parameter (two vector functions are needed as in Example 3).

SOLUTION   We solve for z and x in terms of y. From the equation $y^2 + z^2 = 9$ we have $z^2 = 9 - y^2$ or $z = \pm\sqrt{9 - y^2}$. From the second equation we have:

$$x = y^2 - z^2 + 2 = y^2 - \left(9 - y^2 \right) + 2 = 2y^2 - 7$$

Taking $t = y$ as a parameter, we have $z = \pm\sqrt{9 - t^2}$, $x = 2t^2 - 7$, yielding the following vector parametrization:

$$\mathbf{r}(t) = \left\langle 2t^2 - 7, t, \pm\sqrt{9 - t^2} \right\rangle, \text{ for } -3 \le t \le 3.$$

23. Viviani's Curve C is the intersection of the surfaces (Figure 13)

$$x^2 + y^2 = z^2, \qquad y = z^2$$

(a) Parametrize each of the two parts of C corresponding to $x \ge 0$ and $x \le 0$, taking $t = z$ as parameter.

(b) Describe the projection of C onto the xy-plane.

(c) Show that C lies on the sphere of radius 1 with center $(0, 1, 0)$. This curve looks like a figure eight lying on a sphere [Figure 13(B)].

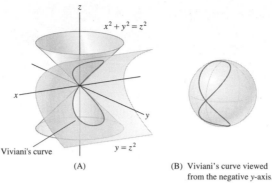

Viviani's curve

$x^2 + y^2 = z^2$

$y = z^2$

(A)

(B) Viviani's curve viewed
from the negative y-axis.

FIGURE 13 Viviani's curve is the intersection of the surfaces $x^2 + y^2 = z^2$ and $y = z^2$.

SOLUTION

(a) We must solve for y and x in terms of z (which is a parameter). We get:

$$y = z^2$$

$$x^2 = z^2 - y^2 \quad \Rightarrow \quad x = \pm\sqrt{z^2 - y^2} = \pm\sqrt{z^2 - z^4}$$

Here, the $\pm$ from $x = \pm\sqrt{z^2 - z^4}$ represents the two parts of the parametrization: $+$ for $x \geq 0$, and $-$ for $x \leq 0$. Substituting the parameter $z = t$ we get:

$$y = t^2, \quad x = \pm\sqrt{t^2 - t^4} = \pm t\sqrt{1 - t^2}.$$

We obtain the following parametrization:

$$\mathbf{r}(t) = \left\langle \pm t\sqrt{1 - t^2}, t^2, t \right\rangle \text{ for } -1 \leq t \leq 1 \tag{1}$$

(b) The projection of the curve onto the xy-plane is the curve on the xy-plane obtained by setting the z-coordinate of $\mathbf{r}(t)$ equal to zero. We obtain the following curve:

$$\left\langle \pm t\sqrt{1 - t^2}, t^2, 0 \right\rangle, \quad -1 \leq t \leq 1$$

We also note that since $x = \pm t\sqrt{1 - t^2}$, then $x^2 = t^2(1 - t^2)$, but also $y = t^2$, so that gives us the equation $x^2 = y(1 - y)$ for the projection onto the xy plane. We rewrite this as follows.

$$x^2 = y(1 - y) \quad \Rightarrow \quad x^2 + y^2 - y = 0$$

$$x^2 + y^2 - y + 1/4 = 1/4$$

$$x^2 + (y - 1/2)^2 = (1/2)^2$$

We can now identify this projection as a circle in the xy plane, with radius $1/2$, centered at the xy point $(0, 1/2)$.

(c) The equation of the sphere of radius 1 with center $(0, 1, 0)$ is:

$$x^2 + (y - 1)^2 + z^2 = 1 \tag{2}$$

To show that $\mathcal{C}$ lies on this sphere, we show that the coordinates of the points on $\mathcal{C}$ (given in (1)) satisfy the equation of the sphere. Substituting the coordinates from (1) into the left side of (2) gives:

$$x^2 + (y - 1)^2 + z^2 = \left(\pm t\sqrt{1 - t^2}\right)^2 + (t^2 - 1)^2 + t^2 = t^2(1 - t^2) + (t^2 - 1)^2 + t^2$$

$$= (t^2 - 1)(t^2 - 1 - t^2) + t^2 = 1$$

We conclude that the curve $\mathcal{C}$ lies on the sphere of radius 1 with center $(0, 1, 0)$.

25. Use sine and cosine to parametrize the intersection of the cylinders $x^2 + y^2 = 1$ and $x^2 + z^2 = 1$ (use two vector-valued functions). Then describe the projections of this curve onto the three coordinate planes.

SOLUTION The circle $x^2 + z^2 = 1$ in the xz-plane is parametrized by $x = \cos t$, $z = \sin t$, and the circle $x^2 + y^2 = 1$ in the xy-plane is parametrized by $x = \cos s$, $y = \sin s$. Hence, the points on the cylinders can be written in the form:

$$x^2 + z^2 = 1: \quad \langle \cos t, y, \sin t \rangle, \quad 0 \leq t \leq 2\pi$$

$$x^2 + y^2 = 1: \quad \langle \cos s, \sin s, z \rangle, \quad 0 \leq t \leq 2\pi$$

The points (x, y, z) on the intersection of the two cylinders must satisfy the following equations:

$$\cos t = \cos s$$

$$y = \sin s$$

$$z = \sin t$$

The first equation implies that $s = \pm t + 2\pi k$. Substituting in the second equation gives $y = \sin(\pm t + 2\pi k) = \sin(\pm t) = \pm \sin t$. Hence, $x = \cos t$, $y = \pm \sin t$, $z = \sin t$. We obtain the following vector parametrization of the intersection:

$$\mathbf{r}(t) = \langle \cos t, \pm \sin t, \sin t \rangle$$

The projection of the curve on the xy-plane is traced by $\langle \cos t, \pm \sin t, 0 \rangle$ which is the unit circle in this plane. The projection of the curve on the xz-plane is traced by $\langle \cos t, 0, \sin t \rangle$ which is the unit circle in the xz-plane. The projection of the curve on the yz-plane is traced by $\langle 0, \pm \sin t, \sin t \rangle$ which is the two segments $z = y$ and $z = -y$ for $-1 \le y \le 1$.

27. Use sine and cosine to parametrize the intersection of the surfaces $x^2 + y^2 = 1$ and $z = 4x^2$ (Figure 14).

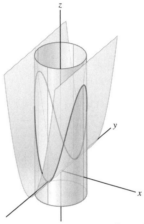

FIGURE 14 Intersection of the surfaces $x^2 + y^2 = 1$ and $z = 4x^2$.

SOLUTION The points on the cylinder $x^2 + y^2 = 1$ and on the parabolic cylinder $z = 4x^2$ can be written in the form:

$$x^2 + y^2 = 1: \quad \langle \cos t, \sin t, z \rangle$$
$$z = 4x^2: \quad \left\langle x, y, 4x^2 \right\rangle$$

The points (x, y, z) on the intersection curve must satisfy the following equations:

$$\begin{array}{l} x = \cos t \\ y = \sin t \\ z = 4x^2 \end{array} \Rightarrow \quad x = \cos t, \ y = \sin t, \ z = 4\cos^2 t$$

We obtain the vector parametrization:

$$\mathbf{r}(t) = \left\langle \cos t, \sin t, 4\cos^2 t \right\rangle, \quad 0 \le t \le 2\pi$$

Using the CAS we obtain the following curve:

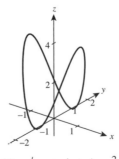

$$\mathbf{r}(t) = \left\langle \cos t, \sin t, 4\cos^2 t \right\rangle$$

In Exercises 28–30, two paths $\mathbf{r}_1(t)$ *and* $\mathbf{r}_2(t)$ *intersect if there is a point P lying on both curves. We say that* $\mathbf{r}_1(t)$ *and* $\mathbf{r}_2(t)$ *collide if* $\mathbf{r}_1(t_0) = \mathbf{r}_2(t_0)$ *at some time* t_0.

29. Determine whether $\mathbf{r}_1$ and $\mathbf{r}_2$ collide or intersect:

$$\mathbf{r}_1(t) = \langle t^2 + 3, t + 1, 6t^{-1} \rangle$$

$$\mathbf{r}_2(t) = \langle 4t, 2t - 2, t^2 - 7 \rangle$$

SOLUTION To determine if the paths collide, we must examine whether the following equations have a solution:

$$\begin{cases} t^2 + 3 = 4t \\ t + 1 = 2t - 2 \\ \dfrac{6}{t} = t^2 - 7 \end{cases}$$

We simplify to obtain:

$$t^2 - 4t + 3 = (t - 3)(t - 1) = 0$$

$$t = 3$$

$$t^3 - 7t - 6 = 0$$

The solution of the second equation is $t = 3$. This is also a solution of the first and the third equations. It follows that $\mathbf{r}_1(3) = \mathbf{r}_2(3)$ so the curves collide. The curves also intersect at the point where they collide. We now check if there are other points of intersection by solving the following equation:

$$\mathbf{r}_1(t) = \mathbf{r}_2(s)$$

$$\left\langle t^2 + 3, t + 1, \frac{6}{t} \right\rangle = \langle 4s, 2s - 2, s^2 - 7 \rangle$$

Equating coordinates we get:

$$\begin{cases} t^2 + 3 = 4s \\ t + 1 = 2s - 2 \\ \dfrac{6}{t} = s^2 - 7 \end{cases}$$

By the second equation, $t = 2s - 3$. Substituting into the first equation yields:

$$(2s - 3)^2 + 3 = 4s$$

$$4s^2 - 12s + 9 + 3 = 4s$$

$$s^2 - 4s + 3 = 0 \quad \rightarrow \quad s_1 = 1, \quad s_2 = 3$$

Substituting $s_1 = 1$ and $s_2 = 3$ into the second equation gives:

$$t_1 + 1 = 2 \cdot 1 - 2 \quad \Rightarrow \quad t_1 = -1$$

$$t_2 + 1 = 2 \cdot 3 - 2 \quad \Rightarrow \quad t_2 = 3$$

The solutions of the first two equations are:

$$t_1 = -1, \quad s_1 = 1; \qquad t_2 = 3, \quad s_2 = 3$$

We check if these solutions satisfy the third equation:

$$\frac{6}{t_1} = \frac{6}{-1} = -6, \quad s_1^2 - 7 = 1^2 - 7 = -6 \quad \Rightarrow \quad \frac{6}{t_1} = s_1^2 - 7$$

$$\frac{6}{t_2} = \frac{6}{3} = 2, \quad s_2^2 - 7 = 3^2 - 7 = 2 \quad \Rightarrow \quad \frac{6}{t_2} = s_2^2 - 7$$

We conclude that the paths intersect at the endpoints of the vectors $\mathbf{r}_1(-1)$ and $\mathbf{r}_1(3)$ (or equivalently $\mathbf{r}_2(1)$ and $\mathbf{r}_2(3)$). That is, at the points $(4, 0, -6)$ and $(12, 4, 2)$.

In Exercises 31–40, find a parametrization of the curve.

31. The vertical line passing through the point $(3, 2, 0)$

SOLUTION The points of the vertical line passing through the point $(3, 2, 0)$ can be written as $(3, 2, z)$. Using $z = t$ as parameter we get the following parametrization:

$$\mathbf{r}(t) = \langle 3, 2, t \rangle, \quad -\infty < t < \infty$$

33. The line through the origin whose projection on the xy-plane is a line of slope 3 and whose projection on the yz-plane is a line of slope 5 (i.e., $\Delta z / \Delta y = 5$)

SOLUTION   We denote by (x, y, z) the points on the line. The projection of the line on the xy-plane is the line through the origin having slope 3, that is the line $y = 3x$ in the xy-plane. The projection of the line on the yz-plane is the line through the origin with slope 5, that is the line $z = 5y$. Thus, the points on the desired line satisfy the following equalities:

$$\begin{matrix} y = 3x \\ z = 5y \end{matrix} \quad \Rightarrow \quad y = 3x, \; z = 5 \cdot 3x = 15x$$

We conclude that the points on the line are all the points in the form $(x, 3x, 15x)$. Using $x = t$ as parameter we obtain the following parametrization:

$$\mathbf{r}(t) = \langle t, 3t, 15t \rangle, \quad -\infty < t < \infty.$$

35. The circle of radius 2 with center $(1, 2, 5)$ in a plane parallel to the yz-plane

SOLUTION   The circle is parallel to the yz-plane and centered at $(1, 2, 5)$, hence the x-coordinates of the points on the circle are $x = 1$. The projection of the circle on the yz-plane is a circle of radius 2 centered at $(2, 5)$. This circle is parametrized by:

$$y = 2 + 2\cos t, \quad z = 5 + 2\sin t$$

We conclude that the points on the required circle can be written as $(1, 2 + 2\cos t, 5 + 2\sin t)$. This gives the following parametrization:

$$\mathbf{r}(t) = \langle 1, 2 + 2\cos t, 5 + 2\sin t \rangle, \quad 0 \le t \le 2\pi.$$

37. The intersection of the plane $y = \frac{1}{2}$ with the sphere $x^2 + y^2 + z^2 = 1$

SOLUTION   Substituting $y = \frac{1}{2}$ in the equation of the sphere gives:

$$x^2 + \left(\frac{1}{2}\right)^2 + z^2 = 1 \quad \Rightarrow \quad x^2 + z^2 = \frac{3}{4}$$

This circle in the horizontal plane $y = \frac{1}{2}$ has the parametrization $x = \frac{\sqrt{3}}{2}\cos t, z = \frac{\sqrt{3}}{2}\sin t$. Therefore, the points on the intersection of the plane $y = \frac{1}{2}$ and the sphere $x^2 + y^2 + z^2 = 1$, can be written in the form $\left(\frac{\sqrt{3}}{2}\cos t, \frac{1}{2}, \frac{\sqrt{3}}{2}\sin t\right)$, yielding the following parametrization:

$$\mathbf{r}(t) = \left\langle \frac{\sqrt{3}}{2}\cos t, \frac{1}{2}, \frac{\sqrt{3}}{2}\sin t \right\rangle, \quad 0 \le t \le 2\pi.$$

39. The ellipse $\left(\frac{x}{2}\right)^2 + \left(\frac{z}{3}\right)^2 = 1$ in the xz-plane, translated to have center $(3, 1, 5)$ [Figure 15(A)]

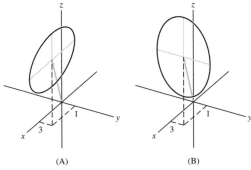

(A)   (B)

FIGURE 15   The ellipses described in Exercises 39 and 40.

SOLUTION   The translated ellipse is in the vertical plane $y = 1$, hence the y-coordinate of the points on this ellipse is $y = 1$. The x and z coordinates satisfy the equation of the ellipse:

$$\left(\frac{x - 3}{2}\right)^2 + \left(\frac{z - 5}{3}\right)^2 = 1.$$

This ellipse is parametrized by the following equations:

$$x = 3 + 2\cos t, \quad z = 5 + 3\sin t.$$

Therefore, the points on the translated ellipse can be written as $(3 + 2\cos t, 1, 5 + 3\sin t)$. This gives the following parametrization:

$$\mathbf{r}(t) = \langle 3 + 2\cos t, 1, 5 + 3\sin t \rangle, \quad 0 \le t \le 2\pi.$$

Further Insights and Challenges

41. Sketch the curve parametrized by $\mathbf{r}(t) = \langle |t| + t, |t| - t \rangle$.

SOLUTION We have:

$$|t| + t = \begin{cases} 0 & t \le 0 \\ 2t & t > 0 \end{cases}; \qquad |t| - t = \begin{cases} 2t & t \le 0 \\ 0 & t > 0 \end{cases}$$

As t increases from $-\infty$ to 0, the x-coordinate is zero and the y-coordinate is positive and decreasing to zero. As t increases from 0 to $+\infty$, the y-coordinate is zero and the x-coordinate is positive and increasing to $+\infty$. We obtain the following curve:

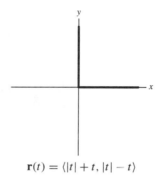

$$\mathbf{r}(t) = \langle |t| + t, |t| - t \rangle$$

43. Let C be the curve obtained by intersecting a cylinder of radius r and a plane. Insert two spheres of radius r into the cylinder above and below the plane, and let F_1 and F_2 be the points where the plane is tangent to the spheres [Figure 16(A)]. Let K be the vertical distance between the equators of the two spheres. Rediscover Archimedes's proof that C is an ellipse by showing that every point P on C satisfies

$$PF_1 + PF_2 = K \qquad \boxed{2}$$

Hint: If two lines through a point P are tangent to a sphere and intersect the sphere at Q_1 and Q_2 as in Figure 16(B), then the segments $\overline{PQ_1}$ and $\overline{PQ_2}$ have equal length. Use this to show that $PF_1 = PR_1$ and $PF_2 = PR_2$.

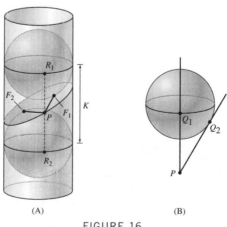

(A) (B)

FIGURE 16

SOLUTION To show that C is an ellipse, we show that every point P on C satisfies:

$$\overline{F_1 P} + \overline{F_2 P} = K$$

We denote the points of intersection of the vertical line through P with the equators of the two spheres by R_1 and R_2 (see figure).

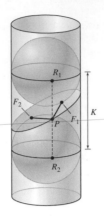

We denote by O_1 and O_2 the centers of the spheres.

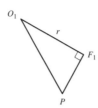

Since F_1 is the tangency point, the radius $\overline{O_1 F_1}$ is perpendicular to the plane of the curve $\mathcal{C}$, and therefore it is orthogonal to the segment $\overline{PF_1}$ on this plane. Hence, $\triangle O_1 F_1 P$ is a right triangle and by Pythagoras' Theorem we have:

$$\overline{O_1 F_1}^2 + \overline{PF_1}^2 = \overline{O_1 P}^2$$

$$r^2 + \overline{PF_1}^2 = \overline{O_1 P}^2 \quad \Rightarrow \quad \overline{PF_1} = \sqrt{\overline{O_1 P}^2 - r^2} \tag{1}$$

$\triangle O_1 R_1 P$ is also a right triangle, hence by Pythagoras' Theorem we have:

$$\overline{O_1 R_1}^2 + \overline{R_1 P}^2 = \overline{O_1 P}^2$$

$$r^2 + \overline{R_1 P}^2 = \overline{O_1 P}^2 \quad \Rightarrow \quad \overline{PR_1} = \sqrt{\overline{O_1 P}^2 - r^2} \tag{2}$$

Combining (1) and (2) we get:

$$\overline{PF_1} = \overline{PR_1} \tag{3}$$

Similarly we have:

$$\overline{PF_2} = \overline{PR_2} \tag{4}$$

We now combine (3), (4) and the equality $\overline{PR_1} + \overline{PR_2} = K$ to obtain:

$$\overline{F_1 P} + \overline{F_2 P} = \overline{PR_1} + \overline{PR_2} = K$$

Thus, the sum of the distances of the points P on $\mathcal{C}$ to the two fixed points F_1 and F_2 is a constant $K > 0$, hence $\mathcal{C}$ is an ellipse.

45. $\boxed{CAS}$ Now reprove the result of Exercise 43 using vector geometry. Assume that the cylinder has equation $x^2 + y^2 = r^2$ and the plane has equation $z = ax + by$.

(a) Show that the upper and lower spheres in Figure 16 have centers

$$C_1 = \left(0, 0, r\sqrt{a^2 + b^2 + 1}\right)$$

$$C_2 = \left(0, 0, -r\sqrt{a^2 + b^2 + 1}\right)$$

(b) Show that the points where the plane is tangent to the sphere are

$$F_1 = \frac{r}{\sqrt{a^2 + b^2 + 1}} \left(a, b, a^2 + b^2\right)$$

$$F_2 = \frac{-r}{\sqrt{a^2 + b^2 + 1}} \left(a, b, a^2 + b^2\right)$$

Hint: Show that $\overline{C_1 F_1}$ and $\overline{C_2 F_2}$ have length r and are orthogonal to the plane.
(c) Verify, with the aid of a computer algebra system, that Eq. (2) holds with

$$K = 2r\sqrt{a^2 + b^2 + 1}$$

To simplify the algebra, observe that since a and b are arbitrary, it suffices to verify Eq. (2) for the point $P = (r, 0, ar)$.

SOLUTION

(a) and (b) Since F_1 is the tangency point of the sphere and the plane, the radius to F_1 is orthogonal to the plane. Therefore to show that the center of the sphere is at C_1 and the tangency point is the given point we must show that:

$$\|\overrightarrow{C_1 F_1}\| = r \tag{1}$$

$$\overrightarrow{C_1 F_1} \text{ is orthogonal to the plane.} \tag{2}$$

We compute the vector $\overrightarrow{C_1 F_1}$:

$$\overrightarrow{C_1 F_1} = \left\langle \frac{ra}{\sqrt{a^2 + b^2 + 1}}, \frac{rb}{\sqrt{a^2 + b^2 + 1}}, \frac{r(a^2 + b^2)}{\sqrt{a^2 + b^2 + 1}} - r\sqrt{a^2 + b^2 + 1} \right\rangle = \frac{r}{\sqrt{a^2 + b^2 + 1}} \langle a, b, -1 \rangle$$

Hence,

$$\|\overrightarrow{C_1 F_1}\| = \frac{r}{\sqrt{a^2 + b^2 + 1}} \| \langle a, b, -1 \rangle \| = \frac{r}{\sqrt{a^2 + b^2 + 1}} \sqrt{a^2 + b^2 + (-1)^2} = r$$

We, thus, proved that (1) is satisfied. To show (2) we must show that $\overrightarrow{C_1 F_1}$ is parallel to the normal vector $\langle a, b, -1 \rangle$ to the plane $z = ax + by$ (i.e., $ax + by - z = 0$). The two vectors are parallel since by (1) $\overrightarrow{C_1 F_1}$ is a constant multiple of $\langle a, b, -1 \rangle$. In a similar manner one can show (1) and (2) for the vector $\overrightarrow{C_2 F_2}$.

(c) This is an extremely challenging problem. As suggested in the book, we use $P = (r, 0, ar)$, and we also use the expressions for F_1 and F_2 as given above. This gives us:

$$P F_1 = \sqrt{\left(1 + 2a^2 + b^2 - 2a\sqrt{1 + a^2 + b^2}\right) r^2}$$

$$P F_2 = \sqrt{\left(1 + 2a^2 + b^2 + 2a\sqrt{1 + a^2 + b^2}\right) r^2}$$

Their sum is not very inspiring:

$$P F_1 + P F_2 = \sqrt{\left(1 + 2a^2 + b^2 - 2a\sqrt{1 + a^2 + b^2}\right) r^2} + \sqrt{\left(1 + 2a^2 + b^2 + 2a\sqrt{1 + a^2 + b^2}\right) r^2}$$

Let us look, instead, at $(PF_1 + PF_2)^2$, and show that this is equal to K^2. Since everything is positive, this will imply that $PF_1 + PF_2 = K$, as desired.

$$(PF_1 + PF_2)^2 = 2r^2 + 4a^2 r^2 + 2b^2 r^2 + 2\sqrt{r^4 + 2b^2 r^4 + b^4 r^4}$$

$$= 2r^2 + 4a^2 r^2 + 2b^2 r^2 + 2(1 + b^2)r^2 = 4r^2(1 + a^2 + b^2) = K^2$$

13.2 Calculus of Vector-Valued Functions (LT Section 14.2)

Preliminary Questions

1. State the three forms of the Product Rule for vector-valued functions.

SOLUTION The Product Rule for scalar multiple $f(t)$ of a vector-valued function $\mathbf{r}(t)$ states that:

$$\frac{d}{dt} f(t)\mathbf{r}(t) = f(t)\mathbf{r}'(t) + f'(t)\mathbf{r}(t)$$

The Product Rule for dot products states that:

$$\frac{d}{dt} \mathbf{r}_1(t) \cdot \mathbf{r}_2(t) = \mathbf{r}_1(t) \cdot \mathbf{r}_2'(t) + \mathbf{r}_1'(t) \cdot \mathbf{r}_2(t)$$

Finally, the Product Rule for cross product is

$$\frac{d}{dt} \mathbf{r}_1(t) \times \mathbf{r}_2(t) = \mathbf{r}_1(t) \times \mathbf{r}_2'(t) + \mathbf{r}_1'(t) \times \mathbf{r}_2(t).$$

In Questions 2–6, indicate whether the statement is true or false, and if it is false, provide a correct statement.

2. The derivative of a vector-valued function is defined as the limit of the difference quotient, just as in the scalar-valued case.

SOLUTION   The statement is true. The derivative of a vector-valued function $\mathbf{r}(t)$ is defined a limit of the difference quotient:

$$\mathbf{r}'(t) = \lim_{t \to 0} \frac{\mathbf{r}(t+h) - \mathbf{r}(t)}{h}$$

in the same way as in the scalar-valued case.

3. There are two Chain Rules for vector-valued functions: one for the composite of two vector-valued functions and one for the composite of a vector-valued and a scalar-valued function.

SOLUTION   This statement is false. A vector-valued function $\mathbf{r}(t)$ is a function whose domain is a set of real numbers and whose range consists of position vectors. Therefore, if $\mathbf{r}_1(t)$ and $\mathbf{r}_2(t)$ are vector-valued functions, the composition "$(\mathbf{r}_1 \cdot \mathbf{r}_2)(t) = \mathbf{r}_1(\mathbf{r}_2(t))$" has no meaning since $\mathbf{r}_2(t)$ is a vector and not a real number. However, for a scalar-valued function $f(t)$, the composition $\mathbf{r}(f(t))$ has a meaning, and there is a Chain Rule for differentiability of this vector-valued function.

4. The terms "velocity vector" and "tangent vector" for a path $\mathbf{r}(t)$ mean one and the same thing.

SOLUTION   This statement is true.

5. The derivative of a vector-valued function is the slope of the tangent line, just as in the scalar case.

SOLUTION   The statement is false. The derivative of a vector-valued function is again a vector-valued function, hence it cannot be the slope of the tangent line (which is a scalar). However, the derivative, $\mathbf{r}'(t_0)$ is the direction vector of the tangent line to the curve traced by $\mathbf{r}(t)$, at $\mathbf{r}(t_0)$.

6. The derivative of the cross product is the cross product of the derivatives.

SOLUTION   The statement is false, since usually,

$$\frac{d}{dt} \mathbf{r}_1(t) \times \mathbf{r}_2(t) \neq \mathbf{r}_1'(t) \times \mathbf{r}_2'(t)$$

The correct statement is the Product Rule for Cross Products. That is,

$$\frac{d}{dt} \mathbf{r}_1(t) \times \mathbf{r}_2(t) = \mathbf{r}_1(t) \times \mathbf{r}_2'(t) + \mathbf{r}_1'(t) \times \mathbf{r}_2(t)$$

7. State whether the following derivatives of vector-valued functions $\mathbf{r}_1(t)$ and $\mathbf{r}_2(t)$ are scalars or vectors:

(a) $\dfrac{d}{dt} \mathbf{r}_1(t)$ **(b)** $\dfrac{d}{dt} \left(\mathbf{r}_1(t) \cdot \mathbf{r}_2(t) \right)$ **(c)** $\dfrac{d}{dt} \left(\mathbf{r}_1(t) \times \mathbf{r}_2(t) \right)$

SOLUTION   (a) vector, (b) scalar, (c) vector.

Exercises

In Exercises 1–6, evaluate the limit.

1. $\lim\limits_{t \to 3} \left\langle t^2, 4t, \dfrac{1}{t} \right\rangle$

SOLUTION   By the theorem on vector-valued limits we have:

$$\lim_{t \to 3} \left\langle t^2, 4t, \frac{1}{t} \right\rangle = \left\langle \lim_{t \to 3} t^2, \lim_{t \to 3} 4t, \lim_{t \to 3} \frac{1}{t} \right\rangle = \left\langle 9, 12, \frac{1}{3} \right\rangle.$$

3. $\lim\limits_{t \to 0} e^{2t} \mathbf{i} + \ln(t+1) \mathbf{j} + 4\mathbf{k}$

SOLUTION   Computing the limit of each component, we obtain:

$$\lim_{t \to 0} \left(e^{2t} \mathbf{i} + \ln(t+1) \mathbf{j} + 4\mathbf{k} \right) = \left(\lim_{t \to 0} e^{2t} \right) \mathbf{i} + \left(\lim_{t \to 0} \ln(t+1) \right) \mathbf{j} + \left(\lim_{t \to 0} 4 \right) \mathbf{k} = e^0 \mathbf{i} + (\ln 1) \mathbf{j} + 4\mathbf{k} = \mathbf{i} + 4\mathbf{k}$$

5. Evaluate $\lim\limits_{h\to 0} \dfrac{\mathbf{r}(t+h)-\mathbf{r}(t)}{h}$ for $\mathbf{r}(t) = \left\langle t^{-1}, \sin t, 4\right\rangle$.

SOLUTION This limit is the derivative $\frac{d\mathbf{r}}{dt}$. Using componentwise differentiation yields:

$$\lim_{h\to 0} \frac{\mathbf{r}(t+h)-\mathbf{r}(t)}{h} = \frac{d\mathbf{r}}{dt} = \left\langle \frac{d}{dt}\left(t^{-1}\right), \frac{d}{dt}\left(\sin t\right), \frac{d}{dt}(4)\right\rangle = \left\langle -\frac{1}{t^2}, \cos t, 0\right\rangle.$$

In Exercises 7–12, compute the derivative.

7. $\mathbf{r}(t) = \left\langle t, t^2, t^3 \right\rangle$

SOLUTION Using componentwise differentiation we get:

$$\frac{d\mathbf{r}}{dt} = \left\langle \frac{d}{dt}(t), \frac{d}{dt}(t^2), \frac{d}{dt}(t^3)\right\rangle = \left\langle 1, 2t, 3t^2 \right\rangle$$

9. $\mathbf{r}(s) = \left\langle e^{3s}, e^{-s}, s^4 \right\rangle$

SOLUTION Using componentwise differentiation we get:

$$\frac{d\mathbf{r}}{ds} = \left\langle \frac{d}{ds}(e^{3s}), \frac{d}{ds}(e^{-s}), \frac{d}{ds}(s^4)\right\rangle = \left\langle 3e^{3s}, -e^{-s}, 4s^3 \right\rangle$$

11. $\mathbf{c}(t) = t^{-1}\mathbf{i} - e^{2t}\mathbf{k}$

SOLUTION Using componentwise differentiation we get:

$$\mathbf{c}'(t) = \left(t^{-1}\right)'\mathbf{i} - \left(e^{2t}\right)'\mathbf{k} = -t^{-2}\mathbf{i} - 2e^{2t}\mathbf{k}$$

13. Calculate $\mathbf{r}'(t)$ and $\mathbf{r}''(t)$ for $\mathbf{r}(t) = \left\langle t, t^2, t^3 \right\rangle$.

SOLUTION We perform the differentiation componentwise to obtain:

$$\mathbf{r}'(t) = \left\langle (t)', (t^2)', (t^3)' \right\rangle = \left\langle 1, 2t, 3t^2 \right\rangle$$

We now differentiate the derivative vector to find the second derivative:

$$\mathbf{r}''(t) = \frac{d}{dt}\left\langle 1, 2t, 3t^2 \right\rangle = \left\langle 0, 2, 6t \right\rangle.$$

15. Sketch the curve $\mathbf{r}_1(t) = \left\langle t, t^2 \right\rangle$ together with its tangent vector at $t = 1$. Then do the same for $\mathbf{r}_2(t) = \left\langle t^3, t^6 \right\rangle$.

SOLUTION Note that $\mathbf{r}_1'(t) = \langle 1, 2t \rangle$ and so $\mathbf{r}_1'(1) = \langle 1, 2 \rangle$. The graph of $\mathbf{r}_1(t)$ satisfies $y = x^2$. Likewise, $\mathbf{r}_2'(t) = \left\langle 3t^2, 6t^5 \right\rangle$ and so $\mathbf{r}_2'(1) = \langle 3, 6 \rangle$. The graph of $\mathbf{r}_2(t)$ also satisfies $y = x^2$. Both graphs and tangent vectors are given here.

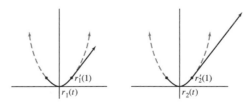

In Exercises 17–20, evaluate the derivative by using the appropriate Product Rule, where

$$\mathbf{r}_1(t) = \left\langle t^2, t^3, t \right\rangle, \qquad \mathbf{r}_2(t) = \left\langle e^{3t}, e^{2t}, e^t \right\rangle$$

17. $\dfrac{d}{dt}\left(\mathbf{r}_1(t)\cdot\mathbf{r}_2(t)\right)$

SOLUTION

$$\frac{d}{dt}(\mathbf{r}_1(t)\cdot\mathbf{r}_2(t)) = \mathbf{r}_1(t)\cdot\mathbf{r}_2'(t) + \mathbf{r}_1'(t)\cdot\mathbf{r}_2(t)$$

$$= \left\langle t^2, t^3, t \right\rangle\cdot\left\langle 3e^{3t}, 2e^{2t}, e^t \right\rangle + \left\langle 2t, 3t^2, 1 \right\rangle\cdot\left\langle e^{3t}, e^{2t}, e^t \right\rangle$$

$$= 3t^2 e^{3t} + 2t^3 e^{2t} + te^t + 2te^{3t} + 3t^2 e^{2t} + e^t$$

$$= (3t^2 + 2t)e^{3t} + (2t^3 + 3t^2)e^{2t} + (t + 1)e^t$$

19. $\dfrac{d}{dt}\big(\mathbf{r}_1(t) \times \mathbf{r}_2(t)\big)$

SOLUTION

$$\frac{d}{dt}(\mathbf{r}_1(t) \times \mathbf{r}_2(t)) = \mathbf{r}_1(t) \times \mathbf{r}_2'(t) + \mathbf{r}_1'(t) \times \mathbf{r}_2(t)$$

$$= \left\langle t^2, t^3, t \right\rangle \times \left\langle 3e^{3t}, 2e^{2t}, e^t \right\rangle + \left\langle 2t, 3t^2, 1 \right\rangle \times \left\langle e^{3t}, e^{2t}, e^t \right\rangle$$

$$= \begin{vmatrix} \mathbf{i} & \mathbf{j} & \mathbf{k} \\ t^2 & t^3 & t \\ 3e^{3t} & 2e^{2t} & e^t \end{vmatrix} + \begin{vmatrix} \mathbf{i} & \mathbf{j} & \mathbf{k} \\ 2t & 3t^2 & 1 \\ e^{3t} & e^{2t} & e^t \end{vmatrix}$$

$$= (t^3 e^t - 2t e^{2t})\mathbf{i} + (3t e^{3t} - t^2 e^t)\mathbf{j} + (2t^2 e^{2t} - 3t^3 e^{3t})\mathbf{k}$$

$$\quad + (3t^2 e^t - e^{2t})\mathbf{i} + (e^{3t} - 2t e^t)\mathbf{j} + (2t e^{2t} - 3t^2 e^{3t})\mathbf{k}$$

$$= [(t^3 + 3t^2)e^t - (2t + 1)e^{2t}]\mathbf{i} + [(3t + 1)e^{3t} - (t^2 + 2t)e^t]\mathbf{j}$$

$$\quad + [(2t^2 + 2t)e^{2t} - (3t^3 + 3t^2)e^{3t}]\mathbf{k}$$

In Exercises 21 and 22, let

$$\mathbf{r}_1(t) = \left\langle t^2, 1, 2t \right\rangle, \qquad \mathbf{r}_2(t) = \left\langle 1, 2, e^t \right\rangle$$

21. Compute $\dfrac{d}{dt}\mathbf{r}_1(t) \cdot \mathbf{r}_2(t)\Big|_{t=1}$ in two ways:

(a) Calculate $\mathbf{r}_1(t) \cdot \mathbf{r}_2(t)$ and differentiate.

(b) Use the Product Rule.

SOLUTION

(a) First we will calculate $\mathbf{r}_1(t) \cdot \mathbf{r}_2(t)$:

$$\mathbf{r}_1(t) \cdot \mathbf{r}_2(t) = \left\langle t^2, 1, 2t \right\rangle \cdot \left\langle 1, 2, e^t \right\rangle$$

$$= t^2 + 2 + 2t e^t$$

And then differentiating we get:

$$\frac{d}{dt}(\mathbf{r}_1(t) \cdot \mathbf{r}_2(t)) = \frac{d}{dt}(t^2 + 2 + 2t e^t) = 2t + 2t e^t + 2e^t$$

$$\frac{d}{dt}(\mathbf{r}_1(t) \cdot \mathbf{r}_2(t))\Big|_{t=1} = 2 + 2e + 2e = 2 + 4e$$

(b) First we differentiate:

$$\mathbf{r}_1(t) = \left\langle t^2, 1, 2t \right\rangle, \qquad \mathbf{r}_1'(t) = \langle 2t, 0, 2 \rangle$$

$$\mathbf{r}_2(t) = \left\langle 1, 2, e^t \right\rangle, \qquad \mathbf{r}_2'(t) = \left\langle 0, 0, e^t \right\rangle$$

Using the Product Rule we see:

$$\frac{d}{dt}(\mathbf{r}_1(t) \cdot \mathbf{r}_2(t)) = \mathbf{r}_1(t) \cdot \mathbf{r}_2'(t) + \mathbf{r}_1'(t) \cdot \mathbf{r}_2(t)$$

$$= \left\langle t^2, 1, 2t \right\rangle \cdot \left\langle 0, 0, e^t \right\rangle + \langle 2t, 0, 2 \rangle \cdot \left\langle 1, 2, e^t \right\rangle$$

$$= 2t e^t + 2t + 2e^t$$

$$\frac{d}{dt}(\mathbf{r}_1(t) \cdot \mathbf{r}_2(t))\Big|_{t=1} = 2e + 2 + 2e = 2 + 4e$$

In Exercises 23–26, evaluate $\dfrac{d}{dt}\mathbf{r}(g(t))$ using the Chain Rule.

23. $\mathbf{r}(t) = \left\langle t^2, 1 - t \right\rangle$, $\quad g(t) = e^t$

SOLUTION We first differentiate the two functions:

$$\mathbf{r}'(t) = \frac{d}{dt}\left\langle t^2, 1 - t \right\rangle = \langle 2t, -1 \rangle$$

$$g'(t) = \frac{d}{dt}(e^t) = e^t$$

Using the Chain Rule we get:

$$\frac{d}{dt}\mathbf{r}(g(t)) = g'(t)\mathbf{r}'(g(t)) = e^t \langle 2e^t, -1\rangle = \langle 2e^{2t}, -e^t\rangle$$

25. $\mathbf{r}(t) = \langle e^t, e^{2t}, 4\rangle, \quad g(t) = 4t + 9$

SOLUTION We first differentiate the two functions:

$$\mathbf{r}'(t) = \frac{d}{dt}\langle e^t, e^{2t}, 4\rangle = \langle e^t, 2e^{2t}, 0\rangle$$

$$g'(t) = \frac{d}{dt}(4t + 9) = 4$$

Using the Chain Rule we get:

$$\frac{d}{dt}\mathbf{r}(g(t)) = g'(t)\mathbf{r}'(g(t)) = 4\langle e^{4t+9}, 2e^{2(4t+9)}, 0\rangle = \langle 4e^{4t+9}, 8e^{8t+18}, 0\rangle$$

27. Let $\mathbf{r}(t) = \langle t^2, 1 - t, 4t\rangle$. Calculate the derivative of $\mathbf{r}(t) \cdot \mathbf{a}(t)$ at $t = 2$, assuming that $\mathbf{a}(2) = \langle 1, 3, 3\rangle$ and $\mathbf{a}'(2) = \langle -1, 4, 1\rangle$.

SOLUTION By the Product Rule for dot products we have

$$\frac{d}{dt}\mathbf{r}(t) \cdot \mathbf{a}(t) = \mathbf{r}(t) \cdot \mathbf{a}'(t) + \mathbf{r}'(t) \cdot \mathbf{a}(t)$$

At $t = 2$ we have

$$\frac{d}{dt}\mathbf{r}(t) \cdot \mathbf{a}(t)\Big|_{t=2} = \mathbf{r}(2) \cdot \mathbf{a}'(2) + \mathbf{r}'(2) \cdot \mathbf{a}(2) \tag{1}$$

We compute the derivative $\mathbf{r}'(2)$:

$$\mathbf{r}'(t) = \frac{d}{dt}\langle t^2, 1 - t, 4t\rangle = \langle 2t, -1, 4\rangle \quad \Rightarrow \quad \mathbf{r}'(2) = \langle 4, -1, 4\rangle \tag{2}$$

Also, $\mathbf{r}(2) = \langle 2^2, 1 - 2, 4 \cdot 2\rangle = \langle 4, -1, 8\rangle$. Substituting the vectors in the equation above, we obtain:

$$\frac{d}{dt}\mathbf{r}(t) \cdot \mathbf{a}(t)\Big|_{t=2} = \langle 4, -1, 8\rangle \cdot \langle -1, 4, 1\rangle + \langle 4, -1, 4\rangle \cdot \langle 1, 3, 3\rangle = (-4 - 4 + 8) + (4 - 3 + 12) = 13$$

The derivative of $\mathbf{r}(t) \cdot \mathbf{a}(t)$ at $t = 2$ is 13.

In Exercises 29–34, find a parametrization of the tangent line at the point indicated.

29. $\mathbf{r}(t) = \langle t^2, t^4\rangle, \quad t = -2$

SOLUTION The tangent line has the following parametrization:

$$\ell(t) = \mathbf{r}(-2) + t\mathbf{r}'(-2) \tag{1}$$

We compute the vectors $\mathbf{r}(-2)$ and $\mathbf{r}'(-2)$:

$$\mathbf{r}(-2) = \langle (-2)^2, (-2)^4\rangle = \langle 4, 16\rangle$$

$$\mathbf{r}'(t) = \frac{d}{dt}\langle t^2, t^4\rangle = \langle 2t, 4t^3\rangle \quad \Rightarrow \quad \mathbf{r}'(-2) = \langle -4, -32\rangle$$

Substituting in (1) gives:

$$\ell(t) = \langle 4, 16\rangle + t \langle -4, -32\rangle = \langle 4 - 4t, 16 - 32t\rangle$$

The parametrization for the tangent line is, thus,

$$x = 4 - 4t, \quad y = 16 - 32t, \quad -\infty < t < \infty.$$

To find a direct relation between y and x, we express t in terms of x and substitute in $y = 16 - 32t$. This gives:

$$x = 4 - 4t \Rightarrow t = \frac{x - 4}{-4}.$$

Hence,

$$y = 16 - 32t = 16 - 32 \cdot \frac{x - 4}{-4} = 16 + 8(x - 4) = 8x - 16.$$

The equation of the tangent line is $y = 8x - 16$.

31. $\mathbf{r}(t) = \langle 1 - t^2, 5t, 2t^3 \rangle$, $t = 2$

SOLUTION The tangent line is parametrized by:

$$\ell(t) = \mathbf{r}(2) + t\mathbf{r}'(2) \tag{1}$$

We compute the vectors in the above parametrization:

$$\mathbf{r}(2) = \langle 1 - 2^2, 5 \cdot 2, 2 \cdot 2^3 \rangle = \langle -3, 10, 16 \rangle$$

$$\mathbf{r}'(t) = \frac{d}{dt}\langle 1 - t^2, 5t, 2t^3 \rangle = \langle -2t, 5, 6t^2 \rangle \quad \Rightarrow \quad \mathbf{r}'(2) = \langle -4, 5, 24 \rangle$$

Substituting the vectors in (1) we obtain the following parametrization:

$$\ell(t) = \langle -3, 10, 16 \rangle + t \langle -4, 5, 24 \rangle = \langle -3 - 4t, 10 + 5t, 16 + 24t \rangle$$

33. $\mathbf{r}(s) = 4s^{-1}\mathbf{i} - \frac{8}{3}s^{-3}\mathbf{k}$, $s = 2$

SOLUTION The tangent line is parametrized by:

$$\ell(s) = \mathbf{r}(2) + s\mathbf{r}'(2) \tag{1}$$

We compute the vectors in the above parametrization:

$$\mathbf{r}(2) = 4(2)^{-1}\mathbf{i} - \frac{8}{3}(2)^{-3}\mathbf{k} = 2\mathbf{i} - \frac{1}{3}\mathbf{k}$$

$$\mathbf{r}'(s) = \frac{d}{ds}\left(4s^{-1}\mathbf{i} - \frac{8}{3}s^{-3}\mathbf{k}\right) = -4s^{-2}\mathbf{i} + 8s^{-4}\mathbf{k} \quad \Rightarrow \quad \mathbf{r}'(2) = -\mathbf{i} + \frac{1}{2}\mathbf{k}$$

Substituting the vectors in (1) we obtain the following parametrization:

$$\ell(t) = \left(2\mathbf{i} - \frac{1}{3}\mathbf{k}\right) + s\left(-\mathbf{i} + \frac{1}{2}\mathbf{k}\right) = (2 - s)\mathbf{i} + \left(\frac{1}{2}s - \frac{1}{3}\right)\mathbf{k}$$

35. Use Example 4 to calculate $\frac{d}{dt}(\mathbf{r} \times \mathbf{r}')$, where $\mathbf{r}(t) = \langle t, t^2, e^t \rangle$.

SOLUTION In Example 4 it is proved that:

$$\frac{d}{dt}\mathbf{r} \times \mathbf{r}' = \mathbf{r} \times \mathbf{r}'' \tag{1}$$

We compute the derivatives $\mathbf{r}'(t)$ and $\mathbf{r}''(t)$:

$$\mathbf{r}'(t) = \frac{d}{dt}\langle t, t^2, e^t \rangle = \langle 1, 2t, e^t \rangle$$

$$\mathbf{r}''(t) = \frac{d}{dt}\langle 1, 2t, e^t \rangle = \langle 0, 2, e^t \rangle$$

Using (1) we get

$$\frac{d}{dt}\mathbf{r} \times \mathbf{r}' = \mathbf{r} \times \mathbf{r}'' = \langle t, t^2, e^t \rangle \times \langle 0, 2, e^t \rangle = \begin{vmatrix} \mathbf{i} & \mathbf{j} & \mathbf{k} \\ t & t^2 & e^t \\ 0 & 2 & e^t \end{vmatrix} = (t^2 e^t - 2e^t)\mathbf{i} - (0 - te^t)\mathbf{j} + (2t - 0)\mathbf{k}$$

$$= (t^2 - 2)e^t\mathbf{i} + te^t\mathbf{j} + 2t\mathbf{k} = \langle (t^2 - 2t)e^t, te^t, 2t \rangle$$

37. Show that the *derivative of the norm* is not equal to the *norm of the derivative* by verifying that $\|\mathbf{r}(t)\|' \neq \|\mathbf{r}'(t)\|$ for $\mathbf{r}(t) = \langle t, 1, 1 \rangle$.

SOLUTION First let us compute $\|\mathbf{r}(t)\|'$ for $\mathbf{r}(t) = \langle t, 1, 1 \rangle$:

$$\|\mathbf{r}(t)\|' = \frac{d}{dt}(\sqrt{t^2 + 2}) = \frac{t}{\sqrt{t^2 + 2}}$$

Now, first let us compute the derivative, $\mathbf{r}'(t)$:

$$\mathbf{r}'(t) = \langle 1, 0, 0 \rangle$$

and then computing the norm:

$$\|\mathbf{r}'(t)\| = \|\langle 1, 0, 0 \rangle\| = \sqrt{1} = 1$$

It is clear in this example, that $\|\mathbf{r}(t)\|' \neq \|\mathbf{r}'(t)\|$.

In Exercises 39–46, evaluate the integrals.

39. $\int_{-1}^{3} \left\langle 8t^2 - t, 6t^3 + t \right\rangle dt$

SOLUTION Vector-valued integration is defined via componentwise integration. Thus, we first compute the integral of each component.

$$\int_{-1}^{3} 8t^2 - t \, dt = \frac{8}{3}t^3 - \frac{t^2}{2}\Big|_{-1}^{3} = \left(72 - \frac{9}{2}\right) - \left(-\frac{8}{3} - \frac{1}{2}\right) = \frac{212}{3}$$

$$\int_{-1}^{3} 6t^3 + t \, dt = \frac{3}{2}t^4 + \frac{t^2}{2}\Big|_{-1}^{3} = \left(\frac{243}{2} + \frac{9}{2}\right) - \left(\frac{3}{2} + \frac{1}{2}\right) = 124$$

Therefore,

$$\int_{-1}^{3} \left\langle 8t^2 - t, 6t^3 + t \right\rangle dt = \left\langle \int_{-1}^{3} 8t^2 - t \, dt, \int_{-1}^{3} 6t^3 + t \, dt \right\rangle = \left\langle \frac{212}{3}, 124 \right\rangle$$

41. $\int_{-2}^{2} \left(u^3 \mathbf{i} + u^5 \mathbf{j} \right) du$

SOLUTION The vector-valued integration is defined via componentwise integration. Thus, we first compute the integral of each component.

$$\int_{-2}^{2} u^3 \, du = \frac{u^4}{4}\Big|_{-2}^{2} = \frac{16}{4} - \frac{16}{4} = 0$$

$$\int_{-2}^{2} u^5 \, du = \frac{u^6}{6}\Big|_{-2}^{2} = \frac{64}{6} - \frac{64}{6} = 0$$

Therefore,

$$\int_{-2}^{2} \left(u^3 \mathbf{i} + u^5 \mathbf{j} \right) du = \left(\int_{-2}^{2} u^3 \, du \right) \mathbf{i} + \left(\int_{2}^{2} u^5 \, du \right) \mathbf{j} = 0\mathbf{i} + 0\mathbf{j}$$

43. $\int_{0}^{1} \left\langle 2t, 4t, -\cos 3t \right\rangle dt$

SOLUTION The vector valued integration is defined via componentwise integration. Therefore,

$$\int_{0}^{1} \left\langle 2t, 4t, -\cos 3t \right\rangle dt = \left\langle \int_{0}^{1} 2t \, dt, \int_{0}^{1} 4t \, dt, \int_{0}^{1} -\cos 3t \, dt \right\rangle = \left\langle t^2\Big|_{0}^{1}, 2t^2\Big|_{0}^{1}, -\frac{\sin 3t}{3}\Big|_{0}^{1} \right\rangle = \left\langle 1, 2, -\frac{\sin 3}{3} \right\rangle$$

45. $\int_{1}^{4} \left(t^{-1}\mathbf{i} + 4\sqrt{t}\,\mathbf{j} - 8t^{3/2}\mathbf{k} \right) dt$

SOLUTION We perform the integration componentwise. Computing the integral of each component we get:

$$\int_{1}^{4} t^{-1} \, dt = \ln t \Big|_{1}^{4} = \ln 4 - \ln 1 = \ln 4$$

$$\int_{1}^{4} 4\sqrt{t} \, dt = 4 \cdot \frac{2}{3}t^{3/2}\Big|_{1}^{4} = \frac{8}{3}\left(4^{3/2} - 1\right) = \frac{56}{3}$$

$$\int_{1}^{4} -8t^{3/2} \, dt = -\frac{16}{5}t^{5/2}\Big|_{1}^{4} = -\frac{16}{5}\left(4^{5/2} - 1\right) = -\frac{496}{5}$$

Hence,

$$\int_{1}^{4} \left(t^{-1}\mathbf{i} + 4\sqrt{t}\mathbf{j} - 8t^{3/2}\mathbf{k} \right) dt = (\ln 4)\,\mathbf{i} + \frac{56}{3}\mathbf{j} - \frac{496}{5}\mathbf{k}$$

In Exercises 47–54, find both the general solution of the differential equation and the solution with the given initial condition.

47. $\dfrac{d\mathbf{r}}{dt} = \langle 1 - 2t, 4t \rangle, \quad \mathbf{r}(0) = \langle 3, 1 \rangle$

SOLUTION   We first find the general solution by integrating $\dfrac{d\mathbf{r}}{dt}$:

$$\mathbf{r}(t) = \int \langle 1 - 2t, 4t \rangle \, dt = \left\langle \int (1 - 2t) \, dt, \int 4t \, dt \right\rangle = \langle t - t^2, 2t^2 \rangle + \mathbf{c} \tag{1}$$

Since $\mathbf{r}(0) = \langle 3, 1 \rangle$, we have:

$$\mathbf{r}(0) = \langle 0 - 0^2, 2 \cdot 0^2 \rangle + \mathbf{c} = \langle 3, 1 \rangle \Rightarrow \mathbf{c} = \langle 3, 1 \rangle$$

Substituting in (1) gives the solution:

$$\mathbf{r}(t) = \langle t - t^2, 2t^2 \rangle + \langle 3, 1 \rangle = \langle -t^2 + t + 3, 2t^2 + 1 \rangle$$

49. $\mathbf{r}'(t) = t^2\mathbf{i} + 5t\mathbf{j} + \mathbf{k}, \quad \mathbf{r}(1) = \mathbf{j} + 2\mathbf{k}$

SOLUTION   We first find the general solution by integrating $\mathbf{r}'(t)$:

$$\mathbf{r}(t) = \int \left(t^2\mathbf{i} + 5t\mathbf{j} + \mathbf{k} \right) dt = \left(\int t^2 \, dt \right)\mathbf{i} + \left(\int 5t \, dt \right)\mathbf{j} + \left(\int 1 \, dt \right)\mathbf{k} = \left(\frac{1}{3}t^3 \right)\mathbf{i} + \left(\frac{5}{2}t^2 \right)\mathbf{j} + t\mathbf{k} + \mathbf{c} \tag{1}$$

The solution which satisfies the initial condition must satisfy:

$$\mathbf{r}(1) = \left(\frac{1}{3} \cdot 1^3 \right)\mathbf{i} + \left(\frac{5}{2} \cdot 1^2 \right)\mathbf{j} + 1 \cdot \mathbf{k} + \mathbf{c} = \mathbf{j} + 2\mathbf{k}$$

That is,

$$\mathbf{c} = -\frac{1}{3}\mathbf{i} - \frac{3}{2}\mathbf{j} + 1\mathbf{k}$$

Substituting in (1) gives the following solution:

$$\mathbf{r}(t) = \left(\frac{1}{3}t^3 \right)\mathbf{i} + \left(\frac{5}{2}t^2 \right)\mathbf{j} + t\mathbf{k} - \frac{1}{3}\mathbf{i} - \frac{3}{2}\mathbf{j} + \mathbf{k} = \left(\frac{1}{3}t^3 - \frac{1}{3} \right)\mathbf{i} + \left(\frac{5t^2}{2} - \frac{3}{2} \right)\mathbf{j} + (t + 1)\mathbf{k}$$

51. $\mathbf{r}''(t) = 16\mathbf{k}, \quad \mathbf{r}(0) = \langle 1, 0, 0 \rangle, \quad \mathbf{r}'(0) = \langle 0, 1, 0 \rangle$

SOLUTION   To find the general solution we first find $\mathbf{r}'(t)$ by integrating $\mathbf{r}''(t)$:

$$\mathbf{r}'(t) = \int \mathbf{r}''(t) \, dt = \int 16\mathbf{k} \, dt = (16t)\mathbf{k} + \mathbf{c}_1 \tag{1}$$

We now integrate $\mathbf{r}'(t)$ to find the general solution $\mathbf{r}(t)$:

$$\mathbf{r}(t) = \int \mathbf{r}'(t) \, dt = \int ((16t)\mathbf{k} + \mathbf{c}_1) \, dt = \left(\int 16(t) \, dt \right)\mathbf{k} + \mathbf{c}_1 t + \mathbf{c}_2 = (8t^2)\mathbf{k} + \mathbf{c}_1 t + \mathbf{c}_2 \tag{2}$$

We substitute the initial conditions in (1) and (2). This gives:

$$\mathbf{r}'(0) = \mathbf{c}_1 = \langle 0, 1, 0 \rangle = \mathbf{j}$$
$$\mathbf{r}(0) = 0\mathbf{k} + \mathbf{c}_1 \cdot 0 + \mathbf{c}_2 = \langle 1, 0, 0 \rangle \quad \Rightarrow \quad \mathbf{c}_2 = \langle 1, 0, 0 \rangle = \mathbf{i}$$

Combining with (2) we obtain the following solution:

$$\mathbf{r}(t) = (8t^2)\mathbf{k} + t\mathbf{j} + \mathbf{i} = \mathbf{i} + t\mathbf{j} + (8t^2)\mathbf{k}$$

53. $\mathbf{r}''(t) = \langle 0, 2, 0 \rangle, \mathbf{r}(3) = \langle 1, 1, 0 \rangle, \mathbf{r}'(3) = \langle 0, 0, 1 \rangle$

SOLUTION   To find the general solution we first find $\mathbf{r}'(t)$ by integrating $\mathbf{r}''(t)$:

$$\mathbf{r}'(t) = \int \mathbf{r}''(t) \, dt = \int \langle 0, 2, 0 \rangle \, dt = \langle 0, 2t, 0 \rangle + \mathbf{c}_1 \tag{1}$$

We now integrate $\mathbf{r}'(t)$ to find the general solution $\mathbf{r}(t)$:

$$\mathbf{r}(t) = \int \mathbf{r}'(t) \, dt = \int (\langle 0, 2t, 0 \rangle + \mathbf{c}_1) \, dt = \langle 0, t^2, 0 \rangle + \mathbf{c}_1 t + \mathbf{c}_2 \tag{2}$$

We substitute the initial conditions in (1) and (2). This gives:

$$\mathbf{r}'(3) = \langle 0, 6, 0 \rangle + \mathbf{c}_1 = \langle 0, 0, 1 \rangle \quad \Rightarrow \quad \mathbf{c}_1 = \langle 0, -6, 1 \rangle$$

$$\mathbf{r}(3) = \langle 0, 9, 0 \rangle + \mathbf{c}_1(3) + \mathbf{c}_2 = \langle 1, 1, 0 \rangle$$

$$\langle 0, 9, 0 \rangle + \langle 0, -18, 3 \rangle + \mathbf{c}_2 = \langle 1, 1, 0 \rangle$$

$$\Rightarrow \quad \mathbf{c}_2 = \langle 1, 10, -3 \rangle$$

Combining with (2) we obtain the following solution:

$$\mathbf{r}(t) = \left\langle 0, t^2, 0 \right\rangle + t \langle 0, -6, 1 \rangle + \langle 1, 10, -3 \rangle$$

$$= \left\langle 1, t^2 - 6t + 10, t - 3 \right\rangle$$

55. Find the location at $t = 3$ of a particle whose path (Figure 8) satisfies

$$\frac{d\mathbf{r}}{dt} = \left\langle 2t - \frac{1}{(t+1)^2}, 2t - 4 \right\rangle, \qquad \mathbf{r}(0) = \langle 3, 8 \rangle$$

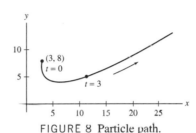

FIGURE 8 Particle path.

SOLUTION To determine the position of the particle in general, we perform integration componentwise on $\mathbf{r}'(t)$ to obtain:

$$\mathbf{r}(t) = \int \mathbf{r}'(t)\, dt$$

$$= \int \left\langle 2t - \frac{1}{(t+1)^2}, 2t - 4 \right\rangle dt$$

$$= \left\langle t^2 + \frac{1}{t+1}, t^2 - 4t \right\rangle + \mathbf{c}_1$$

Using the initial condition, observe the following:

$$\mathbf{r}(0) = \langle 1, 0 \rangle + \mathbf{c}_1 = \langle 3, 8 \rangle$$

$$\Rightarrow \quad \mathbf{c}_1 = \langle 2, 8 \rangle$$

Therefore,

$$\mathbf{r}(t) = \left\langle t^2 + \frac{1}{t+1}, t^2 - 4t \right\rangle + \langle 2, 8 \rangle = \left\langle t^2 + \frac{1}{t+1} + 2, t^2 - 4t + 8 \right\rangle$$

and thus, the location of the particle at $t = 3$ is $\mathbf{r}(3) = \langle 45/4, 5 \rangle = \langle 11.25, 5 \rangle$.

57. A fighter plane, which can shoot a laser beam straight ahead, travels along the path $\mathbf{r}(t) = \langle 5 - t, 21 - t^2, 3 - t^3/27 \rangle$. Show that there is precisely one time t at which the pilot can hit a target located at the origin.

SOLUTION By the given information the laser beam travels in the direction of $\mathbf{r}'(t)$. The pilot hits a target located at the origin at the time t when $\mathbf{r}'(t)$ points towards the origin, that is, when $\mathbf{r}(t)$ and $\mathbf{r}'(t)$ are parallel and point to opposite directions.

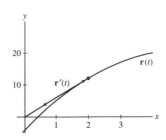

We find $\mathbf{r}'(t)$:

$$\mathbf{r}'(t) = \frac{d}{dt}\left\langle 5 - t, 21 - t^2, 3 - \frac{t^3}{27}\right\rangle = \left\langle -1, -2t, -\frac{t^2}{9}\right\rangle$$

We first find t such that $\mathbf{r}(t)$ and $\mathbf{r}'(t)$ are parallel, that is, we find t such that the cross product of the two vectors is zero. We obtain:

$$\mathbf{0} = \mathbf{r}'(t) \times \mathbf{r}(t) = \begin{vmatrix} \mathbf{i} & \mathbf{j} & \mathbf{k} \\ -1 & -2t & -\frac{t^2}{9} \\ 5 - t & 21 - t^2 & 3 - \frac{t^3}{27} \end{vmatrix}$$

$$= \left(-2t\left(3 - \frac{t^3}{27}\right) + \frac{t^2}{9}(21 - t^2)\right)\mathbf{i} - \left(-\left(3 - \frac{t^3}{27}\right) + \frac{t^2}{9}(5 - t)\right)\mathbf{j} + \left(-(21 - t^2) + 2t(5 - t)\right)\mathbf{k}$$

$$= \left(\frac{-t^4}{27} + \frac{7t^2}{3} - 6t\right)\mathbf{i} - \left(-\frac{2t^3}{27} + \frac{5t^2}{9} - 3\right)\mathbf{j} + (-t^2 + 10t - 21)\mathbf{k}$$

Equating each component to zero we obtain the following equations:

$$-\frac{t^4}{27} + \frac{7}{3}t^2 - 6t = 0$$

$$-\frac{2t^3}{27} + \frac{5t^2}{9} - 3 = 0$$

$$-t^2 + 10t - 21 = -(t - 7)(t - 3) = 0$$

The third equation implies that $t = 3$ or $t = 7$. Only $t = 3$ satisfies the other two equations as well. We now must verify that $\mathbf{r}(3)$ and $\mathbf{r}'(3)$ point in opposite directions. We find these vectors:

$$\mathbf{r}(3) = \left\langle 5 - 3, 21 - 3^2, 3 - \frac{3^3}{27}\right\rangle = \langle 2, 12, 2\rangle$$

$$\mathbf{r}'(3) = \left\langle -1, -2\cdot 3, -\frac{3^2}{9}\right\rangle = \langle -1, -6, -1\rangle$$

Since $\mathbf{r}(3) = -2\mathbf{r}'(3)$, the vectors point in opposite direction. We conclude that only at time $t = 3$ can the pilot hit a target located at the origin.

59. Find all solutions to $\mathbf{r}'(t) = \mathbf{v}$ with initial condition $\mathbf{r}(1) = \mathbf{w}$, where $\mathbf{v}$ and $\mathbf{w}$ are constant vectors in $\mathbf{R}^3$.

SOLUTION   We denote the components of the constant vector $\mathbf{v}$ by $\mathbf{v} = \langle v_1, v_2, v_3\rangle$ and integrate to find the general solution. This gives:

$$\mathbf{r}(t) = \int \mathbf{v}\, dt = \int \langle v_1, v_2, v_3\rangle\, dt = \left\langle \int v_1\, dt, \int v_2\, dt, \int v_3\, dt\right\rangle$$

$$= \langle v_1 t + c_1, v_2 t + c_2, v_3 t + c_3\rangle = t\langle v_1, v_2, v_3\rangle + \langle c_1, c_2, c_3\rangle$$

We let $\mathbf{c} = \langle c_1, c_2, c_3\rangle$ and obtain:

$$\mathbf{r}(t) = t\mathbf{v} + \mathbf{c} = \mathbf{c} + t\mathbf{v}$$

Notice that the solutions are the vector parametrizations of all the lines with direction vector $\mathbf{v}$.

We are also given the initial condition that $\mathbf{r}(1) = \mathbf{w}$, using this information we can determine:

$$\mathbf{r}(1) = (1)\mathbf{v} + \mathbf{c} = \mathbf{w}$$

Therefore $\mathbf{c} = \mathbf{w} - \mathbf{v}$ and we get:

$$\mathbf{r}(t) = (\mathbf{w} - \mathbf{v}) + t\mathbf{v} = (t - 1)\mathbf{v} + \mathbf{w}$$

61. Find all solutions to $\mathbf{r}'(t) = 2\mathbf{r}(t)$ where $\mathbf{r}(t)$ is a vector-valued function in three-space.

SOLUTION   We denote the components of $\mathbf{r}(t)$ by $\mathbf{r}(t) = \langle x(t), y(t), z(t)\rangle$. Then, $\mathbf{r}'(t) = \langle x'(t), y'(t), z'(t)\rangle$. Substituting in the differential equation we get:

$$\langle x'(t), y'(t), z'(t)\rangle = 2\langle x(t), y(t), z(t)\rangle$$

Equating corresponding components gives:

$$\begin{aligned} x'(t) &= 2x(t) & x(t) &= c_1 e^{2t} \\ y'(t) &= 2y(t) &\Rightarrow\quad y(t) &= c_2 e^{2t} \\ z'(t) &= 2z(t) & z(t) &= c_3 e^{2t} \end{aligned}$$

We denote the constant vector by $\mathbf{c} = \langle c_1, c_2, c_3 \rangle$ and obtain the following solutions:

$$\mathbf{r}(t) = \left\langle c_1 e^{2t}, c_2 e^{2t}, c_3 e^{2t} \right\rangle = e^{2t} \langle c_1, c_2, c_3 \rangle = e^{2t} \mathbf{c}$$

63. Prove that the **Bernoulli spiral** (Figure 9) with parametrization $\mathbf{r}(t) = \left\langle e^t \cos 4t, e^t \sin 4t \right\rangle$ has the property that the angle ψ between the position vector and the tangent vector is constant. Find the angle ψ in degrees.

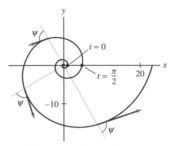

FIGURE 9 Bernoulli spiral.

SOLUTION First, let us compute the tangent vector, $\mathbf{r}'(t)$:

$$\mathbf{r}(t) = \left\langle e^t \cos 4t, e^t \sin 4t \right\rangle, \quad \Rightarrow \quad \mathbf{r}'(t) = \left\langle -4e^t \sin 4t + e^t \cos 4t, 4e^t \cos 4t + e^t \sin 4t \right\rangle$$

Then recall the identity that $\mathbf{a} \cdot \mathbf{b} = \|\mathbf{a}\| \cdot \|\mathbf{b}\| \cos\theta$, where θ is the angle between $\mathbf{a}$ and $\mathbf{b}$, so then,

$$\mathbf{r}(t) \cdot \mathbf{r}'(t) = \left\langle e^t \cos 4t, e^t \sin 4t \right\rangle \cdot \left\langle -4e^t \sin 4t + e^t \cos 4t, 4e^t \cos 4t + e^t \sin 4t \right\rangle$$

$$= -4e^{2t} \sin 4t \cos 4t + e^{2t} \cos^2 4t + 4e^{2t} \sin 4t \cos 4t + e^{2t} \sin^2 4t$$

$$= e^{2t}(\cos^2 4t + \sin^2 4t)$$

$$= e^{2t}$$

Then, computing norms, we get:

$$\|\mathbf{r}(t)\| = \sqrt{e^{2t} \cos^2 4t + e^{2t} \sin^2 4t} = \sqrt{e^{2t}(\cos^2 4t + \sin^2 4t)} = e^t$$

$$\|\mathbf{r}'(t)\| = \sqrt{(-4e^t \sin 4t + e^t \cos 4t)^2 + (4e^t \cos 4t + e^t \sin 4t)^2}$$

$$= \sqrt{16e^{2t} \sin^2 4t - 4e^{2t} \sin 4t \cos 4t + e^{2t} \cos^2 4t + 16e^{2t} \cos^2 4t + 4e^{2t} \sin 4t \cos 4t + e^{2t} \sin^2 4t}$$

$$= \sqrt{16e^{2t}(\sin^2 4t + \cos^2 4t) + e^{2t}(\cos^2 4t + \sin^2 4t)}$$

$$= \sqrt{16e^{2t} + e^{2t}}$$

$$= \sqrt{17} e^t$$

Then using the dot product relation listed above we get:

$$e^{2t} = e^t (\sqrt{17} e^t) \cos\theta = \sqrt{17} e^{2t} \cos\theta$$

Hence

$$\cos\theta = \frac{1}{\sqrt{17}}, \quad \Rightarrow \quad \theta \approx 75.96°$$

Therefore, the angle between the position vector and the tangent vector is constant.

65. Prove that if $\|\mathbf{r}(t)\|$ takes on a local minimum or maximum value at t_0, then $\mathbf{r}(t_0)$ is orthogonal to $\mathbf{r}'(t_0)$. Explain how this result is related to Figure 11. *Hint:* Observe that if $\|\mathbf{r}(t_0)\|$ is a minimum, then $\mathbf{r}(t)$ is tangent at t_0 to the sphere of radius $\|\mathbf{r}(t_0)\|$ centered at the origin.

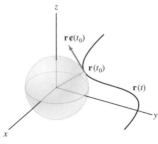

FIGURE 11

SOLUTION   Suppose that $\|\mathbf{r}(t)\|$ takes on a minimum or maximum value at $t = t_0$. Hence, $\|\mathbf{r}(t)\|^2$ also takes on a minimum or maximum value at $t = t_0$, therefore $\frac{d}{dt}\|\mathbf{r}(t)\|^2\big|_{t=t_0} = 0$. Using the Product Rule for dot products we get

$$\frac{d}{dt}\|\mathbf{r}(t)\|^2\bigg|_{t=t_0} = \frac{d}{dt}\mathbf{r}(t) \cdot \mathbf{r}(t)\bigg|_{t=t_0} = \mathbf{r}(t_0) \cdot \mathbf{r}'(t_0) + \mathbf{r}'(t_0) \cdot \mathbf{r}(t_0) = 2\mathbf{r}(t_0) \cdot \mathbf{r}'(t_0) = 0$$

Thus $\mathbf{r}(t_0) \cdot \mathbf{r}'(t_0) = 0$, which implies the orthogonality of $\mathbf{r}(t_0)$ and $\mathbf{r}'(t_0)$. In Figure 11, $\|\mathbf{r}(t_0)\|$ is a minimum and the path intersects the sphere of radius $\|\mathbf{r}(t_0)\|$ at a single point. Therefore, the point of intersection is a tangency point which implies that $\mathbf{r}'(t_0)$ is tangent to the sphere at t_0. We conclude that $\mathbf{r}(t_0)$ and $\mathbf{r}'(t_0)$ are orthogonal.

Further Insights and Challenges

67. Let $\mathbf{r}(t) = \langle x(t), y(t) \rangle$ trace a plane curve $\mathcal{C}$. Assume that $x'(t_0) \neq 0$. Show that the slope of the tangent vector $\mathbf{r}'(t_0)$ is equal to the slope dy/dx of the curve at $\mathbf{r}(t_0)$.

SOLUTION

(a) By the Chain Rule we have

$$\frac{dy}{dt} = \frac{dy}{dx} \cdot \frac{dx}{dt}$$

Hence, at the points where $\frac{dx}{dt} \neq 0$ we have:

$$\frac{dy}{dx} = \frac{\frac{dy}{dt}}{\frac{dx}{dt}} = \frac{y'(t)}{x'(t)}$$

(b) The line $\ell(t) = \langle a, b \rangle + t\mathbf{r}'(t_0)$ passes through (a, b) at $t = 0$. It holds that:

$$\ell(0) = \langle a, b \rangle + 0\mathbf{r}'(t_0) = \langle a, b \rangle$$

That is, (a, b) is the terminal point of the vector $\ell(0)$, hence the line passes through (a, b). The line has the direction vector $\mathbf{r}'(t_0) = \langle x'(t_0), y'(t_0) \rangle$, therefore the slope of the line is $\frac{y'(t_0)}{x'(t_0)}$ which is equal to $\frac{dy}{dx}\big|_{t=t_0}$ by part (a).

69. Verify the Sum and Product Rules for derivatives of vector-valued functions.

SOLUTION   We first verify the Sum Rule stating:

$$(\mathbf{r}_1(t) + \mathbf{r}_2(t))' = \mathbf{r}_1'(t) + \mathbf{r}_2'(t)$$

Let $\mathbf{r}_1(t) = \langle x_1(t), y_1(t), z_1(t) \rangle$ and $\mathbf{r}_2(t) = \langle x_2(t), y_2(t), z_2(t) \rangle$. Then,

$$(\mathbf{r}_1(t) + \mathbf{r}_2(t))' = \frac{d}{dt} \langle x_1(t) + x_2(t), y_1(t) + y_2(t), z_1(t) + z_2(t) \rangle$$

$$= \langle (x_1(t) + x_2(t))', (y_1(t) + y_2(t))', (z_1(t) + z_2(t))' \rangle$$

$$= \langle x_1'(t) + x_2'(t), y_1'(t) + y_2'(t), z_1'(t) + z_2'(t) \rangle$$

$$= \langle x_1'(t), y_1'(t), z_1'(t) \rangle + \langle x_2'(t), y_2'(t), z_2'(t) \rangle = \mathbf{r}_1'(t) + \mathbf{r}_2'(t)$$

The Product Rule states that for any differentiable scalar-valued function $f(t)$ and differentiable vector-valued function $\mathbf{r}(t)$, it holds that:

$$\frac{d}{dt} f(t)\mathbf{r}(t) = f(t)\mathbf{r}'(t) + f'(t)\mathbf{r}(t)$$

To verify this rule, we denote $\mathbf{r}(t) = \langle x(t), y(t), z(t) \rangle$. Then,

$$\frac{d}{df} f(t)\mathbf{r}(t) = \frac{d}{dt} \langle f(t)x(t), f(t)y(t), f(t)z(t) \rangle$$

Applying the Product Rule for scalar functions for each component we get:

$$\frac{d}{dt} f(t)\mathbf{r}(t) = \langle f(t)x'(t) + f'(t)x(t), f(t)y'(t) + f'(t)y(t), f(t)z'(t) + f'(t)z(t) \rangle$$

$$= \langle f(t)x'(t), f(t)y'(t), f(t)z'(t) \rangle + \langle f'(t)x(t), f'(t)y(t), f'(t)z(t) \rangle$$

$$= f(t) \langle x'(t), y'(t), z'(t) \rangle + f'(t) \langle x(t), y(t), z(t) \rangle = f(t)\mathbf{r}'(t) + f'(t)\mathbf{r}(t)$$

71. Verify the Product Rule for cross products [Eq. (5)].

SOLUTION Let $\mathbf{r}_1(t) = \langle a_1(t), a_2(t), a_3(t) \rangle$ and $\mathbf{r}_2(t) = \langle b_1(t), b_2(t), b_3(t) \rangle$. Then (we omit the independent variable t for simplicity):

$$\mathbf{r}_1(t) \times \mathbf{r}_2(t) = \begin{vmatrix} \mathbf{i} & \mathbf{j} & \mathbf{k} \\ a_1 & a_2 & a_3 \\ b_1 & b_2 & b_3 \end{vmatrix} = (a_2 b_3 - a_3 b_2)\mathbf{i} - (a_1 b_3 - a_3 b_1)\mathbf{j} + (a_1 b_2 - a_2 b_1)\mathbf{k}$$

Differentiating this vector componentwise we get:

$$\frac{d}{dt}\mathbf{r}_1 \times \mathbf{r}_2 = \left(a_2 b_3' + a_2' b_3 - a_3 b_2' - a_3' b_2\right)\mathbf{i} - \left(a_1 b_3' + a_1' b_3 - a_3 b_1' - a_3' b_1\right)\mathbf{j} + \left(a_1 b_2' + a_1' b_2 - a_2 b_1' - a_2' b_1\right)\mathbf{k}$$

$$= \left(\left(a_2 b_3' - a_3 b_2'\right)\mathbf{i} - \left(a_1 b_3' - a_3 b_1'\right)\mathbf{j} + \left(a_1 b_2' - a_2 b_1'\right)\mathbf{k}\right)$$
$$+ \left(\left(a_2' b_3 - a_3' b_2\right)\mathbf{i} - \left(a_1' b_3 - a_3' b_1\right)\mathbf{j} + \left(a_1' b_2 - a_2' b_1\right)\mathbf{k}\right)$$

Notice that the vectors in each of the two brackets can be written as the following formal determinants:

$$\frac{d}{dt}\mathbf{r}_1 \times \mathbf{r}_2 = \begin{vmatrix} \mathbf{i} & \mathbf{j} & \mathbf{k} \\ a_1 & a_2 & a_3 \\ b_1' & b_2' & b_3' \end{vmatrix} + \begin{vmatrix} \mathbf{i} & \mathbf{j} & \mathbf{k} \\ a_1' & a_2' & a_3' \\ b_1 & b_2 & b_3 \end{vmatrix} = \langle a_1, a_2, a_3 \rangle \times \langle b_1', b_2', b_3' \rangle + \langle a_1', a_2', a_3' \rangle \times \langle b_1, b_2, b_3 \rangle$$

$$= \mathbf{r}_1 \times \mathbf{r}_2' + \mathbf{r}_1' \times \mathbf{r}_2$$

73. Prove the Substitution Rule (where $g(t)$ is a differentiable scalar function):

$$\int_a^b \mathbf{r}(g(t))g'(t)\,dt = \int_{g^{-1}(a)}^{g^{-1}(b)} \mathbf{r}(u)\,du$$

SOLUTION (Note that an early edition of the textbook had the integral limits as $g(a)$ and $g(b)$; they should actually be $g^{-1}(a)$ and $g^{-1}(b)$.) We denote the components of the vector-valued function by $\mathbf{r}(t)\,dt = \langle x(t), y(t), z(t) \rangle$. Using componentwise integration we have:

$$\int_a^b \mathbf{r}(t)\,dt = \left\langle \int_a^b x(t)\,dt, \int_a^b y(t)\,dt, \int_a^b z(t)\,dt \right\rangle$$

Write $\int_a^b x(t)\,dt$ as $\int_a^b x(s)\,ds$. Let $s = g(t)$, so $ds = g'(t)\,dt$. The substitution gives us $\int_{g^{-1}(a)}^{g^{-1}(b)} x(g(t))g'(t)\,dt$. A similar procedure for the other two integrals gives us:

$$\int_a^b \mathbf{r}(t)\,dt = \left\langle \int_{g^{-1}(a)}^{g^{-1}(b)} x\left(g(t)\right)g'(t)\,dt, \int_{g^{-1}(a)}^{g^{-1}(b)} y\left(g(t)\right)g'(t)\,dt, \int_{g^{-1}(a)}^{g^{-1}(b)} z\left(g(t)\right)g'(t)\,dt \right\rangle$$

$$= \int_{g^{-1}(a)}^{g^{-1}(b)} \left\langle x\left(g(t)\right)g'(t), y\left(g(t)\right)g'(t), z\left(g(t)\right)g'(t) \right\rangle dt$$

$$= \int_{g^{-1}(a)}^{g^{-1}(b)} \left\langle x\left(g(t)\right), y\left(g(t)\right), z\left(g(t)\right) \right\rangle g'(t)\,dt = \int_{g^{-1}(a)}^{g^{-1}(b)} \mathbf{r}\left(g(t)\right)g'(t)\,dt$$

13.3 Arc Length and Speed (LT Section 14.3)

Preliminary Questions

1. At a given instant, a car on a roller coaster has velocity vector $\mathbf{r}' = \langle 25, -35, 10 \rangle$ (in miles per hour). What would the velocity vector be if the speed were doubled? What would it be if the car's direction were reversed but its speed remained unchanged?

SOLUTION The speed is doubled but the direction is unchanged, hence the new velocity vector has the form:

$$\lambda \mathbf{r}' = \lambda \langle 25, -35, 10 \rangle \text{ for } \lambda > 0$$

We use $\lambda = 2$, and so the new velocity vector is $\langle 50, -70, 20 \rangle$. If the direction is reversed but the speed is unchanged, the new velocity vector is:

$$-\mathbf{r}' = \langle -25, 35, -10 \rangle.$$

2. Two cars travel in the same direction along the same roller coaster (at different times). Which of the following statements about their velocity vectors at a given point P on the roller coaster is/are true?

(a) The velocity vectors are identical.

(b) The velocity vectors point in the same direction but may have different lengths.

(c) The velocity vectors may point in opposite directions.

SOLUTION

(a) The length of the velocity vector is the speed of the particle. Therefore, if the speeds of the cars are different the velocities are not identical. The statement is false.

(b) The velocity vector is tangent to the curve. Since the cars travel in the same direction, their velocity vectors point in the same direction. The statement is true.

(c) Since the cars travel in the same direction, the velocity vectors point in the same direction. The statement is false.

3. A mosquito flies along a parabola with speed $v(t) = t^2$. Let $s(t)$ be the total distance traveled at time t.

(a) How fast is $s(t)$ changing at $t = 2$?

(b) Is $s(t)$ equal to the mosquito's distance from the origin?

SOLUTION

(a) By the Arc Length Formula, we have:

$$s(t) = \int_{t_0}^{t} \|\mathbf{r}'(t)\| \, dt = \int_{t_0}^{t} v(t) \, dt$$

Therefore,

$$s'(t) = v(t)$$

To find the rate of change of $s(t)$ at $t = 2$ we compute the derivative of $s(t)$ at $t = 2$, that is,

$$s'(2) = v(2) = 2^2 = 4$$

(b) $s(t)$ is the distance along the path traveled by the mosquito. This distance is usually different from the mosquito's distance from the origin, which is the length of $\mathbf{r}(t)$.

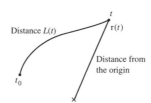

Distance $L(t)$ t $\mathbf{r}(t)$

Distance from the origin

t_0

4. What is the length of the path traced by $\mathbf{r}(t)$ for $4 \leq t \leq 10$ if $\mathbf{r}(t)$ is an arc length parametrization?

SOLUTION   Since $\mathbf{r}(t)$ is an arc length parametrization, the length of the path for $4 \leq t \leq 10$ is equal to the length of the time interval $4 \leq t \leq 10$, which is 6.

Exercises

In Exercises 1–6, compute the length of the curve over the given interval.

1. $\mathbf{r}(t) = \langle 3t, 4t - 3, 6t + 1 \rangle, \quad 0 \leq t \leq 3$

SOLUTION   We have $x(t) = 3t$, $y(t) = 4t - 3$, $z(t) = 6t + 1$ hence

$$x'(t) = 3, \quad y'(t) = 4, \quad z'(t) = 6.$$

We use the Arc Length Formula to obtain:

$$s = \int_0^3 \|\mathbf{r}'(t)\| \, dt = \int_0^3 \sqrt{x'(t)^2 + y'(t)^2 + z'(t)^2} \, dt = \int_0^3 \sqrt{3^2 + 4^2 + 6^2} \, dt = 3\sqrt{61}$$

3. $\mathbf{r}(t) = \langle 2t, \ln t, t^2 \rangle, \quad 1 \leq t \leq 4$

SOLUTION   The derivative of $\mathbf{r}(t)$ is $\mathbf{r}'(t) = \left\langle 2, \frac{1}{t}, 2t \right\rangle$. We use the Arc Length Formula to obtain:

$$s = \int_1^4 \|\mathbf{r}'(t)\| \, dt = \int_1^4 \sqrt{2^2 + \left(\frac{1}{t}\right)^2 + (2t)^2} \, dt = \int_1^4 \sqrt{4t^2 + 4 + \frac{1}{t^2}} \, dt = \int_1^4 \sqrt{\left(2t + \frac{1}{t}\right)^2} \, dt$$

$$= \int_1^4 \left(2t + \frac{1}{t}\right) dt = t^2 + \ln t \Big|_1^4 = (16 + \ln 4) - (1 + \ln 1) = 15 + \ln 4$$

5. $\mathbf{r}(t) = \langle t \cos t, t \sin t, 3t \rangle, \quad 0 \le t \le 2\pi$

SOLUTION The derivative of $\mathbf{r}(t)$ is $\mathbf{r}'(t) = \langle \cos t - t \sin t, \sin t + t \cos t, 3 \rangle$. The length of $\mathbf{r}'(t)$ is, thus,

$$\|\mathbf{r}'(t)\| = \sqrt{(\cos t - t \sin t)^2 + (\sin t + t \cos t)^2 + 9}$$

$$= \sqrt{\cos^2 t - 2t \cos t \sin t + t^2 \sin^2 t + \sin^2 t + 2t \sin t \cos t + t^2 \cos^2 t + 9}$$

$$= \sqrt{\left(\cos^2 t + \sin^2 t\right) + t^2 \left(\sin^2 t + \cos^2 t\right) + 9} = \sqrt{t^2 + 10}$$

Using the Arc Length Formula and the integration formula given in Exercise 6, we obtain:

$$s = \int_0^{2\pi} \|\mathbf{r}'(t)\| \, dt = \int_0^{2\pi} \sqrt{t^2 + 10} \, dt = \frac{1}{2} t \sqrt{t^2 + 10} + \frac{1}{2} \cdot 10 \ln\left(t + \sqrt{t^2 + 10}\right) \Big|_0^{2\pi}$$

$$= \pi \sqrt{4\pi^2 + 10} + 5 \ln\left(2\pi + \sqrt{4\pi^2 + 10}\right) - 5 \ln \sqrt{10} = \pi \sqrt{4\pi^2 + 10} + 5 \ln \frac{2\pi + \sqrt{4\pi^2 + 10}}{\sqrt{10}} \approx 29.3$$

In Exercises 7 and 8, compute the arc length function $s(t) = \int_a^t \|\mathbf{r}'(u)\| \, du$ for the given value of a.

7. $\mathbf{r}(t) = \langle t^2, 2t^2, t^3 \rangle, \quad a = 0$

SOLUTION The derivative of $\mathbf{r}(t)$ is $\mathbf{r}'(t) = \langle 2t, 4t, 3t^2 \rangle$. Hence,

$$\|\mathbf{r}'(t)\| = \sqrt{(2t)^2 + (4t)^2 + (3t^2)^2} = \sqrt{4t^2 + 16t^2 + 9t^4} = t\sqrt{20 + 9t^2}$$

Hence,

$$s(t) = \int_0^t \|\mathbf{r}'(u)\| \, du = \int_0^t u\sqrt{20 + 9u^2} \, du$$

We compute the integral using the substitution $v = 20 + 9u^2$, $dv = 18u \, du$. This gives:

$$s(t) = \frac{1}{18} \int_{20}^{20+9t^2} v^{1/2} \, dv = \frac{1}{18} \cdot \frac{2}{3} v^{3/2} \Big|_{20}^{20+9t^2} = \frac{1}{27} \left((20 + 9t^2)^{3/2} - 20^{3/2}\right).$$

In Exercises 9–12, find the speed at the given value of t.

9. $\mathbf{r}(t) = \langle 2t + 3, 4t - 3, 5 - t \rangle, \quad t = 4$

SOLUTION The speed is the magnitude of the derivative $\mathbf{r}'(t) = \langle 2, 4, -1 \rangle$. That is,

$$v(t) = \|\mathbf{r}'(t)\| = \sqrt{2^2 + 4^2 + (-1)^2} = \sqrt{21} \approx 4.58$$

11. $\mathbf{r}(t) = \langle \sin 3t, \cos 4t, \cos 5t \rangle, \quad t = \frac{\pi}{2}$

SOLUTION The velocity vector is $\mathbf{r}'(t) = \langle 3 \cos 3t, -4 \sin 4t, -5 \sin 5t \rangle$. At $t = \frac{\pi}{2}$ the velocity vector is $\mathbf{r}'\left(\frac{\pi}{2}\right) = \langle 3 \cos \frac{3\pi}{2}, -4 \sin 2\pi, -5 \sin \frac{5\pi}{2} \rangle = \langle 0, 0, -5 \rangle$. The speed is the magnitude of the velocity vector:

$$v\left(\frac{\pi}{2}\right) = \|\langle 0, 0, -5 \rangle\| = 5.$$

13. What is the velocity vector of a particle traveling to the right along the hyperbola $y = x^{-1}$ with constant speed 5 cm/s when the particle's location is $\left(2, \frac{1}{2}\right)$?

SOLUTION The position of the particle is given as $\mathbf{r}(t) = t^{-1}$. The magnitude of the velocity vector $\mathbf{r}'(t)$ is the speed of the particle. Hence,

$$\|\mathbf{r}'(t)\| = 5 \tag{1}$$

The velocity vector points in the direction of motion, hence it is parallel to the tangent line to the curve $y = x^{-1}$ and points to the right. We find the slope of the tangent line at $x = 2$:

$$m = \frac{dy}{dx}\Big|_{x=2} = \frac{d}{dx}(x^{-1})\Big|_{x=2} = -x^{-2}\Big|_{x=2} = -\frac{1}{4}$$

We conclude that the vector $\left\langle 1, -\frac{1}{4} \right\rangle$ is a direction vector of the tangent line at $x = 2$, and for some $\lambda > 0$ we have at the given instance:

$$\mathbf{r}' = \lambda \left\langle 1, -\frac{1}{4} \right\rangle \qquad (2)$$

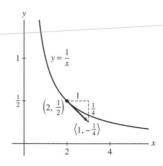

To satisfy (1) we must have:

$$\|\mathbf{r}'\| = \lambda \sqrt{1^2 + \left(-\frac{1}{4}\right)^2} = \lambda \frac{\sqrt{17}}{4} = 5 \quad \Rightarrow \quad \lambda = \frac{20}{\sqrt{17}}$$

Substituting in (2) we obtain the following velocity vector at $\left(2, \frac{1}{2}\right)$:

$$\mathbf{r}' = \frac{20}{\sqrt{17}} \left\langle 1, -\frac{1}{4} \right\rangle = \left\langle \frac{20}{\sqrt{17}}, \frac{-5}{\sqrt{17}} \right\rangle$$

15. Let

$$\mathbf{r}(t) = \left\langle R \cos\left(\frac{2\pi N t}{h}\right), R \sin\left(\frac{2\pi N t}{h}\right), t \right\rangle, \qquad 0 \le t \le h$$

(a) Show that $\mathbf{r}(t)$ parametrizes a helix of radius R and height h making N complete turns.

(b) Guess which of the two springs in Figure 5 uses more wire.

(c) Compute the lengths of the two springs and compare.

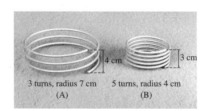

FIGURE 5 Which spring uses more wire?

SOLUTION   We first verify that the projection $\mathbf{p}(t) = \left\langle R \cos\left(\frac{2\pi N t}{h}\right), R \sin\left(\frac{2\pi N t}{h}\right), 0 \right\rangle$ onto the xy-plane describes a point moving around the circle of radius R. We have:

$$x(t)^2 + y(t)^2 = R^2 \cos^2\left(\frac{2\pi N t}{h}\right) + R^2 \sin^2\left(\frac{2\pi N t}{h}\right) = R^2 \left(\cos^2\left(\frac{2\pi N t}{h}\right) + \sin^2\left(\frac{2\pi N t}{h}\right)\right) = R^2$$

This is the equation of the circle of radius R in the xy-plane. As t changes in the interval $0 \le t \le h$ the argument $\frac{2\pi N t}{h}$ changes from 0 to $2\pi N$, that is, it covers N periods of the cos and sin functions. It follows that the projection onto the xy-plane describes a point moving around the circle of radius R, making N complete turns. The height of the helix is the maximum value of the z-component, which is $t = h$.

(a) The second wire seems to use more wire than the first one.

(b) Setting $R = 7, h = 4$ and $N = 3$ in the parametrization in Exercise 15 gives:

$$\mathbf{r}_1(t) = \left\langle 7 \cos\frac{2\pi \cdot 3t}{4}, 7 \sin\frac{2\pi \cdot 3t}{4}, t \right\rangle = \left\langle 7 \cos\frac{3\pi t}{2}, 7 \sin\frac{3\pi t}{2}, t \right\rangle, \qquad 0 \le t \le 4$$

Setting $R = 4, h = 3$ and $N = 5$ in this parametrization we get:

$$\mathbf{r}_2(t) = \left\langle 4 \cos\frac{2\pi \cdot 5t}{3}, 4 \sin\frac{2\pi \cdot 5t}{3}, t \right\rangle = \left\langle 4 \cos\frac{10\pi t}{3}, 4 \sin\frac{10\pi t}{3}, t \right\rangle, \qquad 0 \le t \le 3$$

We find the derivatives of the two vectors and their lengths:

$$\mathbf{r}_1'(t) = \left\langle -\frac{21\pi}{2}\sin\frac{3\pi t}{2}, \frac{21\pi}{2}\cos\frac{3\pi t}{2}, 1 \right\rangle \quad \Rightarrow \quad \|\mathbf{r}_1'(t)\| = \sqrt{\frac{441\pi^2}{4} + 1} = \frac{1}{2}\sqrt{441\pi^2 + 4}$$

$$\mathbf{r}_2'(t) = \left\langle -\frac{40\pi}{3}\sin\frac{10\pi t}{3}, \frac{40\pi}{3}\cos\frac{10\pi t}{3}, 1 \right\rangle \quad \Rightarrow \quad \|\mathbf{r}_2'(t)\| = \sqrt{\frac{1600\pi^2}{9} + 1} = \frac{1}{3}\sqrt{1600\pi^2 + 9}$$

Using the Arc Length Formula we obtain the following lengths:

$$s_1 = \int_0^4 \frac{1}{2}\sqrt{441\pi^2 + 4}\, dt = 2\sqrt{441\pi^2 + 4} \approx 132$$

$$s_2 = \int_0^3 \frac{1}{3}\sqrt{1600\pi^2 + 9}\, dt = \sqrt{1600\pi^2 + 9} \approx 125.7$$

We see that the first spring uses more wire than the second one.

17. The cycloid generated by the unit circle has parametrization

$$\mathbf{r}(t) = \langle t - \sin t, 1 - \cos t \rangle$$

(a) Find the value of t in $[0, 2\pi]$ where the speed is at a maximum.

(b) Show that one arch of the cycloid has length 8. Recall the identity $\sin^2(t/2) = (1 - \cos t)/2$.

SOLUTION One arch of the cycloid is traced as $0 \le t \le 2\pi$. By the Arc Length Formula we have:

$$s = \int_0^{2\pi} \|\mathbf{r}'(t)\|\, dt \tag{1}$$

We compute the derivative and its length:

$$\mathbf{r}'(t) = \langle 1 - \cos t, \sin t \rangle$$

$$\|\mathbf{r}'(t)\| = \sqrt{(1 - \cos t)^2 + (\sin t)^2} = \sqrt{1 - 2\cos t + \cos^2 t + \sin^2 t}$$

$$= \sqrt{2 - 2\cos t} = \sqrt{2(1 - \cos t)} = \sqrt{2 \cdot 2\sin^2 \frac{t}{2}} = 2\left|\sin\frac{t}{2}\right|.$$

For $0 \le t \le 2\pi$, we have $0 \le \frac{t}{2} \le \pi$, so $\sin\frac{t}{2} \ge 0$. Therefore we may omit the absolute value sign and write:

$$\|\mathbf{r}'(t)\| = 2\sin\frac{t}{2}$$

Substituting in (1) and computing the integral using the substitution $u = \frac{t}{2}$, $du = \frac{1}{2}\, dt$, gives:

$$s = \int_0^{2\pi} 2\sin\frac{t}{2}\, dt = \int_0^{\pi} 2\sin u \cdot (2\, du) = 4\int_0^{\pi} \sin u\, du$$

$$= 4(-\cos u)\Big|_0^{\pi} = 4(-\cos\pi - (-\cos 0)) = 4(1 + 1) = 8$$

The length of one arc of the cycloid is $s = 8$. The speed is given by the function:

$$v(t) = \|\mathbf{r}'(t)\| = 2\sin\frac{t}{2}, \quad 0 \le t \le \pi$$

To find the value of t in $[0, 2\pi]$ where the speed is at maximum, we first find the critical point in this interval:

$$v'(t) = 2 \cdot \frac{1}{2}\cos\frac{t}{2} = \cos\frac{t}{2}$$

$$\cos\frac{t}{2} = 0 \quad \Rightarrow \quad \frac{t}{2} = \frac{\pi}{2} \quad \Rightarrow \quad t = \pi$$

Since $v''(t) = -\frac{1}{2}\sin\frac{t}{2}$, we have $v''(\pi) = -\frac{1}{2}\sin\frac{\pi}{2} = -\frac{1}{2} < 0$, hence the speed $v(t)$ has a maximum value at $t = \pi$.

19. Let $\mathbf{r}(t) = \langle 3t + 1, 4t - 5, 2t \rangle$.

(a) Evaluate the arc length integral $s(t) = \int_0^t \|\mathbf{r}'(u)\|\, du$.

(b) Find the inverse $g(s)$ of $s(t)$.

(c) Verify that $\mathbf{r}_1(s) = \mathbf{r}(g(s))$ is an arc length parametrization.

SOLUTION

(a) We differentiate $\mathbf{r}(t)$ componentwise and then compute the norm of the derivative vector. This gives:

$$\mathbf{r}'(t) = \langle 3, 4, 2 \rangle$$

$$\|\mathbf{r}'(t)\| = \sqrt{3^2 + 4^2 + 2^2} = \sqrt{29}$$

We compute $s(t)$:

$$s(t) = \int_0^t \|\mathbf{r}'(u)\|\, du = \int_0^t \sqrt{29}\, du = \sqrt{29}\, u \Big|_0^t = \sqrt{29}\, t$$

(b) We find the inverse $g(s) = t(s)$ by solving $s = \sqrt{29}\, t$ for t. We obtain:

$$s = \sqrt{29}\, t \quad \Rightarrow \quad t = g(s) = \frac{s}{\sqrt{29}}$$

We obtain the following arc length parametrization:

$$\mathbf{r}_1(s) = \mathbf{r}\left(\frac{s}{\sqrt{29}}\right) = \left\langle \frac{3s}{\sqrt{29}} + 1, \frac{4s}{\sqrt{29}} - 5, \frac{2s}{\sqrt{29}} \right\rangle$$

To verify that $\mathbf{r}_1(s)$ is an arc length parametrization we must show that $\|\mathbf{r}_1'(s)\| = 1$. We compute $\mathbf{r}_1'(s)$:

$$\mathbf{r}_1'(s) = \frac{d}{ds}\left\langle \frac{3s}{\sqrt{29}} + 1, \frac{4s}{\sqrt{29}} - 5, \frac{2s}{\sqrt{29}} \right\rangle = \left\langle \frac{3}{\sqrt{29}}, \frac{4}{\sqrt{29}}, \frac{2}{\sqrt{29}} \right\rangle = \frac{1}{\sqrt{29}} \langle 3, 4, 2 \rangle$$

Thus,

$$\|\mathbf{r}_1'(s)\| = \frac{1}{\sqrt{29}} \| \langle 3, 4, 2 \rangle \| = \frac{1}{\sqrt{29}} \sqrt{3^2 + 4^2 + 2^2} = \frac{1}{\sqrt{29}} \cdot \sqrt{29} = 1$$

21. Let $\mathbf{r}(t) = \mathbf{w} + t\mathbf{v}$ be the parametrization of a line.

(a) Show that the arc length function $s(t) = \displaystyle\int_0^t \|\mathbf{r}'(u)\|\, du$ is given by $s(t) = t\|\mathbf{v}\|$. This shows that $\mathbf{r}(t)$ is an arc length parametrizaton if and only if $\mathbf{v}$ is a unit vector.

(b) Find an arc length parametrization of the line with $\mathbf{w} = \langle 1, 2, 3 \rangle$ and $\mathbf{v} = \langle 3, 4, 5 \rangle$.

SOLUTION

(a) Since $\mathbf{r}(t) = \mathbf{w} + t\mathbf{v}$, then $\mathbf{r}'(t) = \mathbf{v}$ and $\|\mathbf{r}'(t)\| = \|\mathbf{v}\|$. Then computing $s(t)$ we get:

$$s(t) = \int_0^t \|\mathbf{r}'(u)\|\, du = \int_0^t \|\mathbf{v}\|\, du = t\|\mathbf{v}\|$$

If we consider $s(t)$,

$$s(t) = t \text{ if and only if } \|\mathbf{v}\| = 1$$

(b) Since $\mathbf{v} = \langle 3, 4, 5 \rangle$, then from part (a) we get:

$$s(t) = t\|\mathbf{v}\| = t\sqrt{3^2 + 4^2 + 5^2} = t\sqrt{50}, \quad \Rightarrow t = g(s) = \frac{s}{\sqrt{50}}$$

Therefore, since we are given $\mathbf{r}(t) = \mathbf{w} + t\mathbf{v}$, the arc length parametrization is:

$$\mathbf{r}_1(s) = \langle 1, 2, 3 \rangle + \frac{s}{\sqrt{50}} \langle 3, 4, 5 \rangle = \left\langle 1 + \frac{3s}{\sqrt{50}}, 2 + \frac{4s}{\sqrt{50}}, 3 + \frac{5s}{\sqrt{50}} \right\rangle$$

23. Find a path that traces the circle in the plane $y = 10$ with radius 4 and center $(2, 10, -3)$ with constant speed 8.

SOLUTION We start with the following parametrization of the circle:

$$\mathbf{r}(t) = \langle 2, 10, -3 \rangle + 4 \langle \cos t, 0, \sin t \rangle = \langle 2 + 4\cos t, 10, -3 + 4\sin t \rangle$$

We need to reparametrize the curve by making a substitution $t = g(s)$, so that the new parametrization $\mathbf{r}_1(s) = \mathbf{r}(g(s))$ satisfies $\|\mathbf{r}_1'(s)\| = 8$ for all s. We find $\mathbf{r}_1'(s)$ using the Chain Rule:

$$\mathbf{r}_1'(s) = \frac{d}{ds} \mathbf{r}(g(s)) = g'(s)\mathbf{r}'(g(s)) \tag{1}$$

Next, we differentiate $\mathbf{r}(t)$ and then replace t by $g(s)$:

$$\mathbf{r}'(t) = \langle -4\sin t, 0, 4\cos t\rangle$$

$$\mathbf{r}'(g(s)) = \langle -4\sin g(s), 0, 4\cos g(s)\rangle$$

Substituting in (1) we get:

$$\mathbf{r}_1'(s) = g'(s)\langle -4\sin g(s), 0, 4\cos g(s)\rangle = -4g'(s)\langle \sin g(s), 0, -\cos g(s)\rangle$$

Hence,

$$\|\mathbf{r}_1'(s)\| = 4|g'(s)|\sqrt{(\sin g(s))^2 + (-\cos g(s))^2} = 4|g'(s)|$$

To satisfy $\|\mathbf{r}_1'(s)\| = 8$ for all s, we choose $g'(s) = 2$. We may take the antiderivative $g(s) = 2 \cdot s$, and obtain the following parametrization:

$$\mathbf{r}_1(s) = \mathbf{r}(g(s)) = \mathbf{r}(2s) = \langle 2 + 4\cos(2s), 10, -3 + 4\sin(2s)\rangle.$$

This is a parametrization of the given circle, with constant speed 8.

25. Find an arc length parametrization of $\mathbf{r}(t) = \langle t^2, t^3\rangle$.

SOLUTION We follow two steps.

Step 1. Find the inverse of the arc length function. The arc length function is the following function:

$$s(t) = \int_0^t \|\mathbf{r}'(u)\|\, du \tag{1}$$

In our case $\mathbf{r}'(t) = \langle 2t, 3t^2\rangle$ hence $\|\mathbf{r}'(t)\| = \sqrt{4t^2 + 9t^4} = \sqrt{4 + 9t^2}\,t$. We substitute in (1) and compute the resulting integral using the substitution $v = 4 + 9u^2$, $dv = 18u\, du$. This gives:

$$s(t) = \int_0^t \sqrt{4 + 9u^2}\,u\, du = \frac{1}{18}\int_4^{4+9t^2} v^{1/2}\, dv = \frac{1}{18}\cdot\frac{2}{3}v^{3/2}\Big|_4^{4+9t^2} = \frac{1}{27}\left((4 + 9t^2)^{3/2} - 4^{3/2}\right)$$

$$= \frac{1}{27}\left(\left(4 + 9t^2\right)^{3/2} - 8\right)$$

We find the inverse of $t = s(t)$ by solving for t in terms of s. This function is invertible for $t \geq 0$ and for $t \leq 0$.

$$s = \frac{1}{27}\left((4 + 9t^2)^{3/2} - 8\right)$$

$$27s + 8 = (4 + 9t^2)^{3/2}$$

$$(27s + 8)^{2/3} - 4 = 9t^2$$

$$t^2 = \frac{1}{9}\left((27s + 8)^{2/3} - 4\right) = \frac{1}{9}(27s + 8)^{2/3} - \frac{4}{9}$$

$$t = \pm\frac{1}{3}\sqrt{(27s + 8)^{2/3} - 4} \tag{2}$$

Step 2. Reparametrize the curve. The arc length parametrization is obtained by replacing t by (2) in $\mathbf{r}(t)$:

$$\mathbf{r}_1(s) = \left\langle \frac{1}{9}(27s + 8)^{2/3} - \frac{4}{9}, \pm\frac{1}{27}\left((27s + 8)^{2/3} - 4\right)^{3/2}\right\rangle$$

27. Find an arc length parametrization of the line $y = mx$ for an arbitrary slope m.

SOLUTION

Step 1. Find the inverse of the arc length function. We are given the line $y = mx$ and a parametrization of this line is $\mathbf{r}(t) = \langle t, mt\rangle$, thus $\mathbf{r}'(t) = \langle 1, m\rangle$ and

$$\|\mathbf{r}'(t)\| = \sqrt{1 + m^2}$$

We then compute $s(t)$:

$$s(t) = \int_0^t \sqrt{1 + m^2}\, du = t\sqrt{1 + m^2}$$

Solving $s = t\sqrt{1 + m^2}$ for t we get:

$$t = \frac{s}{\sqrt{1 + m^2}}$$

Step 2. Reparametrize the curve using the t we just found.

$$\mathbf{r}_1(s) = \left\langle \frac{s}{\sqrt{1+m^2}}, \frac{sm}{\sqrt{1+m^2}} \right\rangle$$

29. The curve known as the **Bernoulli spiral** (Figure 6) has parametrization $\mathbf{r}(t) = \langle e^t \cos 4t, e^t \sin 4t \rangle$.

(a) Evaluate $s(t) = \displaystyle\int_{-\infty}^{t} \|\mathbf{r}'(u)\| \, du$. It is convenient to take lower limit $-\infty$ because $\mathbf{r}(-\infty) = \langle 0, 0 \rangle$.

(b) Use (a) to find an arc length parametrization of $\mathbf{r}(t)$.

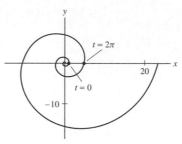

FIGURE 6 Bernoulli spiral.

SOLUTION

(a) We differentiate $\mathbf{r}(t)$ and compute the norm of the derivative vector. This gives:

$$\mathbf{r}'(t) = \langle e^t \cos 4t - 4e^t \sin 4t, e^t \sin 4t + 4e^t \cos 4t \rangle = e^t \langle \cos 4t - 4\sin 4t, \sin 4t + 4\cos 4t \rangle$$

$$\|\mathbf{r}'(t)\| = e^t \sqrt{(\cos 4t - 4\sin 4t)^2 + (\sin 4t + 4\cos 4t)^2}$$

$$= e^t \left(\cos^2 4t - 8\cos 4t \sin 4t + 16\sin^2 4t + \sin^2 4t + 8\sin 4t \cos 4t + 16\cos^2 4t \right)^{1/2}$$

$$= e^t \sqrt{\cos^2 4t + \sin^2 4t + 16(\sin^2 4t + \cos^2 4t)} = e^t \sqrt{1 + 16 \cdot 1} = \sqrt{17}e^t$$

We now evaluate the improper integral:

$$s(t) = \int_{-\infty}^{t} \|\mathbf{r}'(u)\| \, du = \lim_{R \to -\infty} \int_{R}^{t} \sqrt{17}e^u \, du = \lim_{R \to -\infty} \sqrt{17}e^u \Big|_{R}^{t} = \lim_{R \to -\infty} \sqrt{17}(e^t - e^R)$$

$$= \sqrt{17}(e^t - 0) = \sqrt{17}e^t$$

(b) An arc length parametrization of $\mathbf{r}(t)$ is $\mathbf{r}_1(s) = \mathbf{r}(g(s))$ where $t = g(s)$ is the inverse function of $s(t)$. We find $t = g(s)$ by solving $s = \sqrt{17}e^t$ for t:

$$s = \sqrt{17}e^t \quad \Rightarrow \quad e^t = \frac{s}{\sqrt{17}} \Rightarrow t = g(s) = \ln \frac{s}{\sqrt{17}}$$

An arc length parametrization of $\mathbf{r}(t)$ is:

$$\mathbf{r}_1(s) = \mathbf{r}(g(s)) = \left\langle e^{\ln(s/(\sqrt{17}))} \cos \left(4\ln \frac{s}{\sqrt{17}} \right), e^{\ln(s/(\sqrt{17}))} \sin \left(4\ln \frac{s}{\sqrt{17}} \right) \right\rangle$$

$$= \frac{s}{\sqrt{17}} \left\langle \cos \left(4\ln \frac{s}{\sqrt{17}} \right), \sin \left(4\ln \frac{s}{\sqrt{17}} \right) \right\rangle \tag{1}$$

Further Insights and Challenges

31. The unit circle with the point $(-1, 0)$ removed has parametrization (see Exercise 73 in Section 11.1)

$$\mathbf{r}(t) = \left\langle \frac{1-t^2}{1+t^2}, \frac{2t}{1+t^2} \right\rangle, \qquad -\infty < t < \infty$$

Use this parametrization to compute the length of the unit circle as an improper integral. *Hint:* The expression for $\|\mathbf{r}'(t)\|$ simplifies.

SOLUTION We have $x(t) = \frac{1-t^2}{1+t^2}$, $y(t) = \frac{2t}{1+t^2}$. Hence,

$$x^2(t) + y^2(t) = \left(\frac{1-t^2}{1+t^2}\right)^2 + \left(\frac{2t}{1+t^2}\right)^2 = \frac{1 - 2t^2 + t^4 + 4t^2}{(1+t^2)^2} = \frac{1 + 2t^2 + t^4}{(1+t^2)^2} = \frac{\left(1+t^2\right)^2}{(1+t^2)^2} = 1$$

It follows that the path $\mathbf{r}(t)$ lies on the unit circle. We now show that the entire circle is indeed parametrized by $\mathbf{r}(t)$ as t moves from $-\infty$ to ∞. First, note that $x'(t)$ can be written as $\left[-2t(1+t^2) - 2t(1-t^2)\right]/(1+t^2)^2$ which is $-4t/(1+t^2)^2$. So, for t negative, $x(t)$ is an increasing function, $y(t)$ is negative, and since $\lim\limits_{t\to-\infty} x(t) = -1$ and $\lim\limits_{t\to 0} x(t) = 1$, we conclude that $\mathbf{r}(t)$ does indeed parametrize the lower half of the circle for negative t. A similar argument proves that we get the upper half of the circle for positive t. We now compute $\mathbf{r}'(t)$ and its length:

$$\mathbf{r}'(t) = \left\langle \frac{-2t(1+t^2) - 2t(1-t^2)}{(1+t^2)^2}, \frac{2(1+t^2) - 2t\cdot 2t}{(1+t^2)^2} \right\rangle$$

$$= \left\langle -\frac{4t}{(1+t^2)^2}, \frac{2-2t^2}{(1+t^2)^2} \right\rangle = \frac{1}{(1+t^2)^2}\left\langle -4t, 2(1-t^2) \right\rangle$$

$$\|\mathbf{r}'(t)\| = \frac{1}{(1+t^2)^2}\sqrt{16t^2 + 4(1-t^2)^2} = \frac{2}{(1+t^2)^2}\sqrt{t^4 + 2t^2 + 1}$$

$$= \frac{2}{(1+t^2)^2}\sqrt{(t^2+1)^2} = \frac{2(t^2+1)}{(1+t^2)^2} = \frac{2}{1+t^2}$$

That is,

$$\|\mathbf{r}'(t)\| = \frac{2}{1+t^2}$$

We now use the Arc Length Formula to compute the length of the circle:

$$s = \int_{-\infty}^{\infty} \|\mathbf{r}'(t)\|\, dt = 2\int_{-\infty}^{\infty} \frac{dt}{1+t^2} = 2\left(\lim_{R\to\infty} \tan^{-1} R - \lim_{R\to-\infty} \tan^{-1} R \right) = 2\left(\frac{\pi}{2} - \left(-\frac{\pi}{2}\right)\right) = 2\pi$$

33. The curve $\mathbf{r}(t) = \langle t - \tanh t, \operatorname{sech} t\rangle$ is called a **tractrix** (see Exercise 92 in Section 11.1).

(a) Show that $s(t) = \int_0^t \|\mathbf{r}'(u)\|\, du$ is equal to $s(t) = \ln(\cosh t)$.

(b) Show that $t = g(s) = \ln(e^s + \sqrt{e^{2s} - 1})$ is an inverse of $s(t)$ and verify that

$$\mathbf{r}_1(s) = \left\langle \tanh^{-1}\left(\sqrt{1 - e^{-2s}}\right) - \sqrt{1 - e^{-2s}}, e^{-s}\right\rangle$$

is an arc length parametrization of the tractrix.

SOLUTION

(a) We compute the derivative vector and its length:

$$\mathbf{r}'(t) = \left\langle 1 - \operatorname{sech}^2 t, -\operatorname{sech} t \tanh t\right\rangle$$

$$\|\mathbf{r}'(t)\| = \sqrt{(1 - \operatorname{sech}^2 t) + \operatorname{sech}^2 t \tanh^2 t} = \sqrt{1 - 2\operatorname{sech}^2 t + \operatorname{sech}^4 t + \operatorname{sech}^2 t \tanh^2 t}$$

$$= \sqrt{-\operatorname{sech}^2 t(2 - \tanh^2 t) + 1 + \operatorname{sech}^4 t}$$

We use the identity $1 - \tanh^2 t = \operatorname{sech}^2 t$ to write:

$$\|\mathbf{r}'(t)\| = \sqrt{-\operatorname{sech}^2 t(1 + \operatorname{sech}^2 t) + 1 + \operatorname{sech}^4 t} = \sqrt{-\operatorname{sech}^2 t - \operatorname{sech}^4 t + 1 + \operatorname{sech}^4 t}$$

$$= \sqrt{1 - \operatorname{sech}^2 t} = \sqrt{\tanh^2 t} = |\tanh t|$$

For $t \geq 0$, $\tanh t \geq 0$ hence, $\|\mathbf{r}'(t)\| = \tanh t$. We now apply the Arc Length Formula to obtain:

$$s(t) = \int_0^t \|\mathbf{r}'(u)\|\, du = \int_0^t (\tanh u)\, du = \ln(\cosh u)\Big|_0^t = \ln(\cosh t) - \ln(\cosh 0)$$

$$= \ln(\cosh t) - \ln 1 = \ln(\cosh t)$$

That is:

$$s(t) = \ln(\cosh t)$$

(b) We show that the function $t = g(s) = \ln\left(e^s + \sqrt{e^{2s} - 1}\right)$ is an inverse of $s(t)$. First we note that $s'(t) = \tanh t$, hence $s'(t) > 0$ for $t > 0$, which implies that $s(t)$ has an inverse function for $t \geq 0$. Therefore, it suffices to verify that $g(s(t)) = t$. We have:

$$g(s(t)) = \ln\left(e^{\ln(\cosh t)} + \sqrt{e^{2\ln(\cosh t)} - 1}\right) = \ln\left(\cosh t + \sqrt{\cosh^2 t - 1}\right)$$

Since $\cosh^2 t - 1 = \sinh^2 t$ we obtain (for $t \geq 0$):

$$g(s(t)) = \ln\left(\cosh t + \sqrt{\sinh^2 t}\right) = \ln\left(\cosh t + \sinh t\right) = \ln\left(\frac{e^t + e^{-t}}{2} + \frac{e^t - e^{-t}}{2}\right) = \ln\left(e^t\right) = t$$

We thus proved that $t = g(s)$ is an inverse of $s(t)$. Therefore, the arc length parametrization is obtained by substituting $t = g(s)$ in $\mathbf{r}(t) = \langle t - \tanh t, \operatorname{sech} t \rangle$. We compute t, $\tanh t$ and $\operatorname{sech} t$ in terms of s. We have:

$$s = \ln(\cosh t) \quad \Rightarrow \quad e^s = \cosh t \quad \Rightarrow \quad \operatorname{sech} t = e^{-s}$$

Also:

$$\tanh^2 t = 1 - \operatorname{sech}^2 t = 1 - e^{-2s} \quad \Rightarrow \quad \tanh t = \sqrt{1 - e^{-2s}} \quad \Rightarrow \quad t = \tanh^{-1}\sqrt{1 - e^{-2s}}$$

Substituting in $\mathbf{r}(t)$ gives:

$$\mathbf{r}_1(s) = \langle t - \tanh t, \operatorname{sech} t \rangle = \left\langle \tanh^{-1}\sqrt{1 - e^{-2s}} - \sqrt{1 - e^{-2s}}, e^{-s} \right\rangle$$

(c) The tractrix is shown in the following figure:

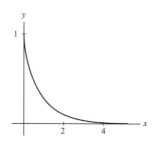

13.4 Curvature   (LT Section 14.4)

Preliminary Questions

1. What is the unit tangent vector of a line with direction vector $\mathbf{v} = \langle 2, 1, -2 \rangle$?

SOLUTION   A line with direction vector $\mathbf{v}$ has the parametrization:

$$\mathbf{r}(t) = \overrightarrow{OP_0} + t\mathbf{v}$$

hence, since $\overrightarrow{OP_0}$ and $\mathbf{v}$ are constant vectors, we have:

$$\mathbf{r}'(t) = \mathbf{v}$$

Therefore, since $\|\mathbf{v}\| = 3$, the unit tangent vector is:

$$\mathbf{T}(t) = \frac{\mathbf{r}'(t)}{\|\mathbf{r}'(t)\|} = \frac{\mathbf{v}}{\|\mathbf{v}\|} = \langle 2/3, 1/3, -2/3 \rangle$$

2. What is the curvature of a circle of radius 4?

SOLUTION   The curvature of a circle of radius R is $\frac{1}{R}$, hence the curvature of a circle of radius 4 is $\frac{1}{4}$.

3. Which has larger curvature, a circle of radius 2 or a circle of radius 4?

SOLUTION   The curvature of a circle of radius 2 is $\frac{1}{2}$, and it is larger than the curvature of a circle of radius 4, which is $\frac{1}{4}$.

4. What is the curvature of $\mathbf{r}(t) = \langle 2 + 3t, 7t, 5 - t \rangle$?

SOLUTION $\mathbf{r}(t)$ parametrizes the line $\langle 2, 0, 5 \rangle + t \langle 3, 7, -1 \rangle$, and a line has zero curvature.

5. What is the curvature at a point where $\mathbf{T}'(s) = \langle 1, 2, 3 \rangle$ in an arc length parametrization $\mathbf{r}(s)$?

SOLUTION The curvature is given by the formula:

$$\kappa(t) = \frac{\|\mathbf{T}'(t)\|}{\|\mathbf{r}'(t)\|}$$

In an arc length parametrization, $\|\mathbf{r}'(t)\| = 1$ for all t, hence the curvature is $\kappa(t) = \|\mathbf{T}'(t)\|$. Using the given information we obtain the following curvature:

$$\kappa = \| \langle 1, 2, 3 \rangle \| = \sqrt{1^2 + 2^2 + 3^2} = \sqrt{14}$$

6. What is the radius of curvature of a circle of radius 4?

SOLUTION The definition of the osculating circle implies that the osculating circles at the points of a circle, is the circle itself. Therefore, the radius of curvature is the radius of the circle, that is, 4.

7. What is the radius of curvature at P if $\kappa_P = 9$?

SOLUTION The radius of curvature is the reciprocal of the curvature, hence the radius of curvature at P is:

$$R = \frac{1}{\kappa_P} = \frac{1}{9}$$

Exercises

In Exercises 1–6, calculate $\mathbf{r}'(t)$ and $\mathbf{T}(t)$, and evaluate $\mathbf{T}(1)$.

1. $\mathbf{r}(t) = \langle 4t^2, 9t \rangle$

SOLUTION We differentiate $\mathbf{r}(t)$ to obtain:

$$\mathbf{r}'(t) = \langle 8t, 9 \rangle \quad \Rightarrow \quad \|\mathbf{r}'(t)\| = \sqrt{(8t)^2 + 9^2} = \sqrt{64t^2 + 81}$$

We now find the unit tangent vector:

$$\mathbf{T}(t) = \frac{\mathbf{r}'(t)}{\|\mathbf{r}'(t)\|} = \frac{1}{\sqrt{64t^2 + 81}} \langle 8t, 9 \rangle$$

For $t = 1$ we obtain the vector:

$$\mathbf{T}(t) = \frac{1}{\sqrt{64 + 81}} \langle 8, 9 \rangle = \left\langle \frac{8}{\sqrt{145}}, \frac{9}{\sqrt{145}} \right\rangle.$$

3. $\mathbf{r}(t) = \langle 3 + 4t, 3 - 5t, 9t \rangle$

SOLUTION We first find the vector $\mathbf{r}'(t)$ and its length:

$$\mathbf{r}'(t) = \langle 4, -5, 9 \rangle \quad \Rightarrow \quad \|\mathbf{r}'(t)\| = \sqrt{4^2 + (-5)^2 + 9^2} = \sqrt{122}$$

The unit tangent vector is therefore:

$$\mathbf{T}(t) = \frac{\mathbf{r}'(t)}{\|\mathbf{r}'(t)\|} = \frac{1}{\sqrt{122}} \langle 4, -5, 9 \rangle = \left\langle \frac{4}{\sqrt{122}}, -\frac{5}{\sqrt{122}}, \frac{9}{\sqrt{122}} \right\rangle$$

We see that the unit tangent vector is constant, since the curve is a straight line.

5. $\mathbf{r}(t) = \langle \cos \pi t, \sin \pi t, t \rangle$

SOLUTION We compute the derivative vector and its length:

$$\mathbf{r}'(t) = \langle -\pi \sin \pi t, \pi \cos \pi t, 1 \rangle$$

$$\|\mathbf{r}'(t)\| = \sqrt{(-\pi \sin \pi t)^2 + (\pi \cos \pi t)^2 + 1^2} = \sqrt{\pi^2(\sin^2 \pi t + \cos^2 \pi t) + 1} = \sqrt{\pi^2 + 1}$$

The unit tangent vector is thus:

$$\mathbf{T}(t) = \frac{\mathbf{r}'(t)}{\|\mathbf{r}'(t)\|} = \frac{1}{\sqrt{\pi^2 + 1}} \langle -\pi \sin \pi t, \pi \cos \pi t, 1 \rangle$$

For $t = 1$ we get:

$$\mathbf{T}(1) = \frac{1}{\sqrt{\pi^2 + 1}} \langle -\pi \sin \pi, \pi \cos \pi, 1 \rangle = \frac{1}{\sqrt{\pi^2 + 1}} \langle 0, -\pi, 1 \rangle = \left\langle 0, -\frac{\pi}{\sqrt{\pi^2 + 1}}, \frac{1}{\sqrt{\pi^2 + 1}} \right\rangle.$$

In Exercises 7–10, use Eq. (3) to calculate the curvature function $\kappa(t)$.

7. $\mathbf{r}(t) = \langle 1, e^t, t \rangle$

SOLUTION   We compute the first and the second derivatives of $\mathbf{r}(t)$:

$$\mathbf{r}'(t) = \langle 0, e^t, 1 \rangle, \quad \mathbf{r}''(t) = \langle 0, e^t, 0 \rangle.$$

Next, we find the cross product $\mathbf{r}'(t) \times \mathbf{r}''(t)$:

$$\mathbf{r}'(t) \times \mathbf{r}''(t) = \begin{vmatrix} \mathbf{i} & \mathbf{j} & \mathbf{k} \\ 0 & e^t & 1 \\ 0 & e^t & 0 \end{vmatrix} = \begin{vmatrix} e^t & 1 \\ e^t & 0 \end{vmatrix} \mathbf{i} - \begin{vmatrix} 0 & 1 \\ 0 & 0 \end{vmatrix} \mathbf{j} + \begin{vmatrix} 0 & e^t \\ 0 & e^t \end{vmatrix} \mathbf{k} = -e^t \mathbf{i} = \langle -e^t, 0, 0 \rangle$$

We need to find the lengths of the following vectors:

$$\|\mathbf{r}'(t) \times \mathbf{r}''(t)\| = |\langle -e^t, 0, 0 \rangle| = e^t$$

$$\|\mathbf{r}'(t)\| = \sqrt{0^2 + (e^t)^2 + 1^2} = \sqrt{1 + e^{2t}}$$

We now use the formula for curvature to calculate $\kappa(t)$:

$$\kappa(t) = \frac{\|\mathbf{r}'(t) \times \mathbf{r}''(t)\|}{\|\mathbf{r}'(t)\|^3} = \frac{e^t}{\left(\sqrt{1 + e^{2t}}\right)^3} = \frac{e^t}{\left(1 + e^{2t}\right)^{3/2}}$$

9. $\mathbf{r}(t) = \langle 4t + 1, 4t - 3, 2t \rangle$

SOLUTION   By Formula (3) we have:

$$\kappa(t) = \frac{\|\mathbf{r}'(t) \times \mathbf{r}''(t)\|}{\|\mathbf{r}'(t)\|^3}$$

We compute $\mathbf{r}'(t)$ and $\mathbf{r}''(t)$:

$$\mathbf{r}'(t) = \langle 4, 4, 2 \rangle, \quad \mathbf{r}''(t) = \langle 0, 0, 0 \rangle$$

Thus $\mathbf{r}'(t) \times \mathbf{r}''(t) = \langle 0, 0, 0 \rangle$, $\|\mathbf{r}'(t) \times \mathbf{r}''(t)\| = 0$, and $\kappa(t) = 0$, as expected.

In Exercises 11–14, use Eq. (3) to evaluate the curvature at the given point.

11. $\mathbf{r}(t) = \langle 1/t, 1/t^2, t^2 \rangle$, $t = -1$

SOLUTION   By the formula for curvature we know:

$$\kappa(t) = \frac{\|\mathbf{r}'(t) \times \mathbf{r}''(t)\|}{\|\mathbf{r}'(t)\|^3}$$

We now find $\mathbf{r}'(t)$, $\mathbf{r}''(t)$ and the cross product. These give:

$$\mathbf{r}'(t) = \left\langle -t^{-2}, -2t^{-3}, 2t \right\rangle, \quad \Rightarrow \mathbf{r}'(-1) = \langle -1, 2, -2 \rangle$$

$$\mathbf{r}''(t) = \left\langle 2t^{-3}, 6t^{-4}, 2 \right\rangle, \quad \Rightarrow \mathbf{r}''(-1) = \langle -2, 6, 2 \rangle$$

$$\mathbf{r}'(-1) \times \mathbf{r}''(-1) = \langle 16, 6, -2 \rangle$$

Now finding the norms, we get:

$$\|\mathbf{r}'(-1)\| = \sqrt{(-1)^2 + 2^2 + (-2)^2} = 3$$

$$\|\mathbf{r}'(-1) \times \mathbf{r}''(-1)\| = \sqrt{16^2 + 6^2 + (-2)^2} = \sqrt{296} = 2\sqrt{74}$$

Therefore,

$$\kappa(-1) = \frac{\|\mathbf{r}'(-1) \times \mathbf{r}''(-1)\|}{\|\mathbf{r}'(-1)\|^3} = \frac{2\sqrt{74}}{3^3} = \frac{2\sqrt{74}}{27}$$

13. $\mathbf{r}(t) = \langle \cos t, \sin t, t^2 \rangle$, $t = \frac{\pi}{2}$

SOLUTION   By the formula for curvature we know:

$$\kappa(t) = \frac{\|\mathbf{r}'(t) \times \mathbf{r}''(t)\|}{\|\mathbf{r}'(t)\|^3}$$

We now find $\mathbf{r}'(t)$, $\mathbf{r}''(t)$ and the cross product. These give:

$$\mathbf{r}'(t) = \langle -\sin t, \cos t, 2t \rangle \quad \Rightarrow \mathbf{r}'(\pi/2) = \langle -1, 0, \pi \rangle$$

$$\mathbf{r}''(t) = \langle -\cos t, -\sin t, 2 \rangle \quad \Rightarrow \mathbf{r}''(\pi/2) = \langle 0, -1, 2 \rangle$$

$$\mathbf{r}'(\pi/2) \times \mathbf{r}''(\pi/2) = \langle \pi, 2, 1 \rangle$$

Now finding norms we get:

$$\|\mathbf{r}'(\pi/2)\| = \sqrt{(-1)^2 + 0^2 + \pi^2} = \sqrt{1 + \pi^2}$$

$$\|\mathbf{r}'(\pi/2) \times \mathbf{r}''(\pi/2)\| = \sqrt{\pi^2 + (-1)^2 + 2^2} = \sqrt{\pi^2 + 5}$$

Therefore,

$$\kappa(\pi/2) = \frac{\|\mathbf{r}'(\pi/2) \times \mathbf{r}''(\pi/2)\|}{\|\mathbf{r}'(\pi/2)\|^3} = \frac{\sqrt{\pi^2 + 5}}{(\sqrt{1 + \pi^2})^3} = \frac{\sqrt{\pi^2 + 5}}{(1 + \pi^2)^{3/2}} \approx 0.108$$

In Exercises 15–18, find the curvature of the plane curve at the point indicated.

15. $y = e^t$, $t = 3$

SOLUTION We use the curvature of a graph in the plane:

$$\kappa(t) = \frac{|f''(t)|}{\left(1 + f'(t)^2\right)^{3/2}}$$

In our case $f(t) = e^t$, hence $f'(t) = f''(t) = e^t$ and we obtain:

$$\kappa(t) = \frac{e^t}{\left(1 + e^{2t}\right)^{3/2}} \quad \Rightarrow \quad \kappa(3) = \frac{e^3}{\left(1 + e^6\right)^{3/2}} \approx 0.0025$$

17. $y = t^4$, $t = 2$

SOLUTION By the curvature of a graph in the plane, we have:

$$\kappa(t) = \frac{|f''(t)|}{\left(1 + f'(t)^2\right)^{3/2}}$$

In this case $f(t) = t^4$, $f'(t) = 4t^3$, $f''(t) = 12t^2$. Hence,

$$\kappa(t) = \frac{12t^2}{\left(1 + \left(4t^3\right)^2\right)^{3/2}} = \frac{12t^2}{\left(1 + 16t^6\right)^{3/2}}$$

At $t = 2$ we obtain the following curvature:

$$\kappa(2) = \frac{12 \cdot 2^2}{(1 + 16 \cdot 2^6)^{3/2}} = \frac{48}{(1025)^{3/2}} \approx 0.0015.$$

19. Find the curvature of $\mathbf{r}(t) = \langle 2 \sin t, \cos 3t, t \rangle$ at $t = \frac{\pi}{3}$ and $t = \frac{\pi}{2}$ (Figure 16).

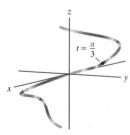

FIGURE 16 The curve $\mathbf{r}(t) = \langle 2 \sin t, \cos 3t, t \rangle$.

SOLUTION   By the formula for curvature we have:

$$\kappa(t) = \frac{\|\mathbf{r}'(t) \times \mathbf{r}''(t)\|}{\|\mathbf{r}'(t)\|^3} \tag{1}$$

We compute the first and second derivatives:

$$\mathbf{r}'(t) = \langle 2\cos t, -3\sin 3t, 1 \rangle, \quad \mathbf{r}''(t) = \langle -2\sin t, -9\cos 3t, 0 \rangle$$

At the points $t = \frac{\pi}{3}$ and $t = \frac{\pi}{2}$ we have:

$$\mathbf{r}'\left(\frac{\pi}{3}\right) = \left\langle 2\cos\frac{\pi}{3}, -3\sin\frac{3\pi}{3}, 1 \right\rangle = \left\langle 2\cos\frac{\pi}{3}, -3\sin\pi, 1 \right\rangle = \langle 1, 0, 1 \rangle$$

$$\mathbf{r}''\left(\frac{\pi}{3}\right) = \left\langle -2\sin\frac{\pi}{3}, -9\cos\frac{3\pi}{3}, 0 \right\rangle = \left\langle -\sqrt{3}, 9, 0 \right\rangle$$

$$\mathbf{r}'\left(\frac{\pi}{2}\right) = \left\langle 2\cos\frac{\pi}{2}, -3\sin\frac{3\pi}{2}, 1 \right\rangle = \langle 0, 3, 1 \rangle$$

$$\mathbf{r}''\left(\frac{\pi}{2}\right) = \left\langle -2\sin\frac{\pi}{2}, -9\cos\frac{3\pi}{2}, 0 \right\rangle = \langle -2, 0, 0 \rangle$$

We compute the cross products required to use (1):

$$\mathbf{r}'\left(\frac{\pi}{3}\right) \times \mathbf{r}''\left(\frac{\pi}{3}\right) = \begin{vmatrix} \mathbf{i} & \mathbf{j} & \mathbf{k} \\ 1 & 0 & 1 \\ -\sqrt{3} & 9 & 0 \end{vmatrix} = \begin{vmatrix} 0 & 1 \\ 9 & 0 \end{vmatrix}\mathbf{i} - \begin{vmatrix} 1 & 1 \\ -\sqrt{3} & 0 \end{vmatrix}\mathbf{j} + \begin{vmatrix} 1 & 0 \\ -\sqrt{3} & 9 \end{vmatrix}\mathbf{k} = -9\mathbf{i} - \sqrt{3}\mathbf{j} + 9\mathbf{k}$$

$$\mathbf{r}'\left(\frac{\pi}{2}\right) \times \mathbf{r}''\left(\frac{\pi}{2}\right) = \begin{vmatrix} \mathbf{i} & \mathbf{j} & \mathbf{k} \\ 0 & 3 & 1 \\ -2 & 0 & 0 \end{vmatrix} = \begin{vmatrix} 3 & 1 \\ 0 & 0 \end{vmatrix}\mathbf{i} - \begin{vmatrix} 0 & 1 \\ -2 & 0 \end{vmatrix}\mathbf{j} + \begin{vmatrix} 0 & 3 \\ -2 & 0 \end{vmatrix}\mathbf{k} = -2\mathbf{j} + 6\mathbf{k}$$

Hence,

$$\left\|\mathbf{r}'\left(\frac{\pi}{3}\right) \times \mathbf{r}''\left(\frac{\pi}{3}\right)\right\| = \sqrt{(-9)^2 + \left(-\sqrt{3}\right)^2 + 9^2} = \sqrt{165}$$

$$\left\|\mathbf{r}'\left(\frac{\pi}{3}\right)\right\| = \sqrt{1^2 + 0^2 + 1^2} = \sqrt{2}$$

At $t = \frac{\pi}{2}$ we have:

$$\left\|\mathbf{r}'\left(\frac{\pi}{2}\right) \times \mathbf{r}''\left(\frac{\pi}{2}\right)\right\| = \sqrt{(-2)^2 + 6^2} = \sqrt{40} = 2\sqrt{10}$$

$$\left\|\mathbf{r}'\left(\frac{\pi}{2}\right)\right\| = \sqrt{0^2 + 3^2 + 1^2} = \sqrt{10}$$

Substituting the values for $t = \frac{\pi}{3}$ and $t = \frac{\pi}{2}$ in (1) we obtain the following curvatures:

$$\kappa\left(\frac{\pi}{3}\right) = \frac{\sqrt{165}}{\left(\sqrt{2}\right)^3} = \frac{\sqrt{165}}{2\sqrt{2}} \approx 4.54$$

$$\kappa\left(\frac{\pi}{2}\right) = \frac{2\sqrt{10}}{\left(\sqrt{10}\right)^3} = \frac{2\sqrt{10}}{10\sqrt{10}} = 0.2$$

21. Show that the tractrix $\mathbf{r}(t) = \langle t - \tanh t, \operatorname{sech} t \rangle$ has the curvature function $\kappa(t) = \operatorname{sech} t$.

SOLUTION   Writing $\mathbf{r}(t) = \langle x(t), y(t) \rangle$, we have $x(t) = t - \tanh t$ and $y(t) = \operatorname{sech} t$. We compute the first and second derivatives of these functions. We use $\tanh^2 t = 1 - \operatorname{sech}^2 t$ to obtain:

$$x'(t) = 1 - \operatorname{sech}^2 t = \tanh^2 t$$

$$x''(t) = -2\operatorname{sech} t\,(-\operatorname{sech} t \tanh t) = 2\operatorname{sech}^2 t \tanh t$$

$$y'(t) = -\operatorname{sech} t \tanh t$$

$$y''(t) = -(-\operatorname{sech} t \tanh^2 t + \operatorname{sech}^3 t) = \operatorname{sech} t(\tanh^2 t - \operatorname{sech}^2 t) = \operatorname{sech} t(1 - 2\operatorname{sech}^2 t)$$

We compute the cross product $\|\mathbf{r}' \times \mathbf{r}''\|$:

$$x'(t)y''(t) - x''(t)y'(t) = \tanh^2 t \operatorname{sech} t (1 - 2 \operatorname{sech}^2 t) + 2 \operatorname{sech}^3 t \tanh^2 t$$

$$= \tanh^2 t \left[\operatorname{sech} t - 2 \operatorname{sech}^3 t + 2 \operatorname{sech}^3 t \right] = \tanh^2 t \operatorname{sech} t$$

We compute the length of $\mathbf{r}'$:

$$x'(t)^2 + y'(t)^2 = \tanh^4 t + \operatorname{sech}^2 t \tanh^2 t = \tanh^2 t (\tanh^2 t + \operatorname{sech}^2 t) = \tanh^2 t$$

Hence

$$\|\mathbf{r}'\|^3 = \left(\tanh^2 t \right)^{3/2} = \tanh^3 t$$

Substituting, we obtain

$$\kappa(t) = \frac{|\operatorname{sech} t \tanh^2 t|}{\tanh^3 t} = \frac{\operatorname{sech} t \tanh^2 t}{\tanh^3 t} = \frac{\operatorname{sech} t}{\tanh t}$$

23. Find the value of α such that the curvature of $y = e^{\alpha x}$ at $x = 0$ is as large as possible.

SOLUTION Using the curvature of a graph in the plane we have:

$$\kappa(x) = \frac{|y''(x)|}{\left(1 + y'(x)^2 \right)^{3/2}} \tag{1}$$

In our case $y'(x) = \alpha e^{\alpha x}$, $y''(x) = \alpha^2 e^{\alpha x}$. Substituting in (1) we obtain

$$\kappa(x) = \frac{\alpha^2 e^{\alpha x}}{\left(1 + \alpha^2 e^{2\alpha x} \right)^{3/2}}$$

The curvature at the origin is thus

$$\kappa(0) = \frac{\alpha^2 e^{\alpha \cdot 0}}{\left(1 + \alpha^2 e^{2\alpha \cdot 0} \right)^{3/2}} = \frac{\alpha^2}{\left(1 + \alpha^2 \right)^{3/2}}$$

Since $\kappa(0)$ and $\kappa^2(0)$ have their maximum values at the same values of α, we may maximize the function:

$$g(\alpha) = \kappa^2(0) = \frac{\alpha^4}{\left(1 + \alpha^2 \right)^3}$$

We find the stationary points:

$$g'(\alpha) = \frac{4\alpha^3 (1 + \alpha^2)^3 - \alpha^4 (3)(1 + \alpha^2)^2 2\alpha}{(1 + \alpha^2)^6} = \frac{2\alpha^3 (1 + \alpha^2)^2 (2 - \alpha^2)}{(1 + \alpha^2)^6} = 0$$

The stationary points are the solutions of the following equation:

$$2\alpha^3 (1 + \alpha^2)^2 (2 - \alpha^2) = 0$$

$$\alpha^3 = 0 \qquad \text{or} \qquad 2 - \alpha^2 = 0$$
$$\alpha = 0 \qquad\qquad\qquad \alpha = \pm\sqrt{2}$$

Since $g(\alpha) \geq 0$ and $g(0) = 0$, $\alpha = 0$ is a minimum point. Also, $g'(\alpha)$ is positive immediately to the left of $\sqrt{2}$ and negative to the right. Hence, $\alpha = \sqrt{2}$ is a maximum point. Since $g(\alpha)$ is an even function, $\alpha = -\sqrt{2}$ is a maximum point as well. Conclusion: $\kappa(x)$ takes its maximum value at the origin when $\alpha = \pm\sqrt{2}$.

25. Show that the curvature function of the parametrization $\mathbf{r}(t) = \langle a \cos t, b \sin t \rangle$ of the ellipse $\left(\frac{x}{a} \right)^2 + \left(\frac{y}{b} \right)^2 = 1$ is

$$\kappa(t) = \frac{ab}{(b^2 \cos^2 t + a^2 \sin^2 t)^{3/2}} \qquad \boxed{9}$$

SOLUTION The curvature is the following function:

$$\kappa(t) = \frac{\|\mathbf{r}'(t) \times \mathbf{r}''(t)\|}{\|\mathbf{r}'(t)\|^3} \tag{1}$$

We compute the derivatives and their cross product:

$$\mathbf{r}'(t) = \langle -a \sin t, b \cos t \rangle, \; \mathbf{r}''(t) = \langle -a \cos t, -b \sin t \rangle$$

$$\mathbf{r}'(t) \times \mathbf{r}''(t) = (-a \sin t\mathbf{i} + b \cos t\mathbf{j}) \times (-a \cos t\mathbf{i} - b \sin t\mathbf{j})$$

$$= ab \sin^2 t\mathbf{k} + ab \cos^2 t\mathbf{k} = ab(\sin^2 t + \cos^2 t)\mathbf{k} = ab\mathbf{k}$$

Thus,

$$\|\mathbf{r}'(t) \times \mathbf{r}''(t)\| = \|ab\mathbf{k}\| = ab$$

$$\|\mathbf{r}'(t)\| = \sqrt{(-a \sin t)^2 + (b \cos t)^2} = \sqrt{a^2 \sin^2 t + b^2 \cos^2 t}$$

Substituting in (1) we obtain the following curvature:

$$\kappa(t) = \frac{ab}{\left(\sqrt{a^2 \sin^2 t + b^2 \cos^2 t}\right)^3} = \frac{ab}{\left(a^2 \sin^2 t + b^2 \cos^2 t\right)^{3/2}}$$

27. In the notation of Exercise 25, assume that $a \geq b$. Show that $b/a^2 \leq \kappa(t) \leq a/b^2$ for all t.

SOLUTION   In Exercise 25 we showed that the curvature of the ellipse $\mathbf{r}(t) = \langle a \cos t, b \sin t \rangle$ is the following function:

$$\kappa(t) = \frac{ab}{\left(b^2 \cos^2 t + a^2 \sin^2 t\right)^{3/2}}$$

Since $a \geq b > 0$ the quotient becomes greater if we replace a by b in the denominator, and it becomes smaller if we replace b by a in the denominator. We use the identity $\cos^2 t + \sin^2 t = 1$ to obtain:

$$\frac{ab}{\left(a^2 \cos^2 t + a^2 \sin^2 t\right)^{3/2}} \leq \kappa(t) \leq \frac{ab}{\left(b^2 \cos^2 t + b^2 \sin^2 t\right)^{3/2}}$$

$$\frac{ab}{\left(a^2(\cos^2 t + \sin^2 t)\right)^{3/2}} \leq \kappa(t) \leq \frac{ab}{\left(b^2(\cos^2 t + \sin^2 t)\right)^{3/2}}$$

$$\frac{ab}{a^3} = \frac{ab}{(a^2)^{3/2}} \leq \kappa(t) \leq \frac{ab}{(b^2)^{3/2}} = \frac{ab}{b^3}$$

$$\frac{b}{a^2} \leq \kappa(t) \leq \frac{a}{b^2}$$

In Exercises 29–32, use Eq. (10) to compute the curvature at the given point.

29. $\langle t^2, t^3 \rangle$, $\quad t = 2$

SOLUTION   For the given parametrization, $x(t) = t^2$, $y(t) = t^3$, hence

$$x'(t) = 2t$$

$$x''(t) = 2$$

$$y'(t) = 3t^2$$

$$y''(t) = 6t$$

At the point $t = 2$ we have

$$x'(2) = 4, \quad x''(2) = 2, \quad y'(2) = 3 \cdot 2^2 = 12, \quad y''(2) = 12$$

Substituting in Eq. (10) we get

$$\kappa(2) = \frac{|x'(2)y''(2) - x''(2)y'(2)|}{\left(x'(2)^2 + y'(2)^2\right)^{3/2}} = \frac{|4 \cdot 12 - 2 \cdot 12|}{(4^2 + 12^2)^{3/2}} = \frac{24}{160^{3/2}} \approx 0.012$$

31. $\langle t \cos t, \sin t \rangle$, $\quad t = \pi$

SOLUTION   We have $x(t) = t \cos t$ and $y(t) = \sin t$, hence:

$$x'(t) = \cos t - t \sin t \quad \Rightarrow \quad x'(\pi) = \cos \pi - \pi \sin \pi = -1$$

$$x''(t) = -\sin t - (\sin t + t \cos t) = -2 \sin t - t \cos t \quad \Rightarrow \quad x''(\pi) = -2 \sin \pi - \pi \cos \pi = \pi$$

$$y'(t) = \cos t \quad \Rightarrow \quad y'(\pi) = \cos \pi = -1$$

$$y''(t) = -\sin t \quad \Rightarrow \quad y''(\pi) = -\sin \pi = 0$$

Substituting in Eq. (10) gives the following curvature:

$$\kappa(\pi) = \frac{|x'(\pi)y''(\pi) - x''(\pi)y'(\pi)|}{\left(x'(\pi)^2 + y'(\pi)^2\right)^{3/2}} = \frac{|-1\cdot 0 - \pi\cdot(-1)|}{\left((-1)^2 + (-1)^2\right)^{3/2}} = \frac{\pi}{2\sqrt{2}} \approx 1.11$$

33. Let $s(t) = \int_{-\infty}^{t} \|\mathbf{r}'(u)\|\, du$ for the Bernoulli spiral $\mathbf{r}(t) = \langle e^t \cos 4t, e^t \sin 4t \rangle$ (see Exercise 29 in Section 13.3). Show that the radius of curvature is proportional to $s(t)$.

SOLUTION The radius of curvature is the reciprocal of the curvature:

$$R(t) = \frac{1}{\kappa(t)}$$

We compute the curvature using the equality given in Exercise 29 in Section 3:

$$\kappa(t) = \frac{|x'(t)y''(t) - x''(t)y'(t)|}{\left(x'(t)^2 + y'(t)^2\right)^{3/2}} \tag{1}$$

In our case, $x(t) = e^t \cos 4t$ and $y(t) = e^t \sin 4t$. Hence:

$$x'(t) = e^t \cos 4t - 4e^t \sin 4t = e^t(\cos 4t - 4\sin 4t)$$

$$x''(t) = e^t(\cos 4t - 4\sin 4t) + e^t(-4\sin 4t - 16\cos 4t) = -e^t(15\cos 4t + 8\sin 4t)$$

$$y'(t) = e^t \sin 4t + 4e^t \cos 4t = e^t(\sin 4t + 4\cos 4t)$$

$$y''(t) = e^t(\sin 4t + 4\cos 4t) + e^t(4\cos 4t - 16\sin 4t) = e^t(8\cos 4t - 15\sin 4t)$$

We compute the numerator in (1):

$$x'(t)y''(t) - x''(t)y'(t) = e^{2t}(\cos 4t - 4\sin 4t)\cdot(8\cos 4t - 15\sin 4t)$$

$$+ e^{2t}(15\cos 4t + 8\sin 4t)\cdot(\sin 4t + 4\cos 4t)$$

$$= e^{2t}\left(68\cos^2 4t + 68\sin^2 4t\right) = 68e^{2t}$$

We compute the denominator in (1):

$$x'(t)^2 + y'(t)^2 = e^{2t}(\cos 4t - 4\sin 4t)^2 + e^{2t}(\sin 4t + 4\cos 4t)^2$$

$$= e^{2t}\left(\cos^2 4t - 8\cos 4t \sin 4t + 16\sin^2 4t + \sin^2 4t + 8\sin 4t \cos 4t + 16\cos^2 4t\right)$$

$$= e^{2t}\left(\cos^2 4t + \sin^2 4t + 16(\sin^2 4t + \cos^2 4t)\right)$$

$$= e^{2t}(1 + 16\cdot 1) = 17e^{2t} \tag{2}$$

Hence

$$\left(x'(t)^2 + y'(t)^2\right)^{3/2} = 17^{3/2}e^{3t}$$

Substituting in (2) we have

$$\kappa(t) = \frac{68e^{2t}}{17^{3/2}e^{3t}} = \frac{4}{\sqrt{17}}e^{-t} \quad\Rightarrow\quad R = \frac{\sqrt{17}}{4}e^t \tag{3}$$

On the other hand, by the Fundamental Theorem and (2) we have

$$s'(t) = \|\mathbf{r}'(t)\| = \sqrt{x'(t)^2 + y'(t)^2} = \sqrt{17e^{2t}} = \sqrt{17}e^t$$

We integrate to obtain

$$s(t) = \int \sqrt{17}\,e^t\, dt = \sqrt{17}\,e^t + C \tag{4}$$

Since $s(t) = \int_{-\infty}^{t} \|\mathbf{r}'(u)\|\, du$, we have $\lim_{t\to-\infty} s(t) = 0$, hence by (4):

$$0 = \lim_{t\to-\infty}\left(\sqrt{17}e^t + C\right) = 0 + C = C.$$

Substituting $C = 0$ in (4) we get:

$$s(t) = \sqrt{17}e^t \tag{5}$$

Combining (3) and (5) gives:

$$R(t) = \frac{1}{4}s(t)$$

which means that the radius of curvature is proportional to $s(t)$.

35. $\boxed{\textsf{CAS}}$ Plot and compute the curvature $\kappa(t)$ of the **clothoid** $\mathbf{r}(t) = \langle x(t), y(t) \rangle$, where

$$x(t) = \int_0^t \sin \frac{u^3}{3}\, du, \qquad y(t) = \int_0^t \cos \frac{u^3}{3}\, du$$

SOLUTION We use the following formula for the curvature (given earlier):

$$\kappa(t) = \frac{|x'(t)y''(t) - x''(t)y'(t)|}{\left(x'(t)^2 + y'(t)^2\right)^{3/2}} \tag{1}$$

We compute the first and second derivatives of $x(t)$ and $y(t)$. Using the Fundamental Theorem and the Chain Rule we get:

$$x'(t) = \sin \frac{t^3}{3}$$

$$x''(t) = \frac{3t^2}{3} \cos \frac{t^3}{3} = t^2 \cos \frac{t^3}{3}$$

$$y'(t) = \cos \frac{t^3}{3}$$

$$y''(t) = \frac{3t^2}{3} \left(-\sin \frac{t^3}{3}\right) = -t^2 \sin \frac{t^3}{3}$$

Substituting in (1) gives the following curvature function:

$$\kappa(t) = \frac{\left|\sin \frac{t^3}{3} \left(-t^2 \sin \frac{t^3}{3}\right) - t^2 \cos \frac{t^3}{3} \cos \frac{t^3}{3}\right|}{\left(\left(\sin \frac{t^3}{3}\right)^2 + \left(\cos \frac{t^3}{3}\right)^2\right)^{3/2}} = \frac{t^2 \left(\sin^2 \frac{t^3}{3} + \cos^2 \frac{t^3}{3}\right)}{1^{3/2}} = t^2$$

That is, $\kappa(t) = t^2$. Here is a plot of the curvature as a function of t:

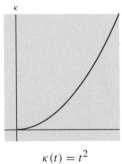

$$\kappa(t) = t^2$$

37. Find the unit normal vector $\mathbf{N}(t)$ to $\mathbf{r}(t) = \langle 4, \sin 2t, \cos 2t \rangle$.

SOLUTION We first find the unit tangent vector:

$$\mathbf{T}(t) = \frac{\mathbf{r}'(t)}{\|\mathbf{r}'(t)\|} \tag{1}$$

We have

$$\mathbf{r}'(t) = \frac{d}{dt} \langle 4, \sin 2t, \cos 2t \rangle = \langle 0, 2\cos 2t, -2\sin 2t \rangle = 2 \langle 0, \cos 2t, -\sin 2t \rangle$$

$$\|\mathbf{r}'(t)\| = 2\sqrt{0^2 + \cos^2 2t + (-\sin 2t)^2} = 2\sqrt{0 + 1} = 2$$

Substituting in (1) gives:

$$T(t) = \frac{2\langle 0, \cos 2t, -\sin 2t\rangle}{2} = \langle 0, \cos 2t, -\sin 2t\rangle$$

The normal vector is the following vector:

$$N(t) = \frac{T'(t)}{\|T'(t)\|} \tag{2}$$

We compute the derivative of the unit tangent vector and its length:

$$T'(t) = \frac{d}{dt}\langle 0, \cos 2t, -\sin 2t\rangle = \langle 0, -2\sin 2t, -2\cos 2t\rangle = -2\langle 0, \sin 2t, \cos 2t\rangle$$

$$\|T'(t)\| = 2\sqrt{0^2 + \sin^2 2t + \cos^2 2t} = 2\sqrt{0+1} = 2$$

Substituting in (2) we obtain:

$$N(t) = \frac{-2\langle 0, \sin 2t, \cos 2t\rangle}{2} = \langle 0, -\sin 2t, -\cos 2t\rangle$$

39. Find the normal vectors to $r(t) = \langle t, \cos t\rangle$ at $t = \frac{\pi}{4}$ and $t = \frac{3\pi}{4}$.

SOLUTION The normal vector to $r(t) = \langle t, \cos t\rangle$ is $T'(t)$, where $T(t) = \frac{r'(t)}{\|r'(t)\|}$ is the unit tangent vector. We have

$$r'(t) = \langle 1, -\sin t\rangle \quad \Rightarrow \quad \|r'(t)\| = \sqrt{1^2 + (\sin t)^2} = \sqrt{1 + \sin^2 t}$$

Hence,

$$T(t) = \frac{1}{\sqrt{1 + \sin^2 t}}\langle 1, -\sin t\rangle$$

We compute the derivative of $T(t)$ to find the normal vector. We use the Product Rule and the Chain Rule to obtain:

$$T'(t) = \frac{1}{\sqrt{1 + \sin^2 t}}\frac{d}{dt}\langle 1, -\sin t\rangle + \left(\frac{1}{\sqrt{1 + \sin^2 t}}\right)'\langle 1, -\sin t\rangle$$

$$= \frac{1}{\sqrt{1 + \sin^2 t}}\langle 0, -\cos t\rangle - \frac{1}{1 + \sin^2 t}\cdot\frac{2\sin t\cos t}{2\sqrt{1 + \sin^2 t}}\langle 1, -\sin t\rangle$$

$$= \frac{1}{\sqrt{1 + \sin^2 t}}\langle 0, \cos t\rangle - \frac{\sin 2t}{2\left(1 + \sin^2 t\right)^{3/2}}\langle 1, -\sin t\rangle$$

At $t = \frac{\pi}{4}$ we obtain the normal vector:

$$T'\left(\frac{\pi}{4}\right) = \frac{1}{\sqrt{1 + \frac{1}{2}}}\left\langle 0, -\frac{1}{\sqrt{2}}\right\rangle - \frac{1}{2\left(1 + \frac{1}{2}\right)^{3/2}}\left\langle 1, -\frac{1}{\sqrt{2}}\right\rangle = \left\langle 0, -\frac{1}{\sqrt{3}}\right\rangle - \left\langle\frac{\sqrt{2}}{3\sqrt{3}}, \frac{-1}{3\sqrt{3}}\right\rangle = \left\langle\frac{-\sqrt{2}}{3\sqrt{3}}, \frac{-2}{3\sqrt{3}}\right\rangle$$

At $t = \frac{3\pi}{4}$ we obtain:

$$T'\left(\frac{3\pi}{4}\right) = \frac{1}{\sqrt{1 + \frac{1}{2}}}\left\langle 0, \frac{1}{\sqrt{2}}\right\rangle - \frac{-1}{2\left(1 + \frac{1}{2}\right)^{3/2}}\left\langle 1, -\frac{1}{\sqrt{2}}\right\rangle = \left\langle 0, \frac{1}{\sqrt{3}}\right\rangle + \left\langle\frac{\sqrt{2}}{3\sqrt{3}}, \frac{-1}{3\sqrt{3}}\right\rangle = \left\langle\frac{\sqrt{2}}{3\sqrt{3}}, \frac{2}{3\sqrt{3}}\right\rangle$$

41. Find the unit normal to the clothoid (Exercise 35) at $t = \pi^{1/3}$.

SOLUTION The Clothoid is the plane curve $r(t) = \langle x(t), y(t)\rangle$ with

$$x(t) = \int_0^t \sin\frac{u^3}{3}\,du, \quad y(t) = \int_0^t \cos\frac{u^3}{3}\,du$$

The unit normal is the following vector:

$$N(t) = \frac{T'(t)}{\|T'(t)\|} \tag{1}$$

We first find the unit tangent vector $\mathbf{T}(t) = \frac{\mathbf{r}'(t)}{\|\mathbf{r}'(t)\|}$. By the Fundamental Theorem we have

$$\mathbf{r}'(t) = \left\langle \sin\frac{t^3}{3}, \cos\frac{t^3}{3} \right\rangle \quad\Rightarrow\quad \|\mathbf{r}'(t)\| = \sqrt{\sin^2\frac{t^3}{3} + \cos^2\frac{t^3}{3}} = \sqrt{1} = 1$$

Hence,

$$\mathbf{T}(t) = \left\langle \sin\frac{t^3}{3}, \cos\frac{t^3}{3} \right\rangle$$

We now differentiate $\mathbf{T}(t)$ using the Chain Rule to obtain:

$$\mathbf{T}'(t) = \left\langle \frac{3t^2}{3}\cos\frac{t^3}{3}, \frac{-3t^2}{3}\sin\frac{t^3}{3} \right\rangle = t^2\left\langle \cos\frac{t^3}{3}, -\sin\frac{t^3}{3} \right\rangle$$

Hence,

$$\|\mathbf{T}'(t)\| = t^2\sqrt{\cos^2\frac{t^3}{3} + \left(-\sin\frac{t^3}{3}\right)^2} = t^2$$

Substituting in (1) we obtain the following unit normal:

$$\mathbf{N}(t) = \left\langle \cos\frac{t^3}{3}, -\sin\frac{t^3}{3} \right\rangle$$

At the point $T = \pi^{1/3}$ the unit normal is

$$\mathbf{N}(\pi^{1/3}) = \left\langle \cos\frac{(\pi^{1/3})^3}{3}, -\sin\frac{(\pi^{1/3})^3}{3} \right\rangle = \left\langle \cos\frac{\pi}{3}, -\sin\frac{\pi}{3} \right\rangle = \left\langle \frac{1}{2}, -\frac{\sqrt{3}}{2} \right\rangle$$

In Exercises 43–48, use Eq. (11) to find $\mathbf{N}$ at the point indicated.

43. $\langle t^2, t^3 \rangle$, $\quad t = 1$

SOLUTION We use the equality

$$\mathbf{N}(t) = \frac{v(t)\mathbf{r}''(t) - v'(t)\mathbf{r}'(t)}{\|v(t)\mathbf{r}''(t) - v'(t)\mathbf{r}'(t)\|}$$

For $\mathbf{r}(t) = \langle t^2, t^3 \rangle$ we have

$$\mathbf{r}'(t) = \langle 2t, 3t^2 \rangle$$

$$\mathbf{r}''(t) = \langle 2, 6t \rangle$$

$$v(t) = \|\mathbf{r}'(t)\| = \sqrt{(2t)^2 + (3t^2)^2} = \sqrt{4t^2 + 9t^4}$$

$$v'(t) = \frac{8t + 36t^3}{2\sqrt{4t^2 + 9t^4}} = \frac{4t + 18t^3}{\sqrt{4t^2 + 9t^4}}$$

At the point $t = 1$ we get

$$\mathbf{r}''(1) = \langle 2, 6 \rangle, \quad v'(1) = \frac{4+18}{\sqrt{4+9}} = \frac{22}{\sqrt{13}},$$

and also

$$\mathbf{r}'(1) = \langle 2, 3 \rangle, \quad v(1) = \sqrt{4+9} = \sqrt{13}$$

Hence,

$$v(1)\mathbf{r}''(1) - v'(1)\mathbf{r}'(1) = \sqrt{13}\langle 2, 6 \rangle - \frac{22}{\sqrt{13}}\cdot\langle 2, 3 \rangle = \left\langle \frac{26-44}{\sqrt{13}}, \frac{78-66}{\sqrt{13}} \right\rangle = \frac{1}{\sqrt{13}}\langle -18, 12 \rangle$$

$$\|v(1)\mathbf{r}''(1) - v'(1)\mathbf{r}'(1)\| = \left\| \frac{1}{\sqrt{13}}\langle -18, 12 \rangle \right\| = \frac{1}{\sqrt{13}}\sqrt{(-18)^2 + 12^2} = \sqrt{\frac{468}{13}} = 6$$

Substituting in (1) gives the following unit normal:

$$\mathbf{N}(1) = \frac{\frac{1}{\sqrt{13}} \langle -18, 12 \rangle}{6} = \frac{1}{\sqrt{13}} \langle -3, 2 \rangle$$

45. $\left\langle t^2/2, t^3/3, t \right\rangle$, $t = 1$

SOLUTION   We use the following equality:

$$\mathbf{N}(t) = \frac{v(t)\mathbf{r}''(t) - v'(t)\mathbf{r}'(t)}{\|v(t)\mathbf{r}''(t) - v'(t)\mathbf{r}'(t)\|}$$

We compute the vectors in the equality above. For $\mathbf{r}(t) = \left\langle t^2/2, t^3/3, t \right\rangle$ we get:

$$\mathbf{r}'(t) = \left\langle t, t^2, 1 \right\rangle$$

$$\mathbf{r}''(t) = \langle 1, 2t, 0 \rangle$$

$$v(t) = \|\mathbf{r}'(t)\| = \sqrt{t^2 + t^4 + 1}$$

$$v'(t) = \frac{1}{2}(t^2 + t^4 + 1)^{-1/2}(4t^3 + 2t) = \frac{4t^3 + 2t}{2\sqrt{t^2 + t^4 + 1}}$$

At the point $t = 1$ we get:

$$\mathbf{r}'(1) = \langle 1, 1, 1 \rangle$$

$$\mathbf{r}''(1) = \langle 1, 2, 0 \rangle$$

$$v'(1) = \frac{6}{2\sqrt{3}} = \frac{3}{\sqrt{3}} = \sqrt{3}$$

$$v(1) = \sqrt{3}$$

Hence,

$$v(1)\mathbf{r}''(1) - v'(1)\mathbf{r}'(1) = \sqrt{3}\,\langle 1, 2, 0 \rangle - \sqrt{3}\,\langle 1, 1, 1 \rangle = \left\langle 0, \sqrt{3}, -\sqrt{3} \right\rangle$$

$$\|v(1)\mathbf{r}''(1) - v'(1)\mathbf{r}'(1)\| = \sqrt{0^2 + (\sqrt{3})^2 + (-\sqrt{3})^2} = \sqrt{6}$$

We now substitute these values in (1) to obtain the following unit normal:

$$\mathbf{N}(1) = \frac{v(1)\mathbf{r}''(1) - v'(1)\mathbf{r}'(1)}{\|v(1)\mathbf{r}''(1) - v'(1)\mathbf{r}'(1)\|} = \frac{\left\langle 0, \sqrt{3}, -\sqrt{3} \right\rangle}{\sqrt{6}} = \left\langle 0, \frac{1}{\sqrt{2}}, -\frac{1}{\sqrt{2}} \right\rangle$$

47. $\left\langle t, e^t, t \right\rangle$, $t = 0$

SOLUTION   We use the equality

$$\mathbf{N}(t) = \frac{v(t)\mathbf{r}''(t) - v'(t)\mathbf{r}'(t)}{\|v(t)\mathbf{r}''(t) - v'(t)\mathbf{r}'(t)\|}$$

For $\mathbf{r}(t) = \left\langle t, e^t, t \right\rangle$ we have

$$\mathbf{r}'(t) = \left\langle 1, e^t, 1 \right\rangle$$

$$\mathbf{r}''(t) = \left\langle 0, e^t, 0 \right\rangle$$

$$v(t) = \|\mathbf{r}'(t)\| = \sqrt{1^2 + (e^t)^2 + 1^2} = \sqrt{e^{2t} + 2}$$

$$v'(t) = \frac{2e^{2t}}{2\sqrt{e^{2t} + 2}} = \frac{e^{2t}}{\sqrt{e^{2t} + 2}}$$

At the point $t = 0$ we have

$$\mathbf{r}'(0 = \langle 1, 1, 1 \rangle, \quad \mathbf{r}''(0) = \langle 0, 1, 0 \rangle, \quad v(0) = \sqrt{3}, \quad v'(0) = \frac{1}{\sqrt{3}},$$

Hence,

$$v(0)\mathbf{r}''(0) - v'(0)\mathbf{r}'(0) = \sqrt{3}\,\langle 0, 1, 0\rangle - \frac{1}{\sqrt{3}}\,\langle 1, 1, 1\rangle$$

$$= \left\langle -\frac{1}{\sqrt{3}}, \frac{2}{\sqrt{3}}, -\frac{1}{\sqrt{3}}\right\rangle$$

$$= \frac{1}{\sqrt{3}}\,\langle -1, 2, -1\rangle$$

$$\|v(0)\mathbf{r}''(0) - v'(0)\mathbf{r}'(0)\| = \frac{1}{\sqrt{3}}\sqrt{1+4+1} = \sqrt{2}$$

Substituting in (1) we obtain the following unit normal:

$$\mathbf{N}(0) = \frac{\frac{1}{\sqrt{3}}\,\langle -1, 2, -1\rangle}{\sqrt{2}} = \frac{1}{\sqrt{6}}\,\langle -1, 2, -1\rangle$$

49. Let $f(x) = x^2$. Show that the center of the osculating circle at (x_0, x_0^2) is given by $\left(-4x_0^3, \frac{1}{2} + 3x_0^2\right)$.

SOLUTION   We parametrize the curve by $\mathbf{r}(x) = \langle x, x^2\rangle$. The center Q of the osculating circle at $x = x_0$ has the position vector

$$\overrightarrow{OQ} = \mathbf{r}(x_0) + \kappa(x_0)^{-1}\mathbf{N}(x_0) \tag{1}$$

We first find the curvature, using the formula for the curvature of a graph in the plane. We have $f'(x) = 2x$ and $f''(x) = 2$, hence,

$$\kappa(x) = \frac{|f''(x)|}{(1 + f'(x)^2)^{3/2}} = \frac{2}{(1 + 4x^2)^{3/2}} \quad \Rightarrow \quad \kappa(x_0)^{-1} = \frac{1}{2}(1 + 4x_0^2)^{3/2}$$

To find the unit normal vector $\mathbf{N}(x_0)$ we use the following considerations:

- The tangent vector is $\mathbf{r}'(x_0) = \langle 1, 2x_0\rangle$, hence the vector $\langle -2x_0, 1\rangle$ is orthogonal to $\mathbf{r}'(x_0)$ (since their dot product is zero). Hence $\mathbf{N}(x_0)$ is one of the two unit vectors $\pm\dfrac{1}{\sqrt{1+4x_0^2}}\,\langle -2x_0, 1\rangle$.

- The graph of $f(x) = x^2$ shows that the unit normal vector points in the positive y-direction, hence, the appropriate choice is:

$$\mathbf{N}(x_0) = \frac{1}{\sqrt{1 + 4x_0^2}}\,\langle -2x_0, 1\rangle \tag{2}$$

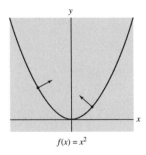

$f(x) = x^2$

We now substitute (2), (3), and $\mathbf{r}(x_0) = \langle x_0, x_0^2\rangle$ in (1) to obtain

$$\overrightarrow{OQ} = \langle x_0, x_0^2\rangle + \frac{1}{2}(1 + 4x_0^2)^{3/2} \cdot \frac{1}{\sqrt{1 + 4x_0^2}}\,\langle -2x_0, 1\rangle = \langle x_0, x_0^2\rangle + \frac{1}{2}(1 + 4x_0^2)\,\langle -2x_0, 1\rangle$$

$$= \langle x_0, x_0^2\rangle + \left\langle -x_0 - 4x_0^3, \frac{1}{2}(1 + 4x_0^2)\right\rangle = \left\langle -4x_0^3, \frac{1}{2} + 3x_0^2\right\rangle$$

The center of the osculating circle is the terminal point of $\overrightarrow{OQ}$, that is,

$$Q = \left(-4x_0^3, \frac{1}{2} + 3x_0^2\right)$$

In Exercises 51–58, find a parametrization of the osculating circle at the point indicated.

51. $\mathbf{r}(t) = \langle \cos t, \sin t \rangle, \quad t = \frac{\pi}{4}$

SOLUTION The curve $\mathbf{r}(t) = \langle \cos t, \sin t \rangle$ is the unit circle. By the definition of the osculating circle, it follows that the osculating circle at each point of the circle is the circle itself. Therefore the osculating circle to the unit circle at $t = \frac{\pi}{4}$ is the unit circle itself.

53. $y = x^2, \quad x = 1$

SOLUTION Let $f(x) = x^2$. We use the parametrization $\mathbf{r}(x) = \langle x, x^2 \rangle$ and proceed by the following steps.

Step 1. Find κ and $\mathbf{N}$. We compute κ using the curvature of a graph in the plane:

$$\kappa(x) = \frac{|f''(x)|}{\left(1 + f'(x)^2\right)^{3/2}}$$

We have $f'(x) = 2x$, $f''(x) = 2$, therefore,

$$\kappa(x) = \frac{2}{(1 + (2x)^2)^{3/2}} = \frac{2}{(1 + 4x^2)^{3/2}} \quad \Rightarrow \quad \kappa(1) = \frac{2}{5^{3/2}} \tag{1}$$

To find $\mathbf{N}(x)$ we notice that the tangent vector is $\mathbf{r}'(x) = \langle 1, 2x \rangle$ hence $\langle -2x, 1 \rangle$ is orthogonal to $\mathbf{r}'(x)$ (their dot product is zero). Therefore, $\mathbf{N}(x)$ is the unit vector in the direction of $\langle -2x, 1 \rangle$ or $- \langle -2x, 1 \rangle$ that points to the "inside" of the curve.

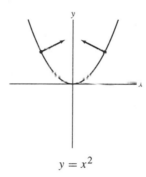

$y = x^2$

As shown in the figure, the unit normal vector points in the positive y-direction, hence:

$$\mathbf{N}(x) = \frac{\langle -2x, 1 \rangle}{\sqrt{4x^2 + 1}} \quad \Rightarrow \quad \mathbf{N}(1) = \frac{1}{\sqrt{5}} \langle -2, 1 \rangle \tag{2}$$

Step 2. Find the center of the osculating circle. The center Q at $\mathbf{r}(1)$ has the position vector

$$\overrightarrow{OQ} = \mathbf{r}(1) + \kappa(1)^{-1} \mathbf{N}(1)$$

Substituting (1), (2) and $\mathbf{r}(1) = \langle 1, 1 \rangle$ we get:

$$\overrightarrow{OQ} = \langle 1, 1 \rangle + \frac{5^{3/2}}{2} \cdot \frac{1}{5^{1/2}} \langle -2, 1 \rangle = \langle 1, 1 \rangle + \frac{5}{2} \langle -2, 1 \rangle = \left\langle -4, \frac{7}{2} \right\rangle$$

Step 3. Parametrize the osculating circle. The osculating circle has radius $R = \frac{1}{\kappa(1)} = \frac{5^{3/2}}{2}$ and it is centered at the point $\left(-4, \frac{7}{2} \right)$, therefore it has the following parametrization:

$$\mathbf{c}(t) = \left\langle -4, \frac{7}{2} \right\rangle + \frac{5^{3/2}}{2} \langle \cos t, \sin t \rangle$$

55. $\langle t - \sin t, 1 - \cos t \rangle, \quad t = \pi$

SOLUTION

Step 1. Find κ and $\mathbf{N}$. In Exercise 44 we found that:

$$\mathbf{N}(\pi) = \langle 0, -1 \rangle \tag{1}$$

To find κ we use the formula for curvature:

$$\kappa(\pi) = \frac{\| \mathbf{r}'(\pi) \times \mathbf{r}''(\pi) \|}{\| \mathbf{r}'(\pi) \|^3} \tag{2}$$

For $\mathbf{r}(t) = \langle t - \sin t, 1 - \cos t \rangle$ we have:

$$\mathbf{r}'(t) = \langle 1 - \cos t, \sin t \rangle \quad \Rightarrow \quad \mathbf{r}'(\pi) = \langle 1 - \cos \pi, \sin \pi \rangle = \langle 2, 0 \rangle$$
$$\mathbf{r}''(t) = \langle \sin t, \cos t \rangle \quad \Rightarrow \quad \mathbf{r}''(\pi) = \langle \sin \pi, \cos \pi \rangle = \langle 0, -1 \rangle$$

Hence,

$$\mathbf{r}'(\pi) \times \mathbf{r}''(\pi) = 2\mathbf{i} \times (-\mathbf{j}) = -2\mathbf{k}$$
$$\|\mathbf{r}'(\pi) \times \mathbf{r}''(\pi)\| = \| -2\mathbf{k} \| = 2 \quad \text{and} \quad \|\mathbf{r}'(\pi)\| = \| \langle 2, 0 \rangle \| = 2$$

Substituting in (2) we get:

$$\kappa(\pi) = \frac{2}{2^3} = \frac{1}{4} \tag{3}$$

Step 2. Find the center of the osculating circle. The center Q of the osculating circle at $\mathbf{r}(\pi) = \langle \pi, 2 \rangle$ has position vector

$$\overrightarrow{OQ} = \mathbf{r}(\pi) + \kappa(\pi)^{-1} N(\pi)$$

Substituting (1), (3) and $\mathbf{r}(\pi) = \langle \pi, 2 \rangle$ we get:

$$\overrightarrow{OQ} = \langle \pi, 2 \rangle + \left(\frac{1}{4} \right)^{-1} \langle 0, -1 \rangle = \langle \pi, 2 \rangle + \langle 0, -4 \rangle = \langle \pi, -2 \rangle$$

Step 3. Parametrize the osculating circle. The osculating circle has radius $R = \frac{1}{\kappa(\pi)}$ and it is centered at $(\pi, -2)$, hence it has the following parametrization:

$$\mathbf{c}(t) = \langle \pi, -2 \rangle + 4 \langle \cos t, \sin t \rangle$$

57. $\mathbf{r}(t) = \langle \cos t, \sin t, t \rangle, \quad t = 0$

SOLUTION   The curvature is the following quotient:

$$\kappa(t) = \frac{\|\mathbf{r}'(t) \times \mathbf{r}''(t)\|}{\|\mathbf{r}'(t)\|^3} \tag{1}$$

We compute the vectors $\mathbf{r}'(t)$ and $\mathbf{r}''(t)$:

$$\mathbf{r}'(t) = \frac{d}{dt} \langle \cos t, \sin t, t \rangle = \langle -\sin t, \cos t, 1 \rangle \tag{2}$$

$$\mathbf{r}''(t) = \frac{d}{dt} \langle -\sin t, \cos t, 1 \rangle = \langle -\cos t, -\sin t, 0 \rangle$$

We now compute the following cross product:

$$\mathbf{r}'(t) \times \mathbf{r}''(t) = \begin{vmatrix} \mathbf{i} & \mathbf{j} & \mathbf{k} \\ -\sin t & \cos t & 1 \\ -\cos t & -\sin t & 0 \end{vmatrix} = \begin{vmatrix} \cos t & 1 \\ -\sin t & 0 \end{vmatrix} \mathbf{i} - \begin{vmatrix} -\sin t & 1 \\ -\cos t & 0 \end{vmatrix} \mathbf{j} + \begin{vmatrix} -\sin t & \cos t \\ -\cos t & -\sin t \end{vmatrix} \mathbf{k}$$
$$= (\sin t)\mathbf{i} - (\cos t)\mathbf{j} + 1 \cdot \mathbf{k} \tag{3}$$

We calculate the norms of the vectors in (1). By (2) and (3) we have:

$$\|\mathbf{r}'(t) \times \mathbf{r}''(t)\| = \sqrt{\sin^2 t + (-\cos t)^2 + 1^2} = \sqrt{1 + 1} = \sqrt{2}$$

$$\|\mathbf{r}'(t)\| = \sqrt{(-\sin t)^2 + \cos^2 t + 1^2} = \sqrt{1 + 1} = \sqrt{2} \tag{4}$$

Substituting (4) in (1) yields the following curvature:

$$\kappa(t) = \frac{\sqrt{2}}{(\sqrt{2})^3} = \frac{1}{2} \quad \Rightarrow \quad \kappa(0) = \frac{1}{2} \tag{5}$$

The unit normal vector is the following vector:

$$N(t) = \frac{\mathbf{T}'(t)}{\|\mathbf{T}'(t)\|} \tag{6}$$

By (2) and (4) we have:

$$\mathbf{T}(t) = \frac{\mathbf{r}'(t)}{\|\mathbf{r}'(t)\|} = \frac{1}{\sqrt{2}} \langle -\sin t, \cos t, 1 \rangle \quad \Rightarrow \quad \mathbf{T}'(t) = \frac{1}{\sqrt{2}} \langle -\cos t, -\sin t, 0 \rangle \tag{7}$$

$$\|\mathbf{T}'(t)\| = \frac{1}{\sqrt{2}} \sqrt{(-\cos t)^2 + (-\sin t)^2 + 0^2} = \frac{1}{\sqrt{2}} \cdot 1 = \frac{1}{\sqrt{2}}$$

Combining (6) and (7) gives:

$$\mathbf{N}(t) = \langle -\cos t, -\sin t, 0 \rangle \quad \Rightarrow \quad \mathbf{N}(0) = \langle -1, 0, 0 \rangle \tag{8}$$

The center of curvature at $t = 0$ is:

$$\overrightarrow{OQ} = \mathbf{r}(0) + \kappa(0)^{-1} \mathbf{N}(0)$$

By (5), (8) and $\mathbf{r}(0) = \langle 1, 0, 0 \rangle$ we get:

$$\overrightarrow{OQ} = \langle 1, 0, 0 \rangle + 2 \langle -1, 0, 0 \rangle = \langle 1, 0, 0 \rangle + \langle -2, 0, 0 \rangle = \langle -1, 0, 0 \rangle$$

Finally, we find a parametrization of the osculating circle at $t = 0$. The osculating circle has radius $R = \frac{1}{\kappa(0)} = 2$ and center $\langle -1, 0, 0 \rangle$, hence it has the following parametrization:

$$\mathbf{c}(t) = \langle -1, 0, 0 \rangle + 2\mathbf{N}(0) \cos t + 2\mathbf{T}(0) \sin t = \langle -1, 0, 0 \rangle + 2\langle -1, 0, 0 \rangle \cos t + \frac{2}{\sqrt{2}} \langle 0, 1, 1 \rangle \sin t$$

$$\mathbf{c}(t) = \left\langle -1 - 2\cos t, \frac{2 \sin t}{\sqrt{2}}, \frac{2 \sin t}{\sqrt{2}} \right\rangle$$

59. Figure 18 shows the graph of the half-ellipse $y = \pm\sqrt{2rx - px^2}$, where r and p are positive constants. Show that the radius of curvature at the origin is equal to r. *Hint:* One way of proceeding is to write the ellipse in the form of Exercise 25 and apply Eq. (9).

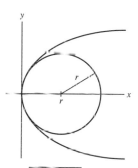

FIGURE 18 The curve $y = \pm\sqrt{2rx - px^2}$ and the osculating circle at the origin.

SOLUTION The radius of curvature is the reciprocal of the curvature. We thus must find the curvature at the origin. We use the following simple variant of the formula for the curvature of a graph in the plane:

$$\kappa(y) = \frac{|x''(y)|}{\left(1 + x'(y)^2\right)^{3/2}} \tag{1}$$

(The traditional formula of $\kappa(x) = \dfrac{|y''(x)|}{\left(1 + y'(x)^2\right)^{3/2}}$ is inappropriate for this problem, as $y'(x)$ is undefined at $x = 0$.) We find x in terms of y:

$$y = \sqrt{2rx - px^2}$$

$$y^2 = 2rx - px^2$$

$$px^2 - 2rx + y^2 = 0$$

We solve for x and obtain:

$$x = \pm\frac{1}{p}\sqrt{r^2 - py^2} + \frac{r}{p}, \quad y \geq 0.$$

We find x' and x'':

$$x' = \pm \frac{-2py}{2p\sqrt{r^2 - py^2}} = \pm \frac{y}{\sqrt{r^2 - py^2}}$$

$$x'' = \pm \frac{1 \cdot \sqrt{r^2 - py^2} - y \cdot \frac{-py}{\sqrt{r^2 - py^2}}}{r^2 - py^2} = \pm \frac{r^2 - py^2 + py^2}{(r^2 - py^2)^{3/2}} = \pm \frac{r^2}{(r^2 - py^2)^{3/2}}$$

At the origin we get:

$$x'(0) = 0, \quad x''(0) = \frac{\pm r^2}{(r^2)^{3/2}} = \frac{\pm 1}{r}$$

Substituting in (1) gives the following curvature at the origin:

$$\kappa(0) = \frac{|x''(0)|}{(1 + x'(0)^2)^{3/2}} = \frac{|\frac{\pm 1}{r}|}{(1 + 0)^{3/2}} = \frac{1}{|r|} = \frac{1}{r}$$

We conclude that the radius of curvature at the origin is

$$R = \frac{1}{\kappa(0)} = r$$

61. The **angle of inclination** at a point P on a plane curve is the angle θ between the unit tangent vector $\mathbf{T}$ and the x-axis (Figure 20). Assume that $\mathbf{r}(s)$ is a arc length parametrization, and let $\theta = \theta(s)$ be the angle of inclination at $\mathbf{r}(s)$. Prove that

$$\kappa(s) = \left| \frac{d\theta}{ds} \right|$$

12

Hint: Observe that $\mathbf{T}(s) = \langle \cos \theta(s), \sin \theta(s) \rangle$.

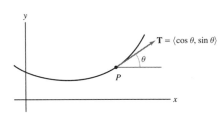

FIGURE 20 The curvature at P is the quantity $|d\theta/ds|$.

SOLUTION Since $\mathbf{T}(t)$ is a unit vector that makes an angle $\theta(t)$ with the positive x-axis, we have

$$\mathbf{T}(t) = \langle \cos \theta(t), \sin \theta(t) \rangle .$$

Differentiating this vector using the Chain Rule gives:

$$\mathbf{T}'(t) = \langle -\theta'(t) \sin \theta(t), \theta'(t) \cos \theta(t) \rangle = \theta'(t) \langle -\sin \theta(t), \cos \theta(t) \rangle$$

We compute the norm of the vector $\mathbf{T}'(t)$:

$$\|\mathbf{T}'(t)\| = \|\theta'(t) \langle -\sin \theta(t), \cos \theta(t) \rangle \| = |\theta'(t)| \sqrt{(-\sin \theta(t))^2 + (\cos \theta(t))^2} = |\theta'(t)| \cdot 1 = |\theta'(t)|$$

When $\mathbf{r}(s)$ is a parametrization by arc length we have:

$$\kappa(s) = \left\| \frac{d\mathbf{T}}{ds} \right\| = \left\| \frac{d\mathbf{T}}{dt} \right\| \left| \frac{dt}{d\theta} \frac{d\theta}{ds} \right| = |\theta'(t)| \frac{1}{|\theta'(t)|} \left| \frac{d\theta}{ds} \right| = \left| \frac{d\theta}{ds} \right|$$

as desired.

63. Let $\theta(x)$ be the angle of inclination at a point on the graph $y = f(x)$ (see Exercise 61).

(a) Use the relation $f'(x) = \tan \theta$ to prove that $\dfrac{d\theta}{dx} = \dfrac{f''(x)}{(1 + f'(x)^2)}$.

(b) Use the arc length integral to show that $\dfrac{ds}{dx} = \sqrt{1 + f'(x)^2}$.

(c) Now give a proof of Eq. (5) using Eq. (12).

SOLUTION

(a) By the relation $f'(x) = \tan\theta$ we have $\theta = \tan^{-1} f'(x)$. Differentiating using the Chain Rule we get:

$$\frac{d\theta}{dx} = \frac{d}{dx}\left(\tan^{-1} f'(x)\right) = \frac{1}{1 + f'(x)^2}\frac{d}{dx}\left(f'(x)\right) = \frac{f''(x)}{1 + f'(x)^2}$$

(b) We use the parametrization $\mathbf{r}(x) = \langle x, f(x)\rangle$. Hence, $\mathbf{r}'(x) = \langle 1, f'(x)\rangle$ and we obtain the following arc length function:

$$S(x) = \int_0^x \|\mathbf{r}'(u)\|\, du = \int_0^x \|\langle 1, f'(u)\rangle\|\, du = \int_0^x \sqrt{1 + f'(u)^2}\, du$$

Differentiating using the Fundamental Theorem gives:

$$\frac{ds}{dx} = \frac{d}{dx}\left(\int_0^x \sqrt{1 + f'(u)^2}\, du\right) = \sqrt{1 + f'(x)^2}$$

(c) By Eq. (12),

$$\kappa(s) = \left|\frac{d\theta}{ds}\right| \tag{1}$$

Using the Chain Rule and the equalities in part (a) and part (b), we obtain:

$$\frac{d\theta}{ds} = \frac{d\theta}{dx}\cdot\frac{dx}{ds} = \frac{d\theta}{dx}\cdot\frac{1}{\frac{ds}{dx}} = \frac{f''(x)}{1 + f'(x)^2}\cdot\frac{1}{\sqrt{1 + f'(x)^2}} = \frac{f''(x)}{\left(1 + f'(x)^2\right)^{3/2}}$$

Combining with (1) we obtain the curvature as the following function of x:

$$\kappa(x) = \frac{|f''(x)|}{\left(1 + f'(x)^2\right)^{3/2}}$$

which proves Eq. (5).

In Exercises 65–67, use Eq. (13) to find the curvature of the curve given in polar form.

65. $f(\theta) = 2\cos\theta$

SOLUTION By Eq. (13):,

$$\kappa(\theta) = \frac{|f(\theta)^2 + 2f'(\theta)^2 - f(\theta)f''(\theta)|}{\left(f(\theta)^2 + f'^2(\theta)\right)^{3/2}}$$

We compute the derivatives $f'(\theta)$ and $f''(\theta)$ and evaluate the numerator of $\kappa(\theta)$. This gives:

$$f'(\theta) = -2\sin\theta$$
$$f''(\theta) = -2\cos\theta$$
$$f(\theta)^2 + 2f'(\theta)^2 - f(\theta)f''(\theta) = 4\cos^2\theta + 2\cdot 4\sin^2\theta - 2\cos\theta(-2\cos\theta)$$
$$= 8\cos^2\theta + 8\sin^2\theta = 8$$

We compute the denominator of $\kappa(\theta)$:

$$\left(f(\theta)^2 + f'(\theta)^2\right)^{3/2} = \left(4\cos^2\theta + 4\sin^2\theta\right)^{3/2} = 4^{3/2} = 8$$

Hence,

$$\kappa(\theta) = \frac{8}{8} = 1$$

67. $f(\theta) = e^\theta$

SOLUTION By Eq. (13) we have the following curvature:

$$\kappa(\theta) = \frac{|f(\theta)^2 + 2f'(\theta)^2 - f(\theta)f''(\theta)|}{\left(f(\theta)^2 + f'^2(\theta)\right)^{3/2}}$$

Since $f(\theta) = e^\theta$ also $f'(\theta) = f''(\theta) = e^\theta$. We compute the numerator and denominator of $\kappa(\theta)$:

$$f(\theta)^2 + 2f'(\theta)^2 - f(\theta)f''(\theta) = e^{2\theta} + 2e^{2\theta} - e^\theta \cdot e^\theta = 2e^{2\theta}$$

$$\left(f(\theta)^2 + f'(\theta)^2\right)^{3/2} = \left(e^{2\theta} + e^{2\theta}\right)^{3/2} = \left(2e^{2\theta}\right)^{3/2} = 2\sqrt{2}e^{3\theta}$$

Substituting in the formula for $\kappa(\theta)$ we obtain:

$$\kappa(\theta) = \frac{2e^{2\theta}}{2\sqrt{2}e^{3\theta}} = \frac{1}{\sqrt{2}}e^{-\theta}$$

69. Show that both $\mathbf{r}'(t)$ and $\mathbf{r}''(t)$ lie in the osculating plane for a vector function $\mathbf{r}(t)$. *Hint:* Differentiate $\mathbf{r}'(t) = v(t)\mathbf{T}(t)$.

SOLUTION   The osculating plane at P is the plane through P determined by the unit tangent $\mathbf{T}$ and the unit normal $\mathbf{N}$ at P. Since $\mathbf{T}(t) = \frac{\mathbf{r}'(t)}{\|\mathbf{r}'(t)\|}$ we have $\mathbf{r}'(t) = v(t)\mathbf{T}(t)$ where $v(t) = \|\mathbf{r}'(t)\|$. That is, $\mathbf{r}'(t)$ is a scalar multiple of $\mathbf{T}(t)$, hence it lies in every plane containing $\mathbf{T}(t)$, in particular in the osculating plane. We now differentiate $\mathbf{r}'(t) = v(t)\mathbf{T}(t)$ using the Product Rule:

$$\mathbf{r}''(t) = v'(t)\mathbf{T}(t) + v(t)\mathbf{T}'(t) \tag{1}$$

By $\mathbf{N}(t) = \frac{\mathbf{T}'(t)}{\|\mathbf{T}'(t)\|}$ we have $\mathbf{T}'(t) = b(t)\mathbf{N}(t)$ for $b(t) = \|\mathbf{T}'(t)\|$. Substituting in (1) gives:

$$\mathbf{r}''(t) = v'(t)\mathbf{T}(t) + v(t)b(t)\mathbf{N}(t)$$

We see that $\mathbf{r}''(t)$ is a linear combination of $\mathbf{T}(t)$ and $\mathbf{N}(t)$, hence $\mathbf{r}''(t)$ lies in the plane determined by these two vectors, that is, $\mathbf{r}''(t)$ lies in the osculating plane.

71. Two vector-valued functions $\mathbf{r}_1(s)$ and $\mathbf{r}_2(s)$ are said to *agree to order 2* at s_0 if

$$\mathbf{r}_1(s_0) = \mathbf{r}_2(s_0), \quad \mathbf{r}_1'(s_0) = \mathbf{r}_2'(s_0), \quad \mathbf{r}_1''(s_0) = \mathbf{r}_2''(s_0)$$

Let $\mathbf{r}(s)$ be an arc length parametrization of a path $\mathcal{C}$, and let P be the terminal point of $\mathbf{r}(0)$. Let $\gamma(s)$ be the arc length parametrization of the osculating circle given in Exercise 70. Show that $\mathbf{r}(s)$ and $\gamma(s)$ agree to order 2 at $s = 0$ (in fact, the osculating circle is the unique circle that approximates $\mathcal{C}$ to order 2 at P).

SOLUTION   The arc length parametrization of the osculating circle at P, described in the xy-coordinate system with P at the origin and the x and y axes in the directions of $\mathbf{T}$ and $\mathbf{N}$ respectively, is given in Exercise 70 by:

$$\gamma(s) = \frac{1}{\kappa}\mathbf{N} + \frac{1}{\kappa}\left((\sin \kappa s)\mathbf{T} - (\cos \kappa s)\mathbf{N}\right)$$

Hence

$$\gamma(0) = \frac{1}{\kappa}\mathbf{N} + \frac{1}{\kappa}\left((\sin 0)\mathbf{T} - (\cos 0)\mathbf{N}\right) = \frac{1}{\kappa}\mathbf{N} + \frac{1}{\kappa}(0 - 1 \cdot \mathbf{N}) = \frac{1}{\kappa}\mathbf{N} - \frac{1}{\kappa}\mathbf{N} = \mathbf{0}$$

$$\mathbf{r}(0) = \overrightarrow{OP} = \mathbf{0}$$

We get:

$$\gamma(0) = \mathbf{r}(0) \tag{1}$$

Differentiating $\gamma(s)$ gives (notice that $\mathbf{N}$, $\mathbf{T}$, and κ are fixed):

$$\gamma'(s) = \frac{1}{\kappa}\left((\kappa \cos \kappa s)\mathbf{T} + (\kappa \sin \kappa s)\mathbf{N}\right) = (\cos \kappa s)\mathbf{T} + (\sin \kappa s)\mathbf{N}$$

Hence:

$$\gamma'(0) = (\cos \kappa \cdot 0)\,\mathbf{T} + (\sin \kappa \cdot 0)\,\mathbf{N} = 1 \cdot \mathbf{T} + 0 \cdot \mathbf{N} = \mathbf{T}$$

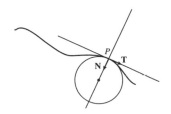

Also, since $\mathbf{r}(s)$ is the arc length parametrization, $\|\mathbf{r}'(s)\| = 1$, hence:

$$\mathbf{T} = \mathbf{T}(0) = \frac{\mathbf{r}'(0)}{\|\mathbf{r}'(0)\|} = \mathbf{r}'(0)$$

We conclude that:

$$\gamma'(0) = \mathbf{r}'(0) \tag{2}$$

We differentiate $\gamma'(s)$ to obtain:

$$\gamma''(s) = (-\kappa \sin \kappa s)\, \mathbf{T} + (\kappa \cos \kappa s)\, \mathbf{N}$$

Hence:

$$\gamma''(0) = (-\kappa \sin 0)\, \mathbf{T} + (\kappa \cos 0)\, \mathbf{N} = 0\mathbf{T} + \kappa \mathbf{N} = \kappa \mathbf{N}$$

For the arc length parametrization $\mathbf{r}(s)$ we have:

$$\mathbf{r}''(s) = \mathbf{T}'(s) = \|\mathbf{T}'(s)\|\mathbf{N}(s) = \|\mathbf{r}'(s)\|\kappa(s)\mathbf{N}(s) = 1 \cdot \kappa(s)\mathbf{N}(s)$$

Hence:

$$\mathbf{r}''(0) = \kappa(0)\mathbf{N}(0) = \kappa \mathbf{N}$$

We conclude that:

$$\gamma''(0) = \mathbf{r}''(0) \tag{3}$$

(1), (2), and (3) show that $\mathbf{r}(s)$ and $\gamma(s)$ agree to order two at $s = 0$.

Further Insights and Challenges

73. Show that the curvature of Viviani's curve, given by $\mathbf{r}(t) = \langle 1 + \cos t, \sin t, 2 \sin(t/2) \rangle$, is

$$\kappa(t) = \frac{\sqrt{13 + 3 \cos t}}{(3 + \cos t)^{3/2}}$$

SOLUTION   We use the formula for curvature:

$$\kappa(t) = \frac{\|\mathbf{r}'(t) \times \mathbf{r}''(t)\|}{\|\mathbf{r}'(t)\|^3} \tag{1}$$

Differentiating $\mathbf{r}(t)$ gives

$$\mathbf{r}'(t) = \left\langle -\sin t, \cos t, 2 \cdot \frac{1}{2} \cos \frac{t}{2} \right\rangle = \left\langle -\sin t, \cos t, \cos \frac{t}{2} \right\rangle$$

$$\mathbf{r}''(t) = \left\langle -\cos t, -\sin t, -\frac{1}{2} \sin \frac{t}{2} \right\rangle$$

We compute the cross product in (1):

$$\mathbf{r}'(t) \times \mathbf{r}''(t) = \begin{vmatrix} \mathbf{i} & \mathbf{j} & \mathbf{k} \\ -\sin t & \cos t & \cos \frac{t}{2} \\ -\cos t & -\sin t & -\frac{1}{2}\sin \frac{t}{2} \end{vmatrix}$$

$$= \left(-\frac{1}{2}\cos t \sin \frac{t}{2} + \sin t \cos \frac{t}{2} \right)\mathbf{i} - \left(\frac{1}{2}\sin t \sin \frac{t}{2} + \cos t \cos \frac{t}{2} \right)\mathbf{j} + \mathbf{k}$$

We find the length of the cross product:

$$\|\mathbf{r}'(t) \times \mathbf{r}''(t)\|^2 = \left(-\frac{1}{2}\cos t \sin \frac{t}{2} + \sin t \cos \frac{t}{2} \right)^2 + \left(\frac{1}{2}\sin t \sin \frac{t}{2} + \cos t \cos \frac{t}{2} \right)^2 + 1$$

$$= \frac{1}{4}\sin^2 \frac{t}{2}\left(\cos^2 t + \sin^2 t \right) + \cos^2 \frac{t}{2}\left(\sin^2 t + \cos^2 t \right) + 1$$

$$= \frac{1}{4}\sin^2 \frac{t}{2} + \cos^2 \frac{t}{2} + 1$$

We use the identities $\sin^2 \frac{t}{2} + \cos^2 \frac{t}{2} = 1$ and $\cos^2 \frac{t}{2} = \frac{1}{2} + \frac{1}{2} \cos t$ to write:

$$\|\mathbf{r}'(t) \times \mathbf{r}''(t)\|^2 = \frac{1}{4} \sin^2 \frac{t}{2} + \cos^2 \frac{t}{2} + 1 = \frac{1}{4} \left(\sin^2 \frac{t}{2} + \cos^2 \frac{t}{2} \right) + \frac{3}{4} \cos^2 \frac{t}{2} + 1$$

$$= \frac{1}{4} + \frac{3}{4} \left(\frac{1}{2} + \frac{1}{2} \cos t \right) + 1 = \frac{3}{8} \cos t + \frac{13}{8}$$

Hence:

$$\|\mathbf{r}'(t) \times \mathbf{r}''(t)\| = \frac{1}{\sqrt{8}} \sqrt{13 + 3 \cos t} \tag{2}$$

We compute the length of $\mathbf{r}'(t)$:

$$\|\mathbf{r}'(t)\|^2 = (-\sin t)^2 + \cos^2 t + \cos^2 \frac{t}{2} = 1 + \cos^2 \frac{t}{2} = 1 + \left(\frac{1}{2} + \frac{1}{2} \cos t \right) = \frac{3}{2} + \frac{1}{2} \cos t$$

Hence,

$$\|\mathbf{r}'(t)\| = \frac{1}{\sqrt{2}} \sqrt{3 + \cos t} \tag{3}$$

Substituting (2) and (3) in (1) gives:

$$\kappa(t) = \frac{\frac{1}{\sqrt{8}} \sqrt{13 + 3 \cos t}}{\left(\frac{1}{\sqrt{2}} \sqrt{3 + \cos t} \right)^3} = \frac{\frac{1}{\sqrt{8}} \sqrt{13 + 3 \cos t}}{\frac{1}{2} \frac{1}{\sqrt{2}} (3 + \cos t)^{3/2}} = \frac{\sqrt{13 + 3 \cos t}}{(3 + \cos t)^{3/2}}$$

In Exercises 75–82, let $\mathbf{B}$ *denote the* **binormal vector** *at a point on a space curve* $\mathcal{C}$, *defined by* $\mathbf{B} = \mathbf{T} \times \mathbf{N}$.

75. Show that $\mathbf{B}$ is a unit vector.

SOLUTION $\mathbf{T}$ and $\mathbf{N}$ are orthogonal unit vectors, therefore the length of their cross product is:

$$\|\mathbf{B}\| = \|\mathbf{T} \times \mathbf{N}\| = \|\mathbf{T}\| \|\mathbf{N}\| \sin \frac{\pi}{2} = 1 \cdot 1 \cdot 1 = 1$$

Therefore $\mathbf{B}$ is a unit vector.

77. Show that if $\mathcal{C}$ is contained in a plane $\mathcal{P}$, then $\mathbf{B}$ is a unit vector normal to $\mathcal{P}$. Conclude that $\tau = 0$ for a plane curve.

SOLUTION   If $\mathcal{C}$ is contained in a plane $\mathcal{P}$, then the unit normal $\mathbf{N}$ and the unit tangent $\mathbf{T}$ are in $\mathcal{P}$. The cross product $\mathbf{B} = \mathbf{T} \times \mathbf{N}$ is orthogonal to $\mathbf{T}$ and $\mathbf{N}$ which are in the plane, hence $\mathbf{B}$ is normal to the plane. Thus, $\mathbf{B}$ is a unit vector normal to the plane. There are only two different unit normal vectors to a plane, one pointing "up" and the other pointing "down". Thus, we can assume (due to continuity) that $\mathbf{B}$ is a constant vector, therefore

$$\frac{d\mathbf{B}}{ds} = 0 \quad \text{or} \quad \tau = 0.$$

79. Use the identity

$$\mathbf{a} \times (\mathbf{b} \times \mathbf{c}) = (\mathbf{a} \cdot \mathbf{c})\mathbf{b} - (\mathbf{a} \cdot \mathbf{b})\mathbf{c}$$

to prove

$$\mathbf{N} \times \mathbf{B} = \mathbf{T}, \qquad \mathbf{B} \times \mathbf{T} = \mathbf{N} \tag{15}$$

SOLUTION   We use the given equality and the definition $\mathbf{B} = \mathbf{T} \times \mathbf{N}$ to write:

$$\mathbf{N} \times \mathbf{B} = \mathbf{N} \times (\mathbf{T} \times \mathbf{N}) = (\mathbf{N} \cdot \mathbf{N}) \mathbf{T} - (\mathbf{N} \cdot \mathbf{T}) \mathbf{N} \tag{1}$$

The unit normal $\mathbf{N}$ and the unit tangent $\mathbf{T}$ are orthogonal unit vectors, hence $\mathbf{N} \cdot \mathbf{N} = \|\mathbf{N}\|^2 = 1$ and $\mathbf{N} \cdot \mathbf{T} = 0$. Therefore, (1) gives:

$$\mathbf{N} \times \mathbf{B} = 1 \cdot \mathbf{T} - 0\mathbf{N} = \mathbf{T}$$

To prove the second equality, we substitute $\mathbf{T} = \mathbf{N} \times \mathbf{B}$ and then use the given equality. We obtain:

$$\mathbf{B} \times \mathbf{T} = \mathbf{B} \times (\mathbf{N} \times \mathbf{B}) = (\mathbf{B} \cdot \mathbf{B}) \mathbf{N} - (\mathbf{B} \cdot \mathbf{N}) \mathbf{B} \tag{2}$$

Now, **B** is a unit vector, hence $\mathbf{B} \cdot \mathbf{B} = \|\mathbf{B}\|^2 = 1$. Also, since $\mathbf{B} = \mathbf{T} \times \mathbf{N}$, **B** is orthogonal to **N** which implies that $\mathbf{B} \cdot \mathbf{N} = 0$. Substituting in (2) we get:

$$\mathbf{B} \times \mathbf{T} = 1\mathbf{N} - 0\mathbf{B} = \mathbf{N}.$$

81. Show that $\mathbf{r}' \times \mathbf{r}''$ is a multiple of **B**. Conclude that

$$\mathbf{B} = \frac{\mathbf{r}' \times \mathbf{r}''}{\|\mathbf{r}' \times \mathbf{r}''\|} \qquad \boxed{17}$$

SOLUTION By the definition of the binormal vector, $\mathbf{B} = \mathbf{T} \times \mathbf{N}$. We denote $a(t) = \frac{1}{\|\mathbf{r}'(t)\|}$ and write:

$$\mathbf{T}(t) = \frac{\mathbf{r}'(t)}{\|\mathbf{r}'(t)\|} = a(t)\mathbf{r}'(t) \qquad (1)$$

We differentiate $\mathbf{T}(t)$ using the Product Rule:

$$\mathbf{T}'(t) = a(t)\mathbf{r}''(t) + a'(t)\mathbf{r}'(t)$$

We denote $b(t) = \|\mathbf{T}'(t)\|$ and obtain:

$$\mathbf{N}(t) = \frac{\mathbf{T}'(t)}{\|\mathbf{T}'(t)\|} = \frac{a(t)}{b(t)}\mathbf{r}''(t) + \frac{a'(t)}{b(t)}\mathbf{r}'(t)$$

For $c_1 = \frac{a(t)}{b(t)}$ and $c_2 = \frac{a'(t)}{b(t)}$ we have:

$$\mathbf{N}(t) - c_1(t)\mathbf{r}''(t) + c_2(t)\mathbf{r}'(t) \qquad (2)$$

We now find **B** as the cross product of $\mathbf{T}(t)$ in (1) and $\mathbf{N}(t)$ in (2). This gives:

$$\mathbf{B}(t) = a(t)\mathbf{r}'(t) \times \big(c_1(t)\mathbf{r}''(t) + c_2(t)\mathbf{r}'(t)\big) = a(t)c_1(t)\mathbf{r}'(t) \times \mathbf{r}''(t) + a(t)c_2(t)\mathbf{r}'(t) \times \mathbf{r}'(t)$$

$$= a(t)c_1(t)\mathbf{r}'(t) \times \mathbf{r}''(t) + \mathbf{0} = a(t)c_1(t)\mathbf{r}'(t) \times \mathbf{r}''(t)$$

We see that **B** is parallel to $\mathbf{r}' \times \mathbf{r}''$. Since **B** is a unit vector we have:

$$\mathbf{B} - \frac{\mathbf{r}' \times \mathbf{r}''}{\|\mathbf{r}' \times \mathbf{r}''\|}.$$

13.5 Motion in Three-Space (LT Section 14.5)

Preliminary Questions

1. If a particle travels with constant speed, must its acceleration vector be zero? Explain.

SOLUTION If the speed of the particle is constant, the tangential component, $a_T(t) = v'(t)$, of the acceleration is zero. However, the normal component, $a_N(t) = \kappa(t)v(t)^2$ is not necessarily zero, since the particle may change its direction.

2. For a particle in uniform circular motion around a circle, which of the vectors $\mathbf{v}(t)$ or $\mathbf{a}(t)$ always points toward the center of the circle?

SOLUTION For a particle in uniform circular motion around a circle, the acceleration vector $\mathbf{a}(t)$ points towards the center of the circle, whereas $\mathbf{v}(t)$ is tangent to the circle.

3. Two objects travel to the right along the parabola $y = x^2$ with nonzero speed. Which of the following statements must be true?

(a) Their velocity vectors point in the same direction.

(b) Their velocity vectors have the same length.

(c) Their acceleration vectors point in the same direction.

SOLUTION

(a) The velocity vector points in the direction of motion, hence the velocities of the two objects point in the same direction.

(b) The length of the velocity vector is the speed. Since the speeds are not necessarily equal, the velocity vectors may have different lengths.

(c) The acceleration is determined by the tangential component $v'(t)$ and the normal component $\kappa(t)v(t)^2$. Since v and v' may be different for the two objects, the acceleration vectors may have different directions.

4. Use the decomposition of acceleration into tangential and normal components to explain the following statement: If the speed is constant, then the acceleration and velocity vectors are orthogonal.

SOLUTION If the speed is constant, $v'(t) = 0$. Therefore, the acceleration vector has only the normal component:

$$\mathbf{a}(t) = a_N(t)\mathbf{N}(t)$$

The velocity vector always points in the direction of motion. Since the vector $\mathbf{N}(t)$ is orthogonal to the direction of motion, the vectors $\mathbf{a}(t)$ and $\mathbf{v}(t)$ are orthogonal.

5. If a particle travels along a straight line, then the acceleration and velocity vectors are (choose the correct description):

(a) Orthogonal   **(b)** Parallel

SOLUTION Since a line has zero curvature, the normal component of the acceleration is zero, hence $\mathbf{a}(t)$ has only the tangential component. The velocity vector is always in the direction of motion, hence the acceleration and the velocity vectors are parallel to the line. We conclude that (b) is the correct statement.

6. What is the length of the acceleration vector of a particle traveling around a circle of radius 2 cm with constant velocity 4 cm/s?

SOLUTION The acceleration vector is given by the following decomposition:

$$\mathbf{a}(t) = v'(t)\mathbf{T}(t) + \kappa(t)v(t)^2\mathbf{N}(t) \tag{1}$$

In our case $v(t) = 4$ is constant hence $v'(t) = 0$. In addition, the curvature of a circle of radius 2 is $\kappa(t) = \frac{1}{2}$. Substituting $v(t) = 4$, $v'(t) = 0$ and $\kappa(t) = \frac{1}{2}$ in (1) gives:

$$\mathbf{a}(t) = \frac{1}{2} \cdot 4^2 N(t) = 8N(t)$$

The length of the acceleration vector is, thus,

$$\|\mathbf{a}(t)\| = 8 \text{ cm/s}^2$$

7. Two cars are racing around a circular track. If, at a certain moment, both of their speedometers read 110 mph. then the two cars have the same (choose one):

(a) a_T **(b)** a_N

SOLUTION The tangential acceleration a_T and the normal acceleration a_N are the following values:

$$a_T(t) = v'(t); \quad a_N(t) = \kappa(t)v(t)^2$$

At the moment where both speedometers read 110 mph, the speeds of the two cars are $v = 110$ mph. Since the track is circular, the curvature $\kappa(t)$ is constant, hence the normal accelerations of the two cars are equal at this moment. Statement (b) is correct.

Exercises

1. Use the table below to calculate the difference quotients $\dfrac{\mathbf{r}(1+h) - \mathbf{r}(1)}{h}$ for $h = -0.2, -0.1, 0.1, 0.2$. Then estimate the velocity and speed at $t = 1$.

$\mathbf{r}(0.8)$	$\langle 1.557, 2.459, -1.970 \rangle$
$\mathbf{r}(0.9)$	$\langle 1.559, 2.634, -1.740 \rangle$
$\mathbf{r}(1)$	$\langle 1.540, 2.841, -1.443 \rangle$
$\mathbf{r}(1.1)$	$\langle 1.499, 3.078, -1.035 \rangle$
$\mathbf{r}(1.2)$	$\langle 1.435, 3.342, -0.428 \rangle$

SOLUTION

$$(h = -0.2)$$

$$\frac{\mathbf{r}(1 - 0.2) - \mathbf{r}(1)}{-0.2} = \frac{\mathbf{r}(0.8) - \mathbf{r}(1)}{-0.2} = \frac{\langle 1.557, 2.459, -1.970 \rangle - \langle 1.540, 2.841, -1.443 \rangle}{-0.2}$$

$$= \frac{\langle 0.017, -0.382, -0.527 \rangle}{-0.2} = \langle -0.085, 1.91, 2.635 \rangle$$

$$(h = -0.1)$$

$$\frac{\mathbf{r}(1 - 0.1) - \mathbf{r}(1)}{-0.1} = \frac{\mathbf{r}(0.9) - \mathbf{r}(1)}{-0.1} = \frac{\langle 1.559, 2.634, -1.740 \rangle - \langle 1.540, 2.841, -1.443 \rangle}{-0.1}$$

$$= \frac{\langle 0.019, -0.207, -0.297 \rangle}{-0.1} = \langle -0.19, 2.07, 2.97 \rangle$$

$$(h = 0.1)$$

$$\frac{\mathbf{r}(1 + 0.1) - \mathbf{r}(1)}{0.1} = \frac{\mathbf{r}(1.1) - \mathbf{r}(1)}{0.1} = \frac{\langle 1.499, 3.078, -1.035 \rangle - \langle 1.540, 2.841, -1.443 \rangle}{0.1}$$

$$= \frac{\langle -0.041, 0.237, 0.408 \rangle}{0.1} = \langle -0.41, 2.37, 4.08 \rangle$$

$$(h = 0.2)$$

$$\frac{\mathbf{r}(1 + 0.2) - \mathbf{r}(1)}{0.2} = \frac{\mathbf{r}(1.2) - \mathbf{r}(1)}{0.2} = \frac{\langle 1.435, 3.342, -0.428 \rangle - \langle 1.540, 2.841, -1.443 \rangle}{0.2}$$

$$= \frac{\langle -0.105, 0.501, 1.015 \rangle}{0.2} = \langle -0.525, 2.505, 5.075 \rangle$$

The velocity vector is defined by:

$$\mathbf{v}(t) = \mathbf{r}'(t) = \lim_{h \to 0} \frac{\mathbf{r}(t + h) - \mathbf{r}(t)}{h}$$

We may estimate the velocity at $t = 1$ by:

$$\mathbf{v}(1) \approx \langle -0.3, 2.2, 3.5 \rangle$$

and the speed by:

$$v(1) = \|\mathbf{v}(1)\| \approx \sqrt{0.3^2 + 2.2^2 + 3.5^2} \cong 4.1$$

In Exercises 3–6, calculate the velocity and acceleration vectors and the speed at the time indicated.

3. $\mathbf{r}(t) = \langle t^3, 1 - t, 4t^2 \rangle$, $t = 1$

SOLUTION In this case $\mathbf{r}(t) = \langle t^3, 1 - t, 4t^2 \rangle$ hence:

$$\mathbf{v}(t) = \mathbf{r}'(t) = \langle 3t^2, -1, 8t \rangle \quad \Rightarrow \quad \mathbf{v}(1) = \langle 3, -1, 8 \rangle$$
$$\mathbf{a}(t) = \mathbf{r}''(t) = \langle 6t, 0, 8 \rangle \quad \Rightarrow \quad \mathbf{a}(1) = \langle 6, 0, 8 \rangle$$

The speed is the magnitude of the velocity vector, that is,

$$v(1) = \|\mathbf{v}(1)\| = \sqrt{3^2 + (-1)^2 + 8^2} = \sqrt{74}$$

5. $\mathbf{r}(\theta) = \langle \sin \theta, \cos \theta, \cos 3\theta \rangle$, $\theta = \frac{\pi}{3}$

SOLUTION Differentiating $\mathbf{r}(\theta) = \langle \sin \theta, \cos \theta, \cos 3\theta \rangle$ gives:

$$\mathbf{v}(\theta) = \mathbf{r}'(\theta) = \langle \cos \theta, -\sin \theta, -3 \sin 3\theta \rangle$$

$$\Rightarrow \mathbf{v}\left(\frac{\pi}{3}\right) = \left\langle \cos \frac{\pi}{3}, -\sin \frac{\pi}{3}, -3 \sin \pi \right\rangle = \left\langle \frac{1}{2}, -\frac{\sqrt{3}}{2}, 0 \right\rangle$$

$$\mathbf{a}(\theta) = \mathbf{r}''(\theta) = \langle -\sin \theta, -\cos \theta, -9 \cos 3\theta \rangle$$

$$\Rightarrow \mathbf{a}\left(\frac{\pi}{3}\right) = \left\langle -\sin \frac{\pi}{3}, -\cos \frac{\pi}{3}, -9 \cos \pi \right\rangle = \left\langle -\frac{\sqrt{3}}{2}, -\frac{1}{2}, 9 \right\rangle$$

The speed is the magnitude of the velocity vector, that is:

$$v\left(\frac{\pi}{3}\right) = \left\| \mathbf{v}\left(\frac{\pi}{3}\right) \right\| = \sqrt{\left(\frac{1}{2}\right)^2 + \left(-\frac{\sqrt{3}}{2}\right)^2 + 0^2} = 1$$

7. Find $\mathbf{a}(t)$ for a particle moving around a circle of radius 8 cm at a constant speed of $v = 4$ cm/s (see Example 4). Draw the path and acceleration vector at $t = \frac{\pi}{4}$.

SOLUTION The position vector is:

$$\mathbf{r}(t) = 8 \langle \cos \omega t, \sin \omega t \rangle$$

Hence,

$$\mathbf{v}(t) = \mathbf{r}'(t) = 8\langle -\omega\sin\omega t, \omega\cos\omega t\rangle = 8\omega\langle -\sin\omega t, \cos\omega t\rangle \tag{1}$$

We are given that the speed of the particle is $v = 4$ cm/s. The speed is the magnitude of the velocity vector, hence:

$$v = 8\omega\sqrt{(-\sin\omega t)^2 + \cos^2\omega t} = 8\omega = 4 \quad\Rightarrow\quad \omega = \frac{1}{2}\ \text{rad/s}$$

Substituting in (2) we get:

$$\mathbf{v}(t) = 4\left\langle -\sin\frac{t}{2}, \cos\frac{t}{2}\right\rangle$$

We now find $\mathbf{a}(t)$ by differentiating the velocity vector. This gives

$$\mathbf{a}(t) = \mathbf{v}'(t) = 4\left\langle -\frac{1}{2}\cos\frac{t}{2}, -\frac{1}{2}\sin\frac{t}{2}\right\rangle = -2\left\langle \cos\frac{t}{2}, \sin\frac{t}{2}\right\rangle$$

The path of the particle is $\mathbf{r}(t) = 8\left\langle\cos\frac{t}{2}, \sin\frac{t}{2}\right\rangle$ and the acceleration vector at $t = \frac{\pi}{4}$ is:

$$\mathbf{a}\left(\frac{\pi}{4}\right) = -2\left\langle\cos\frac{\pi}{8}, \sin\frac{\pi}{8}\right\rangle \approx \langle -1.85, -0.77\rangle$$

The path $\mathbf{r}(t)$ and the acceleration vector at $t = \frac{\pi}{4}$ are shown in the following figure:

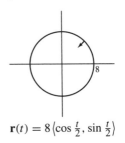

$$\mathbf{r}(t) = 8\left\langle\cos\frac{t}{2}, \sin\frac{t}{2}\right\rangle$$

9. Sketch the path $\mathbf{r}(t) = \langle t^2, t^3\rangle$ together with the velocity and acceleration vectors at $t = 1$.

SOLUTION We compute the velocity and acceleration vectors at $t = 1$:

$$\mathbf{v}(t) = \mathbf{r}'(t) = \langle 2t, 3t^2\rangle \quad\Rightarrow\quad \mathbf{v}(1) = \langle 2, 3\rangle$$

$$\mathbf{a}(t) = \mathbf{v}'(t) = \langle 2, 6t\rangle \quad\Rightarrow\quad \mathbf{a}(1) = \langle 2, 6\rangle$$

The following figure shows the path $\mathbf{r}(t) = \langle t^2, t^3\rangle$ and the vectors $\mathbf{v}(1)$ and $\mathbf{a}(1)$:

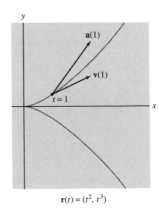

$$\mathbf{r}(t) = (t^2, t^3)$$

In Exercises 11–14, find $\mathbf{v}(t)$ given $\mathbf{a}(t)$ and the initial velocity.

11. $\mathbf{a}(t) = \langle t, 4\rangle, \quad \mathbf{v}(0) = \langle\frac{1}{3}, -2\rangle$

SOLUTION We find $\mathbf{v}(t)$ by integrating $\mathbf{a}(t)$:

$$\mathbf{v}(t) = \int_0^t \mathbf{a}(u)\,du = \int_0^t \langle u, 4\rangle\,du = \left\langle\frac{1}{2}u^2, 4u\right\rangle\Big|_0^t + \mathbf{v}_0 = \left\langle\frac{t^2}{2}, 4t\right\rangle + \mathbf{v}_0$$

The initial condition gives:

$$\mathbf{v}(0) = \langle 0, 0 \rangle + \mathbf{v}_0 = \left\langle \frac{1}{3}, -2 \right\rangle \quad \Rightarrow \quad \mathbf{v}_0 = \left\langle \frac{1}{3}, -2 \right\rangle$$

Hence,

$$\mathbf{v}(t) = \left\langle \frac{t^2}{2}, 4t \right\rangle + \left\langle \frac{1}{3}, -2 \right\rangle = \left\langle \frac{3t^2 + 2}{6}, 4t - 2 \right\rangle$$

13. $\mathbf{a}(t) = \mathbf{k}, \quad \mathbf{v}(0) = \mathbf{i}$

SOLUTION We compute $\mathbf{v}(t)$ by integrating the acceleration vector:

$$\mathbf{v}(t) = \int_0^t \mathbf{a}(u)\, du = \int_0^t \mathbf{k}\, du = \mathbf{k} u \Big|_0^t + \mathbf{v}_0 = t\mathbf{k} + \mathbf{v}_0 \tag{1}$$

Substituting the initial condition gives:

$$\mathbf{v}(0) = 0\mathbf{k} + \mathbf{v}_0 = \mathbf{i} \quad \Rightarrow \quad \mathbf{v}_0 = \mathbf{i}$$

Combining with (1) we obtain:

$$\mathbf{v}(t) = \mathbf{i} + t\mathbf{k}$$

In Exercises 15–18, find $\mathbf{r}(t)$ and $\mathbf{v}(t)$ given $\mathbf{a}(t)$ and the initial velocity and position.

15. $\mathbf{a}(t) = \langle t, 4 \rangle, \quad \mathbf{v}(0) = \langle 3, -2 \rangle, \quad \mathbf{r}(0) = \langle 0, 0 \rangle$

SOLUTION We first integrate $\mathbf{a}(t)$ to find the velocity vector:

$$\mathbf{v}(t) = \int_0^t \langle u, 4 \rangle\, du = \left\langle \frac{u^2}{2}, 4u \right\rangle \Big|_0^t + \mathbf{v}_0 = \left\langle \frac{t^2}{2}, 4t \right\rangle + \mathbf{v}_0 \tag{1}$$

The initial condition $\mathbf{v}(0) = \langle 3, -2 \rangle$ gives:

$$\mathbf{v}(0) = \langle 0, 0 \rangle + \mathbf{v}_0 = \langle 3, -2 \rangle \quad \Rightarrow \quad \mathbf{v}_0 = \langle 3, -2 \rangle$$

Substituting in (1) we get:

$$\mathbf{v}(t) = \left\langle \frac{t^2}{2}, 4t \right\rangle + \langle 3, -2 \rangle = \left\langle \frac{t^2}{2} + 3, 4t - 2 \right\rangle$$

We now integrate the velocity vector to find $\mathbf{r}(t)$:

$$\mathbf{r}(t) = \int_0^t \left\langle \frac{u^2}{2} + 3, 4u - 2 \right\rangle\, du = \left\langle \frac{u^3}{6} + 3u, 2u^2 - 2u \right\rangle \Big|_0^t + \mathbf{r}_0 = \left\langle \frac{t^3}{6} + 3t, 2t^2 - 2t \right\rangle + \mathbf{r}_0$$

The initial condition $\mathbf{r}(0) = \langle 0, 0 \rangle$ gives:

$$\mathbf{r}(0) = \langle 0, 0 \rangle + \mathbf{r}_0 = \langle 0, 0 \rangle \quad \Rightarrow \quad \mathbf{r}_0 = \langle 0, 0 \rangle$$

Hence,

$$\mathbf{r}(t) = \left\langle \frac{t^3}{6} + 3t, 2t^2 - 2t \right\rangle$$

17. $\mathbf{a}(t) = t\mathbf{k}, \quad \mathbf{v}(0) = \mathbf{i}, \quad \mathbf{r}(0) = \mathbf{j}$

SOLUTION Integrating the acceleration vector gives:

$$\mathbf{v}(t) = \int_0^t u\mathbf{k}\, du = \frac{u^2}{2}\mathbf{k} \Big|_0^t + \mathbf{v}_0 = \frac{t^2}{2}\mathbf{k} + \mathbf{v}_0 \tag{1}$$

The initial condition for $\mathbf{v}(t)$ gives:

$$\mathbf{v}(0) = \frac{0^2}{2}\mathbf{k} + \mathbf{v}_0 = \mathbf{i} \quad \Rightarrow \quad \mathbf{v}_0 = \mathbf{i}$$

We substitute in (1):

$$v(t) = \frac{t^2}{2}\mathbf{k} + \mathbf{i} = \mathbf{i} + \frac{t^2}{2}\mathbf{k}$$

We now integrate $\mathbf{v}(t)$ to find $\mathbf{r}(t)$:

$$\mathbf{r}(t) = \int_0^t \left(\mathbf{i} + \frac{u^2}{2}\mathbf{k}\right) du = u\mathbf{i} + \frac{u^3}{6}\mathbf{k}\Big|_0^t + \mathbf{r}_0 = t\mathbf{i} + \frac{t^3}{6}\mathbf{k} + \mathbf{r}_0 \qquad (2)$$

The initial condition for $\mathbf{r}(t)$ gives:

$$\mathbf{r}(0) = 0\mathbf{i} + 0\mathbf{k} + \mathbf{r}_0 = \mathbf{j} \quad \Rightarrow \quad \mathbf{r}_0 = \mathbf{j}$$

Combining with (2) gives the position vector:

$$\mathbf{r}(t) = t\mathbf{i} + \mathbf{j} + \frac{t^3}{6}\mathbf{k}$$

In Exercises 19–24, recall that $g = 9.8$ m/s^2 is the acceleration due to gravity on the earth's surface.

19. A bullet is fired from the ground at an angle of 45°. What initial speed must the bullet have in order to hit the top of a 120-m tower located 180 m away?

SOLUTION   We place the gun at the origin and let $\mathbf{r}(t)$ be the bullet's position vector.

Step 1. Use Newton's Law. The net force vector acting on the bullet is the force of gravity $\mathbf{F} = \langle 0, -gm \rangle = m \langle 0, -g \rangle$. By Newton's Second Law, $\mathbf{F} = m\mathbf{r}''(t)$, hence:

$$m \langle 0, -g \rangle = m\mathbf{r}''(t) \quad \Rightarrow \quad \mathbf{r}''(t) = \langle 0, -g \rangle$$

We compute the position vector by integrating twice:

$$\mathbf{r}'(t) = \int_0^t \mathbf{r}''(u) \, du = \int_0^t \langle 0, -g \rangle \, du = \langle 0, -gt \rangle + \mathbf{v}_0$$

$$\mathbf{r}(t) = \int_0^t \mathbf{r}'(u) \, du = \int_0^t (\langle 0, -gu \rangle + \mathbf{v}_0) \, du = \left\langle 0, -g\frac{t^2}{2} \right\rangle + \mathbf{v}_0 t + \mathbf{r}_0$$

That is,

$$\mathbf{r}(t) = \left\langle 0, \frac{-g}{2}t^2 \right\rangle + \mathbf{v}_0 t + \mathbf{r}_0 \qquad (1)$$

Since the gun is at the origin, $\mathbf{r}_0 = \mathbf{0}$. The bullet is fired at an angle of 45°, hence the initial velocity $\mathbf{v}_0$ points in the direction of the unit vector $\langle \cos 45°, \sin 45° \rangle = \left\langle \frac{\sqrt{2}}{2}, \frac{\sqrt{2}}{2} \right\rangle$ therefore, $\mathbf{v}_0 = v_0 \left\langle \frac{\sqrt{2}}{2}, \frac{\sqrt{2}}{2} \right\rangle$. Substituting these initial values in (1) gives:

$$\mathbf{r}(t) = \left\langle 0, \frac{-g}{2}t^2 \right\rangle + t v_0 \left\langle \frac{\sqrt{2}}{2}, \frac{\sqrt{2}}{2} \right\rangle$$

Step 2. Solve for v_0. The position vector of the top of the tower is $\langle 180, 120 \rangle$, hence at the moment of hitting the tower we have,

$$\mathbf{r}(t) = \left\langle 0, \frac{-g}{2}t^2 \right\rangle + t v_0 \left\langle \frac{\sqrt{2}}{2}, \frac{\sqrt{2}}{2} \right\rangle = \langle 180, 120 \rangle$$

$$\left\langle t v_0 \frac{\sqrt{2}}{2}, \frac{-g}{2}t^2 + \frac{\sqrt{2}}{2} t v_0 \right\rangle = \langle 180, 120 \rangle$$

Equating components, we get the equations:

$$\begin{cases} t v_0 \dfrac{\sqrt{2}}{2} = 180 \\[2mm] -\dfrac{g}{2}t^2 + \dfrac{\sqrt{2}}{2} t v_0 = 120 \end{cases}$$

The first equation implies that $t = \frac{360}{\sqrt{2}v_0}$. We substitute in the second equation and solve for v_0 (we use $g = 9.8$ m/s^2):

$$-\frac{9.8}{2}\left(\frac{360}{\sqrt{2}v_0}\right)^2 + \frac{\sqrt{2}}{2}\left(\frac{360}{\sqrt{2}v_0}\right)v_0 = 120$$

$$-2.45\left(\frac{360}{v_0}\right)^2 + 180 = 120$$

$$\left(\frac{360}{v_0}\right)^2 = \frac{1200}{49} \quad \Rightarrow \quad \frac{360}{v_0} = \sqrt{\frac{1200}{49}} \quad \Rightarrow \quad v_0 = 42\sqrt{3} \approx 72.746 \text{ m/s}$$

The initial speed of the bullet must be $v_0 = 42\sqrt{3}$ m/s ≈ 72.746 m/s.

21. Show that a projectile fired at an angle θ with initial speed v_0 travels a total distance $(v_0^2/g)\sin 2\theta$ before hitting the ground. Conclude that the maximum distance (for a given v_0) is attained for $\theta = 45°$.

SOLUTION We place the gun at the origin and let $\mathbf{r}(t)$ be the projectile's position vector. The net force acting on the projectile is $\mathbf{F} = \langle 0, -mg \rangle = m\langle 0, -g \rangle$. By Newton's Second Law, $\mathbf{F} = m\mathbf{r}''(t)$, hence:

$$m\langle 0, -g \rangle = m\mathbf{r}''(t) \quad \Rightarrow \quad \mathbf{r}''(t) = \langle 0, -g \rangle$$

Integrating twice we get:

$$\mathbf{r}'(t) = \int_0^t \mathbf{r}''(u)\,du = \int_0^t \langle 0, -g \rangle\,du = \langle 0, -gt \rangle + \mathbf{v}_0$$

$$\mathbf{r}(t) = \int_0^t \mathbf{r}'(u)\,du = \int_0^t (\langle 0, -g \cdot u \rangle + \mathbf{v}_0)\,du = \left\langle 0, -\frac{g}{2}t^2 \right\rangle + \mathbf{v}_0 t + \mathbf{r}_0 \tag{1}$$

Since the gun is at the origin, $\mathbf{r}_0 = 0$. The firing was at an angle θ, hence the initial velocity points in the direction of the unit vector $\langle \cos\theta, \sin\theta \rangle$. Hence, $\mathbf{v}_0 = v_0\langle \cos\theta, \sin\theta \rangle$. We substitute the initial vectors in (1) to obtain:

$$\mathbf{r}(t) = \left\langle 0, -\frac{g}{2}t^2 \right\rangle + v_0 t\langle \cos\theta, \sin\theta \rangle \tag{2}$$

The total distance is obtained when the y-component of $\mathbf{r}(t)$ is zero (besides the original moment, that is,

$$-\frac{g}{2}t^2 + (v_0 \sin\theta)t = 0$$

$$t\left(-\frac{g}{2}t + v_0 \sin\theta\right) = 0 \quad \Rightarrow \quad t = 0 \quad \text{or} \quad t = \frac{2v_0 \sin\theta}{g}$$

The appropriate choice is $t = \frac{2v_0 \sin\theta}{g}$. We now find the total distance x_T by substituting this value of t in the x-component of $\mathbf{r}(t)$ in (2). We obtain:

$$x(t) = v_0 t \cos\theta$$

$$x_T = v_0 \cos\theta \cdot \frac{2v_0 \sin\theta}{g} = \frac{2v_0^2 \cos\theta \sin\theta}{g} = \frac{v_0^2 \sin 2\theta}{g}$$

The maximum distance is attained when $\sin 2\theta = 1$, that is $2\theta = 90°$ or $\theta = 45°$.

23. A bullet is fired at an angle $\theta = \frac{\pi}{4}$ at a tower located $d = 600$ m away, with initial speed $v_0 = 120$ m/s. Find the height H at which the bullet hits the tower.

SOLUTION We place the gun at the origin and let $\mathbf{r}(t)$ be the bullet's position vector.

Step 1. Use Newton's Law. The net force vector acting on the bullet is the force of gravity $\mathbf{F} = \langle 0, -gm \rangle = m\langle 0, -g \rangle$. By Newton's Second Law, $\mathbf{F} = m\mathbf{r}''(t)$, hence:

$$m\langle 0, -g \rangle = m\mathbf{r}''(t) \quad \Rightarrow \quad \mathbf{r}''(t) = \langle 0, -g \rangle$$

We compute the position vector by integrating twice:

$$\mathbf{r}'(t) = \int_0^t \mathbf{r}''(u)\,du = \int_0^t \langle 0, -g \rangle\,du = \langle 0, -gt \rangle + \mathbf{v}_0$$

$$\mathbf{r}(t) = \int_0^t \mathbf{r}'(u)\,du = \int_0^t (\langle 0, -gu \rangle + \mathbf{v}_0)\,du = \left\langle 0, -g\frac{t^2}{2} \right\rangle + \mathbf{v}_0 t + \mathbf{r}_0$$

That is,

$$\mathbf{r}(t) = \left\langle 0, \frac{-g}{2}t^2 \right\rangle + \mathbf{v}_0 t + \mathbf{r}_0 \tag{1}$$

Since the gun is at the origin, $\mathbf{r}_0 = \mathbf{0}$. The bullet is fired at an angle of $\pi/4$ radians, hence the initial velocity $\mathbf{v}_0$ points in the direction of the unit vector $\langle \cos \pi/4, \sin \pi/4 \rangle = \left\langle \frac{1}{\sqrt{2}}, \frac{1}{\sqrt{2}} \right\rangle$ therefore, $\mathbf{v}_0 = v_0 \left\langle \frac{1}{\sqrt{2}}, \frac{1}{\sqrt{2}} \right\rangle$. Substituting these initial values in (1) gives:

$$\mathbf{r}(t) = \left\langle 0, \frac{-g}{2}t^2 \right\rangle + t v_0 \left\langle \frac{1}{\sqrt{2}}, \frac{1}{\sqrt{2}} \right\rangle$$

Step 2. Solve for H.

The position vector for the point at which the bullet hits the tower, 600 meters away, is $\langle 600, H \rangle$, hence at the moment of hitting the tower we have,

$$\left\langle 0, \frac{-g}{2}t^2 \right\rangle + t v_0 \left\langle \frac{1}{\sqrt{2}}, \frac{1}{\sqrt{2}} \right\rangle = \langle 600, H \rangle$$

Therefore, for $v_0 = 120$:

$$\frac{t v_0}{\sqrt{2}} = 600 \quad \Rightarrow \quad t = \frac{600\sqrt{2}}{120} = 5\sqrt{2}$$

and

$$-\frac{gt^2}{2} + \frac{t v_0}{\sqrt{2}} = \frac{-9.8(50)}{2} + \frac{5(\sqrt{2})(120)}{\sqrt{2}} = H$$

Hence, $H = 355$ meters. The bullet hits the tower at 355 meters high.

25. A constant force $\mathbf{F} = \langle 5, 2 \rangle$ (in newtons) acts on a 10-kg mass. Find the position of the mass at $t = 10$ s if it is located at the origin at $t = 0$ and has initial velocity $\mathbf{v}_0 = \langle 2, -3 \rangle$ (in meters per second).

SOLUTION   We know that $\mathbf{F} = m\mathbf{a}$ and thus $\langle 5, 2 \rangle = 10\mathbf{a}$ so then $\mathbf{a} = \langle 0.5, 0.2 \rangle$. Using integration we know

$$\mathbf{v}(t) = \int \mathbf{a}(t)\, dt = t\mathbf{a} + \mathbf{c}$$

and we know $\mathbf{v}(0) = \langle 2, -3 \rangle = \mathbf{c}$. Therefore,

$$\mathbf{v}(t) = t\mathbf{a} + \mathbf{v}_0 = t \langle 0.5, 0.2 \rangle + \langle 2, -3 \rangle = \langle 0.5t + 2, 0.2t - 3 \rangle$$

Again, integrating,

$$\mathbf{r}(t) = \int \mathbf{v}(t)\, dt$$

$$= \int t\mathbf{a} + \mathbf{v}_0\, dt$$

$$= \frac{t^2}{2}\mathbf{a} + t\mathbf{v}_0 + \mathbf{c}$$

$$= \frac{t^2}{2}\langle 0.5, 0.2 \rangle + t \langle 2, -3 \rangle$$

$$= \left\langle 0.25t^2 + 2t, 0.1t^2 - 3t \right\rangle + \mathbf{r}_0$$

Using the initial condition $\mathbf{r}(0) = \langle 0, 0 \rangle = \mathbf{c}$, we conclude

$$\mathbf{r}(t) = \left\langle 0.25t^2 + 2t, 0.1t^2 - 3t \right\rangle$$

and hence the position of the mass at $t = 10$ is $\mathbf{r}(10) = \langle 45, -20 \rangle$.

27. A particle follows a path $\mathbf{r}(t)$ for $0 \le t \le T$, beginning at the origin O. The vector $\overline{\mathbf{v}} = \frac{1}{T} \int_0^T \mathbf{r}'(t)\, dt$ is called the **average velocity** vector. Suppose that $\overline{\mathbf{v}} = \mathbf{0}$. Answer and explain the following:

(a) Where is the particle located at time T if $\overline{\mathbf{v}} = \mathbf{0}$?

(b) Is the particle's average speed necessarily equal to zero?

SOLUTION

(a) If the average velocity is 0, then the particle must be back at its original position at time $t = T$. This is perhaps best seen by noting that $\overline{\mathbf{v}} = \frac{1}{T} \int_0^T \mathbf{r}'(t)\, dt = \mathbf{r}(t) \Big|_0^T$.

(b) The average speed need not be zero! Consider a particle moving at constant speed around a circle, with position vector $\mathbf{r}(t) = \langle \cos t, \sin t \rangle$. From 0 to 2π, this has average velocity of 0, but constant average speed of 1.

29. At a certain moment, a particle moving along a path has velocity $\mathbf{v} = \langle 12, 20, 20 \rangle$ and acceleration $\mathbf{a} = \langle 2, 1, -3 \rangle$. Is the particle speeding up or slowing down?

SOLUTION We are asked if the particle is speeding up or slowing down, that is if $\|\mathbf{v}\|$ or $\|\mathbf{v}\|^2$ is increasing or decreasing. We check $\left(\|\mathbf{v}\|^2 \right)'$:

$$\left(\|\mathbf{v}\|^2 \right)' = (\mathbf{v} \cdot \mathbf{v})' = 2\mathbf{v}' \cdot \mathbf{v} = 2 \cdot \mathbf{a} \cdot \mathbf{v} = 2 \langle 2, 1, -3 \rangle \cdot \langle 12, 20, 20 \rangle = 2 \cdot (24 + 20 - 60) = -32 < 0$$

So the speed is decreasing.

In Exercises 30–33, use Eq. (3) to find the coefficients $a_{\mathbf{T}}$ and $a_{\mathbf{N}}$ as a function of t (or at the specified value of t).

31. $\mathbf{r}(t) = \langle t, \cos t, \sin t \rangle$

SOLUTION We find $a_{\mathbf{T}}$ and $a_{\mathbf{N}}$ using the following equalities:

$$a_{\mathbf{T}} = \mathbf{a} \cdot \mathbf{T}, \quad a_{\mathbf{N}} = \frac{\|\mathbf{a} \times \mathbf{v}\|}{\|\mathbf{v}\|}.$$

We compute $\mathbf{v}$ and $\mathbf{a}$ by differentiating $\mathbf{r}$ twice:

$$\mathbf{v}(t) = \mathbf{r}'(t) = \langle 1, -\sin t, \cos t \rangle \quad \Rightarrow \quad \|\mathbf{v}(t)\| = \sqrt{1 + (-\sin t)^2 + \cos^2 t} = \sqrt{2}$$

$$\mathbf{a}(t) = \mathbf{r}''(t) = \langle 0, -\cos t, -\sin t \rangle$$

The unit tangent vector $\mathbf{T}$ is, thus:

$$\mathbf{T}(t) = \frac{\mathbf{v}(t)}{\|\mathbf{v}(t)\|} = \frac{1}{\sqrt{2}} \langle 1, -\sin t, \cos t \rangle$$

Since the speed is constant ($v = \|\mathbf{v}(t)\| = \sqrt{2}$), the tangential component of the acceleration is zero, that is:

$$a_{\mathbf{T}} = 0$$

To find $a_{\mathbf{N}}$ we first compute the following cross product:

$$\mathbf{a} \times \mathbf{v} = \begin{vmatrix} \mathbf{i} & \mathbf{j} & \mathbf{k} \\ 0 & -\cos t & -\sin t \\ 1 & -\sin t & \cos t \end{vmatrix} = \begin{vmatrix} -\cos t & -\sin t \\ -\sin t & \cos t \end{vmatrix} \mathbf{i} - \begin{vmatrix} 0 & -\sin t \\ 1 & \cos t \end{vmatrix} \mathbf{j} + \begin{vmatrix} 0 & -\cos t \\ 1 & -\sin t \end{vmatrix} \mathbf{k}$$

$$= -\left(\cos^2 t + \sin^2 t \right) \mathbf{i} - \sin t \mathbf{j} + \cos t \mathbf{k} = -\mathbf{i} - \sin t \mathbf{j} + \cos t \mathbf{k} = \langle -1, -\sin t, \cos t \rangle$$

Hence,

$$a_{\mathbf{N}} = \frac{\|\mathbf{a} \times \mathbf{v}\|}{\|\mathbf{v}\|} = \frac{\sqrt{(-1)^2 + (-\sin t)^2 + \cos^2 t}}{\sqrt{2}} = \frac{\sqrt{2}}{\sqrt{2}} = 1.$$

33. $\mathbf{r}(t) = \langle e^{2t}, t, e^{-t} \rangle, \quad t = 0$

SOLUTION We will use the following equalities:

$$a_{\mathbf{T}} = \mathbf{a} \cdot \mathbf{T}, \quad a_{\mathbf{N}} = \frac{\|\mathbf{a} \times \mathbf{v}\|}{\|\mathbf{v}\|}.$$

We first find $\mathbf{a}$ and $\mathbf{v}$ by twice differentiating $\mathbf{r}$. We get:

$$\mathbf{v}(t) = \mathbf{r}'(t) = \langle 2e^{2t}, 1, -e^{-t} \rangle$$

$$\mathbf{a}(t) = \mathbf{r}''(t) = \langle 4e^{2t}, 0, e^{-t} \rangle$$

Then evaluating at $t = 0$ we get:

$$\mathbf{v}(0) = \langle 2, 1, -1 \rangle, \quad \Rightarrow \|\mathbf{v}(0)\| = \sqrt{2^2 + 1^2 + (-1)^2} = \sqrt{6}$$

$$\mathbf{a}(0) = \langle 4, 0, 1 \rangle$$

Hence, $\mathbf{T} = \frac{\mathbf{v}}{\|\mathbf{v}\|} = \frac{1}{\sqrt{6}}\langle 2, 1, -1 \rangle$ and we obtain:

$$a_{\mathbf{T}} = \mathbf{a} \cdot \mathbf{T} = \langle 4, 0, 1 \rangle \cdot \frac{1}{\sqrt{6}}\langle 2, 1, -1 \rangle = \frac{1}{\sqrt{6}}(8 + 0 - 1) = \frac{7}{\sqrt{6}}$$

To find $a_{\mathbf{N}}$ we first compute the following cross product:

$$\mathbf{a} \times \mathbf{v} = \begin{vmatrix} \mathbf{i} & \mathbf{j} & \mathbf{k} \\ 4 & 0 & 1 \\ 2 & 1 & -1 \end{vmatrix} = \langle -1, 6, 4 \rangle$$

Therefore,

$$a_{\mathbf{N}} = \frac{\|\mathbf{a} \times \mathbf{v}\|}{\|\mathbf{v}\|} = \frac{\sqrt{(-1)^2 + 6^2 + 4^2}}{\sqrt{6}} = \sqrt{\frac{53}{6}}$$

In Exercise 34–41, find the decomposition of $\mathbf{a}(t)$ into tangential and normal components at the point indicated, as in Example 6.

35. $\mathbf{r}(t) = \left\langle \frac{1}{3}t^3, 1 - 3t \right\rangle$, $t = -2$

SOLUTION   First note here that:

$$\mathbf{v}(t) = \mathbf{r}'(t) = \left\langle t^2, -3 \right\rangle$$

$$\mathbf{a}(t) = \mathbf{r}''(t) = \langle 2t, 0 \rangle$$

At $t = -2$ we have:

$$\mathbf{v} = \mathbf{r}'(-2) = \langle 4, -3 \rangle$$

$$\mathbf{a} = \mathbf{r}''(-2) = \langle -4, 0 \rangle$$

Thus,

$$\mathbf{a} \cdot \mathbf{v} = \langle -4, 0 \rangle \cdot \langle 4, -3 \rangle = -16$$

$$\|\mathbf{v}\| = \sqrt{16 + 9} = 5$$

Recall that we have:

$$\mathbf{T} = \frac{\mathbf{v}}{\|\mathbf{v}\|} = \frac{\langle 4, -3 \rangle}{5} = \left\langle \frac{4}{5}, -\frac{3}{5} \right\rangle$$

$$a_{\mathbf{T}} = \frac{\mathbf{a} \cdot \mathbf{v}}{\|\mathbf{v}\|} = -\frac{16}{5}$$

Next, we compute $a_{\mathbf{N}}$ and $\mathbf{N}$:

$$a_{\mathbf{N}}\mathbf{N} = \mathbf{a} - a_{\mathbf{T}}\mathbf{T} = \langle -4, 0 \rangle + \frac{16}{5}\left\langle \frac{4}{5}, -\frac{3}{5} \right\rangle = \left\langle -\frac{36}{25}, -\frac{48}{25} \right\rangle$$

This vector has length:

$$a_{\mathbf{N}} = \|a_{\mathbf{N}}\mathbf{N}\| = \sqrt{\left(-\frac{36}{25}\right)^2 + \left(-\frac{48}{25}\right)^2} = \frac{60}{25} = \frac{12}{5}$$

and thus,

$$\mathbf{N} = \frac{a_{\mathbf{N}}\mathbf{N}}{a_{\mathbf{N}}} = \frac{\left\langle -\frac{36}{25}, -\frac{48}{25} \right\rangle}{12/5} = \left\langle -\frac{3}{5}, -\frac{4}{5} \right\rangle$$

Finally we obtain the decomposition,

$$\mathbf{a} = \langle -4, 0 \rangle = -\frac{16}{5}\mathbf{T} + \frac{12}{5}\mathbf{N}$$

where $\mathbf{T} = \left\langle \frac{4}{5}, -\frac{3}{5} \right\rangle$ and $\mathbf{N} = \left\langle -\frac{3}{5}, -\frac{4}{5} \right\rangle$.

37. $\mathbf{r}(t) = \left\langle t, \frac{1}{2}t^2, \frac{1}{6}t^3 \right\rangle$, $t = 4$

SOLUTION First note here that:

$$\mathbf{v}(t) = \mathbf{r}'(t) = \left\langle 1, t, \frac{1}{2}t^2 \right\rangle$$

$$\mathbf{a}(t) = \mathbf{r}''(t) = \langle 0, 1, t \rangle$$

At $t = 4$ we have:

$$\mathbf{v} = \mathbf{r}'(4) = \langle 1, 4, 8 \rangle$$

$$\mathbf{a} = \mathbf{r}''(4) = \langle 0, 1, 4 \rangle$$

Thus,

$$\mathbf{a} \cdot \mathbf{v} = \langle 0, 1, 4 \rangle \cdot \langle 1, 4, 8 \rangle = 36$$

$$\|\mathbf{v}\| = \sqrt{1 + 16 + 64} = \sqrt{81} = 9$$

Recall that we have:

$$\mathbf{T} = \frac{\mathbf{v}}{\|\mathbf{v}\|} = \frac{\langle 1, 4, 8 \rangle}{9} = \left\langle \frac{1}{9}, \frac{4}{9}, \frac{8}{9} \right\rangle$$

$$a_{\mathbf{T}} = \frac{\mathbf{a} \cdot \mathbf{v}}{\|\mathbf{v}\|} = \frac{36}{9} = 4$$

Next, we compute $a_{\mathbf{N}}$ and $\mathbf{N}$:

$$a_{\mathbf{N}}\mathbf{N} = \mathbf{a} - a_{\mathbf{T}}\mathbf{T} = \langle 0, 1, 4 \rangle - 4\left\langle \frac{1}{9}, \frac{4}{9}, \frac{8}{9} \right\rangle = \left\langle -\frac{4}{9}, -\frac{7}{9}, \frac{4}{9} \right\rangle$$

This vector has length:

$$a_{\mathbf{N}} = \|a_{\mathbf{N}}\mathbf{N}\| = \sqrt{\frac{16}{81} + \frac{49}{81} + \frac{16}{81}} = 1$$

and thus,

$$\mathbf{N} = \frac{a_{\mathbf{N}}\mathbf{N}}{a_{\mathbf{N}}} = \frac{\left\langle -\frac{4}{9}, -\frac{7}{9}, \frac{4}{9} \right\rangle}{1} = \left\langle -\frac{4}{9}, -\frac{7}{9}, \frac{4}{9} \right\rangle$$

Finally we obtain the decomposition,

$$\mathbf{a} = \langle 0, 1, 4 \rangle = 4\mathbf{T} + (1)\mathbf{N}$$

where $\mathbf{T} = \left\langle \frac{1}{9}, \frac{4}{9}, \frac{8}{9} \right\rangle$ and $\mathbf{N} = \left\langle -\frac{4}{9}, -\frac{7}{9}, \frac{4}{9} \right\rangle$.

39. $\mathbf{r}(t) = \langle t, e^t, te^t \rangle$, $t = 0$

SOLUTION First note here that:

$$\mathbf{v}(t) = \mathbf{r}'(t) = \left\langle 1, e^t, (t+1)e^t \right\rangle$$

$$\mathbf{a}(t) = \mathbf{r}''(t) = \left\langle 0, e^t, (t+2)e^t \right\rangle$$

At $t = 0$ we have:

$$\mathbf{v} = \mathbf{r}'(0) = \langle 1, 1, 1 \rangle$$

$$\mathbf{a} = \mathbf{r}''(0) = \langle 0, 1, 2 \rangle$$

Thus,

$$\mathbf{a} \cdot \mathbf{v} = \langle 0, 1, 2 \rangle \cdot \langle 1, 1, 1 \rangle = 3$$

$$\|\mathbf{v}\| = \sqrt{1 + 1 + 1} = \sqrt{3}$$

Recall that we have:

$$\mathbf{T} = \frac{\mathbf{v}}{\|\mathbf{v}\|} = \frac{1}{\sqrt{3}} \langle 1, 1, 1 \rangle$$

$$a_{\mathbf{T}} = \frac{\mathbf{a} \cdot \mathbf{v}}{\|\mathbf{v}\|} = \frac{3}{\sqrt{3}} = \sqrt{3}$$

Next, we compute a_N and $\mathbf{N}$:

$$a_N\mathbf{N} = \mathbf{a} - a_T\mathbf{T} = \langle 0, 1, 2 \rangle - \sqrt{3}\frac{1}{\sqrt{3}}\langle 1, 1, 1 \rangle = \langle -1, 0, 1 \rangle$$

This vector has length:

$$a_N = \|a_N\mathbf{N}\| = \sqrt{1+1} = \sqrt{2}$$

and thus,

$$\mathbf{N} = \frac{a_N\mathbf{N}}{a_N} = \frac{\langle -1, 0, 1 \rangle}{\sqrt{2}} = \left\langle -\frac{1}{\sqrt{2}}, 0, \frac{1}{\sqrt{2}} \right\rangle$$

Finally we obtain the decomposition,

$$\mathbf{a} = \langle 0, 1, 2 \rangle = \sqrt{3}\mathbf{T} + \sqrt{2}\mathbf{N}$$

where $\mathbf{T} = \frac{1}{\sqrt{3}}\langle 1, 1, 1 \rangle$ and $\mathbf{N} = \left\langle -\frac{1}{\sqrt{2}}, 0, \frac{1}{\sqrt{2}} \right\rangle$.

41. $\mathbf{r}(t) = \langle t, \cos t, t \sin t \rangle, \quad t = \frac{\pi}{2}$

SOLUTION First note here that:

$$\mathbf{v}(t) = \mathbf{r}'(t) = \langle 1, -\sin t, t \cos t + \sin t, \rangle$$

$$\mathbf{a}(t) = \mathbf{r}''(t) = \langle 0, -\cos t, -t \sin t + 2\cos t \rangle$$

At $t = \frac{\pi}{2}$ we have:

$$\mathbf{v} = \mathbf{r}'(\pi/2) = \langle 1, -1, 1 \rangle$$

$$\mathbf{a} = \mathbf{r}''(-2) = \left\langle 0, 0, -\frac{\pi}{2} \right\rangle$$

Thus,

$$\mathbf{a} \cdot \mathbf{v} = \left\langle 0, 0, -\frac{\pi}{2} \right\rangle \cdot \langle 1, -1, 1 \rangle = -\frac{\pi}{2}$$

$$\|\mathbf{v}\| = \sqrt{1+1+1} = \sqrt{3}$$

Recall that we have:

$$\mathbf{T} = \frac{\mathbf{v}}{\|\mathbf{v}\|} = \frac{1}{\sqrt{3}}\langle 1, -1, 1 \rangle$$

$$a_T = \frac{\mathbf{a} \cdot \mathbf{v}}{\|\mathbf{v}\|} = \frac{-\pi/2}{\sqrt{3}} = -\frac{\pi}{2\sqrt{3}}$$

Next, we compute a_N and $\mathbf{N}$:

$$a_N\mathbf{N} = \mathbf{a} - a_T\mathbf{T} = \left\langle 0, 0, -\frac{\pi}{2} \right\rangle + \frac{\pi}{2\sqrt{3}}\frac{1}{\sqrt{3}}\langle 1, -1, 1 \rangle$$

$$= \left\langle 0, 0, -\frac{\pi}{2} \right\rangle + \frac{\pi}{6}\langle 1, -1, 1 \rangle$$

$$= \left\langle \frac{\pi}{6}, -\frac{\pi}{6}, -\frac{\pi}{3} \right\rangle = \frac{\pi}{6}\langle 1, -1, -2 \rangle$$

This vector has length:

$$a_N = \|a_N\mathbf{N}\| = \left\| \frac{\pi}{6}\langle 1, -1, -2 \rangle \right\| = \frac{\pi}{6}\sqrt{1+1+4} = \frac{\pi\sqrt{6}}{6} = \frac{\pi}{\sqrt{6}}$$

and thus,

$$\mathbf{N} = \frac{a_N\mathbf{N}}{a_N} = \frac{\frac{\pi}{6}\langle 1, -1, -2 \rangle}{\frac{\pi}{\sqrt{6}}} = \frac{1}{\sqrt{6}}\langle 1, -1, -2 \rangle$$

Finally we obtain the decomposition,

$$\mathbf{a} = \left\langle 0, 0, -\frac{\pi}{2} \right\rangle = \frac{\pi}{2\sqrt{3}}\mathbf{T} + \frac{\pi}{\sqrt{6}}\mathbf{N}$$

where $\mathbf{T} = \frac{1}{\sqrt{3}}\langle 1, -1, 1 \rangle$ and $\mathbf{N} = \frac{1}{\sqrt{6}}\langle 1, -1, -2 \rangle$.

43. Find the components $a_{\mathbf{T}}$ and $a_{\mathbf{N}}$ of the acceleration vector of a particle moving along a circular path of radius $R = 100$ cm with constant velocity $v_0 = 5$ cm/s.

SOLUTION Since the particle moves with constant speed, we have $v'(t) = 0$, hence:

$$a_{\mathbf{T}} = v'(t) = 0$$

The normal component of the acceleration is $a_{\mathbf{N}} = \kappa(t)v(t)^2$. The curvature of a circular path of radius $R = 100$ is $\kappa(t) = \frac{1}{R} = \frac{1}{100}$, and the velocity is the constant value $v(t) = v_0 = 5$. Hence,

$$a_{\mathbf{N}} = \frac{1}{R}v_0^2 = \frac{25}{100} = 0.25 \text{ cm/s}^2$$

45. Suppose that the Ferris wheel in Example 5 is rotating clockwise and that the point P at angle $45°$ has acceleration vector $\mathbf{a} = \langle 0, -50 \rangle$ m/min² pointing down, as in Figure 11. Determine the speed and tangential acceleration of the Ferris wheel.

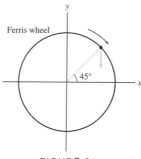

FIGURE 11

SOLUTION The normal and tangential accelerations are both $50/\sqrt{2} \approx 35$ m/min². The normal acceleration is $v^2/R = v^2/30 = 35$, so the speed is

$$v = \sqrt{35(28)} \approx 31.3$$

47. A space shuttle orbits the earth at an altitude 400 km above the earth's surface, with constant speed $v = 28,000$ km/h. Find the magnitude of the shuttle's acceleration (in km/h²), assuming that the radius of the earth is 6378 km (Figure 12).

FIGURE 12 Space shuttle orbit.

SOLUTION The shuttle is in a uniform circular motion, therefore the tangential component of its acceleration is zero, and the acceleration can be written as:

$$\mathbf{a} = \kappa v^2 \mathbf{N} \tag{1}$$

The radius of motion is $6378 + 400 = 6778$ km hence the curvature is $\kappa = \frac{1}{6778}$. Also by the given information the constant speed is $v = 28000$ km/h. Substituting these values in (1) we get:

$$\mathbf{a} = \left(\frac{1}{6778} \cdot 28000^2 \right) \mathbf{N} = (11.5668 \cdot 10^4 \text{ km/h}^2)\mathbf{N}$$

The magnitude of the shuttle's acceleration is thus:

$$\|\mathbf{a}\| = 11.5668 \cdot 10^4 \text{ km/h}^2$$

In units of m/s² we obtain

$$\|\mathbf{a}\| = \frac{11.5668 \cdot 10^4 \cdot 1000}{3600^2} = 8.925 \text{ m/s}^2$$

49. A runner runs along the helix $\mathbf{r}(t) = \langle \cos t, \sin t, t \rangle$. When he is at position $\mathbf{r}(\frac{\pi}{2})$, his speed is 3 m/s and he is accelerating at a rate of $\frac{1}{2}$ m/s². Find his acceleration vector $\mathbf{a}$ at this moment. *Note:* The runner's acceleration vector does not coincide with the acceleration vector of $\mathbf{r}(t)$.

SOLUTION   We have

$$\mathbf{r}'(t) = \langle -\sin t, \cos t, 1 \rangle, \quad \|\mathbf{r}'(t)\| = \sqrt{(-\sin t)^2 + \cos^2 t + 1^2} = \sqrt{2},$$

$$\Rightarrow \quad \mathbf{T} = \frac{1}{\sqrt{2}} \langle -\sin t, \cos t, 1 \rangle$$

By definition, $\mathbf{N}$ is the unit vector in the direction of

$$\frac{d\mathbf{T}}{dt} = \frac{1}{\sqrt{2}} \langle -\cos t, -\sin t, 0 \rangle \qquad \Rightarrow \quad \mathbf{N} = \langle -\cos t, -\sin t, 0 \rangle$$

Therefore $\mathbf{N} = \langle -\cos t, -\sin t, 0 \rangle$. At $t = \pi/2$, we have

$$\mathbf{T} = \frac{1}{\sqrt{2}} \langle -1, 0, 1 \rangle, \qquad \mathbf{N} = \langle 0, -1, 0 \rangle$$

The acceleration vector is

$$\mathbf{a} = v'\mathbf{T} + \kappa v^2 \mathbf{N}$$

We need to find the curvature, which happens to be constant:

$$\kappa = \left\| \frac{d\mathbf{T}}{ds} \right\| = \frac{\|\frac{d\mathbf{T}}{dt}\|}{\|\mathbf{r}'\|} = \frac{\|\frac{1}{\sqrt{2}} \langle -\cos t, -\sin t, 0 \rangle\|}{\sqrt{2}} = \frac{1}{2}$$

Now we have

$$\mathbf{a} = v'\mathbf{T} + \kappa v^2 \mathbf{N} = \left(\frac{1}{2}\right)\mathbf{T} + \left(\frac{1}{2}\right)(3^2)\mathbf{N} = \left(\frac{1}{2}\right)\left(\frac{1}{\sqrt{2}}\right) \langle -1, 0, 1 \rangle + \frac{9}{2} \langle 0, -1, 0 \rangle$$

$$= \left\langle -\frac{1}{2\sqrt{2}}, -\frac{9}{2}, \frac{1}{2\sqrt{2}} \right\rangle$$

51. ✏️ Figure 14 shows acceleration vectors of a particle moving clockwise around a circle. In each case, state whether the particle is speeding up, slowing down, or momentarily at constant speed. Explain.

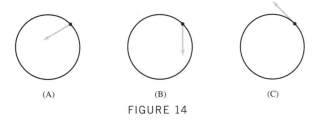

(A)   (B)   (C)

FIGURE 14

SOLUTION   In (A) and (B) the acceleration vector has a nonzero tangential and normal components; these are both possible acceleration vectors. In (C) the normal component of the acceleration toward the inside of the curve is zero, that is, $\mathbf{a}$ is parallel to $\mathbf{T}$, so $\kappa \cdot v(t)^2 = 0$, so either $\kappa = 0$ (meaning our curve is not a circle) or $v(t) = 0$ (meaning our particle isn't moving). Either way, (C) is not a possible acceleration vector.

53. Suppose that $\mathbf{r} = \mathbf{r}(t)$ lies on a sphere of radius R for all t. Let $\mathbf{J} = \mathbf{r} \times \mathbf{r}'$. Show that $\mathbf{r}' = (\mathbf{J} \times \mathbf{r})/\|\mathbf{r}\|^2$. *Hint:* Observe that $\mathbf{r}$ and $\mathbf{r}'$ are perpendicular.

SOLUTION

(a) Solution 1. Since $\mathbf{r} = \mathbf{r}(t)$ lies on the sphere, the vectors $\mathbf{r} = \mathbf{r}(t)$ and $\mathbf{r}' = \mathbf{r}'(t)$ are orthogonal, therefore:

$$\mathbf{r} \cdot \mathbf{r}' = 0 \tag{1}$$

We use the following well-known equality:

$$\mathbf{a} \times (\mathbf{b} \times \mathbf{c}) = (\mathbf{a} \cdot \mathbf{c})\mathbf{b} - (\mathbf{a} \cdot \mathbf{b}) \cdot \mathbf{c}$$

Using this equality and (1) we obtain:

$$\mathbf{J} \times \mathbf{r} = (\mathbf{r} \times \mathbf{r}') \times \mathbf{r} = -\mathbf{r} \times (\mathbf{r} \times \mathbf{r}') = -((\mathbf{r} \cdot \mathbf{r}')\mathbf{r} - (\mathbf{r} \cdot \mathbf{r})\mathbf{r}')$$

$$= -(\mathbf{r} \cdot \mathbf{r}')\mathbf{r} + \|\mathbf{r}\|^2\mathbf{r}' = 0\mathbf{r} + \|\mathbf{r}\|^2\mathbf{r}' = \|\mathbf{r}\|^2\mathbf{r}'$$

Divided by the scalar $\|\mathbf{r}\|^2$ we obtain:

$$\mathbf{r}' = \frac{\mathbf{J} \times \mathbf{r}}{\|\mathbf{r}\|^2}$$

(b) Solution 2. The cross product $\mathbf{J} = \mathbf{r} \times \mathbf{r}'$ is orthogonal to $\mathbf{r}$ and $\mathbf{r}'$. Also, $\mathbf{r}$ and $\mathbf{r}'$ are orthogonal, hence the vectors $\mathbf{r}$, $\mathbf{r}'$ and $\mathbf{J}$ are mutually orthogonal. Now, since $\mathbf{r}'$ is orthogonal to $\mathbf{r}$ and $\mathbf{J}$, the right-hand rule implies that $\mathbf{r}'$ points in the direction of $\mathbf{J} \times \mathbf{r}$. Therefore, for some $\alpha > 0$ we have:

$$\mathbf{r}' = \alpha\mathbf{J} \times \mathbf{r} = \|\mathbf{r}'\| \cdot \frac{\mathbf{J} \times \mathbf{r}}{\|\mathbf{J} \times \mathbf{r}\|} \tag{2}$$

By properties of the cross product and since $\mathbf{J}$, $\mathbf{r}$, and $\mathbf{r}'$ are mutually orthogonal we have:

$$\|\mathbf{J} \times \mathbf{r}\| = \|\mathbf{J}\|\|\mathbf{r}\| = \|\mathbf{r} \times \mathbf{r}'\|\|\mathbf{r}\| = \|\mathbf{r}\|\|\mathbf{r}'\|\|\mathbf{r}\| = \|\mathbf{r}\|^2\|\mathbf{r}'\|$$

Substituting in (2) we get:

$$\mathbf{r}' = \|\mathbf{r}'\| \frac{\mathbf{J} \times \mathbf{r}}{\|\mathbf{r}\|^2\|\mathbf{r}'\|} = \frac{\mathbf{J} \times \mathbf{r}}{\|\mathbf{r}\|^2}$$

Further Insights and Challenges

In Exercises 55–59, we consider an automobile of mass m traveling along a curved but level road. To avoid skidding, the road must supply a frictional force $\mathbf{F} = m\mathbf{a}$, where $\mathbf{a}$ is the car's acceleration vector. The maximum magnitude of the frictional force is μmg, where μ is the coefficient of friction and $g = 9.8$ m/s^2. Let v be the car's speed in meters per second.

55. Show that the car will not skid if the curvature κ of the road is such that (with $R = 1/\kappa$)

$$(v')^2 + \left(\frac{v^2}{R}\right)^2 \leq (\mu g)^2 \qquad \boxed{5}$$

Note that braking ($v' < 0$) and speeding up ($v' > 0$) contribute equally to skidding.

SOLUTION To avoid skidding, the frictional force the road must supply is:

$$\mathbf{F} = m\mathbf{a}$$

where $\mathbf{a}$ is the acceleration of the car. We consider the decomposition of the acceleration $\mathbf{a}$ into normal and tangential directions:

$$\mathbf{a}(t) = v'(t)\mathbf{T}(t) + \kappa v^2(t)\mathbf{N}(t)$$

Since $\mathbf{N}$ and $\mathbf{T}$ are orthogonal unit vectors, $\mathbf{T} \cdot \mathbf{N} = 0$ and $\mathbf{T} \cdot \mathbf{T} = \mathbf{N} \cdot \mathbf{N} = 1$. Thus:

$$\|\mathbf{a}\|^2 = \left(v'\mathbf{T} + \kappa v^2\mathbf{N}\right) \cdot \left(v'\mathbf{T} + \kappa v^2\mathbf{N}\right) = v'^2\mathbf{T} \cdot \mathbf{T} + 2\kappa v^2 v'\mathbf{N} \cdot \mathbf{T} + \kappa^2 v^4\mathbf{N} \cdot \mathbf{N}$$

$$= v'^2 + \kappa^2 v^4 = v'^2 + \frac{v^4}{R^2}$$

Therefore:

$$\|\mathbf{a}\| = \sqrt{(v')^2 + \frac{v^4}{R^2}}$$

Since the maximal fractional force is μmg we obtain that to avoid skidding the curvature must satisfy:

$$m\sqrt{(v')^2 + \frac{v^4}{R^2}} \leq m\mu g.$$

Hence,

$$\left(v'\right)^2 + \frac{v^4}{R^2} \le (\mu g)^2,$$

which becomes:

$$\left(v'\right)^2 + \left(\frac{v^2}{R}\right)^2 \le (\mu g)^2$$

57. Beginning at rest, an automobile drives around a circular track of radius $R = 300$ m, accelerating at a rate of 0.3 m/s^2. After how many seconds will the car begin to skid if the coefficient of friction is $\mu = 0.6$?

SOLUTION   By Exercise 55 the car will begin to skid when:

$$\left(v'\right)^2 + \frac{v^4}{R^2} = \mu^2 g^2 \tag{1}$$

We are given that $v' = 0.3$ and $v_0 = 0$. Integrating gives:

$$v = \int_0^t v'\, dt = \int_0^t 0.3\, dt = 0.3t + v_0 = 0.3t$$

We substitute $v = t$, $v' = 0.3$, $R = 300$, $\mu = 0.6$ and $g = 9.8$ in (1) and solve for t. This gives:

$$(0.3)^2 + \frac{0.3^4 t^4}{300^2} = 0.6^2 \cdot 9.8^2$$

$$t^4 = \frac{300^2 \left(0.6^2 \cdot 9.8^2 - 0.3^2\right)}{0.3^4} = 383{,}160{,}000$$

$$t = 139.91 \text{ s}$$

After 139.91 s or 2.33 minutes, the car will begin to skid.

59. What is the smallest radius R about which an automobile can turn without skidding at 100 km/h if $\mu = 0.75$ (a typical value)?

SOLUTION   In Exercise 55 we showed that the car will not skid if the following inequality holds:

$$\left(v'\right)^2 + \frac{v^4}{R^2} < \mu^2 g^2$$

In case of constant speed, $v' = 0$, so the inequality becomes:

$$\frac{v^4}{R^2} < \mu^2 g^2$$

Solving for R we get:

$$v^4 < \mu^2 g^2 R^2$$

$$\frac{v^4}{\mu^2 g^2} < R^2 \quad \Rightarrow \quad R > \frac{v^2}{\mu g}$$

The smallest radius R in which skidding does not occur is, thus,

$$R \approx \frac{v^2}{\mu g}$$

We substitute $v = 100$ km/h, $\mu = 0.75$, and $g \approx 127{,}008$ km/h^2 to obtain:

$$R \approx \frac{100^2}{0.75 \cdot 127{,}008} = 0.105 \text{ km}.$$

13.6 Planetary Motion According to Kepler and Newton (LT Section 14.6)

Preliminary Questions

1. Describe the relation between the vector $\mathbf{J} = \mathbf{r} \times \mathbf{r}'$ and the rate at which the radial vector sweeps out area.

SOLUTION The rate at which the radial vector sweeps out area equals half the magnitude of the vector $\mathbf{J}$. This relation is expressed in the formula:

$$\frac{dA}{dt} = \frac{1}{2}\|\mathbf{J}\|.$$

2. Equation (1) shows that $\mathbf{r}''$ is proportional to $\mathbf{r}$. Explain how this fact is used to prove Kepler's Second Law.

SOLUTION In the proof of Kepler's Second Law it is shown that the rate at which area is swept out is

$$\frac{dA}{dt} = \frac{1}{2}\|\mathbf{J}\|, \quad \text{where} \quad \mathbf{J} = \mathbf{r}(t) \times \mathbf{r}'(t)$$

To show that $\|\mathbf{J}\|$ is constant, show that $\mathbf{J}$ is constant. This is done using the proportionality of $\mathbf{r}''$ and $\mathbf{r}$ which implies that $\mathbf{r}(t) \times \mathbf{r}''(t) = 0$. Using this we get:

$$\frac{d\mathbf{J}}{dt} = \frac{d}{dt}\left(\mathbf{r} \times \mathbf{r}'\right) = \mathbf{r} \times \mathbf{r}'' + \mathbf{r}' \times \mathbf{r}' = 0 + 0 = 0 \Rightarrow \mathbf{J} = \text{const}$$

3. How is the period T affected if the semimajor axis a is increased four-fold?

SOLUTION Kepler's Third Law states that the period T of the orbit is given by:

$$T^2 = \left(\frac{4\pi^2}{GM}\right)a^3$$

or

$$T = \frac{2\pi}{\sqrt{GM}}a^{3/2}$$

If a is increased four-fold the period becomes:

$$\frac{2\pi}{\sqrt{GM}}(4a)^{3/2} = 8 \cdot \frac{2\pi}{\sqrt{GM}}a^{3/2}$$

That is, the period is increased eight-fold.

Exercises

1. Kepler's Third Law states that T^2/a^3 has the same value for each planetary orbit. Do the data in the following table support this conclusion? Estimate the length of Jupiter's period, assuming that $a = 77.8 \times 10^{10}$ m.

Planet	Mercury	Venus	Earth	Mars
a (10^{10} m)	5.79	10.8	15.0	22.8
T (years)	0.241	0.615	1.00	1.88

SOLUTION Using the given data we obtain the following values of T^2/a^3, where a, as always, is measured not in meters but in 10^{10} m:

Planet	Mercury	Venus	Earth	Mars
T^2/a^3	$2.99 \cdot 10^{-4}$	$3 \cdot 10^{-4}$	$2.96 \cdot 10^{-4}$	$2.98 \cdot 10^{-4}$

The data on the planets supports Kepler's prediction. We estimate Jupiter's period (using the given a) as $T \approx \sqrt{a^3 \cdot 3 \cdot 10^{-4}} \approx 11.9$ years.

3. Ganymede, one of Jupiter's moons discovered by Galileo, has an orbital period of 7.154 days and a semimajor axis of 1.07×10^9 m. Use Exercise 2 to estimate the mass of Jupiter.

SOLUTION By Exercise 2, the mass of Jupiter can be computed using the following equality:

$$M = \frac{4\pi^2}{G}\frac{a^3}{T^2}$$

We substitute the given data $T = 7.154 \cdot 24 \cdot 60^2 = 618,105.6 \, a = 1.07 \times 10^9$ m and $G = 6.67300 \times 10^{-11} \text{m}^3 \text{kg}^{-1} \text{s}^{-1}$, to obtain:

$$M = \frac{4\pi^2 \cdot \left(1.07 \times 10^9\right)^3}{6.67300 \times 10^{-11} \cdot (618,105.6)^2} \approx 1.897 \times 10^{27} \text{ kg}.$$

5. Mass of the Milky Way   The sun revolves around the center of mass of the Milky Way galaxy in an orbit that is approximately circular, of radius $a \approx 2.8 \times 10^{17}$ km and velocity $v \approx 250$ km/s. Use the result of Exercise 2 to estimate the mass of the portion of the Milky Way inside the sun's orbit (place all of this mass at the center of the orbit).

SOLUTION   Write $a = 2.8 \times 10^{20}$ m and $v = 250 \times 10^3$ m/s. The circumference of the sun's orbit (which is assumed circular) is $2\pi a$ m; since the sun's speed is a constant v m/s, its period is $T = \frac{2\pi a}{v}$ s. By Exercise 2, the mass of the portion of the Milky Way inside the sun's orbit is

$$M = \left(\frac{4\pi^2}{G}\right)\left(\frac{a^3}{T^2}\right)$$

Substituting the values of a and T from above, $G = 6.673 \times 10^{-11} \text{ m}^3 \text{kg}^{-1} \text{s}^{-2}$ gives

$$M = \frac{4\pi^2 a^3}{G\left(\frac{4\pi^2 a^2}{v^3}\right)} = \frac{av^2}{G} = \frac{2.8 \cdot 10^{20} \cdot (250 \times 10^3)^2}{6.673 \times 10^{-11}} = 2.6225 \times 10^{41} \text{ kg}.$$

The mass of the sun is 1.989×10^{30} kg, hence M is 1.32×10^{11} times the mass of the sun (132 billions times the mass of the sun).

7. Show that a planet in a circular orbit travels at constant speed. *Hint:* Use that $\mathbf{J}$ is constant and that $\mathbf{r}(t)$ is orthogonal to $\mathbf{r}'(t)$ for a circular orbit.

SOLUTION   It is shown in the proof of Kepler's Second Law that the vector $\mathbf{J} = \mathbf{r}(t) \times \mathbf{r}'(t)$ is constant, hence its length is constant:

$$\|\mathbf{J}\| = \|\mathbf{r}(t) \times \mathbf{r}'(t)\| = \text{const} \tag{1}$$

We consider the orbit as a circle of radius R, therefore, $\mathbf{r}(t)$ and $\mathbf{r}'(t)$ are orthogonal and $\|\mathbf{r}(t)\| = R$. By (1) and using properties of the cross product we obtain:

$$\|\mathbf{r}(t) \times \mathbf{r}'(t)\| = \|\mathbf{r}(t)\| \|\mathbf{r}'(t)\| \sin \frac{\pi}{2} = R \cdot \|\mathbf{r}'(t)\| = \text{const}$$

We conclude that $\|\mathbf{r}'(t)\|$ is constant, that is the speed $v = \|\mathbf{r}'(t)\|$ of the planet is constant.

9. Prove that if a planetary orbit is circular of radius R, then $vT = 2\pi R$, where v is the planet's speed (constant by Exercise 7) and T is the period. Then use Kepler's Third Law to prove that $v = \sqrt{\dfrac{k}{R}}$.

SOLUTION   By the Arc Length Formula and since the speed $v = \|\mathbf{r}'(t)\|$ is constant, the length L of the circular orbit can be computed by the following integral:

$$L = \int_0^T \|\mathbf{r}'(t)\| \, dt = \int_0^T v \, dt = vt \Big|_0^T = vT$$

On the other hand, the length of a circular orbit of radius R is $2\pi R$, so we obtain:

$$vT = 2\pi R \Rightarrow T = \frac{2\pi R}{v} \tag{1}$$

In a circular orbit of radius R, $a = R$, hence by Kepler's Third Law we have:

$$T^2 = \frac{4\pi^2}{GM} R^3 \tag{2}$$

We now substitute (1) in (2) and solve for v. This gives:

$$\left(\frac{2\pi R}{v}\right)^2 = \frac{4\pi^2 R^3}{GM}$$

$$\frac{4\pi^2 R^2}{v^2} = \frac{4\pi^2 R^3}{GM}$$

$$\frac{1}{v^2} = \frac{R}{GM} \Rightarrow v = \sqrt{\frac{GM}{R}}$$

11. A communications satellite orbiting the earth has initial position $\mathbf{r} = \langle 29{,}000, 20{,}000, 0 \rangle$ (in km) and initial velocity $\mathbf{r}' = \langle 1, 1, 1 \rangle$ (in km/s), where the origin is the earth's center. Find the equation of the plane containing the satellite's orbit. *Hint:* This plane is orthogonal to $\mathbf{J}$.

SOLUTION The vectors $\mathbf{r}(t)$ and $\mathbf{r}'(t)$ lie in the plane containing the satellite's orbit, in particular the initial position $\mathbf{r} = \langle 29{,}000, 20{,}000, 0 \rangle$ and the initial velocity $\mathbf{r}' = \langle 1, 1, 1 \rangle$. Therefore, the cross product $\mathbf{J} = \mathbf{r} \times \mathbf{r}'$ is perpendicular to the plane. We compute $\mathbf{J}$:

$$\mathbf{J} = \mathbf{r} \times \mathbf{r}' = \begin{vmatrix} \mathbf{i} & \mathbf{j} & \mathbf{k} \\ 29{,}000 & 20{,}000 & 0 \\ 1 & 1 & 1 \end{vmatrix} = \begin{vmatrix} 20{,}000 & 0 \\ 1 & 1 \end{vmatrix} \mathbf{i} - \begin{vmatrix} 29{,}000 & 0 \\ 1 & 1 \end{vmatrix} \mathbf{j} + \begin{vmatrix} 29{,}000 & 20{,}000 \\ 1 & 1 \end{vmatrix} \mathbf{k}$$

$$= 20{,}000\mathbf{i} - 29{,}000\mathbf{j} + 9000\mathbf{k} = \langle 20{,}000, -29{,}000, 9000 \rangle$$

We now use the vector form of the equation of the plane with $\mathbf{n} = \mathbf{J} = \langle 20{,}000, -29{,}000, 9000 \rangle$ and $\langle x_0, y_0, z_0 \rangle = \mathbf{r} = \langle 29{,}000, 20{,}000, 0 \rangle$, to obtain the following equation:

$$\langle 29{,}000, -20{,}000, 9000 \rangle \cdot \langle x, y, z \rangle = \langle 29{,}000, -20{,}000, 9000 \rangle \cdot \langle 29{,}000, 20{,}000, 9000 \rangle$$

$$1000 \langle 29, -20, 9 \rangle \cdot \langle x, y, z \rangle = 1000 \langle 29, -20, 9 \rangle \cdot \langle 29{,}000, 20{,}000, 9000 \rangle$$

$$29x - 20y + 9z = 841{,}000 - 400{,}000 + 81{,}000 = 0$$

$$29x - 20y + 9z - 522{,}000 = 0$$

The plane containing the satellite's orbit is, thus:

$$\mathcal{P} = \{(x, y, z) : 29x - 20y + 9z - 522{,}000 = 0\}$$

Exercises 13–19: The perihelion and aphelion are the points on the orbit closest to and farthest from the sun, respectively (Figure 8). The distance from the sun at the perihelion is denoted r_{per} and the speed at this point is denoted v_{per}. Similarly, we write r_{ap} and v_{ap} for the distance and speed at the aphelion. The semimajor axis is denoted a.

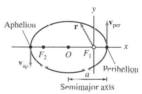

FIGURE 8 $\mathbf{r}$ and $\mathbf{v} = \mathbf{r}'$ are perpendicular at the perihelion and aphelion.

13. Use the polar equation of an ellipse

$$r = \frac{p}{1 + e \cos \theta}$$

to show that $r_{per} = a(1 - e)$ and $r_{ap} = a(1 + e)$. *Hint:* Use the fact that $r_{per} + r_{ap} = 2a$.

SOLUTION We use the polar equation of the elliptic orbit:

$$r = \frac{p}{1 + e \cos \theta} \tag{1}$$

At the perigee, $\theta = 0$ and at the apogee $\theta = \pi$. Substituting these values in (1) gives the distances r_{per} and r_{ap} respectively. That is,

$$r_{per} = \frac{p}{1 + e \cos \theta} = \frac{p}{1 + e} \tag{2}$$

$$r_{ap} = \frac{p}{1 + e \cos \pi} = \frac{p}{1 - e} \tag{3}$$

To obtain the solutions in terms of a rather than p, we notice that:

$$r_{per} + r_{ap} = 2a$$

Hence:

$$2a = \frac{p}{1+e} + \frac{p}{1-e} = \frac{p(1-e) + p(1+e)}{(1+e)(1-e)} = \frac{2p}{(1+e)(1-e)}$$

yielding

$$p = a(1+e)(1-e)$$

Substituting in (2) and (3) we obtain:

$$r_{\text{per}} = \frac{a(1+e)(1-e)}{1+e} = a(1-e)$$

$$r_{\text{ap}} = \frac{a(1+e)(1-e)}{1-e} = a(1+e)$$

15. Use the fact that $\mathbf{J} = \mathbf{r} \times \mathbf{r}'$ is constant to prove

$$v_{\text{per}}(1-e) = v_{\text{ap}}(1+e)$$

Hint: $\mathbf{r}$ is perpendicular to $\mathbf{r}'$ at the perihelion and aphelion.

SOLUTION Since the vector $\mathbf{J}(t) = \mathbf{r}(t) \times \mathbf{r}'(t)$ is constant, it is the same vector at the perigee and at the apogee, hence we may equate the length of $\mathbf{J}(t)$ at these two points. Since at the perigee and at the apogee $\mathbf{r}(t)$ and $\mathbf{r}'(t)$ are orthogonal we have by properties of the cross product:

$$\|\mathbf{r}_{\text{ap}} \times \mathbf{r}'_{\text{ap}}\| = \|\mathbf{r}_{\text{ap}}\| \|\mathbf{r}'_{\text{ap}}\| = r_{\text{ap}} v_{\text{ap}}$$

$$\|\mathbf{r}_{\text{per}} \times \mathbf{r}'_{\text{per}}\| = \|\mathbf{r}_{\text{per}}\| \|\mathbf{r}'_{\text{per}}\| = r_{\text{per}} v_{\text{per}}$$

Equating the two values gives:

$$r_{\text{ap}} v_{\text{ap}} = r_{\text{per}} v_{\text{per}} \tag{1}$$

In Exercise 13 we showed that $r_{\text{per}} = a(1-e)$ and $r_{\text{ap}} = a(1+e)$. Substituting in (1) we obtain:

$$a(1+e)v_{\text{ap}} = a(1-e)v_{\text{per}}$$

$$(1+e)v_{\text{ap}} = (1-e)v_{\text{per}}$$

17. Conservation of Energy The total mechanical energy (kinetic energy plus potential energy) of a planet of mass m orbiting a sun of mass M with position $\mathbf{r}$ and speed $v = \|\mathbf{r}'\|$ is

$$E = \frac{1}{2}mv^2 - \frac{GMm}{\|\mathbf{r}\|} \qquad \boxed{8}$$

(a) Prove the equations

$$\frac{d}{dt}\frac{1}{2}mv^2 = \mathbf{v} \cdot (m\mathbf{a}), \qquad \frac{d}{dt}\frac{GMm}{\|\mathbf{r}\|} = \mathbf{v} \cdot \left(-\frac{GMm}{\|\mathbf{r}\|^3}\mathbf{r}\right)$$

(b) Then use Newton's Law to show that E is conserved—that is, $\dfrac{dE}{dt} = 0$.

SOLUTION We start by observing that since $\|\mathbf{r}\|^2 = \mathbf{r} \cdot \mathbf{r}$, we have (using Eq. (4) in Theorem 3, Section 13.2)

$$\frac{d}{dt}\|\mathbf{r}\|^2 = 2\|\mathbf{r}\|\frac{d}{dt}\|\mathbf{r}\|, \quad \text{and} \quad \frac{d}{dt}\|\mathbf{r}\|^2 = \frac{d}{dt}\mathbf{r} \cdot \mathbf{r} = 2\mathbf{r} \cdot \mathbf{r}'$$

Equating these two expressions gives

$$\frac{d}{dt}\|\mathbf{r}\| = \frac{\mathbf{r} \cdot \mathbf{r}'}{\|\mathbf{r}\|} \tag{1}$$

(a) Applying (1) to $\mathbf{r}'$, we have

$$\frac{d}{dt}\frac{1}{2}mv^2 = \frac{d}{dt}\frac{1}{2}m\|\mathbf{r}'\|^2 = m\|\mathbf{r}'\|\frac{d}{dt}\|\mathbf{r}'\| = m\|\mathbf{r}'\|\frac{\mathbf{r}' \cdot \mathbf{r}''}{\|\mathbf{r}'\|} = \mathbf{r}' \cdot (m\mathbf{r}'') = \mathbf{v} \cdot (m\mathbf{a})$$

proving half of formula 2. For the other half, note that again by (1),

$$\frac{d}{dt}\frac{GMm}{\|\mathbf{r}\|} = GMm\frac{d}{dt}\|\mathbf{r}\|^{-1} = -GMm\|\mathbf{r}\|^{-2}\frac{d}{dt}\|\mathbf{r}\| = -GMm\|\mathbf{r}\|^{-2} \cdot \frac{\mathbf{r} \cdot \mathbf{r}'}{\|r\|}$$

$$= \mathbf{r}' \cdot \left(-\frac{GMm}{\|\mathbf{r}\|^3}\right)\mathbf{r} = \mathbf{v} \cdot \left(-\frac{GMm}{\|\mathbf{r}\|^3}\mathbf{r}\right)$$

(b) We have by part (a)

$$\frac{dE}{dt} = \frac{d}{dt}\left(\frac{1}{2}mv^2\right) - \frac{d}{dt}\left(\frac{GMm}{\|\mathbf{r}\|}\right) = \mathbf{v}\cdot(m\mathbf{a}) + \mathbf{v}\cdot\left(\frac{GMm}{\|\mathbf{r}\|^3}\mathbf{r}\right) = \mathbf{v}\cdot\left(m\mathbf{a} + \frac{GMm}{\|\mathbf{r}\|^3}\mathbf{r}\right) \tag{2}$$

By Newton's Law, formula (1) in the text,

$$\mathbf{r}'' = -\frac{GM}{\|\mathbf{r}\|^2}\mathbf{e}_r = -\frac{GM}{\|\mathbf{r}\|^3}\mathbf{r} \tag{3}$$

Substituting (3) into (2), and noting that $\mathbf{v} = \mathbf{r}'$ and $\mathbf{a} = \mathbf{r}''$ gives

$$\frac{dE}{dt} = \mathbf{r}'\cdot\left(m\mathbf{r}'' + \frac{GMm}{\|\mathbf{r}\|^3}\mathbf{r}\right) = \mathbf{r}'\cdot\left(-\frac{GMm}{\|\mathbf{r}\|^3}\mathbf{r} + \frac{GMm}{\|\mathbf{r}\|^3}\mathbf{r}\right) = 0$$

19. Prove that $v_{\text{per}} = \sqrt{\left(\dfrac{GM}{a}\right)\dfrac{1+e}{1-e}}$ as follows:

(a) Use Conservation of Energy (Exercise 17) to show that

$$v_{\text{per}}^2 - v_{\text{ap}}^2 = 2GM\left(r_{\text{per}}^{-1} - r_{\text{ap}}^{-1}\right)$$

(b) Show that $r_{\text{per}}^{-1} - r_{\text{ap}}^{-1} = \dfrac{2e}{a(1-e^2)}$ using Exercise 13.

(c) Show that $v_{\text{per}}^2 - v_{\text{ap}}^2 = 4\dfrac{e}{(1+e)^2}v_{\text{per}}^2$ using Exercise 15. Then solve for v_{per} using (a) and (b).

SOLUTION

(a) The total mechanical energy of a planet is constant. That is,

$$E = \frac{1}{2}mv^2 - \frac{GMm}{\|\mathbf{r}\|} = \text{const.}$$

Therefore, E has equal values at the perigee and apogee. Hence,

$$\frac{1}{2}mv_{\text{per}}^2 - \frac{GMm}{r_{\text{per}}} = \frac{1}{2}mv_{\text{ap}}^2 - \frac{GMm}{r_{\text{ap}}}$$

$$\frac{1}{2}m\left(v_{\text{per}}^2 - v_{\text{ap}}^2\right) = GMm\left(\frac{1}{r_{\text{per}}} - \frac{1}{r_{\text{ap}}}\right)$$

$$v_{\text{per}}^2 - v_{\text{ap}}^2 = 2GM\left(r_{\text{per}}^{-1} - r_{\text{ap}}^{-1}\right)$$

(b) In Exercise 13 we showed that $r_{\text{per}} = a(1-e)$ and $r_{\text{ap}} = a(1+e)$. Therefore,

$$r_{\text{per}}^{-1} - r_{\text{ap}}^{-1} = \frac{1}{a(1-e)} - \frac{1}{a(1+e)} = \frac{1+e-(1-e)}{a(1-e)(1+e)} = \frac{2e}{a(1-e^2)}$$

(c) In Exercise 15 we showed that

$$v_{\text{per}}(1-e) = v_{\text{ap}}(1+e)$$

Hence,

$$v_{\text{ap}} = \frac{1-e}{1+e}v_{\text{per}}$$

We compute the following difference,

$$v_{\text{per}}^2 - v_{\text{ap}}^2 = v_{\text{per}}^2 - \left(\frac{1-e}{1+e}v_{\text{per}}\right)^2 = v_{\text{per}}^2\left(1 - \left(\frac{1-e}{1+e}\right)^2\right)$$

$$= v_{\text{per}}^2\frac{(1+e)^2 - (1-e)^2}{(1+e)^2} = v_{\text{per}}^2\frac{1+2e+e^2 - (1-2e+e^2)}{(1+e)^2} = 4\frac{e}{(1+e)^2}v_{\text{per}}^2$$

We combine this equality with the equality in part (a) to write

$$\frac{4e}{(1+e)^2}v_{\text{per}}^2 = 2GM\left(r_{\text{per}}^{-1} - r_{\text{ap}}^{-1}\right)$$

Replacing the difference in the right-hand side by $\frac{2e}{a(1-e^2)}$ (from part (b)) and solving for v_{per} we obtain:

$$\frac{4e}{(1+e)^2} v_{\text{per}}^2 = 2GM \cdot \frac{2e}{a(1-e^2)}$$

$$v_{\text{per}}^2 = \frac{4GMe}{a(1-e)(1+e)} \cdot \frac{(1+e)^2}{4e} = \frac{GM(1+e)}{a(1-e)}$$

or,

$$v_{\text{per}} = \sqrt{\frac{GM}{a} \frac{1+e}{1-e}}$$

21. Prove that $v^2 = GM\left(\frac{2}{r} - \frac{1}{a}\right)$ at any point on an elliptical orbit, where $r = \|\mathbf{r}\|$, v is the velocity, and a is the semimajor axis of the orbit.

SOLUTION   The total energy $E = \frac{1}{2}mv^2 - \frac{GMm}{\|\mathbf{r}\|}$ is conserved, and in Exercise 20 we showed that its constant value is $-\frac{GMm}{2a}$. We obtain the following equality:

$$\frac{1}{2}mv^2 - \frac{GMm}{r} = -\frac{GMm}{2a}$$

Algebraic manipulations yield:

$$v^2 = \frac{2GM}{r} - \frac{GM}{a} = GM\left(\frac{2}{r} - \frac{1}{a}\right)$$

Further Insights and Challenges

Exercises 23 and 24 prove Kepler's Third Law. Figure 10 shows an elliptical orbit with polar equation

$$r = \frac{p}{1 + e\cos\theta}$$

where $p = J^2/k$. The origin of the polar coordinates is at F_1. Let a and b be the semimajor and semiminor axes, respectively.

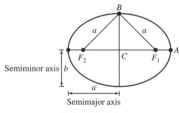

FIGURE 10

23. This exercise shows that $b = \sqrt{pa}$.
(a) Show that $CF_1 = ae$. *Hint:* $r_{\text{per}} = a(1 - e)$ by Exercise 13.
(b) Show that $a = \frac{p}{1 - e^2}$.
(c) Show that $F_1A + F_2A = 2a$. Conclude that $F_1B + F_2B = 2a$ and hence $F_1B = F_2B = a$.
(d) Use the Pythagorean Theorem to prove that $b = \sqrt{pa}$.

SOLUTION
(a) Since $CF_2 = AF_1$, we have:

$$F_2A = CA - CF_2 = 2a - F_1A$$

Therefore,

$$F_1A + F_2A = 2a \tag{1}$$

The ellipse is the set of all points such that the sum of the distances to the two foci F_1 and F_2 is constant. Therefore,

$$F_1A + F_2A = F_1B + F_2B \tag{2}$$

Combining (1) and (2), we obtain:

$$F_1 B + F_2 B = 2a \qquad (3)$$

The triangle $F_2 B F_1$ is isosceles, hence $F_2 B = F_1 B$ and so we conclude that

$$F_1 B = F_2 B = a$$

(b) The polar equation of the ellipse, where the focus F_1 is at the origin is

$$r = \frac{p}{1 + e \cos \theta}$$

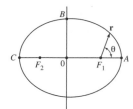

The point A corresponds to $\theta = 0$, hence,

$$F_1 A = \frac{p}{1 + e \cos 0} = \frac{p}{1 + e} \qquad (4)$$

The point C corresponds to $\theta = \pi$ hence,

$$F_1 C = \frac{p}{1 + e \cos \pi} = \frac{p}{1 - e}$$

We now find $F_2 A$. Using the equality $C F_2 = A F_1$ we get:

$$F_2 A = F_2 F_1 + F_1 A = F_2 F_1 + F_2 C = F_1 C = \frac{p}{1 - e}$$

That is,

$$F_2 A = \frac{p}{1 - e} \qquad (5)$$

Combining (1), (4), and (5) we obtain:

$$\frac{p}{1 + e} + \frac{p}{1 - e} = 2a$$

Hence,

$$a = \frac{1}{2} \left(\frac{p}{1 + e} + \frac{p}{1 - e} \right) = \frac{p(1 - e) + p(1 + e)}{2(1 + e)(1 - e)} = \frac{2p}{2 \left(1 - e^2 \right)} = \frac{p}{1 - e^2}$$

(c) We use Pythagoras' Theorem for the triangle $O B F_1$:

$$O B^2 + O F_1^2 = B F_1^2 \qquad (6)$$

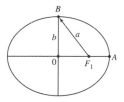

Using (4) we have

$$O F_1 = a - F_1 A = a - \frac{p}{1 + e}$$

Also $O B = b$ and $B F_1 = a$, hence (6) gives:

$$b^2 + \left(a - \frac{p}{1 + e} \right)^2 = a^2$$

We solve for b:

$$b^2 + a^2 - \frac{2ap}{1 + e} + \frac{p^2}{(1 + e)^2} = a^2$$

$$b^2 - \frac{2ap}{1 + e} + \frac{p^2}{(1 + e)^2} = 0$$

In part (b) we showed that $a = \frac{p}{1-e^2}$. We substitute to obtain:

$$b^2 - \frac{2p}{1+e} \cdot \frac{p}{1-e^2} + \frac{p^2}{(1+e)^2} = 0$$

$$b^2 = \frac{2p^2}{(1+e)^2(1-e)} - \frac{p^2}{(1+e)^2} = \frac{2p^2 - p^2(1-e)}{(1+e)^2(1-e)}$$

$$= \frac{p^2(1+e)}{(1+e)^2(1-e)} = \frac{p^2}{1-e^2}$$

Hence,

$$b = \frac{p}{\sqrt{1-e^2}}$$

Since $1 - e^2 = \frac{p}{a}$ we also have

$$b = \frac{p}{\sqrt{\frac{p}{a}}} = \sqrt{ap}$$

25. ✏️ According to Eq. (7) the velocity vector of a planet as a function of the angle θ is

$$\mathbf{v}(\theta) = \frac{k}{J}\mathbf{e}_\theta + \mathbf{c}$$

Use this to explain the following statement: As a planet revolves around the sun, its velocity vector traces out a circle of radius k/J with center $\mathbf{c}$ (Figure 11). This beautiful but hidden property of orbits was discovered by William Rowan Hamilton in 1847.

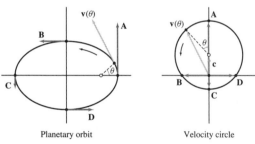

Planetary orbit   Velocity circle

FIGURE 11 The velocity vector traces out a circle as the planet travels along its orbit.

SOLUTION   Recall that $\mathbf{e}_\theta = \langle -\sin\theta, \cos\theta \rangle$, so that

$$\mathbf{v}(\theta) = \frac{k}{J}\langle -\sin\theta, \cos\theta \rangle + \mathbf{c} = \frac{k}{J}\langle \sin(-\theta), \cos(-\theta) \rangle + \mathbf{c}$$

The first term is obviously a clockwise (due to having $-\theta$ instead of θ) parametrization of a circle of radius k/J centered at the origin. It follows that $\mathbf{v}(\theta)$ is a clockwise parametrization of a circle of radius k/J and center $\mathbf{c}$.

CHAPTER REVIEW EXERCISES

1. Determine the domains of the vector-valued functions.

(a) $\mathbf{r}_1(t) = \langle t^{-1}, (t+1)^{-1}, \sin^{-1} t \rangle$

(b) $\mathbf{r}_2(t) = \langle \sqrt{8-t^3}, \ln t, e^{\sqrt{t}} \rangle$

SOLUTION

(a) We find the domain of $\mathbf{r}_1(t) = \langle t^{-1}, (t+1)^{-1}, \sin^{-1} t \rangle$. The function t^{-1} is defined for $t \neq 0$. $(t+1)^{-1}$ is defined for $t \neq -1$ and $\sin^{-1} t$ is defined for $-1 \leq t \leq 1$. Hence, the domain of $\mathbf{r}_1(t)$ is defined by the following inequalities:

$$t \neq 0$$
$$t \neq -1 \quad \Rightarrow \quad -1 < t < 0 \quad \text{or} \quad 0 < t \leq 1$$
$$-1 \leq t \leq 1$$

(b) We find the domain of $\mathbf{r}_2(t) = \langle \sqrt{8 - t^3}, \ln t, e^{\sqrt{t}} \rangle$. The domain of $\sqrt{8 - t^3}$ is $8 - t^3 \geq 0$. The domain of $\ln t$ is $t > 0$ and $e^{\sqrt{t}}$ is defined for $t \geq 0$. Hence, the domain of $\mathbf{r}_2(t)$ is defined by the following inequalities:

$$\begin{aligned} 8 - t^3 &\geq 0 \\ t &> 0 \\ t &\geq 0 \end{aligned} \quad \Rightarrow \quad \begin{aligned} t^3 &\leq 8 \\ t &> 0 \end{aligned} \quad \Rightarrow \quad 0 < t \leq 2$$

3. Find a vector parametrization of the intersection of the surfaces $x^2 + y^4 + 2z^3 = 6$ and $x = y^2$ in $\mathbf{R}^3$.

SOLUTION We need to find a vector parametrization $\mathbf{r}(t) = \langle x(t), y(t), z(t) \rangle$ for the intersection curve. Using $t = y$ as a parameter, we have $x = t^2$ and $y = t$. We substitute in the equation of the surface $x^2 + y^4 + 2z^3 = 6$ and solve for z in terms of t. This gives:

$$t^4 + t^4 + 2z^3 = 6$$

$$2t^4 + 2z^3 = 6$$

$$z^3 = 3 - t^4 \quad \Rightarrow \quad z = \sqrt[3]{3 - t^4}$$

We obtain the following parametrization of the intersection curve:

$$\mathbf{r}(t) = \langle t^2, t, \sqrt[3]{3 - t^4} \rangle.$$

In Exercises 5–10, calculate the derivative indicated.

5. $\mathbf{r}'(t), \quad \mathbf{r}(t) = \langle 1 - t, t^{-2}, \ln t \rangle$

SOLUTION We use the Theorem on Componentwise Differentiation to compute the derivative $\mathbf{r}'(t)$. We get

$$\mathbf{r}'(t) = \langle (1 - t)', (t^{-2})', (\ln t)' \rangle = \left\langle -1, -2t^{-3}, \frac{1}{t} \right\rangle$$

7. $\mathbf{r}'(0), \quad \mathbf{r}(t) = \langle e^{2t}, e^{-4t^2}, e^{6t} \rangle$

SOLUTION We differentiate $\mathbf{r}(t)$ componentwise to find $\mathbf{r}'(t)$:

$$\mathbf{r}'(t) = \langle (e^{2t})', (e^{-4t^2})', (e^{6t})' \rangle = \langle 2e^{2t}, -8te^{-4t^2}, 6e^{6t} \rangle$$

The derivative $\mathbf{r}'(0)$ is obtained by setting $t = 0$ in $\mathbf{r}'(t)$. This gives

$$\mathbf{r}'(0) = \langle 2e^{2 \cdot 0}, -8 \cdot 0e^{-4 \cdot 0^2}, 6e^{6 \cdot 0} \rangle = \langle 2, 0, 6 \rangle$$

9. $\dfrac{d}{dt} e^t \langle 1, t, t^2 \rangle$

SOLUTION Using the Product Rule for differentiation gives

$$\frac{d}{dt} e^t \langle 1, t, t^2 \rangle = e^t \frac{d}{dt} \langle 1, t, t^2 \rangle + (e^t)' \langle 1, t, t^2 \rangle = e^t \langle 0, 1, 2t \rangle + e^t \langle 1, t, t^2 \rangle$$

$$= e^t \left(\langle 0, 1, 2t \rangle + \langle 1, t, t^2 \rangle \right) = e^t \langle 1, 1 + t, 2t + t^2 \rangle$$

In Exercises 11–14, calculate the derivative at $t = 3$, assuming that

$$\mathbf{r}_1(3) = \langle 1, 1, 0 \rangle, \qquad \mathbf{r}_2(3) = \langle 1, 1, 0 \rangle$$

$$\mathbf{r}_1'(3) = \langle 0, 0, 1 \rangle, \qquad \mathbf{r}_2'(3) = \langle 0, 2, 4 \rangle$$

11. $\dfrac{d}{dt} (6\mathbf{r}_1(t) - 4 \cdot \mathbf{r}_2(t))$

SOLUTION Using Differentiation Rules we obtain:

$$\frac{d}{dt} (6\mathbf{r}_1(t) - 4\mathbf{r}_2(t)) \bigg|_{t=3} = 6\mathbf{r}_1'(3) - 4\mathbf{r}_2'(3) = 6 \cdot \langle 0, 0, 1 \rangle - 4 \cdot \langle 0, 2, 4 \rangle$$

$$= \langle 0, 0, 6 \rangle - \langle 0, 8, 16 \rangle = \langle 0, -8, -10 \rangle$$

13. $\dfrac{d}{dt} (\mathbf{r}_1(t) \cdot \mathbf{r}_2(t))$

SOLUTION Using Product Rule for Dot Products we obtain:

$$\frac{d}{dt} \mathbf{r}_1(t) \cdot \mathbf{r}_2(t) = \mathbf{r}_1(t) \cdot \mathbf{r}_2'(t) + \mathbf{r}_1'(t) \cdot \mathbf{r}_2(t)$$

Setting $t = 3$ gives:

$$\frac{d}{dt} \mathbf{r}_1(t) \cdot \mathbf{r}_2(t) \bigg|_{t=3} = \mathbf{r}_1(3) \cdot \mathbf{r}_2'(3) + \mathbf{r}_1'(3) \cdot \mathbf{r}_2(3) = \langle 1, 1, 0 \rangle \cdot \langle 0, 2, 4 \rangle + \langle 0, 0, 1 \rangle \cdot \langle 1, 1, 0 \rangle = 2 + 0 = 2$$

15. Calculate $\int_0^3 \langle 4t + 3, t^2, -4t^3 \rangle \, dt$.

SOLUTION By the definition of vector-valued integration, we have

$$\int_0^3 \langle 4t + 3, t^2, -4t^3 \rangle \, dt = \left\langle \int_0^3 (4t + 3) \, dt, \int_0^3 t^2 \, dt, \int_0^3 -4t^3 \, dt \right\rangle \tag{1}$$

We compute the integrals on the right-hand side:

$$\int_0^3 (4t + 3) \, dt = 2t^2 + 3t \bigg|_0^3 = 2 \cdot 9 + 3 \cdot 3 - 0 = 27$$

$$\int_0^3 t^2 \, dt = \frac{t^3}{3} \bigg|_0^3 = \frac{3^3}{3} = 9$$

$$\int_0^3 -4t^3 \, dt = -t^4 \bigg|_0^3 = -3^4 = -81$$

Substituting in (1) gives the following integral:

$$\int_0^3 \langle 4t + 3, t^2, -4t^3 \rangle \, dt = \langle 27, 9, -81 \rangle$$

17. A particle located at $(1, 1, 0)$ at time $t = 0$ follows a path whose velocity vector is $\mathbf{v}(t) = \langle 1, t, 2t^2 \rangle$. Find the particle's location at $t = 2$.

SOLUTION We first find the path $\mathbf{r}(t)$ by integrating the velocity vector $\mathbf{v}(t)$:

$$\mathbf{r}(t) = \int \langle 1, t, 2t^2 \rangle \, dt = \left\langle \int 1 \, dt, \int t \, dt, \int 2t^2 \, dt \right\rangle = \left\langle t + c_1, \frac{1}{2}t^2 + c_2, \frac{2}{3}t^3 + c_3 \right\rangle$$

Denoting by $\mathbf{c} = \langle c_1, c_2, c_3 \rangle$ the constant vector, we obtain:

$$\mathbf{r}(t) = \left\langle t, \frac{1}{2}t^2, \frac{2}{3}t^3 \right\rangle + \mathbf{c} \tag{1}$$

To find the constant vector $\mathbf{c}$, we use the given information on the initial position of the particle. At time $t = 0$ it is at the point $(1, 1, 0)$. That is, by (1):

$$\mathbf{r}(0) = \langle 0, 0, 0 \rangle + \mathbf{c} = \langle 1, 1, 0 \rangle$$

or,

$$\mathbf{c} = \langle 1, 1, 0 \rangle$$

We substitute in (1) to obtain:

$$\mathbf{r}(t) = \left\langle t, \frac{1}{2}t^2, \frac{2}{3}t^3 \right\rangle + \langle 1, 1, 0 \rangle = \left\langle t + 1, \frac{1}{2}t^2 + 1, \frac{2}{3}t^3 \right\rangle$$

Finally, we substitute $t = 2$ to obtain the particle's location at $t = 2$:

$$\mathbf{r}(2) = \left\langle 2 + 1, \frac{1}{2} \cdot 2^2 + 1, \frac{2}{3} \cdot 2^3 \right\rangle = \left\langle 3, 3, \frac{16}{3} \right\rangle$$

At time $t = 2$ the particle is located at the point

$$\left(3, 3, \frac{16}{3} \right)$$

19. Calculate $\mathbf{r}(t)$ assuming that

$$\mathbf{r}''(t) = \left\langle 4 - 16t, 12t^2 - t \right\rangle, \qquad \mathbf{r}'(0) = \langle 1, 0 \rangle, \qquad \mathbf{r}(0) = \langle 0, 1 \rangle$$

SOLUTION Using componentwise integration we get:

$$\mathbf{r}'(t) = \int \left\langle 4 - 16t, 12t^2 - t \right\rangle dt$$

$$= \left\langle \int 4 - 16t \, dt, \int 12t^2 - t \, dt \right\rangle$$

$$= \left\langle 4t - 8t^2, 4t^3 - \frac{t^2}{2} \right\rangle + \mathbf{c}_1$$

Then using the initial condition $\mathbf{r}'(0) = \langle 1, 0 \rangle$ we get:

$$\mathbf{r}'(0) = \langle 1, 0 \rangle = \mathbf{c}_1$$

so then

$$\mathbf{r}'(t) = \left\langle 4t - 8t^2, 4t^3 - \frac{t^2}{2} \right\rangle + \langle 1, 0 \rangle = \left\langle 4t - 8t^2 + 1, 4t^3 - \frac{t^2}{2} \right\rangle$$

Then integrating componentwise once more we get:

$$\mathbf{r}(t) = \int \left\langle 4t - 8t^2 + 1, 4t^3 - \frac{t^2}{2} \right\rangle dt$$

$$= \left\langle \int 4t - 8t^2 + 1 \, dt, \int 4t^3 - \frac{t^2}{2} \, dt \right\rangle$$

$$= \left\langle 2t^2 - \frac{8}{3}t^3 + t, t^4 - \frac{t^3}{6} \right\rangle + \mathbf{c}_2$$

Using the initial condition $\mathbf{r}(0) = \langle 0, 1 \rangle$ we have:

$$\mathbf{r}(0) = \langle 0, 1 \rangle = \mathbf{c}_2$$

Therefore,

$$\mathbf{r}(t) = \left\langle 2t^2 - \frac{8}{3}t^3 + t, t^4 - \frac{t^3}{6} \right\rangle + \langle 0, 1 \rangle = \left\langle 2t^2 - \frac{8}{3}t^3 + t, t^4 - \frac{t^3}{6} + 1 \right\rangle$$

21. Compute the length of the path

$$\mathbf{r}(t) = \left\langle \sin 2t, \cos 2t, 3t - 1 \right\rangle \quad \text{for } 1 \le t \le 3$$

SOLUTION We use the formula for the arc length:

$$s = \int_1^3 \|\mathbf{r}'(t)\| \, dt \qquad (1)$$

We compute the derivative vector $\mathbf{r}'(t)$ and its length:

$$\mathbf{r}'(t) = \langle 2\cos 2t, -2\sin 2t, 3 \rangle$$

$$\|\mathbf{r}'(t)\| = \sqrt{(2\cos 2t)^2 + (-2\sin 2t)^2 + 3^2} = \sqrt{4\cos^2 2t + 4\sin^2 2t + 9}$$

$$= \sqrt{4\left(\cos^2 2t + \sin^2 2t\right) + 9} = \sqrt{4 \cdot 1 + 9} = \sqrt{13}$$

We substitute in (1) and compute the integral to obtain the following length:

$$s = \int_1^3 \sqrt{13} \, dt = \sqrt{13}t \Big|_1^3 = 2\sqrt{13}.$$

23. Find an arc length parametrization of a helix of height 20 cm that makes four full rotations over a circle of radius 5 cm.

SOLUTION Since the radius is 5 cm and the height is 20 cm, the helix is traced by a parametrization of the form:

$$\mathbf{r}(t) = \langle 5\cos at, 5\sin at, t \rangle, \quad 0 \le t \le 20$$

Since the helix makes exactly 4 full rotations, we have:

$$a \cdot 20 = 4 \cdot 2\pi \quad \Rightarrow \quad a = \frac{2\pi}{5}$$

The parametrization of the helix is, thus:

$$\mathbf{r}(t) = \left\langle 5\cos\frac{2\pi t}{5}, 5\sin\frac{2\pi t}{5}, t \right\rangle, \quad 0 \le t \le 20$$

The helix is shown in the following figure:

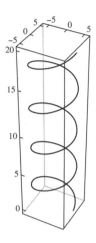

To find the arc length parametrization for the helix, we use:

$$s(t) = \int_0^t \|\mathbf{r}'(u)\| \, du \tag{1}$$

We find $\mathbf{r}'(t)$ and its length:

$$\mathbf{r}'(t) = \left\langle -5 \cdot \frac{2\pi}{5}\sin\frac{2\pi t}{5}, 5 \cdot \frac{2\pi}{5}\cos\frac{2\pi t}{5}, 1 \right\rangle = \left\langle -2\pi\sin\frac{2\pi t}{5}, 2\pi\cos\frac{2\pi t}{5}, 1 \right\rangle$$

$$\|\mathbf{r}'(t)\| = \sqrt{4\pi^2\sin^2\frac{2\pi t}{5} + 4\pi^2\cos^2\frac{2\pi t}{5} + 1} = \sqrt{4\pi^2\left(\sin^2\frac{2\pi t}{5} + \cos^2\frac{2\pi t}{5}\right) + 1} = \sqrt{1 + 4\pi^2}$$

Substituting in (1) we get:

$$s(t) = \int_0^t \sqrt{1 + 4\pi^2} \, du = t\sqrt{1 + 4\pi^2}$$

Therefore, we let $s = t\sqrt{1 + 4\pi^2}$ and thus,

$$t = \frac{s}{\sqrt{1 + 4\pi^2}} = g(s)$$

Thus, we can write

$$\mathbf{r}(s) = \left\langle 5\cos\frac{sa}{\sqrt{1 + 4\pi^2}}, 5\sin\frac{sa}{\sqrt{1 + 4\pi^2}}, \frac{s}{\sqrt{1 + 4\pi^2}} \right\rangle, \quad 0 \le s \le 20\sqrt{1 + 4\pi^2} \approx 127.245$$

25. A projectile fired at an angle of 60° lands 400 m away. What was its initial speed?

SOLUTION Place the projectile at the origin, and let $\mathbf{r}(t)$ be the position vector of the projectile.

Step 1. Use Newton's Law

Gravity exerts a downward force of magnitude mg, where m is the mass of the bullet and $g = 9.8$ m/s^2. In vector form,

$$\mathbf{F} = \langle 0, -mg \rangle = m\langle 0, -g \rangle$$

Newton's Second Law $\mathbf{F} = m\mathbf{r}'(t)$ yields $m \langle 0, -g \rangle = m\mathbf{r}''(t)$ or $\mathbf{r}''(t) = \langle 0, -g \rangle$. We determine $\mathbf{r}(t)$ by integrating twice:

$$\mathbf{r}'(t) = \int_0^t \mathbf{r}''(u)\, du = \int_0^t \langle 0, -g \rangle\, du = \langle 0, -gt \rangle + \mathbf{v}_0$$

$$\mathbf{r}(t) = \int_0^t \mathbf{r}'(u)\, du = \int_0^t (\langle 0, -gu \rangle + \mathbf{v}_0)\, du = \left\langle 0, -\frac{1}{2}gt^2 \right\rangle + t\mathbf{v}_0 + \mathbf{r}_0$$

Step 2. Use the initial conditions

By our choice of coordinates, $\mathbf{r}_0 = \mathbf{0}$. The initial velocity $\mathbf{v}_0$ has unknown magnitude v_0, but we know that it points in the direction of the unit vector $\langle \cos 60°, \sin 60° \rangle$. Therefore,

$$\mathbf{v}_0 = v_0 \langle \cos 60°, \sin 60° \rangle = v_0 \left\langle \frac{1}{2}, \frac{\sqrt{3}}{2} \right\rangle$$

$$\mathbf{r}(t) = \left\langle 0, -\frac{1}{2}gt^2 \right\rangle + t v_0 \left\langle \frac{1}{2}, \frac{\sqrt{3}}{2} \right\rangle$$

Step 3. Solve for v_0.

The projectile hits the point $\langle 400, 0 \rangle$ on the ground if there exists a time t such that $\mathbf{r}(t) = \langle 400, 0 \rangle$; that is,

$$\left\langle 0, -\frac{1}{2}gt^2 \right\rangle + t v_0 \left\langle \frac{1}{2}, \frac{\sqrt{3}}{2} \right\rangle = \langle 400, 0 \rangle$$

Equating components, we obtain

$$\frac{1}{2}t v_0 = 400, \qquad -\frac{1}{2}gt^2 + \frac{\sqrt{3}}{2}t v_0 = 0$$

The first equation yields $t = \frac{800}{v_0}$. Now substitute in the second equation and solve, using $g = 9.8 \text{m/s}^2$:

$$-4.9 \left(\frac{800}{v_0} \right)^2 + \frac{\sqrt{3}}{2} \left(\frac{800}{v_0} \right) v_0 = 0$$

$$\left(\frac{800}{v_0} \right)^2 = \frac{400\sqrt{3}}{4.9}$$

$$\left(\frac{v_0}{800} \right)^2 = \frac{4.9}{400\sqrt{3}} \approx 0.00707$$

$$v_0^2 = 4526.42611, \qquad v_0 \approx 67.279 \text{ m/s}$$

We obtain $v_0 \approx 67.279$ m/s.

27. During a short time interval $[0.5, 1.5]$, the path of an unmanned spy plane is described by

$$\mathbf{r}(t) = \left\langle -\frac{100}{t^2}, 7 - t, 40 - t^2 \right\rangle$$

A laser is fired (in the tangential direction) toward the yz-plane at time $t = 1$. Which point in the yz-plane does the laser beam hit?

SOLUTION Notice first that by differentiating we get the tangent vector:

$$\mathbf{r}'(t) = \left\langle \frac{200}{t^3}, -1, -2t \right\rangle, \qquad \Rightarrow \qquad \mathbf{r}'(1) = \langle 200, -1, -2 \rangle$$

and the tangent line to the path would be:

$$\ell(s) = \mathbf{r}(1) + s\mathbf{r}'(1) = \langle -100, 6, 39 \rangle + s \langle 200, -1, -2 \rangle = \langle -100 + 200s, 6 - s, 39 - 2s \rangle$$

If the laser is fired in the tangential direction toward the yz-plane means that the x-coordinate will be zero - this is when $s = 1/2$. Therefore,

$$\ell(1/2) = \langle 0, 11/2, 38 \rangle$$

Hence, the laser beam will hit the point $(0, 11/2, 38)$.

29. Find the unit tangent vector to $\mathbf{r}(t) = \langle \sin t, t, \cos t \rangle$ at $t = \pi$.

SOLUTION The unit tangent vector at $t = \pi$ is

$$\mathbf{T}(\pi) = \frac{\mathbf{r}'(\pi)}{\|\mathbf{r}'(\pi)\|} \tag{1}$$

We differentiate $\mathbf{r}(t)$ componentwise to obtain:

$$\mathbf{r}'(t) = \langle \cos t, 1, -\sin t \rangle$$

Therefore,

$$\mathbf{r}'(\pi) = \langle \cos \pi, 1, -\sin \pi \rangle = \langle -1, 1, 0 \rangle$$

We compute the length of $\mathbf{r}'(\pi)$:

$$\|\mathbf{r}'(\pi)\| = \sqrt{(-1)^2 + 1^2 + 0^2} = \sqrt{2}$$

Substituting in (1) gives:

$$\mathbf{T}(\pi) = \left\langle \frac{-1}{\sqrt{2}}, \frac{1}{\sqrt{2}}, 0 \right\rangle$$

31. Calculate $\kappa(1)$ for $\mathbf{r}(t) = \langle \ln t, t \rangle$.

SOLUTION Recall,

$$\kappa(t) = \frac{\|\mathbf{r}'(t) \times \mathbf{r}''(t)\|}{\|\mathbf{r}'(t)\|^3}$$

Computing derivatives we get:

$$\mathbf{r}'(t) = \left\langle \frac{1}{t}, 1 \right\rangle, \quad \Rightarrow \quad \mathbf{r}'(1) = \langle 1, 1 \rangle, \quad \Rightarrow \quad \|\mathbf{r}'(1)\| = \sqrt{2}$$

$$\mathbf{r}''(t) = \left\langle -\frac{1}{t^2}, 0 \right\rangle, \quad \Rightarrow \quad \mathbf{r}''(1) = \langle -1, 0 \rangle$$

Computing the cross product we get:

$$\mathbf{r}'(1) \times \mathbf{r}''(1) = \begin{vmatrix} \mathbf{i} & \mathbf{j} & \mathbf{k} \\ 1 & 1 & 0 \\ -1 & 0 & 0 \end{vmatrix} = \langle 0, 0, 1 \rangle$$

and $\|\mathbf{r}'(1) \times \mathbf{r}''(1)\| = 1$. Therefore,

$$\kappa(1) = \frac{\|\mathbf{r}'(1) \times \mathbf{r}''(1)\|}{\|\mathbf{r}'(1)\|^3} = \frac{1}{(\sqrt{2})^3} = \frac{1}{2^{3/2}}$$

In Exercises 33 and 34, write the acceleration vector $\mathbf{a}$ *at the point indicated as a sum of tangential and normal components.*

33. $\mathbf{r}(\theta) = \langle \cos \theta, \sin 2\theta \rangle, \quad \theta = \frac{\pi}{4}$

SOLUTION First note here that:

$$\mathbf{v}(\theta) = \mathbf{r}'(\theta) = \langle -\sin \theta, 2\cos 2\theta \rangle$$

$$\mathbf{a}(\theta) = \mathbf{r}''(\theta) = \langle -\cos \theta, -4\sin 2\theta \rangle$$

At $t = \pi/4$ we have:

$$\mathbf{v} = \mathbf{r}'(\pi/4) = \left\langle -\frac{1}{\sqrt{2}}, 0 \right\rangle$$

$$\mathbf{a} = \mathbf{r}''(\pi/4) = \left\langle -\frac{1}{\sqrt{2}}, -4 \right\rangle$$

Thus,

$$\mathbf{a} \cdot \mathbf{v} = \left\langle -\frac{1}{\sqrt{2}}, -4 \right\rangle \cdot \left\langle -\frac{1}{\sqrt{2}}, 0 \right\rangle = \frac{1}{2}$$

$$\|\mathbf{v}\| = \sqrt{\frac{1}{2} + 0} = \frac{1}{\sqrt{2}}.$$

Recall that we have:

$$\mathbf{T} = \frac{\mathbf{v}}{\|\mathbf{v}\|} = \frac{\left\langle -\frac{1}{\sqrt{2}}, 0 \right\rangle}{1/\sqrt{2}} = \langle -1, 0 \rangle$$

$$a_{\mathbf{T}} = \frac{\mathbf{a} \cdot \mathbf{v}}{\|\mathbf{v}\|} = \frac{1/2}{1/\sqrt{2}} = \frac{1}{\sqrt{2}}$$

Next, we compute $a_{\mathbf{N}}$ and $\mathbf{N}$:

$$a_{\mathbf{N}}\mathbf{N} = \mathbf{a} - a_{\mathbf{T}}\mathbf{T} = \left\langle -\frac{1}{\sqrt{2}}, -4 \right\rangle - \frac{1}{\sqrt{2}} \langle -1, 0 \rangle = \langle 0, -4 \rangle$$

This vector has length:

$$a_{\mathbf{N}} = \|a_{\mathbf{N}}\mathbf{N}\| = 4$$

and thus,

$$\mathbf{N} = \frac{a_{\mathbf{N}}\mathbf{N}}{a_{\mathbf{N}}} = \frac{\langle 0, -4 \rangle}{4} = \langle 0, -1 \rangle$$

Finally, we obtain the decomposition,

$$\mathbf{a} = \left\langle -\frac{1}{\sqrt{2}}, -4 \right\rangle = \frac{1}{\sqrt{2}}\mathbf{T} + 4\mathbf{N}$$

where $\mathbf{T} = \langle\ 1, 0 \rangle$ and $\mathbf{N} = \langle 0, -1 \rangle$.

35. At a certain time t_0, the path of a moving particle is tangent to the y-axis in the positive direction. The particle's speed at time t_0 is 4 m/s, and its acceleration vector is $\mathbf{a} = \langle 5, 4, 12 \rangle$. Determine the curvature of the path at t_0.

SOLUTION We are given that the particle is moving tangent to the y-axis with speed 4 m/s, so then:

$$\mathbf{r}' = \langle 0, 4, 0 \rangle$$

and $\mathbf{a} = \mathbf{r}'' = \langle 5, 4, 12 \rangle$. Recall the formula for curvature:

$$\kappa = \frac{\|\mathbf{r}' \times \mathbf{r}''\|}{\|\mathbf{r}'\|^3}$$

First calculate the cross product:

$$\mathbf{r}' \times \mathbf{r}'' = \begin{vmatrix} \mathbf{i} & \mathbf{j} & \mathbf{k} \\ 0 & 4 & 0 \\ 5 & 4 & 12 \end{vmatrix} = \langle 48, 0, -20 \rangle$$

Then the length of $\mathbf{r}'$ and $\mathbf{r}' \times \mathbf{r}''$:

$$\|\mathbf{r}'\| = 4, \quad \|\mathbf{r}' \times \mathbf{r}''\| = \sqrt{48^2 + 20^2} = \sqrt{2704} = 52$$

so then for curvature we get:

$$\kappa = \frac{\|\mathbf{r}' \times \mathbf{r}''\|}{\|\mathbf{r}'\|^3} = \frac{52}{4^3} = \frac{13}{16}$$

37. Parametrize the osculating circle to $y = \sqrt{x}$ at $x = 4$.

SOLUTION First differentiate twice:

$$f'(x) = \frac{1}{2\sqrt{x}}, \quad f''(x) = -\frac{1}{4x^{3/2}}$$

and at the point $x = 4$ we get:

$$f'(4) = \frac{1}{4}, \quad f''(4) = -\frac{1}{32}$$

Step 1. Find the radius

Then recall the formula for curvature:

$$\kappa(x) = \frac{|f''(x)|}{[1 + (f'(x))^2]^{3/2}}$$

and evaluating at $x = 4$ we have:

$$\kappa(4) = \frac{\frac{1}{32}}{\left[1 + \frac{1}{16}\right]^{3/2}} = \frac{1}{32} \cdot \frac{1}{\left(\frac{17}{16}\right)^{3/2}} = \frac{1}{32} \frac{16^{3/2}}{17^{3/2}} = \frac{2}{17^{3/2}}$$

Therefore the radius of the osculating circle is $R = \frac{17^{3/2}}{2}$.

Step 2. Find N at $x = 4$

First we will parametrize the curve $f(x) = \sqrt{x}$ as:

$$\mathbf{r}(x) = \langle x, \sqrt{x} \rangle, \quad \mathbf{r}(4) = \langle 4, 2 \rangle$$

and differentiate:

$$\mathbf{r}'(x) = \left\langle 1, \frac{1}{2}x^{-1/2} \right\rangle$$

Note here that the vector $\left\langle \frac{1}{2}x^{-1/2}, -1 \right\rangle$ is orthogonal to $\mathbf{r}'(x)$ for all values of x and points in the direction of the bending of the curve $y = \sqrt{x}$.

Computing the unit normal to the curve, using the vector orthogonal to $\mathbf{r}'(x)$ we get:

$$\mathbf{N}(x) = \frac{\left\langle \frac{1}{2}x^{-1/2}, -1 \right\rangle}{\sqrt{\frac{1}{4x} + 1}}, \quad \mathbf{N}(4) = \frac{\left\langle \frac{1}{4}, -1 \right\rangle}{\sqrt{\frac{1}{16} + 1}} = \frac{4}{\sqrt{17}} \left\langle \frac{1}{4}, -1 \right\rangle$$

Step 3. Find the center Q

Now to find the center Q of the osculating circle:

$$\overrightarrow{OQ} = \mathbf{r}(4) + \kappa^{-1}\mathbf{N}(4)$$

$$= \langle 4, 2 \rangle + \frac{17^{3/2}}{2} \frac{4}{\sqrt{17}} \left\langle \frac{1}{4}, -1 \right\rangle$$

$$= \langle 4, 2 \rangle + 34 \left\langle \frac{1}{4}, -1 \right\rangle$$

$$= \langle 4, 2 \rangle + \left\langle \frac{17}{2}, -34 \right\rangle$$

$$= \left\langle \frac{25}{2}, -32 \right\rangle$$

The center of the osculating circle is $Q = (\frac{25}{2}, -32)$.

Step 4. Parametrize the osculating circle

Then parametrizing the osculating circle we get:

$$\mathbf{c}(t) = \left\langle \frac{25}{2}, -32 \right\rangle + \frac{17^{3/2}}{2} \langle \cos t, \sin t \rangle$$

39. Suppose the orbit of a planet is an ellipse of eccentricity $e = c/a$ and period T (Figure 2). Use Kepler's Second Law to show that the time required to travel from A' to B' is equal to

$$\left(\frac{1}{4} + \frac{e}{2\pi} \right) T$$

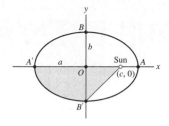

FIGURE 2

SOLUTION By the Law of Equal Areas, the position vector pointing from the sun to the planet sweeps out equal areas in equal times. We denote by S_1 the area swept by the position vector when the planet moves from A' to B', and t is the desired time. Since the position vector sweeps out the whole area of the ellipse (πab) in time T, the Law of Equal Areas implies that:

$$\frac{S_1}{\pi ab} = \frac{t}{T} \quad \Rightarrow \quad t = \frac{T S_1}{\pi ab} \tag{1}$$

We now find the area S_1 as the sum of the area of a quarter of the ellipse and the area of the triangle ODB. That is,

$$S_1 = \frac{\pi ab}{4} + \frac{\overline{OD} \cdot \overline{OB}'}{2} = \frac{\pi ab}{4} + \frac{cb}{2} = \frac{b}{4}(\pi a + 2c)$$

Substituting in (1) we get:

$$t = \frac{Tb(\pi a + 2c)}{4\pi ab} = \frac{T(\pi a + 2c)}{4\pi a} = T\left(\frac{1}{4} + \frac{1}{2\pi}\frac{c}{a}\right) = T\left(\frac{1}{4} + \frac{e}{2\pi}\right)$$

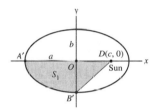

14 DIFFERENTIATION IN SEVERAL VARIABLES

14.1 Functions of Two or More Variables (LT Section 15.1)

Preliminary Questions

1. What is the difference between a horizontal trace and a level curve? How are they related?

SOLUTION A horizontal trace at height c consists of all points (x, y, c) such that $f(x, y) = c$. A level curve is the curve $f(x, y) = c$ in the xy-plane. The horizontal trace is in the $z = c$ plane. The two curves are related in the sense that the level curve is the projection of the horizontal trace on the xy-plane. The two curves have the same shape but they are located in parallel planes.

2. Describe the trace of $f(x, y) = x^2 - \sin(x^3 y)$ in the xz-plane.

SOLUTION The intersection of the graph of $f(x, y) = x^2 - \sin(x^3 y)$ with the xz-plane is obtained by setting $y = 0$ in the equation $z = x^2 - \sin(x^3 y)$. We get the equation $z = x^2 - \sin 0 = x^2$. This is the parabola $z = x^2$ in the xz-plane.

3. Is it possible for two different level curves of a function to intersect? Explain.

SOLUTION Two different level curves of $f(x, y)$ are the curves in the xy-plane defined by equations $f(x, y) = c_1$ and $f(x, y) = c_2$ for $c_1 \neq c_2$. If the curves intersect at a point (x_0, y_0), then $f(x_0, y_0) = c_1$ and $f(x_0, y_0) = c_2$, which implies that $c_1 = c_2$. Therefore, two different level curves of a function do not intersect.

4. Describe the contour map of $f(x, y) = x$ with contour interval 1.

SOLUTION The level curves of the function $f(x, y) = x$ are the vertical lines $x = c$. Therefore, the contour map of f with contour interval 1 consists of vertical lines so that every two adjacent lines are distanced one unit from another.

5. How will the contour maps of

$$f(x, y) = x \quad \text{and} \quad g(x, y) = 2x$$

with contour interval 1 look different?

SOLUTION The level curves of $f(x, y) = x$ are the vertical lines $x = c$, and the level curves of $g(x, y) = 2x$ are the vertical lines $2x = c$ or $x = \frac{c}{2}$. Therefore, the contour map of $f(x, y) = x$ with contour interval 1 consists of vertical lines with distance one unit between adjacent lines, whereas in the contour map of $g(x, y) = 2x$ (with contour interval 1) the distance between two adjacent vertical lines is $\frac{1}{2}$.

Exercises

In Exercises 1–4, evaluate the function at the specified points.

1. $f(x, y) = x + yx^3$, $(2, 2)$, $(-1, 4)$

SOLUTION We substitute the values for x and y in $f(x, y)$ and compute the values of f at the given points. This gives

$$f(2, 2) = 2 + 2 \cdot 2^3 = 18$$
$$f(-1, 4) = -1 + 4 \cdot (-1)^3 = -5$$

3. $h(x, y, z) = xyz^{-2}$, $(3, 8, 2)$, $(3, -2, -6)$

SOLUTION Substituting $(x, y, z) = (3, 8, 2)$ and $(x, y, z) = (3, -2, -6)$ in the function, we obtain

$$h(3, 8, 2) = 3 \cdot 8 \cdot 2^{-2} = 3 \cdot 8 \cdot \frac{1}{4} = 6$$

$$h(3, -2, -6) = 3 \cdot (-2) \cdot (-6)^{-2} = -6 \cdot \frac{1}{36} = -\frac{1}{6}$$

In Exercises 5–12, sketch the domain of the function.

5. $f(x, y) = 12x - 5y$

SOLUTION The function is defined for all x and y, hence the domain is the entire xy-plane.

7. $f(x, y) = \ln(4x^2 - y)$

SOLUTION The function is defined if $4x^2 - y > 0$, that is, $y < 4x^2$. The domain is the region in the xy-plane that is below the parabola $y = 4x^2$.

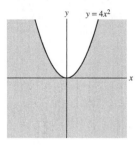

9. $g(y, z) = \dfrac{1}{z + y^2}$

SOLUTION The function is defined if $z + y^2 \neq 0$, that is, $z \neq -y^2$. The domain is the (y, z) plane with the parabola $z = -y^2$ excluded.

$$D = \left\{ (y, z) : z \neq -y^2 \right\}$$

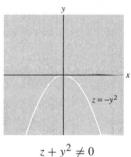

$$z + y^2 \neq 0$$

11. $F(I, R) = \sqrt{IR}$

SOLUTION The function is defined if $IR \geq 0$. Therefore the domain is the first and the third quadrants of the IR plane including both axes.

$$IR \geq 0$$

In Exercises 13–16, describe the domain and range of the function.

13. $f(x, y, z) = xz + e^y$

SOLUTION The domain of f is the entire (x, y, z)-space. Since f takes all the real values, the range is the entire real line.

15. $P(r, s, t) = \sqrt{16 - r^2 s^2 t^2}$

SOLUTION The domain is subset of $\mathbf{R}^3$ where $rst \leq 4$ and the range is $\{w : 0 \leq w \leq 4\}$ because the minimum is 0 and the maximum of P is $\sqrt{16} = 4$.

17. Match graphs (A) and (B) in Figure 21 with the functions

   (i) $f(x, y) = -x + y^2$ (ii) $g(x, y) = x + y^2$

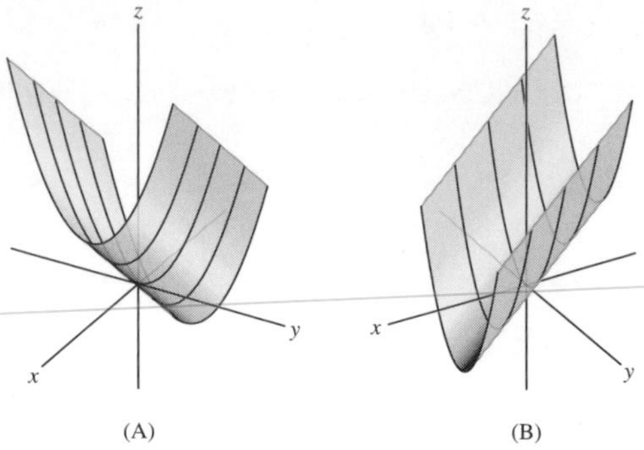

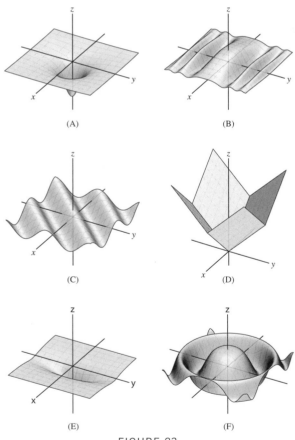

(A)   (B)

FIGURE 21

SOLUTION

(i) The vertical trace for $f(x, y) = -x + y^2$ in the xz-plane ($y = 0$) is $z = -x$. This matches the graph shown in (B).

(ii) The vertical trace for $f(x, y) = x + y^2$ in the xz-plane ($y = 0$) is $z = x$. This matches the graph show in (A).

19. Match the functions (a)–(f) with their graphs (A)–(F) in Figure 23.

(a) $f(x, y) = |x| + |y|$

(b) $f(x, y) = \cos(x - y)$

(c) $f(x, y) = \dfrac{-1}{1 + 9x^2 + y^2}$

(d) $f(x, y) = \cos(y^2)e^{-0.1(x^2+y^2)}$

(e) $f(x, y) = \dfrac{-1}{1 + 9x^2 + 9y^2}$

(f) $f(x, y) = \cos(x^2 + y^2)e^{-0.1(x^2+y^2)}$

(A)   (B)

(C)   (D)

(E)   (F)

FIGURE 23

SOLUTION

(a) $|x| + |y|$. The level curves are $|x| + |y| = c$, $y = c - |x|$, or $y = -c + |x|$. The graph (D) corresponds to the function with these level curves.

(b) $\cos(x - y)$. The vertical trace in the plane $x = c$ is the curve $z = \cos(c - y)$ in the plane $x = c$. These traces correspond to the graph (C).

(c) $\dfrac{-1}{1+9x^2+y^2}$ **(e)** $\dfrac{-1}{1+9x^2+9y^2}$.

The level curves of the two functions are:

$$\frac{-1}{1+9x^2+y^2}=c \qquad\qquad \frac{-1}{1+9x^2+9y^2}=c$$

$$1+9x^2+y^2=-\frac{1}{c} \qquad\qquad 1+9x^2+9y^2=-\frac{1}{c}$$

$$9x^2+y^2=-1-\frac{1}{c} \qquad\qquad 9x^2+9y^2=-1-\frac{1}{c}$$

$$x^2+y^2=-\frac{1+c}{9c}$$

For suitable values of c, the level curves of the function in (c) are ellipses as in (E), and the level curves of the function (e) are circles as in (A).

(d) $\cos(x^2)e^{-1/(x^2+y^2)}$ **(f)** $\cos(x^2+y^2)e^{-1/(x^2+y^2)}$.

The value of $|z|$ is decreasing to zero as x or y are decreasing, hence the possible graphs are (B) and (F).

In (f), z is constant whenever x^2+y^2 is constant, that is, z is constant whenever (x,y) varies on a circle. Hence (f) corresponds to the graph (F) and (d) corresponds to (B).

To summarize, we have the following matching:

$$(a)\leftrightarrow(D) \qquad (b)\leftrightarrow(C) \qquad (c)\leftrightarrow(E)$$

$$(d)\leftrightarrow(B) \qquad (e)\leftrightarrow(A) \qquad (f)\leftrightarrow(F)$$

In Exercises 21–26, sketch the graph and describe the vertical and horizontal traces.

21. $f(x,y)=12-3x-4y$

SOLUTION The graph of $f(x,y)=12-3x-4y$ is shown in the figure:

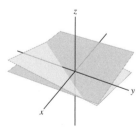

The horizontal trace at height c is the line $12-3x-4y=c$ or $3x+4y=12-c$ in the plane $z=c$.

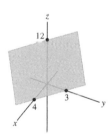

The vertical traces obtained by setting $x=a$ or $y=a$ are the lines $z=(12-3a)-4y$ and $z=-3x+(12-4a)$ in the planes $x=a$ and $y=a$, respectively.

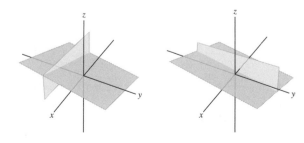

23. $f(x, y) = x^2 + 4y^2$

SOLUTION The graph of the function is shown in the figure:

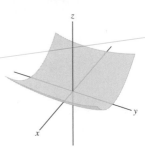

The horizontal trace at height c is the curve $x^2 + 4y^2 = c$, where $c \geq 0$ (if $c = 0$, it is the origin). The horizontal traces are ellipses for $c > 0$.

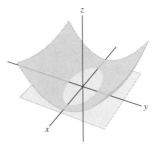

The vertical trace in the plane $x = a$ is the parabola $z = a^2 + 4y^2$, and the vertical trace in the plane $y = a$ is the parabola $z = x^2 + 4a^2$.

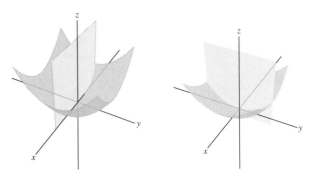

25. $f(x, y) = \sin(x - y)$

SOLUTION The graph of $f(x, y) = \sin(x - y)$ is shown in the figure:

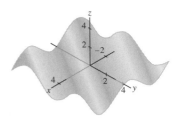

The horizontal trace at the height $z = c$ is $\sin(x - y) = c$ (we could also write $x - y = \sin^{-1}(c)$ or $y = x - \sin^{-1}(c)$). The trace consists of multiple lines all having slope 1, with y-intercepts separated by multiples of 2π.

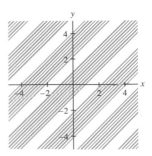

The vertical trace in the plane $x = a$ is $\sin(a - y) = -\sin(y - a) = z$. This curve is a shifted sine curve reflected through the z-axis.

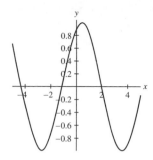

The vertical trace in the plane $y = a$ is $\sin(x - a) = z$. This curve is a shifted sine curve as well.

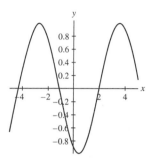

27. Sketch contour maps of $f(x, y) = x + y$ with contour intervals $m = 1$ and 2.

SOLUTION The level curves are $x + y = c$ or $y = c - x$. Using contour interval $m = 1$, we plot $y = c - x$ for various values of c.

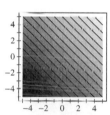

Using contour interval $m = 2$, we plot $y = c - x$ for various values of c.

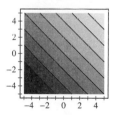

In Exercises 29–36, draw a contour map of $f(x, y)$ with an appropriate contour interval, showing at least six level curves.

29. $f(x, y) = x^2 - y$

SOLUTION The level curves are the parabolas $y = x^2 + c$. We draw a contour plot with contour interval $m = 1$, for $c = 0, 1, 2, 3, 4, 5$:

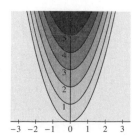

31. $f(x, y) = \dfrac{y}{x}$

SOLUTION   The level curves are $\dfrac{y}{x} = c$ or $y = cx$. We plot $y = cx$ for $c = -2, -1, 0, 1, 2, 3$ using contour interval $m = 1$:

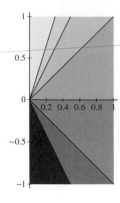

33. $f(x, y) = x^2 + 4y^2$

SOLUTION   The level curves are $x^2 + 4y^2 = c$. These are ellipses centered at the origin in the xy-plane.

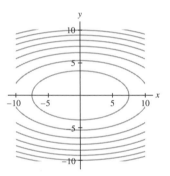

35. $f(x, y) = x^2$

SOLUTION   The level curves are $x^2 = c$. For $c > 0$ these are the two vertical lines $x = \sqrt{c}$ and $x = -\sqrt{c}$ and for $c = 0$ it is the y-axis. We draw a contour map using contour interval $m = 4$ and $c = 0, 4, 8, 12, 16, 20$:

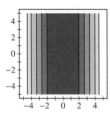

37. ✎   Find the linear function whose contour map (with contour interval $m = 6$) is shown in Figure 25. What is the linear function if $m = 3$ (and the curve labeled $c = 6$ is relabeled $c = 3$)?

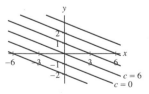

FIGURE 25 Contour map with contour interval $m = 6$

SOLUTION   A linear function has the form $f(x, y) = Ax + By + C$.

Case 1: According to the contour map, the level curve through the origin $(0, 0)$ has equation $f(x, y) = 6$. Therefore

$$f(0, 0) = A(0) + B(0) + C = 6 \quad \Rightarrow \quad C = 6$$

Next, we see from the contour map that the points $(-3, 0) = 0$ and $f(0, -1)$ lie on the level curve $f(x, y) = 0$. Hence

$$f(-3, 0) = A(-3) + B(0) + 6 = 0 \quad \Rightarrow \quad A = 2$$
$$f(0, -1) = A(0) + B(-1) + 6 = 0 \quad \Rightarrow \quad B = 6$$

Therefore $f(x, y) = 2x + 6y + 6$.

Case 1: If $m = 3$, then $(0, 0)$ lies on the level curve $f(x, y) = 3$, and we proceed as before

$$f(0, 0) = A(0) + B(0) + C = 3 \quad \Rightarrow \quad C = 3 f(-3, 0) = A(-3) + B(0) + 3 = 0 \quad \Rightarrow \quad A = 1$$
$$f(0, -1) = A(0) + B(-1) + 3 = 0 \quad \Rightarrow \quad B = 2$$

Therefore $f(x, y) = x + 3y + 3$.

39. Referring to Figure 27, answer the following questions:

(a) At which of (A)–(C) is pressure increasing in the northern direction?

(b) At which of (A)–(C) is pressure increasing in the easterly direction?

(c) In which direction at (B) is pressure increasing most rapidly?

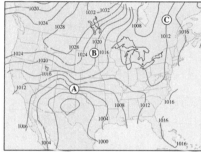

FIGURE 27 Atmospheric Pressure (in millibars) over the continental U.S. on March 26, 2009

SOLUTION

(a) (A) and (B)

(b) (C)

(c) west

In Exercises 40–43, $\rho(S, T)$ is seawater density (kg/m³) as a function of salinity S (ppt) and temperature T (°C). Refer to the contour map in Figure 28.

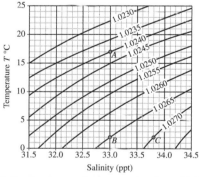

FIGURE 28 Contour map of seawater density $\rho(S, T)$ (kg/m³).

41. Calculate the average rate of change of ρ with respect to S from B to C.

SOLUTION For fixed temperature, the segment $\overline{BC}$ spans one level curve and the level curve of C is to the right of the level curve of B. Therefore, the change in density from B to C is $\Delta\rho = 0.0005$ kg/m³. The salinity at C is greater than the salinity at B and $\Delta S = 0.8$ ppt. Therefore,

$$\text{Average ROC from } B \text{ to } C = \frac{\Delta\rho}{\Delta S} = \frac{0.0005}{0.8} = 0.000625 \text{ kg/m}^3 \cdot \text{ppt}.$$

43. Does water density appear to be more sensitive to a change in temperature at point A or point B?

SOLUTION   The two adjacent level curves are closer to the level curve of A than the corresponding two adjacent level curves are to the level curve of B. This suggests that water density is more sensitive to a change in temperature at A than at B.

In Exercises 44–47, refer to Figure 29.

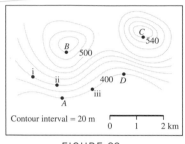

FIGURE 29

45. Estimate the average rate of change from A and B and from A to C.

SOLUTION   The change in elevation from A to B is 140 m. The scale shows that $\overline{AB}$ is approximately 2000 m. Therefore,

$$\text{Average ROC from } A \text{ to } B = \frac{140}{2000} \approx 0.07.$$

The change in elevation from A to C is obtained by multiplying the number of level curves between A and C, which is 8, by the contour interval 20 meters, giving $8 \cdot 20 = 160$ m. Using the scale, we approximate the distance $\overline{AC}$ by 3000 m. Therefore,

$$\text{Average ROC from } A \text{ to } C = \frac{160}{3000} \approx 0.0533.$$

47. Sketch the path of steepest ascent beginning at D.

SOLUTION   Starting at D, we draw a path that everywhere along the way points on the steepest direction, that is, moves as straight as possible from one level curve to the next to end at the point C.

Further Insights and Challenges

49. Let $f(x, y) = \dfrac{x}{\sqrt{x^2 + y^2}}$ for $(x, y) \neq 0$. Write f as a function $f(r, \theta)$ in polar coordinates, and use this to find the level curves of f.

SOLUTION   In polar coordinates $x = r \cos\theta$ and $r = \sqrt{x^2 + y^2}$. Hence,

$$f(r, \theta) = \frac{r \cos\theta}{r} = \cos\theta.$$

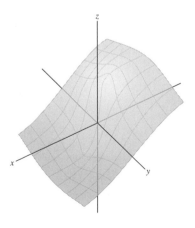

The level curves are the curves $\cos\theta = c$ in the $r\theta$-plane, for $|c| \leq 1$. For $-1 < c < 1$, $c \neq 0$, the level curves $\cos\theta = c$ are the two rays $\theta = \cos^{-1} c$ and $\theta = -\cos^{-1} c$.

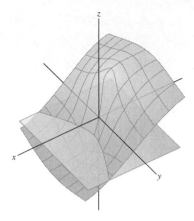

For $c = 0$, the level curve $\cos \theta = 0$ is the y-axis; for $c = 1$ the level curve $\cos \theta = 1$ is the nonnegative x-axis.

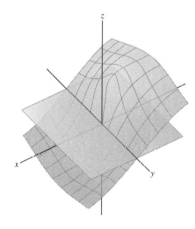

For $c = -1$, the level curve $\cos \theta = -1$ is the negative x-axis.

14.2 Limits and Continuity in Several Variables (LT Section 15.2)

Preliminary Questions

1. What is the difference between $D(P, r)$ and $D^*(P, r)$?

SOLUTION $D(P, r)$ is the open disk of radius r and center (a, b). It consists of all points distanced less than r from P, hence $D(P, r)$ includes the point P. $D^*(P, r)$ consists of all points in $D(P, r)$ other than P itself.

2. Suppose that $f(x, y)$ is continuous at $(2, 3)$ and that $f(2, y) = y^3$ for $y \neq 3$. What is the value $f(2, 3)$?

SOLUTION $f(x, y)$ is continuous at $(2, 3)$, hence the following holds:

$$f(2, 3) = \lim_{(x,y) \to (2,3)} f(x, y)$$

Since the limit exists, we may compute it by approaching $(2, 3)$ along the vertical line $x = 2$. This gives

$$f(2, 3) = \lim_{(x,y) \to (2,3)} f(x, y) = \lim_{y \to 3} f(2, y) = \lim_{y \to 3} y^3 = 3^3 = 27$$

We conclude that $f(2, 3) = 27$.

3. Suppose that $Q(x, y)$ is a function such that $1/Q(x, y)$ is continuous for all (x, y). Which of the following statements are true?

(a) $Q(x, y)$ is continuous for all (x, y).

(b) $Q(x, y)$ is continuous for $(x, y) \neq (0, 0)$.

(c) $Q(x, y) \neq 0$ for all (x, y).

SOLUTION All three statements are true. Let $f(x, y) = \frac{1}{Q(x,y)}$. Hence $Q(x, y) = \frac{1}{f(x,y)}$.

(a) Since f is continuous, Q is continuous whenever $f(x, y) \neq 0$. But by the definition of f it is never zero, therefore Q is continuous at all (x, y).

(b) Q is continuous everywhere including at $(0, 0)$.

(c) Since $f(x, y) = \frac{1}{Q(x,y)}$ is continuous, the denominator is never zero, that is, $Q(x, y) \neq 0$ for all (x, y).

Moreover, there are no points where $Q(x, y) = 0$. (The equality $Q(x, y) = (0, 0)$ is meaningless since the range of Q consists of real numbers.)

4. Suppose that $f(x, 0) = 3$ for all $x \neq 0$ and $f(0, y) = 5$ for all $y \neq 0$. What can you conclude about $\lim_{(x,y)\to(0,0)} f(x, y)$?

SOLUTION We show that the limit $\lim_{(x,y)\to(0,0)} f(x, y)$ does not exist. Indeed, if the limit exists, it may be computed by approaching $(0, 0)$ along the x-axis or along the y-axis. We compute these two limits:

$$\lim_{\substack{(x,y)\to(0,0) \\ \text{along } y=0}} f(x, y) = \lim_{x\to0} f(x, 0) = \lim_{x\to0} 3 = 3$$

$$\lim_{\substack{(x,y)\to(0,0) \\ \text{along } x=0}} f(x, y) = \lim_{y\to0} f(0, y) = \lim_{y\to0} 5 = 5$$

Since the limits are different, $f(x, y)$ does not approach one limit as $(x, y) \to (0, 0)$, hence the limit $\lim_{(x,y)\to(0,0)} f(x, y)$ does not exist.

Exercises

In Exercises 1–8, evaluate the limit using continuity

1. $\lim_{(x,y)\to(1,2)} (x^2 + y)$

SOLUTION Since the function $x^2 + y$ is continuous, we evaluate the limit by substitution:

$$\lim_{(x,y)\to(1,2)} (x^2 + y) = 1^2 + 2 = 3$$

3. $\lim_{(x,y)\to(2,-1)} (xy - 3x^2 y^3)$

SOLUTION The function $xy - 3x^2 y^3$ is continuous everywhere because it is a polynomial, hence we compute the limit by substitution:

$$\lim_{(x,y)\to(2,-1)} (xy - 3x^2 y^3) = 2(-1) - 3(4)(-1)^3 = -2 + 12 = 10$$

5. $\lim_{(x,y)\to(\frac{\pi}{4},0)} \tan x \cos y$

SOLUTION We use the continuity of $\tan x \cos y$ at the point $\left(\frac{\pi}{4}, 0\right)$ to evaluate the limit by substitution:

$$\lim_{(x,y)\to(\frac{\pi}{4},0)} \tan x \cos y = \tan \frac{\pi}{4} \cos 0 = 1 \cdot 1 = 1$$

7. $\lim_{(x,y)\to(1,1)} \dfrac{e^{x^2} - e^{-y^2}}{x + y}$

SOLUTION The function is the quotient of two continuous functions, and the denominator is not zero at the point $(1, 1)$. Therefore, the function is continuous at this point, and we may compute the limit by substitution:

$$\lim_{(x,y)\to(1,1)} \frac{e^{x^2} - e^{-y^2}}{x + y} = \frac{e^{1^2} - e^{-1^2}}{1 + 1} = \frac{e - \frac{1}{e}}{2} = \frac{1}{2}(e - e^{-1})$$

In Exercises 9–12, assume that

$$\lim_{(x,y)\to(2,5)} f(x, y) = 3, \qquad \lim_{(x,y)\to(2,5)} g(x, y) = 7$$

9. $\lim_{(x,y)\to(2,5)} \big(g(x, y) - 2f(x, y)\big)$

SOLUTION

$$\lim_{(x,y)\to(2,5)} \big(g(x, y) - 2f(x, y)\big) = 7 - 2(3) = 1$$

11. $\lim_{(x,y)\to(2,5)} e^{f(x,y)^2 - g(x,y)}$

SOLUTION

$$\lim_{(x,y)\to(2,5)} e^{f(x,y)^2 - g(x,y)} = e^{3^2 - 7} = e^2$$

13. Does $\lim\limits_{(x,y)\to(0,0)} \dfrac{y^2}{x^2+y^2}$ exist? Explain.

SOLUTION This limit does not exist. Consider the following approaches to the point $(x, y) = (0, 0)$ - first along the line $x = 0$ and second, along the line $y = x$.

First along the line $x = 0$ we calculate:

$$\lim\limits_{(x,y)\to(0,0)} \frac{y^2}{x^2+y^2} = \lim\limits_{y\to0} \frac{y^2}{0^2+y^2} = \lim\limits_{y\to0} 1 = 1$$

Second, along the line $y = x$ we calculate:

$$\lim\limits_{(x,y)\to(0,0)} \frac{y^2}{x^2+y^2} = \lim\limits_{x\to0} \frac{x^2}{x^2+x^2} = \lim\limits_{x\to0} \frac{1}{2} = \frac{1}{2}$$

Since these two limits are not equal, the limit in question, $\lim\limits_{(x,y)\to(0,0)} \frac{y^2}{x^2+y^2}$ does not exist.

15. Prove that

$$\lim\limits_{(x,y)\to(0,0)} \frac{x}{x^2+y^2}$$

does not exist by considering the limit along the x-axis.

SOLUTION Compute this limit approaching $(x, y) = (0, 0)$ along the x-axis $(y = 0)$:

$$\lim\limits_{(x,y)\to(0,0)} \frac{x}{x^2+y^2} = \lim\limits_{x\to0} \frac{x}{x^2+0^2} = \lim\limits_{x\to0} \frac{1}{x}$$

This limit is known not to exist (it gets arbitrarily large from the right and arbitrarily small from the left), therefore the limit in question, $\lim\limits_{(x,y)\to(0,0)} \frac{x}{x^2+y^2}$, also does not exist.

17. Use the Squeeze Theorem to evaluate

$$\lim\limits_{(x,y)\to(4,0)} (x^2 - 16) \cos\left(\frac{1}{(x-4)^2+y^2}\right)$$

SOLUTION Consider the following inequalities:

$$-1 \le \cos\left(\frac{1}{(x-4)^2+y^2}\right) \le 1$$

Then for x such that $x \ge 4$ then $x^2 - 16 \ge 0$ and we have:

$$(-1)(x^2 - 16) \le (x^2 \quad 16) \cos\left(\frac{1}{(x-4)^2+y^2}\right) \le (x^2 - 16)$$

$$\lim\limits_{(x,y)\to(4,0)} (-1)(x^2 - 16) \le \lim\limits_{(x,y)\to(4,0)} (x^2 - 16)\cos\left(\frac{1}{(x-4)^2+y^2}\right) \le \lim\limits_{(x,y)\to(4,0)} (x^2 \quad 16)$$

Then the two limits at the ends of the inequality are clearly equal to 0, by the Squeeze Theorem.

Now, if $x < 4$, then $x^2 - 16 < 0$ and we have:

$$(x^2 - 16) \le (x^2 - 16)\cos\left(\frac{1}{(x-4)^2+y^2}\right) \le (-1)(x^2 - 16)$$

$$\lim\limits_{(x,y)\to(4,0)} (x^2 - 16) \le \lim\limits_{(x,y)\to(4,0)} (x^2 - 16)\cos\left(\frac{1}{(x-4)^2+y^2}\right) \le \lim\limits_{(x,y)\to(4,0)} (-1)(x^2 - 16)$$

Then the two limits at the ends of the inequality are clearly equal to 0, by the Squeeze Theorem.

Thus we can conclude

$$\lim\limits_{(x,y)\to(4,0)} (x^2 - 16)\cos\left(\frac{1}{(x-4)^2+y^2}\right) = 0$$

In Exercises 19–32, evaluate the limit or determine that it does not exist.

19. $\lim\limits_{(z,w)\to(-2,1)} \dfrac{z^4 \cos(\pi w)}{e^{z+w}}$

SOLUTION This function is continuous everywhere since the denominator is never equal to 0, therefore, we will evaluate the limit by substitution:

$$\lim\limits_{(z,w)\to(-2,1)} \frac{z^4 \cos(\pi w)}{e^{z+w}} = \frac{(-2)^4 \cos(\pi)}{e^{-2+1}} = \frac{16(-1)}{e^{-1}} = -16e$$

21. $\displaystyle\lim_{(x,y)\to(4,2)} \frac{y-2}{\sqrt{x^2-4}}$

SOLUTION The function is continuous at the point $(4, 2)$, since it is the quotient of two continuous functions and the denominator is not zero at $(4, 2)$. We compute the limit by substitution:

$$\lim_{(x,y)\to(4,2)} \frac{y-2}{\sqrt{x^2-4}} = \frac{2-2}{\sqrt{4^2-4}} = \frac{0}{\sqrt{12}} = 0$$

23. $\displaystyle\lim_{(x,y)\to(3,4)} \frac{1}{\sqrt{x^2+y^2}}$

SOLUTION The function $\dfrac{1}{\sqrt{x^2+y^2}}$ is continuous at the point $(3, 4)$ since it is the quotient of two continuous functions and the denominator is not zero at $(3, 4)$. We compute the limit by substitution:

$$\lim_{(x,y)\to(3,4)} \frac{1}{\sqrt{x^2+y^2}} = \frac{1}{\sqrt{9+16}} = \frac{1}{5}$$

25. $\displaystyle\lim_{(x,y)\to(1,-3)} e^{x-y}\ln(x-y)$

SOLUTION This function $e^{x-y}\ln(x-y)$ is continuous at the point $(1, -3)$ since it is the product of two continuous functions. We can compute the limit by substitution:

$$\lim_{(x,y)\to(1,-3)} e^{x-y}\ln(x-y) = e^{1+3}\ln(1+3) = e^4\ln 4$$

27. $\displaystyle\lim_{(x,y)\to(-3,-2)} (x^2y^3+4xy)$

SOLUTION The function x^2y^3+4xy is continuous everywhere because it is a polynomial. We can compute this limit by substitution:

$$\lim_{(x,y)\to(-3,-2)} (x^2y^3+4xy) = 9(-8)+4(-3)(-2) = -72+24 = -48$$

29. $\displaystyle\lim_{(x,y)\to(0,0)} \tan(x^2+y^2)\tan^{-1}\left(\frac{1}{x^2+y^2}\right)$

SOLUTION Consider the following inequalities:

$$-\frac{\pi}{2} \le \tan^{-1}\left(\frac{1}{x^2+y^2}\right) \le \frac{\pi}{2}$$

$$-\frac{\pi}{2}\cdot\tan(x^2+y^2) \le \tan(x^2+y^2)\cdot\left(\frac{1}{x^2+y^2}\right) \le \frac{\pi}{2}\tan(x^2+y^2)$$

and then taking limits:

$$\lim_{(x,y)\to(0,0)} -\frac{\pi}{2}\cdot\tan(x^2+y^2) \le \lim_{(x,y)\to(0,0)} \tan(x^2+y^2)\cdot\left(\frac{1}{x^2+y^2}\right) \le \lim_{(x,y)\to(0,0)} \frac{\pi}{2}\tan(x^2+y^2)$$

Each of the limits on the endpoints of this inequality is equal to 0, thus we can conclude:

$$\lim_{(x,y)\to(0,0)} \tan(x^2+y^2)\cdot\left(\frac{1}{x^2+y^2}\right) = 0$$

31. $\displaystyle\lim_{(x,y)\to(0,0)} \frac{x^2+y^2}{\sqrt{x^2+y^2+1}-1}$

SOLUTION We rewrite the function by dividing and multiplying it by the conjugate of $\sqrt{x^2+y^2+1}-1$ and using the identity $(a-b)(a+b) = a^2-b^2$. This gives

$$\frac{x^2+y^2}{\sqrt{x^2+y^2+1}-1} = \frac{(x^2+y^2)\left(\sqrt{x^2+y^2+1}+1\right)}{\left(\sqrt{x^2+y^2+1}-1\right)\left(\sqrt{x^2+y^2+1}+1\right)} = \frac{(x^2+y^2)\left(\sqrt{x^2+y^2+1}+1\right)}{(x^2+y^2+1)-1}$$

$$= \frac{(x^2+y^2)\left(\sqrt{x^2+y^2+1}+1\right)}{x^2+y^2} = \sqrt{x^2+y^2+1}+1$$

The resulting function is continuous, hence we may compute the limit by substitution. This gives

$$\lim_{(x,y)\to(0,0)} \frac{x^2+y^2}{\sqrt{x^2+y^2+1}-1} = \lim_{(x,y)\to(0,0)} \left(\sqrt{x^2+y^2+1}+1\right) = \sqrt{0^2+0^2+1}+1 = 2$$

33. Let $f(x, y) = \dfrac{x^3+y^3}{x^2+y^2}$.

(a) Show that

$$|x^3| \le |x|(x^2+y^2), \quad |y^3| \le |y|(x^2+y^2)$$

(b) Show that $|f(x, y)| \le |x| + |y|$.

(c) Use the Squeeze Theorem to prove that $\displaystyle\lim_{(x,y)\to(0,0)} f(x, y) = 0$.

SOLUTION

(a) Since $|x|y^2 \ge 0$, we have

$$|x^3| \le |x^3| + |x|y^2 = |x|^3 + |x|y^2 = |x|(x^2+y^2)$$

Similarly, since $|y|x^2 \ge 0$, we have

$$|y^3| \le |y^3| + |y|x^2 = |y|^3 + |y|x^2 = |y|(x^2+y^2)$$

(b) We use the triangle inequality to write

$$|f(x, y)| = \frac{|x^3+y^3|}{x^2+y^2} \le \frac{|x^3|+|y^3|}{x^2+y^2}$$

We continue using the inequality in part (a):

$$|f(x, y)| \le \frac{|x|(x^2+y^2)+|y|(x^2+y^2)}{x^2+y^2} = \frac{(|x|+|y|)(x^2+y^2)}{x^2+y^2} = |x|+|y|$$

That is,

$$|f(x, y)| \le |x| + |y|$$

(c) In part (b) we showed that

$$|f(x, y)| \le |x| + |y| \tag{1}$$

Let $\epsilon > 0$. Then if $|x| < \frac{\epsilon}{2}$ and $|y| < \frac{\epsilon}{2}$, we have by (1)

$$|f(x, y) - 0| \le |x| + |y| < \frac{\epsilon}{2} + \frac{\epsilon}{2} = \epsilon \tag{2}$$

Notice that if $x^2 + y^2 < \frac{\epsilon^2}{4}$, then $x^2 < \frac{\epsilon^2}{4}$ and $y^2 < \frac{\epsilon^2}{4}$. Hence $|x| < \frac{\epsilon}{2}$ and $|y| < \frac{\epsilon}{2}$, so (1) holds. In other words, using $D^\star\left(\frac{\epsilon}{2}\right)$ to represent the punctured disc of radius $\epsilon/2$ centered at the origin, we have

$$(x, y) \in D^\star\left(\frac{\epsilon}{2}\right) \quad \Rightarrow \quad |x| < \frac{\epsilon}{2}$$

and

$$|y| < \frac{\epsilon}{2} \quad \Rightarrow \quad |f(x, y) - 0| < \epsilon$$

We conclude by the limit definition that

$$\lim_{(x,y)\to(0,0)} f(x, y) = 0$$

35. 📖 Figure 7 shows the contour maps of two functions. Explain why the limit $\displaystyle\lim_{(x,y)\to P} f(x, y)$ does not exist. Does $\displaystyle\lim_{(x,y)\to Q} g(x, y)$ appear to exist in (B)? If so, what is its limit?

(A) Contour map of $f(x, y)$ (B) Contour map of $g(x, y)$

FIGURE 7

SOLUTION As (x, y) approaches arbitrarily close to P, the function $f(x, y)$ takes the values $\pm 1, \pm 3,$ and ± 5. Therefore $f(x, y)$ does not approach one limit as $(x, y) \rightarrow P$. Rather, the limit depends on the contour along which (x, y) is approaching P. This implies that the limit $\lim_{(x,y) \rightarrow P} f(x, y)$ does not exist. In (B) the limit $\lim_{(x,y) \rightarrow Q} g(x, y)$ appears to exist. If it exists, it must be 4, which is the level curve of Q.

Further Insights and Challenges

37. Is the following function continuous?

$$f(x, y) = \begin{cases} x^2 + y^2 & \text{if } x^2 + y^2 < 1 \\ 1 & \text{if } x^2 + y^2 \geq 1 \end{cases}$$

SOLUTION $f(x, y)$ is defined by a polynomial in the domain $x^2 + y^2 < 1$, hence f is continuous in this domain. In the domain $x^2 + y^2 > 1$, f is a constant function, hence f is continuous in this domain also. Thus, we must examine continuity at the points on the circle $x^2 + y^2 = 1$.

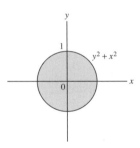

We express $f(x, y)$ using polar coordinates:

$$f(r, \theta) = \begin{cases} r^2 & 0 \leq r < 1 \\ 1 & r \geq 1 \end{cases}$$

Since $\lim_{r \to 1-} f(r, \theta) = \lim_{r \to 1-} r^2 = 1$ and $\lim_{r \to 1+} f(r, \theta) = \lim_{r \to 1+} 1 = 1$, we have $\lim_{r \to 1} f(r, \theta) = 1$. Therefore $f(r, \theta)$ is continuous at $r = 1$, or $f(x, y)$ is continuous on $x^2 + y^2 = 1$. We conclude that f is continuous everywhere on R^2.

39. Prove that the function

$$f(x, y) = \begin{cases} \dfrac{(2^x - 1)(\sin y)}{xy} & \text{if } xy \neq 0 \\ \ln 2 & \text{if } xy = 0 \end{cases}$$

is continuous at $(0, 0)$.

SOLUTION To solve this problem it is necessary to show that $\lim_{(x,y) \to (0,0)} f(x, y) = f(0, 0) = \ln 2$. Consider the following:

$$\lim_{(x,y) \to (0,0)} \frac{(2^x - 1) \sin y}{xy} = \lim_{(x,y) \to (0,0)} \frac{2^x - 1}{x} \cdot \frac{\sin y}{y}$$

$$= \left(\lim_{x \to 0} \frac{2^x - 1}{x} \right) \left(\lim_{y \to 0} \frac{\sin y}{y} \right)$$

$$= \lim_{x \to 0} \frac{(\ln 2) 2^x}{1} \cdot (1) = \ln 2$$

(Using L'Hopital's Rule on the limit in terms of x.) Thus since $\lim_{(x,y) \to (0,0)} f(x, y) = f(0, 0)$, we see that $f(x, y)$ is continuous at $(0, 0)$.

41. The function $f(x, y) = x^2 y/(x^4 + y^2)$ provides an interesting example where the limit as $(x, y) \to (0, 0)$ does not exist, even though the limit along every line $y = mx$ exists and is zero (Figure 8).

(a) Show that the limit along any line $y = mx$ exists and is equal to 0.

(b) Calculate $f(x, y)$ at the points $(10^{-1}, 10^{-2})$, $(10^{-5}, 10^{-10})$, $(10^{-20}, 10^{-40})$. Do not use a calculator.

(c) Show that $\lim\limits_{(x,y)\to(0,0)} f(x, y)$ does not exist. *Hint:* Compute the limit along the parabola $y = x^2$.

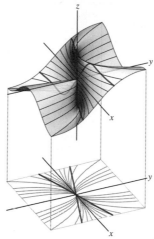

FIGURE 8 Graph of $f(x, y) = \dfrac{x^2 y}{x^4 + y^2}$.

SOLUTION

(a) Substituting $y = mx$ in $f(x, y) = \dfrac{x^2 y}{x^4 + y^2}$, we get

$$f(x, mx) = \frac{x^2 \cdot mx}{x^4 + (mx)^2} = \frac{mx^3}{x^2(x^2 + m^2)} = \frac{mx}{x^2 + m^2}$$

We compute the limit as $x \to 0$ by substitution:

$$\lim_{x\to 0} f(x, mx) = \lim_{x\to 0} \frac{mx}{x^2 + m^2} = \frac{m \cdot 0}{0^2 + m^2} = 0$$

(b) We compute $f(x, y)$ at the given points:

$$f(10^{-1}, 10^{-2}) = \frac{10^{-2} \cdot 10^{-2}}{10^{-4} + 10^{-4}} = \frac{10^{-4}}{2 \cdot 10^{-4}} = \frac{1}{2}$$

$$f(10^{-5}, 10^{-10}) = \frac{10^{-10} \cdot 10^{-10}}{10^{-20} + 10^{-20}} = \frac{10^{-20}}{2 \cdot 10^{-20}} = \frac{1}{2}$$

$$f(10^{-20}, 10^{-40}) = \frac{10^{-40} \cdot 10^{-40}}{10^{-80} + 10^{-80}} = \frac{10^{-80}}{2 \cdot 10^{-80}} = \frac{1}{2}$$

(c) We compute the limit as (x, y) approaches the origin along the parabola $y = x^2$ (by part (b), the limit appears to be $\frac{1}{2}$). We substitute $y = x^2$ in the function and compute the limit as $x \to 0$. This gives

$$\lim_{\substack{(x,y)\to(0,0) \\ \text{along } y=x^2}} f(x, y) = \lim_{x\to 0} f(x, x^2) = \lim_{x\to 0} \frac{x^2 \cdot x^2}{x^4 + (x^2)^2} = \lim_{x\to 0} \frac{x^4}{2x^4} = \lim_{x\to 0} \frac{1}{2} = \frac{1}{2}$$

However, in part (a), we showed that the limit along the lines $y = mx$ is zero. Therefore $f(x, y)$ does not approach one limit as $(x, y) \to (0, 0)$, so the limit $\lim\limits_{(x,y)\to(0,0)} f(x, y)$ does not exist.

14.3 Partial Derivatives (LT Section 15.3)

Preliminary Questions

1. Patricia derived the following *incorrect* formula by misapplying the Product Rule:

$$\frac{\partial}{\partial x}(x^2 y^2) = x^2(2y) + y^2(2x)$$

What was her mistake and what is the correct calculation?

SOLUTION To compute the partial derivative with respect to x, we treat y as a constant. Therefore the Constant Multiple Rule must be used rather than the Product Rule. The correct calculation is:

$$\frac{\partial}{\partial x}(x^2 y^2) = y^2 \frac{\partial}{\partial x}(x^2) = y^2 \cdot 2x = 2xy^2.$$

2. Explain why it is not necessary to use the Quotient Rule to compute $\dfrac{\partial}{\partial x}\left(\dfrac{x+y}{y+1}\right)$. Should the Quotient Rule be used to compute $\dfrac{\partial}{\partial y}\left(\dfrac{x+y}{y+1}\right)$?

SOLUTION In differentiating with respect to x, y is considered a constant. Therefore in this case the Constant Multiple Rule can be used to obtain

$$\frac{\partial}{\partial x}\left(\frac{x+y}{y+1}\right) = \frac{1}{y+1}\frac{\partial}{\partial x}(x+y) = \frac{1}{y+1}\cdot 1 = \frac{1}{y+1}.$$

As for the second part, since y appears in both the numerator and the denominator, the Quotient Rule is indeed needed.

3. Which of the following partial derivatives should be evaluated without using the Quotient Rule?

(a) $\dfrac{\partial}{\partial x}\dfrac{xy}{y^2+1}$ **(b)** $\dfrac{\partial}{\partial y}\dfrac{xy}{y^2+1}$ **(c)** $\dfrac{\partial}{\partial x}\dfrac{y^2}{y^2+1}$

SOLUTION

(a) This partial derivative does not require use of the Quotient Rule, since the Constant Multiple Rule gives

$$\frac{\partial}{\partial x}\left(\frac{xy}{y^2+1}\right) = \frac{y}{y^2+1}\frac{\partial}{\partial x}(x) = \frac{y}{y^2+1}\cdot 1 = \frac{y}{y^2+1}.$$

(b) This partial derivative requires use of the Quotient Rule.

(c) Since y is considered a constant in differentiating with respect to x, we do not need the Quotient Rule to state that

$$\frac{\partial}{\partial x}\left(\frac{y^2}{y^2+1}\right) = 0.$$

4. What is f_x, where $f(x, y, z) = (\sin yz)e^{z^3 - z^{-1}\sqrt{y}}$?

SOLUTION In differentiating with respect to x, we treat y and z as constants. Therefore, the whole expression for $f(x, y, z)$ is treated as constant, so the derivative is zero:

$$\frac{\partial}{\partial x}\left(\sin yz\, e^{z^3 - z^{-1}\sqrt{y}}\right) = 0.$$

5. Assuming the hypotheses of Clairaut's Theorem are satisfied, which of the following partial derivatives are equal to f_{xxy}?

(a) f_{xyx} **(b)** f_{yyx} **(c)** f_{xyy} **(d)** f_{yxx}

SOLUTION f_{xxy} involves two differentiations with respect to x and one differentiation with respect to y. Therefore, if f satisfies the assumptions of Clairaut's Theorem, then

$$f_{xxy} = f_{xyx} = f_{yxx}$$

Exercises

1. Use the limit definition of the partial derivative to verify the formulas

$$\frac{\partial}{\partial x}xy^2 = y^2, \qquad \frac{\partial}{\partial y}xy^2 = 2xy$$

SOLUTION Using the limit definition of the partial derivative, we have

$$\frac{\partial}{\partial x}xy^2 = \lim_{h\to 0}\frac{(x+h)y^2 - xy^2}{h} = \lim_{h\to 0}\frac{xy^2 + hy^2 - xy^2}{h} = \lim_{h\to 0}\frac{hy^2}{h} = \lim_{h\to 0}y^2 = y^2$$

$$\frac{\partial}{\partial y}xy^2 = \lim_{k\to 0}\frac{x(y+k)^2 - xy^2}{k} = \lim_{k\to 0}\frac{x(y^2 + 2yk + k^2) - xy^2}{k} = \lim_{k\to 0}\frac{xy^2 + 2xyk + xk^2 - xy^2}{k}$$

$$= \lim_{k\to 0}\frac{k(2xy + xk)}{k} = \lim_{k\to 0}(2xy + k) = 2xy + 0 = 2xy$$

3. Use the Quotient Rule to compute $\dfrac{\partial}{\partial y}\dfrac{y}{x+y}$.

SOLUTION Using the Quotient Rule we obtain

$$\frac{\partial}{\partial y}\frac{y}{x+y}=\frac{(x+y)\frac{\partial}{\partial y}(y)-y\frac{\partial}{\partial y}(x+y)}{(x+y)^2}=\frac{(x+y)\cdot 1-y\cdot 1}{(x+y)^2}=\frac{x}{(x+y)^2}$$

5. Calculate $f_z(2,3,1)$, where $f(x,y,z)=xyz$.

SOLUTION We first find the partial derivative $f_z(x,y,z)$:

$$f_z(x,y,z)=\frac{\partial}{\partial z}(xyz)=xy$$

Substituting the given point we get

$$f_z(2,3,1)=2\cdot 3=6$$

7. The plane $y=1$ intersects the surface $z=x^4+6xy-y^4$ in a certain curve. Find the slope of the tangent line to this curve at the point $P=(1,1,6)$.

SOLUTION The slope of the tangent line to the curve $z=z(x,1)=x^4+6x-1$, obtained by intersecting the surface $z=x^4+6xy-y^4$ with the plane $y=1$, is the partial derivative $\frac{\partial z}{\partial x}(1,1)$.

$$\frac{\partial z}{\partial x}=\frac{\partial}{\partial x}(x^4+6xy-y^4)=4x^3+6y$$

$$m=\frac{\partial z}{\partial x}(1,1)=4\cdot 1^3+6\cdot 1=10$$

In Exercises 9–12, refer to Figure 8.

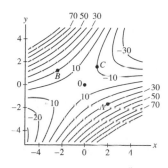

70 50 30

FIGURE 8 Contour map of $f(x,y)$.

9. Estimate f_x and f_y at point A.

SOLUTION To estimate f_x we move horizontally to the next level curve in the direction of growing x, to a point A'. The change in f from A to A' is the contour interval, $\Delta f=40-30=10$. The distance between A and A' is approximately $\Delta x\approx 1.0$. Hence,

$$f_x(A)\approx\frac{\Delta f}{\Delta x}=\frac{10}{1.0}=10$$

To estimate f_y we move vertically from A to a point A'' on the next level curve in the direction of growing y. The change in f from A to A'' is $\Delta f=20-30=-10$. The distance between A and A'' is $\Delta y\approx 0.5$. Hence,

$$f_y(A)\approx\frac{\Delta f}{\Delta y}=\frac{-10}{0.5}\approx -20.$$

11. Starting at point B, in which compass direction (N, NE, SW, etc.) does f increase most rapidly?

SOLUTION The distances between adjacent level curves starting at B are the smallest along the line with slope -1, upward. Therefore, f is increasing most rapidly in the direction of $\theta=135°$ or in the NW direction.

In Exercises 13–40, compute the first-order partial derivatives.

13. $z=x^2+y^2$

SOLUTION We compute $z_x(x,y)$ by treating y as a constant, and we compute $z_y(x,y)$ by treating x as a constant:

$$\frac{\partial}{\partial x}(x^2+y^2)=2x;\quad\frac{\partial}{\partial y}(x^2+y^2)=2y$$

15. $z = x^4 y + xy^{-2}$

SOLUTION We obtain the following partial derivatives:

$$\frac{\partial}{\partial x}(x^4 y + xy^{-2}) = 4x^3 y + y^{-2}$$

$$\frac{\partial}{\partial y}(x^4 y + xy^{-2}) = x^4 + x \cdot (-2y^{-3}) = x^4 - 2xy^{-3}$$

17. $z = \dfrac{x}{y}$

SOLUTION Treating y as a constant we have

$$\frac{\partial}{\partial x}\left(\frac{x}{y}\right) = \frac{1}{y}\frac{\partial}{\partial x}(x) = \frac{1}{y}\cdot 1 = \frac{1}{y}$$

We now find the derivative $z_y(x, y)$, treating x as a constant:

$$\frac{\partial}{\partial y}\left(\frac{x}{y}\right) = x \cdot \frac{\partial}{\partial y}\left(\frac{1}{y}\right) = x \cdot \frac{-1}{y^2} = \frac{-x}{y^2}.$$

19. $z = \sqrt{9 - x^2 - y^2}$

SOLUTION Differentiating with respect to x, treating y as a constant, and using the Chain Rule, we obtain

$$\frac{\partial}{\partial x}\left(\sqrt{9 - x^2 - y^2}\right) = \frac{1}{2\sqrt{9 - x^2 - y^2}}\frac{\partial}{\partial x}(9 - x^2 - y^2) = \frac{-2x}{2\sqrt{9 - x^2 - y^2}} = \frac{-x}{\sqrt{9 - x^2 - y^2}}$$

We now differentiate with respect to y, treating x as a constant:

$$\frac{\partial}{\partial y}\left(\sqrt{9 - x^2 - y^2}\right) = \frac{1}{2\sqrt{9 - x^2 - y^2}}\frac{\partial}{\partial y}(9 - x^2 - y^2) = \frac{-2y}{2\sqrt{9 - x^2 - y^2}} = \frac{-y}{\sqrt{9 - x^2 - y^2}}$$

21. $z = (\sin x)(\sin y)$

SOLUTION We obtain the following partial derivatives:

$$\frac{\partial}{\partial x}(\sin x \sin y) = \sin y \frac{\partial}{\partial x}\sin x = \sin y \cos x$$

$$\frac{\partial}{\partial y}(\sin x \sin y) = \sin x \frac{\partial}{\partial y}\sin y = \sin x \cos y$$

23. $z = \tan \dfrac{x}{y}$

SOLUTION By the Chain Rule,

$$\frac{d}{dx}\tan u = \frac{1}{\cos^2 u}\frac{du}{dx} \quad \text{and} \quad \frac{d}{dy}\tan u = \frac{1}{\cos^2 u}\frac{du}{dy}.$$

(We could also say that the derivative of $\tan u$ is $\sec^2 u$, but of course $\sec^2 u = 1/\cos^2 u$, so it really is the same thing.) We apply this with $u = \frac{x}{y}$ to obtain

$$\frac{\partial}{\partial x}\tan\left(\frac{x}{y}\right) = \frac{1}{\cos^2\left(\frac{x}{y}\right)}\frac{\partial}{\partial x}\left(\frac{x}{y}\right) = \frac{1}{\cos^2\left(\frac{x}{y}\right)}\cdot\frac{1}{y} = \frac{1}{y\cos^2\left(\frac{x}{y}\right)}$$

$$\frac{\partial}{\partial y}\tan\left(\frac{x}{y}\right) = \frac{1}{\cos^2\left(\frac{x}{y}\right)}\frac{\partial}{\partial y}\left(\frac{x}{y}\right) = \frac{1}{\cos^2\left(\frac{x}{y}\right)}\cdot\frac{-x}{y^2} = \frac{-x}{y^2\cos^2\left(\frac{x}{y}\right)}$$

25. $z = \ln(x^2 + y^2)$

SOLUTION Using the Chain Rule we have

$$\frac{\partial z}{\partial x} = \frac{1}{x^2 + y^2}\frac{\partial}{\partial x}(x^2 + y^2) = \frac{1}{x^2 + y^2}\cdot 2x = \frac{2x}{x^2 + y^2}$$

$$\frac{\partial z}{\partial y} = \frac{1}{x^2 + y^2}\frac{\partial}{\partial y}(x^2 + y^2) = \frac{1}{x^2 + y^2}\cdot 2y = \frac{2y}{x^2 + y^2}$$

27. $W = e^{r+s}$

SOLUTION We use the Chain Rule to compute $\frac{\partial W}{\partial r}$ and $\frac{\partial W}{\partial s}$:

$$\frac{\partial W}{\partial r} = e^{r+s} \cdot \frac{\partial}{\partial r}(r+s) = e^{r+s} \cdot 1 = e^{r+s}$$

$$\frac{\partial W}{\partial s} = e^{r+s} \cdot \frac{\partial}{\partial s}(r+s) = e^{r+s} \cdot 1 = e^{r+s}$$

29. $z = e^{xy}$

SOLUTION We use the Chain Rule, $\frac{d}{dx}e^u = e^u\frac{du}{dx}$; $\frac{d}{dy}e^u = e^u\frac{du}{dy}$ with $u = xy$ to obtain

$$\frac{\partial}{\partial x}e^{xy} = e^{xy}\frac{\partial}{\partial x}(xy) = e^{xy}y = ye^{xy}$$

$$\frac{\partial}{\partial y}e^{xy} = e^{xy}\frac{\partial}{\partial y}(xy) = e^{xy}x = xe^{xy}$$

31. $z = e^{-x^2-y^2}$

SOLUTION We use the Chain Rule to find $\frac{\partial z}{\partial x}$ and $\frac{\partial z}{\partial y}$:

$$\frac{\partial z}{\partial x} = e^{-x^2-y^2}\frac{\partial}{\partial x}(-x^2-y^2) = e^{-x^2-y^2} \cdot (-2x) = -2xe^{-x^2-y^2}$$

$$\frac{\partial z}{\partial y} = e^{-x^2-y^2}\frac{\partial}{\partial y}(-x^2-y^2) = e^{-x^2-y^2} \cdot (-2y) = -2ye^{-x^2-y^2}$$

33. $U = \dfrac{e^{-rt}}{r}$

SOLUTION We have

$$\frac{\partial U}{\partial r} = \frac{-te^{-rt} \cdot r - e^{-rt}}{r^2}\ \frac{1}{} = \frac{-(1+rt)e^{-rt}}{r^2}$$

and also

$$\frac{\partial U}{\partial t} = \frac{-re^{-rt}}{r} = -e^{-rt}$$

35. $z = \sinh(x^2 y)$

SOLUTION By the Chain Rule, $\frac{d}{dx}\sinh u = \cosh u\frac{du}{dx}$ and $\frac{d}{dy}\sinh u = \cosh u\frac{du}{dy}$. We use the Chain Rule with $u = x^2 y$ to obtain

$$\frac{\partial}{\partial x}\sinh(x^2 y) = \cosh(x^2 y)\frac{\partial}{\partial x}(x^2 y) = 2xy\cosh(x^2 y)$$

$$\frac{\partial}{\partial y}\sinh(x^2 y) = \cosh(x^2 y)\frac{\partial}{\partial y}(x^2 y) = x^2\cosh(x^2 y)$$

37. $w = xy^2 z^3$

SOLUTION The partial derivatives of w are

$$\frac{\partial w}{\partial x} = y^2 z^3$$

$$\frac{\partial w}{\partial y} = xz^3\frac{\partial}{\partial y}(y^2) = xz^3 \cdot 2y = 2xz^3 y$$

$$\frac{\partial w}{\partial z} = xy^2\frac{\partial}{\partial z}(z^3) = xy^2 \cdot 3z^2 = 3xy^2 z^2$$

39. $Q = \dfrac{L}{M} e^{-Lt/M}$

SOLUTION

$$\frac{\partial Q}{\partial L} = \frac{\partial}{\partial L} \left(\frac{L}{M} e^{-Lt/M} \right)$$

$$= \frac{L}{M} \cdot e^{-Lt/M} \cdot (-t/M) + e^{-Lt/M} \cdot \frac{1}{M}$$

$$= -\frac{Lt}{M^2} e^{-Lt/M} + \frac{e^{-Lt/M}}{M}$$

$$\frac{\partial Q}{\partial M} = \frac{\partial}{\partial M} \left(\frac{L}{M} e^{-Lt/M} \right)$$

$$= \frac{L}{M} \cdot e^{-Lt/M} \cdot \frac{Lt}{M^2} + e^{-Lt/M} \cdot \frac{-L}{M^2}$$

$$= \frac{L^2 t}{M^3} e^{-Lt/M} - \frac{L}{M^2} e^{-Lt/M}$$

$$\frac{\partial Q}{\partial t} = \frac{\partial}{\partial t} \left(\frac{L}{M} e^{-Lt/M} \right)$$

$$= -\frac{L^2}{M^2} e^{-Lt/M}$$

In Exercises 41–44, compute the given partial derivatives.

41. $f(x, y) = 3x^2 y + 4x^3 y^2 - 7xy^5$, $f_x(1, 2)$

SOLUTION Differentiating with respect to x gives

$$f_x(x, y) = 6xy + 12x^2 y^2 - 7y^5$$

Evaluating at $(1, 2)$ gives

$$f_x(1, 2) = 6 \cdot 1 \cdot 2 + 12 \cdot 1^2 \cdot 2^2 - 7 \cdot 2^5 = -164.$$

43. $g(u, v) = u \ln(u + v)$, $g_u(1, 2)$

SOLUTION Using the Product Rule and the Chain Rule we get

$$g_u(u, v) = \frac{\partial}{\partial u} (u \ln(u + v)) = 1 \cdot \ln(u + v) + u \cdot \frac{1}{u + v} = \ln(u + v) + \frac{u}{u + v}$$

At the point $(1, 2)$ we have

$$g_u(1, 2) = \ln(1 + 2) + \frac{1}{1 + 2} = \ln 3 + \frac{1}{3}.$$

Exercises 45 and 46 refer to Example 5.

45. Calculate N for $L = 0.4$, $R = 0.12$, and $d = 10$, and use the linear approximation to estimate ΔN if d is increased from 10 to 10.4.

SOLUTION From the example in the text we have

$$N = \left(\frac{2200R}{Ld} \right)^{1.9}$$

Calculating N for $L = 0.4$, $R = 0.12$, and $d = 10$ we have

$$N = \left(\frac{2200 \cdot 0.12}{0.4 \cdot 10} \right)^{1.9} \approx 2865.058$$

then we will use the derivation

$$\Delta N \approx \frac{\partial N}{\partial d} \Delta d$$

since d is increasing from 10 to 10.4. We need to compute $\partial N/\partial d$, with L and R constant:

$$\frac{\partial N}{\partial d} = \frac{\partial}{\partial d}\left(\frac{2200R}{Ld}\right)^{1.9}$$

$$= \left(\frac{2200R}{L}\right)^{1.9}\frac{\partial}{\partial d}(d^{-1.9})$$

$$= -1.9\left(\frac{2200R}{L}\right)^{1.9}d^{-2.9}$$

we have first

$$\left.\frac{\partial N}{\partial d}\right|_{(L,R,d)=(0.4,0.12,10)} = -1.9\left(\frac{2200\cdot 0.12}{0.4}\right)^{1.9}(10)^{-2.9} \approx -544.361$$

Therefore we can conclude:

$$\Delta N \approx \frac{\partial N}{\partial d}\Delta d \approx (-544.361)(10.4 - 10) = -217.744$$

47. The **heat index** I is a measure of how hot it feels when the relative humidity is H (as a percentage) and the actual air temperature is T (in degrees Fahrenheit). An approximate formula for the heat index that is valid for (T, H) near $(90, 40)$ is

$$I(T, H) = 45.33 + 0.6845T + 5.758H - 0.00365T^2$$

$$- 0.1565HT + 0.001HT^2$$

(a) Calculate I at $(T, H) = (95, 50)$

(b) Which partial derivative tells us the increase in I per degree increase in T when $(T, H) = (95, 50)$. Calculate this partial derivative.

SOLUTION

(a) Let us compute I when $T = 95$ and $H = 50$:

$$I(95, 50) = 45.33 + 0.6845(95) + 5.758(50) - 0.00365(95)^2 - 0.1565(50)(95) + 0.001(50)(95)^2$$

$$= 73.19125$$

(b) The partial derivative we are looking for here is $\partial I/\partial T$:

$$\frac{\partial I}{\partial T} = 0.6845 - 0.00730T - 0.1565H + 0.002HT$$

and evaluating we have:

$$\frac{\partial I}{\partial T}(95, 50) = 0.6845 - 0.00730(95) - 0.1565(50) + 0.002(50)(95) = 1.666$$

49. The volume of a right-circular cone of radius r and height h is $V = \frac{\pi}{3}r^2h$. Suppose that $r = h = 12$ cm. What leads to a greater increase in V, a 1-cm increase in r or a 1-cm increase in h? Argue using partial derivatives.

SOLUTION We obtain the following derivatives:

$$\frac{\partial V}{\partial r} = \frac{\partial}{\partial r}\left(\frac{\pi}{3}r^2h\right) = \frac{\pi h}{3}\frac{\partial}{\partial r}r^2 = \frac{\pi h}{3}\cdot 2r = \frac{2\pi hr}{3}$$

$$\frac{\partial V}{\partial h} = \frac{\partial}{\partial h}\left(\frac{\pi}{3}r^2h\right) = \frac{\pi}{3}r^2$$

An increase $\Delta r = 1$ cm in r leads to an increase of $\frac{\partial V}{\partial r}(12, 12)\cdot 1$ in the volume, and an increase $\Delta h = 1$ cm in h leads to an increase of $\frac{\partial V}{\partial h}(12, 12)\cdot 1$ in V. We compute these values, using the partials computed. This gives

$$\frac{\partial V}{\partial r}(12, 12) = \left.\frac{2\pi hr}{3}\right|_{(12,12)} = \frac{2\pi\cdot 12\cdot 12}{3} = 301.6$$

$$\frac{\partial V}{\partial h}(12, 12) = \frac{\pi}{3}\cdot 12^2 = 150.8$$

We conclude that an increase of 1 cm in r leads to a greater increase in V than an increase of 1 cm in h.

51. Calculate $\partial W/\partial E$ and $\partial W/\partial T$, where $W = e^{-E/kT}$, where k is a constant.

SOLUTION We use the Chain Rule

$$\frac{d}{dE}e^u = e^u\frac{du}{dE} \quad \text{and} \quad \frac{d}{dT}e^u = e^u\frac{du}{dT}$$

with $u = -\frac{E}{kT}$, to obtain

$$\frac{\partial W}{\partial E} = e^{-E/kT}\frac{\partial}{\partial E}\left(-\frac{E}{kT}\right) = e^{-E/kT}\left(-\frac{1}{kT}\right) = -\frac{1}{kT}e^{-E/kT}$$

$$\frac{\partial W}{\partial T} = e^{-E/kT}\frac{\partial}{\partial T}\left(-\frac{E}{kT}\right) = e^{-E/kT}\cdot\left(-\frac{E}{k}\right)\frac{\partial}{\partial T}\left(\frac{1}{T}\right) = e^{-E/kT}\left(-\frac{E}{k}\right)\left(-\frac{1}{T^2}\right) = \frac{E}{kT^2}e^{-E/kT}$$

53. Use the contour map of $f(x, y)$ in Figure 9 to explain the following statements.

(a) f_y is larger at P than at Q, and f_x is smaller (more negative) at P than at Q.

(b) $f_x(x, y)$ is decreasing as a function of y; that is, for any fixed value $x = a$, $f_x(a, y)$ is decreasing in y.

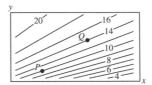

FIGURE 9 Contour interval 2.

SOLUTION

(a) A vertical segment through P meet more level curves than a vertical segment of the same size through Q, so f is increasing more rapidly in the y at P than at Q. Therefore, f_y are both larger at P than at Q.

Similarly, a horizontal segment through P meet more level curves at P than at Q, but f is *decreasing* in the positive x-direction, so f is decreasing more rapidly in the x-direction at P than at Q. Therefore, f_x is more negative at P than at Q.

(b) For any fixed value $x = a$, a horizontal segment meets fewer level curves as we move it vertically upward. This indicates that $f_x(a, y)$ in a decreasing function of y.

55. Over most of the earth, a magnetic compass does not point to true (geographic) north; instead, it points at some angle east or west of true north. The angle D between magnetic north and true north is called the **magnetic declination**. Use Figure 11 to determine which of the following statements is true.

(a) $\left.\dfrac{\partial D}{\partial y}\right|_A > \left.\dfrac{\partial D}{\partial y}\right|_B$ 　　　　**(b)** $\left.\dfrac{\partial D}{\partial x}\right|_C > 0$ 　　　　**(c)** $\left.\dfrac{\partial D}{\partial y}\right|_C > 0$

Note that the horizontal axis increases from right to left because of the way longitude is measured.

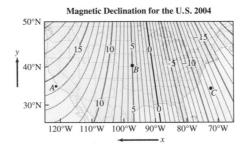

FIGURE 11 Contour interval $1°$.

SOLUTION

(a) To estimate $\left.\frac{\partial D}{\partial y}\right|_A$ and $\left.\frac{\partial D}{\partial y}\right|_B$, we move vertically from A and B to the points on the next level curve in the direction of increasing y (upward). From A, we quickly come to a level curve corresponding to higher value of D; but from B, moving vertically, there is hardly any change as we move along the curve. The statement is thus true.

(b) The derivative $\left.\frac{\partial D}{\partial x}\right|_C$ is estimated by $\frac{\Delta D}{\Delta x}$. Since x varies in the horizontal direction, we move horizontally from C to a point on the next level curve in the direction of increasing x (leftwards). Since the value of D on this level curve is greater than on the level curve of C, $\Delta D = 1$. Also $\Delta x > 0$, hence

$$\left.\frac{\partial D}{\partial x}\right|_C \approx \frac{\Delta D}{\Delta x} = \frac{1}{\Delta x} > 0.$$

The statement is correct.

(c) Moving from C vertically upward (in the direction of increasing y), we come to a point on a level curve with a smaller value of D. Therefore, $\Delta D = -1$ and $\Delta y > 0$, so we obtain

$$\left. \frac{\partial D}{\partial y} \right|_C \approx \frac{\Delta D}{\Delta y} = \frac{-1}{\Delta y} < 0$$

Hence, the statement is false.

In Exercises 57–62, compute the derivatives indicated.

57. $f(x, y) = 3x^2 y - 6xy^4$, $\dfrac{\partial^2 f}{\partial x^2}$ and $\dfrac{\partial^2 f}{\partial y^2}$

SOLUTION We first compute the partial derivatives $\frac{\partial f}{\partial x}$ and $\frac{\partial f}{\partial y}$:

$$\frac{\partial f}{\partial x} = 6xy - 6y^4; \quad \frac{\partial f}{\partial y} = 3x^2 - 6x \cdot 4y^3 = 3x^2 - 24xy^3$$

We now differentiate $\frac{\partial f}{\partial x}$ with respect to x and $\frac{\partial f}{\partial y}$ with respect to y. We get

$$\frac{\partial^2 f}{\partial x^2} = \frac{\partial}{\partial x} f_x = 6y; \quad \frac{\partial^2 f}{\partial y^2} = \frac{\partial}{\partial y} f_y = -24x \cdot 3y^2 = -72xy^2.$$

59. $h(u, v) = \dfrac{u}{u + 4v}$, $h_{vv}(u, v)$

SOLUTION We first note

$$\frac{\partial h}{\partial v} = \frac{-4u}{(u + 4v)^2}$$

so thus

$$\frac{\partial h^2}{\partial v^2} = \frac{\partial}{\partial v} \left(\frac{-4u}{(u + 4v)^2} \right) = \frac{32u}{(u + 4v)^3}$$

61. $f(x, y) = x \ln(y^2)$, $f_{yy}(2, 3)$

SOLUTION We find f_y using the Chain Rule:

$$f_y = \frac{\partial}{\partial y}(x \ln y^2) = x \frac{\partial}{\partial y} \ln y^2 = x \frac{1}{y^2} \cdot 2y = \frac{2x}{y}$$

We now differentiate f_y with respect to y, obtaining

$$f_{yy}(x, y) = \frac{\partial}{\partial y} f_y = 2x \frac{\partial}{\partial y} \left(\frac{1}{y} \right) = \frac{-2x}{y^2}.$$

The derivative at $(2, 3)$ is thus

$$f_{yy}(2, 3) = \frac{-2 \cdot 2}{3^2} = -\frac{4}{9}.$$

63. Compute f_{xyxzy} for

$$f(x, y, z) = y \sin(xz) \sin(x + z) + (x + z^2) \tan y + x \tan \left(\frac{z + z^{-1}}{y - y^{-1}} \right)$$

Hint: Use a well-chosen order of differentiation on each term.

SOLUTION At the points where the derivatives are continuous, the partial derivative f_{xyxzy} may be performed in any order. To simplify the computation we first consider $f(x, y, z)$ as the sum of the following terms:

$$F(x, y, z) = y \sin(xz) \sin(x + z), \quad G(x, y, z) = (x + z^2) \tan y, \quad H(x, y, z) = x \tan \left(\frac{z + z^{-1}}{y - y^{-1}} \right)$$

so that

$$f(x, y, z) = F(x, y, z) + G(x, y, z) + H(x, y, z)$$

We can differentiate each in any order. First, let us work with $F(x, y, z) = y \sin(xz) \sin(x + z)$:

$$F_y(x, y, z) = \frac{\partial}{\partial y}(y \sin(xz) \sin(x + z)) = \sin(xz) \sin(x + z)$$

then

$$F_{yy}(x, y, z) = \frac{\partial}{\partial y}(F_y(x, y, z)) = 0$$

hence,

$$F_{yyxz}(x, y, z) = 0$$

Next, let us work with $G(x, y, z) = (x + z^2) \tan y$:

$$G_x(x, y, z) = \frac{\partial}{\partial x}((x + z^2) \tan y) = \tan y$$

then

$$G_{xx}(x, y, z) = \frac{\partial}{\partial x}(G_x(x, y, z)) = 0$$

Hence

$$G_{xxyyz}(x, y, z) = 0$$

Finally, let us work with $H(x, y, z) = x \tan\left(\dfrac{z + z^{-1}}{y - y^{-1}}\right)$:

$$H_x(x, y, z) = \frac{\partial}{\partial x}\left(x \tan\left(\frac{z + z^{-1}}{y - y^{-1}}\right)\right) = \tan\left(\frac{z + z^{-1}}{y - y^{-1}}\right)$$

then

$$H_{xx}(x, y, z) = \frac{\partial}{\partial x}(H_x(x, y, z)) = 0$$

hence,

$$H_{xxyyz}(x, y, z) = 0$$

Therefore, we can conclude that $f_{xyxzy}(x, y, z) = 0 + 0 + 0 = 0$.

In Exercises 65–72, compute the derivative indicated.

65. $f(u, v) = \cos(u + v^2)$, f_{uuv}

SOLUTION Using the Chain Rule, we have

$$f_u = \frac{\partial}{\partial u}\cos(u + v^2) = -\sin(u + v^2) \cdot \frac{\partial}{\partial u}(u + v^2) = -\sin(u + v^2)$$

$$f_{uu} = \frac{\partial}{\partial u}(-\sin(u + v^2)) = -\cos(u + v^2)$$

$$f_{uuv} = \frac{\partial}{\partial v}(-\cos(u + v^2)) = \sin(u + v^2) \cdot \frac{\partial}{\partial v}(u + v^2) = 2v \sin(u + v^2)$$

67. $F(r, s, t) = r(s^2 + t^2)$, F_{rst}

SOLUTION For $F(r, s, t) = r(s^2 + t^2)$, we have

$$F_r = s^2 + t^2$$

$$F_{rs} = 2s$$

$$F_{rst} = 0$$

69. $F(\theta, u, v) = \sinh(uv + \theta^2)$, $F_{uu\theta}$

SOLUTION We can compute:

$$F_u = v \cdot \cosh(uv + \theta^2)$$

$$F_{uu} = v^2 \cdot \sinh(uv + \theta^2)$$

$$F_{uu\theta} = 2\theta v^2 \cosh(uv + \theta^2)$$

71. $g(x, y, z) = \sqrt{x^2 + y^2 + z^2}$, g_{xyz}

SOLUTION Differentiating with respect to x, using the Chain Rule, we get

$$g_x = \frac{\partial}{\partial x}\sqrt{x^2 + y^2 + z^2} = \frac{1}{2\sqrt{x^2 + y^2 + z^2}}\frac{\partial}{\partial x}(x^2 + y^2 + z^2) = \frac{1}{2\sqrt{x^2 + y^2 + z^2}} \cdot 2x = \frac{x}{\sqrt{x^2 + y^2 + z^2}}$$

We now differentiate g_x with respect to y, using the Chain Rule. This gives

$$g_{xy} = x\frac{\partial}{\partial y}(x^2 + y^2 + z^2)^{-1/2} = x \cdot \left(-\frac{1}{2}\right)(x^2 + y^2 + z^2)^{-3/2} \cdot 2y = \frac{-xy}{(x^2 + y^2 + z^2)^{3/2}}$$

Finally, we differentiate g_{xy} with respect to z, obtaining

$$g_{xyz} = -xy\frac{\partial}{\partial z}(x^2 + y^2 + z^2)^{-3/2} = -xy \cdot \left(-\frac{3}{2}\right)(x^2 + y^2 + z^2)^{-5/2} \cdot 2z = \frac{3xyz}{(x^2 + y^2 + z^2)^{5/2}}$$

73. Find a function such that $\dfrac{\partial f}{\partial x} = 2xy$ and $\dfrac{\partial f}{\partial y} = x^2$.

SOLUTION The function $f(x, y) = x^2y$ satisfies $\frac{\partial f}{\partial y} = x^2$ and $\frac{\partial f}{\partial x} = 2xy$.

75. Assume that f_{xy} and f_{yx} are continuous and that f_{yxx} exists. Show that f_{xyx} also exists and that $f_{yxx} = f_{xyx}$.

SOLUTION Since f_{xy} and f_{yx} are continuous, Clairaut's Theorem implies that

$$f_{xy} = f_{yx} \tag{1}$$

We are given that f_{yxx} exists. Using (1) we get

$$f_{yxx} = \frac{\partial}{\partial x}\frac{\partial}{\partial x}f_y = \frac{\partial}{\partial x}f_{yx} = \frac{\partial}{\partial x}f_{xy} = f_{xyx}$$

Therefore, f_{xyx} also exists and $f_{yxx} = f_{xyx}$.

77. Find all values of A and B such that $f(x, t) = e^{Ax+Bt}$ satisfies Eq. (3).

SOLUTION We compute the following partials, using the Chain Rule:

$$\frac{\partial f}{\partial t} = \frac{\partial}{\partial t}(e^{Ax+Bt}) = e^{Ax+Bt}\frac{\partial}{\partial t}(Ax + Bt) = Be^{Ax+Bt}$$

$$\frac{\partial f}{\partial x} = \frac{\partial}{\partial x}(e^{Ax+Bt}) = e^{Ax+Bt}\frac{\partial}{\partial x}(Ax + Bt) = Ae^{Ax+Bt}$$

$$\frac{\partial^2 f}{\partial x^2} = \frac{\partial}{\partial x}(Ae^{Ax+Bt}) = A\frac{\partial}{\partial x}(e^{Ax+Bt}) = Ae^{Ax+Bt}\frac{\partial}{\partial x}(Ax + Bt) = A^2e^{Ax+Bt}$$

Substituting these partials in the differential equation (3), we get

$$Be^{Ax+Bt} = A^2e^{Ax+Bt}$$

We divide by the nonzero e^{Ax+Bt} to obtain

$$B = A^2$$

We conclude that $f(x, t) = e^{Ax+Bt}$ satisfies equation (5) if and only if $B = A^2$, where A is arbitrary.

*In Exercises 79–82, the **Laplace operator** Δ is defined by $\Delta f = f_{xx} + f_{yy}$. A function $u(x, y)$ satisfying the Laplace equation $\Delta u = 0$ is called **harmonic**.*

79. Show that the following functions are harmonic:

(a) $u(x, y) = x$

(b) $u(x, y) = e^x \cos y$

(c) $u(x, y) = \tan^{-1}\dfrac{y}{x}$

(d) $u(x, y) = \ln(x^2 + y^2)$

SOLUTION

(a) We compute u_{xx} and u_{yy} for $u(x, y) = x$:

$$u_x = \frac{\partial}{\partial x}(x) = 1; \quad u_{xx} = \frac{\partial}{\partial x}(1) = 0$$

$$u_y = \frac{\partial}{\partial y}(x) = 0; \quad u_{yy} = \frac{\partial}{\partial y}(0) = 0$$

Since $u_{xx} + u_{yy} = 0$, u is harmonic.

(b) We compute the partial derivatives of $u(x, y) = e^x \cos y$:

$$u_x = \frac{\partial}{\partial x}\left(e^x \cos y\right) = \cos y \frac{\partial}{\partial x} e^x = (\cos y)e^x$$

$$u_y = \frac{\partial}{\partial y}\left(e^x \cos y\right) = e^x \frac{\partial}{\partial y} \cos y = -e^x \sin y$$

$$u_{xx} = \frac{\partial}{\partial x}\left((\cos y)e^x\right) = \cos y \frac{\partial}{\partial x} e^x = (\cos y)e^x$$

$$u_{yy} = \frac{\partial}{\partial y}\left(-e^x \sin y\right) = -e^x \frac{\partial}{\partial y} \sin y = -e^x \cos y$$

Thus,

$$u_{xx} + u_{yy} = (\cos y)e^x - e^x \cos y = 0$$

Hence $u(x, y) = e^x \cos y$ is harmonic.

(c) We compute the partial derivatives of $u(x, y) = \tan^{-1} \frac{y}{x}$ using the Chain Rule and the formula

$$\frac{d}{dt} \tan^{-1} t = \frac{1}{1 + t^2}$$

We have

$$u_x = \frac{\partial}{\partial x} \tan^{-1} \frac{y}{x} = \frac{1}{1 + (y/x)^2} \frac{\partial}{\partial x} \frac{y}{x} = \frac{1}{1 + (y/x)^2}\left(\frac{-y}{x^2}\right) = -\frac{y}{x^2 + y^2}$$

$$u_y = \frac{\partial}{\partial y} \tan^{-1} \frac{y}{x} = \frac{1}{1 + (y/x)^2} \frac{\partial}{\partial y} \frac{y}{x} = \frac{1}{1 + (y/x)^2}\left(\frac{1}{x}\right) = \frac{x}{x^2 + y^2}$$

$$u_{xx} = \frac{\partial}{\partial x}\left(-\frac{y}{x^2 + y^2}\right) = \frac{2xy}{(x^2 + y^2)^2}$$

$$u_{yy} = \frac{\partial}{\partial y} \frac{x}{x^2 + y^2} = -\frac{2xy}{(x^2 + y^2)^2}$$

Therefore $u_{xx} + u_{xx} = 0$. This shows that $u(x, y) = \tan^{-1} \frac{y}{x}$ is harmonic.

(d) We compute the partial derivatives of $u(x, y) = \ln(x^2 + y^2)$ using the Chain Rule:

$$u_x = \frac{\partial}{\partial x} \ln(x^2 + y^2) = \frac{1}{x^2 + y^2} \cdot 2x = \frac{2x}{x^2 + y^2}$$

$$u_y = \frac{\partial}{\partial y} \ln(x^2 + y^2) = \frac{1}{x^2 + y^2} \cdot 2y = \frac{2y}{x^2 + y^2}$$

We now find u_{xx} and u_{yy} using the Quotient Rule:

$$u_{xx} = \frac{\partial}{\partial x} \frac{2x}{x^2 + y^2} = \frac{2(x^2 + y^2) - 2x \cdot 2x}{(x^2 + y^2)^2} = \frac{2(y^2 - x^2)}{(x^2 + y^2)^2}$$

$$u_{yy} = \frac{\partial}{\partial y} \frac{2y}{x^2 + y^2} = \frac{2(x^2 + y^2) - 2y \cdot 2y}{(x^2 + y^2)^2} = \frac{2(x^2 - y^2)}{(x^2 + y^2)^2}$$

Thus,

$$u_{xx} + u_{yy} = \frac{2(y^2 - x^2)}{(x^2 + y^2)^2} + \frac{2(x^2 - y^2)}{(x^2 + y^2)^2} = 0.$$

Therefore, $u(x, y) = \ln(x^2 + y^2)$ is harmonic.

81. Show that if $u(x, y)$ is harmonic, then the partial derivatives $\partial u/\partial x$ and $\partial u/\partial y$ are harmonic.

SOLUTION We assume that the second-order partials are continuous, hence the partial differentiation may be performed in any order. By the given data, we have

$$u_{xx} + u_{yy} = 0 \tag{1}$$

We must show that

$$(u_x)_{xx} + (u_x)_{yy} = 0 \quad \text{and} \quad (u_y)_{xx} + (u_y)_{yy} = 0$$

We differentiate (1) with respect to x, obtaining

$$0 = (u_{xx})_x + (u_{yy})_x = u_{xxx} + u_{xyy} = (u_x)_{xx} + (u_x)_{yy} \tag{2}$$

We differentiate (1) with respect to y:

$$0 = (u_{xx})_y + (u_{yy})_y = u_{xxy} + u_{yyy} = u_{yxx} + u_{yyy} = (u_y)_{xx} + (u_y)_{yy} \tag{3}$$

Equalities (2) and (3) prove that u_x and u_y are harmonic.

83. Show that $u(x, t) = \text{sech}^2(x - t)$ satisfies the **Korteweg–deVries equation** (which arises in the study of water waves):

$$4u_t + u_{xxx} + 12uu_x = 0$$

SOLUTION In Exercise 72 we found the following derivatives:

$$u_x = -2\,\text{sech}^2(x - t)\tanh(x - t)$$

$$u_{xxx} = 16\,\text{sech}^4(x - t)\tanh(x - t) - 8\,\text{sech}^2(x - t)\tanh^3(x - t)$$

Hence,

$$\begin{aligned}
4u_t + u_{xxx} + 12uu_x &= 8\,\text{sech}^2(x - t)\tanh(x - t) + 16\,\text{sech}^4(x - t)\tanh(x - t) \\
&\quad - 8\,\text{sech}^2(x - t)\tanh^3(x - t) - 24\,\text{sech}^4(x - t)\tanh(x - t) \\
&= 8\,\text{sech}^2(x - t)\{\tanh(x - t) - \tanh^3(x - t)\} - 8\,\text{sech}^4(x - t)\tanh(x - t) \\
&= 8\,\text{sech}^2(x - t)\tanh(x - t)\{1 - \tanh^2(x - t)\} - 8\,\text{sech}^4(x - t)\tanh(x - t) \\
&= 8\,\text{sech}^2(x - t)\tanh(x - t)\{\text{sech}^2(x - t)\} - 8\,\text{sech}^4(x - t)\tanh(x - t) \\
&= 0
\end{aligned}$$

14.4 Differentiability and Tangent Planes (LT Section 15.4)

Preliminary Questions

1. How is the linearization of $f(x, y)$ at (a, b) defined?

SOLUTION The linearization of $f(x, y)$ at (a, b) is the linear function

$$L(x, y) = f(a, b) + f_x(a, b)(x - a) + f_y(a, b)(y - b)$$

This function is the equation of the tangent plane to the surface $z = f(x, y)$ at $(a, b, f(a, b))$.

2. Define local linearity for functions of two variables.

SOLUTION $f(x, y)$ is locally linear at (a, b) if

$$f(x, y) - L(x, y) = \epsilon(x, y)\sqrt{(x - a)^2 + (y - b)^2}$$

for all (x, y) in an open disk D containing (a, b), where $\epsilon(x, y)$ satisfies $\lim\limits_{(x,y)\to(a,b)} \epsilon(x, y) = 0$.

In Exercises 3–5, assume that

$$f(2, 3) = 8, \qquad f_x(2, 3) = 5, \qquad f_y(2, 3) = 7$$

3. Which of (a)–(b) is the linearization of f at $(2, 3)$?
(a) $L(x, y) = 8 + 5x + 7y$
(b) $L(x, y) = 8 + 5(x - 2) + 7(y - 3)$

SOLUTION The linearization of f at $(2, 3)$ is the following linear function:

$$L(x, y) = f(2, 3) + f_x(2, 3)(x - 2) + f_y(2, 3)(y - 3)$$

That is,

$$L(x, y) = 8 + 5(x - 2) + 7(y - 3) = -23 + 5x + 7y$$

The function in (b) is the correct answer.

4. Estimate $f(2, 3.1)$.

SOLUTION We use the linear approximation

$$f(a + h, b + k) \approx f(a, b) + f_x(a, b)h + f_y(a, b)k$$

We let $(a, b) = (2, 3)$, $h = 0$, $k = 3.1 - 3 = 0.1$. Then,

$$f(2, 3.1) \approx f(2, 3) + f_x(2, 3) \cdot 0 + f_y(2, 3) \cdot 0.1 = 8 + 0 + 7 \cdot 0.1 = 8.7$$

We get the estimation $f(2, 3.1) \approx 8.7$.

5. Estimate Δf at $(2, 3)$ if $\Delta x = -0.3$ and $\Delta y = 0.2$.

SOLUTION The change in f can be estimated by the linear approximation as follows:

$$\Delta f \approx f_x(a, b)\Delta x + f_y(a, b)\Delta y$$

$$\Delta f \approx f_x(2, 3) \cdot (-0.3) + f_y(2, 3) \cdot 0.2$$

or

$$\Delta f \approx 5 \cdot (-0.3) + 7 \cdot 0.2 = -0.1$$

The estimated change is $\Delta f \approx -0.1$.

6. Which theorem allows us to conclude that $f(x, y) = x^3 y^8$ is differentiable?

SOLUTION The function $f(x, y) = x^3 y^8$ is a polynomial, hence $f_x(x, y)$ and $f_y(x, y)$ exist and are continuous. Therefore the Criterion for Differentiability implies that f is differentiable everywhere.

Exercises

1. Use Eq. (2) to find an equation of the tangent plane to the graph of $f(x, y) = 2x^2 - 4xy^2$ at $(-1, 2)$.

SOLUTION The equation of the tangent plane at the point $(-1, 2, 18)$ is

$$z = f(-1, 2) + f_x(-1, 2)(x + 1) + f_y(-1, 2)(y - 2) \tag{1}$$

We compute the function and its partial derivatives at the point $(-1, 2)$:

$$f(x, y) = 2x^2 - 4xy^2 \qquad f(-1, 2) = 18$$

$$f_x(x, y) = 4x - 4y^2 \quad \Rightarrow \quad f_x(-1, 2) = -20$$

$$f_y(x, y) = -8xy \qquad f_y(-1, 2) = 16$$

Substituting in (1) we obtain the following equation of the tangent plane:

$$z = 18 - 20(x + 1) + 16(y - 2) = -34 - 20x + 16y$$

That is,

$$z = -34 - 20x + 16y$$

In Exercises 3–10, find an equation of the tangent plane at the given point.

3. $f(x, y) = x^2 y + xy^3$, $(2, 1)$

SOLUTION The equation of the tangent plane at $(2, 1)$ is

$$z = f(2, 1) + f_x(2, 1)(x - 2) + f_y(2, 1)(y - 1) \tag{1}$$

We compute the values of f and its partial derivatives at $(2, 1)$:

$$f(x, y) = x^2 y + xy^3 \qquad f(2, 1) = 6$$

$$f_x(x, y) = 2xy + y^3 \quad \Rightarrow \quad f_x(2, 1) = 5$$

$$f_y(x, y) = x^2 + 3xy^2 \qquad f_y(2, 1) = 10$$

We now substitute these values in (1) to obtain the following equation of the tangent plane:

$$z = 6 + 5(x - 2) + 10(y - 1) = 5x + 10y - 14$$

That is,

$$z = 5x + 10y - 14.$$

5. $f(x, y) = x^2 + y^{-2}$,   (4, 1)

SOLUTION   The equation of the tangent plane at (4, 1) is

$$z = f(4, 1) + f_x(4, 1)(x - 4) + f_y(4, 1)(y - 1) \tag{1}$$

We compute the values of f and its partial derivatives at (4, 1):

$$f(x, y) = x^2 + y^{-2} \qquad f(4, 1) = 17$$
$$f_x(x, y) = 2x \qquad \Rightarrow \quad f_x(4, 1) = 8$$
$$f_y(x, y) = -2y^{-3} \qquad f_y(4, 1) = -2$$

Substituting in (1) we obtain the following equation of the tangent plane:

$$z = 17 + 8(x - 4) - 2(y - 1) = 8x - 2y - 13.$$

7. $F(r, s) = r^2 s^{-1/2} + s^{-3}$,   (2, 1)

SOLUTION   The equation of the tangent plane at (2, 1) is

$$z = f(2, 1) + f_r(2, 1)(r - 2) + f_s(2, 1)(s - 1) \tag{1}$$

We compute f and its partial derivatives at (2, 1):

$$f(r, s) = r^2 s^{-1/2} + s^{-3} \qquad f(2, 1) = 5$$
$$f_r(r, s) - 2rs^{-1/2} \qquad \Rightarrow \quad f_r(2, 1) = 4$$
$$f_s(r, s) = -\frac{1}{2} r^2 s^{-3/2} - 3s^{-4} \qquad f_s(2, 1) = -5$$

We substitute these values in (1) to obtain the following equation of the tangent plane:

$$z = 5 + 4(r - 2) - 5(s - 1) = 4r - 5s + 2.$$

9. $f(x, y) = \operatorname{sech}(x - y)$,   (ln 4, ln 2)

SOLUTION   The equation of the tangent plane at (ln 4, ln 2) is:

$$z = f(\ln 4, \ln 2) + f_x(\ln 4, \ln 2)(x - \ln 4) + f_y(\ln 4, \ln 2)(y - \ln 2)$$

We compute f and its partial derivatives at (ln 4, ln 2):

$$f(x, y) = \operatorname{sech}(x - y), \qquad f(\ln 4, \ln 2) = \operatorname{sech}(\ln 2) = \frac{4}{5}$$

$$f_x(x, y) = -\tanh(x - y)\operatorname{sech}(x - y), \qquad f_x(\ln 4, \ln 2) = -\tanh(\ln 2)\operatorname{sech}(\ln 2) = -\frac{12}{25}$$

$$f_y(x, y) = \tanh(x - y)\operatorname{sech}(x - y), \qquad f_y(\ln 4, \ln 2) = \tanh(\ln 2)\operatorname{sech}(\ln 2) = \frac{12}{25}$$

We substitute these values in the tangent plane equation to obtain:

$$z = \frac{4}{5} - \frac{12}{25}(x - \ln 4) + \frac{12}{25}(x - \ln 2) = -\frac{4}{25}(3x - 3y - 5 - \ln 8)$$

11. Find the points on the graph of $z = 3x^2 - 4y^2$ at which the vector $\mathbf{n} = \langle 3, 2, 2 \rangle$ is normal to the tangent plane.

SOLUTION   The equation of the tangent plane at the point $\left(a, b, f(a, b)\right)$ on the graph of $z = f(x, y)$ is

$$z = f(a, b) + f_x(a, b)(x - a) + f_y(a, b)(y - b)$$

or

$$f_x(a, b)(x - a) + f_y(a, b)(y - b) - z + f(a, b) = 0$$

Therefore, the following vector is normal to the plane:

$$\mathbf{v} = \left\langle f_x(a, b), f_y(a, b), -1 \right\rangle$$

We compute the partial derivatives of the function $f(x, y) = 3x^2 - 4y^2$:

$$f_x(x, y) = 6x \quad \Rightarrow \quad f_x(a, b) = 6a$$
$$f_y(x, y) = -8y \quad \Rightarrow \quad f_y(a, b) = -8b$$

Therefore, the vector $\mathbf{v} = \langle 6a, -8b, -1 \rangle$ is normal to the tangent plane at (a, b). Since we want $\mathbf{n} = \langle 3, 2, 2 \rangle$ to be normal to the plane, the vectors $\mathbf{v}$ and $\mathbf{n}$ must be parallel. That is, the following must hold:

$$\frac{6a}{3} = \frac{-8b}{2} = -\frac{1}{2}$$

which implies that $a = -\frac{1}{4}$ and $b = \frac{1}{8}$. We compute the z-coordinate of the point:

$$z = 3 \cdot \left(-\frac{1}{4}\right)^2 - 4\left(\frac{1}{8}\right)^2 = \frac{1}{8}$$

The point on the graph at which the vector $\mathbf{n} = \langle 3, 2, 2 \rangle$ is normal to the tangent plane is $\left(-\frac{1}{4}, \frac{1}{8}, \frac{1}{8}\right)$.

13. Find the linearization $L(x, y)$ of $f(x, y) = x^2 y^3$ at $(a, b) = (2, 1)$. Use it to estimate $f(2.01, 1.02)$ and $f(1.97, 1.01)$ and compare with values obtained using a calculator.

SOLUTION

(a) We compute the value of the function and its partial derivatives at $(a, b) = (2, 1)$:

$$f(x, y) = x^2 y^3 \qquad f(2, 1) = 4$$
$$f_x(x, y) = 2xy^3 \quad \Rightarrow \quad f_x(2, 1) = 4$$
$$f_y(x, y) = 3x^2 y^2 \qquad f_y(2, 1) = 12$$

The linear approximation is therefore

$$L(x, y) = f(2, 1) + f_x(2, 1)(x - 2) + f_y(2, 1)(y - 1)$$
$$L(x, y) = 4 + 4(x - 2) + 12(y - 1) = -16 + 4x + 12y$$

(b) For $h = x - 2$ and $k = y - 1$ we have the following form of the linear approximation at $(a, b) = (2, 1)$:

$$L(x, y) = f(2, 1) + f_x(2, 1)h + f_y(2, 1)k = 4 + 4h + 12k$$

To approximate $f(2.01, 1.02)$ we set $h = 2.01 - 2 = 0.01, k = 1.02 - 1 = 0.02$ to obtain

$$L(2.01, 1.02) = 4 + 4 \cdot 0.01 + 12 \cdot 0.02 = 4.28$$

The actual value is

$$f(2.01, 1.02) = 2.01^2 \cdot 1.02^3 = 4.2874$$

To approximate $f(1.97, 1.01)$ we set $h = 1.97 - 2 = -0.03, k = 1.01 - 1 = 0.01$ to obtain

$$L(1.97, 1.01) = 4 + 4 \cdot (-0.03) + 12 \cdot 0.01 = 4.$$

The actual value is

$$f(1.97, 1.01) = 1.97^2 \cdot 1.01^3 = 3.998.$$

15. Let $f(x, y) = x^3 y^{-4}$. Use Eq. (4) to estimate the change

$$\Delta f = f(2.03, 0.9) - f(2, 1)$$

SOLUTION We compute the function and its partial derivatives at $(a, b) = (2, 1)$:

$$f(x, y) = x^3 y^{-4} \qquad f(2, 1) = 8$$
$$f_x(x, y) = 3x^2 y^{-4} \quad \Rightarrow \quad f_x(2, 1) = 12$$
$$f_y(x, y) = -4x^3 y^{-5} \qquad f_y(2, 1) = -32$$

Also, $\Delta x = 2.03 - 2 = 0.03$ and $\Delta y = 0.9 - 1 = -0.1$. Therefore,

$$\Delta f = f(2.03, 0.9) - f(2, 1) \approx f_x(2, 1)\Delta x + f_y \Delta y = 12 \cdot 0.03 + (-32) \cdot (-0.1) = 3.56$$
$$\Delta f \approx 3.56$$

17. Use the linear approximation of $f(x, y) = e^{x^2+y}$ at $(0, 0)$ to estimate $f(0.01, -0.02)$. Compare with the value obtained using a calculator.

SOLUTION The linear approximation of f at the point $(0, 0)$ is

$$f(h, k) \approx f(0, 0) + f_x(0, 0)h + f_y(0, 0)k \tag{1}$$

We first must compute f and its partial derivative at the point $(0, 0)$. Using the Chain Rule we obtain

$$f(x, y) = e^{x^2+y} \qquad\qquad f(0, 0) = e^0 = 1$$

$$f_x(x, y) = 2xe^{x^2+y} \quad\Rightarrow\quad f_x(0, 0) = 2 \cdot 0 \cdot e^0 = 0$$

$$f_y(x, y) = e^{x^2+y} \qquad\qquad f_y(0, 0) = e^0 = 1$$

We substitute these values and $h = 0.01$, $k = -0.02$ in (1) to obtain

$$f(0.01, -0.02) \approx 1 + 0 \cdot 0.01 + 1 \cdot (-0.02) = 0.98$$

The actual value is $f(0.01, -0.02) = e^{0.01^2 - 0.02} \approx 0.9803$.

19. Find the linearization of $f(x, y, z) = z\sqrt{x+y}$ at $(8, 4, 5)$.

SOLUTION The linear approximation of f at the point $(8, 4, 5)$ is:

$$f(x, y, z) \approx f(8, 4, 5) + f_x(8, 4, 5)(x - 8) + f_y(8, 4, 5)(y - 4) + f_z(8, 4, 5)(z - 5)$$

We compute the values of f and its partial derivatives at $(8, 4, 5)$:

$$f(x, y, z) = z\sqrt{x+y}, \qquad\qquad f(8, 4, 5) = 5\sqrt{12} = 10\sqrt{3}$$

$$f_x(x, y, z) = \frac{z}{2\sqrt{x+y}}, \qquad\qquad f_x(8, 4, 5) = \frac{5}{2\sqrt{12}} = \frac{5}{4\sqrt{3}}$$

$$f_y(x, y, z) = \frac{z}{2\sqrt{x+y}}, \qquad\qquad f_y(8, 4, 5) = \frac{5}{2\sqrt{12}} = \frac{5}{4\sqrt{3}}$$

$$f_z(x, y, z) = \sqrt{x+y}, \qquad\qquad f_z(8, 4, 5) = \sqrt{12} = 1\sqrt{3}$$

Substituting these values we obtain the linearization:

$$f(x, y, z) \approx 10\sqrt{3} + \frac{5}{1\sqrt{3}}(x - 8) + \frac{5}{4\sqrt{3}}(y - 4) + 4\sqrt{3}(z - 5)$$

$$= \frac{5}{4\sqrt{3}}(x - 8) + \frac{5}{4\sqrt{3}}(y - 4) + 4\sqrt{3}z - 15\sqrt{3}$$

21. Estimate $f(2.1, 3.8)$ assuming that

$$f(2, 4) = 5, \qquad f_x(2, 4) = 0.3, \qquad f_y(2, 4) = -0.2$$

SOLUTION We use the linear approximation of f at the point $(2, 4)$, which is

$$f(2 + h, 4 + k) \approx f(2, 4) + f_x(2, 4)h + f_y(2, 4)k$$

Substituting the given values and $h = 0.1$, $k = -0.2$ we obtain the following approximation:

$$f(2.1, 3.8) \approx 5 + 0.3 \cdot 0.1 + 0.2 \cdot 0.2 = 5.07.$$

In Exercises 23–28, use the linear approximation to estimate the value. Compare with the value given by a calculator.

23. $(2.01)^3(1.02)^2$

SOLUTION The number $(2.01)^3(1.02)^2$ is a value of the function $f(x, y) = x^3 y^2$. We use the li(8, near approximation at $(2, 1)$, which is

$$f(2 + h, 1 + k) \approx f(2, 1) + f_x(2, 1)h + f_y(2, 1)k \tag{1}$$

We compute the value of the function and its partial derivatives at $(2, 1)$:

$$f(x, y) = x^3 y^2 \qquad\qquad f(2, 1) = 8$$

$$f_x(x, y) = 3x^2 y^2 \quad\Rightarrow\quad f_x(2, 1) = 12$$

$$f_y(x, y) = 2x^3 y \qquad\qquad f_y(2, 1) = 16$$

Substituting these values and $h = 0.01$, $k = 0.02$ in (1) gives the approximation

$$(2.01)^3(1.02)^2 \approx 8 + 12 \cdot 0.01 + 16 \cdot 0.02 = 8.44$$

The value given by a calculator is 8.4487. The error is 0.0087 and the percentage error is

$$\text{Percentage error} \approx \frac{0.0087 \cdot 100}{8.4487} = 0.103\%$$

25. $\sqrt{3.01^2 + 3.99^2}$

SOLUTION   This is a value of the function $f(x, y) = \sqrt{x^2 + y^2}$. We use the linear approximation at the point $(3, 4)$, which is

$$f(3 + h, 4 + k) \approx f(3, 4) + f_x(3, 4)h + f_y(3, 4)k \tag{1}$$

Using the Chain Rule gives the following partial derivatives:

$$f(x, y) = \sqrt{x^2 + y^2} \qquad\qquad f(3, 4) = 5$$

$$f_x(x, y) = \frac{2x}{2\sqrt{x^2 + y^2}} = \frac{x}{\sqrt{x^2 + y^2}} \quad \Rightarrow \quad f_x(3, 4) = \frac{3}{5}$$

$$f_y(x, y) = \frac{2y}{2\sqrt{x^2 + y^2}} = \frac{y}{\sqrt{x^2 + y^2}} \qquad f_y(3, 4) = \frac{4}{5}$$

Substituting these values and $h = 0.01$, $k = -0.01$ in (1) gives the following approximation:

$$\sqrt{3.01^2 + 3.99^2} \approx 5 + \frac{3}{5} \cdot 0.01 + \frac{4}{5} \cdot (-0.01) = 4.998$$

The value given by a calculator is $\sqrt{3.01^2 + 3.99^2} \approx 4.99802$. The error is 0.00002 and the percentage error is at most

$$\text{Percentage error} \approx \frac{0.00002 \cdot 100}{4.99802} = 0.0004002\%$$

27. $\sqrt{(1.9)(2.02)(4.05)}$

SOLUTION   We use the linear approximation of the function $f(x, y, z) = \sqrt{xyz}$ at the point $(2, 2, 4)$, which is

$$f(2 + h, 2 + k, 4 + l) \approx f(2, 2, 4) + f_x(2, 2, 4)h + f_y(2, 2, 4)k + f_z(2, 2, 4)l \tag{1}$$

We compute the values of the function and its partial derivatives at $(2, 2, 4)$:

$$f(x, y, z) = \sqrt{xyz} \qquad\qquad f(2, 2, 4) = 4$$

$$f_x(x, y, z) = \frac{yz}{2\sqrt{xyz}} = \frac{1}{2}\sqrt{\frac{yz}{x}} \quad \Rightarrow \quad f_x(2, 2, 4) = 1$$

$$f_y(x, y, z) = \frac{xz}{2\sqrt{xyz}} = \frac{1}{2}\sqrt{\frac{xz}{y}} \qquad f_y(2, 2, 4) = 1$$

$$f_z(x, y, z) = \frac{xy}{2\sqrt{xyz}} = \frac{1}{2}\sqrt{\frac{xy}{z}} \qquad f_z(2, 2, 4) = \frac{1}{2}$$

Substituting these values and $h = -0.1$, $k = 0.02$, $l = 0.05$ in (1) gives the following approximation:

$$\sqrt{(1.9)(2.02)(4.05)} = 4 + 1 \cdot (-0.1) + 1 \cdot 0.02 + \frac{1}{2}(0.05) = 3.945$$

The value given by a calculator is:

$$\sqrt{(1.9)(2.02)(4.05)} \approx 3.9426$$

29. Find an equation of the tangent plane to $z = f(x, y)$ at $P = (1, 2, 10)$ assuming that

$$f(1, 2) = 10, \qquad f(1.1, 2.01) = 10.3, \qquad f(1.04, 2.1) = 9.7$$

SOLUTION   The equation of the tangent plane at the point $(1, 2)$ is

$$z = f(1, 2) + f_x(1, 2)(x - 1) + f_y(1, 2)(y - 2)$$

$$z = 10 + f_x(1, 2)(x - 1) + f_y(1, 2)(y - 2) \tag{1}$$

Since the values of the partial derivatives at $(1, 2)$ are not given, we approximate them as follows:

$$f_x(1, 2) \approx \frac{f(1.1, 2) - f(1, 2)}{0.1} \approx \frac{f(1.1, 2.01) - f(1, 2)}{0.1} = 3$$

$$f_y(1, 2) \approx \frac{f(1, 2.1) - f(1, 2)}{0.1} \approx \frac{f(1.04, 2.1) - f(1, 2)}{0.1} = -3$$

Substituting in (1) gives the following approximation to the equation of the tangent plane:

$$z = 10 + 3(x - 1) - 3(y - 2)$$

That is, $z = 3x - 3y + 13$.

In Exercises 31–34, let $I = W/H^2$ denote the BMI described in Example 5.

31. A boy has weight $W = 34$ kg and height $H = 1.3$ m. Use the linear approximation to estimate the change in I if (W, H) changes to $(36, 1.32)$.

SOLUTION Let $\Delta I = I(36, 1.32) - I(34, 1.3)$ denote the change in I. Using the linear approximation of I at the point $(34, 1.3)$ we have

$$I(34 + h, 1.3 + k) - I(34, 1.3) \approx \frac{\partial I}{\partial W}(34, 1.3)h + \frac{\partial I}{\partial H}(34, 1.3)k$$

For $h = 2, k = 0.02$ we obtain

$$\Delta I \approx \frac{\partial I}{\partial W}(34, 1.3) \cdot 2 + \frac{\partial I}{\partial H}(34, 1.3) \cdot 0.02 \tag{1}$$

We compute the partial derivatives in (1):

$$\frac{\partial I}{\partial W} = \frac{\partial}{\partial W}\frac{W}{H^2} = \frac{1}{H^2} \qquad\qquad \Rightarrow \quad \frac{\partial I}{\partial W}(34, 1.3) = 0.5917$$

$$\frac{\partial I}{\partial H} = W\frac{\partial}{\partial H}H^{-2} = W \cdot (-2H^{-3}) = \frac{-2W}{H^3} \quad \Rightarrow \quad \frac{\partial I}{\partial H}(34, 1.3) = -30.9513$$

Substituting the partial derivatives in (1) gives the following estimation of ΔI:

$$\Delta I \approx 0.5917 \cdot 2 - 30.9513 \cdot 0.02 = 0.5644$$

33. (a) Show that $\Delta I \approx 0$ if $\Delta H / \Delta W \approx H/2W$.
(b) Suppose that $(W, H) = (25, 1.1)$. What increase in H will leave I (approximately) constant if W is increased by 1 kg?

SOLUTION

(a) The linear approximation implies that

$$\Delta I \approx \frac{\partial I}{\partial W}\Delta W + \frac{\partial I}{\partial H}\Delta H$$

Hence, $\Delta I \approx 0$ if

$$\frac{\partial I}{\partial W}\Delta W + \frac{\partial I}{\partial H}\Delta H = 0 \tag{1}$$

We compute the partial derivatives of $I = \dfrac{W}{H^2}$:

$$\frac{\partial I}{\partial W} = \frac{\partial}{\partial W}\left(\frac{W}{H^2}\right) = \frac{1}{H^2}$$

$$\frac{\partial I}{\partial H} = W\frac{\partial}{\partial H}(H^{-2}) = -2WH^{-3} = \frac{-2W}{H^3}$$

We substitute the partial derivatives in (1) to obtain

$$\frac{1}{H^2}\Delta W - \frac{2W}{H^3}\Delta H = 0$$

Hence,

$$\frac{1}{H^2}\Delta W = \frac{2W}{H^3}\Delta H$$

or

$$\frac{\Delta H}{\Delta W} = \frac{1}{H^2} \cdot \frac{H^3}{2W} = \frac{H}{2W}$$

(b) In part (a) we showed that if $\frac{\Delta H}{\Delta W} = \frac{H}{2W}$, then I remains approximately constant. We thus substitute $W = 25$, $H = 1.1$, $\Delta W = 1$, and solve for ΔH. This gives

$$\frac{\Delta H}{1} = \frac{1.1}{50} \quad \Rightarrow \quad \Delta H \approx 0.022 \text{ meters.}$$

That is, an increase of 0.022 meters in H will leave I approximately constant.

35. A cylinder of radius r and height h has volume $V = \pi r^2 h$.

(a) Use the linear approximation to show that

$$\frac{\Delta V}{V} \approx \frac{2\Delta r}{r} + \frac{\Delta h}{h}$$

(b) Estimate the percentage increase in V if r and h are each increased by 2%.

(c) The volume of a certain cylinder V is determined by measuring r and h. Which will lead to a greater error in V: a 1% error in r or a 1% error in h?

SOLUTION

(a) The linear approximation is

$$\Delta V \approx V_r \Delta r + V_h \Delta h \tag{1}$$

We compute the partial derivatives of $V = \pi r^2 h$:

$$V_r = \pi h \frac{\partial}{\partial r} r^2 = 2\pi h r$$

$$V_h = \pi r^2 \frac{\partial}{\partial h} h = \pi r^2$$

Substituting in (1) gives

$$\Delta V \approx 2\pi h r \Delta r + \pi r^2 \Delta h$$

We divide by $V = \pi r^2 h$ to obtain

$$\frac{\Delta V}{V} \approx \frac{2\pi h r \Delta r}{V} + \frac{\pi r^2 \Delta h}{V} = \frac{2\pi h r \Delta r}{\pi r^2 h} + \frac{\pi r^2 \Delta h}{\pi r^2 h} = \frac{2\Delta r}{r} + \frac{\Delta h}{h}$$

That is,

$$\frac{\Delta V}{V} \approx \frac{2\Delta r}{r} + \frac{\Delta h}{h}$$

(b) The percentage increase in V is, by part (a),

$$\frac{\Delta V}{V} \cdot 100 \approx 2\frac{\Delta r}{r} \cdot 100 + \frac{\Delta h}{h} \cdot 100$$

We are given that $\frac{\Delta r}{r} \cdot 100 = 2$ and $\frac{\Delta h}{h} \cdot 100 = 2$, hence the percentage increase in V is

$$\frac{\Delta V}{V} \cdot 100 = 2 \cdot 2 + 2 = 6\%$$

(c) The percentage error in V is

$$\frac{\Delta V}{V} \cdot 100 = 2\frac{\Delta r}{r} \cdot 100 + \frac{\Delta h}{h} \cdot 100$$

A 1% error in r implies that $\frac{\Delta r}{r} \cdot 100 = 1$. Assuming that there is no error in h, we get

$$\frac{\Delta V}{V} \cdot 100 = 2 \cdot 1 + 0 = 2\%$$

A 1% in h implies that $\frac{\Delta h}{h} \cdot 100 = 1$. Assuming that there is no error in r, we get

$$\frac{\Delta V}{V} \cdot 100 = 0 + 1 = 1\%$$

We conclude that a 1% error in r leads to a greater error in V than a 1% error in h.

37. The monthly payment for a home loan is given by a function $f(P, r, N)$, where P is the principal (initial size of the loan), r the interest rate, and N is the length of the loan in months. Interest rates are expressed as a decimal: A 6% interest rate is denoted by $r = 0.06$. If $P = \$100{,}000$, $r = 0.06$, and $N = 240$ (a 20-year loan), then the monthly payment is $f(100{,}000, 0.06, 240) = 716.43$. Furthermore, at these values, we have

$$\frac{\partial f}{\partial P} = 0.0071, \qquad \frac{\partial f}{\partial r} = 5769, \qquad \frac{\partial f}{\partial N} = -1.5467$$

Estimate:

(a) The change in monthly payment per $1000 increase in loan principal.

(b) The change in monthly payment if the interest rate increases to $r = 6.5\%$ and $r = 7\%$.

(c) The change in monthly payment if the length of the loan increases to 24 years.

SOLUTION

(a) The linear approximation to $f(P, r, N)$ is

$$\Delta f \approx \frac{\partial f}{\partial P}\Delta P + \frac{\partial f}{\partial r}\Delta r + \frac{\partial f}{\partial N}\Delta N$$

We are given that $\frac{\partial f}{\partial P} = 0.0071$, $\frac{\partial f}{\partial r} = 5769$, $\frac{\partial f}{\partial N} = -1.5467$, and $\Delta P = 1000$. Assuming that $\Delta r = 0$ and $\Delta N = 0$, we get

$$\Delta f \approx 0.0071 \cdot 1000 = 7.1$$

The change in monthly payment per thousand dollar increase in loan principal is $7.1.

(b) By the given data, we have

$$\Delta f \approx 0.0071\Delta P + 5769\Delta r - 1.5467\Delta N \tag{1}$$

The interest rate 6.5% corresponds to $r = 0.065$, and the interest rate 7% corresponds to $r = 0.07$. In the first case $\Delta r = 0.065 - 0.06 = 0.005$ and in the second case $\Delta r = 0.07 - 0.06 = 0.01$. Substituting in (1), assuming that $\Delta P = 0$ and $\Delta N = 0$, gives

$$\Delta f = 5769 \cdot 0.005 = \$28.845$$

$$\Delta f = 5769 \cdot 0.01 = \$57.69$$

(c) We substitute $\Delta N = (24 - 20) \cdot 12 = 48$ months and $\Delta r = \Delta N = 0$ in (1) to obtain

$$\Delta f \approx -1.5467 \cdot 48 = -74.2416$$

The monthly payment will be reduced by $74.2416.

39. The volume V of a right-circular cylinder is computed using the values 3.5 m for diameter and 6.2 m for height. Use the linear approximation to estimate the maximum error in V if each of these values has a possible error of at most 5%. Recall that $V = \frac{1}{3}\pi r^2 h$.

SOLUTION We denote by d and h the diameter and height of the cylinder, respectively. By the Formula for the Volume of a Cylinder we have

$$V = \pi\left(\frac{d}{2}\right)^2 h = \frac{\pi}{4}d^2 h$$

The linear approximation is

$$\Delta V \approx \frac{\partial V}{\partial d}\Delta d + \frac{\partial V}{\partial h}\Delta h \tag{1}$$

We compute the partial derivatives at $(d, h) = (3.5, 6.2)$:

$$\frac{\partial V}{\partial d}(d, h) = \frac{\pi}{4}h \cdot 2d = \frac{\pi}{2}hd \qquad\qquad \frac{\partial V}{\partial d}(3.5, 6.2) \approx 34.086$$

$$\frac{\partial V}{\partial h}(d, h) = \frac{\pi}{4}d^2 \qquad\Rightarrow\qquad \frac{\partial V}{\partial h}(3.5, 6.2) = 9.621$$

Substituting these derivatives in (1) gives

$$\Delta V \approx 34.086\Delta d + 9.621\Delta h \tag{2}$$

We are given that the errors in the measurements of d and h are at most 5%. Hence,

$$\frac{\Delta d}{3.5} = 0.05 \quad \Rightarrow \quad \Delta d = 0.175$$

$$\frac{\Delta h}{6.2} = 0.05 \quad \Rightarrow \quad \Delta h = 0.31$$

Substituting in (2) we obtain

$$\Delta V \approx 34.086 \cdot 0.175 + 9.621 \cdot 0.31 \approx 8.948$$

The error in V is approximately 8.948 meters. The percentage error is at most

$$\frac{\Delta V \cdot 100}{V} = \frac{8.948 \cdot 100}{\frac{\pi}{4} \cdot 3.5^2 \cdot 6.2} = 15\%$$

Further Insights and Challenges

41. This exercise shows directly (without using Theorem 1) that the function $f(x, y) = 5x + 4y^2$ from Example 1 is locally linear at $(a, b) = (2, 1)$.
(a) Show that $f(x, y) = L(x, y) + e(x, y)$ with $e(x, y) = 4(y - 1)^2$.
(b) Show that

$$0 \le \frac{e(x, y)}{\sqrt{(x - 2)^2 + (y - 1)^2}} \le 4|y - 1|$$

(c) Verify that $f(x, y)$ is locally linear.

SOLUTION   According to Example 1,

$$L(x, y) = -4 + 5x + 8y$$

(a) We compute the difference:

$$f(x, y) - L(x, y) = (5x + 4y^2) - (-4 + 5x + 8y)$$
$$= 4y^2 - 8y + 4 = 4(y - 1)^2$$

Therefore, $f(x, y) = L(x, y) + 4(y - 1)^2$.
(b) For $(x, y) \ne (2, 1)$, we consider

$$\frac{e(x, y)}{\sqrt{(x - 2)^2 + (y - 1)^2}} = \frac{4(y - 1)^2}{\sqrt{(x - 2)^2 + (y - 1)^2}}$$

The following inequality holds

$$\frac{4(y - 1)^2}{\sqrt{(x - 2)^2 + (y - 1)^2}} \le \frac{4(y - 1)^2}{\sqrt{(y - 1)^2}} = 4|y - 1|$$

because we have made the denominator smaller.
(c) We have

$$f(x, y) = L(x, y) + e(x, y)$$

where

$$0 \le \frac{e(x, y)}{\sqrt{(x - 2)^2 + (y - 1)^2}} \le 4|y - 1|$$

We have $\lim\limits_{(x,y) \to (2,1)} 4|y - 1| = 0$, and therefore

$$\lim\limits_{(x,y) \to (2,1)} e(x, y) = 0$$

by the Squeeze Theorem. This proves that $f(x, y)$ is locally linear at $(2, 1)$.

43. Differentiability Implies Continuity Use the definition of differentiability to prove that if f is differentiable at (a, b), then f is continuous at (a, b).

SOLUTION   Suppose that f is differentiable at (a, b), then we know f is locally linear at (a, b), that is

$$f(x, y) = L(x, y) + e(x, y)$$

where $e(x, y)$ satisfies

$$\lim_{(x,y)\to(a,b)} \frac{e(x, y)}{\sqrt{(x - a)^2 + (y - b)^2}} = \lim_{(x,y)\to(a,b)} E(x, y) = 0$$

and

$$L(x, y) = f(a, b) + f_x(a, b)(x - a) + f_y(a, b)(y - b)$$

We would like to show $\lim_{(x,y)\to(a,b)} f(x, y) = f(a, b)$, then f would be continuous at (a, b). Consider the following computation:

$$\lim_{(x,y)\to(a,b)} f(x, y) = \lim_{(x,y)\to(a,b)} L(x, y) + e(x, y)$$

$$= \lim_{(x,y)\to(a,b)} L(x, y) + E(x, y)\sqrt{(x - a)^2 + (y - b)^2}$$

$$= \lim_{(x,y)\to(a,b)} f(a, b) + f_x(a, b)(x - a) + f_y(a, b)(y - b) + E(x, y)\sqrt{(x - a)^2 + (y - b)^2}$$

$$= f(a, b) + 0 + 0 + 0 = f(a, b)$$

Therefore we have shown that if f is differentiable at (a, b) then f is continuous at (a, b).

45. Assumptions Matter Define $g(x, y) = 2xy(x + y)/(x^2 + y^2)$ for $(x, y) \neq 0$ and $g(0, 0) = 0$. In this exercise, we show that $g(x, y)$ is continuous at $(0, 0)$ and that $g_x(0, 0)$ and $g_y(0, 0)$ exist, but $g(x, y)$ is not differentiable at $(0, 0)$.

(a) Show using polar coordinates that $g(x, y)$ is continuous at $(0, 0)$.

(b) Use the limit definitions to show that $g_x(0, 0)$ and $g_y(0, 0)$ exist and that both are equal to zero.

(c) Show that the linearization of $g(x, y)$ at $(0, 0)$ is $L(x, y) = 0$.

(d) Show that if $g(x, y)$ were locally linear at $(0, 0)$, we would have $\lim_{h \to 0} \frac{g(h, h)}{h} = 0$. Then observe that this is not the case because $g(h, h) = 2h$. This shows that $g(x, y)$ is not locally linear at $(0, 0)$ and, hence, not differentiable at $(0, 0)$.

SOLUTION

(a) We would like to show $\lim_{(x,y)\to(0,0)} g(x, y) = g(0, 0)$. Consider the following, using polar coordinates, $x = r \cos\theta$ and $y = r \sin\theta$:

$$\lim_{(x,y)\to(0,0)} \frac{2xy(x + y)}{x^2 + y^2} = \lim_{(r,\theta)\to(0,0)} \frac{2r^2 \cos\theta \sin\theta(r \cos\theta + r \sin\theta)}{r^2 \cos^2\theta + r^2 \sin^2\theta}$$

$$= \lim_{(r,\theta)\to(0,0)} \frac{2r^3 \cos\theta \sin\theta(\cos\theta + \sin\theta)}{r^2}$$

$$= \lim_{(r,\theta)\to(0,0)} 2r \cos\theta \sin\theta(\cos\theta + \sin\theta) = 0 = g(0, 0)$$

Therefore $g(x, y)$ is continuous at $(0, 0)$.

(b) Taking partial derivatives we have:

$$g_x(x, y) = \frac{2y^2(y - x)^2}{(x^2 + y^2)^2}, \quad g_y(x, y) = \frac{2x^2(x - y)^2}{(x^2 + y^2)^2}$$

But we need to use limit definitions for the partial derivatives. Consider the following:

$$g_x(0, 0) = \lim_{h\to 0} \frac{g(h, 0) - g(0, 0)}{h}$$

$$= \lim_{h\to 0} \frac{1}{h}(0 - 0) = 0$$

$$g_y(0, 0) = \lim_{h\to 0} \frac{g(0, h) - g(0, 0)}{h}$$

$$= \lim_{h\to 0} \frac{1}{h}(0 - 0) = 0$$

Thus both partial derivatives exist and $g_x(0, 0) = 0$ and $g_y(0, 0) = 0$.

(c) We know that the linearization of g will be:

$$g(x, y) \approx g(0, 0) + g_x(0, 0)(x - 0) + g_y(0, 0)(y - 0)$$

We are given that $g(0, 0) = 0$. In part (b) we know $g_x(0, 0) = 0$ and $g_y(0, 0) = 0$. Substituting in these values in the linearization we have:

$$g(x, y) \approx 0 + 0 + 0 = 0$$

(d) We know if g were locally linear at $(0, 0)$, we would have:

$$\lim_{h \to 0} \frac{g(h, h)}{h} = 0$$

However, we know:

$$g(h, h) = \frac{2h^2(2h)}{2h^2} = 2h, \quad \frac{g(h, h)}{h} = \frac{2h}{h} = 2$$

This is a contradiction, $g(x, y)$ is not locally linear at $(0, 0)$ and hence, is not differentiable at $(0, 0)$.

14.5 The Gradient and Directional Derivatives   (LT Section 15.5)

Preliminary Questions

1. Which of the following is a possible value of the gradient ∇f of a function $f(x, y)$ of two variables?

(a) 5   **(b)** $\langle 3, 4 \rangle$ **(c)** $\langle 3, 4, 5 \rangle$

SOLUTION   The gradient of $f(x, y)$ is a vector with two components, hence the possible value of the gradient $\nabla f = \left\langle \frac{\partial f}{\partial x}, \frac{\partial f}{\partial y} \right\rangle$ is (b).

2. True or false? A differentiable function increases at the rate $\|\nabla f_P\|$ in the direction of ∇f_P.

SOLUTION   The statement is true. The value $\|\nabla f_P\|$ is the rate of increase of f in the direction ∇f_P.

3. Describe the two main geometric properties of the gradient ∇f.

SOLUTION   The gradient of f points in the direction of maximum rate of increase of f and is normal to the level curve (or surface) of f.

4. You are standing at a point where the temperature gradient vector is pointing in the northeast (NE) direction. In which direction(s) should you walk to avoid a change in temperature?

(a) NE   **(b)** NW   **(c)** SE   **(d)** SW

SOLUTION   The rate of change of the temperature T at a point P in the direction of a unit vector $\mathbf{u}$, is the directional derivative $D_{\mathbf{u}} T(P)$, which is given by the formula

$$D_{\mathbf{u}} T(P) = \|\nabla f_P\| \cos \theta$$

To avoid a change in temperature, we must choose the direction $\mathbf{u}$ so that $D_{\mathbf{u}} T(P) = 0$, that is, $\cos \theta = 0$, so $\theta = \frac{\pi}{2}$ or $\theta = \frac{3\pi}{2}$. Since the gradient at P is pointing NE, we should walk NW or SE to avoid a change in temperature. Thus, the answer is (b) and (c).

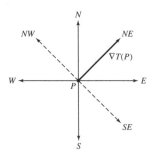

5. What is the rate of change of $f(x, y)$ at $(0, 0)$ in the direction making an angle of $45°$ with the x-axis if $\nabla f(0, 0) = \langle 2, 4 \rangle$?

SOLUTION   By the formula for directional derivatives, and using the unit vector $\langle 1/\sqrt{2}, 1/\sqrt{2} \rangle$, we get $\langle 2, 4 \rangle \cdot \langle 1/\sqrt{2}, 1/\sqrt{2} \rangle = 6/\sqrt{2} = 3\sqrt{2}$.

Exercises

1. Let $f(x, y) = xy^2$ and $\mathbf{c}(t) = \left(\frac{1}{2} t^2, t^3 \right)$.

(a) Calculate ∇f and $\mathbf{c}'(t)$.

(b) Use the Chain Rule for Paths to evaluate $\frac{d}{dt} f(\mathbf{c}(t))$ at $t = 1$ and $t = -1$.

SOLUTION

(a) We compute the partial derivatives of $f(x, y) = xy^2$:

$$\frac{\partial f}{\partial x} = y^2, \quad \frac{\partial f}{\partial y} = 2xy$$

The gradient vector is thus

$$\nabla f = \left\langle y^2, 2xy \right\rangle.$$

Also,

$$\mathbf{c}'(t) = \left\langle \left(\frac{1}{2}t^2\right)', \left(t^3\right)' \right\rangle = \left\langle t, 3t^2 \right\rangle$$

(b) Using the Chain Rule gives

$$\frac{d}{dt} f\left(\mathbf{c}(t)\right) = \frac{d}{dt}\left(\frac{1}{2}t^2 \cdot t^6\right) = \frac{d}{dt}\left(\frac{1}{2}t^8\right) = 4t^7$$

Substituting $x = \frac{1}{2}t^2$ and $y = t^3$, we obtain

$$\frac{d}{dt} f\left(\mathbf{c}(t)\right) = t^6 \cdot t + 2 \cdot \frac{1}{2}t^2 \cdot 3 \cdot t^3 \cdot t^2 = 4t^7$$

At the point $t = 1$ and $t = -1$, we get

$$\frac{d}{dt}\left(f\left(\mathbf{c}(t)\right)\right)\Big|_{t=1} = 4 \cdot 1^7 = 4, \quad \frac{d}{dt}\left(f\left(\mathbf{c}(t)\right)\right)\Big|_{t=-1} = 4 \cdot (-1)^7 = -4.$$

3. Figure 14 shows the level curves of a function $f(x, y)$ and a path $\mathbf{c}(t)$, traversed in the direction indicated. State whether the derivative $\frac{d}{dt} f(\mathbf{c}(t))$ is positive, negative, or zero at points A–D.

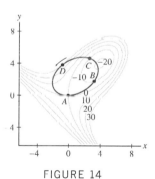

FIGURE 14

SOLUTION At points A and D, the path is (temporarily) tangent to one of the contour lines, which means that along the path $\mathbf{c}(t)$ the function $f(x, y)$ is (temporarily) constant, and so the derivative $\frac{d}{dt} f(\mathbf{c}(t))$ is zero. At point B, the path is moving from a higher contour (of -10) to a lower one (of -20), so the derivative is negative. At the point C, where the path moves from the contour of -10 towards the contour of value 0, the derivative is positive.

In Exercises 5–8, calculate the gradient.

5. $f(x, y) = \cos(x^2 + y)$

SOLUTION We find the partial derivatives using the Chain Rule:

$$\frac{\partial f}{\partial x} = -\sin\left(x^2 + y\right)\frac{\partial}{\partial x}\left(x^2 + y\right) = -2x\sin\left(x^2 + y\right)$$

$$\frac{\partial f}{\partial y} = -\sin\left(x^2 + y\right)\frac{\partial}{\partial y}\left(x^2 + y\right) = -\sin\left(x^2 + y\right)$$

The gradient vector is thus

$$\nabla f = \left\langle \frac{\partial f}{\partial x}, \frac{\partial f}{\partial y} \right\rangle = \left\langle -2x\sin\left(x^2 + y\right), -\sin\left(x^2 + y\right) \right\rangle = -\sin\left(x^2 + y\right)\langle 2x, 1\rangle$$

7. $h(x, y, z) = xyz^{-3}$

SOLUTION We compute the partial derivatives of $h(x, y, z) = xyz^{-3}$, obtaining

$$\frac{\partial h}{\partial x} = yz^{-3}, \quad \frac{\partial h}{\partial y} = xz^{-3}, \quad \frac{\partial h}{\partial z} = xy \cdot \left(-3z^{-4}\right) = -3xyz^{-4}$$

The gradient vector is thus

$$\nabla h = \left\langle \frac{\partial h}{\partial x}, \frac{\partial h}{\partial y}, \frac{\partial h}{\partial z} \right\rangle = \left\langle yz^{-3}, xz^{-3}, -3xyz^{-4} \right\rangle.$$

In Exercises 9–20, use the Chain Rule to calculate $\dfrac{d}{dt} f(\mathbf{c}(t))$.

9. $f(x, y) = 3x - 7y, \quad \mathbf{c}(t) = (\cos t, \sin t), \quad t = 0$

SOLUTION By the Chain Rule for paths, we have

$$\frac{d}{dt} f(\mathbf{c}(t)) = \nabla f_{\mathbf{c}(t)} \cdot \mathbf{c}'(t) \tag{1}$$

We compute the gradient and the derivative $\mathbf{c}'(t)$:

$$\nabla f = \left\langle \frac{\partial f}{\partial x}, \frac{\partial f}{\partial y} \right\rangle = \langle 3, -7 \rangle, \quad \mathbf{c}'(t) = \langle -\sin t, \cos t \rangle$$

We determine these vectors at $t = 0$:

$$\mathbf{c}'(0) = \langle -\sin 0, \cos 0 \rangle = \langle 0, 1 \rangle$$

and since the gradient is a constant vector, we have

$$\nabla f_{\mathbf{c}(0)} = \nabla f_{(1,0)} = \langle 3, -7 \rangle$$

Substituting these vectors in (1) gives

$$\frac{d}{dt} f(\mathbf{c}(t)) \bigg|_{t=0} = \langle 3, -7 \rangle \cdot \langle 0, 1 \rangle = 0 - 7 = -7$$

11. $f(x, y) = x^2 - 3xy, \quad \mathbf{c}(t) = (\cos t, \sin t), \quad t = 0$

SOLUTION By the Chain Rule For Paths we have

$$\frac{d}{dt} f(\mathbf{c}(t)) = \nabla f_{\mathbf{c}(t)} \cdot \mathbf{c}'(t) \tag{1}$$

We compute the gradient and $\mathbf{c}'(t)$:

$$\nabla f = \left\langle \frac{\partial f}{\partial x}, \frac{\partial f}{\partial y} \right\rangle = \langle 2x - 3y, -3x \rangle$$

$$\mathbf{c}'(t) = \langle -\sin t, \cos t \rangle$$

At the point $t = 0$ we have

$$\mathbf{c}(0) = (\cos 0, \sin 0) = (1, 0)$$

$$\mathbf{c}'(0) = \langle -\sin 0, \cos 0 \rangle = \langle 0, 1 \rangle$$

$$\nabla f \bigg|_{\mathbf{c}(0)} = \nabla f_{(1,0)} = \langle 2 \cdot 1 - 3 \cdot 0, -3 \cdot 1 \rangle = \langle 2, -3 \rangle$$

Substituting in (1) we obtain

$$\frac{d}{dt} f(\mathbf{c}(t)) \bigg|_{t=0} = \langle 2, -3 \rangle \cdot \langle 0, 1 \rangle = -3$$

13. $f(x, y) = \sin(xy)$, $\mathbf{c}(t) = (e^{2t}, e^{3t})$, $t = 0$

SOLUTION By the Chain Rule for Paths we have

$$\frac{d}{dt} f(\mathbf{c}(t)) = \nabla f_{\mathbf{c}(t)} \cdot \mathbf{c}'(t) \tag{1}$$

We compute the gradient and $\mathbf{c}'(t)$:

$$\nabla f = \left\langle \frac{\partial f}{\partial x}, \frac{\partial f}{\partial y} \right\rangle = \langle y \cos(xy), x \cos(xy) \rangle$$

$$\mathbf{c}'(t) = \left\langle 2e^{2t}, 3e^{3t} \right\rangle$$

At the point $t = 0$ we have

$$\mathbf{c}(0) = \left(e^0, e^0 \right) = (1, 1)$$

$$\mathbf{c}'(0) = \left\langle 2e^0, 3e^0 \right\rangle = \langle 2, 3 \rangle$$

$$\nabla f_{\mathbf{c}(0)} = \nabla f_{(1,1)} = \langle \cos 1, \cos 1 \rangle$$

Substituting the vectors in (1) we get

$$\frac{d}{dt} f(\mathbf{c}(t)) \bigg|_{t=0} = \langle \cos 1, \cos 1 \rangle \cdot \langle 2, 3 \rangle = 5 \cos 1$$

15. $f(x, y) = x - xy$, $\mathbf{c}(t) = (t^2, t^2 - 4t)$, $t = 4$

SOLUTION We compute the gradient and $\mathbf{c}'(t)$:

$$\nabla f = \left\langle \frac{\partial f}{\partial x}, \frac{\partial f}{\partial y} \right\rangle = \langle 1 - y, -x \rangle$$

$$\mathbf{c}'(t) = (2t, 2t - 4)$$

At the point $t = 4$ we have

$$\mathbf{c}(4) = \left(4^2, 4^2 - 4 \cdot 4 \right) = (16, 0)$$

$$\mathbf{c}'(4) = \langle 2 \cdot 4, 2 \cdot 4 - 4 \rangle = \langle 8, 4 \rangle$$

$$\nabla f_{\mathbf{c}(4)} = \nabla f_{(16,0)} = \langle 1 - 0, -16 \rangle = \langle 1, -16 \rangle$$

We now use the Chain Rule for Paths to compute the following derivative.

$$\frac{d}{dt} f(\mathbf{c}(t)) \bigg|_{t=4} = \nabla f_{\mathbf{c}(4)} \cdot \mathbf{c}'(4) = \langle 1, -16 \rangle \cdot \langle 8, 4 \rangle = 8 - 64 = -56$$

17. $f(x, y) = \ln x + \ln y$, $\mathbf{c}(t) = (\cos t, t^2)$, $t = \frac{\pi}{4}$

SOLUTION We compute the gradient and $\mathbf{c}'(t)$:

$$\nabla f = \left\langle \frac{\partial f}{\partial x}, \frac{\partial f}{\partial y} \right\rangle = \left\langle \frac{1}{x}, \frac{1}{y} \right\rangle$$

$$\mathbf{c}'(t) = \langle -\sin t, 2t \rangle$$

At the point $t = \frac{\pi}{4}$ we have

$$\mathbf{c}\left(\frac{\pi}{4}\right) = \left(\cos \frac{\pi}{4}, \left(\frac{\pi}{4}\right)^2 \right) = \left(\frac{\sqrt{2}}{2}, \frac{\pi^2}{16} \right)$$

$$\mathbf{c}'\left(\frac{\pi}{4}\right) = \left\langle -\sin \frac{\pi}{4}, \frac{2\pi}{4} \right\rangle = \left\langle -\frac{\sqrt{2}}{2}, \frac{\pi}{2} \right\rangle$$

$$\nabla f_{\mathbf{c}\left(\frac{\pi}{4}\right)} = \nabla f_{\left(\frac{\sqrt{2}}{2}, \frac{\pi^2}{16}\right)} = \left\langle \sqrt{2}, \frac{16}{\pi^2} \right\rangle$$

Using the Chain Rule for Paths we obtain the following derivative:

$$\frac{d}{dt} f\left(\mathbf{c}(t)\right)\Big|_{t=\frac{\pi}{4}} = \nabla f_{\mathbf{c}\left(\frac{\pi}{4}\right)} \cdot \mathbf{c}'\left(\frac{\pi}{4}\right) = \left\langle \sqrt{2}, \frac{16}{\pi^2} \right\rangle \cdot \left\langle -\frac{\sqrt{2}}{2}, \frac{\pi}{2} \right\rangle = -1 + \frac{8}{\pi} \approx 1.546$$

19. $g(x, y, z) = xyz^{-1}$, $\mathbf{c}(t) = (e^t, t, t^2)$, $t = 1$

SOLUTION By the Chain Rule for Paths we have

$$\frac{d}{dt} g\left(\mathbf{c}(t)\right) = \nabla g_{\mathbf{c}(t)} \cdot \mathbf{c}'(t) \tag{1}$$

We compute the gradient and $\mathbf{c}'(t)$:

$$\nabla g = \left\langle \frac{\partial g}{\partial x}, \frac{\partial g}{\partial y}, \frac{\partial g}{\partial z} \right\rangle = \left\langle yz^{-1}, xz^{-1}, -xyz^{-2} \right\rangle$$

$$\mathbf{c}'(t) = \left\langle e^t, 1, 2t \right\rangle$$

At the point $t = 1$ we have

$$\mathbf{c}(1) = (e, 1, 1)$$

$$\mathbf{c}'(1) = \langle e, 1, 2 \rangle$$

$$\nabla g_{\mathbf{c}(1)} = \nabla g_{(e,1,1)} = \langle 1, e, -e \rangle$$

Substituting the vectors in (1) gives the following derivative:

$$\frac{d}{dt} g\left(\mathbf{c}(t)\right)\Big|_{t=1} = \langle 1, e, -e \rangle \cdot \langle e, 1, 2 \rangle = e + e - 2e = 0$$

In Exercises 21–30, calculate the directional derivative in the direction of **v** *at the given point. Remember to normalize the direction vector or use Eq. (4).*

21. $f(x, y) = x^2 + y^3$, $\mathbf{v} = \langle 4, 3 \rangle$, $P = (1, 2)$

SOLUTION We first normalize the direction vector **v**:

$$\mathbf{u} = \frac{\mathbf{v}}{\|\mathbf{v}\|} = \frac{\langle 4, 3 \rangle}{\sqrt{4^2 + 3^2}} = \left\langle \frac{4}{5}, \frac{3}{5} \right\rangle$$

We compute the gradient of $f(x, y) = x^2 + y^3$ at the given point:

$$\nabla f = \left\langle \frac{\partial f}{\partial x}, \frac{\partial f}{\partial y} \right\rangle = \left\langle 2x, 3y^2 \right\rangle \quad \Rightarrow \quad \nabla f_{(1,2)} = \langle 2, 12 \rangle$$

Using the Theorem on Evaluating Directional Derivatives, we get

$$D_{\mathbf{u}} f(1, 2) = \nabla f_{(1,2)} \cdot \mathbf{u} = \langle 2, 12 \rangle \cdot \left\langle \frac{4}{5}, \frac{3}{5} \right\rangle = \frac{8}{5} + \frac{36}{5} = \frac{44}{5} = 8.8$$

23. $f(x, y) = x^2 y^3$, $\mathbf{v} = \mathbf{i} + \mathbf{j}$, $P = \left(\frac{1}{6}, 3\right)$

SOLUTION We normalize **v** to obtain a unit vector **u** in the direction of **v**:

$$\mathbf{u} = \frac{\mathbf{v}}{\|\mathbf{v}\|} = \frac{1}{\sqrt{2}} (\mathbf{i} + \mathbf{j}) = \frac{1}{\sqrt{2}} \mathbf{i} + \frac{1}{\sqrt{2}} \mathbf{j}$$

We compute the gradient of $f(x, y) = x^2 y^3$ at the point P:

$$\nabla f = \left\langle \frac{\partial f}{\partial x}, \frac{\partial f}{\partial y} \right\rangle = \left\langle 2xy^3, 3x^2 y^2 \right\rangle \quad \Rightarrow \quad \nabla f_{\left(\frac{1}{6},3\right)} = \left\langle 2 \cdot \frac{1}{6} \cdot 3^3, 3 \cdot \frac{1}{6^2} \cdot 3^2 \right\rangle = \left\langle 9, \frac{3}{4} \right\rangle = 9\mathbf{i} + \frac{3}{4}\mathbf{j}$$

The directional derivative in the direction **v** is thus

$$D_{\mathbf{u}} f\left(\frac{1}{6}, 3\right) = \nabla f_{\left(\frac{1}{6},3\right)} \cdot \mathbf{u} = \left(9\mathbf{i} + \frac{3}{4}\mathbf{j} \right) \cdot \left(\frac{1}{\sqrt{2}}\mathbf{i} + \frac{1}{\sqrt{2}}\mathbf{j} \right) = \frac{9}{\sqrt{2}} + \frac{3}{4\sqrt{2}} = \frac{39}{4\sqrt{2}}$$

25. $f(x, y) = \tan^{-1}(xy)$, $\quad \mathbf{v} = \langle 1, 1 \rangle$, $\quad P = (3, 4)$

SOLUTION We first normalize $\mathbf{v}$ to obtain a unit vector $\mathbf{u}$ in the direction $\mathbf{v}$:

$$\mathbf{u} = \frac{\mathbf{v}}{\|\mathbf{v}\|} = \frac{1}{\sqrt{2}} \langle 1, 1 \rangle$$

We compute the gradient of $f(x, y) = \tan^{-1}(xy)$ at the point $P = (3, 4)$:

$$\nabla f = \left\langle \frac{\partial f}{\partial x}, \frac{\partial f}{\partial y} \right\rangle = \left\langle \frac{y}{1 + (xy)^2}, \frac{x}{1 + (xy)^2} \right\rangle = \frac{1}{1 + x^2 y^2} \langle y, x \rangle$$

$$\nabla f_{(3,4)} = \frac{1}{1 + 3^2 \cdot 4^2} \langle 4, 3 \rangle = \frac{1}{145} \langle 4, 3 \rangle$$

Therefore, the directional derivative in the direction $\mathbf{v}$ is

$$D_{\mathbf{u}} f(3, 4) = \nabla f_{(3,4)} \cdot \mathbf{u} = \frac{1}{145} \langle 4, 3 \rangle \cdot \frac{1}{\sqrt{2}} \langle 1, 1 \rangle = \frac{1}{145\sqrt{2}} (4 + 3) = \frac{7}{145\sqrt{2}} = \frac{7\sqrt{2}}{290}$$

27. $f(x, y) = \ln(x^2 + y^2)$, $\quad \mathbf{v} = 3\mathbf{i} - 2\mathbf{j}$, $\quad P = (1, 0)$

SOLUTION We normalize $\mathbf{v}$ to obtain a unit vector $\mathbf{u}$ in the direction $\mathbf{v}$:

$$\mathbf{u} = \frac{\mathbf{v}}{\|\mathbf{v}\|} = \frac{1}{\sqrt{3^2 + (-2)^2}} (3\mathbf{i} - 2\mathbf{j}) = \frac{1}{\sqrt{13}} (3\mathbf{i} - 2\mathbf{j})$$

We compute the gradient of $f(x, y) = \ln\left(x^2 + y^2\right)$ at the point $P = (1, 0)$:

$$\nabla f = \left\langle \frac{\partial f}{\partial x}, \frac{\partial f}{\partial y} \right\rangle = \left\langle \frac{2x}{x^2 + y^2}, \frac{2y}{x^2 + y^2} \right\rangle = \frac{2}{x^2 + y^2} \langle x, y \rangle$$

$$\nabla f_{(1,0)} = \frac{2}{1^2 + 0^2} \langle 1, 0 \rangle = \langle 2, 0 \rangle = 2\mathbf{i}$$

The directional derivative in the direction $\mathbf{v}$ is thus

$$D_{\mathbf{u}} f(1, 0) = \nabla f_{(1,0)} \cdot \mathbf{u} = 2\mathbf{i} \cdot \frac{1}{\sqrt{13}} (3\mathbf{i} - 2\mathbf{j}) - \frac{6}{\sqrt{13}}$$

29. $g(x, y, z) = xe^{-yz}$, $\quad \mathbf{v} = \langle 1, 1, 1 \rangle$, $\quad P = (1, 2, 0)$

SOLUTION We first compute a unit vector $\mathbf{u}$ in the direction $\mathbf{v}$:

$$\mathbf{u} = \frac{\mathbf{v}}{\|\mathbf{v}\|} = \frac{\langle 1, 1, 1 \rangle}{\sqrt{1^2 + 1^2 + 1^2}} = \frac{1}{\sqrt{3}} \langle 1, 1, 1 \rangle$$

We find the gradient of $f(x, y, z) = xe^{-yz}$ at the point $P = (1, 2, 0)$:

$$\nabla f = \left\langle \frac{\partial f}{\partial x}, \frac{\partial f}{\partial y}, \frac{\partial f}{\partial z} \right\rangle = \left\langle e^{-yz}, -xze^{-yz}, -xye^{-yz} \right\rangle = e^{-yz} \langle 1, -xz, -xy \rangle$$

$$\nabla f_{(1,2,0)} = e^0 \langle 1, 0, -2 \rangle = \langle 1, 0, -2 \rangle$$

The directional derivative in the direction $\mathbf{v}$ is thus

$$D_{\mathbf{u}} f(1, 2, 0) - \nabla f_{(1,2,0)} \cdot \mathbf{u} = \langle 1, 0, -2 \rangle \cdot \frac{1}{\sqrt{3}} \langle 1, 1, 1 \rangle = \frac{1}{\sqrt{3}} (1 + 0 - 2) = -\frac{1}{\sqrt{3}}$$

31. Find the directional derivative of $f(x, y) = x^2 + 4y^2$ at $P = (3, 2)$ in the direction pointing to the origin.

SOLUTION The direction vector is $\mathbf{v} = \overrightarrow{PO} = \langle -3, -2 \rangle$. A unit vector $\mathbf{u}$ in the direction $\mathbf{v}$ is obtained by normalizing $\mathbf{v}$. That is,

$$\mathbf{u} = \frac{\mathbf{v}}{\|\mathbf{v}\|} = \frac{\langle -3, -2 \rangle}{\sqrt{3^2 + 2^2}} = \frac{-1}{\sqrt{13}} \langle 3, 2 \rangle$$

We compute the gradient of $f(x, y) = x^2 + 4y^2$ at the point $P = (3, 2)$:

$$\nabla f = \left\langle \frac{\partial f}{\partial x}, \frac{\partial f}{\partial y} \right\rangle = \langle 2x, 8y \rangle \quad \Rightarrow \quad \nabla f_{(3,2)} = \langle 6, 16 \rangle$$

The directional derivative is thus

$$D_{\mathbf{u}} f(3, 2) = \nabla f_{(3,2)} \cdot \mathbf{u} = \langle 6, 16 \rangle \cdot \frac{-1}{\sqrt{13}} \langle 3, 2 \rangle = \frac{-50}{\sqrt{13}}$$

33. A bug located at $(3, 9, 4)$ begins walking in a straight line toward $(5, 7, 3)$. At what rate is the bug's temperature changing if the temperature is $T(x, y, z) = xe^{y-z}$? Units are in meters and degrees Celsius.

SOLUTION The bug is walking in a straight line from the point $P = (3, 9, 4)$ towards $Q = (5, 7, 3)$, hence the rate of change in the temperature is the directional derivative in the direction of $\mathbf{v} = \overrightarrow{PQ}$. We first normalize $\mathbf{v}$ to obtain

$$\mathbf{v} = \overrightarrow{PQ} = \langle 5 - 3, 7 - 9, 3 - 4 \rangle = \langle 2, -2, -1 \rangle$$

$$\mathbf{u} = \frac{\mathbf{v}}{\|\mathbf{v}\|} = \frac{\langle 2, -2, -1 \rangle}{\sqrt{4 + 4 + 1}} = \frac{1}{3} \langle 2, -2, -1 \rangle$$

We compute the gradient of $T(x, y, z) = xe^{y-z}$ at $P = (3, 9, 4)$:

$$\nabla T = \left\langle \frac{\partial T}{\partial x}, \frac{\partial T}{\partial y}, \frac{\partial T}{\partial z} \right\rangle = \left\langle e^{y-z}, xe^{y-z}, -xe^{y-z} \right\rangle = e^{y-z} \langle 1, x, -x \rangle$$

$$\nabla T_{(3,9,4)} = e^{9-4} \langle 1, 3, -3 \rangle = e^5 \langle 1, 3, -3 \rangle$$

The rate of change of the bug's temperature at the starting point P is the directional derivative

$$D_{\mathbf{u}} f(P) = \nabla T_{(3,9,4)} \cdot \mathbf{u} = e^5 \langle 1, 3, -3 \rangle \cdot \frac{1}{3} \langle 2, -2, -1 \rangle = -\frac{e^5}{3} \approx -49.47$$

The answer is -49.47 degrees Celsius per meter.

35. Suppose that $\nabla f_P = \langle 2, -4, 4 \rangle$. Is f increasing or decreasing at P in the direction $\mathbf{v} = \langle 2, 1, 3 \rangle$?

SOLUTION We compute the derivative of f at P with respect to $\mathbf{v}$:

$$D_{\mathbf{v}} f(P) = \nabla f_P \cdot \mathbf{v} = \langle 2, -4, 4 \rangle \cdot \langle 2, 1, 3 \rangle = 4 - 4 + 12 = 12 > 0$$

Since the derivative is positive, f is increasing at P in the direction of $\mathbf{v}$.

37. Let $f(x, y, z) = \sin(xy + z)$ and $P = (0, -1, \pi)$. Calculate $D_{\mathbf{u}} f(P)$, where $\mathbf{u}$ is a unit vector making an angle $\theta = 30°$ with ∇f_P.

SOLUTION The directional derivative $D_{\mathbf{u}} f(P)$ is the following dot product:

$$D_{\mathbf{u}} f(P) = \nabla f_P \cdot \mathbf{u}$$

Since $\mathbf{u}$ is a unit vector making an angle $\theta = 30°$ with ∇f_P, we have by the properties of the dot product

$$D_{\mathbf{u}} f(P) = \|\nabla f_P\| \cdot \|\mathbf{u}\| \cos 30° = \frac{\sqrt{3}}{2} \|\nabla f_P\| \tag{1}$$

We now must find the gradient at P and its length:

$$\nabla f = \left\langle \frac{\partial f}{\partial x}, \frac{\partial f}{\partial y}, \frac{\partial f}{\partial z} \right\rangle = \langle y \cos(xy + z), x \cos(xy + z), \cos(xy + z) \rangle = \cos(xy + z) \langle y, x, 1 \rangle$$

$$\nabla f_{(0,-1,\pi)} = \cos \pi \langle -1, 0, 1 \rangle = -1 \langle -1, 0, 1 \rangle = \langle 1, 0, -1 \rangle$$

Hence,

$$\|\nabla f_{(0,-1,\pi)}\| = \sqrt{1^2 + 0^2 + (-1)^2} = \sqrt{2}$$

Substituting in (1) we get

$$D_{\mathbf{u}} f(P) = \frac{\sqrt{3}}{2} \sqrt{2} = \frac{\sqrt{6}}{2}.$$

39. Find a vector normal to the surface $x^2 + y^2 - z^2 = 6$ at $P = (3, 1, 2)$.

SOLUTION The gradient ∇f_P is normal to the level curve $f(x, y, z) = x^2 + y^2 - z^2 = 6$ at P. We compute this vector:

$$f_x(x, y, z) = 2x$$
$$f_y(x, y, z) = 2y \quad \Rightarrow \quad \nabla f_P = \nabla f_{(3,1,2)} = \langle 6, 2, -4 \rangle$$
$$f_z(x, y, z) = -2z$$

The vector $\langle 6, 2, -4 \rangle$ is normal to the surface $x^2 + y^2 - z^2 = 6$ at P.

41. Find the two points on the ellipsoid

$$\frac{x^2}{4} + \frac{y^2}{9} + z^2 = 1$$

where the tangent plane is normal to $\mathbf{v} = \langle 1, 1, -2 \rangle$.

SOLUTION The gradient ∇f_P is normal to the level surface $f(x, y, z) = \frac{x^2}{4} + \frac{y^2}{9} + z^2 = 1$. If $\mathbf{v} = \langle 1, 1, -2 \rangle$ is also normal, then ∇f_P and $\mathbf{v}$ are parallel, that is, $\nabla f_P = k\mathbf{v}$ for some constant k. This yields the equation

$$\nabla f_P = \langle \frac{x}{2}, \frac{2y}{9}, 2z \rangle = k \langle 1, 1, -2 \rangle$$

Thus $x = 2k$, $y = 9k/2$, and $z = -k$. To determine k, substitute in the equation of the ellipsoid:

$$\frac{x^2}{4} + \frac{y^2}{9} + z^2 = \frac{(2k)^2}{4} + \frac{(9k/2)^2}{9} + (-k)^2 = 1$$

This yields $k^2 + \frac{9}{4}k^2 + k^2 = 1$ or $k = \pm 2/\sqrt{17}$. The two points are

$$(x, y, z) = (2k, \frac{9}{2}k, -k) = \pm \left(\frac{4}{\sqrt{17}}, \frac{9}{\sqrt{17}}, -\frac{2}{\sqrt{17}} \right)$$

In Exercises 42–45, find an equation of the tangent plane to the surface at the given point.

43. $xz + 2x^2y + y^2z^3 = 11$, $P = (2, 1, 1)$

SOLUTION The equation of the tangent plane at P is

$$\nabla f_P \cdot \langle x - 2, y - 1, z - 1 \rangle = 0 \tag{1}$$

We compute the gradient of $f(x, y, z) = xz + 2x^2y + y^2z^3$ at the point $P = (2, 1, 1)$:

$$\nabla f = \left\langle \frac{\partial f}{\partial x}, \frac{\partial f}{\partial y}, \frac{\partial f}{\partial z} \right\rangle = \langle z + 4xy, 2x^2 + 2yz^3, x + 3y^2z^2 \rangle$$

At the point P we have

$$\nabla f_P = \langle 9, 10, 5 \rangle$$

Substituting in (1) we obtain the following equation of the tangent plane:

$$\langle 9, 10, 5 \rangle \cdot \langle x - 2, y - 1, z - 1 \rangle = 0$$
$$9(x - 2) + 10(y - 1) + 5(z - 1) = 0$$

or

$$9x + 10y + 5z = 33$$

45. $\ln[1 + 4x^2 + 9y^4] - 0.1z^2 = 0$, $P = (3, 1, 6.1876)$

SOLUTION The equation of the tangent plane at P is

$$\nabla f_P \cdot (x - 3, y - 1, z - 6.1876) = 0 \tag{1}$$

We compute the gradient of $f(x, y, z) = \ln(1 + 4x^2 + 9y^4) - 0.1z^2$ at the point P:

$$\nabla f = \left\langle \frac{\partial f}{\partial x}, \frac{\partial f}{\partial y}, \frac{\partial f}{\partial z} \right\rangle = \left\langle \frac{8x}{1 + 4x^2 + 9y^4}, \frac{36y^3}{1 + 4x^2 + 9y^4}, -0.2z \right\rangle$$

At the point $P = (3, 1, 6.1876)$ we have

$$\nabla f_P = \left\langle \frac{24}{1 + 36 + 9}, \frac{36}{46}, -1.2375 \right\rangle = \langle 0.5217, 0.7826, -1.2375 \rangle$$

We substitute in (1) to obtain the following equation of the tangent plane:

$$0.5217(x - 3) + 0.7826(y - 1) - 1.2375(z - 6.1876) = 0$$

or

$$0.5217x + 0.7826y - 1.2375z = -5.309$$

47. $\mathcal{CAS}$ Use a computer algebra system to produce a contour plot of $f(x, y) = x^2 - 3xy + y - y^2$ together with its gradient vector field on the domain $[-4, 4] \times [-4, 4]$.

SOLUTION

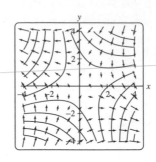

49. Find a function $f(x, y, z)$ such that $\nabla f = \langle 2x, 1, 2 \rangle$.

SOLUTION   The following equality must hold:

$$\nabla f = \left\langle \frac{\partial f}{\partial x}, \frac{\partial f}{\partial y}, \frac{\partial f}{\partial z} \right\rangle = \langle 2x, 1, 2 \rangle$$

Equating corresponding components gives

$$\frac{\partial f}{\partial x} = 2x$$

$$\frac{\partial f}{\partial y} = 1$$

$$\frac{\partial f}{\partial z} = 2$$

One of the functions that satisfies these equalities is $f(x, y, z) = x^2 + y + 2z$.

51. Find a function $f(x, y, z)$ such that $\nabla f = \langle z, 2y, x \rangle$.

SOLUTION $f(x, y, z) = xz + y^2$ is a good choice.

53. Show that there does not exist a function $f(x, y)$ such that $\nabla f = \langle y^2, x \rangle$. *Hint:* Use Clairaut's Theorem $f_{xy} = f_{yx}$.

SOLUTION   Suppose that for some differentiable function $f(x, y)$,

$$\nabla f = \langle f_x, f_y \rangle = \langle y^2, x \rangle$$

That is, $f_x = y^2$ and $f_y = x$. Therefore,

$$f_{xy} = \frac{\partial}{\partial y} f_x = \frac{\partial}{\partial y} y^2 = 2y \quad \text{and} \quad f_{yx} = \frac{\partial}{\partial x} f_y = \frac{\partial}{\partial x} x = 1$$

Since f_{xy} and f_{yx} are both continuous, they must be equal by Clairaut's Theorem. Since $f_{xy} \neq f_{yx}$ we conclude that such a function f does not exist.

55. Use Eq. (8) to estimate

$$\Delta f = f(3.53, 8.98) - f(3.5, 9)$$

assuming that $\nabla f_{(3.5,9)} = \langle 2, -1 \rangle$.

SOLUTION   By Eq. (8),

$$\Delta f \approx \nabla f_P \cdot \Delta \mathbf{v}$$

The vector $\Delta \mathbf{v}$ is the following vector:

$$\Delta \mathbf{v} = \langle 3.53 - 3.5, 8.98 - 9 \rangle = \langle 0.03, -0.02 \rangle$$

Hence,

$$\Delta f \approx \nabla f_{(3,5,9)} \cdot \Delta \mathbf{v} = \langle 2, -1 \rangle \cdot \langle 0.03, -0.02 \rangle = 0.08$$

57. Suppose, in the previous exercise, that a particle located at the point $P = (2, 2, 8)$ travels toward the xy-plane in the direction normal to the surface.

(a) Through which point Q on the xy-plane will the particle pass?

(b) Suppose the axes are calibrated in centimeters. Determine the path $\mathbf{c}(t)$ of the particle if it travels at a constant speed of 8 cm/s. How long will it take the particle to reach Q?

SOLUTION

(a) The particle travels along the line through $P = (2, 2, 8)$ in the direction $(4, 2, -1)$. The vector parametrization of this line is

$$\mathbf{r}(t) = \langle 2, 2, 8 \rangle + t \langle 4, 2, -1 \rangle = \langle 2 + 4t, 2 + 2t, 8 - t \rangle \tag{1}$$

We must find the point where this line intersects the xy-plane. At this point the z-component is zero. Hence,

$$8 - t = 0 \quad \Rightarrow \quad t = 8$$

Substituting $t = 8$ in (1) we obtain

$$\mathbf{r}(8) = \langle 2 + 4 \cdot 8, 2 + 2 \cdot 8, 0 \rangle = \langle 34, 18, 0 \rangle$$

The particle will pass through the point $Q = (34, 18, 0)$ on the xy-plane.

(b) If $\mathbf{v}$ is a direction vector of the line PQ, so that $\|\mathbf{v}\| = 8$, the following parametrization of the line has constant speed 8:

$$\mathbf{c}(t) = \langle 2, 2, 8 \rangle + t\mathbf{v}$$

(This has speed 8 because $\|\mathbf{c}'(t)\| = \|\mathbf{v}\| = 8$). In the previous exercise, we found the unit vector $\mathbf{n} = \frac{1}{\sqrt{21}} \langle 4, 2, -1 \rangle$, therefore we use the direction vector $\mathbf{v} = 8\mathbf{n} = \frac{8}{\sqrt{21}} \langle 4, 2, -1 \rangle$, obtaining the following parametrization of the line:

$$\mathbf{c}(t) = \langle 2, 2, 8 \rangle + t \cdot \frac{8}{\sqrt{21}} \langle 4, 2, -1 \rangle = \left\langle 2 + \frac{32}{\sqrt{21}}t, 2 + \frac{16}{\sqrt{21}}t, 8 - \frac{8t}{\sqrt{21}} \right\rangle$$

To find the time needed for the particle to reach Q if it travels along $\mathbf{c}(t)$, we first compute the distance $\overline{PQ}$:

$$\overline{PQ} = \sqrt{(34 - 2)^2 + (18 - 2)^2 + (0 - 8)^2} = \sqrt{1344} = 8\sqrt{21}$$

The time needed is thus

$$T = \frac{\overline{PQ}}{8} = \frac{8\sqrt{21}}{8} = \sqrt{21} \approx 4.58 \text{ s}$$

59. Suppose that the intersection of two surfaces $F(x, y, z) = 0$ and $G(x, y, z) = 0$ is a curve $\mathcal{C}$, and let P be a point on $\mathcal{C}$. Explain why the vector $\mathbf{v} = \nabla F_P \times \nabla G_P$ is a direction vector for the tangent line to $\mathcal{C}$ at P.

SOLUTION The gradient ∇F_P is orthogonal to all the curves in the level surface $F(x, y, z) = 0$ passing through P. Similarly, ∇G_P is orthogonal to all the curves in the level surface $G(x, y, z) = 0$ passing through P. Therefore, both ∇F_P and ∇G_P are orthogonal to the intersection curve $\mathcal{C}$ at P, hence the cross product $\nabla F_P \times \nabla G_P$ is parallel to the tangent line to $\mathcal{C}$ at P.

61. Let $\mathcal{C}$ be the curve obtained by intersecting the two surfaces $x^3 + 2xy + yz = 7$ and $3x^2 - yz = 1$. Find the parametric equations of the tangent line to $\mathcal{C}$ at $P = (1, 2, 1)$.

SOLUTION The parametric equations of the tangent line to $\mathcal{C}$ at $P = (1, 2, 1)$ are

$$x = 1 + at, \quad y = 2 + bt, \quad z = 1 + ct \tag{1}$$

where $\mathbf{v} = \langle a, b, c \rangle$ is a direction vector for the line. By Exercise 59, $\mathbf{v}$ may be chosen as the cross product:

$$\mathbf{v} = \nabla F_P \times \nabla G_P \tag{2}$$

where $F(x, y, z) = x^3 + 2xy + yz$ and $G(x, y, z) = 3x^2 - yz$. We compute the gradient vectors:

$$F_x(x, y, z) = 3x^2 + 2y \qquad F_x(1, 2, 1) = 7$$
$$F_y(x, y, z) = 2x + z \quad \Rightarrow \quad F_y(1, 2, 1) = 3 \quad \Rightarrow \quad \nabla F_P = \langle 7, 3, 2 \rangle$$
$$F_z(x, y, z) = y \qquad F_z(1, 2, 1) = 2$$

$$G_x(x, y, z) = 6x \qquad G_x(1, 2, 1) = 6$$
$$G_y(x, y, z) = -z \quad \Rightarrow \quad G_y(1, 2, 1) = -1 \quad \Rightarrow \quad \nabla G_P = \langle 6, -1, -2 \rangle$$
$$G_z(x, y, z) = -y \qquad G_z(1, 2, 1) = -2$$

Hence,

$$\mathbf{v} = \langle 7, 3, 2 \rangle \times \langle 6, -1, -2 \rangle = \begin{vmatrix} \mathbf{i} & \mathbf{j} & \mathbf{k} \\ 7 & 3 & 2 \\ 6 & -1 & -2 \end{vmatrix} = -4\mathbf{i} + 26\mathbf{j} - 25\mathbf{k} = \langle -4, 26, -25 \rangle$$

Therefore, $\mathbf{v} = \langle a, b, c \rangle = \langle -4, 26, -25 \rangle$, so we obtain

$$a = -4, \quad b = 26, \quad c = -25.$$

Substituting in (1) gives the following parametric equations of the tangent line:

$$x = 1 - 4t, \quad y = 2 + 26t, \quad z = 1 - 25t.$$

63. Prove the Chain Rule for Gradients (Theorem 1).

SOLUTION   We must show that if $F(t)$ is a differentiable function of t and $f(x, y, z)$ is differentiable, then

$$\nabla F\left(f(x, y, z)\right) = F'\left(f(x, y, z)\right) \nabla f$$

Using the Chain Rule for partial derivatives we get

$$\nabla F\left(f(x, y, z)\right) = \left\langle \frac{\partial}{\partial x} F\left(f(x, y, z)\right), \frac{\partial}{\partial y} F\left(f(x, y, z)\right), \frac{\partial}{\partial z} F\left(f(x, y, z)\right) \right\rangle$$

$$= \left\langle \frac{dF}{dt} \cdot \frac{\partial f}{\partial x}, \frac{dF}{dt} \cdot \frac{\partial f}{\partial y}, \frac{dF}{dt} \cdot \frac{\partial f}{\partial z} \right\rangle = \frac{dF}{dt} \left\langle \frac{\partial f}{\partial x}, \frac{\partial f}{\partial y}, \frac{\partial f}{\partial z} \right\rangle = F'\left(f(x, y, z)\right) \nabla F$$

Further Insights and Challenges

65. Let $\mathbf{u}$ be a unit vector. Show that the directional derivative $D_{\mathbf{u}} f$ is equal to the component of ∇f along $\mathbf{u}$.

SOLUTION   The component of ∇f along $\mathbf{u}$ is $\nabla f \cdot \mathbf{u}$. By the Theorem on Evaluating Directional Derivatives, $D_{\mathbf{u}} f = \nabla f \cdot \mathbf{u}$, which is the component of ∇f along $\mathbf{u}$.

67. Use the definition of differentiability to show that if $f(x, y)$ is differentiable at $(0, 0)$ and

$$f(0, 0) = f_x(0, 0) = f_y(0, 0) = 0$$

then

$$\lim_{(x,y)\to(0,0)} \frac{f(x, y)}{\sqrt{x^2 + y^2}} = 0 \qquad \boxed{9}$$

SOLUTION   If $f(x, y)$ is differentiable at $(0, 0)$, then there exists a function $\epsilon(x, y)$ satisfying $\lim_{(x,y)\to(0,0)} \epsilon(x, y) = 0$ such that

$$f(x, y) = L(x, y) + \epsilon(x, y)\sqrt{x^2 + y^2} \qquad (1)$$

Since $f(0, 0) = 0$, the linear function $L(x, y)$ is

$$L(x, y) = f(0, 0) + f_x(0, 0)x + f_y(0, 0)y = f_x(0, 0)x + f_y(0, 0)y$$

Substituting in (1) gives

$$f(x, y) = f_x(0, 0)x + f_y(0, 0)y + \epsilon(x, y)\sqrt{x^2 + y^2}$$

Therefore,

$$\lim_{(x,y)\to(0,0)} \frac{f(x, y) - f_x(0, 0)x - f_y(0, 0)y}{\sqrt{x^2 + y^2}} = \lim_{(x,y)\to(0,0)} \epsilon(x, y) = 0$$

69. Prove that if $f(x, y)$ is differentiable and $\nabla f_{(x,y)} = \mathbf{0}$ for all (x, y), then f is constant.

SOLUTION   Since $\nabla f = \langle f_x, f_y \rangle = \langle 0, 0 \rangle$ for all (x, y), we have

$$f_x(x, y) = f_y(x, y) = 0 \text{ for all } (x, y) \qquad (1)$$

Let $Q_0 = (x_0, y_0)$ be a fixed point and let $P = (x_1, y_1)$ be any other point. Let $\mathbf{c}(t) = \langle x(t), y(t) \rangle$ be a parametric equation of the line joining Q_0 and P, with $P = \mathbf{c}(t_1)$ and $Q_0 = \mathbf{c}(t_0)$. We define the following function:

$$F(t) = f(x(t), y(t))$$

$F(t)$ is defined for all t, since $f(x, y)$ is defined for all (x, y). By the Chain Rule we have

$$F'(t) = f_x(x(t), y(t))\frac{dx}{dt} + f_y(x(t), y(t))\frac{dy}{dt}$$

Combining with (1) we get $F'(t) = 0$ for all t. We conclude that $F(t) = \text{const}$. That is, f is constant on the line $\mathbf{c}(t)$. In particular, $f(P) = f(Q_0)$. Since P is any point, it follows that $f(x, y)$ is a constant function.

In Exercises 71–73, a path $\mathbf{c}(t) = (x(t), y(t))$ *follows the gradient of a function* $f(x, y)$ *if the tangent vector* $\mathbf{c}'(t)$ *points in the direction of* ∇f *for all t. In other words,* $\mathbf{c}'(t) = k(t)\nabla f_{\mathbf{c}(t)}$ *for some positive function* $k(t)$. *Note that in this case,* $\mathbf{c}(t)$ *crosses each level curve of* $f(x, y)$ *at a right angle.*

71. Show that if the path $\mathbf{c}(t) = (x(t), y(t))$ follows the gradient of $f(x, y)$, then

$$\frac{y'(t)}{x'(t)} = \frac{f_y}{f_x}$$

SOLUTION Since $\mathbf{c}(t)$ follows the gradient of $f(x, y)$, we have

$$\mathbf{c}'(t) = k(t)\nabla f_{\mathbf{c}(t)} = k(t)\langle f_x(\mathbf{c}(t)), f_y(\mathbf{c}(t))\rangle$$

which implies that

$$x'(t) = k(t)f_x(\mathbf{c}(t)) \quad \text{and} \quad y'(t) = k(t)f_y(\mathbf{c}(t))$$

Hence,

$$\frac{y'(t)}{x'(t)} = \frac{k(t)f_y(\mathbf{c}(t))}{k(t)f_x(\mathbf{c}(t))} = \frac{f_y(\mathbf{c}(t))}{f_x(\mathbf{c}(t))}$$

or in short notation,

$$\frac{y'(t)}{x'(t)} = \frac{f_y}{f_x}$$

73. $\mathbf{\mathcal{CAS}}$ Find the curve $y = g(x)$ passing through $(0, 1)$ that crosses each level curve of $f(x, y) = y \sin x$ at a right angle. If you have a computer algebra system, graph $y = g(x)$ together with the level curves of f.

SOLUTION Using $f_x = y \cos x$, $f_y = \sin x$, and $y(0) = 1$, we get

$$\frac{dy}{dx} = \frac{\tan x}{y} \quad \Rightarrow \quad y(0) = 1$$

We solve the differential equation using separation of variables:

$$y\,dy = \tan x\,dx$$

$$\int y\,dy = \int \tan x\,dx$$

$$\frac{1}{2}y^2 = -\ln|\cos x| + k$$

$$y^2 = -2\ln|\cos x| + k = -\ln\left(\cos^2 x\right) + k$$

$$y = \pm\sqrt{-\ln\left(\cos^2 x\right) + k}$$

Since $y(0) = 1 > 0$, the appropriate sign is the positive sign. That is,

$$y = \sqrt{-\ln\left(\cos^2 x\right) + k} \tag{1}$$

We find the constant k by substituting $x = 0$, $y = 1$ and solve for k. This gives

$$1 = \sqrt{-\ln\left(\cos^2 0\right) + k} = \sqrt{-\ln 1 + k} = \sqrt{k}$$

Hence,

$$k = 1$$

Substituting in (2) gives the following solution:

$$y = \sqrt{1 - \ln\left(\cos^2 x\right)} \tag{2}$$

The following figure shows the graph of the curve (3) together with some level curves of f.

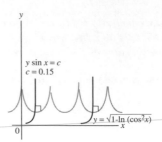

14.6 The Chain Rule   (LT Section 15.6)

Preliminary Questions

1. Let $f(x, y) = xy$, where $x = uv$ and $y = u + v$.

(a) What are the primary derivatives of f?

(b) What are the independent variables?

SOLUTION

(a) The primary derivatives of f are $\frac{\partial f}{\partial x}$ and $\frac{\partial f}{\partial y}$.

(b) The independent variables are u and v, on which x and y depend.

In Questions 2 and 3, suppose that $f(u, v) = ue^v$, where $u = rs$ and $v = r + s$.

2. The composite function $f(u, v)$ is equal to:

(a) rse^{r+s} **(b)** re^s **(c)** rse^{rs}

SOLUTION The composite function $f(u, v)$ is obtained by replacing u and v in the formula for $f(u, v)$ by the corresponding functions $u = rs$ and $v = r + s$. This gives

$$f\big(u(r, s), v(r, s)\big) = u(r, s)e^{v(r,s)} = rse^{r+s}$$

Answer (a) is the correct answer.

3. What is the value of $f(u, v)$ at $(r, s) = (1, 1)$?

SOLUTION We compute $u = rs$ and $v = r + s$ at the point $(r, s) = (1, 1)$:

$$u(1, 1) = 1 \cdot 1 = 1; \quad v(1, 1) = 1 + 1 = 2$$

Substituting in $f(u, v) = ue^v$, we get

$$f(u, v)\Big|_{(r,s)=(1,1)} = 1 \cdot e^2 = e^2.$$

4. According to the Chain Rule, $\partial f / \partial r$ is equal to (choose the correct answer):

(a) $\dfrac{\partial f}{\partial x}\dfrac{\partial x}{\partial r} + \dfrac{\partial f}{\partial x}\dfrac{\partial x}{\partial s}$ **(b)** $\dfrac{\partial f}{\partial x}\dfrac{\partial x}{\partial r} + \dfrac{\partial f}{\partial y}\dfrac{\partial y}{\partial r}$ **(c)** $\dfrac{\partial f}{\partial r}\dfrac{\partial r}{\partial x} + \dfrac{\partial f}{\partial s}\dfrac{\partial s}{\partial x}$

SOLUTION For a function $f(x, y)$ where $x = x(r, s)$ and $y = y(r, s)$, the Chain Rule states that the partial derivative $\frac{\partial f}{\partial r}$ is as given in (b). That is,

$$\frac{\partial f}{\partial x}\frac{\partial x}{\partial r} + \frac{\partial f}{\partial y}\frac{\partial y}{\partial r}$$

5. Suppose that x, y, z are functions of the independent variables u, v, w. Which of the following terms appear in the Chain Rule expression for $\partial f / \partial w$?

(a) $\dfrac{\partial f}{\partial v}\dfrac{\partial x}{\partial v}$ **(b)** $\dfrac{\partial f}{\partial w}\dfrac{\partial w}{\partial x}$ **(c)** $\dfrac{\partial f}{\partial z}\dfrac{\partial z}{\partial w}$

SOLUTION By the Chain Rule, the derivative $\frac{\partial f}{\partial w}$ is

$$\frac{\partial f}{\partial w} = \frac{\partial f}{\partial x}\frac{\partial x}{\partial w} + \frac{\partial f}{\partial y}\frac{\partial y}{\partial w} + \frac{\partial f}{\partial z}\frac{\partial z}{\partial w}$$

Therefore (c) is the only correct answer.

6. With notation as in the previous question, does $\partial x/\partial v$ appear in the Chain Rule expression for $\partial f/\partial u$?

SOLUTION The Chain Rule expression for $\frac{\partial f}{\partial u}$ is

$$\frac{\partial f}{\partial u} = \frac{\partial f}{\partial x}\frac{\partial x}{\partial u} + \frac{\partial f}{\partial y}\frac{\partial y}{\partial u} + \frac{\partial f}{\partial z}\frac{\partial z}{\partial u}$$

The derivative $\frac{\partial x}{\partial v}$ does not appear in differentiating f with respect to the independent variable u.

Exercises

1. Let $f(x, y, z) = x^2 y^3 + z^4$ and $x = s^2$, $y = st^2$, and $z = s^2 t$.

(a) Calculate the primary derivatives $\frac{\partial f}{\partial x}, \frac{\partial f}{\partial y}, \frac{\partial f}{\partial z}$.

(b) Calculate $\frac{\partial x}{\partial s}, \frac{\partial y}{\partial s}, \frac{\partial z}{\partial s}$.

(c) Compute $\frac{\partial f}{\partial s}$ using the Chain Rule:

$$\frac{\partial f}{\partial s} = \frac{\partial f}{\partial x}\frac{\partial x}{\partial s} + \frac{\partial f}{\partial y}\frac{\partial y}{\partial s} + \frac{\partial f}{\partial z}\frac{\partial z}{\partial s}$$

Express the answer in terms of the independent variables s, t.

SOLUTION

(a) The primary derivatives of $f(x, y, z) = x^2 y^3 + z^4$ are

$$\frac{\partial f}{\partial x} = 2xy^3, \quad \frac{\partial f}{\partial y} = 3x^2 y^2, \quad \frac{\partial f}{\partial z} = 4z^3$$

(b) The partial derivatives of x, y, and z with respect to s are

$$\frac{\partial x}{\partial s} = 2s, \quad \frac{\partial y}{\partial s} = t^2, \quad \frac{\partial z}{\partial s} = 2st$$

(c) We use the Chain Rule and the partial derivatives computed in parts (a) and (b) to find the following derivative:

$$\frac{\partial f}{\partial s} = \frac{\partial f}{\partial x}\frac{\partial x}{\partial s} + \frac{\partial f}{\partial y}\frac{\partial y}{\partial s} + \frac{\partial f}{\partial z}\frac{\partial z}{\partial s} = 2xy^3 \cdot 2s + 3x^2 y^2 t^2 + 4z^3 \cdot 2st = 4xy^3 s + 3x^2 y^2 t^2 + 8z^3 st$$

To express the answer in terms of the independent variables s, t we substitute $x = s^2$, $y = st^2$, $z = s^2 t$. This gives

$$\frac{\partial f}{\partial s} = 4s^2 (st^2)^3 s + 3(s^2)^2 (st^2)^2 t^2 + 8(s^2 t)^3 st = 4s^6 t^6 + 3s^6 t^6 + 8s^7 t^4 = 7s^6 t^6 + 8s^7 t^4.$$

In Exercises 3–10, use the Chain Rule to calculate the partial derivatives. Express the answer in terms of the independent variables.

3. $\frac{\partial f}{\partial s}, \frac{\partial f}{\partial r}$; $f(x, y, z) = xy + z^2, x = s^2, y = 2rs, z = r^2$

SOLUTION We perform the following steps:

Step 1. Compute the primary derivatives. The primary derivatives of $f(x, y, z) = xy + z^2$ are

$$\frac{\partial f}{\partial x} = y, \quad \frac{\partial f}{\partial y} = x, \quad \frac{\partial f}{\partial z} = 2z$$

Step 2. Apply the Chain Rule. By the Chain Rule,

$$\frac{\partial f}{\partial s} = \frac{\partial f}{\partial x} \cdot \frac{\partial x}{\partial s} + \frac{\partial f}{\partial y} \cdot \frac{\partial y}{\partial s} + \frac{\partial f}{\partial z} \cdot \frac{\partial z}{\partial s} \tag{1}$$

$$\frac{\partial f}{\partial r} = \frac{\partial f}{\partial x} \cdot \frac{\partial x}{\partial r} + \frac{\partial f}{\partial y} \cdot \frac{\partial y}{\partial r} + \frac{\partial f}{\partial z} \cdot \frac{\partial z}{\partial r} \tag{2}$$

We compute the partial derivatives of x, y, z with respect to s and r:

$$\frac{\partial x}{\partial s} = 2s, \quad \frac{\partial y}{\partial s} = 2r, \quad \frac{\partial z}{\partial s} = 0.$$

$$\frac{\partial x}{\partial r} = 0, \quad \frac{\partial y}{\partial r} = 2s, \quad \frac{\partial z}{\partial r} = 2r.$$

Substituting these derivatives and the primary derivatives computed in step 1 in (1) and (2), we get

$$\frac{\partial f}{\partial s} = y \cdot 2s + x \cdot 2r + 2z \cdot 0 = 2ys + 2xr$$

$$\frac{\partial f}{\partial r} = y \cdot 0 + x \cdot 2s + 2z \cdot 2r = 2xs + 4zr$$

Step 3. Express the answer in terms of r and s. We substitute $x = s^2$, $y = 2rs$, and $z = r^2$ in $\frac{\partial f}{\partial s}$ and $\frac{\partial f}{\partial r}$ in step 2, to obtain

$$\frac{\partial f}{\partial s} = 2rs \cdot 2s + s^2 \cdot 2r = 4rs^2 + 2rs^2 = 6rs^2.$$

$$\frac{\partial f}{\partial r} = 2s^2 \cdot s + 4r^2 \cdot r = 2s^3 + 4r^3.$$

5. $\dfrac{\partial g}{\partial u}, \dfrac{\partial g}{\partial v}$; $g(x, y) = \cos(x - y)$, $x = 3u - 5v$, $y = -7u + 15v$

SOLUTION We use the following steps:

Step 1. Compute the primary derivatives. The primary derivatives of $g(x, y) = \cos(x - y)$ are:

$$\frac{\partial g}{\partial x} = -\sin(x - y), \quad \frac{\partial g}{\partial y} = \sin(x - y)$$

Step 2. Apply the Chain Rule. By the Chain Rule,

$$\frac{\partial g}{\partial u} = \frac{\partial g}{\partial x}\frac{\partial x}{\partial u} + \frac{\partial g}{\partial y}\frac{\partial y}{\partial u} = -\sin(x - y)\frac{\partial x}{\partial u} + \sin(x - y)\frac{\partial y}{\partial u}$$

$$\frac{\partial g}{\partial v} = \frac{\partial g}{\partial x}\frac{\partial x}{\partial v} + \frac{\partial g}{\partial y}\frac{\partial y}{\partial v} = -\sin(x - y)\frac{\partial x}{\partial v} + \sin(x - y)\frac{\partial y}{\partial v}$$

We compute the partial derivatives of x, y with respect to u and v:

$$\frac{\partial x}{\partial u} = 3, \quad \frac{\partial x}{\partial v} = -5$$

$$\frac{\partial y}{\partial u} = -7, \quad \frac{\partial y}{\partial v} = 15$$

substituting in the expressions above we have:

$$\frac{\partial g}{\partial u} = -\sin(x - y)(3) + \sin(x - y)(-7) = -10\sin(x - y)$$

$$\frac{\partial g}{\partial v} = -\sin(x - y)(-5) + \sin(x - y)(15) = 20\sin(x - y)$$

Step 3. Express the answer in terms of u and v. We substitute $x = 3u - 5v$ and $y = -7u + 15v$ in $\partial g/\partial u$ and $\partial g/\partial v$ found in step 2. This gives:

$$\frac{\partial g}{\partial u} = -10\sin(10u - 20v)$$

$$\frac{\partial g}{\partial v} = 20\sin(10u - 20v)$$

7. $\dfrac{\partial F}{\partial y}$; $F(u, v) = e^{u+v}$, $u = x^2$, $v = xy$

SOLUTION We use the following steps:

Step 1. Compute the primary derivatives. The primary derivatives of $F(u, v) = e^{u+v}$ are

$$\frac{\partial f}{\partial u} = e^{u+v}, \quad \frac{\partial f}{\partial v} = e^{u+v}$$

Step 2. Apply the Chain Rule. By the Chain Rule,

$$\frac{\partial F}{\partial y} = \frac{\partial F}{\partial u}\frac{\partial u}{\partial y} + \frac{\partial F}{\partial v}\frac{\partial v}{\partial y} = e^{u+v}\frac{\partial u}{\partial y} + e^{u+v}\frac{\partial v}{\partial y} = e^{u+v}\left(\frac{\partial u}{\partial y} + \frac{\partial v}{\partial y}\right)$$

We compute the partial derivatives of u and v with respect to y:

$$\frac{\partial u}{\partial y} = 0, \quad \frac{\partial v}{\partial y} = x$$

We substitute to obtain

$$\frac{\partial F}{\partial y} = xe^{u+v} \tag{1}$$

Step 3. Express the answer in terms of x and y. We substitute $u = x^2$, $v = xy$ in (1) and (2), obtaining

$$\frac{\partial F}{\partial y} = xe^{x^2+xy}.$$

9. $\dfrac{\partial h}{\partial t_2}; h(x, y) = \dfrac{x}{y}, x = t_1 t_2, y = t_1^2 t_2$

SOLUTION We use the following steps:

Step 1. Compute the primary derivatives. The primary derivatives of $h(x, y) = \frac{x}{y}$ are

$$\frac{\partial h}{\partial x} = \frac{1}{y}, \quad \frac{\partial h}{\partial y} = -\frac{x}{y^2}$$

Step 2. Apply the Chain Rule. By the Chain Rule,

$$\frac{\partial h}{\partial t_2} = \frac{\partial h}{\partial x}\frac{\partial x}{\partial t_2} + \frac{\partial h}{\partial y}\frac{\partial y}{\partial t_2} = \frac{1}{y}\frac{\partial x}{\partial t_2} - \frac{x}{y^2}\frac{\partial y}{\partial t_2}$$

We compute the partial derivatives of x and y with respect to t_2:

$$\frac{\partial x}{\partial t_2} = t_1, \quad \frac{\partial y}{\partial t_2} = t_1^2$$

Hence,

$$\frac{\partial h}{\partial t_2} = \frac{t_1}{y} - \frac{x}{y^2}t_1^2$$

Step 3. Express the answer in terms of t_1 and t_2. We substitute $x = t_1 t_2$, $y = t_1^2 t_2$ in $\frac{\partial h}{\partial t_2}$ computed in step 2, to obtain

$$\frac{\partial h}{\partial t_2} = \frac{t_1}{t_1^2 t_2} - \frac{t_1 t_2 \cdot t_1^2}{(t_1^2 t_2)^2} = \frac{1}{t_1 t_2} - \frac{1}{t_1 t_2} = 0$$

Remark: Notice that $h\big(x(t_1, t_2), y(t_1, t_2)\big) = h(t_1, t_2) = \frac{t_1 t_2}{t_1^2 t_2} = \frac{1}{t_1}$. $h(t_1, t_2)$ is independent of t_2, hence $\frac{\partial h}{\partial t_2} = 0$ (as obtained in our computations).

In Exercises 11–16, use the Chain Rule to evaluate the partial derivative at the point specified.

11. $\partial f/\partial u$ and $\partial f/\partial v$ at $(u, v) = (-1, -1)$, where $f(x, y, z) = x^3 + yz^2$, $x = u^2 + v$, $y = u + v^2$, $z = uv$.

SOLUTION The primary derivatives of $f(x, y, z) = x^3 + yz^2$ are

$$\frac{\partial f}{\partial x} = 3x^2, \quad \frac{\partial f}{\partial y} = z^2, \quad \frac{\partial f}{\partial z} = 2yz$$

By the Chain Rule we have

$$\frac{\partial f}{\partial u} = \frac{\partial f}{\partial x}\frac{\partial x}{\partial u} + \frac{\partial f}{\partial y}\frac{\partial y}{\partial u} + \frac{\partial f}{\partial z}\frac{\partial z}{\partial u} = 3x^2\frac{\partial x}{\partial u} + z^2\frac{\partial y}{\partial u} + 2yz\frac{\partial z}{\partial u} \tag{1}$$

$$\frac{\partial f}{\partial v} = \frac{\partial f}{\partial x}\frac{\partial x}{\partial v} + \frac{\partial f}{\partial y}\frac{\partial y}{\partial v} + \frac{\partial f}{\partial z}\frac{\partial z}{\partial v} = 3x^2\frac{\partial x}{\partial v} + z^2\frac{\partial y}{\partial v} + 2yz\frac{\partial z}{\partial v} \tag{2}$$

We compute the partial derivatives of x, y, and z with respect to u and v:

$$\frac{\partial x}{\partial u} = 2u, \quad \frac{\partial y}{\partial u} = 1, \quad \frac{\partial z}{\partial u} = v$$

$$\frac{\partial x}{\partial v} = 1, \quad \frac{\partial y}{\partial v} = 2v, \quad \frac{\partial z}{\partial v} = u$$

Substituting in (1) and (2) we get

$$\frac{\partial f}{\partial u} = 6x^2 u + z^2 + 2yzv \tag{3}$$

$$\frac{\partial f}{\partial v} = 3x^2 + 2vz^2 + 2yzu \tag{4}$$

We determine (x, y, z) for $(u, v) = (-1, -1)$:

$$x = (-1)^2 - 1 = 0, \quad y = -1 + (-1)^2 = 0, \quad z = (-1) \cdot (-1) = 1.$$

Finally, we substitute $(x, y, z) = (0, 0, 1)$ and $(u, v) = (-1, -1)$ in (3), (4) to obtain the following derivatives:

$$\frac{\partial f}{\partial u}\bigg|_{(u,v)=(-1,-1)} = 6 \cdot 0^2 \cdot (-1) + 1^2 + 2 \cdot 0 \cdot 1 \cdot (-1) = 1$$

$$\frac{\partial f}{\partial v}\bigg|_{(u,v)=(-1,-1)} = 3 \cdot 0^2 + 2 \cdot (-1) \cdot 1^2 + 2 \cdot 0 \cdot 1 \cdot (-1) = -2$$

13. $\partial g/\partial \theta$ at $(r, \theta) = \left(2\sqrt{2}, \frac{\pi}{4}\right)$, where $g(x, y) = 1/(x + y^2)$, $x = r \sin \theta$, $y = r \cos \theta$.

SOLUTION   We compute the primary derivatives of $g(x, y) = \frac{1}{x+y^2}$:

$$\frac{\partial g}{\partial x} = -\frac{1}{(x + y^2)^2}, \quad \frac{\partial g}{\partial y} = -\frac{2y}{(x + y^2)^2}$$

By the Chain Rule we have

$$\frac{\partial g}{\partial \theta} = \frac{\partial g}{\partial x}\frac{\partial x}{\partial \theta} + \frac{\partial g}{\partial y}\frac{\partial y}{\partial \theta} = -\frac{1}{(x + y^2)^2}\frac{\partial x}{\partial \theta} - \frac{2y}{(x + y^2)^2}\frac{\partial y}{\partial \theta} = -\frac{1}{(x + y^2)^2}\left(\frac{\partial x}{\partial \theta} + 2y\frac{\partial y}{\partial \theta}\right)$$

We find the partial derivatives $\frac{\partial x}{\partial \theta}$, $\frac{\partial y}{\partial \theta}$:

$$\frac{\partial x}{\partial \theta} = r \cos \theta, \quad \frac{\partial y}{\partial \theta} = -r \sin \theta$$

Hence,

$$\frac{\partial g}{\partial \theta} = -\frac{r}{(x + y^2)^2}(\cos \theta - 2y \sin \theta) \tag{1}$$

At the point $(r, \theta) = \left(2\sqrt{2}, \frac{\pi}{4}\right)$, we have $x = 2\sqrt{2} \sin \frac{\pi}{4} = 2$ and $y = 2\sqrt{2} \cos \frac{\pi}{4} = 2$. Substituting $(r, \theta) = \left(2\sqrt{2}, \frac{\pi}{4}\right)$ and $(x, y) = (2, 2)$ in (1) gives the following derivative:

$$\frac{\partial g}{\partial \theta}\bigg|_{(r,\theta)=\left(2\sqrt{2}, \frac{\pi}{4}\right)} = \frac{-2\sqrt{2}}{(2 + 2^2)^2}\left(\cos \frac{\pi}{4} - 4 \sin \frac{\pi}{4}\right) = \frac{-\sqrt{2}}{18}\left(\frac{1}{\sqrt{2}} - \frac{4}{\sqrt{2}}\right) = \frac{1}{6}.$$

15. $\partial g/\partial u$ at $(u, v) = (0, 1)$, where $g(x, y) = x^2 - y^2$, $x = e^u \cos v$, $y = e^u \sin v$.

SOLUTION   The primary derivatives of $g(x, y) = x^2 - y^2$ are

$$\frac{\partial g}{\partial x} = 2x, \quad \frac{\partial g}{\partial y} = -2y$$

By the Chain Rule we have

$$\frac{\partial g}{\partial u} = \frac{\partial g}{\partial x} \cdot \frac{\partial x}{\partial u} + \frac{\partial g}{\partial y} \cdot \frac{\partial y}{\partial u} = 2x\frac{\partial x}{\partial u} - 2y\frac{\partial y}{\partial u} \tag{1}$$

We find $\frac{\partial x}{\partial u}$ and $\frac{\partial y}{\partial u}$:

$$\frac{\partial x}{\partial u} = e^u \cos v, \quad \frac{\partial y}{\partial u} = e^u \sin v$$

Substituting in (1) gives

$$\frac{\partial g}{\partial u} = 2xe^u \cos v - 2ye^u \sin v = 2e^u(x \cos v - y \sin v) \tag{2}$$

We determine (x, y) for $(u, v) = (0, 1)$:

$$x = e^0 \cos 1 = \cos 1, \quad y = e^0 \sin 1 = \sin 1$$

Finally, we substitute $(u, v) = (0, 1)$ and $(x, y) = (\cos 1, \sin 1)$ in (2) and use the identity $\cos^2 \alpha - \sin^2 \alpha = \cos 2\alpha$, to obtain the following derivative:

$$\frac{\partial g}{\partial u}\bigg|_{(u,v)=(0,1)} = 2e^0 \left(\cos^2 1 - \sin^2 1\right) = 2 \cdot \cos 2 \cdot 1 = 2 \cos 2$$

17. Jessica and Matthew are running toward the point P along the straight paths that make a fixed angle of θ (Figure 3). Suppose that Matthew runs with velocity v_a m/s and Jessica with velocity v_b m/s. Let $f(x, y)$ be the distance from Matthew to Jessica when Matthew is x meters from P and Jessica is y meters from P.

(a) Show that $f(x, y) = \sqrt{x^2 + y^2 - 2xy \cos \theta}$.

(b) Assume that $\theta = \pi/3$. Use the Chain Rule to determine the rate at which the distance between Matthew and Jessica is changing when $x = 30$, $y = 20$, $v_a = 4$ m/s, and $v_b = 3$ m/s.

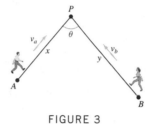

FIGURE 3

SOLUTION

(a) This is a simple application of the Law of Cosines. Connect points A and B in the diagram to form a line segment that we will call f. Then, the Law of Cosines says that $f^2 = x^2 + y^2 - 2xy \cos \theta$. By taking square roots, we find that $f = \sqrt{x^2 + y^2 - 2xy \cos \theta}$.

(b) Using the chain rule,

$$\frac{df}{dt} = \frac{\partial f}{\partial x}\frac{dx}{dt} + \frac{\partial f}{\partial y}\frac{dy}{dt}$$

so we get

$$\frac{df}{dt} = \frac{(x - y \cos \theta)dx/dt}{\sqrt{x^2 + y^2 - 2xy \cos \theta}} + \frac{(y - x \cos \theta)dy/dt}{\sqrt{x^2 + y^2 - 2xy \cos \theta}}$$

and using $x = 30$, $y = 20$, and $dx/dt = 4$, $dy/dt = 3$, we get

$$\frac{df}{dt} = \frac{180 - 170 \cos \theta}{\sqrt{1300 - 1200 \cos \theta}}$$

19. Let $u = u(x, y)$, and let (r, θ) be polar coordinates. Verify the relation

$$\|\nabla u\|^2 = u_r^2 + \frac{1}{r^2}u_\theta^2 \qquad \boxed{8}$$

Hint: Compute the right-hand side by expressing u_θ and u_r in terms of u_x and u_y.

SOLUTION By the Chain Rule we have

$$u_\theta = u_x x_\theta + u_y y_\theta \qquad (1)$$

$$u_r = u_x x_r + u_y y_r \qquad (2)$$

Since $x = r \cos \theta$ and $y = r \sin \theta$, the partial derivatives of x and y with respect to r and θ are

$$x_\theta = -r \sin \theta, \quad y_\theta = r \cos \theta$$

$$x_r = \cos \theta, \quad y_r = \sin \theta$$

Substituting in (1) and (2) gives

$$u_\theta = (-r \sin \theta)u_x + (r \cos \theta)u_y \qquad (3)$$

$$u_r = (\cos \theta)u_x + (\sin \theta)u_y \qquad (4)$$

We now solve these equations for u_x and u_y in terms of u_θ and u_r. Multiplying (3) by $(-\sin\theta)$ and (4) by $r\cos\theta$ and adding the resulting equations gives

$$
\begin{aligned}
(-\sin\theta)u_\theta &= (r\sin^2\theta)u_x - (r\cos\theta\sin\theta)u_y \\
+ \quad r\cos\theta u_r &= (r\cos^2\theta)u_x + (r\cos\theta\sin\theta)u_y \\
\hline
(r\cos\theta)u_r - (\sin\theta)u_\theta &= ru_x
\end{aligned}
$$

or

$$
u_x = (\cos\theta)u_r - \frac{\sin\theta}{r}u_\theta \tag{5}
$$

Similarly, we multiply (3) by $\cos\theta$ and (4) by $r\sin\theta$ and add the resulting equations. We get

$$
\begin{aligned}
(\cos\theta)u_\theta &= (-r\sin\theta\cos\theta)u_x + \left(r\cos^2\theta\right)u_y \\
+ \quad r\sin\theta u_r &= (r\sin\theta\cos\theta)u_x + (r\sin^2\theta)u_y \\
\hline
(\cos\theta)u_\theta + (r\sin\theta)u_r &= ru_y
\end{aligned}
$$

or

$$
u_y = (\sin\theta)u_r + \frac{\cos\theta}{r}u_\theta \tag{6}
$$

We now use (5) and (6) to compute $\|\nabla u\|^2$ in terms of u_r and u_θ. We get

$$
\|\nabla u\|^2 = u_x^2 + u_y^2 = \left((\cos\theta)u_r - \frac{\sin\theta}{r}u_\theta\right)^2 + \left((\sin\theta)u_r + \frac{\cos\theta}{r}u_\theta\right)^2
$$

$$
= \left(\cos^2\theta\right)u_r^2 - \frac{2\cos\theta\sin\theta}{r}u_r u_\theta + \frac{\sin^2\theta}{r^2}u_\theta^2 + \left(\sin^2\theta\right)u_r^2 + \frac{2\sin\theta\cos\theta}{r}u_r u_\theta + \frac{\cos^2\theta}{r^2}u_\theta^2
$$

$$
= \left(\cos^2\theta + \sin^2\theta\right)u_r^2 + \frac{1}{r^2}\left(\sin^2\theta + \cos^2\theta\right)u_\theta^2 = u_r^2 + \frac{1}{r^2}u_\theta^2
$$

That is,

$$
\|\nabla u\|^2 = u_r^2 + \frac{1}{r^2}u_\theta^2
$$

21. Let $x = s+t$ and $y = s-t$. Show that for any differentiable function $f(x,y)$,

$$
\left(\frac{\partial f}{\partial x}\right)^2 - \left(\frac{\partial f}{\partial y}\right)^2 = \frac{\partial f}{\partial s}\frac{\partial f}{\partial t}
$$

SOLUTION   By the Chain Rule we have

$$
\frac{\partial f}{\partial s} = \frac{\partial f}{\partial x}\frac{\partial x}{\partial s} + \frac{\partial f}{\partial y}\frac{\partial y}{\partial s} = \frac{\partial f}{\partial x}\cdot 1 + \frac{\partial f}{\partial y}\cdot 1 = \frac{\partial f}{\partial x} + \frac{\partial f}{\partial y}
$$

$$
\frac{\partial f}{\partial t} = \frac{\partial f}{\partial x}\frac{\partial x}{\partial t} + \frac{\partial f}{\partial y}\frac{\partial y}{\partial t} = \frac{\partial f}{\partial x}\cdot 1 + \frac{\partial f}{\partial y}\cdot (-1) = \frac{\partial f}{\partial x} - \frac{\partial f}{\partial y}
$$

Hence, using the algebraic identity $(a+b)(a-b) = a^2 - b^2$, we get

$$
\frac{\partial f}{\partial s}\cdot\frac{\partial f}{\partial t} = \left(\frac{\partial f}{\partial x} + \frac{\partial f}{\partial y}\right)\cdot\left(\frac{\partial f}{\partial x} - \frac{\partial f}{\partial y}\right) = \left(\frac{\partial f}{\partial x}\right)^2 - \left(\frac{\partial f}{\partial y}\right)^2.
$$

23. Suppose that z is defined implicitly as a function of x and y by the equation $F(x,y,z) = xz^2 + y^2z + xy - 1 = 0$.
(a) Calculate F_x, F_y, F_z.
(b) Use Eq. (7) to calculate $\dfrac{\partial z}{\partial x}$ and $\dfrac{\partial z}{\partial y}$.

SOLUTION
(a) The partial derivatives of F are

$$
F_x = z^2 + y, \quad F_y = 2yz + x, \quad F_z = 2xz + y^2
$$

(b) By Eq. (7) we have

$$\frac{\partial z}{\partial x} = -\frac{F_x}{F_z} = -\frac{z^2 + y}{2xz + y^2}$$

$$\frac{\partial z}{\partial y} = -\frac{F_y}{F_z} = -\frac{2yz + x}{2xz + y^2}$$

In Exercises 25–30, calculate the partial derivative using implicit differentiation.

25. $\dfrac{\partial z}{\partial x}$, $\quad x^2 y + y^2 z + xz^2 = 10$

SOLUTION For $F(x, y, z) = x^2 y + y^2 z + xz^2 = 10$ we have

$$\frac{\partial z}{\partial x} = -\frac{F_x}{F_z} \tag{1}$$

We compute the partial derivatives of F:

$$F_x = 2xy + z^2, \quad F_z = y^2 + 2xz$$

Substituting in (1) gives the following derivative:

$$\frac{\partial z}{\partial x} = -\frac{2xy + z^2}{2xz + y^2}$$

27. $\dfrac{\partial z}{\partial y}$, $\quad e^{xy} + \sin(xz) + y = 0$

SOLUTION We use Eq. (7):

$$\frac{\partial z}{\partial y} = -\frac{F_y}{F_z} \tag{1}$$

The partial derivatives of $F(x, y, z) = e^{xy} + \sin(xz) + y$ are

$$F_y = xe^{xy} + 1, \quad F_z = x\cos(xz)$$

Substituting in (1), we get

$$\frac{\partial z}{\partial y} = -\frac{xe^{xy} + 1}{x\cos(xz)}$$

29. $\dfrac{\partial w}{\partial y}$, $\quad \dfrac{1}{w^2 + x^2} + \dfrac{1}{w^2 + y^2} = 1$ at $(x, y, w) = (1, 1, 1)$

SOLUTION Using the formula obtained by implicit differentiation (Eq. (7)), we have

$$\frac{\partial w}{\partial y} = -\frac{F_y}{F_w} \tag{1}$$

We find the partial derivatives of $F(x, y, w) = \dfrac{1}{w^2 + x^2} + \dfrac{1}{w^2 + y^2} - 1$:

$$F_y = -\frac{2y}{(w^2 + y^2)^2}, \quad F_w = \frac{-2w}{(w^2 + x^2)^2} - \frac{2w}{(w^2 + y^2)^2}$$

We substitute in (1) to obtain

$$\frac{\partial w}{\partial y} = -\frac{\frac{-2y}{(w^2+y^2)^2}}{\frac{-2w}{(w^2+x^2)^2} - \frac{2w}{(w^2+y^2)^2}} = -\frac{y(w^2 + x^2)^2}{w(w^2 + y^2)^2 + w(w^2 + x^2)^2} = \frac{-y(w^2 + x^2)^2}{w\big((w^2 + y^2)^2 + (w^2 + x^2)^2\big)}$$

31. Let $\mathbf{r} = \langle x, y, z \rangle$ and $e_{\mathbf{r}} = \mathbf{r}/\|\mathbf{r}\|$. Show that if a function $f(x, y, z) = F(r)$ depends only on the distance from the origin $r = \|\mathbf{r}\| = \sqrt{x^2 + y^2 + z^2}$, then

$$\nabla f = F'(r)e_{\mathbf{r}} \qquad \boxed{9}$$

SOLUTION The gradient of f is the following vector:

$$\nabla f = \left\langle \frac{\partial f}{\partial x}, \frac{\partial f}{\partial y}, \frac{\partial f}{\partial z} \right\rangle$$

We must express this vector in terms of **r** and r. Using the Chain Rule, we have

$$\frac{\partial f}{\partial x} = F'(r)\frac{\partial r}{\partial x} = F'(r) \cdot \frac{2x}{2\sqrt{x^2 + y^2 + z^2}} = F'(r) \cdot \frac{x}{r}$$

$$\frac{\partial f}{\partial y} = F'(r)\frac{\partial r}{\partial y} = F'(r) \cdot \frac{2y}{2\sqrt{x^2 + y^2 + z^2}} = F'(r) \cdot \frac{y}{r}$$

$$\frac{\partial f}{\partial z} = F'(r)\frac{\partial r}{\partial z} = F'(r) \cdot \frac{2z}{2\sqrt{x^2 + y^2 + z^2}} = F'(r) \cdot \frac{z}{r}$$

Hence,

$$\nabla f = \left\langle F'(r)\frac{x}{r}, F'(r)\frac{y}{r}, F'(r)\frac{z}{r} \right\rangle = \frac{F'(r)}{r}\langle x, y, z \rangle = F'(r)\frac{\mathbf{r}}{\|\mathbf{r}\|} = F'(r)\mathbf{e_r}$$

33. Use Eq. (9) to compute $\nabla\left(\dfrac{1}{r}\right)$.

SOLUTION To compute $\nabla\left(\frac{1}{r}\right)$ using Eq. (9), we let $F(r) = \dfrac{1}{r}$:

$$F'(r) = -\frac{1}{r^2}$$

We obtain

$$\nabla\left(\frac{1}{r}\right) = F'(r)\mathbf{e_r} = -\frac{1}{r^2} \cdot \frac{\mathbf{r}}{\|\mathbf{r}\|} = -\frac{1}{r^3}\mathbf{r}$$

35. Figure 4 shows the graph of the equation

$$F(x, y, z) = x^2 + y^2 - z^2 - 12x - 8z - 4 = 0$$

(a) Use the quadratic formula to solve for z as a function of x and y. This gives two formulas, depending on the choice of sign.
(b) Which formula defines the portion of the surface satisfying $z \geq -4$? Which formula defines the portion satisfying $z \leq -4$?
(c) Calculate $\partial z/\partial x$ using the formula $z = f(x, y)$ (for both choices of sign) and again via implicit differentiation. Verify that the two answers agree.

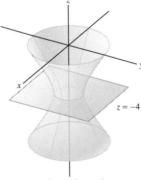

FIGURE 4 Graph of $x^2 + y^2 - z^2 - 12x - 8z - 4 = 0$.

SOLUTION
(a) We rewrite $F(x, y, z) = 0$ as a quadratic equation in the variable z:

$$z^2 + 8z + \left(4 + 12x - x^2 - y^2\right) = 0$$

We solve for z. The discriminant is

$$8^2 - 4\left(4 + 12x - x^2 - y^2\right) = 4x^2 + 4y^2 - 48x + 48 = 4\left(x^2 + y^2 - 12x + 12\right)$$

Hence,

$$z_{1,2} = \frac{-8 \pm \sqrt{4\left(x^2 + y^2 - 12x + 12\right)}}{2} = -4 \pm \sqrt{x^2 + y^2 - 12x + 12}$$

We obtain two functions:

$$z = -4 + \sqrt{x^2 + y^2 - 12x + 12}, \quad z = -4 - \sqrt{x^2 + y^2 - 12x + 12}$$

(b) The formula with the positive root defines the portion of the surface satisfying $z \geq -4$, and the formula with the negative root defines the portion satisfying $z \leq -4$.

(c) Differentiating $z = -4 + \sqrt{x^2 + y^2 - 12x + 12}$ with respect to x, using the Chain Rule, gives

$$\frac{\partial z}{\partial x} = \frac{2x - 12}{2\sqrt{x^2 + y^2 - 12x + 12}} = \frac{x - 6}{\sqrt{x^2 + y^2 - 12x + 12}} \tag{1}$$

Alternatively, using the formula for $\frac{\partial z}{\partial x}$ obtained by implicit differentiation gives

$$\frac{\partial z}{\partial x} = -\frac{F_x}{F_z} \tag{2}$$

We find the partial derivatives of $F(x, y, z) = x^2 + y^2 - z^2 - 12x - 8z - 4$:

$$F_x = 2x - 12, \quad F_z = -2z - 8$$

Substituting in (2) gives

$$\frac{\partial z}{\partial x} = -\frac{2x - 12}{-2z - 8} = \frac{x - 6}{z + 4}$$

This result is the same as the result in (1), since $z = -4 + \sqrt{x^2 + y^2 - 12x + 12}$ implies that

$$\sqrt{x^2 + y^2 - 12x + 12} = z + 4$$

For $z = -4 - \sqrt{x^2 + y^2 - 12x + 12}$, differentiating with respect to x gives

$$\frac{\partial z}{\partial x} = -\frac{2x - 12}{2\sqrt{x^2 + y^2 - 12x + 12}} = \frac{x - 6}{-\sqrt{x^2 + y^2 - 12x + 12}} = \frac{x - 6}{z + 4}$$

which is equal to $-\frac{F_x}{F_z}$ computed above.

37. The pressure P, volume V, and temperature T of a van der Waals gas with n molecules (n constant) are related by the equation

$$\left(P + \frac{an^2}{V^2} \right) (V - nb) = nRT$$

where a, b, and R are constant. Calculate $\partial P / \partial T$ and $\partial V / \partial P$.

SOLUTION Let F be the following function:

$$F(P, V, T) = \left(P + \frac{an^2}{V^2} \right) (V - nb) - nRT$$

By Eq. (7),

$$\frac{\partial P}{\partial T} = -\frac{\frac{\partial F}{\partial T}}{\frac{\partial F}{\partial P}}, \quad \frac{\partial V}{\partial P} = -\frac{\frac{\partial F}{\partial P}}{\frac{\partial F}{\partial V}} \tag{1}$$

We compute the partial derivatives of F:

$$\frac{\partial F}{\partial P} = V - nb$$

$$\frac{\partial F}{\partial T} = -nR$$

$$\frac{\partial F}{\partial V} = -2an^2 V^{-3}(V - nb) + \left(P + \frac{an^2}{V^2} \right) = P + \frac{2an^3 b}{V^3} - \frac{an^2}{V^2}$$

Substituting in (1) gives

$$\frac{\partial P}{\partial T} = -\frac{-nR}{V - nb} = \frac{nR}{V - nb}$$

$$\frac{\partial V}{\partial P} = -\frac{V - nb}{P + \frac{2an^3 b}{V^3} - \frac{an^2}{V^2}} = \frac{nbV^3 - V^4}{PV^3 + 2an^3 b - an^2 V}$$

39. Show that if $f(x)$ is differentiable and $c \neq 0$ is a constant, then $u(x, t) = f(x - ct)$ satisfies the so-called **advection equation**

$$\frac{\partial u}{\partial t} + c\frac{\partial u}{\partial x} = 0$$

SOLUTION   For $s = x - ct$, we have $u(x, t) = f(s)$. We use the Chain Rule to compute $\frac{\partial u}{\partial t}$ and $\frac{\partial u}{\partial x}$:

$$\frac{\partial u}{\partial t} = f'(s)\frac{\partial s}{\partial t} = f'(s) \cdot (-c) = -cf'(s) \tag{1}$$

$$\frac{\partial u}{\partial x} = f'(s)\frac{\partial s}{\partial x} = f'(s) \cdot 1 = f'(s) \tag{2}$$

Equalities (1) and (2) imply that:

$$\frac{\partial u}{\partial t} = -c\frac{\partial u}{\partial x} \quad \text{or} \quad \frac{\partial u}{\partial t} + c\frac{\partial u}{\partial x} = 0$$

Further Insights and Challenges

*In Exercises 40–43, a function $f(x, y, z)$ is called **homogeneous of degree n** if $f(\lambda x, \lambda y, \lambda z) = \lambda^n f(x, y, z)$ for all $\lambda \in \mathbf{R}$.*

41. Prove that if $f(x, y, z)$ is homogeneous of degree n, then $f_x(x, y, z)$ is homogeneous of degree $n - 1$. *Hint:* Either use the limit definition or apply the Chain Rule to $f(\lambda x, \lambda y, \lambda z)$.

SOLUTION   We are given that $f(\lambda x, \lambda y, \lambda z) = \lambda^n f(x, y, z)$ for all λ, and we must show that $f_x(\lambda x, \lambda y, \lambda z) = \lambda^{n-1} f_x(x, y, z)$. We use the limit definition of f_x. Since for all $\lambda \neq 0$, $\lambda h \to 0$ if and only if $h \to 0$, we get

$$f_x(\lambda x, \lambda y, \lambda z) = \lim_{h \to 0} \frac{f(\lambda x + \lambda h, \lambda y, \lambda z) - f(\lambda x, \lambda y, \lambda z)}{\lambda h} = \lim_{h \to 0} \frac{f(\lambda(x + h), \lambda y, \lambda z) - f(\lambda x, \lambda y, \lambda z)}{\lambda h}$$

$$= \lim_{h \to 0} \frac{\lambda^n f(x + h, y, z) - \lambda^n f(x, y, z)}{\lambda h} = \lim_{h \to 0} \frac{\lambda^{n-1} f(x + h, y, z) - \lambda^{n-1} f(x, y, z)}{h}$$

$$= \lambda^{n-1} \lim_{h \to 0} \frac{f(x + h, y, z) - f(x, y, z)}{h} = \lambda^{n-1} f_x(x, y, z)$$

Alternatively, we prove this property using the Chain Rule. We use the Chain Rule to differentiate the following equality with respect to x:

$$f(\lambda x, \lambda y, \lambda z) = \lambda^n f(x, y, z)$$

We get

$$f_x(\lambda x, \lambda y, \lambda z) \cdot \frac{\partial(\lambda x)}{\partial x} + f_y(\lambda x, \lambda y, \lambda z) \cdot \frac{\partial(\lambda y)}{\partial x} + f_z(\lambda x, \lambda y, \lambda z) \cdot \frac{\partial(\lambda z)}{\partial x} = \lambda^n f_x(x, y, z)$$

Since $\frac{\partial(\lambda y)}{\partial x} = \frac{\partial(\lambda z)}{\partial x} = 0$ and $\frac{\partial(\lambda x)}{\partial x} = \lambda$, we obtain for $\lambda \neq 0$,

$$\lambda f_x(\lambda x, \lambda y, \lambda z) = \lambda^n f_x(x, y, z) \quad \text{or} \quad f_x(\lambda x, \lambda y, \lambda z) = \lambda^{n-1} f_x(x, y, z)$$

43. Verify Eq. (11) for the functions in Exercise 40.

SOLUTION   Eq. (11) states that if f is homogeneous of degree n, then

$$x\frac{\partial f}{\partial x} + y\frac{\partial f}{\partial y} + z\frac{\partial f}{\partial z} = nf$$

(a) $f(x, y, z) = x^2 y + xyz$. f is homogeneous of degree $n = 3$. The partial derivatives of f are

$$\frac{\partial f}{\partial x} = 2xy + yz, \quad \frac{\partial f}{\partial y} = x^2 + xz, \quad \frac{\partial f}{\partial z} = xy$$

Hence,

$$x\frac{\partial f}{\partial x} + y\frac{\partial f}{\partial y} + z\frac{\partial f}{\partial z} = x(2xy + yz) + y(x^2 + xz) + zxy = 3x^2 y + 3xyz = 3(x^2 y + xyz) = 3f(x, y, z)$$

(b) $f(x, y, z) = 3x + 2y - 8z$. f is homogeneous of degree $n = 1$. We have

$$x \frac{\partial f}{\partial x} + y \frac{\partial f}{\partial y} + z \frac{\partial f}{\partial z} = x \cdot 3 + y \cdot 2 + z \cdot (-8) = 3x + 2y - 8z = 1 \cdot f(x, y, z)$$

(c) $f(x, y, z) = \ln\left(\frac{xy}{z^2}\right)$. f is homogeneous of degree $n = 0$. The partial derivatives of f are

$$\frac{\partial f}{\partial x} = \frac{\frac{y}{z^2}}{\frac{xy}{z^2}} = \frac{1}{x}, \quad \frac{\partial f}{\partial y} = \frac{\frac{x}{z^2}}{\frac{xy}{z^2}} = \frac{1}{y}, \quad \frac{\partial f}{\partial z} = \frac{-2z^{-3}xy}{xyz^{-2}} = -\frac{2}{z}$$

Hence,

$$x \frac{\partial f}{\partial x} + y \frac{\partial f}{\partial y} + z \frac{\partial f}{\partial z} = x \cdot \frac{1}{x} + y \cdot \frac{1}{y} + z \cdot \left(-\frac{2}{z}\right) = 0 = 0 \cdot f(x, y, z)$$

(d) $f(x, y, z) = z^4$. f is homogeneous of degree $n = 4$. We have

$$x \frac{\partial f}{\partial x} + y \frac{\partial f}{\partial y} + z \frac{\partial f}{\partial z} = x \cdot 0 + y \cdot 0 + z \cdot 4z^3 = 4z^4 = 4f(x, y, z)$$

45. Let $r = \sqrt{x_1^2 + \cdots + x_n^2}$ and let $g(r)$ be a function of r. Prove the formulas

$$\frac{\partial g}{\partial x_i} = \frac{x_i}{r} g_r, \qquad \frac{\partial^2 g}{\partial x_i^2} = \frac{x_i^2}{r^2} g_{rr} + \frac{r^2 - x_i^2}{r^3} g_r$$

SOLUTION By the Chain Rule, we have

$$\frac{\partial g}{\partial x_i} = g'(r) \frac{\partial r}{\partial x_i} = g_r \cdot \frac{2x_i}{2\sqrt{x_1^2 + \cdots + x_n^2}} = g_r \frac{x_i}{r}$$

We differentiate $\frac{\partial g}{\partial x_i}$ with respect to x_i. Using the Product Rule we get

$$\frac{\partial^2 g}{\partial x_i^2} = \frac{\partial}{\partial x_i}(g_r) \cdot \frac{x_i}{r} + g_r \frac{\partial}{\partial x_i}\left(\frac{x_i}{r}\right) \tag{1}$$

We use the Chain Rule to compute $\frac{\partial}{\partial x_i}(g_r)$:

$$\frac{\partial}{\partial x_i}(g_r) = \frac{d}{dr}(g_r) \cdot \frac{\partial r}{\partial x_i} = g_{rr} \cdot \frac{2x_i}{2\sqrt{x_1^2 + \cdots + x_n^2}} = g_{rr} \cdot \frac{x_i}{r} \tag{2}$$

We compute $\frac{\partial}{\partial x_i} \cdot \left(\frac{x_i}{r}\right)$ using the Quotient Rule and the Chain Rule:

$$\frac{\partial}{\partial x_i} \cdot \left(\frac{x_i}{r}\right) = \frac{1 \cdot r - x_i \cdot \frac{\partial r}{\partial x_i}}{r^2} = \frac{r - x_i \cdot \frac{x_i}{r}}{r^2} = \frac{r^2 - x_i^2}{r^3} \tag{3}$$

Substituting (2) and (3) in (1), we obtain

$$\frac{\partial^2 g}{\partial x_i^2} = g_{rr} \cdot \frac{x_i}{r} \cdot \frac{x_i}{r} + g_r \frac{r^2 - x_i^2}{r^3} = \frac{x_i^2}{r^2} g_{rr} + \frac{r^2 - x_i^2}{r^3} g_r$$

*In Exercises 47–51, the **Laplace operator** is defined by $\Delta f = f_{xx} + f_{yy}$. A function $f(x, y)$ satisfying the Laplace equation $\Delta f = 0$ is called **harmonic**. A function $f(x, y)$ is called **radial** if $f(x, y) = g(r)$, where $r = \sqrt{x^2 + y^2}$.*

47. Use Eq. (12) to prove that in polar coordinates (r, θ),

$$\Delta f = f_{rr} + \frac{1}{r^2} f_{\theta\theta} + \frac{1}{r} f_r \qquad \boxed{13}$$

SOLUTION The polar coordinates are $x = r\cos\theta$, $y = r\sin\theta$. Hence,

$$\frac{\partial x}{\partial \theta} = -r\sin\theta, \quad \frac{\partial y}{\partial \theta} = r\cos\theta, \quad \frac{\partial x}{\partial r} = \cos\theta, \quad \frac{\partial y}{\partial r} = \sin\theta,$$

$$\frac{\partial^2 x}{\partial \theta^2} = -r\cos\theta, \quad \frac{\partial^2 y}{\partial \theta^2} = -r\sin\theta, \quad \frac{\partial^2 x}{\partial r^2} = \frac{\partial^2 y}{\partial r^2} = 0$$

By Eq. (12) we have

$$f_{\theta\theta} = f_{xx}\left(\frac{\partial x}{\partial \theta}\right)^2 + f_{yy}\left(\frac{\partial y}{\partial \theta}\right)^2 + 2f_{xy}\left(\frac{\partial x}{\partial \theta}\right)\left(\frac{\partial y}{\partial \theta}\right) + f_x\frac{\partial^2 x}{\partial \theta^2} + f_y\frac{\partial^2 y}{\partial \theta^2}$$

$$= f_{xx}\left(r^2\sin^2\theta\right) + f_{yy}\left(r^2\cos^2\theta\right) - \left(2r^2\sin\theta\cos\theta\right)f_{xy} - (r\cos\theta)f_x - (r\sin\theta)f_y \tag{1}$$

and

$$f_{rr} = f_{xx}\left(\frac{\partial x}{\partial r}\right)^2 + f_{yy}\left(\frac{\partial y}{\partial r}\right)^2 + 2f_{xy}\left(\frac{\partial x}{\partial r}\right)\left(\frac{\partial y}{\partial r}\right) + f_x\frac{\partial^2 x}{\partial r^2} + f_y\frac{\partial^2 y}{\partial r^2}$$

$$= f_{xx}\left(\cos^2\theta\right) + f_{yy}\left(\sin^2\theta\right) + (2\cos\theta\sin\theta)f_{xy} \tag{2}$$

$$f_r = f_x\frac{\partial x}{\partial r} + f_y\frac{\partial y}{\partial r} = f_x(\cos\theta) + f_y(\sin\theta) \tag{3}$$

We now compute the right-hand side of the equality we need to prove. Using (1), (2), and (3), we obtain

$$f_{rr} + \frac{1}{r^2}f_{\theta\theta} + \frac{1}{r}f_r = f_{xx}\left(\cos^2\theta\right) + f_{yy}\left(\sin^2\theta\right) + (2\cos\theta\sin\theta)f_{xy} + f_{xx}\left(\sin^2\theta\right)$$

$$+ f_{yy}\left(\cos^2\theta\right) - (2\sin\theta\cos\theta)f_{xy} - \frac{\cos\theta}{r}f_x - \frac{\sin\theta}{r}f_y + f_x\frac{\cos\theta}{r} + f_y\frac{\sin\theta}{r}$$

$$= f_{xx}\left(\cos^2\theta + \sin^2\theta\right) + f_{yy}\left(\sin^2\theta + \cos^2\theta\right)$$

$$= f_{xx} + f_{yy} = \Delta f$$

We thus showed that

$$\Delta f = f_{rr} + \frac{1}{r^2}f_{\theta\theta} + \frac{1}{r}f_r$$

49. Verify that $f(x, y) = x$ and $f(x, y) = y$ are harmonic using both the rectangular and polar expressions for Δf.

SOLUTION We must show that $\Delta f = 0$.

(a) Using the rectangular expression for Δf:

$$\Delta f = f_{xx} + f_{yy}$$

For $f(x, y) = x$ we have $f_x = 1$, $f_y = 0$, hence, $f_{xx} = 0$, $f_{yy} = 0$. Therefore $\Delta f = f_{xx} + f_{yy} = 0 + 0 = 0$. For $f(x, y) = y$ we have $f_y = 1$, $f_x = 0$, hence, $f_{xx} = 0$, $f_{yy} = 0$, and again, $\Delta f = f_{xx} + f_{yy} = 0 + 0 = 0$.

(b) Using the polar expression for Δf,

$$\Delta f = f_{rr} + \frac{1}{r^2}f_{\theta\theta} + \frac{1}{r}f_r \tag{1}$$

Since $x = r\cos\theta$, we have $f(r, \theta) = x = r\cos\theta$. Hence,

$$f_r = \cos\theta, \quad f_\theta = -r\sin\theta, \quad f_{rr} = 0, \quad f_{\theta\theta} = -r\cos\theta$$

We now show that $\Delta f = 0$:

$$\Delta f = f_{rr} + \frac{1}{r^2}f_{\theta\theta} + \frac{1}{r}f_r = 0 + \frac{1}{r^2}\cdot(-r\cos\theta) + \frac{1}{r}\cos\theta = 0$$

Similarly, since $y = r\sin\theta$, we have $f(r, \theta) = y = r\sin\theta$. Hence,

$$f_r = \sin\theta, \quad f_\theta = r\cos\theta, \quad f_{rr} = 0, \quad f_{\theta\theta} = -r\sin\theta$$

Substituting in (1) gives

$$\Delta f = 0 + \frac{1}{r^2}(-r\sin\theta) + \frac{1}{r}\sin\theta = 0$$

51. Use the Product Rule to show that

$$f_{rr} + \frac{1}{r}f_r = r^{-1}\frac{\partial}{\partial r}\left(r\frac{\partial f}{\partial r}\right)$$

Use this formula to show that if f is a radial harmonic function, then $rf_r = C$ for some constant C. Conclude that $f(x, y) = C\ln r + b$ for some constant b.

SOLUTION We show that $f_{rr} + \frac{1}{r} f_r = r^{-1} \frac{\partial}{\partial r} \left(r \frac{\partial f}{\partial r} \right)$. We use the Product Rule to compute the following derivative:

$$\frac{\partial}{\partial r} \left(r \frac{\partial f}{\partial r} \right) = 1 \cdot \frac{\partial f}{\partial r} + r \frac{\partial}{\partial r} \left(\frac{\partial f}{\partial r} \right) = \frac{\partial f}{\partial r} + r \frac{\partial^2 f}{\partial r^2} = f_r + r f_{rr} = r \left(f_{rr} + \frac{1}{r} f_r \right)$$

Hence,

$$f_{rr} + \frac{1}{r} f_r = r^{-1} \frac{\partial}{\partial r} \left(r \frac{\partial f}{\partial r} \right) \tag{1}$$

Now, suppose that $f(x, y)$ is a radial harmonic function. Since f is radial, $f(x, y) = g(r)$, therefore $f_{\theta\theta} = 0$. Substituting in the polar expressions for Δf gives

$$\Delta f = f_{rr} + \frac{1}{r^2} f_{\theta\theta} + \frac{1}{r} f_r = f_{rr} + \frac{1}{r} f_r = 0$$

Combining with (1), we get

$$r^{-1} \frac{\partial}{\partial r} \left(r \frac{\partial f}{\partial r} \right) = 0 \quad \text{or} \quad \frac{\partial}{\partial r} \left(r \frac{\partial f}{\partial r} \right) = 0$$

yielding

$$r \frac{\partial f}{\partial r} = C \quad \Rightarrow \quad f_r = \frac{C}{r}$$

We now integrate the two sides to obtain

$$\int f_r \, dr = \int \frac{C}{r} \, dr \quad \text{or} \quad f(r) = C \ln r + b.$$

14.7 Optimization in Several Variables (LT Section 15.7)

Preliminary Questions

1. The functions $f(x, y) = x^2 + y^2$ and $g(x, y) = x^2 - y^2$ both have a critical point at $(0, 0)$. How is the behavior of the two functions at the critical point different?

SOLUTION Let $f(x, y) = x^2 + y^2$ and $g(x, y) = x^2 - y^2$. In the domain $\mathbf{R}^2$, the partial derivatives of f and g are

$$f_x = 2x, \quad f_{xx} = 2, \quad f_y = 2y, \quad f_{yy} = 2, \quad f_{xy} = 0$$

$$g_x = 2x, \quad g_{xx} = 2, \quad g_y = -2y, \quad g_{yy} = -2, \quad g_{xy} = 0$$

Therefore, $f_x = f_y = 0$ at $(0, 0)$ and $g_x = g_y = 0$ at $(0, 0)$. That is, the two functions have one critical point, which is the origin. Since the discriminant of f is $D = 4 > 0$, $f_{xx} > 0$, and the discriminant of g is $D = -4 < 0$, f has a local minimum (which is also a global minimum) at the origin, whereas g has a saddle point there. Moreover, since $\lim_{y \to \infty} g(0, y) = -\infty$ and $\lim_{x \to \infty} g(x, 0) = \infty$, g does not have global extrema on the plane. Similarly, f does not have a global maximum but does have a global minimum, which is $f(0, 0) = 0$.

2. Identify the points indicated in the contour maps as local minima, local maxima, saddle points, or neither (Figure 15).

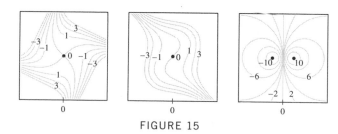

FIGURE 15

SOLUTION If $f(P)$ is a local minimum or maximum, then the nearby level curves are closed curves encircling P. In Figure (C), f increases in all directions emanating from P and decreases in all directions emanating from Q. Hence, f has a local minimum at P and local maximum at Q.

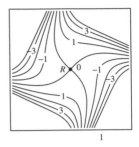

In Figure (A), the level curves through the point R consist of two intersecting lines that divide the neighborhood near R into four regions. f is decreasing in some directions and increasing in other directions. Therefore, R is a saddle point.

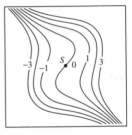

Figure (A)

Point S in Figure (B) is neither a local extremum nor a saddle point of f.

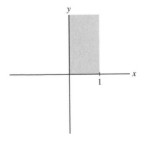

Figure (B)

3. Let $f(x, y)$ be a continuous function on a domain $\mathcal{D}$ in $\mathbf{R}^2$. Determine which of the following statements are true:

(a) If $\mathcal{D}$ is closed and bounded, then f takes on a maximum value on $\mathcal{D}$.

(b) If $\mathcal{D}$ is neither closed nor bounded, then f does not take on a maximum value of $\mathcal{D}$.

(c) $f(x, y)$ need not have a maximum value on the domain $\mathcal{D}$ defined by $0 \leq x \leq 1, 0 \leq y \leq 1$.

(d) A continuous function takes on neither a minimum nor a maximum value on the open quadrant

$$\{(x, y) : x > 0, y > 0\}$$

SOLUTION

(a) This statement is true. It follows by the Theorem on Existence of Global Extrema.

(b) The statement is false. Consider the constant function $f(x, y) = 2$ in the following domain:

$$D = \{(x, y) : 0 < x \leq 1, \ 0 \leq y < \infty\}$$

Obviously f is continuous and D is neither closed nor bounded. However, f takes on a maximum value (which is 2) on D.

(c) The domain $D = \{(x, y) : 0 \le x, y \le 1\}$ is the following rectangle:

$$D = \{(x, y) : 0 \le x, y \le 1\}$$

D is closed and bounded, hence f takes on a maximum value on D. Thus the statement is false.

(d) The statement is false. The constant function $f(x, y) = c$ takes on minimum and maximum values on the open quadrant.

Exercises

1. Let $P = (a, b)$ be a critical point of $f(x, y) = x^2 + y^4 - 4xy$.

(a) First use $f_x(x, y) = 0$ to show that $a = 2b$. Then use $f_y(x, y) = 0$ to show that $P = (0, 0)$, $(2\sqrt{2}, \sqrt{2})$, or $(-2\sqrt{2}, -\sqrt{2})$.

(b) Referring to Figure 16, determine the local minima and saddle points of $f(x, y)$ and find the absolute minimum value of $f(x, y)$.

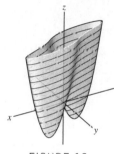

FIGURE 16

SOLUTION

(a) We find the partial derivatives:

$$f_x(x, y) = \frac{\partial}{\partial x}\left(x^2 + y^4 - 4xy\right) = 2x - 4y$$

$$f_y(x, y) = \frac{\partial}{\partial y}\left(x^2 + y^4 - 4xy\right) = 4y^3 - 4x$$

Since $P = (a, b)$ is a critical point, $f_x(a, b) = 0$. That is,

$$2a - 4b = 0 \quad \Rightarrow \quad a = 2b$$

Also $f_y(a, b) = 0$, hence,

$$4b^3 - 4a = 0 \quad \Rightarrow \quad a = b^3$$

We obtain the following equations for the critical points (a, b):

$$\begin{cases} a = 2b \\ a = b^3 \end{cases}$$

Equating the two equations, we get

$$2b = b^3$$

$$b^3 - 2b = b(b^2 - 2) = 0 \quad \Rightarrow \quad \begin{cases} b_1 = 0 \\ b_2 = \sqrt{2} \\ b_3 = -\sqrt{2} \end{cases}$$

Since $a = 2b$, we have $a_1 = 0$, $a_2 = 2\sqrt{2}$, $a_3 = -2\sqrt{2}$. The critical points are thus

$$P_1 = (0, 0), \quad P_2 = \left(2\sqrt{2}, \sqrt{2}\right), \quad P_3 = \left(-2\sqrt{2}, -\sqrt{2}\right)$$

(b) Referring to Figure 14, we see that $P_1 = (0, 0)$ is a saddle point and $P_2 = \left(2\sqrt{2}, \sqrt{2}\right)$, $P_3 = \left(-2\sqrt{2}, -\sqrt{2}\right)$ are local minima. The absolute minimum value of f is -4.

3. Find the critical points of

$$f(x, y) = 8y^4 + x^2 + xy - 3y^2 - y^3$$

Use the contour map in Figure 18 to determine their nature (local minimum, local maximum, or saddle point).

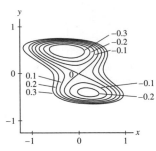

FIGURE 18 Contour map of $f(x, y) = 8y^4 + x^2 + xy - 3y^2 - y^3$.

SOLUTION The critical points are the solutions of $f_x = 0$ and $f_y = 0$. That is,

$$f_x(x, y) = 2x + y = 0$$
$$f_y(x, y) = 32y^3 + x - 6y - 3y^2 = 0$$

The first equation gives $y = -2x$. We substitute in the second equation and solve for x. This gives

$$32(-2x)^3 + x - 6(-2x) - 3(-2x)^2 = 0$$
$$-256x^3 + 13x - 12x^2 = 0$$
$$-x(256x^2 + 12x - 13) = 0$$

Hence $x = 0$ or $256x^2 + 12x - 13 = 0$. Solving the quadratic,

$$x_{1,2} = \frac{-12 \pm \sqrt{12^2 - 4 \cdot 256 \cdot (-13)}}{512} = \frac{-12 \pm 116}{512} \quad \Rightarrow \quad x = \frac{13}{64} \quad \text{or} \quad -\frac{1}{4}$$

Substituting in $y = -2x$ gives the y-coordinates of the critical points. The critical points are thus

$$(0, 0), \quad \left(\frac{13}{64}, -\frac{13}{32}\right), \quad \left(-\frac{1}{4}, \frac{1}{2}\right)$$

We now use the contour map to determine the type of each critical point. The level curves through $(0, 0)$ consist of two intersecting lines that divide the neighborhood near $(0, 0)$ into four regions. The function is decreasing in the y direction and increasing in the x-direction. Therefore, $(0, 0)$ is a saddle point. The level curves near the critical points $\left(\frac{13}{64}, -\frac{13}{32}\right)$ and $\left(-\frac{1}{4}, \frac{1}{2}\right)$ are closed curves encircling the points, hence these are local minima or maxima. The graph shows that both $\left(\frac{13}{64}, -\frac{13}{32}\right)$ and $\left(-\frac{1}{4}, \frac{1}{2}\right)$ are local minima.

5. Let $f(x, y) = y^2x - yx^2 + xy$.
(a) Show that the critical points (x, y) satisfy the equations

$$y(y - 2x + 1) = 0, \qquad x(2y - x + 1) = 0$$

(b) Show that f has three critical points where $x = 0$ or $y = 0$ (or both) and one critical point where x and y are nonzero.
(c) Use the Second Derivative Test to determine the nature of the critical points.

SOLUTION
(a) The critical points are the solutions of the two equations $f_x(x, y) = 0$ and $f_y(x, y) = 0$. That is,

$$f_x(x, y) = y^2 - 2yx + y = 0 \qquad y(y - 2x + 1) = 0$$
$$\qquad\qquad\qquad\qquad\qquad\qquad \Rightarrow$$
$$f_y(x, y) = 2yx - x^2 + x = 0 \qquad x(2y - x + 1) = 0$$

(b) We find the critical points by solving the equations obtained in part (a):

$$y(y - 2x + 1) = 0 \tag{1}$$

$$x(2y - x + 1) = 0 \tag{2}$$

Equation (1) implies that $y = 0$ or $y = 2x - 1$. Substituting $y = 0$ in (2) and solving for x gives

$$x(-x + 1) = 0 \quad \Rightarrow \quad x = 0 \quad \text{or} \quad x = 1$$

We obtain the solutions $(0, 0)$ and $(1, 0)$. We now substitute $y = 2x - 1$ in (2) and solve for x. We get

$$x(4x - 2 - x + 1) = 0$$

$$x(3x - 1) = 0 \quad \Rightarrow \quad x = 0 \quad \text{or} \quad x = \frac{1}{3}$$

We compute the y-coordinate, using $y = 2x - 1$:

$$y = 2 \cdot 0 - 1 = -1$$

$$y = 2 \cdot \frac{1}{3} - 1 = -\frac{1}{3}$$

We obtain the solutions $(0, -1)$ and $\left(\frac{1}{3}, -\frac{1}{3}\right)$. To summarize, the critical points are $(0, 0)$, $(1, 0)$, $(0, -1)$, and $\left(\frac{1}{3}, -\frac{1}{3}\right)$. Three of the critical points have at least one zero coordinate, and one has two nonzero coordinates.

(c) We compute the second-order partial derivatives:

$$f_{xx}(x, y) = \frac{\partial}{\partial x}(y^2 - 2yx + y) = -2y$$

$$f_{yy}(x, y) = \frac{\partial}{\partial y}(2yx - x^2 + x) = 2x$$

$$f_{xy}(x, y) = \frac{\partial}{\partial y}(y^2 - 2yx + y) = 2y - 2x + 1$$

The discriminant is

$$D(x, y) = f_{xx}f_{yy} - f_{xy}^2 = -2y \cdot 2x - (2y - 2x + 1)^2 = -4xy - (2y - 2x + 1)^2$$

We now apply the Second Derivative Test. We first compute the discriminants at the critical points:

$$D(0, 0) = -1 < 0$$

$$D(1, 0) = -1 < 0$$

$$D(0, -1) = -1 < 0$$

$$D\left(\frac{1}{3}, -\frac{1}{3}\right) = -4 \cdot \frac{1}{3}\left(-\frac{1}{3}\right) - \left(-\frac{2}{3} - \frac{2}{3} + 1\right)^2 - \frac{1}{3} > 0,$$

$$f_{xx}\left(\frac{1}{3}, -\frac{1}{3}\right) = -2 \cdot \left(-\frac{1}{3}\right) = \frac{2}{3} > 0$$

The Second Derivative Test implies that the points $(0, 0)$, $(1, 0)$, and $(0, -1)$ are saddle points, and $f\left(\frac{1}{3}, -\frac{1}{3}\right)$ is a local minimum.

In Exercises 7–23, find the critical points of the function. Then use the Second Derivative Test to determine whether they are local minima, local maxima, or saddle points (or state that the test fails).

7. $f(x, y) = x^2 + y^2 - xy + x$

SOLUTION

Step 1. Find the critical points. We set the first-order partial derivatives of $f(x, y) = x^2 + y^2 - xy + x$ equal to zero and solve:

$$f_x(x, y) = 2x - y + 1 = 0 \tag{1}$$

$$f_y(x, y) = 2y - x = 0 \tag{2}$$

Equation (2) implies that $x = 2y$. Substituting in (1) and solving for y gives

$$2 \cdot 2y - y + 1 = 0 \quad \Rightarrow \quad 3y = -1 \quad \Rightarrow \quad y = -\frac{1}{3}$$

The corresponding value of x is $x = 2 \cdot \left(-\frac{1}{3}\right) = -\frac{2}{3}$. The critical point is $\left(-\frac{2}{3}, -\frac{1}{3}\right)$.

Step 2. Compute the Discriminant. We find the second-order partials:

$$f_{xx}(x, y) = 2, \quad f_{yy}(x, y) = 2, \quad f_{xy}(x, y) = -1$$

The discriminant is

$$D(x, y) = f_{xx} f_{yy} - f_{xy}^2 = 2 \cdot 2 - (-1)^2 = 3$$

Step 3. Applying the Second Derivative Test. We have

$$D\left(-\frac{2}{3}, -\frac{1}{3}\right) = 3 > 0 \quad \text{and} \quad f_{xx}\left(-\frac{2}{3}, -\frac{1}{3}\right) = 2 > 0$$

The Second Derivative Test implies that $f\left(-\frac{2}{3}, -\frac{1}{3}\right)$ is a local minimum.

9. $f(x, y) = x^3 + 2xy - 2y^2 - 10x$

SOLUTION

Step 1. Find the critical points. We set the first-order partial derivatives of $f(x, y) = x^3 + 2xy - 2y^2 - 10x$ equal to zero and solve:

$$f_x(x, y) = 3x^2 + 2y - 10 = 0 \tag{1}$$
$$f_y(x, y) = 2x - 4y = 0 \tag{2}$$

Equation (2) implies that $x = 2y$. We substitute in (1) and solve for y. This gives

$$3 \cdot (2y)^2 + 2y - 10 = 0$$
$$12y^2 + 2y - 10 = 0$$
$$6y^2 + y - 5 = 0$$

$$y_{1,2} = \frac{-1 \pm \sqrt{1 - 4 \cdot 6 \cdot (-5)}}{12} = \frac{-1 \pm 11}{12} \quad \Rightarrow \quad y_1 = -1 \quad \text{and} \quad y_2 = \frac{5}{6}$$

We find the x-coordinates using $x = 2y$:

$$x_1 = 2 \cdot (-1) = -2, \quad x_2 = 2 \cdot \frac{5}{6} = \frac{5}{3}$$

The critical points are thus $(-2, -1)$ and $\left(\frac{5}{3}, \frac{5}{6}\right)$.

Step 2. Compute the Discriminant. We find the second-order partials:

$$f_{xx}(x, y) = 6x, \quad f_{yy}(x, y) = -4, \quad f_{xy}(x, y) = 2$$

The discriminant is

$$D(x, y) = f_{xx} f_{yy} - f_{xy}^2 = 6x \cdot (-4) - 2^2 = -24x - 4$$

Step 3. Apply the Second Derivative Test. We have

$$D(-2, -1) = -24 \cdot (-2) - 4 = 44 > 0,$$
$$f_{xx}(-2, -1) = 6 \cdot (-2) = -12 < 0$$
$$D\left(\frac{5}{3}, \frac{5}{6}\right) = -24 \cdot \frac{5}{3} - 4 = -44 < 0$$

We conclude that $f(-2, -1)$ is a local maximum and $\left(\frac{5}{3}, \frac{5}{6}\right)$ is a saddle point.

11. $f(x, y) = 4x - 3x^3 - 2xy^2$

SOLUTION

Step 1. Find the critical points. We set the first-order derivatives of $f(x, y) = 4x - 3x^3 - 2xy^2$ equal to zero and solve:

$$f_x(x, y) = 4 - 9x^2 - 2y^2 = 0 \tag{1}$$
$$f_y(x, y) = -4xy = 0 \tag{2}$$

Equation (2) implies that $x = 0$ or $y = 0$. If $x = 0$, then equation (1) gives

$$4 - 2y^2 = 0 \quad \Rightarrow \quad y^2 = 2 \quad \Rightarrow \quad y = \sqrt{2}, \quad y = -\sqrt{2}$$

If $y = 0$, then equation (1) gives

$$4 - 9x^2 = 0 \quad \Rightarrow \quad 9x^2 = 4 \quad \Rightarrow \quad x = \frac{2}{3}, \quad x = -\frac{2}{3}$$

The critical points are therefore

$$\left(0, \sqrt{2}\right), \quad \left(0, -\sqrt{2}\right), \quad \left(\frac{2}{3}, 0\right), \quad \left(-\frac{2}{3}, 0\right)$$

Step 2. Compute the discriminant. The second-order partials are

$$f_{xx}(x, y) = -18x, \quad f_{yy}(x, y) = -4x, \quad f_{xy} = -4y$$

The discriminant is thus

$$D(x, y) = f_{xx} f_{yy} - f_{xy}^2 = -18x \cdot (-4x) - (-4y)^2 = 72x^2 - 16y^2$$

Step 3. Apply the Second Derivative Test. We have

$$D\left(0, \sqrt{2}\right) = -32 < 0$$

$$D\left(0, -\sqrt{2}\right) = -32 < 0$$

$$D\left(\frac{2}{3}, 0\right) = 72 \cdot \frac{4}{9} = 32 > 0,$$

$$f_{xx}\left(\frac{2}{3}, 0\right) = -18 \cdot \frac{2}{3} = -12 < 0$$

$$D\left(-\frac{2}{3}, 0\right) = 72 \cdot \frac{4}{9} = 32 > 0,$$

$$f_{xx}\left(-\frac{2}{3}, 0\right) = -18 \cdot \left(-\frac{2}{3}\right) = 12 > 0$$

The Second Derivative Test implies that the points $\left(0, \pm\sqrt{2}\right)$ are the saddle points, $f\left(\frac{2}{3}, 0\right)$ is a local maximum, and $f\left(-\frac{2}{3}, 0\right)$ is a local minimum.

13. $f(x, y) = x^4 + y^4 - 4xy$

SOLUTION

Step 1. Find the critical points. We set the first-order derivatives of $f(x, y) = x^4 + y^4 - 4xy$ equal to zero and solve:

$$f_x(x, y) = 4x^3 - 4y = 0, \quad f_y(x, y) = 4y^3 - 4x = 0 \tag{1}$$

Equation (1) implies that $y = x^3$. Substituting in (2) and solving for x, we obtain

$$\left(x^3\right)^3 - x = x^9 - x = x(x^8 - 1) = 0 \quad \Rightarrow \quad x = 0, \quad x = 1, \quad x = -1$$

The corresponding y coordinates are

$$y = 0^3 = 0, \quad y = 1^3 = 1, \quad y = (-1)^3 = -1$$

The critical points are therefore

$$(0, 0), \quad (1, 1), \quad (-1, -1)$$

Step 2. Compute the discriminant. We find the second-order partials:

$$f_{xx}(x, y) = 12x^2, \quad f_{yy}(x, y) = 12y^2, \quad f_{xy}(x, y) = -4$$

The discriminant is thus

$$D(x, y) = f_{xx} f_{yy} - f_{xy}^2 = 12x^2 \cdot 12y^2 - (-4)^2 = 144x^2 y^2 - 16$$

Step 3. Apply the Second Derivative Test. We have

$$D(0, 0) = -16 < 0$$

$$D(1, 1) = 144 - 16 = 128 > 0, \quad f_{xx}(1, 1) = 12 > 0$$

$$D(-1, -1) = 144 - 16 = 128 > 0, \quad f_{xx}(-1, -1) = 12 > 0$$

We conclude that $(0, 0)$ is a saddle point, whereas $f(1, 1)$ and $f(-1, -1)$ are local minima.

15. $f(x, y) = xye^{-x^2 - y^2}$

SOLUTION

Step 1. Find the critical points. We compute the partial derivatives of $f(x, y) = xye^{-x^2 - y^2}$, using the Product Rule and the Chain Rule:

$$f_x(x, y, z) = y\left(1 \cdot e^{-x^2 - y^2} + xe^{-x^2 - y^2} \cdot (-2x)\right) = ye^{-x^2 - y^2}\left(1 - 2x^2\right)$$

$$f_y(x, y, z) = x\left(1 \cdot e^{-x^2 - y^2} + ye^{-x^2 - y^2} \cdot (-2y)\right) = xe^{-x^2 - y^2}\left(1 - 2y^2\right)$$

We set the partial derivatives equal to zero and solve to find the critical points. This gives

$$ye^{-x^2 - y^2}\left(1 - 2x^2\right) = 0$$

$$xe^{-x^2 - y^2}\left(1 - 2y^2\right) = 0$$

Since $e^{-x^2 - y^2} \neq 0$, the first equation gives $y = 0$ or $1 - 2x^2 = 0$, that is, $y = 0$, $x = \frac{1}{\sqrt{2}}$, $x = -\frac{1}{\sqrt{2}}$. We substitute each of these values in the second equation and solve to obtain

$$y = 0: \quad xe^{-x^2} = 0 \quad \Rightarrow \quad x = 0$$

$$x = \frac{1}{\sqrt{2}}: \quad \frac{1}{\sqrt{2}}e^{-\frac{1}{2} - y^2}\left(1 - 2y^2\right) = 0 \quad \Rightarrow \quad 1 - 2y^2 = 0 \quad \Rightarrow \quad y = \pm\frac{1}{\sqrt{2}}$$

$$x = -\frac{1}{\sqrt{2}}: \quad -\frac{1}{\sqrt{2}}e^{-\frac{1}{2} - y^2}\left(1 - 2y^2\right) = 0 \quad \Rightarrow \quad 1 - 2y^2 = 0 \quad \Rightarrow \quad y = \pm\frac{1}{\sqrt{2}}$$

We obtain the following critical points: $(0, 0)$,

$$\left(\frac{1}{\sqrt{2}}, \frac{1}{\sqrt{2}}\right), \quad \left(\frac{1}{\sqrt{2}}, -\frac{1}{\sqrt{2}}\right), \quad \left(-\frac{1}{\sqrt{2}}, \frac{1}{\sqrt{2}}\right), \quad \left(-\frac{1}{\sqrt{2}}, -\frac{1}{\sqrt{2}}\right)$$

Step 2. Compute the second-order partials.

$$f_{xx}(x, y) = y\frac{\partial}{\partial x}\left(e^{-x^2 - y^2}\left(1 - 2x^2\right)\right) = y\left(e^{-x^2 - y^2}(-2x)\left(1 - 2x^2\right) + e^{-x^2 - y^2}(-4x)\right)$$

$$= -2xye^{-x^2 - y^2}\left(3 - 2x^2\right)$$

$$f_{yy}(x, y) = x\frac{\partial}{\partial y}\left(e^{-x^2 - y^2}\left(1 - 2y^2\right)\right) = x\left(e^{-x^2 - y^2}(-2y)\left(1 - 2y^2\right) + e^{-x^2 - y^2}(-4y)\right)$$

$$= -2yxe^{-x^2 - y^2}\left(3 - 2y^2\right)$$

$$f_{xy}(x, y) = \frac{\partial}{\partial y}f_x = \left(1 - 2x^2\right)\frac{\partial}{\partial y}\left(ye^{-x^2 - y^2}\right) = \left(1 - 2x^2\right)\left(1 \cdot e^{-x^2 - y^2} + ye^{-x^2 - y^2}(-2y)\right)$$

$$= e^{-x^2 - y^2}\left(1 - 2x^2\right)\left(1 - 2y^2\right)$$

The discriminant is

$$D(x, y) = f_{xx}f_{yy} - f_{xy}^2$$

Step 3. Apply the Second Derivative Test. We construct the following table:

Critical Point	f_{xx}	f_{yy}	f_{xy}	D	Type
$(0, 0)$	0	0	1	-1	$D < 0$, saddle point
$\left(\frac{1}{\sqrt{2}}, \frac{1}{\sqrt{2}}\right)$	$-\frac{2}{e}$	$-\frac{2}{e}$	0	$\frac{4}{e^2}$	$D > 0$, $f_{xx} < 0$ local maximum
$\left(\frac{1}{\sqrt{2}}, -\frac{1}{\sqrt{2}}\right)$	$\frac{2}{e}$	$\frac{2}{e}$	0	$\frac{4}{e^2}$	$D > 0$, $f_{xx} > 0$ local minimum
$\left(-\frac{1}{\sqrt{2}}, \frac{1}{\sqrt{2}}\right)$	$\frac{2}{e}$	$\frac{2}{e}$	0	$\frac{4}{e^2}$	$D > 0$, $f_{xx} > 0$ local minimum
$\left(-\frac{1}{\sqrt{2}}, -\frac{1}{\sqrt{2}}\right)$	$-\frac{2}{e}$	$-\frac{2}{e}$	0	$\frac{4}{e^2}$	$D > 0$, $f_{xx} < 0$ local maximum

17. $f(x, y) = \sin(x + y) - \cos x$

SOLUTION

Step 1. Find the critical points. We set the first-order derivatives of $f(x, y) = \sin(x + y) - \cos x$ equal to zero and solve:

$$f_x(x, y) = \cos(x + y) + \sin x = 0$$

$$f_y(x, y) = \cos(x + y) = 0$$

First consider the second equation, $\cos(x + y) = 0$ this is when

$$x + y = \frac{(2k + 1)\pi}{2} \rightarrow y = \frac{(2k + 1)\pi}{2} - x \text{ where } k \text{ is an integer}$$

Then setting the two equations equal to one another we gain $\sin x = 0$ which are the values:

$$x = 0, \pm\pi, \pm 2\pi, \cdots = \pm k\pi \text{ where } k \text{ is an integer.}$$

Thus we have:

$$x = k\pi \text{ and } y = \frac{(2n + 1)\pi}{2} \text{ where } n, k \text{ are integers}$$

Step 2. Compute the discriminant. We find the second-order partial derivatives:

$$f_{xx}(x, y) = -\sin(x + y) + \cos x, \quad f_{yy}(x, y) = -\sin(x + y), \quad f_{xy}(x, y) = -\sin(x + y)$$

The discriminant is:

$$D(x, y) = f_{xx}f_{yy} - f_{xy}^2 = (-\sin(x + y) + \cos x)(-\sin(x + y)) - \sin^2(x + y) = -\cos(x)\sin(x + y)$$

Step 3. Apply the Second Derivative Test. We have

$$D = \begin{cases} +1, & \text{if } y = -\dfrac{4n + 3}{2}\pi \\ \\ -1, & y = \dfrac{4n + 1}{2}\pi \end{cases}$$

Therefore, the points $\left(k\pi, \dfrac{4n + 1}{2}\pi\right)$ are saddle points since $D < 0$.

Since $D > 0$ for the points $\left(k\pi, \dfrac{4n + 3}{2}\pi\right)$, we need to examine f_{xx}. The results show:

$$f_{xx} > 0 \text{ if } k \text{ is even and } f_{xx} < 0 \text{ if } k \text{ is odd}$$

Thus:

$$\left(k\pi, \frac{4n + 3}{2}\pi\right) \text{ are local minima if } k \text{ is even}$$

while

$$\left(k\pi, \frac{4n + 3}{2}\pi\right) \text{ are local maxima if } k \text{ is odd}$$

19. $f(x, y) = \ln x + 2\ln y - x - 4y$

SOLUTION

Step 1. Find the critical points. We set the first-order partials of $f(x, y) = \ln x + 2\ln y - x - 4y$ equal to zero and solve:

$$f_x(x, y) = \frac{1}{x} - 1 = 0, \quad f_y(x, y) = \frac{2}{y} - 4 = 0$$

The first equation gives $x = 1$, and the second equation gives $y = \frac{1}{2}$. We obtain the critical point $\left(1, \frac{1}{2}\right)$. Notice that f_x and f_y do not exist if $x = 0$ or $y = 0$, respectively, but these are not critical points since they are not in the domain of f. The critical point is thus $\left(1, \frac{1}{2}\right)$.

Step 2. Compute the discriminant. We find the second-order partials:

$$f_{xx}(x, y) = -\frac{1}{x^2}, \quad f_{yy}(x, y) = -\frac{2}{y^2}, \quad f_{xy}(x, y) = 0$$

The discriminant is

$$D(x, y) = f_{xx}f_{yy} - f_{xy}^2 = \frac{2}{x^2 y^2}$$

Step 3. Apply the Second Derivative Test. We have

$$D\left(1, \frac{1}{2}\right) = \frac{2}{1^2 \cdot \left(\frac{1}{2}\right)^2} = 8 > 0, \quad f_{xx}\left(1, \frac{1}{2}\right) = -\frac{1}{1^2} = -1 < 0$$

We conclude that $f\left(1, \frac{1}{2}\right)$ is a local maximum.

21. $f(x, y) = x - y^2 - \ln(x + y)$

SOLUTION

Step 1. Find the critical points. We set the partial derivatives of $f(x, y) = x - y^2 - \ln(x + y)$ equal to zero and solve.

$$f_x(x, y) = 1 - \frac{1}{x + y} = 0, \quad f_y(x, y) = -2y - \frac{1}{x + y} = 0$$

The first equation implies that $\frac{1}{x+y} = 1$. Substituting in the second equation gives

$$-2y - 1 = 0 \quad \Rightarrow \quad 2y = -1 \quad \Rightarrow \quad y = -\frac{1}{2}$$

We substitute $y = -\frac{1}{2}$ in the first equation and solve for x:

$$1 - \frac{1}{x - \frac{1}{2}} = 0 \quad \Rightarrow \quad x - \frac{1}{2} = 1 \quad \Rightarrow \quad x = \frac{3}{2}$$

We obtain the critical point $\left(\frac{3}{2}, -\frac{1}{2}\right)$. Notice that although f_x and f_y do not exist where $x + y = 0$, these are not critical points since f is not defined at these points.

Step 2. Compute the discriminant. We compute the second-order partial derivatives:

$$f_{xx}(x, y) = \frac{\partial}{\partial x}\left(1 - \frac{1}{x + y}\right) = \frac{1}{(x + y)^2}$$

$$f_{yy}(x, y) = \frac{\partial}{\partial y}\left(-2y - \frac{1}{x + y}\right) = -2 + \frac{1}{(x + y)^2}$$

$$f_{xy}(x, y) = \frac{\partial}{\partial y}\left(1 - \frac{1}{x + y}\right) = \frac{1}{(x + y)^2}$$

The discriminant is

$$D(x, y) = f_{xx} f_{yy} - f_{xy}^2 = \frac{1}{(x + y)^2}\left(-2 + \frac{1}{(x + y)^2}\right) - \frac{1}{(x + y)^4} = \frac{-2}{(x + y)^2}$$

Step 3. Apply the Second Derivative Test. We have

$$D\left(\frac{3}{2}, -\frac{1}{2}\right) = \frac{-2}{\left(\frac{3}{2} - \frac{1}{2}\right)^2} = -2 < 0$$

We conclude that $\left(\frac{3}{2}, -\frac{1}{2}\right)$ is a saddle point.

23. $f(x, y) = (x + 3y)e^{y - x^2}$

SOLUTION

Step 1. Find the critical points. We compute the partial derivatives of $f(x, y) = (x + 3y)e^{y - x^2}$, using the Product Rule and the Chain Rule:

$$f_x(x, y) = 1 \cdot e^{y - x^2} + (x + 3y)e^{y - x^2} \cdot (-2x) = e^{y - x^2}\left(1 - 2x^2 - 6xy\right)$$

$$f_y(x, y) = 3e^{y - x^2} + (x + 3y)e^{y - x^2} \cdot 1 = e^{y - x^2}(3 + x + 3y)$$

We set the partial derivatives equal to zero and solve to find the critical points:

$$e^{y - x^2}\left(1 - 2x^2 - 6xy\right) = 0$$

$$e^{y - x^2}(3 + x + 3y) = 0$$

Since $e^{y-x^2} \neq 0$, we obtain the following equations:

$$1 - 2x^2 - 6xy = 0$$

$$3 + x + 3y = 0$$

The second equation gives $x = -3(1 + y)$. We substitute for x in the first equation and solve for y:

$$1 - 2 \cdot 9(1 + y)^2 + 18(1 + y)y = 0$$

$$1 - 18\left(1 + 2y + y^2\right) + 18\left(y + y^2\right) = 0$$

$$-17 - 18y = 0 \quad \Rightarrow \quad y = -\frac{17}{18}, \quad x = -3\left(1 - \frac{17}{18}\right) = -\frac{1}{6}$$

The critical point is $\left(-\frac{1}{6}, -\frac{17}{18}\right)$.

Step 2. Compute the second-order partials.

$$f_{xx}(x, y) = \frac{\partial}{\partial x} f_x = e^{y-x^2}(-2x)\left(1 - 2x^2 - 6xy\right) + e^{y-x^2}(-4x - 6y) = 2e^{y-x^2}\left(2x^3 + 6x^2 y - 3x - 3y\right)$$

$$f_{yy}(x, y) = \frac{\partial}{\partial y} f_y = e^{y-x^2}(3 + x + 3y) + e^{y-x^2} \cdot 3 = e^{y-x^2}(6 + x + 3y)$$

$$f_{xy}(x, y) = \frac{\partial}{\partial x} f_y = e^{y-x^2}(-2x)(3 + x + 3y) + e^{y-x^2} \cdot 1 = e^{y-x^2}\left(1 - 6xy - 2x^2 - 6x\right)$$

The discriminant is

$$D(x, y) = f_{xx} f_{yy} - f_{xy}^2$$

Step 3. Apply the Second Derivative Test. We obtain the following table:

Critical Point	f_{xx}	f_{yy}	f_{xy}	D	Type
$\left(-\frac{1}{6}, -\frac{17}{18}\right)$	2.4	1.13	0.38	2.57	$D > 0$, $f_{xx} > 0$, local minimum

25. Prove that the function $f(x, y) = \frac{1}{3}x^3 + \frac{2}{3}y^{3/2} - xy$ satisfies $f(x, y) \geq 0$ for $x \geq 0$ and $y \geq 0$.

(a) First, verify that the set of critical points of f is the parabola $y = x^2$ and that the Second Derivative Test fails for these points.

(b) Show that for fixed b, the function $g(x) = f(x, b)$ is concave up for $x > 0$ with a critical point at $x = b^{1/2}$.

(c) Conclude that $f(a, b) \geq f(b^{1/2}, b) = 0$ for all $a, b \geq 0$.

SOLUTION

(a) To find the critical points, we need the first-order partial derivatives, set them equal to zero and solve:

$$f_x(x, y) = x^2 - y = 0, \quad f_y(x, y) = y^{1/2} - x = 0$$

This gives us:

$$y = x^2$$

as the solution set for the critical points.

Now to compute the discriminant, we need the second-order partials

$$f_{xx}(x, y) = 2x, \quad f_{yy}(x, y) = \frac{1}{2}y^{-1/2}, \quad f_{xy}(x, y) = -1$$

Thus the discriminant is

$$D(x, y) = \frac{x}{\sqrt{y}} - 1$$

Since $y = x^2$ is the solution set for the critical points we see:

$$D(x, y) = 1 - 1 = 0$$

Therefore the Second Derivative Test is inconclusive and fails us.

(b) If we fix a value b and consider $g(x) = f(x, b) = \frac{1}{3}x^3 + \frac{2}{3}b^{3/2} - bx$ to find the concavity, we see

$$g'(x) = x^2 - b, \quad g''(x) = 2x$$

Then certainly, for $x > 0$, this function is concave up. The critical point will occur at the point when $x^2 - b = 0$ or $x = b^{1/2}$.

(c) Now, since for fixed b, we know that $g(x) = f(x, b)$ is concave up if $x > 0$, and the critical point is $x = b^{1/2}$. Therefore

$$f(a, b) \geq f(b^{1/2}, b) = 0 \text{ for all } b \geq 0$$

27. *CAS*   Use a computer algebra system to find a numerical approximation to the critical point of

$$f(x, y) = (1 - x + x^2)e^{y^2} + (1 - y + y^2)e^{x^2}$$

Apply the Second Derivative Test to confirm that it corresponds to a local minimum as in Figure 20.

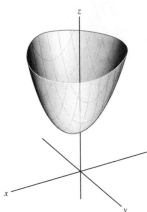

FIGURE 20   Plot of $f(x, y) = (1 - x + x^2)e^{y^2} + (1 - y + y^2)e^{x^2}$.

SOLUTION   The critical points are the solutions of $f_x(x, y) = 0$ and $f_y(x, y) = 0$. We compute the partial derivatives:

$$f_x(x, y) = (-1 + 2x)e^{y^2} + \left(1 - y + y^2\right)e^{x^2} \cdot 2x$$

$$f_y(x, y) = \left(1 - x + x^2\right)e^{y^2} \cdot 2y + (-1 + 2y)e^{x^2}$$

Hence, the critical points are the solutions of the following equations:

$$(2x - 1)e^{y^2} + 2x\left(1 - y + y^2\right)e^{x^2} = 0$$

$$(2y - 1)e^{x^2} + 2y\left(1 - x + x^2\right)e^{y^2} = 0$$

Using a CAS we obtain the following solution: $x = y = 0.27788$, which from the figure is a local minimum.

📖 *In Exercises 29–32, determine the global extreme values of the function on the given set* without using calculus.

29. $f(x, y) = x + y, \quad 0 \leq x \leq 1, \quad 0 \leq y \leq 1$

SOLUTION   The sum $x + y$ is maximum when $x = 1$ and $y = 1$, and it is minimum when $x = 0$ and $y = 0$. Therefore, the global maximum of f on the given set is $f(1, 1) = 1 + 1 = 2$ and the global minimum is $f(0, 0) = 0 + 0 = 0$.

31. $f(x, y) = (x^2 + y^2 + 1)^{-1}, \quad 0 \leq x \leq 3, \quad 0 \leq y \leq 5$

SOLUTION $f(x, y) = \frac{1}{x^2 + y^2 + 1}$ is maximum when x^2 and y^2 are minimum, that is, when $x = y = 0$. f is minimum when x^2 and y^2 are maximum, that is, when $x = 3$ and $y = 5$. Therefore, the global maximum of f on the given set is $f(0, 0) = (0^2 + 0^2 + 1)^{-1} = 1$, and the global minimum is $f(3, 5) = (3^2 + 5^2 + 1)^{-1} = \frac{1}{35}$.

33. Assumptions Matter   Show that $f(x, y) = xy$ does not have a global minimum or a global maximum on the domain

$$\mathcal{D} = \{(x, y) : 0 < x < 1, 0 < y < 1\}$$

Explain why this does not contradict Theorem 3.

SOLUTION   The largest and smallest values of f on the closed square $0 \leq x, y \leq 1$ are $f(1, 1) = 1$ and $f(0, 0) = 0$. However, on the open square $0 < x, y < 1$, f can never attain these maximum and minimum values, since the boundary (and in particular the points $(1, 1)$ and $(-1, -1)$) are not included in the domain. This does not contradict Theorem 3 since the domain is open.

35. Find the maximum of

$$f(x, y) = x + y - x^2 - y^2 - xy$$

on the square, $0 \le x \le 2, 0 \le y \le 2$ (Figure 21).

(a) First, locate the critical point of f in the square, and evaluate f at this point.

(b) On the bottom edge of the square, $y = 0$ and $f(x, 0) = x - x^2$. Find the extreme values of f on the bottom edge.

(c) Find the extreme values of f on the remaining edges.

(d) Find the largest among the values computed in (a), (b), and (c).

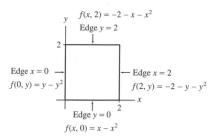

FIGURE 21 The function $f(x, y) = x + y - x^2 - y^2 - xy$ on the boundary segments of the square $0 \le x \le 2, 0 \le y \le 2$.

SOLUTION

(a) To find the critical points, we look at the first-order partial derivatives set equal to zero and solve:

$$f_x(x, y) = 1 - 2x - y = 0, \quad f_y(x, y) = 1 - 2y - x = 0$$

This gives $y = 1 - 2x$ and $x = 1 - 2y$, solving simultaneously we see $y = 1/3$ and $x = 1/3$. The critical point is $(1/3, 1/3)$, subsequently, $f(1/3, 1/3) = 1/3$.

(b) To find the extreme points of $f(x, 0) = x - x^2$ we take the first derivative and set it equal to zero and solve:

$$f'(x, 0) = 1 - 2x = 0 \rightarrow x = 1/2$$

Thus the extreme value on the bottom edge of the square is

$$f(1/2, 0) = 1/4$$

(c) Now to find the extreme values on the other edges of the square.

First, let us use $x = 0$. $f(0, y) = y - y^2$. Taking the first derivative and setting equal to 0 gives us:

$$f'(0, y) = 1 - 2y = 0, \rightarrow y = 1/2$$

Therefore, the extreme value along $x = 0$ is $f(0, 1/2) = 1/4$.

Next, let us use $y = 2$: $f(x, 2) = -x^2 - x - 2$. Take the first derivative and setting equal to 0 gives us:

$$f'(x, 2) = -2x - 1 = 0, \rightarrow x = -1/2$$

Therefore, the extreme value along $y = 2$ is $f(-1/2, 2) = -7/4$.

Finally, let us use $x = 2$: $f(2, y) = -2 - y - y^2$. Take the first derivative and setting equal to 0 gives us:

$$f'(2, y) = -1 - 2y = 0, \rightarrow y = -1/2$$

Therefore, the extreme value along $x = 2$ is $f(2, -1/2) = -7/4$.

(d) Out of all the values we computed in parts (a), (b), and (c), $1/3$ is the largest. This value occurs at the point $(1/3, 1/3)$.

In Exercises 37–43, determine the global extreme values of the function on the given domain.

37. $f(x, y) = x^3 - 2y, \quad 0 \le x \le 1, \quad 0 \le y \le 1$

SOLUTION We use the following steps.

Step 1. Find the critical points. We set the first derivative equal to zero and solve:

$$f_x(x, y) = 3x^2 = 0, \quad f_y(x, y) = -2$$

The two equations have no solutions, hence there are no critical points.

Step 2. Check the boundary. The extreme values occur either at the critical points or at a point on the boundary of the domain. Since there are no critical points, the extreme values occur at boundary points. We consider each edge of the square $0 \le x, y \le 1$ separately.

The segment $\overline{OA}$: On this segment $y = 0, 0 \le x \le 1$, and f takes the values $f(x, 0) = x^3$. The minimum value is $f(0, 0) = 0$ and the maximum value is $f(1, 0) = 1$.

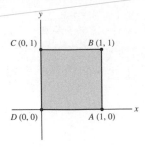

The segment $\overline{AB}$: On this segment $x = 1, 0 \le y \le 1$, and f takes the values $f(1, y) = 1 - 2y$. The minimum value is $f(1, 1) = 1 - 2 \cdot 1 = -1$ and the maximum value is $f(1, 0) = 1 - 2 \cdot 0 = 1$.
The segment $\overline{BC}$: On this segment $y = 1, 0 \le x \le 1$, and f takes the values $f(x, 1) = x^3 - 2$. The minimum value is $f(0, 1) = 0^3 - 2 = -2$ and the maximum value is $f(1, 1) = 1^3 - 2 = -1$.
The segment $\overline{OC}$: On this segment $x = 0, 0 \le y \le 1$, and f takes the values $f(0, y) = -2y$. The minimum value is $f(0, 1) = -2 \cdot 1 = -2$ and the maximum value is $f(0, 0) = -2 \cdot 0 = 0$.

Step 3. Conclusions. The values obtained in the previous steps are

$$f(0, 0) = 0, \quad f(1, 0) = 1, \quad f(1, 1) = -1, \quad f(0, 1) = -2$$

The smallest value is $f(0, 1) = -2$ and it is the global minimum of f on the square. The global maximum is the largest value $f(1, 0) = 1$.

39. $f(x, y) = x^2 + 2y^2, \quad 0 \le x \le 1, \quad 0 \le y \le 1$

SOLUTION The sum $x^2 + 2y^2$ is maximum at the point $(1, 1)$, where x^2 and y^2 are maximum. It is minimum if $x = y = 0$, that is, at the point $(0, 0)$. Hence,

$$\text{Global maximum} = f(1, 1) = 1^2 + 2 \cdot 1^2 = 3$$
$$\text{Global minimum} = f(0, 0) = 0^2 + 2 \cdot 0^2 = 0$$

41. $f(x, y) = x^3 + y^3 - 3xy, \quad 0 \le x \le 1, \quad 0 \le y \le 1$

SOLUTION We use the following steps.
Step 1. Examine the critical points in the interior of the domain. We set the partial derivatives equal to zero and solve:

$$f_x(x, y) = 3x^2 - 3y = 0$$
$$f_y(x, y) = 3y^2 - 3x = 0$$

The first equation gives $y = x^2$. We substitute in the second equation and solve for x:

$$3\left(x^2\right)^2 - 3x = 0$$
$$3x^4 - 3x = 3x\left(x^3 - 1\right) = 0 \quad \Rightarrow \quad x = 0, \quad y = 0^2 = 0$$
$$\text{or} \quad x = 1, \quad y = 1^2 = 1$$

The critical points $(0, 0)$ and $(1, 1)$ are not in the interior of the domain.
Step 2. Find the extreme values on the boundary. We consider each part of the boundary separately.

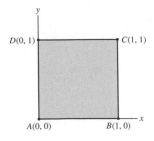

The edge $\overline{AB}$: On this edge, $y = 0$, $0 \le x \le 1$, and $f(x, 0) = x^3$. The maximum value is obtained at $x = 1$ and the minimum value is obtained at $x = 0$. The corresponding extreme points are $(1, 0)$ and $(0, 0)$.

The edge $\overline{BC}$: On this edge $x = 1$, $0 \le y \le 1$, and $f(1, y) = y^3 - 3y + 1$. The critical points are $\frac{d}{dy}\left(y^3 - 3y + 1\right) = 3y^2 - 3 = 0$, that is, $y = \pm 1$. The point in the given domain is $y = 1$. The candidates for extreme values are thus $y = 1$ and $y = 0$, giving the points $(1, 1)$ and $(1, 0)$.

The edge $\overline{DC}$: On this edge $y = 1$, $0 \le x \le 1$, and $f(x, 1) = x^3 - 3x + 1$. Replacing the values of x and y in the previous solutions we get the points $(1, 1)$ and $(0, 1)$.

The edge $\overline{AD}$: On this edge $x = 0$, $0 \le y \le 1$, and $f(0, y) = y^3$. Replacing the values of x and y obtained for the edge $\overline{AB}$, we get $(0, 1)$ and $(0, 0)$.

By Theorem 3, the extreme values occur either at a critical point in the interior of the square or at a point on the boundary of the square. Since there are no critical points in the interior of the square, the candidates for extreme values are the following points:

$$(0, 0), \quad (1, 0), \quad (1, 1), \quad (0, 1)$$

We compute $f(x, y) = x^3 + y^3 - 3xy$ at these points:

$$f(0, 0) = 0^3 + 0^3 - 3 \cdot 0 = 0$$

$$f(1, 0) = 1^3 + 0^3 - 3 \cdot 1 \cdot 0 = 1$$

$$f(1, 1) = 1^3 + 1^3 - 3 \cdot 1 \cdot 1 = -1$$

$$f(0, 1) = 0^3 + 1^3 - 3 \cdot 0 \cdot 1 = 1$$

We conclude that in the given domain, the global maximum is $f(1, 0) = f(0, 1) = 1$ and the global minimum is $f(1, 1) = -1$.

43. $f(x, y) = (4y^2 - x^2)e^{-x^2 - y^2}$, $x^2 + y^2 \le 2$

SOLUTION We use the following steps.

Step 1. Examine the critical points. We compute the partial derivatives of $f(x, y) = \left(4y^2 - x^2\right) e^{-x^2 - y^2}$, set them equal to zero and solve. This gives

$$f_x(x, y) = -2xe^{-x^2 - y^2} + \left(4y^2 - x^2\right) e^{-x^2 - y^2} \cdot (-2x) = -2xe^{-x^2 - y^2}\left(1 + 4y^2 - x^2\right) = 0$$

$$f_y(x, y) = 8ye^{-x^2 - y^2} + \left(4y^2 - x^2\right) e^{-x^2 - y^2} \cdot (-2y) = -2ye^{-x^2 - y^2}\left(-4 + 4y^2 - x^2\right) = 0$$

Since $e^{-x^2 - y^2} \ne 0$, the first equation gives $x = 0$ or $x^2 = 1 + 4y^2$. Substituting $x = 0$ in the second equation gives

$$-2ye^{-y^2}\left(-4 + 4y^2\right) = 0.$$

Since $e^{-y^2} \ne 0$, we get

$$y\left(-1 + y^2\right) = y(y - 1)(y + 1) = 0 \quad \Rightarrow \quad y = 0, \quad y = 1, \quad y = -1$$

We obtain the three points $(0, 0)$, $(0, -1)$, $(0, 1)$. We now substitute $x^2 = 1 + 4y^2$ in the second equation and solve for y:

$$-2ye^{-1 - 5y^2}\left(-4 + 4y^2 - 1 - 4y^2\right) = 0$$

$$-2ye^{-1 - 5y^2} \cdot (-5) = 0 \quad \Rightarrow \quad y = 0$$

The corresponding values of x are obtained from

$$x^2 = 1 + 4 \cdot 0^2 = 1 \quad \Rightarrow \quad x = \pm 1$$

We obtain the solutions $(1, 0)$ and $(-1, 0)$. We conclude that the critical points are

$$(0, 0), \quad (0, -1), \quad (0, 1), \quad (1, 0), \quad \text{and} \quad (-1, 0).$$

All of these points are in the interior $x^2 + y^2 < 2$ of the given disk.

Step 2. Check the boundary. The boundary is the circle $x^2 + y^2 = 2$. On this set $y^2 = 2 - x^2$, hence the function $f(x, y)$ takes the values

$$f(x, y)\Big|_{x^2 + y^2 = 2} = g(x) = \left(4\left(2 - x^2\right) - x^2\right)e^{-2} = \left(-5x^2 + 8\right)e^{-2}$$

That is, $g(x) = -5e^{-2}x^2 + 8e^{-2}$. We determine the interval of x. Since $x^2 + y^2 = 2$, we have $0 \le x^2 \le 2$ or $-\sqrt{2} \le x \le \sqrt{2}$.

We thus must find the extreme values of $g(x) = -5e^{-2}x^2 + 8e^{-2}$ on the interval $-\sqrt{2} \le x \le \sqrt{2}$. With the aid of the graph of $g(x)$, we conclude that the maximum value is $g(0) = 8e^{-2}$ and the minimum value is

$$g\left(-\sqrt{2}\right) = g\left(\sqrt{2}\right) = -5e^{-2}\left(\pm\sqrt{2}\right)^2 + 8e^{-2} = -10e^{-2} + 8e^{-2} = -2e^{-2} \approx -0.271$$

We conclude that the points on the boundary with largest and smallest values of f are

$$f\left(0, \pm\sqrt{2}\right) = 8e^{-2} \approx 1.083, \quad f\left(\pm\sqrt{2}, 0\right) = -2e^{-2} \approx -0.271$$

Step 3. Conclusions. The extreme values either occur at the critical points or at the points on the boundary, found in step 2. We compare the values of f at these points:

$$f(0, 0) = 0$$

$$f(0, -1) = 4e^{-1} \approx 1.472$$

$$f(0, 1) = 4e^{-1} \approx 1.472$$

$$f(1, 0) = -e^{-1} \approx -0.368$$

$$f(-1, 0) = -e^{-1} \approx -0.368$$

$$f\left(0, \pm\sqrt{2}\right) \approx 1.083$$

$$f\left(\pm\sqrt{2}, 0\right) \approx -0.271$$

We conclude that the global minimum is $f(1, 0) = f(-1, 0) = -0.368$ and the global maximum is $f(0, -1) = f(0, 1) = 1.472$.

45. Find the maximum volume of the largest box of the type shown in Figure 22, with one corner at the origin and the opposite corner at a point $P = (x, y, z)$ on the paraboloid

$$z = 1 - \frac{x^2}{4} - \frac{y^2}{9} \quad \text{with } x, y, z \ge 0$$

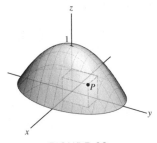

FIGURE 22

SOLUTION To maximize the volume of a rectangular box, start with the relation $V = xyz$ and using the paraboloid equation we see

$$z = 1 - \frac{x^2}{4} - \frac{y^2}{9} \quad \Rightarrow \quad V(x, y) = xy\left(1 - \frac{x^2}{4} - \frac{y^2}{9}\right)$$

Therefore we will consider

$$V(x, y) = xy - \frac{1}{4}x^3y - \frac{1}{9}xy^3$$

First to find the critical points, we take the first-order partial derivatives and set them equal to zero, and solve:

$$V_x(x, y) = y - \frac{3}{4}x^2y - \frac{1}{9}y^3, \quad V_y(x, y) = x - \frac{1}{4}x^3 - \frac{1}{3}xy^2$$

Using the equation $V_y = 0$ we see

$$x - \frac{1}{4}x^3 - \frac{1}{3}xy^2 = 0 \quad \Rightarrow \quad x = 0, \quad y^2 = 3 - \frac{3}{4}x^2 \quad \Rightarrow \quad y = \sqrt{3 - \frac{3}{4}x^2}$$

(Note here, we can ignore the value $x = 0$, since it produces a box having zero volume.)

Using this relation in the first equation, $V_x = 0$, we see:

$$\sqrt{3 - \frac{3}{4}x^2} - \frac{3}{4}x^2\sqrt{3 - \frac{3}{4}x^2} - \frac{1}{9}\left(3 - \frac{3}{4}x^2\right)^{3/2} = 0$$

Factoring we see:

$$\sqrt{3 - \frac{3}{4}x^2}\left[1 - \frac{3}{4}x^2 - \frac{1}{9}\left(3 - \frac{3}{4}x^2\right)\right] = 0$$

and thus

$$3 - \frac{3}{4}x^2 = 0 \quad \Rightarrow \quad x^2 = 4 \quad \Rightarrow \quad x = \pm 2$$

or

$$1 - \frac{3}{4}x^2 - \frac{1}{3} + \frac{1}{12}x^2 = 0 \quad \Rightarrow \quad \frac{2}{3} - \frac{2}{3}x^2 = 0 \quad \Rightarrow \quad x = \pm 1$$

Since the governing equation $f(x, y)$ is a paraboloid, that is symmetric about the z-axis, we need only consider the point when $x = 2$ or $x = 1$.

Therefore, since $y = \sqrt{3 - \frac{3}{4}x^2}$ and $z = 1 - \frac{1}{4}x^2 - \frac{1}{9}y^2$, we have, if $x = 2$

$$y = \sqrt{3 - \frac{3}{4} \cdot 4} = 0 \quad \Rightarrow \quad z = 1 - \frac{1}{4} \cdot 4 - \frac{1}{9} \cdot 0 = 0$$

This will give a box having zero volume - not a maximum volume at all.

Using $x = 1$, and $y = \sqrt{3 - \frac{3}{4}x^2}$, $z = 1 - \frac{1}{4}x^2 - \frac{1}{9}y^2$, we have

$$y = \sqrt{3 - \frac{3}{4}} = \frac{3}{2}, \quad z = 1 - \frac{1}{4} \cdot 1^2 - \frac{1}{9} \cdot \frac{9}{4} = \frac{1}{2}$$

Therefore, the box having maximum volume has dimensions, $x = 1$, $y = 3/2$, and $z = 1/2$ and maximum value for the volume:

$$V = xyz = 1 \cdot \frac{3}{2} \cdot \frac{1}{2} = \frac{3}{4}$$

47. Show that the sum of the squares of the distances from a point $P = (c, d)$ to n fixed points $(a_1, b_1), \dots, (a_n, b_n)$ is minimized when c is the average of the x-coordinates a_i and d is the average of the y-coordinates b_i.

SOLUTION First we must form the sum of the squares of the distances from a point $P(c, d)$ to n fixed points. For instance, the square of the distance from (c, d) to (a_1, b_1) would be:

$$(c - a_1)^2 + (d - b_1)^2$$

using this pattern, the sum in question would be

$$S = \sum_{i=1}^{n}[(c - a_i)^2 + (d - b_i)^2]$$

Using the methods discussed in this section of the text, we want to minimize the sum S. We will examine the first-order partial derivatives with respect to c and d and set them equal to zero and solve:

$$S_c = \sum_{i=1}^{n} 2(c - a_i) = 0, \quad S_d = \sum_{i=1}^{n} 2(d - b_i) = 0$$

Consider first the following:

$$\sum_{i=1}^{n} 2(c - a_i) = 0 \quad \Rightarrow \quad \sum_{i=1}^{n}(c - a_i) = 0 \quad \Rightarrow \quad \sum_{i=1}^{n}c - \sum_{i=1}^{n}a_i = 0$$

Therefore

$$\sum_{i=1}^{n}c = \sum_{i=1}^{n}a_i \quad \Rightarrow \quad n \cdot c = \sum_{i=1}^{n}a_i \quad \Rightarrow \quad c = \frac{1}{n}\sum_{i=1}^{n}a_i$$

Similarly we can examine $S_d = 0$ to see

$$\sum_{i=1}^{n} 2(d - b_i) = 0 \quad \Rightarrow \quad \sum_{i=1}^{n}(d - b_i) = 0 \quad \Rightarrow \quad \sum_{i=1}^{n} d - \sum_{i=1}^{n} b_i = 0$$

and

$$\sum_{i=1}^{n} d = \sum_{i=1}^{n} b_i \quad \Rightarrow \quad n \cdot d = \sum_{i=1}^{n} b_i \quad \Rightarrow \quad d = \frac{1}{n} \sum_{i=1}^{n} b_i$$

Therefore, the sum is minimized when c is the average of the x-coordinates a_i and d is the average of the y-coordinates b_i.

49. Consider a rectangular box B that has a bottom and sides but no top and has minimal surface area among all boxes with fixed volume V.

(a) Do you think B is a cube as in the solution to Exercise 48? If not, how would its shape differ from a cube?

(b) Find the dimensions of B and compare with your response to (a).

SOLUTION

(a) Each of the variables x and y is the length of a side of three faces (for example, x is the length of the front, back, and bottom sides), whereas z is the length of a side of four faces.

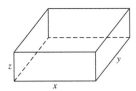

Therefore, the variables x, y, and z do not have equal influence on the surface area. We expect that in the box B with minimal surface area, z is smaller than $\sqrt[3]{V}$, which is the side of a cube with volume V (also we would expect $x = y$).

(b) We must find the dimensions of the box B, with fixed volume V and with smallest possible surface area, when the top is not included.

Step 1. Find a function to be minimized. The surface area of the box with sides lengths x, y, z when the top is not included is

$$S = 2xz + 2yz + xy \tag{1}$$

To express the surface in terms of x and y only, we use the formula for the volume of the box, $V = xyz$, giving $z = \frac{V}{xy}$. We substitute in (1) to obtain

$$S = 2x \cdot \frac{V}{xy} + 2y \cdot \frac{V}{xy} + xy = \frac{2V}{y} + \frac{2V}{x} + xy$$

That is,

$$S = \frac{2V}{y} + \frac{2V}{x} + xy.$$

Step 2. Determine the domain. The variables x, y denote lengths, hence they must be nonnegative. Moreover, S is not defined for $x = 0$ or $y = 0$. Since there are no other limitations on the variables, the domain is

$$D = \{(x, y) : x > 0, y > 0\}$$

We must find the minimum value of S on D. Because this domain is neither closed nor bounded, we are not sure that a minimum value exists. However, it can be proved (in like manner as in Exercise 48) that S does have a minimum value on D. This value occurs at a critical point in D, hence we set the partial derivatives equal to zero and solve. This gives

$$S_x(x, y) = -\frac{2V}{x^2} + y = 0$$

$$S_y(x, y) = -\frac{2V}{y^2} + x = 0$$

The first equation gives $y = \frac{2V}{x^2}$. Substituting in the second equation yields

$$x - \frac{2V}{\frac{4V^2}{x^4}} = x - \frac{x^4}{2V} = x\left(1 - \frac{x^3}{2V}\right) = 0$$

The solutions are $x = 0$ and $x = (2V)^{1/3}$. The solution $x = 0$ is not included in D, so the only solution is $x = (2V)^{1/3}$. We find the value of y using $y = \frac{2V}{x^2}$:

$$y = \frac{2V}{(2V)^{2/3}} = (2V)^{1/3}$$

We conclude that the critical point, which is the point where the minimum value of S in D occurs, is $\left((2V)^{1/3}, (2V)^{1/3}\right)$. We find the corresponding value of z using $z = \frac{V}{xy}$. We get

$$z = \frac{V}{(2V)^{1/3}(2V)^{1/3}} = \frac{V}{2^{2/3}V^{2/3}} = \frac{V^{1/3}}{2^{2/3}} = \left(\frac{V}{4}\right)^{1/3}$$

We conclude that the sizes of the box with minimum surface area are

width: $x = (2V)^{1/3}$;
length: $y = (2V)^{1/3}$;
height: $z = \left(\frac{V}{4}\right)^{1/3}$.

We see that z is smaller than x and y as predicted.

51. The power (in microwatts) of a laser is measured as a function of current (in milliamps). Find the linear least-squares fit (Exercise 50) for the data points.

Current (mA)	1.0	1.1	1.2	1.3	1.4	1.5
Laser power (μW)	0.52	0.56	0.82	0.78	1.23	1.50

SOLUTION By Exercise 50, the coefficients of the linear least-square fit $f(x) = mx + b$ are determined by the following equations:

$$m\sum_{j=1}^{n} x_j + bn = \sum_{j=1}^{n} y_j$$

$$m\sum_{j=1}^{n} x_j^2 + b\sum_{j=1}^{n} x_j = \sum_{j=1}^{n} x_j \cdot y_j \qquad (1)$$

In our case there are $n = 6$ data points:

$$(x_1, y_1) = (1, 0.52), \quad (x_2, y_2) = (1.1, 0.56),$$

$$(x_3, y_3) = (1.2, 0.82), \quad (x_4, y_4) = (1.3, 0.78),$$

$$(x_5, y_5) = (1.4, 1.23), \quad (x_6, y_6) = (1.5, 1.50).$$

We compute the sums in (1):

$$\sum_{j=1}^{6} x_j = 1 + 1.1 + 1.2 + 1.3 + 1.4 + 1.5 = 7.5$$

$$\sum_{j=1}^{6} y_j = 0.52 + 0.56 + 0.82 + 0.78 + 1.23 + 1.50 = 5.41$$

$$\sum_{j=1}^{6} x_j^2 = 1^2 + 1.1^2 + 1.2^2 + 1.3^2 + 1.4^2 + 1.5^2 = 9.55$$

$$\sum_{j=1}^{6} x_j \cdot y_j = 1 \cdot 0.52 + 1.1 \cdot 0.56 + 1.2 \cdot 0.82 + 1.3 \cdot 0.78 + 1.4 \cdot 1.23 + 1.5 \cdot 1.50 = 7.106$$

Substituting in (1) gives the following equations:

$$7.5m + 6b = 5.41$$

$$9.55m + 7.5b = 7.106 \qquad (2)$$

We multiply the first equation by 9.55 and the second by (-7.5), then add the resulting equations. This gives

$$\frac{\begin{aligned}71.625m + 57.3b &= 51.6655\\ + \quad -71.625m - 56.25b &= -53.295\end{aligned}}{1.05b = -1.6295} \quad \Rightarrow \quad b = -1.5519$$

We now substitute $b = -1.5519$ in the first equation in (2) and solve for m:

$$7.5m + 6 \cdot (-1.5519) = 5.41 \qquad \Rightarrow \qquad m = 1.9629$$
$$7.5m = 14.7214$$

The linear least squares fit $f(x) = mx + b$ is thus

$$f(x) = 1.9629x - 1.5519.$$

Further Insights and Challenges

53. In this exercise, we prove that for all $x, y \geq 0$:

$$\frac{1}{\alpha}x^\alpha + \frac{1}{\beta}x^\beta \geq xy$$

where $\alpha \geq 1$ and $\beta \geq 1$ are numbers such that $\alpha^{-1} + \beta^{-1} = 1$. To do this, we prove that the function

$$f(x, y) = \alpha^{-1}x^\alpha + \beta^{-1}y^\beta - xy$$

satisfies $f(x, y) \geq 0$ for all $x, y \geq 0$.

(a) Show that the set of critical points of $f(x, y)$ is the curve $y = x^{\alpha-1}$ (Figure 26). Note that this curve can also be described as $x = y^{\beta-1}$. What is the value of $f(x, y)$ at points on this curve?

(b) Verify that the Second Derivative Test fails. Show, however, that for fixed $b > 0$, the function $g(x) = f(x, b)$ is concave up with a critical point at $x = b^{\beta-1}$.

(c) Conclude that for all $x > 0$, $f(x, b) \geq f(b^{\beta-1}, b) = 0$.

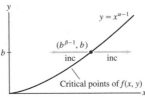

FIGURE 26 The critical points of $f(x, y) = \alpha^{-1}x^\alpha + \beta^{-1}y^\beta - xy$ form a curve $y = x^{\alpha-1}$.

SOLUTION We define the following function:

$$f(x, y) = \frac{1}{\alpha}x^\alpha + \frac{1}{\beta}y^\beta - xy$$

Notice that $f(0, 0) = 0$.

(a) Determine the critical points for $f(x, y) = f(x, y) = \alpha^{-1}x^\alpha + \beta^{-1}y^\beta - xy$. First, take the first-order partial derivatives and set them equal to zero to solve:

$$f_x = \alpha^{-1} \cdot \alpha x^{\alpha-1} - y = x^{\alpha-1} - y = 0, \quad f_y = \beta^{-1} \cdot \beta y^{\beta-1} - x = y^{\beta-1} - x = 0$$

This means that $y = x^{\alpha-1}$ and simultaneously $x = y^{\beta-1}$. Note here that we are guaranteed that the set of points satisfying both equations is nonempty because $1/\alpha + 1/\beta = 1$.

Now to compute the value of $f(x, y)$ at these points:

$$f(x, y) = f(x, x^{\alpha-1}) = \alpha^{-1}x^\alpha + \beta^{-1}(x^{\alpha-1})^\beta - x(x^{\alpha-1}) = \left(\frac{1}{\alpha} - 1\right)x^\alpha + \frac{1}{\beta}x^{\alpha\beta-\beta}$$

But remember that $\alpha^{-1} + \beta^{-1} = 1$ so we can say

$$\frac{1}{\alpha} + \frac{1}{\beta} = 1, \quad \beta + \alpha = \alpha\beta$$

Using these relations we see:

$$f(x, y) = f(x, x^{\alpha-1}) = \left(\frac{1}{\alpha} - 1\right)x^\alpha + \frac{1}{\beta}x^{\alpha\beta - \beta} = -\frac{1}{\beta}x^\alpha + \frac{1}{\beta}x^\alpha = 0$$

or similarly,

$$f(x, y) = f(y^{\beta-1}, y) = \frac{1}{\alpha}y^{\alpha\beta-\alpha} + \left(\frac{1}{\beta} - 1\right)y^\beta = \frac{1}{\alpha}y^\beta - \frac{1}{\alpha}y^\beta = 0$$

(b) Now computing the second-order partial derivatives we get

$$f_{xx} = (\alpha - 1)x^{\alpha-2}, \quad f_{yy} = (\beta - 1)y^{\beta-2}, \quad f_{xy} = -1$$

Therefore we can write the discriminant (while using the relations about α and β above):

$$D = f_{xx}f_{yy} - f_{xy}^2 = (\alpha - 1)(\beta - 1)x^{\alpha-2}y^{\beta-2} - 1 = x^{\alpha-2}y^{\beta-2} - 1$$

Evaluating this expression at the critical points when $y = x^{\alpha-1}$ we see

$$D(x, x^{\alpha-1}) = x^{\alpha-2}(x^{\alpha-1})^{\beta-2} - 1 = x^{\alpha-2}x^{\alpha\beta-\beta-2\alpha+2} - 1 = x^{\alpha-2+\alpha\beta-\beta-2\alpha+2} - 1 = x^0 - 1 = 0$$

Thus the Second Derivative Test is inconclusive and fails.
   Instead, if we fix $b > 0$, consider the function

$$g(x) = f(x, b) = \frac{1}{\alpha}x^\alpha + \frac{1}{\beta}b^\beta - bx$$

Therefore, taking the first derivative and setting it equal to zero to solve, we see

$$g'(x) = x^{\alpha-1} - b = 0 \quad \Rightarrow \quad b = x^{\alpha-1}$$

In order to solve this for x, note here that $(\alpha - 1)(\beta - 1) = 1$ so then $\frac{1}{\alpha-1} = \beta - 1$ and

$$b = x^{\alpha-1} \quad \Rightarrow \quad x = b^{1/(\alpha-1)} \quad \Rightarrow \quad x = b^{\beta-1}$$

Since

$$g''(x) = (\alpha - 1)x^{\alpha-2}, \quad \alpha \geq 1$$

then $g''(x) \geq 0$ for all x. Therefore, $g(x)$ is concave up with critical point $x = b^{\beta-1}$.

(c) From our work in part (b), we can conclude, for all $x > 0$, then

$$f(x, b) \geq f(b^{\beta-1}, b) = 0$$

14.8 Lagrange Multipliers: Optimizing with a Constraint (LT Section 15.8)

Preliminary Questions

1. Suppose that the maximum of $f(x, y)$ subject to the constraint $g(x, y) = 0$ occurs at a point $P = (a, b)$ such that $\nabla f_P \neq 0$. Which of the following statements is true?

(a) ∇f_P is tangent to $g(x, y) = 0$ at P.

(b) ∇f_P is orthogonal to $g(x, y) = 0$ at P.

SOLUTION

(a) Since the maximum of f subject to the constraint occurs at P, it follows by Theorem 1 that ∇f_P and ∇g_P are parallel vectors. The gradient ∇g_P is orthogonal to $g(x, y) = 0$ at P, hence ∇f_P is also orthogonal to this curve at P. We conclude that statement (b) is false (yet the statement can be true if $\nabla f_P = (0, 0)$).

(b) This statement is true by the reasoning given in the previous part.

2. Figure 9 shows a constraint $g(x, y) = 0$ and the level curves of a function f. In each case, determine whether f has a local minimum, a local maximum, or neither at the labeled point.

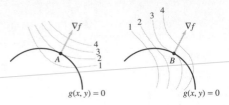

FIGURE 9

SOLUTION The level curve $f(x, y) = 2$ is tangent to the constraint curve at the point A. A close level curve that intersects the constraint curve is $f(x, y) = 1$, hence we may assume that f has a local maximum 2 under the constraint at A. The level curve $f(x, y) = 3$ is tangent to the constraint curve. However, in approaching B under the constraint, from one side f is increasing and from the other side f is decreasing. Therefore, $f(B)$ is neither local minimum nor local maximum of f under the constraint.

3. On the contour map in Figure 10:

(a) Identify the points where $\nabla f = \lambda \nabla g$ for some scalar λ.

(b) Identify the minimum and maximum values of $f(x, y)$ subject to $g(x, y) = 0$.

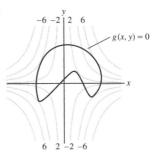

Contour plot of $f(x, y)$
(contour interval 2)

FIGURE 10 Contour map of $f(x, y)$; contour interval 2.

SOLUTION

(a) The gradient ∇g is orthogonal to the constraint curve $g(x, y) = 0$, and ∇f is orthogonal to the level curves of f. These two vectors are parallel at the points where the level curve of f is tangent to the constraint curve. These are the points A, B, C, D, E in the figure:

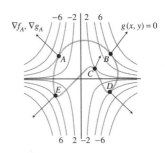

(b) The minimum and maximum occur where the level curve of f is tangent to the constraint curve. The level curves tangent to the constraint curve are

$$f(A) = -4, \quad f(C) = 2, \quad f(B) = 6, \quad f(D) = -4, \quad f(E) = 4$$

Therefore the global minimum of f under the constraint is -4 and the global maximum is 6.

Exercises

In this exercise set, use the method of Lagrange multipliers unless otherwise stated.

1. Find the extreme values of the function $f(x, y) = 2x + 4y$ subject to the constraint $g(x, y) = x^2 + y^2 - 5 = 0$.

(a) Show that the Lagrange equation $\nabla f = \lambda \nabla g$ gives $\lambda x = 1$ and $\lambda y = 2$.

(b) Show that these equations imply $\lambda \neq 0$ and $y = 2x$.

(c) Use the constraint equation to determine the possible critical points (x, y).

(d) Evaluate $f(x, y)$ at the critical points and determine the minimum and maximum values.

SOLUTION

(a) The Lagrange equations are determined by the equality $\nabla f = \lambda \nabla g$. We find them:

$$\nabla f = \langle f_x, f_y \rangle = \langle 2, 4 \rangle, \quad \nabla g = \langle g_x, g_y \rangle = \langle 2x, 2y \rangle$$

Hence,

$$\langle 2, 4 \rangle = \lambda \langle 2x, 2y \rangle$$

or

$$\begin{array}{cc} \lambda(2x) = 2 \\ \lambda(2y) = 4 \end{array} \quad \Rightarrow \quad \begin{array}{c} \lambda x = 1 \\ \lambda y = 2 \end{array}$$

(b) The Lagrange equations in part (a) imply that $\lambda \neq 0$. The first equation implies that $x = \frac{1}{\lambda}$ and the second equation gives $y = \frac{2}{\lambda}$. Therefore $y = 2x$.

(c) We substitute $y = 2x$ in the constraint equation $x^2 + y^2 - 5 = 0$ and solve for x and y. This gives

$$x^2 + (2x)^2 - 5 = 0$$
$$5x^2 = 5$$
$$x^2 = 1 \quad \Rightarrow \quad x_1 = -1, \quad x_2 = 1$$

Since $y = 2x$, we have $y_1 = 2x_1 = -2$, $y_2 = 2x_2 = 2$. The critical points are thus

$$(-1, -2) \quad \text{and} \quad (1, 2).$$

Extreme values can also occur at the points where $\nabla g = \langle 2x, 2y \rangle = \langle 0, 0 \rangle$. However, $(0, 0)$ is not on the constraint.

(d) We evaluate $f(x, y) = 2x + 4y$ at the critical points, obtaining

$$f(-1, -2) = 2 \cdot (-1) + 4 \cdot (-2) = -10$$
$$f(1, 2) = 2 \cdot 1 + 4 \cdot 2 = 10$$

Since f is continuous and the graph of $g = 0$ is closed and bounded, global minimum and maximum points exist. So according to Theorem 1, we conclude that the maximum of $f(x, y)$ on the constraint is 10 and the minimum is -10.

3. Apply the method of Lagrange multipliers to the function $f(x, y) = (x^2 + 1)y$ subject to the constraint $x^2 + y^2 = 5$. *Hint:* First show that $y \neq 0$; then treat the cases $x = 0$ and $x \neq 0$ separately.

SOLUTION We first write out the Lagrange Equations. We have $\nabla f = \langle 2xy, x^2 + 1 \rangle$ and $\nabla g = \langle 2x, 2y \rangle$. Hence, the Lagrange Condition for $\nabla g \neq 0$ is

$$\nabla f = \lambda \nabla g$$
$$\langle 2xy, x^2 + 1 \rangle = \lambda \langle 2x, 2y \rangle$$

We obtain the following equations:

$$\begin{array}{cc} 2xy = \lambda(2x) \\ x^2 + 1 = \lambda(2y) \end{array} \quad \Rightarrow \quad \begin{array}{c} 2x(y - \lambda) = 0 \\ x^2 + 1 = 2\lambda y \end{array} \tag{1}$$

The second equation implies that $y \neq 0$, since there is no real value of x such that $x^2 + 1 = 0$. Likewise, $\lambda \neq 0$. The solutions of the first equation are $x = 0$ and $y = \lambda$.

Case 1: $x = 0$. Substituting $x = 0$ in the second equation gives $2\lambda y = 1$, or $y = \frac{1}{2\lambda}$. We substitute $x = 0$, $y = \frac{1}{2\lambda}$ (recall that $\lambda \neq 0$) in the constraint to obtain

$$0^2 + \frac{1}{4\lambda^2} = 5 \quad \Rightarrow \quad 4\lambda^2 = \frac{1}{5} \quad \Rightarrow \quad \lambda = \pm\frac{1}{\sqrt{20}} = \pm\frac{1}{2\sqrt{5}}$$

The corresponding values of y are

$$y = \frac{1}{2 \cdot \frac{1}{2\sqrt{5}}} = \sqrt{5} \quad \text{and} \quad y = \frac{1}{2 \cdot \left(-\frac{1}{2\sqrt{5}}\right)} = -\sqrt{5}$$

We obtain the critical points:

$$\left(0, \sqrt{5}\right) \quad \text{and} \quad \left(0, -\sqrt{5}\right)$$

Case 2: $x \neq 0$. Then the first equation in (1) implies $y = \lambda$. Substituting in the second equation gives

$$x^2 + 1 = 2\lambda^2 \quad \Rightarrow \quad x^2 = 2\lambda^2 - 1$$

We now substitute $y = \lambda$ and $x^2 = 2\lambda^2 - 1$ in the constraint $x^2 + y^2 = 5$ to obtain

$$2\lambda^2 - 1 + \lambda^2 = 5$$
$$3\lambda^2 = 6$$
$$\lambda^2 = 2 \quad \Rightarrow \quad \lambda = \pm\sqrt{2}$$

The solution (x, y) are thus

$$\lambda = \sqrt{2}: \quad y = \sqrt{2}, \quad x = \pm\sqrt{2 \cdot 2 - 1} = \pm\sqrt{3}$$
$$\lambda = -\sqrt{2}: \quad y = -\sqrt{2}, \quad x = \pm\sqrt{2 \cdot 2 - 1} = \pm\sqrt{3}$$

We obtain the critical points:

$$\left(\sqrt{3}, \sqrt{2}\right), \quad \left(-\sqrt{3}, \sqrt{2}\right), \quad \left(\sqrt{3}, -\sqrt{2}\right), \quad \left(-\sqrt{3}, -\sqrt{2}\right)$$

We conclude that the critical points are

$$\left(0, \sqrt{5}\right), \quad \left(0, -\sqrt{5}\right), \quad \left(\sqrt{3}, \sqrt{2}\right), \quad \left(-\sqrt{3}, \sqrt{2}\right), \quad \left(\sqrt{3}, -\sqrt{2}\right), \quad \left(-\sqrt{3}, -\sqrt{2}\right).$$

We now calculate $f(x, y) = \left(x^2 + 1\right) y$ at the critical points:

$$f\left(0, \sqrt{5}\right) = \sqrt{5} \approx 2.24$$
$$f\left(0, -\sqrt{5}\right) = -\sqrt{5} \approx -2.24$$
$$f\left(\sqrt{3}, \sqrt{2}\right) = f\left(-\sqrt{3}, \sqrt{2}\right) = 4\sqrt{2} \approx 5.66$$
$$f\left(\sqrt{3}, -\sqrt{2}\right) = f\left(-\sqrt{3}, -\sqrt{2}\right) = -4\sqrt{2} \approx -5.66$$

Since the constraint gives a closed and bounded curve, f achieves a minimum and a maximum under it. We conclude that the maximum of $f(x, y)$ on the constraint is $4\sqrt{2}$ and the minimum is $-4\sqrt{2}$.

In Exercises 4–13, find the minimum and maximum values of the function subject to the given constraint.

5. $f(x, y) = x^2 + y^2, \quad 2x + 3y = 6$

SOLUTION   We find the extreme values of $f(x, y) = x^2 + y^2$ under the constraint $g(x, y) = 2x + 3y - 6 = 0$.

Step 1. Write out the Lagrange Equations. The gradients of f and g are $\nabla f = \langle 2x, 2y \rangle$ and $\nabla g = \langle 2, 3 \rangle$. The Lagrange Condition is

$$\nabla f = \lambda \nabla g$$
$$\langle 2x, 2y \rangle = \lambda \langle 2, 3 \rangle$$

We obtain the following equations:

$$2x = \lambda \cdot 2$$
$$2y = \lambda \cdot 3$$

Step 2. Solve for λ in terms of x and y. Notice that if $x = 0$, then the first equation gives $\lambda = 0$, therefore by the second equation also $y = 0$. The point $(0, 0)$ does not satisfy the constraint. Similarly, if $y = 0$ also $x = 0$. We therefore may assume that $x \neq 0$ and $y \neq 0$ and obtain by the two equations:

$$\lambda = x \quad \text{and} \quad \lambda = \frac{2}{3}y.$$

Step 3. Solve for x and y using the constraint. Equating the two expressions for λ gives

$$x = \frac{2}{3}y \quad \Rightarrow \quad y = \frac{3}{2}x$$

We substitute $y = \frac{3}{2}x$ in the constraint $2x + 3y = 6$ and solve for x and y:

$$2x + 3 \cdot \frac{3}{2}x = 6$$

$$13x = 12 \quad \Rightarrow \quad x = \frac{12}{13}, \quad y = \frac{3}{2} \cdot \frac{12}{13} = \frac{18}{13}$$

We obtain the critical point $\left(\frac{12}{13}, \frac{18}{13}\right)$.

Step 4. Calculate f at the critical point. We evaluate $f(x, y) = x^2 + y^2$ at the critical point:

$$f\left(\frac{12}{13}, \frac{18}{13}\right) = \left(\frac{12}{13}\right)^2 + \left(\frac{18}{13}\right)^2 = \frac{468}{169} \approx 2.77$$

Rewriting the constraint as $y = -\frac{2}{3}x + 2$, we see that as $|x| \to +\infty$ then so does $|y|$, and hence $x^2 + y^2$ is increasing without bound on the constraint as $|x| \to \infty$. We conclude that the value $468/169$ is the minimum value of f under the constraint, rather than the maximum value.

7. $f(x, y) = xy, \quad 4x^2 + 9y^2 = 32$

SOLUTION We find the extreme values of $f(x, y) = xy$ under the constraint $g(x, y) = 4x^2 + 9y^2 - 32 = 0$.

Step 1. Write out the Lagrange Equation. The gradient vectors are $\nabla f = \langle y, x \rangle$ and $\nabla g = \langle 8x, 18y \rangle$, hence the Lagrange Condition is

$$\nabla f = \lambda \nabla g$$

$$\langle y, x \rangle = \lambda \langle 8x, 18y \rangle$$

We obtain the following equations:

$$y = \lambda(8x)$$

$$x = \lambda(18y)$$

Step 2. Solve for λ in terms of x and y. If $x = 0$, then the Lagrange equations also imply that $y = 0$ and vice versa. Since the point $(0, 0)$ does not satisfy the equation of the constraint, we may assume that $x \neq 0$ and $y \neq 0$. The two equations give

$$\lambda = \frac{y}{8x} \quad \text{and} \quad \lambda = \frac{x}{18y}$$

Step 3. Solve for x and y using the constraint. We equate the two expressions for λ to obtain

$$\frac{y}{8x} = \frac{x}{18y} \quad \Rightarrow \quad 18y^2 = 8x^2 \quad \Rightarrow \quad y = \pm\frac{2}{3}x$$

We now substitute $y = \pm\frac{2}{3}x$ in the equation of the constraint and solve for x and y:

$$4x^2 + 9 \cdot \left(\pm\frac{2}{3}x\right)^2 = 32$$

$$4x^2 + 9 \cdot \frac{4x^2}{9} = 32$$

$$8x^2 = 32 \quad \Rightarrow \quad x = -2, \quad x = 2$$

We find y by the relation $y = \pm\frac{2}{3}x$:

$$y = \frac{2}{3} \cdot (-2) = -\frac{4}{3}, \quad y = -\frac{2}{3} \cdot (-2) = \frac{4}{3}, \quad y = \frac{2}{3} \cdot 2 = \frac{4}{3}, \quad y = -\frac{2}{3} \cdot 2 = -\frac{4}{3}$$

We obtain the following critical points:

$$\left(-2, -\frac{4}{3}\right), \quad \left(-2, \frac{4}{3}\right), \quad \left(2, \frac{4}{3}\right), \quad \left(2, -\frac{4}{3}\right)$$

Extreme values can also occur at the point where $\nabla g = \langle 8x, 18y \rangle = \langle 0, 0 \rangle$, that is, at the point $(0, 0)$. However, the point does not lie on the constraint.

Step 4. Calculate f at the critical points. We evaluate $f(x, y) = xy$ at the critical points:

$$f\left(-2, -\frac{4}{3}\right) = f\left(2, \frac{4}{3}\right) = \frac{8}{3}$$

$$f\left(-2, \frac{4}{3}\right) = f\left(2, -\frac{4}{3}\right) = -\frac{8}{3}$$

Since f is continuous and the constraint is a closed and bounded set in R^2 (an ellipse), f attains global extrema on the constraint. We conclude that $\frac{8}{3}$ is the maximum value and $-\frac{8}{3}$ is the minimum value.

9. $f(x, y) = x^2 + y^2, \qquad x^4 + y^4 = 1$

SOLUTION We find the extreme values of $f(x, y) = x^2 + y^2$ under the constraint $g(x, y) = x^4 + y^4 - 1 = 0$.

Step 1. Write out the Lagrange Equations. We have $\nabla f = \langle 2x, 2y \rangle$ and $\nabla g = \left\langle 4x^3, 4y^3 \right\rangle$, hence the Lagrange Condition $\nabla f = \lambda \nabla g$ gives

$$\langle 2x, 2y \rangle = \lambda \left\langle 4x^3, 4y^3 \right\rangle$$

or

$$\begin{matrix} 2x = \lambda\left(4x^3\right) \\ 2y = \lambda\left(4y^3\right) \end{matrix} \quad \Rightarrow \quad \begin{matrix} x = 2\lambda x^3 \\ y = 2\lambda y^3 \end{matrix} \qquad (1)$$

Step 2. Solve for λ in terms of x and y. We first assume that $x \neq 0$ and $y \neq 0$. Then the Lagrange equations give

$$\lambda = \frac{1}{2x^2} \quad \text{and} \quad \lambda = \frac{1}{2y^2}$$

Step 3. Solve for x and y using the constraint. Equating the two expressions for λ gives

$$\frac{1}{2x^2} = \frac{1}{2y^2} \quad \Rightarrow \quad y^2 = x^2 \quad \Rightarrow \quad y = \pm x$$

We now substitute $y = \pm x$ in the equation of the constraint $x^4 + y^4 = 1$ and solve for x and y:

$$x^4 + (\pm x)^4 = 1$$
$$2x^4 = 1$$
$$x^4 = \frac{1}{2} \quad \Rightarrow \quad x = \frac{1}{2^{1/4}}, \quad x = -\frac{1}{2^{1/4}}$$

The corresponding values of y are obtained by the relation $y = \pm x$. The critical points are thus

$$\left(\frac{1}{2^{1/4}}, \frac{1}{2^{1/4}}\right), \quad \left(\frac{1}{2^{1/4}}, -\frac{1}{2^{1/4}}\right), \quad \left(-\frac{1}{2^{1/4}}, \frac{1}{2^{1/4}}\right), \quad \left(-\frac{1}{2^{1/4}}, -\frac{1}{2^{1/4}}\right) \qquad (2)$$

We examine the case $x = 0$ or $y = 0$. Notice that the point $(0, 0)$ does not satisfy the equation of the constraint, hence either $x = 0$ or $y = 0$ can hold, but not both at the same time.

Case 1: $x = 0$. Substituting $x = 0$ in the constraint $x^4 + y^4 = 1$ gives $y = \pm 1$. We thus obtain the critical points

$$(0, -1), \quad (0, 1) \qquad (3)$$

Case 2: $y = 0$. We may interchange x and y in the discussion in case 1, and obtain the critical points:

$$(-1, 0), \quad (1, 0) \qquad (4)$$

Combining (2), (3), and (4) we conclude that the critical points are

$$A_1 = \left(\frac{1}{2^{1/4}}, \frac{1}{2^{1/4}}\right), \quad A_2 = \left(\frac{1}{2^{1/4}}, -\frac{1}{2^{1/4}}\right), \quad A_3 = \left(-\frac{1}{2^{1/4}}, \frac{1}{2^{1/4}}\right),$$

$$A_4 = \left(-\frac{1}{2^{1/4}}, -\frac{1}{2^{1/4}}\right), \quad A_5 = (0, -1), \quad A_6 = (0, 1), \quad A_7 = (-1, 0), \quad A_8 = (1, 0)$$

The point where $\nabla g = \left\langle 4x^3, 4y^3 \right\rangle = \langle 0, 0 \rangle$, that is, $(0, 0)$, does not lie on the constraint.

Step 4. Compute f at the critical points. We evaluate $f(x, y) = x^2 + y^2$ at the critical points:

$$f(A_1) = f(A_2) = f(A_3) = f(A_4) = \left(\frac{1}{2^{1/4}}\right)^2 + \left(\frac{1}{2^{1/4}}\right)^2 = \frac{2}{2^{1/2}} = \sqrt{2}$$

$$f(A_5) = f(A_6) = f(A_7) = f(A_8) = 1$$

The constraint $x^4 + y^4 = 1$ is a closed and bounded set in R^2 and f is continuous on this set, hence f has global extrema on the constraint. We conclude that $\sqrt{2}$ is the maximum value and 1 is the minimum value.

11. $f(x, y, z) = 3x + 2y + 4z, \quad x^2 + 2y^2 + 6z^2 = 1$

SOLUTION We find the extreme values of $f(x, y, z) = 3x + 2y + 4z$ under the constraint $g(x, y, z) = x^2 + 2y^2 + 6z^2 - 1 = 0$.

Step 1. Write out the Lagrange Equations. The gradient vectors are $\nabla f = \langle 3, 2, 4 \rangle$ and $\nabla g = \langle 2x, 4y, 12z \rangle$, therefore the Lagrange Condition $\nabla f = \lambda \nabla g$ is:

$$\langle 3, 2, 4 \rangle = \lambda \langle 2x, 4y, 12z \rangle$$

The Lagrange equations are, thus:

$$3 = \lambda(2x) \qquad \frac{3}{2} = \lambda x$$

$$2 = \lambda(4y) \quad \Rightarrow \quad \frac{1}{2} = \lambda y$$

$$4 = \lambda(12z) \qquad \frac{1}{3} = \lambda z$$

Step 2. Solve for λ in terms of x, y, and z. The Lagrange equations imply that $x \neq 0$, $y \neq 0$, and $z \neq 0$. Solving for λ we get

$$\lambda = \frac{3}{2x}, \quad \lambda = \frac{1}{2y}, \quad \lambda = \frac{1}{3z}$$

Step 3. Solve for x, y, and z using the constraint. Equating the expressions for λ gives

$$\frac{3}{2x} = \frac{1}{2y} = \frac{1}{3z} \quad \Rightarrow \quad x = \frac{9}{2}z, \quad y = \frac{3}{2}z$$

Substituting $x = \frac{9}{2}z$ and $y = \frac{3}{2}z$ in the equation of the constraint $x^2 + 2y^2 + 6z^2 = 1$ and solving for z we get

$$\left(\frac{9}{2}z\right)^2 + 2\left(\frac{3}{2}z\right)^2 + 6z^2 = 1$$

$$\frac{123}{4}z^2 = 1 \quad \Rightarrow \quad z_1 = \frac{2}{\sqrt{123}}, z_2 = -\frac{2}{\sqrt{123}}$$

Using the relations $x = \frac{9}{2}z$, $y = \frac{3}{2}z$ we get

$$x_1 = \frac{9}{2} \cdot \frac{2}{\sqrt{123}} = \frac{9}{\sqrt{123}}, \quad y_1 = \frac{3}{2} \cdot \frac{2}{\sqrt{123}} = \frac{3}{\sqrt{123}}, \quad z_1 = \frac{2}{\sqrt{123}}$$

$$x_2 = \frac{9}{2} \cdot \frac{-2}{\sqrt{123}} = -\frac{9}{\sqrt{123}}, \quad y_2 = \frac{3}{2} \cdot \frac{-2}{\sqrt{123}} = -\frac{3}{\sqrt{123}}, \quad z_2 = -\frac{2}{\sqrt{123}}$$

We obtain the following critical points:

$$p_1 = \left(\frac{9}{\sqrt{123}}, \frac{3}{\sqrt{123}}, \frac{2}{\sqrt{123}}\right) \quad \text{and} \quad p_2 = \left(-\frac{9}{\sqrt{123}}, -\frac{3}{\sqrt{123}}, -\frac{2}{\sqrt{123}}\right)$$

Critical points are also the points on the constraint where $\nabla g = 0$. However, $\nabla g = \langle 2x, 4y, 12z \rangle = \langle 0, 0, 0 \rangle$ only at the origin, and this point does not lie on the constraint.

Step 4. Computing f at the critical points. We evaluate $f(x, y, z) = 3x + 2y + 4z$ at the critical points:

$$f(p_1) = \frac{27}{\sqrt{123}} + \frac{6}{\sqrt{123}} + \frac{8}{\sqrt{123}} = \frac{41}{\sqrt{123}} = \sqrt{\frac{41}{3}} \approx 3.7$$

$$f(p_2) = -\frac{27}{\sqrt{123}} - \frac{6}{\sqrt{123}} - \frac{8}{\sqrt{123}} = -\frac{41}{\sqrt{123}} = -\sqrt{\frac{41}{3}} \approx -3.7$$

Since f is continuous and the constraint is closed and bounded in R^3, f has global extrema under the constraint. We conclude that the minimum value of f under the constraint is about -3.7 and the maximum value is about 3.7.

13. $f(x, y, z) = xy + 3xz + 2yz$, $5x + 9y + z = 10$

SOLUTION We show that $f(x, y, z) = xy + 3xz + 2yz$ does not have minimum and maximum values subject to the constraint $g(x, y, z) = 5x + 9y + z - 10 = 0$. First notice that the curve $c_1 : (x, x, 10 - 14x)$ lies on the surface of the constraint since it satisfies the equation of the constraint. On c_1 we have,

$$f(x, y, z) = f(x, x, 10 - 14x) = x^2 + 3x(10 - 14x) + 2x(10 - 14x) = -69x^2 + 50x$$

Since $\lim\limits_{x \to \infty} \left(-69x^2 + 50x\right) = -\infty$, f does not have minimum value on the constraint. Notice that the curve $c_2 :$ $(x, -x, 10 + 4x)$ also lies on the surface of the constraint. The values of f on c_2 are

$$f(x, y, z) = f(x, -x, 10 + 4x) = -x^2 + 3x(10 + 4x) - 2x(10 + 4x) = 3x^2 + 10x$$

The limit $\lim\limits_{x \to \infty} (3x^2 + 10x) = \infty$ implies that f does not have a maximum value subject to the constraint.

15. Find the point (a, b) on the graph of $y = e^x$ where the value ab is as small as possible.

SOLUTION We must find the point where $f(x, y) = xy$ has a minimum value subject to the constraint $g(x, y) = e^x - y = 0$.

Step 1. Write out the Lagrange Equations. Since $\nabla f = \langle y, x \rangle$ and $\nabla g = \langle e^x, -1 \rangle$, the Lagrange Condition $\nabla f = \lambda \nabla g$ is

$$\langle y, x \rangle = \lambda \langle e^x, -1 \rangle$$

The Lagrange equations are thus

$$y = \lambda e^x$$
$$x = -\lambda$$

Step 2. Solve for λ in terms of x and y. The Lagrange equations imply that

$$\lambda = ye^{-x} \quad \text{and} \quad \lambda = -x$$

Step 3. Solve for x and y using the constraint. We equate the two expressions for λ to obtain

$$ye^{-x} = -x \quad \Rightarrow \quad y = -xe^x$$

We now substitute $y = -xe^x$ in the equation of the constraint and solve for x:

$$e^x - (-xe^x) = 0$$
$$e^x(1 + x) = 0$$

Since $e^x \neq 0$ for all x, we have $x = -1$. The corresponding value of y is determined by the relation $y = -xe^x$. That is,

$$y = -(-1)e^{-1} = e^{-1}$$

We obtain the critical point

$$(-1, e^{-1})$$

Step 4. Calculate f at the critical point. We evaluate $f(x, y) = xy$ at the critical point.

$$f(-1, e^{-1}) = (-1) \cdot e^{-1} = -e^{-1}$$

We conclude (see Remark) that the minimum value of xy on the graph of $y = e^x$ is $-e^{-1}$, and it is obtained for $x = -1$ and $y = e^{-1}$.

Remark: Since the constraint is not bounded, we need to justify the existence of a minimum value. The values $f(x, y) = xy$ on the constraint $y = e^x$ are $f(x, e^x) = h(x) = xe^x$. Since $h(x) > 0$ for $x > 0$, the minimum value (if it exists) occurs at a point $x < 0$. Since

$$\lim_{x \to -\infty} xe^x = \lim_{x \to -\infty} \frac{x}{e^{-x}} = \lim_{x \to -\infty} \frac{1}{-e^{-x}} = \lim_{x \to -\infty} -e^x = 0,$$

then for $x <$ some negative number $-R$, we have $|f(x) - 0| < 0.1$, say. Thus, on the bounded region $-R \leq x \leq 0$, f has a minimum value of $-e^{-1} \approx -0.37$, and this is thus a global minimum (for all x).

17. The surface area of a right-circular cone of radius r and height h is $S = \pi r \sqrt{r^2 + h^2}$, and its volume is $V = \frac{1}{3}\pi r^2 h$.
(a) Determine the ratio h/r for the cone with given surface area S and maximum volume V.
(b) What is the ratio h/r for a cone with given volume V and minimum surface area S?
(c) Does a cone with given volume V and maximum surface area S exist?

SOLUTION

(a) Let S_0 denote a given surface area. We must find the ratio $\frac{h}{r}$ for which the function $V(r, h) = \frac{1}{3}\pi r^2 h$ has maximum value under the constraint $S(r, h) = \pi r \sqrt{r^2 + h^2} = \pi\sqrt{r^4 + h^2 r^2} = S_0$.

Step 1. Write out the Lagrange Equation. We have

$$\nabla V = \pi \left\langle \frac{2rh}{3}, \frac{r^2}{3} \right\rangle \quad \text{and} \quad \nabla S = \pi \left\langle \frac{2r^3 + h^2 r}{\sqrt{r^4 + h^2 r^2}}, \frac{hr^2}{\sqrt{r^4 + h^2 r^2}} \right\rangle$$

The Lagrange Condition $\nabla V = \lambda \nabla S$ gives the following equations:

$$\frac{2rh}{3} = \frac{2r^3 + h^2 r}{\sqrt{r^4 + h^2 r^2}}\lambda \quad \Rightarrow \quad \frac{2h}{3} = \frac{2r^2 + h^2}{\sqrt{r^4 + h^2 r^2}}\lambda$$

$$\frac{r^2}{3} = \frac{hr^2}{\sqrt{r^4 + h^2 r^2}}\lambda \quad \Rightarrow \quad \frac{1}{3} = \frac{h}{\sqrt{r^4 + h^2 r^2}}\lambda$$

Step 2. Solve for λ in terms of r and h. These equations yield two expressions for λ that must be equal:

$$\lambda = \frac{2h}{3}\frac{\sqrt{r^4 + h^2 r^2}}{2r^2 + h^2} = \frac{1}{3h}\sqrt{r^4 + h^2 r^2}$$

Step 3. Solve for r and h using the constraint. We have

$$\frac{2h}{3}\frac{\sqrt{r^4 + h^2 r^2}}{2r^2 + h^2} = \frac{1}{3h}\sqrt{r^4 + h^2 r^2}$$

$$2h\frac{1}{2r^2 + h^2} = \frac{1}{h}$$

$$2h^2 = 2r^2 + h^2 \quad \Rightarrow \quad h^2 = 2r^2 \quad \Rightarrow \quad \frac{h}{r} = \sqrt{2}$$

We substitute $h^2 = 2r^2$ in the constraint $\pi r \sqrt{r^2 + h^2} = S_0$ and solve for r. This gives

$$\pi r \sqrt{r^2 + 2r^2} = S_0$$

$$\pi r \sqrt{3r^2} = S_0$$

$$\sqrt{3}\pi r^2 = S_0 \quad \Rightarrow \quad r^2 = \frac{S_0}{\sqrt{3}\pi}, \quad h^2 = 2r^2 = \frac{2S_0}{\sqrt{3}\pi}$$

Extreme values can occur also at points on the constraint where $\nabla S = \left\langle \frac{2r^2 + h^2 r}{\sqrt{r^4 + h^2 r^2}}, \frac{hr^2}{\sqrt{r^4 + h^2 r^2}} \right\rangle = \langle 0, 0 \rangle$, that is, at $(r, h) = (0, h)$, $h \neq 0$. However, since the radius of the cone is positive ($r > 0$), these points are irrelevant. We conclude that for the cone with surface area S_0 and maximum volume, the following holds:

$$\frac{h}{r} = \sqrt{2}, \quad h = \sqrt{\frac{2S_0}{\sqrt{3}\pi}}, \quad r = \sqrt{\frac{S_0}{\sqrt{3}\pi}}$$

For the surface area $S_0 = 1$ we get

$$h = \sqrt{\frac{2}{\sqrt{3}\pi}} \approx 0.6, \quad r = \sqrt{\frac{1}{\sqrt{3}\pi}} = 0.43$$

(b) We now must find the ratio $\frac{h}{r}$ that minimizes the function $S(r, h) = \pi r \sqrt{r^2 + h^2}$ under the constraint

$$V(r, h) = \frac{1}{3}\pi r^2 h = V_0$$

Using the gradients computed in part (a), the Lagrange Condition $\nabla S = \lambda \nabla V$ gives the following equations:

$$\frac{2r^3 + h^2 r}{\sqrt{r^4 + h^2 r^2}} = \lambda\frac{2rh}{3} \qquad \frac{2r^2 + h^2}{\sqrt{r^4 + h^2 r^2}} = \lambda\frac{2h}{3}$$

$$\Rightarrow$$

$$\frac{hr^2}{\sqrt{r^4 + h^2 r^2}} = \lambda\frac{r^2}{3} \qquad \frac{h}{\sqrt{r^4 + h^2 r^2}} = \frac{\lambda}{3}$$

These equations give

$$\frac{\lambda}{3} = \frac{1}{2h} \frac{2r^2 + h^2}{\sqrt{r^4 + h^2 r^2}} = \frac{h}{\sqrt{r^4 + h^2 r^2}}$$

We simplify and solve for $\frac{h}{r}$:

$$\frac{2r^2 + h^2}{2h} = h$$

$$2r^2 + h^2 = 2h^2$$

$$2r^2 = h^2 \quad \Rightarrow \quad \frac{h}{r} = \sqrt{2}$$

We conclude that the ratio $\frac{h}{r}$ for a cone with a given volume and minimal surface area is

$$\frac{h}{r} = \sqrt{2}$$

(c) The constant $V = 1$ gives $\frac{1}{3}\pi r^2 h = 1$ or $h = \frac{3}{\pi r^2}$. As $r \to \infty$, we have $h \to 0$, therefore

$$\lim_{\substack{r \to \infty \\ h \to 0}} S(r, h) = \lim_{\substack{r \to \infty \\ h \to 0}} \pi r \sqrt{r^2 + h^2} = \infty$$

That is, S does not have maximum value on the constraint, hence there is no cone of volume 1 and maximal surface area.

19. Find the point on the ellipse

$$x^2 + 6y^2 + 3xy = 40$$

with largest x-coordinate (Figure 13).

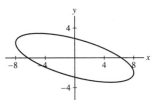

FIGURE 13 Graph of $x^2 + 6y^2 + 3xy = 40$

SOLUTION   We need to maximize $f(x, y) = x$ subject to the constraint

$$g(x, y) = x^2 + 6y^2 + 3xy = 40$$

Step 1. Write out the Lagrange Equations. The gradient vectors are $\nabla f = \langle 1, 0 \rangle$ and $\nabla g = \langle 2x + 3y, 12y + 3x \rangle$, hence the Lagrange Condition $\nabla f = \lambda \nabla g$ gives:

$$\langle 1, 0 \rangle = \lambda \langle 2x + 3y, 12y + 3x \rangle$$

or

$$1 = \lambda(2x + 3y), \qquad 0 = \lambda(12y + 3x)$$

this yields

$$x = -4y$$

Step 2. Solve for x and y using the constraint.

$$x^2 + 6y^2 + 3xy = (-4y)^2 + 6y^2 + 3(-4y)y = (16 + 6 - 12)y^2 = 10y^2 = 40$$

so $y = \pm 2$. If $y = 2$ then $x = -8$ and if $y = -2$ then $x = 8$. The extreme points are $(-8, 2)$ and $(8, -2)$. We conclude that the point with largest x-coordinate is $P = (8, -2)$.

21. Find the point (x_0, y_0) on the line $4x + 9y = 12$ that is closest to the origin.

SOLUTION   Since we are minimizing distance, we can minimize the square of the distance function without loss of generality:

$$f(x, y) = (x - 0)^2 + (y - 0)^2 = x^2 + y^2$$

subject to the constraint $g(x, y) = 4x + 9y - 12$.

Step 1. Write out the Lagrange Equations. The gradient vectors are $\nabla f = \langle 2x, 2y \rangle$ and $\nabla g = \langle 4, 9 \rangle$, hence the Lagrange Condition $\nabla f = \lambda \nabla g$ gives

$$\langle 2x, 2y \rangle = \lambda \langle 4, 9 \rangle$$

or

$$2x = 4\lambda \quad \Rightarrow \quad x = 2\lambda, \quad 2y = 9\lambda$$

Step 2. Solve for λ in terms of x and y. The Lagrange equations give the following two expressions for λ:

$$\lambda = \frac{x}{2}, \quad \lambda = \frac{9}{2}y$$

Equating these two

$$\frac{x}{2} = \frac{9}{2}y \quad \Rightarrow \quad x = 9y$$

Step 3. Solve for x and y using the constraint. We are given $4x + 9y = 12$, therefore we can write:

$$4(9y) + 9y = 12 \quad \Rightarrow \quad 45y = 12 \quad \Rightarrow \quad y = \frac{12}{45} = \frac{4}{15}$$

Since $x = 9y$, then we conclude:

$$y = \frac{4}{15} \quad x = 9 \cdot \frac{4}{15} = \frac{12}{5}$$

Step 4. Conclusions. Therefore the point closest to the origin lying on the plane $4x + 9y = 12$ is the point $(12/5, 4/15)$.

23. Find the maximum value of $f(x, y) = x^a y^b$ for $x \geq 0$, $y \geq 0$ on the line $x + y = 1$, where $a, b > 0$ are constants.

SOLUTION

Step 1. Write the Lagrange Equations. We must find the maximum value of $f(x, y) = x^a y^b$ under the constraints $g(x, y) = x + y - 1$, $x > 0$, $y > 0$. The gradient vectors are $\nabla f = \left\langle ax^{a-1}y^b, bx^a y^{b-1} \right\rangle$ and $\nabla g = \lambda \langle 1, 1 \rangle$, hence the Lagrange Condition $\nabla f = \lambda \nabla g$ is

$$\left\langle ax^{a-1}y^b, bx^a y^{b-1} \right\rangle = \lambda \langle 1, 1 \rangle$$

We obtain the following equations:

$$\begin{aligned} ax^{a-1}y^b &= \lambda \\ bx^a y^{b-1} &= \lambda \end{aligned} \quad \Rightarrow \quad ax^{a-1}y^b = bx^a y^{b-1}$$

Step 2. Solve for x and y using the constraint. We solve the equation in step 1 for y in terms of x. This gives

$$ax^{a-1}y^b = bx^a y^{b-1}$$

$$ay = bx \quad \Rightarrow \quad y = \frac{b}{a}x$$

We now substitute $y = \frac{b}{a}x$ in the constraint $x + y = 1$ and solve for x:

$$x + \frac{b}{a}x = 1$$

$$(a + b)x = a \quad \Rightarrow \quad x = \frac{a}{a + b}$$

We find y using the relation $y = \frac{b}{a}x$:

$$y = \frac{b}{a} \cdot \frac{a}{a + b} = \frac{b}{a + b}$$

The critical point is thus

$$\left(\frac{a}{a + b}, \frac{b}{a + b} \right) \tag{1}$$

Step 3. Conclusions. We compute $f(x, y) = x^a y^b$ at the critical point:

$$f\left(\frac{a}{a+b}, \frac{b}{a+b}\right) = \left(\frac{a}{a+b}\right)^a \left(\frac{b}{a+b}\right)^b = \frac{a^a b^b}{(a+b)^{a+b}}$$

Now, since f is continuous on the segment $x + y = 1$, $x \geq 0$, $y \geq 0$, which is a closed and bounded set in R^2, then f has minimum and maximum values on this segment. The minimum value is 0 (obtained at $(0, 1)$ and $(1, 0)$), therefore the critical point (1) corresponds to the maximum value. We conclude that the maximum value of $x^a y^b$ on $x + y = 1$, $x > 0$, $y > 0$ is

$$\frac{a^a b^b}{(a+b)^{a+b}}$$

25. Find the maximum value of $f(x, y) = x^a y^b$ for $x \geq 0$, $y \geq 0$ on the unit circle, where $a, b > 0$ are constants.

SOLUTION We must find the maximum value of $f(x, y) = x^a y^b$ $(a, b > 0)$ subject to the constraint $g(x, y) = x^2 + y^2 = 1$.

Step 1. Write out the Lagrange Equations. We have $\nabla f = \left\langle ax^{a-1} y^b, bx^a y^{b-1} \right\rangle$ and $\nabla g = \langle 2x, 2y \rangle$. Therefore the Lagrange Condition $\nabla f = \lambda \nabla g$ is

$$\left\langle ax^{a-1} y^b, bx^a y^{b-1} \right\rangle = \lambda \langle 2x, 2y \rangle$$

or

$$ax^{a-1} y^b = 2\lambda x$$
$$bx^a y^{b-1} = 2\lambda y \qquad (1)$$

Step 2. Solve for λ in terms of x and y. If $x = 0$ or $y = 0$, f has the minimum value 0. We thus may assume that $x > 0$ and $y > 0$. The equations (1) imply that

$$\lambda = \frac{ax^{a-2} y^b}{2}, \quad \lambda = \frac{bx^a y^{b-2}}{2}$$

Step 3. Solve for x and y using the constraint. Equating the two expressions for λ and solving for y in terms of x gives

$$\frac{ax^{a-2} y^b}{2} = \frac{bx^a y^{b-2}}{2}$$
$$ax^{a-2} y^b = bx^a y^{b-2}$$
$$ay^2 = bx^2$$
$$y^2 = \frac{b}{a} x^2 \quad \Rightarrow \quad y = \sqrt{\frac{b}{a}} x$$

We now substitute $y = \sqrt{\frac{b}{a}} x$ in the constraint $x^2 + y^2 = 1$ and solve for $x > 0$. We obtain

$$x^2 + \frac{b}{a} x^2 = 1$$
$$(a + b)x^2 = a$$
$$x^2 = \frac{a}{a+b} \quad \Rightarrow \quad x = \sqrt{\frac{a}{a+b}}$$

We find y using the relation $y = \sqrt{\frac{b}{a}} x$:

$$y = \sqrt{\frac{b}{a}} \sqrt{\frac{a}{a+b}} = \sqrt{\frac{ab}{a(a+b)}} = \sqrt{\frac{b}{a+b}}$$

We obtain the critical point:

$$\left(\sqrt{\frac{a}{a+b}}, \sqrt{\frac{b}{a+b}}\right)$$

Extreme points can also occur where $\nabla g = \mathbf{0}$, that is, $\langle 2x, 2y \rangle = \langle 0, 0 \rangle$ or $(x, y) = (0, 0)$. However, this point is not on the constraint.

Step 4. Conclusions. We compute $f(x, y) = x^a y^b$ at the critical point:

$$f\left(\sqrt{\frac{a}{a+b}}, \sqrt{\frac{b}{a+b}}\right) = \left(\frac{a}{a+b}\right)^{a/2}\left(\frac{b}{a+b}\right)^{b/2} = \frac{a^{a/2}b^{b/2}}{(a+b)^{(a+b)/2}} = \sqrt{\frac{a^a b^b}{(a+b)^{a+b}}}$$

The function $f(x, y) = x^a y^b$ is continuous on the set $x^2 + y^2 = 1$, $x \geq 0$, $y \geq 0$, which is a closed and bounded set in R^2, hence f has minimum and maximum values on the set. The minimum value is 0 (obtained at $(0, 1)$ and $(1, 0)$), hence the critical point that we found corresponds to the maximum value. We conclude that the maximum value of $x^a y^b$ on $x^2 + y^2 = 1$, $x > 0$, $y > 0$ is

$$\sqrt{\frac{a^a b^b}{(a+b)^{a+b}}}.$$

27. Show that the minimum distance from the origin to a point on the plane $ax + by + cz = d$ is

$$\frac{|d|}{\sqrt{a^2 + b^2 + c^2}}$$

SOLUTION We want to minimize the distance $P = \sqrt{x^2 + y^2 + z^2}$ subject to $ax + by + cz = d$. Since the square function u^2 is increasing for $u \geq 0$, the square P^2 attains its minimum at the same point where the distance P attains its minimum. Thus, we may minimize the function $f(x, y, z) = x^2 + y^2 + z^2$ subject to the constraint $g(x, y, z) = ax + by + cz = d$.

Step 1. Write out the Lagrange Equations. We have $\nabla f = \langle 2x, 2y, 2z \rangle$ and $\nabla g = \langle a, b, c \rangle$, hence the Lagrange Condition $\nabla f = \lambda \nabla g$ gives the following equations:

$$2x = \lambda a$$
$$2y = \lambda b$$
$$2z = \lambda c$$

Assume for now that $a \neq 0$, $b \neq 0$, $c \neq 0$.

Step 2. Solve for λ in terms of x, y, and z. The Lagrange Equations imply that

$$\lambda = \frac{2x}{a}, \quad \lambda = \frac{2y}{b}, \quad \lambda = \frac{2z}{c}$$

Step 3. Solve for x, y, and z using the constraint. Equating the expressions for λ give the following equations:

$$\begin{aligned} \frac{2x}{a} &= \frac{2z}{c} \\ \frac{2y}{b} &= \frac{2z}{c} \end{aligned} \quad \Rightarrow \quad \begin{aligned} x &= \frac{a}{c}z \\ y &= \frac{b}{c}z \end{aligned} \tag{1}$$

We now substitute $x = \frac{a}{c}z$ and $y = \frac{b}{c}z$ in the equation of the constraint $ax + by + cz = d$ and solve for z. This gives

$$a\left(\frac{a}{c}z\right) + b\left(\frac{b}{c}z\right) + cz = d$$

$$\frac{a^2}{c}z + \frac{b^2}{c}z + cz = d$$

$$\left(a^2 + b^2 + c^2\right)z = dc$$

Since $a^2 + b^2 + c^2 \neq 0$, we get $z = \frac{dc}{a^2+b^2+c^2}$. We now use (1) to compute y and x:

$$x = \frac{a}{c} \cdot \frac{dc}{a^2 + b^2 + c^2} = \frac{ad}{a^2 + b^2 + c^2}, \quad y = \frac{b}{c} \cdot \frac{dc}{a^2 + b^2 + c^2} = \frac{bd}{a^2 + b^2 + c^2}$$

We obtain the critical point:

$$P = \left(\frac{ad}{a^2 + b^2 + c^2}, \frac{bd}{a^2 + b^2 + c^2}, \frac{dc}{a^2 + b^2 + c^2}\right) \tag{2}$$

Step 4. Conclusions. It is clear geometrically that f has a minimum value subject to the constraint, hence the minimum value occurs at the point P. We conclude that the point P is the point on the plane closest to the origin. We now consider the case where $a = 0$. We consider the planes $ax + by + cz = d$, where $a \neq 0$ and $a \to 0$. A continuous change in

a causes a continuous change in the closest point P. Therefore, the point P closest to the origin in case of $a = 0$ can be obtained by computing the limit of P in (2) as $a \to 0$, that is, by substituting $a = 0$. Similar considerations hold for $b = 0$ or $c = 0$. We conclude that the closest point P in (2) holds also for the planes with $a = 0$, $b = 0$, or $c = 0$ (but not all of them 0). The distance P of that point to the origin is

$$P = \sqrt{\frac{(ad)^2 + (bd)^2 + (dc)^2}{(a^2 + b^2 + c^2)^2}} = |d| \sqrt{\frac{a^2 + b^2 + c^2}{(a^2 + b^2 + c^2)^2}} = \frac{|d|}{\sqrt{a^2 + b^2 + c^2}}$$

29. Let Q be the point on an ellipse closest to a given point P outside the ellipse. It was known to the Greek mathematician Apollonius (third century BCE) that $\overline{PQ}$ is perpendicular to the tangent to the ellipse at Q (Figure 15). Explain in words why this conclusion is a consequence of the method of Lagrange multipliers. *Hint:* The circles centered at P are level curves of the function to be minimized.

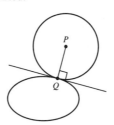

FIGURE 15

SOLUTION   Let $P = (x_0, y_0)$. The distance d between the point P and a point $Q = (x, y)$ on the ellipse is minimum where the square d^2 is minimum (since the square function u^2 is increasing for $u \ge 0$). Therefore, we want to minimize the function

$$f(x, y, z) = (x - x_0)^2 + (y - y_0)^2 + (z - z_0)^2$$

subject to the constraint

$$g(x, y) = \frac{x^2}{a^2} + \frac{y^2}{b^2} = 1$$

The method of Lagrange indicates that the solution Q is the point on the ellipse where $\nabla f = \lambda \nabla g$, that is, the point on the ellipse where the gradients ∇f and ∇g are parallel. Since the gradient is orthogonal to the level curves of the function, ∇g is orthogonal to the ellipse $g(x, y) = 1$, and ∇f is orthogonal to the level curve of f passing through Q. But this level curve is a circle through Q centered at P, hence the parallel vectors ∇g and ∇f are orthogonal to the ellipse and to the circle centered at P respectively. We conclude that the point Q is the point at which the tangent to the ellipse is also the tangent to the circle through Q centered at P. That is, the tangent to the ellipse at Q is perpendicular to the radius $\overline{PQ}$ of the circle.

In Exercises 31 and 32, let V be the volume of a can of radius r and height h, and let S be its surface area (including the top and bottom).

31. Find r and h that minimize S subject to the constraint $V = 54\pi$.

SOLUTION   We see that the surface area of the can is $S = 2\pi r h + 2\pi r^2$ subject to $V = 54\pi = \pi r^2 h$. Let us write the constraint as $V(r, h) = \pi r^2 h - 54\pi$ and use Lagrange Multipliers to solve.

Step 1.   Write out the Lagrange Equations. The gradient vectors are $\nabla S = \langle 2\pi h + 4\pi r, 2\pi r \rangle$ and $\nabla V = \left\langle 2\pi r h, \pi r^2 \right\rangle$. Then using $\nabla S = \lambda \nabla V$, we see

$$\langle 2\pi h + 4\pi r, 2\pi r \rangle = \lambda \left\langle 2\pi r h, \pi r^2 \right\rangle$$

or

$$2\pi h + 4\pi r = 2\pi \lambda r h, \quad 2\pi r = \lambda \pi r^2$$

Consider the second equation, rewriting we have:

$$2\pi r - \lambda \pi r^2 = 0 \quad \Rightarrow \quad \pi r(2 - \lambda r) = 0 \quad \Rightarrow \quad r = 0, \lambda = \frac{2}{r}$$

We can ignore when $r = 0$ since it does not correspond to any point on the constraint curve $54\pi = \pi r^2 h$.
    Using the first equation, rewriting we have:

$$2\pi h + 4\pi r = 2\pi \lambda r h \quad \Rightarrow \quad \lambda = \frac{2\pi h + 4\pi r}{2\pi r h} = \frac{h + 2r}{r h}$$

Step 2. Solve for r, h using the constraint to determine the critical point.

Using the two derived equations for λ we have:

$$\frac{2}{r} = \frac{h + 2r}{rh} \quad\Rightarrow\quad 2rh = hr + 2r^2 \quad r(2h - h - 2r) = 0 \quad\Rightarrow\quad h = 2r$$

Then using the constraint, $54\pi = \pi r^2 h$ we see:

$$54\pi = \pi r^2(2r) \quad\Rightarrow\quad 54 = 2r^3 \quad\Rightarrow\quad r^3 = 27 \quad\Rightarrow\quad r = 3$$

Thus $r = 3$ and $h = 2(3) = 6$.

Step 3. Conclusions. The minimum surface area, given that the volume must be 54π is determined by a can having radius $r = 3$ and height $h = 6$. We know this is the minimum surface area because surface area is an increasing function of r and h.

33. A plane with equation $\dfrac{x}{a} + \dfrac{y}{b} + \dfrac{z}{c} = 1$ $(a, b, c > 0)$ together with the positive coordinate planes forms a tetrahedron of volume $V = \frac{1}{6}abc$ (Figure 18). Find the minimum value of V among all planes passing through the point $P = (1, 1, 1)$.

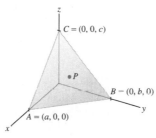

FIGURE 18

SOLUTION The plane is constrained to pass through the point $P = (1, 1, 1)$, hence this point must satisfy the equation of the plane. That is,

$$\frac{1}{a} + \frac{1}{b} + \frac{1}{c} = 1$$

We thus must minimize the function $V(a, b, c) = \frac{1}{6}abc$ subject to the constraint $g(a, b, c) = \frac{1}{a} + \frac{1}{b} + \frac{1}{c} = 1$, $a > 0$, $b > 0$, $c > 0$.

Step 1. Write out the Lagrange Equations. We have $\nabla V = \left\langle \frac{1}{6}bc, \frac{1}{6}ac, \frac{1}{6}ab \right\rangle$ and $\nabla g = \left\langle -\frac{1}{a^2}, -\frac{1}{b^2}, -\frac{1}{c^2} \right\rangle$, hence the Lagrange Condition $\nabla V = \lambda \nabla g$ yields the following equations:

$$\frac{1}{6}bc = -\frac{1}{a^2}\lambda$$

$$\frac{1}{6}ac = -\frac{1}{b^2}\lambda$$

$$\frac{1}{6}ab = -\frac{1}{c^2}\lambda$$

Step 2. Solve for λ in terms of a, b, and c. The Lagrange equations imply that

$$\lambda = -\frac{bca^2}{6}, \quad \lambda = -\frac{acb^2}{6}, \quad \lambda = -\frac{abc^2}{6}$$

Step 3. Solve for a, b, and c using the constraint. Equating the expressions for λ, we obtain the following equations:

$$\begin{array}{ll} bca^2 = acb^2 \\ abc^2 = acb^2 \end{array} \quad\Rightarrow\quad \begin{array}{ll} abc(a - b) = 0 \\ abc(c - b) = 0 \end{array}$$

Since a, b, c are positive numbers, we conclude that $a = b$ and $c = b$. We now substitute $a = b$ and $c = b$ in the equation of the constraint $\frac{1}{a} + \frac{1}{b} + \frac{1}{c} = 1$ and solve for b. This gives

$$\frac{1}{b} + \frac{1}{b} + \frac{1}{b} = 1$$

$$\frac{3}{b} = 1 \quad\Rightarrow\quad b = 3$$

Therefore also $a = b = 3$ and $c = b = 3$. We obtain the critical point $(3, 3, 3)$.

Step 4. Conclusions. If V has a minimum value subject to the constraint then it occurs at the point $(3, 3, 3)$. That is, the plane that minimizes V is

$$\frac{x}{3} + \frac{y}{3} + \frac{z}{3} = 1 \quad \text{or} \quad x + y + z = 3$$

Remark: Since the constraint is not bounded, we need to justify the existence of a minimum value of $V = \frac{1}{6}abc$ under the constraint $\frac{1}{a} + \frac{1}{b} + \frac{1}{c} = 1$. First notice that since a, b, c are nonnegative and the sum of their reciprocals is 1, none of them can tend to zero. In fact, none of a, b, c can be less than 1. Therefore, if $a \to \infty, b \to \infty$, or $c \to \infty$, then $V \to \infty$. This means that we can find a cube that includes the point $\left(\frac{1}{3}, \frac{1}{3}, \frac{1}{3}\right)$ such that, on the part of the constraint that is outside the cube, it holds that $V > V\left(\frac{1}{3}, \frac{1}{3}, \frac{1}{3}\right) = \frac{1}{162}$. On the part of the constraint inside the cube, V has a minimum value m, since it is a closed and bounded set. Clearly m is the minimum of V on the whole constraint.

35. Show that the Lagrange equations for $f(x, y) = x + y$ subject to the constraint $g(x, y) = x + 2y = 0$ have no solution. What can you conclude about the minimum and maximum values of f subject to $g = 0$? Show this directly.

SOLUTION   Using the methods of Lagrange we can write $\nabla f = \lambda \nabla g$ and see

$$\langle 1, 1 \rangle = \lambda \langle 1, 2 \rangle$$

Which gives us the equations:

$$1 = \lambda, \quad 1 = 2\lambda$$

hence, $\lambda = 1$ or $\lambda = 1/2$. This is an inconsistent set of equations, thus the Lagrange method has no solution. What we can conclude from this is that the maximum and minimum values of f subject to $g = 0$ does not exist. This means that $f(x, y)$ increases without an upper bound and decreases without a lower bound.

To show this directly, we can write $y = -1/2x$ from the constraint equation and substitute it into $f(x, y) = f(x, -1/2x) = x - 1/2x = 1/2x$. We know that $y = 1/2x$ is a straight line having slope $1/2$, increasing, with no maximum nor minimum values.

37. Let L be the minimum length of a ladder that can reach over a fence of height h to a wall located a distance b behind the wall.

(a) Use Lagrange multipliers to show that $L = (h^{2/3} + b^{2/3})^{3/2}$ (Figure 19). *Hint:* Show that the problem amounts to minimizing $f(x, y) = (x + b)^2 + (y + h)^2$ subject to $y/b = h/x$ or $xy = bh$.

(b) Show that the value of L is also equal to the radius of the circle with center $(-b, -h)$ that is tangent to the graph of $xy = bh$.

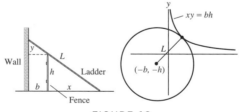

FIGURE 19

SOLUTION

(a) We denote by x and y the lengths shown in the figure, and express the length l of the ladder in terms of x and y.

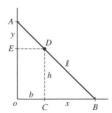

Using the Pythagorean Theorem, we have

$$l = \sqrt{\overline{OA}^2 + \overline{OB}^2} = \sqrt{(y + h)^2 + (x + b)^2} \tag{1}$$

Since the function u^2 is increasing for $u \geq 0$, l and l^2 have their minimum values at the same point. Therefore, we may minimize the function $f(x, y) = l^2(x, y)$, which is

$$f(x, y) = (x + b)^2 + (y + h)^2$$

We now identify the constraint on the variables x and y. (Notice that h, b are constants while x and y are free to change). Using proportional lengths in the similar triangles $\triangle AED$ and $\triangle DCB$, we have

$$\frac{\overline{AE}}{\overline{DC}} = \frac{\overline{ED}}{\overline{CB}}$$

That is,

$$\frac{y}{h} = \frac{b}{x} \quad \Rightarrow \quad xy = bh$$

We thus must minimize $f(x, y) = (x + b)^2 + (y + h)^2$ subject to the constraint $g(x, y) = xy = bh$, $x > 0$, $y > 0$.

Step 1. Write out the Lagrange Equations. We have $\nabla f = \langle 2(x + b), 2(y + h) \rangle$ and $\nabla g = \langle y, x \rangle$, hence the Lagrange Condition $\nabla f = \lambda \nabla g$ gives the following equations:

$$2(x + b) = \lambda y$$

$$2(y + h) = \lambda x$$

Step 2. Solve for λ in terms of x and y. The equation of the constraint implies that $y \neq 0$ and $x \neq 0$. Therefore, the Lagrange equations yield

$$\lambda = \frac{2(x + b)}{y}, \quad \lambda = \frac{2(y + h)}{x}$$

Step 3. Solve for x and y using the constraint. Equating the two expressions for λ gives

$$\frac{2(x + b)}{y} = \frac{2(y + h)}{x}$$

We simplify:

$$x(x + b) = y(y + h)$$

$$x^2 + xb = y^2 + yh$$

The equation of the constraint implies that $y = \frac{bh}{x}$. We substitute and solve for $x > 0$. This gives

$$x^2 + xb = \left(\frac{bh}{x}\right)^2 + \frac{bh}{x} \cdot h$$

$$x^2 + xb = \frac{b^2 h^2}{x^2} + \frac{bh^2}{x}$$

$$x^4 + x^3 b = b^2 h^2 + bh^2 x$$

$$x^4 + bx^3 - bh^2 x - b^2 h^2 = 0$$

$$x^3(x + b) - bh^2(x + b) = 0$$

$$\left(x^3 - bh^2\right)(x + b) = 0$$

Since $x > 0$ and $b > 0$, also $x + b > 0$ and the solution is

$$x^3 - bh^2 = 0 \quad \Rightarrow \quad x = (bh^2)^{1/3}$$

We compute y. Using the relation $y = \frac{bh}{x}$,

$$y = \frac{bh}{(bh^2)^{1/3}} = \frac{bh}{b^{1/3} h^{2/3}} = b^{2/3} h^{1/3} = (b^2 h)^{1/3}$$

We obtain the solution

$$x = \left(bh^2\right)^{1/3}, \quad y = \left(b^2 h\right)^{1/3} \tag{2}$$

Extreme values may also occur at the point on the constraint where $\nabla g = \mathbf{0}$. However, $\nabla g = \langle y, x \rangle = \langle 0, 0 \rangle$ only at the point $(0, 0)$, which is not on the constraint.

Step 4. Conclusions. Notice that on the constraint $y = \frac{bh}{x}$ or $x = \frac{bh}{y}$, as $x \to 0+$ then $y \to \infty$, and as $x \to \infty$, then $y \to 0+$. Also, as $y \to 0+$, $x \to \infty$ and as $y \to \infty$, $x \to 0+$. In either case, $f(x, y)$ is increasing without bound. Using this property and the theorem on the existence of extreme values for a continuous function on a closed and bounded set (for a certain part of the constraint), one can show that f has a minimum value on the constraint. This minimum value occurs at the point (2). We substitute this point in (1) to obtain the following minimum length L:

$$L = \sqrt{\left((b^2h)^{1/3} + h\right)^2 + \left((bh^2)^{1/3} + b\right)^2}$$

$$= \sqrt{(b^2h)^{2/3} + 2h(b^2h)^{1/3} + h^2 + (bh^2)^{2/3} + 2b(bh^2)^{1/3} + b^2}$$

$$= \sqrt{b^{\frac{4}{3}}h^{2/3} + 2h^{\frac{4}{3}}b^{2/3} + h^2 + b^{2/3}h^{\frac{4}{3}} + 2b^{\frac{4}{3}}h^{2/3} + b^2}$$

$$= \sqrt{3b^{\frac{4}{3}}h^{2/3} + 3h^{\frac{4}{3}}b^{2/3} + h^2 + b^2}$$

$$= \sqrt{\left(h^{2/3}\right)^3 + 3\left(h^{2/3}\right)^2 b^{2/3} + 3h^{2/3}\left(b^{2/3}\right)^2 + \left(b^{2/3}\right)^3}$$

Using the identity $(\alpha + \beta)^3 = \alpha^3 + 3\alpha^2\beta + 3\alpha\beta^2 + \beta^3$, we conclude that

$$L = \sqrt{\left(h^{2/3} + b^{2/3}\right)^3} = \left(h^{2/3} + b^{2/3}\right)^{3/2}.$$

(b) The Lagrange Condition states that the gradient vectors ∇f_P and ∇g_P are parallel (where P is the minimizing point). The gradient ∇f_P is orthogonal to the level curve of f passing through P, which is a circle through P centered at $(-b, -h)$. ∇g_P is orthogonal to the level curve of g passing through P, which is the curve of the constraint $xy = bh$. We conclude that the circle and the curve $xy = bh$, both being perpendicular to parallel vectors, are tangent at P. The radius of the circle is the minimum value L, of $f(x, y)$.

39. Find the point lying on the intersection of the plane $x + \frac{1}{2}y + \frac{1}{4}z = 0$ and the sphere $x^2 + y^2 + z^2 = 9$ with the largest z-coordinate.

SOLUTION We will use the method of Lagrange Multipliers with two constraints here. We want to maximize $f(x, y, z) = z$ subject to the two surfaces. Set the first constraint as $g(x, y, z) = x + \frac{1}{2}y + \frac{1}{4}z = 0$ and the second as $h(x, y, z) = x^2 + y^2 + z^2 - 9 = 0$.

Write out the Lagrange equations. We have $\nabla f = \langle 0, 0, 1 \rangle$, $\nabla g = \left\langle 1, \frac{1}{2}, \frac{1}{4} \right\rangle$ and $\nabla g = \langle 2x, 2y, 2z \rangle$, hence the Lagrange condition, $\nabla f = \lambda \nabla g + \mu \nabla h$ yields the following equations:

$$\langle 0, 0, 1 \rangle = \lambda \left\langle 1, \frac{1}{2}, \frac{1}{4} \right\rangle + \mu \langle 2x, 2y, 2z \rangle$$

and

$$0 = \lambda + 2\mu x, \quad 0 = \frac{1}{2}\lambda + 2\mu y, \quad 1 = \frac{1}{4}\lambda + 2\mu z$$

Hence, from the first two equations we see

$$\lambda = -2\mu x, \quad \lambda = -4\mu y$$

Therefore

$$-2\mu x = -4\mu y \quad \Rightarrow \quad x = 2y$$

since $\mu \neq 0$. Using the first constraint equation $x + \frac{1}{2}y + \frac{1}{4}z = 0$ we have

$$2y + \frac{1}{2}y + \frac{1}{4}z = 0 \quad \Rightarrow \quad \frac{5}{2}y + \frac{1}{4}z = 0 \quad \Rightarrow \quad y = -\frac{1}{10}z$$

Finally, we can substitute $y = -1/10z$ and $x = 2y = -1/5z$ into the second constraint equation $x^2 + y^2 + z^2 = 9$ to see

$$\left(-\frac{1}{5}z\right)^2 + \left(-\frac{1}{10}z\right)^2 + z^2 = 9 \quad \Rightarrow \quad \frac{1}{25}z^2 + \frac{1}{100}z^2 + z^2 = 9 \quad \Rightarrow \quad 4z^2 + z^2 + 100z^2 = 900$$

Hence

$$105z^2 = 900 \quad \Rightarrow \quad z^2 = \frac{900}{105} = \frac{60}{7}$$

Therefore $z = \pm\sqrt{\frac{60}{7}} = \pm 2\sqrt{\frac{15}{7}}$. The two critical points are:

$$P\left(-\frac{2}{5}\sqrt{\frac{15}{7}}, -\frac{1}{5}\sqrt{\frac{15}{7}}, 2\sqrt{\frac{15}{7}}\right), \quad Q\left(\frac{2}{5}\sqrt{\frac{15}{7}}, \frac{1}{5}\sqrt{\frac{15}{7}}, -2\sqrt{\frac{15}{7}}\right)$$

The critical point with the largest z-coordinate (the maximum of $f(x, y, z)$) is P with z-coordinate $2\sqrt{\frac{15}{7}} \approx 2.928$.

41. The cylinder $x^2 + y^2 = 1$ intersects the plane $x + z = 1$ in an ellipse. Find the point on that ellipse that is farthest from the origin.

SOLUTION We need to use Lagrange Multipliers with two constraints here. We want to maximize the square of the distance from the origin $f(x, y, z) = x^2 + y^2 + z^2$ subject to $g(x, y, z) = x^2 + y^2 - 1$ and $h(x, y, z) = x + z - 1$. Taking the gradients we have $\nabla f = \langle 2x, 2y, 2z \rangle$, $\nabla g = \langle 2x, 2y, 0 \rangle$, and $\nabla h = \langle 1, 0, 1 \rangle$. Writing the Lagrange condition $\nabla f = \lambda \nabla g + \mu \nabla h$ we have

$$\langle 2x, 2y, 2z \rangle = \lambda \langle 2x, 2y, 0 \rangle + \mu \langle 1, 0, 1 \rangle$$

and

$$2x = 2\lambda x + \mu, \quad 2y = 2\lambda y, \quad 2z = \mu$$

Using the second equation we see:

$$2y - 2\lambda y = 0 \quad \Rightarrow \quad 2y(\lambda - 1) = 0$$

Therefore, either $\lambda = 1$ or $y = 0$.

If $\lambda = 1$ then this implies $\mu = 0$ and $z = 0$. Using the constraint $x + z = 1$ then $x = 1$, and using the constraint $x^2 + y^2 = 1$, then $y = 0$. This gives the critical point

$$(1, 0, 0)$$

If $y = 0$, using the constraint $x^2 + y^2 = 1$, then $x = \pm 1$. If $x = 1$, then $z = 0$, if $x = -1$ then $z = 2$. This gives the critical points

$$(1, 0, 0), \quad (-1, 0, 2)$$

Now we examine $f(x, y, z) = x^2 + y^2 + z^2$ at the two critical points for the maximum value:

$$f(1, 0, 0) = 1, \quad f(-1, 0, 2) = 5$$

Thus, the point farthest from the origin on this ellipse is the point $(-1, 0, 2)$ (at $\sqrt{5}$ units away).

43. Find the minimum value of $f(x, y, z) = x^2 + y^2 + z^2$ subject to two constraints, $x + 2y + z = 3$ and $x - y = 4$.

SOLUTION The constraint equations are

$$g(x, y, z) = x + 2y + z - 3 = 0, \quad h(x, y) = x - y - 4 = 0$$

Step 1. Write out the Lagrange Equations. We have $\nabla f = \langle 2x, 2y, 2z \rangle$, $\nabla g = \langle 1, 2, 1 \rangle$, and $\nabla h = \langle 1, -1, 0 \rangle$, hence the Lagrange Condition is

$$\nabla f = \lambda \nabla g + \mu \nabla h$$

$$\langle 2x, 2y, 2z \rangle = \lambda \langle 1, 2, 1 \rangle + \mu \langle 1, -1, 0 \rangle$$

$$= \langle \lambda + \mu, 2\lambda - \mu, \lambda \rangle$$

We obtain the following equations:

$$2x = \lambda + \mu$$

$$2y = 2\lambda - \mu$$

$$2z = \lambda$$

Step 2. Solve for λ and μ. The first equation gives $\lambda = 2x - \mu$. Combining with the third equation we get

$$2z = 2x - \mu \tag{1}$$

The second equation gives $\mu = 2\lambda - 2y$, combining with the third equation we get $\mu = 4z - 2y$. Substituting in (1) we obtain

$$2z = 2x - (4z - 2y) = 2x - 4z + 2y$$

$$6z = 2x + 2y \quad \Rightarrow \quad z = \frac{x + y}{3} \tag{2}$$

Step 3. Solve for x, y, and z using the constraints. The constraints give x and y as functions of z:

$$x - y = 4 \quad \Rightarrow \quad y = x - 4$$

$$x + 2y + z = 3 \quad \Rightarrow \quad y = \frac{3 - x - z}{2}$$

Combining the two equations we get

$$x - 4 = \frac{3 - x - z}{2}$$

$$2x - 8 = 3 - x - z$$

$$3x = 11 - z \quad \Rightarrow \quad x = \frac{11 - z}{3}$$

We find y using $y = x - 4$:

$$y = \frac{11 - z}{3} - 4 = \frac{-1 - z}{3}$$

We substitute x and y in (2) and solve for z:

$$z = \frac{\frac{11-z}{3} + \frac{-1-z}{3}}{3} = \frac{11 - z - 1 - z}{9} = \frac{10 - 2z}{9}$$

$$9z = 10 - 2z$$

$$11z = 10 \quad \Rightarrow \quad z = \frac{10}{11}$$

We find x and y:

$$y = \frac{-1 - z}{3} = \frac{-1 - \frac{10}{11}}{3} = -\frac{21}{33} = -\frac{7}{11}$$

$$x = \frac{11 - z}{3} = \frac{11 - \frac{10}{11}}{3} = \frac{111}{33} = \frac{37}{11}$$

We obtain the solution

$$P = \left(\frac{37}{11}, -\frac{7}{11}, \frac{10}{11} \right)$$

Step 4. Calculate the critical values. We compute $f(x, y, z) = z^2 + y^2 + z^2$ at the critical point:

$$f(P) = \left(\frac{37}{11} \right)^2 + \left(-\frac{7}{11} \right)^2 + \left(\frac{10}{11} \right)^2 = \frac{1518}{121} = \frac{138}{11} \approx 12.545$$

As x tends to infinity, so also does $f(x, y, z)$ tend to ∞. Therefore f has no maximum value and the given critical point P must produce a minimum. We conclude that the minimum value of f subject to the two constraints is $f(P) = \frac{138}{11} \approx 12.545$.

Further Insights and Challenges

45. Assumptions Matter Consider the problem of minimizing $f(x, y) = x$ subject to $g(x, y) = (x - 1)^3 - y^2 = 0$.

(a) Show, without using calculus, that the minimum occurs at $P = (1, 0)$.

(b) Show that the Lagrange condition $\nabla f_P = \lambda \nabla g_P$ is not satisfied for any value of λ.

(c) Does this contradict Theorem 1?

SOLUTION

(a) The equation of the constraint can be rewritten as

$$(x - 1)^3 = y^2 \quad \text{or} \quad x = y^{2/3} + 1$$

Therefore, at the points under the constraint, $x \geq 1$, hence $f(x, y) \geq 1$. Also at the point $P = (1, 0)$ we have $f(1, 0) = 1$, hence $f(1, 0) = 1$ is the minimum value of f under the constraint.

(b) We have $\nabla f = \langle 1, 0 \rangle$ and $\nabla g = \left\langle 3(x - 1)^2, -2y \right\rangle$, hence the Lagrange Condition $\nabla f = \lambda \nabla g$ yields the following equations:

$$1 = \lambda \cdot 3(x - 1)^2$$

$$0 = -2\lambda y$$

The first equation implies that $\lambda \neq 0$ and $x - 1 = \pm \frac{1}{\sqrt{3\lambda}}$. The second equation gives $y = 0$. Substituting in the equation of the constraint gives

$$(x-1)^3 - y^2 = \left(\frac{\pm 1}{\sqrt{3\lambda}}\right)^3 - 0^2 = \frac{\pm 1}{(3\lambda)^{3/2}} \neq 0$$

We conclude that the Lagrange Condition is not satisfied by any point under the constraint.

(c) Theorem 1 is not violated since at the point $P = (1, 0)$, $\nabla g = \mathbf{0}$, whereas the Theorem is valid for points where $\nabla g_P \neq \mathbf{0}$.

47. Consider the utility function $U(x_1, x_2) = x_1 x_2$ with budget constraint $p_1 x_1 + p_2 x_2 = c$.

(a) Show that the maximum of $U(x_1, x_2)$ subject to the budget constraint is equal to $c^2/(4 p_1 p_2)$.

(b) Calculate the value of the Lagrange multiplier λ occurring in (a).

(c) Prove the following interpretation: λ is the rate of increase in utility per unit increase in total budget c.

SOLUTION

(a) By the earlier exercise, the utility is maximized at a point where the following equality holds:

$$\frac{U_{x_1}}{U_{x_2}} = \frac{p_1}{p_2}$$

Since $U_{x_1} = x_2$ and $U_{x_2} = x_1$, we get

$$\frac{x_2}{x_1} = \frac{p_1}{p_2} \quad \Rightarrow \quad x_2 = \frac{p_1}{p_2} x_1$$

We now substitute x_2 in terms of x_1 in the constraint $p_1 x_1 + p_2 x_2 = c$ and solve for x_1. This gives

$$p_1 x_1 + p_2 \cdot \frac{p_1}{p_2} x_1 = c$$

$$2 p_1 x_1 = c \quad \Rightarrow \quad x_1 = \frac{c}{2 p_1}$$

The corresponding value of x_2 is computed by $x_2 = \frac{p_1}{p_2} x_1$:

$$x_2 = \frac{p_1}{p_2} \cdot \frac{c}{2 p_1} = \frac{c}{2 p_2}$$

That is, $U(x_1, x_2)$ is maximized at the consumption level $x_1 = \frac{c}{2 p_1}$, $x_2 = \frac{c}{2 p_2}$. The maximum value is

$$U\left(\frac{c}{2 p_1}, \frac{c}{2 p_2}\right) = \frac{c}{2 p_1} \cdot \frac{c}{2 p_2} = \frac{c^2}{4 p_1 p_2}$$

(b) The Lagrange condition $\nabla U = \lambda \nabla g$ for $U(x_1, x_2) = x_1 x_2$ and $g(x_1, x_2) = p_1 x_1 + p_2 x_2 - c = 0$ is

$$\langle x_2, x_1 \rangle = \lambda \langle p_1, p_2 \rangle \tag{1}$$

or

$$\begin{aligned} x_2 &= \lambda p_1 \\ x_1 &= \lambda p_2 \end{aligned} \quad \Rightarrow \quad \lambda = \frac{x_2}{p_1} = \frac{x_1}{p_2}$$

In part (a) we showed that at the maximizing point $x_1 = \frac{c}{2 p_1}$, therefore the value of λ is

$$\lambda = \frac{x_1}{p_2} = \frac{c}{2 p_1 p_2}$$

(c) We compute $\frac{dU}{dc}$ using the Chain Rule:

$$\frac{dU}{dc} = \frac{\partial U}{\partial x_1} x_1'(c) + \frac{\partial U}{\partial x_2} x_2'(c) = x_2 x_1'(c) + x_1 x_2'(c) = \langle x_2, x_1 \rangle \cdot \langle x_1'(c), x_2'(c) \rangle$$

Substituting in (1) we get

$$\frac{dU}{dc} = \lambda \langle p_1, p_2 \rangle \cdot \langle x_1'(c), x_2'(c) \rangle = \lambda \left(p_1 x_1'(c) + p_2 x_2'(c) \right) \tag{2}$$

We now use the Chain Rule to differentiate the equation of the constraint $p_1 x_1 + p_2 x_2 = c$ with respect to c:

$$p_1 x_1'(c) + p_2 x_2'(c) = 1$$

Substituting in (2), we get

$$\frac{dU}{dc} = \lambda \cdot 1 = \lambda$$

Using the approximation $\Delta U \approx \frac{dU}{dc} \Delta c$, we conclude that λ is the rate of increase in utility per unit increase of total budget L.

49. Let $B > 0$. Show that the maximum of

$$f(x_1, \ldots, x_n) = x_1 x_2 \cdots x_n$$

subject to the constraints $x_1 + \cdots + x_n = B$ and $x_j \geq 0$ for $j = 1, \ldots, n$ occurs for $x_1 = \cdots = x_n = B/n$. Use this to conclude that

$$(a_1 a_2 \cdots a_n)^{1/n} \leq \frac{a_1 + \cdots + a_n}{n}$$

for all positive numbers $a_1, \ldots, a_n$.

SOLUTION   We first notice that the constraints $x_1 + \cdots + x_n = B$ and $x_j \geq 0$ for $j = 1, \ldots, n$ define a closed and bounded set in the nth dimensional space, hence f (continuous, as a polynomial) has extreme values on this set. The minimum value zero occurs where one of the coordinates is zero (for example, for $n = 2$ the constraint $x_1 + x_2 = B$, $x_1 \geq 0, x_2 \geq 0$ is a triangle in the first quadrant). We need to maximize the function $f(x_1, \ldots, x_n) = x_1 x_2 \cdots x_n$ subject to the constraints $g(x_1, \ldots, x_n) = x_1 + \cdots + x_n - B = 0$, $x_j \geq 0$, $j = 1, \ldots, n$.

Step 1. Write out the Lagrange Equations. The gradient vectors are

$$\nabla f = \langle x_2 x_3 \cdots x_n, x_1 x_3 \cdots x_n, \ldots, x_1 x_2 \cdots x_{n-1} \rangle$$

$$\nabla g = \langle 1, 1, \ldots, 1 \rangle$$

The Lagrange Condition $\nabla f = \lambda \nabla g$ yields the following equations:

$$x_2 x_3 \cdots x_n = \lambda$$

$$x_1 x_3 \cdots x_n = \lambda$$

$$x_1 x_2 \cdots x_{n-1} = \lambda$$

Step 2. Solving for $x_1, x_2, \ldots, x_n$ using the constraint. The Lagrange equations imply the following equations:

$$x_2 x_3 \cdots x_n = x_1 x_2 \cdots x_{n-1}$$

$$x_1 x_3 \cdots x_n = x_1 x_2 \cdots x_{n-1}$$

$$x_1 x_2 x_4 \cdots x_n = x_1 x_2 \cdots x_{n-1}$$

$$\vdots$$

$$x_1 x_2 \cdots x_{n-2} x_n = x_1 x_2 \cdots x_{n-1}$$

We may assume that $x_j \neq 0$ for $j = 1, \ldots, n$, since if one of the coordinates is zero, f has the minimum value zero. We divide each equation by its right-hand side to obtain

$$\frac{x_n}{x_1} = 1 \qquad\qquad x_1 = x_n$$

$$\frac{x_n}{x_2} = 1 \qquad\qquad x_2 = x_n$$

$$\frac{x_n}{x_3} = 1 \quad \Rightarrow \quad x_3 = x_n$$

$$\vdots \qquad\qquad\qquad \vdots$$

$$\frac{x_n}{x_{n-1}} = 1 \qquad\qquad x_{n-1} = x_n$$

Substituting in the constraint $x_1 + \cdots + x_n = B$ and solving for x_n gives

$$\underbrace{x_n + x_n + \cdots + x_n}_{n} = B$$

$$n x_n = B \quad \Rightarrow \quad x_n = \frac{B}{n}$$

Hence $x_1 = \cdots = x_n = \frac{B}{n}$.

Step 3. Conclusions. The maximum value of $f(x_1, \ldots, x_n) = x_1 x_2 \cdots x_n$ on the constraint $x_1 + \cdots + x_n = B$, $x_j \geq 0$, $j = 1, \ldots, n$ occurs at the point at which all coordinates are equal to $\frac{B}{n}$. The value of f at this point is

$$f\left(\frac{B}{n}, \frac{B}{n}, \ldots, \frac{B}{n}\right) = \left(\frac{B}{n}\right)^n$$

It follows that for any point $(x_1, \ldots, x_n)$ on the constraint, that is, for any point satisfying $x_1 + \cdots + x_n = B$ with x_j positive, the following holds:

$$f(x_1, \ldots, x_n) \leq \left(\frac{B}{n}\right)^n$$

That is,

$$x_1 \cdots x_n \leq \left(\frac{x_1 + \cdots + x_n}{n}\right)^n$$

or

$$(x_1 \cdots x_n)^{1/n} \leq \frac{x_1 + \cdots + x_n}{n}.$$

51. Given constants E, E_1, E_2, E_3, consider the maximum of

$$S(x_1, x_2, x_3) = x_1 \ln x_1 + x_2 \ln x_2 + x_3 \ln x_3$$

subject to two constraints:

$$x_1 + x_2 + x_3 = N, \qquad E_1 x_1 + E_2 x_2 + E_3 x_3 = E$$

Show that there is a constant μ such that $x_i = A^{-1} e^{\mu E_i}$ for $i = 1, 2, 3$, where $A = N^{-1}(e^{\mu E_1} + e^{\mu E_2} + e^{\mu E_3})$.

SOLUTION The constraints equations are

$$g(x_1, x_2, x_3) = x_1 + x_2 + x_3 - N = 0$$
$$h(x_1, x_2, x_3) = E_1 x_1 + E_2 x_2 + E_3 x_3 - E = 0$$

We first find the Lagrange equations. The gradient vectors are

$$\nabla S = \left\langle \ln x_1 + x_1 \cdot \frac{1}{x_1}, \ln x_2 + x_2 \cdot \frac{1}{x_2}, \ln x_3 + x_3 \cdot \frac{1}{x_3} \right\rangle = \langle 1 + \ln x_1, 1 + \ln x_2, 1 + \ln x_3 \rangle$$
$$\nabla g = \langle 1, 1, 1 \rangle, \qquad \nabla h = \langle E_1, E_2, E_3 \rangle$$

The Lagrange Condition $\nabla f = \lambda \nabla g + \mu \nabla h$ gives the following equation:

$$\langle 1 + \ln x_1, 1 + \ln x_2, 1 + \ln x_3 \rangle = \lambda \langle 1, 1, 1 \rangle + \mu \langle E_1, E_2, E_3 \rangle = \langle \lambda + \mu E_1, \lambda + \mu E_2, \lambda + \mu E_3 \rangle$$

We obtain the Lagrange equations:

$$1 + \ln x_1 = \lambda + \mu E_1$$
$$1 + \ln x_2 = \lambda + \mu E_2$$
$$1 + \ln x_3 = \lambda + \mu E_3$$

We subtract the third equation from the other equations to obtain

$$\ln x_1 - \ln x_3 = \mu(E_1 - E_3)$$
$$\ln x_2 - \ln x_3 = \mu(E_2 - E_3)$$

or

$$\begin{aligned} \ln \frac{x_1}{x_3} &= \mu(E_1 - E_3) \\ \ln \frac{x_2}{x_3} &= \mu(E_2 - E_3) \end{aligned} \quad \Rightarrow \quad \begin{aligned} x_1 &= x_3 e^{\mu(E_1 - E_3)} \\ x_2 &= x_3 e^{\mu(E_2 - E_3)} \end{aligned} \tag{1}$$

Substituting x_1 and x_2 in the equation of the constraint $g(x_1, x_2, x_3) = 0$ and solving for x_3 gives

$$x_3 e^{\mu(E_1 - E_3)} + x_3 e^{\mu(E_2 - E_3)} + x_3 = N$$

We multiply by $e^{\mu E_3}$:

$$x_3\left(e^{\mu E_1} + e^{\mu E_2} + e^{\mu E_3}\right) = N e^{\mu E_3}$$

$$x_3 = \frac{N e^{\mu E_3}}{e^{\mu E_1} + e^{\mu E_2} + e^{\mu E_3}}$$

Substituting in (1) we get

$$x_1 = \frac{N e^{\mu E_3}}{e^{\mu E_1} + e^{\mu E_2} + e^{\mu E_3}} \cdot e^{\mu(E_1 - E_3)} = \frac{N e^{\mu E_1}}{e^{\mu E_1} + e^{\mu E_2} + e^{\mu E_3}}$$

$$x_2 = \frac{N e^{\mu E_3}}{e^{\mu E_1} + e^{\mu E_2} + e^{\mu E_3}} \cdot e^{\mu(E_2 - E_3)} = \frac{N e^{\mu E_2}}{e^{\mu E_1} + e^{\mu E_2} + e^{\mu E_3}}$$

Letting $A = \frac{e^{\mu E_1} + e^{\mu E_2} + e^{\mu E_3}}{N}$, we obtain

$$x_1 = A^{-1} e^{\mu E_1}, \quad x_2 = A^{-1} e^{\mu E_2}, \quad x_3 = A^{-1} e^{\mu E_3}$$

The value of μ is determined by the second constraint $h(x_1, x_2, x_3) = 0$.

CHAPTER REVIEW EXERCISES

1. Given $f(x, y) = \dfrac{\sqrt{x^2 - y^2}}{x + 3}$:

(a) Sketch the domain of f.

(b) Calculate $f(3, 1)$ and $f(-5, -3)$.

(c) Find a point satisfying $f(x, y) = 1$.

SOLUTION

(a) f is defined where $x^2 - y^2 \geq 0$ and $x + 3 \neq 0$. We solve these two inequalities:

$$x^2 - y^2 \geq 0 \quad \Rightarrow \quad x^2 \geq y^2 \quad \Rightarrow \quad |x| \geq |y|$$

$$x + 3 \neq 0 \quad \Rightarrow \quad x \neq -3$$

Therefore, the domain of f is the following set:

$$D = \{(x, y) : |x| \geq |y|, x \neq -3\}$$

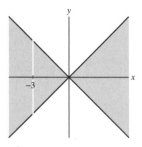

(b) To find $f(3, 1)$ we substitute $x = 3$, $y = 1$ in $f(x, y)$. We get

$$f(3, 1) = \frac{\sqrt{3^2 - 1^2}}{3 + 3} = \frac{\sqrt{8}}{6} = \frac{\sqrt{2}}{3}$$

Similarly, setting $x = -5$, $y = -3$, we get

$$f(-5, -3) = \frac{\sqrt{(-5)^2 - (-3)^2}}{-5 + 3} = \frac{\sqrt{16}}{-2} = -2.$$

(c) We must find a point (x, y) such that

$$f(x, y) = \frac{\sqrt{x^2 - y^2}}{x + 3} = 1$$

We choose, for instance, $y = 1$, substitute and solve for x. This gives

$$\frac{\sqrt{x^2 - 1^2}}{x + 3} = 1$$

$$\sqrt{x^2 - 1} = x + 3$$

$$x^2 - 1 = (x + 3)^2 = x^2 + 6x + 9$$

$$6x = -10 \quad \Rightarrow \quad x = -\frac{5}{3}$$

Thus, the point $\left(-\frac{5}{3}, 1\right)$ satisfies $f\left(-\frac{5}{3}, 1\right) = 1$.

3. Sketch the graph $f(x, y) = x^2 - y + 1$ and describe its vertical and horizontal traces.

SOLUTION The graph is shown in the following figure.

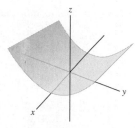

The trace obtained by setting $x = c$ is the line $z = c^2 - y + 1$ or $z = (c^2 + 1) - y$ in the plane $x = c$. The trace obtained by setting $y = c$ is the parabola $z = x^2 - c + 1$ in the plane $y = c$. The trace obtained by setting $z = c$ is the parabola $y = x^2 + 1 - c$ in the plane $z = c$.

5. Match the functions (a)–(d) with their graphs in Figure 1

(a) $f(x, y) = x^2 + y$

(b) $f(x, y) = x^2 + 4y^2$

(c) $f(x, y) = \sin(4xy)e^{-x^2 - y^2}$

(d) $f(x, y) = \sin(4x)e^{-x^2 - y^2}$

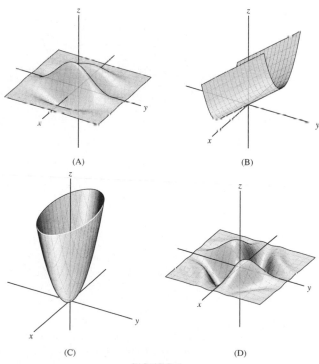

(A) (B) (C) (D)

FIGURE 1

SOLUTION The function $f = x^2 + y$ matches picture (b), as can be seen by taking the $x = 0$ slice. The function $f = x^2 + 4y^2$ matches picture (c), as can be seen by taking $z = c$ slices (giving ellipses). Since $\sin(4xy)e^{-x^2 - y^2}$ is symmetric with respect to x and y, and so also is picture (d), we match $\sin(4xy)e^{-x^2 - y^2}$ with (d). That leaves the third function, $\sin(4x)e^{-x^2 - y^2}$, to match with picture (a).

7. Describe the level curves of:

(a) $f(x, y) = e^{4x-y}$

(b) $f(x, y) = \ln(4x - y)$

(c) $f(x, y) = 3x^2 - 4y^2$

(d) $f(x, y) = x + y^2$

SOLUTION

(a) The level curves of $f(x, y) = e^{4x-y}$ are the curves $e^{4x-y} = c$ in the xy-plane, where $c > 0$. Taking ln from both sides we get $4x - y = \ln c$. Therefore, the level curves are the parallel lines of slope 4, $4x - y = \ln c$, $c > 0$, in the xy-plane.

(b) The level curves of $f(x, y) = \ln(4x - y)$ are the curves $\ln(4x - y) = c$ in the xy-plane. We rewrite it as $4x - y = e^c$ to obtain the parallel lines of slope 4, with negative y-intercepts.

(c) The level curves of $f(x, y) = 3x^2 - 4y^2$ are the hyperbolas $3x^2 - 4y^2 = c$ in the xy plane.

(d) The level curves of $f(x, y) = x + y^2$ arc the curves $x + y^2 = c$ or $x = c - y^2$ in the xy-plane. These are parabolas whose axis is the x-axis.

In Exercises 9–14, evaluate the limit or state that it does not exist.

9. $\lim\limits_{(x,y)\to(1,-3)} (xy + y^2)$

SOLUTION The function $f(x, y) = xy + y^2$ is continuous everywhere because it is a polynomial, therefore we evaluate the limit using substitution:

$$\lim_{(x,y)\to(1,-3)} \left(xy + y^2\right) = 1 \cdot (-3) + (-3)^2 = 6$$

11. $\lim\limits_{(x,y)\to(0,0)} \dfrac{xy + xy^2}{x^2 + y^2}$

SOLUTION We evaluate the limits as (x, y) approaches the origin along the lines $y = x$ and $y = 2x$:

$$\lim_{\substack{(x,y)\to(0,0) \\ \text{along } y=x}} \frac{xy + xy^2}{x^2 + y^2} = \lim_{x\to 0} \frac{x \cdot x + x \cdot x^2}{x^2 + x^2} = \lim_{x\to 0} \frac{x^2 + x^3}{2x^2} = \lim_{x\to 0} \frac{1 + x}{2} = \frac{1}{2}$$

$$\lim_{\substack{x\to(0,0) \\ \text{along } y=2x}} \frac{xy + xy^2}{x^2 + y^2} = \lim_{x\to 0} \frac{x \cdot 2x + x \cdot (2x)^2}{x^2 + (2x)^2} = \lim_{x\to 0} \frac{2x^2 + 4x^3}{5x^2} = \lim_{x\to 0} \frac{2 + 4x}{5} = \frac{2}{5}$$

Since the two limits are different, $f(x, y)$ does not approach one limit as $(x, y) \to (0, 0)$, therefore the limit does not exist.

13. $\lim\limits_{(x,y)\to(1,-3)} (2x + y)e^{-x+y}$

SOLUTION The function $f(x, y) = (2x + y)e^{-x+y}$ is continuous, hence we evaluate the limit using substitution:

$$\lim_{(x,y)\to(1,-3)} (2x + y)e^{-x+y} = (2 \cdot 1 - 3)e^{-1-3} = -e^{-4}$$

15. Let

$$f(x, y) = \begin{cases} \dfrac{(xy)^p}{x^4 + y^4} & (x, y) \neq (0, 0) \\ 0 & (x, y) = (0, 0) \end{cases}$$

Use polar coordinates to show that $f(x, y)$ is continuous at all (x, y) if $p > 2$ but is discontinuous at $(0, 0)$ if $p \leq 2$.

SOLUTION We show using the polar coordinates $x = r\cos\theta$, $y = r\sin\theta$, that the limit of $f(x, y)$ as $(x, y) \to (0, 0)$ is zero for $p > 2$. This will prove that f is continuous at the origin. Since f is a rational function with nonzero denominator for $(x, y) \neq (0, 0)$, f is continuous there. We have

$$\lim_{(x,y)\to(0,0)} f(x, y) = \lim_{r\to 0+} \frac{(r\cos\theta)^p(r\sin\theta)^p}{(r\cos\theta)^4 + (r\sin\theta)^4} = \lim_{r\to 0+} \frac{r^{2p}(\cos\theta\sin\theta)^p}{r^4\left(\cos^4\theta + \sin^4\theta\right)} \tag{1}$$

$$= \lim_{r\to 0+} \frac{r^{2(p-2)}(\cos\theta\sin\theta)^p}{\cos^4\theta + \sin^4\theta}$$

We use the following inequalities:

$$\left| \cos^4 \theta \sin^4 \theta \right| \le 1$$

$$\cos^4 \theta + \sin^4 \theta = \left(\cos^2 \theta + \sin^2 \theta \right)^2 - 2 \cos^2 \theta \sin^2 \theta = 1 - \frac{1}{2} \cdot (2 \cos \theta \sin \theta)^2$$

$$= 1 - \frac{1}{2} \sin^2 2\theta \ge 1 - \frac{1}{2} = \frac{1}{2}$$

Therefore,

$$0 \le \left| \frac{r^{2(p-2)} (\cos \theta \sin \theta)^p}{\cos^4 \theta + \sin^4 \theta} \right| \le \frac{r^{2(p-2)} \cdot 1}{\frac{1}{2}} = 2r^{2(p-2)}$$

Since $p - 2 > 0$, $\lim\limits_{r \to 0+} 2r^{2(p-2)} = 0$, hence by the Squeeze Theorem the limit in (1) is also zero. We conclude that f is continuous for $p > 2$.

We now show that for $p < 2$ the limit of $f(x, y)$ as $(x, y) \to (0, 0)$ does not exist. We compute the limit as (x, y) approaches the origin along the line $y = x$.

$$\lim_{\substack{(x,y) \to (0,0) \\ \text{along } y=x}} f(x, y) = \lim_{x \to 0} \frac{(x^2)^p}{x^4 + x^4} = \lim_{x \to 0} \frac{x^{2p}}{2x^4} = \lim_{x \to 0} \frac{x^{2(p-2)}}{2} = \infty$$

Therefore the limit of $f(x, y)$ as $(x, y) \to (0, 0)$ does not exist for $p < 2$. We now show that the limit $\lim\limits_{(x,y) \to (0,0)} \frac{x^2 y^2}{x^4 + y^4}$ does not exist for $p = 2$ as well. We compute the limits along the line $y = 0$ and $y = x$:

$$\lim_{\substack{(x,y) \to (0,0) \\ \text{along } y=0}} \frac{x^2 y^2}{x^4 + y^4} = \lim_{x \to 0} \frac{x^2 \cdot 0^2}{x^4 + 0^4} = \lim_{x \to 0} \frac{0}{x^4} = 0$$

$$\lim_{\substack{(x,y) \to (0,0) \\ \text{along } y=x}} \frac{x^2 y^2}{x^4 + y^4} = \lim_{x \to 0} \frac{x^2 \cdot x^2}{x^4 + x^4} = \lim_{x \to 0} \frac{x^4}{2x^4} = \frac{1}{2}$$

Since the limits along two paths are different, $f(x, y)$ does not approach one limit as $(x, y) \to (0, 0)$. We thus showed that if $p \le 2$, the limit $\lim\limits_{(x,y) \to (0,0)} f(x, y)$ does not exist, and f is not continuous at the origin for $p \le 2$.

In Exercises 17–20, compute f_x and f_y.

17. $f(x, y) = 2x + y^2$

SOLUTION To find f_x we treat y as a constant, and to find f_y we treat x as a constant. We get

$$f_x = \frac{\partial}{\partial x} \left(2x + y^2 \right) = \frac{\partial}{\partial x} (2x) + \frac{\partial}{\partial x} \left(y^2 \right) = 2 + 0 = 2$$

$$f_y = \frac{\partial}{\partial y} \left(2x + y^2 \right) = \frac{\partial}{\partial y} (2x) + \frac{\partial}{\partial y} \left(y^2 \right) = 0 + 2y = 2y$$

19. $f(x, y) = \sin(xy)e^{-x-y}$

SOLUTION We compute f_x, treating y as a constant and using the Product Rule and the Chain Rule. We get

$$f_x = \frac{\partial}{\partial x} \left(\sin(xy)e^{-x-y} \right) = \frac{\partial}{\partial x} (\sin(xy)) \, e^{-x-y} + \sin(xy) \frac{\partial}{\partial x} e^{-x-y}$$

$$= \cos(xy) \cdot ye^{-x-y} + \sin(xy) \cdot (-1)e^{-x-y} = e^{-x-y} (y \cos(xy) - \sin(xy))$$

We compute f_y similarly, treating x as a constant. Notice that since $f(y, x) = f(x, y)$, the partial derivative f_y can be obtained from f_x by interchanging x and y. That is,

$$f_y = e^{-x-y} (x \cos(yx) - \sin(yx)).$$

21. Calculate f_{xxyz} for $f(x, y, z) = y \sin(x + z)$.

SOLUTION We differentiate f twice with respect to x, once with respect to y, and finally with respect to z. This gives

$$f_x = \frac{\partial}{\partial x} (y \sin(x + z)) = y \cos(x + z)$$

$$f_{xx} = \frac{\partial}{\partial x} (y \cos(x + z)) = -y \sin(x + z)$$

$$f_{xxy} = \frac{\partial}{\partial y}(-y\sin(x+z)) = -\sin(x+z)$$

$$f_{xxyz} = \frac{\partial}{\partial z}(-\sin(x+z)) = -\cos(x+z)$$

23. Find an equation of the tangent plane to the graph of $f(x, y) = xy^2 - xy + 3x^3y$ at $P = (1, 3)$.

SOLUTION   The tangent plane has the equation

$$z = f(1, 3) + f_x(1, 3)(x - 1) + f_y(1, 3)(y - 3) \tag{1}$$

We compute the partial derivatives of $f(x, y) = xy^2 - xy + 3x^3y$:

$$\begin{aligned} f_x(x, y) &= y^2 - y + 9x^2y \\ f_y(x, y) &= 2xy - x + 3x^3 \end{aligned} \Rightarrow \begin{aligned} f_x(1, 3) &= 3^2 - 3 + 9 \cdot 1^2 \cdot 3 = 33 \\ f_y(1, 3) &= 2 \cdot 1 \cdot 3 - 1 + 3 \cdot 1^3 = 8 \end{aligned}$$

Also, $f(1, 3) = 1 \cdot 3^2 - 1 \cdot 3 + 3 \cdot 1^3 \cdot 3 = 15$. Substituting these values in (1), we obtain the following equation:

$$z = 15 + 33(x - 1) + 8(y - 3)$$

or

$$z = 33x + 8y - 42$$

25. Use a linear approximation of $f(x, y, z) = \sqrt{x^2 + y^2 + z}$ to estimate $\sqrt{7.1^2 + 4.9^2 + 69.5}$. Compare with a calculator value.

SOLUTION   The function whose value we want to approximate is

$$f(x, y, z) = \sqrt{x^2 + y^2 + z}$$

We will use the linear approximation at the point $(7, 5, 70)$. Recall that the linear approximation to a surface will be:

$$L(x, y, z) = f(7, 5, 70) + f_x(7, 5, 70)(x - 7) + f_y(7, 5, 70)(y - 5) + f_z(7, 5, 70)(z - 70)$$

We compute the partial derivatives of f:

$$f_x(x, y, z) = \frac{2x}{2\sqrt{x^2 + y^2 + z}} = \frac{x}{\sqrt{x^2 + y^2 + z}} \quad \Rightarrow \quad f_x(7, 5, 70) = \frac{7}{\sqrt{7^2 + 5^2 + 70}} = \frac{7}{12}$$

$$f_y(x, y, z) = \frac{2y}{2\sqrt{x^2 + y^2 + z}} = \frac{y}{\sqrt{x^2 + y^2 + z}} \quad \Rightarrow \quad f_y(7, 5, 70) = \frac{5}{\sqrt{7^2 + 5^2 + 70}} = \frac{5}{12}$$

$$f_z(x, y, z) = \frac{1}{2\sqrt{x^2 + y^2 + z}} \quad \Rightarrow \quad f_z(7, 5, 70) = \frac{1}{2\sqrt{7^2 + 5^2 + 70}} = \frac{1}{24}$$

Also, $f(7, 5, 70) = \sqrt{7^2 + 5^2 + 70} = 12$. Substituting the values in the linear approximation equation we obtain the following approximation:

$$L(x, y, z) = 12 + \frac{7}{12}(x - 7) + \frac{5}{12}(y - 5) + \frac{1}{24}(z - 70)$$

Now we are ready to approximate $\sqrt{7.1^2 + 4.9^2 + 69.5}$. That is, using the linear approximation,

$$L(7.1, 4.9, 69.5) = 12 + \frac{7}{12}(7.1 - 7) + \frac{5}{12}(4.9 - 5) + \frac{1}{24}(69.5 - 70)$$

$$= 12 + \frac{7}{12} \cdot \frac{1}{10} + \frac{5}{12} \cdot -\frac{1}{10} + \frac{1}{24} \cdot -\frac{1}{2}$$

$$= 12 + \frac{7}{120} - \frac{5}{120} - \frac{1}{48}$$

$$= \frac{2879}{240} = 11.9958333$$

The value obtained using a calculator is 11.996667.

27. Figure 4 shows the contour map of a function $f(x, y)$ together with a path $c(t)$ in the counterclockwise direction. The points $c(1)$, $c(2)$, and $c(3)$ are indicated on the path. Let $g(t) = f(c(t))$. Which of statements (i)–(iv) are true? Explain.

(i) $g'(1) > 0$.

(ii) $g(t)$ has a local minimum for some $1 \le t \le 2$.

(iii) $g'(2) = 0$.

(iv) $g'(3) = 0$.

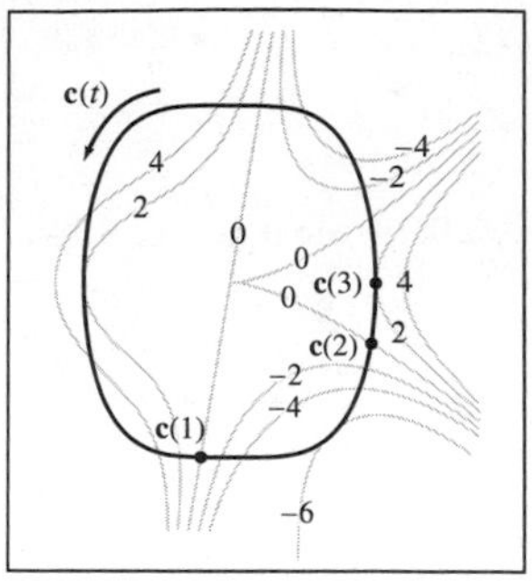

FIGURE 4

SOLUTION   (ii) and (iv) are true

In Exercises 29–32, compute $\dfrac{d}{dt} f(\mathbf{c}(t))$ at the given value of t.

29. $f(x, y) = x + e^y$, $\mathbf{c}(t) = (3t - 1, t^2)$ at $t = 2$

SOLUTION   By the Chain Rule for Paths we have

$$\frac{d}{dt} f(\mathbf{c}(t)) = \nabla f \cdot \mathbf{c}'(t) \tag{1}$$

We evaluate the gradient ∇f and $\mathbf{c}'(t)$:

$$\mathbf{c}'(t) = \langle 3, 2t \rangle$$

$$\nabla f = \langle f_x, f_y \rangle = \langle 1, e^y \rangle \quad \Rightarrow \quad \nabla f_{\mathbf{c}(t)} = \left\langle 1, e^{t^2} \right\rangle$$

Substituting in (1) we get

$$\frac{d}{dt} f(\mathbf{c}(t)) = \left\langle 1, e^{t^2} \right\rangle \cdot \langle 3, 2t \rangle = 3 + 2te^{t^2}$$

At $t = 2$ we have

$$\frac{d}{dt} f(\mathbf{c}(t)) \bigg|_{t=2} = 3 + 2 \cdot 2 \cdot e^{2^2} = 3 + 4e^4 \approx 221.4.$$

31. $f(x, y) = xe^{3y} - ye^{3x}$, $\mathbf{c}(t) = (e^t, \ln t)$ at $t = 1$

SOLUTION   We use the Chain Rule for Paths:

$$\frac{d}{dt} f(\mathbf{c}(t)) = \nabla f_{\mathbf{c}(t)} \cdot \mathbf{c}'(t) \tag{1}$$

We find the ∇f at the point $\mathbf{c}(1)$ and compute $\mathbf{c}'(1)$. We get

$$\nabla f = \langle f_x, f_y \rangle = \left\langle e^{3y} - 3ye^{3x}, 3xe^{3y} - e^{3x} \right\rangle$$

$$\mathbf{c}(1) = \left\langle e^1, \ln 1 \right\rangle = \langle e, 0 \rangle$$

$$\nabla f_{\mathbf{c}(1)} = \left\langle e^{3 \cdot 0} - 3 \cdot 0 e^{3e}, 3ee^{3 \cdot 0} - e^{3e} \right\rangle = \left\langle 1, 3e - e^{3e} \right\rangle \tag{2}$$

$$\mathbf{c}'(t) = \frac{d}{dt} \langle e^t, \ln t \rangle = \left\langle e^t, t^{-1} \right\rangle \quad \Rightarrow \quad \mathbf{c}'(1) = \langle e, 1 \rangle \tag{3}$$

Substituting (2) and (3) in (1) gives

$$\frac{d}{dt} f(\mathbf{c}(t)) \bigg|_{t=1} = \nabla f_{c(1)} \cdot \mathbf{c}'(1) = \left\langle 1, 3e - e^{3e} \right\rangle \cdot \langle e, 1 \rangle = e + 3e - e^{3e} = 4e - e^{3e}$$

In Exercises 33–36, compute the directional derivative at P in the direction of **v.**

33. $f(x, y) = x^3 y^4$, $P = (3, -1)$, $\mathbf{v} = 2\mathbf{i} + \mathbf{j}$

SOLUTION   We first normalize **v** to find a unit vector **u** in the direction of **v**:

$$\mathbf{u} = \frac{\mathbf{v}}{\|\mathbf{v}\|} = \frac{2\mathbf{i} + \mathbf{j}}{\sqrt{2^2 + 1^2}} = \frac{2}{\sqrt{5}}\mathbf{i} + \frac{1}{\sqrt{5}}\mathbf{j}$$

We compute the directional derivative using the following equality:

$$D_{\mathbf{u}} f(3, -1) = \nabla f_{(3,-1)} \cdot \mathbf{u}$$

The gradient vector at the given point is the following vector:

$$\nabla f = \langle f_x, f_y \rangle = \left\langle 3x^2 y^4, 4x^3 y^3 \right\rangle \quad \Rightarrow \quad \nabla f_{(3,-1)} = \langle 27, -108 \rangle$$

Hence,

$$D_{\mathbf{u}} f(3, -1) = \langle 27, -108 \rangle \cdot \left\langle \frac{2}{\sqrt{5}}, \frac{1}{\sqrt{5}} \right\rangle = \frac{54}{\sqrt{5}} - \frac{108}{\sqrt{5}} = -\frac{54}{\sqrt{5}}$$

35. $f(x, y) = e^{x^2 + y^2}$, $\quad P = \left(\dfrac{\sqrt{2}}{2}, \dfrac{\sqrt{2}}{2} \right)$, $\quad \mathbf{v} = \langle 3, -4 \rangle$

SOLUTION   We normalize $\mathbf{v}$ to obtain a vector $\mathbf{u}$ in the direction of $\mathbf{v}$:

$$\mathbf{u} = \frac{\langle 3, -4 \rangle}{\sqrt{3^2 + (-4)^2}} = \left\langle \frac{3}{5}, -\frac{4}{5} \right\rangle$$

We use the following theorem:

$$D_{\mathbf{u}} f(P) = \nabla f_P \cdot \mathbf{u} \tag{1}$$

We find the gradient of f at the given point:

$$\nabla f = \langle f_x, f_y \rangle = \left\langle 2x e^{x^2 + y^2}, 2y e^{x^2 + y^2} \right\rangle = 2 e^{x^2 + y^2} \langle x, y \rangle$$

Hence,

$$\nabla f_P = 2 e^{\left(\frac{\sqrt{2}}{2} \right)^2 + \left(\frac{\sqrt{2}}{2} \right)^2} \left\langle \frac{\sqrt{2}}{2}, \frac{\sqrt{2}}{2} \right\rangle = e\sqrt{2} \langle 1, 1 \rangle$$

Substituting in (1) we get

$$D_{\mathbf{u}} f(P) = \sqrt{2} e \langle 1, 1 \rangle \cdot \left\langle \frac{3}{5}, -\frac{4}{5} \right\rangle = \sqrt{2} e \left(\frac{3}{5} - \frac{4}{5} \right) = -\frac{\sqrt{2} e}{5}$$

37. Find the unit vector $\mathbf{e}$ at $P = (0, 0, 1)$ pointing in the direction along which $f(x, y, z) = xz + e^{-x^2 + y}$ increases most rapidly.

SOLUTION   The gradient vector ∇f_P points in the direction of maximum rate of increase of f. Therefore we need to find a unit vector in the direction of ∇f_P. We first find the gradient of $f(x, y, z) = xz + e^{-x^2 + y}$:

$$\nabla f = \left\langle \frac{\partial f}{\partial x}, \frac{\partial f}{\partial y}, \frac{\partial f}{\partial z} \right\rangle = \left\langle z - 2x e^{-x^2 + y}, e^{-x^2 + y}, x \right\rangle$$

At the point $P = (0, 0, 1)$ we have

$$\nabla f_P = \langle 1, 1, 0 \rangle.$$

We normalize ∇f_P to obtain the unit vector $\mathbf{e}$ at P pointing in the direction of maximum increase of f:

$$\mathbf{e} = \frac{\nabla f_P}{\|\nabla f_P\|} = \left\langle \frac{1}{\sqrt{2}}, \frac{1}{\sqrt{2}}, 0 \right\rangle.$$

39. Let $n \neq 0$ be an integer and r an arbitrary constant. Show that the tangent plane to the surface $x^n + y^n + z^n = r$ at $P = (a, b, c)$ has equation

$$a^{n-1} x + b^{n-1} y + c^{n-1} z = r$$

SOLUTION   The tangent plane to the surface, defined implicitly by $F(x, y, z) = r$ at a point (a, b, c) on the surface, has the following equation:

$$0 = F_x(a, b, c)(x - a) + F_y(a, b, c)(y - b) + F_z(a, b, c)(z - c) \tag{1}$$

The given surface is defined by the function $F(x, y, z) = x^n + y^n + z^n$. We find the partial derivative of F at a point $P = (a, b, c)$ on the surface:

$$F_x(x, y, z) = nx^{n-1} \qquad F_x(a, b, c) = na^{n-1}$$
$$F_y(x, y, z) = ny^{n-1} \quad \Rightarrow \quad F_y(a, b, c) = nb^{n-1}$$
$$F_z(x, y, z) = nz^{n-1} \qquad F_z(a, b, c) = nc^{n-1}$$

Substituting in (1) we get

$$na^{n-1}(x - a) + nb^{n-1}(y - b) + nc^{n-1}(z - c) = 0$$

We divide by n and simplify:

$$a^{n-1}x - a^n + b^{n-1}y - b^n + c^{n-1}z - c^n = 0$$
$$a^{n-1}x + b^{n-1}y + c^{n-1}z = a^n + b^n + c^n \qquad (2)$$

The point $P = (a, b, c)$ lies on the surface, hence it satisfies the equation of the surface. That is,

$$a^n + b^n + c^n = r$$

Substituting in (2) we obtain the following equation of the tangent plane:

$$a^{n-1}x + b^{n-1}y + c^{n-1}z = r$$

41. Let $f(x, y, z) = x^2y + y^2z$. Use the Chain Rule to calculate $\partial f/\partial s$ and $\partial f/\partial t$ (in terms of s and t), where

$$x = s + t, \quad y = st, \quad z = 2s - t$$

SOLUTION We compute the Primary Derivatives:

$$\frac{\partial f}{\partial x} = 2xy, \quad \frac{\partial f}{\partial y} = x^2 + 2yz, \quad \frac{\partial f}{\partial z} = y^2$$

Since $\frac{\partial x}{\partial s} = 1$, $\frac{\partial y}{\partial s} = t$, $\frac{\partial z}{\partial s} = 2$, $\frac{\partial x}{\partial t} = 1$, $\frac{\partial y}{\partial t} = s$, and $\frac{\partial z}{\partial t} = -1$, the Chain Rule gives

$$\frac{\partial f}{\partial s} = \frac{\partial f}{\partial x}\frac{\partial x}{\partial s} + \frac{\partial f}{\partial y}\frac{\partial y}{\partial s} + \frac{\partial f}{\partial z}\frac{\partial z}{\partial s} = 2xy \cdot 1 + \left(x^2 + 2yz\right)t + y^2 \cdot 2$$

$$= 2xy + \left(x^2 + 2yz\right)t + 2y^2$$

$$\frac{\partial f}{\partial t} = \frac{\partial f}{\partial x}\frac{\partial x}{\partial t} + \frac{\partial f}{\partial y}\frac{\partial y}{\partial t} + \frac{\partial f}{\partial z}\frac{\partial z}{\partial t} = 2xy \cdot 1 + \left(x^2 + 2yz\right)s + y^2 \cdot (-1)$$

$$= 2xy + \left(x^2 + 2yz\right)s - y^2$$

We now substitute $x = s + t$, $y = st$, and $z = 2s - t$ to express the answer in terms of the independent variables s, t. We get

$$\frac{\partial f}{\partial s} = 2(s + t)st + \left((s + t)^2 + 2st(2s - t)\right)t + 2s^2t^2$$

$$= 2s^2t + 2st^2 + \left(s^2 + 2st + t^2 + 4s^2t - 2st^2\right)t + 2s^2t^2$$

$$= 3s^2t + 4st^2 + t^3 - 2st^3 + 6s^2t^2$$

$$\frac{\partial f}{\partial t} = 2(s + t)st + \left((s + t)^2 + 2st(2s - t)\right)s - s^2t^2$$

$$= 2s^2t + 2st^2 + \left(s^2 + 2st + t^2 + 4s^2t - 2st^2\right)s - s^2t^2$$

$$= 4s^2t + 3st^2 + s^3 + 4s^3t - 3s^2t^2$$

43. Let $g(u, v) = f(u^3 - v^3, v^3 - u^3)$. Prove that

$$v^2 \frac{\partial g}{\partial u} - u^2 \frac{\partial g}{\partial v} = 0$$

SOLUTION   We are given the function $f(x, y)$, where $x = u^3 - v^3$ and $y = v^3 - u^3$. Using the Chain Rule we have the following derivatives:

$$\frac{\partial g}{\partial u} = \frac{\partial f}{\partial x}\frac{\partial x}{\partial u} + \frac{\partial f}{\partial y}\frac{\partial y}{\partial u}$$

$$\frac{\partial g}{\partial v} = \frac{\partial f}{\partial x}\frac{\partial x}{\partial v} + \frac{\partial f}{\partial y}\frac{\partial y}{\partial v} \tag{1}$$

We compute the following partial derivatives:

$$\frac{\partial x}{\partial u} = 3u^2, \quad \frac{\partial y}{\partial u} = -3u^2$$

$$\frac{\partial x}{\partial v} = -3v^2, \quad \frac{\partial y}{\partial v} = 3v^2$$

Substituting in (1) we obtain

$$\frac{\partial g}{\partial u} = \frac{\partial f}{\partial x} \cdot 3u^2 + \frac{\partial f}{\partial y}\left(-3u^2\right) = 3u^2\left(\frac{\partial f}{\partial x} - \frac{\partial f}{\partial y}\right)$$

$$\frac{\partial g}{\partial v} = \frac{\partial f}{\partial x}\left(-3v^2\right) + \frac{\partial f}{\partial y}\left(3v^2\right) = -3v^2\left(\frac{\partial f}{\partial x} - \frac{\partial f}{\partial y}\right)$$

Therefore,

$$v^2\frac{\partial g}{\partial u} + u^2\frac{\partial g}{\partial v} = 3u^2v^2\left(\frac{\partial f}{\partial x} - \frac{\partial f}{\partial y}\right) - 3u^2v^2\left(\frac{\partial f}{\partial x} - \frac{\partial f}{\partial y}\right) = 0$$

45. Calculate $\partial z/\partial x$, where $xe^z + ze^y = x + y$.

SOLUTION   The function $F(x, y, z) = xe^z + ze^y - x - y = 0$ defines z implicitly as a function of x and y. Using implicit differentiation, the partial derivative of z with respect to x is

$$\frac{\partial z}{\partial x} = -\frac{F_x}{F_z} \tag{1}$$

We compute the partial derivatives F_x and F_z:

$$F_x = e^z - 1$$

$$F_z = xe^z + e^y$$

Substituting in (1) gives

$$\frac{\partial z}{\partial x} = -\frac{e^z - 1}{xe^z + e^y}.$$

In Exercises 47–50, find the critical points of the function and analyze them using the Second Derivative Test.

47. $f(x, y) = x^4 - 4xy + 2y^2$

SOLUTION   To find the critical points, we need the first-order partial derivatives and set them equal to zero to solve for x and y:

$$f_x(x, y) = 4x^3 - 4y = 0, \quad f_y(x, y) = -4x + 4y = 0$$

Looking at the second equation we see $x = y$. Using this in the first equation, then

$$4x^3 - 4x = 0 \quad \Rightarrow \quad 4x(x^2 - 1) = 0 \quad \Rightarrow \quad x = 0, \pm 1$$

Therefore, our critical points are:

$$(0, 0), (1, 1), (-1, -1)$$

Now to find the discriminant, D, we need the second-order partial derivatives:

$$f_{xx}(x, y) = 12x^2, \quad f_{yy}(x, y) = 4, \quad f_{xy}(x, y) = -4$$

Hence,

$$D(x, y) = f_{xx}f_{yy} - f_{xy}^2 = 48x^2 - 16 = 16(3x^2 - 1)$$

Analyzing our three critical points we see:

$$D(0, 0) = -16 < 0, \quad D(1, 1) = 32 > 0, \quad D(-1, -1) = 32 > 0$$

Since the discriminant for $(0, 0)$ is negative, $(0, 0)$ is a saddle point.

Looking at $f_{xx}(1, 1) = 12 > 0$ and $f_{xx}(-1, -1) = 12 > 0$ hence, the points $(1, 1)$ and $(-1, -1)$ are both local minima.

49. $f(x, y) = e^{x+y} - xe^{2y}$

SOLUTION We find the critical point by setting the partial derivatives of $f(x, y) = e^{x+y} - xe^{2y}$ equal to zero and solve. This gives

$$f_x(x, y) = e^{x+y} - e^{2y} = 0$$
$$f_y(x, y) = e^{x+y} - 2xe^{2y} = 0$$

The first equation gives $e^{x+y} = e^{2y}$ and the second equation gives $e^{x+y} = 2xe^{2y}$. Equating the two expressions, dividing by the nonzero function e^{2y}, and solving for x, we obtain

$$e^{2y} = 2xe^{2y} \quad \Rightarrow \quad 1 = 2x \quad \Rightarrow \quad x = \frac{1}{2}$$

We now substitute $x = \frac{1}{2}$ in the first equation and solve for y, to obtain

$$e^{\frac{1}{2}+y} - e^{2y} = 0 \quad \Rightarrow \quad e^{\frac{1}{2}+y} = e^{2y} \quad \Rightarrow \quad \frac{1}{2} + y = 2y \quad \Rightarrow \quad y = \frac{1}{2}$$

There is one critical point, $\left(\frac{1}{2}, \frac{1}{2}\right)$. We examine the critical point using the Second Derivative Test. We compute the second derivatives at this point:

$$f_{xx}(x, y) = e^{x+y} \quad \Rightarrow \quad f_{xx}\left(\frac{1}{2}, \frac{1}{2}\right) = e^{\frac{1}{2}+\frac{1}{2}} = e$$

$$f_{yy}(x, y) = e^{x+y} - 4xe^{2y} \quad \Rightarrow \quad f_{yy}\left(\frac{1}{2}, \frac{1}{2}\right) = e^{\frac{1}{2}+\frac{1}{2}} - 4 \cdot \frac{1}{2}e^{2\cdot\frac{1}{2}} = -e$$

$$f_{xy}(x, y) = e^{x+y} - 2e^{2y} \quad \Rightarrow \quad f_{xy}\left(\frac{1}{2}, \frac{1}{2}\right) = e^{\frac{1}{2}+\frac{1}{2}} - 2e^{2\cdot\frac{1}{2}} = -e$$

Therefore the discriminant at the critical point is

$$D\left(\frac{1}{2}, \frac{1}{2}\right) = f_{xx}f_{yy} - f_{xy}^2 = e \cdot (-e) - (-e)^2 = -2e^2 < 0$$

We conclude that $\left(\frac{1}{2}, \frac{1}{2}\right)$ is a saddle point.

51. Prove that $f(x, y) = (x + 2y)e^{xy}$ has no critical points.

SOLUTION We find the critical points by setting the partial derivatives of $f(x, y) = (x + 2y)e^{xy}$ equal to zero and solving. We get

$$f_x(x, y) = e^{xy} + (x + 2y)ye^{xy} = e^{xy}\left(1 + xy + 2y^2\right) = 0$$

$$f_y(x, y) = 2e^{xy} + (x + 2y)xe^{xy} = e^{xy}\left(2 + x^2 + 2xy\right) = 0$$

We divide the two equations by the nonzero expression e^{xy} to obtain the following equations:

$$1 + xy + 2y^2 = 0$$
$$2 + 2xy + x^2 = 0$$

The first equation implies that $xy = -1 - 2y^2$. Substituting in the second equation gives

$$2 + 2\left(-1 - 2y^2\right) + x^2 = 0$$

$$2 - 2 - 4y^2 + x^2 = 0$$

$$x^2 = 4y^2 \quad \Rightarrow \quad x = 2y \quad \text{or} \quad x = -2y$$

We substitute in the first equation and solve for y:

$$
\begin{array}{ll}
\underline{x = 2y} & \underline{x = -2y} \\
1 + 2y^2 + 2y^2 = 0 & 1 - 2y^2 + 2y^2 = 0 \\
1 + 4y^2 = 0 & 1 = 0 \\
y^2 = -\frac{1}{4} &
\end{array}
$$

In both cases there is no solution. We conclude that there are no solutions for $f_x = 0$ and $f_y = 0$, that is, there are no critical points.

53. Find the global extrema of $f(x, y) = 2xy - x - y$ on the domain $\{y \le 4, y \ge x^2\}$.

SOLUTION   The region is shown in the figure.

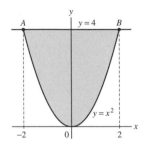

Step 1. Finding the critical points. We find the critical points in the interior of the domain by setting the partial derivatives equal to zero and solving. We get

$$f_x = 2y - 1 = 0$$

$$f_y = 2x - 1 = 0 \quad \Rightarrow \quad x = \frac{1}{2}, \quad y = \frac{1}{2}$$

The critical point is $\left(\frac{1}{2}, \frac{1}{2}\right)$. (It lies in the interior of the domain since $\frac{1}{2} < 4$ and $\frac{1}{2} > \left(\frac{1}{2}\right)^2$).

Step 2. Finding the global extrema on the boundary. We consider the two parts of the boundary separately.

The parabola $y = x^2$, $-2 \le x \le 2$:

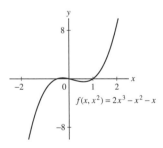

On this curve, $f(x, x^2) = 2 \cdot x \cdot x^2 - x - x^2 = 2x^3 - x^2 - x$. Using calculus in one variable or the graph of the function, we see that the minimum of $f(x, x^2)$ on the interval occurs at $x = -2$ and the maximum at $x = 2$. The corresponding points are $(-2, 4)$ and $(2, 4)$.

The segment $\overline{AB}$: On this segment $y = 4$, $-2 \le x \le 2$, hence $f(x, 4) = 2 \cdot x \cdot 4 - x - 4 = 7x - 4$. The maximum value occurs at $x = 2$ and the minimum value at $x = -2$. The corresponding points on the segment $\overline{AB}$ are $(-2, 4)$ and $(2, 4)$

Step 3. Conclusions. Since the global extrema occur either at critical points in the interior of the domain or on the boundary of the domain, the candidates for global extrema are the following points:

$$\left(\frac{1}{2}, \frac{1}{2}\right), \quad (-2, 4), \quad (2, 4)$$

We compute the values of $f = 2xy - x - y$ at these points:

$$f\left(\frac{1}{2}, \frac{1}{2}\right) = 2 \cdot \frac{1}{2} \cdot \frac{1}{2} - \frac{1}{2} - \frac{1}{2} = -\frac{1}{2}$$

$$f(-2, 4) = 2 \cdot (-2) \cdot 4 + 2 - 4 = -18$$

$$f(2, 4) = 2 \cdot 2 \cdot 4 - 2 - 4 = 10$$

We conclude that the global maximum is $f(2, 4) = 10$ and the global minimum is $f(-2, 4) = -18$.

55. Use Lagrange multipliers to find the minimum and maximum values of $f(x, y) = 3x - 2y$ on the circle $x^2 + y^2 = 4$.

SOLUTION

Step 1. Write out the Lagrange Equations. The constraint curve is $g(x, y) = x^2 + y^2 - 4 = 0$, hence $\nabla g = \langle 2x, 2y \rangle$ and $\nabla f = \langle 3, -2 \rangle$. The Lagrange Condition $\nabla f = \lambda \nabla g$ is thus $\langle 3, -2 \rangle = \lambda \langle 2x, 2y \rangle$. That is,

$$3 = \lambda \cdot 2x$$
$$-2 = \lambda \cdot 2y$$

Note that $\lambda \neq 0$.

Step 2. Solve for x and y using the constraint. The Lagrange equations gives

$$\begin{array}{l} 3 = \lambda \cdot 2x \\ -2 = \lambda \cdot 2y \end{array} \quad \Rightarrow \quad \begin{array}{l} x = \dfrac{3}{2\lambda} \\ y = -\dfrac{1}{\lambda} \end{array} \tag{1}$$

We substitute x and y in the equation of the constraint and solve for λ. We get

$$\left(\frac{3}{2\lambda} \right)^2 + \left(-\frac{1}{\lambda} \right)^2 = 4$$

$$\frac{9}{4\lambda^2} + \frac{1}{\lambda^2} = 4$$

$$\frac{1}{\lambda^2} \cdot \frac{13}{4} = 4 \quad \Rightarrow \quad \lambda = \frac{\sqrt{13}}{4} \quad \text{or} \quad \lambda = -\frac{\sqrt{13}}{4}$$

Substituting in (1), we obtain the points

$$x = \frac{6}{\sqrt{13}}, \quad y = -\frac{4}{\sqrt{13}}$$

$$x = -\frac{6}{\sqrt{13}}, \quad y = \frac{4}{\sqrt{13}}$$

The critical points are thus

$$P_1 = \left(\frac{6}{\sqrt{13}}, -\frac{4}{\sqrt{13}} \right)$$

$$P_2 = \left(-\frac{6}{\sqrt{13}}, \frac{4}{\sqrt{13}} \right)$$

Step 3. Calculate the value at the critical points. We find the value of $f(x, y) = 3x - 2y$ at the critical points:

$$f(P_1) = 3 \cdot \frac{6}{\sqrt{13}} - 2 \cdot \frac{-4}{\sqrt{13}} = \frac{26}{\sqrt{13}}$$

$$f(P_2) = 3 \cdot \frac{-6}{\sqrt{13}} - 2 \cdot \frac{4}{\sqrt{13}} = \frac{-26}{\sqrt{13}}$$

Thus, the maximum value of f on the circle is $\frac{26}{\sqrt{13}}$, and the minimum is $-\frac{26}{\sqrt{13}}$.

57. Find the minimum and maximum values of $f(x, y) = x^2 y$ on the ellipse $4x^2 + 9y^2 = 36$.

SOLUTION   We must find the minimum and maximum values of $f(x, y) = x^2 y$ subject to the constraint $g(x, y) = 4x^2 + 9y^2 - 36 = 0$.

Step 1. Write out the Lagrange Equations. The gradient vectors are $\nabla f = \left\langle 2xy, x^2 \right\rangle$ and $\nabla g = \langle 8x, 18y \rangle$, hence the Lagrange Condition $\nabla f = \lambda \nabla g$ gives

$$\left\langle 2xy, x^2 \right\rangle = \lambda \langle 8x, 18y \rangle = \langle 8\lambda x, 18\lambda y \rangle$$

We obtain the following Lagrange Equations:

$$2xy = 8\lambda x$$
$$x^2 = 18\lambda y$$

Step 2. Solve for λ in terms of x and y. If $x = 0$, the equation of the constraint implies that $y = \pm 2$. The points $(0, 2)$ and $(0, -2)$ satisfy the Lagrange Equations for $\lambda = 0$. If $x \neq 0$, the second Lagrange Equation implies that $y \neq 0$. Therefore the Lagrange Equations give

$$2xy = 8\lambda x \quad \Rightarrow \quad \lambda = \frac{y}{4}$$

$$x^2 = 18\lambda y \quad \Rightarrow \quad \lambda = \frac{x^2}{18y}$$

Step 3. Solve for x and y using the constraint. We equate the two expressions for λ to obtain

$$\frac{y}{4} = \frac{x^2}{18y}$$

$$18y^2 = 4x^2$$

We now substitute $4x^2 = 18y^2$ in the equation of the constraint $4x^2 + 9y^2 = 36$ and solve for y. This gives

$$\begin{aligned} 18y^2 + 9y^2 &= 36 \\ 27y^2 &= 36 \end{aligned} \quad \Rightarrow \quad y^2 = \frac{36}{27} \quad \Rightarrow \quad y_1 = \frac{2}{\sqrt{3}}, \quad y_2 = -\frac{2}{\sqrt{3}}$$

We find the x-coordinates using $x^2 = \frac{9y^2}{2}$:

$$x^2 = \frac{9y^2}{2}$$

$$x^2 = \frac{9}{2} \cdot \frac{4}{3} = 6 \quad \Rightarrow \quad x_1 = \sqrt{6}, \quad x_2 = -\sqrt{6}$$

We obtain the following critical points:

$$P_1 = (0, 2), \quad P_2 = (0, -2), \quad P_3 = \left(\sqrt{6}, \frac{2}{\sqrt{3}}\right)$$

$$P_4 = \left(\sqrt{6}, -\frac{2}{\sqrt{3}}\right), \quad P_5 = \left(-\sqrt{6}, \frac{2}{\sqrt{3}}\right), \quad P_6 = \left(-\sqrt{6}, -\frac{2}{\sqrt{3}}\right)$$

Step 4. Conclusions. We evaluate the function $f(x, y) = x^2 y$ at the critical points:

$$f(P_1) = 0^2 \cdot 2 = 0$$

$$f(P_2) = 0^2 \cdot (-2) = 0$$

$$f(P_3) = f(P_5) = 6 \cdot \frac{2}{\sqrt{3}} = \frac{12}{\sqrt{3}}$$

$$f(P_4) = f(P_5) = 6 \cdot \left(-\frac{2}{\sqrt{3}}\right) = -\frac{12}{\sqrt{3}}$$

Since the min and max of f occur on the ellipse, it must occur at critical points. Thus, we conclude that the maximum and minimum of f subject to the constraint are $\frac{12}{\sqrt{3}}$ and $-\frac{12}{\sqrt{3}}$ respectively.

59. Find the extreme values of $f(x, y, z) = x + 2y + 3z$ subject to the two constraints $x + y + z = 1$ and $x^2 + y^2 + z^2 = 1$.

SOLUTION We must find the extreme values of $f(x, y, z) = x + 2y + 3z$ subject to the constraints $g(x, y, z) = x + y + z - 1 = 0$ and $h(x, y, z) = x^2 + y^2 + z^2 - 1 = 0$.

Step 1. Write out the Lagrange Equations. We have $\nabla f = \langle 1, 2, 3 \rangle$, $\nabla g = \langle 1, 1, 1 \rangle$, $\nabla h = \langle 2x, 2y, 2z \rangle$, hence the Lagrange condition $\nabla f = \lambda \nabla g + \mu \nabla h$ gives

$$\langle 1, 2, 3 \rangle = \lambda \langle 1, 1, 1 \rangle + \mu \langle 2x, 2y, 2z \rangle = \langle \lambda + 2\mu x, \lambda + 2\mu y, \lambda + 2\mu z \rangle$$

or

$$1 = \lambda + 2\mu x$$

$$2 = \lambda + 2\mu y$$

$$3 = \lambda + 2\mu z$$

Step 2. Solve for λ and μ. The Lagrange Equations give

$$1 = \lambda + 2\mu x \qquad \lambda = 1 - 2\mu x$$
$$2 = \lambda + 2\mu y \quad \Rightarrow \quad \lambda = 2 - 2\mu y$$
$$3 = \lambda + 2\mu z \qquad \lambda = 3 - 2\mu z$$

Equating the three expressions for λ, we get the following equations:

$$\begin{aligned} 1 - 2\mu x &= 2 - 2\mu y \\ 1 - 2\mu x &= 3 - 2\mu z \end{aligned} \quad \Rightarrow \quad \begin{aligned} 2\mu(y - x) &= 1 \\ \mu(z - x) &= 2 \end{aligned}$$

The first equation implies that $\mu = \frac{1}{2(y-x)}$, and the second implies that $\mu = \frac{2}{z-x}$. Equating the two expressions for μ, we get

$$\frac{1}{2(y - x)} = \frac{2}{z - x}$$
$$z - x = 4y - 4x \quad \Rightarrow \quad z = 4y - 3x$$

Step 3. Solve for x, y, and z using the constraints. We substitute $z = 4y - 3x$ in the equations of the constraints and solve to find x and y. This gives

$$\begin{aligned} x + y + (4y - 3x) &= 1 \\ x^2 + y^2 + (4y - 3x)^2 &= 1 \end{aligned} \quad \Rightarrow \quad \begin{aligned} y &= \frac{1 + 2x}{5} \\ 10x^2 + 17y^2 - 24xy &= 1 \end{aligned}$$

Substituting in the second equation and solving for x, we get

$$y = \frac{1 + 2x}{5}$$

$$10x^2 + 17\left(\frac{1 + 2x}{5}\right)^2 - 24x \cdot \frac{1 + 2x}{5} = 1$$

$$250x^2 + 17(1 + 2x)^2 - 120x(1 + 2x) - 25$$

$$39x^2 - 26x - 4 = 0$$

$$x_{1,2} = \frac{26 \pm \sqrt{1300}}{78}$$

$$\Rightarrow \quad x_1 = \frac{1}{3} + \frac{5\sqrt{13}}{39} \approx 0.8, \quad x_2 = \frac{1}{3} - \frac{5\sqrt{13}}{39} \approx -0.13$$

We find the y-coordinates using $y = \frac{1+2x}{5}$.

$$y_1 = \frac{1 + 2 \cdot 0.8}{5} = 0.52, \quad y_2 = \frac{1 - 2 \cdot 0.13}{5} = 0.15$$

Finally, we find the z-coordinate using $z = 4y - 3x$:

$$z_1 = 4 \cdot 0.52 - 3 \cdot 0.8 = -0.32, \quad z_2 = 4 \cdot 0.15 + 3 \cdot 0.13 = 0.99$$

We obtain the critical points:

$$P_1 = (0.8, 0.52, -0.32), \quad P_2 = (-0.13, 0.15, 0.99)$$

Step 4. Conclusions. We evaluate the function $f(x, y, z) = x + 2y + 3z$ at the critical points:

$$f(P_1) = 0.8 + 2 \cdot 0.52 - 3 \cdot 0.32 = 0.88$$
$$f(P_2) = -0.13 + 2 \cdot 0.15 + 3 \cdot 0.99 = 3.14 \tag{1}$$

The two constraints determine the common points of the unit sphere $x^2 + y^2 + z^2 = 1$ and the plane $x + y + z = 1$. This set is a circle that is a closed and bounded set in R^3. Therefore, f has a minimum and maximum values on this set. These extrema are given in (1).

61. Use Lagrange multipliers to find the dimensions of a cylindrical can with a bottom but no top, of fixed volume V with minimum surface area.

SOLUTION   We denote the radius of the cylinder by r and the height by h.

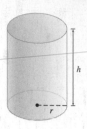

The volume of the cylinder is $g = \pi r^2 h$ and the surface area is

$$f = 2\pi rh + 2\pi r^2$$

We need to minimize $f(r, h) = 2\pi rh + 2\pi r^2$ subject to the constraint $g(r, h) = \pi r^2 h - V = 0$.

Step 1. Write out the Lagrange Equations. We have $\nabla f = \langle 2\pi h + 4\pi r, 2\pi r \rangle = 2\pi \langle h + 2r, r \rangle$ and $\nabla g = \langle 2\pi hr, \pi r^2 \rangle = \pi \langle 2hr, r^2 \rangle$, hence the Lagrange Condition $\nabla f = \lambda \nabla g$ is

$$2\pi \langle h + 2r, r \rangle = \pi \lambda \langle 2hr, r^2 \rangle$$

or

$$2 \langle h + 2r, r \rangle = \lambda \langle 2hr, r^2 \rangle$$

We obtain the following equations:

$$
\begin{array}{ccc}
2(h + 2r) = 2hr\lambda & & h + 2r = hr\lambda \\
& \Rightarrow & \\
2r = \lambda r^2 & & 2r = \lambda r^2
\end{array}
$$

Step 2. Solve for λ in terms of r and h. The equation of the constraint implies that $r \neq 0$ and $h \neq 0$ (we assume that $V > 0$). Therefore, the Lagrange equations give

$$\lambda = \frac{h + 2r}{hr} = \frac{1}{r} + \frac{2}{h}, \quad \lambda = \frac{2}{r}$$

Step 3. Solve for r and h using the constraint. Equating the two expressions for λ gives

$$\frac{1}{r} + \frac{2}{h} = \frac{2}{r}$$

$$\frac{2}{h} = \frac{1}{r} \quad \Rightarrow \quad h = 2r$$

We substitute $h = 2r$ in the equation of the constraint $\pi r^2 h = V$ and solve for r. We obtain

$$\pi r^2 \cdot 2r = V$$

$$2\pi r^3 = V \quad \Rightarrow \quad r = \left(\frac{V}{2\pi}\right)^{1/3}$$

We find h using the relation $h = 2r$:

$$h = 2\left(\frac{V}{2\pi}\right)^{1/3}$$

The critical point is $h = 2\left(\frac{V}{2\pi}\right)^{1/3}, r = \left(\frac{V}{2\pi}\right)^{1/3}$.

Step 4. Conclusions. On the constraint $\pi r^2 h = V$ we have $h = \frac{V}{\pi r^2}$ and $r = \sqrt{\frac{V}{\pi h}}$, hence

$$f\left(r, \frac{V}{\pi r^2}\right) = 2\pi r \cdot \frac{V}{\pi r^2} + 2\pi r^2 = \frac{2V}{r} + 2\pi r^2$$

$$f\left(\sqrt{\frac{V}{\pi h}}, h\right) = 2\pi \sqrt{\frac{V}{\pi h}} h + 2\pi \cdot \frac{V}{\pi h} = 2\sqrt{\pi V}\sqrt{h} + \frac{2V}{h}$$

We see that as $h \to 0+$ or $h \to \infty$, we have $f(r, h) \to \infty$, and as $r \to 0+$ or $r \to \infty$, we have $f(r, h) \to \infty$. Therefore, f has a minimum value on the constraint, which occurs at the critical point. We evaluate $f(r, h) = 2\pi rh + 2\pi r^2 = 2\pi(rh + r^2)$ at the critical point P:

$$f(P) = 2\pi\left(\left(\frac{V}{2\pi}\right)^{1/3} \cdot 2\left(\frac{V}{2\pi}\right)^{1/3} + \left(\frac{V}{2\pi}\right)^{2/3}\right) = 2\pi\left(2\left(\frac{V}{2\pi}\right)^{2/3} + \left(\frac{V}{2\pi}\right)^{2/3}\right) = 6\pi\left(\frac{V}{2\pi}\right)^{2/3}$$

We conclude that the minimum surface area is $6\pi\left(\frac{V}{2\pi}\right)^{2/3}$, and the dimensions of the corresponding cylinder are $r = \left(\frac{V}{2\pi}\right)^{1/3}$, $h = 2\left(\frac{V}{2\pi}\right)^{1/3}$.

63. Given n nonzero numbers $\sigma_1, \ldots, \sigma_n$, show that the minimum value of

$$f(x_1, \ldots, x_n) = x_1^2\sigma_1^2 + \cdots + x_n^2\sigma_n^2$$

subject to $x_1 + \cdots + x_n = 1$ is c, where $c = \left(\sum_{j=1}^{n}\sigma_j^{-2}\right)^{-1}$.

SOLUTION   We must minimize the function $f(x_1, \ldots, x_n) = x_1^2\sigma_1^2 + \cdots + x_n^2\sigma_n^2$ subject to the constraint $g(x_1, \ldots, x_n) = x_1 + \cdots + x_n - 1 = 0$.

Step 1. Write out the Lagrange Equations. We have $\nabla f = \langle 2\sigma_1^2 x_1, \ldots, 2\sigma_n^2 x_n\rangle$ and $\nabla g = \langle 1, \ldots, 1\rangle$, hence the Lagrange Condition $\nabla f = \lambda\nabla g$ gives the following equations:

$$2\sigma_i^2 x_i = \lambda, \quad i = 1, \ldots, n$$

Step 2. Solve for $x_1, \ldots, x_n$ using the constraint. The Lagrange equations imply the following equations:

$$2\sigma_i^2 x_i = 2\sigma_n^2 x_n, \quad x_i = \frac{\sigma_n^2}{\sigma_i^2}x_n; \quad i = 1, \ldots, n-1$$

We substitute these values in the equation of the constraint $x_1 + \cdots + x_n = 1$ and solve for x_n. This gives

$$\frac{\sigma_n^2}{\sigma_1^2}x_n + \frac{\sigma_n^2}{\sigma_2^2}x_n + \cdots + \frac{\sigma_n^2}{\sigma_{n-1}^2}x_n + x_n = 1$$

$$\sigma_n^2\left(\frac{1}{\sigma_1^2} + \frac{1}{\sigma_2^2} + \cdots + \frac{1}{\sigma_{n-1}^2} + \frac{1}{\sigma_n^2}\right)x_n = 1$$

$$\sigma_n^2\left(\sum_{j=1}^{n}\sigma_j^{-2}\right)x_n = 1$$

Denoting $c = \left(\sum_{j=1}^{n}\sigma_j^{-2}\right)^{-1}$, we get $x_n = \frac{c}{\sigma_n^2}$. Using $x_i = \frac{\sigma_n^2}{\sigma_i^2}x_n$ we get

$$x_i = \frac{\sigma_n^2}{\sigma_i^2} \cdot \frac{c}{\sigma_n^2} = \frac{c}{\sigma_i^2}$$

We obtain the following point:

$$P = \left(\frac{c}{\sigma_1^2}, \frac{c}{\sigma_2^2}, \ldots, \frac{c}{\sigma_n^2}\right)$$

Step 3. Conclusions. As $x_i \to \infty$ or $x_i \to -\infty$, for one or more i's the function $f(x_1, \ldots, x_n)$ tends to ∞. f is continuous since it is a polynomial, hence f has a minimum value on the constraint. This minimum occurs at the critical point. We find it:

$$f(P) = \sum_{j=1}^{n}\sigma_j^2\left(\frac{c}{\sigma_j^2}\right)^2 = \sum_{j=1}^{n}\frac{\sigma_j^2 c^2}{\sigma_j^4} = c^2\sum_{j=1}^{n}\sigma_j^{-2} = c^2 \cdot c^{-1} = c$$

15 MULTIPLE INTEGRATION

15.1 Integration in Two Variables (LT Section 16.1)

Preliminary Questions

1. If $S_{8,4}$ is a Riemann sum for a double integral over $\mathcal{R} = [1, 5] \times [2, 10]$ using a regular partition, what is the area of each subrectangle? How many subrectangles are there?

SOLUTION Since the partition is regular, all subrectangles have sides of length

$$\Delta x = \frac{5 - 1}{8} = \frac{1}{2}, \quad \Delta y = \frac{10 - 2}{4} = 2$$

Therefore the area of each subrectangle is $\Delta A = \Delta x \Delta y = \frac{1}{2} \cdot 2 = 1$, and the number of subrectangles is $8 \cdot 4 = 32$.

2. Estimate the double integral of a continuous function f over the small rectangle $\mathcal{R} = [0.9, 1.1] \times [1.9, 2.1]$ if $f(1, 2) = 4$.

SOLUTION Since we are given the value of f at one point in $\mathcal{R}$ only, we can only use the approximation S_{11} for the integral of f over $\mathcal{R}$. For S_{11} we have one rectangle wi th sides

$$\Delta x = 1.1 - 0.9 = 0.2, \quad \Delta y = 2.1 - 1.9 = 0.2$$

Hence, the area of the rectangle is $\Delta A = \Delta x \Delta y = 0.2 \cdot 0.2 = 0.04$. We obtain the following approximation:

$$\iint_{\mathcal{R}} f \, dA \approx S_{1,1} = f(1, 2) \Delta A = 4 \cdot 0.04 = 0.16$$

3. What is the integral of the constant function $f(x, y) = 5$ over the rectangle $[-2, 3] \times [2, 4]$?

SOLUTION The integral of f over the unit square $\mathcal{R} = [-2, 3] \times [2, 4]$ is the volume of the box of base $\mathcal{R}$ and height 5. That is,

$$\iint_{\mathcal{R}} 5 \, dA = 5 \cdot \text{Area}(\mathcal{R}) = 5 \cdot 5 \cdot 2 = 50$$

4. What is the interpretation of $\iint_{\mathcal{R}} f(x, y) \, dA$ if $f(x, y)$ takes on both positive and negative values on $\mathcal{R}$?

SOLUTION The double integral $\iint_{\mathcal{R}} f(x, y) \, dA$ is the signed volume between the graph $z = f(x, y)$ for $(x, y) \in \mathcal{R}$, and the xy-plane. The region below the xy-plane is treated as negative volume.

5. Which of (a) or (b) is equal to $\displaystyle\int_1^2 \int_4^5 f(x, y) \, dy \, dx$?

(a) $\displaystyle\int_1^2 \int_4^5 f(x, y) \, dx \, dy$ **(b)** $\displaystyle\int_4^5 \int_1^2 f(x, y) \, dx \, dy$

SOLUTION The integral $\int_1^2 \int_4^5 f(x, y) \, dy \, dx$ is written with dy preceding dx, therefore the integration is first with respect to y over the interval $4 \le y \le 5$, and then with respect to x over the interval $1 \le x \le 2$. By Fubini's Theorem, we may replace the order of integration over the corresponding intervals. Therefore the given integral is equal to (b) rather than to (a).

6. For which of the following functions is the double integral over the rectangle in Figure 15 equal to zero? Explain your reasoning.

(a) $f(x, y) = x^2 y$ **(b)** $f(x, y) = xy^2$

(c) $f(x, y) = \sin x$ **(d)** $f(x, y) = e^x$

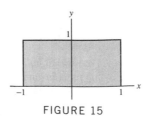

FIGURE 15

SOLUTION The double integral is the signed volume of the region between the graph of $f(x, y)$ and the xy-plane over $\mathcal{R}$. In (b) and (c) the function satisfies $f(-x, y) = -f(x, y)$, hence the region below the xy-plane, where $-1 \leq x \leq 0$ cancels with the region above the xy-plane, where $0 \leq x \leq 1$. Therefore, the double integral is zero. In (a) and (d), the function $f(x, y)$ is always positive on the rectangle, so the double integral is greater than zero.

Exercises

1. Compute the Riemann sum $S_{4,3}$ to estimate the double integral of $f(x, y) = xy$ over $\mathcal{R} = [1, 3] \times [1, 2.5]$. Use the regular partition and upper-right vertices of the subrectangles as sample points.

SOLUTION The rectangle $\mathcal{R}$ and the subrectangles are shown in the following figure:

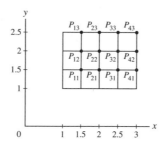

The subrectangles have sides of length

$$\Delta x = \frac{3 - 1}{4} = 0.5, \quad \Delta y = \frac{2.5 - 1}{3} = 0.5 \quad \Rightarrow \quad \Delta A = 0.5 \cdot 0.5 = 0.25$$

The upper right vertices are the following points:

$$
\begin{array}{llll}
P_{11} = (1.5, 1.5) & P_{21} = (2, 1.5) & P_{31} = (2.5, 1.5) & P_{41} = (3, 1.5) \\
P_{12} = (1.5, 2) & P_{22} = (2, 2) & P_{32} = (2.5, 2) & P_{42} = (3, 2) \\
P_{13} = (1.5, 2.5) & P_{23} = (2, 2.5) & P_{33} = (2.5, 2.5) & P_{43} = (3, 2.5)
\end{array}
$$

We compute $f(x, y) = xy$ at these points:

$$
\begin{array}{lll}
f(P_{11}) = 1.5 \cdot 1.5 = 2.25 & f(P_{12}) = 1.5 \cdot 2 = 3 & f(P_{13}) = 3.75 \\
f(P_{21}) = 2 \cdot 1.5 = 3 & f(P_{22}) = 2 \cdot 2 = 4 & f(P_{23}) = 5 \\
f(P_{31}) = 2.5 \cdot 1.5 = 3.75 & f(P_{32}) = 2.5 \cdot 2 = 5 & f(P_{33}) = 6.25 \\
f(P_{41}) = 3 \cdot 1.5 = 4.5 & f(P_{42}) = 3 \cdot 2 = 6 & f(P_{43}) = 7.5
\end{array}
$$

Hence, $S_{4,3}$ is the following sum:

$$S_{4,3} = \sum_{i=1}^{4} \sum_{j=1}^{3} f(P_{ij})\Delta A = 0.25(2.25 + 3 + 3.75 + 4.5 + 3 + 4 + 5 + 6 + 3.75 + 5 + 6.25 + 7.5) = 13.5$$

In Exercises 3–6, compute the Riemann sums for the double integral $\iint_{\mathcal{R}} f(x, y) \, dA$, *where* $\mathcal{R} = [1, 4] \times [1, 3]$, *for the grid and two choices of sample points shown in Figure 16.*

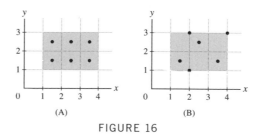

FIGURE 16

3. $f(x, y) = 2x + y$

SOLUTION The subrectangles have sides of length $\Delta x = \frac{4-1}{3} = 1$ and $\Delta y = \frac{3-1}{2} = 1$, and area $\Delta A = \Delta x \Delta y = 1$. We find the sample points in (A) and (B):

(A)

$$P_{11} = (1.5, 1.5) \quad P_{21} = (2.5, 1.5) \quad P_{31} = (3.5, 1.5)$$

$$P_{12} = (1.5, 2.5) \quad P_{22} = (2.5, 2.5) \quad P_{32} = (3.5, 2.5)$$

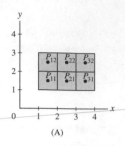

(A)

(B)

$$P_{11} = (1.5, 1.5) \quad P_{21} = (2, 1) \quad P_{31} = (3.5, 1.5)$$

$$P_{21} = (2, 3) \quad P_{22} = (2.5, 2.5) \quad P_{23} = (4, 3)$$

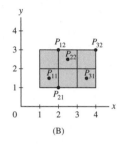

(B)

The Riemann Sum S_{32} is the following estimation of the double integral:

$$\iint_{\mathcal{R}} f(x, y)\, dA \approx S_{32} = \sum_{i=1}^{3} \sum_{j=1}^{2} f(P_{ij}) \Delta A = \sum_{i=1}^{3} \sum_{j=1}^{2} f(P_{ij})$$

We compute S_{32} for the two choices of sample points (A) and (B), and the following function:

$$f(x, y) = 2x + y$$

We compute $f(P_{ij})$ for the sample points computed above:

(A)

$$f(P_{11}) = f(1.5, 1.5) = 2 \cdot 1.5 + 1.5 = 4.5$$

$$f(P_{21}) = f(2.5, 1.5) = 2 \cdot 2.5 + 1.5 = 6.5$$

$$f(P_{31}) = f(3.5, 1.5) = 2 \cdot 3.5 + 1.5 = 8.5$$

$$f(P_{12}) = f(1.5, 2.5) = 2 \cdot 1.5 + 2.5 = 5.5$$

$$f(P_{22}) = f(2.5, 2.5) = 2 \cdot 2.5 + 2.5 = 7.5$$

$$f(P_{32}) = f(3.5, 2.5) = 2 \cdot 3.5 + 2.5 = 9.5$$

Hence,

$$S_{32} = \sum_{i=1}^{3} \sum_{j=1}^{2} f(P_{ij}) \Delta A = 4.5 + 6.5 + 8.5 + 5.5 + 7.5 + 9.5 = 42$$

(B)

$$f(P_{11}) = f(1.5, 1.5) = 2 \cdot 1.5 + 1.5 = 4.5$$

$$f(P_{21}) = f(2, 1) = 2 \cdot 2 + 1 = 5$$

$$f(P_{31}) = f(3.5, 1.5) = 2 \cdot 3.5 + 1.5 = 8.5$$

$$f(P_{21}) = f(2, 3) = 2 \cdot 2 + 3 = 7$$

$$f(P_{22}) = f(2.5, 2.5) = 2 \cdot 2.5 + 2.5 = 7.5$$

$$f(P_{23}) = f(4, 3) = 2 \cdot 4 + 3 = 11$$

Hence,

$$S_{32} = \sum_{i=1}^{3} \sum_{j=1}^{2} f(P_{ij}) \Delta A = 4.5 + 5 + 8.5 + 7 + 7.5 + 11 = 43.5$$

5. $f(x, y) = 4x$

SOLUTION We compute the values of f at the sample points:

(A)

$$f(P_{11}) = f(1.5, 1.5) = 4 \cdot 1.5 = 6$$
$$f(P_{21}) = f(2.5, 1.5) = 4 \cdot 2.5 = 10$$
$$f(P_{31}) = f(3.5, 1.5) = 4 \cdot 3.5 = 14$$
$$f(P_{12}) = f(1.5, 2.5) = 4 \cdot 1.5 = 6$$
$$f(P_{22}) = f(2.5, 2.5) = 4 \cdot 2.5 = 10$$
$$f(P_{32}) = f(3.5, 2.5) = 4 \cdot 3.5 = 14$$
$$\Delta x = \frac{4-1}{3} = 1, \quad \Delta y = \frac{3-1}{2} = 1$$

Hence $\Delta A = \Delta x \cdot \Delta y = 1$ and we get

$$S_{32} = \sum_{i=1}^{3} \sum_{j=1}^{2} f(P_{ij}) \, \Delta A = 6 + 10 + 14 + 6 + 10 + 14 = 60$$

(B)

$$f(P_{11}) = f(1.5, 1.5) = 4 \cdot 1.5 = 6$$
$$f(P_{21}) = f(2, 1) = 4 \cdot 2 = 8$$
$$f(P_{31}) = f(3.5, 1.5) = 4 \cdot 3.5 = 14$$
$$f(P_{12}) = f(2, 3) = 4 \cdot 2 = 8$$
$$f(P_{22}) = f(2.5, 2.5) = 4 \cdot 2.5 = 10$$
$$f(P_{32}) = f(4, 3) = 4 \cdot 4 = 16$$

$\Delta A = 1$. Hence,

$$S_{32} = \sum_{i=1}^{3} \sum_{j=1}^{2} f(P_{ij}) \, \Delta A = 6 + 8 + 14 + 8 + 10 + 16 = 62$$

7. Let $\mathcal{R} = [0, 1] \times [0, 1]$. Estimate $\iint_{\mathcal{R}} (x + y) \, dA$ by computing two different Riemann sums, each with at least six rectangles.

SOLUTION We define the following subrectangles and sample points:

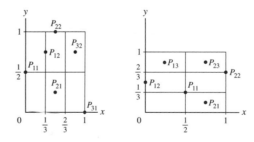

The sample points defined in the two figures are:

(A)

$$P_{11} = \left(0, \tfrac{1}{2}\right) \quad P_{21} = \left(\tfrac{1}{2}, \tfrac{1}{4}\right) \quad P_{31} = (1, 0)$$
$$P_{12} = \left(\tfrac{1}{3}, \tfrac{3}{4}\right) \quad P_{22} = \left(\tfrac{1}{2}, 1\right) \quad P_{32} = \left(\tfrac{5}{6}, \tfrac{3}{4}\right)$$

(B)

$$P_{11} = \left(\tfrac{1}{2}, \tfrac{1}{3}\right) \quad P_{21} = \left(\tfrac{3}{4}, \tfrac{1}{6}\right) \quad P_{12} = \left(0, \tfrac{1}{2}\right)$$
$$P_{22} = \left(1, \tfrac{2}{3}\right) \quad P_{13} = \left(\tfrac{1}{4}, \tfrac{5}{6}\right) \quad P_{23} = \left(\tfrac{3}{4}, \tfrac{5}{6}\right)$$

We compute the values of $f(x, y) = x + y$ at the sample points:

(A)

$$f(P_{11}) = f\left(0, \frac{1}{2}\right) = 0 + \frac{1}{2} = \frac{1}{2}$$

$$f(P_{21}) = f\left(\frac{1}{2}, \frac{1}{4}\right) = \frac{1}{2} + \frac{1}{4} = \frac{3}{4}$$

$$f(P_{31}) = f(1, 0) = 1 + 0 = 1$$

$$f(P_{12}) = f\left(\frac{1}{3}, \frac{3}{4}\right) = \frac{1}{3} + \frac{3}{4} = \frac{13}{12}$$

$$f(P_{22}) = f\left(\frac{1}{2}, 1\right) = \frac{1}{2} + 1 = \frac{3}{2}$$

$$f(P_{32}) = f\left(\frac{5}{6}, \frac{3}{4}\right) = \frac{5}{6} + \frac{3}{4} = \frac{19}{12}$$

Each subrectangle has sides of length $\Delta x = \frac{1}{3}$, $\Delta y = \frac{1}{2}$ and area $\Delta A = \Delta x \Delta y = \frac{1}{3} \cdot \frac{1}{2} = \frac{1}{6}$. We obtain the following Riemann sum:

$$S_{32} = \sum_{i=1}^{3}\sum_{j=1}^{2} f(P_{ij})\,\Delta A = \frac{1}{6}\left(\frac{1}{2} + \frac{3}{4} + 1 + \frac{13}{12} + \frac{3}{2} + \frac{19}{12}\right) = \frac{77}{72} \approx 1.069$$

(B)

$$f(P_{11}) = f\left(\frac{1}{2}, \frac{1}{3}\right) = \frac{1}{2} + \frac{1}{3} = \frac{5}{6}$$

$$f(P_{21}) = f\left(\frac{3}{4}, \frac{1}{6}\right) = \frac{3}{4} + \frac{1}{6} = \frac{11}{12}$$

$$f(P_{12}) = f\left(0, \frac{1}{2}\right) = 0 + \frac{1}{2} = \frac{1}{2}$$

$$f(P_{22}) = f\left(1, \frac{2}{3}\right) = 1 + \frac{2}{3} = \frac{5}{3}$$

$$f(P_{13}) = f\left(\frac{1}{4}, \frac{5}{6}\right) = \frac{1}{4} + \frac{5}{6} = \frac{13}{12}$$

$$f(P_{23}) = f\left(\frac{3}{4}, \frac{5}{6}\right) = \frac{3}{4} + \frac{5}{6} = \frac{19}{12}$$

Each subrectangle has sides of length $\Delta x = \frac{1}{2}$, $\Delta y = \frac{1}{3}$ and area $\Delta A = \Delta x \Delta y = \frac{1}{2} \cdot \frac{1}{3} = \frac{1}{6}$. We obtain the following Riemann sum:

$$S_{23} = \sum_{i=1}^{3}\sum_{j=1}^{2} f(P_{ij})\,\Delta A = \frac{1}{6}\left(\frac{5}{6} + \frac{11}{12} + \frac{1}{2} + \frac{5}{3} + \frac{13}{12} + \frac{19}{12}\right) = \frac{79}{72} \approx 1.097$$

9. Evaluate $\iint_{\mathcal{R}} (15 - 3x)\,dA$, where $\mathcal{R} = [0, 5] \times [0, 3]$, and sketch the corresponding solid region (see Example 2).

SOLUTION   This double integral is the volume V of the solid wedge underneath the graph of $f(x, y) = 15 - 3x$. The triangular face of the wedge has area

$$A = \frac{1}{2} \cdot 5 \cdot 15 = \frac{75}{2}$$

The volume of the wedge is equal to the area A times the length $\ell = 3$; that is

$$V = \ell A = 3\left(\frac{75}{2}\right) = \frac{225}{2}$$

11. The following table gives the approximate height at quarter-meter intervals of a mound of gravel. Estimate the volume of the mound by computing the average of the two Riemann sums $S_{4,3}$ with lower-left and upper-right vertices of the subrectangles as sample points.

0.75	0.1	0.2	0.2	0.15	0.1
0.5	0.2	0.3	0.5	0.4	0.2
0.25	0.15	0.2	0.4	0.3	0.2
0	0.1	0.15	0.2	0.15	0.1
$y \diagdown x$	0	0.25	0.5	0.75	1

SOLUTION Each subrectangle is a square of side 0.25, hence the area of each subrectangle is $\Delta A = 0.25^2 = 0.0625$. By the given data, the lower-left vertex sample points are:

$$\begin{array}{lll}
f(P_{11}) = f(0,0) & f(P_{12}) = f(0,0.25) & f(P_{13}) = f(0,0.50) \\
f(P_{21}) = f(0.25,0) & f(P_{22}) = f(0.25,0.25) & f(P_{23}) = f(0.25,0.50) \\
f(P_{31}) = f(0.50,0) & f(P_{32}) = f(0.50,0.25) & f(P_{33}) = f(0.50,0.50) \\
f(P_{41}) = f(0.75,0) & f(P_{42}) = f(0.75,0.25) & f(P_{43}) = f(0.75,0.50)
\end{array}$$

The Riemann sum $S_{4,3}$ that corresponds to these lower-left vertex sample points is the following sum:

$$S_{4,3} = \sum_{i=1}^{4} \sum_{j=1}^{3} f\left(P_{ij}\right) \Delta A$$

$$= 0.0625(0.1 + 0.15 + 0.2 + 0.15 + 0.2 + 0.3 + 0.2 + 0.4 + 0.5 + 0.15 + 0.3 + 0.4) \approx 0.190625$$

Now by the given data, the upper-right vertex sample points are:

$$\begin{array}{lll}
f(P_{11}) = f(0.25,0.25) & f(P_{12}) = f(0.25,0.50) & f(P_{13}) = f(0.25,0.75) \\
f(P_{21}) = f(0.50,0.25) & f(P_{22}) = f(0.50,0.50) & f(P_{23}) = f(0.50,0.75) \\
f(P_{31}) = f(0.75,0.25) & f(P_{32}) = f(0.75,0.50) & f(P_{33}) = f(0.75,0.75) \\
f(P_{41}) = f(1,0.25) & f(P_{42}) = f(1,0.50) & f(P_{43}) = f(1,0.75)
\end{array}$$

The Riemann sum S'_{43} that corresponds to these upper-right vertex sample points is the following sum:

$$S'_{4,3} = \sum_{i=1}^{4} \sum_{j=1}^{3} f\left(P_{ij}\right) \Delta A$$

$$= 0.0625(0.2 + 0.3 + 0.2 + 0.4 + 0.5 + 0.2 + 0.3 + 0.4 + 0.15 + 0.2 + 0.2 + 0.1) \approx 0.196875$$

Taking the average of the two Riemann sums we have:

$$\text{volume} \approx \frac{S_{4,3} + S'_{4,3}}{2} = \frac{0.190625 + 0.196875}{2} = 0.19375$$

13. CAS Let $S_{N,N}$ be the Riemann sum for $\int_0^1 \int_0^1 e^{x^3 - y^3} \, dy \, dx$ using the regular partition and the lower left-hand vertex of each subrectangle as sample points. Use a computer algebra system to calculate $S_{N,N}$ for $N = 25, 50, 100$.

SOLUTION Using a computer algebra system, we compute $S_{N,N}$ to be 1.0731, 1.0783, and 1.0809.

In Exercises 15–18, use symmetry to evaluate the double integral.

15. $\iint_{\mathcal{R}} x^3 \, dA, \quad \mathcal{R} = [-4, 4] \times [0, 5]$

SOLUTION The double integral is the signed volume of the region between the graph of $f(x, y) = x^3$ and the xy-plane. However, $f(x, y)$ takes opposite values at (x, y) and $(-x, y)$:

$$f(-x, y) = (-x)^3 = -x^3 = -f(x, y)$$

Because of symmetry, the (negative) signed volume of the region below the xy-plane where $-4 \le x \le 0$ cancels with the (positive) signed volume of the region above the xy-plane where $0 \le x \le 4$. The net result is

$$\iint_{\mathcal{R}} x^3 \, dA = 0$$

17. $\iint_{\mathcal{R}} \sin x \, dA, \quad \mathcal{R} = [0, 2\pi] \times [0, 2\pi]$

SOLUTION Since $\sin(\pi + x) = -\sin x$, the region between the graph and the xy-plane where $\pi \le x \le 2\pi$, is below the xy-plane, and it cancels with the region above the xy-plane where $0 \le x \le \pi$. Hence,

$$\iint_{\mathcal{R}} \sin x \, dA = 0$$

In Exercises 19–36, evaluate the iterated integral.

19. $\displaystyle \int_1^3 \int_0^2 x^3 y \, dy \, dx$

SOLUTION We first compute the inner integral, treating x as a constant, then integrate the result with respect to x:

$$\int_1^3 \int_0^2 x^3 y \, dy \, dx = \int_1^3 x^3 \frac{y^2}{2}\bigg|_{y=0}^{2} dx = \int_1^3 x^3 \left(\frac{2^2}{2} - 0\right) dx = \int_1^3 2x^3 \, dx = \frac{x^4}{2}\bigg|_1^3 = 40$$

21. $\displaystyle \int_4^9 \int_{-3}^8 1 \, dx \, dy$

SOLUTION

$$\int_4^9 \int_{-3}^8 1 \, dx \, dy = \int_4^9 1 \left(\int_{-3}^8 1 \, dy\right) dx$$

$$= \int_4^9 1 \left(y\bigg|_{-3}^8\right) dx$$

$$= \int_4^9 11 \, dx$$

$$= 11x\bigg|_4^9$$

$$= 99 - 44 = 55$$

23. $\displaystyle \int_{-1}^1 \int_0^\pi x^2 \sin y \, dy \, dx$

SOLUTION We first evaluate the inner integral, treating x as a constant, then integrate the result with respect to x. This gives

$$\int_{-1}^1 \int_0^\pi x^2 \sin y \, dy \, dx = \int_{-1}^1 x^2(-\cos y)\bigg|_{y=0}^{\pi} dx = \int_{-1}^1 x^2(-\cos \pi + \cos 0) \, dx$$

$$= \int_{-1}^1 x^2(1+1) \, dx = \int_{-1}^1 2x^2 \, dx = \frac{2}{3}x^3\bigg|_{-1}^1 = \frac{2}{3}\left(1^3 - (-1)^3\right) = \frac{4}{3}$$

25. $\displaystyle \int_2^6 \int_1^4 x^2 \, dx \, dy$

SOLUTION We use Iterated Integral of a Product Function to compute the integral as follows:

$$\int_2^6 \int_1^4 x^2 \, dx \, dy = \int_2^6 \int_1^4 x^2 \cdot 1 \, dx \, dy = \left(\int_1^4 x^2 \, dx\right)\left(\int_2^6 1 \, dy\right) = \left(\frac{x^3}{3}\bigg|_1^4\right)\left(y\bigg|_2^6\right)$$

$$= \left(\frac{4^3}{3} - \frac{1^3}{3}\right)(6-2) = 21 \cdot 4 = 84$$

27. $\displaystyle \int_0^1 \int_0^2 (x + 4y^3) \, dx \, dy$

SOLUTION We use additivity of the double integral to write

$$\int_0^1 \int_0^2 \left(x + 4y^3\right) dx \, dy = \int_0^1 \int_0^2 x \, dx \, dy + \int_0^1 \int_0^2 4y^3 \, dx \, dy \tag{1}$$

We now compute each of the double integrals using product of iterated integrals:

$$\int_0^1 \int_0^2 x \, dx \, dy = \left(\int_0^2 x \, dx\right)\left(\int_0^1 1 \, dy\right) = \left(\frac{1}{2}x^2\bigg|_0^2\right)\left(y\bigg|_0^1\right) = 2 \cdot 1 = 2$$

$$\int_0^1 \int_0^2 4y^3 \, dx \, dy = \left(\int_0^1 4y^3 \, dy\right)\left(\int_0^2 1 \, dx\right) = \left(y^4\bigg|_0^1\right)\left(x\bigg|_0^2\right) = 1 \cdot 2 = 2$$

Substituting in (1) gives

$$\int_0^1 \int_0^2 (x + 4y^3)\, dx\, dy = 2 + 2 = 4.$$

29. $\displaystyle\int_0^4 \int_0^9 \sqrt{x + 4y}\, dx\, dy$

SOLUTION We compute the inner integral, treating y as a constant. Then we evaluate the resulting integral with respect to y:

$$\int_0^4 \int_0^9 \sqrt{x + 4y}\, dx\, dy = \int_0^4 \frac{2}{3}(x + 4y)^{3/2}\Big|_{x=0}^9\, dy = \int_0^4 \frac{2}{3}\left((9 + 4y)^{3/2} - (4y)^{3/2}\right) dy$$

$$= \frac{2}{3}\left(\frac{2}{5\cdot 4}(9 + 4y)^{5/2} - \frac{2}{5\cdot 4}(4y)^{5/2}\right)\Big|_0^4$$

$$= \frac{1}{15}(5^5 - 4^5) - \frac{1}{15}(3^5 - 0) \approx 123.8667$$

31. $\displaystyle\int_1^2 \int_0^4 \frac{dy\, dx}{x + y}$

SOLUTION The inner integral is an iterated integral with respect to y. We evaluate it first and then compute the resulting integral with respect to x. This gives

$$\int_1^2 \int_0^4 \frac{dy\, dx}{x + y} = \int_1^2 \left(\int_0^4 \frac{dy}{x + y}\right) dx = \int_1^2 \ln(x + y)\Big|_{y=0}^4\, dx = \int_1^2 (\ln(x + 4) - \ln x)\, dx$$

We use the integral formula:

$$\int \ln(x + a)\, dx = (x + a)(\ln(x + a) - 1) + C$$

We get

$$\int_1^2 \int_0^4 \frac{dy\, dx}{x + y} = (x + 4)(\ln(x + 4) - 1) - x(\ln x - 1)\Big|_1^2$$

$$= 6(\ln 6 - 1) - 2(\ln 2 - 1) - (5(\ln 5 - 1) - (\ln 1 - 1))$$

$$= 6\ln 6 - 2\ln 2 - 5\ln 5 \approx 1.31$$

33. $\displaystyle\int_0^4 \int_0^5 \frac{dy\, dx}{\sqrt{x + y}}$

SOLUTION

$$\int_0^4 \int_0^5 \frac{dy\, dx}{\sqrt{x + y}} = \int_0^4 \left(\int_0^5 \frac{dy}{\sqrt{x + y}}\right) dx$$

$$= \int_0^4 \left(2\sqrt{x + y}\Big|_{y=0}^5\right) dx$$

$$= 2\int_0^4 (\sqrt{x + 5} - \sqrt{x})\, dx$$

$$= 2\left(\frac{2}{3}(x + 5)^{3/2} - \frac{2}{3}x^{3/2}\right)\Big|_0^4$$

$$= 2\left(\frac{2}{3}\cdot 27 - \frac{2}{3}\cdot 8\right) - 2\left(\frac{2}{3}\cdot 5^{3/2} - 0\right)$$

$$= 36 - \frac{32}{3} - \frac{20}{3}\sqrt{5} = \frac{76}{3} - \frac{20}{3}\sqrt{5} \approx 10.426$$

35. $\displaystyle\int_1^2\int_1^3 \frac{\ln(xy)\,dy\,dx}{y}$

SOLUTION

$$\int_1^2\int_1^3 \frac{\ln(xy)\,dy\,dx}{y} = \int_1^2 \left(\frac{1}{2}[\ln(xy)]^2\Big|_1^3\right) dx$$

$$= \frac{1}{2}\int_1^2 [\ln(3x)]^2 - [\ln(x)]^2\,dx$$

$$= \frac{1}{2}\int_1^2 [\ln(3x)]^2\,dx - \frac{1}{2}\int_1^2 [\ln(x)]^2\,dx$$

$$= \frac{1}{2}\left[x(\ln 3x)^2\Big|_1^2 - 2\int_1^2 \ln(3x)\,dx\right] - \frac{1}{2}\left[x(\ln x)^2\Big|_1^2 - 2\int_1^2 \ln x\,dx\right]$$

$$= \frac{1}{2}\left[2(\ln 6)^2 - (\ln 3)^2\right] - \int_1^2 \ln(3x)\,dx - \frac{1}{2}\left[2(\ln 2)^2 - 0\right] + \int_1^2 \ln x\,dx$$

$$= (\ln 6)^2 - \frac{1}{2}(\ln 3)^2 - \left[x\ln(3x) - x\Big|_1^2\right] - (\ln 2)^2 + \left[x\ln x - x\Big|_1^2\right]$$

$$= (\ln 6)^2 - \frac{1}{2}(\ln 3)^2 - (\ln 2)^2 - (2\ln 6 - 2 - \ln 3 + 1) + (2\ln 2 - 2 - 0 + 1)$$

$$= (\ln 6)^2 - \frac{1}{2}(\ln 3)^2 - (\ln 2)^2 - 2\ln 6 + \ln 3 + 1 + 2\ln 2 - 2 + 1$$

$$= (\ln 6)^2 - \frac{1}{2}(\ln 3)^2 - (\ln 2)^2 - 2\ln 6 + \ln 3 + 2\ln 2 \approx 1.028$$

In Exercises 37–42, use Eq. (1) to evaluate the integral.

37. $\displaystyle\iint_{\mathcal{R}} \frac{x}{y}\,dA, \quad \mathcal{R} = [-2, 4] \times [1, 3]$

SOLUTION We compute the double integral as the product of two single integrals:

$$\iint_{\mathcal{R}} \frac{x}{y}\,dA = \int_{-2}^4 \int_1^3 \frac{x}{y}\,dy\,dx = \int_{-2}^4 x\,dx \cdot \int_1^3 \frac{1}{y}\,dy$$

$$= \left(\frac{1}{2}x^2\Big|_{-2}^4\right)\left(\ln y\Big|_1^3\right) = \frac{1}{2}(16 - 4) \cdot (\ln 3 - \ln 1)$$

$$= 6\ln 3$$

39. $\displaystyle\iint_{\mathcal{R}} \cos x \sin 2y\,dA, \quad \mathcal{R} = \left[0, \frac{\pi}{2}\right] \times \left[0, \frac{\pi}{2}\right]$

SOLUTION Since the integrand has the form $f(x, y) = g(x)h(y)$, we may compute the double integral as the product of two single integrals. That is,

$$\iint_{\mathcal{R}} \cos x \sin 2y\,dA = \int_0^{\pi/2}\int_0^{\pi/2} \cos x \sin 2y\,dx\,dy = \left(\int_0^{\pi/2} \cos x\,dx\right)\left(\int_0^{\pi/2} \sin 2y\,dy\right)$$

$$= \left(\sin x\Big|_0^{\pi/2}\right)\left(-\frac{1}{2}\cos 2y\Big|_0^{\pi/2}\right) = \left(\sin\frac{\pi}{2} - \sin 0\right)\left(-\frac{1}{2}\cos\pi + \frac{1}{2}\cos 0\right)$$

$$= (1 - 0)\left(\frac{1}{2} + \frac{1}{2}\right) = 1$$

41. $\displaystyle\iint_{\mathcal{R}} e^x \sin y\,dA, \quad \mathcal{R} = [0, 2] \times \left[0, \frac{\pi}{4}\right]$

SOLUTION We compute the double integral as the product of two single integrals. This can be done since the integrand has the form $f(x, y) = g(x)h(y)$. We get

$$\iint_{\mathcal{R}} e^x \sin y\,dA = \int_0^{\pi/4}\int_0^2 e^x \sin y\,dx\,dy = \left(\int_0^2 e^x\,dx\right)\left(\int_0^{\pi/4} \sin y\,dy\right)$$

$$= \left(e^x\Big|_0^2\right)\left(-\cos y\Big|_0^{\pi/4}\right) = \left(e^2 - e^0\right)\left(-\cos\frac{\pi}{4} + \cos 0\right) = \left(e^2 - 1\right)\left(1 - \frac{\sqrt{2}}{2}\right) \approx 1.87$$

43. Let $f(x, y) = mxy^2$, where m is a constant. Find a value of m such that $\iint_{\mathcal{R}} f(x, y)\, dA = 1$, where $\mathcal{R} = [0, 1] \times [0, 2]$.

SOLUTION Since $f(x, y) = mxy^2$ is a product of a function of x and a function of y, we may compute the double integral as the product of two single integrals. That is,

$$\int_0^2 \int_0^1 mxy^2\, dx\, dy = m \left(\int_0^1 x\, dx \right) \left(\int_0^2 y^2\, dy \right) \tag{1}$$

We compute each integral:

$$\int_0^1 x\, dx = \frac{1}{2}x^2 \Big|_0^1 = \frac{1}{2}\left(1^2 - 0^2\right) = \frac{1}{2}$$

$$\int_0^2 y^2\, dy = \frac{1}{3}y^3 \Big|_0^2 = \frac{1}{3}\left(2^3 - 0^3\right) = \frac{8}{3}$$

We substitute in (1), equate to 1 and solve the resulting equation for m. This gives

$$m \cdot \frac{1}{2} \cdot \frac{8}{3} = 1 \quad \Rightarrow \quad m = \frac{3}{4}$$

45. Evaluate $\int_0^1 \int_0^1 \frac{y}{1 + xy}\, dy\, dx$. *Hint:* Change the order of integration.

SOLUTION Using Fubini's Theorem we change the order of integration, integrating first with respect to x and then with respect to y. This gives

$$\int_0^1 \int_0^1 \frac{y}{1 + xy}\, dy\, dx = \int_0^1 \left(\int_0^1 \frac{y}{1 + xy}\, dx \right) dy = \int_0^1 \left(y \int_0^1 \frac{dx}{1 + xy} \right) dy = \int_0^1 y \cdot \frac{1}{y} \ln(1 + xy) \Big|_{x=0}^1 dy$$

$$= \int_0^1 (\ln(1 + y) - \ln 1)\, dy = \int_0^1 \ln(1 + y)\, dy = (1 + y)(\ln(1 + y) - 1) \Big|_{y=0}^1$$

$$= 2(\ln 2 - 1) - (\ln 1 - 1) = 2 \ln 2 - 1 \approx 0.386$$

47. 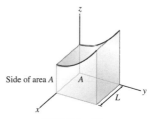 Using Fubini's Theorem, argue that the solid in Figure 18 has volume AL, where A is the area of the front face of the solid.

Side of area A

FIGURE 18

SOLUTION We denote by M the length of the other side of the rectangle in the basis of the solid. The volume of the solid is the double integral of the function $f(x, y) = g(x)$ over the rectangle

$$\mathcal{R} = [0, M] \times [0, L]$$

$$V = \iint_{\mathcal{R}} g(x)\, dA.$$

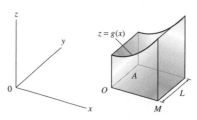

We use Fubini's Theorem to write the double integral as iterated integral, and then compute the resulting integral as the product of two single integrals. This gives

$$V = \iint_{\mathcal{R}} g(x)\, dA = \int_0^L \int_0^M g(x)\, dx\, dy = \left(\int_0^M g(x)\, dx \right) \left(\int_0^L 1\, dy \right) = \left(\int_0^M g(x)\, dx \right) \cdot L \qquad (1)$$

The integral $\int_0^M g(x)\, dx$ is the area A of the region under the graph of $z = g(x)$ over the interval $0 \le x \le M$. Substituting in (1) gives the following volume of the solid:

$$V = \left(\int_0^M g(x)\, dx \right) \cdot L = AL$$

Further Insights and Challenges

49. Let $F(x, y) = x^{-1} e^{xy}$. Show that $\dfrac{\partial^2 F}{\partial x\, \partial y} = y e^{xy}$ and use the result of Exercise 48 to evaluate $\iint_{\mathcal{R}} y e^{xy}\, dA$ for the rectangle $\mathcal{R} = [1, 3] \times [0, 1]$.

SOLUTION   Differentiating $F(x, y) = x^{-1} e^{xy}$ with respect to y gives

$$\frac{\partial F}{\partial y} = \frac{\partial}{\partial y} \left(x^{-1} e^{xy} \right) = x^{-1} x e^{xy} = e^{xy}$$

We now differentiate $\frac{\partial F}{\partial y}$ with respect to x:

$$\frac{\partial^2 F}{\partial x\, \partial y} = \frac{\partial}{\partial x} \left(e^{xy} \right) = y e^{xy}$$

In Exercise 48 we showed that

$$\int_c^d \int_a^b \frac{\partial^2 F}{\partial x\, \partial y}\, dx\, dy = F(b, d) - F(b, c) - F(a, d) + F(a, c)$$

Therefore, for $F(x, y) = x^{-1} e^{xy} = \frac{e^{xy}}{x}$ we obtain

$$\iint_{\mathcal{R}} y e^{xy}\, dA = \int_0^1 \int_1^3 y e^{xy}\, dx\, dy = F(3, 1) - F(3, 0) - F(1, 1) + F(1, 0)$$

$$= \frac{e^3}{3} - \frac{e^0}{3} - \frac{e^1}{1} + \frac{e^0}{1} = \frac{e^3}{3} - \frac{1}{3} - e + 1 = 4.644$$

51. In this exercise, we use double integration to evaluate the following improper integral for $a > 0$ a positive constant:

$$I(a) = \int_0^\infty \frac{e^{-x} - e^{-ax}}{x}\, dx$$

(a) Use L'Hôpital's Rule to show that $f(x) = \dfrac{e^{-x} - e^{-ax}}{x}$, though not defined at $x = 0$, can be made continuous by assigning the value $f(0) = a - 1$.

(b) Prove that $|f(x)| \le e^{-x} + e^{-ax}$ for $x > 1$ (use the triangle inequality), and apply the Comparison Theorem to show that $I(a)$ converges.

(c) Show that $I(a) = \displaystyle\int_0^\infty \int_1^a e^{-xy}\, dy\, dx$.

(d) Prove, by interchanging the order of integration, that

$$I(a) = \ln a - \lim_{T \to \infty} \int_1^a \frac{e^{-Ty}}{y}\, dy \qquad \boxed{2}$$

(e) Use the Comparison Theorem to show that the limit in Eq. (2) is zero. *Hint:* If $a \ge 1$, show that $e^{-Ty}/y \le e^{-T}$ for $y \ge 1$, and if $a < 1$, show that $e^{-Ty}/y \le e^{-aT}/a$ for $a \le y \le 1$. Conclude that $I(a) = \ln a$ (Figure 19).

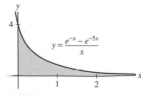

FIGURE 19 The shaded region has area $\ln 5$.

SOLUTION

(a) The function $f(x) = \frac{e^{-x} - e^{-ax}}{x}$, $f(0) = a - 1$ is continuous if $\lim_{x \to 0} f(x) = f(0) = a - 1$. We verify this limit using L'Hôpital's Rule:

$$\lim_{x \to 0} \frac{e^{-x} - e^{-ax}}{x} = \lim_{x \to 0} \frac{-e^{-x} + ae^{-ax}}{1} = -1 + a = a - 1$$

Therefore, f is continuous.

(b) We now show that the following integral converges:

$$I(a) = \int_0^\infty \frac{e^{-x} - e^{-ax}}{x} \, dx \qquad (a > 0)$$

Since $e^{-x} - e^{-ax} < e^{-x} + e^{-ax}$ then also $\frac{e^{-x} - e^{-ax}}{x} < \frac{e^{-x} + e^{-ax}}{x}$ for $x > 0$. Therefore, if $x > 1$ we have

$$\frac{e^{-x} - e^{-ax}}{x} < \frac{e^{-x} + e^{-ax}}{x} < e^{-x} + e^{-ax}$$

That is, for $x > 1$,

$$f(x) < e^{-x} + e^{-ax} \tag{1}$$

Also, since $e^{-ax} - e^{-x} < e^{-ax} + e^{-x}$ we have for $x > 1$,

$$\frac{e^{-ax} - e^{-x}}{x} < \frac{e^{-ax} + e^{-x}}{x} < e^{-ax} + e^{-x}$$

Thus, we get

$$-f(x) < e^{-x} + e^{-ax} \tag{2}$$

Together with (1) we have

$$0 \le |f(x)| < e^{-x} + e^{-ax} \tag{3}$$

We now show that the integral of the right hand-side converges:

$$\int_0^\infty \left(e^{-x} + e^{-ax}\right) dx = \lim_{R \to \infty} \int_0^R \left(e^{-x} + e^{-ax}\right) dx$$

$$= \lim_{R \to \infty} \left(-e^{-x} - \frac{e^{-ax}}{a} \Big|_{x=0}^R\right)$$

$$= \lim_{R \to \infty} \left(-e^{-R} - \frac{e^{-aR}}{a} + e^0 + \frac{e^0}{a}\right)$$

$$= \lim_{R \to \infty} \left(-e^{-R} - \frac{e^{-aR}}{a} + 1 + \frac{1}{a}\right)$$

$$= 1 + \frac{1}{a}$$

Since the integral converges, we conclude by (3) and the Comparison Test for Improper Integrals that

$$\int_0^\infty \frac{e^{-x} - e^{-ax}}{x} \, dx$$

also converges for $a > 0$.

(c) We compute the inner integral with respect to y:

$$\int_1^a e^{-xy} \, dy = -\frac{1}{x} e^{-xy} \Big|_{y=1}^a = -\frac{1}{x} \left(e^{-xa} - e^{-x \cdot 1}\right) = \frac{e^{-x} - e^{-xa}}{x}$$

Therefore,

$$\int_0^\infty \int_1^a e^{-xy} \, dy \, dx = \int_0^\infty \left(\int_1^a e^{-xy} \, dy\right) dx = \int_0^\infty \frac{e^{-x} - e^{-xa}}{x} \, dx = I(a)$$

(d) By the definition of the improper integral,

$$I(a) = \lim_{T \to \infty} \int_0^T \int_1^a e^{-xy} \, dy \, dx \tag{4}$$

We compute the double integral. Using Fubini's Theorem we may compute the iterated integral using reversed order of integration. That is,

$$\int_0^T \int_1^a e^{-xy} \, dy \, dx = \int_1^a \int_0^T e^{-xy} \, dx \, dy = \int_1^a \left(\int_0^T e^{-xy} \, dx \right) dy = \int_1^a \left(-\frac{1}{y} e^{-xy} \Big|_{x=0}^T \right) dy$$

$$= \int_1^a \left(-\frac{1}{y} \left(e^{-Ty} - e^{-0 \cdot y} \right) \right) dy = \int_1^a \frac{1 - e^{-Ty}}{y} \, dy = \int_1^a \frac{dy}{y} - \int_1^a \frac{e^{-Ty}}{y} \, dy$$

$$= \ln y \Big|_1^a - \int_1^a \frac{e^{-Ty}}{y} \, dy = \ln a - \ln 1 - \int_1^a \frac{e^{-Ty}}{y} \, dy = \ln a - \int_1^a \frac{e^{-Ty}}{y} \, dy$$

Combining with (4) we get

$$I(a) = \ln a - \lim_{T \to \infty} \int_1^a \frac{e^{-Ty}}{y} \, dy \tag{5}$$

(e) We now show, using the Comparison Theorem, that

$$\lim_{T \to \infty} \int_1^a \frac{e^{-Ty}}{y} \, dy = 0$$

We consider the following possible cases:

Case 1: $a \geq 1$. Then in the interval of integration $y \geq 1$. Also since $T \to \infty$, we may assume that $T > 0$. Thus,

$$\frac{e^{-Ty}}{y} \leq \frac{e^{-T \cdot 1}}{1} = e^{-T}$$

Hence,

$$0 \leq \int_1^a \frac{e^{-Ty}}{y} \, dy \leq \int_1^a e^{-T} \, dy = e^{-T}(a - 1)$$

By the limit $\lim_{T \to \infty} e^{-T}(a - 1) = 0$ and the Squeeze Theorem we conclude that,

$$\lim_{T \to \infty} \int_1^a \frac{e^{-Ty}}{y} \, dy = 0$$

Case 2: $0 < a < 1$. Then,

$$\int_1^a \frac{e^{-Ty}}{y} \, dy = -\int_a^1 \frac{e^{-Ty}}{y} \, dy$$

and in the interval of integration $a \leq y \leq 1$, therefore

$$\frac{e^{-Ty}}{y} \leq \frac{e^{-Ta}}{a}$$

(the function $\frac{e^{-Ty}}{y}$ is decreasing). Hence,

$$0 \leq \int_a^1 \frac{e^{-Ty}}{y} \, dy \leq \int_a^1 \frac{e^{-Ta}}{a} \, dy = \frac{(1 - a)}{a} e^{-Ta}$$

By the limit $\lim_{T \to \infty} \frac{1-a}{a} e^{-Ta} = 0$ and the Squeeze Theorem we conclude also that

$$\lim_{T \to \infty} \int_1^a \frac{e^{-Ty}}{y} = -\lim_{T \to \infty} \int_a^1 \frac{e^{-Ty}}{y} = 0$$

We thus showed that for all $a > 0$, $\lim_{T \to \infty} \int_1^a \frac{e^{-Ty}}{y} = 0$. Combining with Eq. (5) obtained in part (c), we find that $I(a) = \ln a$.

15.2 Double Integrals over More General Regions (LT Section 16.2)

Preliminary Questions

1. Which of the following expressions do not make sense?

(a) $\displaystyle\int_0^1 \int_1^x f(x, y)\, dy\, dx$

(b) $\displaystyle\int_0^1 \int_1^y f(x, y)\, dy\, dx$

(c) $\displaystyle\int_0^1 \int_x^y f(x, y)\, dy\, dx$

(d) $\displaystyle\int_0^1 \int_x^1 f(x, y)\, dy\, dx$

SOLUTION

(a) This integral is the following iterated integral:

$$\int_0^1 \int_1^x f(x, y)\, dy\, dx = \int_0^1 \left(\int_1^x f(x, y)\, dy \right) dx$$

The inner integral is a function of x and it is integrated with respect to x over the interval $0 \leq x \leq 1$. The result is a number. This integral makes sense.

(b) This integral is the same as

$$\int_0^1 \int_1^y f(x, y)\, dy\, dx = \int_0^1 \left(\int_1^y f(x, y)\, dy \right) dx$$

The inner integral is an integral with respect to y, over the interval $[1, y]$. This does not make sense.

(c) This integral is the following iterated integral:

$$\int_0^1 \left(\int_x^y f(x, y) \right) dy\, dx$$

The inner integral is a function of x and y and it is integrated with respect to y over the interval $x \leq y \leq y$. This does not make sense.

(d) This integral is the following iterated integral:

$$\int_0^1 \left(\int_x^1 f(x, y)\, dy \right) dx$$

The inner integral is a function of x and it is integrated with respect to x. This makes sense.

2. Draw a domain in the plane that is neither vertically nor horizontally simple.

SOLUTION The following region cannot be described in the form $\{a \leq x \leq b, \alpha(x) \leq y \leq \beta(x)\}$ nor in the form $\{c \leq y \leq d, \alpha(y) \leq x \leq \beta(y)\}$, hence it is neither vertically nor horizontally simple.

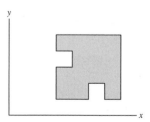

3. Which of the four regions in Figure 18 is the domain of integration for $\displaystyle\int_{-\sqrt{2}/2}^0 \int_{-x}^{\sqrt{1-x^2}} f(x, y)\, dy\, dx$?

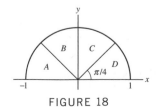

FIGURE 18

SOLUTION   The region B is defined by the inequalities

$$-x \le y \le \sqrt{1-x^2}, \qquad -\frac{\sqrt{2}}{2} \le x \le 0$$

To compute $\int_{-\sqrt{2}/2}^{0} \int_{-x}^{\sqrt{1-x^2}} f(x, y)\, dy\, dx$, we first integrate with respect to y over the interval $-x \le y \le \sqrt{1-x^2}$, and then with respect to x over $-\frac{\sqrt{2}}{2} \le x \le 0$. That is, the domain of integration is B.

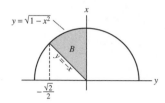

4. Let $\mathcal{D}$ be the unit disk. If the maximum value of $f(x, y)$ on $\mathcal{D}$ is 4, then the largest possible value of $\iint_{\mathcal{D}} f(x, y)\, dA$ is (choose the correct answer):

(a) 4   (b) 4π (c) $\dfrac{4}{\pi}$

SOLUTION   The area of the unit disk is π and the maximum value of $f(x, y)$ on this region is $M = 4$, therefore we have,

$$\iint_{\mathcal{D}} f(x, y)\, dx\, dy \le 4\pi$$

The correct answer is (b).

Exercises

1. Calculate the Riemann sum for $f(x, y) = x - y$ and the shaded domain $\mathcal{D}$ in Figure 19 with two choices of sample points, ● and ○. Which do you think is a better approximation to the integral of f over $\mathcal{D}$? Why?

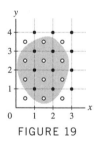

FIGURE 19

SOLUTION   The subrectangles in Figure 17 have sides of length $\Delta x = \Delta y = 1$ and area $\Delta A = 1 \cdot 1 = 1$.

(a) Sample points ●. There are six sample points that lie in the domain $\mathcal{D}$. We compute the values of $f(x, y) = x - y$ at these points:

$$f(1, 1) = 0, \qquad f(1, 2) = -1, \qquad f(1, 3) = -2,$$
$$f(2, 1) = 1, \qquad f(2, 2) = 0, \qquad f(2, 3) = -1$$

The Riemann sum is

$$S_{3,4} = (0 - 1 - 2 + 1 + 0 - 1) \cdot 1 = -3$$

(b) Sample points ○. We compute the values of $f(x, y) = x - y$ at the eight sample points that lie in $\mathcal{D}$:

$$f(1.5, 0.5) = 1, \qquad f(0.5, 1.5) = -1, \qquad f(0.5, 2.5) = -2,$$
$$f(1.5, 3.5) = -2, \qquad f(1.5, 1.5) = 0, \qquad f(1.5, 2.5) = -1,$$
$$f(2.5, 1.5) = 1, \qquad f(2.5, 2.5) = 0.$$

The corresponding Riemann sum is thus

$$S_{34} = (1 - 1 - 2 + 0 - 1 - 2 + 1 + 0) \cdot 1 = -4.$$

3. Express the domain $\mathcal{D}$ in Figure 21 as both a vertically simple region and a horizontally simple region, and evaluate the integral of $f(x, y) = xy$ over $\mathcal{D}$ as an iterated integral in two ways.

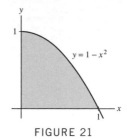

FIGURE 21

SOLUTION The domain $\mathcal{D}$ can be described as a vertically simple region as follows:

$$0 \le x \le 1, \quad 0 \le y \le 1 - x^2 \tag{1}$$

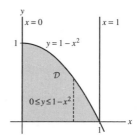

Vertically simple region

The domain $\mathcal{D}$ can also be described as a horizontally simple region. To do this, we must express x in terms of y, for nonnegative values of x. This gives

$$y = 1 - x^2 \quad \Rightarrow \quad x^2 = 1 - y \quad \Rightarrow \quad x = \sqrt{1 - y}$$

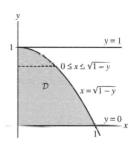

Horizontally simple region

Therefore, we can describe $\mathcal{D}$ by the following inequalities:

$$0 \le y \le 1, \quad 0 \le x \le \sqrt{1 - y} \tag{2}$$

We now compute the integral of $f(x, y) = xy$ over $\mathcal{D}$ first using definition (1) and then using definition (2). We obtain

$$\iint_{\mathcal{D}} xy \, dA = \int_0^1 \int_0^{1-x^2} xy \, dy \, dx = \int_0^1 \frac{xy^2}{2} \Big|_{y=0}^{1-x^2} dx = \int_0^1 \frac{x}{2} \left((1 - x^2)^2 - 0^2 \right) dx = \int_0^1 \frac{x(1 - x^2)^2}{2} dx$$

$$= \frac{1}{2} \int_0^1 (x - 2x^3 + x^5) \, dx = \frac{1}{2} \left(\frac{x^2}{2} - \frac{x^4}{2} + \frac{x^6}{6} \right) \Big|_0^1 = \frac{1}{2} \left(\frac{1}{2} - \frac{1}{2} + \frac{1}{6} \right) = \frac{1}{12}$$

Using definition (2) gives

$$\iint_{\mathcal{D}} xy \, dA = \int_0^1 \int_0^{\sqrt{1-y}} xy \, dx \, dy = \int_0^1 \frac{yx^2}{2} \Big|_{x=0}^{\sqrt{1-y}} dy = \int_0^1 \frac{y}{2} \left(\left(\sqrt{1 - y}\right)^2 - 0^2 \right) dy$$

$$= \int_0^1 \frac{y}{2}(1 - y) \, dy = \int_0^1 \left(\frac{y}{2} - \frac{y^2}{2} \right) dy = \frac{y^2}{4} - \frac{y^3}{6} \Big|_0^1 = \frac{1}{4} - \frac{1}{6} = \frac{1}{12}$$

The answers agree as expected.

In Exercises 5–7, compute the double integral of $f(x, y) = x^2 y$ over the given shaded domain in Figure 22.

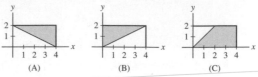

(A)   (B)   (C)

FIGURE 22

5. (A)

SOLUTION

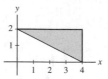

We describe the domain $\mathcal{D}$ as a vertically simple region. We find the equation of the line connecting the points $(0, 2)$ and $(4, 0)$.

$$y - 0 = \frac{2 - 0}{0 - 4}(x - 4) \quad \Rightarrow \quad y = -\frac{1}{2}x + 2$$

Therefore the domain is described as a vertically simple region by the inequalities

$$0 \le x \le 4, \quad -\frac{1}{2}x + 2 \le y \le 2$$

We use Theorem 2 to evaluate the double integral:

$$\iint_{\mathcal{D}} x^2 y \, dA = \int_0^4 \int_{-\frac{x}{2}+2}^2 x^2 y \, dy \, dx = \int_0^4 \left. \frac{x^2 y^2}{2} \right|_{y=-\frac{x}{2}+2}^2 dx = \int_0^4 \frac{x^2}{2}\left(2^2 - \left(-\frac{x}{2} + 2\right)^2\right) dx$$

$$= \int_0^4 \left(x^3 - \frac{x^4}{8}\right) dx = \left. \frac{x^4}{4} - \frac{x^5}{40} \right|_0^4 = \frac{4^4}{4} - \frac{4^5}{40} = \frac{192}{5} = 38.4$$

7. (C)

SOLUTION   The domain in (C) is a horizontally simple region, described by the inequalities

$$0 \le y \le 2, \quad y \le x \le 4$$

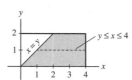

Using Theorem 2 we obtain the following integral:

$$\iint_{\mathcal{D}} x^2 y \, dA = \int_0^2 \int_y^4 x^2 y \, dx \, dy = \int_0^2 \left. \frac{x^3 y}{3} \right|_{x=y}^{x=4} dy = \int_0^2 \frac{y}{3}\left(4^3 - y^3\right) dy = \int_0^2 \left(\frac{64y}{3} - \frac{y^4}{3}\right) dy$$

$$= \left. \frac{32}{3}y^2 - \frac{y^5}{15} \right|_0^2 = \frac{32 \cdot 2^2}{3} - \frac{2^5}{15} = \frac{608}{15} \approx 40.53$$

9. Integrate $f(x, y) = x$ over the region bounded by $y = x^2$ and $y = x + 2$.

SOLUTION   The domain of integration is shown in the following figure:

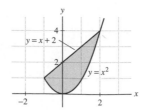

To find the inequalities defining the domain as a vertically simple region we first must find the x-coordinates of the two points where the line $y = x + 2$ and the parabola $y = x^2$ intersect. That is,

$$x + 2 = x^2 \quad \Rightarrow \quad x^2 - x - 2 = (x - 2)(x + 1) = 0$$

$$\Rightarrow \quad x_1 = -1, \quad x_2 = 2$$

We describe the domain by the following inequalities:

$$-1 \le x \le 2, \quad x^2 \le y \le x + 2$$

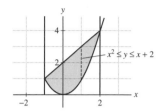

We now evaluate the integral of $f(x, y) = x$ over the vertically simple region $\mathcal{D}$:

$$\iint_{\mathcal{D}} x \, dA = \int_{-1}^{2} \int_{x^2}^{x+2} x \, dy \, dx = \int_{-1}^{2} xy \Big|_{y=x^2}^{x+2} \, dx = \int_{-1}^{2} x \left(x + 2 - x^2 \right) dx$$

$$= \int_{-1}^{2} \left(x^2 + 2x - x^3 \right) dx = \frac{x^3}{3} + x^2 - \frac{x^4}{4} \Big|_{-1}^{2} = \left(\frac{8}{3} + 4 - 4 \right) - \left(-\frac{1}{3} + 1 - \frac{1}{4} \right) = 2\frac{1}{4}$$

11. Evaluate $\displaystyle\iint_{\mathcal{D}} \frac{y}{x} \, dA$, where $\mathcal{D}$ is the shaded part of the semicircle of radius 2 in Figure 23

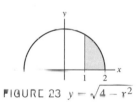

FIGURE 23 $y = \sqrt{4 - x^2}$

SOLUTION The region is defined by the following inequalities:

$$1 \le x \le 2, \quad 0 \le y \le \sqrt{4 - x^2}$$

Therefore, the double integral of f over $\mathcal{D}$ is:

$$\iint_{\mathcal{D}} \frac{y}{x} \, dA = \int_{1}^{2} \int_{0}^{\sqrt{4-x^2}} \frac{y}{x} \, dy \, dx$$

$$= \int_{1}^{2} \frac{1}{x} \left(\frac{1}{2} y^2 \Big|_{0}^{\sqrt{4-x^2}} \right) dx$$

$$= \frac{1}{2} \int_{1}^{2} \frac{1}{x} (4 - x^2) \, dx$$

$$= \frac{1}{2} \int_{1}^{2} \frac{4}{x} - x \, dx$$

$$= \frac{1}{2} \left(4 \ln |x| - \frac{1}{2} x^2 \right) \Big|_{1}^{2}$$

$$= \frac{1}{2} (4 \ln 2 - 2) - \frac{1}{2} \left(0 - \frac{1}{2} \right)$$

$$= 2 \ln 2 - 1 + \frac{1}{4} = 2 \ln 2 - \frac{3}{4} \approx 0.636$$

13. Calculate the double integral of $f(x, y) = x + y$ over the domain $\mathcal{D} = \{(x, y) : x^2 + y^2 \le 4, y \ge 0\}$.

SOLUTION

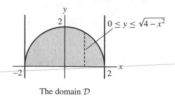

The domain $\mathcal{D}$

The semicircle can be described as a vertically simple region, by the following inequalities:

$$-2 \le x \le 2, \quad 0 \le y \le \sqrt{4 - x^2}$$

We evaluate the double integral by the following iterated integral:

$$\iint_{\mathcal{D}} (x + y)\, dA = \int_{-2}^{2} \int_{0}^{\sqrt{4-x^2}} (x + y)\, dy\, dx = \int_{-2}^{2} xy + \frac{1}{2}y^2 \Big|_{y=0}^{\sqrt{4-x^2}} dx = \int_{-2}^{2} \left(x\sqrt{4 - x^2} + \frac{1}{2}\left(\sqrt{4 - x^2}\right)^2 \right) dx$$

$$= \int_{-2}^{2} x\sqrt{4 - x^2}\, dx + \frac{1}{2} \int_{-2}^{2} (4 - x^2)\, dx = \int_{-2}^{2} x\sqrt{4 - x^2}\, dx + 2x - \frac{x^3}{6} \Big|_{x=-2}^{2}$$

$$= \int_{-2}^{2} x\sqrt{4 - x^2}\, dx + 4 - \frac{8}{6} - \left(-4 - \frac{-8}{6} \right) = \int_{-2}^{2} x\sqrt{4 - x^2}\, dx + \frac{16}{3}$$

The integral of an odd function over an interval that is symmetric with respect to the origin is zero. Hence $\int_{-2}^{2} x\sqrt{4 - x^2}\, dx = 0$, so we get

$$\iint_{\mathcal{D}} (x + y)\, dA = 0 + \frac{16}{3} = \frac{16}{3} \approx 5.33$$

15. Calculate the integral of $f(x, y) = x$ over the region $\mathcal{D}$ bounded above by $y = x(2 - x)$ and below by $x = y(2 - y)$. *Hint:* Apply the quadratic formula to the lower boundary curve to solve for y as a function of x.

SOLUTION   The two graphs are symmetric with respect to the line $y = x$, thus their point of intersection is $(1, 1)$. The region $\mathcal{D}$ is shown in the following figure:

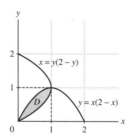

To find the inequalities defining the region $\mathcal{D}$ as a vertically simple region, we first must solve the lower boundary curve for y in terms of x. We get

$$x = y(2 - y) = 2y - y^2$$

$$y^2 - 2y + x = 0$$

We solve the quadratic equation in y:

$$y = 1 \pm \sqrt{1 - x}$$

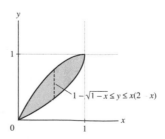

The domain $\mathcal{D}$ lies below the line $y = 1$, hence the appropriate solution is $y = 1 - \sqrt{1-x}$. We obtain the following inequalities for $\mathcal{D}$:

$$0 \le x \le 1, \quad 1 - \sqrt{1-x} \le y \le x(2-x)$$

We now evaluate the double integral of $f(x, y) = x$ over $\mathcal{D}$:

$$\iint_{\mathcal{D}} x \, dA = \int_0^1 \int_{1-\sqrt{1-x}}^{x(2-x)} x \, dy \, dx = \int_0^1 xy \Big|_{y=1-\sqrt{1-x}}^{x(2-x)} dx = \int_0^1 \left(x^2(2-x) - \left(x - x\sqrt{1-x} \right) \right) dx$$

$$= \int_0^1 \left(2x^2 - x^3 - x + x\sqrt{1-x} \right) dx = \frac{2x^3}{3} - \frac{x^4}{4} - \frac{x^2}{2} \Big|_0^1 + \int_0^1 x\sqrt{1-x} \, dx$$

$$= -\frac{1}{12} + \int_0^1 x\sqrt{1-x} \, dx$$

Using the substitution $u = \sqrt{1-x}$ it can be shown that $\int_0^1 x\sqrt{1-x} \, dx = \frac{4}{15}$. Therefore we get

$$\iint_{\mathcal{D}} x \, dA = -\frac{1}{12} + \frac{4}{15} = \frac{11}{60}$$

In Exercises 17–24, compute the double integral of $f(x, y)$ over the domain $\mathcal{D}$ indicated.

17. $f(x, y) = x^2 y; \quad 1 \le x \le 3, \quad x \le y \le 2x + 1$

SOLUTION These inequalities describe $\mathcal{D}$ as a vertically simple region.

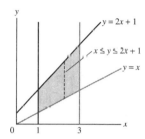

We compute the double integral of $f(x, y) = x^2 y$ on $\mathcal{D}$ by the following iterated integral:

$$\iint_{\mathcal{D}} x^2 y \, dA = \int_1^3 \int_x^{2x+1} x^2 y \, dy \, dx = \int_1^3 \frac{x^2 y^2}{2} \Big|_{y=x}^{2x+1} dx = \int_1^3 \frac{x^2}{2} \left((2x+1)^2 - x^2 \right) dx$$

$$= \int_1^3 \left(\frac{3}{2} x^4 + 2x^3 + \frac{x^2}{2} \right) dx = \frac{3}{10} x^5 + \frac{x^4}{2} + \frac{x^3}{6} \Big|_1^3$$

$$= \frac{3 \cdot 3^5}{10} + \frac{3^4}{2} + \frac{3^3}{6} - \left(\frac{3}{10} + \frac{1}{2} + \frac{1}{6} \right) = \frac{1754}{15} \approx 116.93$$

19. $f(x, y) = x; \quad 0 \le x \le 1, \quad 1 \le y \le e^{x^2}$

SOLUTION We compute the double integral of $f(x, y) = x$ over the vertically simple region $\mathcal{D}$, as the following iterated integral:

$$\iint_{\mathcal{D}} x \, dA = \int_0^1 \int_1^{e^{x^2}} x \, dy \, dx = \int_0^1 xy \Big|_{y-1}^{e^{x^2}} dx = \int_0^1 \left(xe^{x^2} - x \cdot 1 \right) dx$$

$$= \int_0^1 xe^{x^2} \, dx - \int_0^1 x \, dx = \int_0^1 xe^{x^2} \, dx - \frac{x^2}{2} \Big|_0^1 = \int_0^1 xe^{x^2} \, dx - \frac{1}{2} \tag{1}$$

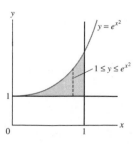

The resulting integral can be computed using the substitution $u = x^2$. The value of this integral is

$$\int_0^1 xe^{x^2}\,dx = \frac{e-1}{2}$$

Combining with (1) we get

$$\iint_{\mathcal{D}} x\,dA = \frac{e-1}{2} - \frac{1}{2} = \frac{e-2}{2} \approx 0.359$$

21. $f(x, y) = 2xy$; bounded by $x = y$, $x = y^2$

SOLUTION The intersection points of the graphs $x = y$ and $x = y^2$ are $(0, 0)$ $(1, 1)$. The horizontally simple region $\mathcal{D}$ is shown in the figure:

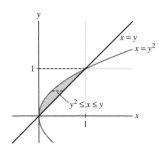

We compute the double integral of $f(x, y) = 2xy$ over $\mathcal{D}$, using Theorem 2. The limits of integration are determined by the inequalities:

$$0 \le y \le 1, \quad y^2 \le x \le y.$$

Defining $\mathcal{D}$, we get

$$\iint_{\mathcal{D}} 2xy\,dA = \int_0^1 \int_{y^2}^y 2xy\,dx\,dy = \int_0^1 x^2 y \Big|_{x=y^2}^y\,dy = \int_0^1 (y^2 \cdot y - y^4 \cdot y)\,dy$$

$$= \int_0^1 (y^3 - y^5)\,dy = \frac{y^4}{4} - \frac{y^6}{6}\Big|_0^1 = \frac{1}{4} - \frac{1}{6} = \frac{1}{12}$$

23. $f(x, y) = e^{x+y}$; bounded by $y = x - 1$, $y = 12 - x$ for $2 \le y \le 4$

SOLUTION The horizontally simple region $\mathcal{D}$ is shown in the figure:

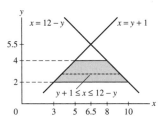

We compute the double integral of $f(x, y) = e^{x+y}$ over $\mathcal{D}$ by evaluating the following iterated integral:

$$\iint_{\mathcal{D}} e^{x+y}\,dA = \int_2^4 \int_{y+1}^{12-y} e^{x+y}\,dx\,dy = \int_2^4 e^{x+y}\Big|_{x=y+1}^{12-y}\,dy = \int_2^4 \left(e^{12-y+y} - e^{y+1+y}\right)\,dy$$

$$= \int_2^4 \left(e^{12} - e^{2y+1}\right)\,dy = e^{12} \cdot y - \frac{1}{2}e^{2y+1}\Big|_2^4 = \left(e^{12} \cdot 4 - \frac{1}{2}e^{2\cdot4+1}\right) - \left(e^{12} \cdot 2 - \frac{1}{2}e^{2\cdot2+1}\right)$$

$$= 4e^{12} - \frac{1}{2}e^9 - 2e^{12} + \frac{1}{2}e^5 = 2e^{12} - \frac{1}{2}e^9 + \frac{1}{2}e^5 \approx 321532.2$$

In Exercises 25–28, sketch the domain of integration and express as an iterated integral in the opposite order.

25. $\int_0^4 \int_x^4 f(x, y)\,dy\,dx$

SOLUTION The limits of integration correspond to the inequalities describing the following domain $\mathcal{D}$:

$$0 \le x \le 4, \quad x \le y \le 4.$$

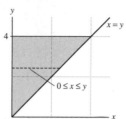

From the sketch of $\mathcal{D}$ we see that $\mathcal{D}$ can also be expressed as a horizontally simple region as follows:

$$0 \le y \le 4, \quad 0 \le x \le y$$

Therefore we can reverse the order of integration as follows:

$$\int_0^4 \int_x^4 f(x, y) \, dy \, dx = \int_0^4 \int_0^y f(x, y) \, dx \, dy.$$

27. $\displaystyle \int_4^9 \int_2^{\sqrt{y}} f(x, y) \, dx \, dy$

SOLUTION The limits of integration correspond to the following inequalities defining the horizontally simple region $\mathcal{D}$:

$$4 < y \le 9, \quad 2 \le x \le \sqrt{y}$$

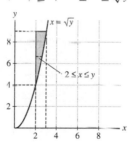

The region $\mathcal{D}$ can also be expressed as a vertically simple region. We first need to write the equation of the curve $x = \sqrt{y}$ in the form $y = x^2$. The corresponding inequalities are

$$2 \le x \le 3, \quad x^2 \le y \le 9$$

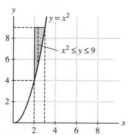

We now can write the iterated integral with reversed order of integration:

$$\int_4^9 \int_2^{\sqrt{y}} f(x, y) \, dx \, dy = \int_2^3 \int_{x^2}^9 f(x, y) \, dy \, dx.$$

29. Sketch the domain $\mathcal{D}$ corresponding to

$$\int_0^4 \int_{\sqrt{y}}^2 \sqrt{4x^2 + 5y}\, dx\, dy$$

Then change the order of integration and evaluate.

SOLUTION   The limits of integration correspond to the following inequalities describing the domain $\mathcal{D}$:

$$0 \le y \le 4, \quad \sqrt{y} \le x \le 2$$

The horizontally simple region $\mathcal{D}$ is shown in the figure:

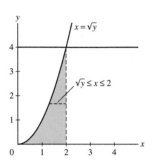

The domain $\mathcal{D}$ can also be described as a vertically simple region. Rewriting the equation $x = \sqrt{y}$ in the form $y = x^2$, we define $\mathcal{D}$ by the following inequalities (see figure):

$$0 \le x \le 2, \quad 0 \le y \le x^2$$

The corresponding iterated integral is

$$\int_0^2 \int_0^{x^2} \sqrt{4x^2 + 5y}\, dy\, dx$$

We evaluate this integral:

$$\int_0^2 \int_0^{x^2} \sqrt{4x^2 + 5y}\, dy\, dx = \int_0^2 \left(\int_0^{x^2} \left(4x^2 + 5y\right)^{1/2} dy \right) dx$$

$$= \int_0^2 \frac{2}{15}\left(4x^2 + 5y\right)^{3/2} \bigg|_{y=0}^{x^2} dx$$

$$= \int_0^2 \left(\frac{2}{15}\left(4x^2 + 5x^2\right)^{3/2} - \frac{2}{15}\left(4x^2 + 0\right)^{3/2} \right) dx$$

$$= \int_0^2 \left(\frac{2}{15}\left(9x^2\right)^{3/2} - \frac{2}{15}\left(4x^2\right)^{3/2} \right) dx$$

$$= \int_0^2 \frac{18}{5}x^3 - \frac{16}{15}x^3\, dx$$

$$= \int_0^2 \frac{38}{15}x^3\, dx$$

$$= \frac{38}{15} \cdot \frac{x^4}{4} \bigg|_0^2 = \frac{38}{15} \cdot \frac{2^4}{4} = \frac{152}{15} \approx 10.133$$

31. Compute the integral of $f(x, y) = (\ln y)^{-1}$ over the domain $\mathcal{D}$ bounded by $y = e^x$ and $y = e^{\sqrt{x}}$. *Hint:* Choose the order of integration that enables you to evaluate the integral.

SOLUTION To express $\mathcal{D}$ as a horizontally simple region, we first must rewrite the equations of the curves $y = e^x$ and $y = e^{\sqrt{x}}$ with x as a function of y. That is,

$$y = e^x \quad \Rightarrow \quad x = \ln y$$

$$y = e^{\sqrt{x}} \quad \Rightarrow \quad \sqrt{x} = \ln y \quad \Rightarrow \quad x = \ln^2 y$$

We obtain the following inequalities:

$$1 \le y \le e, \quad \ln^2 y \le x \le \ln y$$

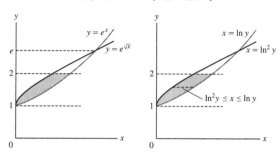

Using Theorem 2, we compute the double integral of $f(x, y) = (\ln y)^{-1}$ over $\mathcal{D}$ as the following iterated integral:

$$\iint_{\mathcal{D}} (\ln y)^{-1} \, dA = \int_1^e \int_{\ln^2 y}^{\ln y} (\ln y)^{-1} \, dx \, dy = \int_1^e (\ln y)^{-1} x \Big|_{x=\ln^2 y}^{\ln y} \, dy = \int_1^e (\ln y)^{-1} \left(\ln y - \ln^2 y \right) \, dy$$

$$= \int_1^e (1 - \ln y) \, dy = \int_1^e 1 \, dy - \int_1^e \ln y \, dy = y \Big|_1^e - y(\ln y - 1) \Big|_1^e$$

$$= (e - 1) - [e(0) - 1(-1)] = e - 2$$

In Exercises 33–36, sketch the domain of integration. Then change the order of integration and evaluate. Explain the simplification achieved by changing the order.

33. $\displaystyle \int_0^1 \int_y^1 \frac{\sin x}{x} \, dx \, dy$

SOLUTION The limits of integration correspond to the following inequalities:

$$0 \le y \le 1, \quad y \le x \le 1$$

The horizontally simple region $\mathcal{D}$ is shown in the figure.

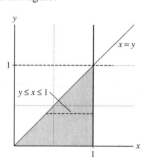

We see that $\mathcal{D}$ can also be described as a vertically simple region, by the following inequalities:

$$0 \le x \le 1, \quad 0 \le y \le x$$

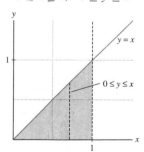

We evaluate the corresponding iterated integral:

$$\int_0^1 \int_0^x \frac{\sin x}{x}\, dy\, dx = \int_0^1 \frac{\sin x}{x} y \Big|_{y=0}^x dx = \int_0^1 \frac{\sin x}{x}(x-0)\, dx = \int_0^1 \sin x\, dx = -\cos x \Big|_0^1 = 1 - \cos 1 \approx 0.46$$

Trying to integrate in reversed order we obtain a complicated integral in the inner integral. That is,

$$\int_0^1 \int_y^1 \frac{\sin x}{x}\, dx\, dy = \int_0^1 \left(\int_y^1 \frac{\sin x}{x}\, dx \right) dy$$

Remark: $f(x, y) = \frac{\sin x}{x}$ is not continuous at the point $(0, 0)$ in $\mathcal{D}$. To make it continuous we need to define $f(0, 0) = 1$.

35. $\int_0^1 \int_{y=x}^1 xe^{y^3}\, dy\, dx$

SOLUTION The limits of integration define a vertically simple region $\mathcal{D}$ by the following inequalities:

$$0 \le x \le 1, \quad x \le y \le 1$$

This region can also be described as a horizontally simple region by the following inequalities (see figure):

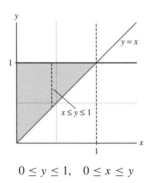

$$0 \le y \le 1, \quad 0 \le x \le y$$

We thus can rewrite the given integral in reversed order of integration as follows:

$$\int_0^1 \int_0^y xe^{y^3}\, dx\, dy = \int_0^1 \frac{x^2}{2} e^{y^3} \Big|_{x=0}^y dy = \int_0^1 e^{y^3}\left(\frac{y^2}{2} - 0 \right) dy = \int_0^1 \frac{1}{2} e^{y^3} y^2\, dy$$

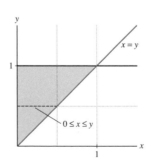

We compute this integral using the substitution $u = y^3$, $du = 3y^2\, dy$. This gives

$$\int_0^1 \int_0^y xe^{y^3}\, dx\, dy = \int_0^1 \frac{1}{2} e^{y^3} y^2\, dy = \int_0^1 e^u \cdot \frac{1}{6}\, du = \frac{e^u}{6} \Big|_0^1 = \frac{e-1}{6} \approx 0.286$$

Trying to evaluate the double integral in the original order of integration, we find that the inner integral is impossible to compute:

$$\int_0^1 \int_x^1 xe^{y^3}\, dy\, dx = \int_0^1 \left(\int_x^1 xe^{y^3}\, dy \right) dx$$

37. Sketch the domain $\mathcal{D}$ where $0 \le x \le 2$, $0 \le y \le 2$, and x *or* y is greater than 1. Then compute $\iint_{\mathcal{D}} e^{x+y} \, dA$.

SOLUTION   The domain $\mathcal{D}$ within the square $0 \le x, y \le 2$ is shown in the figure.

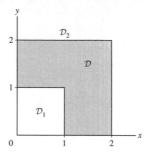

We denote the unit square $0 \le x, y \le 1$ and the square $0 \le x, y \le 2$ by $\mathcal{D}_1$ and $\mathcal{D}_2$ respectively. Then $\mathcal{D}_2$ is the union of $\mathcal{D}_1$ and $\mathcal{D}$, and these two domains do not overlap except on the boundary of $\mathcal{D}_1$. Therefore, by properties of the double integral, we have

$$\iint_{\mathcal{D}_2} e^{x+y} \, dA = \iint_{\mathcal{D}_1} e^{x+y} \, dA + \iint_{\mathcal{D}} e^{x+y} \, dA$$

Hence,

$$\iint_{\mathcal{D}} e^{x+y} \, dA = \iint_{\mathcal{D}_2} e^{x+y} \, dA - \iint_{\mathcal{D}_1} e^{x+y} \, dA = \int_0^2 \int_0^2 e^{x+y} \, dx \, dy - \int_0^1 \int_0^1 e^{x+y} \, dx \, dy$$

$$= \int_0^2 e^{x+y} \Big|_{x=0}^2 \, dy - \int_0^1 e^{x+y} \Big|_{x=0}^1 \, dy = \int_0^2 (e^{2+y} - e^y) \, dy - \int_0^1 (e^{1+y} - e^y) \, dy$$

$$= e^{2+y} - e^y \Big|_{y=0}^2 - (e^{1+y} - e^y) \Big|_{y=0}^1 = e^4 - e^2 - (e^2 - e^0) - \left(e^2 - e - (e - e^0) \right)$$

$$= e^4 - 3e^2 + 2e \approx 37.87$$

In Exercises 39–42, calculate the double integral of $f(x, y)$ over the triangle indicated in Figure 25.

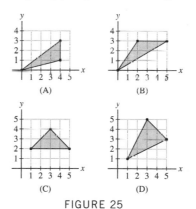

FIGURE 25

39. $f(x, y) = e^{x^2}$,   (A)

SOLUTION   The equations of the lines OA and OB are $y = \frac{3}{4}x$ and $y = \frac{1}{4}x$, respectively. Therefore, the triangle may be expressed as a vertically simple region by the following inequalities:

$$0 \le x \le 4, \quad \frac{x}{4} \le y \le \frac{3x}{4}$$

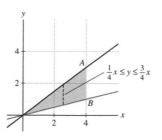

The double integral of $f(x, y) = e^{x^2}$ over the triangle is the following iterated integral:

$$\int_0^4 \int_{x/4}^{3x/4} e^{x^2} \, dy \, dx = \int_0^4 y e^{x^2} \Big|_{y=x/4}^{3x/4} \, dx$$

$$= \int_0^4 e^{x^2} \left(\frac{3x}{4} - \frac{x}{4} \right) dx$$

$$= \frac{1}{2} \int_0^4 x e^{x^2} \, dx$$

$$= \frac{1}{4} e^{x^2} \Big|_0^4 = \frac{1}{4}(e^{16} - 1)$$

41. $f(x, y) = \dfrac{x}{y^2}$, (C)

SOLUTION To find the inequalities defining the triangle as a horizontally simple region, we first find the inequalities of the lines AB and BC:

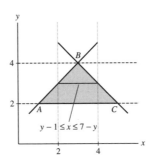

$$AB: \quad y - 2 = \frac{4 - 2}{3 - 1}(x - 1) \quad \Rightarrow \quad y - 2 = x - 1 \quad \Rightarrow \quad x = y - 1$$

$$BC: \quad y - 2 = \frac{4 - 2}{3 - 5}(x - 5) \quad \Rightarrow \quad y - 2 = 5 - x \quad \Rightarrow \quad x = 7 - y$$

We obtain the following inequalities for the triangle:

$$2 \le y \le 4, \quad y - 1 \le x \le 7 - y$$

The double integral of $f(x, y) = \frac{x}{y^2}$ over the triangle is the following iterated integral:

$$\int_2^4 \int_{y-1}^{7-y} \frac{x}{y^2} \, dx \, dy = \int_2^4 \frac{x^2}{2y^2} \Big|_{x=y-1}^{7-y} \, dy = \int_2^4 \frac{(7-y)^2 - (y-1)^2}{2y^2} \, dy = \int_2^4 \left(\frac{24}{y^2} - \frac{6}{y} \right) dy$$

$$= -\frac{24}{y} - 6 \ln y \Big|_2^4 = -\frac{24}{4} - 6 \ln 4 - \left(-\frac{24}{2} - 6 \ln 2 \right) = 6 - 6 \ln 2 = 1.84$$

43. Calculate the double integral of $f(x, y) = \dfrac{\sin y}{y}$ over the region $\mathcal{D}$ in Figure 26.

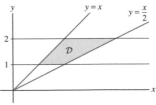

FIGURE 26

SOLUTION To describe $\mathcal{D}$ as a horizontally simple region, we rewrite the equations of the lines with x as a function of y, that is, $x = y$ and $x = 2y$. The inequalities for $\mathcal{D}$ are

$$1 \le y \le 2, \quad y \le x \le 2y$$

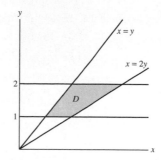

We now compute the double integral of $f(x, y) = \frac{\sin y}{y}$ over $\mathcal{D}$ by the following iterated integral:

$$\iint_{\mathcal{D}} \frac{\sin y}{y} \, dA = \int_1^2 \int_y^{2y} \frac{\sin y}{y} \, dx \, dy = \int_1^2 \frac{\sin y}{y} x \Big|_{x=y}^{2y} \, dy = \int_1^2 \frac{\sin y}{y} (2y - y) \, dy$$

$$= \int_1^2 \frac{\sin y}{y} \cdot y \, dy = \int_1^2 \sin y \, dy = -\cos y \Big|_1^2 = \cos 1 - \cos 2 \approx 0.956$$

45. Find the volume of the region bounded by $z = 40 - 10y$, $z = 0$, $y = 0$, $y = 4 - x^2$.

SOLUTION The volume of the region is the double integral of $f(x, y) = 40 - 10y$ over the domain $\mathcal{D}$ in the xy-plane between the curves $y = 0$ and $y = 4 - x^2$. This is a vertically simple region described by the inequalities:

$$-2 < x \le 2, \quad 0 \le y \le 4 - x^2$$

We compute the double integral as the following iterated integral.

$$V = \iint_{\mathcal{D}} 40 - 10y \, dA = \int_{-2}^2 \int_0^{4-x^2} (40 - 10y) \, dy \, dx$$

$$= \int_{-2}^2 \left(40y - 5y^2 \Big|_0^{4-x^2} \right) dx$$

$$= \int_{-2}^2 40(4 - x^2) - 5(4 - x^2)^2 \, dx = \int_{-2}^2 160 - 40x^2 - 5(16 - 8x^2 + x^4) \, dx$$

$$= \int_{-2}^2 80 - 5x^4 \, dx = 80x - x^5 \Big|_{-2}^2$$

$$\blacksquare (160 - 32) - (-160 + 32) = 256$$

47. Calculate the average value of $f(x, y) = e^{x+y}$ on the square $[0, 1] \times [0, 1]$.

SOLUTION

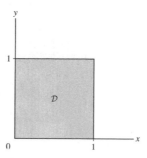

Since the area of the square $\mathcal{D}$ is 1, the average value of $f(x, y) = e^{x+y}$ on $\mathcal{D}$ is the following value:

$$\bar{f} = \frac{1}{\text{Area}(\mathcal{D})} \iint_{\mathcal{D}} f(x, y) \, dA = \frac{1}{1} \int_0^1 \int_0^1 e^{x+y} \, dx \, dy = \int_0^1 e^{x+y} \Big|_{x=0}^1 \, dy = \int_0^1 \left(e^{1+y} - e^{0+y} \right) dy$$

$$= \int_0^1 \left(e^{1+y} - e^y \right) dy = e^{1+y} - e^y \Big|_0^1 = \left(e^2 - e \right) - \left(e^1 - e^0 \right) = e^2 - 2e + 1 \approx 2.95$$

49. Find the average height of the "ceiling" in Figure 28 defined by $z = y^2 \sin x$ for $0 \leq x \leq \pi, 0 \leq y \leq 1$.

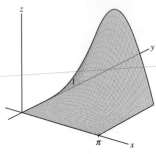

FIGURE 28

SOLUTION

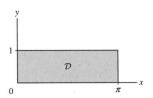

The average height is

$$\overline{H} = \frac{1}{\text{Area}(\mathcal{D})} \iint_{\mathcal{D}} y^2 \sin x \, dA = \frac{1}{\pi \cdot 1} \int_0^1 \int_0^\pi y^2 \sin x \, dx \, dy = \frac{1}{\pi} \int_0^1 y^2 (-\cos x) \Big|_{x=0}^\pi dy$$

$$= \frac{1}{\pi} \int_0^1 y^2 (-\cos \pi + \cos 0) \, dy = \frac{1}{\pi} \int_0^1 2y^2 \, dy = \frac{1}{\pi} \cdot \frac{2}{3} y^3 \Big|_0^1 = \frac{2}{3\pi}$$

51. What is the average value of the linear function

$$f(x, y) = mx + ny + p$$

on the ellipse $\left(\frac{x}{a}\right)^2 + \left(\frac{y}{b}\right)^2 \leq 1$? Argue by symmetry rather than calculation.

SOLUTION   The average value of the linear function $f(x, y) = mx + ny + p$ over the ellipse $\mathcal{D}$ is

$$\overline{f} = \frac{1}{\text{Area}(\mathcal{D})} \iint_{\mathcal{D}} f(x, y) \, dA = \frac{1}{\text{Area}(\mathcal{D})} \iint_{\mathcal{D}} (mx + ny + p) \, dA$$

$$= m \cdot \underbrace{\frac{1}{\text{Area}(\mathcal{D})} \iint_{\mathcal{D}} x \, dA}_{I_1} + n \cdot \underbrace{\frac{1}{\text{Area}(\mathcal{D})} \iint_{\mathcal{D}} y \, dA}_{I_2} + \frac{1}{\text{Area}(\mathcal{D})} \iint_{\mathcal{D}} p \, dA \qquad (1)$$

I_1 and I_2 are the average values of the x and y coordinates of a point in the region enclosed by the ellipse. This region is symmetric with respect to the y-axis, hence $I_1 = 0$. It is also symmetric with respect to the x-axis, hence $I_2 = 0$. We use the formula

$$\iint_{\mathcal{D}} p \, dA = p \cdot \text{Area}(\mathcal{D})$$

to conclude by (1) that

$$\overline{f} = m \cdot 0 + n \cdot 0 + \frac{1}{\text{Area}(\mathcal{D})} \cdot p \cdot \text{Area}(\mathcal{D}) = p$$

53. Let $\mathcal{D}$ be the rectangle $0 \leq x \leq 2, -\frac{1}{8} \leq y \leq \frac{1}{8}$, and let $f(x, y) = \sqrt{x^3 + 1}$. Prove that

$$\iint_{\mathcal{D}} f(x, y) \, dA \leq \frac{3}{2}$$

SOLUTION   Recall that we can write

$$\iint_{\mathcal{D}} f(x, y)\, dA \le M \cdot \text{Area}(\mathcal{D})$$

where M is a constant such that $f(x, y) \le M$. We can see that $\text{Area}(\mathcal{D}) = 2(1/4) = 1/2$. So it remains to show that there is some constant M so that $f(x, y) \le M$. Consider the following:

$$x \le 2 \quad \Rightarrow \quad x^3 + 1 \le 0 \quad \Rightarrow \quad \sqrt{x^3 + 1} \le 3$$

Thus we can let $M = 3$. So then we have

$$\iint_{\mathcal{D}} f(x, y)\, dA \le M \cdot \text{Area}(\mathcal{D}) \quad \Rightarrow \quad \iint_{\mathcal{D}} \sqrt{x^3 + 1}\, dA \le 3 \cdot \frac{1}{2} = \frac{3}{2}$$

55. Prove the inequality $\displaystyle\iint_{\mathcal{D}} \frac{dA}{4 + x^2 + y^2} \le \pi$, where $\mathcal{D}$ is the disk $x^2 + y^2 \le 4$.

SOLUTION   The function $f(x, y) = \frac{1}{4+x^2+y^2}$ satisfies

$$f(x, y) = \frac{1}{4 + x^2 + y^2} \le \frac{1}{4}$$

Also, the area of the disk is

$$\text{Area}(\mathcal{D}) = \pi \cdot 2^2 = 4\pi$$

Therefore, by Theorem 3, we have

$$\iint_{\mathcal{D}} \frac{dA}{4 + x^2 + y^2} \le \frac{1}{4} \cdot 4\pi = \pi.$$

57. Let $\overline{f}$ be the average of $f(x, y) = xy^2$ on $\mathcal{D} = [0, 1] \times [0, 4]$. Find a point $P \in \mathcal{D}$ such that $f(P) = \overline{f}$ (the existence of such a point is guaranteed by the Mean Value Theorem for Double Integrals).

SOLUTION   We first compute the average $\overline{f}$ of $f(x, y) = xy^2$ on $\mathcal{D}$.

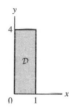

$\overline{f}$ is

$$\overline{f} = \frac{1}{\text{Area}(\mathcal{D})} \iint_{\mathcal{D}} xy^2\, dA = \frac{1}{4 \cdot 1} \int_0^1 \int_0^4 xy^2\, dy\, dx = \frac{1}{4} \int_0^1 \left. \frac{xy^3}{3} \right|_{y=0}^4 dx$$

$$= \frac{1}{4} \int_0^1 \left(\frac{x \cdot 4^3}{3} - \frac{x \cdot 0^3}{3} \right) dx = \int_0^1 \frac{16x}{3}\, dx = \left. \frac{8x^2}{3} \right|_0^1 = \frac{8}{3}$$

We now must find a point $P = (a, b)$ in $\mathcal{D}$ such that

$$f(P) = ab^2 = \frac{8}{3}$$

We choose $b = 2$, obtaining

$$a \cdot 2^2 = \frac{8}{3} \quad \Rightarrow \quad a = \frac{2}{3}$$

The point $P = \left(\frac{2}{3}, 2 \right)$ in the rectangle $\mathcal{D}$ satisfies

$$f(P) = \overline{f} = \frac{8}{3}$$

In Exercises 59 and 60, use (11) to estimate the double integral.

59. The following table lists the areas of the subdomains $\mathcal{D}_j$ of the domain $\mathcal{D}$ in Figure 30 and the values of a function $f(x, y)$ at sample points $P_j \in \mathcal{D}_j$. Estimate $\iint_{\mathcal{D}} f(x, y)\, dA$.

j	1	2	3	4	5	6
Area($\mathcal{D}_j$)	1.2	1.1	1.4	0.6	1.2	0.8
$f(P_j)$	9	9.1	9.3	9.1	8.9	8.8

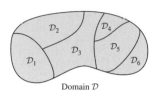

Domain $\mathcal{D}$

FIGURE 30

SOLUTION   By Eq. (11) we have

$$\iint_{\mathcal{D}} f(x, y)\, dA \approx \sum_{j=1}^{6} f\left(P_j\right) \text{Area}\left(\mathcal{D}_j\right)$$

Substituting the data given in the table, we obtain

$$\iint_{\mathcal{D}} f(x, y)\, dA \approx 9 \cdot 1.2 + 9.1 \cdot 1.1 + 9.3 \cdot 1.4 + 9.1 \cdot 0.6 + 8.9 \cdot 1.2 + 8.8 \cdot 0.8 = 57.01$$

Thus,

$$\iint_{\mathcal{D}} f(x, y)\, dA \approx 57.01$$

61. According to Eq. (3), the area of a domain $\mathcal{D}$ is equal to $\iint_{\mathcal{D}} 1\, dA$. Prove that if $\mathcal{D}$ is the region between two curves $y = g_1(x)$ and $y = g_2(x)$ with $g_2(x) \le g_1(x)$ for $a \le x \le b$, then

$$\iint_{\mathcal{D}} 1\, dA = \int_a^b (g_1(x) - g_2(x))\, dx$$

SOLUTION   The region $\mathcal{D}$ is defined by the inequalities

$$a \le x \le b, \quad g_1(x) \le y \le g_2(x)$$

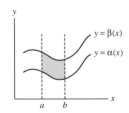

We compute the double integral of $f(x, y) = 1$ on $\mathcal{D}$, using Theorem 2, by evaluating the following iterated integral:

$$\int_{\mathcal{D}} 1\, dA = \int_a^b \int_{g_1(x)}^{g_2(x)} 1\, dy\, dx = \int_a^b \left(\int_{g_1(x)}^{g_2(x)} 1\, dy \right) dx = \int_a^b y \Big|_{y=g_1(x)}^{g_2(x)} dx = \int_a^b (g_2(x) - g_1(x))\, dx$$

Further Insights and Challenges

63. Use the fact that a continuous function on a closed domain $\mathcal{D}$ attains both a minimum value m and a maximum value M, together with Theorem 3, to prove that the average value $\overline{f}$ lies between m and M. Then use the IVT in Exercise 62 to prove the Mean Value Theorem for Double Integrals.

SOLUTION Suppose that $f(x, y)$ is continuous and $\mathcal{D}$ is closed, bounded, and connected. By Theorem 3 in Chapter 15.7 ("Existence of Global Extrema"), $f(x, y)$ takes on a minimum value (call it m) at some point (x_m, y_m) and a maximum value (call it M) at some point (x_M, y_M) in the domain $\mathcal{D}$. Now, by Theorem 3,

$$m \text{ Area}(\mathcal{D}) \le \iint_{\mathcal{D}} f(x, y) \, dA \le M \text{ Area}(\mathcal{D})$$

which can be restated as

$$m \le \frac{1}{\text{Area}(\mathcal{D})} \iint_{\mathcal{D}} f(x, y) \, dA \le M$$

By the IVT in two variables (stated and proved in the previous problem), $f(x, y)$ takes on every value between m and M at some point in $\mathcal{D}$. In particular, f must take on the value $\frac{1}{\text{Area}(\mathcal{D})} \iint_{\mathcal{D}} f(x, y) \, dA$ at some point P. So, $f(P) = \frac{1}{\text{Area}(\mathcal{D})} \iint_{\mathcal{D}} f(x, y) \, dA$, which is rewritten as

$$f(P) \text{ Area}(\mathcal{D}) = \iint_{\mathcal{D}} f(x, y) \, dA$$

15.3 Triple Integrals (LT Section 16.3)

Preliminary Questions

1. Which of (a)–(c) is not equal to $\int_0^1 \int_3^4 \int_6^7 f(x, y, z) \, dz \, dy \, dx$?

(a) $\int_6^7 \int_0^1 \int_3^4 f(x, y, z) \, dy \, dx \, dz$

(b) $\int_3^4 \int_0^1 \int_6^7 f(x, y, z) \, dz \, dx \, dy$

(c) $\int_0^1 \int_3^4 \int_6^7 f(x, y, z) \, dx \, dz \, dy$

SOLUTION The given integral, I, is a triple integral of f over the box $B = [0, 1] \times [3, 4] \times [6, 7]$. In (a) the limits of integration are $0 \le x \le 1$, $3 \le y \le 4$, $6 \le z \le 7$, hence this integral is equal to I. In (b) the limits of integration are $0 \le x \le 1$, $3 \le y \le 4$, $6 \le z \le 7$, hence it is also equal to I. In (c) the limits of integration are $6 \le x \le 7$, $0 \le y \le 1$, $3 \le z \le 4$. This is the triple integral of f over the box $[6, 7] \times [0, 1] \times [3, 4]$, which is different from B. Therefore, the triple integral is usually unequal to I.

2. Which of the following is not a meaningful triple integral?

(a) $\int_0^1 \int_0^x \int_{x+y}^{2x+y} e^{x+y+z} \, dz \, dy \, dx$

(b) $\int_0^1 \int_0^z \int_{x+y}^{2x+y} e^{x+y+z} \, dz \, dy \, dx$

SOLUTION

(a) The limits of integration determine the following inequalities:

$$0 \le x \le 1, \quad 0 \le y \le x, \quad x + y \le z \le 2x + y$$

The integration is over the simple region $\mathcal{W}$, which lies between the planes $z = x + y$ and $z = 2x + y$ over the domain $\mathcal{D}_1 = \{(x, y) : 0 \le x \le 1, \ 0 \le y \le x\}$ in the xy-plane.

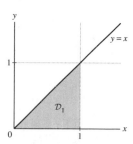

Thus, the integral represents a meaningful triple integral.

(b) Note that the inner integral is with respect to z, but then the middle integral has limits from 0 to z! This makes no sense.

3. Describe the projection of the region of integration $\mathcal{W}$ onto the xy-plane:

(a) $\displaystyle\int_0^1 \int_0^x \int_0^{x^2+y^2} f(x, y, z)\, dz\, dy\, dx$

(b) $\displaystyle\int_0^1 \int_0^{\sqrt{1-x^2}} \int_2^4 f(x, y, z)\, dz\, dy\, dx$

SOLUTION

(a) The region of integration is defined by the limits of integration, yielding the following inequalities:

$$0 \le x \le 1, \quad 0 \le y \le x, \quad 0 \le z \le x^2 + y^2$$

$\mathcal{W}$ is the region between the paraboloid $z = x^2 + y^2$ and the xy-plane which is above the triangle $\mathcal{D} = \{(x, y) : 0 \le x \le 1, 0 \le y \le x\}$ in the xy-plane. This triangle is the projection of $\mathcal{W}$ onto the xy-plane.

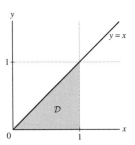

(b) The inequalities determined by the limits of integration are

$$0 \le x \le 1, \quad 0 \le y \le \sqrt{1-x^2}, \quad 2 \le z \le 4$$

This is the region between the planes $z = 2$ and $z = 4$, which is above the region $\mathcal{D} = \left\{(x, y) : 0 \le x \le 1, \ 0 \le y \le \sqrt{1-x^2}\right\}$ in the xy-plane. The projection $\mathcal{D}$ of $\mathcal{W}$ onto the xy-plane is the part of the unit disk in the first quadrant.

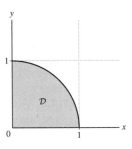

Exercises

In Exercises 1–8, evaluate $\displaystyle\iiint_{\mathcal{B}} f(x, y, z)\, dV$ *for the specified function f and box $\mathcal{B}$.*

1. $f(x, y, z) = z^4; \quad 2 \le x \le 8, \quad 0 \le y \le 5, \quad 0 \le z \le 1$

SOLUTION   We write the triple integral as an iterated integral and compute it to obtain

$$\iiint_{\mathcal{B}} z^4\, dV = \int_2^8 \int_0^5 \int_0^1 z^4\, dz\, dy\, dx = \int_2^8 \int_0^5 \left(\int_0^1 z^4\, dz \right) dy\, dx = \int_2^8 \int_0^5 \frac{1}{5} z^5 \Big|_{z=0}^{1} dy\, dx$$

$$= \int_2^8 \int_0^5 \frac{1}{5}\, dy\, dx = \frac{1}{5} \int_2^8 \int_0^5 dy\, dx = \frac{1}{5} \cdot 6 \cdot 5 = 6$$

3. $f(x, y, z) = xe^{y-2z}; \quad 0 \le x \le 2, \quad 0 \le y \le 1, \quad 0 \le z \le 1$

SOLUTION   We write the triple integral as an iterated integral. Since $f(x, y, z) = xe^y \cdot e^{-2z}$, we may evaluate the iterated integral as the product of three single integrals. We get

$$\iiint_{\mathcal{B}} xe^{y-2z}\, dV = \int_0^2 \int_0^1 \int_0^1 xe^{y-2z}\, dz\, dy\, dx = \left(\int_0^2 x\, dx \right)\left(\int_0^1 e^y\, dy \right)\left(\int_0^1 e^{-2z}\, dz \right)$$

$$= \left(\frac{1}{2} x^2 \Big|_0^2 \right)\left(e^y \Big|_0^1 \right)\left(-\frac{1}{2} e^{-2z} \Big|_0^1 \right) = 2(e-1) \cdot -\frac{1}{2}(e^{-2} - 1) = (e-1)(1 - e^{-2})$$

5. $f(x, y, z) = (x - y)(y - z);$ $[0, 1] \times [0, 3] \times [0, 3]$

SOLUTION   We write the triple integral as an iterated integral and evaluate the inner, middle, and outer integrals successively. This gives

$$\iiint_B (x - y)(y - z) \, dV = \int_0^1 \int_0^3 \int_0^3 (x - y)(y - z) \, dz \, dy \, dx = \int_0^1 \int_0^3 \left(\int_0^3 (x - y)(y - z) \, dz \right) dy \, dx$$

$$= \int_0^1 \int_0^3 (x - y) \left(yz - \frac{1}{2} z^2 \right) \Big|_{z=0}^3 dy \, dx = \int_0^1 \int_0^3 (x - y) \left(3y - \frac{9}{2} \right) dy \, dx$$

$$= \int_0^1 \int_0^3 \left(\left(3x + \frac{9}{2} \right) y - \frac{9}{2} x - 3y^2 \right) dy \, dx = \int_0^1 \left(\frac{3}{2} x + \frac{9}{4} \right) y^2 - \frac{9}{2} xy - y^3 \Big|_{y=0}^3 dx$$

$$= \int_0^1 \left(\left(\frac{3}{2} x + \frac{9}{4} \right) \cdot 9 - \frac{9}{2} x \cdot 3 - 27 \right) dx = \int_0^1 -\frac{27}{4} \, dx = -\frac{27}{4} = -6.75$$

7. $f(x, y, z) = (x + z)^3;$ $[0, a] \times [0, b] \times [0, c]$

SOLUTION   We write the triple integral as an iterated integral and evaluate it to obtain

$$\iiint_B f(x, y, z) \, dV = \int_0^a \int_0^b \int_0^c (x + z)^3 \, dz \, dy \, dx = \int_0^a \int_0^b \frac{(x + z)^4}{4} \Big|_{z=0}^c dy \, dx$$

$$= \int_0^a \int_0^b \left(\frac{(x + c)^4}{4} - \frac{x^4}{4} \right) dy \, dx = \int_0^a \frac{(x + c)^4 - x^4}{4} y \Big|_{y=0}^b dx$$

$$= \int_0^a \frac{b}{4} \left[(x + c)^4 - x^4 \right] dx = \frac{b}{4} \left[\frac{(x + c)^5}{5} - \frac{x^5}{5} \right] \Big|_{x=0}^a$$

$$= \frac{b}{4} \frac{(a + c)^5 - a^5 - c^5}{5} = \frac{b}{20} \left[(a + c)^5 - a^5 - c^5 \right]$$

In Exercises 9–14, evaluate $\iiint_W f(x, y, z) \, dV$ *for the function f and region W specified.*

9. $f(x, y, z) = x + y;$ $W : y \le z \le x,$ $0 \le y \le x,$ $0 \le x \le 1$

SOLUTION W is the region between the planes $z = y$ and $z = x$ lying over the triangle $\mathcal{D}$ in the xy-plane defined by the inequalities $0 \le y \le x, 0 \le x \le 1$.

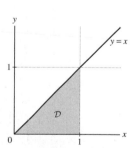

We compute the integral, using Theorem 2, by evaluating the following iterated integral:

$$\iiint_W (x + y) \, dV = \iint_{\mathcal{D}} \left(\int_y^x (x + y) \, dz \right) dA = \iint_{\mathcal{D}} (x + y)z \Big|_{z=y}^x dA = \iint_{\mathcal{D}} (x + y)(x - y) \, dA$$

$$= \iint_{\mathcal{D}} \left(x^2 - y^2 \right) dA = \int_0^1 \int_0^x \left(x^2 - y^2 \right) dy \, dx = \int_0^1 \left(\int_0^x \left(x^2 - y^2 \right) dy \right) dx$$

$$= \int_0^1 x^2 y - \frac{y^3}{3} \Big|_{y=0}^x dx = \int_0^1 \frac{2x^3}{3} \, dx = \frac{2}{12} x^4 \Big|_0^1 = \frac{1}{6}$$

11. $f(x, y, z) = xyz$; $\mathcal{W} : 0 \leq z \leq 1, \quad 0 \leq y \leq \sqrt{1 - x^2}, \quad 0 \leq x \leq 1$

SOLUTION $\mathcal{W}$ is the region between the planes $z = 0$ and $z = 1$, lying over the part $\mathcal{D}$ of the disk in the first quadrant.

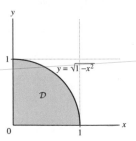

Using Theorem 2, we compute the triple integral as the following iterated integral:

$$\iiint_{\mathcal{W}} xyz \, dV = \iint_{\mathcal{D}} \left(\int_0^1 xyz \, dz \right) dA = \iint_{\mathcal{D}} \left. \frac{xyz^2}{2} \right|_{z=0}^1 dA = \iint_{\mathcal{D}} \frac{xy}{2} \, dA$$

$$= \int_0^1 \left(\int_0^{\sqrt{1-x^2}} \frac{xy}{2} \, dy \right) dx = \int_0^1 \left. \frac{xy^2}{4} \right|_{y=0}^{\sqrt{1-x^2}} dx = \int_0^1 \frac{x(1 - x^2)}{4} \, dx$$

$$= \int_0^1 \frac{x - x^3}{4} \, dx = \left. \frac{x^2}{8} - \frac{x^4}{16} \right|_0^1 = \frac{1}{8} - \frac{1}{16} = \frac{1}{16}$$

13. $f(x, y, z) = e^z$; $\quad \mathcal{W} : x + y + z \leq 1, \quad x \geq 0, \quad y \geq 0, \quad z \geq 0$

SOLUTION   Notice that $\mathcal{W}$ is the tetrahedron under the plane $x + y + z = 1$ above the first quadrant.

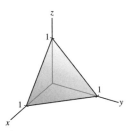

First, we must determine the projection $\mathcal{D}$ of $\mathcal{W}$ onto the xy-plane. The intersection of the plane $x + y + z = 1$ with the xy-plane is obtained by solving

$$\begin{array}{c} x + y + z = 1 \\ z = 0 \end{array} \quad \Rightarrow \quad x + y = 1$$

Therefore, the projection $\mathcal{D}$ of $\mathcal{W}$ onto the xy-plane is the triangle enclosed by the line $x + y = 1$ and the positive axes.

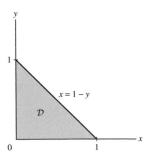

The region $\mathcal{W}$ is the region between the planes $z = 1 - x - y$ and $z = 0$, lying above the triangle $\mathcal{D}$ in the xy-plane. The triple integral can be written as the following iterated integral:

$$\iiint_{\mathcal{W}} e^z \, dV = \iint_{\mathcal{D}} \left(\int_0^{1-x-y} e^z \, dz \right) dA = \iint_{\mathcal{D}} \left. e^z \right|_{z=0}^{1-x-y} dA$$

$$= \iint_{\mathcal{D}} \left(e^{1-x-y} - 1 \right) dA$$

$$= \int_0^1 \left(\int_0^{1-y} \left(e^{1-x-y} - 1 \right) dx \right) dy = \int_0^1 -e^{1-x-y} - x \Big|_{x=0}^{1-y} dy$$

$$= \int_0^1 -e^{1-1+y-y} - (1-y) + e^{1-y} \, dy$$

$$= \int_0^1 e^{1-y} + y - 2 \, dy = -e^{1-y} + \frac{1}{2}y^2 - 2y \Big|_{y=0}^1$$

$$= -1 + \frac{1}{2} - 2 - \left(-e^1\right) = e - \frac{5}{2}$$

15. Calculate the integral of $f(x, y, z) = z$ over the region $\mathcal{W}$ in Figure 10 below the hemisphere of radius 3 and lying over the triangle $\mathcal{D}$ in the xy-plane bounded by $x = 1$, $y = 0$, and $x = y$.

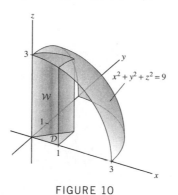

FIGURE 10

SOLUTION

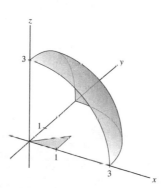

The upper surface is the hemisphere $z = \sqrt{9 - x^2 - y^2}$ and the lower surface is the xy-plane $z = 0$. The projection of $\mathcal{V}$ onto the xy-plane is the triangle $\mathcal{D}$ shown in the figure.

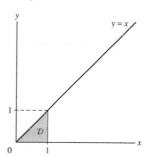

We compute the triple integral as the following iterated integral:

$$\iiint_{\mathcal{V}} z \, dV = \iint_{\mathcal{D}} \left(\int_0^{\sqrt{9-x^2-y^2}} z \, dz \right) dA = \iint_{\mathcal{D}} \frac{z^2}{2} \Big|_0^{\sqrt{9-x^2-y^2}} dA = \iint_{\mathcal{D}} \frac{9 - x^2 - y^2}{2} \, dA$$

$$= \int_0^1 \left(\int_0^x \frac{9 - x^2 - y^2}{2} \, dy \right) dx = \int_0^1 \frac{9y - x^2y - \frac{y^3}{3}}{2} \Big|_{y=0}^x dx = \int_0^1 \left(\frac{9x}{2} - \frac{2x^3}{3} \right) dx$$

$$= \frac{9x^2}{4} - \frac{x^4}{6} \Big|_0^1 = 2\frac{1}{12}$$

17. Integrate $f(x, y, z) = x$ over the region in the first octant ($x \geq 0, y \geq 0, z \geq 0$) above $z = y^2$ and below $z = 8 - 2x^2 - y^2$.

SOLUTION   We first find the projection of the region $\mathcal{W}$ onto the xy-plane. We find the curve of intersection between the upper and lower surfaces, by solving the following equation for $x, y \geq 0$:

$$8 - 2x^2 - y^2 = y^2 \quad \Rightarrow \quad y^2 = 4 - x^2 \quad \Rightarrow \quad y = \sqrt{4 - x^2}, x \geq 0$$

The projection $\mathcal{D}$ of $\mathcal{W}$ onto the xy-plane is the region bounded by the circle $x^2 + y^2 = 4$ and the positive axes.

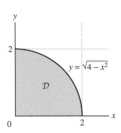

We now compute the triple integral over $\mathcal{W}$ by evaluating the following iterated integral:

$$\iiint_{\mathcal{W}} x \, dV = \iint_{\mathcal{D}} \left(\int_{y^2}^{8-2x^2-y^2} x \, dz \right) dA = \iint_{\mathcal{D}} xz \Big|_{z=y^2}^{8-2x^2-y^2} dA$$

$$= \iint_{\mathcal{D}} x(8 - 2x^2 - y^2 - y^2) \, dA = \iint_{\mathcal{D}} 8x - 2x^3 - 2xy^2 \, dA$$

$$= \int_0^2 \int_0^{\sqrt{4-x^2}} 8x - 2x^3 - 2xy^2 \, dy \, dx = \int_0^2 8xy - 2x^3y - x^2y^2 \Big|_{y=0}^{\sqrt{4-x^2}} dx$$

$$= \int_0^2 8x\sqrt{4 - x^2} \, dx - \int_0^2 2x^3\sqrt{4 - x^2} \, dx - \int_0^2 x^2(4 - x^2) \, dx$$

$$= 8 \int_0^2 x\sqrt{4 - x^2} \, dx - 2 \int_0^2 x^3\sqrt{4 - x^2} \, dx - \int_0^2 4x^2 - x^4 \, dx$$

The first and third integrals are easily computing using u-substitution and term by term integration, respectively. The second integral requires a clever u-substitution, let $u = 4 - x^2$, then $du = -2x \, dx$ and $x^2 = 4 - u$. Using this information we see

$$-2 \int_0^2 x^3\sqrt{4 - x^2} \, dx = -2 \int_0^2 x \cdot x^2\sqrt{4 - x^2} \, dx$$

$$= \int_{u=4}^0 (4 - u)\sqrt{u} \, du$$

$$= \int_4^0 4\sqrt{u} - u^{3/2} \, du$$

$$= \frac{8}{3}u^{3/2} - \frac{2}{5}u^{5/2} \Big|_{u=4}^0$$

$$= -\left(\frac{64}{3} - \frac{64}{5} \right) = -\frac{128}{15}$$

Hence,

$$\iiint_{\mathcal{W}} x \, dV = 8 \int_0^2 x\sqrt{4 - x^2} \, dx - 2 \int_0^2 x^3\sqrt{4 - x^2} \, dx - \int_0^2 4x^2 - x^4 \, dx$$

$$= 8 \int_0^2 x\sqrt{4 - x^2} \, dx - \frac{128}{15} - \int_0^2 4x^2 - x^4 \, dx$$

$$= 8 \cdot -\frac{1}{3}(4 - x^2)^{3/2} \Big|_0^2 - \frac{128}{15} - \left(\frac{4}{3}x^3 - \frac{1}{5}x^5 \Big|_0^2 \right)$$

$$= -\frac{8}{3}(0 - 8) - \frac{128}{15} - \left(\frac{32}{3} - \frac{32}{5} \right) = \frac{128}{15}$$

19. Find the triple integral of the function z over the ramp in Figure 12. Here, z is the height above the ground.

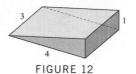

FIGURE 12

SOLUTION   We place the coordinate axes as shown in the figure:

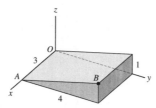

The upper surface is the plane passing through the points $O = (0, 0, 0)$, $A = (3, 0, 0)$, and $B = (3, 4, 1)$. We find a normal to this plane and then determine the equation of the plane. We get

$$\overrightarrow{OA} \times \overrightarrow{AB} = \langle 3, 0, 0 \rangle \times \langle 0, 4, 1 \rangle = \begin{vmatrix} \mathbf{i} & \mathbf{j} & \mathbf{k} \\ 3 & 0 & 0 \\ 0 & 4 & 1 \end{vmatrix} = -3\mathbf{j} + 12\mathbf{k} = 3\,(-\mathbf{j} + 4\mathbf{k})$$

The plane is orthogonal to the vector $\langle 0, -1, 4 \rangle$ and passes through the origin, hence the equation of the plane is

$$0 \cdot x - y + 4z = 0 \quad \Rightarrow \quad z = \frac{y}{4}$$

The projection of the region of integration $\mathcal{W}$ onto the xy-plane is the rectangle $\mathcal{D}$ defined by

$$0 \le x \le 3, \quad 0 \le y \le 4.$$

We now compute the triple integral of f over $\mathcal{W}$, as the following iterated integral:

$$\iiint_{\mathcal{W}} z \, dV = \iint_{\mathcal{D}} \left(\int_0^{y/4} z \, dz \right) dA = \iint_{\mathcal{D}} \frac{z^2}{2} \bigg|_{z=0}^{y/4} dA = \iint_{\mathcal{D}} \frac{y^2}{32} \, dA = \int_0^4 \left(\int_0^3 \frac{y^2}{32} \, dx \right) dy$$

$$= \int_0^4 \frac{y^2 x}{32} \bigg|_{x=0}^{3} dy = \int_0^4 \frac{3y^2}{32} \, dy = \frac{y^3}{32} \bigg|_0^4 = \frac{4^3}{32} = 2$$

21. Find the volume of the solid in the octant $x \ge 0$, $y \ge 0$, $z \ge 0$ bounded by $x + y + z = 1$ and $x + y + 2z = 1$.

SOLUTION   The solid $\mathcal{W}$ is shown in the figure:

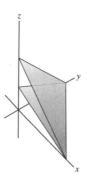

The upper and lower surfaces are the planes $x + y + z = 1$ (or $z = 1 - x - y$) and $x + y + 2z = 1$ (or $z = \frac{1-x-y}{2}$), respectively. The projection of $\mathcal{W}$ onto the xy-plane is the triangle enclosed by the line $AB : y = 1 - x$ and the positive x and y-axes.

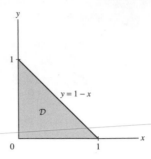

Using the volume of a solid as a triple integral, we have

$$\text{Volume}(\mathcal{W}) = \iiint_{\mathcal{W}} 1\, dV = \iint_{\mathcal{D}} \left(\int_{(1-x-y)/2}^{1-x-y} 1\, dz \right) dA = \iint_{\mathcal{D}} z \Big|_{z=(1-x-y)/2}^{1-x-y} dA$$

$$= \iint_{\mathcal{D}} \left((1-x-y) - \frac{1-x-y}{2} \right) dA = \iint_{\mathcal{D}} \frac{1-x-y}{2}\, dA$$

$$= \int_0^1 \left(\int_0^{1-x} \frac{1-x-y}{2}\, dy \right) dx = \int_0^1 \frac{y - xy - \frac{y^2}{2}}{2} \Big|_{y=0}^{1-x} dx$$

$$= \int_0^1 \frac{1 - x - x(1-x) - \frac{(1-x)^2}{2}}{2}\, dx = \frac{1}{2} \int_0^1 \left(\frac{x^2}{2} - x + \frac{1}{2} \right) dx$$

$$= \frac{1}{2} \left(\frac{x^3}{6} - \frac{x^2}{2} + \frac{1}{2}x \right) \Big|_0^1 = \frac{1}{2} \left(\frac{1}{6} - \frac{1}{2} + \frac{1}{2} \right) = \frac{1}{12}$$

23. Evaluate $\iiint_{\mathcal{W}} xz\, dV$, where $\mathcal{W}$ is the domain bounded by the elliptic cylinder $\dfrac{x^2}{4} + \dfrac{y^2}{9} = 1$ and the sphere $x^2 + y^2 + z^2 = 16$ in the first octant $x \geq 0,\, y \geq 0,\, z \geq 0$ (Figure 13).

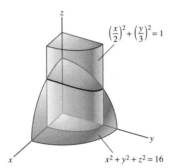

FIGURE 13

SOLUTION

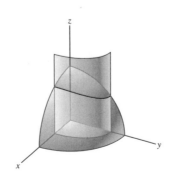

The upper surface is the sphere $x^2 + y^2 + z^2 = 16$, or $z = \sqrt{16 - x^2 - y^2}$, and the lower surface is the xy-plane, $z = 0$. The projection of $\mathcal{W}$ onto the xy-plane is the region in the first quadrant bounded by the ellipse $(x/2)^2 + (y/3)^2 = 1$, or $y = \frac{3}{2}\sqrt{4 - x^2}$.

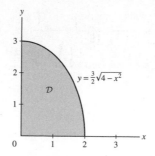

Therefore, the triple integral over $\mathcal{W}$ is equal to the following iterated integral:

$$\iiint_{\mathcal{W}} xz\, dV = \iint_{\mathcal{D}} \left(\int_0^{\sqrt{16-x^2-y^2}} xz\, dz \right) dA = \iint_{\mathcal{D}} \frac{1}{2} xz^2 \Big|_{z=0}^{\sqrt{16-x^2-y^2}}\, dA$$

$$= \frac{1}{2} \iint_{\mathcal{D}} x(16 - x^2 - y^2)\, dA = \frac{1}{2} \int_0^2 \int_0^{\frac{3}{2}\sqrt{4-x^2}} 16x - x^3 - xy^2\, dy\, dx$$

$$= \frac{1}{2} \int_0^2 16xy - x^3 y - \frac{1}{3} xy^3 \Big|_{y=0}^{\frac{3}{2}\sqrt{4-x^2}}\, dx$$

$$= \frac{1}{2} \int_0^2 24x\sqrt{4 - x^2} - \frac{3}{2} x^3 \sqrt{4 - x^2} - \frac{9}{8} x(4 - x^2)^{3/2}\, dx$$

$$= 12 \int_0^2 x\sqrt{4 - x^2}\, dx - \frac{3}{4} \int_0^2 x^3 \sqrt{4 - x^2}\, dx - \frac{9}{16} \int_0^2 x(4 - x^2)^{3/2}\, dx$$

The first and third integrals are simple u-substitution problems. For the second integral, let us use $u = 4 - x^2$ and thus $du = -2x\, dx$ and $x^2 = 4 - u$. Thus, we can write

$$-\frac{3}{4} \int_0^2 x^3 \sqrt{4 - x^2}\, dx = -\frac{3}{4} \int_0^2 x \cdot x^2 \sqrt{4 - x^2}\, dx$$

$$= \frac{3}{8} \int_{u=4}^0 (4 - u)\sqrt{u}\, du$$

$$= \frac{3}{8} \int_4^0 4u^{1/2} - u^{3/2}\, du$$

$$= \frac{3}{8} \left(\frac{8}{3} u^{3/2} - \frac{2}{5} u^{5/2} \Big|_{u=4}^0 \right)$$

$$= -\frac{3}{8} \left(\frac{64}{3} - \frac{64}{5} \right) = -\frac{16}{5}$$

Therefore we have:

$$\iiint_{\mathcal{W}} xz\, dV = 12 \int_0^2 x\sqrt{4 - x^2}\, dx - \frac{3}{4} \int_0^2 x^3 \sqrt{4 - x^2}\, dx - \frac{9}{16} \int_0^2 x(4 - x^2)^{3/2}$$

$$= 12 \int_0^2 x\sqrt{4 - x^2}\, dx - \frac{16}{5} - \frac{9}{16} \int_0^2 x(4 - x^2)^{3/2}\, dx$$

$$= -6 \cdot \frac{2}{3} \left((4 - x^2)^{3/2} \Big|_0^2 \right) - \frac{16}{5} + \frac{9}{32} \cdot \frac{2}{5} \left((4 - x^2)^{5/2} \Big|_0^2 \right)$$

$$= -4 (0 - 8) - \frac{16}{5} + \frac{9}{80} (0 - 32) = \frac{126}{5}$$

25. Describe the domain of integration of the following integral:

$$\int_{-2}^2 \int_{-\sqrt{4-z^2}}^{\sqrt{4-z^2}} \int_1^{\sqrt{5-x^2-z^2}} f(x, y, z)\, dy\, dx\, dz$$

SOLUTION The domain of integration of $\mathcal{W}$ is defined by the following inequalities:

$$-2 \le z \le 2, \quad -\sqrt{4 - z^2} \le x \le \sqrt{4 - z^2}, \quad 1 \le y \le \sqrt{5 - x^2 - z^2}$$

This region is bounded by the plane $y = 1$ and the sphere $y^2 = 5 - x^2 - z^2$ or $x^2 + y^2 + z^2 = 5$, lying over the disk $x^2 + z^2 \leq 4$ in the xz-plane. This is the central cylinder oriented along the y-axis of radius 2 inside a sphere of radius $\sqrt{5}$.

27. In Example 5, we expressed a triple integral as an iterated integral in the three orders

$$dz\,dy\,dx, \quad dx\,dz\,dy, \quad \text{and} \quad dy\,dz\,dx$$

Write this integral in the three other orders:

$$dz\,dx\,dy, \quad dx\,dy\,dz, \quad \text{and} \quad dy\,dx\,dz$$

SOLUTION In Example 5 we considered the triple integral $\iiint_{\mathcal{W}} xyz^2\,dV$, where $\mathcal{W}$ is the region bounded by

$$z = 4 - y^2, \quad z = 0, \quad y = 2x, \quad x = 0.$$

We now write the triple integral as an iterated integral in the orders $dz\,dx\,dy$, $dx\,dy\,dz$, and $dy\,dx\,dz$.

- $dz\,dx\,dy$: The upper surface $z = 4 - y^2$ projects onto the xy-plane on the triangle defined by the lines $y = 2$, $y = 2x$, and $x = 0$.

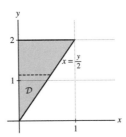

We express the line $y = 2x$ as $x = \frac{y}{2}$ and write the triple integral as

$$\iiint_{\mathcal{W}} xyz^2\,dV = \iint_{\mathcal{D}} \left(\int_0^{4-y^2} xyz^2\,dz \right) dA = \int_0^2 \int_0^{y/2} \int_0^{4-y^2} xyz^2\,dz\,dx\,dy$$

- $dx\,dy\,dz$: The projection of $\mathcal{W}$ onto the yz-plane is the domain T (see Example 5) defined by the inequalities

$$T : 0 \leq y \leq 2, \ 0 \leq z \leq 4 - y^2$$

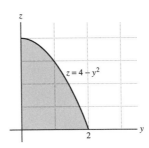

This region can also be expressed as

$$0 \leq z \leq 4, \quad 0 \leq y \leq \sqrt{4-z}$$

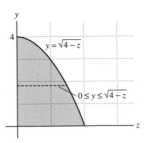

As explained in Example 5, the region $\mathcal{W}$ consists of all points lying between T and the "left-face" $y = 2x$, or $x = \frac{y}{2}$. Therefore, we obtain the following inequalities for $\mathcal{W}$:

$$0 \leq z \leq 4, \quad 0 \leq y \leq \sqrt{4-z}, \quad 0 \leq x \leq \frac{1}{2}y$$

This yields the following iterated integral:

$$\iiint_{\mathcal{W}} xyz^2 \, dV = \iint_{T} \left(\int_{0}^{y/2} xyz^2 \, dx \right) dA = \int_{0}^{4} \int_{0}^{\sqrt{4-z}} \int_{0}^{y/2} xyz^2 \, dx \, dy \, dz$$

- $dy \, dx \, dz$: As explained in Example 5, the projection of $\mathcal{W}$ onto the xz-plane is determined by the inequalities

$$S : 0 \leq x \leq 1, \ 0 \leq z \leq 4 - 4x^2$$

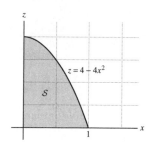

This region can also be described if we write x as a function of z:

$$z = 4 - 4x^2 \quad \Rightarrow \quad 4x^2 = 4 - z \quad \Rightarrow \quad x = \sqrt{1 - \frac{z}{4}}$$

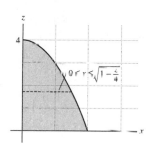

This gives the following inequalities of S:

$$S : 0 \leq z \leq 4, \ 0 \leq x \leq \sqrt{1 - \frac{z}{4}}$$

The upper surface $z = 4 - y^2$ can be described by $y = \sqrt{4 - z}$, hence the limits of y are $2x \leq y \leq \sqrt{4 - z}$. We obtain the following iterated integral.

$$\iiint_{\mathcal{W}} xyz^2 \, dV = \iint_{S} \left(\int_{2x}^{\sqrt{4-z}} xyz^2 \, dy \right) dA = \int_{0}^{4} \int_{0}^{\sqrt{1-\frac{z}{4}}} \int_{2x}^{\sqrt{4-z}} xyz^2 \, dy \, dx \, dz$$

29. Let

$$\mathcal{W} = \left\{ (x, y, z) : \sqrt{x^2 + y^2} \leq z \leq 1 \right\}$$

(see Figure 15). Express $\iiint_{\mathcal{W}} f(x, y, z) \, dV$ as an iterated integral in the order $dz \, dy \, dx$ (for an arbitrary function f).

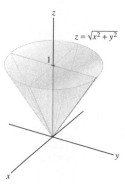

FIGURE 15

SOLUTION To express the triple integral as an iterated integral in order $dz\,dy\,dx$, we must find the projection of $\mathcal{W}$ onto the xy-plane. The upper circle is $\sqrt{x^2 + y^2} = 1$ or $x^2 + y^2 = 1$, hence the projection of $\mathcal{W}$ onto the xy plane is the disk

$$\mathcal{D}: x^2 + y^2 \le 1$$

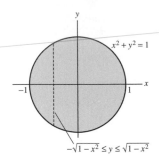

The upper surface is the plane $z = 1$ and the lower surface is $z = \sqrt{x^2 + y^2}$, therefore the triple integral over $\mathcal{W}$ is equal to the following iterated integral:

$$\iiint_{\mathcal{W}} f(x, y, z)\,dV = \iint_{\mathcal{D}} \left(\int_{\sqrt{x^2+y^2}}^{1} f(x, y, z)\,dz \right) dA = \int_{-1}^{1} \int_{-\sqrt{1-x^2}}^{\sqrt{1-x^2}} \int_{\sqrt{x^2+y^2}}^{1} f(x, y, z)\,dz\,dy\,dx$$

31. Let $\mathcal{W}$ be the region bounded by $z = 1 - y^2$, $y = x^2$, and the planes $z = 0$, $y = 1$. Calculate the volume of $\mathcal{W}$ as a triple integral in the order $dz\,dy\,dx$.

SOLUTION $dz\,dy\,dx$:

The projection of $\mathcal{W}$ onto the xy-plane is the region $\mathcal{D}$ bounded by the curve $y = x^2$ and the line $y = 1$. The region $\mathcal{W}$ consists of all points lying between $\mathcal{D}$ and the cylinder $z = 1 - y^2$.

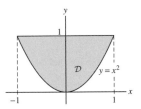

Therefore, $\mathcal{W}$ can be described by the following inequalities:

$$-1 \le x \le 1, \quad x^2 \le y \le 1, \quad 0 \le z \le 1 - y^2$$

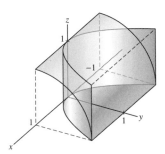

We use the formula for the volume as a triple integral, write the triple integral as an iterated integral, and compute it. We obtain

$$\text{Volume}(\mathcal{W}) = \iiint_{\mathcal{W}} 1\,dV = \int_{-1}^{1} \int_{x^2}^{1} \int_{0}^{1-y^2} 1\,dz\,dy\,dx = \int_{-1}^{1} \int_{x^2}^{1} z \Big|_{z=0}^{1-y^2}\,dy\,dx$$

$$= \int_{-1}^{1} \int_{x^2}^{1} \left(1 - y^2 \right) dy\,dx = \int_{-1}^{1} y - \frac{y^3}{3} \Big|_{y=x^2}^{1}\,dx = \int_{-1}^{1} \left(1 - \frac{1}{3} - \left(x^2 - \frac{x^6}{3} \right) \right) dx$$

$$= 2 \int_{0}^{1} \left(\frac{x^6}{3} - x^2 + \frac{2}{3} \right) dx = 2 \left(\frac{x^7}{21} - \frac{x^3}{3} + \frac{2x}{3} \right) \Big|_{0}^{1} = 2 \left(\frac{1}{21} - \frac{1}{3} + \frac{2}{3} \right) = \frac{16}{21}$$

In Exercises 33–36, compute the average value of $f(x, y, z)$ over the region $\mathcal{W}$.

33. $f(x, y, z) = xy \sin(\pi z);$ $\mathcal{W} = [0, 1] \times [0, 1] \times [0, 1]$

SOLUTION   The volume of the cube is $V = 1$, hence the average of f over the cube is the following value:

$$
\overline{f} = \iiint_{\mathcal{W}} xy \sin(\pi z)\, dV = \int_0^1 \int_0^1 \int_0^1 xy \sin(\pi z)\, dx\, dy\, dz
$$

$$
= \int_0^1 \int_0^1 \frac{1}{2}x^2 y \sin(\pi z)\Big|_{x=0}^1 dy\, dz = \int_0^1 \int_0^1 \frac{1}{2} y \sin(\pi z)\, dy\, dz
$$

$$
= \int_0^1 \frac{y^2}{4} \sin(\pi z)\Big|_{y=0}^1 dz = \int_0^1 \frac{1}{4} \sin(\pi z)\, dz = -\frac{1}{4\pi} \cos(\pi z)\Big|_0^1
$$

$$
= -\frac{1}{4\pi}(\cos \pi - \cos 0) = -\frac{1}{4\pi}(-1 - 1) = \frac{1}{2\pi}
$$

35. $f(x, y, z) = e^y;$ $\mathcal{W} : 0 \le y \le 1 - x^2,$ $0 \le z \le x$

SOLUTION   First we must calculate the volume of $\mathcal{W}$. We will use the symmetry of $y = 1 - x^2$ to write:

$$
V = \iiint_{\mathcal{W}} 1\, dV = 2 \int_0^1 \int_0^{1-x^2} \int_0^x 1\, dz\, dy\, dx
$$

$$
= 2 \int_0^1 \int_0^{1-x^2} z\Big|_{z=0}^x dy\, dx = 2 \int_0^1 \int_0^{1-x^2} x\, dy\, dx
$$

$$
= 2 \int_0^1 xy\Big|_{y=0}^{1-x^2} dx = 2 \int_0^1 x(1 - x^2)\, dx
$$

$$
= 2 \int_0^1 x - x^3\, dx = x^2 - \frac{1}{2}x^4\Big|_0^1 = \frac{1}{2}
$$

Now we can compute the average value for $f(x, y, z) = e^y$:

$$
\overline{f} = \frac{1}{V} \iiint_{\mathcal{W}} e^y\, dV = 2 \cdot \frac{1}{1/2} \int_0^1 \int_0^{1-x^2} \int_0^x e^y\, dz\, dy\, dx
$$

$$
= 4 \int_0^1 \int_0^{1-x^2} z e^y\Big|_{z=0}^x dy\, dx = 4 \int_0^1 \int_0^{1-x^2} x e^y\, dy\, dx
$$

$$
= 4 \int_0^1 x e^y\Big|_{y=0}^{1-x^2} dx = 4 \int_0^1 x e^{1-x^2} - x\, dx
$$

$$
= 4\left(-\frac{1}{2}e^{1-x^2} - \frac{1}{2}x^2\right)\Big|_0^1 = \left(-2e^0 - 2\right) - \left(-2e^1 - 0\right) = 2e - 4
$$

In Exercises 37 and 38, let $I = \int_0^1 \int_0^1 \int_0^1 f(x, y, z)\, dV$ and let $S_{N,N,N}$ be the Riemann sum approximation

$$
S_{N,N,N} = \frac{1}{N^3} \sum_{i=1}^N \sum_{j=1}^N \sum_{k=1}^N f\left(\frac{i}{N}, \frac{j}{N}, \frac{k}{N}\right)
$$

37. CAS   Calculate $S_{N,N,N}$ for $f(x, y, z) = e^{x^2 - y - z}$ for $N = 10, 20, 30$. Then evaluate I and find an N such that $S_{N,N,N}$ approximates I to two decimal places.

SOLUTION   Using a CAS, we get $S_{N,N,N} \approx 0.561, 0.572,$ and 0.576 for $N = 10, 20,$ and 30, respectively. We get $I \approx 0.584$, and using $N = 100$ we get $S_{N,N,N} \approx 0.582$, accurate to two decimal places.

Further Insights and Challenges

39. Use Integration by Parts to verify Eq. (7).

SOLUTION   If $C_n = \int_{-\pi/2}^{\pi/2} \cos^n \theta\, d\theta$, we use integration by parts to show that

$$
C_n = \left(\frac{n-1}{n}\right) C_{n-2}.
$$

We use integration by parts with $u = \cos^{n-1}\theta$ and $V' = \cos\theta$. Hence, $u' = (n-1)\cos^{n-2}\theta(-\sin\theta)$ and $v = \sin\theta$. Thus,

$$C_n = \int_{-\pi/2}^{\pi/2} \cos^n\theta \, d\theta = \int_{-\pi/2}^{\pi/2} \cos^{n-1}\theta\cos\theta \, d\theta = \cos^{n-1}\theta\sin\theta\Big|_{\theta=-\pi/2}^{\pi/2} + \int_{-\pi/2}^{\pi/2}(n-1)\cos^{n-2}\theta\sin^2\theta \, d\theta$$

$$= \cos^{n-1}\frac{\pi}{2}\sin\frac{\pi}{2} - \cos^{n+1}\left(-\frac{\pi}{2}\right)\sin\left(-\frac{\pi}{2}\right) + (n-1)\int_{-\pi/2}^{\pi/2}\cos^{n-2}\theta\sin^2\theta \, d\theta$$

$$= 0 + (n-1)\int_{-\pi/2}^{\pi/2}\cos^{n-2}\theta\left(1 - \cos^2\theta\right)d\theta = (n-1)\int_{-\pi/2}^{\pi/2}\cos^{n-2}\theta \, d\theta - (n-1)\int_{-\pi/2}^{\pi/2}\cos^n\theta \, d\theta$$

$$= (n-1)C_{n-2} - (n-1)C_n$$

We obtain the following equality:

$$C_n = (n-1)C_{n-2} - (n-1)C_n$$

or

$$C_n + (n-1)C_n = (n-1)C_{n-2}$$

$$nC_n = (n-1)C_{n-2}$$

$$C_n = \frac{n-1}{n}C_{n-2}$$

15.4 Integration in Polar, Cylindrical, and Spherical Coordinates (LT Section 16.4)

Preliminary Questions

1. Which of the following represent the integral of $f(x, y) = x^2 + y^2$ over the unit circle?

(a) $\int_0^1 \int_0^{2\pi} r^2 \, dr \, d\theta$

(b) $\int_0^{2\pi} \int_0^1 r^2 \, dr \, d\theta$

(c) $\int_0^1 \int_0^{2\pi} r^3 \, dr \, d\theta$

(d) $\int_0^{2\pi} \int_0^1 r^3 \, dr \, d\theta$

SOLUTION The unit circle is described in polar coordinates by the inequalities

$$0 \le \theta \le 2\pi, \quad 0 \le r \le 1$$

Using double integral in polar coordinates, we have

$$\iint_{\mathcal{D}} f(x, y) \, dA = \int_0^{2\pi} \int_0^1 \left((r\cos\theta)^2 + (r\sin\theta)^2\right) r \, dr \, d\theta = \int_0^{2\pi} \int_0^1 r^2\left(\cos^2\theta + \sin^2\theta\right) r \, dr \, d\theta$$

$$= \int_0^{2\pi} \int_0^1 r^3 \, dr \, d\theta$$

Therefore (d) is the correct answer.

2. What are the limits of integration in $\iiint f(r, \theta, z) r \, dr \, d\theta \, dz$ if the integration extends over the following regions?

(a) $x^2 + y^2 \le 4, \quad -1 \le z \le 2$

(b) Lower hemisphere of the sphere of radius 2, center at origin

SOLUTION

(a) This is a cylinder of radius 2. In the given region the z coordinate is changing between the values -1 and 2, and the angle θ is changing between the values $\theta = 0$ and 2π. Therefore the region is described by the inequalities

$$-1 \le z \le 2, \quad 0 \le \theta < 2\pi, \quad 0 \le r \le 2$$

Using triple integral in cylindrical coordinates gives

$$\int_{-1}^2 \int_0^{2\pi} \int_0^2 f(P) r \, dr \, d\theta \, dz$$

(b) The sphere of radius 2 is $x^2 + y^2 + z^2 = r^2 + z^2 = 4$, or $r = \sqrt{4 - z^2}$.

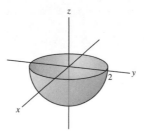

In the lower hemisphere we have $-2 \le z \le 0$ and $0 \le \theta < 2\pi$. Therefore, it has the description

$$-2 \le z \le 0, \quad 0 \le \theta < 2\pi, \quad 0 \le r \le \sqrt{4 - z^2}$$

We obtain the following integral in cylindrical coordinates:

$$\int_{-2}^{0} \int_{0}^{2\pi} \int_{0}^{\sqrt{4-z^2}} r \, dr \, d\theta \, dz$$

3. What are the limits of integration in

$$\iiint f(\rho, \phi, \theta) \rho^2 \sin \phi \, d\rho \, d\phi \, d\theta$$

if the integration extends over the following spherical regions centered at the origin?
(a) Sphere of radius 4
(b) Region between the spheres of radii 4 and 5
(c) Lower hemisphere of the sphere of radius 2

SOLUTION

(a) In the sphere of radius 4, θ varies from 0 to 2π, ϕ varies from 0 to π, and ρ varies from 0 to 4. Using triple integral in spherical coordinates, we obtain the following integral:

$$\int_{0}^{2\pi} \int_{0}^{\pi} \int_{0}^{4} f(P) \rho^2 \sin \phi \, d\rho \, d\phi \, d\theta$$

(b) In the region between the spheres of radii 4 and 5, ρ varies from 4 to 5, ϕ varies from 0 to π, and θ varies from 0 to 2π. We obtain the following integral:

$$\int_{0}^{2\pi} \int_{0}^{\pi} \int_{4}^{5} f(P) \rho^2 \sin \phi \, d\rho \, d\phi \, d\theta$$

(c) The inequalities in spherical coordinates for the lower hemisphere of radius 2 are

$$0 \le \theta \le 2\pi, \quad \frac{\pi}{2} \le \phi \le \pi, \quad 0 \le \rho \le 2$$

Therefore we obtain the following integral:

$$\int_{0}^{2\pi} \int_{\pi/2}^{\pi} \int_{0}^{2} f(P) \rho^2 \sin \phi \, d\rho \, d\phi \, d\theta.$$

4. An ordinary rectangle of sides Δx and Δy has area $\Delta x \, \Delta y$, no matter where it is located in the plane. However, the area of a polar rectangle of sides Δr and $\Delta \theta$ depends on its distance from the origin. How is this difference reflected in the Change of Variables Formula for polar coordinates?

SOLUTION The area ΔA of a small polar rectangle is

$$\Delta A = \frac{1}{2}(r + \Delta r)^2 \Delta \theta - \frac{1}{2} r^2 \Delta \theta = r \, (\Delta r \, \Delta \theta) + \frac{1}{2}(\Delta r)^2 \Delta \theta \approx r \, (\Delta r \, \Delta \theta)$$

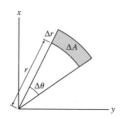

The factor r, due to the distance of the polar rectangle from the origin, appears in $dA = r \, dr \, d\theta$, in the Change of Variables formula.

Exercises

In Exercises 1–6, sketch the region $\mathcal{D}$ indicated and integrate $f(x, y)$ over $\mathcal{D}$ using polar coordinates.

1. $f(x, y) = \sqrt{x^2 + y^2}, \quad x^2 + y^2 \leq 2$

SOLUTION The domain $\mathcal{D}$ is the disk of radius $\sqrt{2}$ shown in the figure:

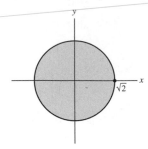

The inequalities defining $\mathcal{D}$ in polar coordinates are

$$0 \leq \theta \leq 2\pi, \quad 0 \leq r \leq \sqrt{2}$$

We describe $f(x, y) = \sqrt{x^2 + y^2}$ in polar coordinates:

$$f(x, y) = \sqrt{x^2 + y^2} = \sqrt{r^2} = r$$

Using change of variables in polar coordinates, we get

$$\iint_{\mathcal{D}} \sqrt{x^2 + y^2} \, dA = \int_0^{2\pi} \int_0^{\sqrt{2}} r \cdot r \, dr \, d\theta = \int_0^{2\pi} \int_0^{\sqrt{2}} r^2 \, dr \, d\theta = \int_0^{2\pi} \frac{r^3}{3} \Big|_{r=0}^{\sqrt{2}} \, d\theta$$

$$= \int_0^{2\pi} \frac{\left(\sqrt{2}\right)^3}{3} \, d\theta = \frac{2\sqrt{2}}{3}\theta \Big|_0^{2\pi} = \frac{4\sqrt{2}\pi}{3}$$

3. $f(x, y) = xy; \quad x \geq 0, \quad y \geq 0, \quad x^2 + y^2 \leq 4$

SOLUTION The domain $\mathcal{D}$ is the quarter circle of radius 2 in the first quadrant.

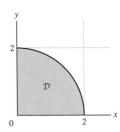

It is described by the inequalities

$$0 \leq \theta \leq \frac{\pi}{2}, \quad 0 \leq r \leq 2$$

We write f in polar coordinates:

$$f(x, y) = xy = (r \cos \theta)(r \sin \theta) = r^2 \cos \theta \sin \theta = \frac{1}{2} r^2 \sin 2\theta$$

Using change of variables in polar coordinates gives

$$\iint_{\mathcal{D}} xy \, dA = \int_0^{\pi/2} \int_0^2 \left(\frac{1}{2} r^2 \sin 2\theta\right) r \, dr \, d\theta = \int_0^{\pi/2} \int_0^2 \frac{1}{2} r^3 \sin 2\theta \, dr \, d\theta = \int_0^{\pi/2} \frac{1}{2} \cdot \frac{r^4}{4} \sin 2\theta \Big|_{r=0}^2 \, d\theta$$

$$= \int_0^{\pi/2} 2 \sin 2\theta \, d\theta = -\cos 2\theta \Big|_0^{\pi/2} = -(\cos \pi - \cos 0) = 2$$

5. $f(x, y) = y(x^2 + y^2)^{-1}; \quad y \geq \frac{1}{2}, \quad x^2 + y^2 \leq 1$

SOLUTION The region $\mathcal{D}$ is the part of the unit circle lying above the line $y = \frac{1}{2}$.

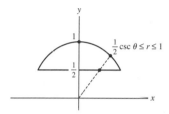

The angle α in the figure is

$$\alpha = \tan^{-1}\frac{\frac{1}{2}}{\frac{\sqrt{3}}{2}} = \tan^{-1}\frac{1}{\sqrt{3}} = \frac{\pi}{6}$$

Therefore, θ varies between $\frac{\pi}{6}$ and $\pi - \frac{\pi}{6} = \frac{5\pi}{6}$. The horizontal line $y = \frac{1}{2}$ has polar equation $r\sin\theta = \frac{1}{2}$ or $r = \frac{1}{2}\csc\theta$. The circle of radius 1 centered at the origin has polar equation $r = 1$. Therefore, r varies between $\frac{1}{2}\csc\theta$ and 1. The inequalities describing $\mathcal{D}$ in polar coordinates are thus

$$\frac{\pi}{6} \le \theta \le \frac{5\pi}{6}, \quad \frac{1}{2}\csc\theta \le r \le 1$$

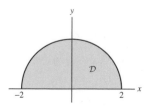

We write f in polar coordinates:

$$f(x, y) = y(x^2 + y^2)^{-1} = (r\sin\theta)(r^2)^{-1} = r^{-1}\sin\theta$$

Using change of variables in polar coordinates, we obtain

$$\iint_{\mathcal{D}} y(x^2+y^2)^{-1}\, dA = \int_{\pi/6}^{5\pi/6}\int_{\frac{1}{2}\csc\theta}^{1} r^{-1}\sin\theta\, r\, dr\, d\theta = \int_{\pi/6}^{5\pi/6}\int_{\frac{1}{2}\csc\theta}^{1} \sin\theta\, dr\, d\theta$$

$$= \int_{\pi/6}^{5\pi/6} r\sin\theta\Big|_{r=\frac{1}{2}\csc\theta}^{1} d\theta = \int_{\pi/6}^{5\pi/6}\left(\sin\theta - \frac{1}{2}\sin\theta\csc\theta\right) d\theta$$

$$= \int_{\pi/6}^{5\pi/6}\left(\sin\theta - \frac{1}{2}\right) d\theta = -\cos\theta - \frac{\theta}{2}\Big|_{\pi/6}^{5\pi/6} = -\cos\frac{5\pi}{6} - \frac{5\pi}{12} - \left(-\cos\frac{\pi}{6} - \frac{\pi}{12}\right)$$

$$= \frac{\sqrt{3}}{2} - \frac{\pi}{3} + \frac{\sqrt{3}}{2} = \sqrt{3} - \frac{\pi}{3} \approx 0.685$$

In Exercises 7–14, sketch the region of integration and evaluate by changing to polar coordinates.

7. $\displaystyle\int_{-2}^{2}\int_{0}^{\sqrt{4-x^2}} (x^2 + y^2)\, dy\, dx$

SOLUTION The domain $\mathcal{D}$ is described by the inequalities

$$\mathcal{D}: -2 \le x \le 2,\ 0 \le y \le \sqrt{4 - x^2}$$

That is, $\mathcal{D}$ is the semicircle $x^2 + y^2 \le 4,\ 0 \le y$.

We describe $\mathcal{D}$ in polar coordinates:

$$\mathcal{D} : 0 \le \theta \le \pi, \ 0 \le r \le 2$$

The function f in polar coordinates is $f(x, y) = x^2 + y^2 = r^2$. We use the Change of Variables Formula to write

$$\int_{-2}^{2} \int_{0}^{\sqrt{4-x^2}} \left(x^2 + y^2\right) dy\, dx = \int_{0}^{\pi} \int_{0}^{2} r^2 \cdot r\, dr\, d\theta = \int_{0}^{\pi} \int_{0}^{2} r^3\, dr\, d\theta = \int_{0}^{\pi} \frac{r^4}{4}\Big|_{r=0}^{2} d\theta = \int_{0}^{\pi} \frac{2^4}{4}\, d\theta = 4\pi$$

9. $\int_{0}^{1/2} \int_{\sqrt{3}x}^{\sqrt{1-x^2}} x\, dy\, dx$

SOLUTION The region of integration is described by the inequalities

$$0 \le x \le \frac{1}{2}, \quad \sqrt{3}x \le y \le \sqrt{1 - x^2}$$

$\mathcal{D}$ is the circular sector shown in the figure.

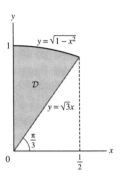

The ray $y = \sqrt{3}x$ in the first quadrant has the polar equation

$$r \sin\theta = \sqrt{3} r \cos\theta \quad \Rightarrow \quad \tan\theta = \sqrt{3} \quad \Rightarrow \quad \theta = \frac{\pi}{3}$$

Therefore, $\mathcal{D}$ lies in the angular sector $\frac{\pi}{3} \le \theta \le \frac{\pi}{2}$. Also, the circle $y = \sqrt{1 - x^2}$ has the polar equation $r = 1$, hence $\mathcal{D}$ can be described by the inequalities

$$\frac{\pi}{3} \le \theta \le \frac{\pi}{2}, \quad 0 \le r \le 1$$

We use change of variables to obtain

$$\int_{0}^{1/2} \int_{\sqrt{3}x}^{\sqrt{1-x^2}} x\, dy\, dx = \int_{\pi/3}^{\pi/2} \int_{0}^{1} r(\cos\theta)r\, dr\, d\theta = \int_{\pi/3}^{\pi/2} \int_{0}^{1} r^2 \cos\theta\, dr\, d\theta = \int_{\pi/3}^{\pi/2} \frac{r^3 \cos\theta}{3}\Big|_{r=0}^{1} d\theta$$

$$= \int_{\pi/3}^{\pi/2} \frac{\cos\theta}{3}\, d\theta = \frac{\sin\theta}{3}\Big|_{\pi/3}^{\pi/2} = \frac{1}{3}\left(\sin\frac{\pi}{2} - \sin\frac{\pi}{3}\right) = \frac{1}{3}\left(1 - \frac{\sqrt{3}}{2}\right) \approx 0.045$$

11. $\int_{0}^{5} \int_{0}^{y} x\, dx\, dy$

SOLUTION

$$\int_{0}^{5} \int_{0}^{y} x\, dx\, dy = \int_{\pi/4}^{\pi/2} \int_{r=0}^{5/\sin\theta} r^2 \cos\theta\, dr\, d\theta = \int_{\pi/4}^{\pi/2} \frac{1}{3} r^3 \cos\theta\Big|_{r=0}^{5/\sin\theta} d\theta$$

$$= \frac{1}{3} \int_{\pi/4}^{\pi/2} \frac{125}{\sin^3\theta} \cos\theta\, d\theta = \frac{125}{3} \int_{\pi/4}^{\pi/2} \frac{\cos\theta}{\sin^3\theta}\, d\theta$$

$$= -\frac{125}{6} \frac{1}{\sin^2\theta}\Big|_{\pi/4}^{\pi/2} = -\frac{125}{6}(1 - 2) = \frac{125}{6}$$

13. $\int_{-1}^{2} \int_{0}^{\sqrt{4-x^2}} \left(x^2 + y^2\right) dy\, dx$

SOLUTION The domain $\mathcal{D}$, shown in the figure, is described by the inequalities

$$-1 \le x \le 2, \quad 0 \le y \le \sqrt{4 - x^2}$$

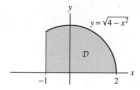

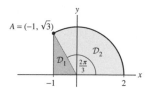

We denote by $\mathcal{D}_1$ and $\mathcal{D}_2$ the triangle and the circular sections, respectively, shown in the figure.

By properties of integrals we have

$$\iint_{\mathcal{D}} \left(x^2 + y^2\right) dA = \iint_{\mathcal{D}_1} \left(x^2 + y^2\right) dA + \iint_{\mathcal{D}_2} \left(x^2 + y^2\right) dA \tag{1}$$

We compute each integral separately, starting with $\mathcal{D}_1$. The vertical line $x = -1$ has polar equation $r\cos\theta = -1$ or $r = -\sec\theta$. The ray OA has polar equation $\theta = \frac{2\pi}{3}$. Therefore $\mathcal{D}_1$ is described by

$$\frac{2\pi}{3} \le \theta \le \pi, \quad 0 \le r \le -\sec\theta$$

Using change of variables gives

$$\iint_{\mathcal{D}_1} \left(x^2 + y^2\right) dA = \int_{2\pi/3}^{\pi} \int_0^{-\sec\theta} r^2 \cdot r \, dr \, d\theta = \int_{2\pi/3}^{\pi} \int_0^{-\sec\theta} r^3 \, dr \, d\theta$$

$$= \int_{2\pi/3}^{\pi} \frac{r^4}{4} \Big|_{r=0}^{-\sec\theta} d\theta = \int_{2\pi/3}^{\pi} \frac{\sec^4\theta}{4} \, d\theta = \frac{1}{4} \int_{2\pi/3}^{\pi} \sec^4\theta \, d\theta \tag{2}$$

We compute the integral (we use substitution $u = \tan\theta$ for the second integral):

$$\int_{2\pi/3}^{\pi} \frac{1}{\cos^4\theta} \, d\theta = \int_{2\pi/3}^{\pi} \frac{\sin^2\theta + \cos^2\theta}{\cos^4\theta} \, d\theta = \int_{2\pi/3}^{\pi} \frac{d\theta}{\cos^2\theta} + \int_{2\pi/3}^{\pi} \tan^2\theta \cdot \frac{1}{\cos^2\theta} \, d\theta$$

$$= \tan\theta \Big|_{\theta=2\pi/3}^{\pi} + \int_{-\sqrt{3}}^{0} u^2 \, du = \tan\pi - \tan\frac{2\pi}{3} + \frac{u^3}{3} \Big|_{-\sqrt{3}}^{0} = \sqrt{3} + \frac{3\sqrt{3}}{3} = 2\sqrt{3}$$

Hence, by (2) we get

$$\iint_{\mathcal{D}_1} \left(x^2 + y^2\right) dA = \frac{\sqrt{3}}{2} \tag{3}$$

$\mathcal{D}_2$ is described by the inequalities

$$0 \le \theta \le \frac{2\pi}{3}, \quad 0 \le r \le 2$$

Hence,

$$\iint_{\mathcal{D}_2} \left(x^2 + y^2\right) dA = \int_0^{2\pi/3} \int_0^2 r^2 \cdot r \, dr \, d\theta = \int_0^{2\pi/3} \int_0^2 r^3 \, dr \, d\theta$$

$$= \int_0^{2\pi/3} \frac{r^4}{4} \Big|_{r=0}^{2} d\theta = \int_0^{2\pi/3} 4 \, d\theta = 4 \cdot \theta \Big|_0^{2\pi/3} = \frac{8\pi}{3} \tag{4}$$

Combining (1), (3) and (4) we obtain the following solution:

$$\iint_{\mathcal{D}} \left(x^2 + y^2\right) dA = \frac{\sqrt{3}}{2} + \frac{8\pi}{3} \approx 9.24$$

In Exercises 15–20, calculate the integral over the given region by changing to polar coordinates.

15. $f(x, y) = (x^2 + y^2)^{-2}$; $x^2 + y^2 \leq 2$, $x \geq 1$

SOLUTION   The region $\mathcal{D}$ lies in the angular sector

$$-\frac{\pi}{4} \leq \theta \leq \frac{\pi}{4}$$

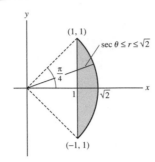

The vertical line $x = 1$ has polar equation $r \cos \theta = 1$ or $r = \sec \theta$. The circle $x^2 + y^2 = 2$ has polar equation $r = \sqrt{2}$. Therefore, $\mathcal{D}$ has the following description:

$$-\frac{\pi}{4} \leq \theta \leq \frac{\pi}{4}, \quad \sec \theta \leq r \leq \sqrt{2}$$

The function in polar coordinates is

$$f(x, y) = (x^2 + y^2)^{-2} = (r^2)^{-2} = r^{-4}.$$

Using change of variables we obtain

$$\iint_{\mathcal{D}} \left(x^2 + y^2\right)^{-2} dA = \int_{-\pi/4}^{\pi/4} \int_{\sec \theta}^{\sqrt{2}} r^{-4} r \, dr \, d\theta = \int_{-\pi/4}^{\pi/4} \int_{\sec \theta}^{\sqrt{2}} r^{-3} \, dr \, d\theta = \int_{-\pi/4}^{\pi/4} \frac{r^{-2}}{-2} \Big|_{\sec \theta}^{\sqrt{2}} \, d\theta$$

$$= \int_{-\pi/4}^{\pi/4} \left(\frac{1}{2\sec^2 \theta} - \frac{1}{4}\right) d\theta = 2 \int_0^{\pi/4} \left(\frac{1}{2}\cos^2 \theta - \frac{1}{4}\right) d\theta$$

$$= \left(\frac{\theta}{2} + \frac{\sin 2\theta}{4}\right) \Big|_0^{\pi/4} - \frac{\theta}{2}\Big|_0^{\pi/4} = \frac{\pi}{8} + \frac{1}{4} - \frac{\pi}{8} = \frac{1}{4}$$

17. $f(x, y) = |xy|$; $x^2 + y^2 \leq 1$

SOLUTION   The unit disk is described in polar coordinates by

$$\mathcal{D} : 0 \leq \theta \leq 2\pi, \ 0 \leq r \leq 1$$

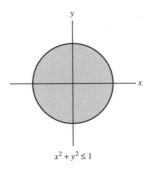

The function is $f(x, y) = |xy| = |r \cos \theta \cdot r \sin \theta| = \frac{1}{2}r^2 |\sin 2\theta|$. Using change of variables we obtain

$$\iint_{\mathcal{D}} |xy| \, dA = \int_0^{2\pi} \int_0^1 \frac{1}{2}r^2 |\sin 2\theta| \cdot r \, dr \, d\theta = \int_0^{2\pi} \int_0^1 \frac{1}{2}r^3 |\sin 2\theta| \, dr \, d\theta$$

$$= \int_0^{2\pi} \frac{r^4}{8} |\sin 2\theta| \Big|_{r=0}^1 \, d\theta = \int_0^{2\pi} \frac{1}{8} |\sin 2\theta| \, d\theta \tag{1}$$

The signs of $\sin 2\theta$ in the interval of integration are

For $0 \le \theta \le \frac{\pi}{2}$ or $\pi \le \theta \le \frac{3\pi}{2}$, $\sin 2\theta \ge 0$, hence $|\sin 2\theta| = \sin 2\theta$.

For $\frac{\pi}{2} \le \theta \le \pi$ or $\frac{3\pi}{2} \le \theta \le 2\pi$, $\sin 2\theta \le 0$, hence $|\sin 2\theta| = -\sin 2\theta$.

Therefore, by (1) we get

$$\iint_{\mathcal{D}} |xy|\, dA = \int_0^{\pi/2} \frac{1}{8} \sin 2\theta\, d\theta - \int_{\pi/2}^{\pi} \frac{1}{8} \sin 2\theta\, d\theta + \int_{\pi}^{3\pi/2} \frac{1}{8} \sin 2\theta\, d\theta - \int_{3\pi/2}^{2\pi} \frac{1}{8} \sin 2\theta\, d\theta$$

$$= -\frac{1}{16} \cos 2\theta \Big|_0^{\pi/2} + \frac{1}{16} \cos 2\theta \Big|_{\pi/2}^{\pi} - \frac{1}{16} \cos 2\theta \Big|_{\pi}^{3\pi/2} + \frac{1}{16} \cos 2\theta \Big|_{3\pi/2}^{2\pi}$$

$$= -\frac{1}{16}(\cos \pi - 1) + \frac{1}{16}(\cos 2\pi - \cos \pi) - \frac{1}{16}(\cos 3\pi - \cos 2\pi) + \frac{1}{16}(\cos 4\pi - \cos 3\pi)$$

$$= \frac{2}{16} + \frac{2}{16} + \frac{2}{16} + \frac{2}{16} = \frac{1}{2}$$

That is,

$$\iint_{\mathcal{D}} |xy|\, dA = \frac{1}{2}$$

19. $f(x, y) = x - y$; $x^2 + y^2 \le 1$, $x + y \ge 1$

SOLUTION As shown in Exercise 24, the region $\mathcal{D}$ is described by the following inequalities:

$$\mathcal{D}: 0 \le \theta \le \frac{\pi}{2}, \quad \frac{1}{\cos \theta + \sin \theta} \le r \le 1$$

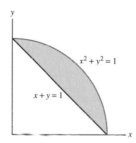

The function in polar coordinates is

$$f(x, y) = x - y = r\cos\theta - r\sin\theta = r(\cos\theta - \sin\theta)$$

Using the Change of Variables Formula we get

$$\iint_{\mathcal{D}} (x - y)\, dA = \int_0^{\pi/2} \int_{\frac{1}{\cos\theta + \sin\theta}}^1 r(\cos\theta - \sin\theta) r\, dr\, d\theta = \int_0^{\pi/2} \int_{\frac{1}{\cos\theta + \sin\theta}}^1 r^2(\cos\theta - \sin\theta)\, dr\, d\theta$$

$$= \int_0^{\pi/2} \frac{r^3(\cos\theta - \sin\theta)}{3} \Big|_{r = \frac{1}{\cos\theta + \sin\theta}}^1 d\theta = \int_0^{\pi/2} \frac{\cos\theta - \sin\theta}{3} \left(1 - \frac{1}{(\cos\theta + \sin\theta)^3}\right) d\theta$$

$$= \int_0^{\pi/2} \frac{\cos\theta - \sin\theta}{3}\, d\theta - \frac{1}{3} \int_0^{\pi/2} \frac{\cos\theta - \sin\theta}{(\cos\theta + \sin\theta)^3}\, d\theta \tag{1}$$

We compute the two integrals:

$$\int_0^{\pi/2} \frac{\cos\theta - \sin\theta}{3}\, d\theta = \frac{\sin\theta + \cos\theta}{3} \Big|_0^{\pi/2} = \frac{(1 + 0) - (0 + 1)}{3} = 0 \tag{2}$$

To compute the second integral we will use u-substitution and let $u = \sin\theta + \cos\theta$:

$$\int_0^{\pi/2} \frac{\cos\theta - \sin\theta}{(\cos\theta + \sin\theta)^3}\, d\theta = \int_{u=1}^1 u^{-3}\, du = 0 \tag{3}$$

Combining (1), (2), and (3) we conclude that

$$\iint_{\mathcal{D}} (x - y)\, dA = 0$$

21. Find the volume of the wedge-shaped region (Figure 17) contained in the cylinder $x^2 + y^2 = 9$, bounded above by the plane $z = x$ and below by the xy-plane.

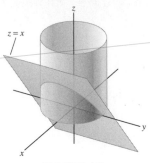

FIGURE 17

SOLUTION

Step 1. Express $\mathcal{W}$ in cylindrical coordinates. $\mathcal{W}$ is bounded above by the plane $z = x$ and below by $z = 0$, therefore $0 \le z \le x$, in particular $x \ge 0$. Hence, $\mathcal{W}$ projects onto the semicircle $\mathcal{D}$ in the xy-plane of radius 3, where $x \ge 0$.

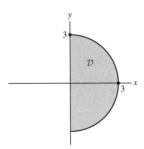

In polar coordinates,

$$\mathcal{D}: -\frac{\pi}{2} \le \theta \le \frac{\pi}{2},\ 0 \le r \le 3$$

The upper surface is $z = x = r \cos\theta$ and the lower surface is $z = 0$. Therefore,

$$\mathcal{W}: -\frac{\pi}{2} \le \theta \le \frac{\pi}{2},\ 0 \le r \le 3,\ 0 \le z \le r \cos\theta$$

Step 2. Set up an integral in cylindrical coordinates and evaluate. The volume of $\mathcal{W}$ is the triple integral $\iiint_{\mathcal{W}} 1\,dV$. Using change of variables in cylindrical coordinates gives

$$\iiint_{\mathcal{W}} 1\,dV = \int_{-\pi/2}^{\pi/2} \int_0^3 \int_0^{r\cos\theta} r\,dz\,dr\,d\theta = \int_{-\pi/2}^{\pi/2} \int_0^3 rz \Big|_{z=0}^{r\cos\theta} dr\,d\theta = \int_{-\pi/2}^{\pi/2} \int_0^3 r^2 \cos\theta\,dr\,d\theta$$

$$= \int_{-\pi/2}^{\pi/2} \frac{r^3}{3} \cos\theta \Big|_{r=0}^3 d\theta = \int_{-\pi/2}^{\pi/2} 9\cos\theta\,d\theta = 9\sin\theta \Big|_{-\pi/2}^{\pi/2} = 9\left(\sin\frac{\pi}{2} - \sin\left(-\frac{\pi}{2}\right)\right) = 18$$

23. Evaluate $\iint_{\mathcal{D}} \sqrt{x^2 + y^2}\,dA$, where $\mathcal{D}$ is the domain in Figure 19. *Hint:* Find the equation of the inner circle in polar coordinates and treat the right and left parts of the region separately.

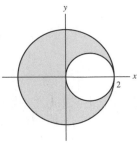

FIGURE 19

SOLUTION We denote by $\mathcal{D}_1$ and $\mathcal{D}_2$ the regions enclosed by the circles $x^2 + y^2 = 4$ and $(x-1)^2 + y^2 = 1$. Therefore,

$$\iint_{\mathcal{D}} \sqrt{x^2 + y^2}\, dx\, dy = \iint_{\mathcal{D}_1} \sqrt{x^2 + y^2}\, dx\, dy - \iint_{\mathcal{D}_2} \sqrt{x^2 + y^2}\, dx\, dy \tag{1}$$

We compute the integrals on the right hand-side.

$\mathcal{D}_1$:

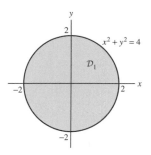

The circle $x^2 + y^2 = 4$ has polar equation $r = 2$, therefore $\mathcal{D}_1$ is determined by the following inequalities:

$$\mathcal{D}_1 : 0 \le \theta \le 2\pi, \; 0 \le r \le 2$$

The function in polar coordinates is $f(x, y) = \sqrt{x^2 + y^2} = r$. Using change of variables in the integral gives

$$\iint_{\mathcal{D}_1} \sqrt{x^2 + y^2}\, dx\, dy = \int_0^{2\pi} \int_0^2 r \cdot r\, dr\, d\theta = \int_0^{2\pi} \int_0^2 r^2\, dr\, d\theta = \int_0^{2\pi} \frac{r^3}{3}\Big|_{r=0}^2 d\theta = \int_0^{2\pi} \frac{8}{3}\, d\theta = \frac{16\pi}{3} \tag{2}$$

$\mathcal{D}_2$:

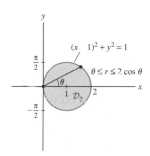

$\mathcal{D}_2$ lies in the angular sector $-\frac{\pi}{2} \le \theta \le \frac{\pi}{2}$. We find the polar equation of the circle $(x-1)^2 + y^2 = 1$:

$$(x-1)^2 + y^2 = x^2 - 2x + 1 + y^2 = x^2 + y^2 - 2x + 1 = 1 \quad \Rightarrow \quad x^2 + y^2 = 2x$$
$$\Rightarrow \quad r^2 = 2r\cos\theta$$
$$\Rightarrow \quad r = 2\cos\theta$$

Thus, the domain $\mathcal{D}_2$ is defined by the following inequalities:

$$\mathcal{D}_2 : -\frac{\pi}{2} \le \theta \le \frac{\pi}{2}, \; 0 \le r \le 2\cos\theta$$

We use the change of variables in the integral and integration table to obtain

$$\iint_{\mathcal{D}_2} \sqrt{x^2 + y^2}\, dx\, dy = \int_{-\pi/2}^{\pi/2} \int_0^{2\cos\theta} r \cdot r\, dr\, d\theta = \int_{-\pi/2}^{\pi/2} \int_0^{2\cos\theta} r^2\, dr\, d\theta = \int_{-\pi/2}^{\pi/2} \frac{r^3}{3}\Big|_{r=0}^{2\cos\theta} d\theta$$

$$= \int_{-\pi/2}^{\pi/2} \frac{8\cos^3\theta}{3}\, d\theta = 2\int_0^{\pi/2} \frac{8\cos^3\theta}{3}\, d\theta = \frac{16}{3}\left(\frac{\cos^2\theta\sin\theta}{3} + \frac{2}{3}\sin\theta\right)\Big|_{\theta=0}^{\pi/2}$$

$$= \frac{16}{3} \cdot \frac{2}{3}\sin\frac{\pi}{2} = \frac{32}{9} \tag{3}$$

Substituting (2) and (3) in (1), we obtain the following solution:

$$\iint_{\mathcal{D}} \sqrt{x^2 + y^2}\, dx\, dy = \frac{16\pi}{3} - \frac{32}{9} = \frac{48\pi - 32}{9} \approx 13.2.$$

Remark: The integral can also be evaluated using the hint as the sum of

$$\int_{\mathcal{D}^*} \iint \sqrt{x^2 + y^2}\, dA \quad \text{and} \quad \int_{\mathcal{D}^{**}} \iint \sqrt{x^2 + y^2}\, dA$$

where $\mathcal{D}^*$ is the left semicircle $x^2 + y^2 = 4$ and $\mathcal{D}^*$ is the right part of $\mathcal{D}$. Since

$$\mathcal{D}^*: \frac{\pi}{2} \le \theta \le \frac{3\pi}{2}, \ 0 \le r \le 2$$

$$\mathcal{D}^{**}: -\frac{\pi}{2} \le \theta \le \frac{\pi}{2}, \ 2\cos\theta \le r \le 2$$

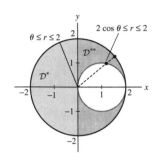

we get

$$\iint_{\mathcal{D}} \sqrt{x^2 + y^2}\, dA = \int_{\pi/2}^{3\pi/2} \int_0^2 r^2\, dr\, d\theta + \int_{-\pi/2}^{\pi/2} \int_{2\cos\theta}^2 r^2\, dr\, d\theta$$

Obviously, computing the integrals leads to the same result.

25. Let $\mathcal{W}$ be the region between the paraboloids $z = x^2 + y^2$ and $z = 8 - x^2 - y^2$.

(a) Describe $\mathcal{W}$ in cylindrical coordinates.

(b) Use cylindrical coordinates to compute the volume of $\mathcal{W}$.

SOLUTION

(a)

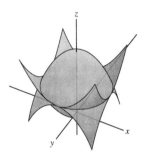

The paraboloids $z = x^2 + y^2$ and $z = 8 - (x^2 + y^2)$ have the polar equations $z = r^2$ and $z = 8 - r^2$, respectively. We find the curve of intersection by solving

$$8 - r^2 = r^2 \quad \Rightarrow \quad 2r^2 = 8 \quad \Rightarrow \quad r = 2$$

Therefore, the projection $\mathcal{D}$ of $\mathcal{W}$ onto the xy-plane is the region enclosed by the circle $r = 2$, and $\mathcal{D}$ has the following description:

$$\mathcal{D} : 0 \le \theta \le 2\pi, \ 0 \le r \le 2$$

The upper and lower boundaries of $\mathcal{W}$ are $z = 8 - r^2$ and $z = r^2$, respectively. Hence,

$$\mathcal{W} : 0 \le \theta \le 2\pi, \ 0 \le r \le 2, \ r^2 \le z \le 8 - r^2$$

(b) Using change of variables in cylindrical coordinates, we get

$$\text{Volume}(\mathcal{W}) = \iiint_{\mathcal{W}} 1\, dV = \int_0^{2\pi} \int_0^2 \int_{r^2}^{8-r^2} r\, dz\, dr\, d\theta = \int_0^{2\pi} \int_0^2 rz \Big|_{z=r^2}^{8-r^2} dr\, d\theta = \int_0^{2\pi} \int_0^2 r\left(8 - 2r^2\right) dr\, d\theta$$

$$= \int_0^{2\pi} \int_0^2 \left(8r - 2r^3\right) dr\, d\theta = \int_0^{2\pi} 4r^2 - \frac{r^4}{2}\Big|_{r=0}^2 d\theta = \int_0^{2\pi} 8\, d\theta = 16\pi$$

In Exercises 27–32, use cylindrical coordinates to calculate $\iiint_{\mathcal{W}} f(x, y, z)\, dV$ for the given function and region.

27. $f(x, y, z) = x^2 + y^2; \quad x^2 + y^2 \le 9, \quad 0 \le z \le 5$

SOLUTION The projection of $\mathcal{W}$ onto the xy-plane is the region inside the circle $x^2 + y^2 = 9$. In polar coordinates,

$$\mathcal{D} : 0 \le \theta \le 2\pi,\ 0 \le r \le 3$$

The upper and lower boundaries are the planes $z = 5$ and $z = 0$, respectively. Therefore, $\mathcal{W}$ has the following description in cylindrical coordinates:

$$\mathcal{W} : 0 \le \theta \le 2\pi,\ 0 \le r \le 3,\ 0 \le z \le 5$$

The integral in cylindrical coordinates is thus

$$\iiint_{\mathcal{W}} (x^2 + y^2)\, dV = \int_0^{2\pi} \int_0^3 \int_0^5 r^2 \cdot r\, dz\, dr\, d\theta = \int_0^{2\pi} \int_0^3 \int_0^5 r^3\, dz\, dr\, d\theta$$

$$= \left(\int_0^{2\pi} 1\, d\theta \right) \left(\int_0^3 r^3\, dr \right) \left(\int_0^5 1\, dz \right) = 2\pi \cdot 5 \cdot \frac{r^4}{4} \Big|_0^3 = \frac{5 \cdot 3^4 \pi}{2} \approx 636.17$$

29. $f(x, y, z) = y; \quad x^2 + y^2 \le 1, \quad x \ge 0, \quad y \ge 0, \quad 0 \le z \le 2$

SOLUTION

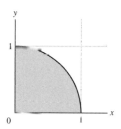

The projection of $\mathcal{W}$ onto the xy plane is the quarter of the unit circle in the first quadrant. It is defined by the following polar equations:

$$\mathcal{D} : 0 \le \theta \le \frac{\pi}{2},\ 0 \le r \le 1$$

The upper and lower boundaries of $\mathcal{W}$ are the planes $z = 2$ and $z = 0$, respectively; hence, $\mathcal{W}$ has the following definition:

$$\mathcal{W} : 0 \le \theta \le \frac{\pi}{2},\ 0 \le r \le 1,\ 0 \le z \le 2$$

The function is $f(x, y, z) = y = r \sin \theta$. The integral in cylindrical coordinates is thus

$$\iiint_{\mathcal{W}} y\, dV = \int_0^{\pi/2} \int_0^1 \int_0^2 (r \sin \theta) r\, dz\, dr\, d\theta = \int_0^{\pi/2} \int_0^1 \int_0^2 r^2 \sin \theta\, dz\, dr\, d\theta$$

$$= \left(\int_0^{\pi/2} \sin \theta\, d\theta \right) \left(\int_0^1 r^2\, dr \right) \left(\int_0^2 1\, dz \right) = \left(-\cos \theta \Big|_0^{\pi/2} \right) \left(\frac{r^3}{3} \Big|_0^1 \right) \left(z \Big|_0^2 \right) = 1 \cdot \frac{1}{3} \cdot 2 = \frac{2}{3}$$

31. $f(x, y, z) = z; \quad x^2 + y^2 \le z \le 9$

SOLUTION

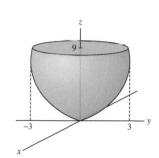

The upper boundary of $\mathcal{W}$ is the plane $z = 9$, and the lower boundary is $z = x^2 + y^2 = r^2$. Therefore, $r^2 \le z \le 9$. The projection $\mathcal{D}$ onto the xy-plane is the circle $x^2 + y^2 = 9$ or $r = 3$. That is,

$$\mathcal{D}: 0 \le \theta \le 2\pi, \ 0 \le r \le 3$$

The inequalities defining $\mathcal{W}$ in cylindrical coordinates are thus

$$\mathcal{W}: 0 \le \theta \le 2\pi, \ 0 \le r \le 3, \ r^2 \le z \le 9$$

Therefore, we obtain the following integral:

$$\iiint_{\mathcal{W}} z \, dV = \int_0^{2\pi} \int_0^3 \int_{r^2}^9 zr \, dz \, dr \, d\theta = \int_0^{2\pi} \int_0^3 \frac{z^2 r}{2} \Big|_{z=r^2}^9 dr \, d\theta = \int_0^{2\pi} \int_0^3 \frac{r(81 - r^4)}{2} \, dr \, d\theta$$

$$= \int_0^{2\pi} \int_0^3 \frac{81r - r^5}{2} \, dr \, d\theta = \int_0^{2\pi} \frac{81r^2}{4} - \frac{r^6}{12} \Big|_0^3 \, d\theta = \int_0^{2\pi} 121.5 \, d\theta = 243\pi$$

In Exercises 33–36, express the triple integral in cylindrical coordinates.

33. $\displaystyle \int_{-1}^1 \int_{y=-\sqrt{1-x^2}}^{y=\sqrt{1-x^2}} \int_{z=0}^4 f(x, y, z) \, dz \, dy \, dx$

SOLUTION   The region of integration is determined by the limits of integration. That is,

$$\mathcal{W}: -1 \le x \le 1, \ -\sqrt{1-x^2} \le y \le \sqrt{1-x^2}, \ 0 \le z \le 4$$

Therefore the projection of $\mathcal{W}$ onto the xy-plane is the disk $x^2 + y^2 \le 1$. This region has the following definition in polar coordinates:

$$\mathcal{D}: 0 \le \theta \le 2\pi, \ 0 \le r \le 1$$

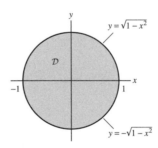

The upper and lower boundaries of $\mathcal{W}$ are the planes $z = 4$ and $z = 0$, respectively. Hence,

$$\mathcal{W}: 0 \le \theta \le 2\pi, \ 0 \le r \le 1, \ 0 \le z \le 4$$

Using change of variables in cylindrical coordinates, we get the integral

$$\int_0^{2\pi} \int_0^1 \int_0^4 f(r \cos\theta, r \sin\theta, z) r \, dz \, dr \, d\theta$$

35. $\displaystyle \int_{-1}^1 \int_{y=0}^{y=\sqrt{1-x^2}} \int_{z=0}^{x^2+y^2} f(x, y, z) \, dz \, dy \, dx$

SOLUTION   The inequalities defining the region of integration are

$$\mathcal{W}: -1 \le x \le 1, \ 0 \le y \le \sqrt{1-x^2}, \ 0 \le z \le x^2 + y^2$$

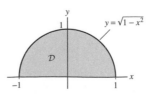

The projection of W onto the xy-plane is the semicircle $x^2 + y^2 = 1$, where $-1 \le x \le 1$. This domain is defined by the polar inequalities

$$\mathcal{D} : 0 \le \theta \le \pi, \, 0 \le r \le 1$$

The lower surface is $z = 0$ and upper surface is $z = x^2 + y^2 = r^2$, hence W has the following description in cylindrical coordinates:

$$W : 0 \le \theta \le \pi, \, 0 \le r \le 1, \, 0 \le z \le r^2$$

We obtain the following integral:

$$\int_0^\pi \int_0^1 \int_0^{r^2} f(r \cos \theta, r \sin \theta, z) r \, dz \, dr \, d\theta$$

37. Find the equation of the right-circular cone in Figure 21 in cylindrical coordinates and compute its volume.

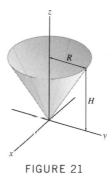

FIGURE 21

SOLUTION   To find the equation of the surface we use proportion in similar triangles.

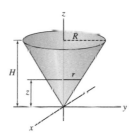

This gives

$$\frac{z}{H} = \frac{r}{R} \quad \Rightarrow \quad z = \frac{H}{R} r$$

The volume of the right circular cone is

$$V = \iiint_W 1 \, dV$$

The projection of W onto the xy-plane is the region $x^2 + y^2 \le R^2$, or in polar coordinates,

$$\mathcal{D} : 0 \le \theta \le 2\pi, \, 0 \le r \le R$$

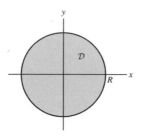

The upper and lower boundaries are the surfaces $z = H$ and $z = \frac{H}{R}r$, respectively. Hence $\mathcal{W}$ is determined by the following cylindrical inequalities:

$$\mathcal{W} : 0 \leq \theta \leq 2\pi, \ 0 \leq r \leq R, \ \frac{H}{R}r \leq z \leq H$$

We compute the volume using the following integral:

$$V = \iiint_{\mathcal{W}} 1 \, dv = \int_0^{2\pi} \int_0^R \int_{\frac{H}{R}r}^H r \, dz \, dr \, d\theta = \int_0^{2\pi} \int_0^R rz \Big|_{z=\frac{Hr}{R}}^H dr \, d\theta = \int_0^{2\pi} \int_0^R r \left(H - \frac{Hr}{R} \right) dr \, d\theta$$

$$= \int_0^{2\pi} \int_0^R \left(rH - \frac{r^2 H}{R} \right) dr \, d\theta = \int_0^{2\pi} \frac{r^2 H}{2} - \frac{r^3 H}{3R} \bigg|_{r=0}^R d\theta = \int_0^{2\pi} \frac{R^2 H}{6} \, d\theta = \frac{R^2 H}{6} \cdot 2\pi = \frac{\pi R^2 H}{3}$$

39. Use cylindrical coordinates to find the volume of a sphere of radius a from which a central cylinder of radius b has been removed where $0 < b < a$.

SOLUTION   Firstly, the equation of the sphere having radius a is

$$x^2 + y^2 + z^2 = a^2 \quad \Rightarrow \quad r^2 + z^2 = a^2$$

in cylindrical coordinates. Next, the equation of the cylinder with radius b that is being removed from the sphere is

$$x^2 + y^2 = b^2 \quad \Rightarrow \quad r^2 = b^2 \quad \Rightarrow \quad r = b$$

in cylindrical coordinates. Thus the region that is remaining can be described by the following inequalities in cylindrical coordinates:

$$0 \leq \theta \leq 2\pi, \quad b \leq r \leq a, \quad -\sqrt{a^2 - r^2} \leq z \leq \sqrt{a^2 - r^2}$$

Thus the volume can be computed:

$$V = \iiint_{\mathcal{W}} 1 \, dV = \iiint_{\mathcal{W}} r \, dz \, dr \, d\theta$$

$$= 2 \int_0^{2\pi} \int_b^a \int_0^{\sqrt{a^2 - r^2}} r \, dz \, dr \, d\theta$$

$$= 2 \int_0^{2\pi} \int_b^a rz \Big|_{z=0}^{\sqrt{a^2 - r^2}} dr \, d\theta$$

$$= 2 \int_0^2 \pi \int_b^a r\sqrt{a^2 - r^2} \, dr \, d\theta$$

$$= 2 \int_0^{2\pi} -\frac{1}{2} \cdot \frac{2}{3} (a^2 - r^2)^{3/2} \bigg|_{r=b}^a d\theta$$

$$= \frac{2}{3} \int_0^{2\pi} (a^2 - b^2)^{3/2} \, d\theta = \frac{2}{3} (a^2 - b^2)^{3/2} \cdot \theta \bigg|_{\theta=0}^{2\pi}$$

$$= \frac{4}{3} \pi (a^2 - b^2)^{3/2}$$

In Exercises 41–46, use spherical coordinates to calculate the triple integral of $f(x, y, z)$ over the given region.

41. $f(x, y, z) = y; \quad x^2 + y^2 + z^2 \leq 1, \quad x, y, z \leq 0$

SOLUTION

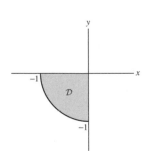

The region inside the unit sphere in the octant $x, y, z \leq 0$ is defined by the inequalities

$$\mathcal{W}: \pi \leq \theta \leq \frac{3\pi}{2}, \ \frac{\pi}{2} \leq \phi \leq \pi, \ 0 \leq \rho \leq 1$$

The function in spherical coordinates is $f(x, y, z) = y = \rho \sin\theta \sin\phi$. Using a triple integral in spherical coordinates, we obtain

$$\iiint_{\mathcal{W}} y \, dV = \int_{\pi}^{3\pi/2} \int_{\pi/2}^{\pi} \int_{0}^{1} (\rho \sin\theta \sin\phi)\rho^2 \sin\phi \, d\rho \, d\phi \, d\theta = \int_{\pi}^{3\pi/2} \int_{\pi/2}^{\pi} \int_{0}^{1} \rho^3 \sin\theta \sin^2\phi \, d\rho \, d\phi \, d\theta$$

$$= \left(\int_{\pi}^{3\pi/2} \sin\theta \, d\theta \right) \left(\int_{\pi/2}^{\pi} \sin^2\phi \, d\phi \right) \left(\int_{0}^{1} \rho^3 d\rho \right) = \left(-\cos\theta \Big|_{\pi}^{3\pi/2} \right) \left(\frac{\theta}{2} - \frac{\sin 2\theta}{4} \Big|_{\pi/2}^{\pi} \right) \left(\frac{\rho^4}{4} \Big|_{0}^{1} \right)$$

$$= (-1) \cdot \left(\frac{\pi}{2} - \frac{\pi}{4} \right) \cdot \frac{1}{4} = -\frac{\pi}{16}$$

43. $f(x, y, z) = x^2 + y^2; \quad \rho \leq 1$

SOLUTION $\mathcal{W}$ is the region inside the unit sphere, therefore it is described by the following inequalities:

$$\mathcal{W}: 0 \leq \theta \leq 2\pi, \ 0 \leq \phi \leq \pi, \ 0 \leq \rho \leq 1$$

The function in spherical coordinates is

$$f(x, y, z) = x^2 + y^2 = (\rho \cos\theta \sin\phi)^2 + (\rho \sin\theta \sin\phi)^2$$
$$= \rho^2 \sin^2\phi \left(\cos^2\theta + \sin^2\theta \right) = \rho^2 \sin^2\phi$$

Using triple integrals in spherical coordinates we get

$$\iiint_{\mathcal{W}} (x^2 + y^2) \, dV = \int_{0}^{2\pi} \int_{0}^{\pi} \int_{0}^{1} (\rho^2 \sin^2\phi)\rho^2 \sin\phi \, d\rho \, d\phi \, d\theta$$

$$= \int_{0}^{2\pi} \int_{0}^{\pi} \int_{0}^{1} \rho^4 \sin^3\phi \, d\rho \, d\phi \, d\theta = \left(\int_{0}^{2\pi} 1 \, d\theta \right) \left(\int_{0}^{\pi} \sin^3\phi \, d\phi \right) \left(\int_{0}^{1} \rho^4 d\rho \right)$$

$$= \left(\theta \Big|_{0}^{2\pi} \right) \left(-\frac{\sin^2\theta \cos\theta}{3} - \frac{2}{3}\cos\theta \Big|_{0}^{\pi} \right) \left(\frac{\rho^5}{5} \Big|_{0}^{1} \right) = 2\pi \cdot \left(\frac{2}{3} + \frac{2}{3} \right) \cdot \frac{1}{5} = \frac{8\pi}{15}$$

45. $f(x, y, z) = \sqrt{x^2 + y^2 + z^2}; \quad x^2 + y^2 + z^2 \leq 2z$

SOLUTION We rewrite the inequality for the region using spherical coordinates:

$$\rho^2 \leq 2\rho \cos\phi \quad \Rightarrow \quad \rho \leq 2\cos\phi$$

Completing the square in $x^2 + y^2 + z^2 = 2z$, we see that this is the equation of the sphere of radius 1 centered at $(0, 0, 1)$. That is,

$$x^2 + y^2 + z - 2z = 0$$
$$x^2 + y^2 + (z - 1)^2 = 1$$

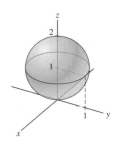

$\mathcal{W}$ is the region inside the sphere, hence θ varies from 0 to 2π, and ϕ varies from 0 to $\frac{\pi}{2}$. The inequalities describing $\mathcal{W}$ in spherical coordinates are thus

$$\mathcal{W}: 0 \leq \theta \leq 2\pi, \ 0 \leq \phi \leq \frac{\pi}{2}, \ 0 \leq \rho \leq 2\cos\phi$$

The function in spherical coordinates is

$$f(x, y, z) = \sqrt{x^2 + y^2 + z^2} = \rho.$$

We obtain the following integral:

$$I = \int_0^{2\pi} \int_0^{\pi/2} \int_0^{2\cos\phi} \rho \cdot \rho^2 \sin\phi \, d\rho \, d\phi \, d\theta = \int_0^{2\pi} \int_0^{\pi/2} \int_0^{2\cos\phi} \rho^3 \sin\phi \, d\rho \, d\phi \, d\theta$$

$$= \int_0^{2\pi} \int_0^{\pi/2} \left. \frac{\rho^4 \sin\phi}{4} \right|_{\rho=0}^{2\cos\phi} d\phi \, d\theta = \int_0^{2\pi} \int_0^{\pi/2} \frac{16 \cos^4\phi \sin\phi}{4} \, d\phi \, d\theta$$

$$= \left(\int_0^{2\pi} 4 \, d\theta \right) \left(\int_0^{\pi/2} \cos^4\phi \sin\phi \, d\phi \right) = 8\pi \int_0^{\pi/2} \cos^4\phi \sin\phi \, d\phi$$

We compute the integral using the substitution $u = \cos\phi$, $du = -\sin\phi \, d\phi$. We obtain

$$I = 8\pi \int_1^0 u^4(-du) = 8\pi \int_0^1 u^4 \, du = 8\pi \left. \frac{u^5}{5} \right|_0^1 = \frac{8\pi}{5}$$

47. Use spherical coordinates to evaluate the triple integral of $f(x, y, z) = z$ over the region

$$0 \leq \theta \leq \frac{\pi}{3}, \qquad 0 \leq \phi \leq \frac{\pi}{2}, \qquad 1 \leq \rho \leq 2$$

SOLUTION   The function in spherical coordinates is $f(x, y, z) = z = \rho \cos\phi$. Using triple integral in spherical coordinates gives

$$\iiint_W z \, dV = \int_0^{\pi/3} \int_0^{\pi/2} \int_1^2 (\rho \cos\phi)\rho^2 \sin\phi \, d\rho \, d\phi \, d\theta = \int_0^{\pi/3} \int_0^{\pi/2} \int_1^2 \rho^3 \cos\phi \sin\phi \, d\rho \, d\phi \, d\theta$$

$$= \left(\int_0^{\pi/3} 1 \, d\theta \right) \left(\int_0^{\pi/2} \frac{1}{2} \sin 2\phi \, d\phi \right) \left(\int_1^2 \rho^3 d\rho \right)$$

$$= \frac{\pi}{3} \cdot \left(-\frac{1}{4} \cos 2\phi \right) \left. \right|_0^{\pi/2} \cdot \left(\frac{\rho^4}{4} \right) \left. \right|_1^2 = \frac{\pi}{3} \cdot \frac{1}{2} \cdot \left(4 - \frac{1}{4} \right) = \frac{5}{8}\pi$$

49. Calculate the integral of

$$f(x, y, z) = z(x^2 + y^2 + z^2)^{-3/2}$$

over the part of the ball $x^2 + y^2 + z^2 \leq 16$ defined by $z \geq 2$.

SOLUTION

The equation of the sphere in spherical coordinates is $\rho^2 = 16$ or $\rho = 4$.

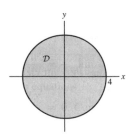

We write the equation of the plane $z = 2$ in spherical coordinates:

$$\rho \cos\phi = 2 \quad \Rightarrow \quad \rho = \frac{2}{\cos\phi}$$

To compute the interval of ϕ, we must find the value of ϕ corresponding to $\rho = 4$ on the plane $z = 2$. We get

$$4 = \frac{2}{\cos\phi} \quad \Rightarrow \quad \cos\phi = \frac{1}{2} \quad \Rightarrow \quad \phi = \frac{\pi}{3}$$

Therefore, ϕ is changing from 0 to $\frac{\pi}{3}$, θ is changing from 0 to 2π, and ρ is changing from $\frac{2}{\cos\phi}$ to 4. We obtain the following description for $\mathcal{W}$:

$$\mathcal{W}: 0 \le \theta \le 2\pi, \ 0 \le \phi \le \frac{\pi}{3}, \ \frac{2}{\cos\phi} \le \rho \le 4$$

The function is

$$f(x, y, z) = z(x^2 + y^2 + z^2)^{-3/2} = \rho\cos\phi \cdot (\rho^2)^{-3/2} = \rho^{-2}\cos\phi$$

We use triple integrals in spherical coordinates to write

$$\iiint_{\mathcal{W}} f(x, y, z)\, dV = \int_0^{2\pi} \int_0^{\pi/3} \int_{2/\cos\phi}^4 (\rho^{-2}\cos\phi)\rho^2 \sin\phi\, d\rho\, d\phi\, d\theta = \int_0^{2\pi} \int_0^{\pi/3} \int_{2/\cos\phi}^4 \frac{\sin 2\phi}{2}\, d\rho\, d\phi\, d\theta$$

$$= \int_0^{2\pi} \int_0^{\pi/3} \frac{\sin 2\phi}{2}\rho \Big|_{\rho=\frac{2}{\cos\phi}}^4 d\phi\, d\theta = \int_0^{2\pi} \int_0^{\pi/3} \left(2\sin 2\phi - \frac{\sin 2\phi}{2} \cdot \frac{2}{\cos\phi} \right) d\phi\, d\theta$$

$$= \int_0^{2\pi} \int_0^{\pi/3} (2\sin 2\phi - 2\sin\phi)\, d\phi\, d\theta = 2\pi \cdot \left(-\cos 2\phi + 2\cos\phi \, \Big|_{\phi=0}^{\pi/3} \right)$$

$$= 2\pi \cdot \left(-\cos\frac{2\pi}{3} + 2\cos\frac{\pi}{3} + 1 - 2 \right) = \pi$$

51. Calculate the volume of the sphere $x^2 + y^2 + z^2 = a^2$, using both spherical and cylindrical coordinates.

SOLUTION Spherical coordinates: In the entire sphere of radius a, we have

$$\mathcal{W}: 0 \le \theta \le 2\pi, \ 0 \le \phi \le \pi, \ 0 \le \rho \le a$$

Using triple integral in spherical coordinates we get

$$V = \iiint_{\mathcal{W}} 1\, dV = \int_0^{2\pi} \int_0^\pi \int_0^a \rho^2 \sin\phi\, d\rho\, d\phi\, d\theta = \left(\int_0^a \rho^2 d\rho \right)\left(\int_0^\pi \sin\phi\, d\phi \right)\left(\int_0^{2\pi} 1\, d\theta \right)$$

$$= \left(\frac{\rho^3}{3} \Big|_0^a \right)\left(-\cos\phi \, \Big|_0^\pi \right)\left(\theta \, \Big|_0^{2\pi} \right) = \frac{a^3}{3} \cdot 2 \cdot 2\pi = \frac{4\pi a^3}{3}$$

Cylindrical coordinates: The projection of $\mathcal{W}$ onto the xy-plane is the circle of radius a, that is,

$$\mathcal{D}: 0 \le \theta \le 2\pi, \ 0 \le r \le a$$

The upper surface is $z = \sqrt{a^2 - (x^2 + y^2)} = \sqrt{a^2 - r^2}$ and the lower surface is $z = -\sqrt{a^2 - r^2}$. Therefore, $\mathcal{W}$ has the following description in cylindrical coordinates:

$$\mathcal{W}: 0 \le \theta \le 2\pi, \ 0 \le r \le a, \ -\sqrt{a^2 - r^2} \le z \le \sqrt{a^2 - r^2}$$

We obtain the following integral:

$$V = \int_0^{2\pi} \int_0^a \int_{-\sqrt{a^2-r^2}}^{\sqrt{a^2-r^2}} r\, dz\, dr\, d\theta = \int_0^{2\pi} \int_0^a rz \Big|_{z=-\sqrt{a^2-r^2}}^{\sqrt{a^2-r^2}} dr\, d\theta = \int_0^{2\pi} \int_0^a 2r\sqrt{a^2-r^2}\, dr\, d\theta \qquad (1)$$

We compute the inner integral using the substitution $u = \sqrt{a^2 - r^2}$, $du = -\frac{r}{u}\, dr$. We get

$$\int_0^a 2r\sqrt{a^2 - r^2}\, dr = \int_a^0 -2u^2\, du = \int_0^a 2u^2\, du = \frac{2u^3}{3} \Big|_0^a = \frac{2a^3}{3}$$

Substituting in (1) gives

$$V = \int_0^{2\pi} \frac{2a^3}{3}\, d\theta = \frac{2a^3}{3}\theta \Big|_0^{2\pi} = \frac{2a^3}{3} \cdot 2\pi = \frac{4\pi a^3}{3}.$$

53. Bell-Shaped Curve One of the key results in calculus is the computation of the area under the bell-shaped curve (Figure 23):

$$I = \int_{-\infty}^{\infty} e^{-x^2} \, dx$$

This integral appears throughout engineering, physics, and statistics, and although e^{-x^2} does not have an elementary antiderivative, we can compute I using multiple integration.

(a) Show that $I^2 = J$, where J is the improper double integral

$$J = \int_{-\infty}^{\infty} \int_{-\infty}^{\infty} e^{-x^2 - y^2} \, dx \, dy$$

Hint: Use Fubini's Theorem and $e^{-x^2 - y^2} = e^{-x^2} e^{-y^2}$.

(b) Evaluate J in polar coordinates.

(c) Prove that $I = \sqrt{\pi}$.

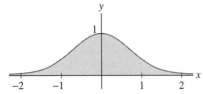

FIGURE 23 The bell-shaped curve $y = e^{-x^2}$.

SOLUTION

(a) We must show that $I^2 = J$. Firstly, consider the following:

$$I^2 = I \cdot I = \int_{-\infty}^{\infty} e^{-x^2} \, dx \cdot \int_{-\infty}^{\infty} e^{-y^2} \, dy = \int_{-\infty}^{\infty} e^{-x^2} \cdot e^{-y^2} \, dx \, dy = \int_{-\infty}^{\infty} e^{-x^2 - y^2} \, dx \, dy$$

This works because each integral after the first equals sign is independent of the other.

(b) The improper integral over the xy-plane can be computed as the limit as $\mathcal{R} \to \infty$ of the double integrals over the disk. $\mathcal{D}_{\mathcal{R}}$ is defined by

$$\mathcal{D}_{\mathcal{R}} : 0 \le \theta \le 2\pi, \ 0 \le r \le \mathcal{R}$$

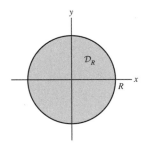

That is,

$$J = \lim_{\mathcal{R} \to \infty} \iint_{\mathcal{D}_{\mathcal{R}}} e^{-(x^2 + y^2)} \, dx \, dy \tag{1}$$

We compute the double integral using polar coordinates. The function is $f(x, y) = e^{-(x^2 + y^2)} = e^{-r^2}$, hence

$$\iint_{\mathcal{D}_{\mathcal{R}}} e^{-(x^2 + y^2)} \, dx \, dy = \int_0^{2\pi} \int_0^{\mathcal{R}} e^{-r^2} r \, dr \, d\theta = \left(\int_0^{2\pi} 1 \, d\theta \right) \left(\int_0^{\mathcal{R}} e^{-r^2} r \, dr \right) = 2\pi \int_0^{\mathcal{R}} e^{-r^2} r \, dr$$

We compute the integral using the substitution $u = r^2$, $du = 2r \, dr$. We get

$$\iint_{\mathcal{D}_{\mathcal{R}}} e^{-(x^2 + y^2)} \, dx \, dy = 2\pi \int_0^{\mathcal{R}^2} e^{-u} \frac{du}{2} = \pi \int_0^{\mathcal{R}^2} e^{-u} \, du = \pi(-e^{-u}) \Big|_0^{\mathcal{R}^2} = \pi(1 - e^{-\mathcal{R}^2}) \tag{2}$$

Combining (1) and (2), we get

$$J = \lim_{\mathcal{R} \to \infty} \left(\pi \left(1 - e^{-\mathcal{R}^2} \right) \right) = \pi$$

On the other hand, using the Iterated Integral of a Product Function, we get

$$\pi = J = \int_{-\infty}^{\infty} \int_{-\infty}^{\infty} e^{-x^2 - y^2} \, dx \, dy = \int_{-\infty}^{\infty} \int_{-\infty}^{\infty} e^{-x^2} \cdot e^{-y^2} \, dx \, dy$$

$$= \left(\int_{-\infty}^{\infty} e^{-x^2} \, dx \right) \left(\int_{-\infty}^{\infty} e^{-y^2} \, dy \right) = I^2$$

(c) That is,

$$I^2 = \pi \quad \Rightarrow \quad I = \sqrt{\pi}$$

Further Insights and Challenges

55. Prove the formula

$$\iint_{\mathcal{D}} \ln r \, dA = -\frac{\pi}{2}$$

where $r = \sqrt{x^2 + y^2}$ and $\mathcal{D}$ is the unit disk $x^2 + y^2 \le 1$. This is an improper integral since $\ln r$ is not defined at $(0, 0)$, so integrate first over the annulus $a \le r \le 1$ where $0 < a < 1$, and let $a \to 0$.

SOLUTION

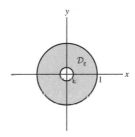

The improper integral I is computed by the limit as $a \to 0^+$ of the integrals over the annulus $\mathcal{D}_a$ defined by

$$\mathcal{D}_a : 0 \le \theta \le 2\pi, \ a \le r \le 1$$

Using double integrals in polar coordinates and integration by parts, we get

$$I_a = \int_0^{2\pi} \int_a^1 (\ln r) \cdot r \, dr \, d\theta = 2\pi \int_a^1 r \ln r \, dr = 2\pi \left(\frac{r^2 \ln r}{2} - \frac{r^2}{4} \Big|_a^1 \right)$$

$$= 2\pi \left(\frac{\ln 1}{2} - \frac{1}{4} - \frac{a^2 \ln a}{2} + \frac{a^2}{4} \right) = \frac{\pi}{2} \left(a^2 - 2a^2 \ln a - 1 \right)$$

We now compute the limit of I_a as $a \to 0^+$. We use L'Hôpital's rule to obtain

$$I = \lim_{a \to 0^+} \frac{\pi}{2} (a^2 - 2a^2 \ln a - 1) = -\frac{\pi}{2} - \pi \lim_{a \to 0^+} a^2 \ln a = -\frac{\pi}{2} - \pi \lim_{a \to 0^+} \frac{\ln a}{a^{-2}}$$

$$= -\frac{\pi}{2} - \pi \lim_{a \to 0^+} \frac{a^{-1}}{-2a^{-3}} = -\frac{\pi}{2} + \frac{\pi}{2} \lim_{a \to 0^+} a^2 = -\frac{\pi}{2}$$

15.5 Applications of Multiple Integrals (LT Section 16.5)

Preliminary Questions

1. What is the mass density $\rho(x, y, z)$ of a solid of volume 5 m^3 with uniform mass density and total mass 25 kg?

SOLUTION Here, recall that

$$\text{total mass} = \iiint_{\mathcal{W}} \rho(x, y, z) \, dV$$

Since we are told that the solid has volume 5, and $\rho(x, y, z)$ is uniform (i.e. constant, let $\rho(x, y, z) = \rho$), we can write:

$$25 = \iiint_{\mathcal{W}} \rho(x, y, z) \, dV = \rho \cdot V(\mathcal{W}) = 5\rho, \quad \Rightarrow \quad \rho = 5 \text{ kg/m}^3$$

2. A domain $\mathcal{D}$ in $\mathbf{R}^2$ with uniform mass density is symmetric with respect to the y-axis. Which of the following are true?

(a) $x_{CM} = 0$ **(b)** $y_{CM} = 0$ **(c)** $I_x = 0$ **(d)** $I_y = 0$

SOLUTION Here, the x-coordinate of the center of mass, $x_{CM} = 0$ (a) since $x_{CM} = \frac{M_y}{M}$ and $M_y = \iint_{\mathcal{D}} x\rho(x, y)\, dA$. Since $\rho(x, y) = \rho(-x, y)$, then we see that $(-x)\rho(-x, y) = -x\rho(x, y)$ and M_y is an integral of an odd function over a symmetric region, hence $M_y = 0$.

3. If $p(x, y)$ is the joint probability density function of random variables X and Y, what does the double integral of $p(x, y)$ over $[0, 1] \times [0, 1]$ represent? What does the integral of $p(x, y)$ over the triangle bounded by $x = 0$, $y = 0$, and $x + y = 1$ represent?

SOLUTION The double integral of $p(x, y)$ over $[0, 1] \times [0, 1]$ represents the probability that both X and Y are between 0 and 1. The integral of $p(x, y)$ over the triangle bounded by $x = 0$, $y = 0$, and $x + y = 1$ represents the probability that both X and Y are nonnegative and $X + Y \leq 1$.

Exercises

1. Find the total mass of the square $0 \leq x \leq 1, 0 \leq y \leq 1$ assuming a mass density of

$$\rho(x, y) = x^2 + y^2$$

SOLUTION

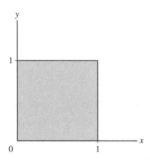

The total mass M is obtained by integrating the mass density $\rho(x, y) = x^2 + y^2$ over the square $\mathcal{D}$ in the xy-plane. This gives

$$M = \iint_{\mathcal{D}} \rho(x, y)\, dA = \int_0^1 \int_0^1 \left(x^2 + y^2\right) dx\, dy = \int_0^1 \frac{x^3}{3} + y^2 x \Big|_{x=0}^{1} dy$$

$$= \int_0^1 \left(\frac{1}{3} + y^2 - 0\right) dy = \int_0^1 \left(\frac{1}{3} + y^2\right) dy = \frac{y}{3} + \frac{y^3}{3} \Big|_0^1 = \frac{1}{3} + \frac{1}{3} = \frac{2}{3}$$

3. Find the total charge in the region under the graph of $y = 4e^{-x^2/2}$ for $0 \leq x \leq 10$ (in centimeters) assuming a charge density of $\rho(x, y) = 10^{-6}xy$ coulombs per square centimeter.

SOLUTION The total charge C of the region is obtained by computing the double integral of charge density $\rho(x, y) = 10^{-6}xy$ over the region defined by the inequalities

$$0 \leq x \leq 10, \quad 0 \leq y \leq 4e^{-x^2/2}$$

Therefore, we compute the double integral

$$C = \iint_{\mathcal{D}} \rho(x, y)\, dA = \int_0^{10} \int_0^{4e^{-x^2/2}} 10^{-6}xy\, dy\, dx = 10^{-6} \int_0^{10} \left(\frac{1}{2}xy^2 \Big|_{y=0}^{4e^{-x^2/2}}\right) dx$$

$$= \frac{1}{2} \cdot 10^{-6} \int_0^{10} 16xe^{-x^2}\, dx = -4 \cdot 10^{-6} \int_0^{10} e^{-x^2}(-2x\, dx) = -4 \cdot 10^{-6} \left(e^{-x^2} \Big|_0^{10}\right)$$

$$= -4 \cdot 10^{-6} \left(e^{-10^2} - 1\right) = \frac{1}{250,000}\left(1 - e^{-100}\right)$$

5. Find the total population within the sector $2|x| \leq y \leq 8$ assuming a population density of $\rho(x, y) = 100e^{-0.1y}$ people per square kilometer.

SOLUTION The total population P of the region is obtained by computing the double integral of population density $\rho(x, y) = 100e^{-0.1y}$ over the region defined by the inequality $2|x| \leq y \leq 8$. This means the region can be split into two vertically simple regions described by the inequalities:

$$0 \leq x \leq 4, \quad 2x \leq y \leq 8$$

and

$$-4 \leq x \leq 4, \quad -2x \leq y \leq 8$$

Now to compute the double integral:

$$\iint_{\mathcal{D}} \rho(x, y)\, dA = \iint_{\mathcal{D}_1} \rho(x, y)\, dA + \iint_{\mathcal{D}_2} \rho(x, y)\, dA$$

$$\iint_{\mathcal{D}_1} \rho(x, y)\, dA + \iint_{\mathcal{D}_2} \rho(x, y)\, dA = \int_0^4 \int_{2x}^8 100e^{-0.1y}\, dy\, dx + \int_{-4}^0 \int_{-2x}^8 100e^{-0.1y}\, dy\, dx$$

$$= \int_0^4 \left.\frac{100}{-0.1}e^{-0.1y}\right|_{y=2x}^8 dx + \int_{-4}^0 \left.\frac{100}{-0.1}e^{-0.1y}\right|_{y=-2x}^8 dx$$

$$= -1000 \int_0^4 e^{-0.8} - e^{-0.2x}\, dx - 1000 \int_{-4}^0 e^{-0.8} - e^{0.2x}\, dx$$

$$= -1000 \left(e^{-0.8}x + 5e^{-0.2x}\Big|_0^4\right) - 1000 \left(e^{-0.8}x - 5e^{0.2x}\Big|_{-4}^0\right)$$

$$= -1000 \left(4e^{-0.8} + 5e^{-0.8} - 5\right) - 1000 \left(-5 + 4e^{-0.8} + 5e^{-0.8}\right)$$

$$= -1000 \left(18e^{-0.8} - 10\right) \approx 1912$$

7. Calculate the total charge of the solid ball $x^2 + y^2 + z^2 \leq 5$ (in centimeters) assuming a charge density (in coulombs per cubic centimeter) of

$$\rho(x, y, z) = (3 \cdot 10^{-8})(x^2 + y^2 + z^2)^{1/2}$$

SOLUTION To calculate total charge, first we consider the solid ball in spherical coordinates:

$$x^2 + y^2 + z^2 \leq 5 \quad \Rightarrow \quad 0 < \rho \leq \sqrt{5}, \quad 0 \leq \theta \leq 2\pi, \quad 0 \leq \phi \leq \pi$$

And the charge density function too, let us rename it $R(x, y, z)$:

$$R(x, y, z) = (3 \cdot 10^{-8})(x^2 + y^2 + z^2)^{1/2} \quad \Rightarrow \quad R(\rho \sin\phi \cos\theta, \rho \sin\phi \sin\theta, \rho \cos\phi) = (3 \cdot 10^{-8})\rho$$

Then integrating to compute the total charge we have:

$$\int_0^{2\pi} \int_0^\pi \int_0^{\sqrt{5}} (3 \cdot 10^{-8})\rho \cdot \rho^2 \sin\phi\, d\rho\, d\phi\, d\theta = 3 \cdot 10^{-8} \int_0^{2\pi} \int_0^\pi \int_0^{\sqrt{5}} \rho^3 \sin\phi\, d\rho\, d\phi\, d\theta$$

$$= 3 \cdot 10^{-8} \int_0^{2\pi} \int_0^\pi \sin\phi \left(\frac{1}{4}\rho^4\Big|_0^{\sqrt{5}}\right) d\phi\, d\theta = 3 \cdot 10^{-8} \cdot \frac{25}{4} \int_0^{2\pi} \int_0^\pi \sin\phi\, d\phi\, d\theta$$

$$= 3 \cdot 10^{-8} \cdot \frac{25}{4} \int_0^{2\pi} -\cos\phi\Big|_0^\pi d\theta = 3 \cdot 10^{-8} \cdot \frac{25}{4} \int_0^{2\pi} 2\, d\theta = 3 \cdot 10^{-8} \cdot 25\pi$$

$$\approx 2.356 \cdot 10^{-6}$$

9. Assume that the density of the atmosphere as a function of altitude h (in km) above sea level is $\rho(h) = ae^{bh}$ kg/km^3, where $a = 1.225 \times 10^9$ and $b = 0.13$. Calculate the total mass of the atmosphere contained in the cone-shaped region $\sqrt{x^2 + y^2} \leq h \leq 3$.

SOLUTION First we must consider the given cone in cylindrical coordinates:

$$\sqrt{x^2 + y^2} \leq z \leq 3 \quad \Rightarrow \quad r \leq z \leq 3$$

while

$$0 \le r \le 3, \quad 0 \le \theta \le 2\pi$$

And the density function as well:

$$\rho(x, y, z) = ae^{-bz} \quad \Rightarrow \quad \rho(r \cos \theta, r \sin \theta, z) = ae^{-bz}$$

Now to compute the total mass of the atmosphere in question:

$$\int_0^{\theta=2\pi} \int_{r=0}^3 \int_{z=r}^3 ae^{-bz} \cdot r \, dz \, dr \, d\theta = \int_0^{2\pi} \int_0^3 \int_r^3 r(ae^{-bz}) \, dz \, dr \, d\theta$$

$$= \int_0^{2\pi} \int_0^3 r \left(-\frac{1}{b} ae^{-bz} \Big|_{z=r}^3 \right) dr \, d\theta$$

$$= -\frac{a}{b} \int_0^{2\pi} \int_0^3 re^{-3b} - re^{-br} \, dr \, d\theta$$

$$= -\frac{a}{b} \int_0^{2\pi} \frac{1}{2} r^2 e^{-3b} \Big|_0^3 - \left(-\frac{1}{b} re^{-br} - \frac{1}{b^2} e^{-br} \Big|_0^3 \right) d\theta$$

$$= -\frac{a}{b} \cdot 2\pi \left(\frac{9}{2} e^{-3b} + \frac{3}{b} e^{-3b} + \frac{1}{b^2} e^{-3b} - \frac{1}{b^2} \right)$$

Now, since $a = 1.225 \times 10^9$ and $b = 0.13$ we have that the total mass is

$$-\frac{a}{b} \cdot 2\pi \left(\frac{9}{2} e^{-3b} + \frac{3}{b} e^{-3b} + \frac{1}{b^2} e^{-3b} - \frac{1}{b^2} \right) \approx 2.593 \times 10^{10}$$

In Exercises 11–14, find the centroid of the given region.

11. Region bounded by $y = 1 - x^2$ and $y = 0$

SOLUTION First we will compute the area of the region:

$$\text{Area}(\mathcal{D}) = \int_{-1}^1 \int_0^{1-x^2} dy \, dx = \int_{-1}^1 1 - x^2 \, dx = x - \frac{1}{3} x^3 \Big|_{-1}^1 = 1 - \frac{1}{3} - \left(-1 + \frac{1}{3} \right) = \frac{4}{3}$$

It is clear from symmetry that $\bar{x} = 0$, and

$$\bar{y} = \frac{1}{\text{Area}(\mathcal{D})} \iint_{\mathcal{D}} y \, dA = \frac{3}{4} \int_{-1}^1 \int_0^{1-x^2} y \, dy \, dx$$

$$= \frac{3}{4} \int_{-1}^1 \frac{1}{2} y^2 \Big|_0^{1-x^2} dx = \frac{3}{8} \int_{-1}^1 (1 - x^2)^2 \, dx$$

$$= \frac{3}{8} \int_{-1}^1 1 - 2x^2 + x^4 \, dx = \frac{3}{8} \left(x - \frac{2}{3} x^3 + \frac{1}{5} x^5 \right) \Big|_{-1}^1$$

$$= \frac{3}{8} \left(1 - \frac{2}{3} + \frac{1}{5} \right) - \frac{3}{8} \left(-1 + \frac{2}{3} - \frac{1}{5} \right) = \frac{2}{5}$$

The centroid has coordinates $(\bar{x}, \bar{y}) = \left(0, \frac{2}{5} \right)$.

13. Quarter circle $x^2 + y^2 \le R^2, x \ge 0, y \ge 0$

SOLUTION

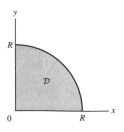

The centroid $P = (\bar{x}, \bar{y})$ has the following coordinates:

$$\bar{x} = \frac{1}{\text{Area}(\mathcal{D})} \iint_{\mathcal{D}} x \, dA = \frac{4}{\pi R^2} \iint_{\mathcal{D}} x \, dA$$

$$\bar{y} = \frac{1}{\text{Area}(\mathcal{D})} \iint_{\mathcal{D}} y \, dA = \frac{4}{\pi R^2} \iint_{\mathcal{D}} y \, dA$$

We compute the integrals using polar coordinates. The domain $\mathcal{D}$ is described in polar coordinates by the inequalities

$$\mathcal{D} : 0 \le \theta \le \frac{\pi}{2}, \ 0 \le r \le R$$

The functions are $x = r \cos\theta$ and $y = r \sin\theta$, respectively. Using the Change of Variables Formula gives

$$\bar{x} = \frac{4}{\pi R^2} \int_0^{\pi/2} \int_0^R r \cos\theta \cdot r \, dr \, d\theta = \frac{4}{\pi R^2} \int_0^{\pi/2} \int_0^R r^2 \cos\theta \, dr \, d\theta = \frac{4}{\pi R^2} \int_0^{\pi/2} \frac{r^3 \cos\theta}{3} \Big|_{r=0}^R d\theta$$

$$= \frac{4}{\pi R^2} \int_0^{\pi/2} \frac{R^3 \cos\theta}{3} \, d\theta = \frac{4R}{3\pi} \sin\theta \Big|_0^{\pi/2} = \frac{4R}{3\pi} \left(\sin\frac{\pi}{2} - \sin 0 \right) = \frac{4R}{3\pi}$$

And,

$$\bar{y} = \frac{4}{\pi R^2} \int_0^{\pi/2} \int_0^R r \sin\theta \cdot r \, dr \, d\theta = \frac{4}{\pi R^2} \int_0^{\pi/2} \int_0^R r^2 \sin\theta \, dr \, d\theta = \frac{4}{\pi R^2} \int_0^{\pi/2} \frac{r^3 \sin\theta}{3} \Big|_{r=0}^R d\theta$$

$$= \frac{4}{\pi R^2} \int_0^{\pi/2} \frac{R^3 \sin\theta}{3} \, d\theta = \frac{4R}{3\pi} (- \cos\theta) \Big|_0^{\pi/2} = \frac{4R}{3\pi} \left(-\cos\frac{\pi}{2} + \cos 0 \right) = \frac{4R}{3\pi}$$

Notice that we can use the symmetry of $\mathcal{D}$ with respect to x and y to conclude that $\bar{y} = \bar{x}$, and save the computation of $\bar{y}$. We obtain the centroid $P = \left(\frac{4R}{3\pi}, \frac{4R}{3\pi} \right)$.

15. $\mathcal{CAS}$ Use a computer algebra system to compute numerically the centroid of the shaded region in Figure 12 bounded by $r^2 = \cos 2\theta$ for $x \ge 0$.

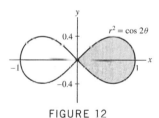

FIGURE 12

SOLUTION Using symmetry, it is easy to see $\bar{y} = 0$. Also, computing the area of the region,

$$\text{Area} = 2 \cdot \frac{1}{2} \int_{-\pi/4}^{\pi/4} r^2 \, d\theta = \int_{-\pi/4}^{\pi/4} \cos 2\theta \, d\theta = \frac{1}{2} \sin 2\theta \Big|_{-\pi/4}^{\pi/4} = 1$$

and we will compute $\bar{x}$ as

$$\bar{x} = \frac{1}{A} \iint_{\mathcal{D}} x \, dA = \int_{\theta=-\pi/4}^{\pi/4} \int_{r=0}^{\sqrt{\cos 2\theta}} r \cos\theta \cdot r \, dr \, d\theta = \frac{\sqrt{2}}{16} \pi \approx 0.278$$

Therefore, we have that the centroid is $(\bar{x}, \bar{y}) = (\sqrt{2}\pi/16, 0)$.

In Exercises 17–19, find the centroid of the given solid region.

17. Hemisphere $x^2 + y^2 + z^2 \le R^2, z \ge 0$

SOLUTION First we need to find the volume of the solid in question. It is a hemisphere, so using geometry, we have

$$\text{Volume} = \frac{1}{2} \cdot \frac{4}{3} \pi R^3 = \frac{2\pi R^3}{3}$$

The centroid is the point P with the following coordinates:

$$\bar{x} = \frac{1}{V} \iiint_{\mathcal{W}} x \, dV, \quad \bar{y} = \frac{1}{V} \iiint_{\mathcal{W}} y \, dV, \quad \bar{z} = \frac{1}{V} \iiint_{\mathcal{W}} z \, dV$$

By symmetry, it is clear that $\bar{x} = \bar{y} = 0$, and using spherical coordinates,

$$\bar{z} = \frac{1}{V} \iiint_{region} z \, dV = \frac{3}{2\pi R^3} \int_0^{2\pi} \int_0^{\pi/2} \int_0^R \rho \cos \phi \cdot \rho^2 \sin \phi \, d\rho \, d\phi \, d\theta$$

$$= \frac{3}{2\pi R^3} \int_0^{2\pi} \int_0^{\pi/2} \int_0^R \rho^3 \cos \phi \sin \phi \, d\rho \, d\phi \, d\theta$$

$$= \frac{3}{2\pi R^3} \int_0^{2\pi} 1 \, d\theta \cdot \int_0^{\pi/2} \cos \phi \sin \phi \, d\phi \cdot \int_0^R \rho^3 \, d\rho$$

$$= \frac{3}{2\pi R^3} \cdot 2\pi \left(\frac{1}{2} \sin^2 \phi \Big|_0^{\pi/2} \right) \left(\frac{1}{4} \rho^4 \Big|_0^R \right) = \frac{3}{2\pi R^3} \cdot 2\pi \cdot \frac{1}{2} \cdot \frac{1}{4} R^4 = \frac{3R}{8}$$

Therefore, the coordinates of the centroid of a hemisphere having radius R, are $(0, 0, 3R/8)$.

19. The "ice cream cone" region $\mathcal{W}$ bounded, in spherical coordinates, by the cone $\phi = \pi/3$ and the sphere $\rho = 2$

SOLUTION First we must find the volume of this solid:

$$V = \int_{\theta=0}^{2\pi} \int_{\phi=0}^{\pi/3} \int_{\rho=0}^{2} \rho^2 \sin \phi \, d\rho \, d\phi \, d\theta$$

$$= 2\pi \left(\int_0^{\pi/3} \sin \phi \, d\phi \right) \left(\int_0^2 \rho^2 \, d\rho \right) = 2\pi \cdot \frac{8}{3} \left(-\cos \phi \Big|_0^{\pi/3} \right)$$

$$= \frac{16\pi}{3} \cdot \frac{1}{2} = \frac{8\pi}{3}$$

And now compute the coordinates of the centroid. By symmetry, it is clear that $\bar{x} = \bar{y} = 0$.

$$\bar{z} = \frac{1}{V} \iiint_{\mathcal{W}} z \, dV = \frac{3}{8\pi} \int_{\theta=0}^{2\pi} \int_{\phi=0}^{\pi/3} \int_{\rho=0}^{2} \rho \cos \phi \cdot \rho^2 \sin \phi \, d\rho \, d\phi \, d\theta$$

$$= \frac{3}{8\pi} \int_0^{2\pi} d\theta \cdot \int_0^{\pi/3} \cos \phi \sin \phi \, d\phi \cdot \int_0^2 \rho^3 \, d\rho$$

$$= \frac{3}{8\pi} \cdot 2\pi \cdot \left(\frac{1}{2} \sin^2 \phi \Big|_0^{\pi/3} \right) \left(\frac{1}{4} \rho^4 \Big|_0^2 \right) = \frac{3}{4} \left(\frac{1}{2} \cdot \frac{3}{4} \right) (4) = \frac{9}{8}$$

The coordinates of the centroid are $(0, 0, 9/8)$.

21. Find the centroid of the region $\mathcal{W}$ lying above the sphere $x^2 + y^2 + z^2 = 6$ and below the paraboloid $z = 4 - x^2 - y^2$ (Figure 15).

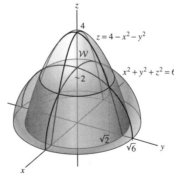

FIGURE 15

SOLUTION The centroid is the point P with the following coordinates:

$$\bar{x} = \frac{1}{V} \iiint_{\mathcal{W}} x \, dV, \quad \bar{y} = \frac{1}{V} \iiint_{\mathcal{W}} y \, dV, \quad \bar{z} = \frac{1}{V} \iiint_{\mathcal{W}} z \, dV$$

In a previous section we showed that the volume of the region is $V = 1.54\pi$. We also showed that $\mathcal{D}$ has the following definition in cylindrical coordinates:

$$0 \le \theta \le 2\pi, \quad 0 \le r \le \sqrt{2}, \quad \sqrt{6 - r^2} \le z \le 4 - r^2$$

Using this information we compute the coordinates of the centroid by the following integrals:

$$\overline{x} = \frac{1}{1.54\pi} \int_0^{2\pi} \int_0^{\sqrt{2}} \int_{\sqrt{6-r^2}}^{4-r^2} (r\cos\theta) r \, dz \, dr \, d\theta = \frac{1}{1.54\pi} \int_0^{2\pi} \int_0^{\sqrt{2}} r^2 \cos\theta \, z \Big|_{z=\sqrt{6-r^2}}^{4-r^2} dr \, d\theta$$

$$= \frac{1}{1.54\pi} \int_0^{2\pi} \int_0^{\sqrt{2}} r^2 \cos\theta \left(4 - r^2 - \sqrt{6-r^2}\right) dr \, d\theta$$

$$= \frac{1}{1.54\pi} \int_0^{2\pi} \cos\theta \int_0^{\sqrt{2}} \left(4r^2 - r^4 - r^2\sqrt{6-r^2}\right) dr \, d\theta \qquad (1)$$

We denote the inner integral by a and compute the second integral to obtain

$$\overline{x} = \frac{1}{1.54\pi} \int_0^{2\pi} \cos\theta \cdot a \, d\theta = \frac{1}{1.54\pi} a \sin\theta \Big|_0^{2\pi} = 0$$

The value $\overline{x} = 0$ is the result of the symmetry of $\mathcal{W}$ with respect to the yz-plane. Similarly, since $\mathcal{W}$ is symmetric with respect to the xz-plane, the average value of the y-coordinate is zero.

$\overline{y} = 0$:

We compute the z-coordinate of the centroid:

$$\overline{z} = \frac{1}{1.54\pi} \int_0^{2\pi} \int_0^{\sqrt{2}} \int_{\sqrt{6-r^2}}^{4-r^2} zr \, dz \, dr \, d\theta = \frac{1}{1.54\pi} \int_0^{2\pi} \int_0^{\sqrt{2}} \frac{z^2 r}{2} \Big|_{z=\sqrt{6-r^2}}^{4-r^2} dr \, d\theta$$

$$= \frac{1}{1.54\pi} \int_0^{2\pi} \int_0^{\sqrt{2}} \frac{r}{2}\left((4-r^2)^2 - \left(\sqrt{6-r^2}\right)^2\right) dr \, d\theta$$

$$= \frac{1}{2 \cdot 1.54\pi} \int_0^{2\pi} \int_0^{\sqrt{2}} (r^5 - 7r^3 + 10r) \, dr \, d\theta$$

$$= \frac{1}{3.08\pi} \int_0^{2\pi} \frac{r^6}{6} - \frac{7r^4}{4} + 5r^2 \Big|_{r=0}^{\sqrt{2}} d\theta = \frac{1}{3.08\pi} \cdot \frac{13}{3} \cdot 2\pi \approx 2.81$$

Therefore the centroid of $\mathcal{W}$ is

$$P = (0, 0, 2.81).$$

In Exercises 23–26, find the center of mass of the region with the given mass density ρ.

23. Region bounded by $y = 4 - x$, $x = 0$, $y = 0$; $\rho(x, y) = x$

SOLUTION The mass of the region is

$$M = \int_0^4 \int_0^{4-x} x \, dy \, dx = \int_0^4 xy \Big|_0^{4-x} dx = \int_0^4 4x - x^2 \, dx = 2x^2 - \frac{1}{3}x^3 \Big|_0^4 = 32 - \frac{64}{3} = \frac{32}{3}$$

and we have

$$M_x = \int_0^4 \int_0^{4-x} yx \, dy \, dx = \int_0^4 \frac{1}{2}xy^2 \Big|_0^{4-x} dx = \frac{1}{2} \int_0^4 16x - 8x^2 + x^3 \, dx$$

$$= \frac{1}{2}\left(8x^2 - \frac{8}{3}x^3 + \frac{1}{4}x^4\right)\Big|_0^4 = \frac{1}{2}\left(128 - \frac{512}{3} + 64\right) = \frac{32}{3}$$

and

$$M_y = \int_0^4 \int_0^{4-x} x^2 \, dy \, dx = \int_0^4 x^2 y \Big|_0^{4-x} dx = \int_0^4 4x^2 - x^3 \, dx$$

$$= \left(\frac{4}{3} x^3 - \frac{1}{4} x^4 \right) \Big|_0^4 = \frac{256}{3} - 64 = \frac{64}{3}$$

and thus the center of mass is

$$\left(\frac{M_y}{M}, \frac{M_x}{M} \right) = \left(\frac{64}{3} \cdot \frac{3}{32}, \frac{32}{3} \cdot \frac{3}{32} \right) = (2, 1)$$

25. Region $|x| + |y| \le 1$; $\rho(x, y) = (x + 1)(y + 1)$

SOLUTION For $x \le 0$, the region is defined by $-1 \le x \le 0$ and $-1 - x \le y \le 1 + x$; for $x \ge 0$, it is parameterized by $0 \le x \le 1$ and $-1 + x \le y \le 1 - x$. The mass of the region is thus

$$M = \int_{-1}^0 \int_{-1-x}^{1+x} (x + 1)(y + 1) \, dy \, dx + \int_0^1 \int_{x-1}^{1-x} (x + 1)(y + 1) \, dy \, dx$$

$$= \frac{1}{2} \left(\int_{-1}^0 (x + 1)(y + 1)^2 \Big|_{y=-1-x}^{1+x} dx + \int_0^1 (x + 1)(y + 1)^2 \Big|_{y=x-1}^{1-x} dx \right)$$

$$= \frac{1}{2} \left(\int_{-1}^0 (x + 1)((x + 2)^2 - (-x)^2) \, dx + \int_0^1 (x + 1)((2 - x)^2 - x^2) \, dx \right)$$

$$= \frac{1}{2} \left(\int_{-1}^0 4(x + 1)^2 \, dx + \int_0^1 4(1 - x^2) \, dx \right)$$

$$= 2 \left(\left(\frac{1}{3}(x + 1)^3 \right) \Big|_{-1}^0 + \left(x - \frac{1}{3} x^3 \right) \Big|_0^1 \right) = 2 \left(\frac{1}{3} + \left(1 - \frac{1}{3} \right) \right) = 2$$

We have

$$M_x = \int_{-1}^0 \int_{-1-x}^{1+x} y(x + 1)(y + 1) \, dy \, dx + \int_0^1 \int_{x-1}^{1-x} y(x + 1)(y + 1) \, dy \, dx$$

$$= \int_{-1}^0 \int_{-1-x}^{1+x} (x + 1)(y^2 + y) \, dy \, dx + \int_0^1 \int_{x-1}^{1-x} (x + 1)(y^2 + y) \, dy \, dx$$

$$= \int_{-1}^0 (x + 1) \left(\frac{1}{3} y^3 + \frac{1}{2} y^2 \right) \Big|_{y=-1-x}^{1+x} dx + \int_0^1 (x + 1) \left(\frac{1}{3} y^3 + \frac{1}{2} y^2 \right) \Big|_{y=x-1}^{1-x} dx$$

$$= \int_{-1}^0 (x + 1) \left(\frac{2}{3} + 2x + 2x^2 + \frac{2}{3} x^3 \right) dx + \int_0^1 (x + 1) \left(\frac{2}{3} - 2x + 2x^2 - \frac{2}{3} x^3 \right) dx$$

$$= \int_{-1}^0 \frac{2}{3} x^4 + \frac{8}{3} x^3 + 4x^2 + \frac{8}{3} x + \frac{2}{3} \, dx + \int_0^1 -\frac{2}{3} x^4 + \frac{4}{3} x^3 - \frac{4}{3} x + \frac{2}{3} \, dx$$

$$= \left(\frac{2}{15} x^5 + \frac{2}{3} x^4 + \frac{4}{3} x^3 + \frac{4}{3} x^2 + \frac{2}{3} x \right) \Big|_{-1}^0 + \left(-\frac{2}{15} x^5 + \frac{1}{3} x^4 - \frac{2}{3} x^3 + \frac{2}{3} x \right) \Big|_0^1$$

$$= \frac{2}{15} + \frac{1}{5} = \frac{1}{3}$$

Since the region and the density function are symmetric in x and y, we must have also $M_y = M_x = \frac{1}{3}$. Then the center of mass is

$$\left(\frac{M_y}{M}, \frac{M_x}{M} \right) = \left(\frac{1}{3} \cdot \frac{1}{2}, \frac{1}{3} \cdot \frac{1}{2} \right) = \left(\frac{1}{6}, \frac{1}{6} \right)$$

27. Find the z-coordinate of the center of mass of the first octant of the unit sphere with mass density $\rho(x, y, z) = y$ (Figure 17).

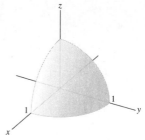

FIGURE 17

SOLUTION We use spherical coordinates:

$$x = \rho \cos \theta \sin \phi, \quad y = \rho \sin \theta \sin \phi, \quad z = \rho \cos \phi$$

$$dV = \rho^2 \sin \phi \, d\rho \, d\phi \, d\theta$$

The octant $\mathcal{W}$ is defined by $0 \le \theta \le \frac{\pi}{2}, 0 \le \phi \le \frac{\pi}{2}, 0 \le \rho \le 1$, so we have

$$M_{xy} = \iiint_{\mathcal{W}} z \, \rho(x, y, z) \, dV = \int_{\theta=0}^{\pi/2} \int_{\phi=0}^{\pi/2} \int_{\rho=0}^{1} (\rho \cos \phi)(\rho \sin \theta \sin \phi) \, \rho^2 \sin \phi \, d\rho \, d\phi \, d\theta$$

$$= \left(\int_{\theta=0}^{\pi/2} \sin \theta \, d\theta \right) \left(\int_{\phi=0}^{\pi/2} \cos \phi \sin^2 \phi \, d\phi \right) \left(\int_{\rho=0}^{1} \rho^4 \, d\rho \right)$$

$$= (1) \left(\frac{1}{3} \sin^3 \phi \Big|_0^{\pi/2} \right) \left(\frac{1}{5} \right) = \frac{1}{15}$$

The total mass M of $\mathcal{W}$ is equal to the integral of the mass density $\rho(x, y, z)$:

$$M = \iiint_{\mathcal{W}} \rho(x, y, z) \, dV = \int_{\theta=0}^{\pi/2} \int_{\phi=0}^{\pi/2} \int_{\rho=0}^{1} (\rho \sin \theta \sin \phi) \, \rho^2 \sin \phi \, d\rho \, d\phi \, d\theta$$

$$= \left(\int_{\theta=0}^{\pi/2} \sin \theta \, d\theta \right) \left(\int_{\phi=0}^{\pi/2} \sin^2 \phi \, d\phi \right) \left(\int_{\rho=0}^{1} \rho^3 \, d\rho \right) = (1) \left(\frac{\pi}{4} \right) \left(\frac{1}{4} \right) = \frac{\pi}{16}$$

We conclude that

$$z_{\text{CM}} = \frac{1}{M} \iiint_{\mathcal{W}} z \, \rho(x, y, z) \, dV = \frac{1/15}{\pi/16} = \frac{16}{15\pi} \approx 0.34$$

29. Let $\mathcal{R}$ be the rectangle $[-a, a] \times [b, -b]$ with uniform density and total mass M. Calculate:
(a) The mass density ρ of $\mathcal{R}$
(b) I_x and I_0
(c) The radius of gyration about the x-axis

SOLUTION

(a) The mass density is simply the mass per unit area since the density is uniform; this is

$$\frac{M}{4ab}$$

(b) We have

$$I_x = \iint_{\mathcal{R}} y^2 \rho(x, y) \, dA = \frac{M}{4ab} \int_{-a}^{a} \int_{-b}^{b} y^2 \, dy \, dx = \frac{2aM}{4ab} \int_{-b}^{b} y^2 \, dy$$

$$= \frac{M}{2b} \cdot \frac{1}{3} y^3 \Big|_{-b}^{b} = \frac{1}{3} M b^2$$

and

$$I_0 = \iint_{\mathcal{R}} (x^2 + y^2) \rho(x, y) \, dA = \frac{M}{4ab} \int_{-a}^{a} \int_{-b}^{b} x^2 + y^2 \, dy \, dx = \frac{M}{4ab} \int_{-a}^{a} x^2 y + \frac{1}{3} y^3 \Big|_{-b}^{b} dx$$

$$= \frac{2M}{4ab} \int_{-a}^{a} x^2 b + \frac{1}{3} b^3 \, dx = \frac{M}{2ab} \left(\frac{b}{3} x^3 + \frac{b^3}{3} x \right) \Big|_{-a}^{a}$$

$$= \frac{M}{2ab} \left(\frac{2}{3} b a^3 + \frac{2}{3} b^3 a \right) = \frac{1}{3} M (a^2 + b^2)$$

(c) The radius of gyration about the x-axis is defined to be

$$\sqrt{\frac{I_x}{M}} = \sqrt{\frac{Mb^2}{3} \cdot \frac{1}{M}} = \frac{b}{\sqrt{3}}$$

31. Calculate I_0 and I_x for the disk $\mathcal{D}$ defined by $x^2 + y^2 \leq 16$ (in meters), with total mass 1000 kg and uniform mass density. *Hint:* Calculate I_0 first and observe that $I_0 = 2I_x$. Express your answer in the correct units.

SOLUTION Note that the area of the disk is $\pi r^2 = 16\pi$ so that the mass density is

$$\rho(x, y) = \frac{1000}{16\pi} = \frac{125}{2\pi}$$

Then using polar coordinates we have

$$I_0 = \iint_{\mathcal{D}} (x^2 + y^2)\frac{125}{2\pi}\, dA = \frac{125}{2\pi} \int_0^4 \int_0^{2\pi} r^2 \cdot r \, d\theta \, dr = 125 \cdot \frac{1}{4}r^4 \Big|_0^4 = 125 \cdot 64 = 8000 \text{ kg-m}^2$$

Since both the region and the mass density are symmetric in x and y, we have $I_x = I_y$. But then $I_0 = I_x + I_y = 2I_x$ so that

$$I_x = 4000 \text{ kg-m}^2$$

In Exercises 33–36, let $\mathcal{D}$ be the triangular domain bounded by the coordinate axes and the line $y = 3 - x$, with mass density $\rho(x, y) = y$. Compute the given quantities.

33. Total mass

SOLUTION The total mass is simply

$$\iint_{\mathcal{D}} \rho(x, y)\, dA = \int_0^3 \int_0^{3-x} y \, dy \, dx = \frac{1}{2} \int_0^3 y^2 \Big|_0^{3-x} dx = \frac{1}{2} \int_0^3 (3 - x)^2 \, dx = -\frac{1}{6}(3 - x)^3 \Big|_0^3 = \frac{27}{6} = \frac{9}{2}$$

35. I_x

SOLUTION

$$I_x = \iint_{\mathcal{D}} y^2 \rho(x, y)\, dA = \int_0^3 \int_0^{3-x} y^3 \, dy \, dx = \frac{1}{4} \int_0^3 (3 - x)^4 \, dx$$

$$= -\frac{1}{20}(3 - x)^5 \Big|_0^3 = \frac{1}{20}3^5 = \frac{243}{20}$$

In Exercises 37–40, let $\mathcal{D}$ be the domain between the line $y = bx/a$ and the parabola $y = bx^2/a^2$ where $a, b > 0$. Assume the mass density is $\rho(x, y) = xy$. Compute the given quantities.

37. Centroid

SOLUTION The curves intersect at $x = 0$ and at $x = a$. The area is

$$A = \iint_{\mathcal{D}} 1 \, dA = \int_0^a \int_{bx^2/a^2}^{bx/a} 1 \, dy \, dx = \int_0^a \frac{bx}{a} - \frac{bx^2}{a^2} \, dx$$

$$= \left(\frac{bx^2}{2a} - \frac{bx^3}{3a^2}\right)\Big|_0^a = \frac{ab}{2} - \frac{ab}{3} = \frac{ab}{6}$$

Then

$$\bar{x} = \frac{1}{A} \iint_{\mathcal{D}} x \, dA = \frac{6}{ab} \int_0^a \int_{bx^2/a^2}^{bx/a} x \, dy \, dx = \frac{6}{ab} \int_0^a x\left(\frac{bx}{a} - \frac{bx^2}{a^2}\right) dx$$

$$= \frac{6}{ab} \int_0^a \frac{b}{a}x^2 - \frac{b}{a^2}x^3 \, dx = \frac{6}{ab}\left(\frac{b}{3a}a^3 - \frac{b}{4a^2}a^4\right)$$

$$= \frac{6}{ab}\left(\frac{a^2 b}{3} - \frac{a^2 b}{4}\right) = \frac{a}{2}$$

and

$$\bar{y} = \frac{1}{A}\iint_{\mathcal{D}} y\,dA = \frac{6}{ab}\int_0^a \int_{bx^2/a^2}^{bx/a} y\,dy\,dx = \frac{6}{2ab}\int_0^a y^2 \Big|_{bx^2/a^2}^{bx/a}\,dx$$

$$= \frac{3}{ab}\int_0^a \frac{b^2}{a^2}x^2 - \frac{b^2}{a^4}x^4\,dx = \frac{3}{ab}\left(\frac{b^2}{3a^2}x^3 - \frac{b^2}{5a^4}x^5\right)\Big|_0^a$$

$$= \frac{3}{ab}\left(\frac{ab^2}{3} - \frac{ab^2}{5}\right) = \frac{2b}{5}$$

39. I_x

SOLUTION The curves intersect at $x = 0$ and at $x = a$, so

$$I_x = \iint_{\mathcal{D}} y^2 \rho(x, y)\,dA = \int_0^a \int_{bx^2/a^2}^{bx/a} xy^3\,dy\,dx = \frac{1}{4}\int_0^a xy^4 \Big|_{bx^2/a^2}^{bx/a}\,dx$$

$$= \frac{1}{4}\int_0^a x\left(\frac{b^4}{a^4}x^4 - \frac{b^4}{a^8}x^8\right)dx = \frac{b^4}{4a^8}\int_0^a a^4 x^5 - x^9\,dx$$

$$= \frac{b^4}{4a^8}\left(\frac{a^4}{6}x^6 - \frac{1}{10}x^{10}\right)\Big|_0^a = \frac{b^4}{4a^8}\left(\frac{a^{10}}{6} - \frac{a^{10}}{10}\right) = \frac{a^2 b^4}{60}$$

41. Calculate the moment of inertia I_x of the disk $\mathcal{D}$ defined by $x^2 + y^2 \le R^2$ (in meters) with total mass M kg. How much kinetic energy (in joules) is required to rotate the disk about the x-axis with angular velocity 10 rad/s?

SOLUTION The area of the disk is πR^2, so its mass density is

$$\rho(x, y) = \frac{M}{\pi R^2}$$

We compute I_x using polar coordinates:

$$I_x = \iint_{\mathcal{D}} y^2 \rho(x, y)\,dA = \frac{M}{\pi R^2}\int_0^{2\pi}\int_0^R (r\sin\theta)^2 r\,dr\,d\theta = \frac{M}{\pi R^2}\int_0^{2\pi}\int_0^R r^3 \sin^2\theta\,dr\,d\theta$$

$$= \frac{M}{\pi R^2}\left(\int_0^{2\pi}\sin^2\theta\,d\theta\right)\left(\int_0^R r^3\,dr\right)$$

$$= \frac{M}{\pi R^2}\cdot \pi \cdot \frac{R^4}{4} = \frac{1}{4}MR^2$$

It follows that the kinetic energy required to rotate the disk about the x-axis with angular velocity 10 rad/s is

$$\frac{1}{2}I_x\omega^2 = \frac{1}{8}MR^2 \cdot 100 = \frac{25}{2}MR^2 \text{ joules}$$

43. Show that the moment of inertia of a sphere of radius R of total mass M with uniform mass density about any axis passing through the center of the sphere is $\frac{2}{5}MR^2$. Note that the mass density of the sphere is $\rho = M/\left(\frac{4}{3}\pi R^3\right)$.

SOLUTION Since the sphere is symmetric under an arbitrary rotation, and since the mass density is uniform, it follows that the moments of inertia of the sphere about all axes passing through its center are equal. Thus it suffices to prove the result for an arbitrary axis; we choose the z-axis. Then, using spherical coordinates, we have

$$I_z = \iiint_{\mathcal{S}} (x^2 + y^2)\rho(x, y, z)\,dA$$

$$= \frac{3M}{4\pi R^3}\int_0^{2\pi}\int_0^\pi \int_0^R \left((r\cos\theta\sin\phi)^2 + (r\sin\theta\sin\phi)^2\right)r^2 \sin\phi\,dr\,d\theta\,d\phi$$

$$= \frac{3M}{4\pi R^3}\int_0^{2\pi}\int_0^\pi \int_0^R r^4 \sin^3\phi\,dr\,d\theta\,d\phi = \frac{3M}{4\pi R^3}\cdot 2\pi\left(\int_0^\pi \sin^3\phi\,d\phi\right)\left(\int_0^R r^4\,dr\right)$$

$$= \frac{3M}{2R^3}\cdot \frac{4}{3}\cdot \frac{1}{5}R^5 = \frac{2}{5}MR^2$$

In Exercises 45 and 46, prove the formula for the right circular cylinder in Figure 18.

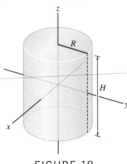

FIGURE 18

45. $I_z = \frac{1}{2}MR^2$

SOLUTION Assuming the cylinder has uniform mass density 1, and using cylindrical coordinates, we have

$$I_z = \iiint_C (x^2 + y^2)\rho(x, y, z)\, dA = \int_0^R \int_0^{2\pi} \int_{-H/2}^{H/2} r^2 \cdot r\, dz\, d\theta\, dr$$

$$= 2\pi H \int_0^R r^3\, dr = \frac{1}{2}\pi R^4 H$$

But the volume of the cylinder, which is equal to its mass, is $\pi R^2 H$, so that

$$I_z = \frac{1}{2}\pi R^4 H = \frac{1}{2}MR^2$$

47. The yo-yo in Figure 19 is made up of two disks of radius $r = 3$ cm and an axle of radius $b = 1$ cm. Each disk has mass $M_1 = 20$ g, and the axle has mass $M_2 = 5$ g.

(a) Use the result of Exercise 45 to calculate the moment of inertia I of the yo-yo with respect to the axis of symmetry. Note that I is the sum of the moments of the three components of the yo-yo.

(b) The yo-yo is released and falls to the end of a 100-cm string, where it spins with angular velocity ω. The total mass of the yo-yo is $m = 45$ g, so the potential energy lost is PE $= mgh = (45)(980)100$ g-cm^2/s^2. Find ω under the assumption that one-third of this potential energy is converted into rotational kinetic energy.

Axle of radius b

r

FIGURE 19

SOLUTION

(a) If the figure is rotated by 90°, it looks like three right circular cylinders oriented as in Exercise 45. The moment of inertia of each of the disks around the axis of rotation is thus

$$\frac{1}{2}MR^2 = \frac{1}{2} \cdot 20 \cdot 3^2 = 90 \text{ g-cm}^2$$

and the moment of inertia of the axle around the axis of rotation is

$$\frac{1}{2}MR^2 = \frac{1}{2} \cdot 5 \cdot 1^2 = \frac{5}{2} \text{ g-cm}^2$$

Thus the total moment of inertia of the yo-yo around its axis of rotation is

$$2 \cdot 90 + \frac{5}{2} = 182.5 \text{ g-cm}^2$$

(b) If one third of the potential energy is converted to kinetic energy, then

$$\frac{1}{3} \cdot 45 \cdot 980 \cdot 100 = \frac{1}{2} \cdot 182.5 \cdot \omega^2 \text{ g-cm}^2$$

so that

$$\omega = \sqrt{\frac{2}{3} \cdot \frac{45 \cdot 980 \cdot 100}{182.5}} \approx 127 \text{ radians/sec} = \frac{127}{2\pi} \approx 20.2 \text{ rotations/sec}$$

49. Calculate $P(0 \le X \le 2; 1 \le Y \le 2)$, where X and Y have joint probability density function

$$p(x, y) = \begin{cases} \frac{1}{72}(2xy + 2x + y) & \text{if } 0 \le x \le 4 \text{ and } 0 \le y \le 2 \\ 0 & \text{otherwise} \end{cases}$$

SOLUTION The region $0 \le X \le 2; 1 \le Y \le 2$ falls into the region where $p(x, y)$ is defined by the first line of the given formula, so that

$$P(0 \le X \le 2; 1 \le Y \le 2) = \int_0^2 \int_1^2 \frac{1}{72}(2xy + 2x + y)\, dy\, dx = \frac{1}{72} \int_0^2 \left(xy^2 + 2xy + \frac{1}{2}y^2 \right) \Big|_1^2 dx$$

$$= \frac{1}{72} \int_0^2 5x + \frac{3}{2}\, dx = \frac{1}{72} \left(\frac{5}{2}x^2 + \frac{3}{2}x \right) \Big|_0^2 = \frac{1}{72} \cdot 13 = \frac{13}{72} \approx 0.18$$

51. The lifetime (in months) of two components in a certain device are random variables X and Y that have joint probability distribution function

$$p(x, y) = \begin{cases} \frac{1}{9216}(48 - 2x - y) & \text{if } x \ge 0, y \ge 0, 2x + y < 48 \\ 0 & \text{otherwise} \end{cases}$$

Calculate the probability that both components function for at least 12 months without failing. Note that $p(x, y)$ is nonzero only within the triangle bounded by the coordinate axes and the line $2x + y = 48$ shown in Figure 20.

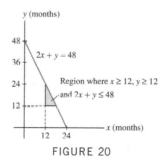

FIGURE 20

SOLUTION Both components function for at least 12 months without failing if $X + Y \ge 12$; however, we must also have $2X + Y \le 48$. Then the region of integration is the shaded triangle in the figure; the lower left corner of that triangle is $(12, 12)$. One of the remaining vertices is the intersection of $x = 12$ and $2x + y = 48$; solving for y we have $y = 24$, so the point is $(12, 24)$. The other vertex is the intersection of $y = 12$ and $2x + y = 48$; solving for x gives $x = 18$, so the point is $(18, 12)$. The region of integration is then $12 \le x \le 18$ and $12 \le y \le 48 - 2x$. Thus the probability is

$$P(X \ge 12, Y \ge 12) = \int_{12}^{18} \int_{12}^{48-2x} \frac{1}{9216}(48 - 2x - y)\, dy\, dx$$

$$= \frac{1}{9216} \int_{12}^{18} \left(48y - 2xy - \frac{1}{2}y^2 \right) \Big|_{12}^{48-2x} dx$$

$$= \frac{1}{9216} \int_{12}^{18} 1800 - 96x - 2x(36 - 2x) - \frac{1}{2}(48 - 2x)^2\, dx$$

$$= \frac{1}{9216} \left(1800x - 48x^2 - 36x^2 + \frac{4}{3}x^3 + \frac{1}{12}(48 - 2x)^3 \right) \Big|_{12}^{18}$$

$$= \frac{144}{9216} = \frac{1}{64} = .015625$$

53. Find a constant C such that

$$p(x, y) = \begin{cases} Cy & \text{if } 0 \le x \le 1 \text{ and } x^2 \le y \le x \\ 0 & \text{otherwise} \end{cases}$$

is a joint probability density function. Then calculate the probability that $Y \ge X^{3/2}$.

SOLUTION $p(x, y)$ is a joint probability density function if

$$1 = \int_0^1 \int_{x^2}^x p(x, y)\, dy\, dx = \int_0^1 \int_{x^2}^x Cy\, dy\, dx$$

$$= \frac{C}{2} \int_0^1 y^2 \Big|_{x^2}^x dx = \frac{C}{2} \int_0^1 x^2 - x^4\, dx = \frac{C}{15}$$

so that we must have $C = 15$. Now, for $0 \le x \le 1$ we have $x^2 \le x^{3/2} \le x$, so that

$$P(Y \ge X^{3/2}) = \int_0^1 \int_{x^{3/2}}^x 15y\, dy\, dx = \frac{15}{2} \int_0^1 y^2 \Big|_{x^{3/2}}^x dx$$

$$= \frac{15}{2} \int_0^1 x^2 - x^3\, dx = \frac{15}{2} \left(\frac{1}{3} - \frac{1}{4} \right) = \frac{15}{24} = \frac{5}{8} = 0.375$$

55. According to quantum mechanics, the x- and y-coordinates of a particle confined to the region $\mathcal{R} = [0, 1] \times [0, 1]$ are random variables with joint probability density function

$$p(x, y) = \begin{cases} C \sin^2(2\pi \ell x) \sin^2(2\pi ny) & \text{if } (x, y) \in \mathcal{R} \\ 0 & \text{otherwise} \end{cases}$$

The integers ℓ and n determine the energy of the particle, and C is a constant.

(a) Find the constant C.

(b) Calculate the probability that a particle with $\ell = 2$, $n = 3$ lies in the region $\left[0, \frac{1}{4}\right] \times \left[0, \frac{1}{8}\right]$.

SOLUTION

(a) We have

$$\int_0^1 \int_0^1 C \sin^2(2\pi \ell x) \sin^2(2\pi ny)\, dx\, dy = C \left(\int_0^1 \sin^2(2\pi \ell x)^2\, dx \right) \left(\int_0^1 \sin^2(2\pi ny)^2\, dy \right)$$

Now, since ℓ is an integer, using the substitution $u = 2\pi \ell x$, $du = 2\pi \ell\, dx$, we have

$$\int_0^1 \sin^2(2\pi \ell x)\, dx = \frac{1}{2\pi \ell} \int_0^{2\pi \ell} \sin^2 u\, du = \frac{1}{2\pi \ell} \left(\frac{1}{2}u - \frac{1}{2} \sin u \cos u \right) \Big|_0^{2\pi \ell}$$

$$= \frac{1}{2\pi \ell} \left(\pi \ell - \frac{1}{2} \sin(2\pi \ell) \cos(2\pi \ell) + \frac{1}{2} \sin 0 \cos 0 \right) = \frac{1}{2}$$

and the same is true of $\int_0^1 \sin^2(2\pi ny)\, dy$. Thus the value of the entire integral is $C \frac{1}{2} \cdot \frac{1}{2} = \frac{C}{4}$. In order for this to be a joint probability density function, then, we must have $C = 4$.

(b) We compute

$$\int_0^{1/4} \int_0^{1/8} 4 \sin^2(2\pi \cdot 2x) \sin^2(2\pi \cdot 3y)\, dy\, dx = 4 \left(\int_0^{1/4} \sin^2(4\pi x)\, dx \right) \left(\int_0^{1/8} \sin^2(6\pi y)\, dy \right)$$

$$= 4 \left(\frac{1}{4\pi} \int_0^\pi \sin^2 u\, du \right) \left(\frac{1}{6\pi} \int_0^{3\pi/4} \sin^2 v\, dv \right)$$

$$= 4 \left(\frac{1}{4\pi} \cdot \frac{\pi}{2} \right) \left(\frac{1}{6\pi} \cdot \left(\frac{3\pi}{8} + \frac{1}{4} \right) \right)$$

$$= 4 \left(\frac{1}{8} \right) \left(\frac{1}{16} + \frac{1}{24\pi} \right) = \frac{1}{32} + \frac{1}{48\pi} \approx 0.03788$$

57. ▱ According to Coulomb's Law, the force between two electric charges of magnitude q_1 and q_2 separated by a distance r is kq_1q_2/r^2 (k is a negative constant). Let F be the net force on a charged particle P of charge Q coulombs located d centimeters above the center of a circular disk of radius R with a uniform charge distribution of density ρ C/m^2 (Figure 21). By symmetry, F acts in the vertical direction.

(a) Let $\mathcal{R}$ be a small polar rectangle of size $\Delta r \times \Delta \theta$ located at distance r. Show that $\mathcal{R}$ exerts a force on P whose vertical component is

$$\left(\frac{k\rho Qd}{(r^2 + d^2)^{3/2}}\right) r \, \Delta r \, \Delta \theta$$

(b) Explain why F is equal to the following double integral, and evaluate:

$$F = k\rho Qd \int_0^{2\pi} \int_0^R \frac{r \, dr \, d\theta}{(r^2 + d^2)^{3/2}}$$

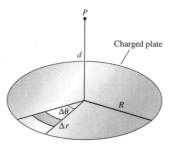

FIGURE 21

SOLUTION

(a) The area of the small polar rectangle $\mathcal{R}$ is

$$\Delta A = \frac{1}{2}(r + \Delta r)^2 \Delta \theta - \frac{1}{2}r^2 \Delta \theta = r(\Delta r \, \Delta \theta) + \frac{1}{2}\Delta r^2 \Delta \theta \approx r(\Delta r \, \Delta \theta)$$

Therefore, the charge on $\mathcal{R}$ is $q_1 = \rho r (\Delta r \, \Delta \theta)$. The distance between P and $\mathcal{R}$ is, by the Pythagorean Law, $\sqrt{r^2 + d^2}$.

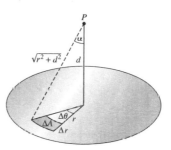

Therefore, the magnitude of the force between P and $\mathcal{R}$ is

$$\frac{kq_1 q_2}{\left(\sqrt{r^2 + d^2}\right)^2} = \frac{k\left(\rho r \Delta r \, \Delta \theta\right) Q}{r^2 + d^2} = \frac{k\rho Q}{r^2 + d^2} r \Delta r \, \Delta \theta$$

The vertical component of this force is obtained by multiplying the force by $\cos\alpha = \dfrac{d}{\sqrt{r^2+d^2}}$. That is,

$$F_{\text{vert}} = \frac{k\rho Q}{r^2 + d^2} \cdot \frac{d}{\sqrt{r^2 + d^2}} r \Delta r \, \Delta \theta = \frac{k\rho Qd}{(r^2 + d^2)^{3/2}} r \Delta r \, \Delta \theta$$

(b) Since F acts in the vertical direction, it is approximated by the Riemann sum of the forces F_{vert} in part (a), over the polar rectangles. This Riemann sum approximates F in higher precision if we let $\Delta\theta \to 0$ and $\Delta r \to 0$. The result is the double integral of $\frac{k\rho Qd}{(r^2+d^2)^{3/2}}$ over the disk. The disk is determined by $0 \le \theta \le 2\pi$ and $0 \le r \le R$. Therefore we get

$$F = \int_0^R \int_0^{2\pi} \frac{k\rho Qd}{(r^2 + d^2)^{3/2}} r \, dr \, d\theta = 2\pi k\rho Qd \int_0^R \int_0^{2\pi} \frac{r \, dr \, d\theta}{(r^2 + d^2)^{3/2}}$$

Using the u-substitution of $u = r^2 + d^2$, $du = 2r \, dr$, we continue

$$F = 2\pi k\rho Qd \int_0^R \frac{r \, dr}{(r^2 + d^2)^{3/2}} = 2\pi k\rho Qd \cdot \frac{1}{2} \int_{d^2}^{R^2 + d^2} u^{-3/2} \, du = -\pi k\rho Qd u^{-1/2} \Big|_{d^2}^{R^2 + d^2}$$

$$= \pi k\rho Qd \left(\frac{1}{d} - \frac{1}{\sqrt{R^2 + d^2}}\right)$$

Further Insights and Challenges

59. Let $\mathcal{D}$ be the domain in Figure 22. Assume that $\mathcal{D}$ is symmetric with respect to the y-axis; that is, both $g_1(x)$ and $g_2(x)$ are even functions.

(a) Prove that the centroid lies on the y-axis—that is, that $\overline{x} = 0$.

(b) Show that if the mass density satisfies $\rho(-x, y) = \rho(x, y)$, then $M_y = 0$ and $x_{\mathrm{CM}} = 0$.

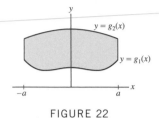

FIGURE 22

SOLUTION

(a) Assume $\mathcal{D}$ has area A. Then the x coordinate of the centroid is

$$\overline{x} = \frac{1}{A} \iint_{\mathcal{D}} x \, dA = \frac{1}{A} \int_{-a}^{a} \int_{g_1(x)}^{g_2(x)} x \, dy \, dx = \frac{1}{A} \int_{-a}^{a} x(g_2(x) - g_1(x)) \, dx$$

Since g_1 and g_2 are both even functions, $x(g_2(x) - g_1(x))$ is an odd function, so its integral over a region symmetric about the x-axis is zero. Thus $\overline{x} = 0$.

(b) Let $R(x, y)$ be an antiderivative of $\rho(x, y)$ with respect to y, i.e. $R(x, y) = \int \rho(x, y) \, dy$. Note that

$$R(-x, y) = \int \rho(-x, y) \, dy = \int \rho(x, y) \, dy = R(x, y)$$

so that R is an even function with respect to x. Now, we have

$$M_y \iint_{\mathcal{D}} x\rho(x, y) \, dA = \int_{-a}^{a} \int_{g_1(x)}^{g_2(x)} x\rho(x, y) \, dy \, dx = \int_{-a}^{a} x(R(x, g_2(x)) - R(x, g_1(x))) \, dx$$

Since g_1, g_2, and R are all even functions of x, we have

$$R(-x, g_2(-x)) - R(-x, g_1(-x)) = R(-x, g_2(x)) - R(-x, g_1(x)) = R(x, g_2(x)) - R(x, g_1(x))$$

so that the second factor in the integrand is an even function of x. But x is an odd function of x, so their product is odd. It follows that the integral over the range $-a \leq x \leq a$ is zero. Thus $M_y = x_{\mathrm{CM}} = 0$.

61. Use Pappus's Theorem in Exercise 60 to show that the torus obtained by revolving a circle of radius b centered at $(0, a)$ about the x-axis (where $b < a$) has volume $V = 2\pi^2 ab^2$.

SOLUTION The centroid of the circle is obviously at the center, $(0, a)$, and the area of the circle is πb^2, so that by Pappus' theorem,

$$V = 2\pi \pi b^2 a = 2\pi ab^2$$

63. Parallel-Axis Theorem Let $\mathcal{W}$ be a region in $\mathbf{R}^3$ with center of mass at the origin. Let I_z be the moment of inertia of $\mathcal{W}$ about the z-axis, and let I_h be the moment of inertia about the vertical axis through a point $P = (a, b, 0)$, where $h = \sqrt{a^2 + b^2}$. By definition,

$$I_h = \iiint_{\mathcal{W}} ((x - a)^2 + (y - b)^2)\rho(x, y, z) \, dV$$

Prove the Parallel-Axis Theorem: $I_h = I_z + Mh^2$.

SOLUTION We have

$$I_h - I_z = \iiint_{\mathcal{W}} ((x - a)^2 + (y - b)^2)\rho(x, y, z) \, dV - \iiint_{\mathcal{W}} (x^2 + y^2)\rho(x, y, z) \, dV$$

$$= \iiint_{\mathcal{W}} (-2ax + a^2 - 2by + b^2)\rho(x, y, z) \, dV$$

$$= (a^2 + b^2) \iiint_{\mathcal{W}} \rho(x, y, z) \, dV - 2a \iiint_{\mathcal{W}} x\rho(x, y, z) \, dV - 2b \iiint_{\mathcal{W}} y\rho(x, y, z) \, dV$$

$$= (a^2 + b^2)M - 2aM_{yz} - 2bM_{xz} = Mh^2$$

since the last two terms are zero because the center of mass of $\mathcal{W}$ is at the origin.

15.6 Change of Variables (LT Section 16.6)

Preliminary Questions

1. Which of the following maps is linear?

(a) (uv, v) **(b)** $(u + v, u)$ **(c)** $(3, e^u)$

SOLUTION

(a) This map is not linear since it does not satisfy the linearity property:

$$\Phi(2u, 2v) = (2u \cdot 2v, 2v) = (4uv, 2v) = 2(2uv, v)$$

$$2\Phi(u, v) = 2(uv, v) \quad \Rightarrow \quad \Phi(2u, 2v) \neq 2\Phi(u, v)$$

(b) This map is linear since it has the form $\Phi(u, v) = (Au + Cv, Bu + Dv)$ where $A = C = 1$, $B = 1$, $D = 0$.

(c) This map is not linear since it does not satisfy the linearity properties. For example,

$$\Phi(2u, 2v) = (3, e^{2u})$$
$$2\Phi(u, v) = 2(3, e^u) \quad \Rightarrow \quad \Phi(2u, 2v) \neq 2\Phi(u, v)$$

2. Suppose that Φ is a linear map such that $\Phi(2, 0) = (4, 0)$ and $\Phi(0, 3) = (-3, 9)$. Find the images of:

(a) $\Phi(1, 0)$ **(b)** $\Phi(1, 1)$ **(c)** $\Phi(2, 1)$

SOLUTION We denote the linear map by $\Phi(u, v) = (Au + Cv, Bu + Dv)$. By the given information we have

$$\Phi(2, 0) = (A \cdot 2 + C \cdot 0, B \cdot 2 + D \cdot 0) = (2A, 2B) = (4, 0)$$

$$\Phi(0, 3) = (A \cdot 0 + C \cdot 3, B \cdot 0 + D \cdot 3) = (3C, 3D) = (-3, 9)$$

Therefore,

$$\begin{aligned} 2A &= 4 \\ 2B &= 0 \\ 3C &= -3 \\ 3D &= 9 \end{aligned} \quad \Rightarrow \quad A = 2, \quad B = 0, \quad C = -1, \quad D = 3$$

The linear map is thus

$$\Phi(u, v) = (2u - v, 3v)$$

We now compute the images:

(a) $\Phi(1, 0) = (2 \cdot 1 - 0, 3 \cdot 0) = (2, 0)$
(b) $\Phi(1, 1) = (2 \cdot 1 - 1, 3 \cdot 1) = (1, 3)$
(c) $\Phi(2, 1) = (2 \cdot 2 - 1, 3 \cdot 1) = (3, 3)$

3. What is the area of $\Phi(\mathcal{R})$ if $\mathcal{R}$ is a rectangle of area 9 and Φ is a mapping whose Jacobian has constant value 4?

SOLUTION

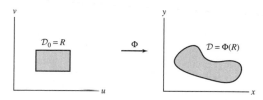

The areas of $\mathcal{D}_0 = \Phi(\mathcal{R})$ and $\mathcal{D} = \mathcal{R}$ are the following integrals:

$$\text{Area}(\mathcal{R}) = 9 = \iint_{\mathcal{D}_0} 1 \, du \, dv$$

$$\text{Area}(\Phi(\mathcal{R})) = \iint_{\mathcal{D}} 1 \, dx \, dy$$

Using the Change of Variables Formula, we have

$$\text{Area}(\Phi(\mathcal{R})) = \iint_{\mathcal{D}} 1 \, dx \, dy = \iint_{\mathcal{D}_0} 1 |\text{Jac}\Phi| \, du \, dv = \iint_{\mathcal{D}_0} 4 \, du \, dv = 4 \iint_{\mathcal{D}_0} 1 \, du \, dv = 4 \cdot 9 = 36$$

The area of $\Phi(\mathcal{R})$ is 36.

4. Estimate the area of $\Phi(\mathcal{R})$, where $\mathcal{R} = [1, 1.2] \times [3, 3.1]$ and Φ is a mapping such that $\text{Jac}(\Phi)(1, 3) = 3$.

SOLUTION

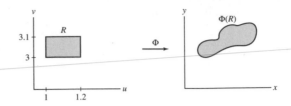

We use the following estimation:

$$\text{Area}\,(\Phi(\mathcal{R})) \approx |\text{Jac}\,(\Phi)\,(P)|\text{Area}(\mathcal{R})$$

The area of the rectangle $\mathcal{R}$ is

$$\text{Area}(\mathcal{R}) = 0.2 \cdot 0.1 = 0.02$$

We choose the sample point $P = (1, 3)$ in $\mathcal{R}$ to obtain the following estimation:

$$\text{Area}\,(\Phi(\mathcal{R})) \approx |\text{Jac}\,(\Phi)\,(1, 3)|\text{Area}(\mathcal{R}) = 3 \cdot 0.02 = 0.06$$

Exercises

1. Determine the image under $\Phi(u, v) = (2u, u + v)$ of the following sets:

(a) The u- and v-axes

(b) The rectangle $\mathcal{R} = [0, 5] \times [0, 7]$

(c) The line segment joining $(1, 2)$ and $(5, 3)$

(d) The triangle with vertices $(0, 1)$, $(1, 0)$, and $(1, 1)$

SOLUTION

(a) The image of the u-axis is obtained by substituting $v = 0$ in $\Phi(u, v) = (2u, u + v)$. That is,

$$\Phi(u, 0) = (2u, u + 0) = (2u, u).$$

The image of the u-axis is the set of points $(x, y) = (2u, u)$, which is the line $y = \frac{1}{2}x$ in the xy-plane. The image of the v-axis is obtained by substituting $u = 0$ in $\Phi(u, v) = (2u, u + v)$. That is,

$$\Phi(0, v) = (0, 0 + v) = (0, v).$$

Therefore, the image of the v-axis is the set $(x, y) = (0, v)$, which is the vertical line $x = 0$ (the y-axis).

(b) Since Φ is a linear map, the segment through points P and Q is mapped to the segment through $\Phi(P)$ and $\Phi(Q)$. We thus must find the images of the vertices of $\mathcal{R}$:

$$\Phi(0, 0) = (2 \cdot 0, 0 + 0) = (0, 0)$$
$$\Phi(5, 0) = (2 \cdot 5, 5 + 0) = (10, 5)$$
$$\Phi(5, 7) = (2 \cdot 5, 5 + 7) = (10, 12)$$
$$\Phi(0, 7) = (2 \cdot 0, 0 + 7) = (0, 7)$$

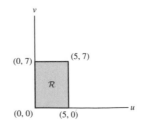

The image of $\mathcal{R}$ is the parallelogram with vertices $(0, 0)$, $(10, 5)$, $(10, 12)$, and $(0, 7)$ in the xy-plane.

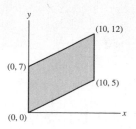

(c) We compute the images of the endpoints of the segment:

$$\Phi(1, 2) = (2 \cdot 1, 1 + 2) = (2, 3)$$

$$\Phi(5, 3) = (2 \cdot 5, 5 + 3) = (10, 8)$$

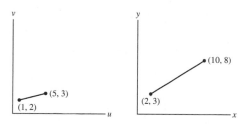

The image is the segment in the xy-plane joining the points $(2, 3)$ and $(10, 8)$.

(d) Since Φ is linear, the image of the triangle is the triangle whose vertices are the images of the vertices of the triangle. We compute these images:

$$\Phi(0, 1) = (2 \cdot 0, 0 + 1) = (0, 1)$$

$$\Phi(1, 0) = (2 \cdot 1, 1 + 0) = (2, 1)$$

$$\Phi(1, 1) = (2 \cdot 1, 1 + 1) = (2, 2)$$

Therefore the image is the triangle in the xy-plane whose vertices are at the points $(0, 1)$, $(2, 1)$, and $(2, 2)$.

3. Let $\Phi(u, v) = (u^2, v)$. Is Φ one-to-one? If not, determine a domain on which Φ is one-to-one. Find the image under Φ of:

(a) The u- and v-axes

(b) The rectangle $\mathcal{R} = [-1, 1] \times [-1, 1]$

(c) The line segment joining $(0, 0)$ and $(1, 1)$

(d) The triangle with vertices $(0, 0)$, $(0, 1)$, and $(1, 1)$

SOLUTION Φ is not one-to-one since for any $u \neq 0$, (u, v) and $(-u, v)$ are two different points with the same image. However, Φ is one-to-one on the domain $\{(u, v) : u \geq 0\}$ and on the domain $\{(u, v) : u \leq 0\}$.

(a) The image of the u-axis is the set of the points

$$(x, y) = \Phi(u, 0) = (u^2, 0) \quad \Rightarrow \quad x = u^2, \quad y = 0$$

That is, the positive x-axis, including the origin. The image of the v-axis is the set of the following points:

$$(x, y) = \Phi(0, v) = (0^2, v) = (0, v) \quad \Rightarrow \quad x = 0, \quad y = v$$

That is, the line $x = 0$, which is the y-axis.

(b) The rectangle $\mathcal{R}$ is defined by

$$|u| \leq 1, \quad |v| \leq 1$$

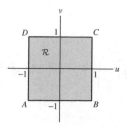

Since $x = u^2$ and $y = v$, we have $u = \pm\sqrt{x}$ and $v = y$ (depending on our choice of domain). Therefore, the inequalities for x and y are

$$|\pm\sqrt{x}| \le 1, \quad |y| \le 1$$

or

$$0 \le x \le 1 \quad \text{and} \quad -1 \le y \le 1.$$

We conclude that the image of $\mathcal{R}$ in the xy-plane is the rectangle $[0, 1] \times [-1, 1]$.

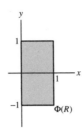

(c) The line segment joining the points $(0, 0)$ and $(1, 1)$ in the uv-plane is defined by

$$0 \le u \le 1, \quad v = u.$$

Substituting $u = \sqrt{x}$ and $v = y$, we get

$$0 \le \sqrt{x} \le 1, \quad y = \sqrt{x}$$

or

$$0 \le x \le 1, \quad y = \sqrt{x}$$

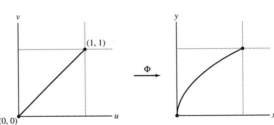

The image is the curve $y = \sqrt{x}$ for $0 \le x \le 1$.

(d) We identify the image of the sides of the triangle OAB.

The image of $\overline{OA}$: This segment is defined by $u = 0$ and $0 \le v \le 1$. That is,

$$\pm\sqrt{x} = 0 \quad \text{and} \quad 0 \le y \le 1$$

or

$$x = 0, \quad 0 \le y \le 1.$$

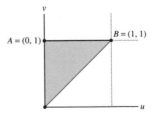

This is the segment joining the points $(0, 0)$ and $(0, 1)$ in the xy-plane.

The image of $\overline{AB}$: This segment is defined by $0 \le u \le 1$ and $v = 1$. That is,

$$0 \le \sqrt{x} \le 1, \quad y = 1$$

or

$$0 \le x \le 1, \quad y = 1.$$

This is the segment joining the points $(0, 1)$ and $(1, 1)$ in the xy-plane.

The image of $\overline{OB}$: In part (c) we showed that the image of the segment is the curve $y = \sqrt{x}$, $0 \le x \le 1$.

Therefore, the image of the triangle is the region shown in the figure:

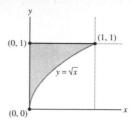

In Exercises 5–12, let $\Phi(u, v) = (2u + v, 5u + 3v)$ *be a map from the uv-plane to the xy-plane.*

5. Show that the image of the horizontal line $v = c$ is the line $y = \frac{5}{2}x + \frac{1}{2}c$. What is the image (in slope-intercept form) of the vertical line $u = c$?

SOLUTION The image of the horizontal line $v = c$ is the set of the following points:

$$(x, y) = \Phi(u, c) = (2u + c, 5u + 3c) \quad \Rightarrow \quad x = 2u + c, \quad y = 5u + 3c$$

The first equation implies $u = \frac{x-c}{2}$. Substituting in the second equation gives

$$y = 5\frac{(x - c)}{2} + 3c = \frac{5x}{2} + \frac{c}{2}$$

Therefore, the image of the line $v = c$ is the line $y = \frac{5x}{2} + \frac{c}{2}$ in the xy-plane.

The image of the vertical line $u = c$ is the set of the following points:

$$(x, y) = \Phi(c, v) = (2c + v, 5c + 3v) \quad \Rightarrow \quad x = 2c + v, \quad y = 5c + 3v$$

By the first equation, $v = x - 2c$. Substituting in the second equation gives

$$y = 5c + 3(x - 2c) = 5c + 3x - 6c = 3x - c$$

Therefore, the image of the line $u = c$ is the line $y = 3x - c$ in the xy-plane.

7. Describe the image of the line $v = 4u$ under Φ in slope-intercept form.

SOLUTION We choose any two points on the line $v = 4u$, for example $(u, v) = (1, 4)$ and $(u, v) = (0, 0)$. By a property of linear maps, the image of the line $v = 4u$ under the linear map $\Phi(u, v) = (2u + v, 5u + 3v)$ is the line in the xy-plane through the points $\Phi(1, 4)$ and $\Phi(0, 0)$. We find these points:

$$\Phi(0, 0) = (2 \cdot 0 + 0, 5 \cdot 0 + 3 \cdot 0) = (0, 0)$$

$$\Phi(1, 4) = (2 \cdot 1 + 4, 5 \cdot 1 + 3 \cdot 4) = (6, 17)$$

We now find the slope-intercept equation of the line in the xy-plane through the points $(0, 0)$ and $(6, 17)$:

$$y - 0 = \frac{17 - 0}{6 - 0}(x - 0) \quad \Rightarrow \quad y = \frac{17}{6}x$$

9. Show that the inverse of Φ is

$$\Phi^{-1}(x, y) = (3x - y, -5x + 2y)$$

Hint: Show that $\Phi(\Phi^{-1}(x, y)) = (x, y)$ and $\Phi^{-1}(\Phi(u, v)) = (u, v)$.

SOLUTION By the definition of the inverse map, we must show that the given maps $\Phi^{-1}(x, y) = (3x - y, -5x + 2y)$ and $\Phi(u, v) = (2u + v, 5u + 3v)$ satisfy $\Phi\left(\Phi^{-1}(x, y)\right) = (x, y)$ and $\Phi^{-1}(\Phi(u, v)) = (u, v)$. We have

$$\Phi\left(\Phi^{-1}(x, y)\right) = \Phi(3x - y, -5x + 2y) = (2(3x - y) + (-5x + 2y), 5(3x - y) + 3(-5x + 2y)) = (x, y)$$

$$\Phi^{-1}(\Phi(u, v)) = \Phi^{-1}(2u + v, 5u + 3v) = (3(2u + v) - (5u + 3v), -5(2u + v) + 2(5u + 3v)) = (u, v)$$

We conclude that Φ^{-1} is the inverse of Φ.

11. Calculate $\text{Jac}(\Phi) = \dfrac{\partial(x, y)}{\partial(u, v)}$.

SOLUTION   The Jacobian of the linear mapping $\Phi(u, v) = (2u + v, 5u + 3v)$ is the following determinant:

$$\text{Jac}(\Phi) = \frac{\partial(x, y)}{\partial(u, v)} = \begin{vmatrix} 2 & 1 \\ 5 & 3 \end{vmatrix} = 2 \cdot 3 - 5 \cdot 1 = 1$$

In Exercises 13–18, compute the Jacobian (at the point, if indicated).

13. $\Phi(u, v) = (3u + 4v, u - 2v)$

SOLUTION   Using the Jacobian of linear mappings we get

$$\text{Jac}(\Phi) = \frac{\partial(x, y)}{\partial(u, v)} = \begin{vmatrix} 3 & 4 \\ 1 & -2 \end{vmatrix} = 3 \cdot (-2) - 1 \cdot 4 = -10$$

15. $\Phi(r, t) = (r \sin t, r - \cos t), \quad (r, t) = (1, \pi)$

SOLUTION   We have $x = r \sin t$ and $y = r - \cos t$. Therefore,

$$\text{Jac}(\Phi) = \frac{\partial(x, y)}{\partial(r, t)} = \begin{vmatrix} \dfrac{\partial x}{\partial r} & \dfrac{\partial x}{\partial t} \\[2mm] \dfrac{\partial y}{\partial r} & \dfrac{\partial y}{\partial t} \end{vmatrix} = \begin{vmatrix} \sin t & r \cos t \\ 1 & \sin t \end{vmatrix} = \sin^2 t - r \cos t$$

At the point $(r, t) = (1, \pi)$ we get

$$\text{Jac}(\Phi)(1, \pi) = \sin^2 \pi - 1 \cdot \cos \pi = 0 - 1 \cdot (-1) = 1$$

17. $\Phi(r, \theta) = (r \cos \theta, r \sin \theta), \quad (r, \theta) = \left(4, \frac{\pi}{6}\right)$

SOLUTION   Since $x = r \cos \theta$ and $y = r \sin \theta$, the Jacobian of Φ is the following determinant:

$$\text{Jac}(\Phi) = \frac{\partial(x, y)}{\partial(r, \theta)} = \begin{vmatrix} \dfrac{\partial x}{\partial r} & \dfrac{\partial x}{\partial \theta} \\[2mm] \dfrac{\partial y}{\partial r} & \dfrac{\partial y}{\partial \theta} \end{vmatrix} = \begin{vmatrix} \cos \theta & -r \sin \theta \\ \sin \theta & r \cos \theta \end{vmatrix}$$

$$= r \cos^2 \theta + r \sin^2 \theta = r(\cos^2 \theta + \sin^2 \theta) = r \cdot 1 = r$$

At the point $(r, \theta) = (4, \pi/6)$ we get:

$$\text{Jac}(\Phi)(4, \pi/6) = 4$$

19. Find a linear mapping Φ that maps $[0, 1] \times [0, 1]$ to the parallelogram in the xy-plane spanned by the vectors $\langle 2, 3 \rangle$ and $\langle 4, 1 \rangle$.

SOLUTION

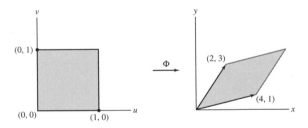

We denote the linear map by

$$\Phi(u, v) = (Au + Cv, Bu + Dv) \tag{1}$$

The image of the unit square $\mathcal{R} = [0, 1] \times [0, 1]$ under the linear map is the parallelogram whose vertices are the images of the vertices of $\mathcal{R}$. Two of vertices of the given parallelogram are $(2, 3)$ and $(4, 1)$. To find $A, B, C,$ and D it suffices to determine four equations. Therefore, we ask that (notice that for linear maps $\Phi(0, 0) = (0, 0)$)

$$\Phi(0, 1) = (2, 3), \quad \Phi(1, 0) = (4, 1)$$

We substitute in (1) and solve for A, B, C, and D:

$$(A \cdot 0 + C \cdot 1, B \cdot 0 + D \cdot 1) = (C, D) = (2, 3)$$
$$(A \cdot 1 + C \cdot 0, B \cdot 1 + D \cdot 0) = (A, B) = (4, 1)$$
$$\Rightarrow \quad C = 2, \quad D = 3$$
$$A = 4, \quad B = 1$$

Substituting in (1) we obtain the following map:

$$\Phi(u, v) = (4u + 2v, u + 3v).$$

21. Let $\mathcal{D}$ be the parallelogram in Figure 13. Apply the Change of Variables Formula to the map $\Phi(u, v) = (5u + 3v, u + 4v)$ to evaluate $\iint_{\mathcal{D}} xy \, dx \, dy$ as an integral over $\mathcal{D}_0 = [0, 1] \times [0, 1]$.

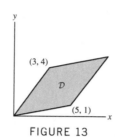

FIGURE 13

SOLUTION

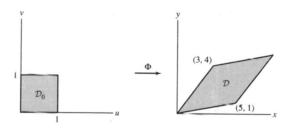

We express $f(x, y) = xy$ in terms of u and v. Since $x = 5u + 3v$ and $y = u + 4v$, we have

$$f(x, y) = xy = (5u + 3v)(u + 4v) = 5u^2 + 12v^2 + 23uv$$

The Jacobian of the linear map $\Phi(u, v) = (5u + 3v, u + 4v)$ is

$$\text{Jac}(\Phi) = \frac{\partial(x, y)}{\partial(u, v)} = \begin{vmatrix} 5 & 3 \\ 1 & 4 \end{vmatrix} = 20 - 3 = 17$$

Applying the Change of Variables Formula we get

$$\iint_{\mathcal{D}} xy \, dA = \iint_{\mathcal{D}_0} f(x, y) \left| \frac{\partial(x, y)}{\partial(u, v)} \right| du \, dv = \int_0^1 \int_0^1 (5u^2 + 12v^2 + 23uv) \cdot 17 \, du \, dv$$

$$= 17 \int_0^1 \frac{5u^3}{3} + 12v^2 u + \frac{23u^2 v}{2} \Big|_{u=0}^1 dv = 17 \int_0^1 \left(\frac{5}{3} + 12v^2 + \frac{23v}{2} \right) dv$$

$$= 17 \left(\frac{5v}{3} + 4v^3 + \frac{23v^2}{4} \Big|_0^1 \right) = 17 \left(\frac{5}{3} + 4 + \frac{23}{4} \right) = \frac{2329}{12} \approx 194.08$$

23. Let $\Phi(u, v) = (3u + v, u - 2v)$. Use the Jacobian to determine the area of $\Phi(\mathcal{R})$ for:
(a) $\mathcal{R} = [0, 3] \times [0, 5]$ **(b)** $\mathcal{R} = [2, 5] \times [1, 7]$

SOLUTION The Jacobian of the linear map $\Phi(u, v) = (3u + v, u - 2v)$ is the following determinant:

$$\text{Jac}\Phi = \frac{\partial(x, y)}{\partial(u, v)} = \begin{vmatrix} 3 & 1 \\ 1 & -2 \end{vmatrix} = -6 - 1 = -7$$

By properties of linear maps, we have

$$\text{Area}(\Phi(\mathcal{R})) = |\text{Jac}\Phi|\text{Area}(\mathcal{R}) = 7 \cdot \text{Area}(\mathcal{R})$$

(a) The area of the rectangle $R = [0, 3] \times [0, 5]$ is $3 \cdot 5 = 15$, therefore the area of $\Phi(\mathcal{R})$ is

$$\text{Area} \left(\Phi(\mathcal{R}) \right) = 7 \cdot 15 = 105$$

(b) The area of the rectangle $\mathcal{R} = [2, 5] \times [1, 7]$ is $3 \cdot 6 = 18$ hence the area of $\Phi(\mathcal{R})$ is

$$\text{Area} \left(\Phi(\mathcal{R}) \right) = 7 \cdot 18 = 126.$$

25. With Φ as in Example 3, use the Change of Variables Formula to compute the area of the image of $[1, 4] \times [1, 4]$.

SOLUTION Let $\mathcal{R}$ represent the rectangle $[1, 4] \times [1, 4]$. We proceed as follows. Jac(Φ) is easily calculated as

$$\text{Jac}(T) = \frac{\partial(x, y)}{\partial(u, v)} = \begin{vmatrix} 1/v & -u/v^2 \\ v & u \end{vmatrix} = 2u/v$$

Now, the area is given by the Change of Variables Formula as

$$\iint_{\Phi(\mathcal{R})} 1 \, dA = \iint_{\mathcal{R}} 1 |\text{Jac}(\Phi)| \, du \, dv = \iint_{\mathcal{R}} 1 |2u/v| \, du \, dv = \int_1^4 \int_1^4 2u/v \, du \, dv$$

$$= \int_1^4 2u \, du \cdot \int_1^4 \frac{1}{v} \, dv = (16 - 1)(\ln 4 - \ln 1) = 15 \ln 4$$

In Exercises 26–28, let $\mathcal{R}_0 = [0, 1] \times [0, 1]$ be the unit square. The translate of a map $\Phi_0(u, v) = (\phi(u, v), \psi(u, v))$ is a map

$$\Phi(u, v) = (a + \phi(u, v), b + \psi(u, v))$$

where a, b are constants. Observe that the map Φ_0 in Figure 15 maps $\mathcal{R}_0$ to the parallelogram $\mathcal{P}_0$ and that the translate

$$\Phi_1(u, v) = (2 + 4u + 2v, 1 + u + 3v)$$

maps $\mathcal{R}_0$ to $\mathcal{P}_1$.

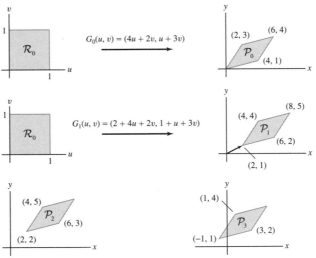

FIGURE 15

27. Sketch the parallelogram $\mathcal{P}$ with vertices $(1, 1), (2, 4), (3, 6), (4, 9)$ and find the translate of a linear mapping that maps $\mathcal{R}_0$ to $\mathcal{P}$.

SOLUTION The parallelogram $\mathcal{P}$ is shown in the figure:

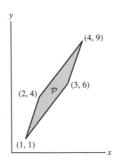

We first translate the parallelogram $\mathcal{P}$ one unit to the left and one unit downward to obtain a parallelogram $\mathcal{P}_0$ with a vertex at the origin.

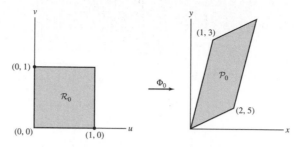

We find a linear map $\Phi_0(u, v) = (Au + Cv, Bu + Dv)$ that maps $\mathcal{R}_0$ to $\mathcal{P}_0$:

$$\Phi_0(0, 1) = (1, 3) \quad \Rightarrow \quad (C, D) = (1, 3) \quad \Rightarrow \quad C = 1, \quad D = 3$$
$$\Phi_0(1, 0) = (2, 5) \quad \Rightarrow \quad (A, B) = (2, 5) \quad \Rightarrow \quad A = 2, \quad B = 5$$

Therefore,

$$\Phi_0(u, v) = (2u + v, 5u + 3v)$$

Now we can determine the translate Φ of Φ_0 that maps $\mathcal{R}_0$ to $\mathcal{P}$. Since $\mathcal{P}$ is obtained by translating $\mathcal{P}_0$ one unit upward and one unit to the right, the map Φ is the following translate of Φ_0:

$$\Phi(u, v) = (1 + 2u + v, 1 + 5u + 3v)$$

29. Let $\mathcal{D} = \Phi(\mathcal{R})$, where $\Phi(u, v) = (u^2, u + v)$ and $\mathcal{R} = [1, 2] \times [0, 6]$. Calculate $\iint_{\mathcal{D}} y \, dx \, dy$. *Note:* It is not necessary to describe $\mathcal{D}$.

SOLUTION

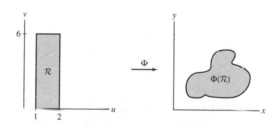

Changing variables, we have

$$\iint_{\mathcal{D}} y \, dA = \iint_{\mathcal{R}} (u + v) \left| \frac{\partial(x, y)}{\partial(u, v)} \right| du \, dv \tag{1}$$

We compute the Jacobian of Φ. Since $x = u^2$ and $y = u + v$, we have

$$\frac{\partial(x, y)}{\partial(u, v)} = \begin{vmatrix} \dfrac{\partial x}{\partial u} & \dfrac{\partial x}{\partial v} \\ \dfrac{\partial y}{\partial u} & \dfrac{\partial y}{\partial v} \end{vmatrix} = \begin{vmatrix} 2u & 0 \\ 1 & 1 \end{vmatrix} = 2u$$

We substitute in (1) and compute the resulting integral:

$$\iint_{\mathcal{D}} y \, dA = \int_0^6 \int_1^2 (u + v) \cdot 2u \, du \, dv = \int_0^6 \int_1^2 (2u^2 + 2uv) \, du \, dv = \int_0^6 \left. \frac{2u^3}{3} + u^2 v \right|_{u=1}^{2} dv$$

$$= \int_0^6 \left(\left(\frac{16}{3} + 4v \right) - \left(\frac{2}{3} + v \right) \right) dv = \int_0^6 \left(3v + \frac{14}{3} \right) dv = \left. \frac{3}{2} v^2 + \frac{14}{3} v \right|_0^6 = 82$$

31. Compute $\iint_{\mathcal{D}} (x + 3y) \, dx \, dy$, where $\mathcal{D}$ is the shaded region in Figure 16. *Hint:* Use the map $\Phi(u, v) = (u - 2v, v)$.

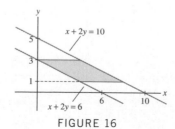

FIGURE 16

SOLUTION   The boundary of $\mathcal{D}$ is defined by the lines $x + 2y = 6$, $x + 2y = 10$, $y = 1$, and $y = 3$.

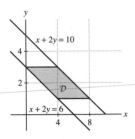

Therefore, $\mathcal{D}$ is mapped to a rectangle $\mathcal{D}_0$ in the uv-plane under the map

$$u = x + 2y, \quad v = y \tag{1}$$

or

$$(u, v) = \Phi^{-1}(x, y) = (x + 2y, y)$$

Since $\mathcal{D}$ is defined by the inequalities $6 \le x + 2y \le 10$ and $1 \le y \le 3$, the corresponding domain in the uv-plane is the rectangle

$$\mathcal{D}_0 : 6 \le u \le 10, \ 1 \le v \le 3$$

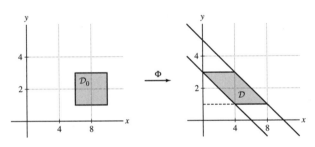

To find $\Phi(u, v)$ we must solve the equations (1) for x and y in terms of u and v. We obtain

$$\begin{array}{l} u = x + 2y \\ v = y \end{array} \Rightarrow \begin{array}{l} x = u - 2v \\ y = v \end{array} \Rightarrow \Phi(u, v) = (u - 2v, v)$$

We compute the Jacobian of the linear mapping Φ:

$$\text{Jac}(\Phi) = \frac{\partial(x, y)}{\partial(u, v)} = \begin{vmatrix} 1 & -2 \\ 0 & 1 \end{vmatrix} = 1 \cdot 1 + 2 \cdot 0 = 1$$

The function $f(x, y) = x + 3y$ expressed in terms of the new variables u and v is

$$f(x, y) = u - 2v + 3v = u + v$$

We now use the Change of Variables Formula to compute the required integral. We get

$$\iint_{\mathcal{D}} f(x, y) \, dx \, dy = \iint_{\mathcal{D}_0} (u + v) \left| \frac{\partial(x, y)}{\partial(u, v)} \right| du \, dv = \int_1^3 \int_6^{10} (u + v) \cdot 1 \, du \, dv$$

$$= \int_1^3 \frac{u^2}{2} + vu \Big|_{u=6}^{10} dv = \int_1^3 \left((50 + 10v) - (18 + 6v) \right) dv$$

$$= \int_1^3 (32 + 4v) \, dv = 32v + 2v^2 \Big|_1^3 = (96 + 18) - (32 + 2) = 80$$

33. Show that $T(u, v) = (u^2 - v^2, 2uv)$ maps the triangle $\mathcal{D}_0 = \{(u, v) : 0 \le v \le u \le 1\}$ to the domain $\mathcal{D}$ bounded by $x = 0$, $y = 0$, and $y^2 = 4 - 4x$. Use T to evaluate

$$\iint_{\mathcal{D}} \sqrt{x^2 + y^2} \, dx \, dy$$

SOLUTION We show that the boundary of $\mathcal{D}_0$ is mapped to the boundary of $\mathcal{D}$.

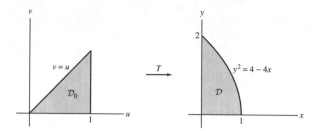

We have

$$x = u^2 - v^2 \quad \text{and} \quad y = 2uv$$

The line $v = u$ is mapped to the following set:

$$(x, y) = (u^2 - u^2, 2u^2) = (0, 2u^2) \quad \Rightarrow \quad x = 0, \quad y \geq 0$$

That is, the image of the line $u = v$ is the positive y-axis. The line $v = 0$ is mapped to the following set:

$$(x, y) = (u^2, 0) \quad \Rightarrow \quad x = u^2, \quad y = 0 \quad \Rightarrow \quad y = 0, \quad x \geq 0$$

Thus, the line $v = 0$ is mapped to the positive x-axis. We now show that the vertical line $u = 1$ is mapped to the curve $y^2 + 4x = 4$. The image of the line $u = 1$ is the following set:

$$(x, y) = (1 - v^2, 2v) \quad \Rightarrow \quad x = 1 - v^2, \quad y = 2v$$

We substitute $v = \frac{y}{2}$ in the equation $x = 1 - v^2$ to obtain

$$x = 1 - \left(\frac{y}{2}\right)^2 = 1 - \frac{y^2}{4} \quad \Rightarrow \quad 4x = 4 - y^2 \quad \Rightarrow \quad y^2 + 4x = 4$$

Since the boundary of $\mathcal{D}_0$ is mapped to the boundary of $\mathcal{D}$, we conclude that the domain $\mathcal{D}_0$ is mapped by T to the domain $\mathcal{D}$ in the xy-plane. We now compute the integral $\iint_{\mathcal{D}} \sqrt{x^2 + y^2} \, dx \, dy$. We express the function $f(x, y) = \sqrt{x^2 + y^2}$ in terms of the new variables u and v:

$$f(x, y) = \sqrt{(u^2 - v^2)^2 + (2uv)^2} = \sqrt{u^4 - 2u^2v^2 + v^4 + 4u^2v^2}$$

$$= \sqrt{u^4 + 2u^2v^2 + v^4} = \sqrt{(u^2 + v^2)^2} = u^2 + v^2$$

We compute the Jacobian of T:

$$\text{Jac}(T) = \frac{\partial(x, y)}{\partial(u, v)} = \begin{vmatrix} \dfrac{\partial x}{\partial u} & \dfrac{\partial x}{\partial v} \\[2mm] \dfrac{\partial y}{\partial u} & \dfrac{\partial y}{\partial v} \end{vmatrix} = \begin{vmatrix} 2u & -2v \\ 2v & 2u \end{vmatrix} = 4u^2 + 4v^2 = 4(u^2 + v^2)$$

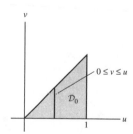

Using the Change of Variables Formula gives

$$\iint_{\mathcal{D}} \sqrt{x^2 + y^2} \, dx \, dy = \iint_{\mathcal{D}_0} (u^2 + v^2) \cdot 4(u^2 + v^2) \, du \, dv = 4 \int_0^1 \int_0^u (u^4 + 2u^2v^2 + v^4) \, dv \, du$$

$$= 4 \int_0^1 u^4 v + \frac{2}{3}u^2v^3 + \frac{v^5}{5} \bigg|_{v=0}^u du = 4 \int_0^1 \left(u^5 + \frac{2}{3}u^5 + \frac{u^5}{5}\right) du$$

$$= 4 \int_0^1 \frac{28}{15}u^5 \, du = \frac{112}{15} \cdot \frac{u^6}{6} \bigg|_0^1 = \frac{56}{45}$$

35. Calculate $\iint_{\mathcal{D}} e^{9x^2+4y^2}\, dx\, dy$, where $\mathcal{D}$ is the interior of the ellipse $\left(\frac{x}{2}\right)^2 + \left(\frac{y}{3}\right)^2 \le 1$.

SOLUTION   We define a map that maps the unit disk $u^2 + v^2 \le 1$ onto the interior of the ellipse. That is,

$$x = 2u, \quad y = 3v \quad \Rightarrow \quad \Phi(u, v) = (2u, 3v)$$

Since $\left(\frac{x}{2}\right)^2 + \left(\frac{y}{3}\right)^2 \le 1$ if and only if $u^2 + v^2 \le 1$, Φ is the map we need.

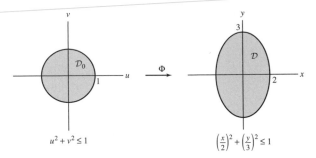

We express the function $f(x, y) = e^{9x^2+4y^2}$ in terms of u and v:

$$f(x, y) = e^{9(2u)^2+4(3v)^2} = e^{36u^2+36v^2} = e^{36(u^2+v^2)}$$

We compute the Jacobian of Φ:

$$\operatorname{Jac}(\Phi) = \begin{vmatrix} \dfrac{\partial x}{\partial u} & \dfrac{\partial x}{\partial v} \\[2mm] \dfrac{\partial y}{\partial u} & \dfrac{\partial y}{\partial v} \end{vmatrix} = \begin{vmatrix} 2 & 0 \\ 0 & 3 \end{vmatrix} = 6$$

Using the Change of Variables Formula gives

$$\iint_{\mathcal{D}} e^{9x^2+4y^2}\, dA = \iint_{\mathcal{D}_0} e^{36(u^2+v^2)} \cdot 6\, du\, dv$$

We compute the integral using polar coordinates $u = r\cos\theta$, $v = r\sin\theta$:

$$\iint_{\mathcal{D}} e^{9x^2+4y^2}\, dA = \int_0^{2\pi} \int_0^1 6e^{36r^2} \cdot r\, dr\, d\theta = \left(6\int_0^{2\pi} d\theta\right)\left(\int_0^1 e^{36r^2} r\, dr\right)$$

$$= 12\pi \left.\frac{e^{36r^2}}{72}\right|_{r=0}^1 = \frac{12\pi(e^{36}-1)}{72} = \frac{\pi(e^{36}-1)}{6}$$

37. Sketch the domain $\mathcal{D}$ bounded by $y = x^2$, $y = \frac{1}{2}x^2$, and $y = x$. Use a change of variables with the map $x = uv$, $y = u^2$ to calculate

$$\iint_{\mathcal{D}} y^{-1}\, dx\, dy$$

This is an improper integral since $f(x, y) = y^{-1}$ is undefined at $(0, 0)$, but it becomes proper after changing variables.

SOLUTION   The domain $\mathcal{D}$ is shown in the figure.

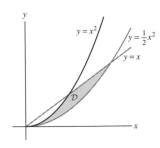

We must identify the domain $\mathcal{D}_0$ in the uv-plane. Notice that Φ is one-to-one, where $u \ge 0$ (or $u \le 0$), since in $\mathcal{D}$, $x \ge 0$, so it also follows by $x = uv$ that $v \ge 0$. Therefore, we search the domain $\mathcal{D}_0$ in the first quadrant of the uv-plane. To do this, we examine the curves that are mapped to the curves defining the boundary of $\mathcal{D}$. We examine each curve separately.

$y = x^2$: Since $x = uv$ and $y = u^2$ we get

$$u^2 = (uv)^2 \quad \Rightarrow \quad 1 = v^2 \quad \Rightarrow \quad v = 1$$

$y = \frac{1}{2}x^2$:

$$u^2 = \frac{1}{2}(uv)^2 \quad \Rightarrow \quad 1 = \frac{1}{2}v^2 \quad \Rightarrow \quad v^2 = 2 \quad \Rightarrow \quad v = \sqrt{2}$$

$y = x$: $u^2 = uv \quad \Rightarrow \quad v = u$. The region $\mathcal{D}_0$ is the region in the first quadrant of the uv-plane enclosed by the curves $v = 1$, $v = \sqrt{2}$, and $v = u$.

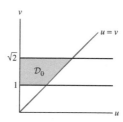

We now use change of variables to compute the integral $\iint_{\mathcal{D}} y^{-1}\, dx\, dy$. The function in terms of the new variables is $f(x, y) = u^{-2}$. We compute the Jacobian of $\Phi(u, v) = (x, y) = (uv, u^2)$:

$$\text{Jac}(\Phi) = \begin{vmatrix} \dfrac{\partial x}{\partial u} & \dfrac{\partial x}{\partial v} \\[2mm] \dfrac{\partial y}{\partial u} & \dfrac{\partial y}{\partial v} \end{vmatrix} = \begin{vmatrix} v & u \\ 2u & 0 \end{vmatrix} = -2u^2$$

Using the Change of Variables Formula gives

$$\iint_{\mathcal{D}} y^{-1}\, dx\, dy = \iint_{\mathcal{D}_0} u^{-2} \cdot 2u^2\, du\, dv = \int_1^{\sqrt{2}} \int_0^v 2\, du\, dv = \int_1^{\sqrt{2}} 2u \Big|_{u=0}^{v} dv = \int_1^{\sqrt{2}} 2v\, dv = v^2 \Big|_1^{\sqrt{2}} = 2 - 1 = 1$$

39. Let Φ be the inverse of the map $F(x, y) = (xy, x^2 y)$ from the xy-plane to the uv-plane. Let $\mathcal{D}$ be the domain in Figure 18. Show, by applying the Change of Variables Formula to the inverse $\Phi = F^{-1}$, that

$$\iint_{\mathcal{D}} e^{xy}\, dx\, dy = \int_{10}^{20} \int_{20}^{40} e^u v^{-1}\, dv\, du$$

and evaluate this result. *Hint: See Example 8.*

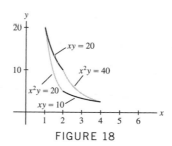

FIGURE 18

SOLUTION The domain $\mathcal{D}$ is defined by the inequalities

$$\mathcal{D}: \ 10 \le xy \le 20, \ 20 \le x^2 y \le 40$$

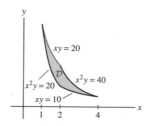

Since $u = xy$ and $v = x^2 y$, the image $\mathcal{D}_0$ of $\mathcal{D}$ (in the uv-plane) under F is the rectangle

$$\mathcal{D}_0: \ 10 \le u \le 20, \ 20 \le v \le 40$$

The function expressed in the new variables is

$$f(x, y) = e^{xy} = e^u$$

To find the Jacobian of the inverse Φ of F, we use the formula for the Jacobian of the inverse mapping. That is,

$$\frac{\partial(x, y)}{\partial(u, v)} = \left(\frac{\partial(u, v)}{\partial(x, y)} \right)^{-1}$$

We find the Jacobian of F. Since $u = xy$ and $v = x^2 y$, we have

$$\text{Jac}(F) = \begin{vmatrix} \dfrac{\partial u}{\partial x} & \dfrac{\partial u}{\partial y} \\[2mm] \dfrac{\partial v}{\partial x} & \dfrac{\partial v}{\partial y} \end{vmatrix} = \begin{vmatrix} y & x \\ 2xy & x^2 \end{vmatrix} = yx^2 - 2x^2 y = -x^2 y$$

Hence,

$$\text{Jac}(\Phi) = -\frac{1}{x^2 y}$$

We now compute the double integral $\displaystyle\iint_{\mathcal{D}} e^{xy}\, dA$ using the Change of Variables Formula. Since $y > 0$ in $\mathcal{D}$, we have $|\text{Jac}(\Phi)| = \left| -\frac{1}{x^2 y} \right| = \frac{1}{x^2 y} = v^{-1}$. Therefore,

$$\iint_{\mathcal{D}} e^{xy}\, dA = \iint_{\mathcal{D}_0} e^u v^{-1}\, dv\, du = \int_{10}^{20} \int_{20}^{40} e^u v^{-1}\, dv\, du = \left(\int_{10}^{20} e^u\, du \right) \left(\int_{20}^{40} v^{-1}\, dv \right)$$

$$= e^u \Big|_{10}^{20} \cdot \ln v \Big|_{20}^{40} = (e^{20} - e^{10})\,(\ln(40) - \ln(20)) = (e^{20} - e^{10})\ln 2$$

41. Let $I = \displaystyle\iint_{\mathcal{D}} (x^2 - y^2)\, dx\, dy$, where

$$\mathcal{D} = \{(x, y) : 2 \le xy \le 4, 0 \le x - y \le 3, x \ge 0, y \ge 0\}$$

(a) Show that the mapping $u = xy$, $v = x - y$ maps $\mathcal{D}$ to the rectangle $\mathcal{R} = [2, 4] \times [0, 3]$.
(b) Compute $\partial(x, y)/\partial(u, v)$ by first computing $\partial(u, v)/\partial(x, y)$.
(c) Use the Change of Variables Formula to show that I is equal to the integral of $f(u, v) = v$ over $\mathcal{R}$ and evaluate.

SOLUTION

(a) The domain $\mathcal{D}$ is defined by the inequalities

$$\mathcal{D} : 2 \le xy \le 4, \ 0 \le x - y \le 3, \ x \ge 0, \ y \ge 0$$

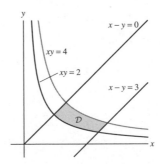

Since $u = xy$ and $v = x - y$, the image of $\mathcal{D}$ under this mapping is the rectangle defined by

$$\mathcal{D}_0 : 2 \le u \le 4, \ 0 \le v \le 3$$

That is, $\mathcal{D}_0 = [2, 4] \times [0, 3]$.

(b) We compute the Jacobian $\frac{\partial(u,v)}{\partial(x,y)}$ and then use the formula for the Jacobian of the inverse mapping to compute $\frac{\partial(x,y)}{\partial(u,v)}$. Since $u = xy$ and $v = x - y$, we have

$$\frac{\partial(u, v)}{\partial(x, y)} = \begin{vmatrix} \dfrac{\partial u}{\partial x} & \dfrac{\partial u}{\partial y} \\[6pt] \dfrac{\partial v}{\partial x} & \dfrac{\partial v}{\partial y} \end{vmatrix} = \begin{vmatrix} y & x \\ 1 & -1 \end{vmatrix} = -y - x = -(x + y)$$

Therefore,

$$\frac{\partial(x, y)}{\partial(u, v)} = \left(\frac{\partial(u, v)}{\partial(x, y)}\right)^{-1} = -\frac{1}{x + y}$$

(c) In $\mathcal{D}$, $x \geq 0$ and $y \geq 0$, hence $\left|\frac{\partial(x,y)}{\partial(u,v)}\right| = \frac{1}{x+y}$. Using the change of variable formula gives:

$$I = \iint_{\mathcal{D}_0} (x^2 - y^2) \cdot \frac{1}{x + y} \, du\, dv = \iint_{\mathcal{D}_0} (x - y) \, du\, dv = \int_0^3 \int_2^4 v \, du\, dv$$

$$= \left(\int_0^3 v \, dv\right)\left(\int_2^4 du\right) = \left(\frac{v^2}{2}\Big|_0^3\right)\left(u\Big|_2^4\right) = \frac{9}{2} \cdot 2 = 9$$

43. Derive formula (9) in Section 15.4 for integration in spherical coordinates from the general Change of Variables Formula.

SOLUTION The spherical coordinates are

$$x = \rho \cos\theta \sin\phi, \quad y = \rho \sin\theta \sin\phi, \quad z = \rho \cos\phi$$

Suppose that a region $\mathcal{W}$ in the (x, y, z)-plane is the image of a region $\mathcal{W}_0$ in the (θ, ϕ, ρ)-space defined by:

$$\mathcal{W}_0 : \theta_1 \leq \theta \leq \theta_2, \ \phi_1 \leq \phi \leq \phi_2, \quad \rho_1(\theta, \phi) \leq \rho \leq \rho_2(\theta, \phi) \tag{1}$$

Then, by the Change of Variables Formula, we have

$$\iiint_{\mathcal{W}} f(x, y, z) = \iiint_{\mathcal{W}_0} f(\rho \cos\theta \sin\phi, \rho \sin\theta \sin\phi, \rho \cos\phi) = \left|\frac{\partial(x, y, z)}{\partial(\theta, \phi, \rho)}\right| d\rho\, d\phi\, d\theta \tag{2}$$

We compute the Jacobian:

$$\frac{\partial(x, y, z)}{\partial(\theta, \phi, \rho)} = \begin{vmatrix} \dfrac{\partial x}{\partial\theta} & \dfrac{\partial x}{\partial\phi} & \dfrac{\partial x}{\partial\rho} \\[6pt] \dfrac{\partial y}{\partial\theta} & \dfrac{\partial y}{\partial\phi} & \dfrac{\partial y}{\partial\rho} \\[6pt] \dfrac{\partial z}{\partial\theta} & \dfrac{\partial z}{\partial\phi} & \dfrac{\partial z}{\partial\rho} \end{vmatrix} = \begin{vmatrix} \rho\sin\theta\sin\phi & \rho\cos\theta\cos\phi & \cos\theta\sin\phi \\ \rho\cos\theta\sin\phi & \rho\sin\theta\cos\phi & \sin\theta\sin\phi \\ 0 & -\rho\sin\phi & \cos\phi \end{vmatrix}$$

$$= -\rho\sin\theta\sin\phi \begin{vmatrix} \rho\sin\theta\cos\phi & \sin\theta\sin\phi \\ -\rho\sin\phi & \cos\phi \end{vmatrix} - \rho\cos\theta\cos\phi \begin{vmatrix} \rho\cos\theta\sin\phi & \sin\theta\sin\phi \\ 0 & \cos\phi \end{vmatrix}$$

$$+ \cos\theta\sin\phi \begin{vmatrix} \rho\cos\theta\sin\phi & \rho\sin\theta\cos\phi \\ 0 & -\rho\sin\phi \end{vmatrix}$$

$$= -\rho\sin\theta\sin\phi(\rho\sin\theta\cos^2\phi + \rho\sin\theta\sin^2\phi) - \rho\cos\theta\cos\phi(\rho\cos\theta\cos\phi\sin\phi - 0)$$

$$+ \cos\theta\sin\phi(-\rho^2\cos\theta\sin^2\phi - 0)$$

$$= -\rho^2\sin^2\theta\sin\phi(\cos^2\phi + \sin^2\phi) - \rho^2\cos^2\theta\cos^2\phi\sin\phi - \rho^2\cos^2\theta\sin^3\phi$$

$$= -\rho^2\sin^2\theta\sin\phi - \rho^2\cos^2\theta\cos^2\phi\sin\phi - \rho^2\cos^2\theta\sin^3\phi$$

$$= -\rho^2\sin\phi(\sin^2\theta + \cos^2\theta\cos^2\phi + \cos^2\theta\sin^2\phi)$$

$$= -\rho^2\sin\phi\left(\sin^2\theta + \cos^2\theta(\cos^2\phi + \sin^2\phi)\right)$$

$$= -\rho^2\sin\phi(\sin^2\theta + \cos^2\theta) = -\rho^2\sin\phi$$

Since $0 \leq \phi \leq \pi$, we have $\sin\phi \geq 0$. Therefore,

$$\left|\frac{\partial(x, y, z)}{\partial(\theta, \phi, \rho)}\right| = \rho^2\sin\phi \tag{3}$$

Combining (1), (2), and (3) gives

$$\iiint_{\mathcal{W}} f(x, y, z)\, dv = \int_{\theta_1}^{\theta_2} \int_{\phi_1}^{\phi_2} \int_{\rho_1(\theta,\phi)}^{\rho_2(\theta,\phi)} f(\rho\cos\theta\sin\phi, \rho\sin\theta\sin\phi, \rho\cos\phi)\rho^2\sin\phi \, d\rho\, d\phi\, d\theta$$

Further Insights and Challenges

45. Use the map

$$x = \frac{\sin u}{\cos v}, \qquad y = \frac{\sin v}{\cos u}$$

to evaluate the integral

$$\int_0^1 \int_0^1 \frac{dx\,dy}{1 - x^2 y^2}$$

This is an improper integral since the integrand is infinite if $x = \pm 1$, $y = \pm 1$, but applying the Change of Variables Formula shows that the result is finite.

SOLUTION   We express the function $f(x, y) = \frac{1}{1 - x^2 y^2}$ in terms of the new variables u and v:

$$1 - x^2 y^2 = 1 - \frac{\sin^2 u}{\cos^2 v} \frac{\sin^2 v}{\cos^2 u} = 1 - \left(\frac{\sin u}{\cos u}\right)^2 \cdot \left(\frac{\sin v}{\cos v}\right)^2 = 1 - \tan^2 u \tan^2 v$$

Hence,

$$f(x, y) = \frac{1}{1 - \tan^2 u \tan^2 v}$$

We compute the Jacobian of the mapping:

$$\frac{\partial(x, y)}{\partial(u, v)} = \begin{vmatrix} \dfrac{\partial x}{\partial u} & \dfrac{\partial x}{\partial v} \\[2mm] \dfrac{\partial y}{\partial u} & \dfrac{\partial y}{\partial v} \end{vmatrix} = \begin{vmatrix} \dfrac{\cos u}{\cos v} & \dfrac{\sin u \sin v}{\cos^2 v} \\[2mm] \dfrac{\sin v \sin u}{\cos^2 u} & \dfrac{\cos v}{\cos u} \end{vmatrix} = \frac{\cos u}{\cos v} \cdot \frac{\cos v}{\cos u} - \frac{\sin u \sin v}{\cos^2 v} \cdot \frac{\sin v \sin u}{\cos^2 u}$$

$$= 1 - \frac{\sin^2 u}{\cos^2 u} \cdot \frac{\sin^2 v}{\cos^2 v} = 1 - \tan^2 u \tan^2 v$$

Now, since $0 \le x \le 1$ and $0 \le y \le 1$, we have $0 \le \frac{\sin u}{\cos v} \cdot \frac{\sin v}{\cos u} \le 1$ or $0 \le \tan u \tan v \le 1$. Therefore, $0 \le \tan^2 u \tan^2 v \le 1$, hence

$$\left| \frac{\partial(x, y)}{\partial(u, v)} \right| = 1 - \tan^2 u \tan^2 v$$

We now identify a domain $\mathcal{D}_0$ in the uv-plane that is mapped by Φ onto $\mathcal{D}$ and Φ is one-to-one on $\mathcal{D}_0$.

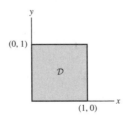

We examine each segment on the boundary of $\mathcal{D}$ separately.

$y = 0$:

$$\frac{\sin v}{\cos u} = 0 \quad \Rightarrow \quad \sin v = 0 \quad \Rightarrow \quad v = \pi k$$

$x = 0$:

$$\frac{\sin u}{\cos v} = 0 \quad \Rightarrow \quad \sin u = 0 \quad \Rightarrow \quad u = \pi k$$

$y = 1$:

$$\frac{\sin v}{\cos u} = 1 \quad \Rightarrow \quad \sin v = \cos u \quad \Rightarrow \quad v + u = \frac{\pi}{2} + 2\pi k \quad \text{or} \quad v - u = \frac{\pi}{2} + 2\pi k \qquad (1)$$

$x = 1$:

$$\frac{\sin u}{\cos v} = 1 \quad \Rightarrow \quad \sin u = \cos v \quad \Rightarrow \quad v + u = \frac{\pi}{2} + 2\pi k \quad \text{or} \quad u - v = \frac{\pi}{2} + 2\pi k \qquad (2)$$

One of the possible regions $\mathcal{D}_0$ is obtained by choosing $k = 0$ in all solutions. We get

$$v = 0, \quad u = 0, \quad \left(v + u = \frac{\pi}{2} \text{ or } v - u = \frac{\pi}{2}\right), \quad \left(u + v = \frac{\pi}{2} \text{ or } u - v = \frac{\pi}{2}\right)$$

The corresponding regions are:

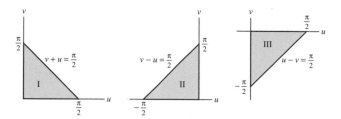

In II, $x = \frac{\sin u}{\cos v} < 0$ and in III $y = \frac{\sin v}{\cos u} < 0$, therefore these regions are not mapped to the unit square in the xy-plane. The appropriate region is I.

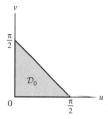

We now use the Change of Variables Formula and the result obtained previously to obtain the following integral:

$$\int_0^1 \int_0^1 \frac{dx\,dy}{1 - x^2 y^2} = \iint_{\mathcal{D}_0} \frac{1}{1 - \tan^2 u \tan^2 v} \cdot (1 - \tan^2 u \tan^2 v)\,du\,dv$$

$$= \iint_{\mathcal{D}_0} 1\,du\,dv = \text{Area}(\mathcal{D}_0) = \frac{\frac{\pi}{2} \cdot \frac{\pi}{2}}{2} = \frac{\pi^2}{8}$$

47. Let P and Q be points in $\mathbf{R}^2$. Show that a linear map $\Phi(u, v) = (Au + Cv, Bu + Dv)$ maps the segment joining P and Q to the segment joining $\Phi(P)$ to $\Phi(Q)$. *Hint:* The segment joining P and Q has parametrization

$$(1 - t)\overrightarrow{OP} + t\overrightarrow{OQ} \quad \text{for} \quad 0 \le t \le 1$$

SOLUTION First let $P(x_0, y_0)$ and $Q(x_1, y_1)$ so that we see if

$$\mathbf{r}(0) = \overrightarrow{OP} = (x_0, y_0), \quad \mathbf{r}(1) = \overrightarrow{OQ} = (x_1, y_1)$$

Then using the linear map we see:

$$\Phi(x_0, y_0) = (Ax_0 + Cy_0, Bx_0 + Dy_0) = \Phi(P)$$

and

$$\Phi(x_1, y_1) = (Ax_1 + Cy_1, Bx_1 + Dy_1) = \Phi(Q)$$

Hence this linear map take the endpoints P and Q to the new endpoints $\Phi(P)$ and $\Phi(Q)$. Now to determine the line segment mapping, consider the following:

$$\Phi(\mathbf{r}(t)) = \Phi((1-t)\overrightarrow{OP} + t\overrightarrow{OQ}) = \Phi((1-t)x_0 + tx_1, (1-t)y_0 + ty_1)$$

$$= (A((1-t)x_0 + tx_1) + C((1-t)y_0 + ty_1), B((1-t)x_0 + tx_1 + D((1-t)y_0 + ty_1))$$

$$= ((1-t)Ax_0 + t(Ax_1) + (1-t)Cy_0 + t(Cy_1), (1-t)Bx_0 + t(Bx_1) + (1-t)Dy_0 + t(Dy_1))$$

$$= ((1-t)(Ax_0 + Cy_0) + t(Ax_1 + Cy_1), (1-t)(Bx_0 + Dy_0) + t(Bx_1 + Dy_1))$$

$$= (1-t)(Ax_0 + Cy_0, Bx_0 + Dy_0) + t(Ax_1 + Cy_1, Bx_1 + Dy_1)$$

$$= (1-t)\Phi(P) + t\Phi(Q)$$

This is a parameterization for the line segment joining $\Phi(P)$ and $\Phi(Q)$. Therefore, the linear map maps the line segment joining P and Q to the line segment joining $\Phi(P)$ and $\Phi(Q)$.

49. The product of 2×2 matrices A and B is the matrix AB defined by

$$\underbrace{\begin{pmatrix} a & b \\ c & d \end{pmatrix}}_{A} \underbrace{\begin{pmatrix} a' & b' \\ c' & d' \end{pmatrix}}_{B} = \underbrace{\begin{pmatrix} aa' + bc' & ab' + bd' \\ ca' + dc' & cb' + dd' \end{pmatrix}}_{AB}$$

The (i, j)-entry of A is the **dot product** of the ith row of A and the jth column of B. Prove that $\det(AB) = \det(A)\det(B)$.

SOLUTION The determinants of A and B are

$$\det(A) = \begin{vmatrix} a & b \\ c & d \end{vmatrix} = ad - bc,$$

$$\det(B) = \begin{vmatrix} a' & b' \\ c' & d' \end{vmatrix} = a'd' - b'c' \tag{1}$$

We now compute the determinant of AB:

$$\det(AB) = \begin{vmatrix} aa' + bc' & ab' + bd' \\ ca' + dc' & cb' + dd' \end{vmatrix} = (aa' + bc')(cb' + dd') - (ab' + bd')(ca' + dc')$$

$$= aa'cb' + aa'dd' + bc'cb' + bc'dd' - ab'ca' - ab'dc' - bd'ca' - bd'dc'$$

$$= (aa'dd' - bd'ca') + (bc'cb' - ab'dc') = a'd'(ad - bc) - b'c'(ad - bc)$$

$$= (ad - bc)(a'd' - b'c') \tag{2}$$

We combine (1) and (2) to conclude

$$\det(AB) = \det(A)\det(B).$$

51. Use Exercise 50 to prove that

$$\mathrm{Jac}(\Phi^{-1}) = \mathrm{Jac}(\Phi)^{-1}$$

Hint: Verify that $\mathrm{Jac}(I) = 1$, where I is the identity map $I(u, v) = (u, v)$.

SOLUTION Since $\Phi^{-1}(\Phi(u, v)) = (u, v)$, we have $(\Phi^{-1} \circ \Phi)(u, v) = (u, v)$. Therefore, $\Phi^{-1} \circ \Phi = I$. Using Exercise 50, we have

$$\mathrm{Jac}(I) = \mathrm{Jac}(\Phi^{-1} \circ \Phi) = \mathrm{Jac}(\Phi^{-1})\mathrm{Jac}(\Phi) \tag{1}$$

The Jacobian of the linear map $I(u, v) = (u, v) = (1 \cdot u + 0 \cdot v, 0 \cdot u + 1 \cdot v)$ is

$$\mathrm{Jac}(I) = \begin{vmatrix} 1 & 0 \\ 0 & 1 \end{vmatrix} = 1 \cdot 1 - 0 \cdot 0 = 1$$

Substituting in (1) gives

$$1 = \mathrm{Jac}(\Phi^{-1})\mathrm{Jac}(\Phi)$$

or

$$\mathrm{Jac}(\Phi^{-1}) = (\mathrm{Jac}(\Phi))^{-1}.$$

CHAPTER REVIEW EXERCISES

1. Calculate the Riemann sum $S_{2,3}$ for $\displaystyle\int_1^4 \int_2^6 x^2 y \, dx \, dy$ using two choices of sample points:

(a) Lower-left vertex

(b) Midpoint of rectangle

Then calculate the exact value of the double integral.

SOLUTION

(a) The rectangle $[2, 6] \times [1, 3]$ is divided into 2×3 subrectangles. The lower-left vertices of the subrectangles are

$$P_{11} = (2, 1) \quad P_{21} = (2, 2) \quad P_{31} = (2, 3)$$
$$P_{12} = (3, 1) \quad P_{22} = (3, 2) \quad P_{32} = (3, 3)$$

Also $\Delta x = \frac{6-2}{2} = 2$, $\Delta y = \frac{4-1}{3} = 1$, hence $\Delta A = 2 \cdot 1 = 2$. The Riemann sum $S_{3,4}$ is the following sum:

$$S_{2,3} = 2\left(2^2 \cdot 1 + 2^2 \cdot 2 + 2^2 \cdot 3 + 3^2 \cdot 1 + 3^2 \cdot 2 + 3^2 \cdot 3\right)$$

$$= 2\,(4 + 8 + 12 + 9 + 18 + 27) = 156$$

(b) The midpoints of the subrectangles are

$$\begin{array}{lll} P_{11} = (3, 3/2) & P_{21} = (3, 5/2) & P_{31} = (3, 7/2) \\ P_{12} = (5, 3/2) & P_{22} = (5, 5/2) & P_{32} = (5, 7/2) \end{array}$$

Also $\Delta x = 2$, $\Delta y = 1$, hence $\Delta A = 2 \cdot 1 = 2$. The Riemann sum $S_{2,3}$ is

$$S_{2,3} = 2\left(3^2 \cdot \frac{3}{2} + 3^2 \cdot \frac{5}{2} + 3^2 \cdot \frac{7}{2} + 5^2 \cdot \frac{3}{2} + 5^2 \cdot \frac{5}{2} + 5^2 \cdot \frac{7}{2}\right)$$

$$= 2\left(\frac{27}{2} + \frac{45}{2} + \frac{63}{2} + \frac{75}{2} + \frac{125}{2} + \frac{175}{2}\right)$$

$$= 510$$

We compute the exact value of the double integral, using an iterated integral of a product function. We get

$$\int_1^4 \int_2^6 x^2 y \, dx \, dy = \left(\int_1^4 y \, dy\right)\left(\int_2^6 x^2 \, dx\right) = \left(\left.\frac{y^2}{2}\right|_1^4\right)\left(\left.\frac{x^3}{3}\right|_2^6\right)$$

$$= \frac{16 - 1}{2} \cdot \frac{216 - 8}{3} = \frac{3120}{6} = 520$$

3. Let $\mathcal{D}$ be the shaded domain in Figure 1.

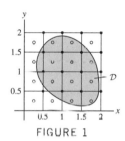

FIGURE 1

Estimate $\displaystyle\iint_{\mathcal{D}} xy \, dA$ by the Riemann sum whose sample points are the midpoints of the squares in the grid.

SOLUTION The subrectangles have sides of length $\Delta x = \Delta y = 0.5$ and area $\Delta A = 0.5^2 = 0.25$. Of sixteen sample points only ten lie in $\mathcal{D}$. The sample points that lie in $\mathcal{D}$ are

$$(0.75, 0.75), \ (0.75, 1.25), \ (0.75, 1.75), \ (1.25, 0.25), \ (1.25, 0.75),$$

$$(1.25, 1.25), \ (1.25, 1.75), \ (1.75, 0.25), \ (1.75, 0.75), \ (1.75, 1.25)$$

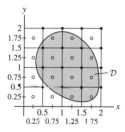

The Riemann sum S_{44} is thus

$$S_{44} = 0.25\,(f(0.75, 0.75) + f(0.75, 1.25) + f(0.75, 1.75) + f(1.25, 0.25) + f(1.25, 0.75)$$

$$+ f(1.25, 1.25) + f(1.25, 1.75) + f(1.75, 0.25) + f(1.75, 0.75) + f(1.75, 1.25))$$

$$= 0.25\left(0.75^2 + 0.75 \cdot 1.25 + 0.75 \cdot 1.75 + 1.25 \cdot 0.25 + 1.25 \cdot 0.75 + 1.25^2\right.$$

$$+ 1.25 \cdot 1.75 + 1.75 \cdot 0.25 + 1.75 \cdot 0.75 + 1.75 \cdot 1.25)$$

$$= 0.25 \cdot 11.75 = 2.9375$$

In Exercises 5–8, evaluate the iterated integral.

5. $\displaystyle\int_0^2 \int_3^5 y(x-y)\,dx\,dy$

SOLUTION   First we evaluate the inner integral treating y as a constant:

$$\int_3^5 y(x-y)\,dx = y\left(\frac{x^2}{2} - yx\right)\Bigg|_{x=3}^5 = y\left(\left(\frac{25}{2} - 5y\right) - \left(\frac{9}{2} - 3y\right)\right) = y(8 - 2y) = 8y - 2y^2$$

Now we integrate this result with respect to y:

$$\int_0^2 (8y - 2y^2)\,dy = 4y^2 - \frac{2}{3}y^3\Bigg|_0^2 = 16 - \frac{16}{3} = \frac{32}{3}$$

Therefore,

$$\int_0^2 \int_3^5 y(x-y)\,dx\,dy = \frac{32}{3}.$$

7. $\displaystyle\int_0^{\pi/3} \int_0^{\pi/6} \sin(x+y)\,dx\,dy$

SOLUTION   We compute the inner integral treating y as a constant:

$$\int_0^{\pi/6} \sin(x+y)\,dx = -\cos(x+y)\Bigg|_{x=0}^{\pi/6} = -\cos\left(\frac{\pi}{6} + y\right) + \cos y = \cos y - \cos\left(y + \frac{\pi}{6}\right)$$

We now integrate the result with respect to y:

$$\int_0^{\pi/3} \int_0^{\pi/6} \sin(x+y)\,dx\,dy = \int_0^{\pi/3} \left(\cos y - \cos\left(y + \frac{\pi}{6}\right)\right)dy = \sin y - \sin\left(y + \frac{\pi}{6}\right)\Bigg|_0^{\pi/3}$$

$$= \sin\frac{\pi}{3} - \sin\left(\frac{\pi}{3} + \frac{\pi}{6}\right) - \left(\sin 0 - \sin\frac{\pi}{6}\right) = \frac{\sqrt{3}}{2} - 1 + \frac{1}{2} = \frac{\sqrt{3}-1}{2}$$

In Exercises 9–14, sketch the domain $\mathcal{D}$ and calculate $\displaystyle\iint_{\mathcal{D}} f(x,y)\,dA$.

9. $\mathcal{D} = \{0 \le x \le 4,\ 0 \le y \le x\},\quad f(x,y) = \cos y$

SOLUTION   The domain $\mathcal{D}$ is shown in the figure:

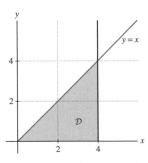

We compute the double integral, considering $\mathcal{D}$ as a vertically simple region. We describe $\mathcal{D}$ by the inequalities

$$0 \le x \le 4, \quad 0 \le y \le x.$$

We now write the double integral as an iterated integral and compute:

$$\iint_{\mathcal{D}} \cos y\,dA = \int_0^4 \int_0^x \cos y\,dy\,dx = \int_0^4 \sin y\Bigg|_{y=0}^x dx$$

$$= \int_0^4 (\sin x - \sin 0)dx = \int_0^4 \sin x\,dx = -\cos x\Bigg|_0^4 = 1 - \cos 4$$

11. $\mathcal{D} = \{0 \le x \le 1,\ 1 - x \le y \le 2 - x\}, \quad f(x, y) = e^{x+2y}$

SOLUTION

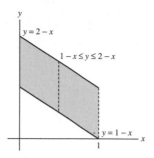

$\mathcal{D}$ is a vertically simple region, hence the double integral over $\mathcal{D}$ is the following iterated integral:

$$\iint_{\mathcal{D}} e^{x+2y}\, dA = \int_0^1 \int_{1-x}^{2-x} e^{x+2y}\, dy\, dx = \int_0^1 \frac{1}{2} e^{x+2y}\Big|_{y=1-x}^{2-x} dx = \int_0^1 \left(\frac{1}{2} e^{x+2(2-x)} - \frac{1}{2} e^{x+2(1-x)} \right) dx$$

$$= \int_0^1 \left(\frac{1}{2} e^{4-x} - \frac{1}{2} e^{2-x} \right) dx = -\frac{1}{2} e^{4-x} + \frac{1}{2} e^{2-x}\Big|_0^1 = -\frac{1}{2} e^3 + \frac{1}{2} e + \frac{1}{2} e^4 - \frac{1}{2} e^2$$

$$= \frac{1}{2} e(e^3 - e^2 - e + 1) = \frac{1}{2} e(e+1)(e-1)^2$$

13. $\mathcal{D} = \{0 \le y \le 1,\ 0.5y^2 \le x \le y^2\}, \quad f(x, y) = y e^{1+x}$

SOLUTION

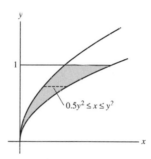

The region is horizontally simple, hence the double integral is equal to the following iterated integral:

$$\iint_{\mathcal{D}} y e^{1+x}\, dA = \int_0^1 \int_{0.5y^2}^{y^2} y e^{1+x}\, dx\, dy = \int_0^1 y e^{1+x}\Big|_{x=0.5y^2}^{y^2} dy$$

$$= \int_0^1 y \left(e^{1+y^2} - e^{1+0.5y^2} \right) dy = \int_0^1 y e^{1+y^2}\, dy - \int_0^1 y e^{1+0.5y^2}\, dy$$

We compute the integrals using the substitutions $u = 1 + y^2$, $du = 2y\, dy$, and $v = 1 + 0.5y^2$, $dv = y\, dy$, respectively. We get

$$\iint_{\mathcal{D}} y e^{1+x}\, dA = \frac{1}{2} \int_1^2 e^u\, du - \int_1^{1.5} e^v\, dv = \frac{1}{2} e^u\Big|_1^2 - e^v\Big|_1^{1.5} = \frac{1}{2} (e^2 - e) - (e^{3/2} - e)$$

$$= \frac{1}{2} e^2 + \frac{1}{2} e - e^{3/2} = 0.5(e^2 - 2e^{1.5} + e)$$

15. Express $\displaystyle \int_{-3}^{3} \int_{0}^{9-x^2} f(x, y)\, dy\, dx$ as an iterated integral in the order $dx\, dy$.

SOLUTION   The limits of integration correspond to the inequalities describing the domain $\mathcal{D}$:

$$-3 \le x \le 3, \quad 0 \le y \le 9 - x^2.$$

A quick sketch verifies that this is the region under the upper part of the parabola $y = 9 - x^2$, that is, the part that is above the x-axis. Therefore, the double integral can be rewritten as the following sum:

$$\int_{-3}^{3} \int_{0}^{9-x^2} f(x, y)\, dy\, dx = \int_{0}^{9} \int_{-\sqrt{9-y}}^{\sqrt{9-y}} f(x, y)\, dx\, dy$$

17. Let $\mathcal{D}$ be the domain between $y = x$ and $y = \sqrt{x}$. Calculate $\iint_{\mathcal{D}} xy \, dA$ as an iterated integral in the order $dx \, dy$ and $dy \, dx$.

SOLUTION In the order $dx \, dy$: The inequalities describing $\mathcal{D}$ as a horizontally simple region are obtained by first rewriting the equations of the curves with x as a function of y, that is, $x = y$ and $x = y^2$, respectively. The points of intersection are found solving the equation

$$y = y^2 \quad \Rightarrow \quad y(1 - y) = 0 \quad \Rightarrow \quad y = 0, \quad y = 1$$

We obtain the following inequalities for $\mathcal{D}$ (see figure):

$$\mathcal{D}: 0 \le y \le 1, \ y^2 \le x \le y$$

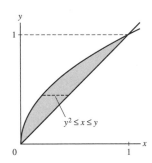

We now compute the double integral as the following iterated integral:

$$\iint_{\mathcal{D}} xy \, dA = \int_0^1 \int_{y^2}^y xy \, dx \, dy = \int_0^1 \frac{x^2 y}{2} \Big|_{x=y^2}^{x=y} dy = \int_0^1 \left(\frac{y \cdot y^2}{2} - \frac{y^4 \cdot y}{2} \right) dy$$

$$= \int_0^1 \left(\frac{y^3}{2} - \frac{y^5}{2} \right) dy = \frac{y^4}{8} - \frac{y^6}{12} \Big|_0^1 = \frac{1}{8} - \frac{1}{12} = \frac{1}{24}$$

In the order $dy \, dx$: $\mathcal{D}$ is described as a vertically simple region by the following inequalities (see figure):

$$\mathcal{D}: 0 \le x \le 1, \ x \le y \le \sqrt{x}$$

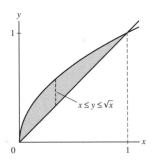

The corresponding iterated integral is

$$\iint_{\mathcal{D}} xy \, dA = \int_0^1 \int_x^{\sqrt{x}} xy \, dy \, dx = \int_0^1 \frac{xy^2}{2} \Big|_{y=x}^{\sqrt{x}} dx = \int_0^1 \left(\frac{x \cdot x}{2} - \frac{x \cdot x^2}{2} \right) dx$$

$$= \int_0^1 \left(\frac{x^2}{2} - \frac{x^3}{2} \right) dx = \frac{x^3}{6} - \frac{x^4}{8} \Big|_0^1 = \frac{1}{6} - \frac{1}{8} = \frac{1}{24}$$

19. Change the order of integration and evaluate $\displaystyle\int_0^9 \int_0^{\sqrt{y}} \frac{x \, dx \, dy}{(x^2 + y)^{1/2}}$.

SOLUTION The region here is described by the inequalities:

$$0 \le x \le \sqrt{y}, \quad 0 \le y \le 9$$

This region can also be described by writing these inequalities:

$$0 \le x \le 3, \quad x^2 \le y \le 9$$

Hence, changing the order of integration and evaluating we get:

$$\int_0^9 \int_0^{\sqrt{y}} \frac{x}{\sqrt{x^2+y}}\, dx\, dy = \int_0^3 \int_{x^2}^9 \frac{x}{\sqrt{x^2+y}}\, dy\, dx = \int_0^3 x \left(2\sqrt{x^2+y}\, \Big|_{x^2}^9 \right) dx$$

$$= 2\int_0^3 x\sqrt{x^2+9} - x\sqrt{x^2+x^2}\, dx = 2\int_0^3 x\sqrt{x^2+9} - x^2\sqrt{2}\, dx$$

$$= 2\left(\frac{1}{3}(x^2+9)^{3/2} - \frac{\sqrt{2}}{3}x^3 \Big|_0^3 \right)$$

$$= \frac{2}{3}\cdot 18^{3/2} - \frac{2\sqrt{2}}{3}\cdot 27 - 2\cdot \frac{1}{3}\cdot 9^{3/2}$$

$$= 36\sqrt{2} - 18\sqrt{2} - 18 = 18\sqrt{2} - 18$$

21. Prove the formula

$$\int_0^1 \int_0^y f(x)\, dx\, dy = \int_0^1 (1-x)f(x)\, dx$$

Then use it to calculate $\int_0^1 \int_0^y \frac{\sin x}{1-x}\, dx\, dy$.

SOLUTION   The region of integration of the double integral $\int_0^1 \int_0^y f(x)\, dx\, dy$ is described as horizontally simple by the inequalities

$$0 \le y \le 1, \quad 0 < x \le y$$

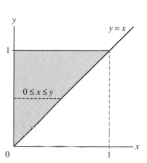

The region can also be described as a vertically simple region, by the inequalities

$$0 \le x \le 1, \quad x \le y \le 1$$

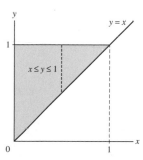

Therefore,

$$\int_0^1 \int_0^y f(x)\, dx\, dy = \int_0^1 \int_x^1 f(x)\, dy\, dx = \int_0^1 f(x)y\, \Big|_{y=x}^1\, dx = \int_0^1 f(x)(1-x)\, dx$$

We use the formula with $f(x) = \frac{\sin x}{1-x}$. We get

$$\int_0^1 \int_0^y \frac{\sin x}{1-x}\, dx\, dy = \int_0^1 (1-x)\cdot \frac{\sin x}{1-x}\, dx = \int_0^1 \sin x\, dx = -\cos x\, \Big|_0^1 = 1 - \cos 1$$

23. Use cylindrical coordinates to compute the volume of the region defined by $4 - x^2 - y^2 \le z \le 10 - 4x^2 - 4y^2$.

SOLUTION

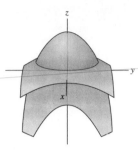

We first find the projection of $\mathcal{W}$ onto the xy-plane. The intersection curve of the upper and lower boundaries of $\mathcal{W}$ is obtained by solving

$$10 - 4x^2 - 4y^2 = 4 - x^2 - y^2$$

$$6 = 3(x^2 + y^2) \quad \Rightarrow \quad x^2 + y^2 = 2$$

Therefore, the projection of $\mathcal{W}$ onto the xy-plane is the circle $x^2 + y^2 \le 2$. The upper surface is $z = 10 - 4(x^2 + y^2)$ or $z = 10 - 4r^2$ and the lower surface is $z = 4 - (x^2 + y^2) = 4 - r^2$. Therefore, the inequalities for $\mathcal{W}$ in cylindrical coordinates are

$$0 \le \theta \le 2\pi, \quad 0 \le r \le \sqrt{2}, \quad 4 - r^2 \le z \le 10 - 4r^2$$

We use the volume as a triple integral and change of variables in cylindrical coordinates to write

$$V = \text{Volume}(\mathcal{W}) = \iiint_{\mathcal{W}} 1 \, dV = \int_0^{2\pi} \int_0^{\sqrt{2}} \int_{4-r^2}^{10-4r^2} r \, dz \, dr \, d\theta = \int_0^{2\pi} \int_0^{\sqrt{2}} rz \Big|_{z=4-r^2}^{10-4r^2} dr \, d\theta$$

$$= \int_0^{2\pi} \int_0^{\sqrt{2}} r\left(10 - 4r^2 - \left(4 - r^2\right)\right) dr \, d\theta = \int_0^{2\pi} \int_0^{\sqrt{2}} \left(6r - 3r^3\right) dr \, d\theta$$

$$= \int_0^{2\pi} 3r^2 - \frac{3}{4}r^4 \Big|_{r=0}^{\sqrt{2}} d\theta = \int_0^{2\pi} (6 - 3) \, d\theta = 6\pi$$

25. Find the volume of the region between the graph of the function $f(x, y) = 1 - (x^2 + y^2)$ and the xy-plane.

SOLUTION

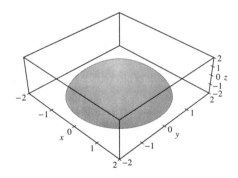

The intersection of the surface $z = 1 - (x^2 + y^2)$ with the xy-plane is obtained by setting $z = 0$. That is, $1 - (x^2 + y^2) = 0$ or $x^2 + y^2 = 1$. Therefore, the projection of the solid onto the xy-plane is the disk $x^2 + y^2 \le 1$. We describe the disk as a vertically simple region:

$$\mathcal{D}: -1 \le x \le 1, \ -\sqrt{1 - x^2} \le y \le \sqrt{1 - x^2}$$

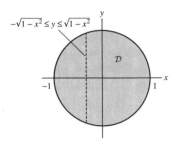

The volume V is the double integral of $z = 1 - (x^2 + y^2)$ over $\mathcal{D}$. That is,

$$V = \iint_{\mathcal{D}} \left(1 - (x^2 + y^2)\right) dA = \int_0^{2\pi} \int_0^1 (1 - r^2) r \, dr \, d\theta = 2\pi \int_0^1 (r - r^3) \, dr = \pi/2$$

27. Calculate $\iiint_{\mathcal{B}} (xy + z) \, dV$, where

$$\mathcal{B} = \left\{0 \le x \le 2, \; 0 \le y \le 1, \; 1 \le z \le 3\right\}$$

as an iterated integral in two different ways.

SOLUTION   The triple integral over the box may be evaluated in any order. For instance,

$$\iiint_{\mathcal{B}} (xy + z) \, dV = \int_0^2 \int_0^1 \int_1^3 (xy + z) \, dz \, dy \, dx = \int_0^1 \int_0^2 \int_1^3 (xy + z) \, dz \, dx \, dy$$

$$= \int_1^3 \int_0^2 \int_0^1 (xy + z) \, dy \, dx \, dz$$

We compute the integral in two of the possible orders:

$$\iiint_{\mathcal{B}} (xy + z) \, dV = \int_0^2 \int_0^1 \int_1^3 (xy + z) \, dz \, dy \, dx = \int_0^2 \int_0^1 xyz + \frac{z^2}{2} \Big|_{z=1}^3 dy \, dx$$

$$= \int_0^2 \int_0^1 \left(\left(3xy + \frac{9}{2}\right) - \left(xy + \frac{1}{2}\right)\right) dy \, dx = \int_0^2 \int_0^1 (2xy + 4) \, dy \, dx$$

$$= \int_0^2 xy^2 + 4y \Big|_{y=0}^1 dx = \int_0^2 (x + 4) \, dx = \frac{x^2}{2} + 4x \Big|_0^2 = \frac{4}{2} + 8 = 10$$

$$\iiint_{\mathcal{B}} (xy + z) \, dV = \int_0^1 \int_0^2 \int_1^3 (xy + z) \, dz \, dx \, dy = \int_0^1 \int_0^2 xyz + \frac{z^2}{2} \Big|_{z=1}^3 dx \, dy$$

$$= \int_0^1 \int_0^2 \left(\left(3xy + \frac{9}{2}\right) - \left(xy + \frac{1}{2}\right)\right) dx \, dy = \int_0^1 \int_0^2 (2xy + 4) \, dx \, dy$$

$$= \int_0^1 x^2 y + 4x \Big|_{x=0}^2 dy = \int_0^1 (4y + 8) \, dy = 2y^2 + 8y \Big|_0^1 = 2 + 8 = 10$$

29. Evaluate $I = \int_{-1}^1 \int_0^{\sqrt{1-x^2}} \int_0^1 (x + y + z) \, dz \, dy \, dx$.

SOLUTION   We compute the triple integral:

$$I_1 = \int_{-1}^1 \int_0^{\sqrt{1-x^2}} \int_0^1 (x + y + z) \, dz \, dy \, dx = \int_{-1}^1 \int_0^{\sqrt{1-x^2}} (x + y)z + \frac{z^2}{2} \Big|_{y=0}^1 dy \, dx$$

$$= \int_{-1}^1 \int_0^{\sqrt{1-x^2}} \left(x + y + \frac{1}{2}\right) dy \, dx = \int_{-1}^1 \left(x + \frac{1}{2}\right) y + \frac{y^2}{2} \Big|_{y=0}^{\sqrt{1-x^2}} dx$$

$$= \int_{-1}^1 \left(x + \frac{1}{2}\right) \sqrt{1 - x^2} + \frac{1 - x^2}{2} \, dx = \int_{-1}^1 x\sqrt{1 - x^2} \, dx + \int_{-1}^1 \frac{1}{2} \sqrt{1 - x^2} \, dx + \int_{-1}^1 \frac{1 - x^2}{2} \, dx \qquad (1)$$

The first integral is zero since the integrand is an odd function. Therefore, using Integration Formulas we get

$$I_1 = \int_0^1 \sqrt{1 - x^2} \, dx + \int_0^1 (1 - x^2) \, dx = \frac{x}{2}\sqrt{1 - x^2} + \frac{1}{2} \sin^{-1} x \Big|_0^1 + \left(x - \frac{x^3}{3}\right) \Big|_0^1$$

$$= \frac{1}{2} \sin^{-1} 1 + \frac{2}{3} = \frac{\pi}{4} + \frac{2}{3}$$

31. Find the volume of the solid contained in the cylinder $x^2 + y^2 = 1$ below the curve $z = (x + y)^2$ and above the curve $z = -(x - y)^2$.

SOLUTION

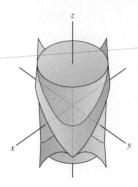

We rewrite the equations of the surfaces using cylindrical coordinates:

$$z = (x + y)^2 = x^2 + y^2 + 2xy = r^2 + 2(r\cos\theta)(r\sin\theta) = r^2(1 + \sin 2\theta)$$

$$z = -(x - y)^2 = -(x^2 + y^2 - 2xy) = -(r^2 - 2r^2\cos\theta\sin\theta) = -r^2(1 - \sin 2\theta)$$

The projection of the solid onto the xy-plane is the unit disk. Therefore, the solid is described by the following inequalities:

$$\mathcal{W}: 0 \le \theta \le 2\pi, \ 0 \le r \le 1, \ -r^2(1 - \sin 2\theta) \le z \le r^2(1 + \sin 2\theta)$$

Expressing the volume as a triple integral and converting the triple integral to cylindrical coordinates, we get

$$V = \text{Volume}(\mathcal{W}) = \iiint_{\mathcal{W}} 1\, dv = \int_0^{2\pi} \int_0^1 \int_{-r^2(1-\sin 2\theta)}^{r^2(1+\sin 2\theta)} r\, dz\, dr\, d\theta$$

$$= \int_0^{2\pi} \int_0^1 rz \Big|_{z=-r^2(1-\sin 2\theta)}^{r^2(1+\sin 2\theta)} dr\, d\theta = \int_0^{2\pi} \int_0^1 r\left(r^2(1 + \sin 2\theta) + r^2(1 - \sin 2\theta)\right) dr\, d\theta$$

$$= \int_0^{2\pi} \int_0^1 r^3 \cdot 2\, dr\, d\theta = \left(\int_0^{2\pi} 2\, d\theta\right) \left(\int_0^1 r^3\, dr\right) = 4\pi \cdot \frac{r^4}{4} \Big|_0^1 = \pi$$

33. Use polar coordinates to calculate $\iint_{\mathcal{D}} \sqrt{x^2 + y^2}\, dA$, where $\mathcal{D}$ is the region in the first quadrant bounded by the spiral $r = \theta$, the circle $r = 1$, and the x-axis.

SOLUTION The region of integration, shown in the figure, has the following description in polar coordinates:

$$\mathcal{D}: 0 \le \theta \le 1, \ \theta \le r \le 1$$

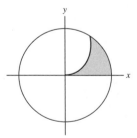

The function is $f(x, y) = \sqrt{x^2 + y^2} = r$. We convert the double integral to polar coordinates and compute to obtain

$$\iint_{\mathcal{D}} \sqrt{x^2 + y^2}\, dA = \int_0^1 \int_\theta^1 r \cdot r\, dr\, d\theta = \int_0^1 \int_\theta^1 r^2\, dr\, d\theta = \int_0^1 \frac{r^3}{3} \Big|_{r=\theta}^1 d\theta$$

$$= \int_0^1 \left(\frac{1}{3} - \frac{\theta^3}{3}\right) d\theta = \frac{\theta}{3} - \frac{\theta^4}{12} \Big|_0^1 = \frac{1}{3} - \frac{1}{12} = \frac{1}{4}$$

35. Express in cylindrical coordinates and evaluate:

$$\int_0^1 \int_0^{\sqrt{1-x^2}} \int_0^{\sqrt{x^2+y^2}} z\, dz\, dy\, dx$$

SOLUTION We evaluate the integral by converting it to cylindrical coordinates. The projection of the region of integration onto the xy-plane, as defined by the limits of integration, is

$$\mathcal{D}: 0 \le x \le 1,\ 0 \le y \le \sqrt{1-x^2}$$

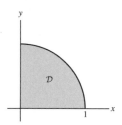

That is, $\mathcal{D}$ is the part of the disk $x^2 + y^2 \le 1$ in the first quadrant. The inequalities defining $\mathcal{D}$ in polar coordinates are

$$\mathcal{D}: 0 \le \theta \le \frac{\pi}{2},\ 0 \le r \le 1$$

The upper surface is $z = \sqrt{x^2 + y^2} = r$ and the lower surface is $z = 0$. Therefore, the inequalities defining the region of integration in cylindrical coordinates are

$$\mathcal{W}: 0 \le \theta \le \frac{\pi}{2},\ 0 \le r \le 1,\ 0 \le z \le r$$

Converting the double integral to cylindrical coordinates gives

$$I = \int_0^{\pi/2} \int_0^1 \int_0^r zr\, dz\, dr\, d\theta = \int_0^{\pi/2} \int_0^1 \frac{z^2 r}{2} \Big|_{z=0}^r dr\, d\theta = \int_0^{\pi/2} \int_0^1 \frac{r^3}{2}\, dr\, d\theta$$

$$= \left(\int_0^{\pi/2} d\theta \right) \left(\int_0^1 \frac{r^3}{2}\, dr \right) = \frac{\pi}{2} \cdot \frac{r^4}{8} \Big|_0^1 = \frac{\pi}{16}$$

37. Convert to spherical coordinates and evaluate:

$$\int_{-2}^2 \int_{-\sqrt{4-x^2}}^{\sqrt{4-x^2}} \int_0^{\sqrt{4-x^2-y^2}} e^{-(x^2+y^2+z^2)^{3/2}}\, dz\, dy\, dx$$

SOLUTION The region of integration as defined by the limits of integration is

$$\mathcal{W}: -2 \le x \le 2,\ -\sqrt{4-x^2} \le y \le \sqrt{4-x^2},\ 0 \le z \le \sqrt{4-x^2-y^2}$$

That is, $\mathcal{W}$ is the region enclosed by the sphere $x^2 + y^2 + z^2 = 4$ and the xy-plane. We see that the region of integration is the upper half-ball $x^2 + y^2 + z^2 \le 4$, hence the inequalities defining $\mathcal{W}$ in spherical coordinates are

$$\mathcal{W}: 0 \le \theta \le 2\pi,\ 0 \le \phi \le \frac{\pi}{2},\ 0 \le \rho \le 2$$

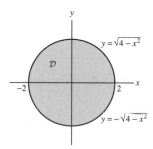

The function is $f(x, y, z) = e^{-(x^2+y^2+z^2)^{3/2}} = e^{-(\rho^2)^{3/2}} = e^{-\rho^3}$, therefore the integral in spherical coordinates is

$$I = \int_0^{2\pi} \int_0^{\pi/2} \int_0^2 e^{-\rho^3} \rho^2 \sin\phi\, d\rho\, d\phi\, d\theta = \left(\int_0^{2\pi} d\theta \right) \left(\int_0^{\pi/2} \sin\phi\, d\phi \right) \left(\int_0^2 e^{-\rho^3} \rho^2\, d\rho \right)$$

$$= 2\pi \left(-\cos\phi \Big|_0^{\pi/2} \right) \int_0^2 e^{-\rho^3} \rho^2\, d\rho = 2\pi \int_0^2 e^{-\rho^3} \rho^2\, d\rho$$

We compute the integral using the substitution $u = \rho^3$, $du = 3\rho^2 d\rho$. We get

$$I = 2\pi \int_0^8 e^{-u} \frac{du}{3} = \frac{2\pi}{3}(-e^{-u})\Big|_0^8 = \frac{2\pi}{3}(-e^{-8} + 1) = \frac{2\pi\left(-1 + e^8\right)}{3e^8}$$

39. Let $\mathcal{W}$ be the ball of radius R in $\mathbf{R}^3$ centered at the origin, and let $P = (0, 0, R)$ be the North Pole. Let $d_P(x, y, z)$ be the distance from P to (x, y, z). Show that the average value of d_P over the sphere $\mathcal{W}$ is equal to $\overline{d} = 6R/5$. *Hint:* Show that

$$\overline{d} = \frac{1}{\frac{4}{3}\pi R^3} \int_{\theta=0}^{2\pi} \int_{\rho=0}^R \int_{\phi=0}^\pi \rho^2 \sin\phi\sqrt{R^2 + \rho^2 - 2\rho R \cos\phi}\, d\phi\, d\rho\, d\theta$$

and evaluate.

SOLUTION We know that the volume of the ball is $\frac{4}{3}\pi R^3$.

In spherical coordinates, the distance from P to a point on the ball is

$$\sqrt{(x-0)^2 + (y-0)^2 + (z-R)^2} = \sqrt{\rho^2 \sin^2\phi \cos^2\theta + \rho^2 \sin^2\phi \sin^2\theta + (\rho\cos\phi - R)^2}$$

$$= \sqrt{\rho^2 \sin^2\phi(\cos^2\theta + \sin^2\theta) + \rho^2\cos^2\phi - 2\rho R\cos\phi + R^2}$$

$$= \sqrt{\rho^2\sin^2\phi + \rho^2\cos^2\phi - 2\rho R\cos\phi + R^2}$$

$$= \sqrt{\rho^2(\sin^2\phi + \cos^2\phi) - 2\rho R\cos\phi + R^2}$$

$$= \sqrt{R^2 + \rho^2 - 2\rho R\cos\phi}$$

Now, to write the average value of d_P we have:

$$d_P = \frac{1}{\frac{4}{3}\pi R^3} \int_{\theta=0}^{2\pi} \int_{\rho=0}^R \int_{\phi=0}^\pi \rho^2\sin\phi\sqrt{R^2 + \rho^2 - 2\rho R\cos\phi}\, d\phi\, d\rho\, d\theta$$

Using substitution, and the fact that $0 \le \rho \le R$,

$$\int_{\phi=0}^\pi \sin\phi\sqrt{R^2 + \rho^2 - 2\rho R\cos\phi}\, d\phi = \frac{2\rho}{3R}(R^2 + \rho^2 - 2\rho R\cos\phi)^{3/2}\Big|_0^\pi$$

$$= \frac{2\rho}{3R}\left((R + \rho^3) - (R - \rho)^3\right) = \frac{2\rho}{3R}(\rho^3 + 3R^2\rho)$$

Now integrate with respect to θ and ρ:

$$d_P = \frac{1}{\frac{4}{3}\pi R^3} \int_0^{2\pi} \int_0^R \rho^2 \cdot \frac{2\rho}{3R}(\rho^3 + 3R^2\rho)\, d\rho\, d\theta$$

$$= \frac{3}{4\pi R^3} \cdot \frac{2}{3R} \int_0^{2\pi} \int_0^R \rho^3(\rho^3 + 3R^2\rho)\, d\rho\, d\theta$$

$$= \frac{1}{2\pi R^4} \int_0^{2\pi} \int_0^R \rho^6 + 3R^2\rho^4\, d\rho\, d\theta$$

$$= \frac{1}{2\pi R^4} \int_0^{2\pi} \frac{1}{7}\rho^7 + \frac{3}{5}R^2\rho^5\Big|_0^R\, d\theta$$

$$= \frac{1}{2\pi R^4} \int_0^{2\pi} \frac{1}{7}R^7 + \frac{3}{5}R^7\, d\theta$$

$$= \frac{1}{2\pi R^4}\left(\frac{26}{35}R^7\right) \cdot 2\pi = \frac{26}{35}R^3$$

to get $8\pi R^4/5$. Dividing by the volume of the sphere gives us $6R/5$

41. Use cylindrical coordinates to find the mass of the solid bounded by $z = 8 - x^2 - y^2$ and $z = x^2 + y^2$, assuming a mass density of $f(x, y, z) = (x^2 + y^2)^{1/2}$.

SOLUTION   The mass of the solid $\mathcal{W}$ is the following integral:

$$M = \iiint_{\mathcal{W}} (x^2 + y^2)^{1/2}\, dV$$

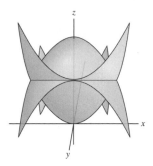

The projection of $\mathcal{W}$ on the xy-plane is obtained by equating the equations of the two surfaces:

$$8 - x^2 - y^2 = x^2 + y^2 \qquad \Rightarrow \qquad x^2 + y^2 = 4$$
$$2(x^2 + y^2) = 8$$

We conclude that the projection is the disk $\mathcal{D}$, $x^2 + y^2 \le 4$.

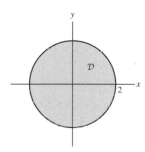

Therefore, $\mathcal{W}$ is described by

$$\mathcal{W}: x^2 + y^2 \le z \le 8 - (x^2 + y^2), \quad (x, y) \in \mathcal{D}$$

Thus,

$$M = \iint_{\mathcal{D}} \int_{x^2+y^2}^{8-(x^2+y^2)} (x^2 + y^2)^{1/2}\, dz\, dx\, dy$$

We convert the integral to cylindrical coordinates. The inequalities for $\mathcal{W}$ are

$$0 \le r \le 2, \quad 0 \le \theta \le 2\pi, \quad r^2 \le z \le 8 - r^2.$$

Also, $(x^2 + y^2)^{1/2} = r$, hence we obtain the following integral:

$$M = \int_0^2 \int_0^{2\pi} \int_{r^2}^{8-r^2} r \cdot r\, dz\, d\theta\, dr = \int_0^2 \int_0^{2\pi} \int_{r^2}^{8-r^2} r^2\, dz\, d\theta\, dr = \int_0^2 \int_0^{2\pi} r^2 z \Big|_{z=r^2}^{8-r^2} d\theta\, dr$$

$$= \int_0^2 \int_0^{2\pi} r^2 (8 - r^2 - r^2)\, d\theta\, dr = \int_0^2 \int_0^{2\pi} (8r^2 - 2r^4)\, d\theta\, dr = \left(\int_0^{2\pi} 1\, d\theta \right) \left(\int_0^2 (8r^2 - 2r^4)\, dr \right)$$

$$= 2\pi \left(\frac{8r^3}{3} - \frac{2}{5}r^5 \Big|_0^2 \right) = \frac{256}{15}\pi \approx 53.62$$

43. Use cylindrical coordinates to find the mass of a cylinder of radius 4 and height 10 if the mass density at a point is equal to the square of the distance from the cylinder's central axis.

SOLUTION

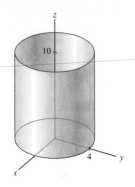

The mass density is $\rho(x, y, z) = x^2 + y^2 = r^2$, hence the mass of the cylinder is

$$M = \iiint_{\mathcal{W}} (x^2 + y^2)\, dV$$

The region $\mathcal{W}$ is described using cylindrical coordinates by the following inequalities:

$$\mathcal{W} : 0 \le \theta \le 2\pi,\ 0 \le r \le 4,\ 0 \le z \le 10$$

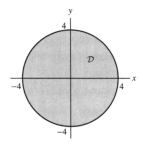

Thus,

$$M = \iiint_{\mathcal{W}} (x^2 + y^2)\, dV = \int_0^{2\pi} \int_0^4 \int_0^{10} r^2 \cdot r\, dz\, dr\, d\theta = \int_0^{2\pi} \int_0^4 \int_0^{10} r^3\, dz\, dr\, d\theta = \int_0^{2\pi} \int_0^4 r^3 z \Big|_{z=0}^{10} dr\, d\theta$$

$$= \int_0^{2\pi} \int_0^4 r^3 \cdot 10\, dr\, d\theta = \int_0^{2\pi} \frac{10 r^4}{4} \Big|_{r=0}^4 d\theta = \int_0^{2\pi} 640\, d\theta = 640 \cdot 2\pi = 1280\pi$$

45. Find the centroid of the solid bounded by the xy-plane, the cylinder $x^2 + y^2 = R^2$, and the plane $x/R + z/H = 1$.

SOLUTION   First to find the volume of this solid. The first equation lends itself well to cylindrical coordinates:

$$x^2 + y^2 = R^2 \quad \Rightarrow \quad r = R, 0 \le \theta \le 2\pi$$

and

$$\frac{x}{R} + \frac{z}{H} = 1 \quad \Rightarrow \quad z = H\left(1 - \frac{x}{R}\right) = H\left(1 - \frac{r\cos\theta}{R}\right)$$

The volume is:

$$Volume = \int_0^{2\pi} \int_0^R \int_0^{H(1 - r\cos\theta/R)} 1\, dz\, dr\, d\theta$$

$$= \int_0^{2\pi} \int_0^R H\left(1 - \frac{r\cos\theta}{R}\right) dr\, d\theta$$

$$= H \int_0^{2\pi} r - \frac{1}{2} \cdot \frac{r^2 \cos\theta}{R} \Big|_{r=0}^R d\theta$$

$$= H \int_0^{2\pi} R - \frac{1}{2} R \cos\theta\, d\theta$$

$$= H\left(R\theta - \frac{1}{2} R \sin\theta \Big|_0^{2\pi}\right)$$

$$= 2\pi H R$$

Now to compute the coordinates of the centroid:

$$\bar{x} = \frac{1}{V} \iiint_{\mathcal{W}} x\, dV = \frac{1}{2\pi HR} \int_0^{2\pi} \int_0^R \int_0^{H(1-r\cos\theta/R)} r\cos\theta\, dz\, dr\, d\theta$$

$$= \frac{1}{2\pi HR} \int_0^{2\pi} \int_0^R r\cos\theta \cdot z \Big|_0^{H(1-r\cos\theta/R)} dr\, d\theta$$

$$= \frac{H}{2\pi HR} \int_0^{2\pi} \int_0^R r\cos\theta \left(1 - \frac{r\cos\theta}{R}\right) dr\, d\theta$$

$$= \frac{1}{2\pi R} \int_0^{2\pi} \int_0^R r\cos\theta - \frac{1}{R} r^2 \cos^2\theta\, dr\, d\theta$$

$$= \frac{1}{2\pi R} \int_0^{2\pi} \frac{1}{2} r^2 \cos\theta - \frac{1}{3R} r^3 \cos^2\theta \Big|_0^R d\theta$$

$$= \frac{1}{2\pi R} \int_0^{2\pi} \frac{1}{2} R^2 \cos\theta - \frac{R^2}{6}(1 + \cos 2\theta)\, d\theta$$

$$= \frac{1}{2\pi R} \left(\frac{1}{2} R^2 \sin\theta - \frac{R^2}{6}\left(\theta + \frac{1}{2}\sin 2\theta\right) \Big|_0^{2\pi}\right) = \frac{1}{2\pi R} \cdot -\frac{R^2}{6}(2\pi) = -\frac{R}{6}$$

$$\bar{y} = \frac{1}{V} \iiint_{\mathcal{W}} y\, dV = \frac{1}{2\pi HR} \int_0^{2\pi} \int_0^R \int_0^{H(1-r\cos\theta/R)} r\sin\theta\, dz\, dr\, d\theta$$

$$- \frac{1}{2\pi HR} \int_0^{2\pi} \int_0^R r\sin\theta \cdot z \Big|_0^{H(1-r\cos\theta/R)} dr\, d\theta$$

$$= \frac{H}{2\pi HR} \int_0^{2\pi} \int_0^R r\sin\theta \left(1 - \frac{r\cos\theta}{R}\right) dr\, d\theta$$

$$= \frac{1}{2\pi R} \int_0^{2\pi} \int_0^R r\sin\theta - \frac{1}{R} r^2 \sin\theta\cos\theta\, dr\, d\theta$$

$$= \frac{1}{2\pi R} \int_0^{2\pi} \frac{1}{2} r^2 \sin\theta - \frac{1}{3R} r^3 \sin\theta\cos\theta \Big|_0^R d\theta$$

$$= \frac{1}{2\pi R} \int_0^{2\pi} \frac{1}{2} R^2 \sin\theta - \frac{R^2}{3}\sin\theta\cos\theta\, d\theta$$

$$= \frac{1}{2\pi R} \left(-\frac{1}{2} R^2 \cos\theta - \frac{R^2}{6}\sin^2\theta \Big|_0^{2\pi}\right) = 0$$

$$\bar{z} = \frac{1}{V} \iiint_{\mathcal{W}} z\, dV = \frac{1}{2\pi HR} \int_0^{2\pi} \int_0^R \int_0^{H(1-r\cos\theta/R)} z\, dz\, dr\, d\theta$$

$$= \frac{1}{2\pi HR} \int_0^{2\pi} \int_0^R \frac{1}{2} z^2 \Big|_0^{H(1-r\cos\theta/R)} dr\, d\theta$$

$$= \frac{H^2}{4\pi HR} \int_0^{2\pi} \int_0^R \left(1 - \frac{r\cos\theta}{R}\right)^2 dr\, d\theta$$

$$- \frac{H}{4\pi R} \int_0^{2\pi} \int_0^R 1 - \frac{2r\cos\theta}{R} + \frac{r^2\cos^2\theta}{R^2}\, dr\, d\theta$$

$$= \frac{H}{4\pi R} \int_0^{2\pi} r - \frac{r^2\cos\theta}{R} + \frac{r^3\cos^2\theta}{3R^2} \Big|_0^R d\theta$$

$$= \frac{H}{4\pi R} \int_0^{2\pi} R - R\cos\theta + \frac{1}{6}R(1 + \cos 2\theta)\, d\theta$$

$$= \frac{H}{4\pi R} \left(R\theta - R\sin\theta + \frac{1}{6}R\left(\theta + \frac{1}{2}\sin 2\theta\right) \Big|_0^{2\pi}\right)$$

$$= \frac{H}{4\pi R} \left(2\pi R + \frac{1}{3}\pi R\right) = \frac{H}{4\pi R} \cdot \frac{7\pi R}{3} = \frac{7}{12} H$$

The coordinates of the centroid are $(-R/6, 0, 7H/12)$.

47. Find the centroid of solid (A) in Figure 4 defined by $x^2 + y^2 \leq R^2$, $0 \leq z \leq H$, and $\frac{\pi}{6} \leq \theta \leq 2\pi$, where θ is the polar angle of (x, y).

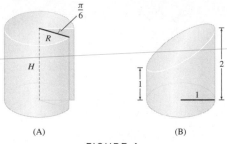

(A)   (B)

FIGURE 4

SOLUTION   Since the mass distribution is uniform, we may assume that $\rho(x, y, z) = 1$, hence the center of mass is

$$x_{CM} = \frac{1}{V} \iiint_{\mathcal{W}} x \, dV, \quad y_{CM} = \frac{1}{V} \iiint_{\mathcal{W}} y \, dV, \quad z_{CM} = \frac{1}{V} \iiint_{\mathcal{W}} z \, dV$$

The inequalities describing $\mathcal{W}$ in cylindrical coordinates are

$$\mathcal{W} : 0 \leq \theta \leq \frac{\pi}{6}, \ 0 \leq r \leq R, \ 0 \leq z \leq H$$

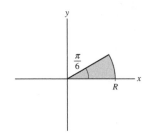

The entire cylinder has a total volume $\pi R^2 H$. The region $\mathcal{W}$ has the fraction $(2\pi - \frac{\pi}{6})/(2\pi)$ of this total volume, so

$$V = \frac{(2\pi - \frac{\pi}{6})}{2\pi}(\pi R^2 H) = \frac{11\pi R^2 H}{12}$$

We use cylindrical coordinates to compute the triple integrals:

$$x_{CM} = \frac{1}{V} \int_{\pi/6}^{2\pi} \int_0^R \int_0^H (r \cos\theta) r \, dz \, dr \, d\theta = \frac{12}{11\pi R^2 H} \left(\int_{\pi/6}^{2\pi} \cos\theta \, d\theta \right) \left(\int_0^R r^2 \, dr \right) \left(\int_0^H dz \right)$$

$$= \frac{12}{11\pi R^2 H} \left(-\frac{1}{2} \right) \left(\frac{R^3}{3} \right) (H) = -\frac{2R}{11\pi}$$

$$y_{CM} = \frac{1}{V} \int_{\pi/6}^{2\pi} \int_0^R \int_0^H (r \sin\theta) r \, dz \, dr \, d\theta = \frac{12}{11\pi R^2 H} \left(\int_{\pi/6}^{2\pi} \sin\theta \, d\theta \right) \left(\int_0^R r^2 \, dr \right) \left(\int_0^H dz \right)$$

$$= \frac{12}{11\pi R^2 H} \left(\frac{-2 + \sqrt{3}}{2} \right) \left(\frac{R^3}{3} \right) (H) = \frac{2R}{11\pi}(\sqrt{3} - 2)$$

$$z_{CM} = \frac{1}{V} \int_{\pi/6}^{2\pi} \int_0^R \int_0^H z r \, dz \, dr \, d\theta = \frac{12}{11\pi R^2 H} \left(\int_{\pi/6}^{2\pi} d\theta \right) \left(\int_0^R r \, dr \right) \left(\int_0^H z \, dz \right)$$

$$= \frac{12}{11\pi R^2 H} \left(\frac{11\pi}{6} \right) \left(\frac{R^2}{2} \right) \left(\frac{H^2}{2} \right) = \frac{H}{2}$$

Therefore, the center of mass is the following point:

$$\left(-\frac{2R}{11\pi}, \frac{2R}{11\pi}(\sqrt{3} - 2), \frac{H}{2} \right)$$

49. Find the center of mass of the cylinder $x^2 + y^2 = 1$ for $0 \le z \le 1$, assuming a mass density of $\rho(x, y, z) = z$.

SOLUTION By symmetry, we can note that the center of mass lies on the z-axis.

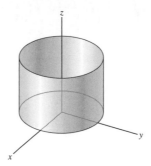

The coordinates of the center of mass are defined as,

$$x_{CM} = \frac{\iiint_{\mathcal{W}} x \left(x^2 + y^2 \right) dV}{M}$$

$$y_{CM} = \frac{\iiint_{\mathcal{W}} y \left(x^2 + y^2 \right) dV}{M} \qquad (1)$$

$$z_{CM} = \frac{\iiint_{\mathcal{W}} z \left(x^2 + y^2 \right) dV}{M}$$

where M is the total mass of $\mathcal{W}$. The cylinder $\mathcal{W}$ is defined by the inequalities

$$-1 \le x \le 1, \quad -\sqrt{1 - x^2} \le y \le \sqrt{1 - x^2}, \quad 0 \le z \le 1$$

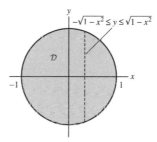

We compute the total mass of $\mathcal{W}$:

$$M = \iiint_{\mathcal{W}} z \, dV = \int_{-1}^{1} \int_{-\sqrt{1-x^2}}^{\sqrt{1-x^2}} \int_{0}^{1} z \, dz \, dy \, dx = \int_{-1}^{1} \int_{-\sqrt{1-x^2}}^{\sqrt{1-x^2}} \frac{1}{2} z^2 \Big|_{z=0}^{1} dy \, dx$$

$$= \frac{1}{2} \int_{-1}^{1} \int_{-\sqrt{1-x^2}}^{\sqrt{1-x^2}} 1 \, dy \, dx = \int_{-1}^{1} \int_{0}^{\sqrt{1-x^2}} 1 \, dy \, dx = \int_{-1}^{1} y \Big|_{y=0}^{\sqrt{1-x^2}} dx$$

$$= \int_{-1}^{1} \sqrt{1 - x^2} \, dx$$

This integral is the area of a half of the unit circle, hence the total mass is

$$\int_{-1}^{1} \sqrt{1 - x^2} \, dx = \frac{\pi}{2} = M$$

We now compute the numerators in (1). Using (2), we get

$$\iiint_{\mathcal{W}} xz \, dV = \int_{-1}^{1} \int_{-\sqrt{1-x^2}}^{\sqrt{1-x^2}} \int_{0}^{1} xz \, dz \, dy \, dx = \int_{-1}^{1} \int_{-\sqrt{1-x^2}}^{\sqrt{1-x^2}} \frac{1}{2} xz^2 \Big|_{z=0}^{1} dy \, dx$$

$$= 2 \int_{-1}^{1} \int_{0}^{\sqrt{1-x^2}} \frac{1}{2} x \, dy \, dx = \int_{-1}^{1} xy \Big|_{y=0}^{\sqrt{1-x^2}} dx$$

$$= \int_{-1}^{1} x\sqrt{1 - x^2} \, dx = -\frac{2}{3} (1 - x^2)^{3/2} \Big|_{-1}^{1} = 0 \qquad (2)$$

Now to compute the next numerator:

$$\iiint_{\mathcal{W}} yz\, dV = \int_{-1}^{1} \int_{-\sqrt{1-x^2}}^{\sqrt{1-x^2}} \int_{0}^{1} yz\, dz\, dy\, dx = \int_{-1}^{1} \int_{-\sqrt{1-x^2}}^{\sqrt{1-x^2}} \frac{1}{2} yz^2 \Big|_{z=0}^{1} dy\, dx$$

$$= 2\int_{-1}^{1} \int_{0}^{\sqrt{1-x^2}} \frac{1}{2} y\, dy\, dx = \int_{-1}^{1} \frac{1}{2} y^2 \Big|_{y=0}^{\sqrt{1-x^2}} dx$$

$$= \int_{-1}^{1} (1-x^2)\, dx = \left(x - x^3 \right) \Big|_{-1}^{1} = 0 \tag{3}$$

Thus far we have:

$$\iiint_{\mathcal{W}} yz\, dV = \iiint_{\mathcal{W}} xz\, dV = 0 \tag{4}$$

We compute the numerator of z_{CM} in (1):

$$\iiint_{\mathcal{W}} z \cdot z\, dV = \int_{-1}^{1} \int_{-\sqrt{1-x^2}}^{\sqrt{1-x^2}} \int_{0}^{1} z^2\, dz\, dy\, dx = \int_{-1}^{1} \int_{-\sqrt{1-x^2}}^{\sqrt{1-x^2}} \frac{1}{3} z^3 \Big|_{z=0}^{1} dy\, dx$$

$$= \int_{-1}^{1} \int_{-\sqrt{1-x^2}}^{\sqrt{1-x^2}} \frac{1}{3}\, dy\, dx = \int_{-1}^{1} \int_{0}^{\sqrt{1-x^2}} \frac{2}{3}\, dy\, dx$$

$$= \int_{-1}^{1} \frac{2}{3} y \Big|_{0}^{\sqrt{1-x^2}} dx$$

$$= \frac{2}{3} \int_{-1}^{1} \sqrt{1 - x^2}\, dx$$

This is $2/3$ times half of the area of the circle, centered at the origin, having radius 1, so the integral is $1/3\pi$.

Finally, we substitute $M = \frac{\pi}{2}$ and the computed integrals for (1) to obtain the following solution:

$$(x_{\text{CM}}, y_{\text{CM}}, z_{\text{CM}}) = \left(0, 0, \frac{\frac{\pi}{3}}{\frac{\pi}{2}} \right) = \left(0, 0, \frac{2}{3} \right).$$

51. Find the center of mass of the first octant of the ball $x^2 + y^2 + z^2 = 1$, assuming a mass density of $\rho(x, y, z) = x$.

SOLUTION

(a) The solid is the part of the unit sphere in the first octant. The inequalities defining the projection of the solid onto the xy-plane are

$$\mathcal{D} : 0 \le y \le 1,\ 0 \le x \le \sqrt{1 - y^2}$$

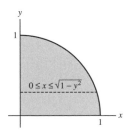

$\mathcal{W}$ is the region bounded by $\mathcal{D}$ and the sphere $z = \sqrt{1 - x^2 - y^2}$, hence $\mathcal{W}$ is defined by the inequalities

$$\mathcal{W} : 0 \le y \le 1,\ 0 \le x \le \sqrt{1 - y^2},\ 0 \le z \le \sqrt{1 - x^2 - y^2} \tag{1}$$

We first must compute the mass of the solid. Using the mass as a triple integral, we have

$$M = \int_{0}^{1} \int_{0}^{\sqrt{1-y^2}} \int_{0}^{\sqrt{1-x^2-y^2}} x\, dz\, dx\, dy = \int_{0}^{1} \int_{0}^{\sqrt{1-y^2}} xz \Big|_{z=0}^{\sqrt{1-x^2-y^2}} dx\, dy$$

$$= \int_{0}^{1} \int_{0}^{\sqrt{1-y^2}} x\sqrt{1 - x^2 - y^2}\, dx\, dy$$

We compute the inner integral using the substitution $u = \sqrt{1 - x^2 - y^2}$, $du = -\frac{x}{u}\,dx$, or $x\,dx = -u\,du$. We get

$$\int_0^{\sqrt{1-y^2}} x\sqrt{1 - x^2 - y^2}\,dx = \int_{\sqrt{1-y^2}}^0 u(-u\,du) = \int_0^{\sqrt{1-y^2}} u^2\,du = \frac{u^3}{3}\Big|_0^{\sqrt{1-y^2}} = \frac{(1-y^2)^{3/2}}{3} \tag{2}$$

We substitute in (2) and compute the resulting integral substituting $y = \sin t$, $dy = \cos t\,dt$:

$$M = \int_0^1 \frac{(1-y^2)^{3/2}}{3}\,dy = \frac{1}{3}\int_0^{\pi/2} (1 - \sin^2 t)^{3/2} \cos t\,dt = \frac{1}{3}\int_0^{\pi/2} \cos^4 t\,dt$$

$$= \frac{1}{3}\left(\frac{\cos^3 t \sin t}{4} + \frac{3}{4}\left(\frac{t}{2} + \frac{\sin 2t}{4}\right)\Big|_0^{\pi/2}\right) = \frac{1}{4} \cdot \frac{\pi}{4} = \frac{\pi}{16}$$

That is, $M = \frac{\pi}{16}$. We now find the coordinates of the center of mass. To compute x_{CM} we use the definition of $\mathcal{D}$ as a vertically simple region to obtain a simpler integral. That is, we write the inequalities for $\mathcal{W}$ as

$$\mathcal{W}: 0 \le x \le 1,\ 0 \le y \le \sqrt{1 - x^2},\ 0 \le z \le \sqrt{1 - x^2 - y^2} \tag{3}$$

Thus,

$$x_{CM} = \frac{1}{M}\iiint_{\mathcal{W}} x\rho\,dV = \frac{16}{\pi}\int_0^1 \int_0^{\sqrt{1-x^2}} \int_0^{\sqrt{1-x^2-y^2}} x^2\,dz\,dy\,dx = \frac{16}{\pi}\int_0^1 \int_0^{\sqrt{1-x^2}} x^2 z\Big|_{z=0}^{\sqrt{1-x^2-y^2}}\,dy\,dx$$

$$= \frac{16}{\pi}\int_0^1 \int_0^{\sqrt{1-x^2}} x^2\sqrt{1 - x^2 - y^2}\,dy\,dx = \frac{16}{\pi}\int_0^1 x^2\left(\int_0^{\sqrt{1-x^2}} \sqrt{1 - x^2 - y^2}\,dy\right)dx \tag{4}$$

We compute the inner integral using Integration Formulas:

$$\int_0^{\sqrt{1-x^2}} \sqrt{1 - x^2 - y^2}\,dy = \frac{y}{2}\sqrt{1 - x^2 - y^2} + \frac{1-x^2}{2}\sin^{-1}\frac{y}{\sqrt{1-x^2}}\Big|_{y=0}^{\sqrt{1-x^2}}$$

$$= \frac{1-x^2}{2}\sin^{-1} 1 = \frac{1-x^2}{2} \cdot \frac{\pi}{2} = \frac{\pi}{4}(1 - x^2)$$

Substituting in (4) gives

$$x_{CM} = \frac{16}{\pi}\int_0^1 x^2 \cdot \frac{\pi}{4}(1 - x^2)\,dx = 4\int_0^1 (x^2 - x^4)\,dx = 4\left(\frac{x^3}{3} - \frac{x^5}{5}\right)\Big|_0^1 = 4\left(\frac{1}{3} - \frac{1}{5}\right) = \frac{8}{15}$$

(b) We compute the y-coordinate of the center of mass, using (1):

$$y_{CM} = \frac{1}{M}\iiint_{\mathcal{W}} y\rho\,dV = \frac{16}{\pi}\int_0^1 \int_0^{\sqrt{1-y^2}} \int_0^{\sqrt{1-x^2-y^2}} yx\,dz\,dx\,dy = \frac{16}{\pi}\int_0^1 \int_0^{\sqrt{1-y^2}} yxz\Big|_{z=0}^{\sqrt{1-x^2-y^2}}\,dx\,dy$$

$$= \frac{16}{\pi}\int_0^1 \int_0^{\sqrt{1-y^2}} yx\sqrt{1 - x^2 - y^2}\,dx\,dy = \frac{16}{\pi}\int_0^1 y\left(\int_0^{\sqrt{1-y^2}} x\sqrt{1 - x^2 - y^2}\,dx\right)dy$$

The inner integral was computed in (2), therefore,

$$y_{CM} = \frac{16}{\pi}\int_0^1 y \cdot \frac{(1-y^2)^{3/2}}{3}\,dy = \frac{16}{3\pi}\int_0^1 y(1 - y^2)^{3/2}\,dy$$

We compute the integral using the substitution $u = 1 - y^2$, $du = -2y\,dy$. We get

$$y_{CM} = \frac{16}{3\pi}\int_1^0 u^{3/2} \cdot \left(-\frac{du}{2}\right) = \frac{8}{3\pi}\int_0^1 u^{3/2}\,du = \frac{8}{3\pi} \cdot \frac{2}{5} \cdot u^{5/2}\Big|_0^1 = \frac{16}{15\pi}$$

Finally we find the z-coordinate of the center of mass, using (1):

$$z_{CM} = \frac{1}{M}\iiint_{\mathcal{W}} z\rho\,dV = \frac{16}{\pi}\int_0^1 \int_0^{\sqrt{1-y^2}} \int_0^{\sqrt{1-x^2-y^2}} zx\,dz\,dx\,dy = \frac{16}{\pi}\int_0^1 \int_0^{\sqrt{1-y^2}} \frac{z^2 x}{2}\Big|_{z=0}^{\sqrt{1-x^2-y^2}}\,dx\,dy$$

$$= \frac{16}{\pi}\int_0^1 \int_0^{\sqrt{1-y^2}} \frac{x}{2}(1 - x^2 - y^2)\,dx\,dy = \frac{8}{\pi}\int_0^1 \int_0^{\sqrt{1-y^2}} (x - x^3 - xy^2)\,dx\,dy$$

$$= \frac{8}{\pi} \int_0^1 \frac{x^2}{2} - \frac{x^4}{4} - \frac{x^2 y^2}{2} \Big|_{x=0}^{\sqrt{1-y^2}} dy = \frac{8}{\pi} \int_0^1 \left(\frac{1-y^2}{2} - \frac{(1-y^2)^2}{4} - \frac{(1-y^2)y^2}{2} \right) dy$$

$$= \frac{2}{\pi} \int_0^1 (y^4 - 2y^2 + 1) \, dy = \frac{2}{\pi} \left(\frac{y^5}{5} - \frac{2y^3}{3} + y \right) \Big|_{y=0}^1 = \frac{2}{\pi} \left(\frac{1}{5} - \frac{2}{3} + 1 \right) = \frac{16}{15\pi}$$

The center mass is the following point:

$$P = \left(\frac{8}{15}, \frac{16}{15\pi}, \frac{16}{15\pi} \right).$$

53. Calculate $P(3X + 2Y \geq 6)$ for the probability density in Exercise 52.

SOLUTION   Previously we found $p(x, y) = \frac{1}{33}(4x - y + 3)$. Then using $P(3X + 2Y \geq 6)$ we want to find $1 - P(3X + 2Y \leq 6)$. Hence we need to integrate the following:

$$P(3X + 2Y \leq 6) = \int_{x=0}^2 \int_{y=0}^{3-3/2x} \frac{1}{33}(4x - y + 3) \, dy \, dx$$

$$= \frac{1}{33} \int_0^2 4xy - \frac{1}{2}y^2 + 3y \Big|_0^{3-3/2x} dx$$

$$= \frac{1}{33} \int_0^2 4x \left(3 - \frac{3}{2}x \right) - \frac{1}{2} \left(3 - \frac{3}{2}x \right)^2 + 3 \left(3 - \frac{3}{2}x \right) dx$$

$$= \frac{1}{33} \int_0^2 -\frac{57}{8}x^2 + 12x + \frac{9}{2} \, dx$$

$$= \frac{1}{33} \left(-\frac{57}{24}x^3 + 6x^2 + \frac{9}{2}x \right) \Big|_0^2$$

$$= \frac{1}{33} (-19 + 24 + 9) = \frac{14}{33}$$

Thus we have:

$$P(3X + 2Y \geq 6) = 1 - P(3X + 2Y \leq 6) = 1 - \frac{14}{33} = \frac{19}{33}$$

55. An insurance company issues two kinds of policies A and B. Let X be the time until the next claim of type A is filed, and let Y be the time (in days) until the next claim of type B is filed. The random variables have joint probability density

$$p(x, y) = 12e^{-4x - 3y}$$

Find the probability that $X \leq Y$.

SOLUTION   We must compute

$$P(X \leq Y) = \int_{x=0}^{\infty} \int_{y=x}^{\infty} p(x, y) \, dy \, dx$$

Now evaluating we get:

$$P(X \leq Y) = \int_{x=0}^{\infty} \int_{y=x}^{\infty} 12e^{-4x - 3y} \, dy \, dx = 12 \int_0^{\infty} \int_x^{\infty} e^{-4x} e^{-3y} \, dy \, dx$$

$$= 12 \int_0^{\infty} e^{-4x} \left(-\frac{1}{3}e^{-3y} \right) \Big|_x^{\infty} dx = -4 \int_0^{\infty} e^{-4x} \left(\lim_{t \to \infty} e^{-3t} - e^{-3x} \right) dx$$

$$= 4 \int_0^{\infty} e^{-4x} \cdot e^{-3x} \, dx = 4 \int_0^{\infty} e^{-7x} \, dx$$

$$= -\frac{4}{7} \left(e^{-7x} \right) \Big|_0^{\infty} = -\frac{4}{7} \lim_{t \to \infty} \left(e^{-7t} - 1 \right) = \frac{4}{7}$$

57. Find a linear mapping $\Phi(u, v)$ that maps the unit square to the parallelogram in the xy-plane spanned by the vectors $\langle 3, -1 \rangle$ and $\langle 1, 4 \rangle$. Then, use the Jacobian to find the area of the image of the rectangle $\mathcal{R} = [0, 4] \times [0, 3]$ under Φ.

SOLUTION   We denote the linear map by

$$G(u, v) = (Au + Cv, \, Bu + Dv) \qquad (1)$$

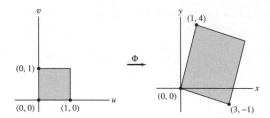

The image of the unit square is the quadrangle whose vertices are the images of the vertices of the square. Therefore we ask that

$$G(0, 0) = (A \cdot 0 + C \cdot 0, B \cdot 0 + D \cdot 0) = (0, 0) \qquad\qquad (0, 0) = (0, 0)$$

$$G(1, 0) = (A \cdot 1 + C \cdot 0, B \cdot 1 + D \cdot 0) = (3, -1) \quad \Rightarrow \quad (A, B) = (3, -1)$$

$$G(0, 1) = (A \cdot 0 + C \cdot 1, B \cdot 0 + D \cdot 1) = (1, 4) \qquad\qquad (C, D) = (1, 4)$$

These equations imply that $A = 3$, $B = -1$, $C = 1$, and $D = 4$. Substituting in (1) we obtain the following map:

$$G(u, v) = (3u + v, -u + 4v)$$

The area of the rectangle $R = [0, 4] \times [0, 3]$ is $4 \cdot 3 = 12$, therefore the transformed area is

$$\text{Area} = |\text{Jac}(G)| \cdot 12$$

The Jacobian of the linear map G is

$$\text{Jac}\,(G) = \begin{vmatrix} A & C \\ B & D \end{vmatrix} = \begin{vmatrix} 3 & 1 \\ -1 & 4 \end{vmatrix} = 12 - (-1) = 13$$

Therefore,

$$\text{Area} = 13 \cdot 12 = 156.$$

59. Let $\mathcal{D}$ be the shaded region in Figure 6, and let F be the map

$$u = y + x^2, \qquad v = y - x^3$$

(a) Show that F maps $\mathcal{D}$ to a rectangle $\mathcal{R}$ in the uv-plane.
(b) Apply Eq. (7) in Section 15.6 with $P = (1, 7)$ to estimate Area($\mathcal{D}$).

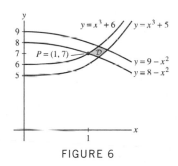

FIGURE 6

SOLUTION

(a) Note that the appropriate map should be $u = y + x^2$ rather than $u = -y + x^2$. We examine the images of the boundary curves of $\mathcal{D}$ under the map $(u, v) = \Phi(x, y) = (x^2 + y, y - x^3)$. The curves $y = x^3 + 6$ and $y = x^3 + 5$ can be rewritten as $y - x^3 = 6$ and $y - x^3 = 5$. Since $v = y - x^3$, these curves are mapped to the horizontal lines $v = 6$ and $v = 5$, respectively. The curves $y = 8 - x^2$ and $y = 9 - x^2$ can be rewritten as $y + x^2 = 8$ and $y + x^2 = 9$. Since $u = y + x^2$, these curves are mapped to the vertical lines $u = 8$ and $u = 9$, respectively. We conclude that $\mathcal{D}$ is mapped to the rectangle $\mathcal{R} = [8, 9] \times [5, 6]$ in the (u, v)-plane.

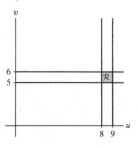

(b) We use Eq. (5) in section 16.5, where this time Φ is a mapping from the (x, y)-plane to the (u, v)-plane, and $P = (1, 7)$ is a point in $\mathcal{D}$:

$$\text{Area}\,\Phi(\mathcal{D}) \approx |\text{Jac}(\Phi)(P)|\text{Area}(\mathcal{D})$$

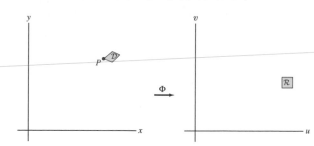

Here, $\text{Area}\,\Phi(\mathcal{D}) = \text{Area}(\mathcal{R}) = 1^2 = 1$, therefore we get

$$1 \approx |\text{Jac}(\Phi)(P)|\text{Area}(\mathcal{D})$$

or

$$\text{Area}(\mathcal{D}) \approx |\text{Jac}(\Phi)(P)|^{-1} \tag{1}$$

We compute the Jacobian of $\Phi(x, y) = (u, v) = (y + x^2, y - x^3)$ at $P = (1, 7)$:

$$\text{Jac}(\Phi) = \begin{vmatrix} \dfrac{\partial u}{\partial x} & \dfrac{\partial u}{\partial y} \\[2mm] \dfrac{\partial v}{\partial x} & \dfrac{\partial v}{\partial y} \end{vmatrix} = \begin{vmatrix} 2x & 1 \\ -3x^2 & 1 \end{vmatrix} = 2x + 3x^2 \quad \Rightarrow \quad \text{Jac}(\Phi)(P) = 2 \cdot 1 + 3 \cdot 7^2 = 149$$

Combining with (1) gives

$$\text{Area}(\mathcal{D}) \approx (149)^{-1} = \frac{1}{149}.$$

61. Sketch the region $\mathcal{D}$ bounded by the curves $y = 2/x$, $y = 1/(2x)$, $y = 2x$, $y = x/2$ in the first quadrant. Let F be the map $u = xy$, $v = y/x$ from the xy-plane to the uv-plane.

(a) Find the image of $\mathcal{D}$ under F.

(b) Let $\Phi = F^{-1}$. Show that $|\text{Jac}(\Phi)| = \dfrac{1}{2|v|}$.

(c) Apply the Change of Variables Formula to prove the formula

$$\iint_{\mathcal{D}} f\left(\frac{y}{x}\right) dx\, dy = \frac{3}{4} \int_{1/2}^{2} \frac{f(v)\, dv}{v}$$

(d) Apply (c) to evaluate $\displaystyle\iint_{\mathcal{D}} \frac{y e^{y/x}}{x} dx\, dy$.

SOLUTION

(a) The region $\mathcal{D}$ is shown in the figure:

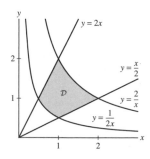

We rewrite the equations of the boundary curves as $xy = 2$, $xy = \frac{1}{2}$, $\frac{y}{x} = 2$, and $\frac{y}{x} = \frac{1}{2}$. These curves are mapped by Φ to the lines $u = 2$, $u = \frac{1}{2}$, $v = 2$, and $v = \frac{1}{2}$. Therefore, the image of $\mathcal{D}$ is the rectangle $\mathcal{R} = \left[\frac{1}{2}, 2\right] \times \left[\frac{1}{2}, 2\right]$ in the (u, v)-plane.

(b) We use the Jacobian of the inverse map:

$$\text{Jac}(F^{-1}) = (\text{Jac}(F))^{-1}$$

We compute the Jacobian of $F(x, y) = (u, v) = \left(xy, \frac{y}{x}\right)$:

$$\text{Jac}(F) = \begin{vmatrix} \dfrac{\partial u}{\partial x} & \dfrac{\partial u}{\partial y} \\ \dfrac{\partial v}{\partial x} & \dfrac{\partial v}{\partial y} \end{vmatrix} = \begin{vmatrix} y & x \\ -\dfrac{y}{x^2} & \dfrac{1}{x} \end{vmatrix} = \frac{y}{x} + \frac{yx}{x^2} = \frac{2y}{x} = 2v$$

(Note that everything is positive, so we don't need absolute values!) Thus,

$$\text{Jac}(F^{-1}) = (\text{Jac}(F))^{-1} = \frac{1}{2v} = \frac{1}{|2v|}.$$

(c) The general change of variables formula is

$$\iint_{\mathcal{D}} f(x, y)\, dA = \iint_{\mathcal{R}} f\left(x(u, v), y(u, v)\right) |\text{Jac}(F^{-1})(u, v)|\, du\, dv$$

Here, $f\left(\frac{y}{x}\right) = f(v)$, $\mathcal{R} = \left[\frac{1}{2}, 2\right] \times \left[\frac{1}{2}, 2\right]$ in the (u, v)-plane and $|\text{Jac}(F^{-1})(u, v)| = |\frac{1}{2v}| = \frac{1}{2v}$ ($v > 0$ in $\mathcal{R}$). Therefore, we have

$$\iint_{\mathcal{D}} f\left(\frac{y}{x}\right) dA = \int_{1/2}^{2} \int_{1/2}^{2} f(v) \cdot \frac{1}{2v}\, du\, dv = \left(\int_{1/2}^{2} 1\, du\right)\left(\int_{1/2}^{2} \frac{f(v)}{2v}\, dv\right) = \frac{3}{4} \int_{1/2}^{2} \frac{f(v)}{v}\, dv$$

(d) We use part (c) with $f\left(\frac{y}{x}\right) = \frac{y}{x} \cdot e^{y/x}$. We have $f(v) = v \cdot e^{v}$, hence

$$\iint_{\mathcal{D}} \frac{y e^{y/x}}{x}\, dx\, dy = \frac{3}{4} \int_{1/2}^{2} \frac{v e^{v}}{v}\, dv = \frac{3}{4} \int_{1/2}^{2} e^{v}\, dv = \frac{3}{4} e^{v} \Big|_{1/2}^{2} = \frac{3}{4}(e^2 - e^{1/2}) = \frac{3}{4}\left(e^2 - \sqrt{e}\right)$$

16 LINE AND SURFACE INTEGRALS

16.1 Vector Fields   (LT Section 17.1)

Preliminary Questions

1. Which of the following is a unit vector field in the plane?

(a) $\mathbf{F} = \langle y, x \rangle$

(b) $\mathbf{F} = \left\langle \dfrac{y}{\sqrt{x^2 + y^2}}, \dfrac{x}{\sqrt{x^2 + y^2}} \right\rangle$

(c) $\mathbf{F} = \left\langle \dfrac{y}{x^2 + y^2}, \dfrac{x}{x^2 + y^2} \right\rangle$

SOLUTION

(a) The length of the vector $\langle y, x \rangle$ is

$$\| \langle y, x \rangle \| = \sqrt{y^2 + x^2}$$

This value is not 1 for all points, hence it is not a unit vector field.

(b) We have

$$\left\| \left\langle \frac{y}{\sqrt{x^2 + y^2}}, \frac{x}{\sqrt{x^2 + y^2}} \right\rangle \right\| = \sqrt{ \left(\frac{y}{\sqrt{x^2 + y^2}} \right)^2 + \left(\frac{x}{\sqrt{x^2 + y^2}} \right)^2 }$$

$$= \sqrt{ \frac{y^2}{x^2 + y^2} + \frac{x^2}{x^2 + y^2} } = \sqrt{ \frac{y^2 + x^2}{x^2 + y^2} } = 1$$

Hence the field is a unit vector field, for $(x, y) \neq (0, 0)$.

(c) We compute the length of the vector:

$$\left\| \left\langle \frac{y}{x^2 + y^2}, \frac{x}{x^2 + y^2} \right\rangle \right\| = \sqrt{ \left(\frac{y}{x^2 + y^2} \right)^2 + \left(\frac{x}{x^2 + y^2} \right)^2 } = \sqrt{ \frac{y^2 + x^2}{(x^2 + y^2)^2} } = \sqrt{ \frac{1}{x^2 + y^2} }$$

Since the length is not identically 1, the field is not a unit vector field.

2. Sketch an example of a nonconstant vector field in the plane in which each vector is parallel to $\langle 1, 1 \rangle$.

SOLUTION   The non-constant vector $\langle x, x \rangle$ is parallel to the vector $\langle 1, 1 \rangle$.

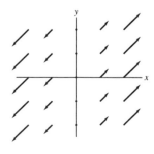

3. Show that the vector field $\mathbf{F} = \langle -z, 0, x \rangle$ is orthogonal to the position vector $\overrightarrow{OP}$ at each point P. Give an example of another vector field with this property.

SOLUTION   The position vector at $P = (x, y, z)$ is $\langle x, y, z \rangle$. We must show that the following dot product is zero:

$$\langle x, y, z \rangle \cdot \langle -z, 0, x \rangle = x \cdot (-z) + y \cdot 0 + z \cdot x = 0$$

Therefore, the vector field $\mathbf{F} = \langle -z, 0, x \rangle$ is orthogonal to the position vector. Another vector field with this property is $\mathbf{F} = \langle 0, -z, y \rangle$, since

$$\langle 0, -z, y \rangle \cdot \langle x, y, z \rangle = 0 \cdot x + (-z) \cdot y + y \cdot z = 0$$

4. Give an example of a potential function for $\langle yz, xz, xy \rangle$ other than $f(x, y, z) = xyz$.

SOLUTION Since any two potential functions of a gradient vector field differ by a constant, a potential function for the given field other than $V(x, y, z) = xyz$ is, for instance, $V_1(x, y, z) = xyz + 1$.

Exercises

1. Compute and sketch the vector assigned to the points $P = (1, 2)$ and $Q = (-1, -1)$ by the vector field $\mathbf{F} = \langle x^2, x \rangle$.

SOLUTION The vector assigned to $P = (1, 2)$ is obtained by substituting $x = 1$ in $\mathbf{F}$, that is,

$$\mathbf{F}(1, 2) = \langle 1^2, 1 \rangle = \langle 1, 1 \rangle$$

Similarly,

$$\mathbf{F}(-1, -1) = \langle (-1)^2, -1 \rangle = \langle 1, -1 \rangle$$

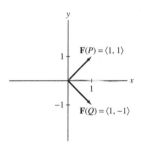

3. Compute and sketch the vector assigned to the points $P = (0, 1, 1)$ and $Q = (2, 1, 0)$ by the vector field $\mathbf{F} = \langle xy, z^2, x \rangle$.

SOLUTION To find the vector assigned to the point $P = (0, 1, 1)$, we substitute $x = 0$, $y = 1$, $z = 1$ in $\mathbf{F} = \langle xy, z^2, x \rangle$. We get

$$\mathbf{F}(P) = \langle 0 \cdot 1, 1^2, 0 \rangle = \langle 0, 1, 0 \rangle$$

Similarly, $\mathbf{F}(Q)$ is obtained by substituting $x = 2$, $y = 1$, $z = 0$ in $\mathbf{F}$. That is,

$$\mathbf{F}(Q) = \langle 2 \cdot 1, 0^2, 2 \rangle = \langle 2, 0, 2 \rangle$$

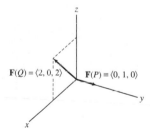

In Exercises 5–12, sketch the following planar vector fields by drawing the vectors attached to points with integer coordinates in the rectangle $-3 \le x \le 3$, $-3 \le y \le 3$. Instead of drawing the vectors with their true lengths, scale them if necessary to avoid overlap.

5. $\mathbf{F} = \langle 1, 0 \rangle$

SOLUTION The constant vector field $\langle 1, 0 \rangle$ is shown in the figure:

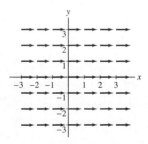

7. F = $x\mathbf{i}$

SOLUTION The vector field $\mathbf{F}(x, y) = x\mathbf{i} = (x, 0)$ is sketched in the following figure:

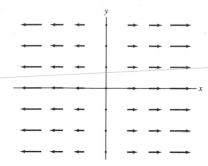

9. F = $\langle 0, x \rangle$

SOLUTION We sketch the vector field $\mathbf{F}(x, y) = \langle 0, x \rangle$:

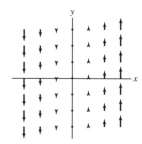

11. F = $\left\langle \dfrac{x}{x^2 + y^2}, \dfrac{y}{x^2 + y^2} \right\rangle$

SOLUTION

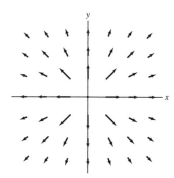

In Exercises 13–16, match each of the following planar vector fields with the corresponding plot in Figure 10.

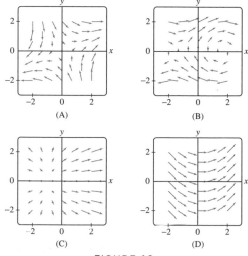

FIGURE 10

13. $\mathbf{F} = \langle 2, x \rangle$

SOLUTION The x coordinate of the vector field $\langle 2, x \rangle$ is always 2. This matches only with Plot (D).

15. $\mathbf{F} = \langle y, \cos x \rangle$

SOLUTION We compute the images of the point $(0, 2)$, for instance, and identify the corresponding graph accordingly:

$$\mathbf{F}(x, y) = \langle y, \cos x \rangle \quad \Rightarrow \quad \mathbf{F}(0, 2) = \langle 2, 1 \rangle \quad \Rightarrow \quad \text{Plot(B)}$$

In Exercises 17–20, match each three-dimensional vector field with the corresponding plot in Figure 11.

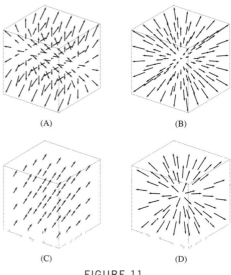

(A)　　　　　　　　　　(B)

(C)　　　　　　　　　　(D)

FIGURE 11

17. $\mathbf{F} = \langle 1, 1, 1 \rangle$

SOLUTION The constant vector field $\langle 1, 1, 1 \rangle$ is shown in plot (C).

19. $\mathbf{F} = \langle x, y, z \rangle$

SOLUTION $\langle x, y, z \rangle$ is shown in plot (B). Note that the vectors are pointing away from the origin and are of increasing magnitude.

21. Find (by inspection) a potential function for $\mathbf{F} = \langle x, 0 \rangle$ and prove that $\mathbf{G} = \langle y, 0 \rangle$ is not conservative.

SOLUTION For $f(x, y) = \frac{1}{2}x^2$ we have $\nabla f = \langle x, 0 \rangle$.

$$\frac{\partial G_1}{\partial y} = 1 \neq \frac{\partial G_2}{\partial x} = 0$$

Thus $\mathbf{G}$ is not conservative.

In Exercises 23–26, find a potential function for the vector field $\mathbf{F}$ by inspection.

23. $\mathbf{F} = \langle x, y \rangle$

SOLUTION We must find a function $\varphi(x, y)$ such that $\frac{\partial \varphi}{\partial x} = x$ and $\frac{\partial \varphi}{\partial y} = y$. We choose the following function:

$$\varphi(x, y) = \frac{1}{2}x^2 + \frac{1}{2}y^2.$$

25. $\mathbf{F} = \langle yz^2, xz^2, 2xyz \rangle$

SOLUTION We choose a function $\varphi(x, y, z)$ such that

$$\frac{\partial \varphi}{\partial x} = yz^2, \quad \frac{\partial \varphi}{\partial y} = xz^2, \quad \frac{\partial \varphi}{\partial z} = xyz$$

The function $\varphi(x, y, z) = xyz^2$ is a potential function for the given field.

27. Find potential functions for $\mathbf{F} = \dfrac{\mathbf{e}_r}{r^3}$ and $\mathbf{G} = \dfrac{\mathbf{e}_r}{r^4}$ in $\mathbf{R}^3$. *Hint:* See Example 6.

SOLUTION We use the gradient of r, $\nabla r = \mathbf{e}_r$, and the Chain Rule for Gradients to write

$$\nabla\left(-\frac{1}{2}r^{-2}\right) = r^{-3}\nabla r = r^{-3}\mathbf{e}_r = \frac{\mathbf{e}_r}{r^3} = \mathbf{F}$$

$$\nabla\left(-\frac{1}{3}r^{-3}\right) = r^{-4}\nabla r = r^{-4}\mathbf{e}_r = \frac{\mathbf{e}_r}{r^4} = \mathbf{G}$$

Therefore $\varphi_1(r) = -\dfrac{1}{2r^2}$ and $\varphi_2(r) = -\dfrac{1}{3r^3}$ are potential functions for $\mathbf{F}$ and $\mathbf{G}$, respectively.

29. Let $\varphi = \ln r$, where $r = \sqrt{x^2 + y^2}$. Express $\nabla\varphi$ in terms of the unit radial vector $\mathbf{e}_r$ in $\mathbf{R}^2$.

SOLUTION Since $r = (x^2 + y^2 + z^2)^{1/2}$, we have $\varphi = \ln(x^2 + y^2 + z^2)^{1/2} = \frac{1}{2}\ln(x^2 + y^2 + z^2)$. We compute the partial derivatives:

$$\frac{\partial\varphi}{\partial x} = \frac{1}{2}\frac{2x}{x^2 + y^2 + z^2} = \frac{x}{r^2}$$

$$\frac{\partial\varphi}{\partial y} = \frac{1}{2}\frac{2y}{x^2 + y^2 + z^2} = \frac{y}{r^2}$$

$$\frac{\partial\varphi}{\partial z} = \frac{1}{2}\frac{2z}{x^2 + y^2 + z^2} = \frac{z}{r^2}$$

Therefore, the gradient of φ is the following vector:

$$\nabla\varphi = \left\langle\frac{\partial\varphi}{\partial x}, \frac{\partial\varphi}{\partial y}, \frac{\partial\varphi}{\partial z}\right\rangle = \left\langle\frac{x}{r^2}, \frac{y}{r^2}, \frac{z}{r^2}\right\rangle = \frac{1}{r}\left\langle\frac{x}{r}, \frac{y}{r}, \frac{z}{r}\right\rangle = \frac{\mathbf{e}_r}{r}$$

31. Which of (A) or (B) in Figure 12 is the contour plot of a potential function for the vector field $\mathbf{F}$? Recall that the gradient vectors are perpendicular to the level curves.

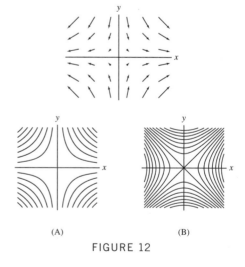

(A)   (B)

FIGURE 12

SOLUTION By the equality $\nabla\varphi = \mathbf{F}$ and since the gradient vectors are perpendicular to the level curves, it follows that the vectors $\mathbf{F}$ are perpendicular to the corresponding level curves of φ. This property is satisfied in (B) and not satisfied in (A). Therefore (B) is the contour plot of φ.

33. Match each of these descriptions with a vector field in Figure 14:

(a) The gravitational field created by two planets of equal mass located at P and Q.

(b) The electrostatic field created by two equal and opposite charges located at P and Q (representing the force on a negative test charge; opposite charges attract and like charges repel).

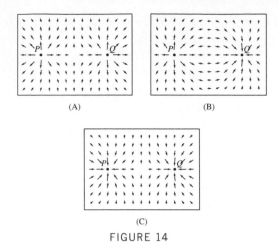

(A)   (B)

(C)

FIGURE 14

SOLUTION

(a) There will be an equilibrium point half way between the two planets. The vector field should pull objects near one planet toward that planet. (C)

(b) A test charge at the midpoint between the two charges will be drawn by one, and repelled by the other. Therefore no equilibrium. (B)

Further Insights and Challenges

35. Show that any vector field of the form

$$\mathbf{F} = \langle f(x), g(y), h(z) \rangle$$

has a potential function. Assume that f, g, and h are continuous.

SOLUTION   Let $F(x)$, $G(y)$, and $H(z)$ be antiderivatives of $f(x)$, $g(y)$, and $h(z)$, respectively. That is, $F'(x) = f(x)$, $G'(y) = g(y)$, and $H'(y) = h(z)$. We define the function

$$\varphi(x, y, z) = F(x) + G(y) + H(z)$$

Then,

$$\frac{\partial \varphi}{\partial x} = F'(x) = f(x), \quad \frac{\partial \varphi}{\partial x} = G'(y) = g(y), \quad \frac{\partial \varphi}{\partial z} = H'(z) = h(z)$$

Therefore, $\nabla \varphi = \mathbf{F}$, or φ is a potential function for $\mathbf{F}$.

16.2 Line Integrals   (LT Section 17.2)

Preliminary Questions

1. What is the line integral of the constant function $f(x, y, z) = 10$ over a curve C of length 5?

SOLUTION   Since the length of C is the line integral $\int_C 1 \, ds = 5$, we have

$$\int_C 10 \, ds = 10 \int_C 1 \, ds = 10 \cdot 5 = 50$$

2. Which of the following have a zero line integral over the vertical segment from $(0, 0)$ to $(0, 1)$?

(a) $f(x, y) = x$ **(b)** $f(x, y) = y$
(c) $\mathbf{F} = \langle x, 0 \rangle$ **(d)** $\mathbf{F} = \langle y, 0 \rangle$
(e) $\mathbf{F} = \langle 0, x \rangle$ **(f)** $\mathbf{F} = \langle 0, y \rangle$

SOLUTION   The vertical segment from $(0, 0)$ to $(0, 1)$ has the parametrization

$$\mathbf{c}(t) = (0, t), \quad 0 \le t \le 1$$

Therefore, $\mathbf{c}'(t) = \langle 0, 1 \rangle$ and $\|\mathbf{c}'(t)\| = 1$. The line integrals are thus computed by

$$\int_C f(x, y) \, ds = \int_0^1 f(\mathbf{c}(t)) \, \|\mathbf{c}'(t)\| \, dt \tag{1}$$

$$\int_C \mathbf{F} \cdot d\mathbf{s} = \int_0^1 \mathbf{F}(\mathbf{c}(t)) \cdot \mathbf{c}'(t) \, dt \tag{2}$$

(a) We have $f(\mathbf{c}(t)) = x = 0$. Therefore by (1) the line integral is zero.

(b) By (1), the line integral is

$$\int_C f(x, y)\, ds = \int_0^1 t \cdot 1\, dt = \frac{1}{2}t^2 \Big|_0^1 = \frac{1}{2} \neq 0$$

(c) This vector line integral is computed using (2). Since $\mathbf{F}(\mathbf{c}(t)) = \langle x, 0 \rangle = \langle 0, 0 \rangle$, the vector line integral is zero.

(d) By (2) we have

$$\int_C \mathbf{F} \cdot d\mathbf{s} = \int_0^1 \langle t, 0 \rangle \cdot \langle 0, 1 \rangle\, dt = \int_0^1 0\, dt = 0$$

(e) The vector integral is computed using (2). Since $\mathbf{F}(\mathbf{c}(t)) = \langle 0, x \rangle = \langle 0, 0 \rangle$, the line integral is zero.

(f) For this vector field we have

$$\int_C \mathbf{F} \cdot d\mathbf{s} = \int_0^1 \mathbf{F}(\mathbf{c}(t)) \cdot \mathbf{c}'(t)\, dt = \int_0^1 \langle 0, t \rangle \cdot \langle 0, 1 \rangle\, dt = \int_0^1 t\, dt = \frac{t^2}{2}\Big|_0^1 = \frac{1}{2} \neq 0$$

So, we conclude that (a), (c), (d), and (e) have an integral of zero.

3. State whether each statement is true or false. If the statement is false, give the correct statement.

(a) The scalar line integral does not depend on how you parametrize the curve.

(b) If you reverse the orientation of the curve, neither the vector line integral nor the scalar line integral changes sign.

SOLUTION

(a) True: It can be shown that any two parametrizations of the curve yield the same value for the scalar line integral, hence the statement is true.

(b) False: For the definition of the scalar line integral, there is no need to specify a direction along the path, hence reversing the orientation of the curve does not change the sign of the integral. However, reversing the orientation of the curve changes the sign of the vector line integral.

4. Suppose that C has length 5. What is the value of $\int_C \mathbf{F} \cdot d\mathbf{s}$ if:

(a) $\mathbf{F}(P)$ is normal to C at all points P on C?

(b) $\mathbf{F}(P)$ is a unit vector pointing in the negative direction along the curve?

SOLUTION

(a) The vector line integral is the integral of the tangential component of the vector field along the curve. Since $\mathbf{F}(P)$ is normal to C at all points P on C, the tangential component is zero, hence the line integral $\int_C \mathbf{F} \cdot d\mathbf{s}$ is zero.

(b) In this case we have

$$\mathbf{F}(P) \cdot \mathbf{T}(P) = \mathbf{T}(P) \cdot \mathbf{T}(P) = \|\mathbf{T}(P)\|^2 = 1$$

Therefore,

$$\int_C \mathbf{F} \cdot d\mathbf{s} = \int_C (\mathbf{F} \cdot \mathbf{T})\, ds = \int_C 1\, ds = \text{Length of } C = 5.$$

Exercises

1. Let $f(x, y, z) = x + yz$, and let C be the line segment from $P = (0, 0, 0)$ to $(6, 2, 2)$.

(a) Calculate $f(\mathbf{c}(t))$ and $ds = \|\mathbf{c}'(t)\|\, dt$ for the parametrization $\mathbf{c}(t) = (6t, 2t, 2t)$ for $0 \leq t \leq 1$.

(b) Evaluate $\int_C f(x, y, z)\, ds$.

SOLUTION

(a) We substitute $x = 6t$, $y = 2t$, $z = 2t$ in the function $f(x, y, z) = x + yz$ to find $f(\mathbf{c}(t))$:

$$f(\mathbf{c}(t)) = 6t + (2t)(2t) = 6t + 4t^2$$

We differentiate the vector $c(t)$ and compute the length of the derivative vector:

$$\mathbf{c}'(t) = \frac{d}{dt}\langle 6t, 2t, 2t \rangle = \langle 6, 2, 2 \rangle \quad \Rightarrow \quad \mathbf{c}'(t) = \sqrt{6^2 + 2^2 + 2^2} = \sqrt{44} = 2\sqrt{11}$$

Hence,

$$ds = \|\mathbf{c}'(t)\|\, dt = 2\sqrt{11}\, dt$$

(b) Computing the scalar line integral, we obtain

$$\int_C f(x, y, z)\, ds = \int_0^1 f\left(\mathbf{c}(t)\right) \|\mathbf{c}'(t)\|\, dt = \int_0^1 (6t + 4t^2) \cdot 2\sqrt{11}\, dt$$

$$= 2\sqrt{11}\left(3t^2 + \frac{4}{3}t^3\right)\Big|_0^1 = 2\sqrt{11}\left(3 + \frac{4}{3}\right) = \frac{26\sqrt{11}}{3}$$

3. Let $\mathbf{F} = \langle y^2, x^2 \rangle$, and let C be the curve $y = x^{-1}$ for $1 \le x \le 2$, oriented from left to right.

(a) Calculate $\mathbf{F}(\mathbf{c}(t))$ and $d\mathbf{s} = \mathbf{c}'(t)\, dt$ for the parametrization of C given by $\mathbf{c}(t) = (t, t^{-1})$.

(b) Calculate the dot product $\mathbf{F}(\mathbf{c}(t)) \cdot \mathbf{c}'(t)\, dt$ and evaluate $\displaystyle\int_C \mathbf{F} \cdot d\mathbf{s}$.

SOLUTION

(a) We calculate $\mathbf{F}\left(\mathbf{c}(t)\right)$ by substituting $x = t$, $y = t^{-1}$ in $\mathbf{F} = \langle y^2, x^2 \rangle$. We get

$$\mathbf{F}(\mathbf{c}(t)) = \left\langle \left(t^{-1}\right)^2, t^2 \right\rangle = \left\langle t^{-2}, t^2 \right\rangle$$

We compute $\mathbf{c}'(t)$:

$$\mathbf{c}'(t) = \frac{d}{dt}\langle t, t^{-1} \rangle = \langle 1, -t^{-2} \rangle \quad \Rightarrow \quad d\mathbf{s} = \langle 1, -t^{-2} \rangle\, dt$$

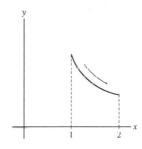

(b) We compute the dot product:

$$\mathbf{F}(\mathbf{c}(t)) \cdot \mathbf{c}'(t) = \langle t^{-2}, t^2 \rangle \cdot \langle 1, -t^{-2} \rangle = t^{-2} \cdot 1 + t^2 \,(-t^{-2}) = t^{-2} - 1$$

Computing the vector line integral, we obtain

$$\int_C \mathbf{F} \cdot d\mathbf{s} = \int_1^2 \mathbf{F}\left(\mathbf{c}(t)\right) \cdot \mathbf{c}'(t)\, dt = \int_1^2 (t^{-2} - 1)\, dt = -t^{-1} - t\Big|_1^2 = \left(-\frac{1}{2} - 2\right) - (-1 - 1) = -\frac{1}{2}$$

In Exercises 5–8, compute the integral of the scalar function or vector field over $\mathbf{c}(t) = (\cos t, \sin t, t)$ *for* $0 \le t \le \pi$.

5. $f(x, y, z) = x^2 + y^2 + z^2$

SOLUTION

Step 1. Compute $\|\mathbf{c}'(t)\|$. We differentiate $\mathbf{c}(t)$:

$$\mathbf{c}'(t) = \frac{d}{dt}\,\langle \cos t, \sin t, t \rangle = \langle -\sin t, \cos t, 1 \rangle$$

Hence,

$$\|\mathbf{c}'(t)\| = \sqrt{(-\sin t)^2 + \cos^2 t + 1^2} = \sqrt{\sin^2 t + \cos^2 t + 1} = \sqrt{2}$$

$$ds = \|\mathbf{c}'(t)\|\, dt = \sqrt{2}\, dt$$

Step 2. Write out $f\left(\mathbf{c}(t)\right)$. We substitute $x = \cos t$, $y = \sin t$, $z = t$ in $f(x, y, z) = x^2 + y^2 + z^2$ to obtain

$$f\left(\mathbf{c}(t)\right) = \cos^2 t + \sin^2 t + t^2 = 1 + t^2$$

Step 3. Compute the line integral. Using the Theorem on Scalar Line Integrals we obtain

$$\int_C (x^2 + y^2 + z^2)\, ds = \int_0^\pi f\left(\mathbf{c}(t)\right) \|\mathbf{c}'(t)\|\, dt = \int_0^\pi (1 + t^2)\sqrt{2}\, dt = \sqrt{2}\left(t + \frac{t^3}{3}\right)\Big|_0^\pi = \sqrt{2}\left(\pi + \frac{\pi^3}{3}\right)$$

7. $\mathbf{F} = \langle x, y, z \rangle$

SOLUTION

Step 1. Calculate the integrand. We write out the vectors:

$$\mathbf{c}(t) = (\cos t, \sin t, t)$$

$$\mathbf{F}(\mathbf{c}(t)) = \langle x, y, z \rangle = \langle \cos t, \sin t, t \rangle$$

$$\mathbf{c}'(t) = \langle -\sin t, \cos t, 1 \rangle$$

The integrand is the dot product:

$$\mathbf{F}(\mathbf{c}(t)) \cdot \mathbf{c}'(t) = \langle \cos t, \sin t, t \rangle \cdot \langle -\sin t, \cos t, 1 \rangle = -\cos t \sin t + \sin t \cos t + t = t$$

Step 2. Evaluate the integral. We use the Theorem on Vector Line Integrals to evaluate the integral:

$$\int_C \mathbf{F} \, d\mathbf{s} = \int_0^\pi \mathbf{F}(\mathbf{c}(t)) \cdot \mathbf{c}'(t) \, dt = \int_0^\pi t \, dt = \frac{1}{2} t^2 \Big|_0^\pi = \frac{\pi^2}{2}$$

In Exercises 9–16, compute $\int_C f \, ds$ for the curve specified.

9. $f(x, y) = \sqrt{1 + 9xy}, \quad y = x^3$ for $0 \le x \le 1$

SOLUTION The curve is parametrized by $\mathbf{c}(t) = \left(t, t^3\right)$ for $0 \le t \le 1$

Step 1. Compute $\|\mathbf{c}'(t)\|$. We have

$$\mathbf{c}'(t) = \frac{d}{dt}\left(t, t^3\right) = \left(1, 3t^2\right) \quad \Rightarrow \quad \|\mathbf{c}'(t)\| = \sqrt{1 + 9t^4}$$

Step 2. Write out $f(\mathbf{c}(t))$. We substitute $x = t$, $y = t^3$ in $f(x, y) = \sqrt{1 + 9xy}$ to obtain

$$f(\mathbf{c}(t)) = \sqrt{1 + 9t \cdot t^3} = \sqrt{1 + 9t^4}$$

Step 3. Compute the line integral. We use the Theorem on Scalar Line Integrals to write

$$\int_C f(x, y) \, ds = \int_0^1 f(\mathbf{c}(t)) \|\mathbf{c}'(t)\| \, dt = \int_0^1 \sqrt{1 + 9t^4} \sqrt{1 + 9t^4} \, dt = \int_0^1 \left(1 + 9t^4\right) dt$$

$$= t + \frac{9t^5}{5} \Big|_0^1 = \frac{14}{5} = 2.8$$

11. $f(x, y, z) = z^2, \quad \mathbf{c}(t) = (2t, 3t, 4t)$ for $0 \le t \le 2$

SOLUTION

Step 1. Compute $\|\mathbf{c}'(t)\|$ We have

$$\mathbf{c}'(t) = \frac{d}{dt}\langle 2t, 3t, 4t \rangle = \langle 2, 3, 4 \rangle \quad \Rightarrow \quad \|\mathbf{c}'(t)\| = \sqrt{2^2 + 3^2 + 4^2} = \sqrt{29}$$

Step 2. Write out $f(\mathbf{c}(t))$ We substitute $z = 4t$ in $f(x, y, z) = z^2$ to obtain:

$$f(\mathbf{c}(t)) = 16t^2$$

Step 3. Compute the line integral. By the Theorem on Scalar Line Integrals we have

$$\int_C f(x, y, z) \, ds = \int_0^2 f(\mathbf{c}(t)) \|\mathbf{c}'(t)\| \, dt = \int_0^2 16t^2 \cdot \sqrt{29} \, dt = \sqrt{29} \cdot \frac{16}{3} t^3 \Big|_0^2 = \frac{128\sqrt{29}}{3} \approx 229.8$$

13. $f(x, y, z) = xe^{z^2},$ piecewise linear path from $(0, 0, 1)$ to $(0, 2, 0)$ to $(1, 1, 1)$

SOLUTION Let C_1 be the segment joining the points $(0, 0, 1)$ and $(0, 2, 0)$ and C_2 be the segment joining the points $(0, 2, 0)$ and $(1, 1, 1)$. We parametrize C_1 and C_2 by the following parametrization:

$$C_1 : \mathbf{c}_1(t) = (0, 2t, 1 - t), \ 0 \le t \le 1$$

$$C_2 : \mathbf{c}_2(t) = (t, 2 - t, t), \ 0 \le t \le 1$$

For $\mathcal{C} = \mathcal{C}_1 + \mathcal{C}_1$ we have

$$\int_{\mathcal{C}} f(x, y, z) \, ds = \int_{\mathcal{C}_1} f(x, y, z) \, ds + \int_{\mathcal{C}_2} f(x, y, z) \, ds \tag{1}$$

We compute the integrals on the right hand side.

- The integral over $\mathcal{C}_1$: We have

$$\mathbf{c}_1'(t) = \frac{d}{dt} \langle 0, 2t, 1 - t \rangle = \langle 0, 2, -1 \rangle \quad \Rightarrow \quad \|\mathbf{c}_1'(t)\| = \sqrt{0 + 4 + 1} = \sqrt{5}$$

$$f(\mathbf{c}(t)) = xe^{z^2} = 0 \cdot e^{(1-t)^2} = 0$$

Hence,

$$\int_{\mathcal{C}_1} f(x, y, z) \, ds = \int_0^1 f(\mathbf{c}_1(t)) \|\mathbf{c}_1'(t)\| \, dt = \int_0^1 0 \, dt = 0 \tag{2}$$

- The integral over $\mathcal{C}_2$: We have

$$\mathbf{c}_2'(t) = \frac{d}{dt} \langle t, 2 - t, t \rangle = \langle 1, -1, 1 \rangle \quad \Rightarrow \quad \|\mathbf{c}_2'(t)\| = \sqrt{1 + 1 + 1} = \sqrt{3}$$

$$f(\mathbf{c}_2(t)) = xe^{z^2} = te^{t^2}$$

Hence,

$$\int_{\mathcal{C}_2} f(x, y, z) \, ds = \int_0^1 te^{t^2} \sqrt{3} \, dt \tag{3}$$

Using the substitution $u = t^2$ we find that

$$\int_{\mathcal{C}_2} f(x, y, z) \, ds = \int_0^1 \frac{\sqrt{3}}{2} e^u \, du = \frac{\sqrt{3}}{2}(e - 1) \approx 1.488$$

Hence,

$$\int_{\mathcal{C}} f(x, y, z) \, ds \approx 1.488$$

15. $f(x, y, z) = 2x^2 + 8z$, $\quad \mathbf{c}(t) = (e^t, t^2, t)$, $\quad 0 \le t \le 1$

SOLUTION

Step 1. Compute $\|\mathbf{c}'(t)\|$.

$$\mathbf{c}'(t) = \frac{d}{dt} \left\langle e^t, t^2, t \right\rangle = \left\langle e^t, 2t, 1 \right\rangle \quad \Rightarrow \quad \|\mathbf{c}'(t)\| = \sqrt{e^{2t} + 4t^2 + 1}$$

Step 2. Write out $f(\mathbf{c}(t))$. We substitute $x = e^t$, $y = t^2$, $z = t$ in $f(x, y, z) = 2x^2 + 8z$ to obtain:

$$f(\mathbf{c}(t)) = 2e^{2t} + 8t$$

Step 3. Compute the line integral. We have

$$\int_{\mathcal{C}} f(x, y, z) \, ds = \int_0^1 f(\mathbf{c}(t)) \|\mathbf{c}'(t)\| \, dt = \int_0^1 (2e^{2t} + 8t)\sqrt{e^{2t} + 4t^2 + 1} \, dt$$

We compute the integral using the substitution $u = e^{2t} + 4t^2 + 1$, $du = 2e^{2t} + 8t \, dt$. We get:

$$\int_{\mathcal{C}} f(x, y, z) \, ds = \int_2^{e^2 + 5} u^{1/2} \, du = \frac{2}{3} u^{3/2} \Big|_2^{e^2 + 5} = \frac{2}{3} \left((e^2 + 5)^{3/2} - 2^{3/2} \right)$$

17. Calculate $\int_{\mathcal{C}} 1 \, ds$, where the curve $\mathcal{C}$ is parametrized by $\mathbf{c}(t) = (4t, -3t, 12t)$ for $2 \le t \le 5$. What does this integral represent?

SOLUTION Compute $\|\mathbf{c}'(t)\|$.

$$\mathbf{c}'(t) = \frac{d}{dt} < 4t, -3t, 12t > = < 4, -3, 12 > \quad \Rightarrow \quad \|\mathbf{c}'(t)\| = \sqrt{4^2 + (-3)^2 + (12)^2} = 13$$

Compute the line integral. We have

$$\int_{\mathcal{C}} 1 \, ds = \int_2^5 \|\mathbf{c}'(t)\| \, dt = \int_2^5 13 \, dt = 13(5 - 2) = 39$$

This represents the distance from the point $(8, -6, 24)$ to the point $(20, -15, 60)$.

In Exercises 19–26, compute $\int_C \mathbf{F} \cdot d\mathbf{s}$ for the oriented curve specified.

19. $\mathbf{F} = \langle x^2, xy \rangle$, line segment from $(0, 0)$ to $(2, 2)$

SOLUTION The oriented line segment is parametrized by

$$\mathbf{c}(t) = (t, t), \quad t \text{ varies from 0 to 2.}$$

Therefore,

$$\mathbf{F}(\mathbf{c}(t)) = \langle x^2, xy \rangle = \langle t^2, t \cdot t \rangle = \langle t^2, t^2 \rangle$$

$$\mathbf{c}'(t) = \frac{d}{dt} \langle t, t \rangle = \langle 1, 1 \rangle$$

The integrand is the dot product:

$$\mathbf{F}(\mathbf{c}(t)) \cdot \mathbf{c}'(t) = \langle t^2, t^2 \rangle \cdot \langle 1, 1 \rangle = t^2 + t^2 = 2t^2$$

We now use the Theorem on vector line integral to compute $\int_C \mathbf{F} \cdot d\mathbf{s}$:

$$\int_C \mathbf{F} \cdot d\mathbf{s} = \int_0^2 \mathbf{F}(\mathbf{c}(t)) \cdot \mathbf{c}'(t)\, dt = \int_0^2 2t^2\, dt = \frac{2t^3}{3}\bigg|_0^2 = \frac{16}{3}$$

21. $\mathbf{F} = \langle x^2, xy \rangle$, part of circle $x^2 + y^2 = 9$ with $x \le 0, y \ge 0$, oriented clockwise

SOLUTION

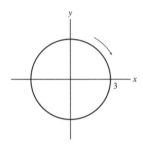

The oriented path is parametrized by

$$\mathbf{c}(t) = (-3\cos t, 3\sin t); \quad t \text{ is changing from 0 to } \frac{\pi}{2}.$$

Note: $\mathbf{c}(0) = (-3, 0)$ and $\mathbf{c}\left(\frac{\pi}{2}\right) = (0, 3)$. $\cos t$ and $\sin t$ are both positive in this range, so $x = -3\cos t \le 0$ and $y = 3\sin t \ge 0$. We compute the integrand:

$$\mathbf{F}(\mathbf{c}(t)) = \langle x^2, xy \rangle = \langle 9\cos^2 t, -9\cos t \sin t \rangle$$

$$\mathbf{c}'(t) = \langle 3\sin t, 3\cos t \rangle$$

$$\mathbf{F}(\mathbf{c}(t)) \cdot \mathbf{c}'(t) = \langle 9\cos^2 t, -9\cos t \sin t \rangle \cdot \langle 3\sin t, 3\cos t \rangle = 27\cos^2 t \sin t - 27\cos^2 t \sin t = 0$$

Hence,

$$\int_C \mathbf{F} \cdot d\mathbf{s} = \int_0^{\frac{\pi}{2}} \mathbf{F}(\mathbf{c}(t)) \cdot \mathbf{c}'(t)\, dt = \int_0^{\frac{\pi}{2}} 0\, dt = 0$$

23. $\mathbf{F} = \langle 3zy^{-1}, 4x, -y \rangle$, $\mathbf{c}(t) = (e^t, e^t, t)$ for $-1 \le t \le 1$

SOLUTION

Step 1. Calculate the integrand. We write out the vectors and compute the integrand:

$$\mathbf{c}(t) = \left(e^t, e^t, t\right)$$

$$\mathbf{F}(\mathbf{c}(t)) = \left\langle 3zy^{-1}, 4x, -y \right\rangle = \left\langle 3te^{-t}, 4e^t, -e^t \right\rangle$$

$$\mathbf{c}'(t) = \left\langle e^t, e^t, 1 \right\rangle$$

The integrand is the dot product:

$$\mathbf{F}(\mathbf{c}(t)) \cdot \mathbf{c}'(t) = \left\langle 3te^{-t}, 4e^t, -e^t \right\rangle \cdot \left\langle e^t, e^t, 1 \right\rangle = 3te^{-t} \cdot e^t + 4e^t \cdot e^t - e^t \cdot 1 = 3t + 4e^{2t} - e^t$$

Step 2. Evaluate the integral. The vector line integral is:

$$\int_C \mathbf{F} \cdot d\mathbf{s} = \int_{-1}^{1} \mathbf{F}\left(\mathbf{c}(t)\right) \cdot \mathbf{c}'(t)\, dt = \int_{-1}^{1} \left(3t + 4e^{2t} - e^t\right) dt = 0 + \int_{-1}^{1} \left(4e^{2t} - e^t\right) dt = 2e^{2t} - e^t \Big|_{-1}^{1}$$

$$= \left(2e^2 - e\right) - \left(2e^{-2} - e^{-1}\right) = 2\left(e^2 - e^{-2}\right) - \left(e - e^{-1}\right) \approx 12.157$$

25. $\mathbf{F} = \left\langle \dfrac{1}{y^3 + 1}, \dfrac{1}{z + 1}, 1 \right\rangle$, $\quad \mathbf{c}(t) = (t^3, 2, t^2)$ for $0 \le t \le 1$

SOLUTION

Step 1. Calculate the integrand. We have

$$\mathbf{c}(t) = \left(t^3, 2, t^2\right)$$

$$\mathbf{F}\left(\mathbf{c}(t)\right) = \left\langle \frac{1}{y^3 + 1}, \frac{1}{z + 1}, 1 \right\rangle = \left\langle \frac{1}{2^3 + 1}, \frac{1}{t^2 + 1}, 1 \right\rangle = \left\langle \frac{1}{9}, \frac{1}{t^2 + 1}, 1 \right\rangle$$

$$\mathbf{c}'(t) = \left\langle 3t^2, 0, 2t \right\rangle$$

Hence,

$$\mathbf{F}\left(\mathbf{c}(t)\right) \cdot \mathbf{c}'(t) = \left\langle \frac{1}{9}, \frac{1}{t^2 + 1}, 1 \right\rangle \cdot \left\langle 3t^2, 0, 2t \right\rangle = \frac{3t^2}{9} + 0 + 2t$$

Step 2. Evaluate the integral. Using the Theorem on vector line integrals we obtain:

$$\int_C \mathbf{F} \cdot d\mathbf{s} = \int_0^1 \mathbf{F}\left(\mathbf{c}(t)\right) \cdot \mathbf{c}'(t)\, dt = \int_0^1 \frac{t^2}{3}\, dt + \int_0^1 2t\, dt$$

$$= \frac{t^3}{9} \Big|_0^1 + t^2 \Big|_0^1 = \frac{10}{9}$$

In Exercises 27–32, evaluate the line integral.

27. $\displaystyle\int_C y\, dx - x\, dy,$ parabola $y = x^2$ for $0 \le x \le 2$

SOLUTION

Step 1. Calculate the integrand.

$$\mathbf{c}(t) = (t, t^2)$$

$$\mathbf{F}\left(\mathbf{c}(t)\right) = \langle y, -x \rangle = \left\langle t^2, -t \right\rangle$$

$$\mathbf{c}'(t) = \langle 1, 2t \rangle$$

The integrand is the dot product

$$\mathbf{F}\left(\mathbf{c}(t)\right) \cdot \mathbf{c}'(t) = \left\langle t^2, -t \right\rangle \cdot \langle 1, 2t \rangle = t^2 - 2t^2 = -t^2$$

Step 2.

$$\int_C y\, dx - x\, dy = \int_0^2 -t^2\, dt = -\frac{t^3}{3} \Big|_0^2 = -\frac{8}{3}$$

29. $\displaystyle\int_C (x - y)\, dx + (y - z)\, dy + z\, dz,$ line segment from $(0, 0, 0)$ to $(1, 4, 4)$

SOLUTION The oriented line segment from $(0, 0, 0)$ to $(1, 4, 4)$ has the parametrization:

$$\mathbf{c}(t) = (t, 4t, 4t),\ 0 \le t \le 1$$

Step 1. Calculate the integrand. We have

$$\mathbf{F}\left(\mathbf{c}(t)\right) = \langle x - y, y - z, z \rangle = \langle t - 4t, 4t - 4t, 4t \rangle = \langle -3t, 0, 4t \rangle$$

$$\mathbf{c}'(t) = \frac{d}{dt}\langle t, 4t, 4t \rangle = \langle 1, 4, 4 \rangle$$

The integrand is the dot product:

$$\mathbf{F}\left(\mathbf{c}(t)\right) \cdot \mathbf{c}'(t) = \langle -3t, 0, 4t \rangle \cdot \langle 1, 4, 4 \rangle = -3t \cdot 1 + 0 \cdot 4 + 4t \cdot 4 = 13t$$

Step 2. Evaluate the integral. The vector line integral is:

$$\int_C \mathbf{F} \cdot d\mathbf{s} = \int_0^1 \mathbf{F}\left(\mathbf{c}(t)\right) \cdot \mathbf{c}'(t) \, dt = \int_0^1 13t \, dt = \left. \frac{13}{2} t^2 \right|_0^1 = 6.5$$

31. $\displaystyle\int_C \frac{-y \, dx + x \, dy}{x^2 + y^2}$, segment from $(1, 0)$ to $(0, 1)$.

SOLUTION

Step 1. Calculate the integrand.

$$\mathbf{c}(t) = (1 - t, t) \quad (0 \leq t \leq 1)$$

$$\mathbf{F}\left(\mathbf{c}(t)\right) = \frac{1}{x^2 + y^2} \langle -y, x \rangle = \frac{1}{(1 - t)^2 + t^2} \langle -t, 1 - t \rangle$$

$$\mathbf{c}'(t) = \langle -1, 1 \rangle$$

The integrand is the dot product

$$\mathbf{F}\left(\mathbf{c}(t)\right) \cdot \mathbf{c}'(t) = \frac{1}{(1 - t)^2 + t^2} \langle -t, 1 - t \rangle \cdot \langle -1, 1 \rangle = \frac{t + 1 - t}{(1 - t)^2 + t^2} = \frac{1}{2t^2 - 2t + 1}$$

Step 2.

$$\int_C \frac{-y \, dx + x \, dy}{x^2 + y^2} = \int_0^1 \frac{dt}{2t^2 - 2t + 1} = \frac{1}{2} \int_0^1 \frac{dt}{\left(t - \frac{1}{2}\right)^2 + \frac{1}{4}}$$

We use the trigonometric substitution $t = \frac{1}{2} + \frac{1}{2} \tan \theta \Rightarrow dt = \frac{1}{2} \sec^2 \theta \, d\theta$.

$$= \frac{1}{2} \int_{-\frac{\pi}{4}}^{\frac{\pi}{4}} \frac{\frac{1}{2} \sec^2 \theta \, d\theta}{\frac{1}{4} (\tan^2 \theta + 1)} = \int_{-\frac{\pi}{4}}^{\frac{\pi}{4}} d\theta = \frac{\pi}{2}$$

33. CAS Let $f(x, y, z) = x^{-1} yz$, and let C be the curve parametrized by $\mathbf{c}(t) = (\ln t, t, t^2)$ for $2 \leq t \leq 4$. Use a computer algebra system to calculate $\displaystyle\int_C f(x, y, z) \, ds$ to four decimal places.

SOLUTION Note that $\mathbf{c}'(t) = \langle 1/t, 1, 2t \rangle$, so $\|\mathbf{c}'(t)\| = \sqrt{1/t^2 + 1 + 4t^2}$. Our line integral is

$$\int_2^4 f(\ln t, t, t^2) \sqrt{1/t^2 + 1 + 4t^2} \, dt,$$

which we calculate to be 339.5587.

In Exercises 35 and 36, calculate the line integral of $\mathbf{F} = \left\langle e^z, e^{x-y}, e^y \right\rangle$ over the given path.

35. The blue path from P to Q in Figure 14

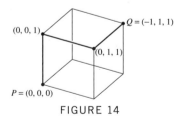

FIGURE 14

SOLUTION

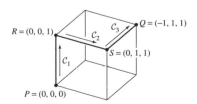

Let C_1, C_2, C_3 denote the oriented segments from P to R, from R to S and S to Q respectively. These paths have the following parametrizations (see figure):

$$C_1: \quad \mathbf{c}_1(t) = (0,0,t) \quad 0 \le t \le 1 \qquad\qquad \mathbf{c}_1'(t) = \langle 0,0,1\rangle$$

$$C_2: \quad \mathbf{c}_2(t) = (0,t,1) \quad 0 \le t \le 1 \quad \Rightarrow \quad \mathbf{c}_2'(t) = \langle 0,1,0\rangle$$

$$C_3: \quad \mathbf{c}_3(t) = (-t,1,1) \quad 0 \le t \le 1 \qquad\qquad \mathbf{c}_3'(t) = \langle -1,0,0\rangle$$

Since $C = C_1 + C_2 + C_3$ we have

$$\int_C \mathbf{F} \cdot d\mathbf{s} = \int_{C_1} \mathbf{F} \cdot d\mathbf{s} + \int_{C_2} \mathbf{F} \cdot d\mathbf{s} + \int_{C_3} \mathbf{F} \cdot d\mathbf{s} \tag{1}$$

We compute each integral on the right hand side separately.

$$\int_{C_1} \mathbf{F} \cdot d\mathbf{s} = \int_0^1 \mathbf{F}(\mathbf{c}_1(t)) \cdot \mathbf{c}_1'(t)\, dt = \int_0^1 \left\langle e^t, e^{0-0}, e^0\right\rangle \cdot \langle 0,0,1\rangle\, dt = \int_0^1 1\, dt = 1$$

$$\int_{C_2} \mathbf{F} \cdot d\mathbf{s} = \int_0^1 \mathbf{F}(\mathbf{c}_2(t)) \cdot \mathbf{c}_2'(t)\, dt = \int_0^1 \left\langle e^1, e^{0-t}, e^t\right\rangle \cdot \langle 0,1,0\rangle\, dt = \int_0^1 e^{-t}\, dt = -e^{-t}\Big|_0^1 = 1 - e^{-1}$$

$$\int_{C_3} \mathbf{F} \cdot d\mathbf{s} = \int_0^1 \mathbf{F}(\mathbf{c}_3(t)) \cdot \mathbf{c}_3'(t)\, dt = \int_0^1 \left\langle e^1, e^{t-1}, e^1\right\rangle \cdot \langle -1,0,0\rangle\, dt = \int_0^1 -e\, dt = -e$$

Substituting these integrals in (1) gives

$$\int_C \mathbf{F} \cdot d\mathbf{s} = 1 + (1 - e^{-1}) - e = 2 - e^{-1} - e$$

In Exercises 37 and 38, C is the path from P to Q in Figure 16 that traces C_1, C_2, and C_3 in the orientation indicated, and $\mathbf{F}$ is a vector field such that

$$\int_C \mathbf{F} \cdot d\mathbf{s} = 5, \qquad \int_{C_1} \mathbf{F} \cdot d\mathbf{s} = 8, \qquad \int_{C_3} \mathbf{F} \cdot d\mathbf{s} = 8$$

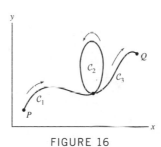

FIGURE 16

37. Determine:

(a) $\displaystyle \int_{-C_3} \mathbf{F} \cdot d\mathbf{s}$
(b) $\displaystyle \int_{C_2} \mathbf{F} \cdot d\mathbf{s}$
(c) $\displaystyle \int_{-C_1-C_3} \mathbf{F} \cdot d\mathbf{s}$

SOLUTION

(a) If the orientation of the path is reversed, the line integral changes sign, thus:

$$\int_{-C_3} \mathbf{F} \cdot d\mathbf{s} = -\int_{C_3} \mathbf{F} \cdot d\mathbf{s} = -8$$

(b) By additivity of line integrals, we have

$$\int_{C} \mathbf{F} \cdot d\mathbf{s} = \int_{C_1} \mathbf{F} \cdot d\mathbf{s} + \int_{C_2} \mathbf{F} \cdot d\mathbf{s} + \int_{C_3} \mathbf{F} \cdot d\mathbf{s}$$

Substituting the given values we obtain

$$5 = 8 + \int_{C_2} \mathbf{F} \cdot d\mathbf{s} + 8$$

or

$$\int_{C_2} \mathbf{F} \cdot d\mathbf{s} = 5 - 16 = -11$$

(c) Using properties of line integrals gives

$$\int_{-C_1-C_3} \mathbf{F} \cdot d\mathbf{s} = \int_{-C_1} \mathbf{F} \cdot d\mathbf{s} + \int_{-C_3} \mathbf{F} \cdot d\mathbf{s} = -\int_{C_1} \mathbf{F} \cdot d\mathbf{s} - \int_{C_3} \mathbf{F} \cdot d\mathbf{s} = -8 - 8 = -16$$

39. The values of a function $f(x, y, z)$ and vector field $\mathbf{F}(x, y, z)$ are given at six sample points along the path ABC in Figure 17. Estimate the line integrals of f and $\mathbf{F}$ along ABC.

Point	$f(x, y, z)$	$\mathbf{F}(x, y, z)$
$\left(1, \frac{1}{6}, 0\right)$	3	$\langle 1, 0, 2 \rangle$
$\left(1, \frac{1}{2}, 0\right)$	3.3	$\langle 1, 1, 3 \rangle$
$\left(1, \frac{5}{6}, 0\right)$	3.6	$\langle 2, 1, 5 \rangle$
$\left(1, 1, \frac{1}{6}\right)$	4.2	$\langle 3, 2, 4 \rangle$
$\left(1, 1, \frac{1}{2}\right)$	4.5	$\langle 3, 3, 3 \rangle$
$\left(1, 1, \frac{5}{6}\right)$	4.2	$\langle 5, 3, 3 \rangle$

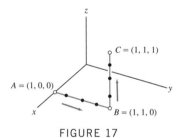

FIGURE 17

SOLUTION

We write the integrals as sum of integrals and estimate each integral by a Riemann Sum. That is,

$$\int_{ABC} f(x, y, z)\, ds = \int_{AB} f(x, y, z)\, ds + \int_{BC} f(x, y, z)\, ds \approx \sum_{i=1}^{3} f(P_i)\, \Delta s_i + \sum_{i=4}^{6} f(P_i) \Delta s_i \tag{1}$$

$$\int_{ABC} \mathbf{F} \cdot d\mathbf{s} = \int_{AB} \mathbf{F} \cdot d\mathbf{s} + \int_{BC} \mathbf{F} \cdot d\mathbf{s} = \int_{AB} (\mathbf{F} \cdot \mathbf{T})\, ds + \int_{BC} (\mathbf{F} \cdot \mathbf{T})\, ds$$

On AB, the unit tangent vector is $\mathbf{T} = \langle 0, 1, 0 \rangle$, hence $\mathbf{F} \cdot \mathbf{T} = F_2$. On BC, the unit tangent vector is $\mathbf{T} = \langle 0, 0, 1 \rangle$, hence $\mathbf{F} \cdot \mathbf{T} = F_3$. Therefore,

$$\int_{ABC} \mathbf{F}\, ds = \int_{AB} F_1\, ds + \int_{BC} F_3\, ds \approx \sum_{i=1}^{3} F_1(P_i)\, \Delta s_i + \sum_{i=4}^{6} F_3(P_i)\, \Delta s_i \tag{2}$$

We consider the partitions of AB and BC to three subarcs with equal length $\Delta s_i = \frac{1}{3}$, therefore (1) and (2) give

$$\int_{ABC} f(x, y, z)\, ds \approx \frac{1}{3}\left(f\left(P_1\right) + f\left(P_2\right) + f\left(P_3\right) + f\left(P_4\right) + f\left(P_5\right) + f\left(P_6\right)\right)$$

$$\int_{ABC} \mathbf{F}\, d\mathbf{s} \approx \frac{1}{3}\left(F_2\left(P_1\right) + F_2\left(P_2\right) + F_2\left(P_3\right) + F_3\left(P_4\right) + F_3\left(P_5\right) + F_3\left(P_6\right)\right)$$

We now substitute the values of the functions at the sample points to obtain the following approximations:

$$\int_{ABC} f(x, y, z)\, ds \approx \frac{1}{3}(3 + 3.3 + 3.6 + 4.2 + 4.5 + 4.2) = 7.6$$

$$\int_{ABC} \mathbf{F} \cdot d\mathbf{s} \approx \frac{1}{3}(0 + 1 + 1 + 4 + 3 + 3) = 4$$

41. Determine whether the line integrals of the vector fields around the circle (oriented counterclockwise) in Figure 19 are positive, negative, or zero.

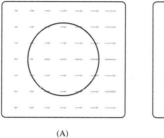
(A)

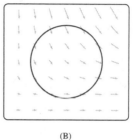
(B)

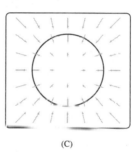
(C)

FIGURE 19

SOLUTION The vector line integral of F is the integral of the tangential component of F along the curve. The positive direction of a curve is counterclockwise.

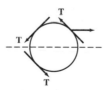

For the vector field in (A), the line integral around the circle is zero because the contribution of the negative tangential components from the upper part of the circle is the same as the contribution of the positive tangential components from the lower part. For the vector in (B) the contribution of the negative tangential component appear to dominate over the positive contribution, hence the line integral is negative. In (C), the vector field is orthogonal to the unit tangent vector at each point, hence the line integral is zero.

43. Calculate the total mass of a circular piece of wire of radius 4 cm centered at the origin whose mass density is $\rho(x, y) = x^2$ g/cm.

SOLUTION The total mass is the following integral:

$$M = \int_C x^2\, ds$$

We use the following parametrization of the wire:

$$\mathbf{c}(t) = (4\cos t, 4\sin t), \quad 0 \le t \le 2\pi$$

Hence,

$$\mathbf{c}'(t) = \langle -4\sin t, \ 4\cos t \rangle \quad \Rightarrow \quad \|\mathbf{c}'(t)\| = \sqrt{(-4\sin t)^2 + (4\cos t)^2} = 4$$

We compute the line integral using the Theorem on Scalar Line Integrals. We get

$$M = \int_0^{2\pi} \rho\left(\mathbf{c}(t)\right) \|\mathbf{c}'(t)\| \, dt = \int_0^{2\pi} (4\cos t)^2 \cdot 4 \, dt$$

$$= 64 \int_0^{2\pi} \cos^2 t \, dt = 64 \left(\frac{t}{2} + \frac{\sin 2t}{4} \right) \Big|_0^{2\pi} = 64 \cdot \frac{2\pi}{2} = 64\pi \, \text{g}$$

45. Find the total charge on the curve $y = x^{4/3}$ for $1 \le x \le 8$ (in cm) assuming a charge density of $\rho(x,y) = x/y$ (in units of 10^{-6} C/cm).

SOLUTION   We parametrize the curve by $\mathbf{c}(t) = (t, t^{\frac{4}{3}})$ $(1 \le t \le 8)$. Then

$$\mathbf{c}'(t) = \left\langle 1, \frac{4}{3} t^{\frac{1}{3}} \right\rangle \quad \Rightarrow \quad \|\mathbf{c}'(t)\| = \sqrt{1 + \frac{16}{9} t^{\frac{2}{3}}}$$

$$\rho(\mathbf{c}(t)) = \frac{x}{y} = \frac{t}{t^{\frac{4}{3}}}$$

Therefore the total charge will be

$$\int_C \frac{x}{y} \, ds = \int_1^8 \frac{t}{t^{\frac{4}{3}}} \sqrt{1 + \frac{16}{9} t^{\frac{2}{3}}} \, dt = \int_1^8 \sqrt{1 + \frac{16}{9} t^{\frac{2}{3}}} \, t^{-\frac{1}{3}} \, dt$$

Using the substitution $u = 1 + \frac{16}{9} t^{\frac{2}{3}} \Rightarrow du = \frac{32}{27} t^{-\frac{1}{3}} \, dt$, we calculate the total charge as

$$\int_{\frac{25}{9}}^{\frac{73}{9}} \sqrt{u} \, \frac{27}{32} \, du = \frac{27}{32} \cdot \frac{2}{3} u^{\frac{3}{2}} \Big|_{\frac{25}{9}}^{\frac{73}{9}} = \frac{1}{48} \left(73^{\frac{3}{2}} - 25^{\frac{3}{2}} \right) \approx 10.39$$

Thus the total charge is 10.39×10^{-6} C.

In Exercises 47–50, use Eq. (6) to compute the electric potential $V(P)$ at the point P for the given charge density (in units of 10^{-6} C).

47. Calculate $V(P)$ at $P = (0, 0, 12)$ if the electric charge is distributed along the quarter circle of radius 4 centered at the origin with charge density $\rho(x, y, z) = xy$.

SOLUTION   We parametrize the curve by $\mathbf{c}(t) = (4\cos t, 4\sin t, 0)$, $(0 \le t \le \frac{\pi}{2})$. Then $\mathbf{c}'(t) = (-4\sin t, 4\cos t, 0) \Rightarrow \|\mathbf{c}'(t)\| = 4$. The distance from the point $(0, 0, 12)$ to $\mathbf{c}(t)$ is

$$r_P(t) = \sqrt{(0 - 4\cos t)^2 + (0 - 4\sin t)^2 + (12 - 0)^2} = \sqrt{16 + 144} = 4\sqrt{10}$$

while the charge density along the curve is

$$\rho(\mathbf{c}(t)) = xy = 4\cos t \, 4\sin t = 16\sin t \cos t = 8\sin 2t$$

Therefore

$$V(P) = k \int_C \frac{\rho}{r_P} \, ds = k \int_0^{\frac{\pi}{2}} \frac{8\sin 2t}{4\sqrt{10}} 4 \, dt = \frac{8k}{\sqrt{10}} \cdot \frac{-\cos 2t}{2} \Big|_0^{\frac{\pi}{2}}$$

$$= \frac{4k}{\sqrt{10}} (-\cos \pi + \cos 0) = \frac{8k}{\sqrt{10}}$$

Thus the electric potential is $\frac{8k}{\sqrt{10}} \times 10^{-6}$ C ≈ 22743.1 volts

49. Calculate $V(P)$ at $P = (2, 0, 2)$ if the negative charge is distributed along the y-axis for $1 \le y \le 3$ with charge density $\rho(x, y, z) = -y$.

SOLUTION   A parametrization for the curve is $\mathbf{c}(t) = (0, t, 0)$ $(1 \le t \le 3)$. Then $\mathbf{c}'(t) = (0, 1, 0) \quad \Rightarrow \quad \|\mathbf{c}'(t)\| = 1$, and the charge density along the curve is $\rho(\mathbf{c}(t)) = -y = -t$. The distance from the origin to $\mathbf{c}(t)$ is

$$r_P(t) = \sqrt{(2 - 0)^2 + (0 - t)^2 + (2 - 0)^2} = \sqrt{8 + t^2}$$

Therefore,

$$V(P) = k \int_C \frac{\rho}{r_P} \, ds = k \int_1^3 \frac{-t}{\sqrt{8 + t^2}} \cdot 1 \, dt$$

Using the substitution $u = 8 + t^2 \Rightarrow du = 2t \, dt$, we have

$$V(P) = -k \int_9^{17} u^{-\frac{1}{2}} \frac{1}{2} \, du = -\frac{k}{2} \cdot 2u^{\frac{1}{2}} \Big|_9^{17} = -\frac{k}{2} \left(17^{\frac{1}{2}} - 9^{\frac{1}{2}} \right)$$

Thus the electric potential is $-\frac{k}{2} \left(17^{\frac{1}{2}} - 9^{\frac{1}{2}} \right) \times 10^{-6} \, C \approx -10097$ volts

51. Calculate the work done by a field $\mathbf{F} = \langle x + y, x - y \rangle$ when an object moves from $(0, 0)$ to $(1, 1)$ along each of the paths $y = x^2$ and $x = y^2$.

SOLUTION   We calculate the work done by $\mathbf{F} = \langle x + y, x - y \rangle$ along the path $y = x^2$ from $(0, 0)$ to $(1, 1)$. We use the parametrization:

$$\mathbf{c}_1(t) = (t, t^2), \quad 0 \le t \le 1$$

We have

$$\mathbf{F}(\mathbf{c}_1(t)) = \left\langle t + t^2, t - t^2 \right\rangle$$

$$\mathbf{c}_1'(t) = \langle 1, 2t \rangle$$

$$\mathbf{F}(\mathbf{c}(t)) \cdot \mathbf{c}'(t) = \left\langle t + t^2, t - t^2 \right\rangle \cdot \langle 1, 2t \rangle = t + t^2 + 2t^2 - 2t^3 = -2t^3 + 3t^2 + t$$

The work is the following line integral:

$$W = \int_{\mathcal{C}_1} \mathbf{F} \cdot d\mathbf{s} = \int_0^1 \mathbf{F}(\mathbf{c}_1(t)) \cdot \mathbf{c}_1'(t) \, dt = \int_0^1 \left(-2t^3 + 3t^2 + t \right) dt = -\frac{1}{2} t^4 + t^3 + \frac{1}{2} t^2 \Big|_0^1 = 1$$

We now compute the work along the path $x = y^2$. We parametrize the path by:

$$\mathbf{c}_2(t) = (t^2, t), \quad 0 \le t \le 1$$

Then

$$\mathbf{F}(\mathbf{c}_2(t)) = \left\langle t^2 + t, t^2 - t \right\rangle$$

$$\mathbf{c}_2'(t) = \langle 2t, 1 \rangle$$

$$\mathbf{F}(\mathbf{c}_2(t)) \cdot \mathbf{c}_2'(t) = \left\langle t^2 + t, t^2 - t \right\rangle \cdot \langle 2t, 1 \rangle = 2t^3 + 2t^2 + t^2 - t = 2t^3 + 3t^2 - t$$

The work is the line integral

$$W = \int_{\mathcal{C}_2} \mathbf{F} \cdot d\mathbf{s} = \int_0^1 \mathbf{F}(\mathbf{c}_2(t)) \cdot \mathbf{c}_2'(t) \, dt = \int_0^1 \left(2t^3 + 3t^2 - t \right) dt = \frac{1}{2} t^4 + t^3 - \frac{1}{2} t^2 \Big|_0^1 = \frac{1}{2} + 1 - \frac{1}{2} = 1$$

We obtain the same work along the two paths.

53. Figure 21 shows a force field $\mathbf{F}$.

(a) Over which of the two paths, ADC or ABC, does $\mathbf{F}$ perform less work?

(b) If you have to work against $\mathbf{F}$ to move an object from C to A, which of the paths, CBA or CDA, requires less work?

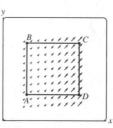

FIGURE 21

SOLUTION

(a) Since x is constant on $\overline{AB}$ and $\overline{DC}$, $\mathbf{F}(x, y) = \langle x, x \rangle$ is also constant on these segments.

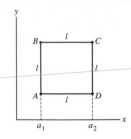

Let a_1 and a_2 denote the constant values of x on the segments $\overline{AB}$ and $\overline{DC}$ respectively, and l denote the lengths of these segments. By Exercise 55 we have

$$\int_{AB} \mathbf{F} \cdot d\mathbf{s} = \langle a_1, a_1 \rangle \cdot \langle 0, l \rangle = a_1 \cdot 0 + a_1 \cdot l = a_1 l$$

$$\int_{DC} \mathbf{F} \cdot d\mathbf{s} = \langle a_2, a_2 \rangle \cdot \langle 0, l \rangle = a_2 \cdot 0 + a_2 \cdot l = a_2 l$$

Since $a_1 < a_2$ we have $\int_{AB} \mathbf{F} \cdot d\mathbf{s} < \int_{DC} \mathbf{F} \cdot d\mathbf{s}$.

(b) We compute the integral over BC. This segment is parametrized by:

$$\mathbf{c}(t) = (a_1 + lt, b), \ 0 \le t \le 1.$$

Hence,

$$\mathbf{F}(\mathbf{c}(t)) = \langle x, x \rangle = \langle a_1 + lt, a_1 + lt \rangle, \ \mathbf{c}'(t) = \langle l, 0 \rangle$$

$$\mathbf{F}(\mathbf{c}(t)) \cdot \mathbf{c}'(t) = \langle a_1 + lt, a_1 + lt \rangle \cdot \langle l, 0 \rangle = a_1 l + l^2 t$$

Thus,

$$\int_{BC} \mathbf{F} \cdot d\mathbf{s} = \int_0^1 \left(a_1 l + l^2 t \right) dt = a_1 lt + \frac{l^2 t^2}{2} \bigg|_{t=0}^1 = a_1 l + \frac{l^2}{2}$$

We see that the line integral does not depend on b, therefore,

$$\int_{AD} \mathbf{F} \cdot d\mathbf{s} = \int_{BC} \mathbf{F} \cdot d\mathbf{s} \tag{1}$$

In part (a) we showed that:

$$\int_{AB} \mathbf{F} \cdot d\mathbf{s} < \int_{DC} \mathbf{F} \cdot d\mathbf{s} \tag{2}$$

Combining (1) and (2) gives:

$$\int_{ABC} \mathbf{F} \cdot d\mathbf{s} = \int_{AB} \mathbf{F} \cdot d\mathbf{s} + \int_{BC} \mathbf{F} \cdot d\mathbf{s} < \int_{DC} \mathbf{F} \cdot d\mathbf{s} + \int_{AD} \mathbf{F} \cdot d\mathbf{s} = \int_{ADC} \mathbf{F} \cdot d\mathbf{s}$$

55. Show that work performed by a constant force field $\mathbf{F}$ over any path C from P to Q is equal to $\mathbf{F} \cdot \overrightarrow{PQ}$.

SOLUTION We denote by $\mathbf{c}(t) = (x(t), y(t), c(t)), t_0 \le t \le t_1$ a parametrization of the oriented path from P to Q (then $\mathbf{c}(t_0) = P$ and $\mathbf{c}(t_1) = Q$). Let $\mathbf{F} = \langle a, b, c \rangle$ be a constant vector field. Then,

$$\mathbf{F}(\mathbf{c}(t)) \cdot \mathbf{c}'(t) = \langle a, b, c \rangle \cdot \langle x'(t), y'(t), z'(t) \rangle = ax'(t) + by'(t) + cz'(t)$$

The vector line integral is, thus,

$$\int_C \mathbf{F} \cdot d\mathbf{s} = \int_{t_0}^{t_1} \mathbf{F}(\mathbf{c}(t)) \cdot \mathbf{c}'(t) \, dt = \int_{t_0}^{t_1} \left(ax'(t) + by'(t) + cz'(t) \right) dt$$

$$= a \int_{t_0}^{t_1} x'(t) \, dt + b \int_{t_0}^{t_1} y'(t) \, dt + c \int_{t_0}^{t_1} z'(t) \, dt = ax(t) \big|_{t=t_0}^{t_1} + by(t) \big|_{t=t_0}^{t_1} + cz(t) \big|_{t=t_0}^{t_1}$$

$$= a \left(x(t_1) - x(t_0) \right) + b \left(y(t_1) - y(t_0) \right) + c \left(z(t_1) - z(t_0) \right)$$

$$= \langle a, b, c \rangle \cdot \langle x(t_1) - x(t_0), y(t_1) - y(t_0), z(t_1) - z(t_0) \rangle$$

Since $P = \langle x(t_0), y(t_0), z(t_0) \rangle$ and $Q = \langle x(t_1), y(t_1), z(t_1) \rangle$ we conclude that,

$$\int_C \mathbf{F} \cdot d\mathbf{s} = \langle a, b, c \rangle \cdot \overrightarrow{PQ} = \mathbf{F} \cdot \overrightarrow{PQ}.$$

57. Charge is distributed along the spiral with polar equation $r = \theta$ for $0 \le \theta \le 2\pi$. The charge density is $\rho(r, \theta) = r$ (assume distance is in centimeters and charge in units of 10^{-6} C/cm). Use the result of Exercise 56(a) to compute the total charge.

SOLUTION Following Exercise 56(a), $f(\theta) = \theta$, and $f'(\theta) = 1$. Thus $\|\mathbf{c}'(\theta)\| = \sqrt{\theta^2 + 1}$. The total charge will be

$$\int_C \rho \, ds = \int_0^{2\pi} \theta \sqrt{\theta^2 + 1} \, d\theta$$

Substituting $u = \theta^2 + 1 \Rightarrow du = 2\theta \, d\theta$, we have

$$\int_C \rho \, ds = \int_1^{4\pi^2 + 1} \sqrt{u} \, \frac{1}{2} \, du = \frac{1}{2} \cdot \frac{2}{3} u^{\frac{3}{2}} \Big|_1^{4\pi^2 + 1}$$

$$= \frac{1}{2} \left((4\pi^2 + 1)^{\frac{3}{2}} - 1 \right) \approx 85.5$$

Thus the total charge is 85.5×10^{-6} C.

In Exercises 58–61, let **F** *be the* **vortex field** *(so-called because it swirls around the origin as in Figure 23):*

$$\mathbf{F} = \left\langle \frac{-y}{x^2 + y^2}, \frac{x}{x^2 + y^2} \right\rangle$$

FIGURE 23 Vortex field $\mathbf{F} = \left\langle \dfrac{-y}{x^2 + y^2}, \dfrac{x}{x^2 + y^2} \right\rangle$.

59. Show that the value of $\displaystyle\int_{C_R} \mathbf{F} \cdot d\mathbf{s}$, where C_R is the circle of radius R centered at the origin and oriented counterclockwise, does not depend on R

SOLUTION We parametrize C_R by:

$$\mathbf{c}(t) = (R \cos t, R \sin t), \quad 0 \le t < 2\pi.$$

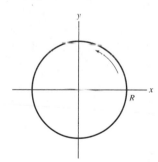

Step 1. Calculate the integrand:

$$\mathbf{F}(\mathbf{c}(t)) = \left\langle -\frac{y}{x^2 + y^2}, \frac{x}{x^2 + y^2} \right\rangle = \left\langle -\frac{R \sin t}{R^2}, \frac{R \cos t}{R^2} \right\rangle = \frac{1}{R} \langle -\sin t, \cos t \rangle$$

$$\mathbf{c}'(t) = \frac{d}{dt} \langle R \cos t, R \sin t \rangle = \langle -R \sin t, R \cos t \rangle = R \langle -\sin t, \cos t \rangle$$

The integrand is the dot product:

$$\mathbf{F}(\mathbf{c}(t)) \cdot \mathbf{c}'(t) = \frac{1}{R} \langle -\sin t, \cos t \rangle \cdot R \langle -\sin t, \cos t \rangle = \sin^2 t + \cos^2 t = 1$$

Step 2. Evaluate the integral.

$$\int_C \mathbf{F} \cdot d\mathbf{s} = \int_0^{2\pi} \mathbf{F}(\mathbf{c}(t)) \cdot \mathbf{c}'(t) \, dt = \int_0^{2\pi} 1 \, dt = 2\pi$$

61. Let $\mathcal{C}$ be a curve in polar form $r = f(\theta)$ for $\theta_1 \le \theta \le \theta_2$ [Figure 24(B)], parametrized by $\mathbf{c}(\theta) = (f(\theta)\cos\theta, f(\theta)\sin\theta))$ as in Exercise 56.

(a) Show that the vortex field in polar coordinates is written $\mathbf{F} = r^{-1}\langle -\sin\theta, \cos\theta\rangle$.

(b) Show that $\mathbf{F} \cdot \mathbf{c}'(\theta)\, d\theta = d\theta$.

(c) Show that $\displaystyle\int_{\mathcal{C}} \mathbf{F} \cdot d\mathbf{s} = \theta_2 - \theta_1$.

SOLUTION

(a) Letting $x = r\cos(\theta)$ and $y = r\sin(\theta)$ we have

$$\mathbf{F} = \left\langle \frac{-y}{x^2+y^2}, \frac{x}{x^2+y^2} \right\rangle = \left\langle \frac{-r\sin(\theta)}{(r\cos(\theta))^2 + (r\sin(\theta))^2}, \frac{r\cos(\theta)}{(r\cos(\theta))^2 + (r\sin(\theta))^2} \right\rangle$$

$$= \frac{r}{r^2}\left\langle \frac{-\sin(\theta)}{\cos^2(\theta) + \sin^2(\theta)}, \frac{\cos(\theta)}{\cos^2(\theta) + \sin^2(\theta)} \right\rangle = r^{-1}\langle -\sin\theta, \cos\theta\rangle$$

(b) From the solution to Exercise 56(a) we have

$$\mathbf{c}'(\theta) = \left\langle f'(\theta)\cos(\theta) - f(\theta)\sin(\theta),\ f'(\theta)\sin(\theta) + f(\theta)\cos(\theta) \right\rangle$$

Substituting $r = f(\theta)$ into the previous part, we have

$$\mathbf{F} \cdot \mathbf{c}'(\theta)\, d\theta = \frac{1}{f(\theta)}\langle -\sin(\theta), \cos(\theta)\rangle \cdot \left\langle f'(\theta)\cos(\theta) - f(\theta)\sin(\theta),\ f'(\theta)\sin(\theta) + f(\theta)\cos(\theta)\right\rangle d\theta$$

$$= \frac{1}{f(\theta)}\left(-f'(\theta)\sin(\theta)\cos(\theta) + f(\theta)\sin^2(\theta) + f'(\theta)\cos(\theta)\sin(\theta) + f(\theta)\cos^2(\theta)\right) d\theta$$

$$= \frac{1}{f(\theta)} f(\theta)(\sin^2(\theta) + \cos^2(\theta))\, d\theta = d\theta$$

(c)

$$\int_{\mathcal{C}} \mathbf{F} \cdot d\mathbf{s} = \int_{\theta_1}^{\theta_2} \mathbf{F} \cdot \mathbf{c}'(\theta)\, d\theta = \int_{\theta_1}^{\theta_2} d\theta = \theta_2 - \theta_1$$

In Exercises 62–65, use Eq. (10) to calculate the flux of the vector field across the curve specified.

63. $\mathbf{F} = \langle x^2, y^2\rangle$;   segment from $(3, 0)$ to $(0, 3)$, oriented upward

SOLUTION   The curve is parametrized by $\mathbf{c}(t) = (3 - t, t)$ $(0 \le t \le 3) \Rightarrow \mathbf{c}'(t) = \langle -1, 1\rangle$. Then

$$\mathbf{F}(\mathbf{c}(t)) = \langle x^2, y^2\rangle = \left\langle (3 - t)^2, t^2\right\rangle$$

Therefore the flux is

$$\int_{\mathcal{C}} F_1\, dy - F_2\, dx = \int_0^3 (3 - t)^2(1) - (t^2)(-1)\, dt$$

$$= \int_0^3 2t^2 - 6t + 9\, dt = \left. \frac{2t^3}{3} - 3t^2 + 9t \right|_0^3 = 18$$

65. $\mathbf{v} = \langle e^y, 2x - 1\rangle$;   parabola $y = x^2$ for $0 \le x \le 1$, oriented left to right

SOLUTION   The curve is parametrized by $\mathbf{c}(t) = (t, t^2)$ $(0 \le t \le 1) \Rightarrow \mathbf{c}'(t) = \langle 1, 2t\rangle$. Then

$$\mathbf{v}(\mathbf{c}(t)) = \langle e^y, 2x - 1\rangle = \left\langle e^{t^2}, 2t - 1\right\rangle$$

Therefore the flux is

$$\int_{\mathcal{C}} v_1\, dy - v_2\, dx = \int_0^1 \left(e^{t^2}\right)(2t) - (2t - 1)(1)\, dt = \left. e^{t^2} - t^2 + t \right|_0^1 = e - 1$$

Further Insights and Challenges

67. Let $\mathbf{F} = \langle x, 0 \rangle$. Prove that if $\mathcal{C}$ is any path from (a, b) to (c, d), then

$$\int_{\mathcal{C}} \mathbf{F} \cdot d\mathbf{s} = \frac{1}{2}(c^2 - a^2)$$

SOLUTION

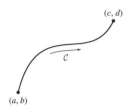

We denote the parametrization of the path by,

$$\mathbf{c}(t) = (x(t), y(t)), \quad t_0 \le t \le t_1, \quad \mathbf{c}(t_0) = (a, b), \quad \mathbf{c}(t_1) = (c, d)$$

By the Theorem on vector line integrals we have,

$$\int_{\mathbf{c}} \mathbf{F} \cdot d\mathbf{s} = \int_{t_0}^{t_1} \mathbf{F}(\mathbf{c}(t)) \cdot \mathbf{c}'(t) \, dt = \int_{t_0}^{t_1} \langle x(t), 0 \rangle \cdot \langle x'(t), y'(t) \rangle \, dt = \int_{t_0}^{t_1} x(t) x'(t) \, dt$$

We use the hint to compute the integral, obtaining

$$\int_{\mathbf{c}} \mathbf{F} \cdot d\mathbf{s} = \frac{1}{2} x(t)^2 \Big|_{t_0}^{t_1} = \frac{1}{2} \left(x(t_1)^2 - x(t_0)^2 \right) = \frac{1}{2} \left(c^2 - a^2 \right)$$

Proof of the hint: By the Chain Rule for differentiation we have

$$\frac{d}{dt} f^2(t) = 2 f(t) f'(t) \quad \Rightarrow \quad f(t) f'(t) = \frac{1}{2} \frac{d}{dt} f^2(t)$$

Applying the Fundamental Theorem of calculus we obtain

$$\int_{t_0}^{t_1} f(t) f'(t) \, dt = \frac{1}{2} \int_{t_0}^{t_1} \frac{d}{dt} \left(f^2(t) \right) dt = \frac{1}{2} \left(f^2(t_1) - f^2(t_0) \right)$$

Alternatively we can evaluate the integral $\int f(t) f'(t) \, dt$ using the substitution $u = f(t)$, $du = f'(t) \, dt$. We get

$$\int f(t) f'(t) \, dt = \int u \, du = \frac{1}{2} u^2 + c = \frac{1}{2} f^2(t) + c.$$

69. We wish to define the **average value** Av(f) of a continuous function f along a curve $\mathcal{C}$ of length L. Divide $\mathcal{C}$ into N consecutive arcs $\mathcal{C}_1, \ldots, \mathcal{C}_N$, each of length L/N, and let P_i be a sample point in $\mathcal{C}_i$ (Figure 25). The sum

$$\frac{1}{N} \sum_{i=1}^{N} f(P_i)$$

may be considered an approximation to Av(f), so we define

$$\mathrm{Av}(f) = \lim_{N \to \infty} \frac{1}{N} \sum_{i=1}^{N} f(P_i)$$

Prove that

$$\mathrm{Av}(f) = \frac{1}{L} \int_{\mathcal{C}} f(x, y, z) \, ds \qquad \boxed{11}$$

Hint: Show that $\dfrac{L}{N} \sum_{i=1}^{N} f(P_i)$ is a Riemann sum approximation to the line integral of f along $\mathcal{C}$.

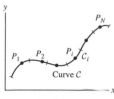

FIGURE 25

SOLUTION   The Riemann sum approximation to the line integral is:

$$\sum_{i=1}^{N} f(P_i) \, \Delta S_i$$

If the consecutive arcs $C_1, \ldots, C_2$ have equal lengths $\frac{L}{N}$, the corresponding Riemann sum is,

$$\sum_{i=1}^{N} f(P_i) \cdot \frac{L}{N} = \frac{L}{N} \sum_{i=1}^{N} f(P_i)$$

We let $N \to \infty$,

$$\int_C f(x, y, z) \, ds = \lim_{N \to \infty} \frac{L}{N} \sum_{i=1}^{N} f(P_i) = L \lim_{N \to \infty} \frac{1}{N} \sum_{i=1}^{N} f(P_i) = L\text{Av}(f)$$

That is,

$$\text{Av}(f) = \frac{1}{L} \int_C f(x, y, z) \, ds.$$

71. Use Eq. (11) to calculate the average value of $f(x, y) = x$ along the curve $y = x^2$ for $0 \le x \le 1$.

SOLUTION   The average value is

$$\text{Av}(f) = \frac{1}{L} \int_{\mathbf{c}} x \, ds \tag{1}$$

We parametrize the curve by the parametrization,

$$\mathbf{c}(t) = \left(t, t^2 \right), \quad 0 \le t \le 1.$$

Hence,

$$\mathbf{c}'(t) = \langle 1, 2t \rangle \quad \Rightarrow \quad \|\mathbf{c}'(t)\| = \sqrt{1 + 4t^2}$$

We first must calculate the length of the path. That is,

$$L = \int_{\mathbf{c}} \|\mathbf{c}'(t)\| \, dt = \int_0^1 \sqrt{1 + 4t^2} \, dt = \frac{1}{2} t \sqrt{1 + 4t^2} + \frac{1}{4} \ln\left(2t + \sqrt{1 + 4t^2} \right) \Big|_0^1$$

$$= \frac{\sqrt{5}}{2} + \frac{1}{4} \ln\left(2 + \sqrt{5} \right) = \frac{2\sqrt{5} + \ln\left(2 + \sqrt{5} \right)}{4}$$

We compute the line integral in (1):

$$\int_{\mathbf{c}} x \, ds = \int_0^1 t \|\mathbf{c}'(t)\| \, dt = \int_0^1 t \sqrt{1 + 4t^2} \, dt$$

We compute the integral using the substitution $u = 1 + 4t^2$, $du = 8t \, dt$.

$$\int_{\mathbf{c}} x \, ds = \int_0^1 \sqrt{1 + 4t^2} \cdot t \, dt = \int_1^5 u^{1/2} \cdot \frac{du}{8}$$

$$= \frac{1}{8} \cdot \frac{2}{3} u^{\frac{3}{2}} \Big|_1^5 = \frac{1}{12} \left(5^{\frac{3}{2}} - 1 \right)$$

Combining gives the following solution:

$$\text{Av}(f) = \frac{4}{2\sqrt{5} + \ln\left(2 + \sqrt{5} \right)} \cdot \frac{5^{\frac{3}{2}} - 1}{12} = \frac{5\sqrt{5} - 1}{\left(6\sqrt{5} + 3 \ln\left(2 + \sqrt{5} \right) \right)}$$

73. The value of a scalar line integral does not depend on the choice of parametrization (because it is defined without reference to a parametrization). Prove this directly. That is, suppose that $\mathbf{c}_1(t)$ and $\mathbf{c}(t)$ are two parametrizations such that $\mathbf{c}_1(t) = \mathbf{c}(\varphi(t))$, where $\varphi(t)$ is an increasing function. Use the Change of Variables Formula to verify that

$$\int_c^d f(\mathbf{c}_1(t)) \|\mathbf{c}_1'(t)\| \, dt = \int_a^b f(\mathbf{c}(t)) \|\mathbf{c}'(t)\| \, dt$$

where $a = \varphi(c)$ and $b = \varphi(d)$.

SOLUTION We compute the integral $\int_a^b f(\mathbf{c}(t)) \|\mathbf{c}'(t)\| \, dt$ using the substitution $t = \varphi(u)$, $a = \varphi(c)$, $b = \varphi(d)$. We get:

$$\int_a^b f(\mathbf{c}_1(t)) \|\mathbf{c}'(t)\| \, dt = \int_{\varphi^{-1}(a)}^{\varphi^{-1}(b)} f(\mathbf{c}(\varphi(t))) \|\mathbf{c}'(\varphi(t))\| \|\varphi'(u)\| \, du \tag{1}$$

Since φ is an increasing function, $\varphi'(u) > 0$ for all u, therefore:

$$\|\mathbf{c}'(\varphi(u))\| \varphi'(u) = \|\mathbf{c}'(\varphi(u)) \varphi'(u)\| \tag{2}$$

By the Chain Rule for vector valued functions, we have,

$$\frac{d}{du} \mathbf{c}(\varphi(u)) = \varphi'(u)\mathbf{c}'(\varphi(u)) \tag{3}$$

Combining (2) and (3) gives:

$$\|\mathbf{c}'(\varphi(u))\| \varphi'(u) = \left\| \frac{d}{du} \mathbf{c}(\varphi(u)) \right\| = \left\| \frac{d}{du} \mathbf{c}_1(u) \right\| = \|\mathbf{c}_1'(u)\| \tag{4}$$

We substitute (4) in (1) to obtain:

$$\int_a^b f(\mathbf{c}(t)) \|\mathbf{c}'(t)\| \, dt = \int_c^d f(\mathbf{c}_1(u)) \|\mathbf{c}_1'(u)\| \, du = \int_c^d f(\mathbf{c}_1(t)) \|\mathbf{c}_1'(t)\| \, dt$$

The last step is simply replacing the dummy variable of integration u by t.

16.3 Conservative Vector Fields (LT Section 17.3)

Preliminary Questions

1. The following statement is false. *If $\mathbf{F}$ is a gradient vector field, then the line integral of $\mathbf{F}$ along every curve is zero.* Which single word must be added to make it true?

SOLUTION The missing word is "closed" (curve). The line integral of a gradient vector field along every closed curve is zero.

2. Which of the following statements are true for all vector fields, and which are true only for conservative vector fields?
(a) The line integral along a path from P to Q does not depend on which path is chosen.
(b) The line integral over an oriented curve C does not depend on how C is parametrized.
(c) The line integral around a closed curve is zero.
(d) The line integral changes sign if the orientation is reversed.
(e) The line integral is equal to the difference of a potential function at the two endpoints.
(f) The line integral is equal to the integral of the tangential component along the curve.
(g) The cross-partials of the components are equal.

SOLUTION
(a) This statement is true only for conservative vector fields.
(b) This statement is true for all vector fields.
(c) This statement holds only for conservative vector fields.
(d) This is a property of all vector fields.
(e) Only conservative vector fields have a potential function, and the line integral is computed by using the potential function as stated.
(f) All vector fields' line integrals share this property.
(g) The cross-partial of the components of a conservative field are equal. For other fields, the cross-partial of the components may or may not equal.

3. Let $\mathbf{F}$ be a vector field on an open, connected domain $\mathcal{D}$. Which of the following statements are always true, and which are true under additional hypotheses on $\mathcal{D}$?
(a) If $\mathbf{F}$ has a potential function, then $\mathbf{F}$ is conservative.
(b) If $\mathbf{F}$ is conservative, then the cross-partials of $\mathbf{F}$ are equal.
(c) If the cross-partials of $\mathbf{F}$ are equal, then $\mathbf{F}$ is conservative.

SOLUTION
(a) This statement is always true, since every gradient vector field is conservative.
(b) If $\mathbf{F}$ is conservative on a connected domain $\mathcal{D}$, then $\mathbf{F}$ has a potential function $\mathcal{D}$ and consequently the cross partials of $\mathbf{F}$ are equal in $\mathcal{D}$.
(c) If the cross partials of $\mathbf{F}$ are equal in a simply-connected region $\mathcal{D}$, then $\mathbf{F}$ is a gradient vector field in $\mathcal{D}$.

4. Let $\mathcal{C}$, $\mathcal{D}$, and $\mathcal{E}$ be the oriented curves in Figure 16 and let $\mathbf{F} = \nabla V$ be a gradient vector field such that $\displaystyle\int_{\mathcal{C}} \mathbf{F} \cdot d\mathbf{s} = 4$. What are the values of the following integrals?

(a) $\displaystyle\int_{\mathcal{D}} \mathbf{F} \cdot d\mathbf{s}$

(b) $\displaystyle\int_{\mathcal{E}} \mathbf{F} \cdot d\mathbf{s}$

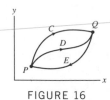

FIGURE 16

SOLUTION   Since $\mathbf{F}$ is a gradient vector field the integrals over closed paths are zero. Therefore, by the equivalent conditions for path independence we have:

(a) $\int_{\mathcal{D}} \mathbf{F} \cdot d\mathbf{s} = \int_{\mathcal{C}} \mathbf{F} \cdot d\mathbf{s} = 4$

(b) $\int_{\mathcal{E}} \mathbf{F} \cdot d\mathbf{s} = \int_{-\mathcal{C}} \mathbf{F} \cdot d\mathbf{s} = -\int_{\mathcal{C}} \mathbf{F} \cdot d\mathbf{s} = -4$

Exercises

1. Let $V(x, y, z) = xy \sin(yz)$ and $\mathbf{F} = \nabla V$. Evaluate $\displaystyle\int_{\mathbf{c}} \mathbf{F} \cdot d\mathbf{s}$, where $\mathbf{c}$ is any path from $(0, 0, 0)$ to $(1, 1, \pi)$.

SOLUTION   By the Fundamental Theorem for Gradient Vector Fields, we have:

$$\int_{\mathbf{c}} \nabla V \cdot d\mathbf{s} = V(1, 1, \pi) - V(0, 0, 0) = 1 \cdot 1 \sin \pi - 0 = 0$$

In Exercises 3–6, verify that $\mathbf{F} = \nabla V$ and evaluate the line integral of $\mathbf{F}$ over the given path.

3. $\mathbf{F} = \langle 3, 6y \rangle$, $V(x, y, z) = 3x + 3y^2$; $\mathbf{c}(t) = (t, 2t^{-1})$ for $1 \le t \le 4$

SOLUTION   The gradient of $V = 3x + 3y^2$ is:

$$\nabla V = \left\langle \frac{\partial V}{\partial x}, \frac{\partial V}{\partial y} \right\rangle = \langle 3, 6y \rangle = \mathbf{F}$$

Using the Fundamental Theorem for Gradient Vector Fields, we have:

$$\int_{\mathbf{c}} \mathbf{F} \cdot d\mathbf{s} = V\left(\mathbf{c}(4)\right) - V\left(\mathbf{c}(1)\right) = V\left(4, \frac{1}{2}\right) - V(1, 2) = \left(3 \cdot 4 + 3 \cdot \frac{1}{4}\right) - (3 \cdot 1 + 3 \cdot 4) = -\frac{9}{4}$$

5. $\mathbf{F} = ye^z\mathbf{i} + xe^z\mathbf{j} + xye^z\mathbf{k}$, $V(x, y, z) = xye^z$; $\mathbf{c}(t) = (t^2, t^3, t - 1)$ for $1 \le t \le 2$

SOLUTION   We verify that $\mathbf{F}$ is the gradient of V:

$$\nabla V = \left\langle \frac{\partial V}{\partial x}, \frac{\partial V}{\partial y}, \frac{\partial V}{\partial z} \right\rangle = \left\langle ye^z, xe^z, xye^z \right\rangle = \mathbf{F}$$

We use the Fundamental Theorem for Gradient Vectors with the initial point $\mathbf{c}(1) = (1, 1, 0)$ and terminal point $\mathbf{c}(2) = (4, 8, 1)$, to obtain:

$$\int_{\mathbf{c}} \mathbf{F} \cdot d\mathbf{s} = V(4, 8, 1) - V(1, 1, 0) = 32e - 1$$

In Exercises 7–16, find a potential function for $\mathbf{F}$ or determine that $\mathbf{F}$ is not conservative.

7. $\mathbf{F} = \langle z, 1, x \rangle$

SOLUTION   We check whether the vector field $\mathbf{F} = \langle z, 1, x \rangle$ satisfies the cross partials condition:

$$\begin{aligned} \frac{\partial F_1}{\partial y} &= \frac{\partial}{\partial y}(z) = 0 \\ \frac{\partial F_2}{\partial x} &= \frac{\partial}{\partial x}(1) = 0 \end{aligned} \quad \Rightarrow \quad \frac{\partial F_1}{\partial y} = \frac{\partial F_2}{\partial x}$$

$$\frac{\partial F_2}{\partial z} = \frac{\partial}{\partial z}(1) = 0$$
$$\frac{\partial F_3}{\partial y} = \frac{\partial}{\partial y}(x) = 0 \quad \Rightarrow \quad \frac{\partial F_2}{\partial z} = \frac{\partial F_3}{\partial y}$$

$$\frac{\partial F_3}{\partial x} = \frac{\partial}{\partial x}(x) = 1$$
$$\frac{\partial F_1}{\partial z} = \frac{\partial}{\partial z}(z) = 1 \quad \Rightarrow \quad \frac{\partial F_3}{\partial x} = \frac{\partial F_1}{\partial z}$$

$\mathbf{F}$ satisfies the cross partials condition everywhere. Hence, $\mathbf{F}$ is conservative. We find a potential function $V(x, y, z)$.

Step 1. Use the condition $\frac{\partial V}{\partial x} = F_1$. V is an antiderivative of $F_1 = z$ when y and z are fixed, therefore:

$$V(x, y, z) = \int z \, dx = zx + g(y, z) \tag{1}$$

Step 2. Use the condition $\frac{\partial V}{\partial y} = F_2$. By (1) we have:

$$\frac{\partial}{\partial y}(zx + g(y, z)) = 1$$

$$g_y(y, z) = 1$$

Integrating with respect to y, while holding z fixed, gives:

$$g(y, z) = \int 1 \, dy = y + h(z)$$

We substitute in (1) to obtain:

$$V(x, y, z) = zx + y + h(z) \tag{2}$$

Step 3. Use the condition $\frac{\partial V}{\partial z} = F_3$. Using (2) we get:

$$\frac{\partial}{\partial z}(zx + y + h(z)) = x$$

$$x + h'(z) = x$$

$$h'(z) = 0 \quad \Rightarrow \quad h(z) = c$$

Substituting in (2) gives the following potential functions:

$$V(x, y, z) = zx + y + c.$$

One of the potential functions is obtained by choosing $c = 0$:

$$V(x, y, z) = zx + y$$

9. $\mathbf{F} = y^2 \mathbf{i} + (2xy + e^z) \mathbf{j} + y e^z \mathbf{k}$

SOLUTION We examine whether $\mathbf{F}$ satisfies the cross partials condition:

$$\frac{\partial F_1}{\partial y} = \frac{\partial}{\partial y}\left(y^2\right) = 2y$$
$$\frac{\partial F_2}{\partial x} = \frac{\partial}{\partial x}\left(2xy + e^z\right) = 2y \quad \Rightarrow \quad \frac{\partial F_1}{\partial y} = \frac{\partial F_2}{\partial x}$$

$$\frac{\partial F_2}{\partial z} = \frac{\partial}{\partial z}\left(2xy + e^z\right) = e^z$$
$$\frac{\partial F_3}{\partial y} = \frac{\partial}{\partial y}\left(y e^z\right) = e^z \quad \Rightarrow \quad \frac{\partial F_2}{\partial z} = \frac{\partial F_3}{\partial y}$$

$$\frac{\partial F_3}{\partial x} = \frac{\partial}{\partial x}\left(y e^z\right) = 0$$
$$\frac{\partial F_1}{\partial z} = \frac{\partial}{\partial z}\left(y^2\right) = 0 \quad \Rightarrow \quad \frac{\partial F_3}{\partial x} = \frac{\partial F_1}{\partial z}$$

We see that $\mathbf{F}$ satisfies the cross partials condition everywhere, hence $\mathbf{F}$ is conservative. We find a potential function for $\mathbf{F}$.

Step 1. Use the condition $\frac{\partial V}{\partial x} = F_1$. V is an antiderivative of $F_1 = y^2$ when y and z are fixed. Hence:

$$V(x, y, z) = \int y^2 \, dx = y^2 x + g(y, z) \tag{1}$$

Step 2. Use the condition $\frac{\partial V}{\partial y} = F_2$. By (1) we have:

$$\frac{\partial}{\partial y} \left(y^2 x + g(y, z) \right) = 2xy + e^z$$

$$2yx + g_y(y, z) = 2xy + e^z \quad \Rightarrow \quad g_y(y, z) = e^z$$

We integrate with respect to y, holding z fixed:

$$g(y, z) = \int e^z \, dy = e^z y + h(z)$$

Substituting in (1) gives:

$$V(x, y, z) = y^2 x + e^z y + h(z) \tag{2}$$

Step 3. Use the condition $\frac{\partial V}{\partial z} = F_3$. By (2), we get:

$$\frac{\partial}{\partial z} \left(y^2 x + e^z y + h(z) \right) = y e^z$$

$$e^z y + h'(z) = y e^z \quad \Rightarrow \quad h'(z) = 0$$

Therefore $h(z) = c$. Substituting in (2) we get:

$$V(x, y, z) = y^2 x + e^z y + c$$

The potential function corresponding to $c = 0$ is:

$$V(x, y, z) = y^2 x + e^z y.$$

11. $\mathbf{F} = \langle \cos(xz), \sin(yz), xy \sin z \rangle$

SOLUTION Since $\frac{\partial F_2}{\partial z} = \frac{\partial}{\partial z}(\sin(yz)) = y \cos(yz)$ and $\frac{\partial F_3}{\partial y} = \frac{\partial}{\partial y}(xy \sin z) = x \sin z$, we have $\frac{\partial F_2}{\partial z} \neq \frac{\partial F_3}{\partial y}$. The cross partials condition is not satisfied, therefore the vector field is not conservative.

13. $\mathbf{F} = \langle z \sec^2 x, z, y + \tan x \rangle$

SOLUTION

Step 1. Use the condition $\frac{\partial V}{\partial x} = F_1$. $V(x, y, z)$ is an antiderivative of $F_1 = z \sec^2 x$ when y and z are fixed, therefore:

$$V(x, y, z) = \int z \sec^2 x \, dx = z \tan x + g(y, z) \tag{1}$$

Step 2. Use the condition $\frac{\partial V}{\partial y} = F_2$. Using (1) we get:

$$\frac{\partial}{\partial y} (z \tan x + g(y, z)) = z$$

$$g_y(y, z) = z$$

We integrate with respect to y, holding z fixed:

$$g(y, z) = \int z \, dy = yz + h(z)$$

Substituting in (1) gives

$$V(x, y, z) = z \tan x + yz + h(z) \tag{2}$$

Step 3. Use the condition $\frac{\partial V}{\partial z} = F_3$. By (2) we have

$$\frac{\partial}{\partial z} (z \tan x + yz + h(z)) = y + \tan x$$

$$\tan x + y + h'(z) = y + \tan x$$

$$h'(z) = 0 \quad \Rightarrow \quad h(z) = c$$

Substituting in (2) we obtain the general potential function:

$$V(x, y, z) = z \tan x + yz + c$$

Choosing $c = 0$ gives the potential function:

$$V(x, y, z) = z \tan x + yz$$

15. $\mathbf{F} = \langle 2xy + 5, x^2 - 4z, -4y \rangle$

SOLUTION We find a potential function $V(x, y, z)$ for $\mathbf{F}$, using the following steps.

Step 1. Use the condition $\frac{\partial V}{\partial x} = F_1$. V is an antiderivative of $F_1 = 2xy + 5$ when y and z are fixed, therefore,

$$V(x, y, z) = \int (2xy + 5)\, dx = x^2 y + 5x + g(y, z) \tag{1}$$

Step 2. Use the condition $\frac{\partial V}{\partial y} = F_2$. We have,

$$\frac{\partial}{\partial y}\left(x^2 y + 5x + g(y, z) \right) = x^2 - 4z$$

$$x^2 + g_y(y, z) = x^2 - 4z \quad \Rightarrow \quad g_y(y, z) = -4z$$

We integrate with respect to y, holding z fixed:

$$g(y, z) = \int -4z\, dy = -4zy + h(z)$$

Combining with (1) gives:

$$V(x, y, z) = x^2 y + 5x - 4zy + h(z) \tag{2}$$

Step 3. Use the condition $\frac{\partial V}{\partial z} = F_3$. We have,

$$\frac{\partial}{\partial z}\left(x^2 y + 5x - 4zy + h(z) \right) = -4y$$

$$-4y + h'(z) = -4y$$

$$h'(z) = 0 \quad \Rightarrow \quad h(z) = c$$

Substituting in (2) we obtain the general potential function:

$$V(x, y, z) = x^2 y + 5x - 4zy + c$$

To compute the line integral we need one of the potential functions. We choose $c = 0$ to obtain the function,

$$V(x, y, z) = x^2 y + 5x - 4zy$$

17. Evaluate

$$\int_{\mathbf{c}} 2xyz\, dx + x^2 z\, dy + x^2 y\, dz$$

over the path $\mathbf{c}(t) = (t^2, \sin(\pi t/4), e^{t^2 - 2t})$ for $0 \le t \le 2$.

SOLUTION A potential function is

$$V(x, y, z) = x^2 yz$$

The path begins at $\mathbf{c}(0) = (0, 0, 1)$ and ends at $\mathbf{c}(2) = (4, 1, 1)$ so the line integral is

$$V(4, 1, 1) - V(0, 0, 1) = 16 - 0 = 16$$

19. A vector field $\mathbf{F}$ and contour lines of a potential function for $\mathbf{F}$ are shown in Figure 17. Calculate the common value of $\int_{\mathcal{C}} \mathbf{F} \cdot d\mathbf{s}$ for the curves shown in Figure 17 oriented in the direction from P to Q.

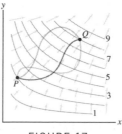

FIGURE 17

SOLUTION

$$\int_{\mathcal{C}} \mathbf{F} \cdot d\mathbf{s} = \int_{\mathcal{C}} \nabla V \cdot d\mathbf{s} = V(Q) - V(P) = 8 - 2 = 6$$

21. Calculate the work expended when a particle is moved from O to Q along segments $\overline{OP}$ and $\overline{PQ}$ in Figure 19 in the presence of the force field $\mathbf{F} = \langle x^2, y^2 \rangle$. How much work is expended moving in a complete circuit around the square?

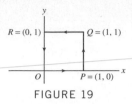

FIGURE 19

SOLUTION

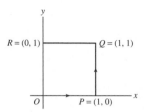

Since $\frac{\partial F_1}{\partial y} = \frac{\partial}{\partial y}(x^2) = 0$ and $\frac{\partial F_2}{\partial x} = \frac{\partial}{\partial x}(y^2) = 0$, we have $\frac{\partial F_1}{\partial y} = \frac{\partial F_2}{\partial x}$. That is, $\mathbf{F}$ satisfies the cross partials condition, therefore $\mathbf{F}$ is conservative. We choose the function $\frac{x^3}{3} + \frac{y^3}{3}$, such that $\mathbf{F}$ is the gradient of the function. The potential energy is, thus, $V = -\frac{x^3}{3} - \frac{y^3}{3}$. The work done against $\mathbf{F}$ is computed by the Fundamental Theorem for Gradient vectors:

$$\text{Work against } \mathbf{F} = -\int_C \mathbf{F} \cdot d\mathbf{s} = V(Q) - V(O) = V(1, 1) - V(0) = -\frac{2}{3} - 0 = -\frac{2}{3}$$

(The negative sign is to be expected, as our force field is actually helping us move along OP and OQ. The line integral of a conservative field along a closed curve is zero, therefore the integral of $\mathbf{F}$ along the complete square is zero, and we get:

$$W = -\int_{OPQR} \mathbf{F} \cdot d\mathbf{s} = 0$$

23. Compute the work W against the earth's gravitational field required to move a satellite of mass $m = 1000$ kg along any path from an orbit of altitude 4000 km to an orbit of altitude 6000 km.

SOLUTION Work against gravity is calculated with the integral

$$W = -\int_C m\mathbf{F} \cdot d\mathbf{s} = 1000 \int_C \nabla V \cdot d\mathbf{s} = 1000(V(r_2) - V(r_1))$$

Since r_1 and r_2 are measured from the center of the earth,

$$r_1 = 4 \times 10^6 + 6.4 \times 10^6 = 10.4 \times 10^6 \text{ meters}$$

$$r_2 = 6 \times 10^6 + 6.4 \times 10^6 = 12.4 \times 10^6 \text{ meters}$$

$$V(r) = -\frac{k}{r} \quad \Rightarrow \quad W = -\frac{1000k}{10^6}\left(\frac{1}{12.4} - \frac{1}{10.4}\right) \approx 6.2 \times 10^9 \text{ J}$$

25. On the surface of the earth, the gravitational field (with z as vertical coordinate measured in meters) is $\mathbf{F} = \langle 0, 0, -g \rangle$.
(a) Find a potential function for $\mathbf{F}$.
(b) Beginning at rest, a ball of mass $m = 2$ kg moves under the influence of gravity (without friction) along a path from $P = (3, 2, 400)$ to $Q = (-21, 40, 50)$. Find the ball's velocity when it reaches Q.

SOLUTION
(a) By inspection $\mathbf{F} = -\nabla V$ for $V(x, y, z) = gz$.
(b) The force of gravity is $m\mathbf{F} = \langle 0, 0, -mg \rangle$, therefore $m\mathbf{F} = -\nabla V$ for $V(x, y, z) = mgz$. The work performed moving the ball from P to Q is the line integral of $m\mathbf{F}$ over the path. Since $m\mathbf{F}$ is conservative, the energy is independent of the path connecting the two points. Using the Fundamental Theorem for Gradient Vector Fields we have:

$$W = -\int_c m\mathbf{F} \cdot d\mathbf{s} = V(-21, 40, 50) - V(3, 2, 400) = 2 \cdot 9.8(50 - 400) = -6860 \text{ joules}$$

By conservation of energy, the kinetic energy of the ball will be 6860 joules, so

$$\frac{mv^2}{2} = 6860 \Rightarrow v = \sqrt{\frac{2 \cdot 6860}{2}} \approx 82.8 \text{ m/s}$$

27. Let $\mathbf{F} = \left\langle \dfrac{-y}{x^2 + y^2}, \dfrac{x}{x^2 + y^2} \right\rangle$ be the vortex field. Determine $\displaystyle\int_c \mathbf{F} \cdot d\mathbf{s}$ for each of the paths in Figure 20.

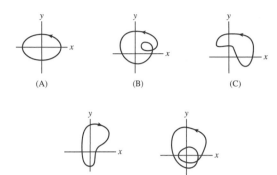

(A)   (B)   (C)

(D)   (E)

FIGURE 20

SOLUTION Since the cross partials of $\mathbf{F}$ are equal, $\mathbf{F}$ has the property,

$$\int_c \mathbf{F} \cdot d\mathbf{s} = 2\pi n$$

where $\mathbf{c}$ is a closed curve not passing through the origin, and n is the number of times $\mathbf{c}$ winds around the origin (n is negative if n winds in the clockwise direction). We use this property to compute the line integrals of $\mathbf{F}$ over the paths in Figure 18:

(A) The path (A) winds around the origin one time in the counterclockwise direction hence the line integral is $2\pi \cdot 1 = 2\pi$.

(B) The point (B) winds around the origin one time in the counterclockwise direction hence the line integral is $2\pi \cdot 1 = 2\pi$.

(C) The path (C) does not encounter the origin, hence the line integral is $2\pi \cdot 0 = 0$. Notice that there exists a simply connected domain $\mathcal{D}$, not including the origin, so that the path c and the region inside $\mathbf{c}$ are in $\mathcal{D}$. Therefore, Theorem 4 applies in $\mathcal{D}$ and $\mathbf{F}$ is a gradient vector in $\mathcal{D}$. Consequently, the line integral of $\mathbf{F}$ over $\mathbf{c}$ is zero.

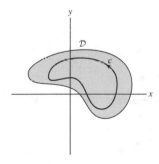

(D) This path winds around the origin one time in the clockwise direction, hence $\int_c \mathbf{F} \cdot d\mathbf{s} = 2\pi \cdot (-1) = -2\pi$.

(E) The path winds around the origin twice in the counterclockwise direction, hence the line integral is $2\pi \cdot 2 = 4\pi$.

Further Insights and Challenges

29. Suppose that $\mathbf{F}$ is defined on $\mathbf{R}^3$ and that $\oint_c \mathbf{F} \cdot d\mathbf{s} = 0$ for all closed paths $\mathbf{c}$ in $\mathbf{R}^3$. Prove:

(a) $\mathbf{F}$ is path-independent; that is, for any two paths $\mathbf{c}_1$ and $\mathbf{c}_2$ in $\mathcal{D}$ with the same initial and terminal points,

$$\int_{c_1} \mathbf{F} \cdot d\mathbf{s} = \int_{c_2} \mathbf{F} \cdot d\mathbf{s}$$

(b) $\mathbf{F}$ is conservative.

SOLUTION

(a) Choose two distinct points P and Q, and let c_1 and c_2 be paths from P to Q. We construct a path from P to P by first using c_1 to reach Q, then using c_2 with its orientation reversed to return to P. (This reversed path is designated $-c_2$.) Such a closed path c can be represented as a difference $c = c_1 - c_2$. (See figure below)

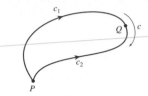

The closed loop c is represented as $c_1 - c_2$.

Thus,

$$\oint_{\mathbf{c}} \mathbf{F} \cdot d\mathbf{s} = \int_{\mathbf{c}_1} \mathbf{F} \cdot d\mathbf{s} + \int_{-\mathbf{c}_2} \mathbf{F} \cdot d\mathbf{s}$$

$$= \int_{\mathbf{c}_1} \mathbf{F} \cdot d\mathbf{s} - \int_{\mathbf{c}_2} \mathbf{F} \cdot d\mathbf{s}$$

Since the problem states that the integral around any closed path is zero, we have

$$\int_{\mathbf{c}_1} \mathbf{F} \cdot d\mathbf{s} - \int_{\mathbf{c}_2} \mathbf{F} \cdot d\mathbf{s} = 0 \quad \Rightarrow \quad \int_{\mathbf{c}_1} \mathbf{F} \cdot d\mathbf{s} = \int_{\mathbf{c}_2} \mathbf{F} \cdot d\mathbf{s}$$

(b) Since $\mathbf{F}$ is defined for all of $\mathbb{R}^3$, it is certainly defined in a simply connected domain $\mathcal{D}$. Since we have just established that $\mathbf{F}$ is also path independent, $\mathbf{F}$ is conservative by Theorem 2.

16.4 Parametrized Surfaces and Surface Integrals (LT Section 17.4)

Preliminary Questions

1. What is the surface integral of the function $f(x, y, z) = 10$ over a surface of total area 5?

SOLUTION Using Surface Integral and Surface Area we have:

$$\iint_S f(x, y, z)\, dS = \iint_{\mathcal{D}} f\left(\Phi(u, v)\right) \|\mathbf{n}(u, v)\|\, du\, dv = \iint_{\mathcal{D}} 10 \|\mathbf{n}(u, v)\|\, du\, dv$$

$$= 10 \iint_{\mathcal{D}} \|\mathbf{n}(u, v)\|\, du\, dv = 10\, \text{Area}(S) = 10 \cdot 5 = 50$$

2. What interpretation can we give to the length $\|\mathbf{n}\|$ of the normal vector for a parametrization $G(u, v)$?

SOLUTION The approximation:

$$\text{Area}\left(S_{ij}\right) \approx \|\mathbf{n}\left(u_{ij}, v_{ij}\right)\| \text{Area}\left(R_{ij}\right)$$

tells that $\|\mathbf{n}\|$ is a distortion factor that indicates how much the area of a small rectangle R_{ij} is altered under the map ϕ.

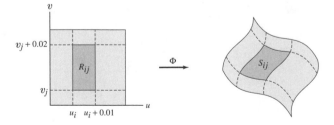

3. A parametrization maps a rectangle of size 0.01×0.02 in the uv-plane onto a small patch $\mathcal{S}$ of a surface. Estimate Area$(\mathcal{S})$ if $\mathbf{T}_u \times \mathbf{T}_v = \langle 1, 2, 2 \rangle$ at a sample point in the rectangle.

SOLUTION We use the estimation

$$\text{Area}(S) \approx \|\mathbf{n}(u, v)\| \text{Area}(R)$$

where $\mathbf{n}(u, v) = \mathbf{T}_u \times \mathbf{T}_v$ at a sample point in R. We get:

$$\text{Area}(S) \approx \| \langle 1, 2, 2 \rangle \| \cdot 0.01 \cdot 0.02 = \sqrt{1^2 + 2^2 + 2^2} \cdot 0.0002 = 0.0006$$

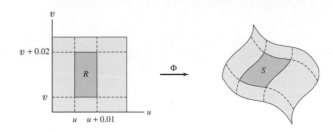

4. A small surface S is divided into three small pieces, each of area 0.2. Estimate $\iint_S f(x, y, z)\, dS$ if $f(x, y, z)$ takes the values 0.9, 1, and 1.1 at sample points in these three pieces.

SOLUTION We use the approximation obtained by the Riemann Sum:

$$\iint_S f(x, y, z)\, dS \approx \sum_{ij} f\left(P_{ij}\right) \text{Area}\left(S_{ij}\right) = 0.9 \cdot 0.2 + 1 \cdot 0.2 + 1.1 \cdot 0.2 = 0.6$$

5. A surface S has a parametrization whose domain is the square $0 \le u, v \le 2$ such that $\|\mathbf{n}(u, v)\| = 5$ for all (u, v). What is Area(S)?

SOLUTION Writing the surface area as a surface integral where $\mathcal{D}$ is the square $[0, 2] \times [0, 2]$ in the uv-plane, we have:

$$\text{Area}(S) = \iint_{\mathcal{D}} \|\mathbf{n}(u, v)\|\, du\, dv = \iint_{\mathcal{D}} 5\, du\, dv = 5 \iint_{\mathcal{D}} 1\, du\, dv = 5\text{Area}(\mathcal{D}) = 5 \cdot 2^2 = 20$$

6. What is the outward-pointing unit normal to the sphere of radius 3 centered at the origin at $P = (2, 2, 1)$?

SOLUTION The outward-pointing normal to the sphere of radius $R = 3$ centered at the origin is the following vector:

$$\langle \cos\theta \sin\phi, \sin\theta \sin\phi, \cos\phi \rangle \tag{1}$$

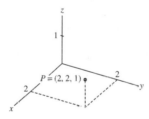

We compute the values in (1) corresponding to $P = (2, 2, 1)$: $x = y = 2, z = 1$ hence $0 \le \theta \le \frac{\pi}{2}$ and $0 < \phi < \frac{\pi}{2}$. We get:

$$\cos\phi = \frac{z}{\rho} = \frac{1}{3} \quad \Rightarrow \quad \sin\phi = \sqrt{1 - \left(\frac{1}{3}\right)^2} = \frac{2\sqrt{2}}{3}$$

$$\cos\theta = \frac{x}{\rho \sin\phi} = \frac{2}{3 \cdot \frac{2\sqrt{2}}{3}} = \frac{1}{\sqrt{2}} \quad \Rightarrow \quad \sin\theta = \sqrt{1 - \frac{1}{2}} = \frac{1}{\sqrt{2}}$$

Substituting in (1) we get the following unit normal:

$$\left\langle \frac{1}{\sqrt{2}} \cdot \frac{2\sqrt{2}}{3}, \frac{1}{\sqrt{2}} \cdot \frac{2\sqrt{2}}{3}, \frac{1}{3} \right\rangle = \left\langle \frac{2}{3}, \frac{2}{3}, \frac{1}{3} \right\rangle$$

Exercises

1. Match each parametrization with the corresponding surface in Figure 16.

(a) $(u, \cos v, \sin v)$

(b) $(u, u + v, v)$

(c) (u, v^3, v)

(d) $(\cos u \sin v, 3 \cos u \sin v, \cos v)$

(e) $(u, u(2 + \cos v), u(2 + \sin v))$

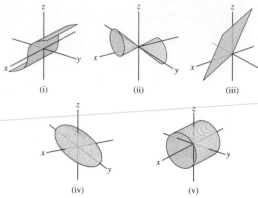

(i)   (ii)   (iii)

(iv)   (v)

FIGURE 16

SOLUTION   (a) = (v), because the y and z coordinates describe a circle with fixed radius.

(b) = (iii), because the coordinates are all linear in u and v.

(c) = (i), because the parametrization gives $y = z^3$.

(d) = (iv), an ellipsoid.

(e) = (ii), because the y and z coordinates describe a circle with varying radius.

3. Show that $G(u, v) = (2u + 1, u - v, 3u + v)$ parametrizes the plane $2x - y - z = 2$. Then:

(a) Calculate $\mathbf{T}_u$, $\mathbf{T}_v$, and $\mathbf{n}(u, v)$.

(b) Find the area of $\mathcal{S} = G(\mathcal{D})$, where $\mathcal{D} = \{(u, v) : 0 \leq u \leq 2, 0 \leq v \leq 1\}$.

(c) Express $f(x, y, z) = yz$ in terms of u and v, and evaluate $\iint_\mathcal{S} f(x, y, z)\,dS$.

SOLUTION   We show that $x = 2u + 1$, $y = u - v$, and $z = 3u + v$ satisfy the equation of the plane,

$$2x - y - z = 2(2u + 1) - (u - v) - (3u + v) = 4u + 2 - u + v - 3u - v = 2$$

Moreover, for any x, y, z satisfying $2x - y - z = z$, there are values of u and v such that $x = 2u + 1$, $y = u - v$, and $z = 3u + v$, since the following equations can be solved for u and v:

$$\begin{aligned} x &= 2u + 1 \\ y &= u - v \\ z &= 3u + v \\ 2x - y - z &= 2 \end{aligned} \quad \Rightarrow \quad u = \frac{x - 1}{2}, \quad v = \frac{x - 1}{2} - y$$

We conclude that $\Phi(u, v)$ parametrizes the whole plane $2x - y - z = 2$.

(a) The tangent vectors $\mathbf{T}_u$ and $\mathbf{T}_v$ are:

$$\mathbf{T}_u = \frac{\partial \phi}{\partial u} = \frac{\partial}{\partial u}(2u + 1, u - v, 3u + v) = \langle 2, 1, 3 \rangle$$

$$\mathbf{T}_v = \frac{\partial \phi}{\partial v} = \frac{\partial}{\partial v}(2u + 1, u - v, 3u + v) = \langle 0, -1, 1 \rangle$$

The normal vector is the following cross product:

$$\mathbf{n}(u, v) = \mathbf{T}_u \times \mathbf{T}_v = \begin{vmatrix} \mathbf{i} & \mathbf{j} & \mathbf{k} \\ 2 & 1 & 3 \\ 0 & -1 & 1 \end{vmatrix} = \begin{vmatrix} 1 & 3 \\ -1 & 1 \end{vmatrix}\mathbf{i} - \begin{vmatrix} 2 & 3 \\ 0 & 1 \end{vmatrix}\mathbf{j} + \begin{vmatrix} 2 & 1 \\ 0 & -1 \end{vmatrix}\mathbf{k}$$

$$= 4\mathbf{i} - 2\mathbf{j} - 2\mathbf{k} = \langle 4, -2, -2 \rangle$$

(b) That area of $S = \Phi(\mathcal{D})$ is the following surface integral:

$$\text{Area}(S) = \iint_\mathcal{D} \|\mathbf{n}(u, v)\|\,du\,dv = \iint_\mathcal{D} \| \langle 4, -2, -2 \rangle \|\,du\,dv = \sqrt{24} \iint_\mathcal{D} 1\,du\,dv$$

$$= \sqrt{24}\,\text{Area}(\mathcal{D}) = \sqrt{24} \cdot 2 \cdot 1 = 4\sqrt{6}$$

(c) We express $f(x, y, z) = yz$ in terms of the parameters u and v:

$$f(\phi(u, v)) = (u - v)(3u + v) = 3u^2 - 2uv - v^2$$

Using the Theorem on Surface Integrals we have:

$$\iint_S f(x, y, z)\, dS = \iint_D f\left(\phi(u, v)\right) \|\mathbf{n}(u, v)\|\, du\, dv = \iint_D \left(3u^2 - 2uv - v^2\right) \|\langle 4, -2, -2\rangle\|\, du\, dv$$

$$= \sqrt{24} \int_0^1 \int_0^2 \left(3u^2 - 2uv - v^2\right) du\, dv = \sqrt{24} \int_0^1 \left(u^3 - u^2 v - v^2 u\right)\Big|_{u=0}^2 dv$$

$$= \sqrt{24} \int_0^1 \left(8 - 4v - 2v^2\right) dv = \sqrt{24}\left(8v - 2v^2 - \frac{2}{3}v^3\right)\Big|_0^1 = \frac{32\sqrt{6}}{3}$$

5. Let $G(x, y) = (x, y, xy)$.

(a) Calculate $\mathbf{T}_x$, $\mathbf{T}_y$, and $\mathbf{n}(x, y)$.

(b) Let S be the part of the surface with parameter domain $\mathcal{D} = \{(x, y) : x^2 + y^2 \le 1, x \ge 0, y \ge 0\}$. Verify the following formula and evaluate using polar coordinates:

$$\iint_S 1\, dS = \iint_D \sqrt{1 + x^2 + y^2}\, dx\, dy$$

(c) Verify the following formula and evaluate:

$$\iint_S z\, dS = \int_0^{\pi/2} \int_0^1 (\sin\theta \cos\theta) r^3 \sqrt{1 + r^2}\, dr\, d\theta$$

SOLUTION

(a) The tangent vectors are:

$$\mathbf{T}_x = \frac{\partial\phi}{\partial x} = \frac{\partial}{\partial x}(x, y, xy) = \langle 1, 0, y\rangle$$

$$\mathbf{T}_y = \frac{\partial\phi}{\partial y} = \frac{\partial}{\partial y}(x, y, xy) = \langle 0, 1, x\rangle$$

The normal vector is the cross product:

$$\mathbf{n}(x, y) = \mathbf{T}_x \times \mathbf{T}_y = \begin{vmatrix} \mathbf{i} & \mathbf{j} & \mathbf{k} \\ 1 & 0 & y \\ 0 & 1 & x \end{vmatrix} = \begin{vmatrix} 0 & y \\ 1 & x \end{vmatrix}\mathbf{i} - \begin{vmatrix} 1 & y \\ 0 & x \end{vmatrix}\mathbf{j} + \begin{vmatrix} 1 & 0 \\ 0 & 1 \end{vmatrix}\mathbf{k}$$

$$= -y\mathbf{i} - x\mathbf{j} + \mathbf{k} = \langle -y, -x, 1\rangle$$

(b) Using the Theorem on evaluating surface integrals we have:

$$\iint_S 1\, dS = \iint_D \|\mathbf{n}(x, y)\|\, dx\, dy = \iint_D \|\langle -y, -x, 1\rangle\|\, dx\, dy = \iint_D \sqrt{y^2 + x^2 + 1}\, dx\, dy$$

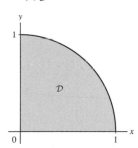

We convert the integral to polar coordinates $x = r\cos\theta$, $y = r\sin\theta$. The new region of integration is:

$$0 \le r \le 1, \quad 0 \le \theta \le \frac{\pi}{2}.$$

We get:

$$\iint_S 1\, dS = \int_0^{\pi/2} \int_0^1 \sqrt{r^2 + 1} \cdot r\, dr\, d\theta = \int_0^{\pi/2}\left(\int_0^1 \sqrt{r^2 + 1} \cdot r\, dr\right) d\theta$$

$$= \int_0^{\pi/2}\left(\int_1^2 \frac{\sqrt{u}}{2}\, du\right) d\theta = \int_0^{\pi/2} \frac{2\sqrt{2} - 1}{3}\, d\theta = \frac{\left(2\sqrt{2} - 1\right)\pi}{6}$$

(c) The function z expressed in terms of the parameters x, y is $f\left(\Phi(x, y)\right) = xy$. Therefore,

$$\iint_S z\, dS = \iint_D xy \cdot \|\mathbf{n}(x, y)\|\, dx\, dy = \iint_D xy\sqrt{1 + x^2 + y^2}\, dx\, dy$$

We compute the double integral by converting it to polar coordinates. We get:

$$\iint_S z \, dS = \int_0^{\pi/2} \int_0^1 (r\cos\theta)(r\sin\theta)\sqrt{1+r^2} \cdot r \, dr \, d\theta = \int_0^{\pi/2} \int_0^1 (\sin\theta\cos\theta) r^3 \sqrt{1+r^2} \, dr \, d\theta$$

$$= \left(\int_0^{\pi/2} (\sin\theta\cos\theta) \, d\theta \right) \left(\int_0^1 r^3\sqrt{1+r^2} \, dr \right) \tag{1}$$

We compute each integral in (1). Using the substitution $u = 1 + r^2$, $du = 2r \, dr$ we get:

$$\int_0^1 r^3\sqrt{1+r^2} \, dr = \int_0^1 r^2\sqrt{1+r^2} \cdot r \, dr = \int_1^2 \left(u^{3/2} - u^{1/2} \right) \frac{du}{2} = \frac{u^{5/2}}{5} - \frac{u^{3/2}}{3} \Big|_1^2 = \frac{2\left(\sqrt{2}+1\right)}{15}$$

Also,

$$\int_0^{\pi/2} \sin\theta\cos\theta \, d\theta = \int_0^{\pi/2} \frac{\sin 2\theta}{2} \, d\theta = -\frac{\cos 2\theta}{4} \Big|_0^{\pi/2} = \frac{1}{2}$$

We substitute the integrals in (1) to obtain the following solution:

$$\iint_S z \, dS = \frac{1}{2} \cdot \frac{2\left(\sqrt{2}+1\right)}{15} = \frac{\sqrt{2}+1}{15}$$

In Exercises 7–10, calculate $\mathbf{T}_u$, $\mathbf{T}_v$, and $\mathbf{n}(u, v)$ for the parametrized surface at the given point. Then find the equation of the tangent plane to the surface at that point.

7. $G(u, v) = (2u + v, u - 4v, 3u);$ $u = 1,$ $v = 4$

SOLUTION The tangent vectors are the following vectors,

$$\mathbf{T}_u = \frac{\partial \Phi}{\partial u} = \frac{\partial}{\partial u}(2u + v, u - 4v, 3u) = \langle 2, 1, 3 \rangle$$

$$\mathbf{T}_v = \frac{\partial \Phi}{\partial v} = \frac{\partial}{\partial v}(2u + v, u - 4v, 3u) = \langle 1, -4, 0 \rangle$$

The normal is the cross product:

$$\mathbf{n}(u, v) = \mathbf{T}_u \times \mathbf{T}_v = \begin{vmatrix} \mathbf{i} & \mathbf{j} & \mathbf{k} \\ 2 & 1 & 3 \\ 1 & -4 & 0 \end{vmatrix} = \begin{vmatrix} 1 & 3 \\ -4 & 0 \end{vmatrix} \mathbf{i} - \begin{vmatrix} 2 & 3 \\ 1 & 0 \end{vmatrix} \mathbf{j} + \begin{vmatrix} 2 & 1 \\ 1 & -4 \end{vmatrix} \mathbf{k}$$

$$= 12\mathbf{i} + 3\mathbf{j} - 9\mathbf{k} = 3\langle 4, 1, -3 \rangle$$

The equation of the plane passing through the point $P : \Phi(1, 4) = (6, -15, 3)$ with the normal vector $\langle 4, 1, -3 \rangle$ is:

$$\langle x - 6, y + 15, z - 3 \rangle \cdot \langle 4, 1, -3 \rangle = 0$$

or

$$4(x - 6) + y + 15 - 3(z - 3) = 0$$

$$4x + y - 3z = 0$$

9. $G(\theta, \phi) = (\cos\theta\sin\phi, \sin\theta\sin\phi, \cos\phi);$ $\theta = \frac{\pi}{2},$ $\phi = \frac{\pi}{4}$

SOLUTION We compute the tangent vectors:

$$\mathbf{T}_\theta = \frac{\partial \Phi}{\partial \theta} = \frac{\partial}{\partial \theta}(\cos\theta\sin\phi, \sin\theta\sin\phi, \cos\phi) = \langle -\sin\theta\sin\phi, \cos\theta\sin\phi, 0 \rangle$$

$$\mathbf{T}_\phi = \frac{\partial \Phi}{\partial \phi} = \frac{\partial}{\partial \phi}(\cos\theta\sin\phi, \sin\theta\sin\phi, \cos\phi) = \langle \cos\theta\cos\phi, \sin\theta\cos\phi, -\sin\phi \rangle$$

The normal vector is the cross product:

$$\mathbf{n}(\theta, \phi) = \mathbf{T}_\theta \times \mathbf{T}_\phi = \begin{vmatrix} \mathbf{i} & \mathbf{j} & \mathbf{k} \\ -\sin\theta\sin\phi & \cos\theta\sin\phi & 0 \\ \cos\theta\cos\phi & \sin\theta\cos\phi & -\sin\phi \end{vmatrix}$$

$$= \left(-\cos\theta\sin^2\phi \right)\mathbf{i} - \left(\sin\theta\sin^2\phi \right)\mathbf{j} + \left(-\sin^2\theta\sin\phi\cos\phi - \cos^2\theta\cos\phi\sin\phi \right)\mathbf{k}$$

$$= -\left(\cos\theta\sin^2\phi \right)\mathbf{i} - \left(\sin\theta\sin^2\phi \right)\mathbf{j} - (\sin\phi\cos\phi)\mathbf{k}$$

The tangency point and the normal at this point are,

$$P = \Phi\left(\frac{\pi}{2}, \frac{\pi}{4}\right) = \left(\cos\frac{\pi}{2}\sin\frac{\pi}{4}, \sin\frac{\pi}{2}\sin\frac{\pi}{4}, \cos\frac{\pi}{4}\right) = \left(0, \frac{\sqrt{2}}{2}, \frac{\sqrt{2}}{2}\right)$$

$$\mathbf{n}\left(\frac{\pi}{2}, \frac{\pi}{4}\right) = -\frac{1}{2}\mathbf{j} - \frac{1}{2}\mathbf{k} = -\frac{1}{2}(\mathbf{j} + \mathbf{k}) = -\frac{1}{2}\langle 0, 1, 1\rangle$$

The equation of the plane orthogonal to the vector $\langle 0, 1, 1\rangle$ and passing through $P = \left(0, \frac{\sqrt{2}}{2}, \frac{\sqrt{2}}{2}\right)$ is:

$$\left\langle x, y - \frac{\sqrt{2}}{2}, z - \frac{\sqrt{2}}{2}\right\rangle \cdot \langle 0, 1, 1\rangle = 0$$

or

$$y - \frac{\sqrt{2}}{2} + z - \frac{\sqrt{2}}{2} = 0$$

$$y + z = \sqrt{2}$$

11. Use the normal vector computed in Exercise 8 to estimate the area of the small patch of the surface $G(u, v) = (u^2 - v^2, u + v, u - v)$ defined by

$$2 \le u \le 2.1, \qquad 3 \le v \le 3.2$$

SOLUTION   We denote the rectangle $\mathcal{D} = \{(u, v) : 2 \le u \le 2.1, 3 \le v \le 3.2\}$. Using the sample point corresponding to $u = 2$, $v = 3$ we obtain the following estimation for the area of $S = \Phi(\mathcal{D})$:

$$\text{Area}(S) \approx \|\mathbf{n}(2, 3)\|\text{Area}(\mathcal{D}) = \|\mathbf{n}(2, 3)\| \cdot 0.1 \cdot 0.2 = 0.02\|\mathbf{n}(2, 3)\| \tag{1}$$

In Exercise 8 we found that $\mathbf{n}(2, 3) = 2\langle -1, -1, 5\rangle$. Therefore,

$$\|\mathbf{n}(2, 3)\| = 2\sqrt{1^2 + 1^2 + 5^2} = 2\sqrt{27}$$

Substituting in (1) gives the following estimation:

$$\text{Area}(S) \approx 0.02 \cdot 2 \cdot \sqrt{27} \approx 0.2078.$$

In Exercises 13–26, calculate $\displaystyle\iint_S f(x, y, z)\, dS$ *for the given surface and function.*

13. $G(u, v) = (u\cos v, u\sin v, u)$, $0 \le u \le 1$, $0 \le v \le 1$; $f(x, y, z) = z(x^2 + y^2)$

SOLUTION

Step 1. Compute the tangent and normal vectors. We have:

$$\mathbf{T}_u = \frac{\partial\Phi}{\partial u} = \frac{\partial}{\partial u}(u\cos v, u\sin v, u) = \langle\cos v, \sin v, 1\rangle$$

$$\mathbf{T}_v = \frac{\partial\Phi}{\partial v} = \frac{\partial}{\partial v}(u\cos v, u\sin v, u) = \langle -u\sin v, u\cos v, 0\rangle$$

The normal vector is the cross product:

$$\mathbf{n} = \mathbf{T}_u \times \mathbf{T}_v = \begin{vmatrix} \mathbf{i} & \mathbf{j} & \mathbf{k} \\ \cos v & \sin v & 1 \\ -u\sin v & u\cos v & 0 \end{vmatrix}$$

$$= (-u\cos v)\mathbf{i} - (u\sin v)\mathbf{j} + \left(u\cos^2 v + u\sin^2 v\right)\mathbf{k}$$

$$= (-u\cos v)\mathbf{i} - (u\sin v)\mathbf{j} + u\mathbf{k} = \langle -u\cos v, -u\sin v, u\rangle$$

We compute the length of $\mathbf{n}$:

$$\|\mathbf{n}\| = \sqrt{(-u\cos v)^2 + (-u\sin v)^2 + u^2} = \sqrt{u^2\left(\cos^2 v + \sin^2 v + 1\right)} = \sqrt{u^2 \cdot 2} = \sqrt{2}|u| = \sqrt{2}u$$

Notice that in the region of integration $u \ge 0$, therefore $|u| = u$.

Step 2. Calculate the surface integral. We express the function $f(x, y, z) = z\left(x^2 + y^2\right)$ in terms of the parameters u, v:

$$f(\Phi, (u, v)) = u\left(u^2\cos^2 v + u^2\sin^2 v\right) = u \cdot u^2 = u^3$$

We obtain the following integral:

$$\iint_S f(x, y, z) \, dS = \int_0^1 \int_0^1 f(\Phi, (u, v)) \, \|\mathbf{n}\| \, du \, dv = \int_0^1 \int_0^1 u^3 \cdot \sqrt{2} u \, du \, dv$$

$$= \left(\int_0^1 \sqrt{2} \, dv \right) \left(\int_0^1 u^4 \, du \right) = \sqrt{2} \cdot \frac{u^5}{5} \bigg|_0^1 = \frac{\sqrt{2}}{5}$$

15. $y = 9 - z^2, \quad 0 \le x \le 3, 0 \le z \le 3; \qquad f(x, y, z) = z$

SOLUTION We use the formula for the surface integral over a graph $y = g(x, z)$:

$$\iint_S f(x, y, z) \, dS = \iint_{\mathcal{D}} f(x, g(x, z), z) \sqrt{1 + g_x^2 + g_z^2} \, dx \, dz \qquad (1)$$

Since $y = g(x, z) = 9 - z^2$, we have $g_x = 0$, $g_z = -2z$, hence:

$$\sqrt{1 + g_x^2 + g_z^2} = \sqrt{1 + 4z^2}$$

$$f(x, g(x, z), z) = z$$

The domain of integration is the square $[0, 3] \times [0, 3]$ in the xz-plane. By (1) we get:

$$\iint_S f(x, y, z) \, dS = \int_0^3 \int_0^3 z\sqrt{1 + 4z^2} \, dz \, dx = \left(\int_0^3 1 \, dx \right) \left(\int_0^3 z\sqrt{1 + 4z^2} \, dz \right) = 3 \int_0^3 z\sqrt{1 + 4z^2} \, dz$$

We use the substitution $u = 1 + 4z^2$, $du = 8z \, dz$ to compute the integral. This gives:

$$\iint_S f(x, y, z) \, dS = 3 \int_0^3 z\sqrt{1 + 4z^2} \, dz = 3 \int_1^{37} \frac{u^{1/2}}{8} \, du = \frac{37\sqrt{37} - 1}{4} \approx 56$$

17. $x^2 + y^2 + z^2 = 1, \quad x, y, z \ge 0; \quad f(x, y, z) = x^2$.

SOLUTION The octant of the unit sphere centered at the origin, where $x, y, z \ge 0$ has the following parametrization in spherical coordinates:

$$\Phi(\theta, \phi) = (\cos\theta \sin\phi, \sin\theta \sin\phi, \cos\phi), \quad 0 \le \theta \le \frac{\pi}{2}, \quad 0 \le \phi \le \frac{\pi}{2}$$

The length of the normal vector is:

$$\|\mathbf{n}\| = \sin\phi$$

The function x^2 expressed in terms of the parameters is $\cos^2\theta \sin^2\phi$. Using the theorem on computing surface integrals we obtain,

$$\iint_S x^2 \, dS = \int_0^{\pi/2} \int_0^{\pi/2} \left(\cos^2\theta \sin^2\phi \right) (\sin\phi) \, d\phi \, d\theta = \int_0^{\pi/2} \int_0^{\pi/2} \cos^2\theta \sin^3\phi \, d\phi \, d\theta$$

$$= \left(\int_0^{\pi/2} \cos^2\theta \, d\theta \right) \left(\int_0^{\pi/2} \sin^3\phi \, d\phi \right) = \left(\frac{\theta}{2} + \frac{\sin 2\theta}{4} \right) \bigg|_{\theta=0}^{\pi/2} \cdot \left(-\frac{\sin^2\phi \cos\phi}{3} - \frac{2}{3}\cos\phi \right) \bigg|_{\phi=0}^{\pi/2}$$

$$= \frac{\pi}{4} \cdot \frac{2}{3} = \frac{\pi}{6}$$

19. $x^2 + y^2 = 4, \quad 0 \le z \le 4; \qquad f(x, y, z) = e^{-z}$

SOLUTION The cylinder has the following parametrization in cylindrical coordinates:

$$\Phi(\theta, z) = (2\cos\theta, 2\sin\theta, z), \quad 0 \le \theta \le 2\pi, 0 \le z \le 4$$

Step 1. Compute the tangent and normal vectors. The tangent vectors are the partial derivatives:

$$\mathbf{T}_\theta = \frac{\partial \Phi}{\partial \theta} = \frac{\partial}{\partial \theta} (2\cos\theta, 2\sin\theta, z) = \langle -2\sin\theta, 2\cos\theta, 0 \rangle$$

$$\mathbf{T}_z = \frac{\partial}{\partial z} (2\cos\theta, 2\sin\theta, z) = \langle 0, 0, 1 \rangle$$

The normal vector is their cross product:

$$\mathbf{n}(\theta, z) = \mathbf{T}_\theta \times \mathbf{T}_z = \begin{vmatrix} \mathbf{i} & \mathbf{j} & \mathbf{k} \\ -2\sin\theta & 2\cos\theta & 0 \\ 0 & 0 & 1 \end{vmatrix} = (2\cos\theta)\mathbf{i} + (2\sin\theta)\mathbf{j} = \langle 2\cos\theta, 2\sin\theta, 0 \rangle$$

The length of the normal vector is thus

$$\|\mathbf{n}(\theta, z)\| = \sqrt{(2\cos\theta)^2 + (2\sin\theta)^2 + 0} = \sqrt{4\left(\cos^2\theta + \sin^2\theta\right)} = \sqrt{4} = 2$$

Step 2. Calculate the surface integral. The surface integral equals the following double integral:

$$\iint_S f(x, y, z)\, dS = \iint_D f\left(\Phi(\theta, z)\right) \|\mathbf{n}\|\, d\theta\, dz = \int_0^{2\pi} \int_0^4 e^{-z} \cdot 2\, d\theta\, dz$$

$$= \left(\int_0^{2\pi} 2\, d\theta\right)\left(\int_0^4 e^{-z}\, dz\right) = 4\pi \cdot (-e^{-z})\Big|_0^4 = 4\pi\left(1 - e^{-4}\right)$$

21. Part of the plane $x + y + z = 1$, where $x, y, z \ge 0$; $\quad f(x, y, z) = z$

SOLUTION We let $z = g(x, y) = 1 - x - y$ and use the formula for the surface integral over the graph of $z = g(x, y)$, where $\mathcal{D}$ is the parameter domain in the xy-plane. That is:

$$\iint_S f(x, y, z)\, dS = \iint_{\mathcal{D}} f\left(x, y, g(x, y)\right) \sqrt{1 + g_x^2 + g_y^2}\, dx\, dy \tag{1}$$

We have, $g_x = -1$ and $g_y = -1$ therefore:

$$\sqrt{1 + g_x^2 + g_y^2} = \sqrt{1 + (-1)^2 + (-1)^2} = \sqrt{3}$$

We express the function $f(x, y, z) = z$ in terms of the parameters x and y:

$$f\left(x, y, g(x, y)\right) = z = 1 - x - y$$

The domain of integration is the triangle $\mathcal{D}$ in the xy-plane shown in the figure.

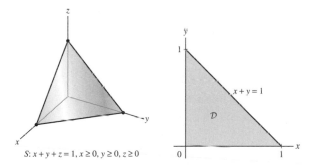

$S: x + y + z = 1, x \ge 0, y \ge 0, z \ge 0$

By (1) we get:

$$\iint_S f(x, y, z)\, dS = \int_0^1 \int_0^{1-y} (1 - x - y)\sqrt{3}\, dx\, dy = \sqrt{3} \int_0^1 x - \frac{x^2}{2} - yx \Big|_{x=0}^{1-y}\, dy$$

$$= \sqrt{3} \int_0^1 \left((1 - y)^2 - \frac{(1 - y)^2}{2}\right) dy = \frac{\sqrt{3}}{2} \int_0^1 \left(1 - 2y + y^2\right) dy$$

$$= \frac{\sqrt{3}}{2}\left(y - y^2 + \frac{y^3}{3}\right)\Big|_0^1 = \frac{\sqrt{3}}{6}$$

23. $x^2 + y^2 + z^2 = 4, 1 \le z \le 2;$ $\quad f(x, y, z) = z^2(x^2 + y^2 + z^2)^{-1}$

SOLUTION We use spherical coordinates to parametrize the cap S.

$$\Phi(\theta, \phi) = (2\cos\theta\sin\phi, 2\sin\theta\sin\phi, 2\cos\phi)$$

$$\mathcal{D} : 0 \le \theta \le 2\pi, 0 \le \phi \le \phi_0$$

The angle ϕ_0 is determined by $\cos \phi_0 = \frac{1}{2}$, that is, $\phi_0 = \frac{\pi}{3}$. The length of the normal vector in spherical coordinates is:

$$\|\mathbf{n}\| = R^2 \sin \phi = 4 \sin \phi$$

We express the function $f(x, y, z) = z^2 \left(x^2 + y^2 + z^2\right)^{-1}$ in terms of the parameters:

$$f\left(\Phi(\theta, \phi)\right) = (2 \cos \phi)^2 4^{-1} = \cos^2 \phi$$

Using the theorem on computing the surface integral we get:

$$\iint_S f(x, y, z) \, dS = \iint_{\mathcal{D}} f\left(\Phi(\theta, \phi)\right) \|\mathbf{n}\| \, d\phi \, d\theta = \int_0^{2\pi} \int_0^{\pi/3} \left(\cos^2 \phi\right) \cdot 4 \sin \phi \, d\phi \, d\theta$$

$$= \left(\int_0^{2\pi} 4 \, d\theta\right) \left(\int_0^{\pi/3} \cos^2 \phi \sin \phi \, d\phi\right) = 8\pi \left(-\frac{\cos \phi}{3}\right) \Big|_0^{\pi/3}$$

$$= \frac{8\pi}{3} \left(-\left(\frac{1}{2}\right)^3 - (-1)\right) = \frac{8\pi}{3} \cdot \frac{7}{8} = \frac{7\pi}{3}$$

25. Part of the surface $z = x^3$, where $0 \le x \le 1, 0 \le y \le 1$; $f(x, y, z) = z$

SOLUTION   Use the formula for the surface integral over the graph of $z = g(x, y)$. We have, $g_x = 3x^2$ and $g_y = 0$ therefore:

$$\|\mathbf{n}\| = \sqrt{1 + g_x^2 + g_y^2} = \sqrt{1 + (3x^2)^2 + (0)^2} = \sqrt{1 + 9x^4}$$

The integral then is

$$\iint_S z \, dS = \int_0^1 \int_0^1 x^3 \sqrt{1 + 9x^4} \, dx \, dy = \left(\int_0^1 dy\right) \left(\int_0^1 x^3 \sqrt{1 + 9x^4} \, dx\right)$$

Substituting $u = 1 + 9x^4$, $du = 36x^3 \, dx$

$$= 1 \cdot \int_1^{10} u^{\frac{1}{2}} \frac{du}{36} = \frac{1}{36} \cdot \frac{2}{3} \cdot u^{\frac{3}{2}} \Big|_1^{10} = \frac{1}{54}(10\sqrt{10} - 1)$$

27. A surface S has a parametrization $G(u, v)$ with domain $0 \le u \le 2, 0 \le v \le 4$ such that the following partial derivatives are constant:

$$\frac{\partial G}{\partial u} = \langle 2, 0, 1 \rangle, \qquad \frac{\partial G}{\partial v} = \langle 4, 0, 3 \rangle$$

What is the surface area of S?

SOLUTION   Since the partial derivatives are constant, the normal vector is also constant. We find it by computing the cross product:

$$\mathbf{n} = \mathbf{T}_u \times \mathbf{T}_v = \frac{\partial \Phi}{\partial u} \times \frac{\partial \Phi}{\partial v} = \begin{vmatrix} \mathbf{i} & \mathbf{j} & \mathbf{k} \\ 2 & 0 & 1 \\ 4 & 0 & 3 \end{vmatrix} = -2\mathbf{j} = \langle 0, -2, 0 \rangle \quad \Rightarrow \quad \|\mathbf{n}\| = 2$$

We denote the rectangle $\mathcal{D} = \{(u, v) : 0 \le u \le 2, \ 0 \le v \le 4\}$, and use the surface area to compute the area of $S = \Phi(\mathcal{D})$. We obtain:

$$\text{Area}(S) = \iint_{\mathcal{D}} \|\mathbf{n}\| \, du \, dv = \iint_{\mathcal{D}} 2 \, du \, dv = 2 \iint_{\mathcal{D}} 1 \, du \, dv = 2 \cdot \text{Area}(\mathcal{D}) = 2 \cdot 2 \cdot 4 = 16$$

29. Calculate $\iint_S (xy + e^z) \, dS$, where S is the triangle in Figure 18 with vertices $(0, 0, 3)$, $(1, 0, 2)$, and $(0, 4, 1)$.

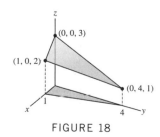

FIGURE 18

SOLUTION We find the equation of the plane through the points $A = (0, 0, 3)$, $B = (0, 4, 1)$ and $C = (1, 0, 2)$.

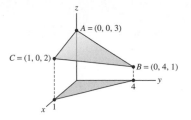

A normal to the plane is the cross product:

$$\overrightarrow{AB} \times \overrightarrow{AC} = \langle 0, 4, -2 \rangle \times \langle 1, 0, -1 \rangle = \begin{vmatrix} \mathbf{i} & \mathbf{j} & \mathbf{k} \\ 0 & 4 & -2 \\ 1 & 0 & -1 \end{vmatrix} = -4\mathbf{i} - 2\mathbf{j} - 4\mathbf{k} = -2 \langle 2, 1, 2 \rangle$$

The equation of the plane passing through $A = (0, 0, 3)$ and perpendicular to the vector $\langle 2, 1, 2 \rangle$ is:

$$\langle x - 0, y - 0, z - 3 \rangle \cdot \langle 2, 1, 2 \rangle = 0$$

$$2x + y + 2(z - 3) = 0$$

$$2x + y + 2z = 6$$

or

$$z - g(x, y) = -x - \frac{1}{2}y + 3$$

We compute the surface integral of $f(x, y, z) = xy + e^z$ over the triangle ABC using the formula for the surface integral over a graph. The parameter domain $\mathcal{D}$ is the projection of the triangle ABC onto the xy-plane (see figure).

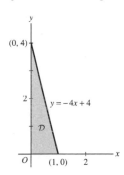

We have:

$$\iint_S f(x, y, z)\, dS = \iint_{\mathcal{D}} f\big(x, y, g(x, y)\big) \sqrt{1 + g_x^2 + g_z^2}\, dx\, dy \tag{1}$$

We compute the functions in the integrand. Since $z = g(x, y) = -x - \frac{y}{2} + 3$, we have:

$$g_x = -1,\ g_y = -\frac{1}{2} \quad \Rightarrow \quad \sqrt{1 + g_x^2 + g_z^2} = \sqrt{1 + (-1)^2 + \left(-\frac{1}{2}\right)^2} = \frac{3}{2}$$

$$f\big(x, y, g(x, y, z)\big) = xy + e^z = xy + e^{-x - \frac{y}{2} + 3}$$

Substituting in (1) gives:

$$\iint_S f(x, y, z)\, dS = \int_0^1 \int_0^{-4x+4} \left(xy + e^{-x - y/2 + 3} \right) \cdot \frac{3}{2}\, dy\, dx = \int_0^1 \left. \frac{3xy^2}{4} - 3e^{-x - y/2 + 3} \right|_{y=0}^{-4x+4} dx$$

$$= \int_0^1 \left(\frac{3x(-4x + 4)^2}{4} - 3e^{-x - (-4x+4)/2 + 3} + 3e^{-x+3} \right) dx$$

$$= \int_0^1 \left(12x^3 - 24x^2 + 12x - 3e^{x+1} + 3e^{-x+3} \right) dx = \left. 3x^4 - 8x^3 + 6x^2 - 3e^{x+1} - 3e^{-x+3} \right|_0^1$$

$$= (1 - 3e^2 - 3e^2) - (-3e - 3e^3) = 3e^3 - 6e^2 + 3e + 1 \approx 25.08$$

31. Use cylindrical coordinates to compute the surface area of a sphere of radius R.

SOLUTION As $z = \pm\sqrt{R^2 - (x^2 + y^2)}$ we may parametrize the upper hemisphere by the map

$$G(r, \theta) = (r\cos\theta, r\sin\theta, \sqrt{R^2 - r^2})$$

To compute the surface area of the hemisphere S, we first must find the tangent vectors and the normal vector. That is,

$$\mathbf{T}_r = \frac{\partial G}{\partial r} = \frac{\partial}{\partial r}\left\langle r\cos\theta, r\sin\theta, \sqrt{R^2 - r^2}\right\rangle = \left\langle\cos\theta, \sin\theta, -\frac{r}{\sqrt{R^2 - r^2}}\right\rangle$$

$$\mathbf{T}_\theta = \frac{\partial G}{\partial\theta} = \frac{\partial}{\partial\theta}\left\langle r\cos\theta, r\sin\theta, \sqrt{R^2 - r^2}\right\rangle = \langle -r\sin\theta, r\cos\theta, 0\rangle$$

The normal vector is the cross product:

$$\mathbf{n} = \mathbf{T}_r \times \mathbf{T}_\theta = \begin{vmatrix} \mathbf{i} & \mathbf{j} & \mathbf{k} \\ \cos\theta & \sin\theta & \frac{-r}{\sqrt{R^2-r^2}} \\ -r\sin\theta & r\cos\theta & 0 \end{vmatrix}$$

$$= \left(\frac{r^2\cos\theta}{\sqrt{R^2 - r^2}}\right)\mathbf{i} + \left(\frac{r^2\sin\theta}{\sqrt{R^2 - r^2}}\right)\mathbf{j} + \left(r\cos^2\theta + r\sin^2\theta\right)\mathbf{k}$$

$$= \left(\frac{r^2\cos\theta}{\sqrt{R^2 - r^2}}\right)\mathbf{i} + \left(\frac{r^2\sin\theta}{\sqrt{R^2 - r^2}}\right)\mathbf{j} + r\mathbf{k}$$

The length of the normal vector is thus

$$\|\mathbf{n}\| = \sqrt{\frac{r^4\cos^2\theta}{R^2 - r^2} + \frac{r^4\sin^2\theta}{R^2 - r^2} + r^2} = \sqrt{\frac{r^4}{R^2 - r^2}\left(\cos^2\theta + \sin^2\theta\right) + r^2} = \sqrt{\frac{r^4}{R^2 - r^2} + r^2} = \frac{rR}{\sqrt{R^2 - r^2}}$$

We now compute the surface area as the following surface integral:

$$\text{Area}(S) = \iint_{\mathcal{D}} \|\mathbf{n}\|\, dr\, d\theta = \int_0^{2\pi}\int_0^R \frac{rR}{\sqrt{R^2 - r^2}}\, dr\, d\theta$$

$$= \left(\int_0^{2\pi} R\, d\theta\right)\left(\int_0^R \frac{r}{\sqrt{R^2 - r^2}}\, dr\right) = 2\pi R\int_0^R \frac{r}{\sqrt{R^2 - r^2}}\, dr$$

We compute the integral using the substitution $t = R^2 - r^2$, $dt = -2r\, dr$. We get:

$$\text{Area}(S) = 2\pi R\int_{R^2}^0 \frac{-1}{2t^{1/2}}\, dt = 2\pi R^2$$

The area of the entire sphere is twice this or $4\pi R^2$.

33. $\boxed{\textbf{CAS}}$ Let S be the surface $z = \ln(5 - x^2 - y^2)$ for $0 \le x \le 1, 0 \le y \le 1$. Using a computer algebra system:

(a) Calculate the surface area of S to four decimal places.

(b) Calculate $\iint_S x^2y^3\, dS$ to four decimal places.

SOLUTION

(a) Using that $z_x = -2x/(5 - x^2 - y^2)$ and $z_y = -2y/(5 - x^2 - y^2)$, we calculate $\|n\|$ to be

$$\|n\| = \sqrt{1 + (z_x)^2 + (z_y)^2} = \frac{\sqrt{(5 - x^2 - y^2)^2 + 4x^2 + 4y^2}}{5 - x^2 - y^2}$$

Thus, the surface area is

$$\text{Area}(S) = \int_0^1\int_0^1 \frac{\sqrt{(5 - x^2 - y^2)^2 + 4x^2 + 4y^2}}{5 - x^2 - y^2}\, dx\, dy \approx 1.078$$

(b) We calculate $\iint_S x^2y^3\, dS$ as follows:

$$\iint_S x^2y^3\, dS = \int_0^1\int_0^1 x^2y^3\frac{\sqrt{(5 - x^2 - y^2)^2 + 4x^2 + 4y^2}}{5 - x^2 - y^2}\, dx\, dy \approx 0.09814$$

35. What is the area of the portion of the plane $2x + 3y + 4z = 28$ lying above the domain $\mathcal{D}$ in the xy-plane in Figure 19 if Area$(\mathcal{D}) = 5$?

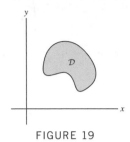

FIGURE 19

SOLUTION We rewrite the equation of the plane as:

$$z = g(x, y) = -\frac{x}{2} - \frac{3}{4}y + 7$$

Hence:

$$\sqrt{1 + g_x^2 + g_y^2} = \sqrt{1 + \left(-\frac{1}{2}\right)^2 + \left(-\frac{3}{4}\right)^2} = \frac{\sqrt{29}}{4}$$

We use the integral for surface area and the surface integral over a graph to write:

$$\text{Area}(S) = \iint_S 1 \, dS = \iint_{\mathcal{D}} \sqrt{1 + g_x^2 + g_y^2} \, dx \, dy = \iint_{\mathcal{D}} \frac{\sqrt{29}}{4} \, dx \, dy$$

$$= \frac{\sqrt{29}}{4} \iint_{\mathcal{D}} 1 \, dx \, dy = \frac{\sqrt{29}}{4} \text{Area}(\mathcal{D}) = \frac{\sqrt{29}}{4} \cdot 5 = \frac{5\sqrt{29}}{4} \approx 6.73$$

37. Find the surface area of the portion S of the cone $z^2 = x^2 + y^2$, where $z \geq 0$, contained within the cylinder $y^2 + z^2 \leq 1$.

SOLUTION We rewrite the equation of the cone as $x = \pm\sqrt{z^2 - y^2}$. The projection of the cone onto the yz-plane is obtained by setting $x = 0$ in the equation of the cone, that is,

$$x = 0 = \sqrt{z^2 - y^2} \quad \Rightarrow \quad z = \pm y$$

Since on S, $z > 0$, we get $z = |y|$. We conclude that the projection of the upper part of the cone $x^2 + y^2 = z^2$ onto the yz-plane is the region between the lines $z = y$ and $z = -y$ on the upper part of the yz-plane. Therefore, the projection $\mathcal{D}$ of S onto the yz-plane is the region shown in the figure:

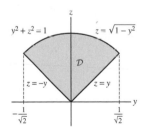

There are two identical portions of the surface parametrized by this region—one for $x \geq 0$, and one for $x \leq 0$. Therefore the area of S is twice the integral over the domain $\mathcal{D}$:

$$\text{Area}(S) = \iint_S dS = 2 \iint_{\mathcal{D}} \sqrt{1 + g_y^2 + g_z^2} \, dy \, dz$$

We compute the integral using a surface integral over a graph. Since $x = g(y, z) = \pm\sqrt{z^2 - y^2}$ we have,

$$g_z = \pm\frac{z}{\sqrt{z^2 - y^2}}, \quad g_y = \pm\frac{y}{\sqrt{z^2 - y^2}}$$

Hence, (notice that $z \geq 0$ on S):

$$\sqrt{1 + g_y^2 + g_z^2} = \sqrt{1 + \frac{z^2}{z^2 - y^2} + \frac{y^2}{z^2 - y^2}} = \sqrt{\frac{2z^2}{z^2 - y^2}} = \frac{z\sqrt{2}}{\sqrt{z^2 - y^2}}$$

We obtain the following integral:

$$\text{Area}(S) = 2 \iint_{\mathcal{D}} \sqrt{1 + g_y^2 + g_z^2} \, dy \, dz = 2 \iint_{\mathcal{D}} \frac{z\sqrt{2}}{\sqrt{z^2 - y^2}} \, dz \, dy$$

Using symmetry gives:

$$\text{Area}(S) = 4 \int_0^{1/(\sqrt{2})} \int_y^{\sqrt{1-y^2}} \frac{z\sqrt{2}}{\sqrt{z^2 - y^2}} \, dz \, dy = 4\sqrt{2} \int_0^{1/(\sqrt{2})} \left(\int_y^{\sqrt{1-y^2}} \frac{z \, dz}{\sqrt{z^2 - y^2}} \right) dy \tag{1}$$

We compute the inner integral using the substitution $u = \sqrt{z^2 - y^2}$, $du = \frac{z}{u} \, dz$. We get:

$$\int_y^{\sqrt{1-y^2}} \frac{z \, dz}{\sqrt{z^2 - y^2}} = \int_0^{\sqrt{1-y^2}} \frac{u \, du}{u} = \int_0^{\sqrt{1-2y^2}} du = \sqrt{1 - 2y^2}$$

We substitute in (1) and compute the resulting integral using the substitution $t = \sqrt{2}y$. We get:

$$\text{Area}(S) = 4\sqrt{2} \int_0^{1/(\sqrt{2})} \sqrt{1 - 2y^2} \, dy = 4\sqrt{2} \int_0^1 \sqrt{1 - t^2} \frac{dt}{\sqrt{2}} = 4 \int_0^1 \sqrt{1 - t^2} \, dt = 4 \cdot \frac{\pi}{4} = \pi$$

39. Calculate $\iint_G x^2 z \, dS$, where G is the cylinder (including the top and bottom) $x^2 + y^2 = 4$, $0 \leq z \leq 3$.

SOLUTION We calculate the surface integral for each of the three surfaces. We begin with the bottom.

$$S_1 : \phi(x, y) = (x, y, 0)$$

$$\iint_{S_1} x^2 z \, dS_1 = \iint_{\mathcal{D}} x^2(0) \|\mathbf{n}_1\| \, dx \, dy = 0$$

Then the top

$$S_2 : \phi(x, y) = (x, y, 3)$$

$$\mathbf{T}_x = \langle 1, 0, 0 \rangle, \mathbf{T}_y \langle 0, 1, 0 \rangle \Rightarrow \mathbf{n}_2 = \langle 0, 0, 1 \rangle$$

$$\iint_{S_2} x^2 z \, dS_2 = \iint_{\mathcal{D}} x^2(3) \|\mathbf{n}_2\| \, dx \, dy = 3 \iint_{\mathcal{D}} x^2 \, dx \, dy$$

The domain, $\mathcal{D}$, is the disk of radius 2. Changing to polar coordinates,

$$= 3 \int_0^{2\pi} \int_0^2 (r \cos \theta)^2 \, r \, dr \, d\theta = 3 \int_0^{2\pi} \cos^2 \theta \, d\theta \cdot \int_0^2 r^3 \, dr$$

$$= 3 \left(\frac{1}{2} + \frac{\sin 2\theta}{2} \right) \Big|_0^{2\pi} \cdot \frac{r^4}{4} \Big|_0^2 = 12\pi$$

Finally the side,

$$S_3 : \phi(r, \theta) = (2 \cos \theta, 2 \sin \theta, z)$$

$$\mathbf{T}_\theta = \langle -2 \sin \theta, 2 \cos \theta, 0 \rangle, \mathbf{T}_z \langle 0, 0, 1 \rangle$$

$$\Rightarrow \quad \mathbf{n}_3 = \langle 2 \cos \theta, 2 \sin \theta, 0 \rangle \Rightarrow \quad \|\mathbf{n}_3\| = 2$$

$$\iint_{S_3} x^2 z \, dS_2 = \int_0^{2\pi} \int_0^3 (2 \cos \theta)^2 \, z \, 2 \, dz \, d\theta$$

$$= 8 \int_0^{2\pi} \cos^2 \theta \, d\theta \cdot \int_0^3 z \, dz = 8 \left(\frac{1}{2} + \frac{\sin 2\theta}{2} \right) \Big|_0^{2\pi} \cdot \frac{z^2}{2} \Big|_0^3 = 36\pi$$

The total surface integral is thus

$$\iint_G x^2 z \, dS = 0 + 12\pi + 36\pi = 48\pi$$

41. Prove a famous result of Archimedes: The surface area of the portion of the sphere of radius R between two horizontal planes $z = a$ and $z = b$ is equal to the surface area of the corresponding portion of the circumscribed cylinder (Figure 22).

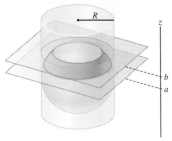

FIGURE 22

SOLUTION We compute the area of the portion of the sphere between the planes a and b. The portion S_1 of the sphere has the parametrization,

$$\Phi(\theta, \phi) = (r \cos \theta \sin \phi, r \sin \theta \sin \phi, r \cos \phi)$$

where,

$$\mathcal{D}_1 : 0 \le \theta \le 2\pi, \ \phi_0 \le \phi \le \phi_1$$

If we assume $0 < a < b$, then the angles ϕ_0 and ϕ_1 are determined by,

$$\cos \phi_0 = \frac{b}{r} \quad \Rightarrow \quad \phi_0 = \cos^{-1} \frac{b}{r}$$

$$\cos \phi_1 = \frac{a}{r} \quad \Rightarrow \quad \phi_1 = \cos^{-1} \frac{a}{r}$$

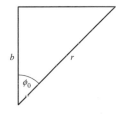 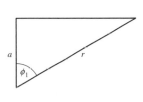

The length of the normal vector is $\|\mathbf{n}\| = r^2 \sin \phi$. We obtain the following integral:

$$\text{Area} (S_1) = \iint_{\mathcal{D}_1} \|\mathbf{n}\| d\phi \, d\theta = \int_0^{2\pi} \int_{\phi_0}^{\phi_1} r^2 \sin \phi \, d\phi \, d\theta = \left(\int_0^{2\pi} r^2 d\theta \right) \left(\int_{\phi_1}^{\phi_2} \sin \phi \, d\phi \right)$$

$$= 2\pi r^2 \left(- \cos \phi \Big|_{\phi = \cos^{-1} \frac{b}{r}}^{\cos^{-1} \frac{a}{r}} \right) = 2\pi r^2 \left(-\frac{a}{r} + \frac{b}{r} \right) = 2\pi r (b - a)$$

The area of the part S_2 of the cylinder of radius r between the planes $z = a$ and $z = b$ is:

$$\text{Area} (S_2) = 2\pi r \cdot (b - a)$$

We see that the two areas are equal:

$$\text{Area} (S_1) = \text{Area} (S_2)$$

Further Insights and Challenges

43. Use Eq. (14) to compute the surface area of $z = 4 - y^2$ for $0 \le y \le 2$ rotated about the z-axis.

SOLUTION Since $g(y) = 4 - y^2$, we have $g'(y) = -2y$. By Eq. (14) we obtain the following integral,

$$\text{Area}(S) = 2\pi \int_0^2 |y| \sqrt{1 + (-2y)^2} \, dy = 2\pi \int_0^2 y \cdot \sqrt{1 + 4y^2} \, dy$$

We compute the integral using the substitution $u = 1 + 4y^2$, $du = 8y \, dy$. We get:

$$\text{Area}(S) = 2\pi \int_1^{17} u^{1/2} \cdot \frac{du}{8} = 2\pi \frac{2}{3} \cdot \frac{u^{3/2}}{8} \Big|_1^{17} = \frac{\pi}{6} \left(17\sqrt{17} - 1 \right) \approx 36.18$$

45. Area of a Torus Let T be the torus obtained by rotating the circle in the yz-plane of radius a centered at $(0, b, 0)$ about the z-axis (Figure 24). We assume that $b > a > 0$.

(a) Use Eq. (14) to show that

$$\text{Area(T)} = 4\pi \int_{b-a}^{b+a} \frac{ay}{\sqrt{a^2 - (b-y)^2}}\, dy$$

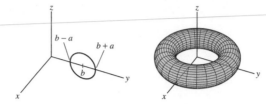

FIGURE 24 The torus obtained by rotating a circle of radius a.

(b) Show that $\text{Area(T)} = 4\pi^2 ab$.

SOLUTION

(a) Using symmetry, the area of the surface obtained by rotating the upper part of the circle is half the area of the torus.

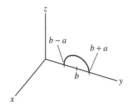

The rotated graph is $z = g(y) = \sqrt{a^2 - (y-b)^2}$, $b - a \le y \le b + a$. So, we have,

$$g'(y) = \frac{-2(y-b)}{2\sqrt{a^2 - (y-b)^2}} = -\frac{y-b}{\sqrt{a^2 - (y-b)^2}}$$

$$\sqrt{1 + g'(y)^2} = \sqrt{1 + \frac{(y-b)^2}{a^2 - (y-b)^2}} = \sqrt{\frac{a^2 - (y-b)^2 + (y-b)^2}{a^2 - (y-b)^2}} = \frac{a}{\sqrt{a^2 - (y-b)^2}}$$

We now use symmetry and Eq. (14) to obtain the following area of the torus (we assume that $b - a > 0$, hence $y > 0$):

$$\text{Area (T)} = 2 \cdot 2\pi \int_{b-a}^{b+a} |y|\sqrt{1 + g'(y)^2}\, dy = 4\pi \int_{b-a}^{b+a} \frac{ay}{\sqrt{a^2 - (y-b)^2}}\, dy \tag{1}$$

(b) We compute the integral using the substitution $u = \frac{y-b}{a}$, $du = \frac{1}{a}\, dy$. We get:

$$\int_{b-a}^{b+a} \frac{ay}{\sqrt{a^2 - (y-b)^2}}\, dy = \int_{-1}^{1} \frac{a^2 u + ab}{\sqrt{a^2 - a^2 u^2}} a\, du = \int_{-1}^{1} \frac{a^2 u + ab}{\sqrt{1 - u^2}}\, du = \int_{-1}^{1} \frac{a^2 u}{\sqrt{1 - u^2}}\, du + \int_{-1}^{1} \frac{ab}{\sqrt{1 - u^2}}\, du$$

The first integral is zero since the integrand is an odd function. We get:

$$\int_{b-a}^{b+a} \frac{ay}{\sqrt{a^2 - (y-b)^2}}\, dy = 2 \int_{0}^{1} \frac{ab}{\sqrt{1 - u^2}}\, du = 2ab \sin^{-1} u \Big|_{0}^{1} = 2ab \left(\frac{\pi}{2} - 0\right) = \pi ab$$

Substituting in (1) gives the following area:

$$\text{Area (T)} = 4\pi \cdot \pi ab = 4\pi^2 ab$$

47. Compute the surface area of the torus in Exercise 45 using Pappus's Theorem.

SOLUTION The generating curve is the circle of radius a in the (y, z)-plane centered at the point $(0, b, 0)$. The length of the generating curve is $L = \pi a$.

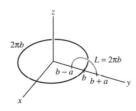

The center of mass of the circle is at the center $(\bar{y}, \bar{z}) = (b, 0)$, and it traverses a circle of radius b centered at the origin. Therefore, the center of mass makes a distance of $2\pi b$. Using Pappus' Theorem, the area of the torus is:

$$L \cdot 2\pi a = 2\pi a \cdot 2\pi b = 4\pi^2 ab.$$

49. Calculate the gravitational potential V for a hemisphere of radius R with uniform mass distribution.

SOLUTION In Exercise 48(b) we expressed the potential φ for a sphere of radius R. To find the potential for a hemisphere of radius R, we need only to modify the limits of the angle ϕ to $0 \le \phi \le \frac{\pi}{2}$. This gives the following integral:

$$\varphi(0, 0, r) = \varphi(r) = -\frac{Gm}{4\pi} \int_0^{\pi/2} \int_0^{2\pi} \frac{\sin\phi \, d\theta \, d\phi}{\sqrt{R^2 + r^2 - 2Rr\cos\phi}} = -\frac{Gm}{4\pi} \cdot 2\pi \int_0^{\pi/2} \frac{\sin\phi \, d\phi}{\sqrt{R^2 + r^2 - 2Rr\cos\phi}}$$

$$= -\frac{Gm}{2} \int_0^{\pi/2} \frac{\sin\phi \, d\phi}{\sqrt{R^2 + r^2 - 2Rr\cos\phi}}$$

We compute the integral using the substitution $u = R^2 + r^2 - 2Rr\cos\phi$, $du = 2Rr\sin\phi \, d\phi$. We obtain:

$$\varphi(r) = -\frac{Gm}{2} \int_{(R-r)^2}^{R^2+r^2} \frac{\frac{du}{2Rr}}{\sqrt{u}} = -\frac{Gm}{4Rr} \int_{(R-r)^2}^{R^2+r^2} u^{-1/2} \, du = -\frac{Gm}{4Rr} \cdot 2u^{1/2} \Big|_{u=(R-r)^2}^{R^2+r^2}$$

$$= -\frac{Gm}{2Rr} \left(\left(R^2 + r^2\right)^{1/2} - \left((R-r)^2\right)^{1/2} \right) = -\frac{Gm}{2Rr} \left(\sqrt{R^2 + r^2} - |R - r| \right)$$

51. Let S be the part of the graph $z = g(x, y)$ lying over a domain $\mathcal{D}$ in the xy-plane. Let $\phi = \phi(x, y)$ be the angle between the normal to S and the vertical. Prove the formula

$$\text{Area}(S) = \iint_{\mathcal{D}} \frac{dA}{|\cos\phi|}$$

SOLUTION

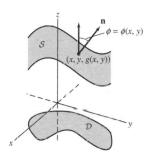

Using the Surface Integral over a Graph we have:

$$\text{Area}(S) = \iint_S 1 \, dS = \iint_{\mathcal{D}} \sqrt{1 + g_x^2 + g_y^2} \, dA \qquad (1)$$

In parametrizing the surface by $\phi(x, y) = (x, y, g(x, y))$, $(x, y) = \mathcal{D}$, we have:

$$\mathbf{T}_x = \frac{\partial \Phi}{\partial x} = \langle 1, 0, g_x \rangle$$

$$\mathbf{T}_y = \frac{\partial \Phi}{\partial y} = \langle 0, 1, g_y \rangle$$

Hence,

$$\mathbf{n} = \mathbf{T}_x \times \mathbf{T}_y = \begin{vmatrix} \mathbf{i} & \mathbf{j} & \mathbf{k} \\ 1 & 0 & g_x \\ 0 & 1 & g_y \end{vmatrix} = -g_x\mathbf{i} - g_y\mathbf{j} + \mathbf{k} = \langle -g_x, -g_y, 1 \rangle$$

$$\|\mathbf{n}\| = \sqrt{g_x^2 + g_y^2 + 1}$$

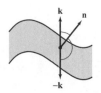

There are two adjacent angles between the normal **n** and the vertical, and the cosines of these angles are opposite numbers. Therefore we take the absolute value of $\cos\phi$ to obtain a positive value for Area(S). Using the Formula for the cosine of the angle between two vectors we get:

$$|\cos\phi| = \frac{|\mathbf{n}\cdot\mathbf{k}|}{\|\mathbf{n}\|\|\mathbf{k}\|} = \frac{|\langle -g_x, -g_y, 1\rangle \cdot \langle 0, 0, 1\rangle|}{\sqrt{1 + g_x^2 + g_y^2}\cdot 1} = \frac{1}{\sqrt{1 + g_x^2 + g_y^2}}$$

Substituting in (1) we get:

$$\text{Area}(S) = \iint_{\mathcal{D}} \frac{dA}{|\cos\phi|}$$

16.5 Surface Integrals of Vector Fields   (LT Section 17.5)

Preliminary Questions

1. Let **F** be a vector field and $G(u, v)$ a parametrization of a surface S, and set $\mathbf{n} = \mathbf{T}_u \times \mathbf{T}_v$. Which of the following is the normal component of **F**?

(a) $\mathbf{F}\cdot\mathbf{n}$ **(b)** $\mathbf{F}\cdot\mathbf{e_n}$

SOLUTION   The normal component of **F** is $\mathbf{F}\cdot\mathbf{e}_n$ rather than $\mathbf{F}\cdot\mathbf{n}$.

2. The vector surface integral $\iint_S \mathbf{F}\cdot d\mathbf{S}$ is equal to the scalar surface integral of the function (choose the correct answer):

(a) $\|\mathbf{F}\|$

(b) $\mathbf{F}\cdot\mathbf{n}$, where **n** is a normal vector

(c) $\mathbf{F}\cdot\mathbf{e_n}$, where $\mathbf{e_n}$ is the unit normal vector

SOLUTION   The vector surface integral $\iint_S \mathbf{F}\cdot d\mathbf{S}$ is defined as the scalar surface integral of the normal component of **F** on the oriented surface. That is, $\iint_S \mathbf{F}\cdot d\mathbf{S} = \iint_S (\mathbf{F}\cdot\mathbf{e}_n)\,dS$ as stated in (c).

3. $\iint_S \mathbf{F}\cdot d\mathbf{S}$ is zero if (choose the correct answer):

(a) **F** is tangent to S at every point.

(b) **F** is perpendicular to S at every point.

SOLUTION   Since $\iint_S \mathbf{F}\cdot d\mathbf{S}$ is equal to the scalar surface integral of the normal component of **F** on S, this integral is zero when the normal component is zero at every point, that is, when **F** is tangent to S at every point as stated in (a).

4. If $\mathbf{F}(P) = \mathbf{e_n}(P)$ at each point on S, then $\iint_S \mathbf{F}\cdot d\mathbf{S}$ is equal to (choose the correct answer):

(a) Zero   **(b)** Area(S)   **(c)** Neither

SOLUTION   If $\mathbf{F}(P) = \mathbf{e}_n(P)$ at each point on S, then,:

$$\iint_S \mathbf{F}\cdot d\mathbf{S} = \iint_S (\mathbf{e}_n\cdot\mathbf{e}_n)\,dS = \iint_S \|\mathbf{e}_n\|^2\,dS = \iint_S 1\,dS = \text{Area}(S)$$

Therefore, (b) is the correct answer.

5. Let S be the disk $x^2 + y^2 \le 1$ in the xy-plane oriented with normal in the positive z-direction. Determine $\iint_S \mathbf{F}\cdot d\mathbf{S}$ for each of the following vector constant fields:

(a) $\mathbf{F} = \langle 1, 0, 0\rangle$ **(b)** $\mathbf{F} = \langle 0, 0, 1\rangle$ **(c)** $\mathbf{F} = \langle 1, 1, 1\rangle$

SOLUTION   The unit normal vector to the oriented disk is $\mathbf{e}_n = \langle 0, 0, 1\rangle$.

(a) Since $\mathbf{F}\cdot\mathbf{e}_n = \langle 1, 0, 0\rangle\cdot\langle 0, 0, 1\rangle = 0$, **F** is perpendicular to the unit normal vector at every point on S, therefore $\iint_S \mathbf{F}\cdot d\mathbf{S} = 0$.

(b) Since $\mathbf{F} = \mathbf{e}_n$ at every point on S, we have:

$$\iint_S \mathbf{F}\cdot d\mathbf{S} = \iint_S (\mathbf{e}_n\cdot\mathbf{e}_n)\,dS = \iint_S \|\mathbf{e}_n\|^2\,dS = \iint_S 1\,dS = \text{Area}(S) = \pi$$

(c) For $\mathbf{F} = \langle 1, 1, 1\rangle$ we have:

$$\iint_S \mathbf{F}\cdot d\mathbf{S} = \iint_S (\mathbf{F}\cdot\mathbf{e}_n)\,dS = \iint_S \langle 1, 1, 1\rangle\cdot\langle 0, 0, 1\rangle\,dS = \iint_S 1\,dS = \text{Area}(S) = \pi$$

6. Estimate $\iint_S \mathbf{F} \cdot d\mathbf{S}$, where S is a tiny oriented surface of area 0.05 and the value of $\mathbf{F}$ at a sample point in S is a vector of length 2 making an angle $\frac{\pi}{4}$ with the normal to the surface.

SOLUTION

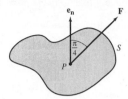

Since S is a tiny surface, we may assume that the dot product $\mathbf{F} \cdot \mathbf{e}_n$ on S is equal to the dot product at the sample point. This gives the following approximation:

$$\iint_S \mathbf{F} \cdot d\mathbf{S} = \iint_S (\mathbf{F} \cdot \mathbf{e}_n) \, dS \approx \iint_S (\mathbf{F}(P) \cdot \mathbf{e}_n(P)) \, dS = \mathbf{F}(P) \cdot \mathbf{e}_n(P) \iint_S 1 dS = \mathbf{F}(P) \cdot \mathbf{e}_n \text{Area}(S)$$

That is,

$$\iint_S \mathbf{F} \cdot d\mathbf{S} \approx \mathbf{F}(P) \cdot \mathbf{e}_n(P) \text{Area}(S) \tag{1}$$

We are given that Area$(S) = 0.05$. We compute the dot product:

$$\mathbf{F}(P) \cdot \mathbf{e}_n(P) = \|\mathbf{F}(P)\|\|\mathbf{e}_n(P)\| \cos \frac{\pi}{4} = 2 \cdot 1 \cdot \frac{1}{\sqrt{2}} = \sqrt{2}$$

Combining with (1) gives the following estimation:

$$\iint_S \mathbf{F} \cdot d\mathbf{S} \approx 0.05\sqrt{2} \approx 0.0707.$$

7. A small surface S is divided into three pieces of area 0.2. Estimate $\iint_S \mathbf{F} \cdot d\mathbf{S}$ if $\mathbf{F}$ is a unit vector field making angles of $85°$, $90°$, and $95°$ with the normal at sample points in these three pieces.

SOLUTION

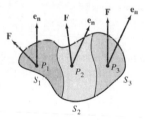

We estimate the vector surface integral by the following sum:

$$\iint_S \mathbf{F} \cdot d\mathbf{S} = \mathbf{F}(P_1) \cdot \mathbf{e}_n(P_1) \text{Area}(S_1) + \mathbf{F}(P_2) \cdot \mathbf{e}_n(P_2) \text{Area}(S_2) + \mathbf{F}(P_3) \cdot \mathbf{e}_n(P_3) \text{Area}(S_3)$$

$$= 0.2 \left(\mathbf{F}(P_1) \cdot \mathbf{e}_n(P_1) + \mathbf{F}(P_2) \cdot \mathbf{e}_n(P_2) + \mathbf{F}(P_3) \cdot \mathbf{e}_n(P_3) \right)$$

We compute the dot product. Since $\mathbf{F}$ and $\mathbf{e}_n$ are unit vectors, we have:

$$\mathbf{F}(P_1) \cdot \mathbf{e}_n(P_1) = \cos 85° \approx 0.0872$$

$$\mathbf{F}(P_2) \cdot \mathbf{e}_n(P_2) = \cos 90° = 0$$

$$\mathbf{F}(P_3) \cdot \mathbf{e}_n(P_3) = \cos 95° \approx -0.0872$$

Substituting gives the following estimation:

$$\iint_S \mathbf{F} \cdot d\mathbf{S} \approx 0.2(0.0872 + 0 - 0.0872) = 0.$$

Exercises

1. Let $\mathbf{F} = \langle z, 0, y \rangle$ and let S be the oriented surface parametrized by $G(u, v) = (u^2 - v, u, v^2)$ for $0 \le u \le 2$, $-1 \le v \le 4$. Calculate:

(a) $\mathbf{n}$ and $\mathbf{F} \cdot \mathbf{n}$ as functions of u and v

(b) The normal component of $\mathbf{F}$ to the surface at $P = (3, 2, 1) = G(2, 1)$

(c) $\iint_S \mathbf{F} \cdot d\mathbf{S}$

SOLUTION

(a) The tangent vectors are,

$$\mathbf{T}_u = \frac{\partial G}{\partial u} = \frac{\partial}{\partial u}\left(u^2 - v, u, v^2\right) = \langle 2u, 1, 0 \rangle$$

$$\mathbf{T}_v = \frac{\partial G}{\partial v} = \frac{\partial}{\partial v}\left(u^2 - v, u, v^2\right) = \langle -1, 0, 2v \rangle$$

The normal vector is their cross product:

$$\mathbf{n} = \mathbf{T}_u \times \mathbf{T}_v = \begin{vmatrix} \mathbf{i} & \mathbf{j} & \mathbf{k} \\ 2u & 1 & 0 \\ -1 & 0 & 2v \end{vmatrix} = v\mathbf{i} - 4uv\mathbf{j} + \mathbf{k} = \langle 2v, -4uv, 1 \rangle$$

We write $\mathbf{F} = \langle z, 0, y \rangle$ in terms of the parameters $x = u^2 - v$, $y = u$, $z = v^2$ and then compute $\mathbf{F} \cdot \mathbf{n}$:

$$\mathbf{F}\left(\Phi(u, v)\right) = \langle z, 0, y \rangle = \left\langle v^2, 0, u \right\rangle$$

$$\mathbf{F}\left(\Phi(u, v)\right) \cdot \mathbf{n}(u, v) = \left\langle v^2, 0, u \right\rangle \cdot \langle 2v, -4uv, 1 \rangle$$

$$= 2v^3 + u$$

(b) At the point $P = (3, 2, 1) = \Phi(2, 1)$ we have:

$$\mathbf{F}(P) = \langle 1, 0, 2 \rangle$$

$$\mathbf{n}(P) = \langle 2, -8, 1 \rangle$$

$$\mathbf{e}_n(P) = \frac{\mathbf{n}(P)}{\|\mathbf{n}(P)\|} = \frac{\langle 2, -8, 1 \rangle}{\sqrt{4 + 64 + 1}} = \frac{1}{\sqrt{69}}\langle 2, -8, 1 \rangle$$

Hence, the normal component of $\mathbf{F}$ to the surface at P is the dot product:

$$\mathbf{F}(P) \cdot \mathbf{e}_n(P) = \langle 1, 0, 2 \rangle \cdot \frac{1}{\sqrt{69}}\langle 2, -8, 1 \rangle = \frac{4}{\sqrt{69}}$$

(c) Using the definition of the vector surface integral and the dot product in part (a), we have:

$$\iint_S \mathbf{F} \cdot d\mathbf{S} = \iint_D \mathbf{F}\left(\phi(u, v)\right) \cdot \mathbf{n}(u, v)\, du\, dv = \int_0^2 \int_{-1}^4 \left(2v^3 + u\right) dv\, du$$

$$= \int_0^2 \frac{2v^4}{4} + uv \Big|_{v=-1}^4 du$$

$$= \int_0^2 \left(128 - \frac{1}{2}\right) + (4 - (-1))u\, du$$

$$= \int_0^2 \frac{255}{2} + 5u\, du = \frac{255u}{2} + \frac{5u^2}{2}\Big|_0^2 = 265$$

3. Let S be the unit square in the xy-plane shown in Figure 14, oriented with the normal pointing in the positive z-direction. Estimate

$$\iint_S \mathbf{F} \cdot d\mathbf{S}$$

where $\mathbf{F}$ is a vector field whose values at the labeled points are

$$\mathbf{F}(A) = \langle 2, 6, 4 \rangle, \qquad \mathbf{F}(B) = \langle 1, 1, 7 \rangle$$

$$\mathbf{F}(C) = \langle 3, 3, -3 \rangle, \qquad \mathbf{F}(D) = \langle 0, 1, 8 \rangle$$

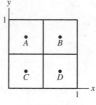

FIGURE 14

SOLUTION The unit normal vector to S is $\mathbf{e}_n = \langle 0, 0, 1 \rangle$. We estimate the vector surface integral $\iint_S \mathbf{F} \cdot d\mathbf{S}$ using the division and sample points given in Figure 14.

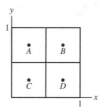

Each subsquare has area $\frac{1}{4}$, therefore we obtain the following estimation:

$$\iint_S \mathbf{F} \cdot d\mathbf{S} \approx (\mathbf{F}(A) \cdot \mathbf{e}_n + \mathbf{F}(B) \cdot \mathbf{e}_n + \mathbf{F}(C) \cdot \mathbf{e}_n + \mathbf{F}(D) \cdot \mathbf{e}_n) \cdot \frac{1}{4}$$

$$= (\langle 2, 6, 4 \rangle \cdot \langle 0, 0, 1 \rangle + \langle 1, 1, 7 \rangle \cdot \langle 0, 0, 1 \rangle + \langle 3, 3, -3 \rangle \cdot \langle 0, 0, 1 \rangle + \langle 0, 1, 8 \rangle \cdot \langle 0, 0, 1 \rangle) \cdot \frac{1}{4}$$

$$= (4 + 7 - 3 + 8) \cdot \frac{1}{4} = 4$$

In Exercises 5–17, compute $\iint_S \mathbf{F} \cdot d\mathbf{S}$ *for the given oriented surface.*

5. $\mathbf{F} = \langle y, z, x \rangle$, plane $3x - 4y + z = 1, 0 \le x \le 1, 0 \le y \le 1$, upward-pointing normal

SOLUTION We rewrite the equation of the plane as $z = 1 - 3x + 4y$, and parametrize the plane by:

$$\Phi(x, y) = (x, y, 1 - 3x + 4y)$$

Here, the parameter domain is the square $\mathcal{D} = \{(x, y) : 0 \le x, y \le 1\}$ in the xy-plane.

Step 1. Compute the tangent and normal vectors.

$$\mathbf{T}_x = \frac{\partial \Phi}{\partial x} = \frac{\partial}{\partial x}(x, y, 1 - 3x + 4y) = \langle 1, 0, -3 \rangle$$

$$\mathbf{T}_y = \frac{\partial \Phi}{\partial y} = \frac{\partial}{\partial y}(x, y, 1 - 3x + 4y) = \langle 0, 1, 4 \rangle$$

$$\mathbf{T}_x \times \mathbf{T}_y = \begin{vmatrix} \mathbf{i} & \mathbf{j} & \mathbf{k} \\ 1 & 0 & -3 \\ 0 & 1 & 4 \end{vmatrix} = 3\mathbf{i} - 4\mathbf{j} + \mathbf{k} = \langle 3, -4, 1 \rangle$$

Since the plane is oriented with upward pointing normal, the normal vector $\mathbf{n}$ is:

$$\mathbf{n} = \langle 3, -4, 1 \rangle$$

Step 2. Evaluate the dot product $\mathbf{F} \cdot \mathbf{n}$. We write $\mathbf{F}$ in terms of the parameters:

$$\mathbf{F}(\Phi(x, y)) = \langle y, z, x \rangle = \langle y, 1 - 3x + 4y, x \rangle$$

The dot product $\mathbf{F} \cdot \mathbf{n}$ is thus

$$\mathbf{F}(\Phi(x, y)) \cdot \mathbf{n} = \langle y, 1 - 3x + 4y, x \rangle \cdot \langle 3, -4, 1 \rangle = 3y - 4(1 - 3x + 4y) + x = 13x - 13y - 4$$

Step 3. Evaluate the surface integral. The surface integral is equal to the following double integral:

$$\iint_S \mathbf{F} \cdot d\mathbf{S} = \iint_{\mathcal{D}} \mathbf{F}(\Phi(x, y)) \cdot \mathbf{n}(x, y) \, dx \, dy = \int_0^1 \int_0^1 (13x - 13y - 4) \, dx \, dy$$

$$= \int_0^1 \frac{13x^2}{2} - 13yx - 4x \Big|_{x=0}^1 \, dy = \int_0^1 \left(\frac{13}{2} - 13y - 4 \right) dy = \frac{5y}{2} - \frac{13y^2}{2} \Big|_0^1 = -4$$

7. $\mathbf{F} = \langle 0, 3, x \rangle$, part of sphere $x^2 + y^2 + z^2 = 9$, where $x \geq 0, y \geq 0, z \geq 0$ outward-pointing normal

SOLUTION We parametrize the octant S by:

$$\Phi(\theta, \phi) = (3\cos\theta\sin\phi, 3\sin\theta\sin\phi, 3\cos\phi), \ 0 \leq \theta \leq \frac{\pi}{2}, \ 0 \leq \phi \leq \frac{\pi}{2}$$

Step 1. Compute the normal vector. As seen in the text, the normal vector that points to the outside of the sphere is:

$$\mathbf{n} = \mathbf{T}_\phi \times \mathbf{T}_\theta = 9\sin\phi \, \langle \cos\theta\sin\phi, \sin\theta\sin\phi, \cos\phi \rangle$$

For $0 \leq \theta \leq \frac{\pi}{2}$, $0 \leq \phi \leq \frac{\pi}{2}$, all trigonometric functions are positive. Therefore all components of $\mathbf{n}$ are positive, so $\mathbf{n}$ points to the outside of the sphere.

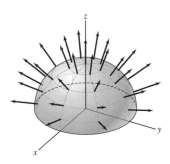

Step 2. Evaluate the dot product $\mathbf{F} \cdot \mathbf{n}$. We express the vector field in terms of the parameters:

$$\mathbf{F}(\Phi(\theta, \phi)) = \langle 0, 3, x \rangle = \langle 0, 3, 3\cos\theta\sin\phi \rangle$$

Hence:

$$\mathbf{F}(\Phi(\theta, \phi)) \cdot \mathbf{n}(\theta, \phi) = \langle 0, 3, 3\cos\theta\sin\phi \rangle \cdot 9\sin\phi \, \langle \cos\theta\sin\phi, \sin\theta\sin\phi, \cos\phi \rangle$$
$$= 27\sin\theta\sin^2\phi + 27\cos\theta\sin^2\phi\cos\phi$$

Step 3. Evaluate the surface integral. The surface integral is equal to the following double integral:

$$\iint_S \mathbf{F} \cdot d\mathbf{S} = \iint_{\mathcal{D}} \mathbf{F}(\Phi(\theta, \phi)) \cdot \mathbf{n}(\theta, \phi) \, d\theta \, d\phi$$

$$= \int_0^{\frac{\pi}{2}} \int_0^{\frac{\pi}{2}} \left(27\sin\theta\sin^2\phi + 27\cos\theta\sin^2\phi\cos\phi \right) d\theta \, d\phi$$

$$= 27 \left(\int_0^{\frac{\pi}{2}} \sin\theta \, d\theta \cdot \int_0^{\frac{\pi}{2}} \sin^2\phi \, d\phi + \int_0^{\frac{\pi}{2}} \cos\theta \, d\theta \cdot \int_0^{\frac{\pi}{2}} \sin^2\phi\cos\phi \, d\phi \right)$$

$$= 27 \left(-\cos\theta \Big|_0^{\frac{\pi}{2}} \cdot \left(\frac{\phi}{2} - \frac{\sin 2\phi}{4} \right) \Big|_0^{\frac{\pi}{2}} + \sin\theta \Big|_0^{\frac{\pi}{2}} \cdot \frac{\sin^3\phi}{3} \Big|_0^{\frac{\pi}{2}} \right)$$

$$= 27 \left(\frac{\pi}{4} \cdot 1 + \frac{1}{3} \cdot 1 \right) = \frac{27}{12}(3\pi + 4)$$

9. $\mathbf{F} = \langle z, z, x \rangle$, $z = 9 - x^2 - y^2, x \geq 0, y \geq 0, z \geq 0$ upward-pointing normal

SOLUTION

Step 1. Find a parametrization. We use x and y as parameters and parametrize the surface by:

$$\Phi(x, y) = \left(x, y, 9 - x^2 - y^2 \right)$$

The parameter domain $\mathcal{D}$ is determined by the conditions $z = 9 - x^2 - y^2 \geq 0 \Rightarrow x^2 + y^2 \leq 9$ and $x, y \geq 0$. That is:

$$\mathcal{D} = \left\{ (x, y) : x^2 + y^2 \leq 9, \ x, y \geq 0 \right\}$$

$\mathcal{D}$ is the portion of the disk of radius 3 in the first quadrant.

Step 2. Compute the tangent and normal vectors. We have:

$$\mathbf{T}_x = \frac{\partial \Phi}{\partial x} = \frac{\partial}{\partial x}\left(x, y, 9 - x^2 - y^2\right) = \langle 1, 0, -2x \rangle$$

$$\mathbf{T}_y = \frac{\partial \Phi}{\partial y} = \frac{\partial}{\partial y}\left(x, y, 9 - x^2 - y^2\right) = \langle 0, 1, -2y \rangle$$

We compute the cross product of the tangent vectors:

$$\mathbf{T}_x \times \mathbf{T}_y = \begin{vmatrix} \mathbf{i} & \mathbf{j} & \mathbf{k} \\ 1 & 0 & -2x \\ 0 & 1 & -2y \end{vmatrix} = (2x)\mathbf{i} + (2y)\mathbf{j} + \mathbf{k} = \langle 2x, 2y, 1 \rangle$$

Since the z-component is positive, the vector points upward, and we have:

$$\mathbf{n} = \langle 2x, 2y, 1 \rangle$$

Step 3. Evaluate the dot product $\mathbf{F} \cdot \mathbf{n}$. We first express the vector field in terms of the parameters x and y, by setting $z = 9 - x^2 - y^2$. We get:

$$\mathbf{F}\left(\Phi(x, y)\right) = \langle z, z, x \rangle = \left\langle 9 - x^2 - y^2, 9 - x^2 - y^2, x \right\rangle$$

We now compute the dot product:

$$\mathbf{F}\left(\Phi(x, y)\right) \cdot \mathbf{n}(x, y) = \left\langle 9 - x^2 - y^2, 9 - x^2 - y^2, x \right\rangle \cdot \langle 2x, 2y, 1 \rangle$$

$$= 2x(9 - x^2 - y^2) + 2y(9 - x^2 - y^2) + x$$

$$= 19x + 18y - 2xy(x^2 + y^2)$$

Step 4. Evaluate the surface integral.

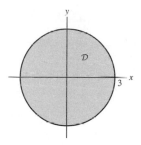

The surface integral is equal to the following double integral:

$$\iint_S \mathbf{F} \cdot d\mathbf{S} = \iint_{\mathcal{D}} \mathbf{F}\left(\Phi(x, y)\right) \cdot \mathbf{n}(x, y)\, dx\, dy = \iint_{\mathcal{D}} \left(19x + 18y - 2xy(x^2 + y^2)\right) dx\, dy$$

We convert the integral to polar coordinates and use the identity $\sin 2\theta = 2\cos\theta\sin\theta$ to obtain:

$$\iint_S \mathbf{F} \cdot d\mathbf{S} = \int_0^3 \int_0^{\pi/2} \left(19r\cos\theta + 18r\sin\theta - 2r^2\cos\theta\sin\theta\right) r\, d\theta\, dr$$

$$= \left(\int_0^3 r^2\, dr\right) \cdot \left(\int_0^{\pi/2} 19\cos\theta + 18\sin\theta\, d\theta\right) + \left(\int_0^3 -r^3\, dr\right) \cdot \left(\int_0^{\pi/2} \sin 2\theta\, d\theta\right)$$

$$= \left(\frac{r^3}{3}\Big|_0^3\right) \cdot \left(19\sin\theta - 18\cos\theta\Big|_0^{\pi/2}\right) + \left(-\frac{r^4}{4}\Big|_0^3\right) \cdot \left(-\frac{1}{2}\cos 2\theta\Big|_0^{\pi/2}\right)$$

$$= 9 \cdot 37 - \frac{81}{4} \cdot (1) = 312.75$$

11. $\mathbf{F} = y^2\mathbf{i} + 2\mathbf{j} - x\mathbf{k}$, portion of the plane $x + y + z = 1$ in the octant $x, y, z \geq 0$, upward-pointing normal

SOLUTION

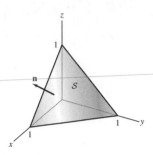

We parametrize the surface by:

$$\Phi(x, y) = (x, y, 1 - x - y),$$

using the parameter domain $\mathcal{D}$ shown in the figure.

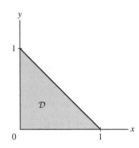

Step 1. Compute the tangent and normal vectors. We have:

$$\mathbf{T}_x = \frac{\partial \Phi}{\partial x} = \frac{\partial}{\partial x}(x, y, 1 - x - y) = \langle 1, 0, -1 \rangle$$

$$\mathbf{T}_y = \frac{\partial \Phi}{\partial y} = \frac{\partial}{\partial y}(x, y, 1 - y) = \langle 0, 1, -1 \rangle$$

$$\mathbf{n} = \mathbf{T}_x \times \mathbf{T}_y = \begin{vmatrix} \mathbf{i} & \mathbf{j} & \mathbf{k} \\ 1 & 0 & -1 \\ 0 & 1 & -1 \end{vmatrix} = \mathbf{i} + \mathbf{j} + \mathbf{k} = \langle 1, 1, 1 \rangle$$

Note that $\mathbf{n}$ points upward.

Step 2. Evaluate the dot product $\mathbf{F} \cdot \mathbf{n}$.

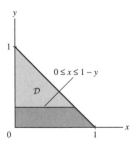

We compute the dot product:

$$\mathbf{F}(\Phi(x, y)) \cdot \mathbf{n} = \left\langle y^2, 2, -x \right\rangle \cdot \langle 1, 1, 1 \rangle = y^2 + 2 - x$$

Step 3. Evaluate the surface integral. The surface integral is equal to the following double integral:

$$\iint_S \mathbf{F} \cdot d\mathbf{S} = \iint_{\mathcal{D}} \mathbf{F}(\Phi(x, y)) \cdot \mathbf{n} \, dx \, dy = \int_0^1 \int_0^{1-y} \left(y^2 + 2 - x \right) dx \, dy = \int_0^1 y^2 x + 2x - \frac{x^2}{2} \bigg|_{x=0}^{1-y} dy$$

$$= \int_0^1 \left(y^2(1 - y) + 2(1 - y) - \frac{(1 - y)^2}{2} \right) dy = \int_0^1 \left(y^2 - y^3 + 2(1 - y) - \frac{(y - 1)^2}{2} \right) dy$$

$$= \frac{y^3}{3} - \frac{y^4}{4} - (1 - y)^2 - \frac{(y - 1)^3}{6} \bigg|_0^1 = \left(\frac{1}{3} - \frac{1}{4} \right) + \left(1 - \frac{1}{6} \right) = \frac{11}{12}$$

13. $\mathbf{F} = \langle xz, yz, z^{-1} \rangle$, disk of radius 3 at height 4 parallel to the xy-plane, upward-pointing normal

SOLUTION

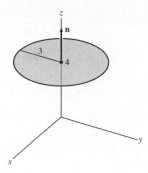

We parametrize the surface S by:

$$\Phi(\theta, r) = (r \cos \theta, r \sin \theta, 4)$$

with the parameter domain:

$$\mathcal{D} = \{(\theta, r) : 0 \le \theta \le 2\pi, 0 \le r \le 3\}$$

Step 1. Compute the tangent and normal vectors. We have:

$$\mathbf{T}_\theta = \frac{\partial \Phi}{\partial \theta} = \frac{\partial}{\partial \theta}(r \cos \theta, r \sin \theta, 4) = \langle -r \sin \theta, r \cos \theta, 0 \rangle$$

$$\mathbf{T}_r = \frac{\partial \Phi}{\partial r} = \frac{\partial}{\partial r}(r \cos \theta, r \sin \theta, 4) = \langle \cos \theta, \sin \theta, 0 \rangle$$

$$\mathbf{T}_\theta \times \mathbf{T}_r = \begin{vmatrix} \mathbf{i} & \mathbf{j} & \mathbf{k} \\ -r \sin \theta & r \cos \theta & 0 \\ \cos \theta & \sin \theta & 0 \end{vmatrix} = \left(-r \sin^2 \theta - r \cos^2 \theta \right) \mathbf{k} = -r \mathbf{k} = \langle 0, 0, -r \rangle$$

Since the orientation of S is with an upward pointing normal, the z-coordinate of $\mathbf{n}$ must be positive. Hence:

$$\mathbf{n} = \langle 0, 0, r \rangle$$

Step 2. Evaluate the dot product $\mathbf{F} \cdot \mathbf{n}$. We first express $\mathbf{F}$ in terms of the parameters:

$$\mathbf{F}(\Phi(\theta, r)) = \langle xz, yz, z^{-1} \rangle = \left\langle r \cos \theta \cdot 4, r \sin \theta \cdot 4, 4^{-1} \right\rangle = \left\langle 4r \cos \theta, 4r \sin \theta, \frac{1}{4} \right\rangle$$

We now compute the dot product:

$$\mathbf{F}(\Phi(\theta, r)) \cdot \mathbf{n}(\theta, r) = \left\langle 4r \cos \theta, 4r \sin \theta, \frac{1}{4} \right\rangle \cdot \langle 0, 0, r \rangle = \frac{r}{4}$$

Step 3. Evaluate the surface integral. The surface integral is equal to the following double integral:

$$\iint_S \mathbf{F} \cdot d\mathbf{S} = \iint_\mathcal{D} \mathbf{F}(\Phi(\theta, r)) \cdot \mathbf{n}(\theta, r) \, dr \, d\theta = \int_0^{2\pi} \int_0^3 \frac{r}{4} \, dr \, d\theta = 2\pi \int_0^3 \frac{r}{4} \, dr = 2\pi \cdot \left. \frac{r^2}{8} \right|_0^3 = \frac{9\pi}{4}$$

15. $\mathbf{F} = \langle 0, 0, e^{y+z} \rangle$, boundary of unit cube $0 \le x \le 1, 0 \le y \le 1, 0 \le z \le 1$, outward-pointing normal

SOLUTION

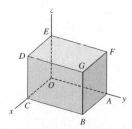

We denote the faces of the cube by:

$$S_1 = \text{Face } OABC \quad S_2 = \text{Face } DGEF \quad S_3 = \text{Face } ABGF$$

$$S_4 = \text{Face } OCDE \quad S_5 = \text{Face } BCDG \quad S_6 = \text{Face } OAFE$$

- On S_1

$$\Phi_1(x, y) = (x, y, 0)$$

and $\mathbf{n}_1 = \langle 0, 0, -1 \rangle$. Thus,

$$\mathbf{F}(\Phi_1(x, y)) \cdot \mathbf{n}_1 = \langle 0, 0, e^y \rangle \cdot \langle 0, 0, -1 \rangle = -e^y$$

- On S_2

$$\Phi_2(x, y) = (x, y, 1)$$

and $\mathbf{n}_2 = \langle 0, 0, 1 \rangle$. Thus,

$$\mathbf{F}(\Phi_2(x, y)) \cdot \mathbf{n}_2 = \langle 0, 0, e^{y+1} \rangle \cdot \langle 0, 0, 1 \rangle = e^{y+1}$$

- On any other surface S_i, $3 \le i \le 6$, we have

$$\mathbf{F}(\Phi_1(x, y)) \cdot \mathbf{n}_i = 0,$$

because the z-component of $\mathbf{n}_i = 0$ and the x, y components of $\mathbf{F}$ equal 0. Thus,

$$\iint_S \mathbf{F} \cdot d\mathbf{S} = \iint_{S_1} \mathbf{F} \cdot d\mathbf{S} + \iint_{S_2} \mathbf{F} \cdot d\mathbf{S} = \int_0^1 \int_0^1 -e^y \, dx \, dy + \int_0^1 \int_0^1 e^{y+1} \, dx \, dy$$

$$= \int_0^1 \int_0^1 \left(e^{y+1} - e^y \right) dx \, dy = \int_0^1 \left(e^{y+1} - e^y \right) dy$$

$$= \int_0^1 e^y (e - 1) \, dy = (e - 1)e^y \Big|_0^1 = (e - 1)^2$$

17. $\mathbf{F} = \langle y, z, 0 \rangle$, $\quad G(u, v) = (u^3 - v, u + v, v^2)$, $0 \le u \le 2, 0 \le v \le 3$, $\quad$ downward-pointing normal

SOLUTION

Step 1. Compute the tangent and normal vectors. We have,

$$\mathbf{T}_u = \frac{\partial \Phi}{\partial u} = \frac{\partial}{\partial u} \left(u^3 - v, u + v, v^2 \right) = \langle 3u^2, 1, 0 \rangle$$

$$\mathbf{T}_v = \frac{\partial \Phi}{\partial v} = \frac{\partial}{\partial v} \left(u^3 - v, u + v, v^2 \right) = \langle -1, 1, 2v \rangle$$

$$\mathbf{T}_u \times \mathbf{T}_v = \begin{vmatrix} \mathbf{i} & \mathbf{j} & \mathbf{k} \\ 3u^2 & 1 & 0 \\ -1 & 1 & 2v \end{vmatrix} = (2v)\mathbf{i} - \left(6u^2 v \right) \mathbf{j} + \left(3u^2 + 1 \right) \mathbf{k} = \langle 2v, -6u^2 v, 3u^2 + 1 \rangle$$

Since the normal is pointing downward, the z-coordinate is negative, hence,

$$\mathbf{n} = \langle -2v, 6u^2 v, -3u^2 - 1 \rangle$$

Step 2. Evaluate the dot product $\mathbf{F} \cdot \mathbf{n}$. We first express $\mathbf{F}$ in terms of the parameters:

$$\mathbf{F}(\Phi(u, v)) = \langle y, z, 0 \rangle = \langle u + v, v^2, 0 \rangle$$

We compute the dot product:

$$\mathbf{F}(\Phi(u, v)) \cdot \mathbf{n}(u, v) = \langle u + v, v^2, 0 \rangle \cdot \langle -2v, 6u^2 v, -3u^2 - 1 \rangle$$

$$= -2v(u + v) + 6u^2 v \cdot v^2 + 0 = -2vu - 2v^2 + 6u^2 v^3$$

Step 3. Evaluate the surface integral. The surface integral is equal to the following double integral:

$$\iint_S \mathbf{F} \cdot d\mathbf{S} = \iint_{\mathcal{D}} \mathbf{F}(\Phi(u, v)) \cdot \mathbf{n}(u, v) \, du \, dv = \int_0^3 \int_0^2 \left(-2uv - 2v^2 + 6u^2 v^3 \right) du \, dv$$

$$= \int_0^3 -u^2 v - 2v^2 u + 2u^3 v^3 \Big|_{u=0}^2 \, dv = \int_0^3 \left(16v^3 - 4v^2 - 4v \right) dv = 4v^4 - \frac{4}{3}v^3 - 2v^2 \Big|_0^3 = 270$$

19. Let $\mathbf{e_r} = \langle x/r, y/r, z/r \rangle$ be the unit radial vector, where $r = \sqrt{x^2 + y^2 + z^2}$. Calculate the integral of $\mathbf{F} = e^{-r}\mathbf{e_r}$ over:

(a) The upper hemisphere of $x^2 + y^2 + z^2 = 9$, outward-pointing normal.

(b) The octant $x \geq 0$, $y \geq 0$, $z \geq 0$ of the unit sphere centered at the origin.

SOLUTION

(a) We parametrize the upper-hemisphere by,

$$\Phi : x = 3\cos\theta\sin\phi, \ y = 3\sin\theta\sin\phi, \ z = 3\cos\phi$$

with the parameter domain:

$$\mathcal{D} = \left\{ (\theta, \phi) : 0 \leq \theta < 2\pi, 0 \leq \phi < \frac{\pi}{2} \right\}$$

The outward pointing normal is (see Eq. (4) in Section 16.4):

$$\mathbf{n} = 9\sin\phi\,\mathbf{e_r}$$

We compute the dot product $\mathbf{F} \cdot \mathbf{n}$ on the sphere. On the sphere $r = 3$, hence,

$$\mathbf{F} \cdot \mathbf{n} = e^{-r}\,\mathbf{e_r} \cdot \mathbf{n} = e^{-3}\mathbf{e_r} \cdot 9\sin\phi\,\mathbf{e_r} = 9e^{-3}\sin\phi\,\mathbf{e_r} \cdot \mathbf{e_r} = 9\,e^{-3}\sin\phi$$

We obtain the following integral:

$$\iint_S \mathbf{F} \cdot d\mathbf{S} = \iint_{\mathcal{D}} (\mathbf{F} \cdot \mathbf{n})\,d\phi\,d\theta = \int_0^{2\pi} \int_0^{\pi/2} 9e^{-3}\sin\phi\,d\phi\,d\theta$$

$$= 18\pi e^{-3} \int_0^{\pi/2} \sin\phi\,d\phi = 18\pi e^{-3}\left(-\cos\phi \Big|_0^{\pi/2}\right) = 18\pi e^{-3}$$

(b) We parametrize the first octant of the sphere by,

$$\Phi : x = \cos\theta\sin\phi, \ y = \sin\theta\sin\phi, \ z = \cos\phi$$

with the parameter domain:

$$\mathcal{D} = \left\{ (\theta, \phi) : 0 \leq \theta < \frac{\pi}{2}, 0 \leq \phi < \frac{\pi}{2} \right\}$$

The outward pointing normal is (as seen above):

$$\mathbf{n} = 1\sin\phi\,\mathbf{e_r}$$

We compute the dot product $\mathbf{F} \cdot \mathbf{n}$ on the sphere. On the sphere $r = 1$, hence,

$$\mathbf{F} \cdot \mathbf{n} = e^{-r}\,\mathbf{e_r} \cdot \mathbf{n} = e^{-1}\mathbf{e_r} \cdot \sin\phi\,\mathbf{e_r} = e^{-1}\sin\phi\,\mathbf{e_r} \cdot \mathbf{e_r} = e^{-1}\sin\phi$$

We obtain the following integral:

$$\iint_S \mathbf{F} \cdot d\mathbf{S} = \iint_{\mathcal{D}} (\mathbf{F} \cdot \mathbf{n})\,d\phi\,d\theta = \int_0^{\pi/2} \int_0^{\pi/2} e^{-1}\sin\phi\,d\phi\,d\theta$$

$$= \frac{\pi}{2}e^{-1} \int_0^{\pi/2} \sin\phi\,d\phi = \frac{\pi}{2}e^{-1}\left(-\cos\phi \Big|_0^{\pi/2}\right) = \frac{\pi}{2}e^{-1}$$

21. The electric field due to a point charge located at the origin in $\mathbf{R}^3$ is $\mathbf{E} = k\dfrac{\mathbf{e_r}}{r^2}$, where $r = \sqrt{x^2 + y^2 + z^2}$ and k is a constant. Calculate the flux of $\mathbf{E}$ through the disk D of radius 2 parallel to the xy-plane with center $(0, 0, 3)$.

SOLUTION Let $r = \sqrt{x^2 + y^2 + z^2}$ and $\hat{r} = \sqrt{x^2 + y^2}$. We parametrize the disc by:

$$\Phi(\hat{r}, \theta) = (\hat{r}\cos\theta, \hat{r}\sin\theta, 3)$$

$$\mathbf{T}_{\hat{r}} = \frac{\partial\Phi}{\partial\hat{r}} = \langle\cos\theta, \sin\theta, 0\rangle$$

$$\mathbf{T}_\theta = \frac{\partial\Phi}{\partial\theta} = \langle-\hat{r}\sin\theta, \hat{r}\cos\theta, 0\rangle$$

$$\mathbf{n} = \mathbf{T}_{\hat{r}} \times \mathbf{T}_\theta = \begin{vmatrix} \mathbf{i} & \mathbf{j} & \mathbf{k} \\ \cos\theta & \sin\theta & 0 \\ -\hat{r}\sin\theta & \hat{r}\cos\theta & 0 \end{vmatrix} = \langle 0, 0, \hat{r}\rangle$$

Now,

$$\mathbf{E} \cdot \mathbf{n} = k\frac{\mathbf{e}_r}{r^2} \cdot \langle 0, 0, \hat{r} \rangle = \frac{k\hat{r}}{r^3} \langle x, y, z \rangle \cdot \langle 0, 0, 1 \rangle = \frac{zk\hat{r}}{r^3}$$

Since on the disk $z = 3$, we get:

$$\mathbf{E} \cdot \mathbf{n} = 3k\frac{\hat{r}}{r^3} \text{ and } r = \sqrt{\hat{r}^2 + 9}$$

so $\mathbf{E} \cdot \mathbf{n} = 3k\dfrac{\hat{r}}{\left(\sqrt{\hat{r}^2+9}\right)^3}$.

$$\iint_{\mathcal{D}} \mathbf{E} \cdot d\mathbf{S} = \int_0^{2\pi} \int_0^2 \frac{3k\hat{r}}{(\hat{r}^2+9)^{3/2}} \, d\hat{r} \, d\theta = 6\pi k \int_0^2 \frac{\hat{r}}{(\hat{r}^2+9)^{3/2}} \, d\hat{r}$$

Substituting $u = \hat{r}^2 + 9$ and $\frac{1}{2} du = \hat{r} \, d\hat{r}$, we get:

$$\iint_{\mathcal{D}} \mathbf{E} \cdot d\mathbf{S} = 3\pi k \int_9^{13} \frac{du}{u^{3/2}} = -6\pi k u^{-1/2} \Big|_9^{13} = \left(2 - \frac{6}{\sqrt{13}}\right) \pi k$$

23. Let $\mathbf{v} = z\mathbf{k}$ be the velocity field (in meters per second) of a fluid in $\mathbf{R}^3$. Calculate the flow rate (in cubic meters per second) through the upper hemisphere ($z \geq 0$) of the sphere $x^2 + y^2 + z^2 = 1$.

SOLUTION   We use the spherical coordinates:

$$x = \cos\theta \sin\phi, \quad y = \sin\theta \sin\phi, \quad z = \cos\phi$$

with the parameter domain

$$0 \leq \theta < 2\pi, \quad 0 \leq \phi \leq \frac{\pi}{2}$$

The normal vector is (see Eq. (4) in Section 16.4):

$$\mathbf{n} = \mathbf{T}_\phi \times \mathbf{T}_\theta = \sin\phi \, \langle \cos\theta \sin\phi, \sin\theta \sin\phi, \cos\phi \rangle$$

We express the function in terms of the parameters:

$$\mathbf{v} = \langle 0, 0, z \rangle = \langle 0, 0, \cos\phi \rangle$$

Hence,

$$\mathbf{v} \cdot \mathbf{n} = \langle 0, 0, \cos\phi \rangle \cdot \sin\phi \, \langle \cos\theta \sin\phi, \sin\theta \sin\phi, \cos\phi \rangle = \sin\phi \cos^2\phi$$

The flow rate of the fluid through the upper hemisphere S is equal to the flux of the velocity vector through S. That is,

$$\iint_S \mathbf{v} \cdot d\mathbf{S} = \int_0^{\frac{\pi}{2}} \int_0^{2\pi} \sin\phi \cos^2\phi \, d\theta \, d\phi$$

$$= \int_0^{2\pi} d\theta \cdot \int_0^{\frac{\pi}{2}} \sin\phi \cos^2\phi \, d\phi = 2\pi \cdot \frac{-\cos^3\phi}{3} \Big|_0^{\frac{\pi}{2}}$$

$$= \frac{2\pi}{3} \text{m}^3/\text{s}$$

In Exercises 25–26, let $\mathcal{T}$ be the triangular region with vertices $(1, 0, 0)$, $(0, 1, 0)$, and $(0, 0, 1)$ oriented with upward-pointing normal vector (Figure 16). Assume distances are in meters.

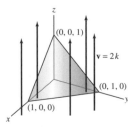

FIGURE 16

25. A fluid flows with constant velocity field $\mathbf{v} = 2\mathbf{k}$ (m/s). Calculate:

(a) The flow rate through $\mathcal{T}$

(b) The flow rate through the projection of $\mathcal{T}$ onto the xy-plane [the triangle with vertices $(0, 0, 0)$, $(1, 0, 0)$, and $(0, 1, 0)$]

SOLUTION

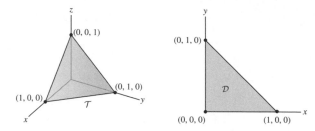

The equation of the plane through the three vertices is $x + y + z = 1$, hence the upward pointing normal vector is:

$$\mathbf{n} = \langle 1, 1, 1 \rangle$$

and the unit normal is:

$$\mathbf{e_n} = \left\langle \frac{1}{\sqrt{3}}, \frac{1}{\sqrt{3}}, \frac{1}{\sqrt{3}} \right\rangle$$

We compute the dot product $\mathbf{v} \cdot \mathbf{e_n}$:

$$\mathbf{v} \cdot \mathbf{e_n} = \langle 0, 0, 2 \rangle \cdot \left\langle \frac{1}{\sqrt{3}}, \frac{1}{\sqrt{3}}, \frac{1}{\sqrt{3}} \right\rangle = \frac{2}{\sqrt{3}}$$

The flow rate through $\mathcal{T}$ is equal to the flux of $\mathbf{v}$ through $\mathcal{T}$. That is,

$$\iint_S \mathbf{v} \cdot d\mathbf{S} = \iint_S (\mathbf{v} \cdot \mathbf{e_n}) \, dS = \iint_S \frac{2}{\sqrt{3}} \, dS = \frac{2}{\sqrt{3}} \iint_S 1 \, dS = \frac{2}{\sqrt{3}} \cdot \text{Area}(S)$$

The area of the equilateral triangle $\mathcal{T}$ is $\dfrac{\left(\sqrt{2}\right)^2 \cdot \sqrt{3}}{4} = \dfrac{\sqrt{3}}{2}$. Therefore,

$$\iint_S \mathbf{v} \cdot d\mathbf{S} = \frac{2}{\sqrt{3}} \cdot \frac{\sqrt{3}}{2} = 1$$

Let $\mathcal{D}$ denote the projection of $\mathcal{T}$ onto the xy-plane. Then the upward pointing normal is $\mathbf{n} = \langle 0, 0, 1 \rangle$. We compute the dot product $\mathbf{v} \cdot \mathbf{n}$:

$$\mathbf{v} \cdot \mathbf{n} = \langle 0, 0, 2 \rangle \cdot \langle 0, 0, 1 \rangle = 2$$

The flow rate through $\mathcal{D}$ is equal to the flux of $\mathbf{v}$ through $\mathcal{D}$. That is,

$$\iint_{\mathcal{D}} \mathbf{v} \cdot d\mathbf{S} = \iint_{\mathcal{D}} (\mathbf{v} \cdot \mathbf{n}) \, dS = \iint_{\mathcal{D}} 2 \, dS = 2 \iint_{\mathcal{D}} 1 \, dS = 2 \cdot \text{Area}(\mathcal{D}) = 2 \cdot \frac{1 \cdot 1}{2} = 1$$

27. Prove that if $\mathcal{S}$ is the part of a graph $z = g(x, y)$ lying over a domain $\mathcal{D}$ in the xy-plane, then

$$\iint_S \mathbf{F} \cdot d\mathbf{S} = \iint_{\mathcal{D}} \left(-F_1 \frac{\partial g}{\partial x} - F_2 \frac{\partial g}{\partial y} + F_3 \right) dx \, dy$$

SOLUTION

Step 1. Find a parametrization. We parametrize the surface by

$$\Phi(x, y) = (x, y, g(x, y)), \, (x, y) \in \mathcal{D}$$

Step 2. Compute the tangent and normal vectors. We have,

$$\mathbf{T}_x = \frac{\partial \Phi}{\partial x} = \frac{\partial}{\partial x}(x, y, g(x, y)) = \left\langle 1, 0, \frac{\partial g}{\partial x} \right\rangle$$

$$\mathbf{T}_y = \frac{\partial \Phi}{\partial y} = \frac{\partial}{\partial y}(x, y, g(x, y)) = \left\langle 0, 1, \frac{\partial g}{\partial y} \right\rangle$$

$$\mathbf{n} = \mathbf{T}_x \times \mathbf{T}_y = \begin{vmatrix} \mathbf{i} & \mathbf{j} & \mathbf{k} \\ 1 & 0 & \frac{\partial g}{\partial x} \\ 0 & 1 & \frac{\partial g}{\partial y} \end{vmatrix} = -\frac{\partial g}{\partial x}\mathbf{i} - \frac{\partial g}{\partial x}\mathbf{j} + \mathbf{k} = \left\langle -\frac{\partial g}{\partial x}, -\frac{\partial g}{\partial y}, 1 \right\rangle$$

Step 3. Evaluate the dot product $\mathbf{F} \cdot \mathbf{n}$.

$$\mathbf{F} \cdot \mathbf{n} = \langle F_1, F_2, F_3 \rangle \cdot \left\langle -\frac{\partial g}{\partial x}, -\frac{\partial g}{\partial y}, 1 \right\rangle = -F_1 \frac{\partial g}{\partial x} - F_2 \frac{\partial g}{\partial y} + F_3$$

Step 4. Evaluate the surface integral. The surface integral is equal to the following double integral:

$$\iint_S \mathbf{F} \cdot d\mathbf{S} = \iint_D (\mathbf{F} \cdot \mathbf{n})\, dx\, dy = \iint_D \left(-F_1 \frac{\partial g}{\partial x} - F_2 \frac{\partial g}{\partial y} + F_3 \right) dx\, dy$$

In Exercises 28–29, a varying current $i(t)$ flows through a long straight wire in the xy-plane as in Example 5. The current produces a magnetic field $\mathbf{B}$ whose magnitude at a distance r from the wire is $B = \dfrac{\mu_0 i}{2\pi r}$ T, where $\mu_0 = 4\pi \cdot 10^{-7}$ T-m/A. Furthermore, $\mathbf{B}$ points into the page at points P in the xy-plane.

29. Assume that $i = 10e^{-0.1t}$ A (t in seconds). Calculate the flux $\Phi(t)$, at time t, of $\mathbf{B}$ through the isosceles triangle of base 12 cm and height 6 cm whose bottom edge is 3 cm from the wire, as in Figure 17. Assume the triangle is oriented with normal vector pointing out of the page. Use Faraday's Law to determine the voltage drop around the triangular loop (the boundary of the triangle) at time t.

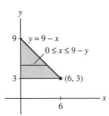

FIGURE 17

SOLUTION The magnetic field is $\mathbf{B} = \dfrac{-\mu_0 i}{2\pi r}\mathbf{k}$ and the unit normal on the triangle points out of the page, hence $\mathbf{n} = \mathbf{e_n} = \mathbf{k}$.

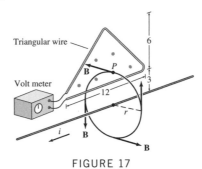

Also, the distance from a point $P = (x, y)$ in $\mathcal{R}$ to the wire is $r = y$. Hence:

$$\mathbf{B} \cdot \mathbf{n} = \frac{-\mu_0 i}{2\pi y}\mathbf{k} \cdot \mathbf{k} = \frac{-\mu_0 i}{2\pi y}$$

The flux $\Phi(t)$ of $\mathbf{B}$ through $\mathcal{R}$ is the following integral:

$$\Phi(t) = \iint_{\mathcal{R}} \mathbf{B} \cdot \mathbf{n}\, dx\, dy = \frac{-\mu_0 i}{2\pi} \iint_{\mathcal{R}} \frac{1}{y}\, dx\, dy$$

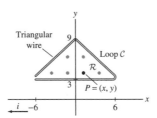

Using symmetry we have:

$$\Phi(t) = \frac{-\mu_0 i}{\pi} \int_3^9 \int_0^{9-y} \frac{1}{y}\, dx\, dy = \frac{-\mu_0 i}{\pi} \int_3^9 \frac{x}{y}\Big|_{x=0}^{9-y} dy = \frac{-\mu_0 i}{\pi} \int_3^9 \frac{9-y}{y}\, dy$$

$$= \frac{-\mu_0 i}{\pi} \int_3^9 \left(\frac{9}{y} - 1 \right) dy = \frac{-\mu_0 i}{\pi} \left(9\ln y - y \Big|_{y=3}^9 \right) = \frac{-\mu_0 i}{\pi} \left((9\ln 9 - 9) - (9\ln 3 - 3) \right)$$

$$= \frac{-\mu_0 i}{\pi}(9\ln 3 - 6) = \frac{-4\pi \cdot 10^{-7}}{\pi}(9\ln 3 - 6)i = -1.56 \cdot 10^{-6} i = -1.56 \cdot 10^{-5} \cdot e^{-0.1t}$$

Using Faraday's Law, the voltage drop around the triangular loop $\mathcal{C}$ (oriented counterclockwise):

$$\int_{\mathcal{C}} \mathbf{E} \cdot d\mathbf{S} = -\frac{d\phi}{dt} = -\frac{d}{dt}\left(-1.56 \cdot 10^{-5} \cdot e^{-0.1t}\right) = -1.56 \cdot 10^{-6} \cdot e^{-0.1t} \text{ Volts}$$

Further Insights and Challenges

In Exercises 31 and 32, let S be the surface with parametrization

$$G(u, v) = \left(\left(1 + v\cos\frac{u}{2}\right)\cos u, \left(1 + v\cos\frac{u}{2}\right)\sin u, v\sin\frac{u}{2}\right)$$

for $0 \le u \le 2\pi$, $-\frac{1}{2} \le v \le \frac{1}{2}$.

31. *CAS* Use a computer algebra system.
(a) Plot S and confirm visually that S is a Möbius strip.
(b) The intersection of S with the xy-plane is the unit circle $G(u, 0) = (\cos u, \sin u, 0)$. Verify that the normal vector along this circle is

$$\mathbf{n}(u, 0) = \left\langle \cos u \sin\frac{u}{2}, \sin u \sin\frac{u}{2}, -\cos\frac{u}{2}\right\rangle$$

(c) As u varies from 0 to 2π, the point $G(u, 0)$ moves once around the unit circle, beginning and ending at $G(0, 0) = G(2\pi, 0) = (1, 0, 0)$. Verify that $\mathbf{n}(u, 0)$ is a unit vector that varies continuously but that $\mathbf{n}(2\pi, 0) = -\mathbf{n}(0, 0)$. This shows that S is not orientable—that is, it is not possible to choose a nonzero normal vector at each point on S in a continuously varying manner (if it were possible, the unit normal vector would return to itself rather than to its negative when carried around the circle).

SOLUTION

(a) We use a computer algebra system to graph the plot of S, and it is indeed a Möbius strip.
(b) To find the normal vector along the unit circle, we use our computer to first find the cross product $\frac{\partial \mathbf{n}}{\partial u} \times \frac{\partial \mathbf{n}}{\partial v}$ and simplify, we get the very ugly expression

$$\mathbf{n}(u, v) = \frac{\partial \mathbf{n}}{\partial u} \times \frac{\partial \mathbf{n}}{\partial v} = \left\langle \frac{1}{2}\left(-v\cos\left(\frac{u}{2}\right) + 2\cos u + v\cos\left(\frac{3u}{2}\right)\right)\sin\left(\frac{u}{2}\right),\right.$$
$$\frac{1}{4}\left(v + 2\cos\left(\frac{u}{2}\right) + 2v\cos(u) - 2\cos\left(\frac{3u}{2}\right) - v\cos(2u)\right),$$
$$\left. -\cos\left(\frac{u}{2}\right)\left(1 + v\cos\left(\frac{u}{2}\right)\right)\right\rangle$$

(Different computer algebra systems may produce different simplifications.) When we replace v with 0 and simplify, we find that:

$$\mathbf{n}(u, 0) = \left\langle \cos u \sin\frac{u}{2}, \frac{1}{2}\left(\cos\frac{u}{2} - \cos\frac{3u}{2}\right), -\cos\frac{u}{2}\right\rangle$$

This is almost, but not quite, what we want. Let's examine that middle term a bit more.

$$\frac{1}{2}\left(\cos\frac{u}{2} - \cos\frac{3u}{2}\right) = \frac{1}{2}\left(\cos\frac{u}{2} - \left(\cos u \cos\frac{u}{2} - \sin u \sin\frac{u}{2}\right)\right) = \frac{1}{2}\left(\cos\frac{u}{2}(1 - \cos u) + \sin u \sin\frac{u}{2}\right)$$
$$= \frac{1}{2}\left(\cos\frac{u}{2} \cdot 2\sin^2\frac{u}{2} + \sin u \sin\frac{u}{2}\right) = \frac{1}{2}\sin\frac{u}{2}\left(2\sin\frac{u}{2}\cos\frac{u}{2} + \sin u\right)$$
$$= \frac{1}{2}\sin\frac{u}{2}(\sin u + \sin u) = \sin u \sin\frac{u}{2}$$

which is what we expect. Thus, we see that

$$\mathbf{n}(u, 0) = \left\langle \cos u \sin\frac{u}{2}, \sin u \sin\frac{u}{2}, -\cos\frac{u}{2}\right\rangle$$

(c) To verify that $\mathbf{n}(u, 0)$ is a unit vector, we note that

$$\|\mathbf{n}(u, 0)\| = \sqrt{\left(\cos u \sin\frac{u}{2}\right)^2 + \left(\sin u \sin\frac{u}{2}\right)^2 + \left(\cos\frac{u}{2}\right)^2}$$
$$= \sqrt{\cos^2 u \sin^2\frac{u}{2} + \sin^2 u \sin^2\frac{u}{2} + \cos^2\frac{u}{2}} = \sqrt{\sin^2\frac{u}{2} + \cos^2\frac{u}{2}} = \sqrt{1} = 1$$

It is clear that $\mathbf{n}(u, 0)$ varies continuously with u, as each of its three components are non-constant continuous functions of u. Finally, we note that $\mathbf{n}(0, 0) = \langle 0, 0, -1\rangle$ but $\mathbf{n}(2\pi, 0) = \langle 0, 0, 1\rangle$, so indeed $\mathbf{n}(2\pi, 0) = -\mathbf{n}(0, 0)$.

CHAPTER REVIEW EXERCISES

1. Compute the vector assigned to the point $P = (-3, 5)$ by the vector field:

(a) $\mathbf{F} = \langle xy, y - x \rangle$

(b) $\mathbf{F} = \langle 4, 8 \rangle$

(c) $\mathbf{F} = \langle 3^{x+y}, \log_2(x + y) \rangle$

SOLUTION

(a) Substituting $x = -3$, $y = 5$ in $\mathbf{F} = \langle xy, y - x \rangle$ we obtain:

$$\mathbf{F} = \langle -3 \cdot 5, 5 - (-3) \rangle = \langle -15, 8 \rangle$$

(b) The constant vector field $\mathbf{F} = \langle 4, 8 \rangle$ assigns the vector $\langle 4, 8 \rangle$ to all the vectors. Thus:

$$\mathbf{F}(-3, 5) = \langle 4, 8 \rangle$$

(c) Substituting $x = -3$, $y = 5$ in $\mathbf{F} = \langle 3^{x+y}, \log_2(x + y) \rangle$ we obtain

$$\mathbf{F} = \langle 3^{-3+5}, \log_2(-3 + 5) \rangle = \langle 3^2, \log_2(2) \rangle = \langle 9, 1 \rangle$$

In Exercises 3–6, sketch the vector field.

3. $\mathbf{F}(x, y) = \langle y, 1 \rangle$

SOLUTION Notice that the vector field is constant along horizontal lines.

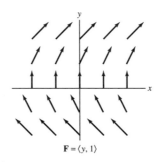

$$\mathbf{F} = \langle y, 1 \rangle$$

5. ∇V, where $V(x, y) = x^2 - y$

SOLUTION The gradient of $V(x, y) = x^2 - y$ is the following vector:

$$\mathbf{F}(x, y) = \left\langle \frac{\partial V}{\partial x}, \frac{\partial V}{\partial y} \right\rangle = \langle 2x, -1 \rangle$$

This vector is sketched in the following figure:

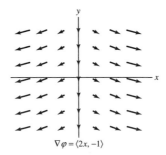

$$\nabla \varphi = \langle 2x, -1 \rangle$$

In Exercises 7–15, determine whether the vector field is conservative, and if so, find a potential function.

7. $\mathbf{F}(x, y) = \langle x^2 y, y^2 x \rangle$

SOLUTION If $\mathbf{F}$ is conservative, the cross partials must be equal. We compute the cross partials:

$$\frac{\partial F_1}{\partial y} = \frac{\partial}{\partial y} \left(x^2 y \right) = x^2$$

$$\frac{\partial F_2}{\partial x} = \frac{\partial}{\partial x} \left(y^2 x \right) = y^2$$

Since the cross-partials are not equal, $\mathbf{F}$ is not conservative.

9. $\mathbf{F}(x, y, z) = \langle \sin x, e^y, z \rangle$

SOLUTION We examine the cross partials of $\mathbf{F}$. Since $F_1 = \sin x$, $F_2 = e^y$, $F_3 = z$ we have:

$$\frac{\partial F_1}{\partial y} = 0 \quad \frac{\partial F_2}{\partial z} = 0 \quad \frac{\partial F_3}{\partial x} = 0$$

$$\frac{\partial F_2}{\partial x} = 0 \quad \frac{\partial F_3}{\partial y} = 0 \quad \frac{\partial F_1}{\partial z} = 0$$

$$\Rightarrow \quad \frac{\partial F_1}{\partial y} = \frac{\partial F_2}{\partial x}, \quad \frac{\partial F_2}{\partial z} = \frac{\partial F_3}{\partial y}, \quad \frac{\partial F_3}{\partial x} = \frac{\partial F_1}{\partial z}$$

Since the cross partials are equal, $\mathbf{F}$ is conservative. We denote the potential field by $V(x, y, z)$. So we have:

$$V_x = \sin x \quad V_y = e^y \quad V_z = z$$

By integrating we get:

$$V(x, y, z) = \int \sin x \, dx = -\cos x + C(y, z)$$

$$V_y = C_y = e^y \quad \Rightarrow \quad C(y, z) = e^y + D(z)$$

$$V(x, y, z) = -\cos x + e^y + D(z)$$

$$V_z = D_z = z \quad \Rightarrow \quad D(z) = \frac{z^2}{2}$$

We conclude that $V(x, y, z) = -\cos x + e^y + \frac{z^2}{2}$. Indeed:

$$\nabla V = \left\langle \frac{\partial V}{\partial x}, \frac{\partial V}{\partial y}, \frac{\partial V}{\partial z} \right\rangle = \langle \sin x, e^y, z \rangle = \mathbf{F}$$

11. $\mathbf{F}(x, y, z) = \langle xyz, \frac{1}{2}x^2 z, 2z^2 y \rangle$

SOLUTION No. We show that the cross partials for x and z are not equal. Since the equality of the cross partials is a necessary condition for a field to be a gradient vector field, we conclude that $\mathbf{F}$ is not a gradient field. We have:

$$\frac{\partial F_1}{\partial z} = \frac{\partial}{\partial z}(xyz) = xy$$

$$\frac{\partial F_3}{\partial x} = \frac{\partial}{\partial x}(2z^2 y) = 0$$

$$\Rightarrow \quad \frac{\partial F_1}{\partial z} \neq \frac{\partial F_3}{\partial x}$$

Therefore the cross partials condition is not satisfied, hence $\mathbf{F}$ is not a gradient vector field.

13. $\mathbf{F}(x, y, z) = \left\langle \frac{y}{1 + x^2}, \tan^{-1} x, 2z \right\rangle$

SOLUTION We examine the cross partials of $\mathbf{F}$. Since $F_1 = \frac{y}{1+x^2}$, $F_2 = \tan^{-1} x$, $F_3 = 2z$ we have:

$$\frac{\partial F_1}{\partial y} = \frac{1}{1 + x^2}$$

$$\frac{\partial F_2}{\partial x} = \frac{1}{1 + x^2}$$

$$\Rightarrow \quad \frac{\partial F_1}{\partial y} = \frac{\partial F_2}{\partial x}$$

$$\frac{\partial F_2}{\partial z} = 0$$

$$\frac{\partial F_3}{\partial y} = 0$$

$$\Rightarrow \quad \frac{\partial F_2}{\partial z} = \frac{\partial F_3}{\partial y}$$

$$\frac{\partial F_3}{\partial x} = 0$$

$$\frac{\partial F_1}{\partial z} = 0$$

$$\Rightarrow \quad \frac{\partial F_3}{\partial x} = \frac{\partial F_1}{\partial z}$$

Since the cross partials are equal, $\mathbf{F}$ is conservative. We denote the potential function by $V(x, y, z)$. We have:

$$V_x = \frac{y}{1 + x^2}, \quad V_y = \tan^{-1}(x), \quad V_z = 2z$$

By integrating we get:

$$V(x, y, z) = \int \frac{y}{1 + x^2} \, dx = y \tan^{-1}(x) + c(y, z)$$

$$V_y = \tan^{-1}(x) + c_y(y, z) = \tan^{-1}(x) \quad \Rightarrow \quad c_y(y, z) = 0 \quad \Rightarrow \quad c(y, z) = c(z)$$

Hence $V(x, y, z) = y \tan^{-1}(x) + c(z)$. $V_z = c'(z) = 2z \Rightarrow c(z) = z^2$. We conclude that $V(x, y, z) = y \tan^{-1}(x) + z^2$. Indeed:

$$\nabla V = \left\langle \frac{\partial V}{\partial x}, \frac{\partial V}{\partial y}, \frac{\partial V}{\partial z} \right\rangle = \left\langle \frac{y}{1 + x^2}, \tan^{-1} x, 2z \right\rangle = \mathbf{F}$$

15. $\mathbf{F}(x, y, z) = \langle xe^{2x}, ye^{2z}, ze^{2y} \rangle$

SOLUTION   We have:

$$\frac{\partial F_3}{\partial y} = \frac{\partial}{\partial y}\left(ze^{2y}\right) = 2ze^{2y}$$

$$\frac{\partial F_2}{\partial z} = \frac{\partial}{\partial z}\left(ye^{2z}\right) = 2ye^{2y}$$

Since $\frac{\partial F_3}{\partial y} \neq \frac{\partial F_2}{\partial z}$, the cross-partials condition is not satisfied , hence $\mathbf{F}$ is not conservative.

In Exercises 17–20, compute the line integral $\int_C f(x, y)\, ds$ for the given function and path or curve.

17. $f(x, y) = xy$,   the path $\mathbf{c}(t) = (t, 2t - 1)$ for $0 \leq t \leq 1$

SOLUTION

Step 1. Compute $ds = \|\mathbf{c}'(t)\|\, dt$. We differentiate $\mathbf{c}(t) = (t, 2t - 1)$ and compute the length of the derivative vector:

$$\mathbf{c}'(t) = \langle 1, 2 \rangle \quad \Rightarrow \quad \|\mathbf{c}'(t)\| = \sqrt{1^2 + 2^2} = \sqrt{5}$$

Hence,

$$ds = \|\mathbf{c}'(t)\|\, dt = \sqrt{5}\, dt$$

Step 2. Write out $f(\mathbf{c}(t))$ and evaluate the line integral. We have:

$$f(\mathbf{c}(t)) = xy = t(2t - 1) = 2t^2 - t$$

Using the Theorem on Scalar Line Integral we have:

$$\int_C f(x, y)\, ds = \int_0^1 f(\mathbf{c}(t)) \|\mathbf{c}'(t)\|\, dt = \int_0^1 \left(2t^2 - t\right) \sqrt{5}\, dt = \sqrt{5}\left(\frac{2}{3}t^3 - \frac{1}{2}t^2\right)\Big|_0^1 = \sqrt{5}\left(\frac{2}{3} - \frac{1}{2}\right) = \frac{\sqrt{5}}{6}$$

19. $f(x, y, z) = e^x - \frac{y}{2\sqrt{2z}}$,   the path $\mathbf{c}(t) = \left(\ln t, \sqrt{2}t, \frac{1}{2}t^2\right)$ for $1 \leq t \leq 2$

SOLUTION

Step 1. Compute $ds = \|\mathbf{c}'(t)\|\, dt$. We have:

$$\mathbf{c}'(t) = \frac{d}{dt}\left\langle \ln t, \sqrt{2}t, \frac{1}{2}t^2 \right\rangle = \left\langle \frac{1}{t}, \sqrt{2}, t \right\rangle$$

$$\|\mathbf{c}'(t)\| = \sqrt{\left(\frac{1}{t}\right)^2 + \left(\sqrt{2}\right)^2 + t^2} = \sqrt{\frac{1}{t^2} + 2 + t^2} = \sqrt{\left(\frac{1}{t} + t\right)^2} = \frac{1}{t} + t$$

Hence:

$$ds = \|\mathbf{c}'(t)\|\, dt = \left(t + \frac{1}{t}\right) dt$$

Step 2. Write out $f(\mathbf{c}(t))$ and evaluate the integral.

$$f(\mathbf{c}(t)) = e^x - \frac{y}{2\sqrt{2z}} = e^{\ln t} - \frac{\sqrt{2}t}{2\sqrt{2} \cdot \frac{1}{2}t^2} = t - \frac{1}{t}$$

We use the Theorem on Scalar Line Integrals to compute the line integral:

$$\int_C f(x, y)\, ds = \int_1^2 f(\mathbf{c}(t)) \|\mathbf{c}'(t)\|\, dt = \int_1^2 \left(t - \frac{1}{t}\right)\left(t + \frac{1}{t}\right) dt$$

$$= \int_1^2 \left(t^2 - \frac{1}{t^2}\right) dt = \frac{t^3}{3} + \frac{1}{t}\Big|_1^2 = \left(\frac{8}{3} + \frac{1}{2}\right) - \left(\frac{1}{3} + 1\right) = \frac{11}{6}$$

21. Find the total mass of an L-shaped rod consisting of the segments $(2t, 2)$ and $(2, 2 - 2t)$ for $0 \le t \le 1$ (length in centimeters) with mass density $\rho(x, y) = x^2 y$ g/cm.

SOLUTION

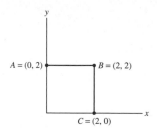

The total mass of the rod is the following sum:

$$M = \int_{AB} x^2 y \, ds + \int_{BC} x^2 y \, ds \tag{1}$$

The segment $\overline{AB}$ is parametrized by $\mathbf{c}_1(t) = (2t, 2), 0 \le t \le 1$. Hence

$$\mathbf{c}_1'(t) = \langle 2, 0 \rangle , \ \|\mathbf{c}_1'(t)\| = 2$$

and

$$f(\mathbf{c}_1(t)) = x^2 y = (2t)^2 \cdot 2 = 8t^2.$$

The segment $\overline{BC}$ is parametrized by $\mathbf{c}_2(t) = (2, 2 - 2t), 0 \le t \le 1$. Hence

$$\mathbf{c}_2'(t) = \langle 0, -2 \rangle , \ \|\mathbf{c}_2'(t)\| = 2$$

and

$$f(\mathbf{c}_2(t)) = x^2 y = 2^2 (2 - 2t) = 8 - 8t.$$

Using these values, the Theorem on Scalar Line Integrals and (1) we get:

$$M = \int_0^1 8t^2 \cdot 2 \, dt + \int_0^1 (8 - 8t) \cdot 2 \, dt = \left. \frac{16t^3}{3} \right|_0^1 + \left. 16t - 8t^2 \right|_0^1 = \frac{40}{3} = 13\frac{1}{3}$$

23. Calculate $\displaystyle \int_{\mathcal{C}_1} y \, dx + x^2 y \, dy$, where $\mathcal{C}_1$ is the oriented curve in Figure 1(A).

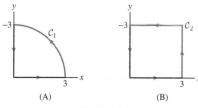

(A)   (B)

FIGURE 1

SOLUTION   We compute the line integral as the sum of the line integrals over the segments $\overline{AO}, \overline{OB}$ and the circular arc BA.

The vector field is $\mathbf{F} = \left(y, x^2 y \right)$. We have:

$$\int_{\mathcal{C}_1} \mathbf{F} \cdot d\mathbf{s} = \int_{AO} \mathbf{F} \cdot d\mathbf{s} + \int_{OB} \mathbf{F} \cdot d\mathbf{s} + \int_{\text{arc } BA} \mathbf{F} \cdot d\mathbf{s} \tag{1}$$

We compute each integral separately.

- The line integral over $\overline{AO}$. The segment $\overline{AO}$ is parametrized by $\mathbf{c}(t) = (0, -t)$, $-3 \leq t \leq 0$. Hence:

$$\mathbf{F}(\mathbf{c}(t)) = \left\langle y, x^2 y \right\rangle = \langle -t, 0 \rangle$$

$$\mathbf{c}'(t) = \langle 0, -1 \rangle$$

$$\mathbf{F}(\mathbf{c}(t)) \cdot \mathbf{c}'(t) = \langle -t, 0 \rangle \cdot \langle 0, -1 \rangle = 0$$

Therefore:

$$\int_{\overline{AO}} \mathbf{F} \cdot d\mathbf{s} = \int_{-3}^{0} \mathbf{F}(\mathbf{c}(t)) \cdot \mathbf{c}'(t)\, dt = 0 \tag{2}$$

- The line integral over $\overline{OB}$. We parametrize the segment $\overline{OB}$ by $\mathbf{c}(t) = (t, 0)$, $0 \leq t \leq 3$. Hence:

$$\mathbf{F}(\mathbf{c}(t)) = \left\langle y, x^2 y \right\rangle = \langle 0, 0 \rangle$$

$$\mathbf{c}'(t) = \langle 1, 0 \rangle$$

$$\mathbf{F}(\mathbf{c}(t)) \cdot \mathbf{c}'(t) = 0$$

Therefore:

$$\int_{\overline{OB}} \mathbf{F} \cdot d\mathbf{s} = \int_{0}^{3} \mathbf{F}(\mathbf{c}(t)) \cdot \mathbf{c}'(t)\, dt = 0 \tag{3}$$

- The line integral over the circular arc BA. We parametrize the circular arc by $\mathbf{c}(t) = (3\cos t, 3\sin t)$, $0 \leq t \leq \frac{\pi}{2}$. Then $\mathbf{c}'(t) = \langle -3\sin t, 3\cos t \rangle$ and $\mathbf{F}(\mathbf{c}(t)) = \left\langle y, x^2 y \right\rangle = \left\langle 3\sin t, 27\cos^2 t \sin t \right\rangle$. We compute the dot product:

$$\mathbf{F}(\mathbf{c}(t)) \cdot \mathbf{c}'(t) = \left\langle 3\sin t, 27\cos^2 t \sin t \right\rangle \cdot \langle -3\sin t, 3\cos t \rangle = -9\sin^2 t + 81\cos^3 t \sin t$$

We obtain the integral:

$$\int_{\text{arc } BA} \mathbf{F} \cdot d\mathbf{s} = \int_{0}^{\pi/2} -9\sin^2 t + 81\cos^3 t \sin t \, dt$$

$$= -9\left(\frac{t}{2} - \frac{\sin 2t}{4} \right) - 81\left(\frac{\cos^4 t}{4} \right) \Big|_{0}^{\pi/2}$$

$$= -\frac{9\pi}{4} + \frac{81}{4} = \frac{81 - 9\pi}{4}$$

Combining (1), (2), (3), and (4) gives:

$$\int_{C_1} \mathbf{F} \cdot d\mathbf{s} = 0 + 0 + \frac{81 - 9\pi}{4} \approx 13.18$$

In Exercises 25–28, compute the line integral $\int_{\mathbf{c}} \mathbf{F} \cdot d\mathbf{s}$ *for the given vector field and path.*

25. $\mathbf{F}(x, y) = \left\langle \dfrac{2y}{x^2 + 4y^2}, \dfrac{x}{x^2 + 4y^2} \right\rangle$,

the path $\mathbf{c}(t) = \left(\cos t, \frac{1}{2}\sin t \right)$ for $0 \leq t \leq 2\pi$

SOLUTION

Step 1. Calculate the integrand $\mathbf{F}(\mathbf{c}(t)) \cdot \mathbf{c}'(t)$.

$$\mathbf{c}(t) = \left(\cos t, \frac{1}{2}\sin t \right)$$

$$\mathbf{F}(\mathbf{c}(t)) = \left\langle \frac{2y}{x^2 + 4y^2}, \frac{x}{x^2 + 4y^2} \right\rangle = \left\langle \frac{2 \cdot \frac{1}{2} \cdot \sin t}{\cos^2 t + 4 \cdot \frac{1}{4}\sin^2 t}, \frac{\cos t}{\cos^2 t + 4 \cdot \frac{1}{4}\sin^2 t} \right\rangle$$

$$= \left\langle \frac{\sin t}{\cos^2 t + \sin^2 t}, \frac{\cos t}{\cos^2 t + \sin^2 t} \right\rangle = \langle \sin t, \cos t \rangle$$

$$\mathbf{c}'(t) = \left\langle -\sin t, \frac{1}{2}\cos t \right\rangle$$

The integrand is the dot product:

$$\mathbf{F}(\mathbf{c}(t)) \cdot \mathbf{c}'(t) = \langle \sin t, \cos t \rangle \cdot \left\langle -\sin t, \frac{1}{2}\cos t \right\rangle = -\sin^2 t + \frac{1}{2}\cos^2 t = \frac{1}{2}\cos 2t - \frac{1}{2}\sin^2 t$$

Step 2. Evaluate the line integral.

$$\int_{\mathcal{C}} \mathbf{F} \cdot d\mathbf{s} = \int_0^{2\pi} \mathbf{F}(\mathbf{c}(t)) \cdot \mathbf{c}'(t)\, dt = \int_0^{2\pi} \left(\frac{1}{2}\cos 2t - \frac{1}{2}\sin^2 t \right) dt = \frac{\sin 2t}{4} - \frac{t}{4} + \frac{\sin 2t}{8} \Big|_0^{2\pi} = -\frac{\pi}{2}$$

27. $\mathbf{F}(x, y) = \langle x^2 y, y^2 z, z^2 x \rangle$,   the path $\mathbf{c}(t) = \left(e^{-t}, e^{-2t}, e^{-3t} \right)$ for $0 \le t < \infty$

SOLUTION

Step 1. Calculate the integrand $\mathbf{F}(\mathbf{c}(t)) \cdot \mathbf{c}'(t)$.

$$\mathbf{c}(t) = \left(e^{-t}, e^{-2t}, e^{-3t} \right)$$

$$\mathbf{c}'(t) = \left\langle e^{-t}, -2e^{-2t}, -3e^{-3t} \right\rangle$$

$$\mathbf{F}(\mathbf{c}(t)) = \left\langle x^2 y, y^2 z, z^2 x \right\rangle = \left\langle e^{-2t} \cdot e^{-2t}, e^{-4t} \cdot e^{-3t}, e^{-6t} \cdot e^{-t} \right\rangle = \left\langle e^{-4t}, e^{-7t}, e^{-7t} \right\rangle$$

The integrand is the dot product:

$$\mathbf{F}(\mathbf{c}(t)) \cdot \mathbf{c}'(t) = \left\langle e^{-4t}, e^{-7t}, e^{-7t} \right\rangle \cdot \left\langle e^{-t}, -2e^{-2t}, -3e^{-3t} \right\rangle = -e^{-5t} - 2e^{-9t} - 3e^{-10t}$$

Step 2. Evaluate the line integral.

$$\int_{\mathcal{C}} \mathbf{F} \cdot d\mathbf{s} = \int_0^\infty \mathbf{F}(\mathbf{c}(t)) \cdot \mathbf{c}'(t)\, dt = \int_0^\infty \left(-e^{-5t} - 2e^{-9t} - 3e^{-10t} \right) dt$$

$$= \lim_{R \to \infty} \left(\frac{1}{5}e^{-5R} + \frac{2}{9}e^{-9R} + \frac{3}{10}e^{-10R} \right) - \left(\frac{1}{5} + \frac{2}{9} + \frac{3}{10} \right) = 0 - \frac{13}{18} = -\frac{13}{18}$$

29. Consider the line integrals $\int_{\mathbf{c}} \mathbf{F} \cdot d\mathbf{s}$ for the vector fields $\mathbf{F}$ and paths $\mathbf{c}$ in Figure 2. Which two of the line integrals appear to have a value of zero? Which of the other two appears to have a negative value?

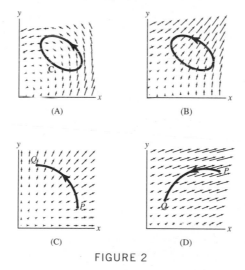

(A)   (B)

(C)   (D)

FIGURE 2

SOLUTION   In (A), the line integral around the ellipse appears to be positive, because the negative tangential components from the lower part of the curve appears to be smaller than the positive contribution of the tangential components from the upper part.

In (B), the line integral around the ellipse appears to be zero, since $\mathbf{F}$ is orthogonal to the ellipse at all points except for two points where the tangential components of $\mathbf{F}$ cancel each other.

In (C), $\mathbf{F}$ is orthogonal to the path, hence the tangential component is zero at all points on the curve. Therefore the line integral $\int_{\mathcal{C}} \mathbf{F} \cdot d\mathbf{s}$ is zero.

In (D), the direction of $\mathbf{F}$ is opposite to the direction of the curve. Therefore the dot product $\mathbf{F} \cdot \mathbf{T}$ is negative at each point along the curve, resulting in a negative line integral.

31. Find constants a, b, c such that

$$G(u, v) = (u + av, bu + v, 2u - c)$$

parametrizes the plane $3x - 4y + z = 5$. Calculate $\mathbf{T}_u$, $\mathbf{T}_v$, and $\mathbf{n}(u, v)$.

SOLUTION   We substitute $x = u + av$, $y = bu + v$ and $z = 2u - c$ in the equation of the plane $3x - 4y + z = 5$, to obtain:

$$5 = 3x - 4y + z = 3(u + av) - 4(bu + v) + 2u - c = (5 - 4b)u + (3a - 4)v - c$$

or

$$(5 - 4b)u + (3a - 4)v - (5 + c) = 0$$

This equation must be satisfied for all u and v, therefore the following must hold:

$$5 - 4b = 0 \qquad b = \frac{5}{4}$$

$$3a - 4 = 0 \quad \Rightarrow \quad a = \frac{4}{3}$$

$$5 + c = 0 \qquad c = -5$$

We obtain the following parametrization for the plane $3x - 4y + z = 5$:

$$\phi(u, v) = \left(u + \frac{4}{3}v, \frac{5}{4}u + v, 2u + 5 \right)$$

We compute the tangent vectors $\mathbf{T}_u$ and $\mathbf{T}_v$:

$$\mathbf{T}_u = \frac{\partial \phi}{\partial u} = \left\langle 1, \frac{5}{4}, 2 \right\rangle; \mathbf{T}_v = \frac{\partial \phi}{\partial v} = \left\langle \frac{4}{3}, 1, 0 \right\rangle$$

The normal vector is their cross product:

$$\mathbf{n} = \mathbf{T}_u \times \mathbf{T}_v = \begin{vmatrix} \mathbf{i} & \mathbf{j} & \mathbf{k} \\ 1 & \frac{5}{4} & 2 \\ \frac{4}{3} & 1 & 0 \end{vmatrix} = \begin{vmatrix} \frac{5}{4} & 2 \\ 1 & 0 \end{vmatrix} \mathbf{i} - \begin{vmatrix} 1 & 2 \\ \frac{4}{3} & 0 \end{vmatrix} \mathbf{j} + \begin{vmatrix} 1 & \frac{5}{4} \\ \frac{4}{3} & 1 \end{vmatrix} \mathbf{k}$$

$$= -2\mathbf{i} + \frac{8}{3}\mathbf{j} + \left(1 - \frac{5}{3} \right) \mathbf{k} = \left\langle -2, \frac{8}{3}, -\frac{2}{3} \right\rangle$$

33. Let $\mathcal{S}$ be the surface parametrized by

$$G(u, v) = \left(2u \sin \frac{v}{2}, 2u \cos \frac{v}{2}, 3v \right)$$

for $0 \le u \le 1$ and $0 \le v \le 2\pi$.

(a) Calculate the tangent vectors $\mathbf{T}_u$ and $\mathbf{T}_v$ and the normal vector $\mathbf{n}(u, v)$ at $P = G(1, \frac{\pi}{3})$.

(b) Find the equation of the tangent plane at P.

(c) Compute the surface area of $\mathcal{S}$.

SOLUTION

(a) The tangent vectors are the partial derivatives:

$$\mathbf{T}_u = \frac{\partial G}{\partial u} = \frac{\partial}{\partial u} \left\langle 2u \sin \frac{v}{2}, 2u \cos \frac{v}{2}, 3v \right\rangle = \left\langle 2 \sin \frac{v}{2}, 2 \cos \frac{v}{2}, 0 \right\rangle$$

$$\mathbf{T}_v = \frac{\partial G}{\partial v} = \frac{\partial}{\partial v} \left\langle 2u \sin \frac{v}{2}, 2u \cos \frac{v}{2}, 3v \right\rangle = \left\langle u \cos \frac{v}{2}, -u \sin \frac{v}{2}, 3 \right\rangle$$

The normal vector is their cross-product:

$$\mathbf{n} = \mathbf{T}_u \times \mathbf{T}_v = \begin{vmatrix} \mathbf{i} & \mathbf{j} & \mathbf{k} \\ 2 \sin \frac{v}{2} & 2 \cos \frac{v}{2} & 0 \\ u \cos \frac{v}{2} & -u \sin \frac{v}{2} & 3 \end{vmatrix} = \begin{vmatrix} 2 \cos \frac{v}{2} & 0 \\ -u \sin \frac{v}{2} & 3 \end{vmatrix} \mathbf{i} - \begin{vmatrix} 2 \sin \frac{v}{2} & 0 \\ u \cos \frac{v}{2} & 3 \end{vmatrix} \mathbf{j} + \begin{vmatrix} 2 \sin \frac{v}{2} & 2 \cos \frac{v}{2} \\ u \cos \frac{v}{2} & -u \sin \frac{v}{2} \end{vmatrix} \mathbf{k}$$

$$= \left(6 \cos \frac{v}{2} \right) \mathbf{i} - \left(6 \sin \frac{v}{2} \right) \mathbf{j} + \left(-2u \sin^2 \frac{v}{2} - 2u \cos^2 \frac{v}{2} \right) \mathbf{k}$$

$$= \left(6 \cos \frac{v}{2} \right) \mathbf{i} - \left(6 \sin \frac{v}{2} \right) \mathbf{j} - 2u\mathbf{k} = \left\langle 6 \cos \frac{v}{2}, -6 \sin \frac{v}{2}, -2u \right\rangle$$

At the point $P = G\left(1, \frac{\pi}{3}\right)$, $u = 1$ and $v = \frac{\pi}{3}$. The tangents and the normal vector at this point are,

$$\mathbf{T}_u\left(1, \frac{\pi}{3}\right) = \left\langle 2\sin\frac{\pi}{6}, 2\cos\frac{\pi}{6}, 0 \right\rangle = \left\langle 1, \sqrt{3}, 0 \right\rangle$$

$$\mathbf{T}_v\left(1, \frac{\pi}{3}\right) = \left\langle 1 \cdot \cos\frac{\pi}{6}, -1 \cdot \sin\frac{\pi}{6}, 3 \right\rangle = \left\langle \frac{\sqrt{3}}{2}, -\frac{1}{2}, 3 \right\rangle$$

$$\mathbf{n}\left(1, \frac{\pi}{3}\right) = \left\langle 6\cos\frac{\pi}{6}, -6\sin\frac{\pi}{6}, -2\cdot 1 \right\rangle = \left\langle 3\sqrt{3}, -3, -2 \right\rangle$$

(b) A normal to the plane is $\mathbf{n}\left(1, \frac{\pi}{3}\right) = \left\langle 3\sqrt{3}, -3, -2 \right\rangle$ found in part (a). We find the tangency point:

$$P = \phi\left(1, \frac{\pi}{3}\right) = \left\langle 2\cdot 1\sin\frac{\pi}{6}, 2\cdot 1\cos\frac{\pi}{6}, 3\cdot\frac{\pi}{3} \right\rangle = \left\langle 1, \sqrt{3}, \pi \right\rangle$$

The equation of the tangent plane is, thus,

$$\left\langle x - 1, y - \sqrt{3}, z - \pi \right\rangle \cdot \left\langle 3\sqrt{3}, -3, -2 \right\rangle = 0$$

or

$$3\sqrt{3}(x - 1) - 3\left(y - \sqrt{3}\right) - 2(z - \pi) = 0$$

$$3\sqrt{3}x - 3y - 2z + 2\pi = 0$$

(c) In part (a) we found the normal vector:

$$\mathbf{n} = \left\langle 6\cos\frac{v}{2}, -6\sin\frac{v}{2}, -2u \right\rangle$$

We compute the length of $\mathbf{n}$:

$$\|\mathbf{n}\| = \sqrt{36\cos^2\frac{v}{2} + 36\sin^2\frac{v}{2} + 4u^2} = \sqrt{36 + 4u^2} = 2\sqrt{9 + u^2}$$

Using the Integral for the Surface Area we get:

$$\text{Area}(S) = \iint_{\mathcal{D}} \|n(u, v)\|\, du\, dv = \int_0^{2\pi}\int_0^1 2\sqrt{9 + u^2}\, du\, dv = 4\pi\int_0^1 \sqrt{9 + u^2}\, du$$

$$= 4\pi\left(\frac{u}{2}\sqrt{u^2 + 9} + \frac{9}{2}\ln\left(u + \sqrt{9 + u^2}\right)\Big|_{u=0}^1\right) = 4\pi\left(\frac{1}{2}\sqrt{10} + \frac{9}{2}\ln\left(1 + \sqrt{10}\right) - \frac{9}{2}\ln 3\right)$$

$$= 2\sqrt{10}\pi + 18\pi\ln\left(1 + \sqrt{10}\right) - 18\pi\ln 3 = 2\sqrt{10}\pi + 18\pi\ln\frac{1 + \sqrt{10}}{3} \approx 38.4$$

35. CAS Express the surface area of the surface $z = 10 - x^2 - y^2$ for $1 \le x \le 1, -3 \le y \le 3$ as a double integral. Evaluate the integral numerically using a CAS.

SOLUTION We use the Surface Integral over a graph. Let $g(x, y) = 10 - x^2 - y^2$. Then $g_x = -2x$, $g_y = -2y$ hence $\sqrt{1 + g_x^2 + g_y^2} = \sqrt{1 + 4x^2 + 4y^2}$. The area at the surface is the following integral which we compute using a CAS:

$$\text{Area}(S) = \iint_{\mathcal{D}} \sqrt{1 + g_x^2 + g_y^2}\, dx\, dy = \int_{-3}^3\int_{-1}^1 \sqrt{1 + 4x^2 + 4y^2}\, dx\, dy \approx 41.8525$$

37. Calculate $\iint_{\mathcal{S}} \left(x^2 + y^2\right)e^{-z}\, dS$, where $\mathcal{S}$ is the cylinder with equation $x^2 + y^2 = 9$ for $0 \le z \le 10$.

SOLUTION We parametrize the cylinder $\mathcal{S}$ by,

$$G(\theta, z) = (3\cos\theta, 3\sin\theta, z)$$

with the parameter domain:

$$0 \le \theta \le 2\pi, \quad 0 \le z \le 10.$$

We compute the tangent and normal vectors:

$$\mathbf{T}_\theta = \frac{\partial\phi}{\partial\theta} = \frac{\partial}{\partial\theta}(3\cos\theta, 3\sin\theta, z) = \langle -3\sin\theta, 3\cos\theta, 0 \rangle$$

$$\mathbf{T}_z = \frac{\partial\phi}{\partial\theta} = \frac{\partial}{\partial\theta}(3\cos\theta, 3\sin\theta, z) = \langle 0, 0, 1 \rangle$$

The normal vector is their cross product:

$$\mathbf{n} = \mathbf{T}_\theta \times \mathbf{T}_z = \begin{vmatrix} \mathbf{i} & \mathbf{j} & \mathbf{k} \\ -3\sin\theta & 3\cos\theta & 0 \\ 0 & 0 & 1 \end{vmatrix} = \begin{vmatrix} 3\cos\theta & 0 \\ 0 & 1 \end{vmatrix}\mathbf{i} - \begin{vmatrix} -3\sin\theta & 0 \\ 0 & 1 \end{vmatrix}\mathbf{j} + \begin{vmatrix} -3\sin\theta & 3\cos\theta \\ 0 & 0 \end{vmatrix}\mathbf{k}$$

$$= (3\cos\theta)\mathbf{i} + (3\sin\theta)\mathbf{j} = 3\langle\cos\theta, \sin\theta, 0\rangle$$

We compute the length of the normal vector:

$$\|\mathbf{n}\| = 3\sqrt{\cos^2\theta + \sin^2\theta + 0} = 3$$

We now express the function $f(x, y, z) = \left(x^2 + y^2\right)e^{-z}$ in terms of the parameters:

$$f(\phi(\theta, z)) = \left(x^2 + y^2\right)e^{-z} = \left(9\cos^2\theta + 9\sin^2\theta\right)e^{-z} = 9e^{-z}$$

Using the Theorem on Surface Integrals, we obtain:

$$\iint_S \left(x^2 + y^2\right)e^{-z}\,dS = \int_0^{10}\int_0^{2\pi} 9e^{-z}3\,d\theta\,dz = 27\cdot 2\pi\int_0^{10} e^{-z}\,dz = 54\pi\left(-e^{-z}\right)\Big|_{z=0}^{10}$$

$$= 54\pi\left(-e^{-10} + 1\right) \approx 54\pi$$

39. Let S be a small patch of surface with a parametrization $G(u, v)$ for $0 \le u \le 0.1$, $0 \le v \le 0.1$ such that the normal vector $\mathbf{n}(u, v)$ for $(u, v) = (0, 0)$ is $\mathbf{n} = \langle 2, -2, 4\rangle$. Use Eq. (3) in Section 16.4 to estimate the surface area of S.

SOLUTION

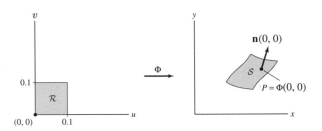

We use Eq. (3) in section 16.4 with $\left(u_{ij}, v_{ij}\right) = (0, 0)$, $\mathcal{R}_{ij} = \mathcal{R} = [0, 0.1] \times [0, 0.1]$ in the (u, v)-plane and $\mathcal{S}_{ij} = \mathcal{S} = G(\mathcal{R})$, in the (x, y)-plane to obtain the following estimation for the area of $\mathcal{S}$:

$$\text{Area}(\mathcal{S}) \approx \|\mathbf{n}(0, 0)\|\text{Area}(\mathcal{R})$$

That is:

$$\text{Area}(\mathcal{S}) \approx \|\langle 2, -2, 4\rangle\|0.1^2 = \sqrt{2^2 + (-2)^2 + 4^2}\cdot(0.1)^2 = 0.02\sqrt{6} \approx 0.049$$

In Exercises 41–46, compute $\iint_S \mathbf{F}\cdot d\mathbf{S}$ *for the given oriented surface or parametrized surface.*

41. $\mathbf{F}(x, y, z) = \langle y, x, e^{xz}\rangle$, $\quad x^2 + y^2 = 9$, $x \ge 0$, $y \ge 0$, $\quad -3 \le z \le 3$, $\quad$ outward-pointing normal

SOLUTION The part of the cylinder is parametrized by:

$$G(\theta, z) = (3\cos\theta, 3\sin\theta, z), \quad 0 \le \theta \le \frac{\pi}{2}, \quad -3 \le z \le 3$$

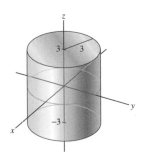

Step 1. Compute the tangent and normal vectors.

$$\mathbf{T}_\theta = \frac{\partial G}{\partial \theta} = \frac{\partial}{\partial \theta} \langle 3\cos\theta, 3\sin\theta, z \rangle = \langle -3\sin\theta, 3\cos\theta, 0 \rangle$$

$$\mathbf{T}_z = \frac{\partial G}{\partial z} = \frac{\partial}{\partial z} \langle 3\cos\theta, 3\sin\theta, z \rangle = \langle 0, 0, 1 \rangle$$

We compute the cross product:

$$\mathbf{T}_\theta \times \mathbf{T}_z = ((-3\sin\theta)\mathbf{i} + (3\cos\theta)\mathbf{j}) \times \mathbf{k} = (3\sin\theta)\mathbf{j} + (3\cos\theta)\mathbf{i} = \langle 3\cos\theta, 3\sin\theta, 0 \rangle$$

The outward pointing normal is (when $\theta = 0$, the x-component must be positive):

$$\mathbf{n} = \langle 3\cos\theta, 3\sin\theta, 0 \rangle$$

Step 2. Evaluate the dot product $\mathbf{F} \cdot \mathbf{n}$. We write $\mathbf{F}(x, y, z) = \langle y, x, e^{xz} \rangle$ in terms of the parameters by substituting $x = 3\cos\theta$, $y = 3\sin\theta$. We get:

$$\mathbf{F}(G(\theta, z)) = \left\langle 3\sin\theta, 3\cos\theta, e^{3z\cos\theta} \right\rangle$$

Hence:

$$\mathbf{F}(G(\theta, z)) \cdot \mathbf{n} = \left\langle 3\sin\theta, 3\cos\theta, e^{3z\cos\theta} \right\rangle \cdot \langle 3\cos\theta, 3\sin\theta, 0 \rangle$$

$$= 18\sin\theta\cos\theta$$

Step 3. Evaluate the surface integral. The surface integral is equal to the following double integral (we use the trigonometric identities $\sin\theta\cos\theta = \frac{\sin 2\theta}{2}$ and $\sin^2 2\theta = \frac{1}{2}(1 - \cos 4\theta)$):

$$\iint_S \mathbf{F} \cdot d\mathbf{S} = \int_0^{\pi/2} \int_{-3}^3 \mathbf{F}(G(\theta, z)) \cdot \mathbf{n}(\theta, z) \, dz \, d\theta = \int_0^{\pi/2} \int_{-3}^3 18\sin\theta\cos\theta \, d\theta$$

$$= 18 \int_0^{\pi/2} \frac{\sin 2\theta}{2} \, d\theta \cdot \int_{-3}^3 dz = -9\frac{\cos 2\theta}{2}\bigg|_0^{\pi/2} \cdot z \bigg|_{-3}^3$$

$$= -\frac{9}{2}(-1 - 1) \cdot (3 - (-3)) = 54$$

43. $\mathbf{F}(x, y, z) = \langle 0, 0, x^2 + y^2 \rangle$, $x^2 + y^2 + z^2 = 4$, $z \geq 0$, outward-pointing normal

SOLUTION The upper hemisphere is parametrized by:

$$G(\theta, \phi) = \langle 2\cos\theta\sin\phi, 2\sin\theta\sin\phi, 2\cos\phi \rangle, \quad 0 \leq \theta \leq 2\pi, \quad 0 \leq \phi \leq \frac{\pi}{2}$$

As seen in Section 17.4, since $0 \leq \phi \leq \frac{\pi}{2}$ then the outward-pointing normal is:

$$\mathbf{n} = 4\sin\phi \langle \cos\theta\sin\phi, \sin\theta\sin\phi, \cos\psi \rangle$$

We express $\mathbf{F}$ in terms of the parameters:

$$\mathbf{F}(G(\theta, \phi)) = \left\langle 0, 0, x^2 + y^2 \right\rangle = \left\langle 0, 0, 4\sin^2\phi \left(\cos^2\theta + \sin^2\theta\right) \right\rangle$$

$$= \left\langle 0, 0, 4\sin^2\phi \right\rangle$$

The dot product $\mathbf{F} \cdot \mathbf{n}$ is thus

$$\mathbf{F}(G(\theta, \phi)) \cdot \mathbf{n}(\theta, \phi) = 16\sin^3\phi\cos\phi$$

We obtain the following integral:

$$\iint_S \mathbf{F} \cdot d\mathbf{s} = \iint_D \mathbf{F}(G(\theta, \phi)) \cdot \mathbf{n}(\theta, \phi) \, d\theta \, d\phi$$

$$= \int_0^{\pi/2} \int_0^{2\pi} 16\sin^3\phi\cos\phi \, d\theta \, d\phi = 16 \int_0^{2\pi} d\theta \cdot \int_0^{\pi/2} \sin^3\phi\cos\phi \, d\phi$$

$$= 16 \cdot 2\pi \cdot \frac{\sin^4\phi}{4}\bigg|_0^{\pi/2} = 8\pi$$

45. $\mathbf{F}(x, y, z) = \langle 0, 0, xze^{xy} \rangle$, $\quad z = xy$, $\quad 0 \le x \le 1, 0 \le y \le 1$, $\quad$ upward-pointing normal

SOLUTION We parametrize the surface by:

$$G(x, y) = (x, y, xy)$$

Where the parameter domain is the square:

$$\mathcal{D} = \{(x, y) : 0 \le x \le 1, 0 \le y \le 1\}$$

Step 1. Compute the tangent and normal vectors.

$$\mathbf{T}_x = \frac{\partial G}{\partial x} = \frac{\partial}{\partial x} \langle x, y, xy \rangle = \langle 1, 0, y \rangle$$

$$\mathbf{T}_y = \frac{\partial G}{\partial y} = \frac{\partial}{\partial y} \langle x, y, xy \rangle = \langle 0, 1, x \rangle$$

$$\mathbf{T}_x \times \mathbf{T}_y = \begin{vmatrix} \mathbf{i} & \mathbf{j} & \mathbf{k} \\ 1 & 0 & y \\ 0 & 1 & x \end{vmatrix} = \begin{vmatrix} 0 & y \\ 1 & x \end{vmatrix} \mathbf{i} - \begin{vmatrix} 1 & y \\ 0 & x \end{vmatrix} \mathbf{j} + \begin{vmatrix} 1 & 0 \\ 0 & 1 \end{vmatrix} \mathbf{k} = -y\mathbf{i} - x\mathbf{j} + \mathbf{k} = \langle -y, -x, 1 \rangle$$

Since the normal points upwards, the z-coordinate is positive. Therefore the normal vector is:

$$\mathbf{n} = \langle -y, -x, 1 \rangle$$

Step 2. Evaluate the dot product $\mathbf{F} \cdot \mathbf{n}$. We express $\mathbf{F}$ in terms of x and y:

$$\mathbf{F}(G(x, y)) = \langle 0, 0, xze^{xy} \rangle = \langle 0, 0, x(xy)e^{xy} \rangle = \langle 0, 0, x^2 ye^{xy} \rangle$$

Hence:

$$\mathbf{F}(G(x, y)) \cdot \mathbf{n}(x, y) = \langle 0, 0, x^2 ye^{xy} \rangle \cdot \langle -y, -x, 1 \rangle = x^2 ye^{xy}$$

Step 3. Evaluate the surface integral. The surface integral is equal to the following double integral:

$$\iint_S \mathbf{F} \cdot d\mathbf{s} = \iint_{\mathcal{D}} \mathbf{F}(G(x, y)) \cdot \mathbf{n}(x, y) \, dx \, dy$$

$$= \int_0^1 \int_0^1 x^2 ye^{xy} \, dy \, dx = \int_0^1 x^2 \left(\int_0^1 ye^{xy} \, dy \right) dx \tag{1}$$

We evaluate the inner integral using integration by parts:

$$\int_0^1 ye^{xy} \, dy = \frac{y}{x} e^{xy} \Big|_{y=0}^1 - \int_0^1 \frac{1}{x} e^{xy} \, dy = \frac{e^x}{x} - \frac{1}{x^2} e^{xy} \Big|_{y=0}^1 = \frac{e^x}{x} - \frac{1}{x^2} \left(e^x - 1 \right)$$

Substituting this integral in (1) gives:

$$\iint_S \mathbf{F} \cdot d\mathbf{s} = \int_0^1 \left(xe^x - (e^x - 1) \right) dx = \int_0^1 xe^x \, dx - \int_0^1 (e^x - 1) \, dx$$

$$= \int_0^1 xe^x \, dx - (e^x - x) \Big|_0^1 = \int_0^1 xe^x \, dx - (e - 2)$$

Using integration by parts we have:

$$\iint_S F \cdot dS = xe^x - e^x \Big|_0^1 - (e - 2) = 1 - (e - 2) = 3 - e$$

47. Calculate the total charge on the cylinder

$$x^2 + y^2 = R^2, \qquad 0 \le z \le H$$

if the charge density in cylindrical coordinates is $\rho(\theta, z) = Kz^2 \cos^2 \theta$, where K is a constant.

SOLUTION The total change on the surface S is $\iint_S \rho \, dS$. We parametrize the surface by,

$$G(\theta, z) = (R \cos \theta, R \sin \theta, Hz)$$

with the parameter domain,

$$0 \le \theta \le 2\pi, \ 0 \le z \le 1.$$

We compute the tangent and normal vectors:

$$\mathbf{T}_\theta = \frac{\partial G}{\partial \theta} = \frac{\partial}{\partial \theta} \langle R\cos\theta, R\sin\theta, Hz\rangle = \langle -R\sin\theta, R\cos\theta, 0\rangle$$

$$\mathbf{T}_z = \frac{\partial G}{\partial z} = \frac{\partial}{\partial z} \langle R\cos\theta, R\sin\theta, Hz\rangle = \langle 0, 0, H\rangle$$

The normal vector is their cross product:

$$\mathbf{n} = \mathbf{T}_\theta \times \mathbf{T}_z = \begin{vmatrix} \mathbf{i} & \mathbf{j} & \mathbf{k} \\ -R\sin\theta & R\cos\theta & 0 \\ 0 & 0 & H \end{vmatrix}$$

$$= \begin{vmatrix} R\cos\theta & 0 \\ 0 & H \end{vmatrix} \mathbf{i} - \begin{vmatrix} -R\sin\theta & 0 \\ 0 & H \end{vmatrix} \mathbf{j} + \begin{vmatrix} -R\sin\theta & R\cos\theta \\ 0 & 0 \end{vmatrix} \mathbf{k}$$

$$= (RH\cos\theta)\mathbf{i} + (RH\sin\theta)\mathbf{j} = RH\langle \cos\theta, \sin\theta, 0\rangle$$

We find the length of $\mathbf{n}$:

$$\|\mathbf{n}\| = RH\sqrt{\cos^2\theta + \sin^2\theta} = RH$$

We compute $\rho\,(G(\theta, z))$:

$$\rho\,(G(\theta, z)) = K(Hz)^2\cos^2\theta = KH^2z^2\cos^2\theta$$

Using the Theorem on Surface Integrals we obtain:

$$\iint_S \rho \cdot dS = \iint_D \rho\,(G(\theta, z)) \cdot \|\mathbf{n}(\theta, z)\|\, dz\, d\theta = \int_0^{2\pi} \int_0^1 KH^2z^2\cos^2\theta \cdot HR\, dz\, d\theta$$

$$= \left(\int_0^1 KH^3Rz^2\,dz\right)\left(\int_0^{2\pi}\cos^2\theta\, d\theta\right) = \left(\frac{KH^3Rz^3}{3}\bigg|_0^1\right)\left(\frac{\theta}{2} + \frac{\sin 2\theta}{4}\bigg|_0^{2\pi}\right)$$

$$= \frac{KH^3R}{3}\cdot\pi = \frac{\pi}{3}KH^3R$$

49. With $\mathbf{v}$ as in Exercise 48, calculate the flow rate across the part of the elliptic cylinder $\dfrac{x^2}{4} + y^2 = 1$ where $x \geq 0, y \geq 0$, and $0 \leq z \leq 4$.

SOLUTION The flow rate of a fluid with velocity field $\mathbf{v} = \langle 2x, y, xy\rangle$ through the elliptic cylinder S is the surface integral:

$$\iint_S \mathbf{v}\cdot d\mathbf{S} \tag{1}$$

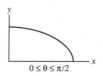

To compute this integral, we parametrize S by,

$$G(\theta, z) = (2\cos\theta, \sin\theta, z), \quad 0 \leq \theta \leq \frac{\pi}{2}, \quad 0 \leq z \leq 4$$

$0 \leq \theta \leq \pi/2$

Step 1. Compute the tangent and normal vectors.

$$\mathbf{T}_\theta = \frac{\partial G}{\partial \theta} = \frac{\partial}{\partial \theta} \langle 2\cos\theta, \sin\theta, z \rangle = \langle -2\sin\theta, \cos\theta, 0 \rangle$$

$$\mathbf{T}_z = \frac{\partial G}{\partial z} = \frac{\partial}{\partial z} \langle 2\cos\theta, \sin\theta, z \rangle = \langle 0, 0, 1 \rangle$$

$$\mathbf{n} = \mathbf{T}_\theta \times \mathbf{T}_z = \begin{vmatrix} \mathbf{i} & \mathbf{j} & \mathbf{k} \\ -2\sin\theta & \cos\theta & 0 \\ 0 & 0 & 1 \end{vmatrix} = (\cos\theta)\mathbf{i} + (2\sin\theta)\mathbf{j} = \langle \cos\theta, 2\sin\theta, 0 \rangle$$

Step 2. Compute the dot product $\mathbf{v} \cdot \mathbf{n}$

$$\mathbf{v}\,(G(\theta, z)) \cdot \mathbf{n} = \langle 4\cos\theta, \sin\theta, 2\cos\theta\sin\theta \rangle \cdot \langle \cos\theta, 2\sin\theta, 0 \rangle = 4\cos^2\theta + 2\sin^2\theta$$

$$= 2\cos^2\theta + 2\left(\cos^2\theta + \sin^2\theta\right) = 2\cos^2\theta + 2$$

Step 3. Evaluate the flux of $\mathbf{v}$. The flux of $\mathbf{v}$ in (1) is equal to the following double integral (we use the equality $2\cos^2\theta = 1 + \cos 2\theta$ in our calculation):

$$\iint_S \mathbf{v} \cdot d\mathbf{S} = \iint_D \mathbf{v}\,(G(\theta, z)) \cdot \mathbf{n}\, d\theta\, dz = \int_0^4 \int_0^{\pi/2} \left(2\cos^2\theta + 2\right) d\theta\, dz$$

$$= 4\int_0^{\pi/2} \left(2\cos^2\theta + 2\right) d\theta = 4\int_0^{\pi/2} (3 + \cos 2\theta)\, d\theta = 4\left(3\theta + \frac{\sin 2\theta}{2}\Big|_{\theta=0}^{\pi/2}\right) = 6\pi$$

17 FUNDAMENTAL THEOREMS OF VECTOR ANALYSIS

17.1 Green's Theorem   (LT Section 18.1)

Preliminary Questions

1. Which vector field $\mathbf{F}$ is being integrated in the line integral $\oint x^2\,dy - e^y\,dx$?

SOLUTION The line integral can be rewritten as $\oint -e^y\,dx + x^2\,dy$. This is the line integral of $\mathbf{F} = \left\langle -e^y, x^2 \right\rangle$ along the curve.

2. Draw a domain in the shape of an ellipse and indicate with an arrow the boundary orientation of the boundary curve. Do the same for the annulus (the region between two concentric circles).

SOLUTION The orientation on $\mathcal{C}$ is counterclockwise, meaning that the region enclosed by $\mathcal{C}$ lies to the left in traversing $\mathcal{C}$.

For the annulus, the inner boundary is oriented clockwise and the outer boundary is oriented counterclockwise. The region between the circles lies to the left while traversing each circle.

3. The circulation of a conservative vector field around a closed curve is zero. Is this fact consistent with Green's Theorem? Explain.

SOLUTION Green's Theorem asserts that

$$\int_{\mathcal{C}} \mathbf{F}\cdot d\mathbf{s} = \int_{\mathcal{C}} P\,dx + Q\,dy = \iint_{\mathcal{D}} \left(\frac{\partial Q}{\partial x} - \frac{\partial P}{\partial y} \right) dA \tag{1}$$

If $\mathbf{F}$ is conservative, the cross partials are equal, that is,

$$\frac{\partial P}{\partial y} = \frac{\partial Q}{\partial x} \quad \Rightarrow \quad \frac{\partial Q}{\partial x} - \frac{\partial P}{\partial y} = 0 \tag{2}$$

Combining (1) and (2) we obtain $\int_{\mathcal{C}} \mathbf{F}\cdot d\mathbf{s} = 0$. That is, Green's Theorem implies that the integral of a conservative vector field around a simple closed curve is zero.

4. Indicate which of the following vector fields possess the following property: For every simple closed curve $\mathcal{C}$, $\int_{\mathcal{C}} \mathbf{F}\cdot d\mathbf{s}$ is equal to the area enclosed by $\mathcal{C}$.

(a) $\mathbf{F} = \langle -y, 0 \rangle$ **(b)** $\mathbf{F} = \langle x, y \rangle$ **(c)** $\mathbf{F} = \left\langle \sin(x^2), x + e^{y^2} \right\rangle$

SOLUTION By Green's Theorem,

$$\int_{\mathcal{C}} \mathbf{F}\cdot d\mathbf{s} = \iint_{\mathcal{D}} \left(\frac{\partial Q}{\partial x} - \frac{\partial P}{\partial y} \right) dx\,dy \tag{1}$$

(a) Here, $P = -y$ and $Q = 0$, hence $\frac{\partial Q}{\partial x} - \frac{\partial P}{\partial y} = 0 - (-1) = 1$. Therefore, by (1),

$$\int_C \mathbf{F} \cdot d\mathbf{s} = \iint_D 1\, dx\, dy = \text{Area}(D)$$

(b) We have $P = x$ and $Q = y$, therefore $\frac{\partial Q}{\partial x} - \frac{\partial P}{\partial y} = 0 - 0 = 0$. By (1) we get

$$\int_C \mathbf{F} \cdot d\mathbf{s} = \iint_D 0\, dx\, dy = 0 \neq \text{Area}(D)$$

(c) In this vector field we have $P = \sin(x^2)$ and $Q = x + e^{y^2}$. Therefore,

$$\frac{\partial Q}{\partial x} - \frac{\partial P}{\partial y} = 1 - 0 = 1.$$

By (1) we obtain

$$\int_C \mathbf{F} \cdot d\mathbf{s} = \iint_D 1\, dx\, dy = \text{Area}(D).$$

Exercises

1. Verify Green's Theorem for the line integral $\oint_C xy\, dx + y\, dy$, where C is the unit circle, oriented counterclockwise.

SOLUTION

Step 1. Evaluate the line integral. We use the parametrization $\gamma(\theta) = \langle \cos\theta, \sin\theta \rangle$, $0 \le \theta \le 2\pi$ of the unit circle. Then

$$dx = -\sin\theta\, d\theta, \quad dy = \cos\theta\, d\theta$$

and

$$xy\, dx + y\, dy = \cos\theta \sin\theta(-\sin\theta\, d\theta) + \sin\theta \cos\theta\, d\theta = \left(-\cos\theta \sin^2\theta + \sin\theta \cos\theta\right) d\theta$$

The line integral is thus

$$\int_C xy\, dx + y\, dy = \int_0^{2\pi} \left(-\cos\theta \sin^2\theta + \sin\theta \cos\theta\right) d\theta$$

$$= \int_0^{2\pi} -\cos\theta \sin^2\theta\, d\theta + \int_0^{2\pi} \sin\theta \cos\theta\, d\theta = -\frac{\sin^3\theta}{3}\bigg|_0^{2\pi} - \frac{\cos 2\theta}{4}\bigg|_0^{2\pi} = 0 \qquad (1)$$

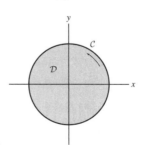

Step 2. Evaluate the double integral. Since $P = xy$ and $Q = y$, we have

$$\frac{\partial Q}{\partial x} - \frac{\partial P}{\partial y} = 0 - x = -x.$$

We compute the double integral in Green's Theorem:

$$\iint_D \left(\frac{\partial Q}{\partial x} - \frac{\partial P}{\partial y}\right) dx\, dy = \iint_D -x\, dx\, dy = -\iint_D x\, dx\, dy$$

The integral of x over the disk D is zero, since by symmetry the positive and negative values of x cancel each other. Therefore,

$$\iint_D \left(\frac{\partial Q}{\partial x} - \frac{\partial P}{\partial y}\right) dx\, dy = 0 \qquad (2)$$

Step 3. Compare. The line integral in (1) is equal to the double integral in (2), as stated in Green's Theorem.

In Exercises 3–10, use Green's Theorem to evaluate the line integral. Orient the curve counterclockwise unless otherwise indicated.

3. $\oint_C y^2\,dx + x^2\,dy$, where C is the boundary of the unit square $0 \le x \le 1, 0 \le y \le 1$

SOLUTION

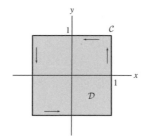

We have $P = y^2$ and $Q = x^2$, therefore

$$\frac{\partial Q}{\partial x} - \frac{\partial P}{\partial y} = 2x - 2y$$

Using Green's Theorem we obtain

$$\int_C y^2\,dx + x^2\,dy = \iint_D \frac{\partial Q}{\partial x} - \frac{\partial P}{\partial y}\,dA = \iint_D (2x - 2y)\,dx\,dy = 2\iint_D x\,dx\,dy - 2\iint_D y\,dx\,dy$$

By symmetry, the positive and negative values of x cancel each other in the first integral, so this integral is zero. The second double integral is zero by similar reasoning. Therefore,

$$\int_C y^2\,dx + x^2\,dy = 0 - 0 = 0$$

5. $\oint_C x^2\,y\,dx$, where C is the unit circle centered at the origin

SOLUTION

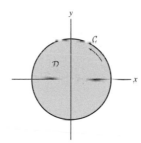

In this function $P = x^2 y$ and $Q = 0$. Therefore,

$$\frac{\partial Q}{\partial x} - \frac{\partial P}{\partial y} = 0 - x^2 = -x^2$$

We obtain the following integral:

$$I = \int_C x^2 y\,dx = \iint_D \left(\frac{\partial Q}{\partial x} - \frac{\partial P}{\partial y}\right) dA = \iint_D -x^2\,dA$$

We convert the integral to polar coordinates. This gives

$$I = \int_0^{2\pi} \int_0^1 -r^2 \cos^2\theta \cdot r\,dr\,d\theta = \int_0^{2\pi} \int_0^1 -r^3 \cos^2\theta\,dr\,d\theta$$

$$= \left(\int_0^{2\pi} \cos^2\theta\,d\theta\right)\left(\int_0^1 -r^3\,dr\right) = \left(\frac{\theta}{2} + \frac{\sin 2\theta}{4}\bigg|_{\theta=0}^{2\pi}\right)\left(-\frac{r^4}{4}\bigg|_{r=0}^1\right) = \pi \cdot \left(-\frac{1}{4}\right) = -\frac{\pi}{4}$$

7. $\oint_C \mathbf{F} \cdot d\mathbf{s}$, where $\mathbf{F} = \langle x^2, x^2 \rangle$ and C consists of the arcs $y = x^2$ and $y = x$ for $0 \leq x \leq 1$

SOLUTION   By Green's Theorem,

$$I = \int_C \mathbf{F} \cdot d\mathbf{s} = \iint_D \left(\frac{\partial Q}{\partial x} - \frac{\partial P}{\partial y} \right) dA$$

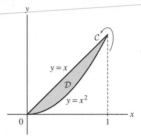

We have $P = Q = x^2$, therefore

$$\frac{\partial Q}{\partial x} - \frac{\partial P}{\partial y} = 2x - 0 = 2x$$

Hence,

$$I = \iint_D 2x \, dA = \int_0^1 \int_{x^2}^x 2x \, dy \, dx = \int_0^1 2xy \Big|_{y=x^2}^x dx = \int_0^1 2x(x - x^2) \, dx = \int_0^1 (2x^2 - 2x^3) \, dx$$

$$= \frac{2}{3}x^3 - \frac{1}{2}x^4 \Big|_0^1 = \frac{2}{3} - \frac{1}{2} = \frac{1}{6}$$

9. The line integral of $\mathbf{F} = \langle e^{x+y}, e^{x-y} \rangle$ along the curve (oriented clockwise) consisting of the line segments by joining the points $(0, 0)$, $(2, 2)$, $(4, 2)$, $(2, 0)$, and back to $(0, 0)$ (note the orientation).

SOLUTION   Consider $\mathbf{F} = \langle e^{x+y}, e^{x-y} \rangle$. Here, $P = e^{x+y}$ and $Q = e^{x-y}$, hence

$$\frac{\partial Q}{\partial x} - \frac{\partial P}{\partial y} = e^{x-y} - e^{x+y} = e^x(e^{-y} - e^y).$$

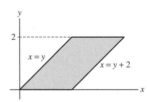

Using Green's Theorem we obtain

$$\int_C \mathbf{F} \cdot d\mathbf{s} = \iint_D e^x(e^{-y} - e^y) \, dx \, dy = \int_0^2 \int_y^{y+2} e^x(e^{-y} - e^y) \, dx \, dy = \int_0^2 e^x(e^{-y} - e^y) \Big|_{x=y}^{y+2} dy$$

$$= \int_0^2 (e^{y+2} - e^y)(e^{-y} - e^y) \, dy = \int_0^2 (e^2 - 1)(1 - e^{2y}) \, dy = (e^2 - 1) \left(y - \frac{e^{2y}}{2} \right) \Big|_{y=0}^2$$

$$= (e^2 - 1) \left(2 - \frac{e^4}{2} - \left(-\frac{1}{2} \right) \right) = \frac{(e^2 - 1)(5 - e^4)}{2}$$

11. Let $\mathbf{F} = \langle 2xe^y, x + x^2 e^y \rangle$ and let C be the quarter-circle path from A to B in Figure 17. Evaluate $I = \oint_C \mathbf{F} \cdot d\mathbf{s}$ as follows:

(a) Find a function $V(x, y)$ such that $\mathbf{F} = \mathbf{G} + \nabla V$, where $\mathbf{G} = \langle 0, x \rangle$.

(b) Show that the line integrals of $\mathbf{G}$ along the segments $\overline{OA}$ and $\overline{OB}$ are zero.

(c) Evaluate I. *Hint:* Use Green's Theorem to show that

$$I = V(B) - V(A) + 4\pi$$

B = (0, 4)

A = (4, 0)

O

FIGURE 17

SOLUTION

(a) We need to find a potential function $V(x, y)$ for the difference

$$\mathbf{F} - \mathbf{G} = \left\langle 2xe^y, x + x^2e^y \right\rangle - \langle 0, x \rangle = \left\langle 2xe^y, x^2e^y \right\rangle$$

We let $V(x, y) = x^2e^y$.

(b) We use the parametrizations $\overline{AO} : \langle t, 0 \rangle, 0 \le t \le 4$ and $\overline{OB} : \langle 0, t \rangle, 0 \le t \le 4$ to evaluate the integrals of $\mathbf{G} = \langle 0, x \rangle$. We get

$$\int_{\overline{OA}} \mathbf{G} \cdot d\mathbf{s} = \int_0^4 \langle 0, t \rangle \cdot \langle 1, 0 \rangle \, dt = \int_0^4 0 \, dt = 0$$

$$\int_{\overline{OB}} \mathbf{G} \cdot d\mathbf{s} = \int_0^4 \langle 0, 0 \rangle \cdot \langle 0, 1 \rangle \, dt = \int_0^4 0 \, dt = 0$$

B = (0, 4)

A = (4, 0)

O

(c) Since $\mathbf{F} - \mathbf{G} = \nabla V$, we have

$$\int_C (\mathbf{F} - \mathbf{G}) \cdot d\mathbf{s} = V(B) - V(A) = \int_C \mathbf{F} \cdot d\mathbf{s} - \int_C \mathbf{G} \cdot d\mathbf{s} = I - \int_C \mathbf{G} \cdot d\mathbf{s}$$

That is,

$$I = V(B) - V(A) + \int_C \mathbf{G} \cdot d\mathbf{s} \tag{1}$$

To compute the line integral on the right-hand side, we rewrite it as

$$\int_C \mathbf{G} \cdot d\mathbf{s} = \int_{\overline{BO} + \overline{OA} + C} \mathbf{G} \cdot d\mathbf{s} - \int_{\overline{BO}} \mathbf{G} \cdot d\mathbf{s} - \int_{\overline{OA}} \mathbf{G} \cdot d\mathbf{s}$$

Using part (b) we may write

$$\int_C \mathbf{G} \cdot d\mathbf{s} = \int_{\overline{BO} + \overline{OA} + C} \mathbf{G} \cdot d\mathbf{s} \tag{2}$$

We now use Green's Theorem. Since $\mathbf{G} = \langle 0, x \rangle$, we have $P = 0$ and $Q = x$, hence $\frac{\partial Q}{\partial x} - \frac{\partial P}{\partial y} = 1 - 0 = 1$. Thus,

$$\int_{\overline{BO} + \overline{OA} + C} \mathbf{G} \cdot d\mathbf{s} = \iint_{\mathcal{D}} 1 \, dA = \text{Area}(\mathcal{D}) = \frac{\pi \cdot 4^2}{4} = 4\pi \tag{3}$$

Combining (1), (2), and (3), we obtain

$$I = V(B) - V(A) + 4\pi$$

Since $V(x, y) = x^2e^y$, we conclude that

$$I = V(0, 4) - V(4, 0) + 4\pi = 0 - 4^2e^0 + 4\pi = 4\pi - 16.$$

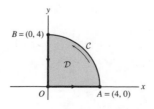

13. Evaluate $I = \int_C (\sin x + y)\, dx + (3x + y)\, dy$ for the nonclosed path $ABCD$ in Figure 19. Use the method of Exercise 12.

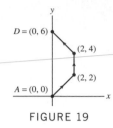

FIGURE 19

SOLUTION

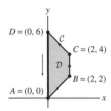

Let $\mathbf{F} = \langle \sin x + y, 3x + y \rangle$, hence $P = \sin x + y$ and $Q = 3x + y$. We denote by $\mathcal{C}_1$ the closed path determined by $\mathcal{C}$ and the segment $\overline{DA}$. Then by Green's Theorem,

$$\int_{\mathcal{C}_1} P\, dx + Q\, dy = \iint_{\mathcal{D}} \left(\frac{\partial Q}{\partial x} - \frac{\partial P}{\partial y} \right) dA = \iint_{\mathcal{D}} (3-1)\, dA = 2 \iint_{\mathcal{D}} dA = 2\,\text{Area}(\mathcal{D}) \qquad (1)$$

The area of $\mathcal{D}$ is the area of the trapezoid $ABCD$, that is,

$$\text{Area}(\mathcal{D}) = \frac{\left(\overline{BC} + \overline{AD}\right) h}{2} = \frac{(2+6)\cdot 2}{2} = 8.$$

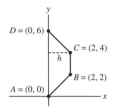

Combining with (1) we get

$$\int_{\mathcal{C}_1} P\, dx + Q\, dy = 2 \cdot 8 = 16$$

Using properties of line integrals, we have

$$\int_{\mathcal{C}} P\, dx + Q\, dy + \int_{\overline{DA}} P\, dx + Q\, dy = 16 \qquad (2)$$

We compute the line integral over $\overline{DA}$, using the parametrization

$$\overline{DA} : x = 0, \ y = t, t \text{ varies from 6 to 0.}$$

We get

$$\int_{\overline{DA}} P\, dx + Q\, dy = \int_6^0 F(0, t) \cdot \frac{d}{dt} \langle 0, t \rangle \, dt = \int_6^0 \langle \sin 0 + t, 3 \cdot 0 + t \rangle \cdot \langle 0, 1 \rangle \, dt$$

$$= \int_6^0 \langle t, t \rangle \cdot \langle 0, 1 \rangle \, dt = \int_6^0 t\, dt = \left. \frac{t^2}{2} \right|_{t=6}^{0} = -18$$

We substitute in (2) and solve for the required integral:

$$\int_{\mathcal{C}} P\, dx + Q\, dy - 18 = 16 \quad \text{or} \quad \int_{\mathcal{C}} P\, dx + Q\, dy = 34.$$

In Exercises 15–18, use Eq. (6) to calculate the area of the given region.

15. The circle of radius 3 centered at the origin

SOLUTION   By Eq. (6), we have

$$A = \frac{1}{2} \int_C x \, dy - y \, dx$$

We parametrize the circle by $x = 3\cos\theta$, $y = 3\sin\theta$, hence,

$$x \, dy - y \, dx = 3\cos\theta \cdot 3\cos\theta \, d\theta - 3\sin\theta(-3\sin\theta) \, d\theta = (9\cos^2\theta + 9\sin^2\theta) \, d\theta = 9 \, d\theta$$

Therefore,

$$A = \frac{1}{2} \int_C x \, dy - y \, dx = \frac{1}{2} \int_0^{2\pi} 9 \, d\theta = \frac{9}{2} \cdot 2\pi = 9\pi.$$

17. The region between the x-axis and the cycloid parametrized by $\mathbf{c}(t) = (t - \sin t, 1 - \cos t)$ for $0 \le t \le 2\pi$ (Figure 20)

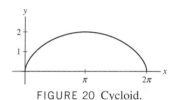

FIGURE 20 Cycloid.

SOLUTION   By Eq. (6), the area is the following integral:

$$A = \frac{1}{2} \int_C x \, dy - y \, dx$$

where C is the closed curve determined by the segment OA and the cycloid Γ.

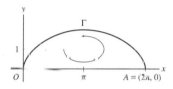

Therefore,

$$A = \frac{1}{2} \int_{OA} x \, dy - y \, dx + \frac{1}{2} \int_{\Gamma} x \, dy - y \, dx \tag{1}$$

We compute the two integrals. The segment OA is parametrized by $\langle t, 0 \rangle$, $t = 0$ to $t = 2\pi$. Hence, $x = t$ and $y = 0$. Therefore,

$$x \, dy - y \, dx = t \cdot 0 \, dt - 0 \cdot dt = 0$$

$$\int_{OA} x \, dy - y \, dx = 0 \tag{2}$$

On Γ we have $x = t - \sin t$ and $y = 1 - \cos t$, therefore

$$x \, dy - y \, dx = (t - \sin t)\sin t \, dt - (1 - \cos t)(1 - \cos t) \, dt$$

$$= (t\sin t - \sin^2 t - 1 + 2\cos t - \cos^2 t) \, dt = (t\sin t + 2\cos t - 2) \, dt$$

Hence,

$$\int_{\Gamma} x \, dy - y \, dx = \int_{2\pi}^{0} (t\sin t + 2\cos t - 2) \, dt = \int_0^{2\pi} (2 - 2\cos t - t\sin t) \, dt$$

$$= 2t - 2\sin t + t\cos t - \sin t \Big|_0^{2\pi} = 2t - 3\sin t + t\cos t \Big|_0^{2\pi} = 6\pi \tag{3}$$

Substituting (2) and (3) in (1) we get

$$A = \frac{1}{2} \cdot 0 + \frac{1}{2} \cdot 6\pi = 3\pi.$$

19. Let $x^3 + y^3 = 3xy$ be the **folium of Descartes** (Figure 21).

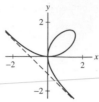

FIGURE 21 Folium of Descartes.

(a) Show that the folium has a parametrization in terms of $t = y/x$ given by

$$x = \frac{3t}{1+t^3}, \qquad y = \frac{3t^2}{1+t^3} \quad (-\infty < t < \infty) \quad (t \neq -1)$$

(b) Show that

$$x\,dy - y\,dx = \frac{9t^2}{(1+t^3)^2}\,dt$$

Hint: By the Quotient Rule,

$$x^2\,d\left(\frac{y}{x}\right) = x\,dy - y\,dx$$

(c) Find the area of the loop of the folium.

SOLUTION

(a) We show that $x = \frac{3t}{1+t^3}$, $y = \frac{3t^2}{1+t^3}$ satisfy the equation $x^3 + y^3 - 3xy = 0$ of the folium:

$$x^3 + y^3 - 3xy = \left(\frac{3t}{1+t^3}\right)^3 + \left(\frac{3t^2}{1+t^3}\right)^3 - 3 \cdot \frac{3t}{1+t^3} \cdot \frac{3t^2}{1+t^3}$$

$$= \frac{27t^3 + 27t^6}{(1+t^3)^3} - \frac{27t^3(1+t^3)}{(1+t^3)^3} = \frac{27t^3\left(1 + t^3 - (1+t^3)\right)}{(1+t^3)^3} = \frac{0}{(1+t^3)^3} = 0$$

This proves that the curve parametrized by $x = \frac{3t}{1+t^3}$, $y = \frac{3t^2}{1+t^3}$ lies on the folium of Descartes. This parametrization parametrizes the whole folium since the two equations can be solved for t in terms of x and y. That is,

$$\begin{matrix} x = \dfrac{3t}{1+t^3} \\[2mm] y = \dfrac{3t^2}{1+t^3} \end{matrix} \quad \Rightarrow \quad t = \frac{y}{x}$$

A glance at the graph of the folium shows that any line $y = tx$, with slope t, intersects the folium exactly once. Thus, there is a one-to-one relationship between the values of t and the points on the graph.

(b) We differentiate the two sides of $t = \frac{y}{x}$ with respect to t. Using the Quotient Rule gives

$$1 = \frac{x\frac{dy}{dt} - y\frac{dx}{dt}}{x^2}$$

or

$$x\frac{dy}{dt} - y\frac{dx}{dt} = x^2 = \left(\frac{3t}{1+t^3}\right)^2$$

This equality can be written in the form

$$x\,dy - y\,dx = \frac{9t^2}{(1+t^3)^2}\,dt$$

(c) We use the formula for the area enclosed by a closed curve and the result of part (b) to find the required area. That is,

$$A = \frac{1}{2}\int_C x\,dy - y\,dx = \frac{1}{2}\int_0^\infty \frac{9t^2}{(1+t^3)^2}\,dt$$

From our earlier discussion on the parametrization of the folium, we see that the loop is traced when the parameter t is increasing along the interval $0 \le t < \infty$. We compute the improper integral using the substitution $u = 1 + t^3$, $du = 3t^2 \, dt$. This gives

$$A = \frac{1}{2} \lim_{R \to \infty} \int_0^R \frac{9t^2}{(1 + t^3)^2} \, dt = \frac{1}{2} \lim_{R \to \infty} \int_1^{1+R^3} \frac{3 \, du}{u^2} = \frac{3}{2} \lim_{R \to \infty} \left. -\frac{1}{u} \right|_{u=1}^{1+R^3}$$

$$= \frac{3}{2} \lim_{R \to \infty} \left(1 - \frac{1}{1 + R^3} \right) = \frac{3}{2}(1 - 0) = \frac{3}{2}$$

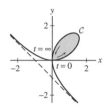

21. The Centroid via Boundary Measurements The centroid (see Section 15.5) of a domain $\mathcal{D}$ enclosed by a simple closed curve $\mathcal{C}$ is the point with coordinates $(\overline{x}, \overline{y}) = (M_y/M, M_x/M)$, where M is the area of $\mathcal{D}$ and the moments are defined by

$$M_x = \iint_{\mathcal{D}} y \, dA, \qquad M_y = \iint_{\mathcal{D}} x \, dA$$

Show that $M_x = \oint_{\mathcal{C}} xy \, dy$. Find a similar expression for M_y.

SOLUTION Consider the moment $M_x = \iint_{\mathcal{D}} y \, dA$, we know from Green's Theorem:

$$\iint_{\mathcal{D}} \left(\frac{\partial F_2}{\partial x} - \frac{\partial F_1}{\partial y} \right) dA = \oint_{\mathcal{C}} F_1 dx + F_2 dy$$

So then we need

$$\frac{\partial F_2}{dx} - \frac{\partial F_1}{\partial y} = y$$

If we set $F_2 = xy$ and $F_1 = 0$, then $\frac{\partial F_2}{dx} - \frac{\partial F_1}{\partial y} = y$ and

$$M_x = \iint_{\mathcal{D}} y \, dA = \oint_{\mathcal{C}} xy \, dy$$

Similarly, consider the moment $M_y = \iint_{\mathcal{D}} x \, dA$. We will now use Green's Theorem, stated above. Here we need

$$\frac{\partial F_2}{dx} - \frac{\partial F_1}{\partial y} = x$$

If we set $F_1 = -xy$ and $F_2 = 0$ then $\frac{\partial F_2}{\partial x} - \frac{\partial F_1}{\partial y} = x$ and

$$M_y = \iint_{\mathcal{D}} x \, dA = \oint_{\mathcal{C}} -xy \, dx$$

23. Let $\mathcal{C}_R$ be the circle of radius R centered at the origin. Use the general form of Green's Theorem to determine $\oint_{\mathcal{C}_2} \mathbf{F} \cdot d\mathbf{s}$, where $\mathbf{F}$ is a vector field such that $\oint_{\mathcal{C}_1} \mathbf{F} \cdot d\mathbf{s} = 9$ and $\frac{\partial F_2}{\partial x} - \frac{\partial F_1}{\partial y} = x^2 + y^2$ for (x, y) in the annulus $1 \le x^2 + y^2 \le 4$.

SOLUTION We use Green's Theorem for the annulus $\mathcal{D}$ between the circles $\mathcal{C}_1$ and $\mathcal{C}_2$ oriented as shown in the figure.

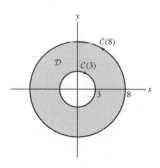

That is,

$$\int_{C_2} \mathbf{F} \cdot d\mathbf{s} - \int_{C_1} \mathbf{F} \cdot d\mathbf{s} = \iint_{D} \left(\frac{\partial F_2}{\partial x} - \frac{\partial F_1}{\partial y} \right) dx\, dy$$

Substituting the given information, we get

$$\int_{C_2} \mathbf{F} \cdot d\mathbf{s} - 9 = \iint_{D} (x^2 + y^2)\, dx\, dy$$

or

$$\int_{C_2} \mathbf{F} \cdot d\mathbf{s} = 9 + \iint_{D} (x^2 + y^2)\, dx\, dy$$

We compute the double integral by converting it to polar coordinates:

$$\int_{C_2} \mathbf{F} \cdot d\mathbf{s} = 9 + \int_{0}^{2\pi} \int_{1}^{2} r^2 \cdot r\, dr\, d\theta = 9 + 2\pi \int_{1}^{2} r^3\, dr = 9 + 2\pi \cdot \frac{r^4}{4} \Big|_{1}^{2} = 9 + 2\pi \left(\frac{2^4 - 1^4}{4} \right) = 9 + \frac{15\pi}{2}$$

25. Referring to Figure 24, suppose that

$$\oint_{C_2} \mathbf{F} \cdot d\mathbf{s} = 3\pi, \qquad \oint_{C_3} \mathbf{F} \cdot d\mathbf{s} = 4\pi$$

Use Green's Theorem to determine the circulation of $\mathbf{F}$ around C_1, assuming that $\dfrac{\partial F_2}{\partial x} - \dfrac{\partial F_1}{\partial x} = 9$ on the shaded region.

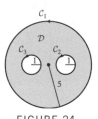

FIGURE 24

SOLUTION   We must calculate $\int_{C_1} \mathbf{F} \cdot d\mathbf{s}$. We use Green's Theorem for the region D between the three circles C_1, C_2, and C_3. Because of orientation, the line integrals $\int_{-C_2} \mathbf{F} \cdot d\mathbf{s} = -\int_{C_2} \mathbf{F} \cdot d\mathbf{s}$ and $\int_{-C_3} \mathbf{F} \cdot d\mathbf{s} = -\int_{C_3} \mathbf{F} \cdot d\mathbf{s}$ must be used in applying Green's Theorem. That is,

$$\int_{C_1} \mathbf{F} \cdot d\mathbf{s} - \int_{C_2} \mathbf{F} \cdot d\mathbf{s} - \int_{C_3} \mathbf{F} \cdot d\mathbf{s} = \iint_{D} \text{curl}(\mathbf{F})\, dA$$

We substitute the given information to obtain

$$\int_{C_1} \mathbf{F} \cdot d\mathbf{s} - 3\pi - 4\pi = \iint_{D} 9\, dA = 9 \iint_{D} 1 \cdot dA = 9\, \text{Area}(D) \tag{1}$$

The area of D is computed as the difference of areas of discs. That is,

$$\text{Area}(D) = \pi \cdot 5^2 - \pi \cdot 1^2 - \pi \cdot 1^2 = 23\pi$$

We substitute in (1) and compute the desired circulation:

$$\int_{C_1} \mathbf{F} \cdot d\mathbf{s} - 7\pi = 9 \cdot 23\pi$$

or

$$\int_{C_1} \mathbf{F} \cdot d\mathbf{s} = 214\pi.$$

In Exercises 27–30, refer to the Conceptual Insight that discusses the curl, defined by

$$\text{curl}_z(\mathbf{F}) = \frac{\partial F_2}{\partial x} - \frac{\partial F_1}{\partial y}$$

27. For the vector fields (A)–(D) in Figure 26, state whether the curl$_z$ at the origin appears to be positive, negative, or zero.

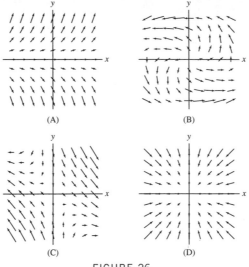

FIGURE 26

SOLUTION The vector field (A) does not have spirals, nor is it a "shear flow." Therefore, the curl appears to be zero. The vector field (B) rotates in the counterclockwise direction, hence we expect the curl to be positive. The vector field (C) rotates in a clockwise direction about the origin—we expect the curl to be negative. Finally, in the vector field (D), the fluid flows straight toward the origin without spiraling. We expect the curl to be zero.

29. Estimate $\oint_{\mathcal{C}} \mathbf{F} \cdot d\mathbf{s}$, where $\mathbf{F} = \left\langle x + 0.1y^2, y - 0.1x^2 \right\rangle$ and $\mathcal{C}$ encloses a small region of area 0.25 containing the point $P = (1, 1)$.

SOLUTION Use the following estimation:

$$\mathbf{F} \cdot d\mathbf{s} = \oint_{\mathcal{C}} F_1 \, dx + F_2 \, dy \approx \text{curl}_z(\mathbf{F})(P) \cdot \text{Area}(\mathcal{D})$$

First computing curl$\mathbf{F}$ we have:

$$\text{curl}\mathbf{F} = \begin{vmatrix} \mathbf{i} & \mathbf{j} & \mathbf{k} \\ \partial x & \partial y & \partial z \\ x + 0.1y^2 & y - 0.1x^2 & 0 \end{vmatrix} = \langle 0, 0, -0.2x - 0.2y \rangle$$

Thus curl$_z(\mathbf{F}) = -0.2x - 0.2y$ and curl$_z(\mathbf{F})(1, 1) = -0.2 - 0.2 = -0.4$. Also, we are given the area of the region is 0.25. Hence, we see:

$$\oint \mathbf{F} \cdot d\mathbf{s} \approx (-0.4)(0.25) = -0.10$$

31. Let $\mathcal{C}_R$ be the circle of radius R centered at the origin. Use Green's Theorem to find the value of R that maximizes $\oint_{\mathcal{C}_R} y^3 \, dx + x \, dy$.

SOLUTION Using Green's Theorem we can write:

$$\oint_{\mathcal{C}_R} y^3 \, dx + x \, dy = \iint_{\mathcal{D}} \frac{\partial F_2}{\partial x} - \frac{\partial F_1}{\partial y} \, dA$$

$$= \iint_{\mathcal{D}} 1 - 3y^2 \, dA$$

Then we have the following, using polar coordinates:

$$\oint_{\mathcal{C}_R} y^3 \, dx + x \, dy = \iint_{\mathcal{D}} \frac{\partial F_2}{\partial x} - \frac{\partial F_1}{\partial y} \, dA$$

$$= \iint_{\mathcal{D}} 1 - 3y^2 \, dA$$

$$= \int_0^{2\pi} \int_0^R (1 - 3r^2 \sin^2 \theta)(r) \, dr \, d\theta$$

$$= \int_0^{2\pi} \int_0^R r - 3r^3 \sin^2 \theta \, dr \, d\theta$$

$$= \int_0^{2\pi} \frac{1}{2}r^2 - \frac{3}{4}r^4 \sin^2 \theta \Big|_0^R \, d\theta$$

$$= \int_0^{2\pi} \frac{1}{2}R^2 - \frac{3}{8}R^4(1 - \cos 2\theta) \, dr \, d\theta$$

$$= \int_0^{2\pi} \frac{3R^4}{8}(\cos 2\theta) + \frac{1}{2}R^2 - \frac{3}{8}R^4 \, d\theta$$

$$= \frac{3R^4}{16} \sin 2\theta + \left(\frac{1}{2}R^2 - \frac{3}{8}R^4\right) \theta \Big|_0^{2\pi}$$

$$= 0 + 2\pi \left(\frac{1}{2}R^2 - \frac{3}{8}R^4\right) = \pi \left(R^2 - \frac{3}{4}R^4\right)$$

Now to maximize this quantity, we need to let $f(R) = \pi(R^2 - 3/4R^4)$ and take the first derivative.

$$f'(R) = \pi(2R - 3R^3) = 0 \quad \Rightarrow R = 0, \pm\sqrt{2/3}$$

This quantity is maximized when $R = \pm\sqrt{\frac{2}{3}}$ (that is, $R = 0$ is a minimum).

33. Use the result of Exercise 32 to compute the areas of the polygons in Figure 27. Check your result for the area of the triangle in (A) using geometry.

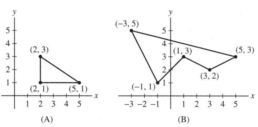

(A)   (B)

FIGURE 27

SOLUTION

(a) The vertices of the triangle are

$$(x_1, y_1) = (x_4, y_4) = (2, 1), \quad (x_2, y_2) = (5, 1), \quad (x_3, y_3) = (2, 3)$$

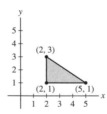

Using the formula obtained in Exercise 28, the area of the triangle is the following sum:

$$A = \frac{1}{2}\big((x_1y_2 - x_2y_1) + (x_2y_3 - x_3y_2) + (x_3y_1 - x_1y_3)\big)$$

$$= \frac{1}{2}\big((2 \cdot 1 - 5 \cdot 1) + (5 \cdot 3 - 2 \cdot 1) + (2 \cdot 1 - 2 \cdot 3)\big) = \frac{1}{2}(-3 + 13 - 4) = 3$$

We verify our result using the formula for the area of a triangle:

$$A = \frac{1}{2}bh = \frac{1}{2} \cdot (5 - 2) \cdot (3 - 1) = 3$$

(b) The vertices of the polygon are

$$(x_1, y_1) = (x_6, y_6) = (-1, 1)$$

$$(x_2, y_2) = (1, 3)$$

$$(x_3, y_3) = (3, 2)$$

$$(x_4, y_4) = (5, 3)$$

$$(x_5, y_5) = (-3, 5)$$

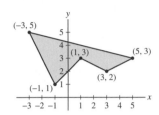

Using the formula in part (a), the area of the polygon is the following sum:

$$A = \frac{1}{2}\left((x_1y_2 - x_2y_1) + (x_2y_3 - x_3y_2) + (x_3y_4 - x_4y) + (x_4y_5 - x_5y_4) + (x_5y_1 - x_1y_5)\right)$$

$$= \frac{1}{2}\left((-1 \cdot 3 - 1 \cdot 1) + (1 \cdot 2 - 3 \cdot 3) + (3 \cdot 3 - 5 \cdot 2) + \left(5 \cdot 5 - (-3) \cdot 3\right) + \left(-3 \cdot 1 - (-1) \cdot 5\right)\right)$$

$$= \frac{1}{2}(-4 - 7 - 1 + 34 + 2) = 12$$

Exercises 34–39: In Section 16.2, we defined the flux of **F** *across a curve* $\mathcal{C}$ *(Figure 28) as the integral of the normal component of* **F** *along* $\mathcal{C}$, *and we showed that if* $\mathbf{c}(t) = (x(t), y(t))$ *is a parametrization of* C *for* $a \le t \le b$, *then the flux is equal to*

$$\int_a^b \mathbf{F}(\mathbf{c}(t)) \cdot \mathbf{n}(t)\, dt$$

where $\mathbf{n}(t) = \langle y'(t), -x'(t) \rangle$.

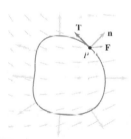

FIGURE 28 The flux of **F** is the integral of the normal component **F** · **n** around the curve.

35. Define $\operatorname{div}(\mathbf{F}) = \dfrac{\partial P}{\partial x} + \dfrac{\partial Q}{\partial y}$. Use Green's Theorem to prove that for any simple closed curve $\mathcal{C}$,

$$\text{Flux across } \mathcal{C} = \iint_{\mathcal{D}} \operatorname{div}(\mathbf{F})\, dA \qquad \boxed{12}$$

where $\mathcal{D}$ is the region enclosed by $\mathcal{C}$. This is a two-dimensional version of the **Divergence Theorem** discussed in Section 17.3.

SOLUTION Since $\mathbf{F} = \langle P, Q \rangle$ and $\mathbf{F}^* = \langle -Q, P \rangle$, we have

$$\operatorname{div}(\mathbf{F}) = \frac{\partial P}{\partial x} + \frac{\partial Q}{\partial y}$$

$$\operatorname{curl}(\mathbf{F}^*) = \frac{\partial P}{\partial x} - \frac{\partial}{\partial y}(-Q) = \frac{\partial P}{\partial x} + \frac{\partial Q}{\partial y}$$

Therefore,

$$\operatorname{div}(\mathbf{F}) = \operatorname{curl}(\mathbf{F}^*) \qquad (1)$$

Using Exercise 33, the flux of $\mathbf{F}$ across $\mathcal{C}$ is

$$\text{flux of } \mathbf{F} \text{ across } \mathcal{C} = \int_{\mathcal{C}} \mathbf{F}^* \cdot d\mathbf{s} \tag{2}$$

Green's Theorem and (1) imply that

$$\int_{\mathcal{C}} \mathbf{F}^* \cdot d\mathbf{s} = \iint_{\mathcal{D}} \text{curl}(\mathbf{F}^*)\, dA = \iint_{\mathcal{D}} \text{div}(\mathbf{F})\, dA \tag{3}$$

Combining (2) and (3) we have

$$\text{flux of } \mathbf{F} \text{ across } \mathcal{C} = \iint_{\mathcal{D}} \text{div}(\mathbf{F})\, dA.$$

37. Use Eq. (12) to compute the flux of $\mathbf{F} = \langle \cos y, \sin y \rangle$ across the square $0 \le x \le 2, 0 \le y \le \frac{\pi}{2}$.

SOLUTION Using the result:

$$\text{flux} = \iint_{\mathcal{D}} \text{div}(\mathbf{F})\, dA$$

we can compute the divergence:

$$\text{div}(\mathbf{F}) = \frac{\partial P}{\partial x} + \frac{\partial Q}{\partial y} = 0 + \cos y$$

Therefore,

$$\begin{aligned} \text{flux} &= \iint_{\mathcal{D}} \text{div}(\mathbf{F})\, dA \\ &= \iint_{\mathcal{D}} (0 + \cos y)\, dA \\ &= \int_0^2 \int_0^{\pi/2} \cos y \, dy \, dx \\ &= (2 - 0)\left(\sin y \Big|_0^{\pi/2} \right) = 2 \end{aligned}$$

39. A buffalo (Figure 29) stampede is described by a velocity vector field $\mathbf{F} = \langle xy - y^3, x^2 + y \rangle$ km/h in the region $\mathcal{D}$ defined by $2 \le x \le 3, 2 \le y \le 3$ in units of kilometers (Figure 30). Assuming a density is $\rho = 500$ buffalo per square kilometer, use Eq. (12) to determine the net number of buffalo leaving or entering $\mathcal{D}$ per minute (equal to ρ times the flux of $\mathbf{F}$ across the boundary of $\mathcal{D}$).

FIGURE 29 Buffalo stampede.

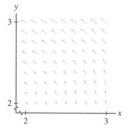

FIGURE 30 The vector field $\mathbf{F} = \langle xy - y^3, x^2 + y \rangle$.

SOLUTION The flux of $\mathbf{F}$ across the boundary ∂D has units of area per second. We multiply the buffalo density to obtain the number of buffalo per second crossing the boundary. Using Green's Theorem:

$$\begin{aligned} \text{flux of buffalo} &= \int_{\partial D} 500\mathbf{F}\, d\mathbf{s} \\ &= 500 \int_{\partial D} \langle xy - y^3, x^2 + y \rangle\, d\mathbf{s} \\ &= 500 \int_2^3 \int_2^3 \text{div}(\mathbf{F})\, dy\, dx \end{aligned}$$

$$= 500 \int_2^3 \int_2^3 (y+1)\, dy\, dx$$

$$= 500 \int_2^3 dx \cdot \int_2^3 (y+1)\, dy$$

$$= 500(3-2)\left(\frac{1}{2}y^2 + y\right)\Big|_2^3$$

$$= 500(1)\left(\frac{9}{2} + 3 - 2 - 2\right)$$

$$= 500(3.5)$$

$$= 1750 \text{ buffalos per second}$$

Further Insights and Challenges

In Exercises 40–43, the **Laplace operator** Δ *is defined by*

$$\Delta\varphi = \frac{\partial^2\varphi}{\partial x^2} + \frac{\partial^2\varphi}{\partial y^2} \qquad \boxed{13}$$

For any vector field $\mathbf{F} = \langle F_1, F_2 \rangle$, *define the conjugate vector field* $\mathbf{F}^* = \langle -F_2, F_1 \rangle$.

41. Let $\mathbf{n}$ be the outward-pointing unit normal vector to a simple closed curve $\mathcal{C}$. The **normal derivative** of a function φ, denoted $\frac{\partial\varphi}{\partial\mathbf{n}}$, is the directional derivative $D_\mathbf{n}(\varphi) = \nabla\varphi \cdot \mathbf{n}$. Prove that

$$\oint_\mathcal{C} \frac{\partial\varphi}{\partial\mathbf{n}}\, ds = \iint_\mathcal{D} \Delta\varphi\, dA$$

where $\mathcal{D}$ is the domain enclosed by a simple closed curve $\mathcal{C}$. *Hint:* Let $\mathbf{F} = \nabla\varphi$. Show that $\frac{\partial\varphi}{\partial\mathbf{n}} = \mathbf{F}^* \cdot \mathbf{T}$ where $\mathbf{T}$ is the unit tangent vector, and apply Green's Theorem.

SOLUTION In Exercise 34 we showed that for any vector field $\mathbf{F}$, $\mathbf{F}^*$ is a rotation of $\mathbf{F}$ by $\frac{\pi}{2}$ counterclockwise. The unit tangent $\mathbf{e}_n$ is a rotation of $\mathbf{n}$ by $\frac{\pi}{2}$ counterclockwise.

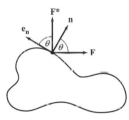

These properties imply that the angle θ between $\mathbf{F}$ and $\mathbf{n}$ is equal to the angle between $\mathbf{F}^*$ and $\mathbf{e}_n$, and $\|\mathbf{F}\| = \|\mathbf{F}^*\|$. Therefore,

$$\mathbf{F} \cdot \mathbf{n} = \|\mathbf{F}\|\|\mathbf{n}\| \cos\theta = \|\mathbf{F}\| \cos\theta$$
$$\mathbf{F}^* \cdot \mathbf{e}_n = \|\mathbf{F}^*\| \|\mathbf{e}_n\| \cos\theta = \|\mathbf{F}\| \cos\theta \qquad \Rightarrow \qquad \mathbf{F} \cdot \mathbf{n} = \mathbf{F}^* \cdot \mathbf{e}_n$$

Now, if $\mathbf{F} = \nabla\varphi$, then

$$\frac{\partial\varphi}{\partial\mathbf{n}} = \nabla\varphi \cdot \mathbf{n} = \mathbf{F} \cdot \mathbf{n} = \mathbf{F}^* \cdot \mathbf{e}_n$$

By the definition of the vector line integral $\int_\mathcal{C} \mathbf{F}^* \cdot d\mathbf{s} = \int_\mathcal{C} (\mathbf{F}^* \cdot \mathbf{e}_n)\, ds$. Therefore,

$$\int_\mathcal{C} \frac{\partial\varphi}{\partial\mathbf{n}}\, ds = \int_\mathcal{C} (\mathbf{F}^* \cdot \mathbf{e}_n)\, ds = \int_\mathcal{C} \mathbf{F}^* \cdot d\mathbf{s}$$

Using Green's Theorem and the equality $\operatorname{curl}(\mathbf{F}^*) = \Delta\varphi$ obtained in Exercise 40, we get

$$\int_\mathcal{C} \frac{\partial\varphi}{\partial\mathbf{n}}\, ds = \int_\mathcal{C} \mathbf{F}^* \cdot d\mathbf{s} = \iint_\mathcal{D} \operatorname{curl}(\mathbf{F}^*)\, dA = \iint_\mathcal{D} \Delta\varphi\, dA.$$

43. Prove that $m(r) \le I_\varphi(r) \le M(r)$, where $m(r)$ and $M(r)$ are the minimum and maximum values of φ on C_r. Then use the continuity of φ to prove that $\lim\limits_{r \to 0} I_\varphi(r) = \varphi(P)$.

SOLUTION $I_\varphi(r)$ is defined by

$$I_\varphi(r) = \frac{1}{2\pi} \int_0^{2\pi} \varphi(a + r\cos\theta, b + r\sin\theta)\, d\theta$$

The points on C_r have the form $(a + r\cos\theta, b + r\sin\theta)$. Therefore, since $m(r)$ and $M(r)$ are the minimum and maximum values of φ on C_r, we have for all $0 \le \theta \le 2\pi$,

$$m(r) \le \varphi(a + r\cos\theta, b + r\sin\theta) \le M(r)$$

Using properties of integrals (Eq. (6) in Section 5.2), we conclude that

$$2\pi m(r) \le \int_0^{2\pi} \varphi(a + r\cos\theta + b + r\sin\theta) \le 2\pi M(r)$$

Dividing by 2π we obtain

$$m(r) \le I_\varphi(r) \le M(r) \tag{1}$$

Now, since φ is continuous and the functions $\sin\theta$ and $\cos\theta$ are bounded for all $0 \le \theta \le 2\pi$, the following holds:

$$\lim_{r \to 0} \varphi(a + r\cos\theta, b + r\sin\theta) = \varphi\left(\lim_{r \to 0} (a + r\cos\theta, b + r\sin\theta)\right) = \varphi(a, b)$$

which means that for $\epsilon > 0$ there exists $\delta > 0$ so that

$$|\varphi(a + r\cos\theta, b + r\sin\theta) - \varphi(a, b)| < \epsilon$$

for all $0 \le \theta \le 2\pi$, whenever $0 < r < \delta$. Hence also

$$\lim_{r \to 0} m(r) = \lim_{r \to 0} M(r) = \varphi(a, b) \tag{2}$$

Combining (1), (2), and the Squeeze Theorem, we obtain the following conclusion:

$$\lim_{r \to 0} I_\varphi(r) = \varphi(a, b).$$

In Exercises 44 and 45, let $\mathcal{D}$ be the region bounded by a simple closed curve C. A function $\varphi(x, y)$ on $\mathcal{D}$ (whose second-order partial derivatives exist and are continuous) is called **harmonic** *if $\Delta\varphi = 0$, where $\Delta\varphi$ is the Laplace operator defined in Eq. (13).*

45. Show that $f(x, y) = x^2 - y^2$ is harmonic. Verify the mean-value property for $f(x, y)$ directly [expand $f(a + r\cos\theta, b + r\sin\theta)$ as a function of θ and compute $I_\varphi(r)$]. Show that $x^2 + y^2$ is not harmonic and does not satisfy the mean-value property.

SOLUTION We show that the function $f(x, y) = x^2 - y^2$ is harmonic by showing that $\Delta f = \frac{\partial^2 f}{\partial x^2} + \frac{\partial^2 f}{\partial y^2} = 0$. We have

$$\frac{\partial f}{\partial x} = 2x, \quad \frac{\partial f}{\partial y} = -2y$$

$$\frac{\partial^2 f}{\partial x^2} = 2, \quad \frac{\partial^2 f}{\partial y^2} = -2$$

Hence,

$$\Delta f = \frac{\partial^2 f}{\partial x^2} + \frac{\partial^2 f}{\partial y^2} = 2 - 2 = 0$$

We now verify the mean-value property for f. That is, we show that for all r,

$$I_f(r) = \frac{1}{2\pi} \int_0^{2\pi} f(a + r\cos\theta, b + r\sin\theta)\, d\theta = f(a, b)$$

We compute the integrand:

$$f(a + r\cos\theta, b + r\sin\theta) = x^2 - y^2 = (a + r\cos\theta)^2 - (b + r\sin\theta)^2$$

$$= a^2 + 2ar\cos\theta + r^2\cos^2\theta - \left(b^2 + 2br\sin\theta + r^2\sin^2\theta\right)$$

$$= a^2 - b^2 + 2r(a\cos\theta - b\sin\theta) + r^2\cos 2\theta$$

We compute the integral:

$$2\pi I_f(r) = \int_0^{2\pi} \left(a^2 - b^2 + 2r(a\cos\theta - b\sin\theta) + r^2 \cos 2\theta \right) d\theta$$

$$= (a^2 - b^2)\theta + 2ar\sin\theta + 2br\cos\theta + \frac{r^2}{2}\sin 2\theta \Big|_{\theta=0}^{2\pi} = 2\pi(a^2 - b^2)$$

Hence,

$$I_f(r) = a^2 - b^2$$

However, we have $f(a, b) = a^2 - b^2$. Hence, for all r, $I_f(r) = f(a, b)$, which proves the mean-value property for f. For $g(x, y) = x^2 + y^2$ we have

$$g_{xx} = 2, \quad g_{yy} = 2, \quad \text{and} \quad \Delta g = 2 + 2 = 4 \neq 0.$$

We check the mean value property:

$$I_g(r) = \frac{1}{2\pi}\int_0^{2\pi} g(a + r\cos\theta, b + r\sin\theta)\, d\theta = \frac{1}{2\pi}\int_0^{2\pi} (a + r\cos\theta)^2 + (b + r\sin\theta)^2 \, d\theta$$

$$= \frac{1}{2\pi}\int_0^{2\pi} a^2 + b^2 + 2r(a\cos\theta + b\sin\theta) + r^2 \, d\theta = a^2 + b^2 + r^2 \neq a^2 + b^2 = \varphi(a, b)$$

The mean value property does not hold for g.

17.2 Stokes' Theorem (LT Section 18.2)

Preliminary Questions

1. Indicate with an arrow the boundary orientation of the boundary curves of the surfaces in Figure 14, oriented by the outward-pointing normal vectors.

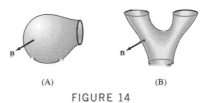

(A)   (B)

FIGURE 14

SOLUTION The indicated orientation is defined so that if the normal vector is moving along the boundary curve, the surface lies to the left. Since the surfaces are oriented by the outward-pointing normal vectors, the induced orientation is as shown in the figure:

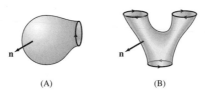

(A)   (B)

2. Let $\mathbf{F} = \text{curl}(\mathbf{A})$. Which of the following are related by Stokes' Theorem?

(a) The circulation of $\mathbf{A}$ and flux of $\mathbf{F}$.

(b) The circulation of $\mathbf{F}$ and flux of $\mathbf{A}$.

SOLUTION Stokes' Theorem states that the circulation of $\mathbf{A}$ is equal to the flux of $\mathbf{F}$. The correct answer is (a).

3. What is the definition of a vector potential?

SOLUTION A vector field $\mathbf{A}$ such that $\mathbf{F} = \text{curl}(\mathbf{A})$ is a vector potential for $\mathbf{F}$.

4. Which of the following statements is correct?

(a) The flux of $\text{curl}(\mathbf{A})$ through every oriented surface is zero.

(b) The flux of $\text{curl}(\mathbf{A})$ through every closed, oriented surface is zero.

SOLUTION Statement (b) is the correct statement. The flux of $\text{curl}(\mathbf{F})$ through an oriented surface is not necessarily zero, unless the surface is closed.

5. Which condition on $\mathbf{F}$ guarantees that the flux through $\mathcal{S}_1$ is equal to the flux through $\mathcal{S}_2$ for any two oriented surfaces $\mathcal{S}_1$ and $\mathcal{S}_2$ with the same oriented boundary?

SOLUTION   If $\mathbf{F}$ has a vector potential $\mathbf{A}$, then by a corollary of Stokes' Theorem,

$$\iint_{\mathcal{S}} \mathbf{F} \cdot d\mathbf{s} = \int_{\mathcal{C}} \mathbf{A} \cdot d\mathbf{s}$$

Therefore, if two oriented surfaces $\mathcal{S}_1$ and $\mathcal{S}_2$ have the same oriented boundary curve, $\mathcal{C}$, then

$$\iint_{\mathcal{S}_1} \mathbf{F} \cdot d\mathbf{s} = \int_{\mathcal{C}} \mathbf{A} \cdot d\mathbf{s} \quad \text{and} \quad \iint_{\mathcal{S}_2} \mathbf{F} \cdot d\mathbf{s} = \int_{\mathcal{C}} \mathbf{A} \cdot d\mathbf{s}$$

Hence,

$$\iint_{\mathcal{S}_1} \mathbf{F} \cdot d\mathbf{s} = \iint_{\mathcal{S}_2} \mathbf{F} \cdot d\mathbf{s}$$

Exercises

In Exercises 1–4, calculate curl($\mathbf{F}$).

1. $\mathbf{F} = \langle z - y^2, x + z^3, y + x^2 \rangle$

SOLUTION   We have

$$\text{curl}(\mathbf{F}) = \begin{vmatrix} \mathbf{i} & \mathbf{j} & \mathbf{k} \\ \dfrac{\partial}{\partial x} & \dfrac{\partial}{\partial y} & \dfrac{\partial}{\partial z} \\ z - y^2 & x + z^3 & y + x^2 \end{vmatrix} = (1 - 3z^2)\mathbf{i} - (2x - 1)\mathbf{j} + (1 + 2y)\mathbf{k} = \langle 1 - 3z^2, 1 - 2x, 1 + 2y \rangle$$

3. $\mathbf{F} = \langle e^y, \sin x, \cos x \rangle$

SOLUTION   We have

$$\text{curl}(\mathbf{F}) = \begin{vmatrix} \mathbf{i} & \mathbf{j} & \mathbf{k} \\ \dfrac{\partial}{\partial x} & \dfrac{\partial}{\partial y} & \dfrac{\partial}{\partial z} \\ e^y & \sin x & \cos x \end{vmatrix} = 0\mathbf{i} - (-\sin x)\mathbf{j} + (\cos x - e^y)\mathbf{k} = \langle 0, \sin x, \cos x - e^y \rangle$$

In Exercises 5–8, verify Stokes' Theorem for the given vector field and surface, oriented with an upward-pointing normal.

5. $\mathbf{F} = \langle 2xy, x, y + z \rangle$,   the surface $z = 1 - x^2 - y^2$ for $x^2 + y^2 \leq 1$

SOLUTION   We must show that

$$\int_{\mathcal{C}} \mathbf{F} \cdot d\mathbf{s} = \iint_{\mathcal{S}} \text{curl}(\mathbf{F}) \cdot d\mathbf{S}$$

Step 1. Compute the line integral around the boundary curve. The boundary curve $\mathcal{C}$ is the unit circle oriented in the counterclockwise direction. We parametrize $\mathcal{C}$ by

$$\gamma(t) = (\cos t, \sin t, 0), \quad 0 \leq t \leq 2\pi$$

Then,

$$\mathbf{F}(\gamma(t)) = \langle 2\cos t \sin t, \cos t, \sin t \rangle$$

$$\gamma'(t) = \langle -\sin t, \cos t, 0 \rangle$$

$$\mathbf{F}(\gamma(t)) \cdot \gamma'(t) = \langle 2\cos t \sin t, \cos t, \sin t \rangle \cdot \langle -\sin t, \cos t, 0 \rangle = -2\cos t \sin^2 t + \cos^2 t$$

We obtain the following integral:

$$\int_C \mathbf{F} \, ds = \int_0^{2\pi} \left(-2\cos t \, \sin^2 t + \cos^2 t\right) dt = -\frac{2 \sin^3 t}{3} + \frac{t}{2} + \frac{\sin 2t}{4} \bigg|_0^{2\pi} = \pi \qquad (1)$$

Step 2. Compute the flux of the curl through the surface. We parametrize the surface by

$$\Phi(\theta, t) = \left(t \cos\theta, t \sin\theta, 1 - t^2\right), \quad 0 \le t \le 1, \quad 0 \le \theta \le 2\pi$$

We compute the normal vector:

$$\mathbf{T}_\theta = \frac{\partial \Phi}{\partial \theta} = \langle -t \sin\theta, t \cos\theta, 0 \rangle$$

$$\mathbf{T}_t = \frac{\partial \Phi}{\partial t} = \langle \cos\theta, \sin\theta, -2t \rangle$$

$$\mathbf{T}_\theta \times \mathbf{T}_t = \begin{vmatrix} \mathbf{i} & \mathbf{j} & \mathbf{k} \\ -t \sin\theta & t \cos\theta & 0 \\ \cos\theta & \sin\theta & -2t \end{vmatrix} = (-2t^2 \cos\theta)\mathbf{i} - (2t^2 \sin\theta)\mathbf{j} - t(\sin^2\theta + \cos^2\theta)\mathbf{k}$$

$$= (-2t^2 \cos\theta)\mathbf{i} - (2t^2 \sin\theta)\mathbf{j} - t\mathbf{k}$$

Since the normal is always supposed to be pointing upward, the z-coordinate of the normal vector must be positive. Therefore, the normal vector is

$$\mathbf{n} = \langle 2t^2 \cos\theta, 2t^2 \sin\theta, t \rangle$$

We compute the curl:

$$\text{curl}(\mathbf{F}) = \begin{vmatrix} \mathbf{i} & \mathbf{j} & \mathbf{k} \\ \dfrac{\partial}{\partial x} & \dfrac{\partial}{\partial y} & \dfrac{\partial}{\partial z} \\ 2xy & x & y + z \end{vmatrix} = \mathbf{i} + (1 - 2x)\mathbf{k} = \langle 1, 0, 1 - 2x \rangle$$

We compute the curl in terms of the parameters:

$$\text{curl}(\mathbf{F}) = \langle 1, 0, 1 - 2t \cos\theta \rangle$$

Hence,

$$\text{curl}(\mathbf{F}) \cdot \mathbf{n} = \langle 1, 0, 1 - 2t \cos\theta \rangle \cdot \langle 2t^2 \cos\theta, 2t^2 \sin\theta, t \rangle = 2t^2 \cos\theta + t - 2t^2 \cos\theta = t$$

The surface integral is thus

$$\iint_S \text{curl}(\mathbf{F}) \cdot d\mathbf{S} = \int_0^{2\pi} \int_0^1 t \, dt \, d\theta = 2\pi \int_0^1 t \, dt = 2\pi \cdot \frac{t^2}{2} \bigg|_0^1 = \pi \qquad (2)$$

The values of the integrals in (1) and (2) are equal, as stated in Stokes' Theorem.

7. $\mathbf{F} = \langle e^{y-z}, 0, 0 \rangle$, the square with vertices $(1, 0, 1)$, $(1, 1, 1)$, $(0, 1, 1)$, and $(0, 0, 1)$

SOLUTION

Step 1. Compute the integral around the boundary curve. The boundary consists of four segments C_1, C_2, C_3, and C_4 shown in the figure:

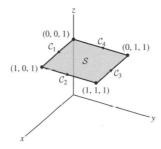

We parametrize the segments by

$$C_1 : \gamma_1(t) = (t, 0, 1), \quad 0 \le t \le 1$$

$$C_2 : \gamma_2(t) = (1, t, 1), \quad 0 \le t \le 1$$

$$C_3 : \gamma_3(t) = (1 - t, 1, 1), \quad 0 \le t \le 1$$

$$C_4 : \gamma_4(t) = (0, 1 - t, 1), \quad 0 \le t \le 1$$

We compute the following values:

$$\mathbf{F}\left(\gamma_1(t)\right) = \left\langle e^{y-z}, 0, 0 \right\rangle = \left\langle e^{-1}, 0, 0 \right\rangle$$

$$\mathbf{F}\left(\gamma_2(t)\right) = \left\langle e^{y-z}, 0, 0 \right\rangle = \left\langle e^{t-1}, 0, 0 \right\rangle$$

$$\mathbf{F}\left(\gamma_3(t)\right) = \left\langle e^{y-z}, 0, 0 \right\rangle = \langle 1, 0, 0 \rangle$$

$$\mathbf{F}\left(\gamma_4(t)\right) = \left\langle e^{y-z}, 0, 0 \right\rangle = \left\langle e^{-t-1}, 0, 0 \right\rangle$$

Hence,

$$\mathbf{F}\left(\gamma_1(t)\right) \cdot \gamma_1'(t) = \left\langle e^{-1}, 0, 0 \right\rangle \cdot \langle 1, 0, 0 \rangle = e^{-1}$$

$$\mathbf{F}\left(\gamma_2(t)\right) \cdot \gamma_2'(t) = \left\langle e^{t-1}, 0, 0 \right\rangle \cdot \langle 0, 1, 0 \rangle = 0$$

$$\mathbf{F}\left(\gamma_3(t)\right) \cdot \gamma_3'(t) = \langle 1, 0, 0 \rangle \cdot \langle -1, 0, 0 \rangle = -1$$

$$\mathbf{F}\left(\gamma_4(t)\right) \cdot \gamma_4'(t) = \left\langle e^{-t-1}, 0, 0 \right\rangle \cdot \langle 0, -1, 0 \rangle = 0$$

We obtain the following integral:

$$\int_C \mathbf{F} \cdot d\mathbf{s} = \sum_{i=1}^{4} \int_{C_i} \mathbf{F} \cdot d\mathbf{s} = \int_0^1 e^{-1}\,dt + 0 + \int_0^1 (-1)\,dt + 0 = e^{-1} - 1$$

Step 2. Compute the curl.

$$\text{curl}(\mathbf{F}) = \begin{vmatrix} \mathbf{i} & \mathbf{j} & \mathbf{k} \\ \dfrac{\partial}{\partial x} & \dfrac{\partial}{\partial y} & \dfrac{\partial}{\partial z} \\ e^{y-z} & 0 & 0 \end{vmatrix} = -e^{y-z}\,\mathbf{j} - e^{y-z}\,\mathbf{k} = \left\langle 0, -e^{y-z}, -e^{y-z} \right\rangle$$

Step 3. Compute the flux of the curl through the surface. We parametrize the surface by

$$\Phi(x, y) = (x, y, 1), \quad 0 \le x,\ y \le 1$$

The upward pointing normal is $\mathbf{n} = \langle 0, 0, 1 \rangle$. We express $\text{curl}(\mathbf{F})$ in terms of the parameters x and y:

$$\text{curl}(\mathbf{F})\,(\Phi(x, y)) = \left\langle 0, -e^{y-1}, -e^{y-1} \right\rangle$$

Hence,

$$\text{curl}(\mathbf{F}) \cdot \mathbf{n} = \left\langle 0, -e^{y-1}, -e^{y-1} \right\rangle \cdot \langle 0, 0, 1 \rangle = -e^{y-1}$$

The surface integral is thus

$$\iint_S \text{curl}(\mathbf{F}) \cdot d\mathbf{S} = \iint_D -e^{y-1}\,dA = \int_0^1 \int_0^1 -e^{y-1}\,dy\,dx = \int_0^1 -e^{y-1}\,dy = -e^{y-1}\Big|_0^1$$

$$= -1 + e^{-1} = e^{-1} - 1 \tag{1}$$

We see that the integrals in (1) and (2) are equal.

In Exercises 9 and 10, calculate $\text{curl}(\mathbf{F})$ *and then use Stokes' Theorem to compute the flux of* $\text{curl}(\mathbf{F})$ *through the given surface as a line integral.*

9. $\mathbf{F} = \left\langle e^{z^2} - y, e^{z^3} + x, \cos(xz) \right\rangle$, the upper hemisphere $x^2 + y^2 + z^2 = 1, z \ge 0$ with outward-pointing normal

SOLUTION

Step 1. Compute the curl.

$$\text{curl}(\mathbf{F}) = \begin{vmatrix} \mathbf{i} & \mathbf{j} & \mathbf{k} \\ \partial x & \partial y & \partial z \\ e^{z^2} - y & e^{z^3} + x & \cos(xz) \end{vmatrix} = \left\langle -3z^2 e^{z^3}, 2ze^{z^3} + z\sin(xz), 2 \right\rangle$$

Step 2. Compute the flux of the curl through the surface. We will use Stokes' Theorem and compute the line integral around the boundary curve. The boundary curve is the unit circle oriented in the counterclockwise direction. We use the parametrization $\gamma(t) = \langle \cos t, \sin t, 0 \rangle$, $0 \le t \le 2\pi$. Then

$$\mathbf{F}(\gamma(t)) \cdot \gamma'(t) = \left\langle e^0 - \sin t, e^0 + \cos t, \cos(0) \right\rangle \cdot \langle -\sin t, \cos t, 0 \rangle$$

$$= \langle 1 - \sin t, 1 + \cos t, 1 \rangle \cdot \langle -\sin t, \cos t, 0 \rangle$$

$$= -\sin t(1 - \sin t) + \cos t(1 + \cos t) + 0$$

$$= -\sin t + \sin^2 t + \cos t + \cos^2 t$$

$$= 1 - \sin t + \cos t$$

The line integral is:

$$\int_{\mathcal{C}} \mathbf{F} \cdot d\mathbf{s} = \int_0^{2\pi} (1 - \sin t + \cos t)\, dt = t + \cos t + \sin t \Big|_0^{2\pi} = 2\pi$$

11. Let $\mathcal{S}$ be the surface of the cylinder (not including the top and bottom) of radius 2 for $1 \le z \le 6$, oriented with outward-pointing normal (Figure 16).

(a) Indicate with an arrow the orientation of $\partial \mathcal{S}$ (the top and bottom circles).

(b) Verify Stokes' Theorem for $\mathcal{S}$ and $\mathbf{F} = \langle yz^2, 0, 0 \rangle$.

FIGURE 16

SOLUTION

(a) The induced orientation is defined so that as the normal vector travels along the boundary curve, the surface lies to its left. Therefore, the boundary circles on top and bottom have opposite orientations, which are shown in the figure.

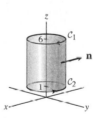

(b) We verify Stokes' Theorem for $\mathcal{S}$ and $\mathbf{F} = \langle yz^2, 0, 0 \rangle$.

Step 1. Compute the integral around the boundary circles. We use the following parametrizations:

$$\mathcal{C}_1 : \gamma_1(t) = (2\cos t, 2\sin t, 6), \quad t \text{ from } 2\pi \text{ to } 0$$

$$\mathcal{C}_2 : \gamma_2(t) = (2\cos t, 2\sin t, 1), \quad t \text{ from } 0 \text{ to } 2\pi$$

We compute the following values:

$$\mathbf{F}(\gamma_1(t)) = \left\langle yz^2, 0, 0 \right\rangle = \langle 72 \sin t, 0, 0 \rangle,$$

$$\gamma_1'(t) = \langle -2\sin t, 2\cos t, 0 \rangle$$

$$\mathbf{F}(\gamma_1(t)) \cdot \gamma_1'(t) = \langle 72 \sin t, 0, 0 \rangle \cdot \langle -2\sin t, 2\cos t, 0 \rangle = -144 \sin^2 t$$

$$\mathbf{F}(\gamma_2(t)) = \left\langle yz^2, 0, 0 \right\rangle = \langle 2\sin t, 0, 0 \rangle,$$

$$\gamma_2'(t) = \langle -2\sin t, 2\cos t, 0 \rangle$$

$$\mathbf{F}(\gamma_2(t)) \cdot \gamma_2'(t) = \langle 2\sin t, 0, 0 \rangle \cdot \langle -2\sin t, 2\cos t, 0 \rangle = -4 \sin^2 t$$

The line integral is thus

$$\int_{\mathcal{C}} \mathbf{F} \cdot d\mathbf{s} = \int_{\mathcal{C}_1} \mathbf{F} \cdot d\mathbf{s} + \int_{\mathcal{C}_2} \mathbf{F} \cdot d\mathbf{s} = \int_{2\pi}^{0} (-144\sin^2 t)\, dt + \int_{0}^{2\pi} (-4\sin^2 t)\, dt$$

$$= \int_{0}^{2\pi} 140\sin^2 t\, dt = 140 \int_{0}^{2\pi} \frac{1 - \cos 2t}{2}\, dt = 70 \cdot 2\pi - \left. \frac{70\sin 2t}{2} \right|_{0}^{2\pi} = 140\pi$$

Step 2. Compute the curl

$$\text{curl}(\mathbf{F}) = \begin{vmatrix} \mathbf{i} & \mathbf{j} & \mathbf{k} \\ \dfrac{\partial}{\partial x} & \dfrac{\partial}{\partial y} & \dfrac{\partial}{\partial z} \\ yz^2 & 0 & 0 \end{vmatrix} = (2yz)\mathbf{j} - z^2\mathbf{k} = \left\langle 0, 2yz, -z^2 \right\rangle$$

Step 3. Compute the flux of the curl through the surface. We parametrize $\mathcal{S}$ by

$$\Phi(\theta, z) = (2\cos\theta, 2\sin\theta, z), \quad 0 \le \theta \le 2\pi, \quad 1 \le z \le 6$$

In Example 2 in the text, it is shown that the outward pointing normal is

$$\mathbf{n} = \langle 2\cos\theta, 2\sin\theta, 0 \rangle$$

We compute the dot product:

$$\text{curl}(\mathbf{F})\,(\Phi(\theta, z)) \cdot \mathbf{n} = \left\langle 0, 4z\sin\theta, -z^2 \right\rangle \cdot \langle 2\cos\theta, 2\sin\theta, 0 \rangle = 8z\sin^2\theta$$

We obtain the following integral (and use the integral we computed before):

$$\iint_{\mathcal{S}} \text{curl}(\mathbf{F}) \cdot d\mathbf{S} = \int_{1}^{6} \int_{0}^{2\pi} 8z\sin^2\theta\, d\theta\, dz = \left(\int_{1}^{6} 8z\, dz \right) \left(\int_{0}^{2\pi} \sin^2\theta\, d\theta \right) = 4z^2 \Big|_{1}^{6} \cdot \pi = 140\pi$$

The line integral and the flux have the same value. This verifies Stokes' Theorem.

13. Let I be the flux of $\mathbf{F} = \left\langle e^y, 2xe^{x^2}, z^2 \right\rangle$ through the upper hemisphere $\mathcal{S}$ of the unit sphere.

(a) Let $\mathbf{G} = \left\langle e^y, 2xe^{x^2}, 0 \right\rangle$. Find a vector field $\mathbf{A}$ such that $\text{curl}(\mathbf{A}) = \mathbf{G}$.
(b) Use Stokes' Theorem to show that the flux of $\mathbf{G}$ through $\mathcal{S}$ is zero. *Hint:* Calculate the circulation of $\mathbf{A}$ around $\partial\mathcal{S}$.
(c) Calculate I. *Hint:* Use (b) to show that I is equal to the flux of $\left\langle 0, 0, z^2 \right\rangle$ through $\mathcal{S}$.

SOLUTION
(a) We search for a vector field $\mathbf{A}$ so that $\mathbf{G} = \text{curl}(\mathbf{A})$. That is,

$$\left\langle \frac{\partial A_3}{\partial y} - \frac{\partial A_2}{\partial z}, \frac{\partial A_1}{\partial z} - \frac{\partial A_3}{\partial x}, \frac{\partial A_2}{\partial x} - \frac{\partial A_1}{\partial y} \right\rangle = \left\langle e^y, 2xe^{x^2}, 0 \right\rangle$$

We note that the third coordinate of this curl vector must be zero; this can be satisfied if $A_1 = 0$ and $A_2 = 0$. With this in mind, we let $\mathbf{A} = \left\langle 0, 0, e^y - e^{x^2} \right\rangle$. The vector field $\mathbf{A} = \left\langle 0, 0, e^y - e^{x^2} \right\rangle$ satisfies this equality. Indeed,

$$\frac{\partial A_3}{\partial y} - \frac{\partial A_2}{\partial z} = e^y, \qquad \frac{\partial A_1}{\partial z} - \frac{\partial A_3}{\partial x} = 2xe^{x^2}, \qquad \frac{\partial A_2}{\partial x} - \frac{\partial A_1}{\partial y} = 0$$

(b) We found that $\mathbf{G} = \text{curl}(\mathbf{A})$, where $\mathbf{A} = \left\langle 0, 0, e^y - e^{x^2} \right\rangle$. We compute the flux of $\mathbf{G}$ through $\mathcal{S}$. By Stokes' Theorem,

$$\iint_{\mathcal{S}} \mathbf{G} \cdot d\mathbf{S} = \iint_{\mathcal{S}} \text{curl}(\mathbf{A}) \cdot d\mathbf{S} = \int_{\mathcal{C}} \mathbf{A} \cdot d\mathbf{s}$$

The boundary $\mathcal{C}$ is the circle $x^2 + y^2 = 1$, parametrized by

$$\gamma(t) = (\cos t, \sin t, 0), \quad 0 \le t \le 2\pi$$

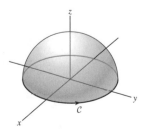

We compute the following values:

$$\mathbf{A}\left(\gamma(t)\right) = \left\langle 0, 0, e^{y} - e^{x^2} \right\rangle = \left\langle 0, 0, e^{\sin t} - e^{\cos^2 t} \right\rangle$$

$$\gamma'(t) = \langle -\sin t, \cos t, 0 \rangle$$

$$\mathbf{A}\left(\gamma(t)\right) \cdot \gamma'(t) = \left\langle 0, 0, e^{\sin t} - e^{\cos^2 t} \right\rangle \cdot \langle -\sin t, \cos t, 0 \rangle = 0$$

Therefore,

$$\int_{C} \mathbf{A} \cdot d\mathbf{s} = \int_{0}^{2\pi} 0 \, dt = 0$$

(c) We rewrite the vector field $\mathbf{F} = \left\langle e^{y}, 2xe^{x^2}, z^2 \right\rangle$ as

$$\mathbf{F} = \left\langle e^{y}, 2xe^{x^2}, z^2 \right\rangle = \left\langle e^{y}, 2xe^{x^2}, 0 \right\rangle + \left\langle 0, 0, z^2 \right\rangle = \text{curl}(\mathbf{A}) + \left\langle 0, 0, z^2 \right\rangle$$

Therefore,

$$\iint_{S} \mathbf{F} \cdot d\mathbf{S} = \iint_{S} \text{curl}(\mathbf{A}) \cdot d\mathbf{S} + \iint_{S} \left\langle 0, 0, z^2 \right\rangle \cdot d\mathbf{S} \tag{1}$$

In part (b) we showed that the first integral on the right-hand side is zero. Therefore,

$$\iint_{S} \mathbf{F} \cdot d\mathbf{S} = \iint_{S} \left\langle 0, 0, z^2 \right\rangle \cdot d\mathbf{S} \tag{2}$$

The upper hemisphere is parametrized by

$$\Phi(\theta, \phi) = (\cos\theta \sin\phi, \sin\theta \sin\phi, \cos\phi), \quad 0 \le \theta \le 2\pi, \quad 0 \le \phi \le \frac{\pi}{2}.$$

with the outward pointing normal

$$\mathbf{n} = \sin\phi \, \langle \cos\theta \sin\phi, \sin\theta \sin\phi, \cos\phi \rangle$$

See Example 4, Section 16.4. We have

$$\left\langle 0, 0, \cos^2\phi \right\rangle \cdot \mathbf{n} = \sin\phi \cos^3\phi$$

Therefore,

$$\iint_{S} \left\langle 0, 0, z^2 \right\rangle \cdot d\mathbf{S} = \int_{0}^{2\pi} \int_{0}^{\pi/2} \sin\phi \cos^3\phi \, d\phi \, d\theta = 2\pi \int_{0}^{\pi/2} \sin\phi \cos^3\phi \, d\phi$$

$$= 2\pi \frac{-\cos^4\phi}{4} \Big|_{0}^{\pi/2} = -\frac{\pi}{2}(0 - 1) = \frac{\pi}{2}$$

Combining with (2) we obtain the solution

$$\iint_{S} \mathbf{F} \cdot d\mathbf{S} = \frac{\pi}{2}.$$

15. Let $\mathbf{A}$ be the vector potential and $\mathbf{B}$ the magnetic field of the infinite solenoid of radius R in Example 6. Use Stokes' Theorem to compute:

(a) The flux of $\mathbf{B}$ through a circle in the xy-plane of radius $r < R$

(b) The circulation of $\mathbf{A}$ around the boundary C of a surface lying outside the solenoid

SOLUTION

(a) In Example 6 it is shown that $\mathbf{B} = \text{curl}(\mathbf{A})$, where

$$\mathbf{A} = \begin{cases} \dfrac{1}{2} R^2 B \left\langle -\dfrac{y}{r^2}, \dfrac{x}{r^2}, 0 \right\rangle & \text{if} \quad r > R \\[2mm] \dfrac{1}{2} B \langle -y, x, 0 \rangle & \text{if} \quad r < R \end{cases} \tag{1}$$

Therefore, using Stokes' Theorem, we have (S is the disk of radius r in the xy-plane)

$$\iint_{S} \mathbf{B} \cdot d\mathbf{S} = \iint_{S} \text{curl}(\mathbf{A}) \cdot d\mathbf{S} = \int_{\partial S} \mathbf{A} \cdot d\mathbf{s} \tag{2}$$

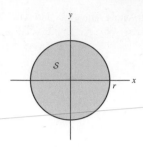

We parametrize the circle $\mathcal{C} = \partial\mathcal{S}$ by $\mathbf{c}(t) = \langle r\cos t, r\sin t, 0\rangle$, $0 \le t \le 2\pi$. Then

$$\mathbf{c}'(t) = \langle -r\sin t, r\cos t, 0\rangle$$

By (1) for $r < R$,

$$\mathbf{A}\left(\mathbf{c}(t)\right) = \frac{1}{2}B\,\langle -r\sin t, r\cos t, 0\rangle$$

Hence,

$$\mathbf{A}\left(\mathbf{c}(t)\right) \cdot \mathbf{c}'(t) = \frac{1}{2}B\,\langle -r\sin t, r\cos t, 0\rangle \cdot \langle -r\sin t, r\cos t, 0\rangle = \frac{1}{2}B\left(r^2\sin^2 t + r^2\cos^2 t\right) = \frac{1}{2}r^2 B$$

Now, by (2) we get

$$\iint_{\mathcal{S}}\mathbf{B}\cdot d\mathbf{S} = \int_{\partial\mathcal{S}}\mathbf{A}\cdot d\mathbf{S} = \int_0^{2\pi}\frac{1}{2}r^2\mathbf{B}\,dt = \frac{1}{2}r^2\mathbf{B}\int_0^{2\pi}dt = r^2\mathbf{B}\pi$$

(b) Outside the solenoid $\mathbf{B}$ is the zero field, hence $\mathbf{B} = \mathbf{0}$ on every domain lying outside the solenoid. Therefore, Stokes' Theorem implies that

$$\int_{\partial\mathcal{S}}\mathbf{A}\cdot d\mathbf{S} = \iint_{\mathcal{S}}\operatorname{curl}(\mathbf{A})\cdot d\mathbf{S} = \iint_{\mathcal{S}}\mathbf{B}\cdot d\mathbf{S} = \iint_{\mathcal{S}}\mathbf{0}\cdot d\mathbf{S} = 0.$$

17. A uniform magnetic field $\mathbf{B}$ has constant strength b in the z-direction [that is, $\mathbf{B} = \langle 0, 0, b\rangle$].
(a) Verify that $\mathbf{A} = \frac{1}{2}\mathbf{B} \times \mathbf{r}$ is a vector potential for $\mathbf{B}$, where $\mathbf{r} = \langle x, y, 0\rangle$.
(b) Calculate the flux of $\mathbf{B}$ through the rectangle with vertices A, B, C, and D in Figure 19.

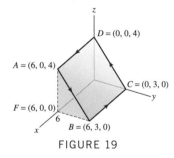

FIGURE 19

SOLUTION

(a) We compute the vector $\mathbf{A} = \frac{1}{2}\mathbf{B} \times \mathbf{r}$. Since $\mathbf{B} = b\mathbf{k}$ and $\mathbf{r} = x\mathbf{i} + y\mathbf{j}$, we have

$$\mathbf{A} = \frac{1}{2}\mathbf{B} \times \mathbf{r} = \frac{1}{2}b\mathbf{k} \times (x\mathbf{i} + y\mathbf{j}) = \frac{1}{2}b(x\mathbf{k}\times\mathbf{i} + y\mathbf{k}\times\mathbf{j}) = \frac{1}{2}b(x\mathbf{j} - y\mathbf{i}) = \left\langle -\frac{by}{2}, \frac{bx}{2}, 0\right\rangle$$

We now show that $\operatorname{curl}(\mathbf{A}) = \mathbf{B}$. We compute the curl of $\mathbf{A}$:

$$\operatorname{curl}(\mathbf{A}) = \begin{vmatrix} \mathbf{i} & \mathbf{j} & \mathbf{k} \\ \dfrac{\partial}{\partial x} & \dfrac{\partial}{\partial y} & \dfrac{\partial}{\partial z} \\ -\dfrac{by}{2} & \dfrac{bx}{2} & 0 \end{vmatrix} = \left\langle 0, 0, \frac{b}{2} + \frac{b}{2}\right\rangle = \langle 0, 0, b\rangle = \mathbf{B}$$

Therefore, $\mathbf{A}$ is a vector potential for $\mathbf{B}$.

(b) Let $\mathcal{S}$ be the rectangle $\square ABCD$ and let $\mathcal{C}$ be the boundary of $\mathcal{S}$. Since $\mathbf{B} = \text{Curl}(\mathbf{A})$, we see that $\mathbf{B}$ has a vector potential. It follows, as explained in this section, that the flux of $\mathbf{B}$ through rectangle $\mathcal{S}$ is equal to the flux of $\mathbf{B}$ through any surface with the same boundary $\mathcal{C}$. Let $\mathcal{S}'$ be the wedge-shaped box with four sides and open top. Since the boundary of $\mathcal{S}'$ is also $\mathcal{C}$, we have

$$\iint_{\mathcal{S}} \mathbf{B} \cdot d\mathbf{S} = \iint_{\mathcal{S}'} \mathbf{B} \cdot d\mathbf{S}$$

The vector field $\mathbf{B}$ points in the $\mathbf{k}$ direction, so it has zero flux through the three vertical sides of $\mathcal{S}'$. On the other hand, the unit normal vector to the bottom face of $\mathcal{S}'$ is $\mathbf{k}$, so the normal component of $\mathbf{B}$ along the bottom face is equal to b. We obtain

$$\iint_{\mathcal{S}'} \mathbf{B} \cdot d\mathbf{S} = \iint_{\text{Bottom Face of } \mathcal{S}'} b \, dA$$

$$= b(\text{Area of Bottom Face of } \mathcal{S}') = 18b$$

19. Let $\mathbf{F} = \langle y^2, 2z + x, 2y^2 \rangle$. Use Stokes' Theorem to find a plane with equation $ax + by + cz = 0$ (where a, b, c are not all zero) such that $\oint_{\mathcal{C}} \mathbf{F} \cdot d\mathbf{s} = 0$ for every closed $\mathcal{C}$ lying in the plane. *Hint:* Choose a, b, c so that curl($\mathbf{F}$) lies in the plane.

SOLUTION   Since we are interested in $\oint_{\mathcal{C}} \mathbf{F} \cdot d\mathbf{s}$, we can also consider $\iint \text{curl}\mathbf{F} \cdot d\mathbf{S}$, by Stokes' Theorem. The curl is $\langle 4y - 2, 0, 1 - 2y \rangle$ and the normal to the plane is $\mathbf{n} = \langle a, b, c \rangle$. They are orthogonal if

$$\langle 4y - 2, 0, 1 - 2y \rangle \cdot \langle a, b, c \rangle = a(4y - 2) + c(1 - 2y) = 0$$

which means:

$$4ay - 2a + c - 2cy = 0 \quad \Rightarrow \quad (4a - 2c) = 0, (c - 2a) = 0$$

This yields $c = 2a$ and b is arbitrary.

21. Let $\mathbf{F} = \langle y^2, x^2, z^2 \rangle$. Show that

$$\int_{\mathcal{C}_1} \mathbf{F} \cdot d\mathbf{s} = \int_{\mathcal{C}_2} \mathbf{F} \cdot d\mathbf{s}$$

for any two closed curves lying on a cylinder whose central axis is the z-axis (Figure 21).

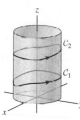

FIGURE 21

SOLUTION   We denote by $\mathcal{S}$ the part of the cylinder for which $\mathcal{C}_1$ and $\mathcal{C}_2$ are boundary curves. Using Stokes' Theorem (notice that $\mathcal{C}_1$ and $\mathcal{C}_2$ have the same orientations), we have

$$\int_{\mathcal{C}_1} \mathbf{F} \cdot d\mathbf{s} - \int_{\mathcal{C}_2} \mathbf{F} \cdot d\mathbf{s} = \iint_{\mathcal{S}} \text{curl}(\mathbf{F}) \cdot d\mathbf{S} \tag{1}$$

We compute the curl:

$$\text{curl}(\mathbf{F}) = \left\langle \frac{\partial F_3}{\partial y} - \frac{\partial F_2}{\partial z}, \frac{\partial F_1}{\partial z} - \frac{\partial F_3}{\partial x}, \frac{\partial F_2}{\partial x} - \frac{\partial F_1}{\partial y} \right\rangle = \langle 0, 0, 2x - 2y \rangle$$

We parametrize $\mathcal{S}$ by

$$\Phi(\theta, z) = \langle R\cos\theta, R\sin\theta, z \rangle$$

where (θ, z) varies in a certain parameter domain $\mathcal{D}$. The outward-pointing normal is

$$\mathbf{n} = \langle R\cos\theta, R\sin\theta, 0 \rangle$$

We compute $\text{curl}(\mathbf{F})$ in terms of the parameters:

$$\text{curl}(\mathbf{F}) = \langle 0, 0, 2x - 2y \rangle = \langle 0, 0, 2R\cos\theta - 2R\sin\theta \rangle$$

We compute the dot product:

$$\text{curl}(\mathbf{F}) \cdot \mathbf{n} = 2R \langle 0, 0, \cos\theta - \sin\theta \rangle \cdot R \langle \cos\theta, \sin\theta, 0 \rangle = 2R^2(0 + 0 + 0) = 0$$

Combining with (1) gives

$$\int_{\mathcal{C}_1} \mathbf{F} \cdot d\mathbf{s} - \int_{\mathcal{C}_2} \mathbf{F} \cdot d\mathbf{s} = \iint_{\mathcal{S}} \text{curl}(\mathbf{F}) \cdot d\mathbf{S} = \iint_{\mathcal{D}} 0\, d\theta\, dr = 0$$

or

$$\int_{\mathcal{C}_1} \mathbf{F} \cdot d\mathbf{s} = \int_{\mathcal{C}_2} \mathbf{F} \cdot d\mathbf{s}.$$

23. You know two things about a vector field $\mathbf{F}$:

(i) $\mathbf{F}$ has a vector potential $\mathbf{A}$ (but $\mathbf{A}$ is unknown).

(ii) The circulation of $\mathbf{A}$ around the unit circle (oriented counterclockwise) is 25.

Determine the flux of $\mathbf{F}$ through the surface $\mathcal{S}$ in Figure 22, oriented with upward pointing normal.

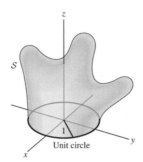

FIGURE 22 Surface $\mathcal{S}$ whose boundary is the unit circle.

SOLUTION Since $\mathbf{F}$ has a vector potential—that is, $\mathbf{F}$ is the curl of a vector field—the flux of $\mathbf{F}$ through a surface depends only on the boundary curve $\mathcal{C}$. Now, the surface $\mathcal{S}$ and the unit disc $\mathcal{S}_1$ in the xy-plane share the same boundary $\mathcal{C}$. Therefore,

$$\iint_{\mathcal{S}} \mathbf{F} \cdot d\mathbf{S} = \iint_{\mathcal{S}_1} \mathbf{F} \cdot d\mathbf{S} \qquad (1)$$

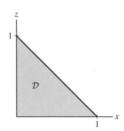

We compute the flux of $\mathbf{F}$ through $\mathcal{S}_1$, using the parametrization

$$\mathcal{S}_1 : \quad \Phi(r, \theta) = (r\cos\theta, r\sin\theta, 0), \quad 0 \le r \le 1, \quad 0 \le \theta \le 2\pi$$

$$\mathbf{n} = \langle 0, 0, 1 \rangle$$

By the given information, we have

$$\mathbf{F}(\Phi(r, \theta)) = \mathbf{F}(r\cos\theta, r\sin\theta, 0) = \langle 0, 0, 1 \rangle$$

Hence,

$$\mathbf{F}(\Phi(r, \theta)) \cdot \mathbf{n} = \langle 0, 0, 1 \rangle \cdot \langle 0, 0, 1 \rangle = 1$$

We obtain the following integral:

$$\iint_{\mathcal{S}_1} \mathbf{F} \cdot d\mathbf{S} = \int_0^{2\pi} \int_0^1 \mathbf{F}(\Phi(r, \theta)) \cdot \mathbf{n}\, dr\, d\theta = \int_0^{2\pi} \int_0^1 1\, dr\, d\theta = 2\pi$$

Combining with (1) we obtain

$$\iint_{\mathcal{S}} \mathbf{F} \cdot d\mathbf{S} = 2\pi$$

25. Prove that $\operatorname{curl}(f\mathbf{a}) = \nabla f \times \mathbf{a}$, where f is a differentiable function and $\mathbf{a}$ is a constant vector.

SOLUTION Let us first write $\mathbf{a}$ as a constant vector $\mathbf{a} = \langle a_1, a_2, a_3 \rangle$ and $f = f(x, y, z)$. Then consider the following:

$$\operatorname{curl}(f\mathbf{a}) = \operatorname{curl}(f(x, y, z)\langle a_1, a_2, a_3 \rangle) = \begin{vmatrix} \mathbf{i} & \mathbf{j} & \mathbf{k} \\ \partial/\partial x & \partial/\partial y & \partial/\partial z \\ a_1 f(x, y, z) & a_2 f(x, y, z) & a_3 f(x, y, z) \end{vmatrix}$$

$$= \left\langle \frac{\partial}{\partial y}(a_3 f) - \frac{\partial}{\partial z}(a_2 f), -\frac{\partial}{\partial z}(a_1 f) + \frac{\partial}{\partial x}(a_3 f), \frac{\partial}{\partial x}(a_2 f) - \frac{\partial}{\partial y}(a_1 f) \right\rangle$$

$$= \langle a_3 f_y - a_2 f_z, a_3 f_x - a_1 f_z, a_2 f_x - a_1 f_y \rangle$$

And now consider the following:

$$\nabla f \times \mathbf{a} = \langle f_x, f_y, f_z \rangle \times \langle a_1, a_2, a_3 \rangle$$

$$= \begin{vmatrix} \mathbf{i} & \mathbf{j} & \mathbf{k} \\ f_x & f_y & f_z \\ a_1 & a_2 & a_3 \end{vmatrix}$$

$$= \langle a_3 f_y - a_2 f_z, a_3 f_x - a_1 f_z, a_2 f_x - a_1 f_y \rangle$$

Since the two expressions above are equal, we conclude

$$\operatorname{curl}(f\mathbf{a}) = \nabla f \times \mathbf{a}$$

27. Prove the following Product Rule:

$$\operatorname{curl}(f\mathbf{F}) = f\operatorname{curl}(\mathbf{F}) + \nabla f \times \mathbf{F}$$

SOLUTION We evaluate the curl of $f\mathbf{F}$. Since $f\mathbf{F} = \langle fF_1, fF_2, fF_3 \rangle$, using the Product Rule for scalar functions we have

$$\operatorname{curl}(f\mathbf{F}) = \left\langle \frac{\partial}{\partial y}(fF_3) - \frac{\partial}{\partial z}(fF_2), \frac{\partial}{\partial z}(fF_1) - \frac{\partial}{\partial x}(fF_3), \frac{\partial}{\partial x}(fF_2) - \frac{\partial}{\partial y}(fF_1) \right\rangle$$

$$= \left\langle \frac{\partial f}{\partial y}F_3 + f\frac{\partial F_3}{\partial y} - \frac{\partial f}{\partial z}F_2 - f\frac{\partial F_2}{\partial z}, \frac{\partial f}{\partial z}F_1 + f\frac{\partial F_1}{\partial z} - \frac{\partial f}{\partial x}F_3 - f\frac{\partial F_3}{\partial x}, \right.$$

$$\left. \frac{\partial f}{\partial x}F_2 + f\frac{\partial F_2}{\partial x} - \frac{\partial f}{\partial y}F_1 - f\frac{\partial F_1}{\partial y} \right\rangle$$

$$= f \left\langle \frac{\partial F_3}{\partial y} - \frac{\partial F_2}{\partial z}, \frac{\partial F_1}{\partial z} - \frac{\partial F_3}{\partial x}, \frac{\partial F_2}{\partial x} - \frac{\partial F_1}{\partial y} \right\rangle$$

$$+ \left\langle \frac{\partial f}{\partial y}F_3 - \frac{\partial f}{\partial z}F_2, \frac{\partial f}{\partial z}F_1 - \frac{\partial f}{\partial x}F_3, \frac{\partial f}{\partial x}F_2 - \frac{\partial f}{\partial y}F_1 \right\rangle \tag{1}$$

The vector in the first term is $\operatorname{curl}(\mathbf{F})$. We show that the second term is the cross product $\nabla f \times \mathbf{F}$. We compute the cross product:

$$\nabla f \times \mathbf{F} = \begin{vmatrix} \mathbf{i} & \mathbf{j} & \mathbf{k} \\ \dfrac{\partial f}{\partial x} & \dfrac{\partial f}{\partial y} & \dfrac{\partial f}{\partial z} \\ F_1 & F_2 & F_3 \end{vmatrix} = \left(\frac{\partial f}{\partial y}F_3 - \frac{\partial f}{\partial z}F_2 \right)\mathbf{i} - \left(\frac{\partial f}{\partial x}F_3 - \frac{\partial f}{\partial z}F_1 \right)\mathbf{j} + \left(\frac{\partial f}{\partial x}F_2 - \frac{\partial f}{\partial y}F_1 \right)\mathbf{k}$$

$$= \left\langle \frac{\partial f}{\partial y}F_3 - \frac{\partial f}{\partial z}F_2, \frac{\partial f}{\partial z}F_1 - \frac{\partial f}{\partial x}F_3, \frac{\partial f}{\partial x}F_2 - \frac{\partial f}{\partial y}F_1 \right\rangle$$

Therefore, (1) gives

$$\operatorname{curl}(f\mathbf{F}) = f\operatorname{curl}(\mathbf{F}) + \nabla f \times \mathbf{F}$$

29. Verify that $\mathbf{B} = \operatorname{curl}(\mathbf{A})$ for $r > R$ in the setting of Example 6.

SOLUTION As observed in the example,

$$\operatorname{curl}(\langle f, g, 0 \rangle) = \langle -g_z, f_z, g_x - f_y \rangle$$

and recall $r = x^2 + y^2$. For $r > R$, this yields

$$\operatorname{curl}(\mathbf{A}) = \frac{1}{2}R^2 B \left\langle 0, 0, \frac{\partial}{\partial x}(xr^{-2}) - \frac{\partial}{\partial y}(-yr^{-2}) \right\rangle$$

The z-component on the right is also zero:

$$\frac{\partial}{\partial x}(xr^{-2}) + \frac{\partial}{\partial y}(yr^{-2}) = \frac{\partial}{\partial x}\left(\frac{x}{x^2+y^2}\right) + \frac{\partial}{\partial y}\left(\frac{y}{x^2+y^2}\right)$$

$$= \frac{(x^2+y^2)-x(2x)}{(x^2+y^2)^2} + \frac{(x^2+y^2)-y(2y)}{(x^2+y^2)^2}$$

$$= 0$$

Thus, $\text{curl}(\mathbf{A}) = \mathbf{0}$ when $r > R$ as required.

Further Insights and Challenges

31. In this exercise, we use the notation of the proof of Theorem 1 and prove

$$\oint_C F_3(x, y, z)\mathbf{k} \cdot d\mathbf{s} = \iint_S \text{curl}(F_3(x, y, z)\mathbf{k}) \cdot d\mathbf{S} \qquad \boxed{11}$$

In particular, S is the graph of $z = f(x, y)$ over a domain $\mathcal{D}$, and C is the boundary of S with parametrization $(x(t), y(t), f(x(t), y(t)))$.

(a) Use the Chain Rule to show that

$$F_3(x, y, z)\mathbf{k} \cdot d\mathbf{s} = F_3(x(t), y(t), f(x(t), y(t)))\left(f_x(x(t), y(t))x'(t) + f_y(x(t), y(t))y'(t)\right) dt$$

and verify that

$$\oint_C F_3(x, y, z)\mathbf{k} \cdot d\mathbf{s} = \oint_{C_0} \langle F_3(x, y, z)f_x(x, y), F_3(x, y, z)f_y(x, y)\rangle \cdot d\mathbf{s}$$

where C_0 has parametrization $(x(t), y(t))$.

(b) Apply Green's Theorem to the line integral over C_0 and show that the result is equal to the right-hand side of Eq. (11).

SOLUTION Let $(x(t), y(t))$, $a \le t \le b$ be a parametrization of the boundary curve C_0 of the domain $\mathcal{D}$.

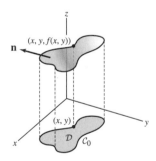

The boundary curve C of S projects on C_0 and has the parametrization

$$\gamma(t) = (x(t), y(t), f(x(t), y(t))), \quad a \le t \le b$$

Let

$$\mathbf{F} = \langle 0, 0, F_3(x, y, z)\rangle$$

We must show that

$$\int_C \mathbf{F} \cdot d\mathbf{s} = \iint_S \text{curl}(\mathbf{F}) \cdot d\mathbf{S} \qquad (1)$$

We first compute the surface integral, using the parametrization

$$S : \Phi(x, y) = (x, y, f(x, y))$$

The normal vector is

$$\mathbf{n} = \frac{\partial \Phi}{\partial x} \times \frac{\partial \Phi}{\partial y} = \langle 1, 0, f_x(x, y)\rangle \times \langle 0, 1, f_y(x, y)\rangle = (\mathbf{i} + f_x(x, y)\mathbf{k}) \times (\mathbf{j} + f_y(x, y)\mathbf{k})$$

$$= -f_y(x, y)\mathbf{j} - f_x(x, y)\mathbf{i} + \mathbf{k} = \langle -f_x(x, y), -f_y(x, y), 1\rangle$$

We compute the curl of **F**:

$$\text{curl}(\mathbf{F}) = \begin{vmatrix} \mathbf{i} & \mathbf{j} & \mathbf{k} \\ \dfrac{\partial}{\partial x} & \dfrac{\partial}{\partial y} & \dfrac{\partial}{\partial z} \\ 0 & 0 & F_3(x, y, z) \end{vmatrix} = \left\langle \frac{\partial F_3(x, y, z)}{\partial y}, -\frac{\partial F_3(x, y, z)}{\partial x}, 0 \right\rangle$$

Hence,

$$\text{curl}(\mathbf{F})\,(\Phi(x, y)) \cdot \mathbf{n} = \left\langle \frac{\partial F_3}{\partial y}(x, y, f(x, y)) - \frac{\partial F_3}{\partial x}(x, y, f(x, y)), 0 \right\rangle \cdot \langle -f_x(x, y), -f_y(x, y), 1 \rangle$$

$$= -\frac{\partial F_3(x, y, f(x, y))}{\partial y} f_x(x, y) + \frac{\partial F_3(x, y, f(x, y))}{\partial x} f_y(x, y)$$

The surface integral is thus

$$\iint_S \text{curl}(\mathbf{F}) \cdot d\mathbf{S} = \iint_D \left(-\frac{\partial F_3(x, y, f(x, y))}{\partial y} f_x(x, y) + \frac{\partial F_3(x, y, f(x, y))}{\partial x} f_y(x, y) \right) dx\, dy \qquad (2)$$

We now evaluate the line integral in (1). We have

$$\mathbf{F}(\gamma(t)) \cdot \gamma'(t) = \left\langle 0, 0, F_3\big(x(t), y(t), f\big(x(t), y(t)\big)\big) \right\rangle \cdot \left\langle x'(t), y'(t), \frac{d}{dt} f\big(x(t), y(t)\big) \right\rangle$$

$$= F_3\big(x(t), y(t), f\big(x(t), y(t)\big)\big) \frac{d}{dt} f\big(x(t), y(t)\big) \qquad (3)$$

Using the Chain Rule gives

$$\frac{d}{dt} f\big(x(t), y(t)\big) = f_x\big(x(t), y(t)\big) x'(t) + f_y\big(x(t), y(t)\big) y'(t)$$

Substituting in (3), we conclude that the line integral is

$$\int_C \mathbf{F} \cdot d\mathbf{s} = \int_a^b \left(F_3\big(x(t), y(t), f\big(x(t), y(t)\big)\big) \cdot \big(f_x\big(x(t), y(t)\big)x'(t) + f_y\big(x(t), y(t)\big)y'(t)\big) \right) dt \qquad (4)$$

We consider the following vector field:

$$\mathbf{G}(x, y) = \left\langle F_3\big(x, y, f(x, y)\big) f_x(x, y),\ F_3\big(x, y, f(x, y)\big) f_y(x, y) \right\rangle$$

Then the integral in (4) is the line integral of the planar vector field **G** over C_0. That is,

$$\int_C \mathbf{F} \cdot d\mathbf{s} = \int_{C_0} \mathbf{G} \cdot d\mathbf{s}$$

Therefore, we may apply Green's Theorem and write

$$\int_C \mathbf{F} \cdot d\mathbf{s} = \int_{C_0} \mathbf{G} \cdot d\mathbf{s} = \iint_D \left(\frac{\partial}{\partial x}\big(F_3(x, y, f(x, y)) f_y(x, y)\big) - \frac{\partial}{\partial y}\big(F_3(x, y, f(x, y)) f_x(x, y)\big) \right) dx\, dy \qquad (5)$$

We use the Product Rule to evaluate the integrand:

$$\frac{\partial F_3}{\partial x}(x, y, f(x, y))\, f_y(x, y) + F_3\big(x, y, f(x, y)\big) f_{yx}(x, y) - \frac{\partial F_3}{\partial y}(x, y, f(x, y))\, f_x(x, y) - F_3\big(x, y, f(x, y)\big) f_{xy}(x, y)$$

$$= \frac{\partial F_3}{\partial x}(x, y, f(x, y))\, f_y(x, y) - \frac{\partial F_3}{\partial y}(x, y, f(x, y))\, f_x(x, y)$$

Substituting in (5) gives

$$\int_C \mathbf{F} \cdot d\mathbf{s} = \iint_D \left(\frac{\partial F_3(x, y, f(x, y))}{\partial x} f_y(x, y) - \frac{\partial F_3(x, y, f(x, y))}{\partial y} f_x(x, y) \right) dx\, dy \qquad (6)$$

Equations (2) and (6) give the same result, hence

$$\int_C \mathbf{F} \cdot d\mathbf{s} = \iint_S \text{curl}(\mathbf{F}) \cdot d\mathbf{s}$$

for

$$\mathbf{F} = \langle 0, 0, F_3(x, y, z) \rangle$$

17.3 Divergence Theorem   (LT Section 18.3)

Preliminary Questions

1. What is the flux of $\mathbf{F} = \langle 1, 0, 0 \rangle$ through a closed surface?

SOLUTION   The divergence of $\mathbf{F} = \langle 1, 0, 0 \rangle$ is $\text{div}(\mathbf{F}) = \frac{\partial P}{\partial x} + \frac{\partial Q}{\partial y} + \frac{\partial R}{\partial z} = 0$, therefore the Divergence Theorem implies that the flux of $\mathbf{F}$ through a closed surface $\mathcal{S}$ is

$$\iint_{\mathcal{S}} \mathbf{F} \cdot d\mathbf{S} = \iiint_{\mathcal{W}} \text{div}(\mathbf{F}) \, dV = \iiint_{\mathcal{W}} 0 \, dV = 0$$

2. Justify the following statement: The flux of $\mathbf{F} = \langle x^3, y^3, z^3 \rangle$ through every closed surface is positive.

SOLUTION   The divergence of $\mathbf{F} = \langle x^3, y^3, z^3 \rangle$ is

$$\text{div}(\mathbf{F}) = 3x^2 + 3y^2 + 3z^2$$

Therefore, by the Divergence Theorem, the flux of $\mathbf{F}$ through a closed surface $\mathcal{S}$ is

$$\iint_{\mathcal{S}} \mathbf{F} \cdot d\mathbf{S} = \iiint_{\mathcal{W}} (3x^2 + 3y^2 + 3z^2) \, dV$$

Since the integrand is positive for all $(x, y, z) \neq (0, 0, 0)$, the triple integral, hence also the flux, is positive.

3. Which of the following expressions are meaningful (where $\mathbf{F}$ is a vector field and f is a function)? Of those that are meaningful, which are automatically zero?

(a) $\text{div}(\nabla f)$ **(b)** $\text{curl}(\nabla f)$ **(c)** $\nabla \text{curl}(f)$

(d) $\text{div}(\text{curl}(\mathbf{F}))$ **(e)** $\text{curl}(\text{div}(\mathbf{F}))$ **(f)** $\nabla(\text{div}(\mathbf{F}))$

SOLUTION

(a) The divergence is defined on vector fields. The gradient is a vector field, hence $\text{div}(\nabla \varphi)$ is defined. It is not automatically zero since for $\varphi = x^2 + y^2 + z^2$ we have

$$\text{div}(\nabla \varphi) = \text{div} \langle 2x, 2y, 2z \rangle = 2 + 2 + 2 = 6 \neq 0$$

(b) The curl acts on vector valued functions, and $\nabla \varphi$ is such a function. Therefore, $\text{curl}(\nabla \varphi)$ is defined. Since the gradient field $\nabla \varphi$ is conservative, the cross partials of $\nabla \varphi$ are equal, or equivalently, $\text{curl}(\nabla \varphi)$ is the zero vector.

(c) The curl is defined on vector fields rather than on scalar functions. Therefore, $\text{curl}(\varphi)$ is undefined. Obviously, $\nabla \text{curl}(\varphi)$ is also undefined.

(d) The curl is defined on the vector field $\mathbf{F}$ and the divergence is defined on the vector field $\text{curl}(\mathbf{F})$. Therefore the expression $\text{div}(\text{curl}(\mathbf{F}))$ is meaningful. We show that this vector is automatically zero:

$$\text{div}(\text{curl}(\mathbf{F})) = \text{div} \left\langle \frac{\partial F_3}{\partial y} - \frac{\partial F_2}{\partial z}, \frac{\partial F_1}{\partial z} - \frac{\partial F_3}{\partial x}, \frac{\partial F_2}{\partial x} - \frac{\partial F_1}{\partial y} \right\rangle$$

$$= \frac{\partial}{\partial x} \left(\frac{\partial F_3}{\partial y} - \frac{\partial F_2}{\partial z} \right) + \frac{\partial}{\partial y} \left(\frac{\partial F_1}{\partial z} - \frac{\partial F_3}{\partial x} \right) + \frac{\partial}{\partial z} \left(\frac{\partial F_2}{\partial x} - \frac{\partial F_1}{\partial y} \right)$$

$$= \frac{\partial^2 F_3}{\partial x \partial y} - \frac{\partial^2 F_2}{\partial x \partial z} + \frac{\partial^2 F_1}{\partial y \partial z} - \frac{\partial^2 F_3}{\partial y \partial x} + \frac{\partial^2 F_2}{\partial z \partial x} - \frac{\partial^2 F_1}{\partial z \partial y}$$

$$= \left(\frac{\partial^2 F_3}{\partial x \partial y} - \frac{\partial^2 F_3}{\partial y \partial x} \right) + \left(\frac{\partial^2 F_2}{\partial z \partial x} - \frac{\partial^2 F_2}{\partial x \partial z} \right) + \left(\frac{\partial^2 F_1}{\partial y \partial z} - \frac{\partial^2 F_1}{\partial z \partial y} \right)$$

$$= 0 + 0 + 0 = 0$$

(e) The curl acts on vector valued functions, whereas $\text{div}(\mathbf{F})$ is a scalar function. Therefore the expression $\text{curl}(\text{div}(\mathbf{F}))$ has no meaning.

(f) $\text{div}(\mathbf{F})$ is a scalar function, hence $\nabla(\text{div}\,\mathbf{F})$ is meaningful. It is not necessarily the zero vector as shown in the following example:

$$\mathbf{F} = \langle x^2, y^2, z^2 \rangle$$

$$\text{div}(\mathbf{F}) = 2x + 2y + 2z$$

$$\nabla(\text{div}\,\mathbf{F}) = \langle 2, 2, 2 \rangle \neq \langle 0, 0, 0 \rangle$$

4. Which of the following statements is correct (where $\mathbf{F}$ is a continuously differentiable vector field defined everywhere)?

(a) The flux of $\text{curl}(\mathbf{F})$ through all surfaces is zero.

(b) If $\mathbf{F} = \nabla \varphi$, then the flux of $\mathbf{F}$ through all surfaces is zero.

(c) The flux of $\text{curl}(\mathbf{F})$ through all closed surfaces is zero.

SOLUTION

(a) This statement holds only for conservative fields. If $\mathbf{F}$ is not conservative, there exist closed curves such that $\int_C \mathbf{F} \cdot d\mathbf{s} \neq 0$, hence by Stokes' Theorem $\iint_S \text{curl}(\mathbf{F}) \cdot d\mathbf{S} \neq 0$.

(b) This statement is false. Consider the unit sphere S in the three-dimensional space and the function $\varphi(x, y, z) = x^2 + y^2 + z^2$. Then $\mathbf{F} = \nabla\varphi = \langle 2x, 2y, 2z \rangle$ and div $(\mathbf{F}) = 2 + 2 + 2 = 6$. Using the Divergence Theorem, we have ($\mathcal{W}$ is the unit ball in R^3)

$$\iint_S \mathbf{F} \cdot d\mathbf{S} = \iiint_{\mathcal{W}} \text{div}(\mathbf{F}) \, dV = \iiint_{\mathcal{W}} 6 \, dV = 6 \iiint_{\mathcal{W}} dV = 6 \, \text{Vol}(\mathcal{W})$$

(c) This statement is correct, as stated in the corollary of Stokes' Theorem in section 17.3.

5. How does the Divergence Theorem imply that the flux of $\mathbf{F} = \langle x^2, y - e^z, y - 2zx \rangle$ through a closed surface is equal to the enclosed volume?

SOLUTION By the Divergence Theorem, the flux is

$$\iint_S \mathbf{F} \cdot d\mathbf{S} = \iiint_{\mathcal{W}} \text{div}(\mathbf{F}) \, dV = \iiint_{\mathcal{W}} (2x + 1 - 2x) \, dV = \iiint_{\mathcal{W}} 1 \, dV = \text{Volume}(\mathcal{W})$$

Therefore the statement is true.

Exercises

In Exercises 1–4, compute the divergence of the vector field.

1. $\mathbf{F} = \langle xy, yz, y^2 - x^3 \rangle$

SOLUTION The divergence of $\mathbf{F}$ is

$$\text{div}(\mathbf{F}) = \frac{\partial}{\partial x}(xy) + \frac{\partial}{\partial y}(yz) + \frac{\partial}{\partial z}(y^2 - x^3) = y + z + 0 = y + z$$

3. $\mathbf{F} = \langle x - 2zx^2, z - xy, z^2x^2 \rangle$

SOLUTION

$$\text{div}(\mathbf{F}) = \frac{\partial}{\partial x}(x - 2zx^2) + \frac{\partial}{\partial y}(z - xy) + \frac{\partial}{\partial z}(z^2x^2) = (1 - 4zx) + (-x) + (2zx^2) = 1 - 4zx - x + 2zx^2$$

5. Find a constant c for which the velocity field

$$\mathbf{v} = (cx - y)\mathbf{i} + (y - z)\mathbf{j} + (3x + 4cz)\mathbf{k}$$

of a fluid is incompressible [meaning that $\text{div}(\mathbf{v}) = 0$].

SOLUTION We compute the divergence of $\mathbf{v}$:

$$\text{div}(\mathbf{v}) = \frac{\partial}{\partial x}(cx - y) + \frac{\partial}{\partial y}(y - z) + \frac{\partial}{\partial z}(3x + 4cz) = c + 1 + 4c = 5c + 1$$

Therefore, $\text{div}(\mathbf{v}) = 0$ if $5c + 1 = 0$ or $c = -\frac{1}{5}$.

In Exercises 7–10, verify the Divergence Theorem for the vector field and region.

7. $\mathbf{F} = \langle z, x, y \rangle$, the box $[0, 4] \times [0, 2] \times [0, 3]$

SOLUTION Let S be the surface of the box and $\mathcal{R}$ the region enclosed by S.

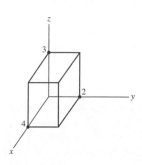

We first compute the surface integral in the Divergence Theorem:

$$\iint_{\mathcal{S}} \mathbf{F} \cdot d\mathbf{S} = \iiint_{\mathcal{R}} \operatorname{div}(\mathbf{F}) \, dV \tag{1}$$

We denote by $\mathcal{S}_i$, $i = 1, \ldots, 6$, the faces of the box, starting at the face on the xz-plane and moving counterclockwise, then moving to the bottom and the top. We use parametrizations

$$\mathcal{S}_1 : \ \Phi_1(x, z) = (x, 0, z), \quad 0 \le x \le 4, \quad 0 \le z \le 3$$
$$\mathbf{n} = \langle 0, -1, 0 \rangle$$
$$\mathcal{S}_2 : \ \Phi_2(y, z) = (0, y, z), \quad 0 \le y \le 2, \quad 0 \le z \le 3$$
$$\mathbf{n} = \langle -1, 0, 0 \rangle$$
$$\mathcal{S}_3 : \ \Phi_3(x, z) = (x, 2, z), \quad 0 \le x \le 4, \quad 0 \le z \le 3$$
$$\mathbf{n} = \langle 0, 1, 0 \rangle$$
$$\mathcal{S}_4 : \ \Phi_4(y, z) = (4, y, z), \quad 0 \le y \le 2, \quad 0 \le z \le 3$$
$$\mathbf{n} = \langle 1, 0, 0 \rangle$$
$$\mathcal{S}_5 : \ \Phi_5(x, y) = (x, y, 0), \quad 0 \le x \le 4, \quad 0 \le y \le 2$$
$$\mathbf{n} = \langle 0, 0, -1 \rangle$$
$$\mathcal{S}_6 : \ \Phi_6(x, y) = (x, y, 3), \quad 0 \le x \le 4, \quad 0 \le y \le 2$$
$$\mathbf{n} = \langle 0, 0, 1 \rangle$$

Then,

$$\iint_{\mathcal{S}_1} \mathbf{F} \cdot d\mathbf{S} = \int_0^3 \int_0^4 \mathbf{F}(\Phi_1(x, z)) \cdot \langle 0, -1, 0 \rangle \, dx \, dz = \int_0^3 \int_0^4 \langle z, x, 0 \rangle \cdot \langle 0, -1, 0 \rangle \, dx \, dz$$

$$= \int_0^3 \int_0^4 -x \, dx \, dz = 3 \frac{-x^2}{2} \Big|_0^4 = -24$$

$$\iint_{\mathcal{S}_2} \mathbf{F} \cdot d\mathbf{S} = \int_0^3 \int_0^2 \mathbf{F}(\Phi_2(y, z)) \cdot \langle -1, 0, 0 \rangle \, dy \, dz = \int_0^3 \int_0^2 \langle z, 0, y \rangle \cdot \langle -1, 0, 0 \rangle \, dy \, dz$$

$$= \int_0^3 \int_0^2 -z \, dy \, dz = 2 \cdot \frac{-z^2}{2} \Big|_0^3 = -9$$

$$\iint_{\mathcal{S}_3} \mathbf{F} \cdot d\mathbf{S} = \int_0^3 \int_0^4 \mathbf{F}(\Phi_3(x, z)) \cdot \langle 0, 1, 0 \rangle \, dx \, dz = \int_0^3 \int_0^4 \langle z, x, 2 \rangle \cdot \langle 0, 1, 0 \rangle \, dx \, dz$$

$$= \int_0^3 \int_0^4 x \, dx \, dz = 3 \cdot \frac{x^2}{2} \Big|_0^4 = 24$$

$$\iint_{\mathcal{S}_4} \mathbf{F} \cdot d\mathbf{S} = \int_0^3 \int_0^2 \mathbf{F}(\Phi_4(y, z)) \cdot \langle 1, 0, 0 \rangle \, dy \, dz = \int_0^3 \int_0^2 \langle z, 4, y \rangle \cdot \langle 1, 0, 0 \rangle \, dy \, dz$$

$$= \int_0^3 \int_0^2 z \, dy \, dz = 2 \cdot \frac{z^2}{2} \Big|_0^3 = 9$$

$$\iint_{\mathcal{S}_5} \mathbf{F} \cdot d\mathbf{S} = \int_0^2 \int_0^4 \mathbf{F}(\Phi_5(x, y)) \cdot \langle 0, 0, -1 \rangle \, dx \, dy = \int_0^2 \int_0^4 \langle 0, x, y \rangle \cdot \langle 0, 0, -1 \rangle \, dx \, dy$$

$$= \int_0^2 \int_0^4 -y \, dx \, dy = 4 \cdot \frac{-y^2}{2} \Big|_0^2 = -8$$

$$\iint_{\mathcal{S}_6} \mathbf{F} \cdot d\mathbf{S} = \int_0^2 \int_0^4 \mathbf{F}(\Phi_6(x, y)) \cdot \mathbf{n} \, dx \, dy = \int_0^2 \int_0^4 \langle 3, x, y \rangle \cdot \langle 0, 0, 1 \rangle \, dx \, dy$$

$$= \int_0^2 \int_0^4 y \, dx \, dy = 4 \cdot \frac{y^2}{2} \Big|_0^2 = 8$$

We add the integrals to obtain the surface integral

$$\iint_{\mathcal{S}} \mathbf{F} \cdot d\mathbf{S} = \sum_{i=1}^{6} \iint_{\mathcal{S}_i} \mathbf{F} \cdot d\mathbf{S} = -24 - 9 + 24 + 9 - 8 + 8 = 0 \tag{2}$$

We now evaluate the triple integral in (1). We compute the divergence of $\mathbf{F} = \langle z, x, y \rangle$:

$$\text{div}(\mathbf{F}) = \frac{\partial}{\partial x}(z) + \frac{\partial}{\partial y}(x) + \frac{\partial}{\partial z}(y) = 0$$

Hence,

$$\iiint_{\mathcal{R}} \text{div}(\mathbf{F})\, dV = \iiint_{\mathcal{R}} 0\, dV = 0 \tag{3}$$

The equality of the integrals in (2) and (3) verifies the Divergence Theorem.

9. $\mathbf{F} = \langle 2x, 3z, 3y \rangle$,   the region $x^2 + y^2 \le 1, 0 \le z \le 2$

SOLUTION

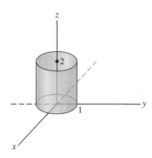

Let $\mathcal{S}$ be the surface of the cylinder and $\mathcal{R}$ the region enclosed by $\mathcal{S}$. We compute the two sides of the Divergence Theorem:

$$\iint_{\mathcal{S}} \mathbf{F} \cdot d\mathbf{S} = \iiint_{\mathcal{R}} \text{div}(\mathbf{F})\, dV \tag{1}$$

We first calculate the surface integral.

Step 1. Integral over the side of the cylinder. The side of the cylinder is parametrized by

$$\Phi(\theta, z) = (\cos\theta, \sin\theta, z), \quad 0 \le \theta \le 2\pi, \quad 0 \le z \le 2$$
$$\mathbf{n} = \langle \cos\theta, \sin\theta, 0 \rangle$$

Then,

$$\mathbf{F}\left(\Phi(\theta, z)\right) \cdot \mathbf{n} = \langle 2\cos\theta, 3z, 3\sin\theta \rangle \cdot \langle \cos\theta, \sin\theta, 0 \rangle = 2\cos^2\theta + 3z\sin\theta$$

We obtain the integral

$$\iint_{\text{side}} \mathbf{F} \cdot d\mathbf{S} = \int_0^2 \int_0^{2\pi} \left(2\cos^2\theta + 3z\sin\theta\right) d\theta\, dz = 4\int_0^{2\pi} \cos^2\theta\, d\theta + \left(\int_0^2 3z\, dz\right)\left(\int_0^{2\pi} \sin\theta\, d\theta\right)$$

$$= 4 \cdot \left(\frac{\theta}{2} + \frac{\sin 2\theta}{4}\bigg|_0^{2\pi}\right) + 0 = 4\pi$$

Step 2. Integral over the top of the cylinder. The top of the cylinder is parametrized by

$$\Phi(x, y) = (x, y, 2)$$

with parameter domain $\mathcal{D} = \left\{(x, y) : x^2 + y^2 \le 1\right\}$. The upward pointing normal is

$$\mathbf{n} = \mathbf{T}_x \times \mathbf{T}_y = \langle 1, 0, 0 \rangle \times \langle 0, 1, 0 \rangle = \mathbf{i} \times \mathbf{j} = \mathbf{k} = \langle 0, 0, 1 \rangle$$

Also,

$$\mathbf{F}\left(\Phi(x, y)\right) \cdot \mathbf{n} = \langle 2x, 6, 3y \rangle \cdot \langle 0, 0, 1 \rangle = 3y$$

Hence,

$$\iint_{\text{top}} \mathbf{F} \cdot d\mathbf{S} = \iint_{\mathcal{D}} 3y\, dA = 0$$

The last integral is zero due to symmetry.

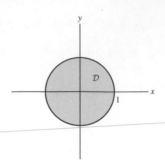

Step 3. Integral over the bottom of the cylinder. We parametrize the bottom by

$$\Phi(x, y) = (x, y, 0), \quad (x, y) \in \mathcal{D}$$

The downward pointing normal is $\mathbf{n} = \langle 0, 0, -1 \rangle$. Then

$$\mathbf{F}\left(\Phi(x, y)\right) \cdot \mathbf{n} = \langle 2x, 0, 3y \rangle \cdot \langle 0, 0, -1 \rangle = -3y$$

We obtain the following integral, which is zero due to symmetry:

$$\iint_{\text{bottom}} \mathbf{F} \cdot d\mathbf{S} = \iint_{\mathcal{D}} -3y \, dA = 0$$

Adding the integrals we get

$$\iint_{\mathcal{S}} \mathbf{F} \cdot d\mathbf{S} = \iint_{\text{side}} \mathbf{F} \cdot d\mathbf{S} + \iint_{\text{top}} \mathbf{F} \cdot d\mathbf{S} + \iint_{\text{bottom}} \mathbf{F} \cdot d\mathbf{S} = 4\pi + 0 + 0 = 4\pi \tag{2}$$

Step 4. Compare with integral of divergence.

$$\text{div}(\mathbf{F}) = \text{div} \langle 2x, 3z, 3y \rangle = \frac{\partial}{\partial x}(2x) + \frac{\partial}{\partial y}(3z) + \frac{\partial}{\partial z}(3y) = 2$$

$$\iiint_{\mathcal{R}} \text{div}(\mathbf{F}) \, dV = \iiint_{\mathcal{R}} 2 \, dV = 2 \iiint_{\mathcal{R}} dV = 2 \, \text{Vol}(\mathcal{R}) = 2 \cdot \pi \cdot 2 = 4\pi \tag{3}$$

The equality of (2) and (3) verifies the Divergence Theorem.

In Exercises 11–18, use the Divergence Theorem to evaluate the flux $\iint_{\mathcal{S}} \mathbf{F} \cdot d\mathbf{S}$.

11. $\mathbf{F} = \left\langle 0, 0, z^3/3 \right\rangle$, $\mathcal{S}$ is the sphere $x^2 + y^2 + z^2 = 1$.

SOLUTION We compute the divergence of $\mathbf{F} = \left\langle 0, 0, z^3/3 \right\rangle$:

$$\text{div}\mathbf{F} = \frac{\partial}{\partial x}(0) + \frac{\partial}{\partial y}(0) + \frac{\partial}{\partial z}(z^3/3) = z^2$$

Hence, by the Divergence Theorem ($\mathcal{W}$ is the unit ball),

$$\iint_{\mathcal{S}} \mathbf{F} \cdot d\mathbf{S} = \iiint_{\mathcal{W}} \text{div}(\mathbf{F}) \, dV = \iiint_{\mathcal{W}} z^2 \, dV$$

Computing this integral we see:

$$\iiint_{\mathcal{W}} z^2 \, dV = \int_0^{2\pi} \int_0^{\pi} \int_0^1 \rho^2 \cos^2 \phi \cdot \rho^2 \sin \phi \, d\rho \, d\phi \, d\theta$$

$$= \int_0^{2\pi} d\theta \cdot \int_0^{\pi} \cos^2 \phi \sin \phi \, d\phi \cdot \int_0^1 \rho^4 \, d\rho$$

$$= (2\pi) \cdot \left(-\frac{\cos^3 \phi}{3} \Big|_0^{\pi} \right) \cdot \left(\frac{\rho^5}{5} \Big|_0^1 \right)$$

$$= 2\pi \left(-\frac{1}{3}(-1 - 1) \right) \left(\frac{1}{5} \right)$$

$$= 2\pi \left(\frac{2}{3} \right) \left(\frac{1}{5} \right) = \frac{4\pi}{15}$$

13. $\mathbf{F} = \langle x^3, 0, z^3 \rangle$, $\mathcal{S}$ is the octant of the sphere $x^2 + y^2 + z^2 = 4$, in the first octant $x \geq 0$, $y \geq 0$, $z \geq 0$.

SOLUTION We compute the divergence of $\mathbf{F} = \langle x^3, 0, z^3 \rangle$:

$$\text{div}(\mathbf{F}) = \frac{\partial}{\partial x}(x^3) + \frac{\partial}{\partial y}(0) + \frac{\partial}{\partial z}(z^3) = 3x^2 + 3z^2 = 3(x^2 + z^2)$$

Using the Divergence Theorem we obtain ($\mathcal{W}$ is the region inside the sphere)

$$\iint_{\mathcal{S}} \mathbf{F} \cdot d\mathbf{S} = \iiint_{\mathcal{W}} \text{div}(\mathbf{F}) \, dV = \iiint_{\mathcal{W}} 3(x^2 + z^2) \, dV$$

We convert the integral to spherical coordinates. We have

$$x^2 + z^2 = \rho^2 \cos^2 \theta \sin^2 \phi + \rho^2 \cos^2 \phi = \rho^2 \cos^2 \theta \sin^2 \phi + \rho^2(1 - \sin^2 \phi)$$
$$= -\rho^2 \sin^2 \phi(1 - \cos^2 \theta) + \rho^2 = -\rho^2 \sin^2 \phi \sin^2 \theta + \rho^2 = \rho^2(1 - \sin^2 \phi \sin^2 \theta)$$

We obtain the following integral:

$$\iint_{\mathcal{S}} \mathbf{F} \cdot d\mathbf{S} = 3 \int_0^{2\pi} \int_0^{\pi/2} \int_0^2 \rho^2(1 - \sin^2 \phi \sin^2 \theta) \cdot \rho^2 \sin \phi \, d\rho \, d\phi \, d\theta$$

$$= 3 \int_0^{2\pi} \int_0^{\pi/2} \int_0^2 \rho^4(\sin \phi - \sin^3 \phi \sin^2 \theta) d\rho \, d\phi \, d\theta$$

$$= 3 \int_0^{2\pi} \int_0^{\pi/2} \int_0^2 \rho^4 \sin \phi \, d\rho \, d\phi \, d\theta - 3 \int_0^{2\pi} \int_0^{\pi/2} \int_0^2 \rho^4 \sin^3 \phi \sin^2 \theta \, d\rho \, d\phi \, d\theta$$

$$= 6\pi \left(\int_0^{\pi/2} \sin \phi \, d\phi \right) \left(\int_0^2 \rho^4 \, d\rho \right) - 3 \left(\int_0^{2\pi} \sin^2 \theta \, d\theta \right) \left(\int_0^{\pi/2} \sin^3 \phi \, d\phi \right) \left(\int_0^2 \rho^4 \, d\rho \right)$$

$$= 6\pi \left(-\cos \phi \Big|_{\phi=0}^{\pi/2} \right) \left(\frac{\rho^5}{5} \Big|_{\rho=0}^2 \right) \left(-3\frac{\theta}{2} - \frac{\sin 2\theta}{4} \Big|_{\theta=0}^{2\pi} \right) \cdot \left(-\frac{\sin^2 \phi \cos \phi}{3} - \frac{2}{3} \cos \phi \Big|_{\phi=0}^{\pi/2} \right) \left(\frac{\rho^5}{5} \Big|_{\rho=0}^2 \right)$$

$$= 6\pi \cdot \frac{32}{5} - 3\pi \cdot \frac{2}{3} \cdot \frac{32}{5} = \frac{128\pi}{5}$$

15. $\mathbf{F} = \langle x, y^2, z + y \rangle$, $\mathcal{S}$ is the boundary of the region contained in the cylinder $x^2 + y^2 = 4$ between the planes $z = x$ and $z = 8$.

SOLUTION Let $\mathcal{W}$ be the region enclosed by $\mathcal{S}$.

We compute the divergence of $\mathbf{F} = \langle x, y^2, z + y \rangle$:

$$\text{div}(\mathbf{F}) = \frac{\partial}{\partial x}(x) + \frac{\partial}{\partial y}(y^2) + \frac{\partial}{\partial z}(z + y) = 1 + 2y + 1 = 2 + 2y.$$

By the Divergence Theorem we have

$$\iint_{\mathcal{S}} \mathbf{F} \cdot d\mathbf{S} = \iiint_{\mathcal{W}} \text{div}(\mathbf{F}) \, dV = \iiint_{\mathcal{W}} (2 + 2y) \, dV$$

We compute the triple integral. Denoting by $\mathcal{D}$ the disk $x^2 + y^2 \leq 4$ in the xy-plane, we have

$$\iint_{\mathcal{S}} \mathbf{F} \cdot d\mathbf{S} = \iint_{\mathcal{D}} \int_x^8 (2 + 2y) \, dz \, dx \, dy = \iint_{\mathcal{D}} (2 + 2y)z \Big|_{z=x}^8 dx \, dy = \iint_{\mathcal{D}} (2 + 2y)(8 - x) \, dx \, dy$$

We convert the integral to polar coordinates:

$$\iint_S \mathbf{F} \cdot d\mathbf{S} = \int_0^{2\pi} \int_0^2 (2 + 2r \sin\theta)(8 - r\cos\theta) r \, dr \, d\theta$$

$$= \int_0^{2\pi} \int_0^2 \left(16r + 2r^2 (8\sin\theta - \cos\theta) - r^3 \sin 2\theta \right) dr \, d\theta$$

$$= \int_0^{2\pi} 8r^2 + \frac{2}{3} r^3 (8\sin\theta - \cos\theta) - \frac{r^4}{4}\sin 2\theta \Big|_{r=0}^2 \, d\theta$$

$$= \int_0^{2\pi} \left(32 + \frac{16}{3}(8\sin\theta - \cos\theta) - 4\sin 2\theta \right) d\theta$$

$$= 64\pi + \frac{128}{3} \int_0^{2\pi} \sin\theta \, d\theta - \frac{16}{3} \int_0^{2\pi} \cos\theta \, d\theta - \int_0^{2\pi} 4\sin 2\theta \, d\theta = 64\pi$$

17. $\mathbf{F} = \langle x + y, z, z - x \rangle$, S is the boundary of the region between the paraboloid $z = 9 - x^2 - y^2$ and the xy-plane.

SOLUTION We compute the divergence of $\mathbf{F} = \langle x + y, z, z - x \rangle$,

$$\text{div}(\mathbf{F}) = \frac{\partial}{\partial x}(x + y) + \frac{\partial}{\partial y}(z) + \frac{\partial}{\partial z}(z - x) = 1 + 0 + 1 = 2.$$

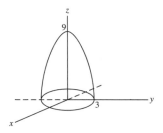

Using the Divergence Theorem we have

$$\iint_S \mathbf{F} \cdot d\mathbf{S} = \iiint_{\mathcal{W}} \text{div}(\mathbf{F}) \, dV = \iiint_{\mathcal{W}} 2 \, dV$$

We compute the triple integral:

$$\iint_S \mathbf{F} \cdot d\mathbf{S} = \iiint_{\mathcal{W}} 2 \, dV = \iint_{\mathcal{D}} \int_0^{9-x^2-y^2} 2 \, dz \, dx \, dy = \iint_{\mathcal{D}} 2z \Big|_0^{9-x^2-y^2} dx \, dy$$

$$= \iint_{\mathcal{W}} 2(9 - x^2 - y^2) \, dx \, dy$$

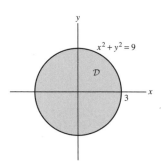

We convert the integral to polar coordinates:

$$x = r\cos\theta, \quad y = r\sin\theta, \quad 0 \le r \le 3, \quad 0 \le \theta \le 2\pi$$

$$\iint_S \mathbf{F} \cdot d\mathbf{S} = \int_0^{2\pi} \int_0^3 2\left(9 - r^2\right) r \, dr \, d\theta = 4\pi \int_0^3 (9r - r^3) \, dr = 4\pi \left(\frac{9r^2}{2} - \frac{r^4}{4} \Big|_0^3 \right) = 81\pi$$

19. Calculate the flux of the vector field $\mathbf{F} = 2xy\mathbf{i} - y^2\mathbf{j} + \mathbf{k}$ through the surface S in Figure 18. *Hint:* Apply the Divergence Theorem to the closed surface consisting of S and the unit disk.

SOLUTION From the diagram in the book, S is the surface in question bounded by the unit circle. Let T be the union of S and the unit disk D. Then T is a closed surface, and we may apply the Divergence Theorem:

$$\iint_S \mathbf{F} \cdot d\mathbf{S} + \iint_D \mathbf{F} \cdot d\mathbf{S} = \iint_T \mathbf{F} \cdot d\mathbf{S} = \iiint_{\mathcal{W}} \text{div}(\mathbf{F}) \cdot d\mathbf{S}$$

where $\mathcal{W}$ is the region enclosed by T. Now we observe that $\mathbf{F} = \left\langle 2xy, -y^2, 1 \right\rangle$ and we compute the divergence of $\mathbf{F}$:

$$\operatorname{div}(\mathbf{F}) = \frac{\partial}{\partial x}(2xy) + \frac{\partial}{\partial y}(-y^2) + \frac{\partial}{\partial z}(1) = 2y - 2y + 0 = 0$$

Therefore, the triple integral is zero and we obtain:

$$\iint_{\mathcal{S}} \mathbf{F} \cdot d\mathbf{S} = -\iint_{D} \mathbf{F} \cdot d\mathbf{S} \tag{1}$$

where D is oriented with a downward pointing normal. Let $\Phi(r, \theta) = (r\cos\theta, r\sin\theta, 0)$ be the parametrization of D with polar coordinates. Then

$$\mathbf{F}(\Phi(r, \theta)) = \left\langle 2r^2 \cos\theta \sin\theta, -r^2 \sin^2\theta, 1 \right\rangle$$

Furthermore,

$$\Phi_r(r, \theta) = \langle \cos\theta, \sin\theta, 0 \rangle, \quad \Phi_\theta(r, \theta) = \langle -r\sin\theta, r\cos\theta, 0 \rangle$$

and $\Phi_r \times \Phi_\theta = \langle 0, 0, r \rangle$ is an upward pointing normal. Finally,

$$\mathbf{F} \cdot d\mathbf{S} = \mathbf{F}(\Phi(r, \theta)) \cdot (\Phi_r \times \Phi_\theta) \, dr \, d\theta = r \, dr \, d\theta$$

The integral on the right in (1) uses a downward pointing normal, so we may drop the minus sign and use the upward-pointing normal to obtain:

$$\iint_{\mathcal{S}} \mathbf{F} \cdot d\mathbf{S} = \int_0^{2\pi} \int_0^1 r \, dr \, d\theta = \pi$$

21. Let $\mathcal{S}$ be the half-cylinder $x^2 + y^2 = 1, x \geq 0, 0 \leq z \leq 1$. Assume that $\mathbf{F}$ is a horizontal vector field (the z component is zero) such that $\mathbf{F}(0, y, z) = zy^2\mathbf{i}$. Let $\mathcal{W}$ be the solid region enclosed by $\mathcal{S}$, and assume that

$$\iiint_{\mathcal{W}} \operatorname{div}(\mathbf{F}) \, dV = 4$$

Find the flux of $\mathbf{F}$ through the curved side of $\mathcal{S}$.

SOLUTION The flux through the top and bottom of the surface are zero. The flux through the flat side (with outward normal $-\mathbf{i}$) is

$$-\int_{z=0}^1 \int_{y=-1}^1 zy^2 \, dy \, dz = -\frac{1}{2}(\frac{2}{3}) = -\frac{1}{3}$$

The flux through the curved side is $4 + \frac{1}{3} = \frac{13}{3}$.

23. Use Eq. (10) to calculate the volume of the unit ball as a surface integral over the unit sphere.

SOLUTION Let $\mathcal{S}$ be the unit sphere and $\mathcal{W}$ is the unit ball. By Eq. (10) we have

$$\operatorname{Volume}(\mathcal{W}) = \frac{1}{3} \iint_{\mathcal{S}} \mathbf{F} \cdot d\mathbf{S}, \quad \mathbf{F} = \langle x, y, z \rangle$$

To compute the surface integral, we parametrize $\mathcal{S}$ by

$$\Phi(\theta, \phi) = (\cos\theta \sin\phi, \sin\theta \sin\phi, \cos\phi), \quad 0 \leq \theta \leq 2\pi, \quad 0 \leq \phi \leq \pi$$

$$\mathbf{n} = \sin\phi \, \langle \cos\theta \sin\phi, \sin\theta \sin\phi, \cos\phi \rangle$$

Then

$$\mathbf{F}(\Phi(\theta, \phi)) \cdot \mathbf{n} = \langle \cos\theta \sin\phi, \sin\theta \sin\phi, \cos\phi \rangle \cdot \left\langle \cos\theta \sin^2\phi, \sin\theta \sin^2\phi, \cos\phi \sin\phi \right\rangle$$

$$= \cos^2\theta \sin^3\phi + \sin^2\theta \sin^3\phi + \cos^2\phi \sin\phi = \sin^3\phi(\cos^2\theta + \sin^2\theta) + \cos^2\phi \sin\phi$$

$$= \sin^3\phi + \cos^2\phi \sin\phi = \sin^3\phi + (1 - \sin^2\phi)\sin\phi = \sin\phi$$

We obtain the following integral:

$$\operatorname{Volume}(\mathcal{W}) = \frac{1}{3} \int_0^{2\pi} \int_0^{\pi} \sin\phi \, d\phi \, d\theta = \frac{1}{3} \cdot 2\pi \int_0^{\pi} \sin\phi \, d\phi = \frac{2\pi}{3} \left(-\cos\phi \Big|_0^{\pi} \right) = \frac{2\pi}{3}(1 + 1) = \frac{4\pi}{3}$$

25. Let $\mathcal{W}$ be the region in Figure 19 bounded by the cylinder $x^2 + y^2 = 4$, the plane $z = x + 1$, and the xy-plane. Use the Divergence Theorem to compute the flux of $\mathbf{F} = \langle z, x, y + z^2 \rangle$ through the boundary of $\mathcal{W}$.

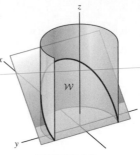

FIGURE 19

SOLUTION We compute the divergence of $\mathbf{F} = \langle z, x, y + z^2 \rangle$:

$$\text{div}(\mathbf{F}) = \frac{\partial}{\partial x}(z) + \frac{\partial}{\partial y}(x) + \frac{\partial}{\partial z}(y + z^2) = 2z$$

By the Divergence Theorem we have

$$\iint_S \mathbf{F} \cdot d\mathbf{S} = \iiint_{\mathcal{W}} \text{div}(\mathbf{F})\, dV = \iiint_{\mathcal{W}} 2\, dV$$

To compute the triple integral, we identify the projection $\mathcal{D}$ of the region on the xy-plane. $\mathcal{D}$ is the region in the xy plane enclosed by the circle $x^2 + y^2 = 4$ and the line $0 = x + 1$ or $x = -1$. We obtain the following integral:

$$\iint_S \mathbf{F} \cdot d\mathbf{S} = \iiint_{\mathcal{W}} 2z\, dV = \iint_{\mathcal{D}} \int_0^{x+1} 2z\, dz\, dx\, dy = \iint_{\mathcal{D}} z^2 \Big|_{z=0}^{x+1} dx\, dy = \iint_{\mathcal{D}} (x+1)^2\, dx\, dy$$

We compute the double integral as the difference of two integrals: the integral over the disk $\mathcal{D}_2$ of radius 2, and the integral over the part $\mathcal{D}_1$ of the disk, shown in the figure.

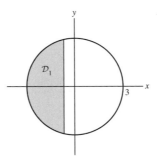

We obtain

$$\iint_S \mathbf{F} \cdot d\mathbf{S} = \iint_{\mathcal{D}_2} (x+1)^2\, dx\, dy - \iint_{\mathcal{D}_1} (x+1)^2\, dx\, dy$$

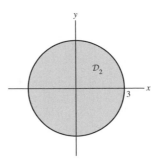

We compute the first integral, converting to polar coordinates:

$$\iint_{\mathcal{D}_2} (x+1)^2\, dx\, dy = \int_0^{2\pi} \int_0^2 (r\cos\theta + 1)^2 r\, dr\, d\theta$$

$$= \int_0^{2\pi} \int_0^2 r^3 \cos^2\theta + 2r^2 \cos\theta + r\, dr\, d\theta$$

$$= \int_0^{2\pi} \frac{r^4}{4} \cos^2\theta + \frac{2}{3}r^3 \cos\theta + \frac{1}{2}r^2 \Big|_0^2 \, d\theta$$

$$= \int_0^{2\pi} 4\cos^2\theta + \frac{16}{3}\cos\theta + 2 \, d\theta$$

$$= \int_0^{2\pi} 2\cos 2\theta + \frac{16}{3}\cos\theta + 4 \, d\theta$$

$$= \sin 2\theta + \frac{16}{3}\sin\theta + 4\theta \Big|_0^{2\pi} = 8\pi$$

We compute the second integral over the upper part of $\mathcal{D}_1$. Due to symmetry, this integral is equal to half of the integral over $\mathcal{D}_1$.

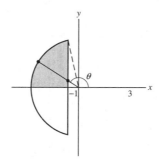

We describe the region in polar coordinates:

$$\frac{2\pi}{3} \le \theta \le \pi, \quad \frac{-1}{\cos\theta} \le r \le 2$$

Then

$$\iint_{\mathcal{D}_1} (x+1)^2 \, dx \, dy = 2 \int_{2\pi/3}^{\pi} \int_{-1/\cos\theta}^{2} (r\cos\theta + 1)^2 r \, dr \, d\theta$$

$$= \int_{2\pi/3}^{\pi} \int_{-1/\cos\theta}^{2} (r^3 \cos^2\theta + 2r^2 \cos\theta + r) \, dr \, d\theta$$

$$= \int_{2\pi/3}^{\pi} \frac{r^4}{4}\cos^2\theta + \frac{2}{3}r^3 \cos\theta + \frac{1}{2}r^2 \Big|_{r=\frac{-1}{\cos\theta}}^{2} \, d\theta$$

$$= 2\int_{2\pi/3}^{\pi} \left(4\cos^2\theta + \frac{16}{3}\cos\theta + 2\right) - \left(\frac{\cos^2\theta}{4\cos^4\theta} - \frac{2\cos\theta}{3\cos^3\theta} + \frac{1}{2\cos^2\theta}\right) d\theta$$

$$= 2\int_{2\pi/3}^{\pi} 2\cos 2\theta + 4 + \frac{16}{3}\cos\theta - \frac{1}{4}\sec^2\theta + \frac{2}{3}\sec^2\theta - \frac{1}{2}\sec^2\theta \, d\theta$$

$$= 2\int_{2\pi/3}^{\pi} 2\cos 2\theta + 4 + \frac{16}{3}\cos\theta - \frac{1}{12}\sec^2\theta \, d\theta$$

$$= \sin 2\theta + 4\theta + \frac{16}{3}\sin\theta - \frac{1}{12}\tan\theta \Big|_{2\pi/3}^{\pi}$$

$$= 2(4\pi) - 2\left(\sin\frac{4\pi}{3} + \frac{8\pi}{3} + \frac{16}{3}\sin\frac{2\pi}{3} - \frac{1}{12}\tan\frac{2\pi}{3}\right)$$

$$= 8\pi - 2\left(-\frac{\sqrt{3}}{2} + \frac{8\pi}{3} + \frac{16\sqrt{3}}{6} + \frac{\sqrt{3}}{12}\right)$$

$$= 8\pi + \sqrt{3} - \frac{16\pi}{3} - \frac{16\sqrt{3}}{3} - \frac{\sqrt{3}}{6}$$

$$= \frac{8\pi}{3} + \sqrt{3}\left(1 - \frac{16}{3} - \frac{1}{6}\right)$$

$$= \frac{8\pi}{3} - \frac{9}{2}\sqrt{3}$$

so we have

$$\iint_S \mathbf{F} \cdot d\mathbf{S} = 8\pi - \iint_{\mathcal{D}_1} (x+1)^2 \, dx \, dy \approx 8\pi - \left(\frac{8\pi}{3} - \frac{9}{2}\sqrt{3} \right) = \frac{16\pi}{3} + \frac{9}{2}\sqrt{3} \approx 24.550.$$

27. The velocity field of a fluid $\mathbf{v}$ (in meters per second) has divergence $\operatorname{div}(\mathbf{v})(P) = 3$ at the point $P = (2, 2, 2)$. Estimate the flow rate out of the sphere of radius 0.5 centered at P.

SOLUTION

$$\text{flow rate through the box} \approx \operatorname{div}(\mathbf{v})(P) \cdot \left(\frac{4}{3}\pi (0.5)^3 \right) = \frac{\pi}{2} \approx 1.57 \ \text{m}^3/\text{s}$$

29. The electric field due to a unit electric dipole oriented in the $\mathbf{k}$-direction is $\mathbf{E} = \nabla(z/r^3)$, where $r = (x^2 + y^2 + z^2)^{1/2}$ (Figure 20). Let $\mathbf{e}_r = r^{-1} \langle x, y, z \rangle$.
(a) Show that $\mathbf{E} = r^{-3}\mathbf{k} - 3zr^{-4}\mathbf{e}_r$.
(b) Calculate the flux of $\mathbf{E}$ through a sphere centered at the origin.
(c) Calculate $\operatorname{div}(\mathbf{E})$.
(d) Can we use the Divergence Theorem to compute the flux of $\mathbf{E}$ through a sphere centered at the origin?

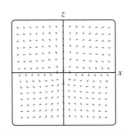

FIGURE 20 The dipole vector field restricted to the xz-plane.

SOLUTION
(a) We first compute the partial derivatives of r:

$$\frac{\partial r}{\partial x} = \frac{1}{2}(x^2 + y^2 + z^2)^{-1/2} \cdot 2x = \frac{x}{r}$$

$$\frac{\partial r}{\partial y} = \frac{1}{2}(x^2 + y^2 + z^2)^{-1/2} \cdot 2y = \frac{y}{r}$$

$$\frac{\partial r}{\partial z} = \frac{1}{2}(x^2 + y^2 + z^2)^{-1/2} \cdot 2z = \frac{z}{r} \tag{1}$$

We compute the partial derivatives of $\frac{z}{r^3}$, using the Chain Rule and the partial derivatives in (1):

$$\frac{\partial}{\partial x}\left(\frac{z}{r^3} \right) = z\frac{\partial}{\partial x}(r^{-3}) = z \cdot (-3)r^{-4}\frac{\partial r}{\partial x} = -3z \cdot r^{-4}\frac{x}{r} = -\frac{3zx}{r^5} = -3zr^{-5}x$$

$$\frac{\partial}{\partial y}\left(\frac{z}{r^3} \right) = z\frac{\partial}{\partial y}(r^{-3}) = z \cdot (-3)r^{-4}\frac{\partial r}{\partial y} = -3z \cdot r^{-4}\frac{y}{r} = -3zr^{-5}y$$

$$\frac{\partial}{\partial z}\left(\frac{z}{r^3} \right) = \frac{\partial}{\partial z}(z \cdot r^{-3}) = 1 \cdot r^{-3} + z \cdot (-3)r^{-4}\frac{\partial r}{\partial z} = r^{-3} - 3z \cdot r^{-4} \cdot \frac{z}{r} = r^{-3} - 3z^2r^{-5}$$

Therefore,

$$\mathbf{E} = \nabla\left(\frac{z}{r^3} \right) = -3zr^{-5}x\mathbf{i} - 3zr^{-5}y\mathbf{j} + (r^{-3} - 3z^2r^{-5})\mathbf{k}$$

$$= r^{-3}\mathbf{k} - 3zr^{-4} \cdot r^{-1}(x\mathbf{i} + y\mathbf{j} + z\mathbf{k}) = r^{-3}\mathbf{k} - 3zr^{-4}\mathbf{e}_r$$

(b) To compute the flux $\iint_S \mathbf{E} \cdot d\mathbf{S}$ we use the parametrization $\Phi(\theta, \phi) = (R\cos\theta \sin\phi, R\sin\theta \sin\phi, R\cos\phi)$, $0 \le \theta \le 2\pi$, $0 \le \phi \le \pi$:

$$\mathbf{n} = R^2 \sin\phi \, \mathbf{e}_r$$

We compute $\mathbf{E}(\Phi(\theta, \phi)) \cdot \mathbf{n}$. Since $r = R$ on $\mathcal{S}$, we get

$$\mathbf{E}(\Phi(\theta, \phi)) \cdot \mathbf{n} = \left(R^{-3}\mathbf{k} - 3zR^{-4}\mathbf{e}_r \right) \cdot R^2 \sin\phi \, \mathbf{e}_r = R^{-1}\sin\phi \, \mathbf{k} \cdot \mathbf{e}_r - 3zR^{-2}\sin\phi$$

$$= R^{-1}\sin\phi \, \mathbf{k} \cdot R^{-1}(x\mathbf{i} + y\mathbf{j} + z\mathbf{k}) - 3zR^{-2}\sin\phi$$

$$= R^{-2}z\sin\phi - 3zR^{-2}\sin\phi = -2zR^{-2}\sin\phi$$

$$= -2R\cos\phi \cdot R^{-2}\sin\phi = -R^{-1}\sin 2\phi$$

Hence,

$$\iint_{\mathcal{S}} \mathbf{E} \cdot d\mathbf{S} = \int_0^{2\pi} \int_0^{\pi} -R^{-1} \sin 2\phi \, d\phi \, d\theta = -\frac{2\pi}{R} \int_0^{\pi} \sin 2\phi \, d\phi = \frac{\pi}{R} \cos 2\phi \Big|_{\phi=0}^{\pi} = 0$$

(c) We use part (a) to write the vector **E** componentwise:

$$\mathbf{E} = r^{-3}\mathbf{k} - 3zr^{-4}\mathbf{e}_r = r^{-3}\mathbf{k} - 3zr^{-4}r^{-1}\langle x, y, z \rangle = \left\langle -3zr^{-5}x, -3zr^{-5}y, -3z^2r^{-5} + r^{-3} \right\rangle$$

To find div(**E**) we compute the following derivatives, using (1) and the laws of differentiation. This gives

$$\frac{\partial}{\partial x}(-3zr^{-5}x) = -3z\frac{\partial}{\partial x}(r^{-5}x) = -3z\left(-5r^{-6}\frac{\partial r}{\partial x}x + r^{-5} \cdot 1\right)$$

$$= -3z\left(-5r^{-6}x\frac{x}{r} + r^{-5}\right) = 3zr^{-7}(5x^2 - r^2)$$

Similarly,

$$\frac{\partial}{\partial y}(-3zr^{-5}y) = 3zr^{-7}(5y^2 - r^2)$$

and

$$\frac{\partial}{\partial z}(-3z^2r^{-5} + r^{-3}) = -6zr^{-5} - 3z^2(-5)r^{-6}\frac{\partial r}{\partial z} - 3r^{-4}\frac{\partial r}{\partial z}$$

$$= -6zr^{-5} + 15z^2r^{-6}\frac{z}{r} - 3r^{-4}\frac{z}{r} = 3zr^{-7}(5z^2 - 3r^2)$$

Hence,

$$\text{div}(\mathbf{E}) = 3zr^{-7}(5x^2 - r^2 + 5y^2 - r^2 + 5z^2 - 3r^2) = 15zr^{-7}(x^2 + y^2 + z^2 - r^2)$$

$$= 15zr^{-7}(r^2 - r^2) = 0$$

(d) Since **E** is not defined at the origin, which is inside the ball $\mathcal{W}$, we cannot use the Divergence Theorem to compute the flux of **E** through the sphere.

31. Let $\mathcal{W}$ be the region between the sphere of radius 4 and the cube of side 1, both centered at the origin. What is the flux through the boundary $\mathcal{S} = \partial\mathcal{W}$ of a vector field **F** whose divergence has the constant value div(**F**) $= -4$?

SOLUTION Recall,

$$\text{flux} = \iiint_{\mathcal{W}} \text{div}(\mathbf{F}) dV$$

Using this fact we see:

$$\text{flux} = \iiint_{\mathcal{W}} (-4)dV = -4 \cdot V(\mathcal{W}) = (-4)\left(\frac{256\pi}{3} - 1\right)$$

33. Find and prove a Product Rule expressing div($f\mathbf{F}$) in terms of div(**F**) and ∇f.

SOLUTION Let $\mathbf{F} = \langle P, Q, R \rangle$. We compute div($f\mathbf{F}$):

$$\text{div}(f\mathbf{F}) = \text{div} \langle fP, fQ, fR \rangle = \frac{\partial}{\partial x}(fP) + \frac{\partial}{\partial y}(fQ) + \frac{\partial}{\partial z}(fR)$$

Applying the product rule for scalar functions we obtain

$$\text{div}(f\mathbf{F}) = \left(f\frac{\partial P}{\partial x} + \frac{\partial f}{\partial x}P\right) + \left(f\frac{\partial Q}{\partial y} + \frac{\partial f}{\partial y}Q\right) + \left(f\frac{\partial R}{\partial z} + \frac{\partial f}{\partial z}R\right)$$

$$= f\left(\frac{\partial P}{\partial x} + \frac{\partial Q}{\partial y} + \frac{\partial R}{\partial z}\right) + \frac{\partial f}{\partial x}P + \frac{\partial f}{\partial y}Q + \frac{\partial f}{\partial z}R = f\text{div}(\mathbf{F}) + \mathbf{F} \cdot \nabla f$$

We thus proved the following identity:

$$\text{div}(f\mathbf{F}) = f\text{div}(\mathbf{F}) + \mathbf{F} \cdot \nabla f$$

35. Prove that $\operatorname{div}(\nabla f \times \nabla g) = 0$.

SOLUTION We compute the cross product:

$$\nabla f \times \nabla g = \langle f_x, f_y, f_z \rangle \times \langle g_x, g_y, g_z \rangle = \begin{vmatrix} \mathbf{i} & \mathbf{j} & \mathbf{k} \\ f_x & f_y & f_z \\ g_x & g_y & g_z \end{vmatrix}$$

$$= \langle f_y g_z - f_z g_y, f_z g_x - f_x g_z, f_x g_y - f_y g_x \rangle$$

We now compute the divergence of this vector. Using the Product Rule for scalar functions and the equality of the mixed partials, we obtain

$$\operatorname{div}(\nabla f \times \nabla g) = \frac{\partial}{\partial x}(f_y g_z - f_z g_y) + \frac{\partial}{\partial y}(f_z g_x - f_x g_z) + \frac{\partial}{\partial z}(f_x g_y - f_y g_x)$$

$$= f_{yx} g_z + f_y g_{zx} - f_{zx} g_y - f_z g_{yx} + f_{zy} g_x + f_z g_{xy} - f_{xy} g_z - f_x g_{zy} + f_{xz} g_y + f_x g_{yz}$$

$$\qquad - f_{yz} g_x - f_y g_{xz}$$

$$= (f_{yx} - f_{xy})g_z + (g_{zx} - g_{xz})f_y + (f_{xz} - f_{zx})g_y + (g_{xy} - g_{yx})f_z$$

$$\qquad + (f_{zy} - f_{yz})g_x + (g_{yz} - g_{zy})f_x = 0$$

In Exercises 36–38, Δ denotes the Laplace operator defined by

$$\Delta \varphi = \frac{\partial^2 \varphi}{\partial x^2} + \frac{\partial^2 \varphi}{\partial y^2} + \frac{\partial^2 \varphi}{\partial z^2}$$

37. A function φ satisfying $\Delta \varphi = 0$ is called **harmonic**.

(a) Show that $\Delta \varphi = \operatorname{div}(\nabla \varphi)$ for any function φ.

(b) Show that φ is harmonic if and only if $\operatorname{div}(\nabla \varphi) = 0$.

(c) Show that if $\mathbf{F}$ is the gradient of a harmonic function, then $\operatorname{curl}(F) = 0$ and $\operatorname{div}(F) = 0$.

(d) Show that $\mathbf{F} = \left\langle xz, -yz, \frac{1}{2}(x^2 - y^2) \right\rangle$ is the gradient of a harmonic function. What is the flux of $\mathbf{F}$ through a closed surface?

SOLUTION

(a) We compute the divergence of $\nabla \varphi$:

$$\operatorname{div}(\nabla \varphi) = \operatorname{div}\left(\left\langle \frac{\partial \varphi}{\partial x}, \frac{\partial \varphi}{\partial y}, \frac{\partial \varphi}{\partial z} \right\rangle \right) = \frac{\partial^2 \varphi}{\partial x^2} + \frac{\partial^2 \varphi}{\partial y^2} + \frac{\partial^2 \varphi}{\partial z^2} = \Delta \varphi$$

(b) In part (a) we showed that $\Delta \varphi = \operatorname{div}(\nabla \varphi)$. Therefore $\Delta \varphi = 0$ if and only if $\operatorname{div}(\nabla \varphi) = 0$. That is, φ is harmonic if and only if $\nabla \varphi$ is divergence free.

(c) We are given that $\mathbf{F} = \nabla \varphi$, where $\Delta \varphi = 0$. In part (b) we showed that

$$\operatorname{div}(\mathbf{F}) = \operatorname{div}(\nabla \varphi) = 0$$

We now show that $\operatorname{curl}(\mathbf{F}) = 0$. We have

$$\operatorname{curl}(\mathbf{F}) = \operatorname{curl}(\nabla \varphi) = \operatorname{curl}\langle \varphi_x, \varphi_y, \varphi_z \rangle = \begin{vmatrix} \mathbf{i} & \mathbf{j} & \mathbf{k} \\ \dfrac{\partial}{\partial x} & \dfrac{\partial}{\partial y} & \dfrac{\partial}{\partial z} \\ \varphi_x & \varphi_y & \varphi_z \end{vmatrix}$$

$$= \langle \varphi_{zy} - \varphi_{yz}, \varphi_{xz} - \varphi_{zx}, \varphi_{yx} - \varphi_{xy} \rangle = \langle 0, 0, 0 \rangle = \mathbf{0}$$

The last equality is due to the equality of the mixed partials.

(d) We first show that $\mathbf{F} = \left\langle xz, -yz, \frac{x^2 - y^2}{2} \right\rangle$ is the gradient of a harmonic function. We let $\varphi = \frac{x^2 z}{2} - \frac{y^2 z}{2}$ such that $\mathbf{F} = \nabla \varphi$. Indeed,

$$\nabla \varphi = \left\langle \frac{\partial \varphi}{\partial x}, \frac{\partial \varphi}{\partial y}, \frac{\partial \varphi}{\partial z} \right\rangle = \left\langle xz, -yz, \frac{x^2 - y^2}{2} \right\rangle = \mathbf{F}$$

We show that φ is harmonic, that is, $\Delta\varphi = 0$. We compute the partial derivatives:

$$\frac{\partial\varphi}{\partial x} = xz \quad\Rightarrow\quad \frac{\partial^2\varphi}{\partial x^2} = z$$

$$\frac{\partial\varphi}{\partial y} = -yz \quad\Rightarrow\quad \frac{\partial^2\varphi}{\partial y^2} = -z$$

$$\frac{\partial\varphi}{\partial z} = \frac{x^2 - y^2}{2} \quad\Rightarrow\quad \frac{\partial^2\varphi}{\partial z^2} = 0$$

Therefore,

$$\Delta\varphi = \frac{\partial^2\varphi}{\partial x^2} + \frac{\partial^2\varphi}{\partial y^2} + \frac{\partial^2\varphi}{\partial z^2} = z - z + 0 = 0$$

Since $\mathbf{F}$ is the gradient of a harmonic function, we know by part (c) that $\text{div}(\mathbf{F}) = 0$. Therefore, by the Divergence Theorem, the flux of $\mathbf{F}$ through a closed surface is zero:

$$\iint_S \mathbf{F}\cdot d\mathbf{S} = \iiint_{\mathcal{W}} \text{div}(\mathbf{F})\, dV = \iiint_{\mathcal{W}} 0\, dV = 0$$

Further Insights and Challenges

39. Let S be the boundary surface of a region $\mathcal{W}$ in $\mathbf{R}^3$ and let $D_{\mathbf{e_n}}\varphi$ denote the directional derivative of φ, where $\mathbf{e_n}$ is the outward unit normal vector. Let Δ be the Laplace operator defined earlier.

(a) Use the Divergence Theorem to prove that

$$\iint_S D_{\mathbf{e_n}}\varphi\, dS = \iiint_{\mathcal{W}} \Delta\varphi\, dV$$

(b) Show that if φ is a harmonic function (defined in Exercise 37), then

$$\iint_S D_{\mathbf{e_n}}\varphi\, dS = 0$$

SOLUTION

(a) By the theorem on evaluating directional derivatives, $D_{\mathbf{e}_n}\varphi = \nabla\varphi \cdot \mathbf{e}_n$, hence,

$$\iint_S D_{\mathbf{e}_n}\varphi\, dS = \iint_S \nabla\varphi \cdot \mathbf{e}_n\, dS \tag{1}$$

By the definition of the vector surface integral, we have

$$\iint_S \nabla\varphi \cdot d\mathbf{S} = \iint_S (\nabla\varphi \cdot \mathbf{e}_n)\, dS$$

Combining with (1) gives

$$\iint_S D_{\mathbf{e}_n}\varphi\, dS = \iint_S \nabla\varphi \cdot d\mathbf{S}$$

We now apply the Divergence Theorem and the identity $\text{div}(\nabla\varphi) = \Delta\varphi$ shown in part (a) of Exercise 27, to write

$$\iint_S D_{\mathbf{e}_n}\varphi\, dS = \iint_S \nabla\varphi \cdot d\mathbf{S} = \iiint_{\mathcal{W}} \text{div}(\nabla\varphi)\, dV = \iiint_{\mathcal{W}} \Delta\varphi\, dV$$

(b) If φ is harmonic, then $\Delta\varphi = 0$; therefore, by the equality of part (a) we have

$$\iint_S D_{\mathbf{e}_n}\varphi\, dS = \iiint_{\mathcal{W}} \Delta\varphi \cdot dV = \iiint_{\mathcal{W}} 0\, dV = 0.$$

41. Let $\mathbf{F} = \langle P, Q, R \rangle$ be a vector field defined on $\mathbf{R}^3$ such that $\text{div}(\mathbf{F}) = 0$. Use the following steps to show that $\mathbf{F}$ has a vector potential.

(a) Let $\mathbf{A} = \langle f, 0, g \rangle$. Show that

$$\text{curl}(\mathbf{A}) = \left\langle \frac{\partial g}{\partial y}, \frac{\partial f}{\partial z} - \frac{\partial g}{\partial x}, -\frac{\partial f}{\partial y} \right\rangle$$

(b) Fix any value y_0 and show that if we define

$$f(x, y, z) = -\int_{y_0}^{y} R(x, t, z)\,dt + \alpha(x, z)$$

$$g(x, y, z) = \int_{y_0}^{y} P(x, t, z)\,dt + \beta(x, z)$$

where α and β are any functions of x and z, then $\partial g/\partial y = P$ and $-\partial f/\partial y = R$.

(c) It remains for us to show that α and β can be chosen so $Q = \partial f/\partial z - \partial g/\partial x$. Verify that the following choice works (for any choice of z_0):

$$\alpha(x, z) = \int_{z_0}^{z} Q(x, y_0, t)\,dt, \qquad \beta(x, z) = 0$$

Hint: You will need to use the relation $\mathrm{div}(\mathbf{F}) = 0$.

SOLUTION

(a) If $\mathbf{A} = \langle f, 0, g \rangle$, then the curl of $\mathbf{A}$ is the following vector field:

$$\mathrm{curl}(\mathbf{A}) = \begin{vmatrix} \mathbf{i} & \mathbf{j} & \mathbf{k} \\ \dfrac{\partial}{\partial x} & \dfrac{\partial}{\partial y} & \dfrac{\partial}{\partial z} \\ f & 0 & g \end{vmatrix} = \left(\frac{\partial g}{\partial y} - 0 \right)\mathbf{i} - \left(\frac{\partial g}{\partial x} - \frac{\partial f}{\partial z} \right)\mathbf{j} + \left(0 - \frac{\partial f}{\partial y} \right)\mathbf{k} = \left\langle \frac{\partial g}{\partial y}, \frac{\partial f}{\partial z} - \frac{\partial g}{\partial x}, -\frac{\partial f}{\partial y} \right\rangle$$

(b) Using the Fundamental Theorem of Calculus, we have

$$\frac{\partial g}{\partial y}(x, y, z) = \frac{\partial}{\partial y}\int_{y_0}^{y} P(x, t, z)\,dt + \frac{\partial}{\partial y}\beta(x, z) = P(x, y, z) + 0 = P(x, y, z)$$

$$-\frac{\partial f}{\partial y}(x, y, z) = \frac{\partial}{\partial y}\int_{y_0}^{y} R(x, t, z)\,dt + \frac{\partial}{\partial y}\alpha(x, z) = R(x, y, z) + 0 = R(x, y, z)$$

(c) We verify that the functions

$$\alpha(x, z) = \int_{z_0}^{z} Q(x, y_0, t)\,dt, \qquad \beta(x, z) = 0$$

satisfy the equality

$$Q = \frac{\partial f}{\partial z} - \frac{\partial g}{\partial x}$$

We differentiate to obtain

$$\frac{\partial f}{\partial z} - \frac{\partial g}{\partial x} = -\int_{y_0}^{y} R_z(x, t, z)\,dt + \alpha_z(x, z) - \int_{y_0}^{y} P_x(x, t, z)\,dz - \beta_x(x, z)$$

$$= -\int_{y_0}^{y} \left(P_x(x, t, z) + R_z(x, t, z) \right) dt + \alpha_z(x, z) \tag{1}$$

By the Fundamental Theorem of Calculus,

$$\alpha_z(x, z) = \frac{\partial}{\partial z}\int_{z_0}^{z} Q(x, y_0, t)\,dt = Q(x, y_0, z) \tag{2}$$

Also, since $\mathrm{div}(\mathbf{F}) = 0$, we have

$$\mathrm{div}(\mathbf{F}) = P_x + Q_y + R_z = 0 \quad \Rightarrow \quad P_x + R_z = -Q_y \tag{3}$$

Substituting (2) and (3) in (1) gives

$$\frac{\partial f}{\partial z} - \frac{\partial g}{\partial x} = \int_{y_0}^{y} Q_y(x, t, z)\,dt + Q(x, y_0, z) = Q(x, y, z) - Q(x, y_0, z) + Q(x, y_0, z) = Q(x, y, z)$$

Parts (a)–(c) prove that $\mathbf{F} = \mathrm{curl}(\mathbf{A})$, or $\mathbf{A}$ is a vector potential for $\mathbf{F}$.

43. Show that

$$\mathbf{F} = \langle 2ye^z - xy, y, yz - z \rangle$$

has a vector potential and find one.

SOLUTION   As shown in Exercise 41, if $\mathbf{F}$ is divergence free, then $\mathbf{F}$ has a vector potential. We show that $\text{div}(\mathbf{F}) = 0$:

$$\text{div}(\mathbf{F}) = \frac{\partial}{\partial x}(2ye^z - xy) + \frac{\partial}{\partial y}(y) + \frac{\partial}{\partial z}(yz - z) = -y + 1 + y - 1 = 0$$

We find a vector potential $\mathbf{A}$, using the result in Exercise 41:

$$\mathbf{A} = \langle f, 0, g \rangle \tag{1}$$

Using $z_0 = 0$, we have

$$f(x, y, z) = -\int_{y_0}^{y} R(x, t, z)\, dt + \int_{0}^{z} Q(x, y_0, t)\, dt$$

$$g(x, y, z) = \int_{y_0}^{y} P(x, t, z)\, dt$$

Hence, $P(x, y, z) = 2ye^z - xy$, $Q(x, y, z) = y$, and $R(x, y, z) = yz - z$. We choose $y_0 = 0$ and compute the functions f and g:

$$f(x, y, z) = -\int_{0}^{y}(tz - z)\, dt + \int_{0}^{z} 0\, dt = -\left(\frac{t^2 z}{2} - zt\right)\Bigg|_{t=0}^{y} = zy - \frac{y^2 z}{2} = z\left(y - \frac{y^2}{2}\right)$$

$$g(x, y, z) = \int_{0}^{y}(2te^z - xt)\, dt = t^2 e^z - \frac{xt^2}{2}\Bigg|_{t=0}^{y} = y^2 e^z - \frac{xy^2}{2} = y^2\left(e^z - \frac{x}{2}\right)$$

Substituting in (1) we obtain

$$\mathbf{A} = \left\langle z\left(y - \frac{y^2}{2}\right), 0, y^2\left(e^z - \frac{x}{2}\right)\right\rangle$$

CHAPTER REVIEW EXERCISES

1. Let $\mathbf{F}(x, y) = \langle x + y^2, x^2 - y\rangle$ and let C be the unit circle, oriented counterclockwise. Evaluate $\oint_C \mathbf{F}\cdot d\mathbf{s}$ directly as a line integral and using Green's Theorem

SOLUTION   We parametrize the unit circle by $\mathbf{c}(t) = (\cos t, \sin t)$, $0 \le t \le 2\pi$. Then, $\mathbf{c}'(t) = \langle -\sin t, \cos t\rangle$ and $\mathbf{F}(\mathbf{c}(t)) = (\cos t + \sin^2 t, \cos^2 t - \sin t)$. We compute the dot product:

$$\mathbf{F}(\mathbf{c}(t))\cdot\mathbf{c}'(t) = \left\langle \cos t + \sin^2 t, \cos^2 t - \sin t\right\rangle\cdot\langle -\sin t, \cos t\rangle$$

$$= (-\sin t)(\cos t + \sin^2 t) + \cos t(\cos^2 t - \sin t)$$

$$= \cos^3 t - \sin^3 t - 2\sin t\cos t$$

The line integral is thus

$$\int_C \mathbf{F}(\mathbf{c}(t))\cdot\mathbf{c}'(t)\, dt = \int_0^{2\pi}\left(\cos^3 t - \sin^3 t - 2\sin t\cos t\right)dt$$

$$= \int_0^{2\pi}\cos^3 t\, dt - \int_0^{2\pi}\sin^3 t\, dt - \int_0^{2\pi}\sin 2t\, dt$$

$$= \frac{\cos^2 t\sin t}{3} + \frac{2\sin t}{3}\Bigg|_0^{2\pi} + \left(\frac{\sin^2 t\cos t}{3} + \frac{2\cos t}{3}\right)\Bigg|_0^{2\pi} + \frac{\cos 2t}{2}\Bigg|_0^{2\pi} = 0$$

We now compute the integral using Green's Theorem. We compute the curl of $\mathbf{F}$. Since $P = x + y^2$ and $Q = x^2 - y$, we have

$$\frac{\partial Q}{\partial x} - \frac{\partial P}{\partial y} = 2x - 2y$$

Thus,

$$\int_C \mathbf{F}\cdot d\mathbf{s} = \iint_{\mathcal{D}}(2x - 2y)\, dx\, dy$$

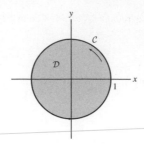

We compute the double integral by converting to polar coordinates. We get

$$\int_C \mathbf{F} \cdot d\mathbf{s} = \int_0^{2\pi} \int_0^1 (2r\cos\theta - 2r\sin\theta)r\,dr\,d\theta = \int_0^{2\pi} \int_0^1 2r^2(\cos\theta - \sin\theta)\,dr\,d\theta$$

$$= \left(\int_0^1 2r^2\,dr\right)\left(\int_0^{2\pi} (\cos\theta - \sin\theta)\,d\theta\right) = \left(\frac{2}{3}r^3\Big|_0^1\right)\left(\sin\theta + \cos\theta\Big|_0^{2\pi}\right) = \frac{2}{3}(1-1) = 0$$

In Exercises 3–6, use Green's Theorem to evaluate the line integral around the given closed curve.

3. $\oint_C xy^3\,dx + x^3y\,dy$, where C is the rectangle $-1 \le x \le 2$, $-2 \le y \le 3$, oriented counterclockwise.

SOLUTION

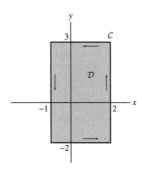

Since $P = xy^3$, $Q = x^3y$ the curl of $\mathbf{F}$ is

$$\frac{\partial Q}{\partial x} - \frac{\partial P}{\partial y} = 3x^2y - 3xy^2$$

By Green's Theorem we obtain

$$\int_C xy^3\,dx + x^3y\,dy = \iint_D (3x^2y - 3xy^2)\,dx\,dy = \int_{-2}^3 \int_{-1}^2 (3x^2y - 3xy^2)\,dx\,dy$$

$$= \int_{-2}^3 x^3y - \frac{3x^2y^2}{2}\Big|_{x=-1}^2 dy = \int_{-2}^3 \left((8y - 6y^2) - \left(-y - \frac{3y^2}{2}\right)\right)dy$$

$$= \int_{-2}^3 \left(-\frac{9y^2}{2} + 9y\right)dy = -\frac{3y^3}{2} + \frac{9y^2}{2}\Big|_{-2}^3 = \left(-\frac{81}{2} + \frac{81}{2}\right) - (12 + 18) = -30$$

5. $\oint_C y^2\,dx - x^2\,dy$, where C consists of the arcs $y = x^2$ and $y = \sqrt{x}$, $0 \le x \le 1$, oriented clockwise.

SOLUTION We compute the curl of $\mathbf{F}$.

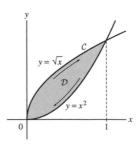

We have $P = y^2$ and $Q = -x^2$, hence

$$\frac{\partial Q}{\partial x} - \frac{\partial P}{\partial y} = -2x - 2y$$

We now compute the line integral using Green's Theorem. Since the curve is oriented clockwise, we consider the negative of the double integrals. We get

$$\int_C y^2 \, dx - x^2 \, dy = -\iint_D (-2x - 2y) \, dA = -\int_0^1 \int_{x^2}^{\sqrt{x}} (-2x - 2y) \, dy \, dx$$

$$= \int_0^1 2xy + y^2 \Big|_{y=x^2}^{\sqrt{x}} \, dx = \int_0^1 \left((2x\sqrt{x} + x) - (2x \cdot x^2 + x^4) \right) dx$$

$$= \int_0^1 (-x^4 - 2x^3 + 2x^{3/2} + x) \, dx = -\frac{x^5}{5} - \frac{x^4}{2} + \frac{4x^{5/2}}{5} + \frac{x^2}{2} \Big|_0^1$$

$$= -\frac{1}{5} - \frac{1}{2} + \frac{4}{5} + \frac{1}{2} = \frac{3}{5}$$

7. Let $\mathbf{c}(t) = \left(t^2(1-t), t(t-1)^2\right)$.

(a) $\boxed{\text{GU}}$ Plot the path $\mathbf{c}(t)$ for $0 \le t \le 1$.

(b) Calculate the area A of the region enclosed by $\mathbf{c}(t)$ for $0 \le t \le 1$ using the formula $A = \frac{1}{2} \oint_C (x \, dy - y \, dx)$.

SOLUTION

(a) The path $\mathbf{c}(t)$ for $0 \le t \le 1$ is shown in the figure:

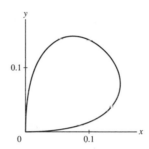

Note that the path is traced out clockwise as t goes from 0 to 1.

(b) We use the formula for the area enclosed by a closed curve,

$$A = \frac{1}{2} \int_C (x \, dy - y \, dx)$$

We compute the line integral. Since $x = t^2(1-t)$ and $y = t(t-1)^2$, we have

$$dx = \left(2t(1-t) - t^2\right) dt = \left(2t - 3t^2\right) dt$$

$$dy = (t-1)^2 + t \cdot 2(t-1) = (t-1)(3t-1) \, dt$$

Therefore,

$$x \, dy - y \, dx = t^2(1-t) \cdot (t-1)(3t-1) \, dt - t(t-1)^2 \cdot (2t - 3t^2) \, dt = t^2(t-1)^2 \, dt$$

We obtain the following integral (note that the path must be counterclockwise):

$$A = \frac{1}{2} \int_1^0 -t^2(t-1)^2 \, dt = \frac{1}{2} \int_0^1 (t^4 - 2t^3 + t^2) \, dt = \frac{1}{2} \left(\frac{t^5}{5} - \frac{t^4}{2} + \frac{t^3}{3} \Big|_0^1 \right) = \frac{1}{60}$$

In Exercises 9–12, calculate the curl and divergence of the vector field.

9. $\mathbf{F} = y\mathbf{i} - z\mathbf{k}$

SOLUTION We compute the curl of the vector field,

$$\text{curl}(\mathbf{F}) = \begin{vmatrix} \mathbf{i} & \mathbf{j} & \mathbf{k} \\ \dfrac{\partial}{\partial x} & \dfrac{\partial}{\partial y} & \dfrac{\partial}{\partial z} \\ y & 0 & -z \end{vmatrix}$$

$$= \left(\frac{\partial}{\partial y}(-z) - \frac{\partial}{\partial z}(0) \right) \mathbf{i} - \left(\frac{\partial}{\partial x}(-z) - \frac{\partial}{\partial z}(y) \right) \mathbf{j} + \left(\frac{\partial(0)}{\partial x} - \frac{\partial(y)}{\partial y} \right) \mathbf{k}$$

$$= 0\mathbf{i} + 0\mathbf{j} - 1\mathbf{k} = -\mathbf{k}$$

The divergence of **F** is

$$\text{div}(\mathbf{F}) = \frac{\partial}{\partial x}(y) + \frac{\partial}{\partial y}(0) + \frac{\partial}{\partial z}(-z) = 0 + 0 - 1 = -1.$$

11. $\mathbf{F} = \nabla(e^{-x^2 - y^2 - z^2})$

SOLUTION In Exercise 8 we proved the identity $\text{curl}(\nabla \varphi) = \mathbf{0}$. Here, $\varphi = e^{-x^2 - y^2 - z^2}$, and we have $\text{curl}\left(\nabla\left(e^{-x^2 - y^2 - z^2}\right)\right) = \mathbf{0}$. To compute div **F**, we first write **F** explicitly:

$$\mathbf{F} = \nabla\left(e^{-x^2 - y^2 - z^2}\right) = \left\langle -2xe^{-x^2 - y^2 - z^2}, -2ye^{-x^2 - y^2 - z^2}, -2ze^{-x^2 - y^2 - z^2}\right\rangle = \langle P, Q, R \rangle$$

$$\text{div}(\mathbf{F}) = \frac{\partial P}{\partial x} + \frac{\partial Q}{\partial y} + \frac{\partial R}{\partial z}$$

$$= \left(-2e^{-x^2 - y^2 - z^2} + 4x^2 e^{-x^2 - y^2 - z^2}\right) + \left(-2e^{-x^2 - y^2 - z^2} + 4y^2 e^{-x^2 - y^2 - z^2}\right)$$

$$+ \left(-2e^{-x^2 - y^2 - z^2} + 4z^2 e^{-x^2 - y^2 - z^2}\right)$$

$$= 2e^{-x^2 - y^2 - z^2}\left(2(x^2 + y^2 + z^2) - 3\right)$$

13. Recall that if F_1, F_2, and F_3 are differentiable functions of one variable, then

$$\text{curl}\left(\langle F_1(x), F_2(y), F_3(z)\rangle\right) = \mathbf{0}$$

Use this to calculate the curl of

$$\mathbf{F} = \left\langle x^2 + y^2, \ln y + z^2, z^3 \sin(z^2)e^{z^3}\right\rangle$$

SOLUTION We use the linearity of the curl and the property mentioned in the exercise to compute the curl of **F**:

$$\text{curl } \mathbf{F} = \text{curl}\left(\left\langle x^2 + y^2, \ln y + z^2, z^3 \sin\left(z^2\right)e^{z^3}\right\rangle\right) = \text{curl}\left(\left\langle x^2, \ln y, z^3 \sin(z^2)e^{z^3}\right\rangle\right) + \text{curl}\left(\left\langle y^2, z^2, 0\right\rangle\right)$$

$$= 0 + \text{curl}\left\langle y^2, z^2, 0\right\rangle = \left\langle \frac{\partial}{\partial y}(0) - \frac{\partial}{\partial z}z^2, \frac{\partial}{\partial z}y^2 - \frac{\partial}{\partial x}(0), \frac{\partial}{\partial x}z^2 - \frac{\partial}{\partial y}y^2\right\rangle = \langle -2z, 0, -2y\rangle$$

15. Verify the identities of Exercises 6 and 34 in Section 17.3 for the vector fields $\mathbf{F} = \langle xz, ye^x, yz\rangle$ and $\mathbf{G} = \langle z^2, xy^3, x^2y\rangle$.

SOLUTION We first show $\text{div}(\text{curl}(\mathbf{F})) = 0$. Let $\mathbf{F} = \langle P, Q, R\rangle = \langle xz, ye^x, yz\rangle$. We compute the curl of **F**:

$$\text{curl}(\mathbf{F}) = \begin{vmatrix} \mathbf{i} & \mathbf{j} & \mathbf{k} \\ \dfrac{\partial}{\partial x} & \dfrac{\partial}{\partial y} & \dfrac{\partial}{\partial z} \\ P & Q & R \end{vmatrix} = \left\langle \frac{\partial R}{\partial y} - \frac{\partial Q}{\partial z}, \frac{\partial P}{\partial z} - \frac{\partial R}{\partial x}, \frac{\partial Q}{\partial x} - \frac{\partial P}{\partial y}\right\rangle$$

Substituting in the appropriate values for P, Q, R and taking derivatives, we get

$$\text{curl}(\mathbf{F}) = \langle z - 0, x - 0, ye^x - 0\rangle$$

Thus,

$$\text{div}(\text{curl}(\mathbf{F})) = (z)_x + (x)_y + (ye^x)_z = 0 + 0 + 0 = 0.$$

Likewise, for $\mathbf{G} = \langle P, Q, R\rangle = \langle z^2, xy^3 x^2y\rangle$, we compute the curl of **G**:

$$\text{curl}(\mathbf{G}) = \begin{vmatrix} \mathbf{i} & \mathbf{j} & \mathbf{k} \\ \dfrac{\partial}{\partial x} & \dfrac{\partial}{\partial y} & \dfrac{\partial}{\partial z} \\ P & Q & R \end{vmatrix} = \left\langle \frac{\partial R}{\partial y} - \frac{\partial Q}{\partial z}, \frac{\partial P}{\partial z} - \frac{\partial R}{\partial x}, \frac{\partial Q}{\partial x} - \frac{\partial P}{\partial y}\right\rangle$$

Substituting in the appropriate values for P, Q, R and taking derivatives, we get

$$\text{curl}(\mathbf{G}) = \left\langle x^2 - 0, 2z - 2xy, y^3 - 0\right\rangle$$

Thus,

$$\text{div}\,(\text{curl}(\mathbf{G})) = (x^2)_x + (2z - 2xy)_y + (y^3)_z = 2x - 2x = 0.$$

We now work on the second identity. For $\mathbf{F} = \langle xz, ye^x, yz \rangle$ and $\mathbf{G} = \langle z^2, xy^3, x^2y \rangle$, it is easy to calculate

$$\mathbf{F} \times \mathbf{G} = \langle x^2y^2e^x - xy^4z, yz^3 - x^3yz, x^2y^3z - yz^2e^x \rangle$$

Thus,

$$\text{div}(\mathbf{F} \times \mathbf{G}) = (2xy^2e^x + x^2y^2e^x - y^4z) + (z^3 - x^3z) + (x^2y^3 - 2yze^x)$$

On the other hand, from our work above,

$$\text{curl}(\mathbf{F}) = \langle z, x, ye^x \rangle$$

$$\text{curl}(\mathbf{G}) = \left\langle x^2, 2z - 2xy, y^3 \right\rangle$$

So, we calculate

$$\mathbf{G} \cdot \text{curl}(\mathbf{F} - \mathbf{F}) \cdot \text{curl}(\mathbf{G}) = z^2 \cdot z + xy^3 \cdot x + x^2y \cdot ye^x - xz \cdot x^2 - ye^x \cdot (2z - 2xy) - yz \cdot y^3$$

$$= z^3 + x^2y^3 + x^2y^2e^x + 2xy^2e^x - x^3z - 2yze^x - y^4z$$

$$= (2xy^2e^x + x^2y^2e^x - y^4z) + (z^3 - x^3z) + (x^2y^3 - 2yze^x) = \text{div}(\mathbf{F} \times \mathbf{G})$$

17. Prove that if $\mathbf{F}$ is a gradient vector field, then the flux of $\text{curl}(\mathbf{F})$ through a smooth surface $\mathcal{S}$ (whether closed or not) is equal to zero.

SOLUTION If $\mathbf{F}$ is a gradient vector field, then $\mathbf{F}$ is conservative; therefore the line integral of $\mathbf{F}$ over any closed curve is zero. Combining with Stokes' Theorem yields

$$\iint_{\mathcal{S}} \text{curl}(\mathbf{F}) \cdot d\mathbf{S} = \int_{\partial \mathcal{S}} \mathbf{F} \cdot d\mathbf{s} = 0$$

19. Let $\mathbf{F} = \langle z^2, x + z, y^2 \rangle$ and let $\mathcal{S}$ be the upper half of the ellipsoid

$$\frac{x^2}{4} + y^2 + z^2 = 1$$

oriented by outward-pointing normals. Use Stokes' Theorem to compute $\iint_{\mathcal{S}} \text{curl}(\mathbf{F}) \cdot d\mathbf{S}$.

SOLUTION We compute the curl of $\mathbf{F} = \langle z^2, x + z, y^2 \rangle$:

$$\text{curl}(\mathbf{F}) = \begin{vmatrix} \mathbf{i} & \mathbf{j} & \mathbf{k} \\ \frac{\partial}{\partial x} & \frac{\partial}{\partial y} & \frac{\partial}{\partial z} \\ z^2 & x + z & y^2 \end{vmatrix} = (2y - 1)\mathbf{i} - (0 - 2z)\mathbf{j} + (1 - 0)\mathbf{k} = \langle 2y - 1, 2z, 1 \rangle$$

Let $\mathcal{C}$ denote the boundary of $\mathcal{S}$, that is, the ellipse $\frac{x^2}{4} + y^2 = 1$ in the xy-plane, oriented counterclockwise. Then by Stoke's Theorem we have

$$\iint_{\mathcal{S}} \text{curl}(\mathbf{F}) \cdot d\mathbf{S} = \int_{\mathcal{C}} \mathbf{F} \cdot d\mathbf{s} \tag{1}$$

We parametrize $\mathcal{C}$ by

$$\mathcal{C} : r(t) = (2\cos t, \sin t, 0), \quad 0 \le t \le 2\pi$$

Then

$$\mathbf{F}(r(t)) \cdot r'(t) = \left\langle 0, 2\cos t, \sin^2 t \right\rangle \cdot \langle -2\sin t, \cos t, 0 \rangle = 2\cos^2 t$$

Combining with (1) gives

$$\iint_{\mathcal{S}} \text{curl}(\mathbf{F}) \cdot d\mathbf{s} = \int_0^{2\pi} 2\cos^2 t \, dt = t + \frac{\sin 2t}{2} \Big|_0^{2\pi} = 2\pi$$

21. Let S be the side of the cylinder $x^2 + y^2 = 4$, $0 \le z \le 2$ (not including the top and bottom of the cylinder). Use Stokes' Theorem to compute the flux of $\mathbf{F} = \langle 0, y, -z \rangle$ through S (with outward pointing normal) by finding a vector potential $\mathbf{A}$ such that $\mathrm{curl}(\mathbf{A}) = \mathbf{F}$.

SOLUTION   We can write $\mathbf{F} = \mathrm{curl}(\mathbf{A})$ where $\mathbf{A} = \langle yz, 0, 0 \rangle$. The flux of $\mathbf{F}$ through S is equal to the line integral of A around the oriented boundary which consists of two circles of radius 2 with center on the z-axis (one at height $z = 0$ and one at height $z = 2$).

However, the line integrals of A about both circles are zero. This is clear for the circle at $z = 0$ because then $A = 0$, but it is also true at $z = 2$ because the vector field $A = \langle 2y, 0, 0 \rangle$ integrates to zero around the circle.

In Exercises 23–26, use the Divergence Theorem to calculate $\iint_S \mathbf{F} \cdot d\mathbf{S}$ for the given vector field and surface.

23. $\mathbf{F} = \langle xy, yz, x^2 z + z^2 \rangle$, S is the boundary of the box $[0, 1] \times [2, 4] \times [1, 5]$.

SOLUTION

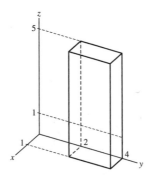

We compute the divergence of $\mathbf{F} = \langle xy, yz, x^2 z + z^2 \rangle$:

$$\mathrm{div}(\mathbf{F}) = \frac{\partial}{\partial x} xy + \frac{\partial}{\partial y} yz + \frac{\partial}{\partial z}(x^2 z + z^2) = y + z + x^2 + 2z = x^2 + y + 3z$$

The Divergence Theorem gives

$$\iint_S \langle xy, yz, x^2 z + z^2 \rangle \cdot d\mathbf{S} = \int_1^5 \int_2^4 \int_0^1 (x^2 + y + 3z)\, dx\, dy\, dz = \int_1^5 \int_2^4 \frac{x^3}{3} + (y + 3z)x \Big|_{x=0}^1 dy\, dz$$

$$= \int_1^5 \int_2^4 \left(\frac{1}{3} + y + 3z \right) dy\, dz = \int_1^5 \frac{1}{3} y + \frac{1}{2} y^2 + 3zy \Big|_{y=2}^4 dz$$

$$= \int_1^5 \left(\left(\frac{4}{3} + \frac{16}{2} + 12z \right) - \left(\frac{2}{3} + 2 + 6z \right) \right) dz = \int_1^5 \left(\frac{20}{3} + 6z \right) dz$$

$$= \frac{20z}{3} + \frac{3z^2}{2} \Big|_1^5 = \left(75 + \frac{100}{3} \right) - \left(3 + \frac{20}{3} \right) = \frac{296}{3}$$

25. $\mathbf{F} = \langle xyz + xy, \frac{1}{2} y^2 (1 - z) + e^x, e^{x^2 + y^2} \rangle$, S is the boundary of the solid bounded by the cylinder $x^2 + y^2 = 16$ and the planes $z = 0$ and $z = y - 4$.

SOLUTION   We compute the divergence of $\mathbf{F}$:

$$\mathrm{div}(\mathbf{F}) = \frac{\partial}{\partial x}(xyz + xy) + \frac{\partial}{\partial y} \left(\frac{y^2}{2} (1 - z) + e^x \right) + \frac{\partial}{\partial z}(e^{x^2 + y^2}) = yz + y + y(1 - z) = 2y$$

Let S denote the surface of the solid $\mathcal{W}$. The Divergence Theorem gives

$$\iint_S \mathbf{F} \cdot d\mathbf{S} = \iiint_{\mathcal{W}} \mathrm{div}(\mathbf{F})\, dV = \iiint_{\mathcal{W}} 2y\, dV = \iint_{\mathcal{D}} \int_{y-4}^0 2y\, dz\, dx\, dy$$

$$= \iint_{\mathcal{D}} 2yz \Big|_{z=y-4}^0 dx\, dy = \iint_{\mathcal{D}} 2y\, (0 - (y - 4))\, dx\, dy = \iint_{\mathcal{D}} (8y - 2y^2)\, dx\, dy$$

We convert the integral to polar coordinates:

$$\iint_S \mathbf{F} \cdot d\mathbf{S} = \int_0^{2\pi} \int_0^4 (8r\cos\theta - 2r^2\cos^2\theta) r \, dr \, d\theta$$

$$= 8 \left(\int_0^4 r^2 \, dr \right) \left(\int_0^{2\pi} \cos\theta \, d\theta \right) - \left(\int_0^4 r^3 \, dr \right) \left(\int_0^{2\pi} 2\cos^2\theta \, d\theta \right)$$

$$= 0 - \left(\frac{r^4}{4} \Big|_0^4 \right) \left(\theta + \frac{\sin 2\theta}{2} \Big|_0^{2\pi} \right) = -\frac{4^4}{4} \cdot 2\pi = -128\pi$$

27. Find the volume of a region $\mathcal{W}$ if

$$\iint_{\partial \mathcal{W}} \left\langle x + xy + z, x + 3y - \frac{1}{2}y^2, 4z \right\rangle \cdot d\mathbf{S} = 16$$

SOLUTION   Let $\mathbf{F} = \left\langle x + xy + z, x + 3y - \frac{1}{2}y^2, 4z \right\rangle$. We compute the divergence of $\mathbf{F}$:

$$\text{div}(\mathbf{F}) = \frac{\partial}{\partial x}(x + xy + z) + \frac{\partial}{\partial y}\left(x + 3y - \frac{1}{2}y^2\right) + \frac{\partial}{\partial z}(4z) = 1 + y + 3 - y + 4 = 8$$

Using the Divergence Theorem and the given information, we obtain

$$16 = \iint_S \mathbf{F} \cdot d\mathbf{S} = \iint_{\mathcal{W}} \text{div}(\mathbf{F}) \, dV = \iint_{\mathcal{W}} 8 \, dV = 8 \iint_{\mathcal{W}} 1 \, dV = 8 \, \text{Volume}\,(\mathcal{W})$$

That is,

$$16 = 8\,\text{Volume}\,(\mathcal{W})$$

or

$$\text{Volume}\,(\mathcal{W}) = 2$$

In Exercises 29–32, let $\mathbf{F}$ be a vector field whose curl and divergence at the origin are

$$\text{curl}(\mathbf{F})(0, 0, 0) = \langle 2, -1, 4 \rangle, \qquad \text{div}(\mathbf{F})(0, 0, 0) = -2$$

29. Estimate $\oint_C \mathbf{F} \cdot d\mathbf{s}$, where C is the circle of radius 0.03 in the xy-plane centered at the origin.

SOLUTION   We use the estimation

$$\int_C \mathbf{F} \cdot d\mathbf{s} \approx (\text{curl}(\mathbf{F})(0) \cdot \mathbf{e}_n) \, \text{Area}(\mathcal{R})$$

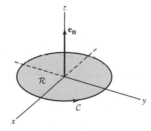

The unit normal vector to the disk $\mathcal{R}$ is $\mathbf{e}_n = \mathbf{k} = \langle 0, 0, 1 \rangle$. The area of the disk is

$$\text{Area}\,(\mathcal{R}) = \pi \cdot 0.03^2 = 0.0009\pi.$$

Using the given curl at the origin, we have

$$\int_C \mathbf{F} \cdot d\mathbf{s} \approx \langle 2, -1, 4 \rangle \cdot \langle 0, 0, 1 \rangle \cdot 0.0009\pi = 4 \cdot 0.0009\pi \approx 0.0113$$

31. Suppose that $\mathbf{v}$ is the velocity field of a fluid and imagine placing a small paddle wheel at the origin. Find the equation of the plane in which the paddle wheel should be placed to make it rotate as quickly as possible.

SOLUTION   The paddle wheel has the maximum spin when the circulation of the velocity field $\mathbf{v}$ around the wheel is maximum. The maximum circulation occurs when $\mathbf{e}_n$, and the curl of $\mathbf{v}$ at the origin (i.e., the vector $\langle 2, -1, 4 \rangle$) point in the same direction. Therefore, the plane in which the paddle wheel should be placed is the plane through the origin with the normal $\langle 2, -1, 4 \rangle$. This plane has the equation, $2x - y + 4z = 0$.

33. The velocity vector field of a fluid (in meters per second) is

$$\mathbf{F} = \langle x^2 + y^2, 0, z^2 \rangle$$

Let $\mathcal{W}$ be the region between the hemisphere

$$\mathcal{S} = \{(x, y, z) : x^2 + y^2 + z^2 = 1, \quad x, y, z \geq 0\}$$

and the disk $\mathcal{D} = \{(x, y, 0) : x^2 + y^2 \leq 1\}$ in the xy-plane. Recall that the flow rate of a fluid across a surface is equal to the flux of $\mathbf{F}$ through the surface.

(a) Show that the flow rate across $\mathcal{D}$ is zero.

(b) Use the Divergence Theorem to show that the flow rate across $\mathcal{S}$, oriented with outward-pointing normal, is equal to $\iiint_{\mathcal{W}} \mathrm{div}(\mathbf{F})\, dV$. Then compute this triple integral.

SOLUTION

(a) To show that no fluid flows across $\mathcal{D}$, we show that the normal component of $\mathbf{F}$ at each point on $\mathcal{D}$ is zero. At each point $P = (x, y, 0)$ on the xy-plane,

$$\mathbf{F}(P) = \langle x^2 + y^2, 0, 0^2 \rangle = \langle x^2 + y^2, 0, 0 \rangle.$$

Moreover, the unit normal vector to the xy-plane is $\mathbf{e}_n = (0, 0, 1)$. Therefore,

$$\mathbf{F}(P) \cdot \mathbf{e}_n = \langle x^2 + y^2, 0, 0 \rangle \cdot \langle 0, 0, 1 \rangle = 0.$$

Since $\mathcal{D}$ is contained in the xy-plane, we conclude that the normal component of $\mathbf{F}$ at each point on $\mathcal{D}$ is zero. Therefore, no fluid flows across $\mathcal{D}$.

(b) By the Divergence Theorem and the linearity of the flux we have

$$\iint_{\mathcal{S}} \mathbf{F} \cdot d\mathbf{S} + \iint_{\mathcal{D}} \mathbf{F} \cdot d\mathbf{S} = \iiint_{\mathcal{W}} \mathrm{div}(\mathbf{F})\, dV$$

Since the flux through the disk $\mathcal{D}$ is zero, we have

$$\iint_{\mathcal{S}} \mathbf{F} \cdot d\mathbf{S} = \iiint_{\mathcal{W}} \mathrm{div}(\mathbf{F})\, dV \tag{1}$$

To compute the triple integral, we first compute $\mathrm{div}(\mathbf{F})$:

$$\mathrm{div}(\mathbf{F}) = \frac{\partial}{\partial x}(x^2 + y^2) + \frac{\partial}{\partial y}(0) + \frac{\partial}{\partial z}(z^2) = 2x + 2z = 2(x + z).$$

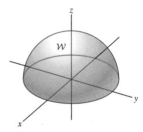

Using spherical coordinates we get

$$\iiint_{\mathcal{W}} \mathrm{div}(\mathbf{F})\, dV = 2 \int_0^{\pi/2} \int_0^{2\pi} \int_0^1 (\rho \sin\phi \cos\theta + \rho \cos\phi)\rho^2 \sin\phi \, d\rho \, d\phi$$

$$= 2 \int_0^1 \rho^3 d\rho \left(\left(\int_0^{\pi/2} \sin^2\phi \, d\phi \right) \left(\int_0^{2\pi} \cos\theta \, d\theta \right) + 2\pi \int_0^{\pi/2} \cos\phi \sin\phi \, d\rho \right)$$

$$= \frac{1}{2} \left(0 + \pi \int_0^{\pi/2} \sin 2\phi \, d\phi \right) = \frac{\pi}{2} \left(-\frac{\cos 2\phi}{2} \right) \Big|_0^{\pi/2} = -\frac{\pi}{4}(-1 - 1) = \frac{\pi}{2}$$

Combining with (1) we obtain the flux:

$$\iint_{\mathcal{S}} \mathbf{F} \cdot d\mathbf{S} = \frac{\pi}{2}$$

35. Let $V(x, y) = x + \dfrac{x}{x^2 + y^2}$. The vector field $\mathbf{F} = \nabla V$ (Figure 5) provides a model in the plane of the velocity field of an incompressible, irrotational fluid flowing past a cylindrical obstacle (in this case, the obstacle is the unit circle $x^2 + y^2 = 1$).

(a) Verify that $\mathbf{F}$ is irrotational [by definition, $\mathbf{F}$ is irrotational if $\mathrm{curl}(\mathbf{F}) = 0$].

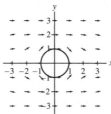

FIGURE 5 The vector field ∇V for $V(x, y) = x + \dfrac{x}{x^2 + y^2}$.

(b) Verify that $\mathbf{F}$ is tangent to the unit circle at each point along the unit circle except $(1, 0)$ and $(-1, 0)$ (where $\mathbf{F} = 0$).

(c) What is the circulation of $\mathbf{F}$ around the unit circle?

(d) Calculate the line integral of $\mathbf{F}$ along the upper and lower halves of the unit circle separately.

SOLUTION

(a) In Exercise 8, we proved the identity $\mathrm{curl}(\nabla \varphi) = 0$. Since $\mathbf{F}$ is a gradient vector field, it is irrotational; that is, $\mathrm{curl}(\mathbf{F}) = 0$ for $(x, y) \neq (0, 0)$, where $\mathbf{F}$ is defined.

(b) We compute $\mathbf{F}$ explicitly:

$$\mathbf{F} = \nabla \varphi = \left\langle \frac{\partial \varphi}{\partial x}, \frac{\partial \varphi}{\partial y} \right\rangle = \left\langle 1 + \frac{y^2 - x^2}{(x^2 + y^2)^2}, -\frac{2xy}{(x^2 + y^2)^2} \right\rangle$$

Now, using $x = \cos t$ and $y = \sin t$ as a parametrization of the circle, we see that

$$\mathbf{F} = \left\langle 1 + \sin^2 t - \cos^2 t, -2 \cos t \sin t \right\rangle = \left\langle 2 \sin^2 t, -2 \cos t \sin t \right\rangle,$$

and so

$$\mathbf{F} = 2 \sin t \, \langle \sin t, -\cos t \rangle = 2 \sin t \, \langle y, -x \rangle,$$

which is clearly perpendicular to the radial vector $\langle x, y \rangle$ for the circle.

(c) We use our expression of $\mathbf{F}$ from Part (b):

$$\mathbf{F} = \nabla \psi = \left\langle 1 + \frac{y^2 - x^2}{(x^2 + y^2)^2}, -\frac{2xy}{(x^2 + y^2)^2} \right\rangle$$

Now, $\mathbf{F}$ is not defined at the origin and therefore we cannot use Green's Theorem to compute the line integral along the unit circle. We thus compute the integral directly, using the parametrization

$$\mathbf{c}(t) = (\cos t, \sin t), \quad 0 \leq t \leq 2\pi.$$

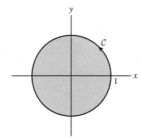

Then,

$$\mathbf{F}(\mathbf{c}(t)) \cdot \mathbf{c}'(t) = \left\langle 1 + \frac{\sin^2 t - \cos^2 t}{(\cos^2 t + \sin^2 t)^2}, -\frac{2 \cos t \sin t}{(\cos^2 t + \sin^2 t)^2} \right\rangle \cdot \langle -\sin t, \cos t \rangle$$

$$= \left\langle 1 + \sin^2 t - \cos^2 t, -2 \cos t, \sin t \right\rangle \cdot \langle -\sin t, \cos t \rangle = \left\langle 2 \sin^2 t, -2 \cos t \sin t \right\rangle \cdot \langle -\sin t, \cos t \rangle$$

$$= -2 \sin^3 t - 2 \cos^2 t \sin t = -2 \sin t (\sin^2 t + \cos^2 t) = -2 \sin t$$

Hence,

$$\int_C \mathbf{F} \cdot d\mathbf{s} = \int_0^{2\pi} -2 \sin t \, dt = 0$$

(d) We denote by C_1 and C_2 the upper and lower halves of the unit circle. Using part (c) we have

$$\int_{C_1} \mathbf{F} \cdot d\mathbf{s} + \int_{C_2} \mathbf{F} \cdot d\mathbf{s} = 0 \quad \Rightarrow \quad \int_{C_2} \mathbf{F} \cdot d\mathbf{s} = -\int_{C_1} \mathbf{F} \cdot d\mathbf{s} \tag{1}$$

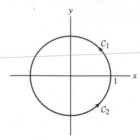

To compute the circulation along C_1, we compute the integral as in part (c), only that the limits of integration are now $t = 0$ and $t = \pi$. Using the computations in part (c) we obtain

$$\int_{C_1} \mathbf{F} \cdot d\mathbf{s} = \int_0^\pi -2\sin^2 t \, dt = -4$$

Therefore, by (1),

$$\int_{C_2} \mathbf{F} \cdot d\mathbf{s} = 4.$$

37. In Section 17.1, we showed that if C is a simple closed curve, oriented counterclockwise, then the line integral is

$$\text{Area enclosed by } C = \frac{1}{2} \oint_C x \, dy - y \, dx \qquad \boxed{1}$$

Suppose that C is a path from P to Q that is not closed but has the property that every line through the origin intersects C in at most one point, as in Figure 7. Let $\mathcal{R}$ be the region enclosed by C and the two radial segments joining P and Q to the origin. Show that the line integral in Eq. (1) is equal to the area of $\mathcal{R}$. *Hint:* Show that the line integral of $\mathbf{F} = \langle -y, x \rangle$ along the two radial segments is zero and apply Green's Theorem.

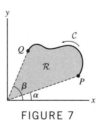

FIGURE 7

SOLUTION

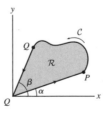

Let $\mathbf{F} = \langle -y, x \rangle$. Then $P = -y$ and $Q = x$, and $\frac{\partial Q}{\partial x} - \frac{\partial P}{\partial y} = 2$. By Green's Theorem, we have

$$\int_C -y \, dx + x \, dy + \int_{QO} -y \, dx + x \, dy + \int_{OP} -y \, dx + x \, dy = \iint_{\mathcal{R}} 2 \, dA = 2 \iint_{\mathcal{R}} dA$$

Denoting by A the area of the region $\mathcal{R}$, we obtain

$$A = \frac{1}{2} \int_C -y \, dx + x \, dy + \frac{1}{2} \int_{QO} -y \, dx + x \, dy + \frac{1}{2} \int_{OP} -y \, dx + x \, dy \tag{1}$$

We parametrize the two segments by

$$\overline{QO} : \mathbf{c}(t) = (t, t\tan\beta) \qquad \mathbf{c}'(t) = \langle 1, \tan\beta\rangle$$
$$\overline{OP} : \mathbf{d}(t) = (t, t\tan\alpha) \quad\Rightarrow\quad \mathbf{d}'(t) = \langle 1, \tan\alpha\rangle$$

Then,

$$\mathbf{F}\,(\mathbf{c}(t)) \cdot \mathbf{c}'(t) = \langle -t\tan\beta, t\rangle \cdot \langle 1, \tan\beta\rangle = -t\tan\beta + t\tan\beta = 0$$

$$\mathbf{F}\,(\mathbf{d}(t)) \cdot \mathbf{d}'(t) = \langle -t\tan\alpha, t\rangle \cdot \langle 1, \tan\alpha\rangle = -t\tan\alpha + t\tan\alpha = 0$$

Therefore,

$$\int_{\overline{QO}} \mathbf{F}\cdot d\mathbf{s} = \int_{\overline{OP}} \mathbf{F}\cdot d\mathbf{s} = 0.$$

Combining with (1) gives

$$A = \frac{1}{2}\int_{\mathcal{C}} -y\,dx + x\,dy.$$

39. Prove the following generalization of Eq. (1). Let $\mathcal{C}$ be a simple closed curve in the plane (Figure 8)

$$\mathcal{S}: \quad ax + by + cz + d = 0$$

Then the area of the region R enclosed by $\mathcal{C}$ is equal to

$$\frac{1}{2\|\mathbf{n}\|}\oint_{\mathcal{C}} (bz - cy)\,dx + (cx - az)\,dy + (ay - bx)\,dz$$

where $\mathbf{n} = \langle a, b, c\rangle$ is the normal to $\mathcal{S}$, and $\mathcal{C}$ is oriented as the boundary of R, (relative to the normal vector $\mathbf{n}$). *Hint:* Apply Stokes' Theorem to $\mathbf{F} = \langle bz - cy, cx - az, ay - bx\rangle$.

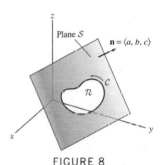

FIGURE 8

SOLUTION   By Stokes' Theorem,

$$\iint_{\mathcal{S}} \mathrm{curl}(\mathbf{F})\cdot d\mathbf{S} = \iint_{\mathcal{S}} (\mathrm{curl}(\mathbf{F})\cdot \mathbf{e}_n)\,dS = \int_{\mathcal{C}} \mathbf{F}\cdot d\mathbf{s} \tag{1}$$

We compute the curl of $\mathbf{F}$:

$$\mathrm{curl}(\mathbf{F}) = \begin{vmatrix} \mathbf{i} & \mathbf{j} & \mathbf{k} \\ \dfrac{\partial}{\partial x} & \dfrac{\partial}{\partial y} & \dfrac{\partial}{\partial z} \\ bz - cy & cx - az & ay - bx \end{vmatrix} = 2a\mathbf{i} + 2b\mathbf{j} + 2c\mathbf{k} = 2\langle a, b, c\rangle$$

The unit normal to the plane $ax + by + cz + d = 0$ is

$$\mathbf{e}_n = \frac{\langle a, b, c\rangle}{\sqrt{a^2 + b^2 + c^2}}$$

Therefore,

$$\mathrm{curl}(\mathbf{F}) \cdot \mathbf{e}_n = 2\langle a, b, c\rangle \cdot \frac{1}{\sqrt{a^2 + b^2 + c^2}}\langle a, b, c\rangle$$

$$= \frac{2}{\sqrt{a^2 + b^2 + c^2}}(a^2 + b^2 + c^2) = 2\sqrt{a^2 + b^2 + c^2}$$

Hence,

$$\iint_{\mathcal{S}} \text{curl}(\mathbf{F}) \cdot d\mathbf{S} = \iint_{\mathcal{S}} \text{curl}(\mathbf{F}) \cdot \mathbf{e}_n \, dS = \iint_{\mathcal{S}} 2\sqrt{a^2+b^2+c^2} \, dS = 2\sqrt{a^2+b^2+c^2} \iint_{\mathcal{S}} 1 \, dS \qquad (2)$$

The sign of $\iint_{\mathcal{S}} 1 \, d\mathbf{S}$ is determined by the orientation of $\mathcal{S}$. Since the area is a positive value, we have

$$\left| \iint_{\mathcal{S}} 1 \, ds \right| = \text{Area}\,(\mathcal{S})$$

Therefore, (2) gives

$$\left| \iint_{\mathcal{S}} \text{curl}(\mathbf{F}) \cdot d\mathbf{S} \right| = 2\sqrt{a^2+b^2+c^2}\,\text{Area}(\mathcal{S})$$

Combining with (1) we obtain

$$2\sqrt{a^2+b^2+c^2}\,\text{Area}(\mathcal{S}) = \left| \int_{\mathcal{C}} \mathbf{F} \cdot ds \right|$$

or

$$\text{Area}(\mathcal{S}) = \frac{1}{2\sqrt{a^2+b^2+c^2}} = \frac{1}{2\|\mathbf{n}\|} \cdot \left| \int_{\mathcal{C}} (bz-cy)\,dx + (cx-az)\,dy + (ay-bx)\,dz \right|$$

41. Show that $G(\theta, \phi) = (a\cos\theta\sin\phi, b\sin\theta\sin\phi, c\cos\phi)$ is a parametrization of the ellipsoid

$$\left(\frac{x}{a}\right)^2 + \left(\frac{y}{b}\right)^2 + \left(\frac{z}{c}\right)^2 = 1$$

Then calculate the volume of the ellipsoid as the surface integral of $\mathbf{F} = \frac{1}{3}\langle x, y, z\rangle$ (this surface integral is equal to the volume by the Divergence Theorem).

SOLUTION For the given parametrization,

$$x = a\cos\theta\sin\phi, \quad y = b\sin\theta\sin\phi, \quad z = c\cos\phi \qquad (1)$$

We show that it satisfies the equation of the ellipsoid

$$\left(\frac{x}{a}\right)^2 + \left(\frac{y}{b}\right)^2 + \left(\frac{z}{c}\right)^2 = \left(\frac{a\cos\theta\sin\phi}{a}\right)^2 + \left(\frac{b\sin\theta\sin\phi}{b}\right)^2 + \left(\frac{c\cos\phi}{c}\right)^2$$

$$= \cos^2\theta\sin^2\phi + \sin^2\theta\sin^2\phi + \cos^2\phi$$

$$= \sin^2\phi(\cos^2\theta + \sin^2\theta) + \cos^2\phi$$

$$= \sin^2\phi + \cos^2\phi = 1$$

Conversely, for each (x, y, z) on the ellipsoid, there exists θ and ϕ so that (1) holds. Therefore $\Phi(\theta, \phi)$ parametrizes the whole ellipsoid. Let $\mathcal{W}$ be the interior of the ellipsoid $\mathcal{S}$. Then by Eq. (10):

$$\text{Volume}(\mathcal{W}) = \frac{1}{3} \iint_{\mathcal{S}} \mathbf{F} \cdot d\mathbf{S}, \quad \mathbf{F} = \langle x, y, z\rangle$$

We compute the surface integral, using the given parametrization. We first compute the normal vector:

$$\frac{\partial\Phi}{\partial\theta} = \langle -a\sin\theta\sin\phi, b\cos\theta\sin\phi, 0\rangle$$

$$\frac{\partial\Phi}{\partial\phi} = \langle a\cos\theta\cos\phi, b\sin\theta\cos\phi, -c\sin\phi\rangle$$

$$\frac{\partial\Phi}{\partial\theta} \times \frac{\partial\Phi}{\partial\phi} = -ab\sin^2\theta\sin\phi\cos\phi\mathbf{k} - ac\sin\theta\sin^2\phi\mathbf{j} - ab\cos^2\theta\sin\phi\cos\phi\mathbf{k} - bc\cos\theta\sin^2\phi\mathbf{i}$$

$$= \left\langle -bc\cos\theta\sin^2\phi, -ac\sin\theta\sin^2\phi, -ab\sin\phi\cos\phi\right\rangle$$

Hence, the outward pointing normal is

$$\mathbf{n} = \left\langle bc\cos\theta\sin^2\phi,\, ac\sin\theta\sin^2\phi,\, ab\sin\phi\cos\phi \right\rangle$$

$$\mathbf{F}\left(\Phi(\theta,\phi)\right)\cdot\mathbf{n} = \langle a\cos\theta\sin\phi,\, b\sin\theta\sin\phi,\, c\cos\phi\rangle \cdot \left\langle bc\cos\theta\sin^2\phi,\, ac\sin\theta\sin^2\phi,\, ab\sin\phi\cos\phi \right\rangle$$

$$= abc\cos^2\theta\sin^3\phi + abc\sin^2\theta\sin^3\phi + abc\sin\phi\cos^2\phi$$

$$= abc\sin^3\phi(\cos^2\theta + \sin^2\theta) + abc\sin\phi\cos^2\phi$$

$$= abc\sin^3\phi + abc\sin\phi\cos^2\phi = abc\sin^3\phi + abc\sin\phi(1 - \sin^2\phi)$$

$$= abc\sin\phi$$

We obtain the following integral:

$$\text{Volume}(\mathcal{W}) = \frac{1}{3}\iint_{\mathcal{S}}\mathbf{F}\cdot d\mathbf{S} = \frac{1}{3}\int_0^{2\pi}\int_0^{\pi} abc\sin\phi\, d\phi\, d\theta$$

$$= \frac{2\pi abc}{3}\int_0^{\pi}\sin\phi\, d\varphi = \frac{2\pi abc}{3}\left(-\cos\phi\Big|_0^{\pi}\right) = \frac{4\pi abc}{3}$$